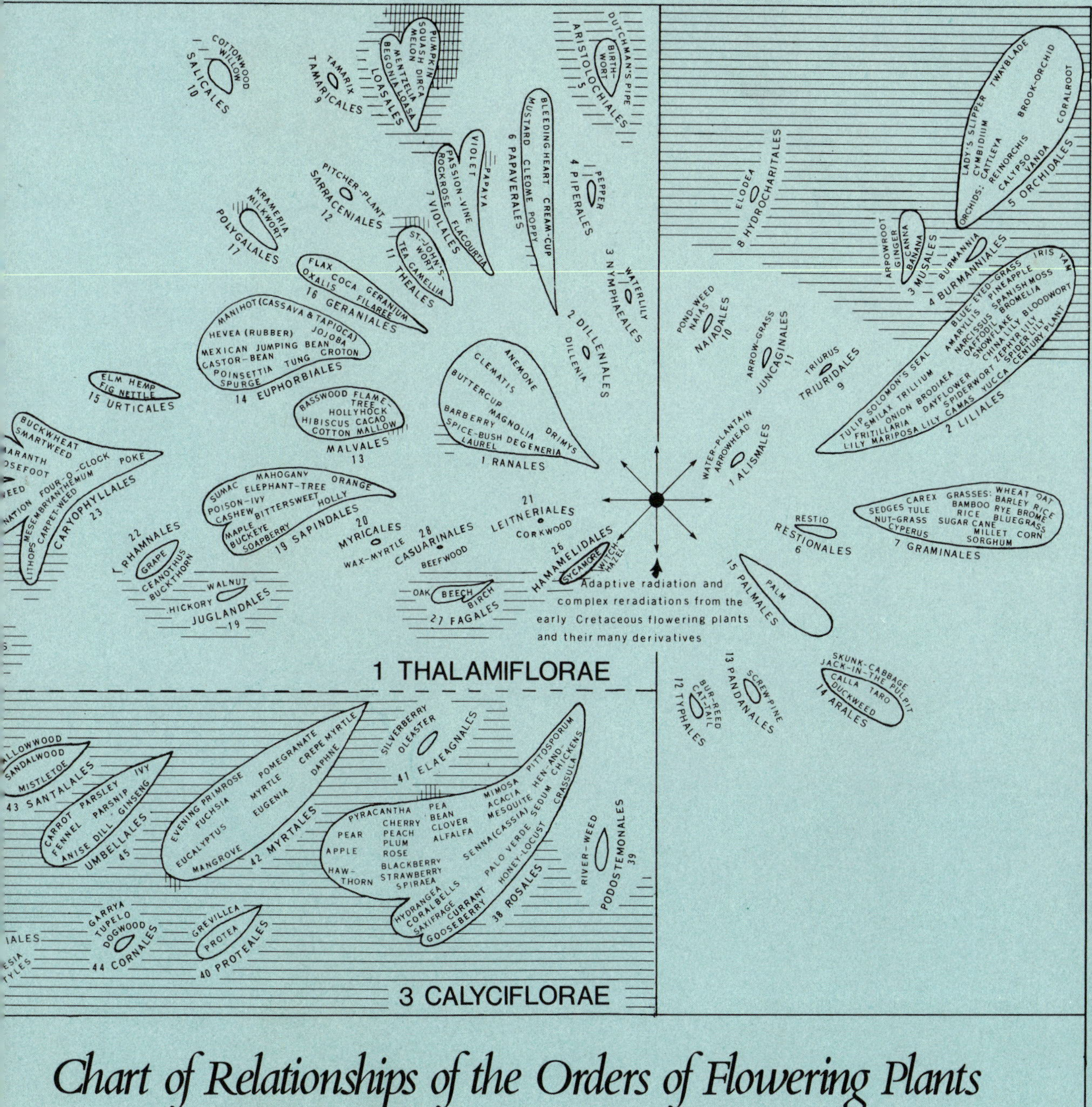

Chart of Relationships of the Orders of Flowering Plants

Plant Classification

Asa Gray, 1810 -1888

DR. GRAY, Fisher Professor of Natural History, Harvard College, Harvard University, 1842–1888, was essentially the father of botany in North America. His textbooks, such as *Lessons in Botany* and *How Plants Grow,* made botany a popular science in the United States and Canada, and they still appear in many private libraries and in the classroom. Gray's numerous technical publications spanned more than half a century. Among them were a *Flora of North America* (with John Torrey), a *Manual of Botany* (covering northeastern North America), and a *Synoptical Flora of North America* (incomplete), as well as a large section of a flora of California. Shorter papers dealt with plants of the entire Continent and even of Japan. Gray promoted development of all fields of botany, and he was the first to sponsor in North America the concept of evolution, as interpreted by his close friend and correspondent, Charles Darwin. The Gray Herbarium at Harvard preserves his plant collection, and his work is continued there. (Photograph by Frank B. Salisbury from an old print at Pomona College, showing Dr. Gray with his plant press at La Veta Pass, Colorado.)

Plant Classification

SECOND EDITION

Lyman Benson

Emeritus Professor of Botany
Chairman of the Botany Department,
Director of the Herbarium
Pomona College, Claremont, California

Principal Plant Dissections and Illustrations by the late
Jerome D. Laudermilk
Research Associate in Paleobotany
Pomona College

D.C. HEATH AND COMPANY
Lexington, Massachusetts • Toronto

Designed by Carmela M. Ciampa

Published simultaneously in Canada.

Printed in the United States of America.

International Standard Book Number: 0–669–01489–3

Library of Congress Catalog Card Number: 78–61856

Preface to Second Edition

A number of ideas have been revised during the 22 years since the first edition of *Plant Classification* was published, especially concerning the origin and relationships of the major taxa of vascular plants other than the flowering plants, or Angiospermae.

Classification of the flowering plants has been outlined in four new systems, some basically similar. Although these systems are not accepted as complete for reasons discussed in Chapter 16, each one contributes good ideas derived from the research of many individuals, and some of them have been incorporated into the system in this book. The recent angiosperm classification systems are considered toward the end of Chapter 16. The basis for the system employed here is presented in Chapter 17.

The most conspicuous change is final abandonment of the Amentiferae, or catkin flowers, as a partly natural and partly artificial group. This was foreshadowed in the first edition, where the group was described as more artificial than natural, while the other four groups of dicotyledons were the reverse. The need for retention for a time of the Amentiferae arose from prevalence of the group in the numerous older manuals and floras arranged according to the Engler and Prantl system, in which the Amentiferae were considered the original and therefore the key group of angiosperms. Since the older manuals have been partly replaced by other books arranged according to different systems, the need for retention of the Amentiferae in *Plant Classification* for practical purposes has declined. Present-day students are much less likely than those of 1957 to encounter the Amentiferae in their manuals and floras of local plants used (after finding the order and family in *Plant Classification*) for identification of the genus, species, and variety.

The system of classification of units needed for understanding plant geography according to natural **floras** and included **floristic associations** has been developed in a number of publications (see the opening page of Chapter 24) since it appeared in the first edition, and it is much revised for this edition. The system and its relationship to ecological classifications of vegetation are discussed further in additions to the **Preface to the First Edition** and in Chapters 24 and 25.

L.B.

Preface to First Edition (updated)

[To the Teacher]

During the first half of the Twentieth Century plant classification has been modified by data from many sources. Because the factors to be taken into account have changed, the entire field of taxonomy requires redescription and revaluation. Two books on the vascular plants are needed, as follows.

1. *An elementary textbook* intended to open up the new world of living plants to college students and the educated public. The objective is development of a more than superficial appreciation of nature through precise and effective methods of study based upon (1) acquiring for use as a tool an adequate *vocabulary describing the characteristics of each plant group,* (2) applying the use of keys and descriptions to the *process of identification,* (3) gaining knowledge of plant taxa through *preparation and preservation of specimens* forming an ordered collection, (4) developing an understanding of the *basis for classification* of plant groups, (5) gaining an appreciation of the *association of species* in natural vegetation. An introductory book must deal primarily with the higher taxa—that is, the divisions, classes, subclasses, orders, and families.

2. *An advanced textbook* with primary emphasis upon the principles underlying classifying, naming, and describing botanical taxa. Assuming the student has learned the outlines of the higher taxa and something of their classification by studying the elementary book, emphasis falls upon application of principles of taxonomy and nomenclature to genera, species, and varieties. Major items are (1) exploration for data (from field and herbarium studies, microscopic morphology and development, plant physiology and ecology, cytogenetics,* and experimental studies* combining various fields), (2) classification or delimitation of taxa, (3) choice of scientific names, (4) description, and (5) documentation.†

Plant Classification is an *elementary text* with the objectives listed above for a book of that scope. It is designed for a college course *without prerequisite.* The earlier chapters are very elementary, the later ones more and more advanced. For courses with a prerequisite of general botany or biology some chapters (e.g., 1–5, or 13) may be omitted or reviewed rapidly.

* Application of both cytogenetic and experimental investigations to plant taxonomy is restricted so far almost wholly to study of taxa of lower ranks; consequently, these "new methods" are appropriate to only the advanced text.

† *Plant Taxonomy, Methods and Principles,* Ronald Press Co., New York, 1962; republished by John Wiley & Sons, New York, 1978.

The Keys and Descriptions

The keys and descriptions in this book are designed to be followed until the family is determined; then the keys in the local flora or manual should be used to identify the genus and species. All flowering plant families represented in North America north of Mexico by native or introduced species growing without the intentional aid of man and all gymnosperm and spore plant families are covered in the keys. This includes nearly all the families of flowering plants in cultivation. Both these and the remaining flowering plant families are described and discussed.

The keys and descriptions offer the following features differing from those in most manuals and floras.

1. *A simplified vocabulary.* However, the keys and descriptions are complete, and the terms are technical.

2. *Arrangement according to a new system of classification,* combining the best features of the older systems with interpretations based upon new data.

3. *Provision of natural keys leading to the orders and then to their included families.* As far as possible the keys are arranged according to natural relationships, and the characters of orders and the distinctions between related families are emphasized. Consequently, the broad outlines of classification are evident and readily remembered. The orders of dicotyledons are arranged in four groups recognized so easily that after identification of a few plants the first divisions of the key may be covered from memory. These features are lost in elongated artificial keys ignoring the orders and segregating the families directly.

The Divisions, Classes, and Subclasses of Vascular Plants

The system of classification of the higher taxa of vascular plants is new, and it includes revisions of taxa at all levels from division to family. The changes from the classical system involve problems of both *classification* and *nomenclature.*

CLASSIFICATION

Because of the accumulation of information accentuating the differences and the gaps between components of certain classical groupings, the major taxa of vascular plants are reclassified as follows.

The Divisions of the Spore Plants. The four major groups of vascular spore plants are connected more clearly than other high ranking taxa by series of fossil plants. The fern, *Psilotum,* horsetail, and club moss lines of development are plausible series radiating from the Silurian and Devonian Rhyniophyta and culminating in or near the living plants of each series.

This raises a question of principle in classification—are the limits of taxa to be set in accordance with the past as well as the present? Presumably the divergence of any two taxa from a common ancestor may have occurred as shown by the circles in Figure P-1, each successive pair representing the state of isolation of the two taxa in a later geological unit of time. Classification of the taxa indicated by the circles could not be the same for any two long periods. At first they were a single unit, later a less homogeneous unit, then two units difficult to segregate, finally two clearly different units. Presumably this is the course of history of most pairs or groups of taxa. Consequently, their limits must have

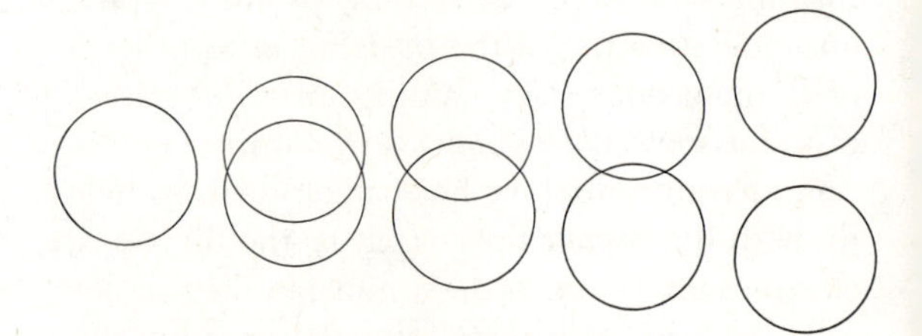

Figure P-1. Diagram illustrating the gradual divergence of taxa from a common ancestor.

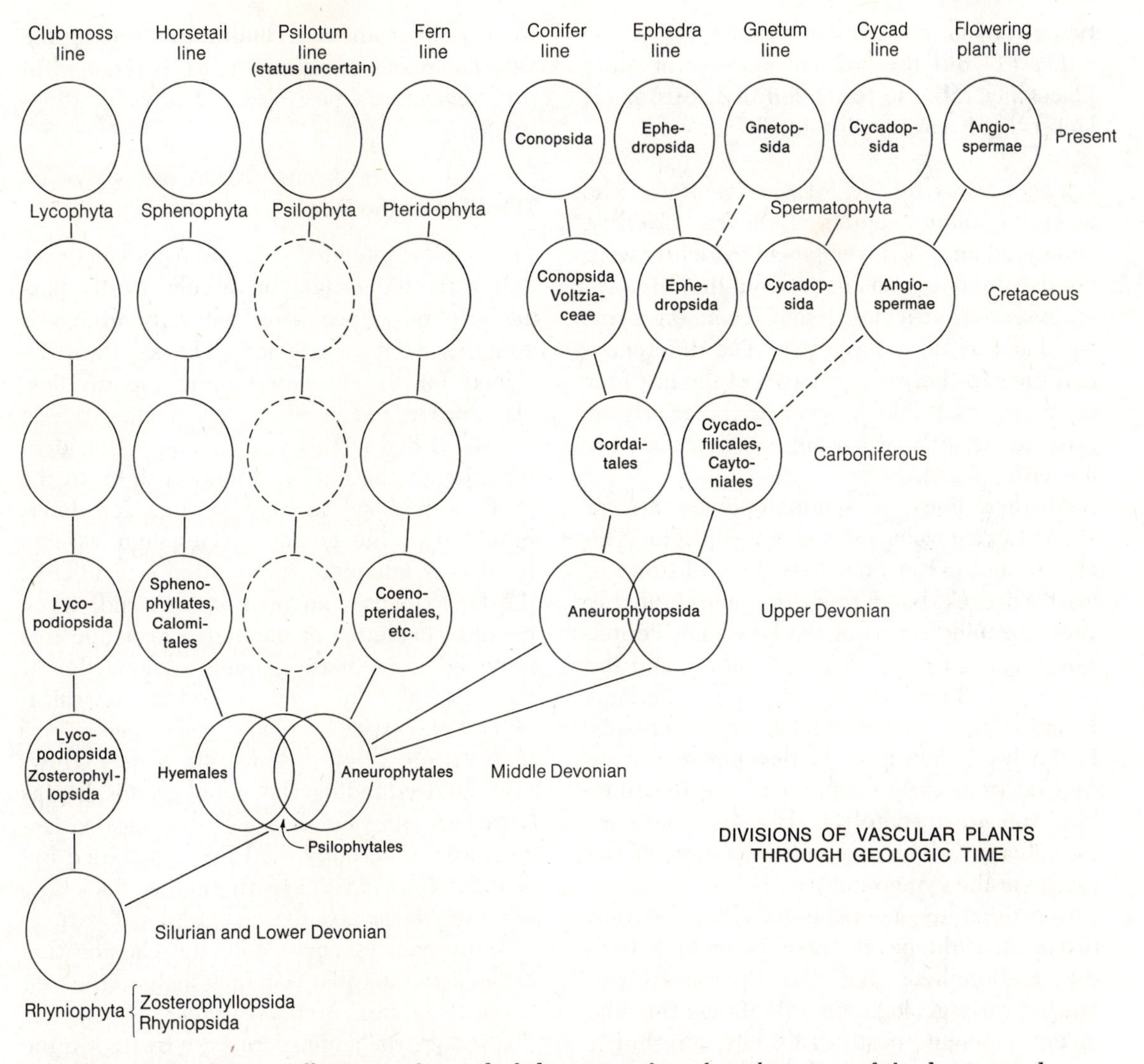

Figure P-2. Diagram illustrating the gradual divergence through geologic time of the divisions and classes of the vascular plants. (The nature and status of the Psilophyta are in dispute.)

changed with geologic time, and a classification of present plants must represent them as they are now. Taxa known only from fossils (as for example the orders Hyeniales, Lepidodendrales, and Psilophytales) belonged to the temporary larger units (divisions and classes) of past geologic periods. Fossil records may indicate, as with the spore plants, the history and relationships of the living taxa, but they do not bring about the amalgamation or determine the limits of the taxa formed by their modern descendants. The accompanying diagram indicates the known spore plants and their possible classification from Silurian and early Devonian time to the present. According to this interpretation the ancient division Rhyniophyta has become divided into five modern divisions.

In this instance connecting series of fossils are known; in others they are not, probably being rare or nonexistent because the plants either (1) did not grow in marshes or other places favorable to formation of fossils or (2) lacked hard parts readily fossilized.

The Classes of the Spermatophytes. The flowering plant, conifer, *Ephedra, Gnetum,* and cycad lines of development evidently were not derived one from another, but they are separate series from a more remote common ancestry than has been supposed. The differences and the gaps between any two of the last four (gymnosperm) lines appear to be nearly as great as those between any of these and the flowering plants.

All five lines of spermatophytes are of ancient (Paleozoic and Mesozoic) origin. The conifer and cycad lines have been distinct at least since Carboniferous time and probably since the middle part of the Devonian Period. The origins of the *Ephedra, Gnetum,* and the angiosperm lines during Mesozoic or perhaps Paleozoic time are not well known from fossils. Probably all five lines of development were derived from early members of the fern line, but data are incomplete. However, new information is improving understanding of the origins of the gymnosperms.

According to one point of view, the fern division should be enlarged to include both the Pteridophyta and the Spermatophyta. During early geologic time doubtless this line of development constituted a division, but it has divided into two modern units as significant and clearly distinct as each of the three other entire surviving lines of development from the Rhyniophyta.

NOMENCLATURE

In general the names of the taxa have been adjusted to correspond with Recommendation 16A of the International Code of Botanical Nomenclature adopted at Stockholm in 1950 (revision of 1972). However, a few names seem to warrant conservation in the form of their usage in the older literature. Among the vascular plants "Angiospermae" is retained in preference to the possible alternative "Anthopsida."

The Plan of the Book

This book is intended to be used for classification of the higher taxa of vascular plants, particularly those occurring naturally in North America north of Mexico. The keys are designed for use in determining the division, class, order, and family of any plant native to or established in this vast region. After identification of the family, attention shifts to the local or regional manual or flora for determination of the genus, species, and variety. Insofar as adequate information is available, the book focuses on relationships and, when possible, evolution of the higher taxa, and it is intended to facilitate gaining a knowledge of the principles involved in general classification of the vascular plants and especially of the intricate array of flowering plants. However, understanding the plant groups begins with becoming familiar with the local or regional flora, accomplished by a combined use of *Plant Classification* with another book covering the lesser taxa of a geographical area.

Many courses cover only the classification of the flowering plants, which include the vast majority of taxa of the vascular plants. For this reason, the angiosperms are treated in the earlier chapters, 1 to 17, and the other vascular plants appear in the later chapters, 18 to 23. The introductory material for classification of the other vascular plants covers the earliest vascular plants and the beginnings of their evolution into the modern higher taxa (see the inside front cover). This precedes the treatment of the taxa other than those of the Angiospermae. The flowering plants may be studied nearly as well without this introduction, because their origin from extinct taxa is not well understood at present. Aside from a tie, at least a vague one, with the extinct seed ferns

(Cycadofilicales) and the related Caytoniales, there is little of known evolutionary history to be discussed. In courses covering all the vascular plants or all but the flowering plants, the key is the material on the origin and early evolution of the other vascular plants. Although knowledge of these taxa is far from complete, some plausible developmental series have become known during recent years, as indicated in the chart in that section.

General biological or botanical principles are discussed in many parts of the book. The subject of evolution and its application to classification of plants receives special attention in Chapters 13 and 14; the history of development of botanical classification systems for the flowering plants in particular is in Chapters 15 and 16; preparation of herbarium specimens and their significance is discussed in Chapter 12. All these subjects, as well as the introduction to roots, stems, and leaves (Chapters 1 to 3), appear under the flowering plants so that they may be studied by beginning students without much reference to the other vascular plants, but they are applicable to the other classes and divisions studied mostly by more advanced students.

Chapters 24 and 25 deal with relationships of plants to the environment and their aggregation into the formations studied as vegetation types in ecology on the one hand, and more especially to their occurrence as parts of natural floras on the other. Vegetation types follow the climatic belts around the world at given latitudes, but the similar formations occurring on different land masses are similar only in being able to cope with like physical environments and in the forms taken by the dominant species. The included taxa, such as genera, species, and varieties, commonly are wholly different at the same latitudes on different continents, and they were derived from different continental or regional floras. Floristically they are different, even though they constitute ecologically similar forests, woodlands, brushlands, or grasslands. For the taxonomist, the floristic composition is of primary importance. Hence the plants of the groups of ecosystems are classified into natural **floras** composed of **floristic associations** developed in different climatic regions. For the most part, the floristic associations correspond with ecological formations occurring around the world, but only with those parts of each formation colonized by plants of a particular flora. Thus, the chaparral of California is not associated with the similar plant formation in the Mediterranean region or with that in Cape Province of South Africa, in which not even one species is the same and most of the genera and even many families are different.

L.B.

Acknowledgments

The late Mr. Jerome D. Laudermilk of Claremont prepared most of the more detailed drawings, as indicated by initials. Before making many of these drawings Mr. Laudermilk made minute and careful dissections of original plant material. The writer wishes to express deep appreciation for Mr. Laudermilk's interest in the work, his originality, and his painstaking accuracy. It is regretted that Mr. Laudermilk's death in January, 1956, prevented him from seeing his illustrations in print. He clung to life for several months with this object in view.

A few other drawings were prepared by Lucretia Breazeale Hamilton, the others by the author as indicated by initials.

The author is grateful for use of the many illustrations acknowledged in the legends.

An early version of the manuscript prepared in mimeographed form for the first edition was sent to a number of botanists for criticism and suggestions, and a later version of the chapter on evolution was circulated among several botanists and zoölogists. The writer is grateful to each person in the following list for suggestions and notation of errors and to several of them for keen and penetrating criticisms of the original manuscript. Several of those listed are deceased. Special gratitude for thorough reading and deep insight into the problems involved in writing the book is due to the late Dr. Sherff and for several chapters to Dr. Grant.

The criticisms and suggestions for the first edition of Dr. Reed C. Rollins, Asa Gray Professor of Botany and Emeritus Director of the Gray Herbarium of Harvard University, are much appreciated. He read the preliminary version for the author and the final manuscript for the publisher.

Dr. Yost U. Amrein, Pomona College
Dr. Clair A. Brown, Louisiana State University
Dr. R. W. Darland, University of Minnesota, Duluth Branch
Dr. Ray J. Davis, Idaho State College, Pocatello
Dr. Graham DuShane, Stanford University
Prof. Joseph Ewan, Tulane University
Mr. Mason G. Fenwick, University of Minnesota, Duluth Branch
Dr. Helen M. Gilkey, Oregon State College
Dr. George J. Goodman, University of Oklahoma
Dr. Verne Grant, University of Texas, Austin
Dr. Dorothy R. Harvey, San Diego State College, California
Dr. William A. Hilton, Pomona College

Dr. C. Leo Hitchcock,
University of Washington
Dr. A. H. Holmgren,
Utah State Agricultural College
Dr. Olga Lakela,
University of Minnesota, Duluth,
and University of South Florida
Dr. Miles D. McCarthy,
California State University, Fullerton
Dr. Ernst Mayr, Harvard University
Dr. Alden H. Miller,
University of California, Berkeley
Mr. C. V. Morton,
United States National Herbarium,
Smithsonian Institution
Dr. Gerald Ownbey, University of Minnesota
Dr. Marion Ownbey,
Washington State College
Dr. Reed C. Rollins,
Gray Herbarium, Harvard University
Dr. Dwight L. Ryerson, Pomona College
Dr. Earl E. Sherff,
Chicago Natural History Museum
Dr. A. C. Smith, University of Hawaii
Dr. Albert N. Steward, Oregon State College
Dr. G. Ledyard Stebbins, Jr.,
University of California, Davis
Dr. Jason R. Swallen,
United States National Herbarium,
Smithsonian Institution
Dr. Victor C. Twitty, Stanford University
Dr. Edgar T. Wherry,
University of Pennsylvania
Dr. Robert E. Woodson, Jr.,
Washington University
Dr. Truman G. Yuncker, De Pauw University

Contents

Gymnosperms

Ferns

Psilophytes and Horsetails

APPENDIX

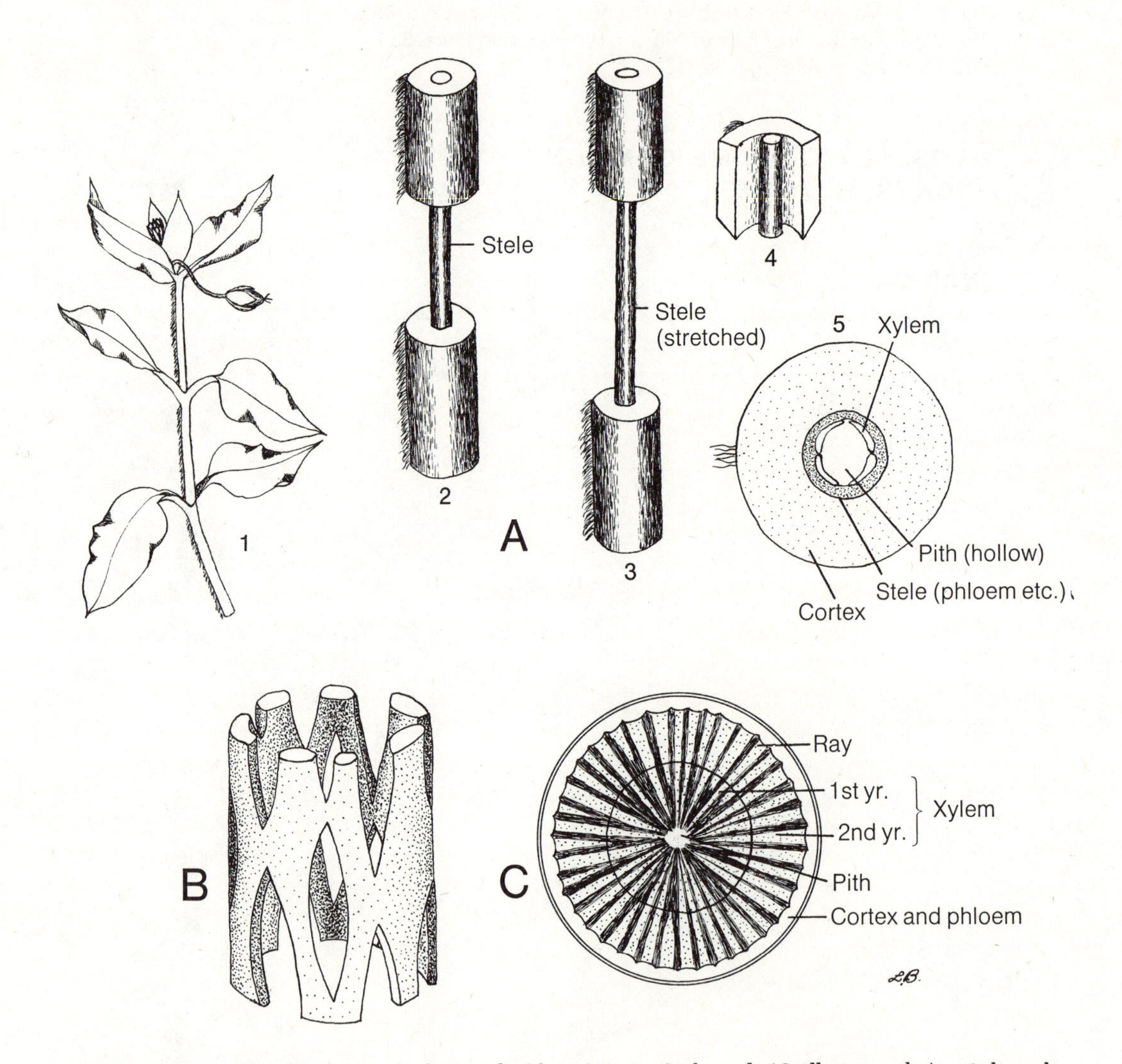

Figure Int.-1. Vascular tissues (xylem and phloem): **A** chickweed (*Stellaria media*), **1** branch with leaves, a fruit, and a flower, **2** portion of the stem with all but the vascular tissue (stele) removed, **3** the same with the elastic vascular tissues stretched by pulling, **4** portion of the stem with the outer tissues split away, exposing the stele, **5** cross section of the stem; **B** xylem (wood) remaining after decay of the fleshy tissues of a cholla (cylindroidal-stemmed cactus, *Opuntia Parryi*), showing the network formed by the "vascular bundles"; **C** diagrammatic cross section of the woody stem of a two-year-old branch of the coast live oak (*Quercus agrifolia*).

Introduction

In the 1880's the Apaches were on the warpath. No white man in the vast network of mountains and high plains of Arizona and New Mexico was safe. Even the lowland cities like Phoenix and Tucson were under military protection.

Two botanists, Mr. and Mrs. J. G. Lemmon of Berkeley, California, wished to collect plants in the Huachuca Mountains in the heart of the Apache country. They rode southeastward from Tucson a day's journey, expecting to spend the night at the ranch of some friends, but the house had been burned and the settlers scalped. And so the Lemmons camped overnight, and the next day they went on toward the Huachuca Mountains, planning to spend the night with friends at a ranch farther on. But these friends had been scalped. And so the Lemmons camped again, and the next day they went on toward the Huachuca Mountains.

Soon they met a small band of Apaches in war paint. The Lemmons were stopped and searched; the plant presses were opened and the specimens examined. Finally the chief came up with the main band of Apaches. He inspected the plant presses gravely, then tapped his forehead. And so the Lemmons went on to the Huachuca Mountains.

If plant collecting is a form of insanity, it is a mild and pleasant one associated with a love of the out-of-doors, fascination with the infinite variety of patterns of plant characteristics, and the challenge of problems arising from their study. The beginning is like solving puzzles—determining the identity of each unknown plant by a process of elimination. Choice after choice is made in a key between alternative characteristics, only one of which is present in the plant in question. Finally, if every decision is correct, the identity of the plant is learned.

As the vocabulary describing the characteristics of a plant group becomes better known, as *the process of identification* becomes easier, and as the *preparation and preservation of specimens* result in an orderly collection, the chief center of interest shifts from these primarily mechanical processes to *the basis for classification.* Many plants are seen to have characters in common yet to differ in varying degrees from each other, and new problems arising from the complexities of classification gradually replace the gamelike process of identification. Ultimately classification of plants results in yet another field of interest—the *association of species* in the natural vegetation of the region.

Units of the Plant Kingdom

Plant identification is an orderly process resulting in assignment of each individual to a descending series of groups of related plants, as judged by characteristics in common. A rose is placed first among the plants with seeds, then among the seed plants having flowers, and so on through subordinate group after group until it is associated finally with its nearest relatives—the other kinds of roses.

The Plant Kingdom is made up of classically recognized major groups as follows:

The Vascular Plants (or Tracheophytes). These include the flowering plants, as illustrated by many figures in especially Chapters 10 and 11; the gymnosperms (e.g., pines, firs, etc.), as illustrated in Chapters 18 to 20; and the ferns, horsetails, and club mosses (the producing plants), as illustrated in Chapters 21 to 23. Parts are illustrated in Chapters 2 to 9.

The Mosses and Liverworts. The mosses in particular are common chiefly in moist places, including the shady sides of rocks and trees, and the liverworts occur mostly in similar situations, though often on moist soil.

The Algae and Fungi. The algae include the seaweeds and the common essentially microscopic algae or pond scums, which to the naked eye may resemble green hairs. Examples of the fungi are mushrooms and bread mold.

This book is concerned with only the vascular plants, that is, the flowering plants, gymnosperms, and pteridophytes. These are the large plants of the Earth. Their bodies are differentiated into stems, leaves, and roots, and internally these organs contain **vascular** (conducting) **tissues** known technically as xylem and phloem. In older stems of trees the *xylem and associated tissues* occur as a thick inner cylinder of wood surrounding a small cylinder of pith; the *phloem and associated tissues* form a thin outer hollow cylinder of bark surrounding the wood. In young stems and in older plants which are not markedly woody these tissues form an internal network within the stem, and the strands of the network are spoken of as "vascular bundles." Vascular traces from the stem lead to the veins of the leaves. Young roots contain a solid core of xylem and phloem. The vascular plants are the only ones in the Plant Kingdom having xylem and phloem.

The vascular plants include the five **divisions** listed below:

Spermatophyta or **Seed Plants.** Literally "seed plants."

Pteridophyta or **Ferns.** Literally "fern plants."

Sphenophyta or **Horsetails.** The technical name means "wedge plants." It is based upon *Sphenophyllum* ("wedge-leaf"), an extinct plant known only from fossils.

Division	*Class*	*English Name*	*Group Name*
Spermatophyta	Angiospermae (covered seeds)		
	Subclass		
	Dicotyledoneae (2 seed leaves)	Diocotyledons	Flowering Plants
	Monocotyledoneae (1 seed leaf)	Monocotyledons	Flowering Plants
	Conopsida (cone)	Conifers	Gymnosperms
	Ephedropsida (*Ephedra*)	*Ephedra*	Gymnosperms
	Gnetopsida (*Gnetum*)	*Gnetum, Welwitschia*	Gymnosperms
	Cycadopsida (cycad)	Cycads	Gymnosperms
Pteridophyta	Pteropsida (feather)	Ferns	Pteridophytes
Sphenophyta	Sphenopsida (wedge)	Horsetails	Pteridophytes
Lycophyta	Lycopsida (wolf)	Club mosses	Pteridophytes
Psilophyta	Psilopsida (bareness)	*Psilotum,* etc.	Pteridophytes

NOTE: The ending *-opsida* for classes means "with the appearance of" the object or quality in parentheses.

Lycophyta or **Club Mosses.** The technical name means "wolf plant." It is based upon *Lycopodium,* club moss, or literally "wolf-foot," derived from a fancied resemblance to a wolf's foot or track.

Psilophyta. The name is derived from that of *Psilotum,* a largely tropical plant with no leaves, i.e., with bare branches.

Each division of the vascular plants is composed of subordinate groups (**taxa,** singular **taxon**). Arranged in order of rank, these are as follows: class, order, family, genus, species, and variety (see table at bottom of p. 2).

The divisions of vascular plants are composed of the following classes and (in the flowering plants) subclasses:

The Angiospermae are composed of two subclasses: (1) the Dicotyledoneae (dicotyledons or dicots), the larger group including, for example, the broad-leaved trees, roses, peas, sunflowers, and buttercups; and (2) the Monocotyledoneae (monocotyledons or monocots), including, for example, the grasses, lilies, orchids, irises, palms, and cannas. The Dicotyledoneae include forty-nine **orders.** One of these is the Rosales (rose order), which includes a number of **families,** among them the rose, pea, saxifrage, stonecrop, and witch-hazel families. A family is composed of one or more **genera** (singular **genus**—from the same Latin origin as general). In the rose family (Rosaceae) an example is the genus *Rosa,* which includes all the roses. Each genus is made up of **species** (singular as well as plural—from the same Latin origin as specific), such as *Rosa setigera,* climbing rose. The basic **name** of the plant is a combination of a word designating the genus with a word designating the species. Species are made up of **varieties.*** The name of one of these repeats the designation of the species, e.g., *Rosa setigera* var. *setigera,* but often this is omitted because it is considered to be understood. The other varieties have epithets (adjectives used as nouns) similar to those of species, e.g., *Rosa setigera* var. *tomentosa.* The botanical variety is not to be confused with the horticultural "variety," which is not a taxon but a minor variant or hybrid of economic or aesthetic significance now officially a cultivar.

The term "subspecies" is used by some authors in exactly the same sense as variety. Others employ it as a designation for a group of higher rank than variety but lower than species. The majority of botanists have not used the term, but it is in common usage in zoölogy. The arguments for or against its use in botany are relatively complex, and the uniform use of one category or the other would outweigh the advantages of either. A taxon of lower rank than variety is forma. It is used by some authors but not by others.

To summarize, the list of categories (taxa) is as follows:

Taxon	*Example*	*English Name*
Division	Spermatophyta	Seed plants
Class	Angiospermae	Flowering plants
Subclass	Dicotyledoneae	Dicots
Order	Rosales	Rose order
Family	Rosaceae	Rose family
Genus	Rosa	Rose
Species	*setigera	Wild climbing rose
Variety	tomentosa	A special climbing rose

Pronunciation of scientific names may be either Latinized or Anglicized. Few American botanists use a "Latin" pronunciation. Unfortunately, agreement upon English pronunciation is by no means complete.

Placing of the accent in the names of orders and families follows a simple rule. It is on the first syllable of the termination, "-ales" (pronounced ā′lēs) for order and "-aceae" (pronounced ā′sē-ē) for family. Thus it is Rosā′les, Ranā′les, Rhamnā′les, etc., and Rosā′ceae, Ranunculā′ceae, Rhamnā′ceae, etc.

* Variety is the singular of the English word, but, strictly speaking, according to the International Code of Botanical Nomenclature, the botanical word is not *variety* but varietas.

* Specific "names" (epithets) formed from those of persons or deities or the adopted names of genera or aboriginal names used for species may be capitalized or, according to preference, all specific epithets may be uncapitalized.

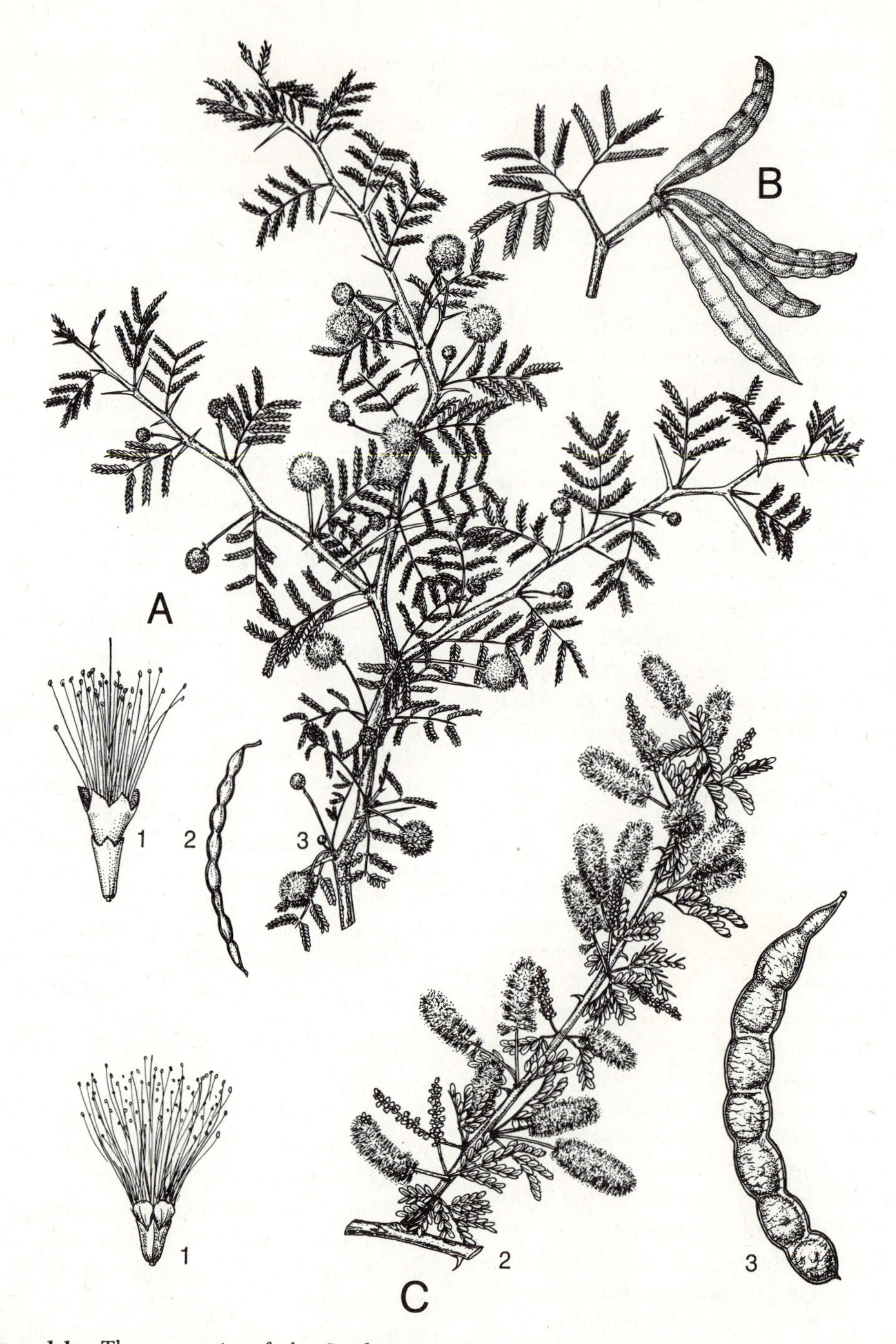

Figure 1-1. Thorny acacias of the Southwestern Deserts: **A** white thorn (*Acacia constricta*), **1** flower, **2** fruit (pod or legume), **3** flowering branch showing the pairs of spines at the nodes (joints) of the stem; **B** huisache (*Acacia Farnesiana*), fruiting branch; **C** cat-claw (*Acacia Greggii*), **1** flower, **2** flowering branch showing the prickles irregularly arranged along the stem, **3** fruit. By permission from *Trees and Shrubs of the Southwestern Deserts*, LYMAN BENSON and ROBERT A. DARROW, Tucson: University of Arizona Press, copyright 1979.

1

Identification of Vascular Plants

Keys and Their Use

Unknown plants are identified by means of keys. Fundamentally these are like outlines, but they are arranged so that the taxa of which the plant is a member may be found by a process of elimination. This is much like the parlor game which begins with the question, "Is it in the vegetable, animal, or mineral kingdom?" and then proceeds through a series of other questions to determine the identity of an object by eliminating first one possibility then another. The following is an illustration of a simple key to the species of a genus.

A Key to the Species of Acacia Occurring in the Southwestern Deserts of the United States

1. Cluster of flowers cylindroidal or ovoid (egg-shaped).
    2. Flower cluster cylindroidal; seeds broader than long.
        3. Stem bearing prickles similar to rose prickles or a cat's claws.

            1. *Acacia Greggii.*

        3. Stem not bearing prickles.

            2. *Acacia millefolia.*

    2. Flower cluster ovoid (egg-shaped); seeds longer than broad.

        3. *Acacia angustissima.*

1. Cluster of flowers spheroidal.
    2'. Pod flattened, splitting immediately at maturity, constricted between the seeds.

    4. *Acacia constricta.*

    2'. Pod cylindroidal, splitting tardily (long after maturity), not constricted between the seeds.

    5. *Acacia Farnesiana.*

The key above is used as illustrated below. First suppose that a botanist is visiting the Baboquivari Mountains in southern Arizona and that he finds an *Acacia* with remarkable sweet-smelling flowers. He is impressed not only by the balls of bright golden yellow flowers but also by the smooth, thick, rodlike pods, which evidently have persisted for several months without splitting open. Identification by means of the key involves use of these characters, as follows:

1. The two "lines" (**leads**) of the key beginning at the extreme left margin and numbered "1" are opposed to each other. The plant "fits" one but not the other. Since the flowers are in "balls," the plant must match the lower lead.

2. The two subordinate leads under the lower "1" are set over a little from the left margin and numbered "2'." They are opposed to each other. Since the pods are "rodlike" and "smooth" *instead of flat and constricted* and since they "evidently have persisted for several months without splitting open," the plant must be the one described by the lower number "2'," that is, *Acacia Farnesiana*, or, translated into English, the Farnesian acacia. It is known also as huisache, and it is cultivated in southern Europe to yield "cassie flowers" used for manufacturing perfumes.

The botanist leaves the mountains and follows one of the "washes" in the desert where he encounters another *Acacia*. This one has paler flowers in clusters about the size and shape of his little finger. He leans over to smell them, and straightening up, he tears his shirt and scratches his back and shoulders on the branches above. At the same time he causes several pods to release their seeds, some of which fall into his shirt pocket. As he takes these out of his pocket, he is impressed by their shape because their breadth exceeds their length. Which species is it?

In the key to the species of *Acacia*, recognition of the genus was taken for granted. Identification of an unknown plant requires use of a key to the divisions of the Plant Kingdom, then of one to the classes of the proper division and of keys leading on to the order, family, genus, and species and often the variety, as illustrated on page 99. Before these series of keys can be used effectively, it is necessary to acquire an understanding of the characteristics of at least the major plant group to be dealt with, e.g., the flowering plants, and of the words used as tools in identification.

Key to the Divisions of Vascular Plants

NOTE: The Divisions **Pteridophyta, Psilophyta, Sphenophyta,** and **Lycophyta** discussed in the section of the text on pp. 655–712.

1. Seeds present; plant producing flowers or woody or sometimes pulpy or scaly seed cones or rarely solitary large seeds; plant not producing spores visible to the naked eye, the spores enclosed within the young flower or cone, the sporelike pollen normally *not* germinating on moist ground.

   Division ***SPERMATOPHYTA***, cf. key below.

1. Seeds none; plant never producing flowers or woody seed cones, sometimes bearing soft cones with spores but not seeds; plant reproducing by spores which germinate ordinarily on moist ground.
   2. Plant with roots (these rarely lacking in aquatic plants) and leaves (each leaf with at least a midrib and often with other veins).
      3. Stems commonly but not always below ground, leaves large and usually divided or compound—1 to only a few or up to about 50 on a plant; sporangia never in conelike strobili, produced usually in dense clusters on the backs of the leaves (sometimes near or along the margins, but clearly dorsal), sometimes on specialized nongreen leaves or leaf-segments.

         Division ***PTERIDOPHYTA***, page 660.

      3. Stems largely above ground; leaves small and scalelike, numerous, either thickly clothing the stems and giving the plant a mosslike appearance or coalescent in a circle at each joint (node) of the stem; sporangia produced either on the upper side of an ordinary or sometimes a specialized yellowish or brownish scalelike leaf or around the inner margin of an umbrellalike scale, the specialized leaves (sporophylls) often in more or less conelike structures (strobili).
         4. Stems markedly jointed, each joint capable of being pulled apart and put back together, the coalescent leaves several to many in a circle around each joint; sporangia several on the inner side of each horizontal umbrellalike sporophyll; sporophylls always in a strobilus at the apex of the stem.

            Division ***SPHENOPHYTA***, page 691.

         4. Stems not jointed, the leaves clothing the stem thickly like shingles on a roof; sporangia produced singly on the upper sides of ordinary or specialized but not umbrellalike leaves; sporophylls often in a strobilus at the apex of the stem.

            Division ***LYCOPHYTA***, page 701.

   2. Plant with no roots or true leaves (the North American genus with scales on the stem, the

scales without midribs); stem forking repeatedly (dichotomous, that is, with two usually equal divisions at each point of forking); plants terrestrial, *not* aquatic. Chapters 21, pp. 679–680, and 22, pp. 686–689.

Division ***PSILOPHYTA***, page 684.

Key to the Classes of the Spermatophyta

NOTE: The class **Angiospermae** is discussed in the text in the section dealing with **Flowering Plants**, the other four classes in the section dealing with **Gymnosperms.**

1. Flowers present, including stamens or pistils or both and usually sepals and petals; seeds enclosed in an ovary, which (with the enclosed seeds and sometimes associated structures) becomes a fruit at maturity. (**Flowering Plants.**)

Class ***ANGIOSPERMAE***, pages 95, 99.*

1. Flowers none; seeds not enclosed in an ovary, the ovules and seeds naked, the seeds wedged between the scales of a woody or sometimes pulpy or scaly cone (rarely enclosed in the scales) or sometimes solitary or in pairs or on the margins of reduced specialized leaves; pollen produced in soft cones. (**Gymnosperms.**)

2. Leaves simple.

3. Trees or sometimes shrubs; resin (pitch) or sometimes mucilage produced in the stems and leaves; micropyle of the ovule *not* surrounded by a tubular outgrowth; leaves alternate or, if opposite or whorled, then green and usually scalelike; plants usually monoecious but rarely dioecious.

Class ***CONOPSIDA***, page 616.*

3. Shrubs, woody vines, or bulbous plants; resin none; micropyle surrounded by an elongated tube, this resembling the style of a flowering plant pistil; leaves opposite or in whorls of 3, if scalelike, then not green; plants nearly always dioecious.

4. Leaves very small, without chlorophyll, brown or tan, dry, scalelike, opposite or in whorls of 3; shrubs, the stems branched diffusely; ovules solitary or in pairs in the strobilus.

Class ***EPHEDROPSIDA***, page 629.*

4. Leaves of moderate to large size, with chlorophyll, not scalelike, opposite (sometimes frayed into long ribbons); ovules several to each strobilus.

Class ***GNETOPSIDA***, page 629.*

2. Leaves pinnate or bipinnate; small trees or sometimes shrubs or bulbous plants.

Class ***CYCADOPSIDA***, page 631.*

* The page reference after each class refers to the place where the next key is to be found.

Check List of Small Spermatophytes and Pteridophytes Readily Confused with Nonvascular Plants
(for reference purposes)

The following list should be consulted if the status of a specimen as a vascular plant is in doubt. The plant bodies may be small and undifferentiated or the sexual reproductive parts obscure or rarely formed. The Eighth Edition of *Gray's Manual of Botany* by M. L. Fernald (1950) keys out northeastern dicotyledons (p. lii) and monocotyledons (p. xxxiv) having an obvious flowering plant body but with the flowers undeveloped or abortive.

The plant must correspond with every character in the combination. If the specimen does not coincide with one character, even the first listed, the combination may be eliminated from further consideration. The characters in **boldface type** should be checked first to determine whether the combination requires further inspection.

1. **Plant attached to rocks** (stones) **in running water, the plant body in ribbonlike segments** (resembling a seaweed, moss, or liverwort) thalluslike, without well-differentiated and evident stems or leaves, small, flowering only occasionally. *Podostemonaceae* (riverweed family).

Order **Podostemonales**, DICOTYLEDONEAE, page 313.

2. **Plant floating, the plant body thalluslike, without well-differentiated and apparent stems and leaves, small, circular or elongated,** budding asexually, the individuals sometimes remaining attached to each other. *Lemnaceae* (duckweed family).

Order **Arales**, MONOCOTYLEDONEAE, page 413.

3. **Floating** (**or sometimes stranded**) **mosslike plants about 1–3 cm. long,** often reddish; the slender branching stems with no roots; bearing small, deeply 2-lobed leaves arranged in two rows; sporangia usually not present but occasionally appearing in abundance, enclosed in soft sporocarps (closed sacs) about 1–5 mm. in diameter, these borne on the ends of the leaf lobes. *Azolla, Salviniaceae* (floating fern family).

Order **Filicales**, PTERIDOPHYTA, page 663.

4. Plant as in number 3 above but the plant body and the sporocarps larger and not red; leaves 1–1.5 cm. long, simple. *Salvinia, Salviniaceae* (floating fern family).

Order **Filicales**, PTERIDOPHYTA, page 663.

5. **Plants appearing to be four-leaf clovers, growing in mud or under water;** each leaf with a long petiole (or the petiole shorter when not submerged) and four cloverlike leaflets; stems rhizomatous, often elongated and branched; sporangia in hard subterranean sporocarps about 2–4 mm. long. *Marsilea, Marsileaceae* (*Marsilea* family).

Order **Filicales**, PTERIDOPHYTA, page 663.

6. **Plants resembling young grass, growing in mud of vernal pools;** leaves about 2.5 cm. long, slender, needlelike; stems rhizomatous, short; **sporangia in hard subterranean sporocarps about 2 mm. in diameter.** *Pilularia* (pillwort), *Marsileaceae* (*Marsilea* family).

Order **Filicales**, PTERIDOPHYTA, page 663.

7. **Plant resembling young grass (but the leaves somewhat fleshy), growing in mud or under water;** stem minute, not visible (without dissection) to the naked eye, enclosed in a basal tuft of elongate leaves; **sporangia imbedded in the swollen subterranean bases of the leaves,** the spores of some sporangia minute and powdery, of others larger and granular. *Isoëtes* (quillwort), *Isoëtaceae* (quillwort family).

Order **Isoëtales**, LYCOPHYTA, page 703.

8. **Plant mosslike; leaves small, scalelike, clothing the stem densely, overlapping each other like shingles;** sporangia spheroidal, appearing in the leaf axils on some branches (or on some plants), the spores of two sizes. *Selaginella, Selaginellaceae* (*Selaginella* family).

Order **Selaginellales**, LYCOPHYTA, page 702.

Regional and Local Floras and Manuals for Identification of Genera, Species, and Varieties

The following manuals and floras are suggested for identification of the lower taxa of vascular plants after the division, class, subclass, order, and family have been determined by the keys in this book. The local manuals listed below are mostly those relating to areas not well covered by more general works. The name of the family may be found in the index of the manual or flora.

ABRAMS, LEROY. *An Illustrated Flora of the Pacific States, Washington, Oregon, and California.* 1: 1923. 2: 1944. 3: 1951. 4: 1960 (with R. S. FERRIS). Stanford University Press, Stanford, California.

ANDERSON, J. P. *Flora of Alaska and Adjacent Parts of Canada.* 1959. Iowa State University Press, Ames. (See WELSH.)

BARKLEY, THEODORE M. *Manual of the Flowering Plants of Kansas.* 1968. Kansas State University Endowment Association, Manhattan.

BEATLEY, JANICE C. *Vascular Plants of the Nevada Test Site and Central-Southern Nevada.* 1976. National Technical Information Service, U.S. Department of Commerce, Springfield, Virginia.

BENSON, LYMAN. *The Cacti of Arizona.* Eds. 1–3, 1940, 1950, 1969. University of Arizona Press, Tucson.

———. *The Native Cacti of California.* 1969. Stanford University Press, Stanford, California.

———. *The Cacti of the United States and Canada.* In press. Stanford University Press, Stanford, California.

NOTE: *Systems of Measurement.* Except in some English-speaking countries the common system of measurement is based upon the meter, which is a little more than a yard (39.37 inches). The meter (m.) is divided into decimeters (dm.; tenths; about 4 inches), centimeters (cm.; hundredths; about ⅖ of an inch), and millimeters (mm.; thousandths; about 1/25 of an inch). This system is used universally in the natural sciences, and **a metric system rule** (obtainable in stationery, hardware, or "dime" stores) **is indispensable to use of the keys.**

In some botanical books the English System is used, and the inch may be divided into lines. A line is 1/12 inch and therefore almost exactly 2 millimeters. It is about the height of the small letter "o" on a standard pica typewriter. Other measures may be worked out by the individual, e.g., a man's clenched fist is about 10 cm. (1 dm.) broad, his little fingernail about 10 mm. (1 cm.) broad. **(See inside back cover for illustration.)**

———, and ROBERT A. DARROW. *A Manual of Southwestern Desert Trees and Shrubs.* Ed. 1. University of Arizona Biological Science Bulletin (6): 1–411. University of Arizona, Tucson. Ed. 2. (Title changed to) *The Trees and Shrubs of the Southwestern Deserts.* 1954. University of Arizona Press and University of New Mexico Press, Albuquerque. Ed. 3 (in press).

BRITTON, NATHANIAL LORD. *Manual of the Flora of the Northern States and Canada.* Ed. 3. 1907. New York City.

———, and ADDISON BROWN. *An Illustrated Flora of the Northern United States, Canada, and the British Possessions from Newfoundland to the Parallel of the Southern Boundary of Virginia, and from the Atlantic Ocean Westward to the 102d Meridian.* Ed. 2. 1–3: 1913. Charles Scribner's Sons, New York.

———. *The New Britton and Brown Illustrated Flora of the Northeastern United States and Adjacent Canada.* By H. A. Gleason, 1–3: 1952. New York Botanical Garden, New York.

BUDD, A. C. *Wild Plants of the Canadian Prairies.* 1957. Experimental Farms Service, Canadian Department of Agriculture, Swift Current. Edmond Cloutier, Queen's Printer, Ottawa.

CLOKEY, IRA W. *Flora of the Charleston Mountains, Clark County, [southern] Nevada.* 1951. University of California Press, Berkeley and Los Angeles.

Contributions toward a Flora of Nevada. Issued by the Division of Plant Exploration and Introduction, Bureau of Plant Industry, etc., U.S. Department of Agriculture, Washington, D.C. (1–): 1940– . (Issued in mimeographed form with a separate number for each plant family. Incomplete.)

CORRELL, DONOVAN S., and HELEN B. CORRELL. *Aquatic and Wetland Plants of the Southwestern United States.* 1972. Stanford University Press, Stanford, California.

———, and MARSHALL CONRING JOHNSTON. *Manual of the Vascular Plants of Texas.* 1970. Texas Research Foundation, Renner.

COULTER, JOHN M. *Botany of Western Texas. A Manual of the Phanerogams and Pteridophytes of Western Texas.* Contributions from the United States National Herbarium 2: 1891–94. U.S. Government Printing Office, Washington, D.C.

COULTER, JOHN M. and AVEN NELSON. *New Manual of Botany of the Central Rocky Mountains (Vascular Plants).* 1909. American Book Company, New York.

CRONQUIST, ARTHUR, ARTHUR H. HOLMGREN, NOEL H. HOLMGREN, JAMES L. REVEAL and PATRICIA K. HOLMGREN. *Intermountain Flora.* 1972. Vol. 1. Hafner Press, New York. 1977. Vol. 6. Columbia University Press, New York.

DAVIS, RAY J. *Flora of Idaho.* 1952. William C. Brown Company, Dubuque, Iowa.

DEAM, CHARLES C. *Flora of Indiana.* 1940. Buford Printing Company, Indianapolis. State of Indiana Department of Conservation. Distributed by the State Forester, State Library, Indianapolis, Indiana.

FASSETT, NORMAN C. *A Manual of Aquatic Plants.* 1940. McGraw-Hill Book Company. New York.

GARRETT, A. O. *The Spring Flora of the Wasatch Region.* Ed. 5. 1936. Stevens & Wallis, Salt Lake City.

GLEASON, H. A. *Plants of the Vicinity of New York.* 1935. New York Botanical Garden, New York.

———, and ARTHUR CRONQUIST. *Manual of the Vascular Plants of the Northeastern United States and Adjacent Canada.* 1963. D. Van Nostrand Company, Princeton, New Jersey.

GRAY, ASA. *Gray's New Manual of Botany. A Handbook of the Flowering Plants and Ferns of the Central and Northeastern United States and Adjacent Canada.* Seventh Edition by Benjamin Lincoln Robinson and Merritt Lyndon Fernald. 1908. American Book Company, New York.

———. *Gray's Manual of Botany. A Handbook of the Flowering Plants and Ferns of the Central and Northeastern United States and Adjacent Canada.* Eighth Edition by Merritt Lyndon Fernald. 1950. American Book Company, New York. Corrected printing, 1970. Van Nostrand Reinhold Company, Princeton, New Jersey.

HARRINGTON, H. D. *Manual of the Plants of Colorado.* 1954. Sage Books, Denver.

HENRY, JOSEPH K. *Flora of Southern British Columbia and Vancouver Island with Many References to Alaska and Northern Species.* 1915. W. J. Gage Company Ltd., Toronto.

HILLEBRAND, W. *Flora of the Hawaiian Islands.* 1888. B. Waterman Company, New York.

HITCHCOCK, ALBERT SPEAR. *Manual of the Grasses of the United States.* Ed. 2. Revised by AGNES CHASE. 1950. U.S. Government Printing Office, Washington, D.C. 1971. Reprinted by Dover Books, New York.

HITCHCOCK, C. LEO, ARTHUR CRONQUIST, MARION OWNBEY, and J. WILLIAM THOMPSON. *Vascu-*

lar Plants of the Pacific Northwest. 1–5: 1955–69. University of Washington Press, Seattle.

———, and ARTHUR CRONQUIST. *Flora of the Pacific Northwest.* 1973. University of Washington Press, Seattle.

HOLMGREN, ARTHUR H. *Handbook of the Vascular Plants of the Northern Wasatch.* 1948. Lithotype Process Company, San Francisco, California. Copyright by the author, Utah State College, Logan, Utah.

HULTÉN, ERIC. *Flora of Alaska and Yukon.* Lunds Universitets Arsskrift. N. F. Avd. 2. 37(1): 1941; 38(1): 1942; 39(1): 1943; 40(1): 1944; 41(1): 1945; 42(1): 1946; 43(1): 1947; 44(1): 1948; 45(1): 1949; 46(1): 1950. Kungl. Fysiografiska Sällskapets Handlingar. N. F. 52(1): 1941; 53(1): 1942; 54(1): 1943; 55(1): 1944; 56(1): 1945; 57(1): 1946; 58(1): 1947; 59(1): 1948; 60(1): 1949; 61(1): 1950. (Issued in two ways each in 10 parts.) 1941–50. Lund, Sweden.

———. *Flora of Alaska and Neighboring Territories.* 1968. Stanford University Press, Stanford, California.

JEPSON, WILLIS LINN. *A Flora of California.* (3 vols. issued in parts; vols. 1, 3 incomplete.) 1909– . Berkeley, California.

———. *A Manual of the Flowering Plants of California.* 1923–25. Associated Students' Store, University of California, Berkeley.

JONES, GEORGE NEVILLE. *Flora of Illinois; Containing Keys for Identification of the Flowering Plants and Ferns.* 1945. American Midland Naturalist Monograph Series; University of Notre Dame, Notre Dame, Indiana. Ed. 2. 1950. Ed. 3. 1963.

KEARNEY, THOMAS H., and ROBERT H. PEEBLES. *Flowering Plants and Ferns of Arizona.* 1942. U.S. Department of Agriculture Miscellaneous Publication (423). U.S. Government Printing Office, Washington, D.C.

———, and Collaborators. *Arizona Flora.* 1951; supplement, 1960. University of California Press, Berkeley and Los Angeles.

LAKELA, OLGA. *Flora of Northeastern Minnesota.* 1965. University of Minnesota Press, Minneapolis.

LITTLE, ELBERT L., JR. *Southwestern Trees* (Arizona and New Mexico). 1950. U.S. Department of Agriculture. Superintendent of Documents, Washington, D.C.

LONG, ROBERT W., and OLGA LAKELA. *A Flora of Tropical Florida.* 1971. University of Miami Press, Coral Gables, Florida.

LOUIS-MARIE, P. *Flore-Manuel de la Province de Quebec.* 1967. Centre de Psychologie et de Pedagogie, Montreal, Quebec.

LUNDELL, CYRUS LONGWORTH, and Collaborators. *Flora of Texas.* 1–3: 1942–70. Texas Research Foundation, Renner.

MACOUN, JOHN. *Catalog of Canadian Plants. Geological Survey of Canada.* 1883–88. Dawson Brothers, Montreal.

MARIE-VICTORIN, FRÈRE. *Flore Laurentienne; Illustrée de 22 cartes et de 2800 dessins par frère Alexandre.* 1935. Inprimerie de LaSalle, 949 rue Côté, 949, Montreal. Reprint with a Supplement by Ernest Rouleau, 1947.

MASON, HERBERT L. *A Flora of the California Marshes.* 1969. University of California Press, Berkeley.

MOHLENBROCK, ROBERT H. *The Illustrated Flora of Illinois.* 1967, 1970, 1972, 1976, 1978 (issued in parts). Southern Illinois University Press, Carbondale.

———, and JOHN W. VOIGT. *A Flora of Southern Illinois.* 1959. Southern Illinois University Press, Carbondale.

MOHR, CHARLES. *Plant Life of Alabama. An Account of the Distribution, Modes of Association, and Adaptations of the Flora of Alabama, Together with a Systematic Catalogue of the Plants Growing in the State.* Contributions from the United States National Herbarium. 6: 1901. U.S. Government Printing Office, Washington, D.C.

MOSS, E. H. *Flora of Alberta.* 1959. University of Toronto Press, Toronto, Ontario.

MUENSCHER, WALTER CONRAD. *Aquatic Plants of the United States.* 1944. Comstock Publishing Company, Inc., Cornell University, Ithaca, New York.

MUNZ, PHILIP A. *A Manual of Southern California Botany.* 1935. Distributed by J. W. Stacey Incorporated, San Francisco, California.

———. *A Flora of Southern California.* 1974. University of California Press, Berkeley.

———, and DAVID D. KECK. *A California Flora.* 1959; supplement, 1968. University of California Press, Berkeley.

PECK, MORTON E. *A Manual of the Higher Plants of Oregon.* 1941. Binfords and Mort, Portland, Oregon. 1961. Oregon State University, Corvallis.

PETERSEN, N. F. *Flora of Nebraska. A List of the Conifers and Flowering Plants of the State with Keys for Their Determination.* 1912. Published by the author through the State Printing Company, Lincoln, Nebraska.

PIPER, CHARLES VANCOUVER. *Flora of the State of Washington.* Contributions from the United

States National Herbarium. 11: 1906. U.S. Government Printing Office, Washington, D.C.

———, and R. KENT BEATTIE. *Flora of the Northwest Coast, Including the Area West of the Summit of the Cascade Mountains, from the Forty-Ninth Parallel South to the Calapooia Mountains on the South Border of Lane County, Oregon.* 1915. Published by the authors through the Press of the New Era Printing Company, Lancaster, Pennsylvania.

POLUNIN, NICHOLAS. *Botany of the Canadian Eastern Arctic. Part I, Pteridophyta and Spermatophyta.* National Museum of Canada Bulletin (92): 1940. Canada, Department of Mines and Resources, Ottawa.

PORSILD, A. ERLING. *Botany of the Southern Yukon Adjacent to the Canol Road.* 1951. Department of Mineral Resources and Development, Ottawa.

———. *Illustrated Flora of the Canadian Arctic Archipelago.* 1957. Minister of Northern Affairs and Natural Resources, Ottawa.

———. *Rocky Mountain Wildflowers* [Canada]. Undated, about 1970. National Museums of Canada, Ottawa.

PORTER, C. L. *Contributions Toward a Flora of Wyoming.* (Issued in numbered leaflets reproduced by ditto machine.) Rocky Mountain Herbarium, University of Wyoming, Laramie. (Incomplete.)

RADFORD, A. E., HARRY E. AHLES, and C. RITCHIE BELL. *Guide to the Vascular Flora of the Carolinas.* 1964. The Book Exchange, University of North Carolina, Chapel Hill.

———. *Manual of the Vascular Flora of the Carolinas.* 1968. University of North Carolina Press, Chapel Hill.

REEVES, R. G., and D. C. BAIN. *A Flora of South Central Texas.* 1946. College Station, Texas.

ROLAND, A. E. *The Flora of Nova Scotia.* Proceedings of the Nova Scotian Institute of Science. 21(3): 1944–45.

RYDBERG, P. A. *Flora of Colorado.* 1906. Agricultural Experiment Station of the Colorado College of Agriculture, Fort Collins, Colorado.

———. *Flora of the Prairies and Plains of Central North America.* 1932. New York Botanical Garden, New York.

———. *Flora of the Rocky Mountains and Adjacent Plains, Colorado, Utah, Wyoming, Idaho, Montana, Saskatchewan, Alberta, and Neighboring Parts of Nebraska, South Dakota, North Dakota, and British Columbia.* Ed. 2. 1922. New York.

ST. JOHN, HAROLD. *Flora of Southeastern Washington and of Adjacent Idaho.* 1937. Students' Book Corp., Pullman, Washington.

SCHAFFNER, J. H. *Field Manual of the Flora of Ohio and Adjacent Territories.* 1928. Columbus, Ohio.

SCOGGAN, H. J. *The Flora of Bic and the Gaspé Peninsula.* National Museum of Canada Bulletin (115): 1950. Department of Mines and Resources, Ottawa.

———. *Flora of Manitoba.* National Museum of Canada Bulletin (140): 1957. Department of Northern Affairs and Natural Resources, Ottawa.

SHINNERS, LLOYD H. *Spring Flora of the Dallas–Fort Worth Area, Texas.* 1958. Southern Methodist University, Dallas. 1972. Ed. 2. W. E. MAHLER, editor. Prestige Press, Fort Worth.

SHREVE, FORREST, and IRA L. WIGGINS. *Vegetation and Flora of the Sonoran Desert.* 1964. Stanford University Press, Stanford, California.

SMALL, JOHN KUNKEL. *Flora of the Southeastern United States, Being Descriptions of the Seed-Plants, Ferns, and Fern-Allies Growing Naturally in North Carolina, South Carolina, Georgia, Florida, Tennessee, Alabama, Mississippi, Arkansas, Louisiana, and the Indian Territory and in Oklahoma and Texas East of the 100th Meridian.* 1903. Published by the author, New York, through the New Era Printing Company, Lancaster, Pennsylvania.

———. *Manual of the Southeastern Flora, Being Descriptions of the Seed Plants Growing Naturally in Florida, Alabama, Mississippi, Eastern Louisiana, Tennessee, North Carolina, South Carolina, and Georgia.* 1935. Published by the author, New York, through the Science Press Printing Company, Lancaster, Pennsylvania.

STANDLEY, PAUL C. *Flora of Glacier National Park, Montana.* Contributions from the United States National Herbarium 22(5): 1921. U.S. Government Printing Office, Washington, D.C.

STEMEN, THOMAS RAY, and W. STANLEY MYERS. *Oklahoma Flora* (Illustrated). 1937. Harlow Publishing Corporation, Oklahoma City.

STEVENS, A. O. *Handbook of North Dakota Plants.* 1950. North Dakota Agricultural College, Fargo. North Dakota Institute for Regional Studies.

STEYERMARK, JULIAN A. *Spring Flora of Missouri.* 1940. Missouri Botanical Garden, St. Louis, and Field Museum of Natural History, Chicago, Illinois. 1963. Iowa State University Press, Ames.

STRASBAUGH, P. D., and EARL L. CORE. *Flora of West Virginia.* 1964. West Virginia University Bulletin, Morgantown.

TAYLOR, THOMAS M. C. *Pacific Northwest Ferns and Their Allies.* 1970. University of Toronto Press, Toronto, Ontario, and Buffalo, New York.

TIDESTROM, IVAR. *Flora of Utah and Nevada.* Contributions from the United States National Herbarium. 25: 1925. U.S. Government Printing Office, Washington, D.C.

———, and SISTER TERESITA KITTELL. *A Flora of Arizona and New Mexico.* 1941. Catholic University of America Press, Washington, D.C.

VIERECK, LESLIE A., and ELBERT L. LITTLE, JR. *Alaska Trees and Shrubs.* 1972. Forest Service, U.S. Department of Agriculture. Superintendent of Documents, Washington, D.C.

VOSS, EDWARD G. *Michigan Flora.* Part I. *Gymnosperms and Monocotyledons.* 1972. Cranbrook Institute of Science and University of Michigan Herbarium, Ann Arbor.

WEBER, WILLIAM A. *Handbook of Plants of the Colorado Front Range.* 1953. University of Colorado Press, Boulder.

———. *Rocky Mountain Flora.* 1967. University of Colorado Press, Boulder.

WELSH, STANLEY L. *Anderson's Flora of Alaska and Adjacent Parts of Canada.* 1974. Brigham Young University Press, Provo, Utah. (See ANDERSON.)

WIGGINS, IRA L., and JOHN HUNTER THOMAS. *A Flora of the Alaskan Arctic Slope.* 1962. University of Toronto Press, Toronto, Ontario.

WIGGINS (see SHREVE and WIGGINS).

WOOTON, E. O., and PAUL C. STANDLEY. *Flora of New Mexico.* Contributions from the United States National Herbarium. 19: 1915. U.S. Government Printing Office, Washington, D.C. 1972. Wheldon & Wesley, Ltd., Codicote, Herts, England, and Stechert-Hafner Service Agency, New York.

The flowering plants are the predominant part of the vegetation of the Earth today. They originated at least 135,000,000 years ago (before the middle of Mesozoic time, cf. p. 459), and they have been the dominant plants of at least the last 60,000,000 or 65,000,000 years (Cenozoic time). There are at least 170,000 species, and they include most of the large well-known plants. Nearly all the food plants of man and most other macroscopic animals as well as almost all textile, medicinal, ornamental, beverage, oil, and perfume plants are angiosperms. The flowering plants are important sources of fine cabinet woods, but in the North Temperate Zone they are less important for lumber than the cone-bearing trees.

As pointed out in Chapter 1, the angiosperms, like the gymnosperms, are vascular plants characterized by seeds (Spermatophyta), but they differ from the gymnosperms in the following characters:

1. Presence of flowers.
2. Enclosure of the seeds in fruits formed by development of the pistil (p. 54) of the flower.

Other highly technical distinguishing characters are seen only on prepared slides by means of a compound microscope.

Flowering Plants

Division *SPERMATOPHYTA*
Class *ANGIOSPERMAE*

Part One

THE VOCABULARY DESCRIBING FLOWERING PLANT CHARACTERISTICS

Part Two

THE PROCESS OF IDENTIFICATION

Part Three

PREPARATION AND PRESERVATION OF SPECIMENS

Part Four

THE BASIS FOR CLASSIFICATION

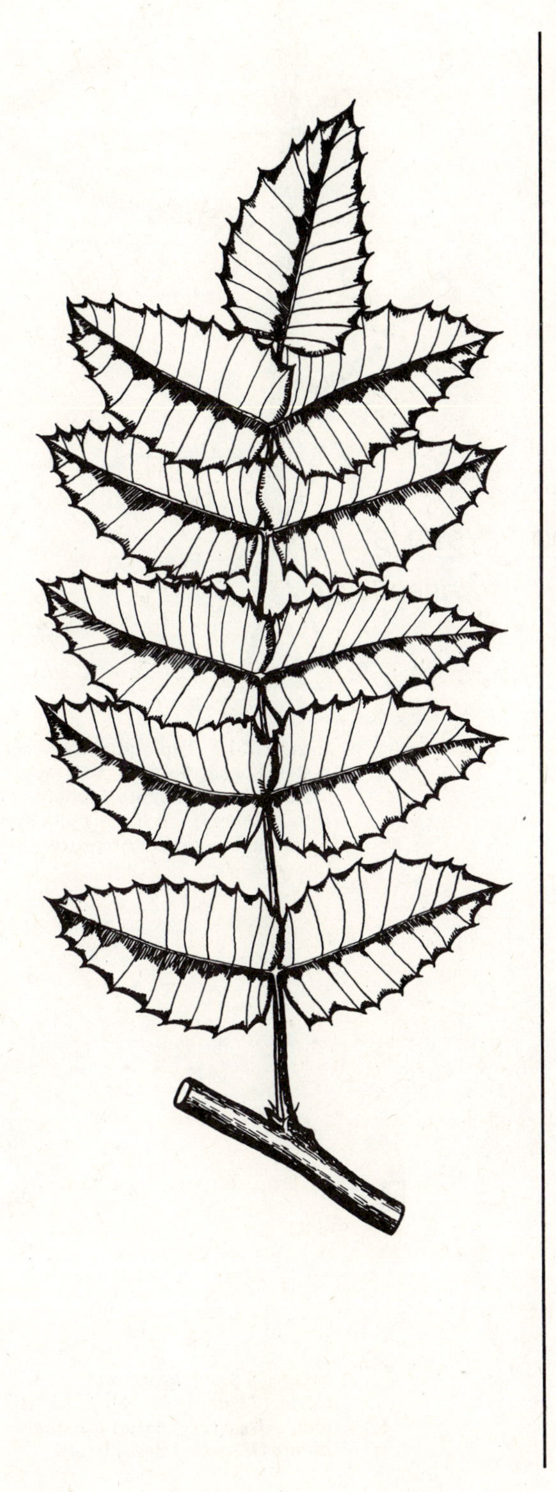

Part One

THE VOCABULARY DESCRIBING FLOWERING PLANT CHARACTERISTICS

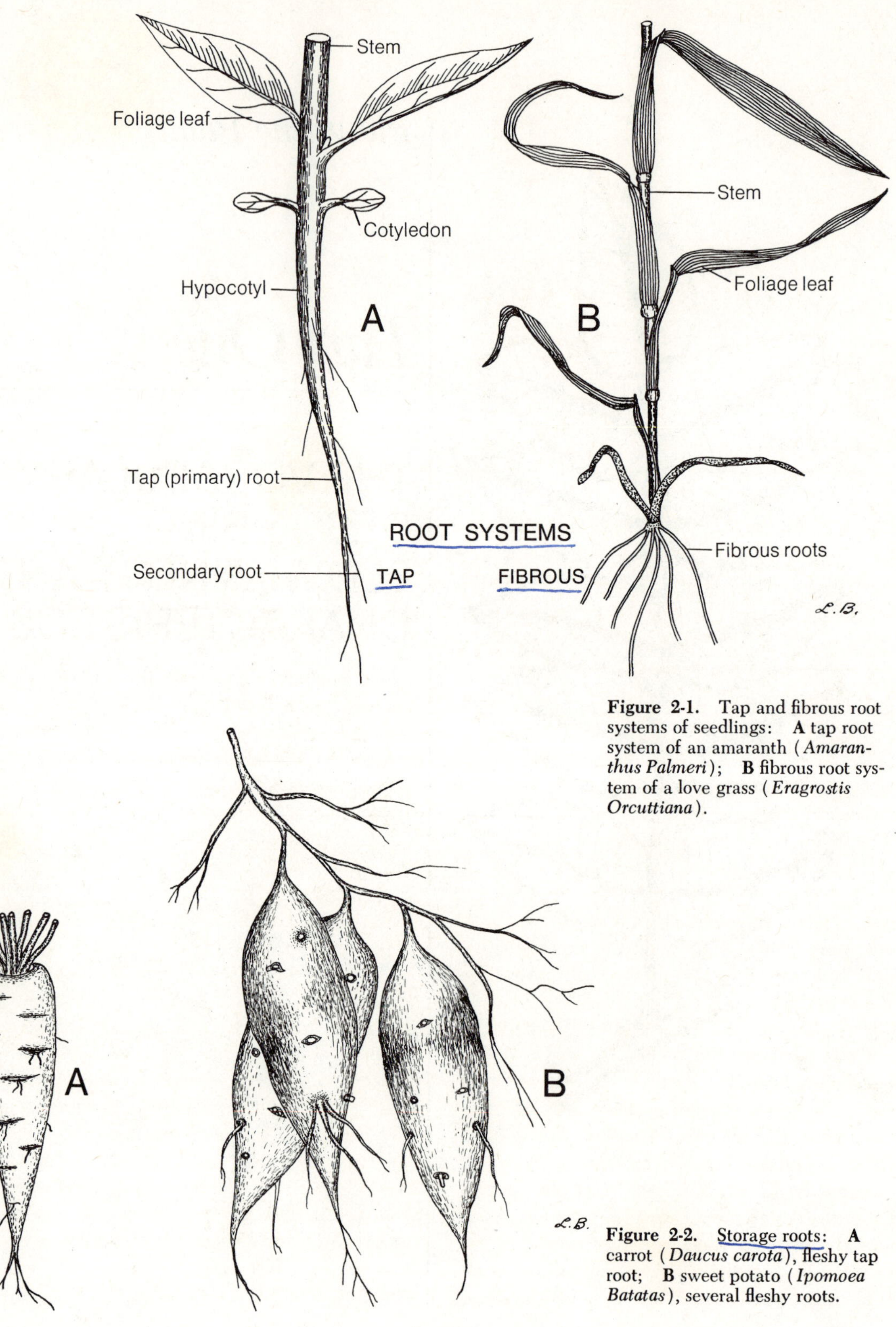

Figure 2-1. Tap and fibrous root systems of seedlings: **A** tap root system of an amaranth (*Amaranthus Palmeri*); **B** fibrous root system of a love grass (*Eragrostis Orcuttiana*).

Figure 2-2. Storage roots: **A** carrot (*Daucus carota*), fleshy tap root; **B** sweet potato (*Ipomoea Batatas*), several fleshy roots.

2

Roots, Stems, and Buds

ROOTS

Root Systems

When a seed germinates, the first structure to emerge through the seed coat is the **root tip,** or **radicle.** This first root is forced outside the seed coat by elongation of the hypocotyl with which it is continuous; it becomes the **primary root.** Soon it develops branch roots known as **secondary roots.** For practical purposes, all roots but the primary one may be referred to as "secondary roots." The primary root of a dicotyledonous flowering plant usually is much larger than the secondary roots, and it is called a **taproot.** Such an arrangement of roots is called a **taproot system.** In a few dicotyledons and all monocotyledons several or many secondary roots of the mature plant are as large as the primary root and not readily distinguishable from it, but in other dicotyledons the primary root may degenerate, being replaced by a group of secondary roots. Such an arrangement is called a **fibrous root system.**

The outermost layer of cells in the stele (central conducting region) of the young root is called the pericycle. The cells of this tissue remain capable of division, and at any point they may form a hump of cells which grows rapidly and secretes substances which dissolve a pathway through the outer tissues of the root. The rapidly dividing cells form a branch root structurally similar to the one from which it was formed. This is the normal method by which branch roots are developed. Branch roots not formed in this way are known as **adventitious.** The best known **adventitious roots** are those formed from stems or leaves. The branches of many plants are capable of developing roots, provided they are kept moist either naturally, as they may lie in the mud, or with the aid of man, as he places cut branches (cuttings) in moist sand or soil. Often the capacity to produce adventitious roots occurs in some species and not others of even the same genus, and this is a useful character in classification. For example, in the genus *Ranunculus* (buttercup) several only distantly

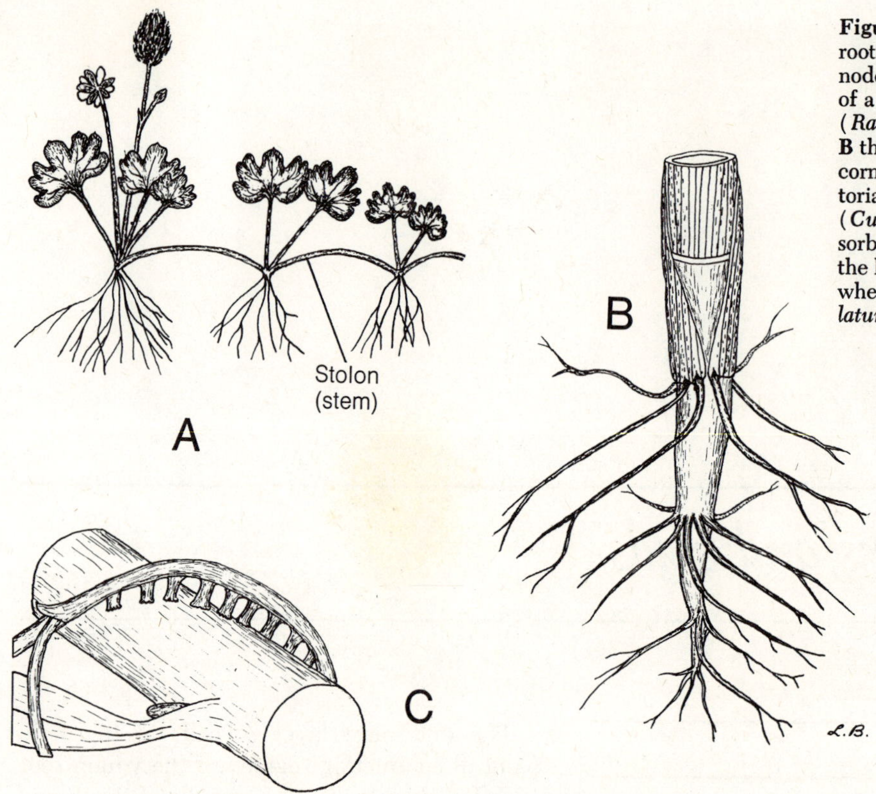

Figure 2-3. Adventitious roots: **A** roots from the nodes of a stolon (runner) of a buttercup or crowfoot (*Ranunculus Cymbalaria*); **B** the prop roots of Indian corn (*Zea Mays*); **C** haustorial roots of a dodder (*Cuscuta californica*) absorbing food and water from the host plant, a wild buckwheat (*Eriogonum fasciculatum* var. *foliolosum*).

related species or species groups may be separated each from its close relatives by presence or absence of the capacity to produce adventitious roots (in combination with certain other characters). Special adventitious root systems serve for support of the stem, as, for example, the prop roots of corn, sugarcane, or the banyan-tree.

Storage Roots

Often, as in the fibrous roots of a sweet potato or a dahlia, each root may be greatly enlarged because of development of an internal storage tissue. Sometimes such a cluster of roots is referred to as a **fascicled root system.** Often a taproot is enlarged greatly, forming a storage organ, which may be **spheroidal,** as in some varieties of beets; top-shaped, or **turbinate,** as in a turnip; **obconical,** as in most carrots or parsnips; or spindle-shaped (**fusiform**), as in the sweet potato. Storage roots are tuberous roots.

Aërial Roots

In the tropical rain forests some species of plants live perched on trees, and many of them, especially orchids or bromeliads (members of the *Bromelia* family), produce aërial root systems.

The common English ivy, poison ivy, poison-oak, and trumpet-vines have rootlets which serve as tendrils do in climbing.

Roots of Parasites

Many kinds of plants become parasitized by the slender, twining yellow branches of dodder, a leafless parasitic relative of the morning-glory. This plant produces short adventitious roots penetrating the stems and branches of its

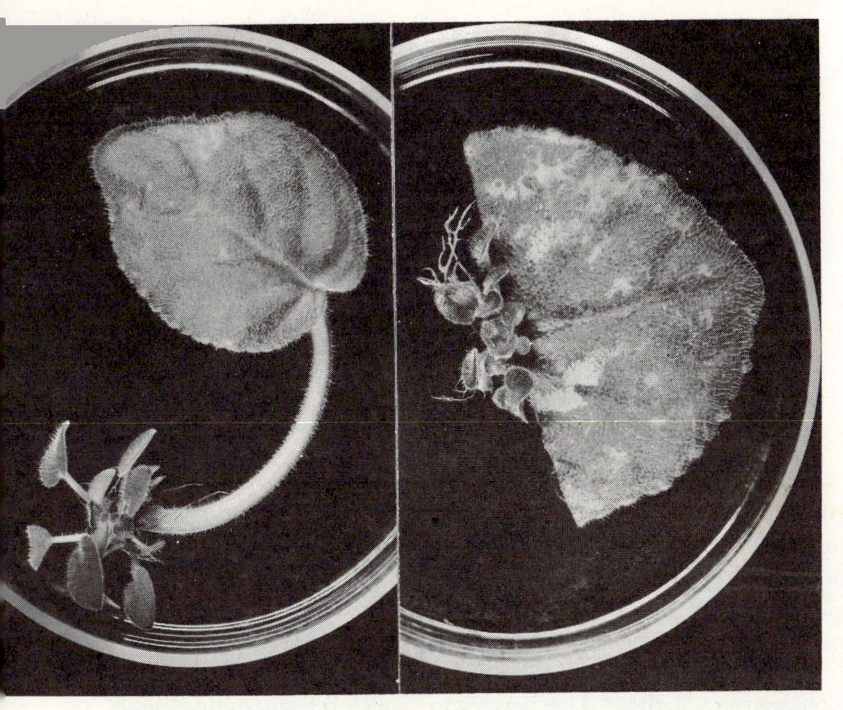

Figure 2-4. Adventitious roots from the petiole (stalk) and from the midrib region of a leaf: African violet (*Saintpaulia ionantha*). From ERNEST L. STOVER, *An Introduction to the Anatomy of Seed Plants.* D. C. Heath and Company.

Figure 2-5. The aërial roots of an orchid (*Cattleya*). Photograph by Ladislaus Cutak.

Figure 2-6. Shrub, arborescent plant, and tree.

host plant and absorbing food and water. Other plants growing on host plants include the mistletoes, some of which are green and dependent for only water and salts while others are yellowish and dependent for carbohydrates as well.

Other plants, including some species of the Indian-paintbrush (*Castilleja*), the Indian-warrior of California (*Pedicularis*), the pinesap (*Monotropa*), the sand-food (*Ammobroma*) of the Southwestern Deserts, the broom-rape (*Orobanche*), and beech-drops (*Epifagus*), have roots attached to those of other plants, and they absorb food and water from the host.

STEMS

The Principal Kinds of Stems

The classical division of plants, used by authors from the time of the ancient Greek botanist Theophrastus until the Seventeenth Century and by the layman even today, is into **herbs, shrubs,** and **trees,** that is, according to the kind of stem. An herb is a plant with a soft stem which does not live over from year to year. Its cambium, if any, ordinarily does not become very active. **Shrubs** or **trees** are woody plants with stems and branches which live over from year to year, becoming a little thicker each growing season as the cambium adds new layers of xylem and phloem. A shrub has no main trunk but a number of major branches arising from ground level. A tree has a single main trunk with branches on only the upper part of the plant. Ordinarily shrubs are smaller than trees.

There are many intermediate plants between herbs and shrubs and between shrubs and trees. The adjectives in the following list do not classify all plants, but they cover most of the cases.

Herbaceous, soft, i.e., not woody. Herbs develop soft new stems each year. After the flowering season either the plant dies or the portion above ground dies back about to ground level and sprouts up again from the underground parts at the end of the cold winter in the North or the dry summer, spring, or fall in the Southwest.

Suffrutescent, with the lower part of the stem just above ground level slightly woody and usually living over from year to year.

Suffruticose, the next degree above suffrutescent, that is, with decidedly woody permanent stems extending up a few inches above ground level. The branches from them are herbaceous, and they die at the end of each season.

Fruticose, that is, distinctly woody, living over from year to year and attaining considerable size, the plant being a shrub.

Arborescent, the plant large and almost treelike, but the main trunk relatively short.

Arboreous, with a well-developed trunk, the plant being a tree.

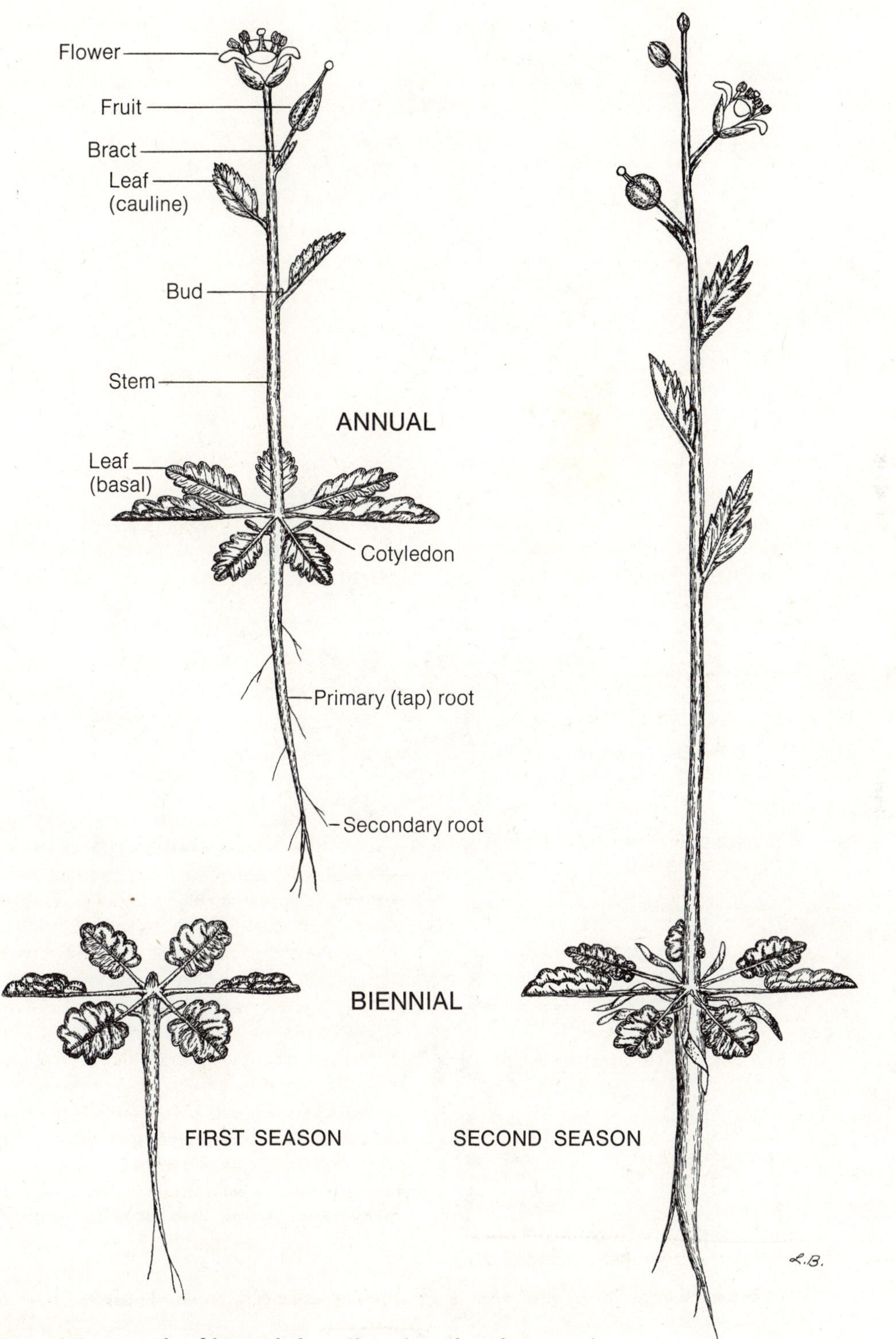

Figure 2-7. Annual and biennial plants (hypothetical similar species).

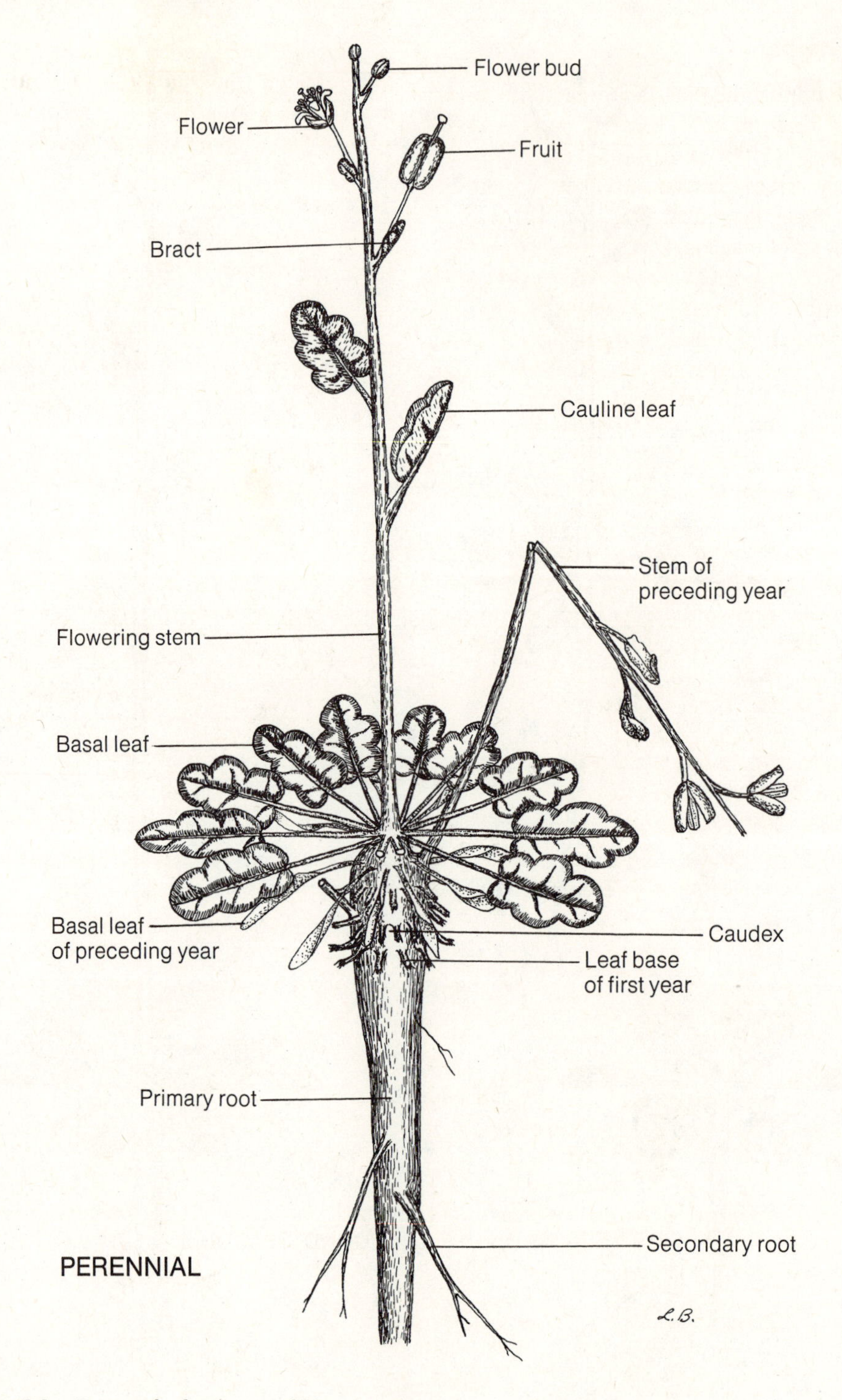

Figure 2-8. Perennial plant in its fifth year of growth (hypothetical species similar to those in **Figure 2-7**).

Kinds of Herbs

The most important classification of herbs is according to duration. **Annuals** live through the northern winter or the southwestern dry season only as seeds, coming up each year as seedlings, then flowering, fruiting, and dying. **Biennials,** such as the sugar beet, produce only basal leaves the first season and leaves, stems, flowers, and fruits the second; they live only two growing seasons altogether. **Perennials** survive the winter or dry season not only by seeds but also by the persistence of underground parts. They may be distinguished by the presence of thickened underground structures (in extreme forms **tubers** or tuberous structures), by a woody crown just below ground level, or by the persistence of dead stems from preceding seasons. Some perennials produce stems and flowers the first season, while others produce only leaves and do not flower or fruit until the second year of growth or later. Both usually continue to flower and fruit for several years.

The distinctions between annuals, biennials, and perennials are not hard and fast; and there are many intergradations, particularly in accordance with climate. "Annuals," like the castor-bean, become trees in the mild winter climate of California, reaching a height of thirty feet or more and a trunk diameter of up to a foot. Along the southern margin of the United States tomato plants growing in favored locations during a series of mild winters may become three or four years old, and geraniums are grown outdoors as perennial plants. Sugar beets are grown in the desert regions of Arizona and adjacent southeastern California for seed production because the long growing season enables the farmer to produce a crop of seeds in a single year instead of two.

Direction of Stems

The following series of adjectives describes direction or method of growth of stems.

Diffuse, spreading in all directions.
Declined, bending over in one direction.
Decumbent, resting on the ground, the stem not having sufficient internal strengthening tissue to stand erect (except at the tip, where only a little weight is to be supported).
Procumbent or **prostrate,** flat on the ground.
Creeping or **repent,** lying on the ground and producing adventitious roots, especially at the nodes, as in some plants used for lawns or in many weeds growing in lawns.

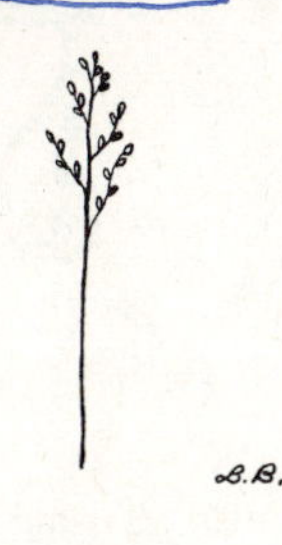
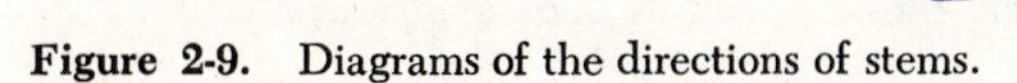

Figure 2-9. Diagrams of the directions of stems.

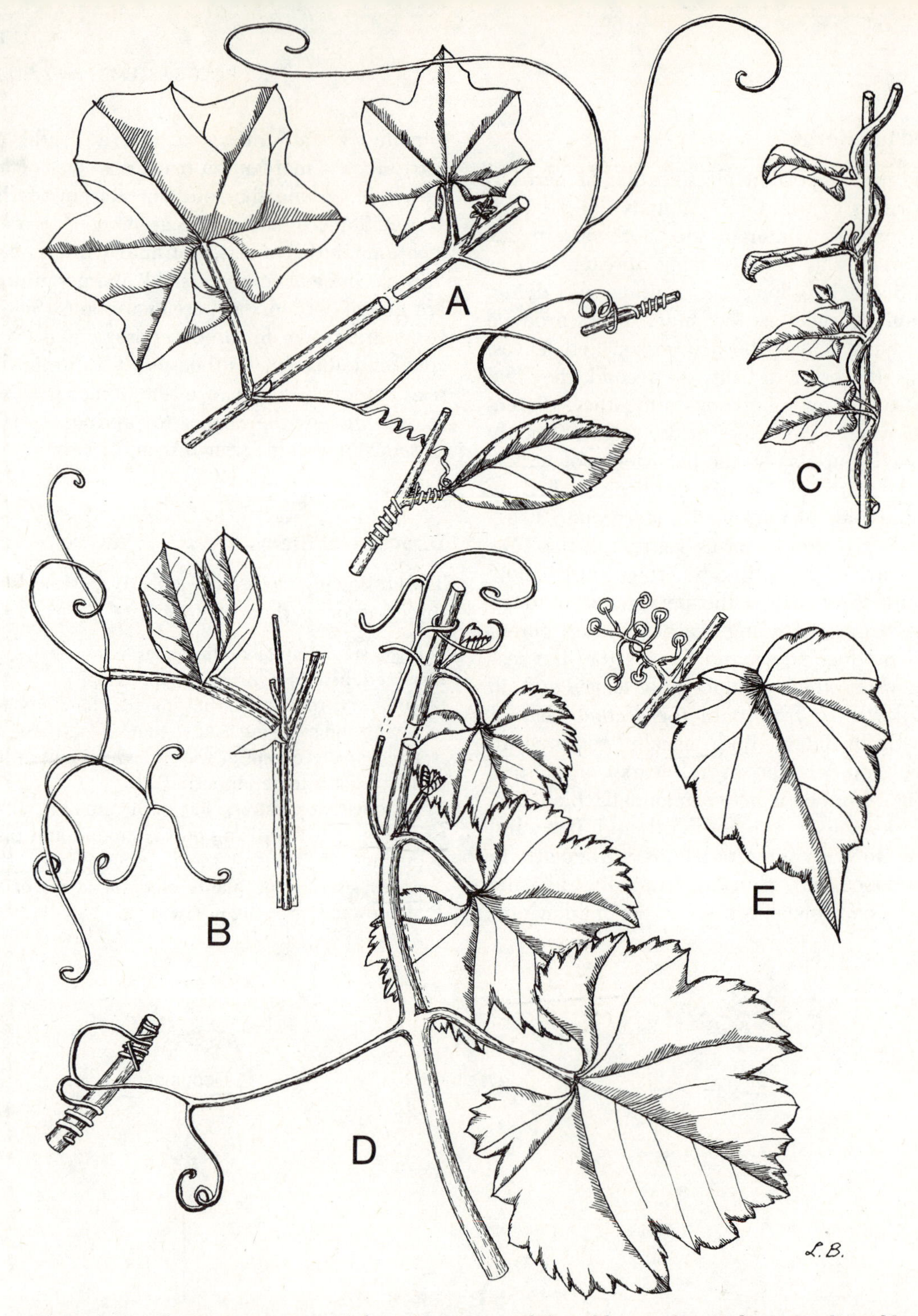

Figure 2-10. Twining structures: **A** leaves forming tendrils, wild cucumber (*Echinocystis* [*Marah*] *macrocarpa*); **B** leaflets forming tendrils, sweet pea (*Lathyrus odoratus*); **C** twining stem, bindweed (known in some sections as morning-glory) (*Convolvulus arvensis*); **D** branchlets forming tendrils, wild grape (*Vitis Girdiana*); **E** branchlets forming tendrils ending in attachment discs, Boston ivy (*Parthenocissus tricuspidata*).

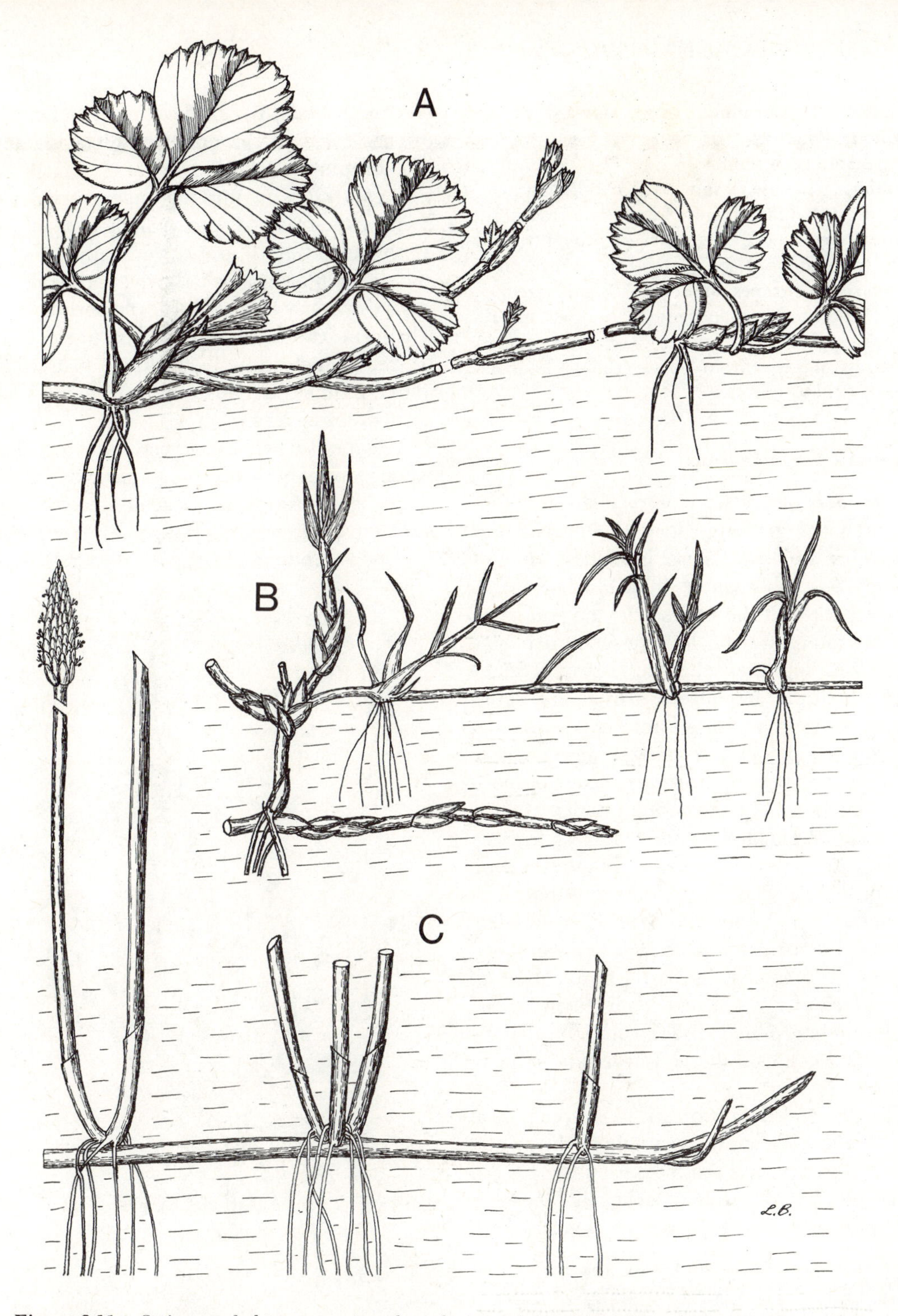

Figure 2-11. Stolons and rhizomes: **A** stolon of a strawberry (*Fragaria,* hybrid developed at the Rancho Santa Ana Botanic Garden); **B** stolon and (lower level) rhizome of Bermuda grass (*Cynodon dactylon*); **C** rhizome of a spike-rush (*Eleocharis mamillata*).

Assurgent or **ascending,** rising upward at an oblique angle.

Erect, standing upright.

Climbing or **scandent,** supported by clinging, i.e., either by twining stems (*Clematis* or morning-glory), by special roots (ivy, poison ivy, and the trumpet-vine), or by tendrils formed from branchlets (grapes and the Virginia creeper, the tendrils being in these cases perhaps reduced flowering branches), or from leaflets (garden pea), or from leaves (the wild cucumbers, *Echinocystis*).

Special Types of Stems

The **runners** of the strawberry plant and similar branches growing along the ground are known as **stolons.** These branches develop adventitious roots where they touch the soil (in most cases at the nodes). In the course of time the internodes die, leaving separate plants which develop the following season. Some plants multiply by **offsets,** which are short stolons each with a single internode. An underground more or less horizontal stem is a **rhizome or rootstock.** Rhizomes occur in most ferns and in many flowering plants. Often they may be elongated and slender as in Bermuda grass (devil grass) or Johnson grass. Except that they appear rootlike and that they are underground, such rhizomes are much like stolons. Even this distinction is arbitrary because some rhizomes lie partly on or close to the surface of the soil and some stolons grow partly underground. Fragmentation of rhizomes is a common method of asexual reproduction.*

The thickened portion of a rhizome is a **tuber.** The best known example is the white potato; this species produces slender underground branches forming thickened tubers at their ends. Like other stems the tuber is covered with buds, which develop in the centers of the "eyes" of the potato. Thickened roots are not tubers.

A **caudex** is an upright underground stem living over from year to year; it is much like a rhizome, but it grows vertically. Usually such a stem grows very slowly, and each year it gives rise to leaves and to flowering stems. The term "caudex" is applied also to the trunks of trees which grow in the same manner but above ground, as, for example, the trunk of a palm or a tree fern.

A **scape** is a leafless flowering stem. Plants having no leafy stems above ground but only scapes are referred to as **acaulescent** ("without stems," discounting the scape). Frequently a scape is produced from a caudex, the caudex

Figure 2-12. Bulb of the Humboldt lily (*Lilium Humboldtii*), showing the fleshy, scaly leaves.

* Plants with rhizomes (e.g., quack grass, Johnson grass, and Bermuda grass) are among the worst weeds, because hoeing simply increases their number by cutting up the rhizomes into new plants. Species with no rhizomes but only roots underground usually may be eliminated quickly, because ordinarily the roots do not reproduce. However, there are exceptions, for example, the dandelion, Canada thistle, and bindweeds (*Convolvulus*, known in some sections as morning-glory).

living over from year to year and the scape being regrown each year, as in the coral-bells (*Heuchera*) of gardens.

Some stems (e.g., those of *Brodiaea* and *Crocus*) form bulblike structures underground. These tuberous enlargements of the stem are called corms, or solid bulbs. Many corms are covered with thin membranous coats. A true **bulb** is a short thickened stem bearing many fleshy or scalelike leaves; it is really an underground bud. **Bulbils** are small bulbs formed above ground either in the axils of the leaves or, as with some century-plants and garden onions, in the inflorescence in place of flowers. When these fall to the ground they develop roots and grow into new plants. **Bulblets** are small bulbs produced underground. Some authors use the terms as synonyms for small bulbs of any kind or for those above ground.

The stems of plants in several families are remarkably thickened and modified structures. The best American examples are afforded by various cacti, in which the stems may be large and leafless, and the cirio or boogum-tree, *Idria columnaris,* of Baja California and Sonora, which resembles a gigantic carrot turned upside down and growing above ground.

Thorns, Prickles, and Spines*

Many plants, especially shrubs, have sharp-pointed woody branches known as **thorns.** Nearly all the branchlets of the foothill palo verde and the three desert crucifixion-thorns of the Southwest are thorns, as are many branchlets of pyracantha (*Cotoneaster Pyracantha*). **Prickles** are not developed from whole branches but from the superficial tissues of the stem. Rose prickles are the best known examples, but similar prickles occur in many other plants, including the cat-claw (*Acacia Greggii*) and various species of *Mimosa.* Since it has no connection with the vascular tissues of the stem, a prickle may be dislodged easily by pushing it sidewise. **Spines** are sharp structures formed from the apex, margin, stipules, or other parts of the leaf. They may be composed of even the whole leaf. Examples include the following: leaf apex, *Yucca;* leaf apex and margin, many century-plants (*Agave*) and holly (*Ilex*); leaflet margins, Oregon-grape and related species (*Berberis*). Spines may be distinguished from thorns by their relative position on the stem because in the seed plants a leaf, whatever its form, always stands just below a branch or bud (embryonic branch).

* As defined in many books, thorn and spine are synonymous terms covering sharp-pointed structures formed from either branches or leaves.

BUDS

The Relative Positions of Buds

The cells at the extreme apex of a stem or branch are capable of dividing (meristematic). A bud is the meristematic, or growing, region together with its enclosing immature or special protective leaves. Down the sides of the stem are larger and larger mounds of tissue projecting laterally; these are embryonic leaves each having (except when it is very young) a small mound of meristematic tissue in its **axil** (the angle above it). This is a potential **lateral** or **axillary** bud. At this early stage of development the potential leaves within the bud are called **leaf primordia,** and the young potential axillary buds are called **branch primordia.**

The form of the plant is determined by the relative activity of the terminal bud as opposed to the lateral ones. If the terminal bud is very active the main stem will be elongated, whereas, if the axillary buds are more active the plant tends to become bushy. Whether a woody plant is a shrub or a tree depends, in large measure, upon which buds are relatively more active during the first years of growth. The form of a tree is determined by a relative activity of terminal and axillary buds after formation of the first several feet of the trunk. In many trees (e.g., oaks, walnuts, and elms) the lateral buds become very active, and the

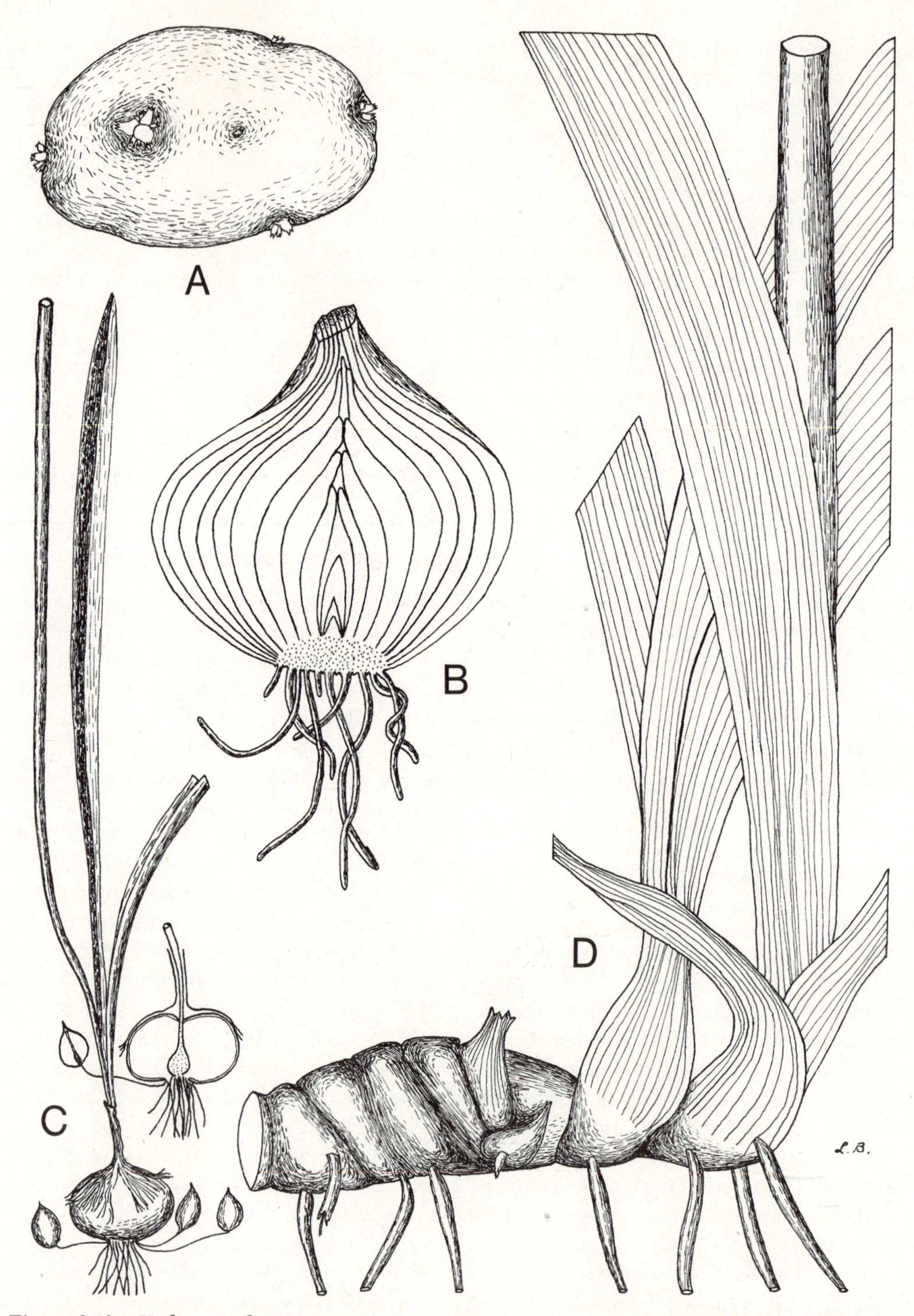

Figure 2-13. Underground storage structures: **A** white potato (*Solanum tuberosum*), storage branch with buds ("eyes"); **B** onion (*Allium Cepa*) bulb, showing the fleshy stem base and the large fleshy leaves; **C** brodiaea (*Brodiaea* [*Dichelostemma*] *pulchella*), corm and young (offset) corms produced at the ends of slender rhizomes, longitudinal section in inset; **D** the fleshy rhizome of an iris (*Iris*).

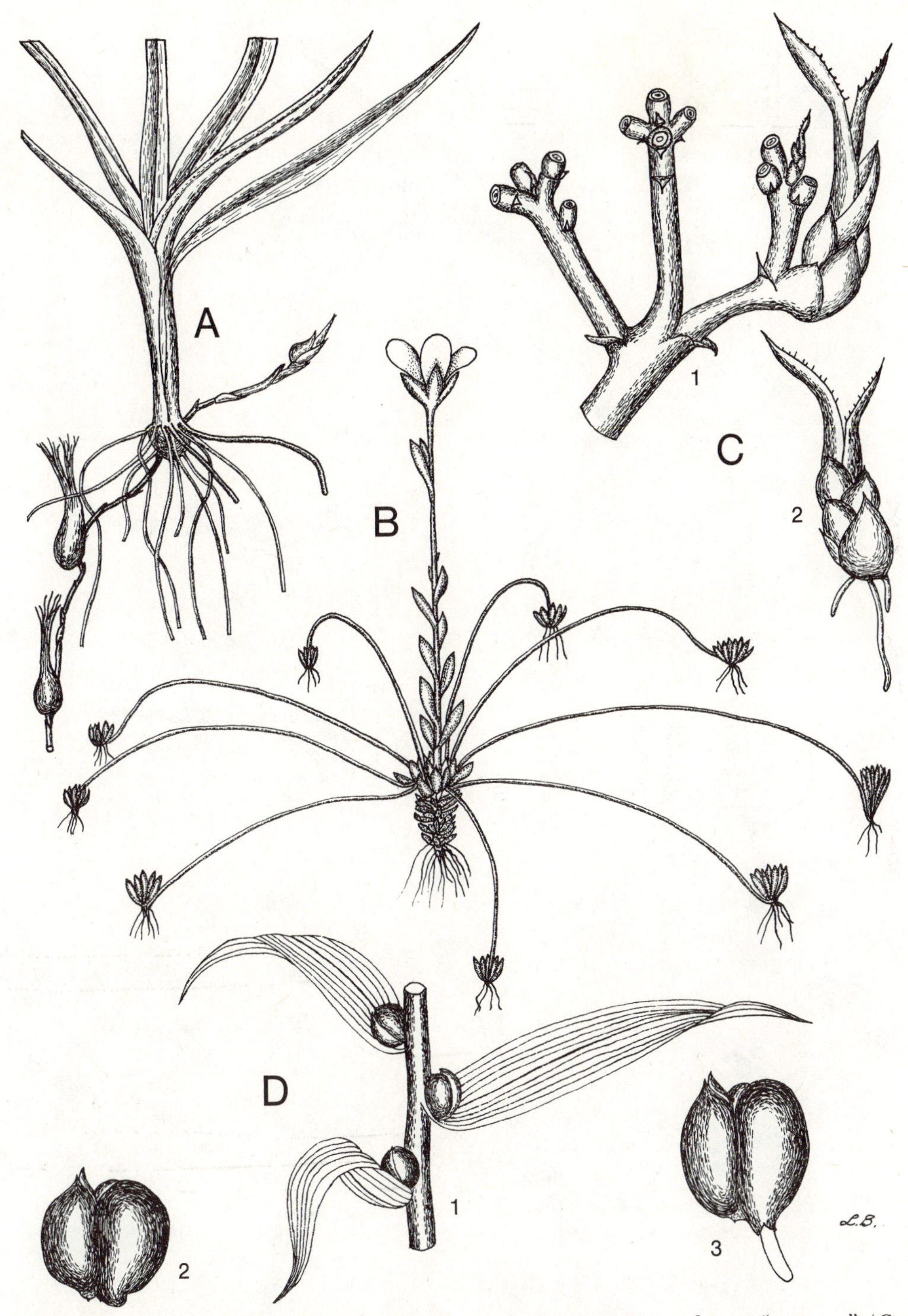

Figure 2-14. Vegetative reproductive structures: **A** small tubers of a sedge or "nut-grass" (*Cyperus rotundus*); **B** offsets of an arctic saxifrage (*Saxifraga flagellaris*); **C** bulbils formed in the places of flowers of a century-plant (*Agave*) (flowers having dropped off without forming fruits), **1** small branch, **2** bulbil; **D** bulbils formed from buds in the angles (axils) above the leaves, tiger-lily (*Lilium tigrinum*), **1** stem with bulbils, **2** bulbil, **3** bulbil germinating.

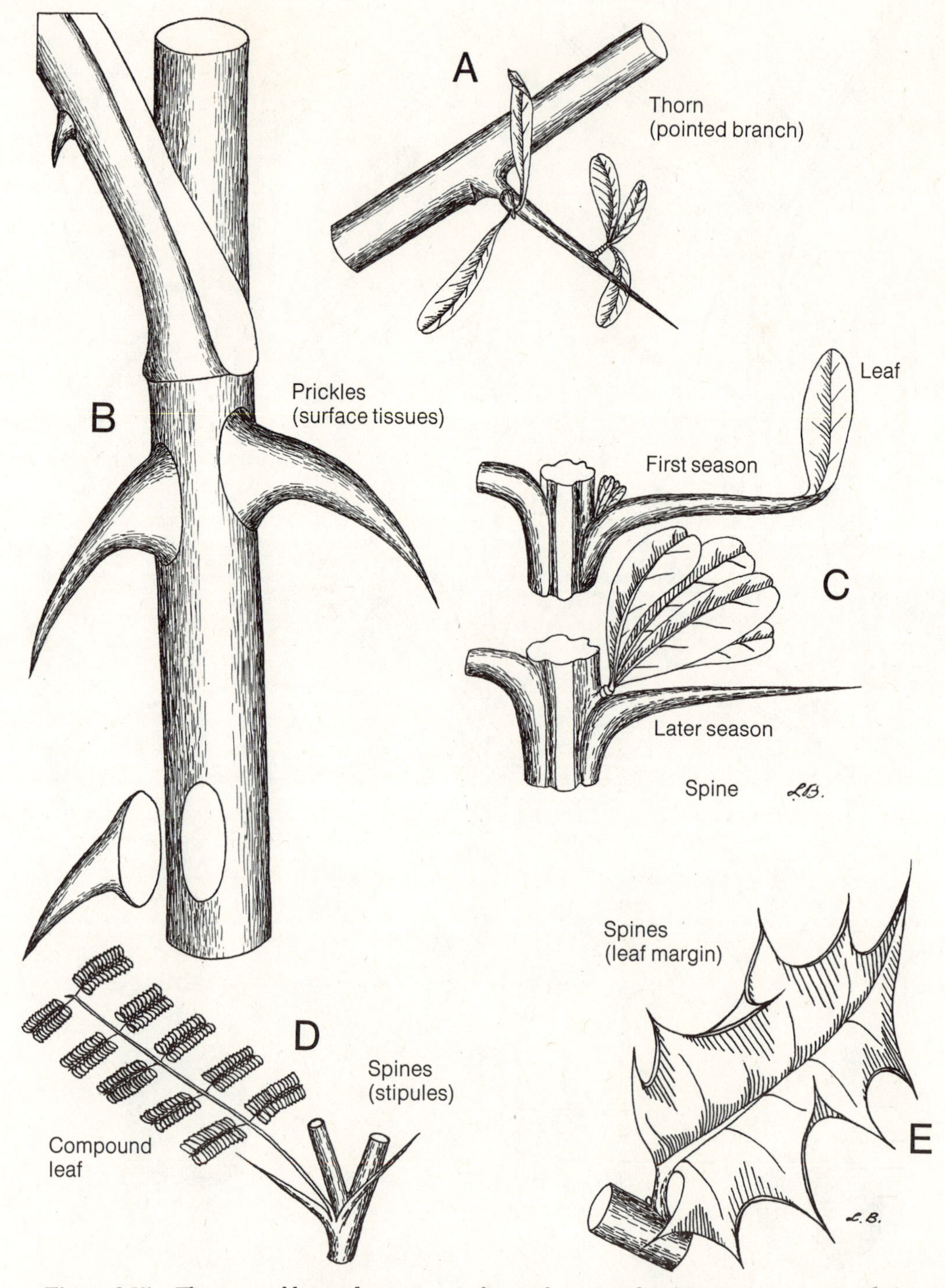

Figure 2-15. Thorns, prickles, and spines: **A** thorn of pyracantha (*Cotoneaster Pyracantha*); **B** prickles of a rose (*Rosa*, hybrid); **C** spine formed from the midrib and petiole of a leaf, the soft tissues of the blade falling away after the first season, ocotillo (*Fouquieria splendens* of the Southwestern Deserts); **D** spines formed from stipules (appendages of the base of the compound leaf), white thorn (*Acacia constricta*, cf. **Figure 1-1**); **E** spines on the margins of the leaf, holly (*Ilex*).

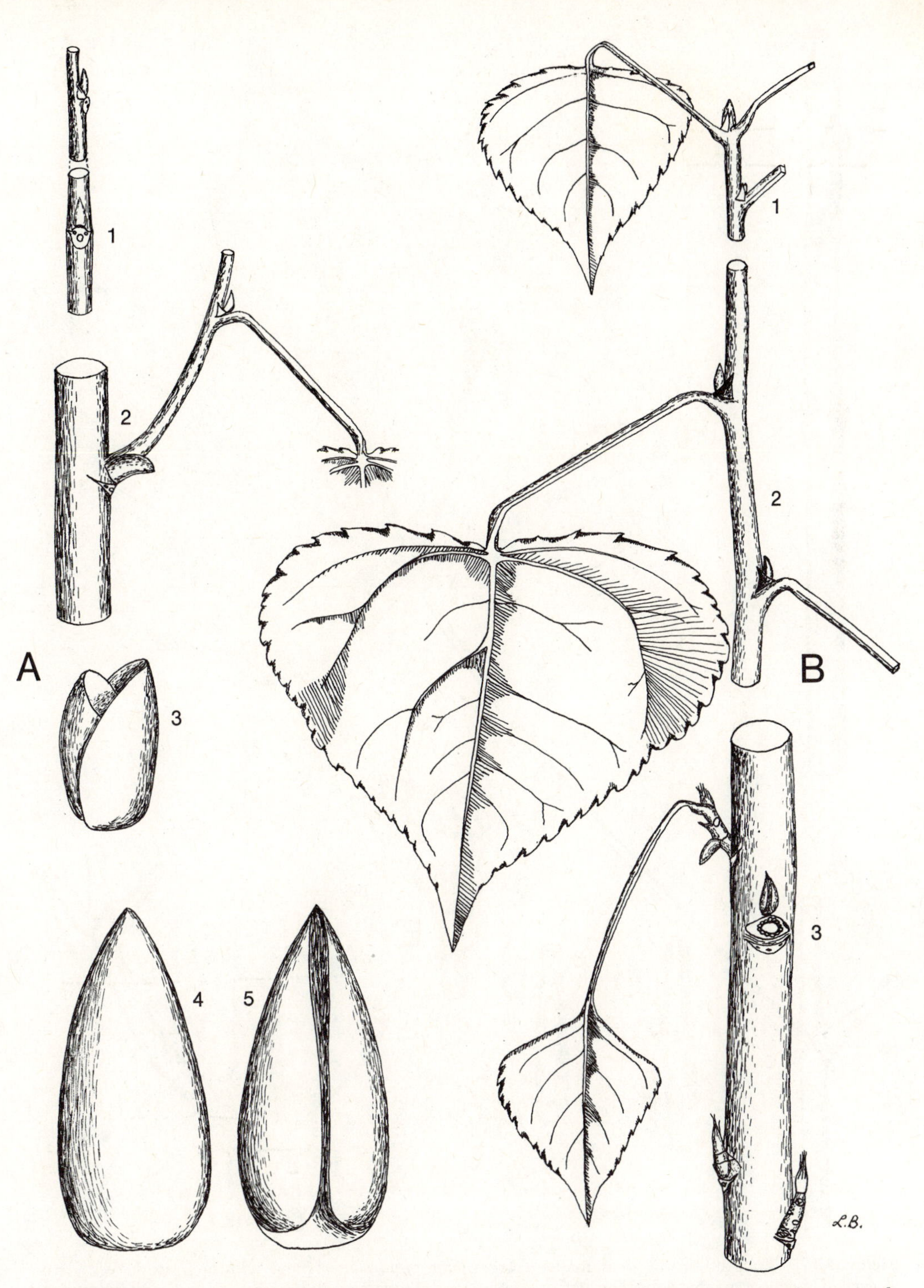

Figure 2-16. Buds of Lombardy poplar (*Populus nigra*), a plant with alternate leaves (one at each node or joint of the stem): **A** winter condition, persistent in the spring, **1** twig with a resting bud above each leaf scar, **2** lower portion of the same branch where growth has started, a bud scar with a new branch above it and with one persistent bud scale, **3, 4,** and **5** views of bud scales; **B** late spring condition, **1, 2,** and **3** portions of the same newly formed branch, the lower lateral (side) or axillary buds starting to enlarge.

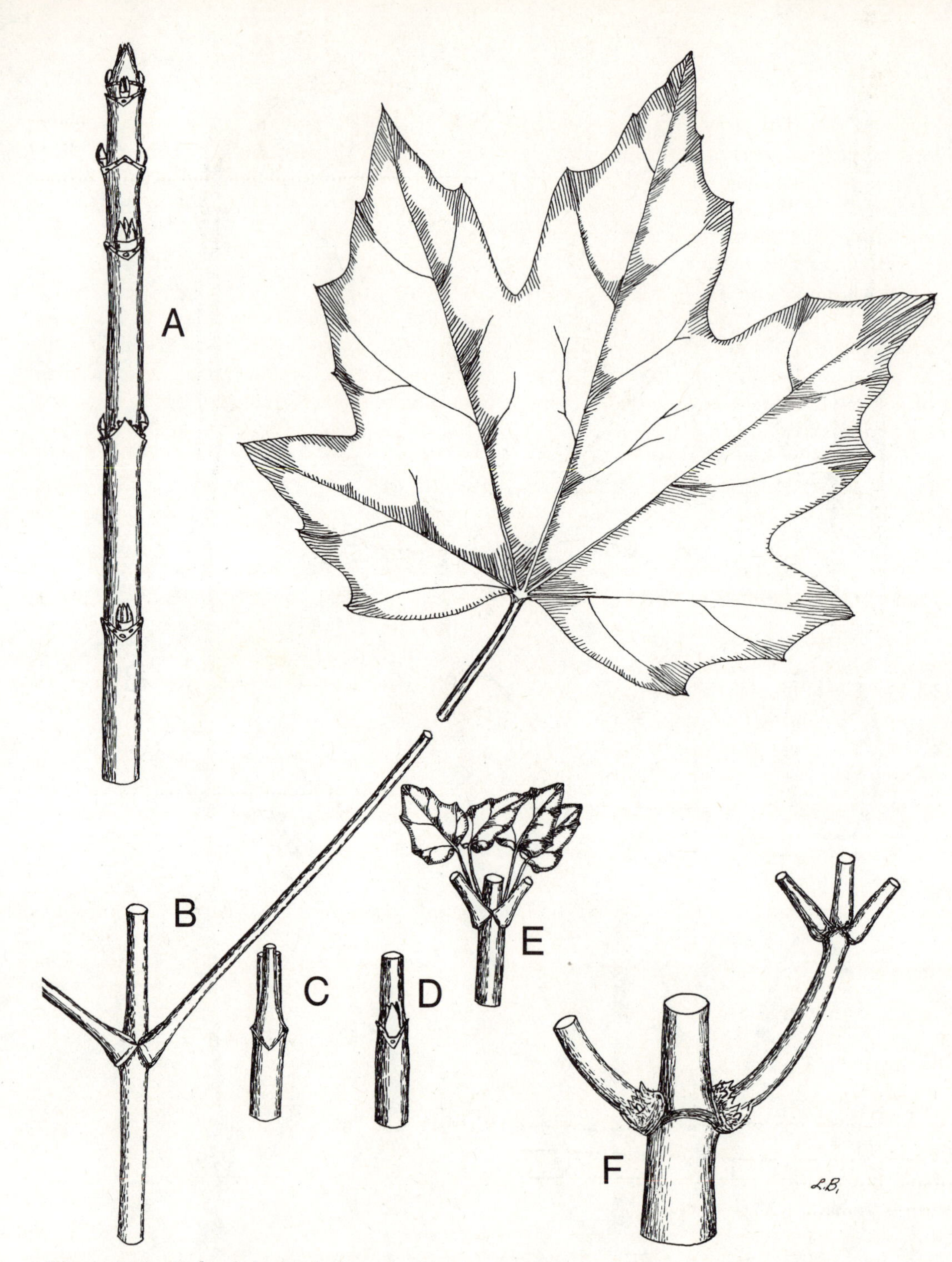

Figure 2-17. Buds of the big-leaf maple (*Acer macrophyllum*), a plant with opposite leaves (two at each node): **A** winter condition with a resting bud above each leaf scar; **B** young shoot in late spring, showing the opposite leaves, the bases of which cover the young buds; **C** another view; **D** the view in **C** with the leaf removed, exposing the bud; **E** young leaves developed from buds still obscured by the leaf bases; **F** young branches with bud scales persistent around their bases, the subtending leaves having fallen.

tree forms a rounded crown; in others (particularly coniferous trees) the terminal bud is far more active than any of the lateral ones, and consequently the trunk is elongated and straight while the lateral branches are relatively small. For this reason the conifers are the principal lumber trees of North Temperate regions.

Not every bud is produced in the usual position at the apex of a stem or branch or in the axil of a leaf. If a plant is wounded, a great many new buds may arise on or near the spot. These are called **adventitious buds.** In some plants adventitious buds may arise from leaves or roots.

Variation in the Contents of Buds

The buds described so far produce only branches, which in turn bear leaves. They are known as **foliage buds.** Some buds contain the primordia of flower or other reproductive parts, and these are **reproductive buds.** In the angiosperms they are **flower buds** or **floral buds.** Some buds on, for example, an apple tree, contain both branch and leaf primordia and rudimentary flowers. These are called **mixed buds.**

Inactive Buds

In cool weather buds of herbaceous plants may become inactive and each one may be protected by a tight "head" of leaves. Two buds well known for their great size as well as for being edible are the heads of lettuce and cabbage. In warm regions often lettuce and cabbage fail to form heads because the temperatures remain too high during the winter for the plants to become partly dormant, and the stem elongates instead of forming an enlarged bud. Buds having only the normal young leaves as protection are called **naked buds.** In the dormant season water loss from the buds of trees and shrubs is reduced or eliminated by growth of specialized leaves covering the dormant buds. These specialized leaves are small, scalelike, and usually brown. The epidermis of each scale leaf is covered more heavily with wax, resin, or other waterproof material than is an ordinary foliage leaf. In some plants (e.g., poplars and cottonwoods) the resinous covering is conspicuous. **Resting buds** with this extra protection are called **protected buds.** They are in marked contrast to the **active buds** of the growing season. Many plants occurring in the South do not develop resting buds, and in the Southwestern Deserts resting buds may be developed at the onset of a dry season rather than of a cold one.

Some buds never develop actively or may not do so for many years; these are known as **latent buds.** If all the buds of a tree or shrub were to develop in any one year there would be a branch produced at every point where there had been a leaf the preceding season. Instead, ordinarily only a few of the axillary buds develop; most buds remain latent, the vast majority of latent buds never developing into any structures at all.

Active Buds

At the beginning of a new growing season resting buds open up and become active with amazing rapidity. This is because their internal structure is already fairly well formed as a result of cell divisions which occurred before or during the apparent resting stage and because food is available from the enormous reserve in the other parts of the plant. When the bud opens, ordinarily the scale leaves fall away, but they may persist for a time. Usually the terminal bud continues its growth when the new growing season comes, but in some instances it dies; and then it is replaced by one or more axillary buds. The age of a branch may be determined by inspection of the scars left at the attachment points of the scale

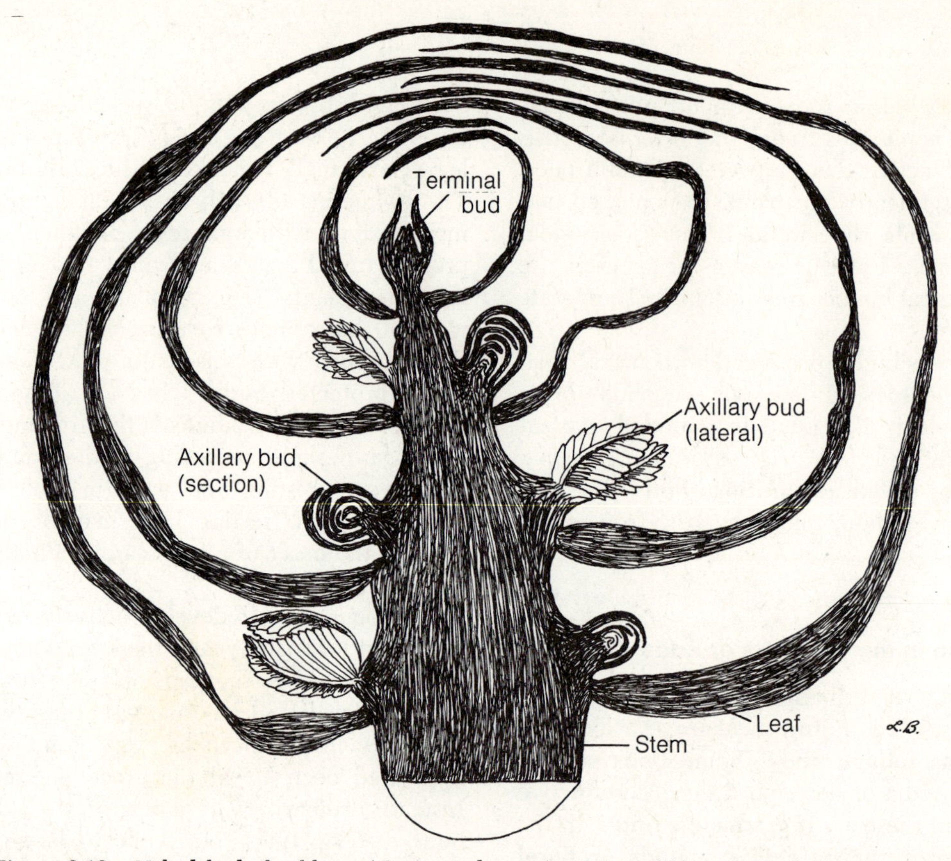

Figure 2-18. Naked bud of cabbage (*Brassica oleracea*) a large edible inactive bud with a smaller bud in the axil of each leaf. There are no bud scales, and protection is from the outer, more mature, leaves.

leaves; each area of such scars indicates a point where the branch was inactive between growing seasons. Eventually, though, as the branch becomes large, the scars of the scale leaves are obliterated. In some plants, as, for example, pears and the pitted fruits, the flower buds may become active before the foliage buds. In the elm even the fruits may have fallen away before the vegetative buds have developed and the leaves have appeared.

SUGGESTED FIELD TRIP

"Scavenger Hunt." Students, working singly or in pairs, according to instructions, should bring in examples of the items in the following list. This should be at a prearranged time or at a signal.

Roots: Tap, fibrous, adventitious, storage, and (if possible) those of a parasite.
Stems: 1. The roots and stems of an annual, a biennial, and a perennial herb.
2. A suffrutescent and a suffruticose plant.
3. Illustrations of fruticose, arborescent, and arboreous plants. Obviously these can not be brought in, but the student should be able to point out examples.
4. Diffuse, declined, decumbent, procumbent, and scandent stems of herbs or woody plants.
5. Rhizomes, caudices (plural of caudex), stolons, scapes.
6. Thorns, prickles, spines (of various types), and tendrils.

Buds: Terminal, lateral (alternately and oppositely arranged), naked, protected.

3

Leaves

The Makeup of a Leaf

Most flowering plant leaves are composed of a thickened **leaf base** (attached to the stem), a slender **petiole** or stalk, and a flat **blade.** In some families or genera a small scale (**stipule**) may be produced on each side of the leaf base. Ordinarily stipules are brown scales with no function, but they may be glands, spines, or flat green photosynthetic (food-manufacturing) structures.

The leaf blade consists of veins with intervening softer tissues. Usually the cells of the soft tissues of the upper side of the leaf are arranged like the cells of a honeycomb, forming a photosynthetic tissue; those of the underside are more or less spheroidal and loosely packed, forming an aërated tissue. Pores (stomata) in the lower surface of the leaf permit gas exchange with the atmosphere. In most leaves the middle vein, or **midrib,** is larger than the rest and is surrounded by more supporting tissue. The arrangement of smaller veins varies. In dicotyledons usually the veins are branched and rebranched, the smaller ones finally rejoining each other and forming a network—the leaves are **netted veined.** In the monocotyledons usually the veins (except the minute ones) lie parallel to each other—the leaves are **parallel veined.** In a few monocotyledon leaves (e.g., bananas) the veins diverge from the midrib but lie parallel to each other. Some dicotyledon leaves may appear to be parallel-veined because they are narrow or thick and the netted vein system is obscure.

The terms **palmate** and **pinnate,** described below, are applicable also to the arrangement of the main veins of a dicotyledon leaf.

Complexity of Leaves

A leaf with the blade in a single segment is **simple;** one with the blade divided into several **leaflets** or sections is **compound.** A compound leaf with all the leaflets arising from one point at the end of the petiole somewhat like the fingers on a human hand is **palmate** (**digitate**). A compound leaf with the leaflets arranged

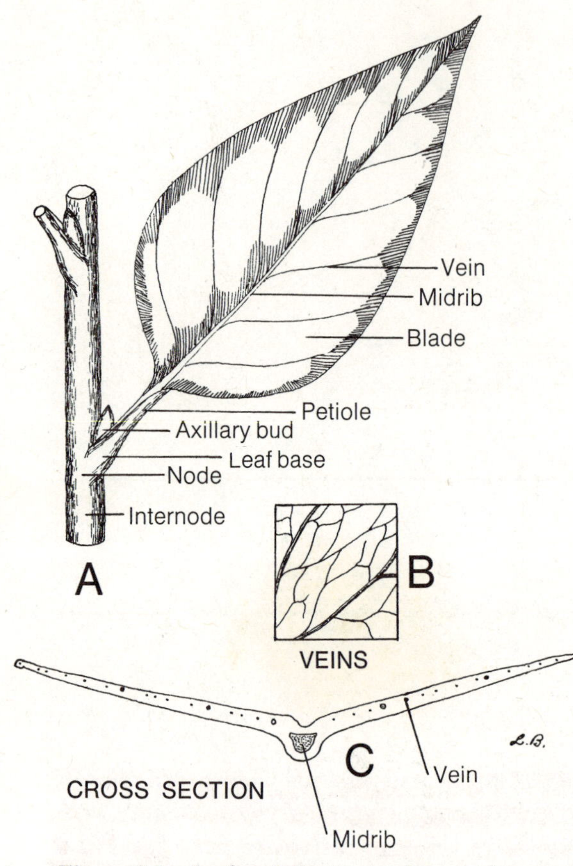

Figure 3-1. Leaf: **A** the parts of a simple, entire leaf; **B** enlargement of a small portion of the blade, showing veins; **C** cross section as seen with a hand lens.

along both sides of the axis (actually the midrib) is **pinnate** (like a feather).

In pinnate leaves typically the leaflets are in opposite pairs along the axis, or **rachis** (**rhachis**), and there is a terminal solitary leaflet. Since there is an odd number of leaflets, such a leaf is **odd-pinnate** (imparipinnate). However, the term is used for any pinnate leaf with a terminal leaflet regardless of whether the lateral leaflets are in pairs or whether the number is odd. Since theoretically when there is no terminal leaflet the number is even, a leaf lacking one is called **even-pinnate.** Since the leaf blade is terminated abruptly, an alternative term is **abruptly pinnate.**

Pinnate leaves may be still more complex because in some the leaflets are pinnate again (the leaf being **bipinnate**), that is, composed of secondary leaflets arranged pinnately on a minor axis (**rachilla**). The leaflets of the larger degree are called **primary leaflets** or **pinnae** (singular **pinna**); the minor ones are **secondary leaflets** or **pinnules. Tripinnate leaves** are composed of primary, secondary, and **tertiary leaflets.** The term pinnule is transferred then to the tertiary leaflets because they are the ones of the smallest degree. Rachilla is applied to the *axes* of both the primary and secondary or lesser leaflets. In short, a **rachis** is the axis of a pinnate leaf; a **rachilla** is the axis of a pinnate leaflet. Leaves pinnate to the fourth degree are **quadripinnate,** and the smallest leaflets (pinnules) are **quaternary leaflets.**

Indentation of Leaves

Leaves may be **entire** (with a simple, smooth margin having no indentation), **toothed** (indented only slightly along the margins), or **lobed** (with deep indentations).

Characters of the Leaf Margins. The margins of leaves or leaflets may be described more specifically by the following list of adjectives.

Undulate (**wavy** or **repand**), with the margin winding gradually in and out or up and down, that is, more or less wavy.

Sinuate, with the margin winding strongly inward and outward.

Crisped, ruffled, that is, with the same kind of winding as in sinuate but in the vertical plane instead of the horizontal.

Revolute, with the margin rolled under.

Incised, cut, or **jagged,** with the margins indented irregularly by deep sharp incisions.

Crenate, with broad, rounded teeth and narrow,

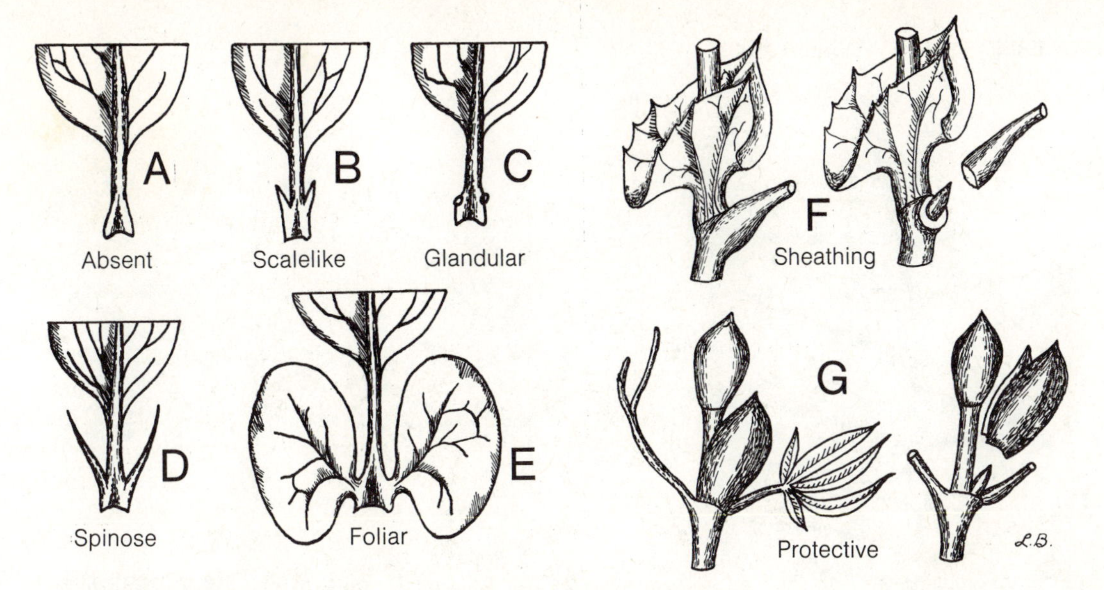

Figure 3-2. Stipules. In **G** the stipules protect the axillary buds.

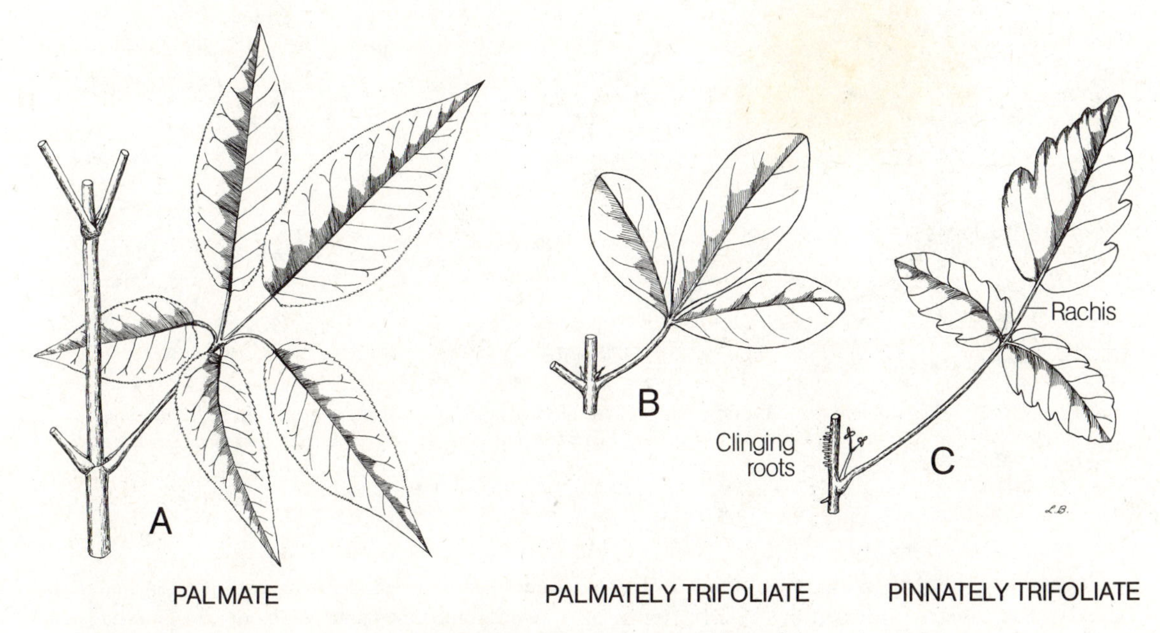

Figure 3-3. Compound leaves: **A** California buckeye (*Aesculus californica*); **B** Mexican orange (*Choisya ternata*); **C** poison-oak of the Pacific coast (*Rhus diversiloba*).

open spaces (**sinuses**) between them. The diminutive, indicating very small teeth, is **crenulate.**

Serrate, with the margin cut into sharp forward-projecting teeth resembling saw teeth. The diminutive is **serrulate.**

Retrorsely toothed, similar to serrate but with the teeth projecting backward toward the base of the leaf.

Dentate, with angular teeth projecting approximately at right angles to the leaf margin. The diminutive is **denticulate.**

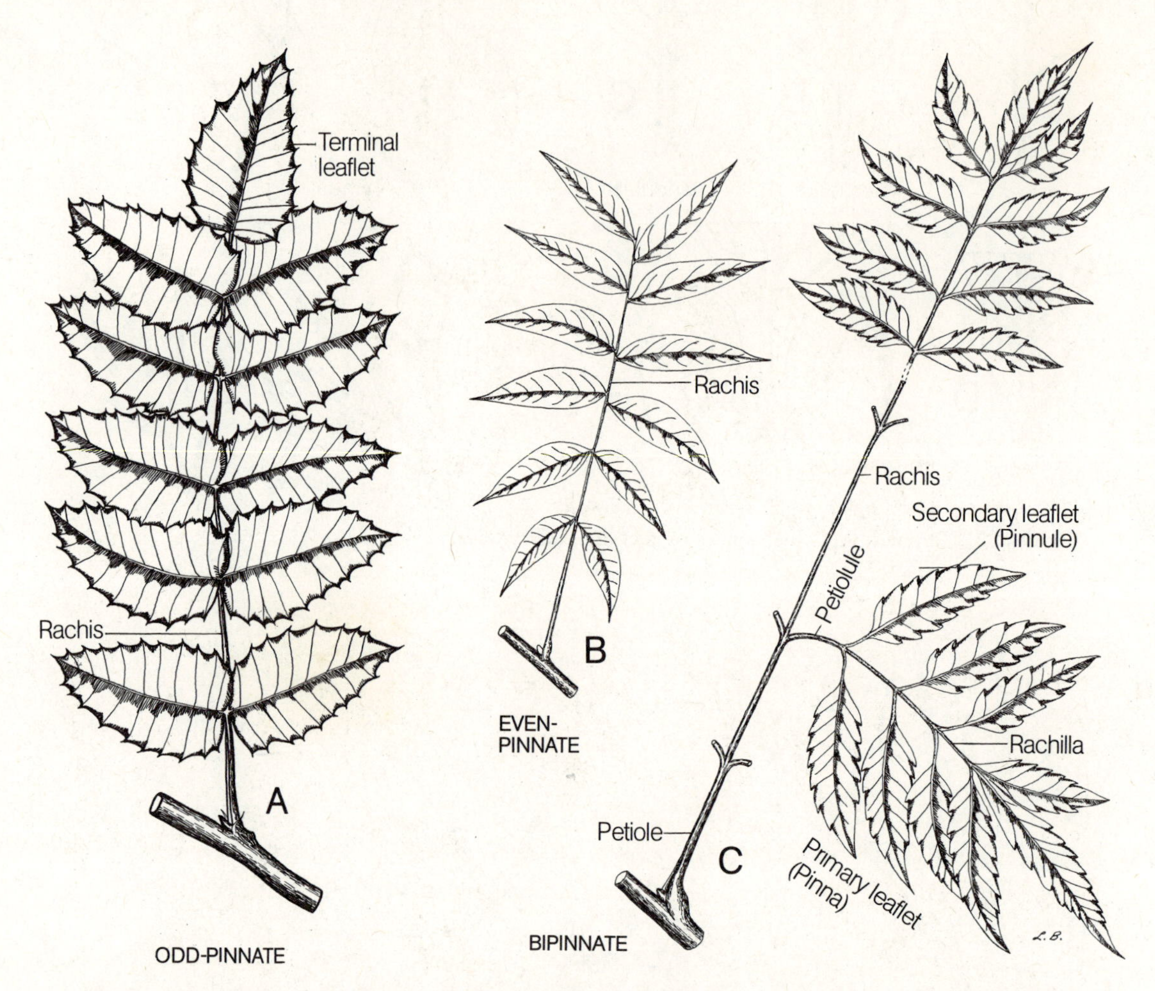

Figure 3-4. Pinnate leaves: **A** odd-pinnate leaf of Oregon-grape (*Berberis Aquifolium*); **B** even- or abruptly pinnate leaf of Chinese pistachio (*Pistachia chinensis*); **C** bipinnate leaf of the China-berry (*Melia Azedarach*).

Degrees of Leaf Lobing. Lobing, in the broad sense, means indentation of any depth from perhaps a quarter of the way to all the way to the base or midrib of the leaf. Lobing may be either palmate or pinnate according to the direction of the sinuses or spaces between the lobes. In **palmately lobed** leaves the indentation is toward the base; in **pinnately lobed** leaves it is toward the midrib. Degrees of lobing are termed arbitrarily as follows:

Lobed (restricted sense), with the indentations extending one-quarter to nearly one-half the distance to the base or midrib.

Cleft, with the indentations running halfway or a little more than halfway.

Parted, with the indentations running decidedly more than halfway to nearly all the way.

Divided, with the indentations running practically all the way or all the way.

Leaves palmately cleft or parted but not divided and, therefore, not quite palmate, are

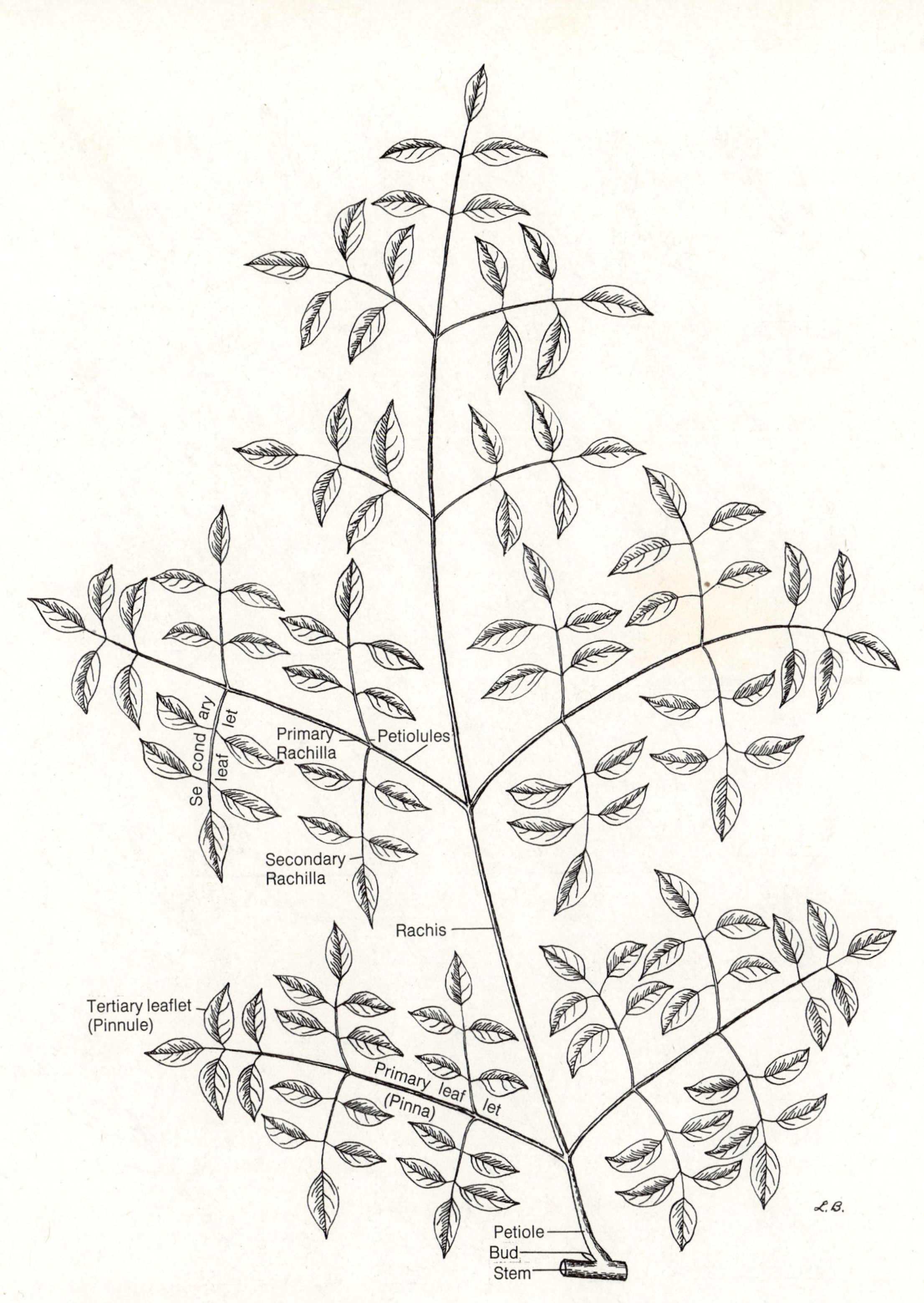

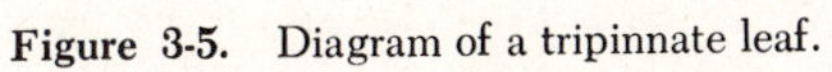
Figure 3-5. Diagram of a tripinnate leaf.

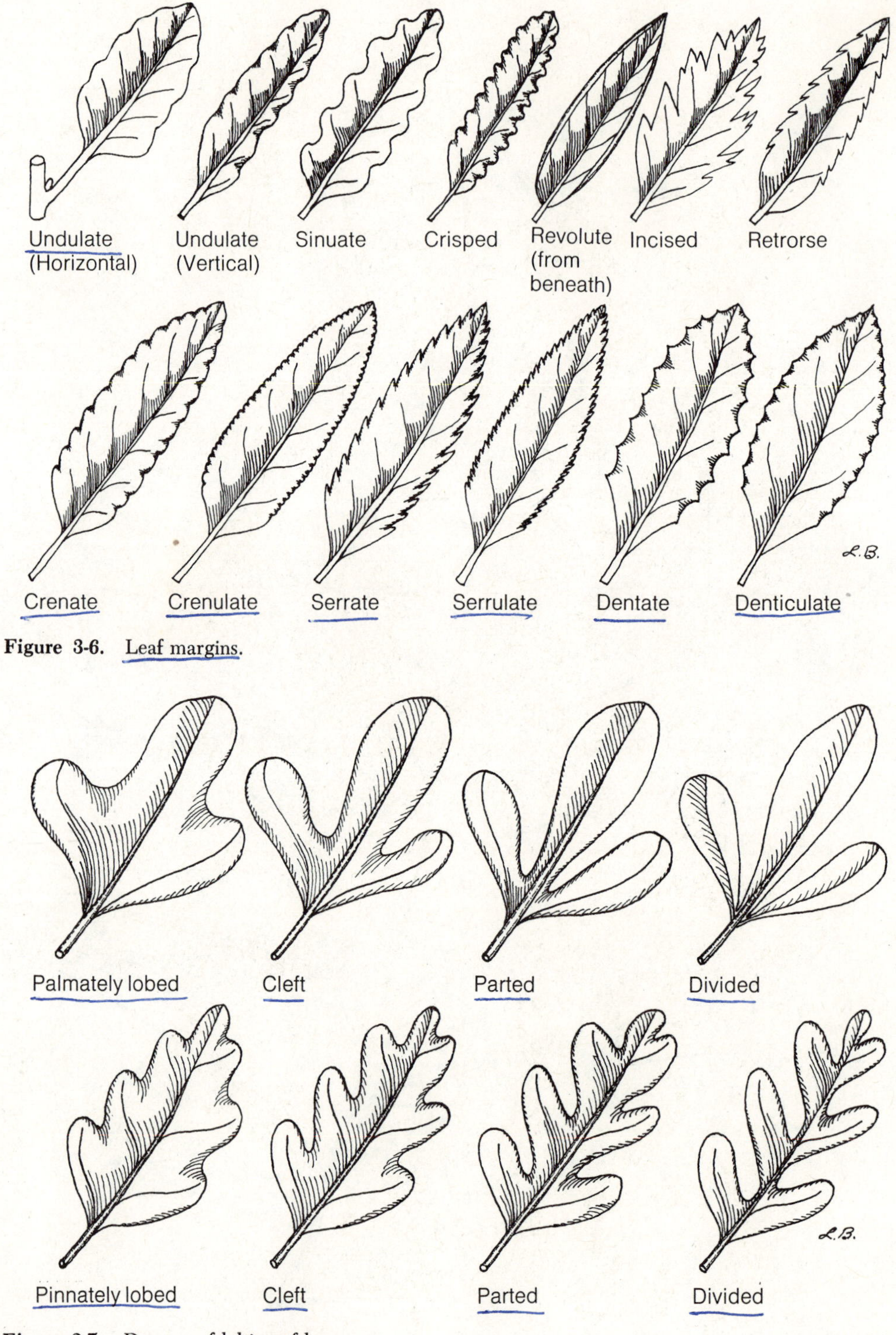

Figure 3-6. Leaf margins.

Figure 3-7. Degrees of lobing of leaves.

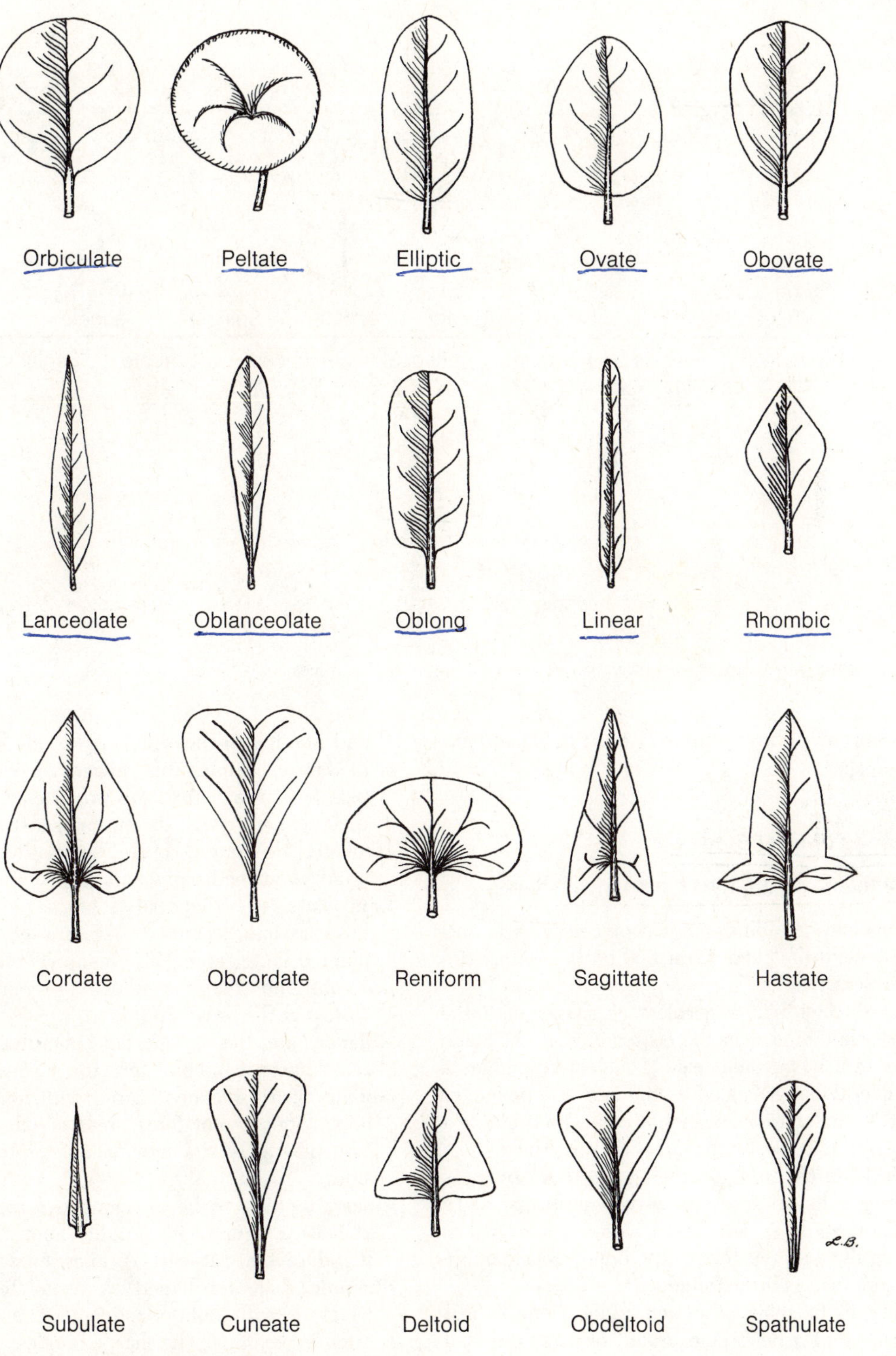

Figure 3-8. Shapes of leaf blades.

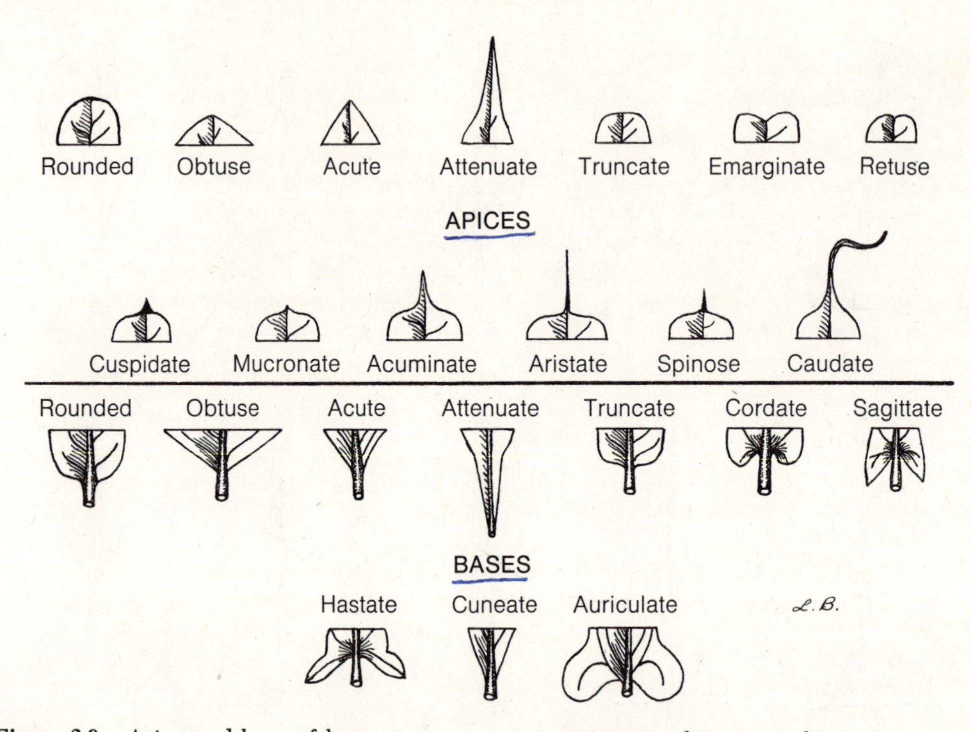

Figure 3-9. Apices and bases of leaves; two upper rows, apices; two lower rows, bases.

palmatifid. Those pinnately cleft or parted are **pinnatifid.**

Shapes of Leaf Blades

The following are the common leaf shapes.

Orbiculate or **rotund,** circular or nearly so in outline with the petiole attached on the edge. Also **orbicular.**

Peltate, shieldlike; regardless of shape, with the petiole attached in the center of the leaf blade as in a nasturtium leaf. The leaf resembles a silver dollar balanced on the end of a pencil, or it is similar to a mushroom or an umbrella, except that typically the edges do not droop.

Elliptic, in the form of an ellipse. About twice as long as broad and with curved sides and ends. The petiole is attached at one end. **Oval** is broadly elliptic, the width being considerably more than half the length.

Ovate, in the shape of a longitudinal (lengthwise) section of a hen's egg. About one and one-half times as long as broad, with curved sides and ends, and with the widest point *not* at the middle, as in elliptic, but toward one end. The petiole is attached at the broad end. **Ovoid** is the same but three-dimensional instead of flat.

Obovate, the same as ovate, but with the petiole attached at the narrow end.

Lanceolate, the shape of a lance. Four to six times as long as broad, with curved sides and pointed ends (especially the apical end), larger at one end than the other. The petiole is attached at the broad end.

Oblanceolate, the same as lanceolate but with the attachment of the petiole at the narrow end.

Oblong, nearly rectangular but with rounded corners, about two or three times as long as broad. The petiole is attached on one of the shorter sides.

Linear, tending to be an elongated rectangle or at least a figure with parallel long sides. The length several (at least 8) times the width.

Rhombic, diamond-shaped. An equilateral parallelogram with oblique angles. The petiole is attached to one of the sharper angles.

Cordate, the classical heart shape, the length being greater than the width. The petiole is attached in the indentation at the base. The apex is acute.

Obcordate, cordate but with the attachment of the petiole at the sharp point of the "heart."

Reniform, kidney-shaped or bean-shaped, with a rounded apex and with the width greater than the length. The petiole is attached in the broad basal indentation of the blade.

Sagittate, arrowhead-shaped, a narrow isosceles triangle with an angular indentation in the short side. The petiole is attached in the indentation.

Hastate, somewhat in the form of an arrowhead, but with basal divergent lobes.

Subulate, slender and tapering from the base upward, awl-shaped.

Cuneate or **cuneiform,** wedge-shaped. Essentially a narrow isosceles triangle with the distal corners (those away from the petiole) rounded off. The petiole is attached at the sharp angle.

Deltoid, in the shape of the Greek letter Delta. An equilateral triangle. The petiole is attached in the middle of one of the sides.

Obdeltoid, deltoid but with the attachment of the petiole at one of the angles instead of on a side of the triangle.

Spathulate or **spatulate,** resembling a spatula. Rounded above but tapering into an elongated, narrow, gradually diminished lower portion.

The apex of the leaf blade is described by the following terms.

Rounded, that is, gently curved.

Obtuse, forming an obtuse angle, that is, more than a right angle.

Acute, forming an acute angle, that is, less than a right angle.

Attenuate, drawn out into an elongated tapering point.

Truncate, ending abruptly as if chopped off, that is, with the apical end running at right angles to the midrib.

Notched, with an indentation at the apex.

Emarginate, with a deep, rather broad indentation at the apex.

Retuse, with a shallow and rather narrow indentation at the apex.

Cuspidate, with an abrupt, sharp, firm point (**cusp**) at the apex.

Mucronate, with an apical short abrupt point (**mucro**). The diminutive is **mucronulate.**

Acuminate, with the summit more or less prolonged into a tapering point.

Aristate, bristle-pointed, that is, with a terminal **awn,** which is a bristlelike structure.

Spinose, with a spine at the tip. The same adjective is used to indicate presence of marginal spines, as in **spinose-dentate.**

Caudate, the apex elongated and taillike.

The following terms are applied to the bases of leaf blades.

Rounded, obtuse, acute, attenuate, and **truncate** may be applied to bases in the same way as to apices.

Cordate, with the leaf base indented like the base of a "heart," regardless of the shape of the rest of the leaf.

Auriculate, with a pair of basal divergent rounded lobes, each like the lobe of a human ear.

The Petiole

Usually the leaf blade is supported by a petiole, but some leaves do not have one (i.e., they are **sessile**). The leaves of some monocotyledonous groups (e.g., grasses, irises, jonquils, or daffodils) show no distinct petioles. In the grasses the basal portion of the leaf is a **sheath** surrounding the stem (this perhaps corresponding to a petiole). Where the blade is attached to the sheath, there is a small scale called the **ligule.**

The black acacia has leaflike structures (**phyllodia**) developed by broadening of the petiole, and ordinarily there is no leaf blade. Suckers and branches in the vicinity of wounds have leaves with somewhat broadened petioles and with portions of a bipinnate blade. There are all transitions between the pinnate type of foliage typical of many of the acacias and the special flattened petioles of the black acacia.

In some plants the base of a sessile leaf encircles the stem. Such a leaf is **perfoliate.** In some honeysuckles and the miner's-lettuce (*Claytonia perfoliata*) a pair of joined leaves

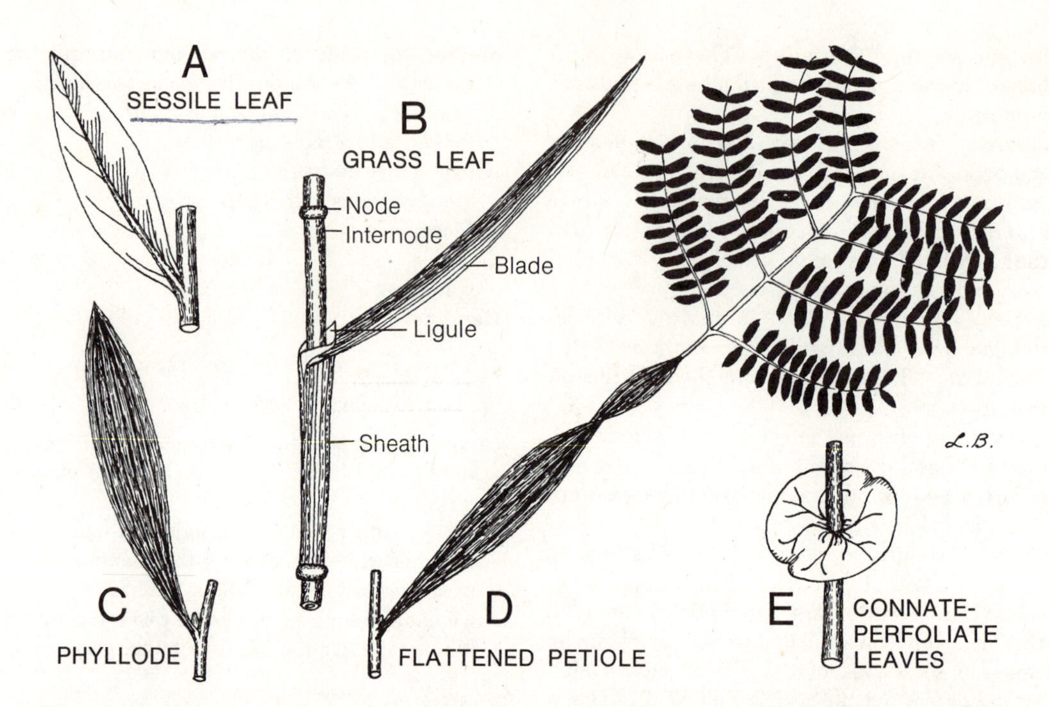

Figure 3-10. Absence or specialization of petioles: **A** sessile leaf; **B** leaf with a basal sheath instead of a petiole, brome-grass (lower leaf of *Bromus rigidus*); **C** phyllode (phyllodium) (i.e., flat petiole) of the black acacia (*Acacia melanoxylon*); **D** bipinnate leaf of a seedling of black acacia, indicating the ancestral condition; **E** connate-perfoliate pair of leaves of chaparral honeysuckle (*Lonicera interrupta*).

from the same node may be perfoliate. Leaves of this type are **connate-perfoliate. Connate** means grown together from the beginning of development.

Relative Positions of Leaves

In the seed plants the **primary leaves** are the cotyledons. **Secondary leaves** may be classified according to their position on the plant.

In herbaceous plants commonly there is a basal whorl or rosette of leaves at ground level just above the cotyledons (or, after these have fallen, their attachment scars). These are known as **radical** or preferably as **basal leaves** because the latter term does not carry the connotation of coming from the root or radicle. Frequently the basal leaves differ from the **cauline** leaves (*caulis,* stem) higher on the plant. Sometimes there is a gradual transition of leaf types from the base of the stem upward. The uppermost cauline leaves are **bracts,** that is, leaves subtending flowers (or other reproductive parts). Usually these are simpler in form and smaller than the lower leaves, and they tend to be sessile. They may not be green.

Number of Leaves at a Node

Ordinarily there is only one leaf at a node. From node to node up the stem the leaves are in a gradually ascending spiral, but especially in pressed specimens they appear to alternate on the two sides of the stem, and they are referred to as **alternate.** In many plants there are two leaves at each node, and this arrangement is **opposite.** In a few plants there are three or more leaves at a node, and these are described as **whorled, cyclic,** or **verticillate.** The noun forms are **whorl, cycle,** and **verticil.**

Special Kinds of Leaves

The variety of leaves is infinite. The more striking special types include bud scales, functionless scale leaves of desert plants, spines, tendrils, insect-capturing leaves, and succulent storage leaves such as those of the century-plants (*Agave*), stonecrops (*Sedum*), echeverias (*Echeveria*), and the garden hen-and-chickens (*Sempervivum*).

SUGGESTED CAMPUS CLASS EXERCISE

About 12 to 20 trees or shrubs will be marked by tags bearing their names and assigning them numbers. A small leafy branch should be collected from each. A key employing leaf characters is to be worked out. This key should enable another person to determine the identity of any of these plants if the tags should be removed. However, the key is for segregation of only these plants from each other; it does not apply to other species.

In preparation of the key, it is best first to sort out the branches into two groups distinguishable on the basis of some clear cut character(s). For example, one pile of specimens may include all having simple leaves, the other compound. Next the compound leaves may be separated into those which are palmate and those which are pinnate, and so on.

The following is an example of a key of this type.

1. Leaves simple.
    2. Leaves palmately lobed.
        3. Leaves opposite. 1. *Big-leaf maple.*
        3. Leaves alternate. 2. *English ivy.*
    2. Leaves entire or toothed.
        3'. Leaves spinose-dentate. 5. *Holly.*
        3'. Leaves *not* spinose-dentate.
            4. Leaves obovate, entire. 8. *Pittosporum.*
            4. Leaves ovate, crenate-serrate. 10. *Viburnum.*

1. Leaves compound.
    2'. Leaves palmate.
        3''. Leaves opposite. 4. *Mexican orange.*
        3''. Leaves alternate. 7. *Scotch broom.*
    2'. Leaves pinnately compound.
        3'''. Leaves simply pinnate.
            4'. Leaves even-pinnate; leaflets entire. 3. *Carob.*
            4'. Leaves odd-pinnate; leaflets serrate. 6. *Cape honeysuckle.*
        3'''. Leaves more than once pinnate.
            4''. Leaves bipinnate.
                5. Rachis and rachillae with prickles; leaflets oblong. 11. *Caesalpinia.*

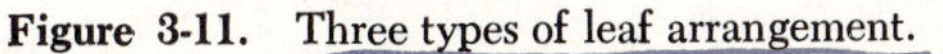
Figure 3-11. Three types of leaf arrangement.

5. Rachis and rachillae without prickles; leaflets linear. 9. *Acacia.*

4''. Leaves partly tri- and partly quadripinnate (or to be interpreted as 3 or 4 times ternate). 12. *Nandina.*

Each descriptive statement in a key begins with a noun, and the opposed descriptive statement should begin with the same noun. The modifiers follow the nouns. Whenever there is more than one descriptive statement in a lead of the key, the separation is by semicolons (cf. leads 4' and 5 above).

A good key *never* has more than two opposed leads. A little thought will reveal ways of avoiding this common error in key construction.

Such terms as cordate and ovate are not necessarily usable. Often the leaf is intermediate, i.e., cordate-ovate or ovate-cordate. These are compounds, but the emphasis is upon the latter term in the compound. Similar compounds are ovate-lanceolate, dentate-serrate, elliptic-oblong, obovate-oblanceolate, etc.

A *scavenger hunt* for leaf types in the lists in the text is suggested also.

4

Color and Surface Characters

Plastid Pigments

The characteristic coloring matter of plants is the green pigment **chlorophyll.** Through the presence of this material, living green cells are able to absorb the energy of light, and this makes possible a series of reactions producing, ultimately, various sugars and other carbohydrates. Later other organic compounds are formed. Since most plants contain chlorophyll, they are green at least in their vegetative parts. The chlorophyll is contained in special ellipsoidal solid bodies called **plastids** (in this case **chloroplasts,** or green bodies) and chlorophyll is said to be a **plastid pigment.** Plastid pigments are insoluble in water, and therefore they remain within the plastid instead of being dissolved in the sap of the cell.

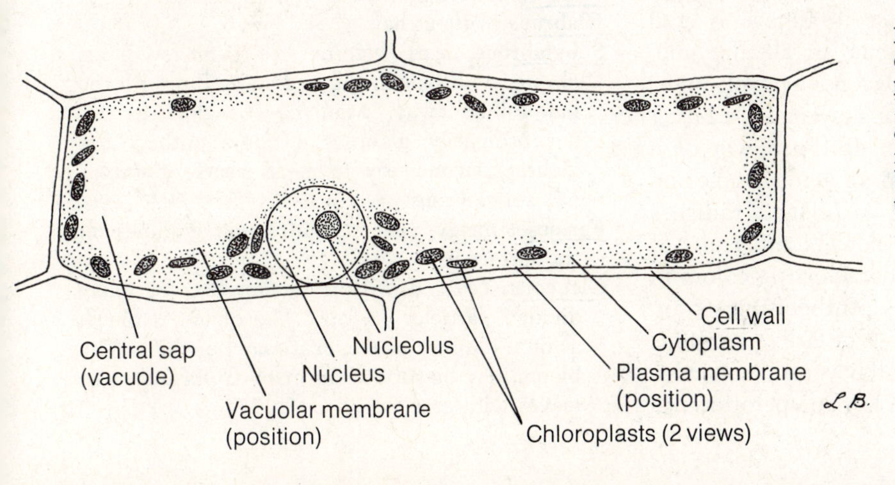

Figure 4-1. A typical plant cell as seen under the compound microscope; from the leaf of elodea (*Elodea densa*), a water plant common in fish ponds and fish bowls. The chloroplasts are to be noted.

The term "chlorophyll" is used inclusively to cover mixtures of associated pigments, the best known being chlorophyll A, chlorophyll B, xanthophyll, and carotene (carotin). The green color of the plant may be modified according to the proportions of these pigments. A higher percentage of chlorophyll A produces a tendency toward blue-green, a higher percentage of chlorophyll B toward yellow-green. The relative amounts of xanthophyll (deep yellow or orange) and carotene (light yellow) increase or decrease the tendency toward yellow-green.

Certain reds, yellows, and other colors may be due to other plastid pigments which, like the chlorophyll group, are insoluble in water. As explained in Chapter 12, pigments insoluble in water are both relatively stable in pressed specimens and often relatively reliable for use as key characters for plant classification.

Water-Soluble Pigments

Many of the pigments characteristic of plants are not located in plastids but are water soluble and dissolved in the cell sap. The most common of these are the **anthocyanins.** Usually these pigments are in the series of colors between blue and red, and they vary from lavender to purple according to the acidity or alkalinity of the cell sap. The resulting color, like that of litmus paper, contains a higher proportion of red as opposed to blue in an acid medium and the reverse in an alkaline medium. If acid is added to lavender-purple juice, the juice turns red; if drops of alkali (a base) are added, it turns blue. If the flower of a purple *Iris* is dipped into an acid, it takes on a reddish color. Red cabbage, radishes (roots), *Coleus,* and the flowers and fruits of many plants owe their characteristic colors to anthocyanin pigments. Anthocyanins are formed only in the presence of large quantities of sugar because the anthocyanin molecules contain sugar. Light and low temperature are favorable also to formation of anthocyanins.

Colors due to water-soluble pigments are destroyed or changed in pressing specimens. They are variable under all conditions and, except for their presence as opposed to other types of pigments, relatively unreliable for purposes of distinguishing plant groups. Where anthocyanins are concerned distinctions of fine shades of color frequently are meaningless in classification.

Autumn Colors

In regions with cold winters, brilliant autumn colors are characteristic. With the gradual coming of autumn, but not necessarily a frost, the chlorophylls disintegrate, leaving the characteristic yellow of carotene and xanthophyll. Furthermore, bright cool days favor accumulation of sugar in leaves, and both low temperature and presence of sugar promote the formation of anthocyanins, which yield red coloring.

Surface Characters of Leaves and Stems

The following are some of the more common terms applied to the surfaces of the leaf or stem and their minute appendages. It is to be noted that plant hairs (**trichomes**) do not correspond with those of mammals.

Smooth, not roughened.

Glabrous, without hair.

Subglabrous, nearly glabrous, i.e., with few hairs.

Glabrate, with hair in the beginning but with the hair falling away. Man may be glabrate, but not ordinarily glabrous. (Some authors use glabrate erroneously to mean nearly glabrous, i.e., subglabrous.)

Farinose, mealy, that is, with small granules on the surface.

Glaucous, covered with a bloom, that is, finely divided particles of wax. The bloom is bluish as on a plum or prune, or it may be white. The bloom may be rubbed off many fruits and some leaves with ease.

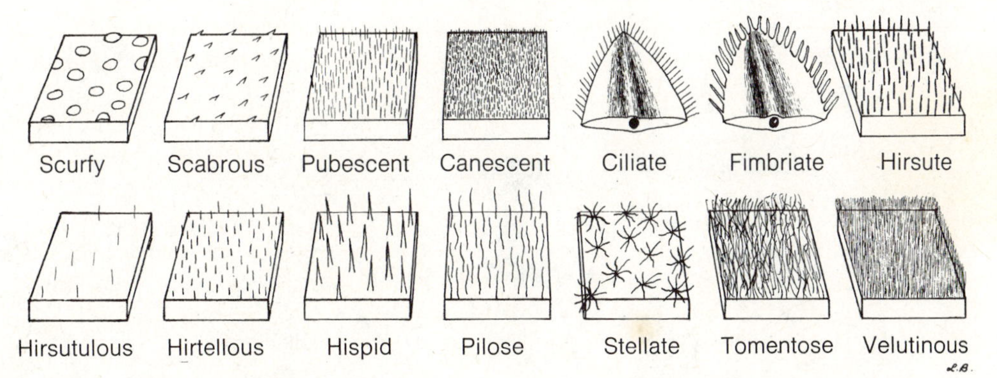

Figure 4-2. Surface features (including hairs) of leaves.

Glaucescent, slightly glaucous, or bluish gray.

Scurfy, covered with dandrufflike scales, scalelike hair, or fringed scales. Scales are characteristic of the goosefoot family and scaly hairs of the oleaster family.

Scabrous, with minute, rough projections; therefore rough to the touch.

Pubescent, hairy or downy, usually with fine soft hairs. Commonly the term pubescent is used to indicate hairiness of a generalized instead of a specialized type, and the term is used loosely to cover any kind of hair. The noun is **pubescence.**

The following terms are used for specific types of hairiness.

Canescent, grayish white or hoary, due to the dense covering of the surface with fine, short, white hairs.

Ciliate, with hairs along the margin of the structure, after the manner of the eyelashes on the human eyelid.

Fimbriate, with a fringe along the margin, that is, resembling the fringe on the sleeve of an early American buckskin shirt.

Hirsute, hairy, with more or less stiff hairs.

Hirsutulous, slightly hirsute.

Hirtellous, minutely hirsute.

Hispid, bristly, covered with markedly stiff hairs. The diminutive form is **hispidulous.**

Pilose, with soft, elongated, slender hairs.

Pubescent (see above). The diminutive is **puberulent** or **puberulous.**

Stellate, descriptive of hairs branched so that each one appears like a small star. However, the term is applied to hairs with only two or three branches, and these may not be suggestive of a star.

Tomentose, covered with matted woolly hairs, these forming a **tomentum.**

Velutinous, velvety, that is, with a pile like that of velvet.

Many plants are characterized by surface glands. These are small cellular organs which secrete oils or other aromatic substances, resins, tarlike substances, or inert materials. Their form and structure vary greatly. Sometimes they appear on the surface of the plant as small projections of various shapes; sometimes they are sunken, occurring in pits or depressions; or sometimes they are elevated on stalks or produced at the ends of hairs or bristles. Almost any kind of swelling is likely to be referred to as a gland whether it is actually glandular or not, that is, whether or not it produces any secretion. Some glands are so active that their secretions cover nearly the entire surface of the plant. This is true, for example, of the creosote-bush of the Southwestern Deserts.

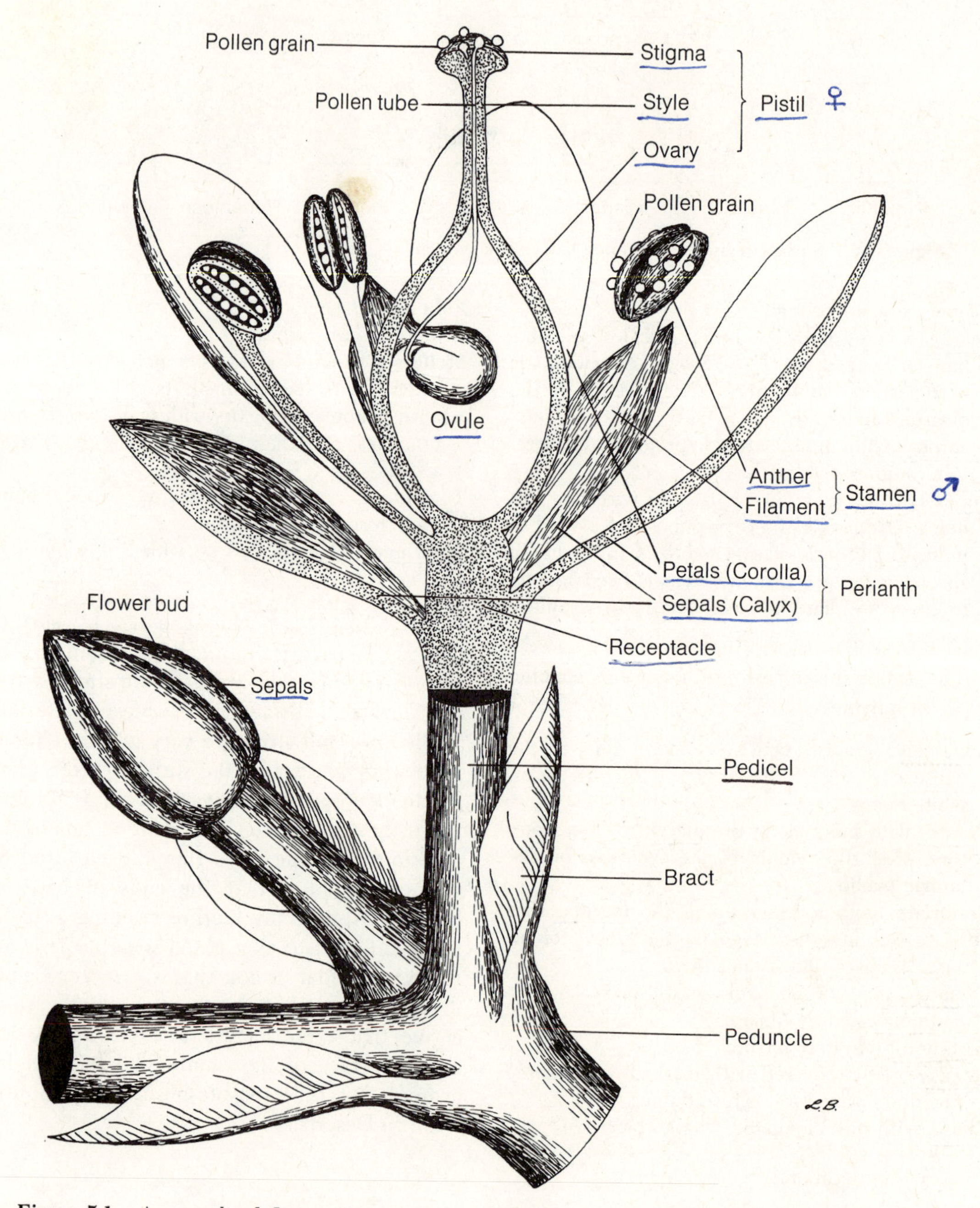

Figure 5-1. A generalized flower (longitudinal section of one in a cluster [umbel] of three), showing the parts. Semidiagrammatic.

5

The Flower

The Structure of a Generalized Flower

A **flower** is a branch with special appendages. The end of the branch which supports the flower parts is the **receptacle** (**thalamus,** or **torus**), consisting of several or many crowded nodes separated by very short internodes. The essential parts of the flower grow one or more from each level of the receptacle. The first internode of the branch below the receptacle is the **pedicel.** The second internode below a single flower may be called the **peduncle,** though this term is used more often to designate a larger branch bearing a cluster of flowers, each on a pedicel or sessile. At the base of the pedicel usually there is a **bract** (specialized leaf) subtending the flower, and there may be another at the base of the peduncle subtending the cluster of flowers (or subtending the single flower if the peduncle bears only one pedicel).

The four kinds of stem appendages which are the usual parts of a flower are as follows.

1. The **sepals,** known collectively as the **calyx.** In most flowers the sepals are green, although in some they are brown and scalelike or in others highly colored and petallike. Usually they are of approximately the same structure as leaves. Commonly they serve as a cover preventing rapid evaporation from the developing inner flower parts; however, when the sepals are colored they may attract insects* as petals do.

2. The **petals,** known collectively as the **corolla.** Commonly the petals are colored or white, and they attract insects to nectar glands on or near their bases. Usually the insects carry pollen, thus effecting **pollination** (transfer of pollen from stamen to stigma, cf. below) of the flower. A collective name for both sepals and petals is **perianth,** or **floral envelope.** The distinction between sepals and petals is not always clear. In many monocotyledonous flowering plants, such as the lily family, petals and sepals usually look alike, and

* Although most animals are not capable of seeing color and they live in a world of white, black, and various grays resembling a black-on-white photograph, birds (except owls) and many insects have color vision.

they are referred to simply as the perianth or all the parts are called **tepals.** In cactus and water-lily flowers there is a gradual transition from sepals to petals. In such cases the term perianth is used sometimes, or the words **sepaloids** and **petaloids** are employed to denote structures appearing like sepals and petals but of different origins.

3. The **stamens,** known collectively as the **androecium.** Each stamen consists of an **anther** or pollen-bearing portion and a **filament** or stalk. The anther contains usually four microsporangia (spore cases) or **pollen sacs** or **chambers.** Most commonly the pollen liberated by the anthers appears upon magnification as a group of minute, yellow spheres; however, it may have various colors and shapes, and it may be marked with peculiar surface patterns. Usually it is carried by insects or wind to the stigma (cf. below). Since the male gametes (sex cells) are developed in the pollen grains and these are developed within the anthers, the stamens are the male organs of the flower.

4. The **carpels,** known collectively as the **gynoecium.** If there is only one carpel, it is the **pistil;** if there are two or more separate carpels, each is a pistil; if there are two or more united carpels, they together are the pistil. A pistil consists of a basal **ovary,** a median **style,** and a terminal **stigma.** The ovary contains **ovules** (each a megasporangium surrounded by an integument) which, after fertilization, develop into **seeds.** The **stigma** is a collector of pollen, and the style beneath it is a passageway through which a tubular outgrowth from the pollen grain (the pollen tube) may grow into the ovary cavity. The pistil is a female organ because the ovule contains a female cell or gamete which unites with a male cell in fertilization.

The Nature of the Flower Parts

Each of the four principal kinds of structures noted above is an appendage of the stem, and each flower part is thought by many botanists to be fundamentally a specialized leaf. One source of evidence is from abnormal flowers in which there is development of each of the parts into a leaflike structure. Evidence from other sources is presented in Chapter 17.

Sepals. In even normal flowers similarity of the sepals to leaves seems obvious because usually these structures are broad and green and with ordinarily three leaf (vascular) traces from the stele of the stem,* as in the foliage leaves of the plant.

Stamens and Petals. The possible relationship of stamens to leaves is less obvious in flowering plants because ordinarily there is no blade. There is only one vascular trace. The filament may represent a petiole and the connective between the pollen sacs a leaf midrib. Some evidence indicates origin of stamens from branchlets, but the most primitive known flowering plants have stamens evidently derived from leaves (cf. pp. 562–563).

Most petals are derived from stamens, and they have also only one vascular bundle. In some garden plants and more rarely in wild ones some of the stamens have been transformed by genetic mutation into petals, forming a "double flower" as in the more highly modified roses and camellias. This transformation is due to broadening of the filaments or anthers of all or most of the stamens. Broadening of the filament may be accompanied by ultimate nondevelopment or opening, broadening, and modification of the anther and assumption by the stamen of the form and color of a petal. Gradual transition from stamens to petals occurs naturally in the evening-star (*Mentzelia laevicaulis*), and normally the five outermost stamens have somewhat broadened filaments. In some flowers these and others of the stamens may be partly or wholly modified into petals.

In the flowers of the spicebush (*Calycanthus*) the petals shade off gradually into sepals. The flower of the water-lily displays gradual transition from sepals to petals and from petals to stamens, emphasizing the interrelationships of these three floral parts in at least some

* I.e., in this case, the receptacle.

Figure 5-2. Transition from stamens to petals in a rose (*Rosa,* hybrid); degrees of expansion of the pollen sacs of the anther into the blade of a petal.

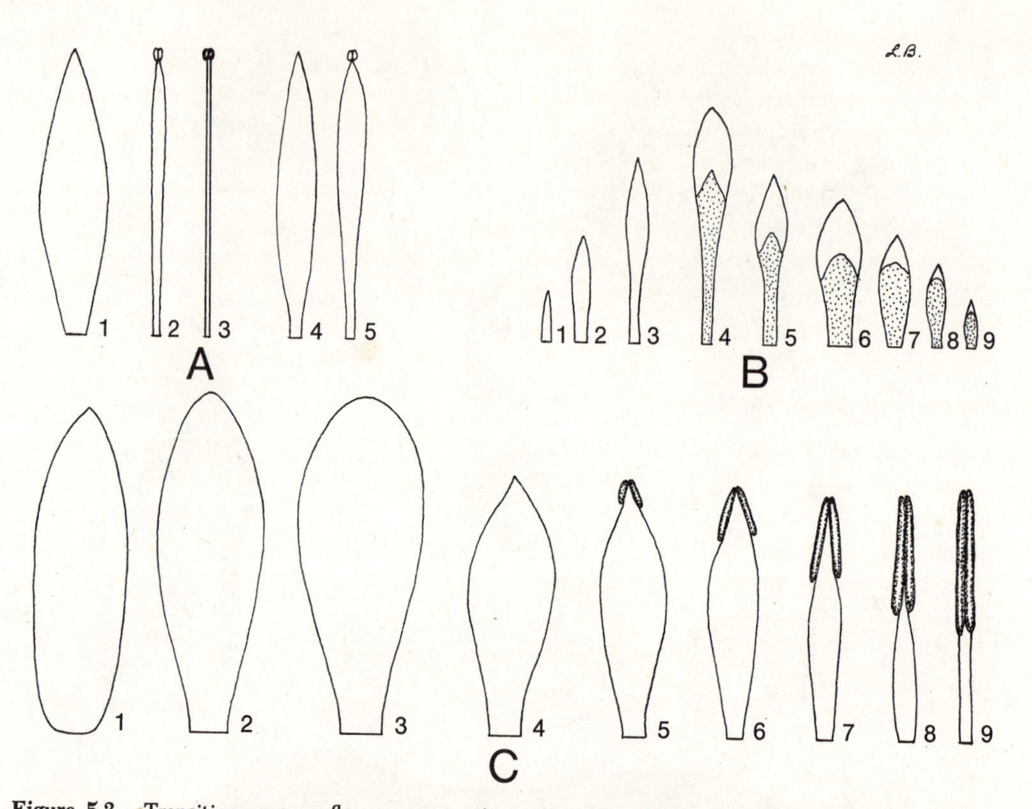

Figure 5-3. Transitions among flower parts: **A** evening- (or blazing-) star (*Mentzelia laevicaulis*), **1** petal, **2** one of the five outermost stamens with a broadened filament, **3** typical stamen, **4** and **5** stamens among the outer five from flowers in which these organs are transformed partly or completely into petals; **B** spicebush (*Calycanthus occidentalis*), transition from the outermost sepals (**1**) to the innermost petals (**9**); **C** cultivated hybrid water-lily (*Nuphar*), transition from sepals (**1**) to petals (**2**) and to stamens (**9**).

flowers. In prickly pears and chollas (cacti) the gradation of petaloids into sepaloids is continued into vegetative leaves.

Pistils. A specialized leaf which forms either all or part of a pistil is a **carpel**. That this is a specialized leaf is indicated by developmental series among the most primitive known flowering plants (cf. p. 555). The vascular trace number is usually three, as in the leaves and the sepals.

A pea pod is a pistil formed from a single carpel. The stigma is developed from crests of the leaf margin near the tip, and the area just basal to the stigma is elongated and folded into a slender tube, the style. The rest of the blade is folded into an ovary with the upper (ventral) surface inside and the lower (dorsal) surface outside. The edges of the leaf are coherent with each other through the entire length of the carpel. The **ovules,** which develop eventually into **seeds,** are in the internal cavity. The ovules are developed on a fleshy structure, the **placenta,** which runs along both sides of the **suture,** i.e., the fused margins of the leaf. The lower edge of the carpel is the **midrib,** which is the line of folding. In the pea pod the ovules alternate in points of attachment near the left and right margins of the leaf, so that the apparently single row of ovules or seeds actually is made up of two rows meshed into one, that is, a double row.

In some plants, such as the perennial larkspurs (*Delphinium*), there is more than one pistil in a flower, and each pistil is formed from one carpel; in other flowers, as those of the orange, several carpels (as many as there are sections later in the fruit) are united into a single pistil. Union of carpels will be discussed in the next chapter.

The Flower and Seed Development

The Pollen Grain. A flowering plant pollen grain consists of two cells, one (the generative cell) enclosed within the other (the tube cell). When a pollen grain is placed in the proper kind of sugar solution, simulating the exudate on the stigma, the tube cell grows rapidly into a long tube, and the generative cell produces two male nuclei (gametes) which correspond to the antherozoids (male cells) of the lower plants or the sperms of animals. The pollen tube may attain various lengths, ranging from a minute fraction of an inch as in wheat to several inches as in corn, in which the styles (silks) are very long and flexible. After the pollen tube enters the ovary, it grows downward into the opening (micropyle) in one of the ovules and its male gametes are discharged.

The Ovule. The outer layers of a flowering plant ovule include one or two integuments and a megasporangium (known as the nucellus), the two or three layers being grown together. A minute opening in the integument(s) (the micropyle) exposes a portion of the nucellus. The pollen tube usually grows through the micropyle and digests its way through the nucellus. A group of seven special cells is enclosed by the nucellus; five are unimportant, but one is the egg or female gamete, and another is the primary endosperm cell.

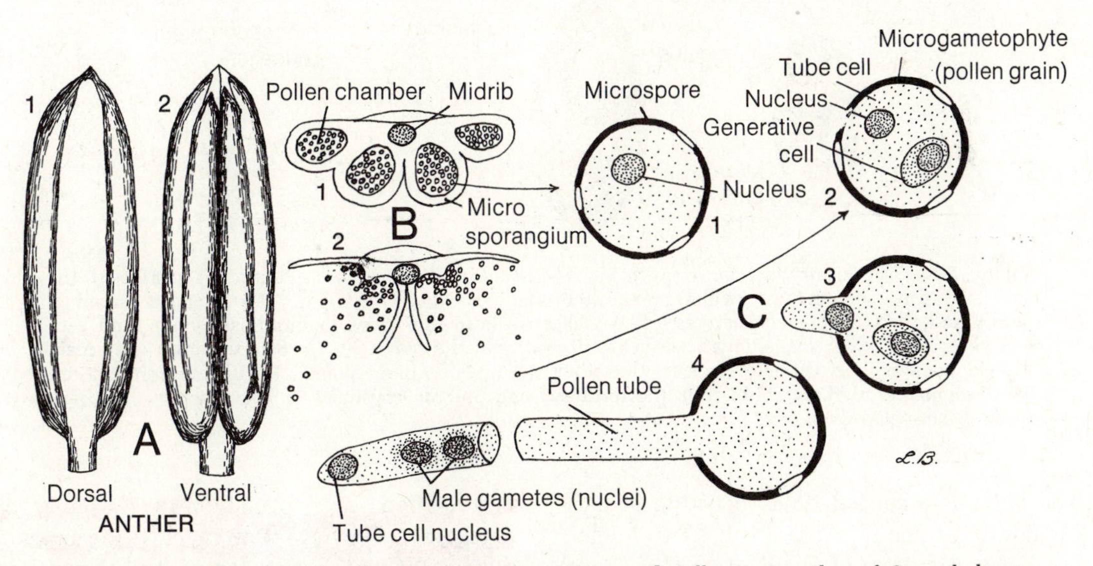

Figure 5-4. Life history of a flowering plant I—anthers and pollen: **A** anther of Spanish broom (*Spartium junceum*), **1** ventral view (side toward the stem axis, i.e., the upper side), **2** dorsal view (side away from the axis, the underside); **B** cross sections of the same anther, **1** before dehiscence, **2** after dehiscence; **C** flowering plant pollen grain development, **1** microspore developing into a pollen grain, **2** pollen grain composed of two cells, one within the other, **3** pollen grain germinating in a sugar solution, the pollen tube emerging, **4** pollen tube with the male gametes.

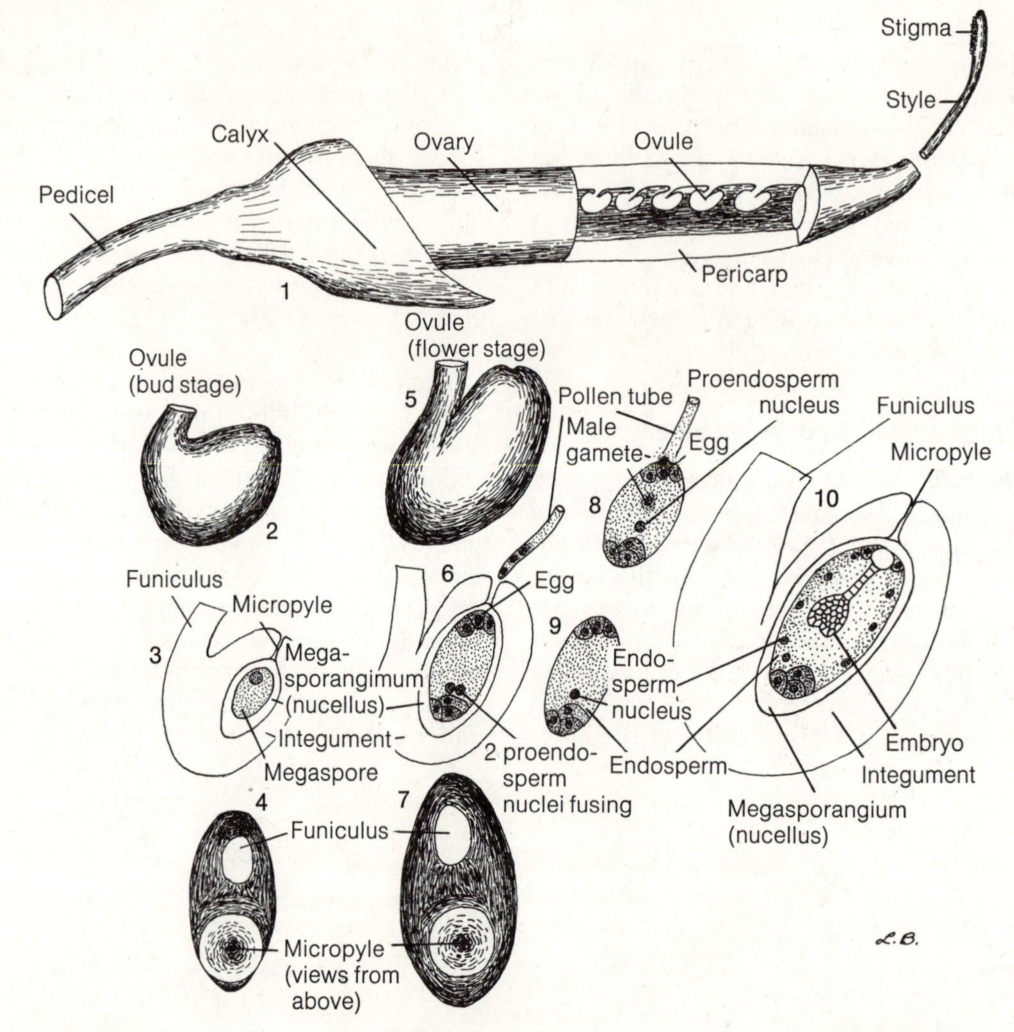

Figure 5-5. Life history of a flowering plant II—ovule, fertilization, and embryo: **1** pistil with the side cut away, showing the ovules (Spanish broom, *Spartium junceum*), **2–9** composite series of ovules and their included structures, **2–4** young ovule in side view, longitudinal section, and ventral view, **5–7** older ovule (just before fertilization) in the same views, **8** fertilization (one male gamete nucleus joins the egg nucleus, the other the proendosperm nucleus), **9** after fertilization, **10** development of the embryo from the fertilized egg and development of the endosperm. (This ovule is campylotropous.)

The latter is a special type occurring in only the flowering plants. At first it has two nuclei; but before the time of fertilization these fuse, forming a single nucleus.

Fertilization. The pollen tube bursts open, freeing the two male gamete nuclei. One of these unites with the egg nucleus effecting fertilization, i.e., formation of the first cell (zygote) of a new individual. This union is followed by rapid development of the embryo within the ovule. The other male gamete unites with the nucleus of the primary endosperm cell, thus giving rise to a special tissue, the **endosperm.** The endosperm absorbs food from surrounding tissues, especially the spo-

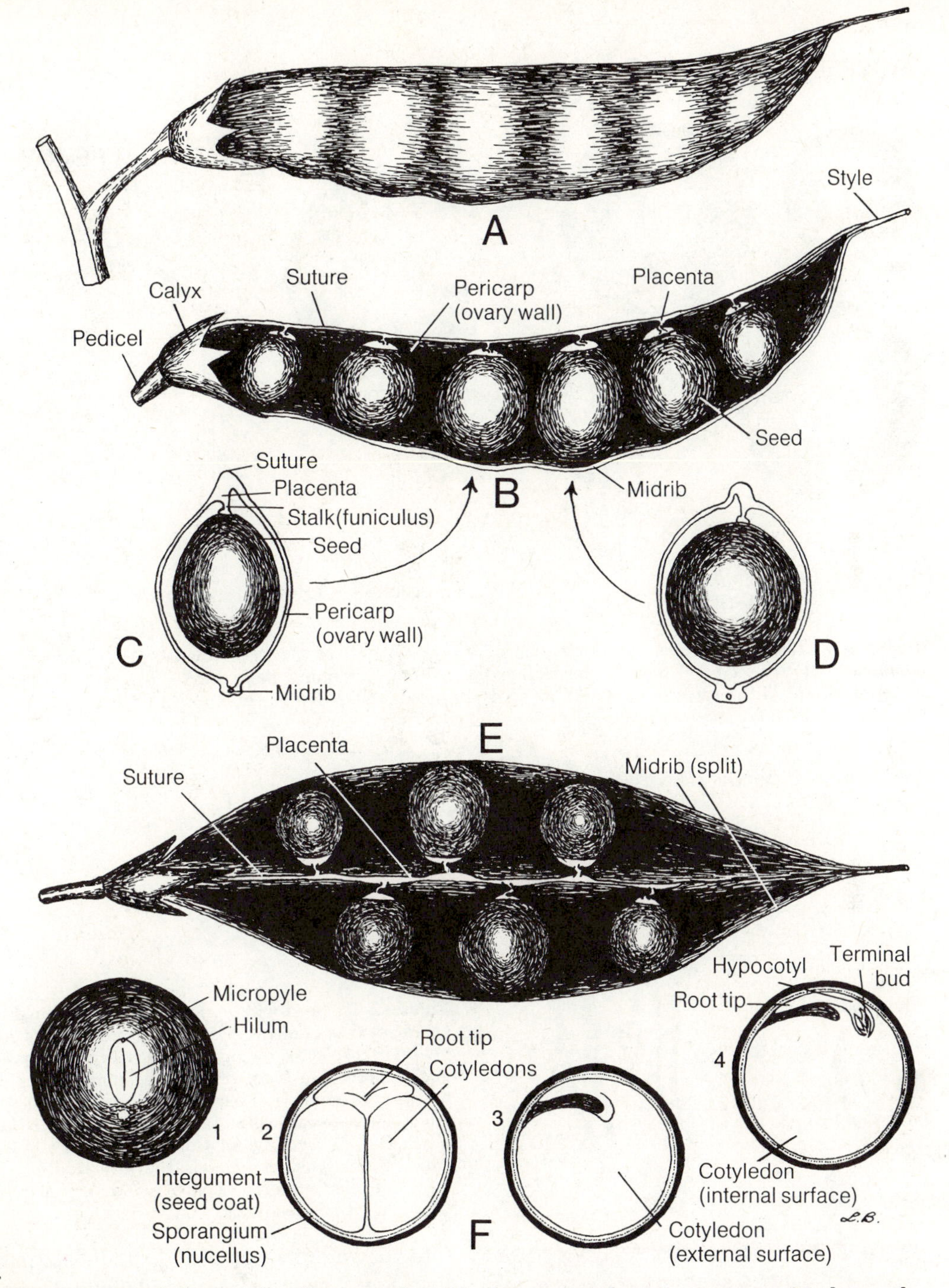

Figure 5-6. Life history of a flowering plant III—fruit and seeds of the sweet pea (*Lathyrus odoratus*): **A** fruit (a pod or legume); **B** fruit with half of the ovary wall removed; **C, D** cross sections at the points indicated, showing the alternate attachments of ovules on the left and right sides of the fruit; **E** dorsal view of the fruit opened along the midrib; **F** seed, **1** external view, **2** sectional view with the same orientation, **3** sectional view, the seed rotated 90° and with the seed coat or integument removed on the near side, the embryo shown whole (i.e., not sectioned), **4** strictly sectional view in the same plane, showing the embryo.

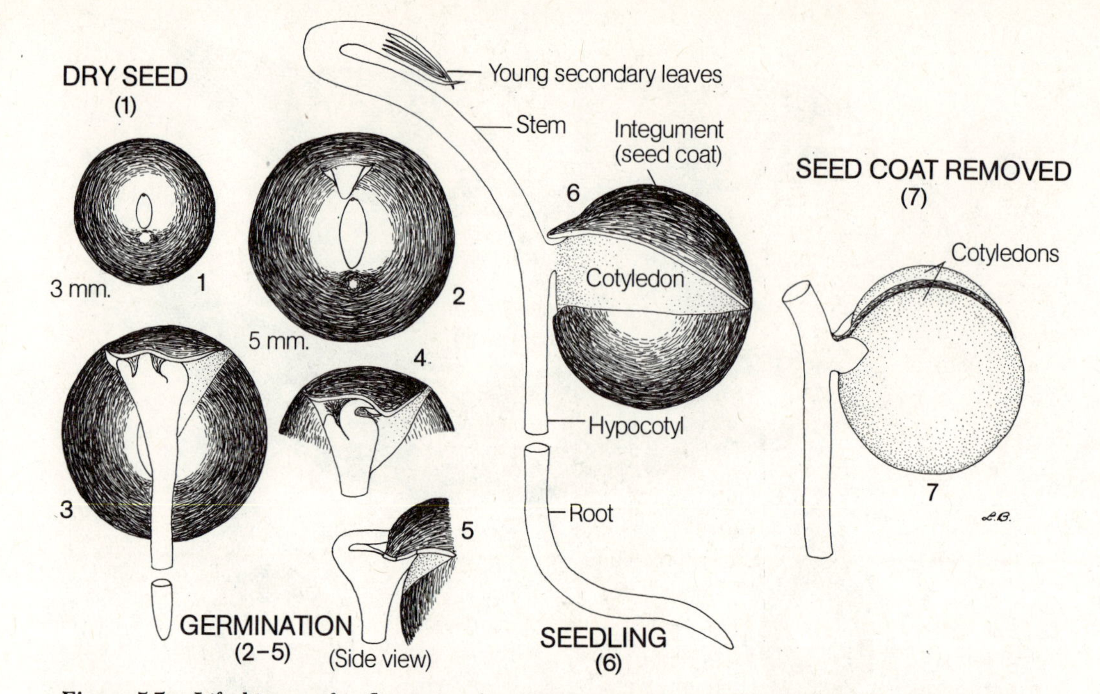

Figure 5-7. Life history of a flowering plant IV—germination of the seed, sweet pea (*Lathyrus odoratus*): **1** dry seed, **2** seed after soaking, the root-tip beginning to emerge, **3–4** later stages in germination, **5** the same but rotated 90°, **6** advanced stage, **7** the same with the seed coat removed.

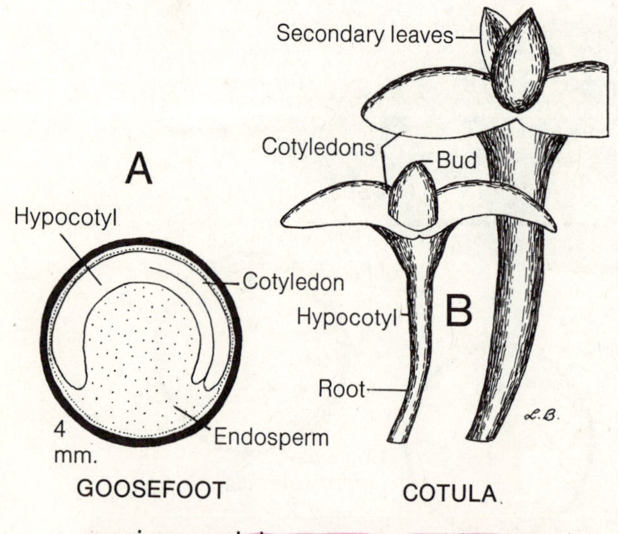

Figure 5-8. Seed with endosperm and seedlings with emergent cotyledons: **A** seed of a goosefoot (*Chenopodium murale*), showing the endosperm (not present in the sweet pea seeds of the preceding figures) and the curved embryo characteristic of the goosefoot family and several related families, cf. the order Caryophyllales; **B** seedlings of *Cotula australis,* a plant in which the cotyledons emerge from the seed coat, enlarge, and become green (photosynthetic), as they do *not* in sweet pea seedlings.

rangium, and becomes a nutritive reservoir around the developing embryo, which draws upon it before or during germination or both.

The Embryo. During the resting stage before germination of the seed, the embryo consists of a hypocotyl with the tip differentiated into the rudiment of a root (radicle), one or two (or rarely three) **cotyledons (primary leaves),** and the epicotyl (stem tip), which is enclosed by very young **secondary leaves.** When the embryo germinates, the cotyledons may either remain inside the seed coat or emerge and become green.

6

Variation in Flower Structure

Incomplete Flowers

The flower discussed in the preceding chapter is **complete,** that is, with all the four usual series of parts—sepals, petals, stamens, and pistils—but many are **incomplete.** Incomplete flowers may lack all or part of the perianth or either stamens or pistils.

The perianth may be wholly lacking, as in the flowers of willows (*Salix*) or the lizard's-tail (*Saururus*), but usually only the petals are absent, and the calyx is present. Petals may be present and sepals absent, but this is very rare. Flowers lacking petals and with or without sepals are **apetalous** (*a*-, without). Examples include the following: grasses, sedges, willows, poplars, cottonwoods, walnuts, hickories, pecans, hazelnuts, birches, alders, beeches, oaks, elms, nettles, mulberries, mistletoes, buckwheat, smartweeds, and Russian thistle. In some apetalous plants, e.g., *Anemone* and *Clematis,* the sepals are large and white or colored, and they attract insects.

List of Symbols Used in Flower Illustrations

An	Anther	Ps	Pistil
Br	Bract	Pt(s)	Petal(s)
Ca	Calyx	Ovl	Ovule
Cr	Corolla	Ovr	Ovary
Ds	Disc	Sp(s)	Sepal(s)
Gl	Gland	St(s)	Stamen(s)
Hyp	Hypanthium	Stg(s)	Stigma(s)
Pd	Pedicel	Sty	Style
Pr	Perianth	Tb	Tube

Flowers having both stamens and pistils are **perfect** or **bisexual** (or in the older works, hermaphrodite). Flowers lacking *either* stamens *or* pistils are **unisexual,** as distinguished from the usual bisexual or perfect type. Those having only pistils are **pistillate;** those having only stamens are **staminate.** If both types are produced on the same individual, the plants are **monoecious** (*monos*-, one; *oikos*-, household); if the staminate and pistillate flowers

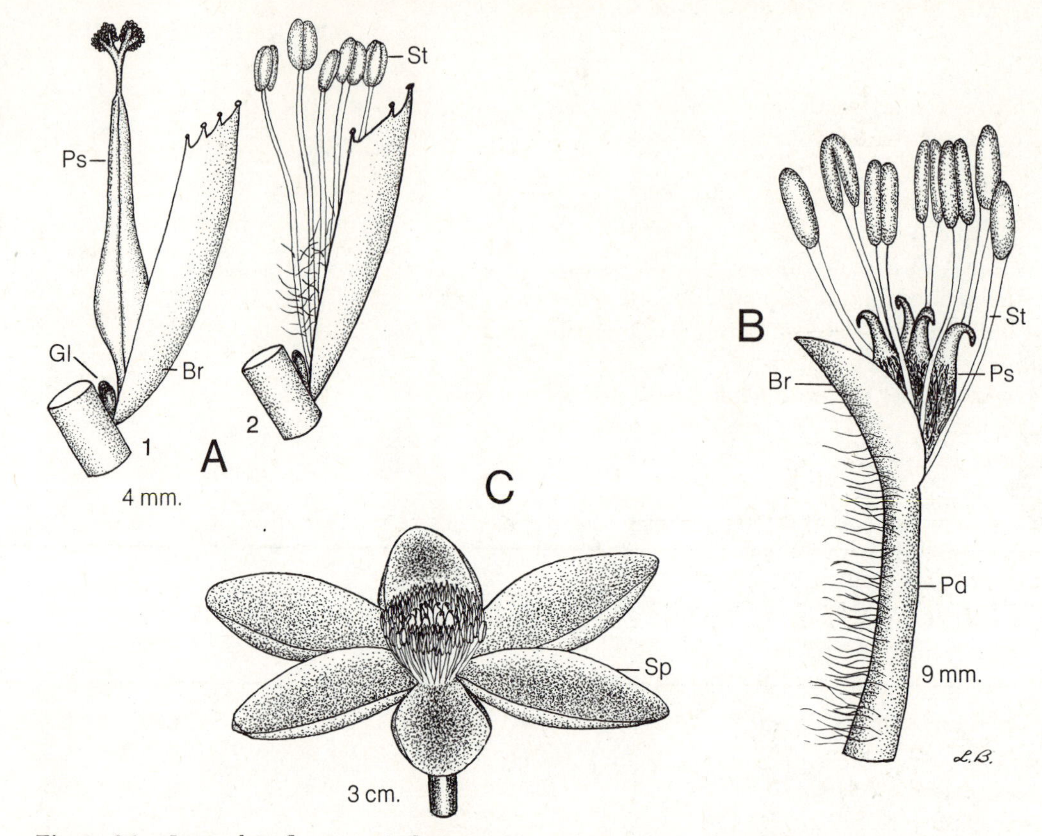

Figure 6-1. Incomplete flowers: **A** flowers composed of only the reproductive structures, red willow (*Salix laevigata*), **1** pistillate flower consisting of only a pistil and a gland, **2** staminate flower consisting of only stamens and a gland; **B** flower consisting of only stamens and pistils, lizard's-tail (*Saururus cernuus*); **C** flower consisting of sepals, stamens, and pistils but no petals, marsh marigold (*Caltha palustris*).

are on different plants, the plants are **dioecious** (*di-*, two; *oikos*). If perfect, pistillate, and staminate flowers are on the same individual, the plant is **polygamous** or, strictly speaking, **polygamomonoecious.** If one individual has perfect and staminate flowers and another perfect and pistillate, the species is **polygamodioecious.**

The tendency toward unisexuality varies among species. In some plants, such as the Peruvian pepper-tree, both stamens and pistils are present in every flower, but in some flowers the stamens are underdeveloped and in others the pistils are nonfunctional. In many others, such as the walnut, birch, and willow, each flower includes only pistils or stamens with no macroscopic sterile rudiments.

Union of Flower Parts

If the parts of a flower are separate from each other, they are described as **distinct.** When the members of a particular series of flower parts, such as the petals, are united to each other, they are said to be **coalescent.** The noun is **coalescence.** If the members of one series are united with those of another, the two series are **adnate.** The noun is **adnation.**

COALESCENCE

Sepals. If the sepals of a flower are coalescent edge to edge, the calyx is **synsepalus,** and the resulting structure is referred to as a **calyx tube,** even though its form may be cuplike or saucerlike.

Petals. If the petals are coalescent, the corolla is **sympetalous** (*sym-*, together, this being the same in meaning as *syn-*), and the resulting structure is a **corolla tube.** Examples of sympetalous flowers are those of the Jimsonweed (*Datura*), the morning-glory (*Ipomoea*), and the bindweed (*Convolvulus*). An older term indicating coalescent petals is **gamopetalous** ("with petals united").

Flowers having separate petals are **choripetalous** (*chori-*, apart). Examples include the mustards, radishes, roses, buttercups, cherries, apples, and geraniums. An older term still in fairly wide usage is polypetalous ("with many petals"), but this is not accurate because the word employed should mean "with separate petals."

Three of the above terms relating to the corolla are used repeatedly in classification of flowering plants. These are summarized as follows:

Apetalous, without petals.
Choripetalous, with separate petals.
Sympetalous, with coalescent petals.

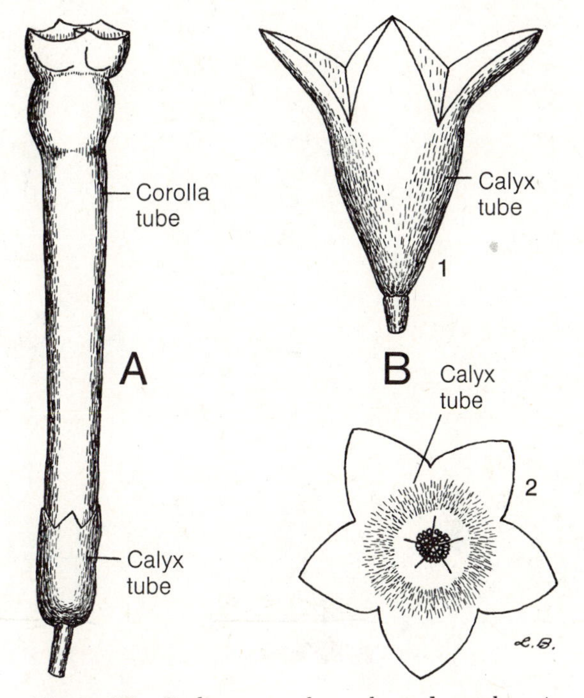

Figure 6.2 Coalescence of sepals and petals: **A** coalescent sepals and petals of the tree tobacco (*Nicotiana glauca*), **B** coalescent sepals of *Sterculia*, **1** side view, **2** top view of the flower, which has no petals.

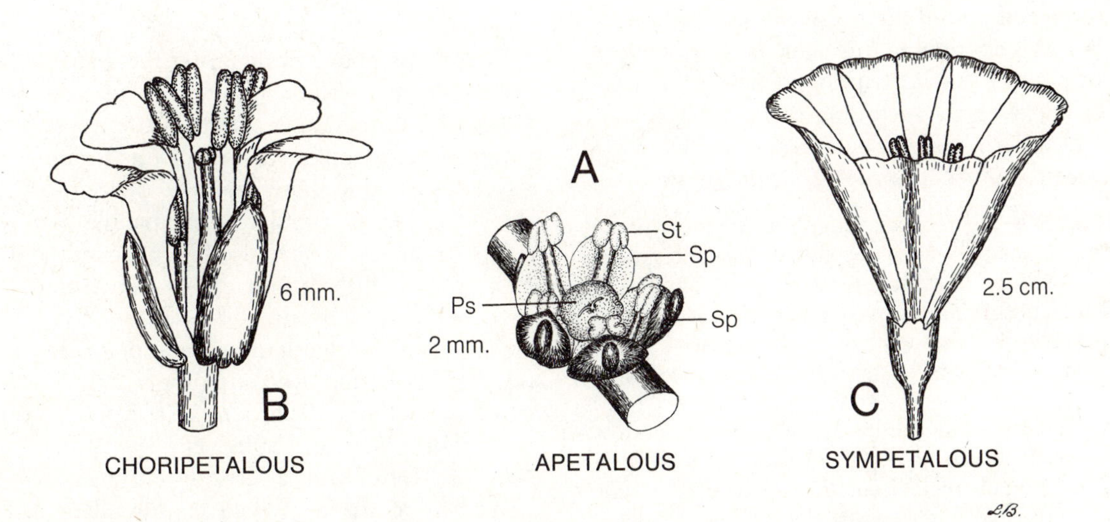

Figure 6-3. Apetalous, choripetalous, and sympetalous flowers: **A** view of the apetalous flower of a goosefoot (*Chenopodium murale*), showing the sepals and stamens and the top of the pistil (cf. also the flowers in **Figure 6-1**); **B** side view of the choripetalous flower of a mustard (*Brassica Rapa* [*campestris*]), showing the sepals, separate petals, stamens, and pistil; **C** sympetalous flower of a bindweed (in some places known as morning-glory) (*Convolvulus arvensis*) (cf. also **Figure 6-2 A**).

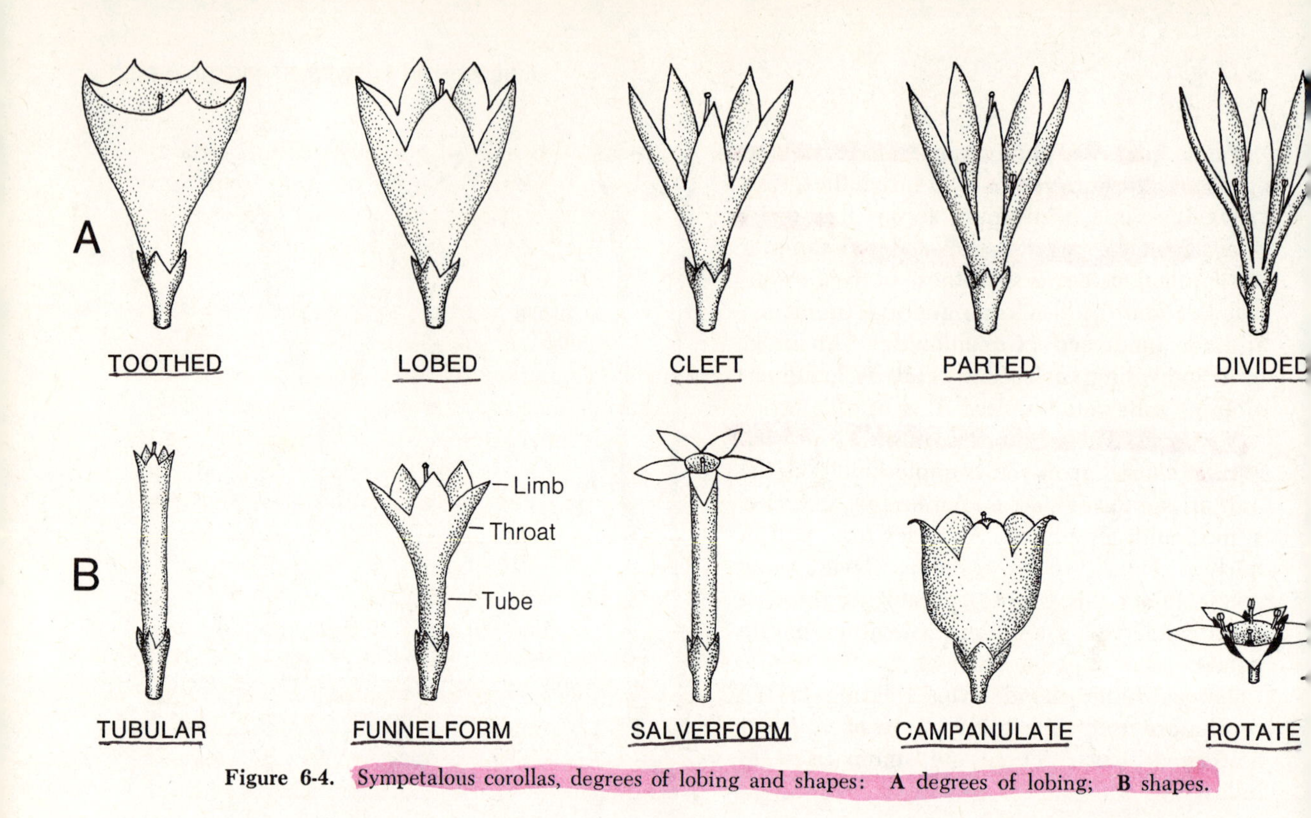

Figure 6-4. Sympetalous corollas, degrees of lobing and shapes: **A** degrees of lobing; **B** shapes.

The degree of coalescence of petals varies; but if there is *any basal* coalescence of *all* the petals, the flower is counted as sympetalous and as having a corolla tube. If only two or three and *not all* the petals are coalescent, the flower is counted in this book as choripetalous. In practice the description of borderline cases in books varies according to the author.

The forms of sympetalous corollas vary as described by the series of terms below.

Tubular. Forming essentially a hollow cylinder, but usually with independent projection of the tips of the individual petals.

Funnelform. The shape of a funnel.

Salverform. The shape of a salver, that is, a shallow bowl-shaped serving dish mounted on a pedestal. Actually a salverform corolla has a flat, spreading border at the end of a narrow, tubular basal portion; and the border diverges at a right angle from the slender basal tube. There may be a constriction at the upper end of the tubular portion of the corolla.

Campanulate. The shape of a church bell. The corolla has a wide base, and it continues to widen gradually toward the distal end. Usually there is at least a slight flare where the petals finally become separated from each other.

Rotate. Literally "wheel-shaped," but actually spreading like a saucer, the corolla being without a basal tube.

The terms above designating the shapes of sympetalous corollas apply also to certain types of choripetalous corollas—for example, rotate or campanulate. Some of this terminology is applicable more rarely to sepals—e.g., the calyx may be tubular or campanulate.

The form of a sympetalous corolla is altered also by the depth of indentation between any two component petals. The series of terms used to describe depth of indentation of leaves (cf. p. 38) is employed also for sympetalous corollas. If the free portions of the petals are very short, they are called **corolla teeth.** If they are significant in length but less than half as long as the whole corolla, they are called **corolla lobes,** and the corolla is **lobed.** If the **lobes** are approximately half or a little more than half as long as the corolla as a whole, the corolla is **cleft.** If the separate portions (**parts**) of the petals extend much more than halfway to the base or nearly to the base, the corolla is **parted.** If they barely do not reach to the base, the corolla is **divided,** and the segments are **divisions.** However, the term **lobes** may be used in a broader sense to include lobes, parts, or divisions. The more or less tubular portion of a sympetalous corolla is the **tube,** and the spreading upper portion is the **limb.** If the upper portion of the tube is dilated and more or less distinguished from both the lower part and the limb, it is known as the **throat.**

Stamens. If all the stamens of the flower are coalescent by their filaments into a common tube, they are **monadelphous,** meaning "in one brotherhood." If the stamens are coalescent into two groups, they are **diadelphous.** Most frequently this term is used to describe a condition occurring in some members of the pea family, having ten stamens, nine coalescent into one group or brotherhood, the other remaining independent and forming a "brotherhood" of its own. In other plants, such as *Camellia,* the stamens may be coalescent into several groups, i.e., **polyadelphous.** In the sunflower family usually the stamens are coalescent by only the anthers, the filaments being distinct.

Carpels. If the carpels of the flower are coalescent, they are **syncarpous,** and the resulting structure is a **compound** pistil. If the carpels are coalescent by their margins, there is no partition across the central cavity of the ovary. A double row of ovules (pp. 54–55) is produced along each of the vertical sutures on the walls of the ovary. If the carpels are coalescent by more than their margins so that the edges meet at the middle of the ovary, radiating vertical partitions are formed. The result is a two- to several-chambered ovary with a double row of ovules along the suture in each **chamber** (**locule** or **cell**). The violets and the lilies are examples of plants with the pistil formed from three carpels—the violets with the ovaries not divided into chambers and the lilies with the ovaries divided into chambers. A similar structural arrangement

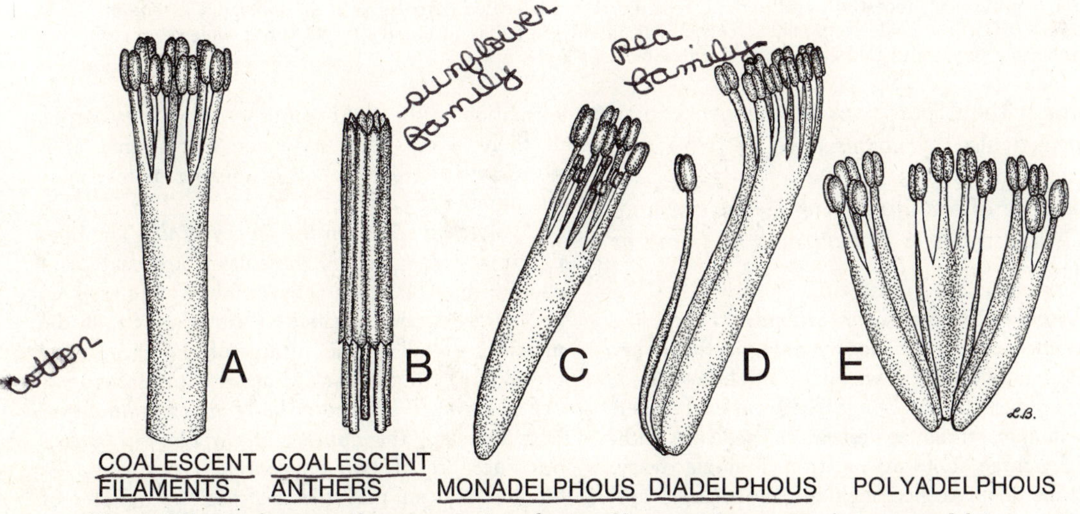

Figure 6-5. Coalescence of stamens: **A** coalescent filaments, the stamens being monadelphous; **B** coalescent anthers, as in most plants of the sunflower family (Compositae); **C** coalescence of two unequal sets of stamens as in the lupines (*Lupinus,* pea family), the stamens being monadelphous; **D** coalescence of only 9 of the 10 stamens, as in the sweet pea, garden pea, and many other members of the pea family, the stamens being diadelphous; **E** coalescence of the stamens into more than two groups, the stamens being polyadelphous.

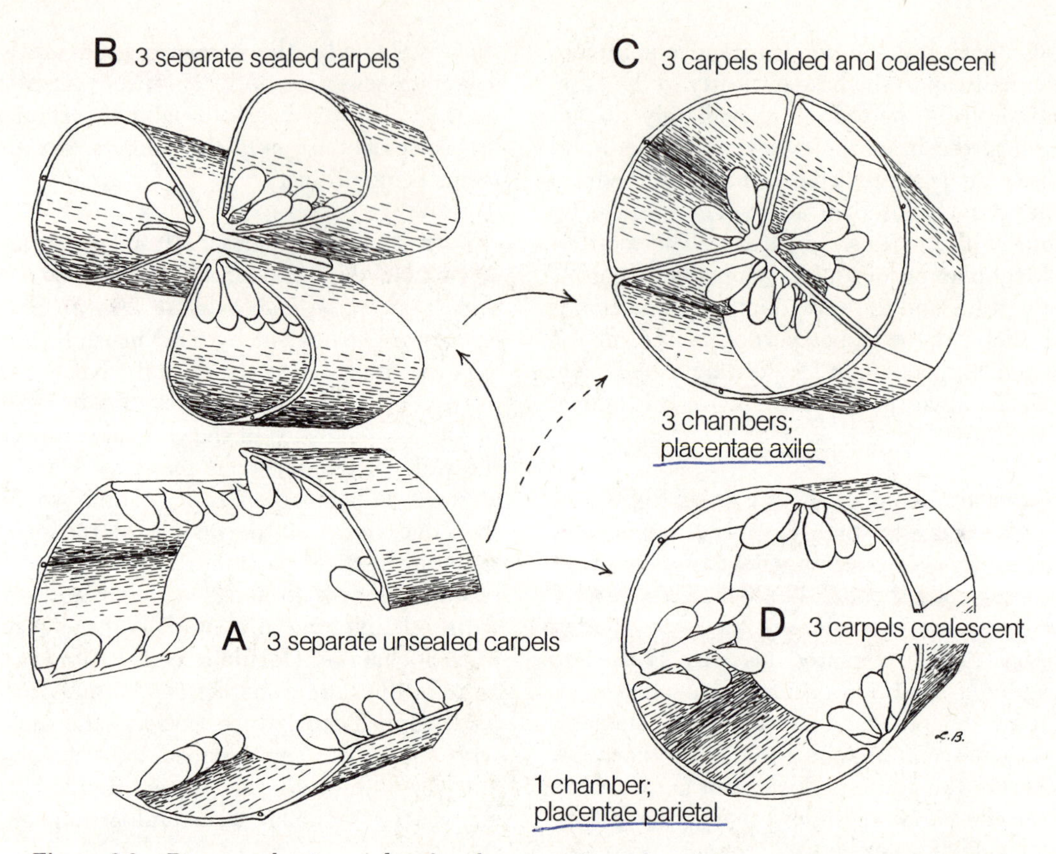

Figure 6-6. Diagrams showing modes of coalescence of carpels: **A** three separate unsealed carpels (a hypothetical ancestral condition); **B** three separate sealed carpels, as in the larkspurs (*Delphinium*); **C** three coalescent, sealed carpels, as in the camellias (*Camellia*); **D** three coalescent (by their margins), unsealed carpels, as in the violets (*Viola*).

with or without partitions in the ovary may be formed also by coalescence of two, four, five, or more carpels.

Evidence of the number of carpels forming the pistil of a flower or fruit may be found as follows.

1. ***Number of Styles or Stigmas.*** Often the coalescence of carpels occurs only at the ovary level (or only on the lower parts of the ovaries), the styles and stigmas being free from each other. For example, three coalescent carpels are indicated by three styles arising from a single ovary. Sometimes coalescence includes the ovaries and styles but not the stigmas, three stigmas arising from a single style and ovary indicating three coalescent carpels. On the other hand, coalescence of carpels may be complete, and there may be only one ovary, one style, and one stigma, despite formation of the pistil from more than one carpel. If there is only one style and one stigma, it is necessary to resort to other means to determine the number of carpels.

2. ***Vertical Lines on the Ovary Wall.*** The lines or sutures along which the carpels are coalescent usually may be seen. The number of lines is an indication of the number of carpels, but, if the midribs as well as the sutures of the carpels are prominent, the number of lines is double the number of carpels. Thus, ordinarily an odd number of lines indicates the number of carpels, but, unless the lines are of two alternating types, an even number may or may not indicate the number of carpels.

3. ***Internal Structure of the Ovary.*** Except in pistils producing very few seeds, usually the most reliable method of determining the number of carpels is examination of a cross section of the ovary

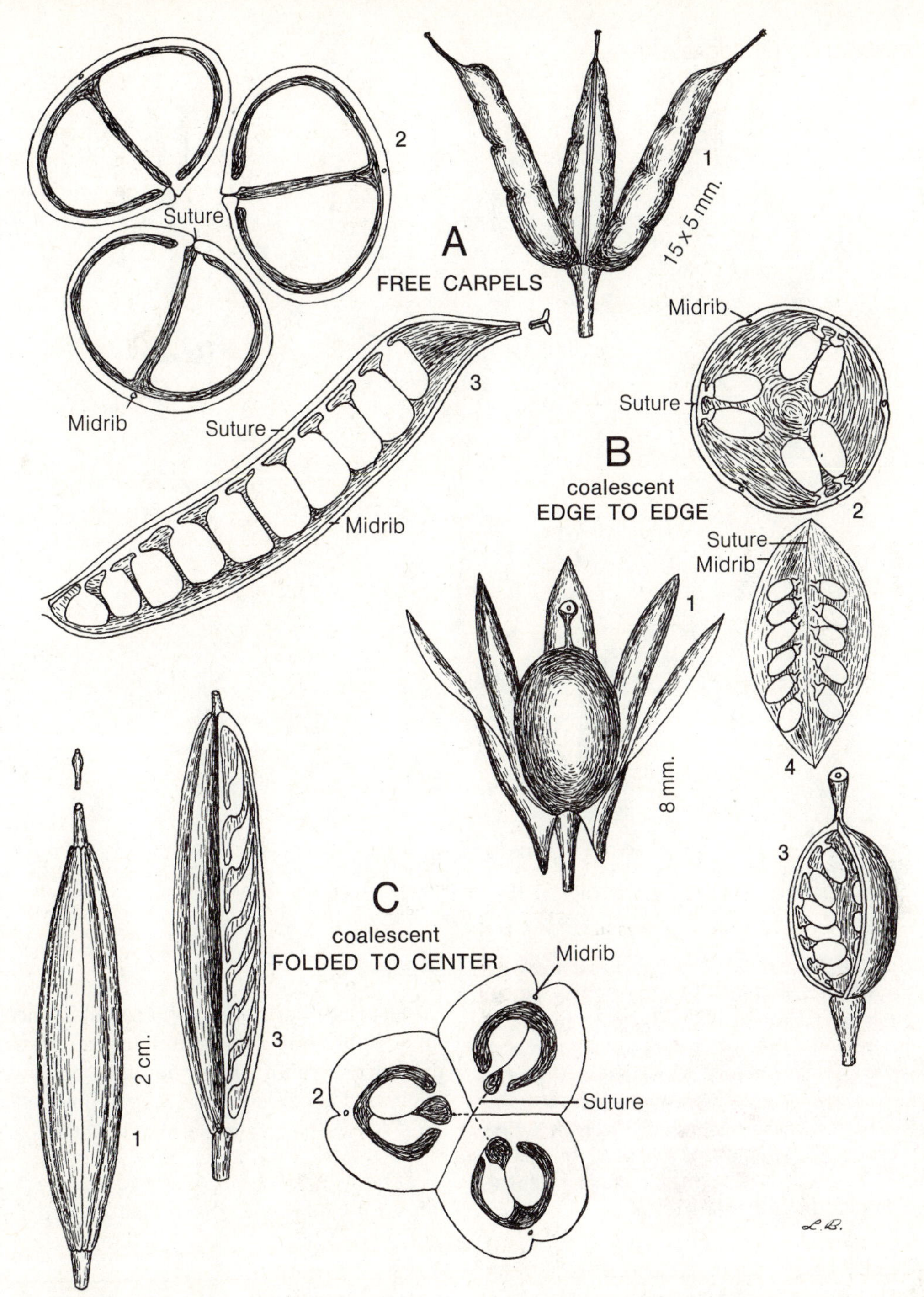

Figure 6-7. Coalescence of carpels: **A** the three separate carpels (fruits) of the red larkspur (*Delphinium cardinale*), **1** external view, **2** in cross section, **3** one carpel in longitudinal section; **B** three carpels coalescent edge to edge, a violet (*Viola*), **1** external view of the fruit and the persistent sepals surrounding it, **2** lower portion of the fruit with the top removed, **3** fruit with the back of the ovary wall removed, **4** interior view of the adjacent halves of two carpels with seeds along the line of coalescence; **C** three coalescent individually sealed carpels, African lily (*Agapanthus africanus*), **1** external view of the fruit, **2** the same with a portion of the ovary wall removed showing the seeds, **3** cross section showing the arrangement of the seeds.

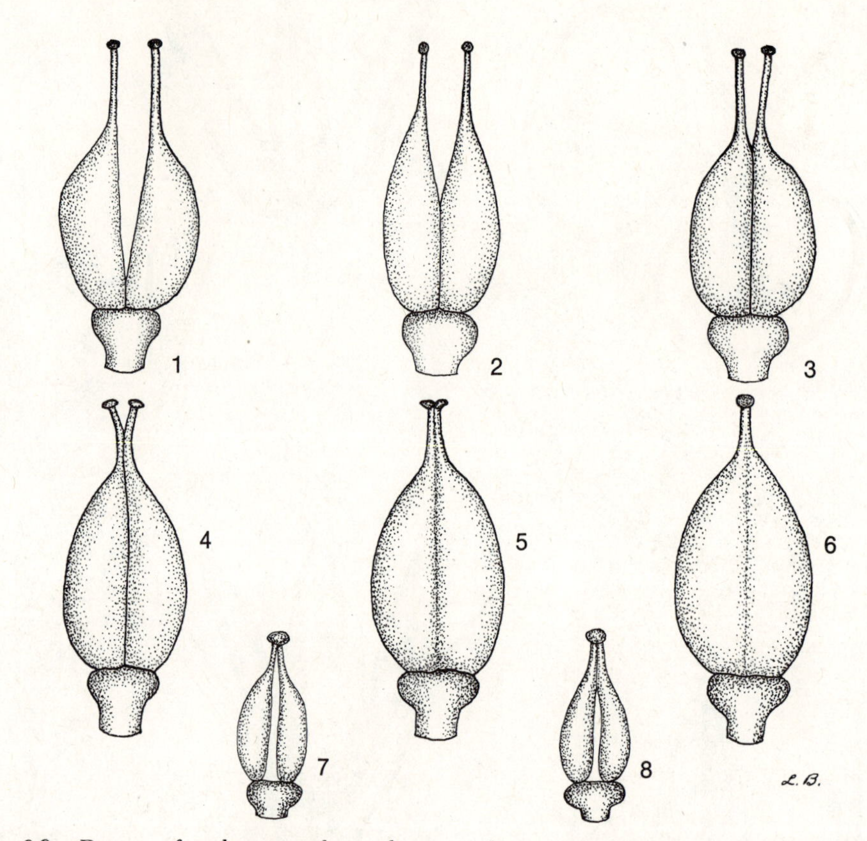

Figure 6-8. Degrees of coalescence of carpels: **1–3** degrees of coalescence of the ovaries, **4–5** degrees of coalescence of styles, as well, **6** coalescence of ovaries, styles, and stigmas, **7** coalescence of stigmas alone, **8** coalescence of the styles and stigmas but not the ovaries.

to determine the number of double rows of ovules, because these reveal the number of sutures and consequently the number of carpels. Usually, too, the number of partitions across the ovary and consequently the number of seed chambers indicates the number of carpels. **Cross sections may be made to best advantage through the fruit rather than through the young pistil because the fruit is larger and easier to see and the technical difficulties of cutting the section are less.**

ADNATION

Adnation is less common than coalescence, but in various forms it is characteristic of a number of orders and many families of flowering plants. The following are examples of adnation between various series of flower parts.

Sepals and petals. Edge to edge adnation of the members of the perianth occurs in at

Figure 6-9. Adnation of flower parts: **A** edge-to-edge coalescence of the sepals and petals forming a perianth tube (with the additional adnation of the stamens), this being characteristic of many monocotyledons, African lily (*Agapanthus africanus*), **1** external view, **2** perianth laid open; **B** similar adnation, but the perianth tube also adnate with the ovary, the ovary being inferior and the flower epigynous, amaryllis (*Amaryllis vittata*); **C** adnation of sepals and stamens, sand-verbena (*Abronia villosa*), **1** external view of the apetalous flower with the conspicuous coalescent petaloid sepals forming a calyx tube, **2** internal view of the calyx tube and of the adnate stamens; **D** probable adnation of sepals and ovary, the floral tube adherent with the ovary probably formed by coalescence of the bases of the sepals, the

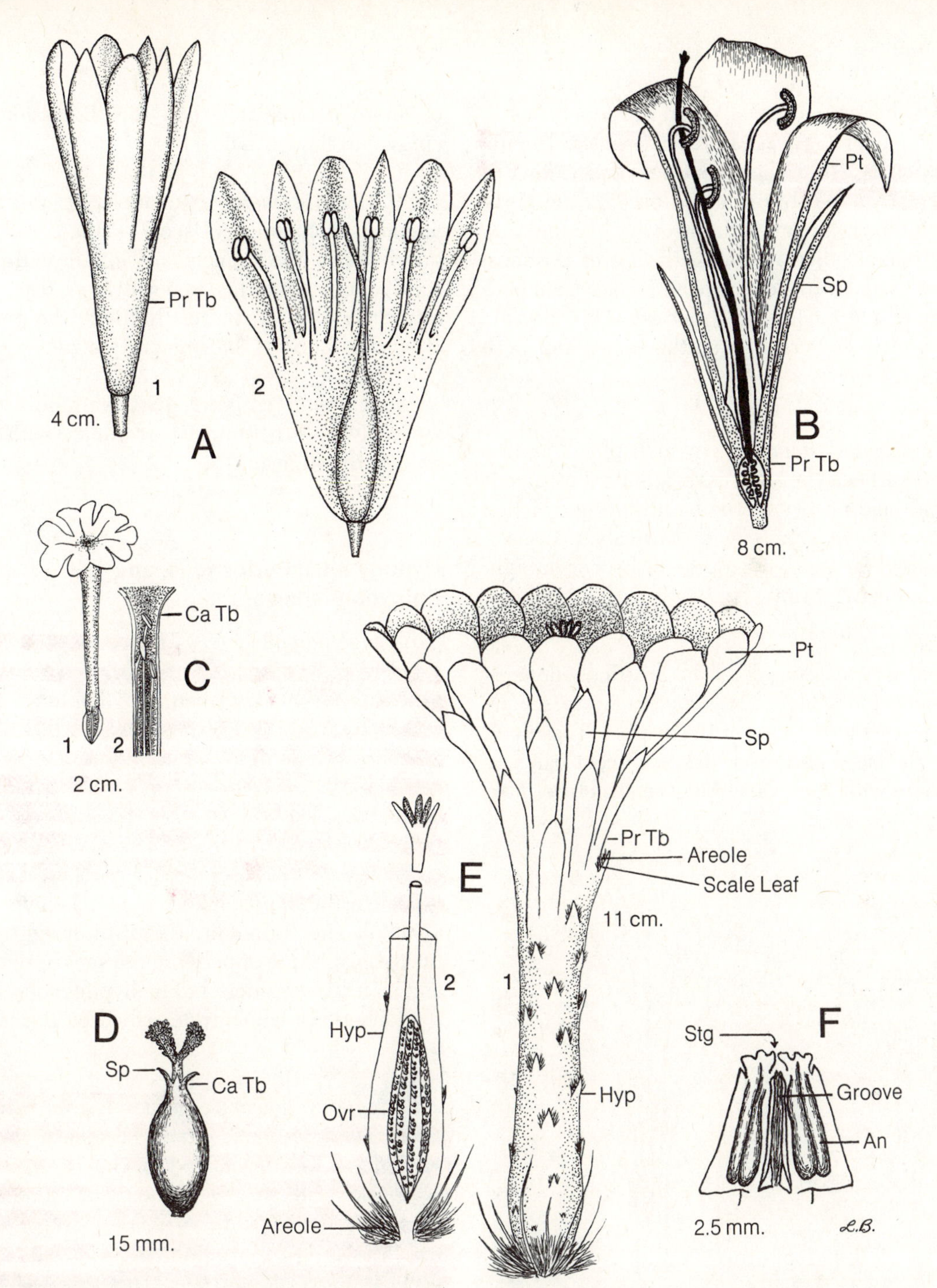

separate sepals at the top probably representing only the tips of the original sepals, California black walnut (*Juglans californica*); **E** adnation of the dorsal and ventral surfaces of sepals and petals, flower of the saguaro (giant cactus) (*Cereus giganteus*), **1** external view of the flower, showing the hypanthium adnate with the ovary, this continuous as a tube with the perianth tube (stem) above, the perianth tube consisting of numerous gradually intergrading sepaloids, the stamens attached to the inner side of this tube, the scalelike leaves on the hypanthium grading into the sepaloids higher up, **2** longitudinal section of the hypanthium and the base of the perianth tube and an internal view of the ovary showing the rows of seeds; **F** stamens adnate with the stigma, milkweed (*Asclepias eriocarpa*).

least some members of several orders of monocotyledons. The sepals and petals alternate in position and the resulting structure is a **perianth tube.** In irises and orchids the perianth tube is adnate with the ovary. In the cactus family (dicotyledons), the lower portions of the sepals and petals are both coalescent and adnate into a common structure. This may be seen after flowering when the sepals and petals fall away as a unit.

Sepals and stamens. In such plants as the sand-verbenas, which have adnate (colored) sepals and no petals, the basal portions of the filaments are adnate with the calyx. Consequently, the stamens appear to be borne high up in the calyx tube.

Sepals and carpels. The pistillate flowers of oak and walnut trees have minute sepals appearing at the tops of the ovaries. Presumably the basal portion of the coalescent calyx is adnate with the ovary, but the structure may be more complicated (cf. the discussion of epigynous flowers below).

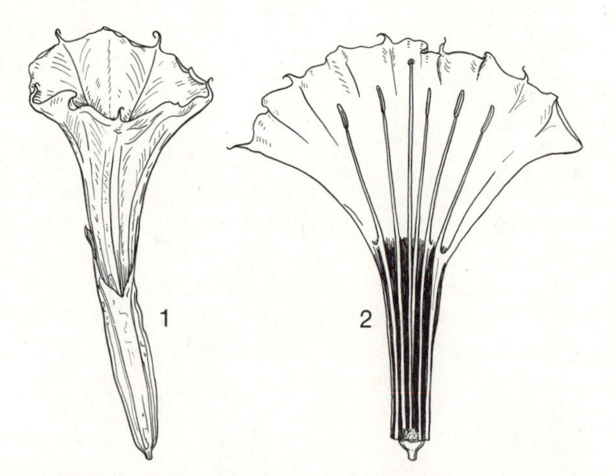

Figure 6-10. Adnation of stamens and petals, flower of a thorn apple (*Datura meteloides,* related to the Jimson-weed): **1** external view of the flower, showing the coalescent sepals and petals, **2** internal view of the flower, showing the adnation of the stamens and the corolla tube. By Lucretia Breazeale Hamilton.

Petals and stamens. Adnation of petals and filaments is common. Usually if the corolla is sympetalous, the stamens are adnate with the corolla tube, but this is not always true. In many members of the heath family the petals are coalescent but the stamens are free.

Stamens and carpels. In the orchids and the milkweeds the stamens are adnate with the style or the stigma.

Hypogynous, Perigynous, and Epigynous Flowers

Hypogynous flowers. The floral parts (the sepals, petals, stamens, and pistils) of the generalized flower described in Chapter 5 are attached directly to the receptacle, which may be conical, spheroidal, ovoid, or discoid. Since the sepals, petals, and stamens are attached at points below the ovary attachment, the ovary is **superior** to them in the sense that it stands above them. The other parts are below the pistil, or **hypogynous** (*hypo-*, below; *gynoecium*—i.e., in botany the pistil or aggregation of pistils). Not only are the sepals, petals, and stamens considered to be hypogynous, but the flower as a whole is described by the same term.

Perigynous flowers. In flowers like those of a cherry, peach, or plum, the sepals, petals, and stamens are produced on the border of a saucerlike, cuplike, or tubular structure, the **floral cup** or **floral tube**; consequently, they are in a position around the ovary or even slightly above it, and they are described as **perigynous** (*peri-*, around). This term is used not only to describe the position of the sepals, petals, and stamens, but also to describe the flower as a whole. The ovary is *superior* despite the fact that a perigynous flower is inter-

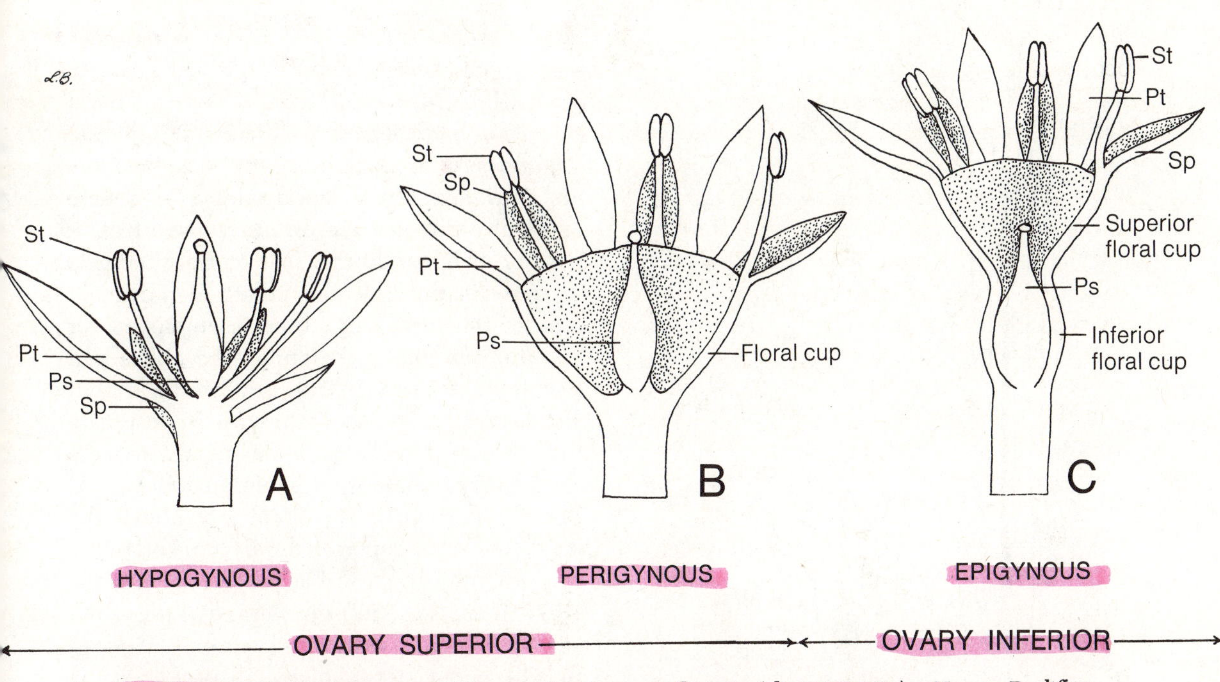

Figure 6-11. Hypogynous, perigynous, and epigynous flowers (diagrammatic). NOTE: **B** *differs from* **A** in **presence of a floral cup,** and **C** *differs from* **B** in **adnation of the floral cup with the ovary.**

mediate between hypogynous and epigynous (cf. below).

Epigynous flowers. Flowers like those of currants or gooseberries differ from perigynous flowers in only one respect—**the floral cup is adnate with the ovary.*** The two have grown fast together in the course of early development; consequently, the ovary appears to be sunken into tissues below the flower, and it is spoken of as **inferior.** The sepals, petals, and stamens of a flower having an inferior ovary are said to be **epigynous** (*epi-*, upon) because they appear to be produced upon the pistil (i.e., in this case the ovary). This term is applied also to the flower as a whole.

An inferior ovary is not within direct reach of pollinating beetles or birds; consequently, the ovules are less likely to be eaten.

* Some authors consider a flower perigynous when only the basal portion of the floral cup is adnate with the ovary. In such instances the ovary is only partly inferior. As the terms are defined here, such flowers are epigynous.

To summarize the preceding paragraphs, the sepals, petals, and stamens may be described as (1) **hypogynous,** below the pistil, (2) **perigynous,** around the pistil, or (3) **epigynous,** upon the pistil (i.e., ovary). In a hypogynous flower there is no floral cup but usually only a simple receptacle; in both perigynous and epigynous flowers there is a floral cup. In a perigynous flower the cup is free from the ovary; in an epigynous flower it is adnate with the ovary. The ovary is considered to be superior unless it is adnate with the floral cup; therefore hypogynous and perigynous flowers have superior ovaries, whereas epigynous flowers alone have inferior ovaries.

In some flowers the floral cup is formed by a coalescence of the tissues of the sepals, petals, and stamens, the bulk of the tissue being derived from the sepals and once the whole structure was known as the "calyx tube" (in contrast to a true calyx tube composed only of coalescent sepals). In other flowers the floral cup is an outgrowth of the receptacle

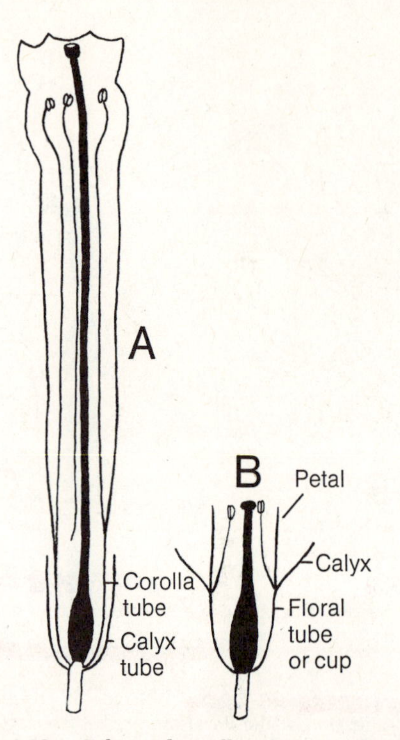

Figure 6-12. Calyx and corolla tubes vs. a floral tube or cup (diagrams): **A** longitudinal section of a flower with coalescent sepals (calyx tube in the correct sense) and coalescent petals forming an elongate corolla tube (cf. **Figure 6-2 A**); **B** longitudinal section of a perigynous flower. Note that both the sepals and the petals in **A** are attached directly to the receptacle but in **B** both are attached to the rim of the floral cup. The flower in **A** is hypogynous.

tissue, and it is called a **hypanthium.** Microscopic examination of a floral cup in longitudinal section reveals its nature through the arrangement of the vascular traces leaving the vascular cylinder (stele) of the receptacle and pedicel. In most perigynous and epigynous flowers the traces leaving the stele of the receptacle pass through the entire floral cup, showing that this organ (a "calyx tube") was formed by coalescence of the bases of the sepals, petals, and stamens. In other flowers the distorted stele cylinder of the stem extends to the margin of the floral cup, showing that the cup is an extension of the edge of the receptacle and is therefore a **hypanthium** and not formed by coalescence of flower parts. Hypanthia are rare. Since determination depends upon detailed and laborious microscopic study, it has been proposed that the term **floral tube*** be used to cover both calyx tube and hypanthium. This term, or the usually more appropriate variation **floral cup,** is used here as a substitute for the exclusive use of calyx tube or of hypanthium. It is to be noted that in some manuals of the plants of particular regions the term calyx tube is employed for only the portion of the floral tube above and free from the ovary. The same term is used also in a strict and accurate sense to designate the coalescent portions of the sepals in a coalescent (**synsepalous,** or **gamosepalous**) calyx. For the portion of the floral cup above the ovary the term **superior floral cup** (or **tube**) was proposed (ed. 1). The part attached to the ovary, then, is the **inferior floral cup** (or **tube**).

In accordance with another point of view, an alternative terminology is preferred by some authors. They propose to abandon use of "calyx tube" in any but its correct sense, i.e., a tube formed by coalescence of the sepals alone, and to extend the use of "hypanthium" to include all other cases, i.e., to be synonymous with floral cup, or floral tube as employed here. If this policy is adopted, it is necessary only to make the substitution of "hypanthium" wherever the other terminology has been employed.

In the monocotyledons epigynous flowers are formed by adnation of the perianth tube with the ovary.

The distinction between a corolla tube and a floral cup or tube is of primary importance. A corolla tube is formed by coalescence of only the petals. If the flower is hypogynous the sepals and petals are attached directly to the receptacle independently of each other.

---

* Benson, Lyman and Robert A. Darrow. *A Manual of Southwestern Desert Trees and Shrubs.* University of Arizona Biological Science Bulletin (6): 35. 1945. Ed. 2 (*The Trees and Shrubs of the Southwestern Deserts*). 25–26. 1954. In French a similar suggestion for use of the term "coupe floral" was made in 1928 by Bonne (Bonne, Gabrielle. *Recherches sur le pédicelle et la fleur des* Rosacées. 1928). See also, Benson, Lyman. *The Cacti of Arizona.* Ed. 2. 3–4. 1950.

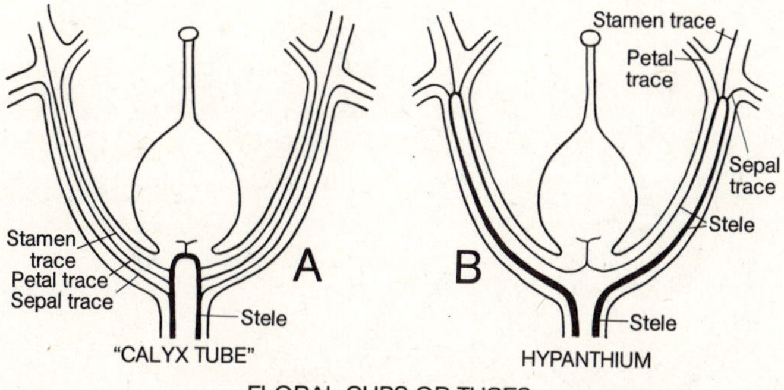

Figure 6-13. Distinction of a "calyx tube" and a hypanthium, both being considered in this book as floral cups (or tubes) (diagrammatic longitudinal sections): **A** "calyx tube," formed from the coalescence and adnation of the bases of the sepals, petals, and stamens, the veins of each of these being connected to the stele of the receptacle (stem) by one or more separate vascular traces; **B** hypanthium, an outgrowth of the margin of the receptacle, the stele of the receptacle being distorted and extending into the cup of the hypanthium and the vascular traces of the sepals, petals, and stamens joining it near the rim of the cup. By Lucretia Breazeale Hamilton.

Flowers with Discs. Some flowers, as those of poison ivy, poison-oak, the sumacs, and the Peruvian pepper-tree, have sepals and petals attached directly to the receptacle, but the stamens (or very rarely the stamens and petals) are borne on an expanded disc produced from the receptacle below the pistil and above the attachment levels of the sepals (and usually the petals). Such a disc is called a **hypogynous disc.** In this book flowers with hypogynous discs are considered as hypogynous and *not* perigynous. Perigynous flowers are restricted to include only those with a floral cup for the following reasons: (1) There is rarely any close relationship between plant groups with flowers having a hypogynous disc and those with perigynous flowers, except when the flower has *both* a floral cup and a disc. (2) The origins of the disc and of the floral cup are different. For example, in the citrus family the disc is formed from vestigial stamens. In some manuals the stamens of flowers with hypogynous discs as well as the flowers themselves are described as perigynous. In perigynous or epigynous flowers an accessory disc may be described as a **perigynous** or an **epigynous** disc.

Good examples of the four types of flowers described above are as follows.

Hypogynous. Geranium, orange, hollyhock, buttercup, violet, poppy, snapdragon, mint, morning-glory, any plant in the lily family or the mustard family.

Perigynous (with a floral cup). Prune, plum, cherry, peach, apricot, nectarine, almond, and other pitted or stone fruits; rose, strawberry, blackberry.

Epigynous. Apple, pear, quince, pomegranate, banana, *Iris,* loquat, honeysuckle, elderberry, cucumber, watermelon, pumpkin, huckleberry, blueberry, cranberry, evening-primrose, evening-star (*Mentzelia*), *Fuchsia,* and *Eucalyptus.*

With a Hypogynous Disc. Poison ivy, poison-oak, sumac, soapberry, bittersweet, burning-bush, citrus fruit trees, spindle-tree, and the Peruvian pepper-tree (*Schinus*).

Arrangement of Flower Parts

Flower parts may be either **spiral** or **cyclic.** Spiral or irregular arrangement occurs with large indefinite numbers of parts. It may be seen most frequently in the stamens and carpels. For example, in the head of fruits of

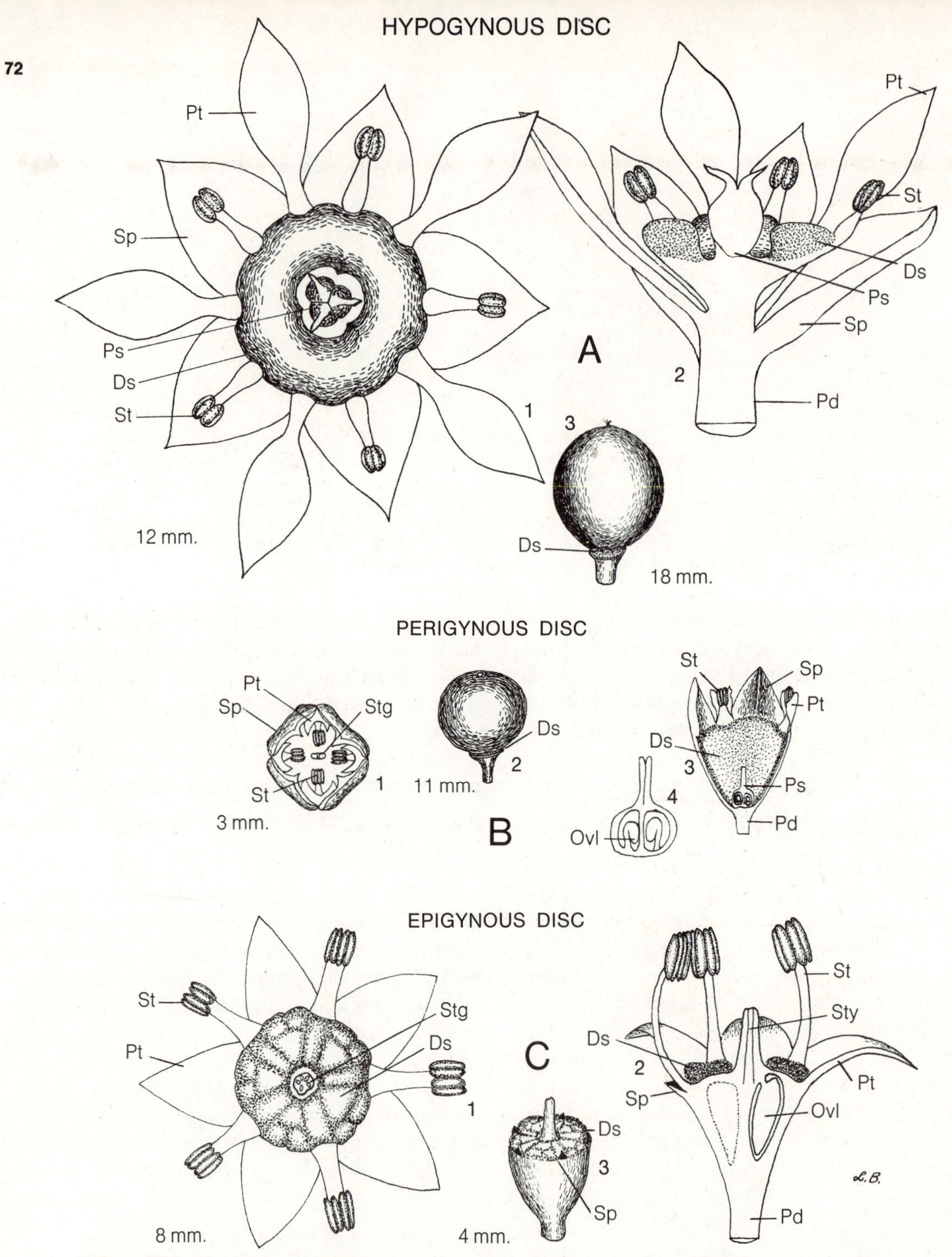

Figure 6-14. Hypogynous, perigynous, and epigynous discs: **A** hypogynous disc of the usual type (hypothetical), **1** top view of the flower, **2** longitudinal section of the flower, **3** fruit (reduced scale); **B** perigynous disc of a coffee berry (*Rhamnus californica*), **1** top view of the flower, **2** fruit subtended by a remnant of the disc and the floral cup, **3** longitudinal section of the flower showing the disc lining the floral cup, **4** enlargement of the ovary section in **3**; **C** epigynous disc of Algerian ivy (*Hedera canariensis*), **1** top view of the flower, **2** longitudinal section of the flower, **3** fruit (reduced scale).

Magnolia the spiral arrangement of the stamen attachment scars on the receptacle and of the carpels (of course, still present as fruits) is clear. Microscopic examination of the development of flower buds may show spiral arrangement to be present also in the sepals and the petals. In the California buttercup, *Ranunculus californicus,* the petals, stamens, and pistils are all indefinite in number and arranged spirally or irregularly. Basically the vascular bundles of the stamens are in clusters yielding irregularity in contrast to the preciseness of the cyclic plan. The sepals are definitely five, and they appear to be arranged at a single level. Actually they are in a spiral which makes only one turn; however, this is not visible readily to the naked eye. The flowers of most buttercups, such as *Ranunculus acris,* common as a weed in the Northern States and Canada though a native of Europe, have only five petals arranged in a single turn of a spiral.

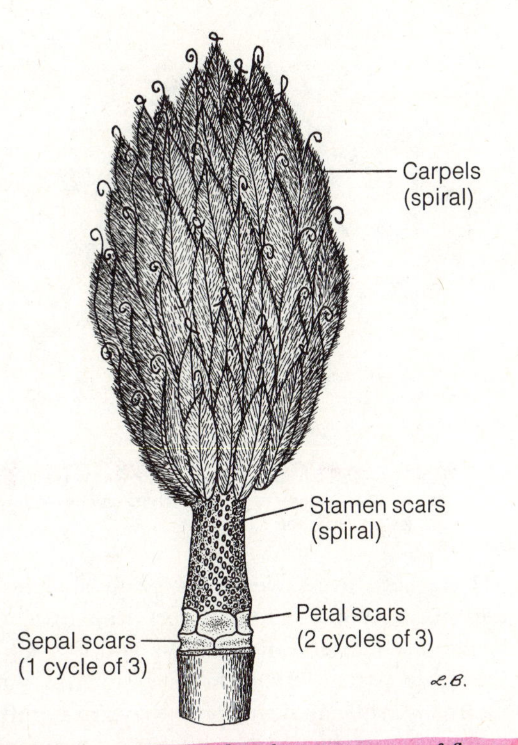

Figure 6-15. Spiral and cyclic arrangement of flower parts, the head of fruits formed from a magnolia (*Magnolia grandiflora*) flower. As shown by the scars, the sepals and petals were arranged in cycles of three; the stamens were arranged spirally; the carpels (still present) are arranged spirally.

The Numerical Arrangement of the Flower

The parts of most flowers are not in spirals but in cycles. A flower built on a plan of five or multiples of five, that is, with five parts in each series—five sepals, five petals, one or more series of five stamens, and five carpels (often coalescent with each other)—is **pentamerous** or **5-merous.** The members of the different series alternate with each other; that is, each of the five petals stands between and inward from two sepals; each stamen of the outer series has the same relationship to two petals; each stamen of succeeding series to two of the next outer series. Often this plan is followed by the carpels also with respect to the inner series of stamens. A flower usually is **3-, 4-,** or **5-merous; 2-merous** flowers are relatively rare. Occasionally there are higher numbers than five. Other variations on this plan will be discussed with respect to the particular orders and families in which they are characteristic.

Symmetry of the Flower

Ordinarily flowers are **radially symmetrical,** that is, their parts radiate regularly from a common center and each within any given series (e.g., the petals) is of the same shape and size as every other part in the same series. Such flowers are known also as **regular** (**actinomorphic**). If they are divided on *any* diameter through the middle of one part, the two segments are mirror images.

Other flowers are **bilaterally symmetrical,** that is, if such a flower were divided in half along the proper line (*and only that line*), the two parts would be mirror images of each other. Such flowers are known also as **irregular** (**zygomorphic**). The most common bilateral (two-sided) symmetry occurs in the corolla, the petals being of different shapes or

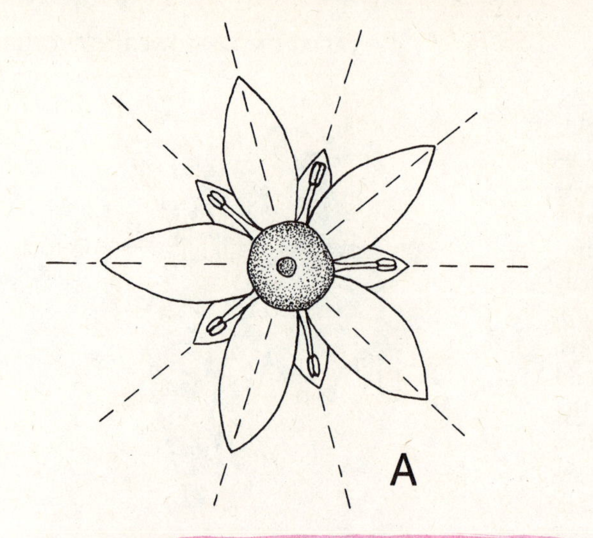

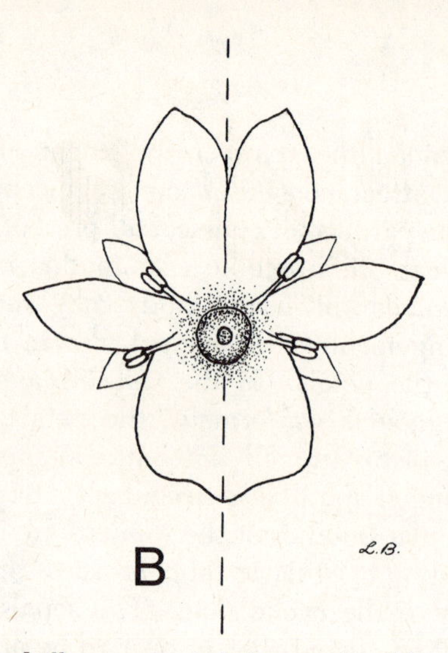

Figure 6-16. Radial and bilateral symmetry (diagrams): **A** radially symmetrical flower, divisible into mirror images on more than one axis; **B** bilaterally symmetrical flower, divisible into mirror images on only one axis.

sizes or both, as in sweet peas, orchids, violets, and snapdragons. A few flowers are most conspicuously bilateral in the calyx and less so in the other parts. Others have a radially symmetrical corolla and calyx, but they are slightly (or rarely conspicuously) bilaterally symmetrical in the stamens or carpels; often this is discounted in the writing of keys, although, strictly speaking, it should be taken into account. An example of a minor departure from symmetry is typical of the mustard family, wherein all but a few species have four long inner stamens arranged in a square and two short outer ones standing opposite and outside two sides of the square.

The following are some special types of bilaterally symmetrical flowers. In the sweet pea one large petal (the **banner** or **standard**) stands above the others; two others (**wings**) are at the sides; the remaining two are more or less enclosed by the wings and are coalescent with each other along the lower margins, forming a boatlike structure (the **keel**). The flower as a whole is called butterflylike or **papilionaceous.** In all orchids one petal (really the upper one but apparently the lower because the base of the flower is twisted) is markedly bilaterally symmetrical and decidedly different from the other two. In violets the petals are somewhat unequal, and the lowest one of the five has a spur (an elongated backward-protruding sac) at the base. In the monkey-flowers, snapdragons, and mints, the two upper petals stand erect, and they attract insects; the three lower petals spread horizontally, and they form a landing place. Such a corolla is **2-lipped** or **bilabiate.**

In all four kinds of plants discussed in the preceding paragraph there is some degree of bilateral symmetry in the stamens as well as the corolla. In the sweet pea there are ten stamens, the nine lower ones being coalescent by their filaments and the single upper one being free from the others. In addition all the stamens are turned more or less upward at their outer ends. In most orchids, although the flower is 3-merous, there are only one or two stamens, and in some there is only half a stamen. The stamen is coalescent with the style and the combined structure stands above and opposite the specialized petal. In violets the two lower stamens bear spurs which project into the spur of the lower petal. In the monkey-flowers, snapdragons, and most mints there are four stamens despite presence of five sepals and five petals, and in some other plants of the same family, including *Penstemon*, the fifth stamen is developed more or less as a sterile structure sometimes highly specialized and resembling a toothbrush.

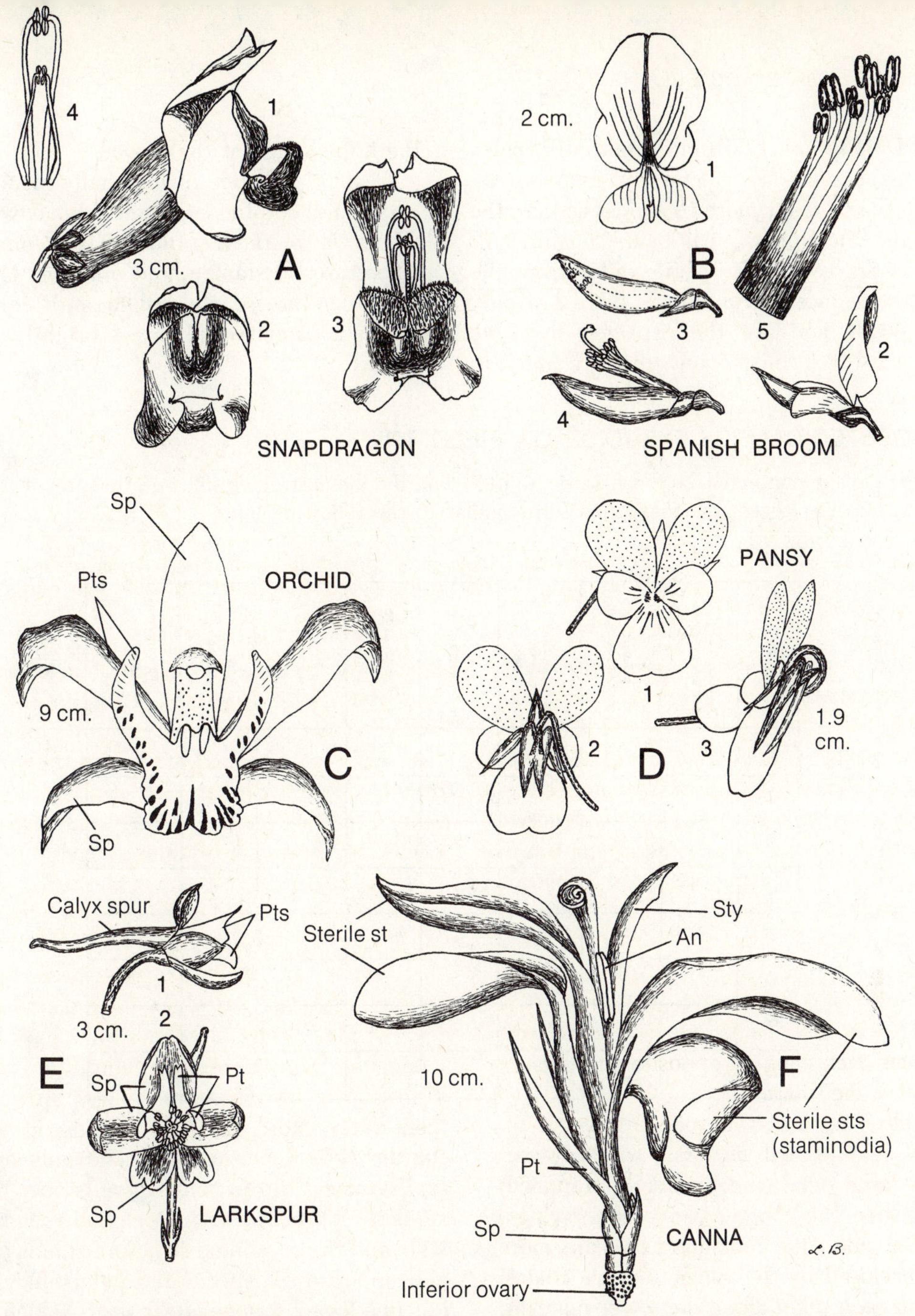

Figure 6-17. Bilaterally symmetrical flowers: **A** the bilaterally symmetrical corolla (**1–3**) and stamens (**3–4**) of a snapdragon (*Antirrhinum majus*), the flower being pulled open in **3**; **B** the bilaterally symmetrical calyx (**2–4**), corolla (**1–2**), and stamens (**4–5**) of the Spanish broom (*Spartium junceum*), this being typical of many members of the pea family, the petals other than the lower coalescent pair (keel) being removed in **3** and **4**, the upper petal (cf. **1**) being known as the banner (standard) and the two lateral petals as wings; **C** the bilaterally symmetrical flower of an orchid (*Cymbidium*), showing the high degree of differentiation of one petal from the other two; **D** the bilaterally symmetrical corolla of a pansy (*Viola*); **E** the bilaterally symmetrical calyx and corolla of the red larkspur (*Delphinium cardinale*); **F** the bilaterally symmetrical stamens (including petal-like, sterile stamens or staminodia) of a canna (*Canna*, garden hybrid), the calyx and corolla being radially symmetrical or nearly so.

In *Delphinium* (larkspur) both the calyx and the corolla are bilaterally symmetrical, but the calyx is far more conspicuous than the corolla. The upper sepal bears a spur. Although the two upper petals differ from the two lower ones and although there are only four petals and not the expected five, the corolla is too inconspicuous to figure much in the appearance of the flower.

Despite presence of a radially symmetrical calyx and corolla in *Canna,* the flower as a whole is markedly bilaterally symmetrical. There are six stamens but only one is fertile; the others are sterile, unlike, and highly specialized, each resembling a brightly colored petal.

SUGGESTED CAMPUS EXERCISE OR FIELD TRIP

For each plant numbered in the field or pointed out by the instructor, fill out the proper squares in an enlarged mimeographed form similar to the following table.

PLANT NO.	HYPO-, PERI-, OR EPIGYNOUS	A-, CHORI-, OR SYMPETALOUS	COALESCENCE OF PARTS	ADNATION OF PARTS	PARTS CYCLIC OR SPIRAL	NUMERICAL PLAN OF FLOWER	SYMMETRY OF FLOWER	COMPLETENESS OF FLOWER
1								
2								
3								
4								
5								
6								
7								
8								

7

Arrangement of Flowers on the Plant

Terminal and Axillary Flowers

Terminal (solitary) flowers occur singly at the apex of a main stem or a branch of some importance. An **axillary flower** terminates a small lateral branch, even though the branch may be so short as to be microscopic. In the flowering plants each branch develops fundamentally in the axil of a leaf, and the basic arrangement of axillary flowers is with one flower terminating each short branch along the stem or the main branches.

The leaves subtending the flowers may be large or be reduced to **bracts** (cf. p. 51) or **bractlets** or **bracteoles** (mere scales). This may result in an elaborate **inflorescence** (cluster of flowers*) composed of many flowers often closely associated with each other.

The next sections describe the common types of inflorescences.

* Referring particularly to the mode of arrangement of flowers within the cluster.

Racemes, Spikes, Corymbs, and Panicles

(indeterminate inflorescences)

An **indeterminate inflorescence** is one in which the apical bud continues to grow for an indefinite period, sometimes as long as conditions remain favorable.

A **raceme** consists of an elongated axis with pedicelled flowers along it. Since a raceme is indeterminate, there is a constant supply of new flower buds at the tip, and the oldest flowers or fruits are at the bottom.

A **spike** is similar to raceme, but the flowers are **sessile,** that is, practically without pedicels. The diminutive form is **spikelet.** The highly specialized spikelets of the grasses and sedges are illustrated in Figures 11-10 to 11-12. A specialized fleshy spike often with the flowers embedded or partly so in the thick axis is a **spadix.** Commonly it is enveloped at least at first by a single large bract—a **spathe.**

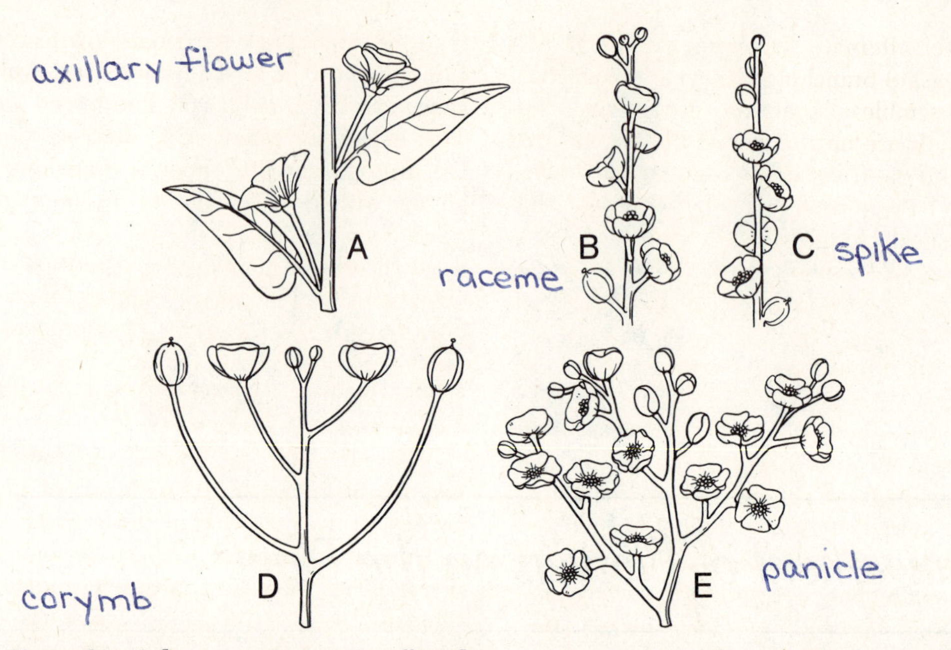

Figure 7-1. Inflorescence types: **A** axillary flower; **B** raceme; **C** spike; **D** corymb (note buds on the inside [apical], fruits outside); **E** panicle (in this case of racemes but often of spikes or sometimes of corymbs). By Lucretia Breazeale Hamilton.

A soft scaly raceme, spike, or spikelike inflorescence of small apetalous unisexual flowers is a **catkin,** or **ament** (Figures 10-16 and 10-54 on pp. 172 and 245). Usually such an inflorescence falls as a unit after flowering. Catkins occur only on trees and shrubs. In the birch the plants are monoecious and both kinds of flowers are in cymules (cf. below) in dense catkins, the floral structures being more or less obscured by bracts. The willow is dioecious, and both staminate and pistillate flowers are in catkins, each staminate flower consisting of two or more stamens in a cluster subtended by a scalelike bract and each pistillate flower consisting of a single pistil subtended by a bract. In the walnut and oak the staminate flowers are in catkins, but the pistillate flowers are solitary or two or three together.

A **corymb** differs from a raceme in having a gradation in the lengths of the pedicels, those of the lower flowers being much longer than those of the upper. The result at the height of flowering is a flat-topped cluster of buds, flowers, and fruits. The buds are at the center of the inflorescence near the terminal bud of the stem which forms the axis. The flowers are in a spiral around the buds, and the fruits are at the outside of the cluster.

A **panicle** is a compound inflorescence; that is, typically it is a cluster of associated racemes, spikes, or corymbs, although the term is applied to other complex clusters.

The adjective forms of the nouns for the inflorescences described above are **racemose, spicate, corymbose,** and **paniculate.**

Cymes

(determinate inflorescences)

In a determinate inflorescence the terminal bud becomes a flower; and since no further elongation can occur, theoretically the growth of the inflorescence is predetermined.

A cyme may have either of two common forms (cf. Figure 7-2 **A** & **B**), depending upon whether the branching in the inflorescence is apparently alternate or opposite. On plants with alternate branching the cyme (monochasium) resembles a corymb, except that the flower buds are on the rim of the flat-topped cluster and the fruits in the center. The flower formed at the termination of the axis of the inflorescence blooms first, and the flowers formed on successively lower pedicels bloom later. The lower pedicels are elongated as they are in a corymb.

In plants with opposite branching in the inflorescence a cyme (dichasium) develops as illustrated in Figure 7-2 **A.** The terminal bud of the stem forms a flower, and the two axillary branches at the first node below the flower elongate. Each of these forms a flower, and again the two axillary buds at the next node of each branch elongate, forming flowers and continuing the procedure. The result is a more or less flat-topped cluster of buds, flowers, and fruits with the oldest fruit in the center.

Since the terminal bud forms a flower, both types of cymes are determinate inflorescences. Consequently, at least in theory, the inflorescence can reach only a predetermined length. This is true for cymes with alternate branching, but cymes with opposite branching may go on with development to an unrestricted size (diameter but not length) because there is no theoretical limit to the repetition of development of the pair of axillary buds into flowers with accompanying development of a pair of subordinate axillary buds.

The diminutive of cyme is **cymule**; the adjective form is **cymose.**

A very dense single cyme is a **glomerule.** It may be distinguished from a head (cf. below) by blooming of the central flower first.

A densely congested or compact panicle is a **thyrse.** The main axis is indeterminate, but typically the lateral branches are determinate. This is a mixed type of inflorescence. In some books it is defined as a congested cyme. The adjective is **thyrsoid.**

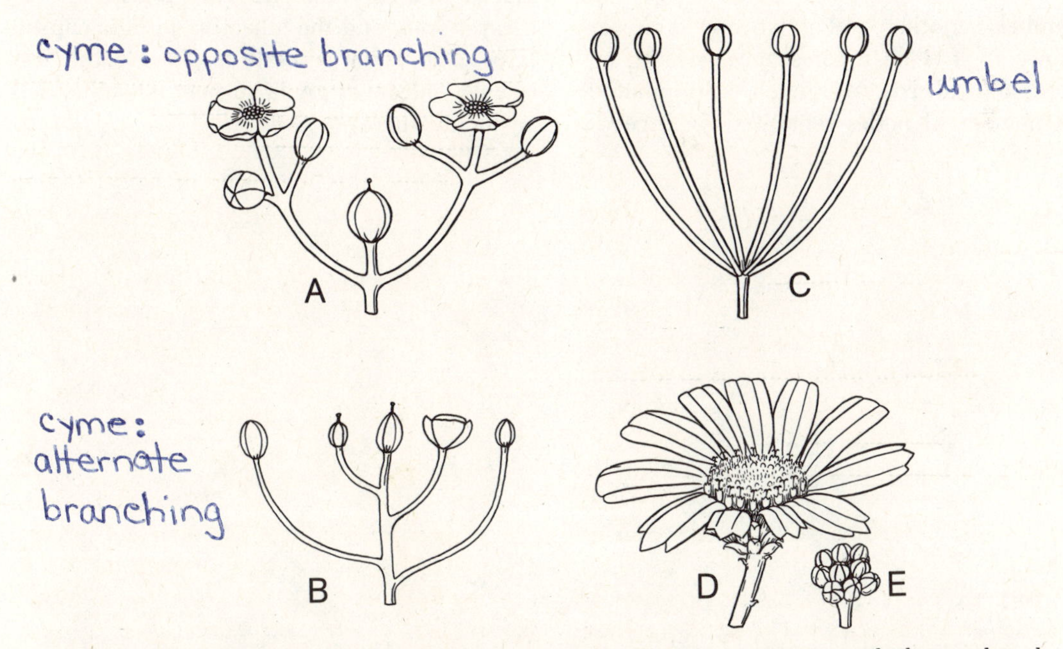

Figure 7-2. Inflorescence types: **A** cyme with opposite branching; **B** cyme with alternate branching (note fruits on the inside, buds outside [basal]); **C** umbel (in bud); **D** head of ray flowers (external and petal-like) and disc flowers (internal) typical of the sunflower family (Compositae, Order Asterales); **E** head of a generalized type (in bud). By Lucretia Breazeale Hamilton.

A **helicoid cyme** (Figures 10-67 and 10-68) (often referred to as scorpioid) is similar to the ordinary cyme with opposite inflorescence branches, but the branches are produced singly instead of in opposite pairs. The stem ends in a flower and a single branch grows from the last node. The branch ends in a flower and a subordinate branch grows out in the same direction. This procedure is repeated several or many times, as in most species of buttercups. In the borage and waterleaf families the helicoid cymes are compact, one-sided, and without bracts; they are curved like the circinate young leaf of a fern or like a "fiddle neck." Commonly they are referred to as "scorpioid spikes" or "scorpioid racemes," but actually they are helicoid cymes. In a truly scorpioid cyme, the alternate branches which successively replace each other grow out in opposite directions, and the inflorescence is not one-sided.

Umbels and Heads

An **umbel** is another kind of flat-topped cluster of flowers. It is an inflorescence without an obvious central axis. A branch is terminated by a number of nodes separated by exceedingly short internodes so that apparently there is no axis at all within the inflorescence. The pedicels of many flowers appear to arise from the same point. Usually these pedicels are of about equal length. Often a ring of bracts (an **involucre**) subtends the umbel.

The definition and the illustration (Figure 7-2 **C**) fit the most commonly appearing definition, but, strictly, in a true umbel the buds are in the center (at the apex of the stem) and the flowers are intermediate and the fruits outside. An umbel is really a shortened raceme, that is, one with only a very short axis. The fundamentally lower flowers bloom first.

A **compound umbel** is an umbel of umbels, each **ray** (peduncle) bearing an umbel instead of a single flower. Each secondary umbel is called an **umbellet,** and its bracts form an **involucel.**

A **head** is similar to an umbel, but the flowers are **sessile;** that is, they lack pedicels or at least appear to because the pedicels are very short. In a head the marginal flowers bloom first as in a corymb. A head is known also as a **capitulum,** and the adjective form is **capitulate.** The term **capitate** is used most frequently, however. A head may be subtended by an involucre.

8

Fruits and Seeds

What Is a Fruit?

The most characteristic feature of flowering plants is the pistil, which encloses the ovules within its ovary wall. After flowering time (**anthesis**) the ovary grows and changes in texture (becoming hard, leathery, fleshy, etc.) as the ovules mature into seeds, and ultimately it becomes a fruit. Since man is likely to think in the terms of the fruits most useful to him, in popular language the word is applied largely to fleshy fruits. Technically the term encompasses many types, some not at all fleshy but dry or grainlike and known to the layman as "pods" or "seeds."

A fruit is a matured ovary (or this and the adnate floral cup) with its enclosed seeds; a vegetable is a vegetative part of the plant, that is, a root, stem, or leaf or some structure developed from one or more of them. These definitions are applied strictly in the botanical sense. From a culinary point of view the term vegetable is used loosely because squashes, string beans, and tomatoes are really fruits.

The Receptacle at Fruiting Time

Usually the receptacle enlarges relatively little after flowering, but in some plants it elongates or expands greatly, becoming fleshy (as in the strawberry) or dry. In the garden geranium or in the filarees of the Southwest the receptacle may be as long as the elongate style of the fruit, the five carpels being lightly coalescent around it. In *Nelumbo,* a member of the water-lily family often cultivated in ponds and pools, each pistil occupies a pocket or pouch in the top of the greatly enlarged receptacle. In a few plants a portion of the receptacle may become greatly elongated so that the ovary is elevated far above the former attachment point of the sepals, petals, and stamens (cf. Figure 10-9).

The Floral Cup

Although a floral cup is present in fewer than half the flowering plants, some of the principal variations in fruits are related to it.

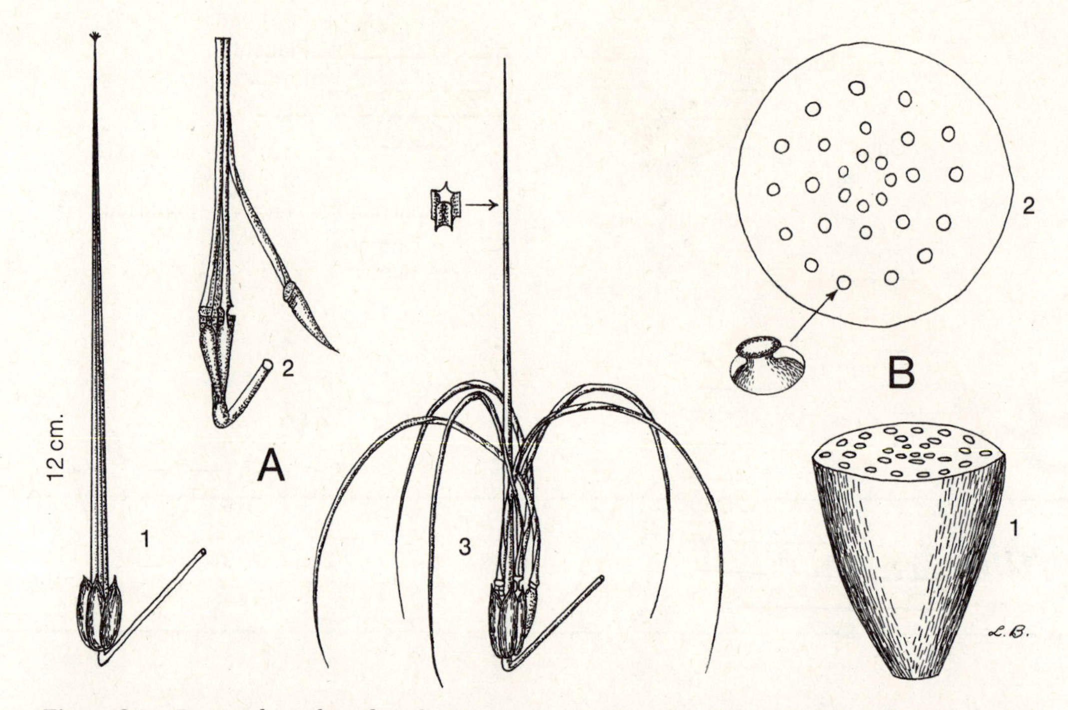

Figure 8-1. Receptacles enlarged at fruiting time: **A** filaree or alfilaria (*Erodium botrys*, a relative of the garden geranium), **1** fruit surrounded by the calyx, the elongate compound style protruding, **2** carpels separating from each other and from the axis (receptacle), **3** carpels separated and nearly ready to fall away, exposing the elongate protruding receptacle; **B** *Nelumbo lutea* (water-lily family, Nymphaeaceae), **1** enlarged fruiting receptacle with cavities in the apical area, **2** apical area with enlargement of a fruit protruding slightly from a cavity. (Cf. also **Figure 8-4 A.**)

The following are examples of structures associated with fruits developed from perigynous flowers. Both strawberries and blackberries have a persistent green floral cup as a basal husk. At fruiting time this still bears the sepals and dried stamens. The fruit of a cherry, plum, peach, apricot, or almond is subtended by a dried, sometimes partly detached, floral cup still bearing the dried sepals and stamens. The small, dry, hairy true fruits of the rose are concealed completely by the floral cup or "hip" which becomes large, fleshy, and usually colored.

An example of a fruit developed from an epigynous flower is illustrated in Figure 8-2 **B.** In all these there is distinction of a "stem end" and a "blossom end." The inferior ovary appears to be an expanded portion of the stem, and the pedicel below it usually is much more slender. At fruiting time the usually persistent sepals and dried stamens at the apical end of the fruit are dwarfed by the great growth of the ovary and the portion of the floral cup covering it. The edible portion of an apple or a pear is developed chiefly from the floral cup (in this case at least largely a hypanthium), and the core is the enclosed ovary. Only a little of the outer tissue of the ovary is fleshy and edible, and this is distinguishable from the surrounding cup tissue only in sections studied under a compound microscope. The fruits of melons and gourds, including watermelons, cantaloupes, cucumbers, and pumpkins, are similar to apples and pears in that there is the same distinction of a stem end and a blossom end. The fruits of many other

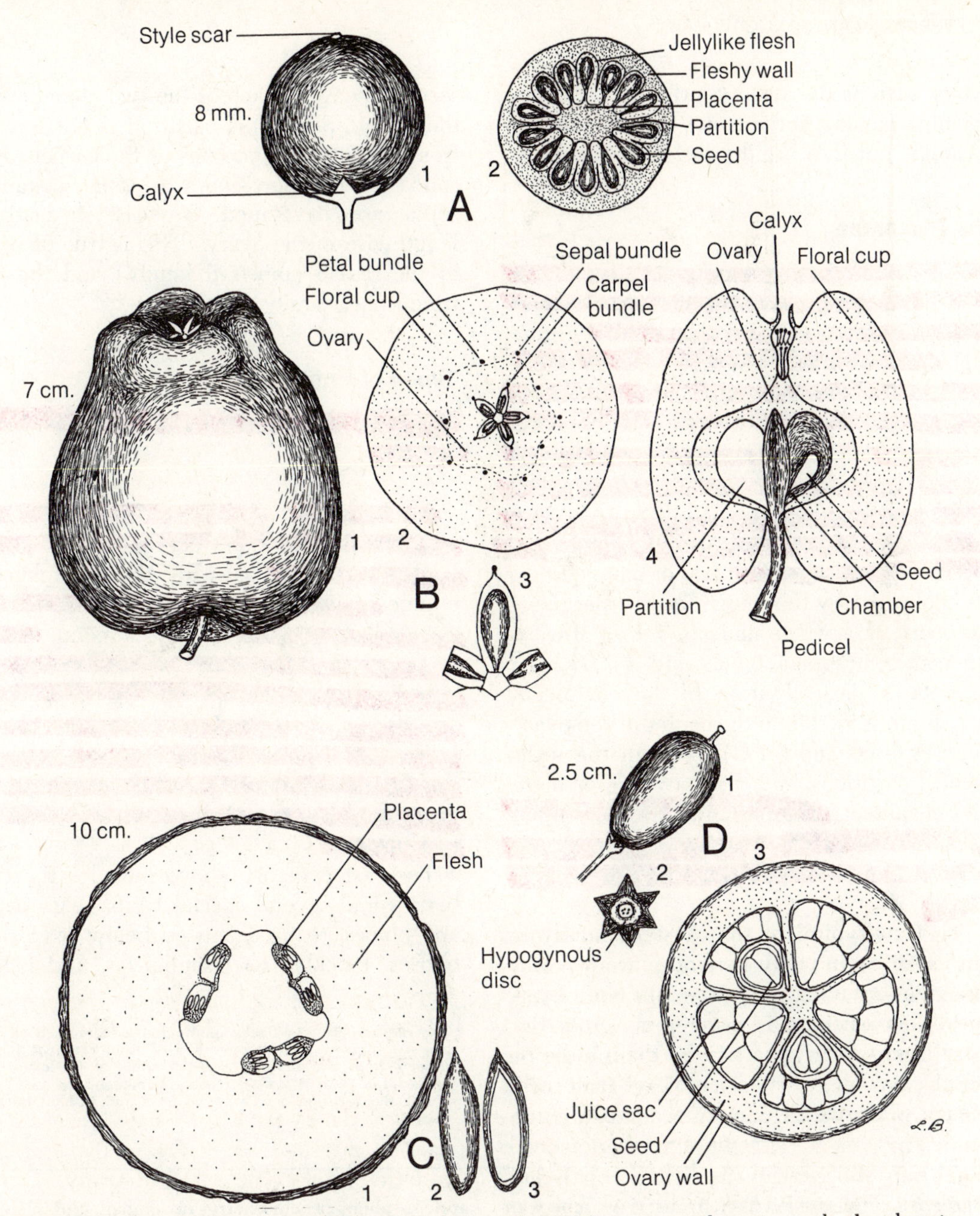

Figure 8-2. Berries and berrylike fruits: **A** berry formed from two coalescent carpels, the placenta central (axile) and fleshy, a nightshade (*Solanum Douglasii*): **1** external view, **2** cross section; **B** pome formed from five coalescent carpels and the adnate floral cup, apple (*Pyrus Malus*), **1** external view, **2** cross section, **3** enlargement of a portion of the ovary showing the attachment of the seeds, **4** longitudinal section; **C** pepo formed from 3 coalescent carpels and the adnate floral cup, muskmelon (*Cucumis Melo*), **1** cross section, **2** seed, external view, **3** seed, longitudinal section, the embryo omitted; **D** citrus fruit formed from five coalescent carpels, kumquat (*Fortunella margarita*), **1** external view, **2** top view of the calyx and the hypogynous disc after removal of the fruit, **3** cross section of the fruit. **In B (inferior ovary) note the persistent sepals at the end opposite the pedicel, in A and D (superior ovary) the sepals at the pedicel end and only the style at the opposite end.**

plants, such as the evening-primroses, carrots, parsnips, parsley, and dill, are the same fundamentally, but they are dry instead of fleshy.

The Placentae

The position of the placenta or placentae is determined by the number and arrangement of carpels. If there is a single carpel (Figure 5-6), there is a single placenta. At least potentially, this follows along the suture. Since the suture is at one side of the ovary, the placenta is **parietal.** In some ovaries formed from a single carpel there may be only one or a few seeds, and these may be so low in the ovary that the placenta is considered to be **basal** or so high that it is **apical.** If two or more carpels are coalescent by their edges at the margin of the ovary as a whole and not folded toward the center (Figures 6-6 **D** and 6-7 **B**), the placenta is parietal also. If the coalescent carpels are folded in to the center of the ovary (Figures 6-6 **C** and 6-7 **C**), parts of the walls become partitions, and there are two or more seed chambers. The placentae are developed at the center of the ovary along the joined margins of each carpel, and they are **axile** or **central.**

The fruit typical of the mustard family is unusual in having both parietal placentae and a partition dividing the ovary into two chambers. This partition is developed across the ovary from what appears to be the middle of one placental area to the middle of the other. In each placental area one of the two rows of seeds appears in each of the two chambers of the ovary. Since this partition is not formed from the infolded margins of the carpels, it is called a **false partition.*** In some instances septa are developed crosswise instead of lengthwise in the ovary. This is true of fruits of the radish (mustard family) and the tick trefoil (pea family).

Types of Fruits

There are two principal types of fruits: **fleshy** and **dry.**

Fleshy Fruits. In fleshy fruits the ovary wall often is juicy and edible, but it may be leathery instead, as with the pumpkin or orange, or at maturity it may dry up and become a leathery husk, as in the almond. Fleshy fruits are of two principal kinds: **berries** and berrylike fruits, characterized by pulpy interiors, no stony layer, and usually but not always several to many seeds, and **drupes,** characterized by a stony inner layer of the pericarp, and most often but not always one seed chamber and one seed.

The first category above includes the true berries and several special fruits (e.g., pepos and pomes, to be discussed below). (Strawberries, blackberries, raspberries, and logan-

* The mustard fruit has been interpreted as composed of four carpels. According to Eames and Wilson, the partition is formed from two carpels, and the walls of the ovary are formed from two, the ovules being formed by the two composing the partition, which are reduced and without internal cavities.

Figure 8-3. One-seeded fleshy fruits: **A** drupe with one stone, formed from a single carpel, apricot (*Prunus Armeniaca*), **1** external view showing the persistent style and the suture, **2** stone (pit) formed by the hard endocarp, the fleshy exocarp removed, **3** longitudinal section of the stone exposing the seed, **4** seed in longitudinal section, embryo (except cotyledons), pointed end, **5** embryo enlarged; **B** drupe with four stones, formed from four coalescent carpels, holly (*Ilex*), **1** side view, **2** basal view showing the calyx, **3** cross section, the embryos being central in the cross sections of the stones, the endosperm surrounding them; **C** aggregation of drupelets each similar to an apricot, plum, peach, cherry, or almond, this plant a Boysenberry (*Rubus,* hybrid): **1** external view, the calyx ("cap") and the floral cup persistent, basal view, **2** the same, side view, **3** longitudinal section, **4** single fruit, side view, showing the persistent style, **5** stone (pit), **6** longitudinal section of the stone, showing the relatively large embryo and the small amount of endosperm, **7** embryo removed from the stone, showing the two large cotyledons; **D** berry- or drupelike fruit of the avocado (*Persea americana*), **1** longitudinal section including the seed and showing the embryo, **2** external view of the large seed, **3** enlargement of a portion of the edge of the seed in **1.**

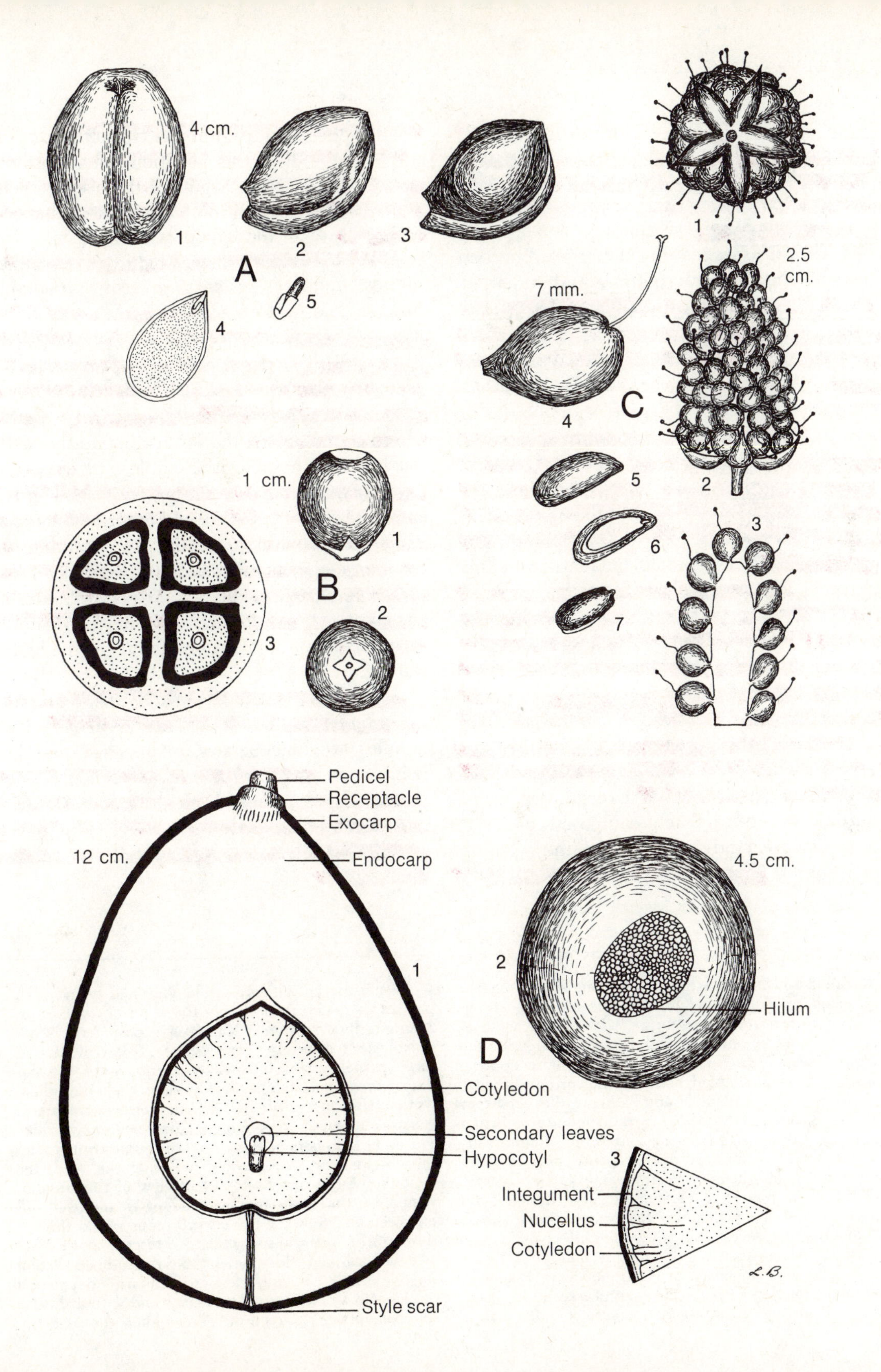
4 cm.
1
2
3
A
5
4
1
2.5
cm.
7 mm.
4
C
2
5
6
3
7
1 cm.
1
B
2
3
Pedicel
Receptacle
Exocarp
12 cm.
Endocarp
4.5 cm.
1
2
Hilum
D
Cotyledon
Secondary leaves
Hypocotyl
3
Integument
Nucellus
Cotyledon
L.B.
Style scar

berries are not berries, cf. below.) The true berries have a fleshy or leathery outer wall and fleshy pulp (in superior ovaries, the ovary wall; in inferior ovaries, both the ovary wall and the attached floral cup). Examples are gooseberries, currants, grapes, huckleberries, cranberries, blueberries, elderberries, and tomatoes. The fruits (**pepos**) formed from the inferior ovaries of the gourd family have a hard or leathery rind on the outside and a fleshy inner layer. Watermelons, cantaloupes, squashes, pumpkins, and cucumbers are examples of this family. Although bananas have a similar outer rind, they are not called pepos. In the fruits (**pomes**) of the apple and its relatives the principal fleshy edible tissue is on the outside, and it is formed largely from the floral cup (hypanthium), as pointed out earlier in this chapter. The enclosed ovary wall may be of firmer texture, and the seeds are not embedded in pulp. Oranges, lemons, grapefruits, and other citrus fruits are berries (many-seeded fleshy fruits) with leathery pericarps. Each of the sections of a citrus fruit with the adjacent part of the rind is a carpel. The seeds, if they are present, are on an axile placenta at the inner edge of the carpel. Exceedingly large hairlike structures, each composed of very large cells, extend inward from the wall of each carpel. The cells are the juice sacs, which attain their greatest size in grapefruit.

Berries may be formed from either inferior or superior ovaries. Usually they have more than one carpel, but in some instances there is only one, as in the baneberry.

The pitted fruits or stone fruits, such as the plums, cherries, peaches, apricots, almonds, prunes, and nectarines, are in the second category. The matured ovary wall (**pericarp**) of these fruits is composed of two layers, an outer fleshy layer (**exocarp**) and an inner stony layer or pit (**endocarp**). Ordinarily the outer fleshy part is eaten, but in the almond the seed enclosed in the pit is the edible part; the exocarp dries into a husk. Technically these fruits are known as **drupes**. Each is formed from a single carpel which ordinarily contains only one seed, although occasionally there are two. In some instances drupes may have several seeds or even several hard-walled sections or chambers, as in the ginseng family or the holly family.

Sometimes there are many small drupes (**drupelets**) produced by a single flower. Examples are blackberries, raspberries, and loganberries. Each of the many pistils of the flower develops into a drupelet identical in structure with a plum or a cherry. These are arranged spirally on an elongated or enlarged receptacle.

Figure 8-4. Dry and indehiscent one-seeded fruits: **A** achenes, each formed from a single carpel, strawberry (*Fragaria vesca,* hybrid), **1** external view of the calyx and the enlarged red receptacle, showing the fruits ("seeds") each in a depression in the receptacle, **2** single achene enlarged, showing the surrounding depression in the receptacle and showing the persistent style, **3** achene in side view; **B** achene formed from a single carpel, a species of buttercup (*Ranunculus inamoenus*), **1** old head of fruits, with some achenes persistent (note their spiral arrangement characteristic of the order Ranales), **2** single achene showing the persistent style, which becomes an achene beak, **3** longitudinal section showing the thin-coated seed, with much endosperm and a minute embryo, **4** embryo enlarged, showing the two cotyledons; **C** achene formed from two coalescent carpels and the floral cup, dandelion (*Taraxacum vulgare* or *officinale*), **1** achene, the apical portion elongate and bearing the pappus, a structure probably formed from the calyx, **2** longitudinal section of the fruit exposing the seed; **D** achene formed by coalescence of the ovaries of three carpels, wild buckwheat (*Eriogonum fasciculatum* var. *foliolosum*), **1** side view, **2** longitudinal section showing seed; **E** achene formed from two coalescent carpels, goosefoot (*Chenopodium murale*) (in other members of the same family the fruit similar but the pericarp not closely investing the seed, the fruit therefore a utricle), **1** top view of the fruit invested by the calyx removed, **2** calyx removed, **3** cross section, showing the curved embryo and the endosperm; **F** caryopsis, wheat (*Triticum aestivum*), **1–2** ventral and side views, **3** longitudinal section (**a–d** embryo, **a** cotyledon [scutellum], **b** epicotyl or stem tip among secondary leaves and with a sheath, the coleoptile, **c** hypocotyl, **d** radicle or root tip,

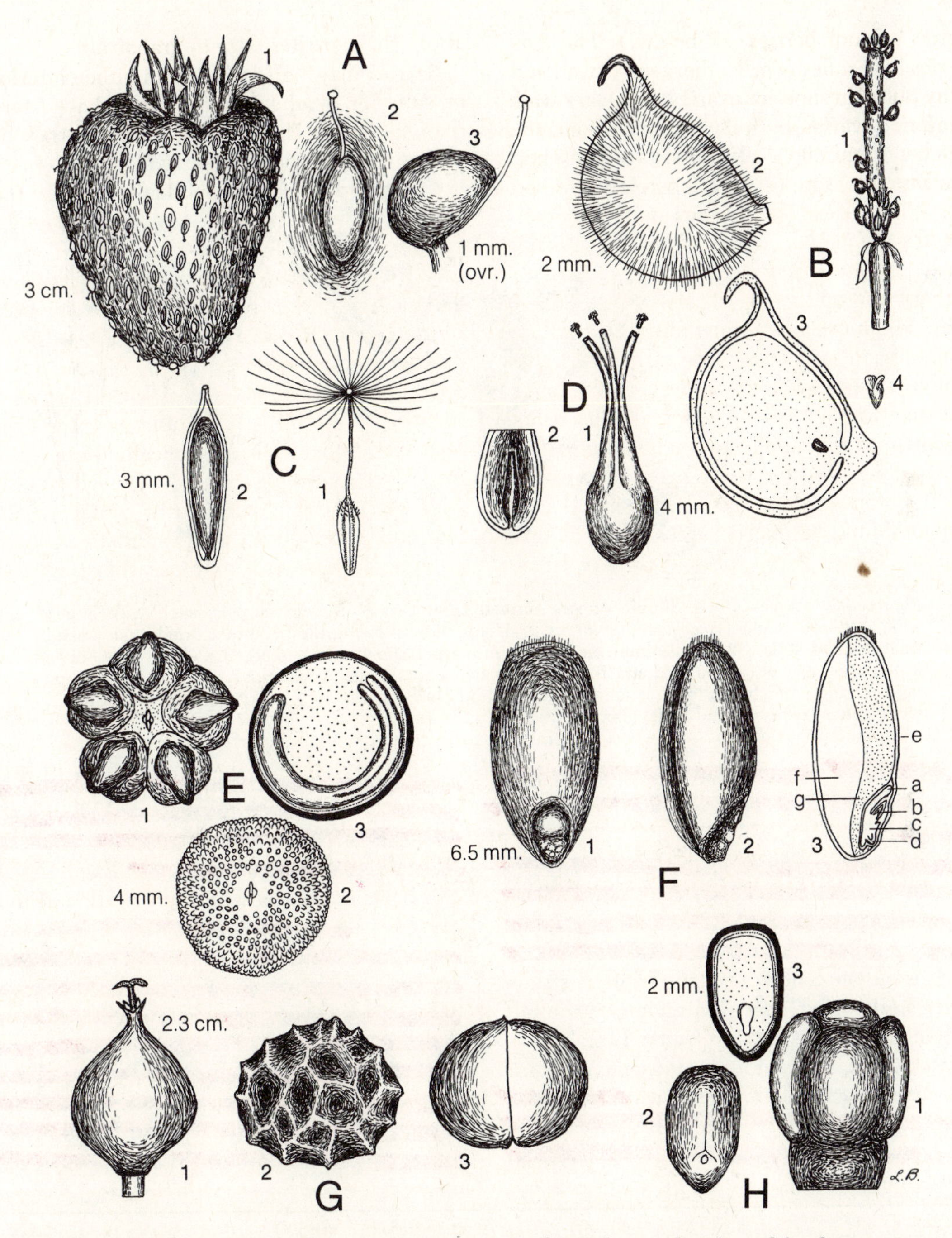

e pericarp (sectional view), **f** surface view of pericarp along the groove on the dorsal side of the fruit, **g** endosperm (aleurone); **G** nut formed from two carpels and invested in the first fleshy, then dry floral cup, southern California black walnut (*Juglans californica*), **1** side view of a young fruit, **2** a mature fruit with dried floral cup (husk), **3** floral cup removed; **H** four nutlets formed by deep constriction between the halves of the ovaries of two carpels and between the carpels, bee-sage (*Salvia apiana*), **1** the four nutlets in side view, **2** single nutlet in ventral view, **3** longitudinal section showing the large endosperm area and the small embryo.

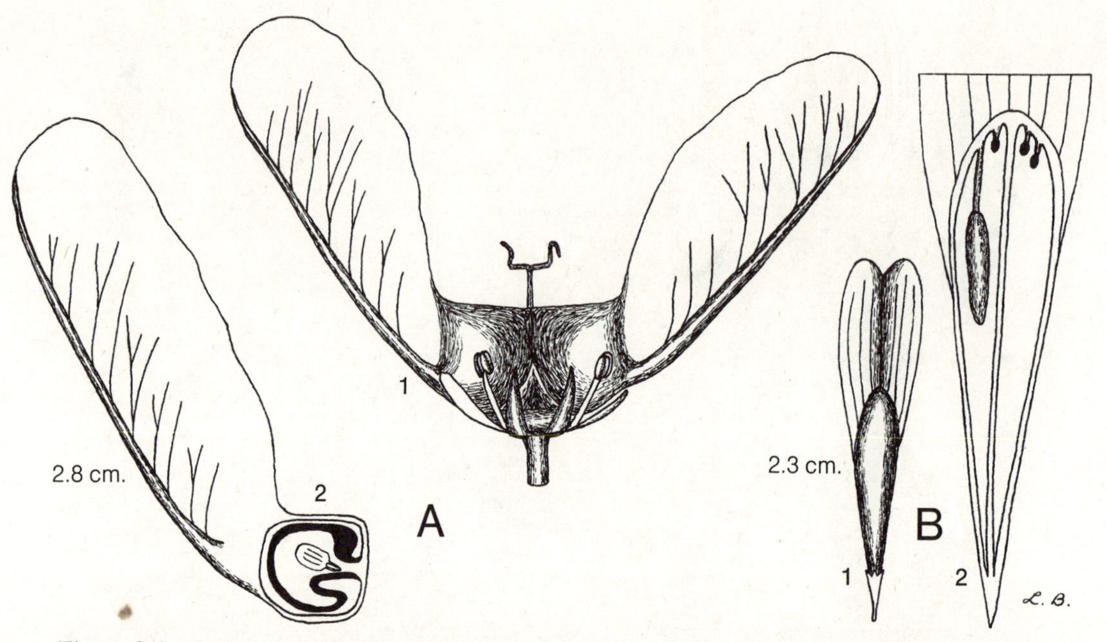

Figure 8-5. Samaras: **A** double samara formed from two coalescent carpels, each with a large wing, big-leaf maple (*Acer macrophyllum*), **1** side view showing the persistent dried sepals, petals, stamens, and style, **2** longitudinal section of one carpel showing the seed, the embryo suspended in endosperm; **B** (single) samara formed from two carpels, ash (*Fraxinus*), **1** side view, showing the single notched wing, **2** longitudinal section showing the two ovary chambers and the attachments of one seed and three abortive ovules.

Dry Fruits. Dry fruits are divided into groups according to the number of seeds, as follows:

Indehiscent, one-seeded, dry fruits. An **achene** is a one-seeded dry fruit with a firm, close-fitting pericarp which does not split open, at least along regular established lines (i.e., is **indehiscent**). The true fruits of strawberries (the hard "seeds" in niches on the red edible receptacle) and the fruits of the buttercups, nettles, and the numerous members of the sunflower family are achenes. A **utricle** is similar to an achene, but the ovary wall is relatively thin and somewhat inflated so that it fits only loosely around the seed. At maturity it opens either irregularly or along a horizontal circular line and exposes the seed. Sometimes the upper part falls off like a lid. Examples are the fruits of the goosefoot, pigweed, and their relatives. A **caryopsis** is similar to an achene also, but the pericarp is adherent to the seed, the two often being indistinguishable, as in a grain of wheat or corn. This fruit type is restricted practically to the grass family. A **nut** is similar to an achene but usually much larger and with a very large seed and a hard and usually thick wall (shell). The diminutive is **nutlet.** A **samara** usually, but not always, falls

Figure 8-6. Dry and dehiscent several-seeded fruits: **A** capsule formed from three individually sealed coalescent carpels and the adnate floral tube, iris (*Iris*), **1** side view showing the persistent perianth above the inferior ovary, **2** cross section showing the double row of seeds in each carpel, **3** loculicidally dehiscent capsule (splitting along the midrib of each carpel); **B** capsule formed from five coalescent carpels, St.-John's-wort (*Hypericum*), **1** side view with one sepal removed, the stigmas shown detached and part of the style omitted, **2** cross section, **3** septicidally dehiscent capsule (splitting through the partitions of the chambers of the ovary); **C** pyxis formed by coalescence of five carpels, scarlet pimpernel (*Anagallis*

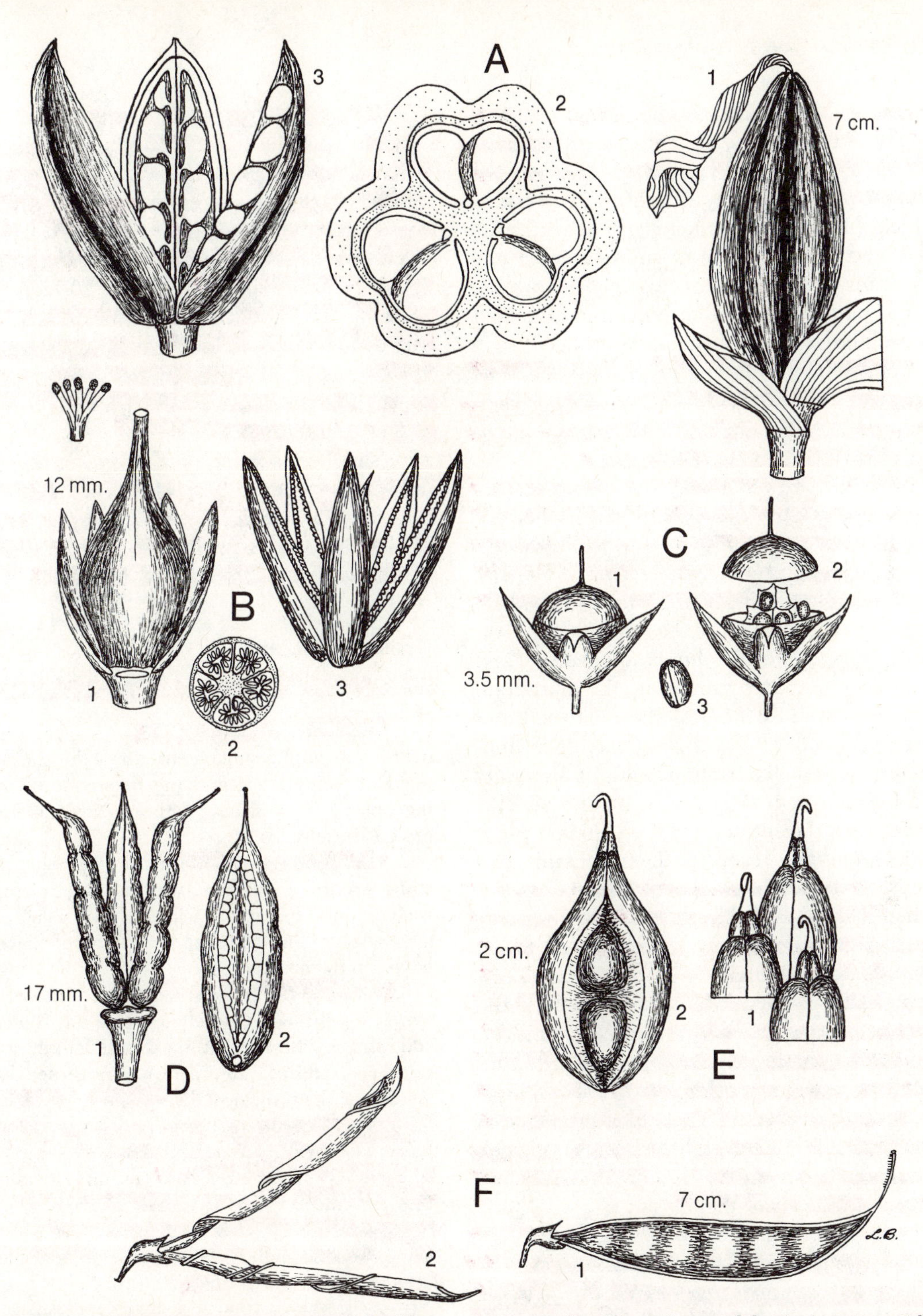

arvensis), **1** side view before circumscissile dehiscence, **2** after dehiscence, showing the central placenta, the fruit one-chambered, **3** seed; **D** three separate follicles, larkspur (*Delphinium cardinale*), **1** side view, **2** dehiscence along the suture between the members of the double row of seeds; **E** fruits of a magnolia (*Magnolia grandiflora*), these essentially folicular but each dehiscent along the midrib of the carpel (**2**) (cf. **Figure 6-15**); **F** legume, sweet pea (*Lathyrus splendens*), the carpel dehiscent along both margins, **1** side view before dehiscence, **2** after dehiscence.

within the one-seeded classification. It is distinguished by a wing, that is, a long, thin, flat structure developed from a portion of the ovary wall. Examples are ash and elm fruits, which flutter to the ground, often being carried away by the wind. In the maple the samara is double, each of two carpels forming a wing and the two often remaining attached to each other for a considerable period.

Dehiscent, several–to many-seeded dry fruits (some types rarely with a single seed). These fruits may be classified according to the number of carpels entering into them.

Either of two kinds may be formed from *a single carpel*. A **follicle** splits open (**dehisces**) at maturity along only the suture, although commonly the term is extended to include any similar fruit splitting along only the midrib. Examples of follicles are the fruits of the larkspurs (*Delphinium*), columbines, peonies, and a number of other groups in the buttercup family, as well as those of the milkweed and dogbane families. The individual fruits composing the conelike fruiting heads of *Magnolia* are folliclelike, but they split on the midribs rather than the sutures. In the common *Magnolia grandiflora* each of them contains two seeds. A **legume** is similar to a follicle, but it is **dehiscent** along both margins. Legumes are restricted to the pea family, but some members of the family have highly specialized indehiscent fruits. Peanuts are an example. In the latter, the flowers are borne near ground level; and after flowering, each supporting structure turns downward, pushing the fruit underground. Each fruit develops usually two seeds surrounded by a nonedible indehiscent shell (pericarp) well known from peanuts roasted with the "shells" on them.

Many-seeded dry fruits formed from *more than one carpel* are **capsules.** These, as well as follicles and legumes, are known also as **pods.** Capsules are characterized by lengthwise dehiscence along regular lines. If the capsule is dehiscent along the midrib of each carpel, it is **loculicidal.** If it divides through the partitions, that is, between the carpels, it is **septicidal.** Loculicidal means "cutting into the cells [chambers or locules]"; septicidal means "cutting through the partitions." A few capsules are dehiscent in both ways. The segments into which the capsule wall splits are called **valves.** In the mustard family (Figure 10-12 **A**) the fruit opens from the base upward; the two valves become detached, leaving the false partition (see p. 84) still attached to the receptacle. This special type of capsule is called a **silique.** A short broad silique (e.g., the fruit of a shepherd's-purse or a pepper-grass) is a **silicle.** A **pyxis** is a special type of capsule which opens around a horizontal ring, the top of the fruit falling away like a lid. This mode of dehiscence is termed **circumscissile.**

Types of Ovules

Ovules are of three principal types, as follows:

1. ***Orthotropous or straight.*** An orthotropous ovule is a simple, stalked structure, the stalk (funiculus) being basal and the micropyle or opening apical. The ovules of the buckwheat family are good examples.
2. ***Campylotropous or curved.*** The body of the ovule is curved in such a way that the micropyle is close to the funiculus. The chickweed and other members of the pink family (Caryophyllaceae) have campylotropous ovules.
3. ***Anatropous or inverted.*** The body of the ovule is bent back closely against the funiculus and adnate with it along the whole length of one side. As a result, the micropyle is turned backward near the funiculus. This is the extreme form. In a less extreme type, often described as ***amphitropous,*** the micropyle is not bent clear back to the edge of the funiculus (example, mallow). The extreme form, that is, an anatropous ovule, is the prevailing type among the flowering plants. The others occur in a few or all families of each of several unrelated orders. (Cf. p. 569.)

Ovules vary in length of the funiculus and in number of integuments. The funiculus may be elongated or nearly obsolete. There may be one integument or two or rarely none.

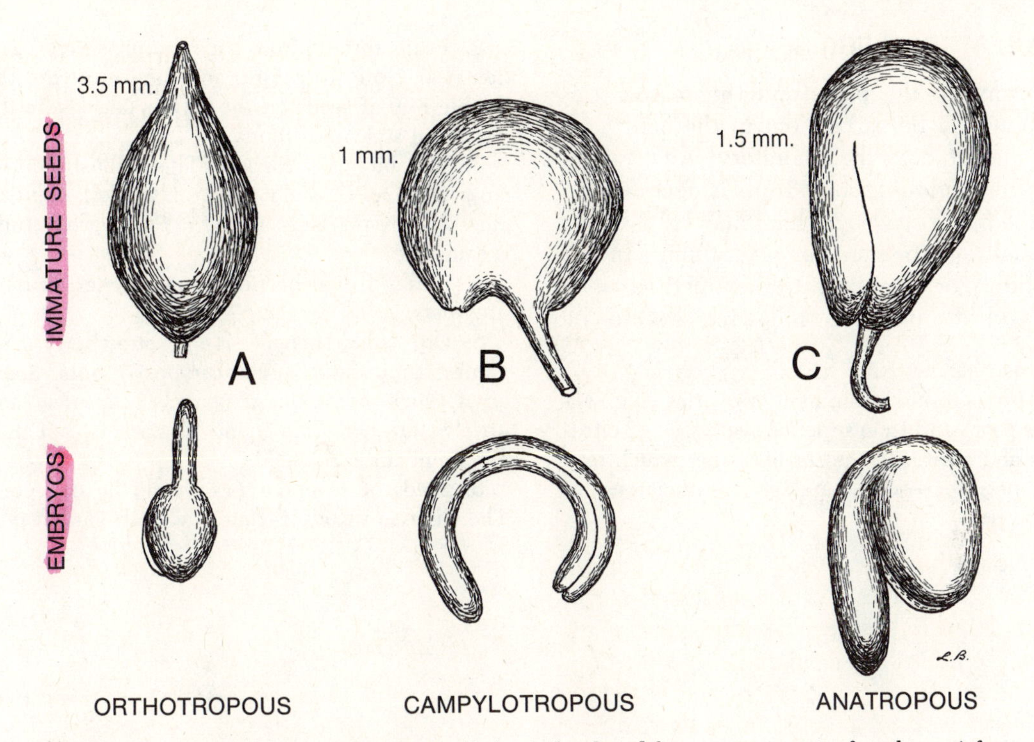

Figure 8-7. Seeds (above) and embryos (below) developed from various types of ovules: **A** from an orthotropous ovule, wild buckwheat (*Eriogonum fasciculatum* var. *foliolosum*); **B** from a campylotropous ovule, chickweed (*Stellaria media*); **C** from an anatropous ovule, shepherd's purse (*Capsella Bursa-pastoris*).

Correlation of the number of integuments with the major outlines of classification is not consistent.

The positions of ovules within the ovary vary greatly. Some stand out horizontally from the placenta; others ascend obliquely, stand erect, or hang downward. According to position the ovules are described as **horizontal, ascending, erect, pendulous,** or **suspended** (hanging perpendicularly from the very summit of the seed chamber of the ovary).

Seeds

The seed coat of a detached seed has two well-marked points—a scar, the **hilum** (marking the point at which the funiculus was attached), and the micropyle (still distinguishable at this stage as a pinpricklike hole). Usually the outermost part of the megasporangium or nucellus persists as a thin papery layer between the seed coat and the endosperm. The endosperm varies greatly, and it may be characteristic of families or genera or even orders. It may be mealy (with starch grains), oily, bony, or pulpy, often according to the chemical nature of the food reserve.

Seeds vary in shape and size and in many other ways. The surface of the seed coat may be either smooth or sculptured. Sometimes part of it forms a thin flat wing; sometimes it is covered with hair as in cotton; sometimes it has a single tuft of hairs at one end as in the milkweed or the fireweed; or sometimes it has a tuft of hairs at each end as in the dogbane family. In some seeds there is usually a fleshy outgrowth (**aril**) from the funiculus or the placenta. It is conspicuous in the burning-bush after the fruits open in Fall. In some water-lilies the aril is a bag enclosing the seed.

SUGGESTED EXERCISE

According to the place and the season, this exercise may be carried out in the field, at a fruit market, or in the laboratory. An arrangement for sectioning some fruits is necessary.

In the field a scavenger hunt may be possible, especially in the summer or autumn. In the laboratory or at a market other methods may be worked out. The following should be noted:

1. Fruits formed from inferior ovaries (e.g., apple or pear) and from superior ovaries (e.g., citrus fruits and especially the smaller types which may have persistent sepals and styles—tomatoes, or garden peas).

2. Fruits formed from a single carpel, e.g., garden pea. Note the midrib and the suture and the double row of seeds on the suture side. Note the persistent style and stigma.

3. The joining of carpels. This is illustrated by cross sections of such fruits as apples, oranges, etc. The by-products are edible only after study is complete.

4. Fleshy fruits: berries, pepos, pomes, drupes, drupelets.

5. Dry fruits; achenes (e.g., strawberry), legumes, caryopses (singular caryopsis), nuts. Some types, such as siliques, samaras, capsules, and utricles, are not often found in markets, but they are abundant in the field.

6. Seeds of several types should be dissected. The easiest to study is that of a fresh garden pea.

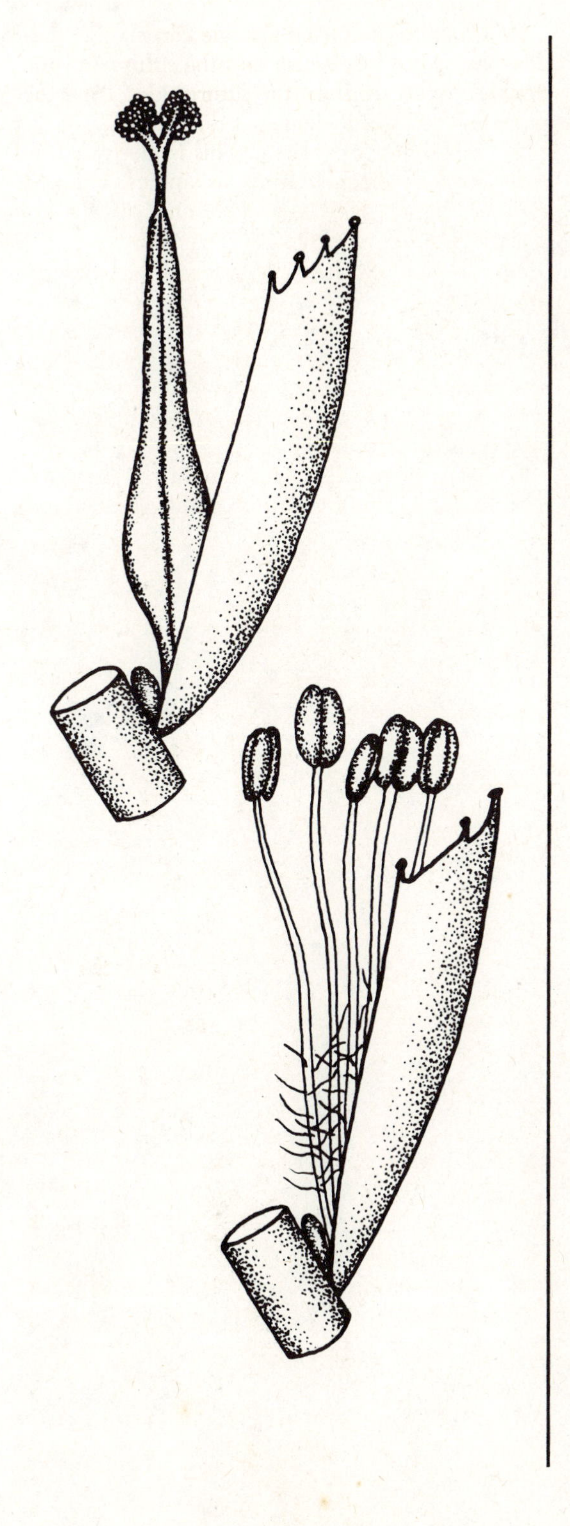

Flowering Plants

Part Two

THE PROCESS OF IDENTIFICATION

9

Identification of Dicotyledons and Monocotyledons

NOTE: The divisions of the Plant Kingdom are segregated in a key on pages 6–7 in Chapter 1.

Differentiation of the Subclasses

The flowering plants include two subclasses, the Dicotyledoneae and Monocotyledoneae. These are clearly distinct from each other, despite the fact that no single character occurs in only one and not the other. A complex of pairs of opposed characters, each usually but not always reliable, marks off one group from the other. Six of these are seen readily, and they are illustrated on pp. 96–98.

Nearly all the flowering trees are dicotyledons; only a few of the monocotyledons are trees or shrubs. These include the palms, *Dracaena* (the dragon-tree), a few large canelike grasses (such as bamboo), century-plants, and yuccas. The century-plants and most yuccas may be interpreted as large herbs, but some yuccas, like the palmillas and the Joshua-tree, are clearly arborescent or arboreous. The rarity of woody plants among the monocotyledons is due to absence of the cambium which in the dicotyledons and the cone-bearing trees (gymnosperms) adds new layers ("rings") of wood (primarily xylem) on the inside and new layers of bark (primarily phloem) on the outside. Stems of nearly all monocotyledons attain their full diameter upon final enlargement of cells formed by divisions accompanying growth of the bud. A few of the woody monocotyledons have a cambiumlike cell layer, which divides, forming new pithy tissue and whole vascular bundles both internally and externally. This is in no way homologous to the cambium of a dicotyledon.

NOTE: **The keys to angiosperm orders and families are based specifically upon only the groups occurring in North America north of Mexico.*** In general they may be used in other re-

* **Orders and families not native or introduced in this area are referred to as "extraterritorial."**

gions and especially North Temperate ones; and the descriptions of the orders and families are on a world-wide basis. Although keys applicable to any flowering plant anywhere in the world would be desirable, the enormous task of preparation of workable universal keys has not been accomplished by any author. It is a project for many years.

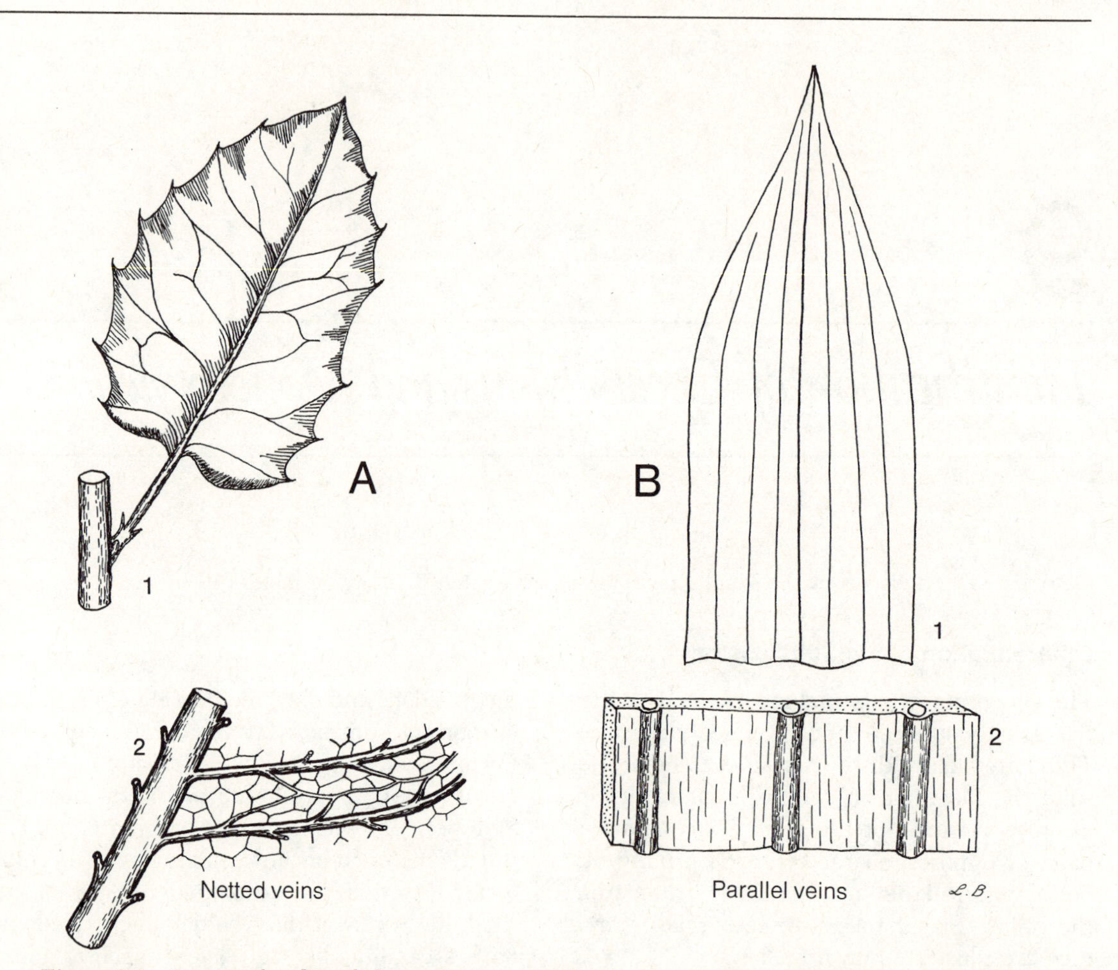

Figure 9-1. Leaves of a dicotyledon and a monocotyledon: **A** evergreen cherry (plant from a hybrid swarm of the holly-leaved cherry, *Prunus ilicifolia,* and the Catalina cherry, *Prunus ilicifolia* var. *occidentalis*), **1** leaf, **2** enlargement of the netted veins; **B** Bermuda (or devil) grass (*Cynodon dactylon*), **1** apical portion of leaf showing the parallel veins, **2** enlargement of a segment.

Dicotyledons

1. Principal veins of the leaves branching out from the midrib or from the base of it, not parallel, forming a distinct network (e.g., sunflower, maple, orange, or apple leaves).

Monocotyledons

1. Principal veins of the leaves parallel to each other (e.g., corn or other grass leaves).

Figure 9-2. Flower patterns of a dicotyledon and a monocotyledon: **A** diagram of a 5-merous dicotyledonous flower; **B** a 3-merous monocotyledonous flower, a yucca (*Yucca Whipplei*).

Dicotyledons

2. Sepals, petals, and ordinarily the other parts of the flower arranged in twos, fours (e.g., evening-primrose), or usually fives (e.g., wild rose, plum, cherry, most buttercups, apple, or geranium flowers).

Monocotyledons

2. Sepals, petals, and usually the other flower parts in threes (e.g., lily, wake-robin or *Trillium, Yucca,* Solomon's-seal, or dogtooth-violet flowers).

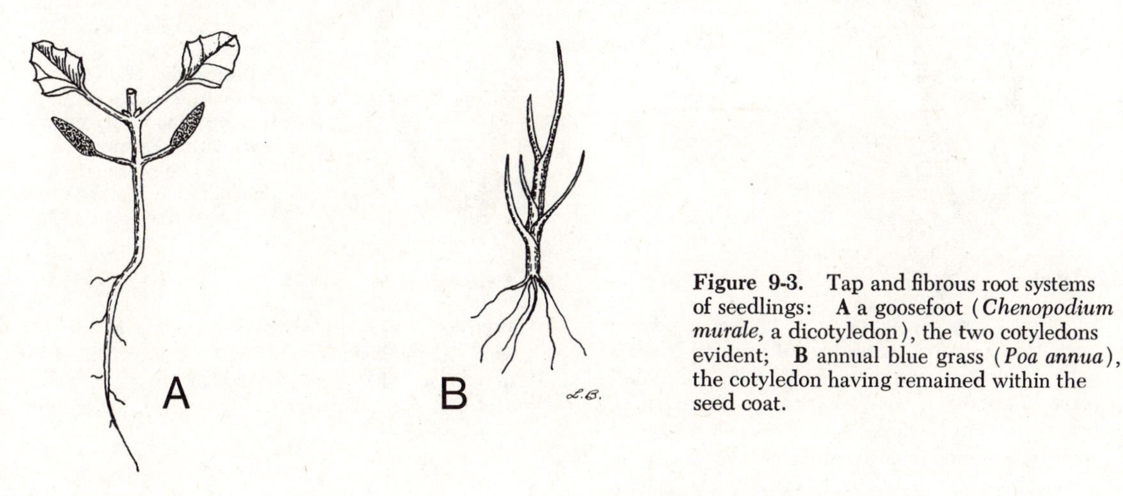

Figure 9-3. Tap and fibrous root systems of seedlings: **A** a goosefoot (*Chenopodium murale,* a dicotyledon), the two cotyledons evident; **B** annual blue grass (*Poa annua*), the cotyledon having remained within the seed coat.

Dicotyledons

3. Root system characterized by a taproot, that is, a large primary root with branch roots growing from it (e.g., sunflower, mallow, pigweed, or lamb's-quarter roots).

Monocotyledons

3. Root system fibrous, that is, without a principal or taproot, the evident roots being of about the same size (e.g., grass roots).

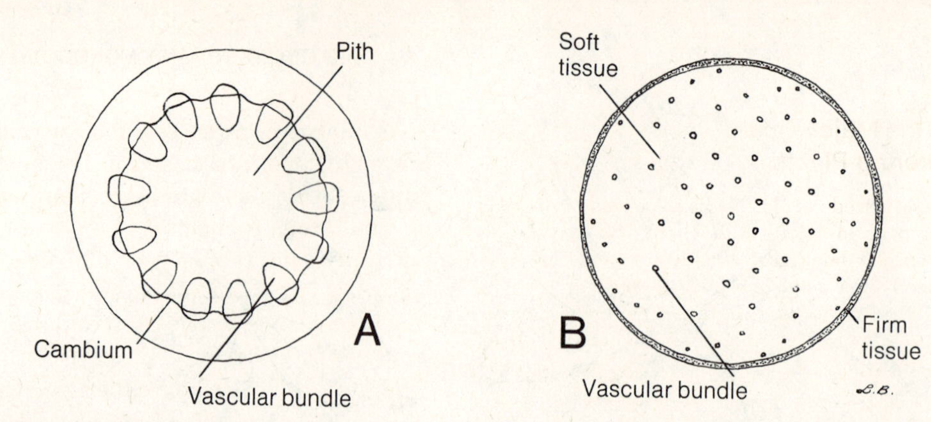

Figure 9-4. Arrangement of vascular tissues in stems, cross sections: **A** typical annual dicotyledon (diagram); **B** typical monocotyledon (garden asparagus, *Asparagus officinalis*). Note the cambium in **A** but not in **B**.

Dicotyledons	*Monocotyledons*
4. Stem with the vascular bundles in a single cylinder ("ring," as seen in cross section) (e.g., sunflower, mallow, pigweed, or lamb's-quarter stems).	4. Stem with the vascular bundles scattered apparently irregularly through pithy tissue (e.g., a corn, *Yucca,* or asparagus stalk).
5. The cambium adding a new cylindroidal layer ("ring") of wood each year or each growing season.	5. Stem and root without a cambium, not increasing in girth by the formation of annual cylindroidal layers of wood.

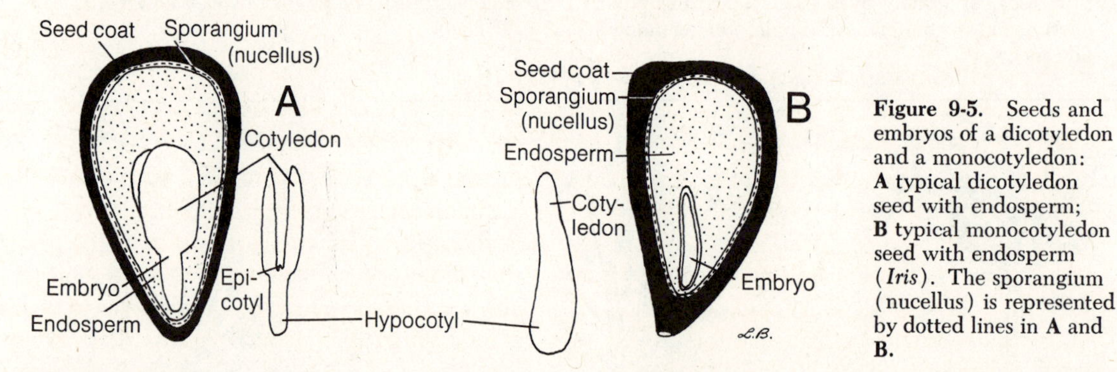

Figure 9-5. Seeds and embryos of a dicotyledon and a monocotyledon: **A** typical dicotyledon seed with endosperm; **B** typical monocotyledon seed with endosperm (*Iris*). The sporangium (nucellus) is represented by dotted lines in **A** and **B**.

Dicotyledons	*Monocotyledons*
6. Cotyledons (seed leaves) two. (A bean, pea, or peanut seed is divided readily into two large cotyledons which may form the bulk of the embryonic plant. In some plants—as for example, peanuts, beans, or cotton—these seed leaves enlarge, become green, and emerge from the seed coat at germination time; in others, as for example, garden peas, they remain within the seed coat.)	6. Cotyledon (seed leaf) one. (A corn or wheat "seed" [really a one-seeded fruit or caryopsis], for example, is not divided readily into two conspicuous parts or cotyledons). This is true also of lily or *Iris* seeds.

Examples. prunes, peaches, apricots, nectarines, cherries, almonds, manzanitas, walnuts, oaks, sunflowers, dandelions, tobaccos, tomatoes, nightshades, mints, buttercups, larkspurs, poppies, haws, sumacs, sycamores, mustards, mesquites, and palos verdes.	Examples. lilies of all types (including yuccas), century-plants, palms, irises, orchids, grasses of all kinds (including corn and bamboo), sedges, rushes, tules, cannas, bananas, and pineapples.

Key to the Major Taxa of Flowering Plants

1. Leaves netted-veined; taproot usually present; flower parts in whorls of 2, 4, or 5 (or very rarely 3); vascular bundles or woody tissue of the stem arranged in a single cylinder ("ring" as seen in cross section); cambium present, adding layers of wood (primarily xylem) internally and of bark (primarily phloem) externally each growing season; cotyledons 2 (or rarely 3 or more).

   Subclass 1. ***DICOTYLEDONEAE,*** page 123.

1. Leaves with the main veins close together and parallel to the midrib or rarely diverging from it but parallel and close to each other; taproot not present, the root system of several or many approximately equal members (fibrous); flower parts in whorls of 3 (or very rarely 2 or 4 or more); vascular bundles of the stem scattered apparently irregularly through pithy tissue; cambium none or in a few woody plants adding whole vascular bundles and pithy tissue both internally and externally; cotyledon 1.

   Subclass 2. ***MONOCOTYLEDONEAE,*** page 357.

Series of Keys

Identification of an unknown plant is accomplished by checking its characteristics against the leads of a key to the divisions of vascular plants (Chapter 1), then of keys to the classes of the proper division and to the subclasses (when this taxonomic rank is employed), orders, families, genera, and species.

The following is an example employing a series of keys for identification of a flowering plant. Because buttercups are conspicuous early-blooming plants almost throughout North America, species of this group are chosen for illustration. The characters of several species are illustrated in order to represent at least one in each major region of North America.

Identification begins with the "Key to the Divisions of Vascular Plants" on page 6. This key includes two leads numbered 1, and obviously the buttercup (Figure 10-4) fits the upper one because each pistil becomes a fruit (achene) containing a seed, which may be seen by holding the single young fruit against a bright light, and because there are flowers.

As indicated in the key, the Spermatophyta are segregated into classes in the next key on page 7. Since there are flowers and each seed of a buttercup is enclosed in an ovary, the upper lead is the proper one. This leads to the Class Angiospermae and to the segregation of Monocotyledoneae and Dicotyledoneae in this chapter (cf. key above).

This short key separates the Dicotyledoneae from the Monocotyledoneae by means of the six character pairs discussed above. These apply to the buttercup as follows: (1) The leaves are not parallel-veined but netted-veined. (2) The root system is fibrous. (Note that this character contradicts the others.) (3) The flower is largely 5-merous because the sepals and usually the petals are 5. The numbers of stamens and carpels are indefinite, and their arrangement is spiral, these characters occurring in the primitive members of either subclass. (4) The vascular bundles are in a single cylinder. (5) The stem grows for only one season, and its capacity to produce a cambium is left undetermined. (6) The seed contains two cotyledons (primary leaves). All the determinable characters but one are those of dicotyledons. The fibrous root system usually is a monocotyledon feature. Such a discrepancy is not uncommon among members of the primitive order of dicotyledons to which the buttercup belongs. Since the preponderance of characters points to the dicotyledons, that group is chosen.

The next step requires the "Key to the Groups," page 124. Because, as shown by the illustration, the flower is hypogynous, and the upper lead 2 is chosen. Because the petals are separate, the plant corresponds to the upper lead 3, and it is placed in the Thalamiflorae, page 129. The results should be checked with the arrangement of the major groups of dicotyledons according to their characters as expressed in the table on page 123. This summarizes the data in the "Key to the Groups."

Under the Thalamiflorae the first step is checking the plant with the "Preliminary List for Elimination of Exceptional Character Combinations" in accordance with the instructions at the beginning of the list. Item 1 starts with **"Aquatic herbs . . ."** in boldface type. This eliminates consideration of number 1 in relation to most buttercups. In the unlikely event that an aquatic species has been chosen, the character combinations a and b must be eliminated. Item 2 begins with **"Desert shrubs . . ."** and it is eliminated; item 3 begins with **"Shrubs or trees . . . ,"** and so on. One by one the special combinations are eliminated at a glance, and use of the "Key to the Orders" on page 133 is the next step.

There are many pistils in each flower, so the upper lead 1 is chosen, and all the rest of the key subordinate to the lower lead 1 is eliminated. The plant fits upper, not lower, lead 2, and it has the characters in upper lead 3. The order is Ranales. This order should be located in the large chart of relationships on the inside front cover. **This should be done whenever the order of a flowering plant is determined so that the relationships of the orders may be learned.**

The Ranales are discussed on page 136, and there is a "Key to the Families." The stamens are separate and numerous, as stated in lower lead 1. Because the carpels are sealed completely and not open on the ventral margins, the lower lead 2 is chosen; because there are numerous separate carpels in the flower and the anthers split along their full length, the upper lead 3; because the plant is an herb, the lower lead 4 and the upper lead 5'; because the stamens are numerous and spirally arranged and the leaves are not peltate, the upper lead 6'. The family is Ranunculaceae (buttercup family).

Determination of the genus and species requires use of the keys in a regional or local manual (cf. list on pp. 8 to 12). Although the principles are the same, details of procedure and arrangement vary considerably from book to book. The following examples are chosen to represent major geographical regions and the text arrangements used in different local manuals. The species to be used as illustrations for use of the keys are shown in detail in Figures 9-6 to 9-12.

1. ALASKA AND ADJACENT CANADA

Eric Hultén, *Flora of Alaska and Neighboring Territories,* is chosen for the example. The Ranunculaceae are keyed on page 12. The buttercup in Figure 9-6 corresponds with the leads in the key to the genera as follows, the leads not being numbered:

The fruits are achenes in dense heads (lower lead).

The achenes are in heads (lower lead).

There are both sepals and petals (upper lead at top of page 13).

The plants are large (at maturity the stems are 6 to many inches long) (lower subordinate lead at the top of the page).

The genus is *Ranunculus,* and the key to the species appears on page 467.

The plant is terrestrial, rather than aquatic or in wet marshes (lower lead).

The flowers are yellow and the sepals are deciduous after flowering (lower lead).

The stems are not scapose (lower lead).

The stems or at least their bases are decumbent, and they develop roots at the nodes in contact with moist soil (upper lead marked by a black square and opposed by the lead similarly marked at the top of the next page).

The head of achenes is spheroidal; the petals are 6–10 mm. long.

The species is *Ranunculus repens* L., number 26 on page 481. It is introduced in meadows and along streams and even in lawns in subarctic and temperate Alaska from the Yukon River southward.

The initial "L." following the name of the species is discussed under the flora of the Pacific Northwest, as identified in Hitchcock and Cronquist, *Flora of the Pacific Northwest.*

Figure 9-6. *Ranuculus repens,* an introduced species from Europe occurring from Alaska to the Atlantic Coast and southward through the cooler and moister parts of Canada and the United States: **1** plant with elongate stolon, leaves, a flower, and fruits in heads, **2** petal, showing the glossy upper portion and dull basal portion and the nectary scale above the nectary, **3** achene. Modifications from LYMAN BENSON, *Plant Taxonomy, Methods and Principles.* John Wiley & Sons, New York. Used by permission.

2. NORTHWEST COAST

Morton E. Peck, *A Manual of the Higher Plants of Oregon,* is chosen for the example. The Ranunculaceae are described on page 299. The buttercup in Figure 9-7 corresponds with the leads in the key to the genera as follows:

The fruits are achenes, and there are many in each head (upper lead A).

The petals are well developed, and the nectar gland on the base of the blade of each petal is covered by a scale (lower lead).

The plant is caulescent, i.e., it has leafy stems (upper lead).

The genus is *Ranunculus,* number 1 in the list, page 300, and the key to the species appears on that page. The plant illustrated is keyed out as follows.

The petals are yellow; the achenes do not have crosswise ridges resembling rough mountain ranges (lower lead).

The wall of the achene is thick and firm at maturity (when dry), and it fits closely around the seed (lower lead).

The plant is terrestrial (lower lead).

The leaves are lobed to parted (lower lead).

Figure 9-7. *Ranunculus occidentalis,* a buttercup common west of the Cascade Mountains in the Pacific Northwest: plants with flowers and fruits and an enlargement of an achene. The nectary scale is a flap as in **9-8, 4.** Reprinted from *Illustrated Flora of the Pacific States,* Volume II, by LeRoy Abrams, with the permission of the publishers, Stanford University Press. Copyright 1944 by the Board of Trustees of Leland Stanford Junior University. (*Ranunculus* by LYMAN BENSON) 2: 201. fig. 1826. 1944.

The plant is a perennial, as shown by the old stems from the preceding year (upper lead A).

The stems and leaves are hirsute (lower lead).

The leaves are palmately lobed or parted (upper lead B).

The stems do not root at the nodes (lower lead in each of the next two cases).

The flowers are conspicuous, the petals being more than 5 mm. long (lower lead).

The plant is 1–5 dm. tall; the petals are narrowly obovate (lower lead).

There are 5 petals (lower lead).

The species is *Ranunculus occidentalis,* number 17 in the list, described on page 304. In the Pacific Northwest the prevailing variety is the "typical" one, *Ranunculus occidentalis* Nutt. var. *occidentalis.* The achene beak is curved and 1–1.5 mm. long, longer than in var. *Eisenii.*

The first paragraph of description applies to both the species as a whole and the "typical" variety *occidentalis.* The varietal epithet ("name") is considered to be understood. The special characters of three other varieties appear in the next short paragraphs. Following each description there is a brief statement of the habitat or at least of the geographical distribution of the variety.

The name of the species is followed by the abbreviation "Nutt." for Thomas Nuttall, who named it and wrote the first description. The abbreviation constitutes a reference to the original place of publication, and it facilitates finding this description and the information included with it, such as a reference to the original collection of the species. This was by Nuttall on "Plains of the Oregon [Columbia] River . . . ," and the specimen is preserved at the British Museum of Natural History in London. This (type) specimen establishes permanently the identity of the plant named *Ranunculus occidentalis.*

3. PACIFIC NORTHWEST

The Pacific Northwest, as whole (including all of Washington, the northern half of Oregon, northern Idaho, the mountainous western part of Montana, and an indefinite part of southern British Columbia [as defined in the *Flora*]). C. Leo Hitchcock and Arthur Cronquist, *Flora of the Pacific Northwest* (see also the larger work by Hitchcock, Cronquist, Ownbey, and Thompson, *Vascular Plants of the Pacific*

Northwest, for a comprehensive, though more expensive, coverage of the same area), is chosen for the example. The buttercup in Figure 9-6 corresponds with the leads of the key to the genera as follows, the leads of the key being in numbered pairs, but according to a system somewhat different from that in this book:

The flowers are not bilaterally symmetrical (lead 1b).

The petals do not have spurs or sacs; the numerous fruits are achenes (lead 3b).

The plants do not produce scapes; the flowers are conspicuous; the fruits are achenes (lead 4b).

The pistils, and later the achenes, are numerous (lead 5b).

There is one ovule per pistil, and the fruit is an achene (lead 6a).

The petals are conspicuous (lead 7a).

There is a nectary at the base of each petal blade, and a minute scale projects over it; the petals are yellow (lead 8b).

The genus is *Ranunculus.* The genera are arranged alphabetically, and *Ranunculus* is described on page 133. The key begins on page 134.

The leaves are lobed deeply (lead 1b).

The stems are leafy and not scapose (lead 2b).

The achenes are not bristly, spiny, or papillate (lead 5b).

The achenes are not cross-corrugated; the petals are yellow; the plants are terrestrial, though sometimes growing in wet places; the nectary scale is well developed (lead 6b).

The plants are not aquatic; each achene has a conspicuous beak developed from the style (lead 7b).

The achenes are compressed and more or less discoid to obovate and glabrous; the margins of the nectary scale are free or mostly free from the petal (lead 8a; proceed to Group VI, leads 37, page 139).

The achene beak is both curved apically and less than 2.5 mm. long (lead 37b).

The petals are much larger than the sepals, and they are 5–18 mm. broad; the stems produce roots at any node on wet ground (lead 38a).

The stems root at any node in contact with the ground (lead 39a).

The species is *Ranunculus repens* L. var. *repens,* which the authors interpret to include var. *glabratus,* not mentioned. There is no description of the species or statement of its geographical distribution. It is native in Eurasia, but it is common throughout the Pacific Northwest as an introduction occurring in meadows and near streams, as well as in lawns and gardens.

The name of the species is followed by the initial "L." for Linnaeus, who named it in his book *Species Plantarum,* and consultation of technical indices will expand the reference to include "Sp. Pl. 554. 1753." The first edition of *Species Plantarum* (1753) was the first great comprehensive book on the flora of the world, and it is the starting point for the naming of plants. In it the plant is said to occur "*. . . in Europae cultis,*" an unfortunately vague statement.

4. CALIFORNIA

W. L. Jepson, *A Manual of the Flowering Plants of California,* is chosen as the first example. The Ranunculaceae are described on page 372. The buttercup in Figure 9-8 corresponds with the leads in the key to the genera, as follows:

The ovary has but one ovule, and the fruit is an achene (lower lead B).

Some leaves are alternate (and some basal); the flowers are complete (upper lead).

Petals are present (lower lead).

The sepals do not have spurs (deep pockets) (lower lead).

Each petal has a nectar gland on the upper side of the base of the blade near the claw (stalk); the sepals are greenish yellow (upper lead).

The petals are yellow; the achene wall is thick and firm at maturity (when it is dry) (upper lead).

The genus is *Ranunculus,* number 13 in the list, page 383, and the key to the species ap-

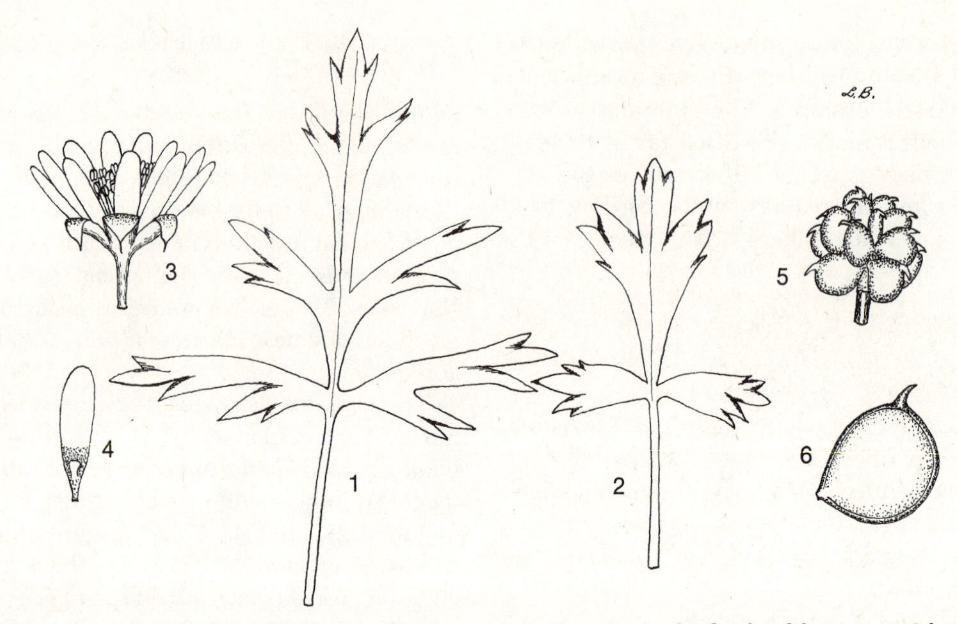

Figure 9-8. *Ranunculus californicus,* a buttercup common in the lowlands of California: **1–2** basal leaves, **3** flower, **4** petal, showing the flaplike scale over the nectary (nectar gland), **5** head of achenes, **6** achene.

pears on that page. The plant illustrated is Californian, and it is listed in Jepson's Manual. It is keyed out as follows.

The nectar gland at the base of the blade (not on the claw) of the petal is covered by a scale (upper lead A).

The coat (pericarp) of the achene is firm and close around the seed (upper lead i.e., *1*).

The plant is terrestrial (lower lead).

The achenes are smooth and with no spines or hooked hairs on the surfaces (upper lead).

The leaves are lobed to parted (lower lead).

The plant is hirsute, though sparingly so in places; the achenes are thin and flat (lower lead).

The achene beaks are curved and shorter than the bodies (upper lead).

The stems are more or less erect or ascending, and they do not root at the nodes (rooting is the characteristic feature of the "creeping or reclining" stems of the lower lead); the sepals are reflexed (upper lead).

The corolla is relatively large, rather than "very small" (i.e., in this case the petals *not* 2–3 mm. long); the beak of the achene is not strongly hooked (lower lead).

The achene beak is not deltoid (lower lead).

There are 9–16 petals (lower lead).

The species is *Ranunculus californicus,* number 12 in the list and described on page 387. In interior parts of northern California living plants with 5 or 6 petals, with usually less dissected leaves, and with longer achene beaks (*Ranunculus occidentalis*) may have been chosen. This is a closely related species keyed out in the lower lead instead of the upper in the last step of the key.

The name of the species is followed by the abbreviation Benth. for George Bentham, who named it and wrote the first description. The abbreviation constitutes a reference to the original place of publication, and it facilitates finding this description and the information included with it, such as a reference to the original collection of the species. This was by Hartweg "...in sylvis prope Monterey," in

1846–47, and the specimen is preserved at the Royal Botanic Gardens at Kew near London. This (type) specimen establishes permanently the identity of the plant named *Ranunculus californicus*.

The first description in the paragraph on page 387 applies both to the species as a whole and to the "typical" variety, which is technically *Ranunculus californicus* var. *californicus*, the varietal epithet (adjective used as a name) being considered to be understood. Three other varieties are listed and described below the first description. The plant illustrated answers the description of none of these. It is var. *californicus*. Following the description there is a brief statement of the habitat in which the plant grows and of its geographical distribution.

Philip A. Munz and David D. Keck, *A California Flora*, is chosen as a second example. The Ranunculaceae are described on page 78. The buttercup in Figure 9-8 corresponds with the leads in the key to the genera as follows (the key is divided between pp. 78 and 79):

Each pistil contains a single ovule, and the fruit is an achene (lower lead, on p. 79).

The leaves of the stem are alternate (lower lead).

There are petals (upper lead).

The petals do not have hollow spurs; the stems bear leaves (lower lead).

The genus is *Ranunculus*, number 6 in the sequence, described on page 92.

The petals are yellow (lead AA and again BB).

The plants are perennial (lead CC).

The stems obviously bear hairs, being usually hirsute (lead D).

The stems do not root at the nodes (lead EE).

The basal leaves are broad and with 3–7 lobes or leaflets (lead FF).

The petals are 7–18 mm. long (lead GG).

The receptacle (as seen at fruiting time) bears hairs (lead H).

Each sepal is reflexed near the base; the petals are 9–16 (lead II).

The species is *Ranunculus californicus* Benth. See discussion under the first example for California, Jepson's *Manual of the Flowering Plants of California*.

The description of the species, a statement of distribution, and a key to the varieties and descriptions of them appear on page 95. Here the key to the varieties of the species retains the old method of designating the typical variety as "the species," and it appears as "*R. californicus*," instead of var. *californicus*. The plant is keyed as follows:

The stems (not shown in Figure 9-8) are erect; the leaf lobes are narrow and sharply acute (lower lead).

The stems are hirsute (not shown in the figure) (lower lead).

The petals are about 8–15 mm. long; however, the leaves are 3–5-lobed to mostly divided (not with 3–5 leaflets as indicated erroneously by "3-foliate") (upper lead).

The plant is *Ranunculus californicus* Benth. var. *californicus*. According to the present rules, the name of the typical variety repeats the specific epithet without an author's name as a reference. Var. *californicus* does not appear in the list below the key, and the description of the species above the key applies to both it and the species as a whole, following a practice carried over from earlier texts in botany.

5. COLUMBIA AND GREAT BASINS AND NORTHERN ROCKY MOUNTAIN REGION

Ray J. Davis, *Flora of Idaho*, is chosen as the example. The Ranunculaceae appear on page 292. The buttercup in Figure 9-9 corresponds with the leads in the key to the genera as follows:

The fruit is an achene; there is but one ovule (upper lead).

Figure 9-9. *Ranunculus glaberrimus,* a buttercup common and the first flower of spring in the Columbia River Basin, the Great Basin, and the northern Rocky Mountains: plant with flowers and enlargements of a head of achenes and a single achene. The nectary scale forms a pocket. By Jeanne Russel Janish. Reprinted from *Illustrated Flora of the Pacific States,* Volume II, by LeRoy Abrams, with the permission of the publishers, Stanford University Press. Copyright 1944 by the Board of Trustees of Leland Stanford Junior University. (*Ranunculus* by LYMAN BENSON) 2: 209. fig. 1842. 1944.

Petals are present; the sepals are not petallike (lower lead).

The sepals are not spurred (lower lead).

Each petal has a nectar gland on the basal portion of the blade next to the claw (stalk).

The genus is *Ranunculus,* described on page 301 (genera arranged alphabetically). The buttercup illustrated occurs throughout the Columbia Basin, the northern Great Basin, and the Northern Rocky Mountain Region. It is chosen because it is the first flower of spring through much of its range and therefore frequently studied in classes. Another species flowering in early summer and occurring through a wider area is used as an alternative illustration also, cf. Figure 9-10. The plant mentioned first, Figure 9-9, is keyed out as follows:

The petals are yellow; the achenes do not have cross ridges (upper lead 1).

The sepals fall away immediately after flowering; the achenes are not inflated and thin-walled (upper lead 2).

The pericarp is smooth, and it is thick and firm at fruiting time (upper lead 3).

The style matures into an achene beak; the pericarp is not corky; the nectary on the blade of the petal is covered by a scale which forms a pocket (upper lead 4).

The basal leaves of this species may be entire or lobed, but the cauline ones are almost unfailingly lobed or divided (upper lead 5).

The nectary scale is not free laterally; it forms a pocket; the achenes are plump and not disclike (lower lead 6 near the top of page 302).

The achenes are numerous, usually 75 or more, and they are in a globose head 10–20 mm. in diameter at maturity, etc. (lower lead 7).

The species is *Ranunculus glaberrimus,* described on page 311 (species arranged alphabetically). The abbreviation following the scientific name of the plant is for Sir W. J. Hooker, who named the species and wrote the original description. The abbreviation constitutes a reference to the place of original publication, and it facilitates finding this description and the information included with it, such as a reference to the original collection of the species. The species was indicated to be "Common on the mountains around the Kettle Falls [Columbia River in Washington] . . . *Douglas.*" A specimen in the herbarium of Hooker at the Royal Botanic Gardens, at Kew near London, is labeled as follows: "Open pine woods near Kettle Falls on the Columbia. 1826, *Douglas.*" This type specimen (technically a lectotype) collected by David Douglas establishes permanently the identity of the plant given the name *Ranunculus glaberrimus* Hook.

The name of the species is followed by a description and by a statement of the habitat and the region in which the plant grows.

Ranunculus glaberrimus is composed of two varieties occurring in Idaho and of another occurring elsewhere. In the *Flora of Idaho* the first description applies to the general characters of the species and to the "typical" variety, *Ranunculus glaberrimus* Hook. var. *glaberrimus* (understood). The paragraph below applies to *Ranunculus glaberrimus* Hook. var. *ellipticus* Greene insofar as it differs from var. *glaberrimus*.

In the summer or in other parts of the Rocky Mountain System, the buttercup in Figure 9-10, may be keyed out, but the discussion of *Ranunculus glaberrimus* above should be read carefully. On page 301 the upper leads 1, 2, 3, 4, and 5 are chosen for the reasons given under *R. glaberrimus* above. Under the upper lead 5, the key is used as follows:

The nectary scale is free on its sides, i.e., it does not form a pocket; the dorsal-to-ventral measurement of the achene is more than 3 times the lateral, i.e., the achene is not plump but more or less disclike (upper lead 6).

The achene is smooth (upper lead 7).

The beak of the achene is curved and not more than 2 mm. long (upper lead 8).

After flowering, the receptacle (which is hairy) grows until it is much larger than at flowering time; the head of achenes is ovoid (lower lead 9).

The petals are of approximately the length of the sepals or somewhat longer (lower lead 10).

(NOTE: The relative length of sepals and petals is reversed in the key.)

The species is *Ranunculus Macounii* Britt. The abbreviation stands for Dr. N. L. Britton who first named and described the plant, cf. discussion under *Ranunculus glaberrimus* above. This species is based upon a single unit not divisible into varieties.

6. CENTRAL ROCKY MOUNTAIN REGION

H. D. Harrington, *Manual of the Plants of Colorado,* is chosen for the example. The buttercup in Figure 9-10 is keyed out as follows, beginning with the key to the genera of the Ranunculaceae on page 236:

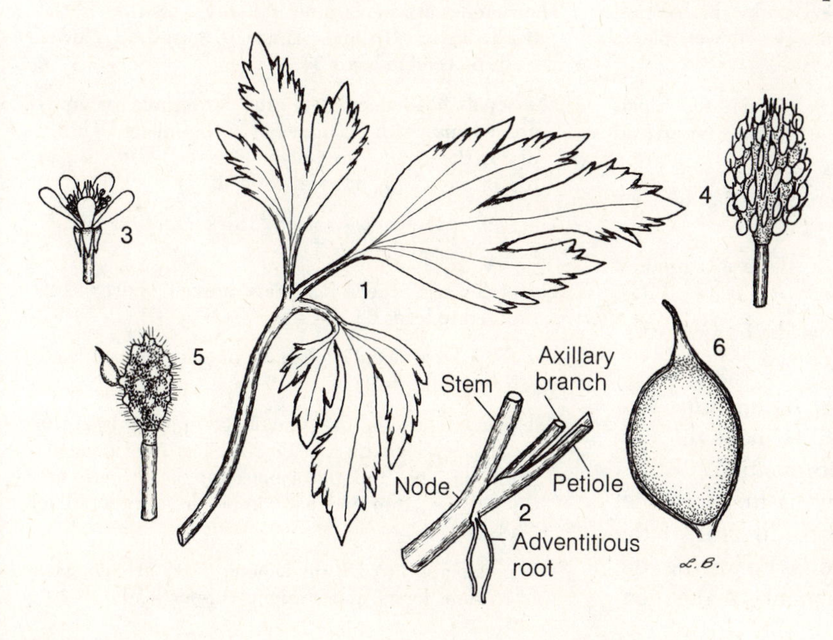

Figure 9-10. *Ranunculus Macounii,* a summer-flowering buttercup common in the Great Basin and Rocky Mountain regions (as well as from Alaska to James Bay): **1** basal leaf, **2** adventitious roots formed often from the lower nodes of stems lying on mud, **3** flower, **4** head of achenes, **5** receptacle with the last persistent achene, **6** achene.

The flower is regular (bilaterally symmetrical) (lower lead 1).

The sepals have no spurs (lower lead 3; under the system used in this book this lead would be 2').

The petals have no spurs; the fruits are achenes (lower lead 4).

Petals are present (lower lead 5).

The fruits are dry (achenes); the flowers are not in racemes (lower lead 11).

The fruits are achenes; the sepals are less than 1 cm. long; the petals are yellow (lower lead 12).

The genus is number 14, *Ranunculus*, described on page 245. The species may be determined as follows from the key:

The petals are yellow and glossy; the achenes are smooth; the plants occur in wet places, but they are not aquatic (lower lead 1).

The sepals are deciduous at the end of anthesis (lower lead 4).

The pericarp of the achene is not striate (lower lead 5).

The leaves are toothed to divided (lower lead 8).

The achenes are not spiny (lower lead 11).

The achene beaks (persistent styles) are long and obvious (lower lead 12).

The achenes do not have marginal corky thickenings; the plants are terrestrial, though of wet places (lower lead 16).

The achenes are compressed (flattened); the sepals are not tinged with lavender or purple (lower lead 18).

The achene beaks are not more than 2 mm. long (lower lead 34).

If the plants grow in wet places, the stems produce adventitious roots from the nodes (upper lead 35).

The petals are only 5–7 mm. long; the fruiting receptacle is ovoid or oblong (lower lead 36).

The species is number 28, *Ranunculus Macounii,* discussed and described near the bottom of page 250. The abbreviation, "Britt." following the scientific name is for Nathaniel Lord Britton, for many years director of the New York Botanical Garden. Following the description is a brief statement of the geographical range of the species. Cf. discussion under *Ranunculus glaberrimus* in section 5 (*Columbia and Great Basins and Northern Rocky Mountain Region*) above.

7. ARIZONA AND NEW MEXICO

Thomas H. Kearney and Robert H. Peebles, *Arizona Flora,* is chosen as the example. The Ranunculaceae are described on page 304. The key is written without indentation, but fundamentally it is similar to other keys. The buttercup in Figure 9-11 corresponds with the leads in the key to the genera, as follows:

Each carpel has but one ovule; the fruit is an achene (upper lead 1). The number (7) indicates proceeding to the pair of leads numbered 7.

Petals are present (upper lead; proceed to leads 8).

The sepals do not have spurs (very deep pockets); the petals are conspicuous; the stems bear leaves (lower lead).

The genus is *Ranunculus,* number 11 in the list, and the key to the species appears on page 314. The buttercup illustrated occurs in southern Arizona and in southern Texas. It is keyed out as follows.

The achenes are smooth, not roughly cross-ridged; the petals are glossy; the plants are terrestrial (lower lead; proceed to leads 3).

The sepals fall immediately after flowering; the fruits are achenes with no tendency to be inflated (utricular); the petals are yellow ("lower" lead, i.e., at top of page 315; proceed to leads 4).

The wall of the fruit is smooth, thick, and firm (upper lead, proceed to leads 5).

The styles and achene beaks are present (lower lead; proceed to leads 6).

The leaves are deeply lobed to pinnate (lower lead; proceed to leads 9).

The nectary scale on the base of the upper side of the petal is covered by a scale which is free along its sides (i.e., not forming a pocket); the achenes are discoid, i.e., thin and flat (upper lead; proceed to leads 10).

The petals are 8 to 18; the achene beaks are elongate (3–5 mm. long) and straight (upper lead).

Figure 9-11. *Ranunculus macranthus*, a species common in the higher hills in southeastern Arizona, in southern Texas, and through most of Mexico: **1** plant with a flower and a head of achenes, **2** petal, showing the flaplike nectary scale, **3** achene.

The species is *Ranunculus macranthus* Scheele, number 9 in the list, described on page 317. The personal name (or in many cases an initial or an abbreviation) following the scientific name of the plant is that of the man who named the species and wrote the original description. The name constitutes a reference to the place of original publication, and it facilitates finding this description and the information included with it, such as a reference to the original collection of the species. The original specimen was collected near New Braunfels, Texas, by Römer. Such original or type specimens are preserved in herbaria (collections of pressed plants), and they establish permanently the identity of the plant given a particular name as a species or variety.

In most books the name of the species is followed by a description, but this is not true in Kearney and Peebles' *Arizona Flora.* There is a detailed statement of the habitat and geographical distribution of the species.

Ranunculus macranthus Scheele, as it occurs in the United States, is a single unit not divisible into varieties. However, the next species above it on the page, *Ranunculus hydrocharoides,* A. Gray, is composed of two varieties. One, the "typical" var. *hydrocharoides,* is not named because the varietal epithet (adjective used as a name) is taken to be understood; the other has a special epithet (var. *stolonifer*) different from that of the species. Both varieties are described in the second paragraph.

8. TEXAS

Donovan S. Correll and Marshall C. Johnston, *Manual of the Vascular Plants of Texas,* is chosen for the example. The buttercup in Figure 9-11 is keyed out as follows, beginning with the key to the genera of the Ranunculaceae on page 634:

The sepals and petals do not have spurs (lower lead 1, the "(3)" indicates going to leads 3).

In the pair of leads numbered 2, the upper 1 is indicated by "3(1)," showing that this lead was reached from lead 1, this being a means of retracing steps in this nonindented key, which is a form difficult to follow.

The leaves are *not* linear or entire; the flowers are large and not in spikes (there being an inadvertance for *Myosurus,* in which the flower is terminal and solitary and the *achenes* are in a spikelike cluster); the sepals do not have spurs (lower lead 3; proceed to leads 4).

The conspicuous part of the flower is the sepals and petals (lower lead 4; proceed to leads 5).

The stems are not woody, and the leaves are not in an apical cluster; the flowers are yellow; there are no sterile stamens (lower lead 5; proceed to leads 6).

The leaves on the stem are alternate; most of the leaves are basal (lower lead 6; proceed to leads 8).

The petals are yellow (lower lead 8; proceed to leads 9).

The petals are yellow; the achenes are numerous (lower lead 9).

The genus is *Ranunculus,* number 6 in the sequence, described on page 641. The species may be determined as follows from the key:

The achenes are smooth and not with cross ridges on the faces; the petals are yellow and glossy (lower lead 1; proceed to leads 3).

The wall of the achene is not striate or nerved (lower lead 3; proceed to leads 4).

The leaves are with lobes, parts, and divisions (lower lead 4; proceed to leads 6).

The achenes are smooth (lower lead 6; proceed to leads 10).

The style and achene beaks are present; the body of the achene has no specially thickened area; the nectary scale projects over the gland (lower lead 10; proceed to leads 11).

The nectary scale does not form a pocket, its margins are free from the petal; the measurement from the dorsal (back) side of the fruit to the ventral (front) is 3–15 times that from side to side; at fruiting time the receptacle is up to only 3 times its length during flowering (lower lead 11; proceed to leads 12).

The achene beaks are straight and elongate, 2–4 mm. long, as long as or longer than the bodies (lower lead 12; proceed to leads 14).

There are 35 or commonly 40 to 130 achenes from one flower; the sepals are 6 or usually 8–10 mm. long; there are 8 to 18 petals (upper lead 14).

The species is *Ranunculus macranthus* Scheele, number 14 in the list and described on page 647, and the distributional range is indicated. See the discussion of the species in section 7, *Arizona and New Mexico,* above, Kearney and Peebles, *Arizona Flora.*

9. EASTERN CANADA AND THE MIDDLE WESTERN AND EASTERN UNITED STATES

Gray's Manual of Botany (8th ed. by M. L. Fernald) is chosen for the example. The Ranunculaceae are described on page 642. The buttercup in Figure 9-12 corresponds with the (lettered instead of numbered) leads in the key to the genera, as follows:

The fruits are achenes occurring in dense heads (upper lead a).

The sepals are 5, and they overlap each other (are imbricated) in the bud; some leaves are basal, the rest alternate (upper lead b).

Petals are present and well developed (upper lead c).

The sepals do not have spurs or sacs (deep pockets) (upper unlettered lead).

The genus is *Ranunculus,* number 1 in the list. The key to the species appears on page 643. The buttercup illustrated occurs throughout eastern North America. It corresponds with the leads in the key to the species, as follows:

The achenes are not with rough cross ridges; the petals are yellow (lower lead a).

The surface of the achene is smooth, not ridged lengthwise (lower lead d).

The achenes are glabrous, flattened, and with beaks (persistent styles); there are 3 sepals and 5 petals ("lower" lead e, top of page 644).

The plants are terrestrial (lower lead f).

The leaves are cleft or parted ("lower" lead i, top of page 645).

The achenes are not harshly hairy (upper lead l).

The achenes are markedly flattened (lower lead m).

The petals are only 2–5 mm. long, about the length of the sepals (upper lead s).

The stems do not root; they are not stoloniferous; the petals are 1–3 mm. broad (upper lead t).

The plant is perennial, and it fits the other characters of the upper unlettered lead (top of page 646).

The species is *Ranunculus recurvatus,* number 25 in the list, page 652. The name of the species is followed by the abbreviation Poir. for Poiret, the man who named and first de-

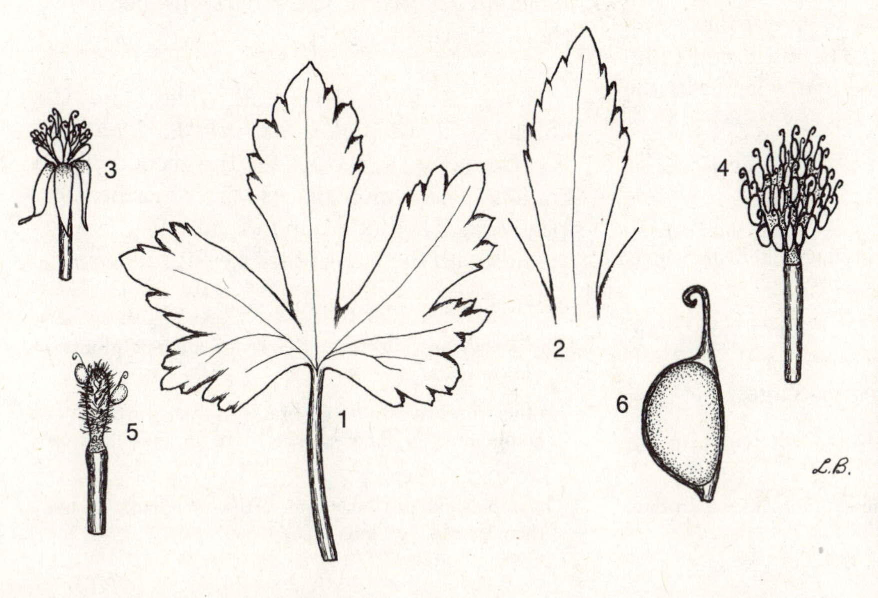

Figure 9-12. *Ranunculus recurvatus,* a species common throughout the temperate regions of the eastern half of North America: **1** basal leaf, **2** the middle lobe of a somewhat different upper cauline leaf, **3** flower, **4** head of achenes, **5** receptacle with the last persistent achenes, **6** achene.

scribed it. The abbreviation constitutes a reference to the original place of publication, and it facilitates finding the basic description and the information associated with it, including a reference to the original collection of the species. The original collection was made in the vicinity of New York prior to 1804, and it is preserved in the collection of Lamarck, now a part of the herbarium of the Muséum National d'Histoire Naturelle in Paris. This (type) specimen establishes permanently the identity of the plant named *Ranunculus recurvatus.*

Ranunculus recurvatus, as it is defined in *Gray's Manual,* is composed of two varieties. The first description combines the characters of the species as a whole and those of the "typical" variety, *Ranunculus recurvatus* var. *recurvatus,* the varietal epithet being omitted as understood. The other variety (*adpressipilis*) is described in the next short paragraph. Habitat data and the geographical distribution of the varieties follow the descriptions.

10. SOUTHERN STATES

J. K. Small, *Manual of the Southeastern Flora,* is chosen for the first example. The Ranunculaceae appear on page 510. A preliminary key divides the family into five tribes, consisting of related genera. The buttercup (Figure 9-12) corresponds with the leads in the key as follows:

The fruit is an achene; there is but one ovule in each carpel (lower lead).

The sepals overlap in the bud (are imbricated); some leaves are basal, the others alternate (upper lead).

The tribe is number IV, Anemoneae, and the key to the genera composing the tribe is below the middle of the same page.

The flowers are subtended by solitary bracts, *not* by involucres (lower lead).

The leaves are simple or pinnate, *not* more than once compound (upper lead).

The sepals do not have spurs (deep pockets) (lower lead).

Petals are present (lower lead).

Each petal has a nectar gland on the upper side of the base of the blade (upper lead).

The achene does not have crosswise ridges resembling mountain ranges; the petals are yellow (lower lead).

The genus is *Ranunculus,* number 18 in the list, described on page 519. The buttercup illustrated occurs throughout eastern North America. It corresponds with the leads in the preliminary key to the species groups as follows.

The plant is terrestrial (lower lead).

The leaf blades are cleft to parted (lower lead).

The achenes are flattened (lower lead).

The achenes are smooth (upper lead).

The achene beak is hooked (upper lead).

The species group is *Recurvati.* This appears as number IV on page 520. The single species is *Ranunculus recurvatus,* number 12 in the whole list, described at the top of page 522. The discussion of this species above (section 9, *Eastern Canada and the Middle Western and Eastern United States*) should be consulted. Only one variety, the "typical" *Ranunculus recurvatus* var. *recurvatus,* occurs in the Southern States.

A. E. Radford, Harry E. Ahles, and C. Ritchie Bell, *Manual of the Vascular Plants of the Carolinas,* is chosen for the second illustration. The Ranunculaceae are described on page 452. The buttercup in Figure 9-12 corresponds with the leads of the key as follows:

The fruit is an achene (lower lead).

The leaves are alternate; the styles are not plumose (lower lead).

Neither the leaves nor the bracts are more than once compound; the inflorescence is not an umbel (lower lead).

The sepals do not bear spurs; the receptacle is less than 1 cm. long (lower lead).

Petals are present; the sepals and the filaments are inconspicuous (upper lead).

The genus is *Ranunculus,* genus number 13 on page 462. The description and the key to the species are there.

The achenes are smooth (lower lead).

The achenes are smooth; the petals are yellow (lower lead).

The achenes are flattened, and each has a distinctly differentiated margin (lower lead).

The achene beaks are hooked; the petals are no longer than the sepals (upper lead).

The species is *Ranunculus recurvatus,* number 14 on page 415. The same species is discussed above under Small, *Manual of the Southeastern Flora,* and in section 9, *Eastern Canada and the Middle Western and Eastern United States,* Fernald, *Gray's Manual of Botany,* 8th. ed. The discussions there should be consulted.

11. TROPICAL SOUTHERN FLORIDA

Robert W. Long and Olga Lakela, *A Flora of Tropical Florida,* is chosen for the example. Inasmuch as *Ranunculus* does not occur in the tropical part of Florida, the cultivated lemon, a tropical species that has escaped from cultivation there and which is available everywhere in landscaping and in horticulture, will be used to illustrate the key. It is illustrated in Figures 10-32 and 10-33. In Long and Lakela the Rutaceae, or rue family, is described on pages 508 and 510, and the key to the genera is on page 510.

The leaves appear to be simple, and they are persistent for more than one season (evergreen) (upper lead 1, this one going directly to the genus).

The genus is *Citrus,* described on the same page. It includes all the citrus fruits, such as oranges, grapefruit, limes, mandarins, tangerines, and citrons. The key to the species begins at the bottom of page 510.

There is a joint in the leaf where the petiole and the blade join; the rind of the fruit is thin (upper lead 1; proceed to the leads 2 on page 511).

The rind is attached firmly to the inner part of the fruit and not easily peeled away as in tangerines (upper lead 2).

The flower bud is tinged with red on the outside; the fruit is ellipsoid [rather than ovoid]; there may be surface projections, but commercial lemons are smooth (lower lead 3).

The species is *Citrus Limon* (L.) Burm. f. It is number 5 in the list and the description is on page 512. The first reference is Linnaeus, "(L.)." This appears in parentheses because Linnaeus supplied the epithet, as *Citrus Medica* L. var. *Limon* L. The combination, *Citrus Limon,* was published by Nicolas Laurens Burman (1734–1779). The initial "f." for *filius* distinguishes him from his father, Johannes Burman (1706–1779), a botanist as well. The citrus fruits are native in tropical and subtropical Asia and the Malay Archipelago; they are introduced in the Americas, including Florida.

Analysis of Floral Characteristics

The basis for the system of classification of flowering plants employed here is presented and discussed in Chapter 17. The four major groups of Dicotyledoneae (Thalamiflorae, Corolliflorae, Calyciflorae, and Ovariflorae) are mostly but not necessarily natural. They include mostly orders more nearly related to each other than to those in other groups. The four principal groups are not taxa. They are adopted partly for convenience, i.e., to permit ready reference of dicotyledons to a few major categories. Wholly natural categories of this rank would be preferable, but, at least for the present, their outlines are not clear. The orders, on the other hand, are thought to be largely natural, although the limitations of knowledge and of the experience of an individual make their completely natural alignment or definition improbable. In some cases their interrelationships are clear, in others open to question as indicated on the detailed chart of relationships of the flowering plant order (inside front cover). The Monocotyledoneae form a single, more compact, much smaller unit, and it is not necessary to divide them into major groups.

As presented in Chapter 17, the order **Ranales** (buttercups, magnolias, etc.) is considered to include plants retaining more primitive features than any other order among the Dicotyledoneae. Similarly, the **Alismales** are considered to have retained more primitive features than any other order of Monocotyledoneae. The two orders have similar flower structure, and they are distinguished from each other by only about two-thirds of the characters segregating most dicots and monocots. The Ranales include among their less specialized common temperate members the buttercups (used above as an example for employing a series of keys), anemones, peonies, *Clematis,* marsh-marigolds, magnolias, and tulip-trees; the Alismales include such marsh plants as the arrowheads (*Sagittaria*), water-plantains (*Alisma*), and burheads (*Echinodorus*).

The floral characters of both the unspecialized Ranales and the Alismales are summarized as follows:

1. All the flower parts are **free** from each other, that is, *no* parts are grown together (coalescent or adnate).
2. The parts of any one series are **equal** and therefore the flower is radially symmetrical. In particular, the petals are the same shape and size.
3. The *stamens* and *carpels* are **numerous.** Sometimes this is true of the petals and even the sepals, as well. However, Stebbins has presented evidence that the numbers of stamens and carpels may be increased and that large numbers are not necessarily primitive, though they are almost universal in the Ranales and Alismales. (See Chapter 17.)
4. The *stamens* and *pistils* (separate carpels) are attached to the receptacle apparently in a **spiral or irregular arrangement.** This is not always evident in the sepals and petals because often each of these series includes only one turn of a spiral. If a cyclic arrangement of stamens and carpels is not discernible and these parts are numerous, it may be assumed, as a rule of thumb, that the basic arrangement is spiral, or at least not cyclic. However, in most cases the stamens are not fundamentally spiral in their arrangement; they are in groups of several, each group developed from an internal branching structure (see Chapter 17).
5. The flowers are **hypogynous.**

The list above should be memorized. The key words are **free, equal, numerous, spiral or irregular** (at least not being cyclic), and **hypogynous.** This series may be remembered as follows: All men are said to be created free and equal, and they are numerous, but not spiral or hypogynous. These features may be seen clearly in the flowers of the buttercup (Figures 8-4, **B,** 9-6 to 12, and 10-4), magnolia (see Figures 6-15, 10-1, and 10-2), or arrowhead (Figure 11-1), or in a head of fruits of some buttercup species or the magnolia or arrowhead. In the very large fruiting head of a magnolia the attachment point of each fallen sepal, petal, and stamen is shown clearly by a conspicuous scar, and the enlarged pistils remain as fruits (Figure 6-15).

The five characters above serve as a summary of the floral characters of the unspecialized and, in some ways, primitive members of the Ranales and Alismales, and study of the departure from them in various respects provides a practical means of becoming familiar with the flowering plants and with problems of relationships and classification. For the most part deviations from the characters of the Ranales and Alismales are related to evolutionary divergence from the primitive flowering plants that were antecedents of the living orders. Probably these plants disappeared sometime during Mesozoic. The Ranales, especially the primitive woody members of the order, have retained more characters of the ancestral plants than have other orders, though not all the characters of the Ranales are primitive. To a certain extent the Ranales are a guide to the interpretation of primitive flowering plants, but departures from their characters are not simple or regular in occurrence. Many new characters have arisen more than once, in different orders, and they occur in various combinations with other characters. Thus, the characters *free, equal, numerous, spiral* or *irregular*, and *hypogynous* are not to be taken as necessarily always primitive. They form only a guide intended to bring some order to classification of flowering plants, and they call attention to occurrence of characters which must be considered in formulating ideas concerning the relationships of the various groups of flowering plants. The lines of evolution in the angiosperms are too little known to be worked out in detail, and the evolutionary problems are complex and not even close to solution. Much more needs to be learned concerning the part each character plays in the relationship of each order, family, genus, species, and variety to the environment before its significance in the evolution of any taxon can be evaluated.

Departure from any of the five characters listed above often represents some degree of specialization, and usually it is associated with limiting insect movement in pollination. The flowers of the Ranales (except in certain highly specialized genera, e.g., those of the larkspur, columbine, and monkshood) and the Alismales have no structural arrangement compelling insect visitors to follow a special pathway leading past both the stigmas and the anthers on the way to the nectaries. In fact, the insect is free to crawl in any direction, and the chances of touching both pistils and stamens are not high. The numerous stamens of the Ranales produce a vast quantity of pollen; some of it is effective in pollination, but most is wasted. Often all the pistils are pollinated, and usually they develop seeds; but frequently there are many abortive carpels, apparently due to lack of pollination.

In contrast to the prevailing inefficient and wasteful pollinating mechanism in many Ranales, pollen transfer is amazingly efficient in some orders of flowering plants. Precision correlates well with the presence of special structural arrangements limiting the pathway of the insect. Examples include the following:

Scrophulariales. In the snapdragon family usually the free portions of especially the two upper petals form a color beacon which attracts insects to the flower, the free tips of the three lower petals spread horizontally forming a landing field, and the corolla tube forms a hangar. When an insect follows the beam of color to the runway, it encounters first the pair of stigmas, which is receptive to pollen. After the insect touches them they may fold tightly together for a time, excluding other pollen. When the insect leaves the flower, no pollen is deposited because now the stigmas have closed together. Therefore self-pollination does not occur ordinarily, at least within the same flower, though it may occur among the flowers on the same plant. The structural arrangement assures efficient transfer of the small amount of pollen from the few stamens to the stigma of a different flower. Little pollen is wasted and self-pollination is reduced. Cross-pollination promotes genetic variability within the species, and it is likely that some individuals will be adaptable to changing conditions.

The precise pollinating mechanism of the Scrophulariales is the result of departures from three of the five floral characters of the Ranales and Alismales (*free, equal,* and *spiral or irregular*). The sepals, petals, and carpels are each coalescent, and the stamens are adnate with the corolla tube; the corolla is bilaterally symmetrical; the stamens and carpels are in precise cycles. Departure from these characters produces a structural arrangement in which few stamens and carpels are necessary because precision in pollination insures seed production with relatively little pollen and fewer pistils. Consequently departure

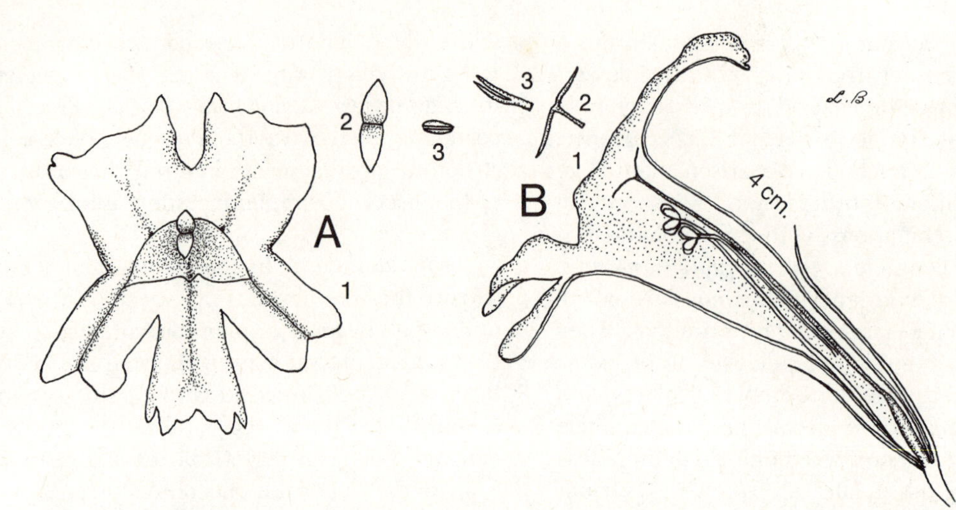

Figure 9-13. Pollinating mechanism in some Scrophulariaceae, as described in the text, *Mimulus* (*Diplacus*) *longiflorus,* a bush monkey-flower: **A** flower, face view, showing the protruding 2-lobed stigma, this open in **1** and **2** and closed (after being touched) in **3**; **B** flower opened lengthwise, showing the pathway to be followed by an insect, passing first the stigma, then the stamens, the stigma open in **1** and **2**, closed in **3**.

from a fourth character (*numerous*) is made possible.

Lamiales. Mint flower structure is similar to that in the Scrophulariales, but in one special group of mints, the sages (*Salvia*), pollination is even more precise. There are only two stamens, each with a short filament and with a long connective (longer than the filament) between the two potential pollen sacs of the anther. In most species the half anther farthest down the corolla tube is missing; in others it is present but poorly developed and usually without pollen. In most sages the stamens are attached to the upper side of the corolla tube; consequently, when the insect crawls under the fertile portion of the anther, it bumps its head against the sterile part—the fertile part patting the insect on the back at just the right spot to have it deposit the pollen on the stigma of the next flower.

Asterales. This group consists of a single large family, the Compositae or sunflower family. What appears to be a single flower (Figure 10-88) is actually an aggregation of many, usually of two kinds—**ray flowers** arranged around the margin of the head and simulating petals and **disc flowers** arranged in the center and simulating slightly the inner flower parts. The ray flowers attract insects; the more numerous disc flowers are so constructed that the insect or its proboscis (tube formed from mouth parts) is guided into a narrow channel passing first the stigma and then the stamens. The flowers depart from all five floral characters of the Ranales and Alismales, for (1) the petals, the anthers, and the carpels are each coalescent, and the filaments of the stamens are adnate with the petals; (2) the corollas of the ray flowers are bilaterally symmetrical; (3) there are only five stamens and two carpels; (4) all the parts are arranged in cycles rather than spirals; and (5) the flowers are epigynous.

The structure of disc and ray flowers not only insures precision in guiding the insect or its proboscis along a definite pathway, but also the numerous small flowers are arranged in such a way that a single butterfly visiting one head may pollinate many flowers without fly-

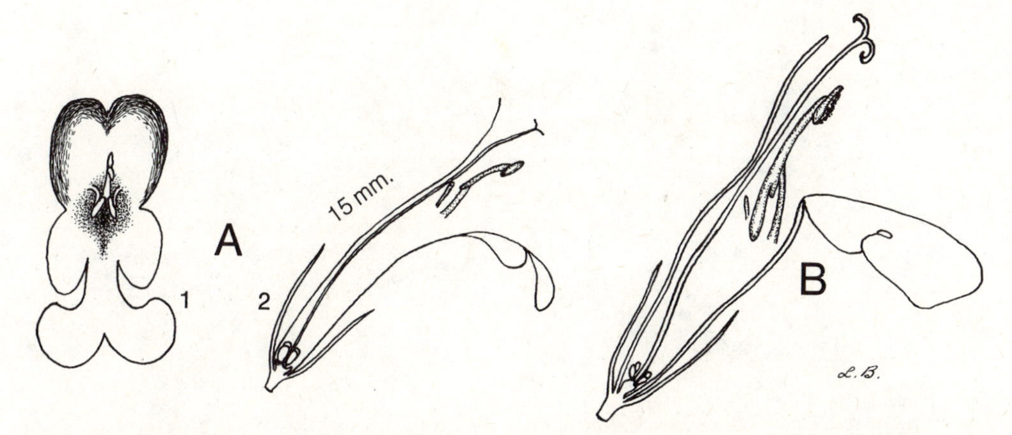

Figure 9-14. Pollinating mechanism in *Salvia,* sage: **A** *Salvia mellifera,* black sage, **1** flower, face view, showing the protruding stigma and two half-anthers, **2** longitudinal view, showing the filament and connective and the single half-anther of a functional stamen as well as a staminodium (rudiment above), the connective not prolonged in this species; **B** *Salvia microphylla,* a species with the sterile end of the connective prolonged and the connective hinged to the filament as described in the text, the sterile portion of the connective serving as a trigger and forcing the half-anther against the back of the insect.

ing. At any time during the period of flowering one or more complete turns of the spiral of disc flowers is open, and the visiting butterfly goes around the whole circle. Thus, within limits, assembly line methods are introduced into flower pollination.

Gentianales. The flowers of the milkweeds depart in many respects from the floral characters of the Ranales. Among other departures, the five anthers are adnate with the edge of the thick stigma. The top of the stigma is pentagonal, and at each corner of the pentagon a groove leads down the side between the adjacent pollen sacs of two anthers. The pollen grains are embedded in waxy masses, and near the top of the stigma a narrow wax bridge runs across the groove between the adjacent anther halves. Thus, the total mass of pollen (**pollinium**) from two adjacent half-anthers has the shape of a pair of saddle bags. The relatively large pentagonal apex of the stigma is shiny and slippery. If an insect alights there, its foot slides down the groove between two anthers, and one of the hooks on the tarsus (leg) carries away a pair of saddlebags full of pollen as the leg rises.

Orchidales. All forms of the highly specialized and diverse orchid flower structure compel the insect or its proboscis to follow a narrow and often tortuous pathway. In some genera one petal resembles a Dutch wooden shoe. The insect enters this petal and proceeds along a definite route leading first past the stigma at the entrance, where pollen from previous flowers may be deposited. After visiting the nectar gland inside the petal, the insect proceeds toward the light, past a waxy mass of pollen produced by the half-anther. This arrangement usually insures delivery of the small amount of pollen to the stigma of another flower. The orchid flower departs from the five characters of the Alismales as follows: (1) the carpels are coalescent, and the stamen(s) and style are adnate; (2) one petal is markedly different from the others; (3) there are only ½ to 3 stamens and 3 carpels; (4) all parts are cyclic; (5) the flower is epigynous. These departures are associated with some of the most effective pollination methods.

Graminales. Aside from insects, wind is the most important carrier of flowering plant

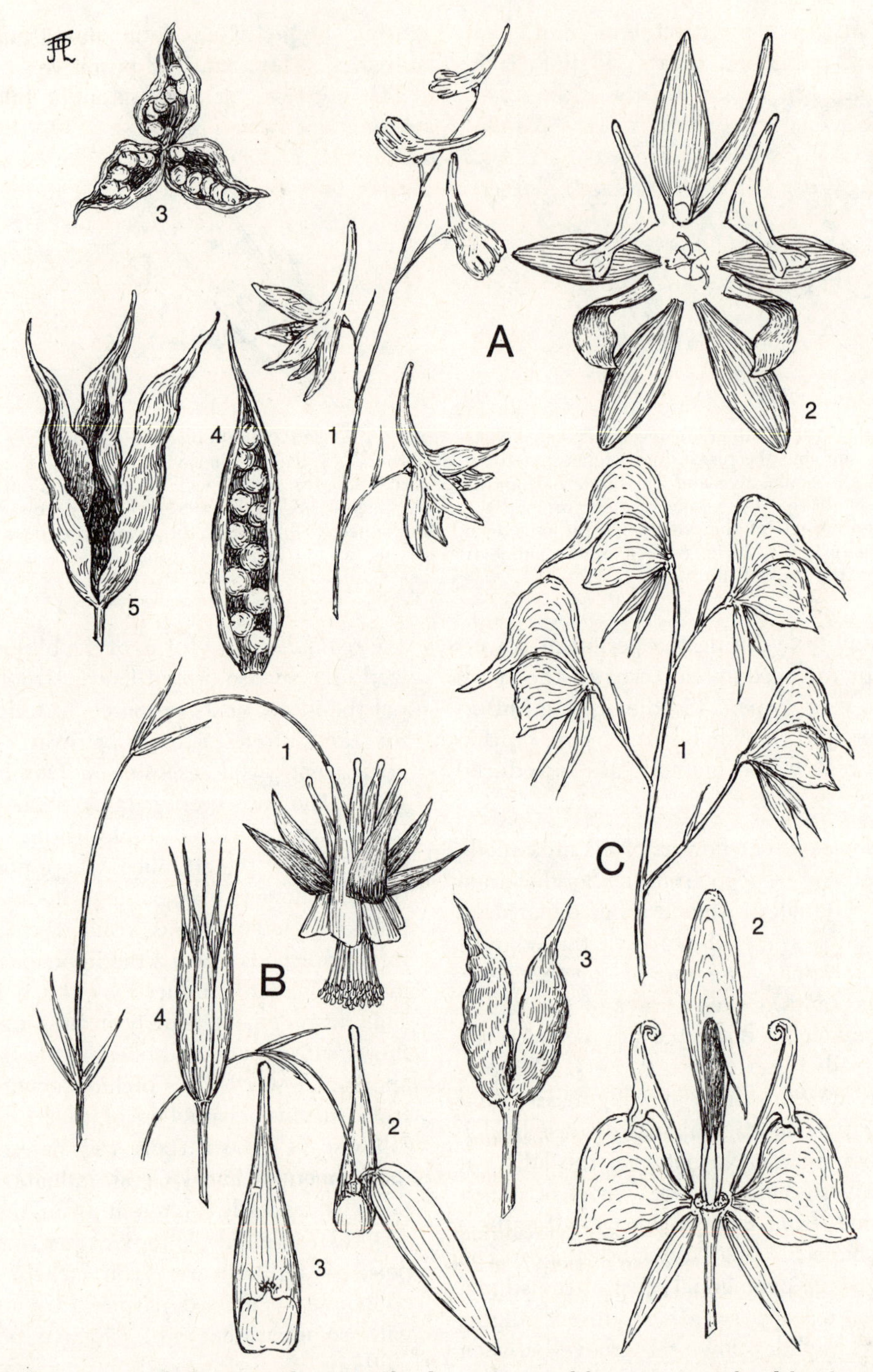

Figure 9-15. Specialized flowers and unspecialized fruits (separate carpels forming follicles) among members of the Ranales (Ranunculaceae): **A** larkspur (*Delphinium*), **1** upper portion of the inflorescence, **2** the bilaterally symmetrical calyx and corolla, the upper sepal bearing a spur lined by the upper petal, **3** the three follicles dehiscing, **4** one of these enlarged, **5** three empty follicles after falling of the seeds; **B** columbine (*Aquilegia truncata*), **1** flower, **2** one sepal and one petal, **3** petal developed almost

pollen. Wind is not a respecter of color, and usually wind-pollinated plants do not have showy corollas. In the grasses the anthers are shaken readily by air currents, for they dangle on long flexible filaments (Figure 11-12 **A**). The stigmas are elongated, feathery brushes ideal for catching the pollen drifting past. These flowers are as highly specialized and as effectively pollinated by wind as others are by insects.

Few would argue that the highly specialized flowers described above are those of primitive flowering plants. By this line of reasoning, a high percentage of the angiosperms may be eliminated from consideration as possibly primitive, and the orders with the most primitive insect pollination are the Ranales and Alismales. However, the wind-pollinated Dicotyledons and Pandanales have a simple flower structure suggesting a primitive type. The carrying of spores by wind existed at least one hundred fifty million years before pollen is known to have been transferred by insects. Although it is evident that some wind-pollinated flowering plants, particularly the grasses, have arisen from insect-pollinated ancestors, others may be primitive.

The presence of such strongly bilaterally symmetrical flowers as those of the larkspur, columbine, and monkshood in the Ranales indicates that specialization in features correlated with the pollinating mechanism may be reached in an order where specialization is not the rule. Even some buttercups (*Ranunculus*) in the order Ranales, despite flower structure corresponding perfectly with all five floral characters of the supposedly primitive flowering plants, have a remarkable method of pollination. Most species have petals which are glossy on the outer (distal) two-thirds of their length; this area sheds water readily, but the portions of the petals toward the center of the flower do not. Hagerup has shown that raindrops accumulate only in the center of the flower, the water rising to the level determined by the gloss on the petals, i.e., just above the stigmas. Pollen from the flower and that brought in by insects is floated on the surface and deposited on the stigmas as the water recedes.

The following analyses according to the floral characters of the Ranales for dicotyledons and the Alismales for monocotyledons indicate some trends of floral evolution in each group of Dicotyledoneae.

Group 1. ***THALAMIFLORAE.*** RECEPTACLE FLOWERS

(Hypogynous; choripetalous or apetalous)

DEPARTURE FROM THE FLORAL CHARACTERS OF THE RANALES

Characters	***Departure from the Characters***
Free	Stamens occasionally coalescent; carpels, except in Ranales, coalescent or rarely solitary.
Equal	Rarely bilaterally symmetrical.
Numerous	Stamens from numerous to few.
Spiral or irregular	Groups near the Ranales with the stamens and carpels spiral, others cyclic.
Hypogynous	

---

wholly as a spur, **4** the five follicles; **C** monkshood (*Aconitum columbianum*), **1** upper portion of the inflorescence, **2** the bilaterally symmetrical calyx and the two specialized upper petals, the lower three petals being lacking or rudimentary, **3** the two follicles. Specialization of the flower parts is in bilateral symmetry and presence of spurs or pouches (on a sepal in *Delphinium* or *Aconitum*, on the petals in *Aquilegia*).

Group 2. ***COROLLIFLORAE.*** COROLLA FLOWERS

(Hypogynous; sympetalous)

The Corolliflorae constitute a more homogeneous group than any of the others except the Ovariflorae. The only significant variability from order to order is in radial or bilateral symmetry of the flower and in the numbers of stamens and carpels; otherwise, the distinction of one order from another is based upon characters not included in the list of floral characters of the Ranales and Alismales.

DEPARTURE FROM THE FLORAL CHARACTERS OF THE RANALES

Characters	***Departure from the Characters***
Free	Sepals, petals, and carpels each coalescent; stamens adnate with the petals.
Equal	Corollas of the orders nearer the Ranales radially symmetrical, of the others bilaterally symmetrical.
Numerous	Stamens usually 5 or fewer, but rarely as many as 20; carpels 2–5.
Spiral or irregular	Stamens and carpels cyclic.
Hypogynous	

Group 3. ***CALYCIFLORAE.*** CUP FLOWERS

(Perigynous or epigynous; choripetalous or apetalous)

DEPARTURE FROM THE FLORAL CHARACTERS OF THE RANALES

Characters	***Departure from the Characters***
Free	Carpels free or coalescent.
Equal	
Numerous	Variable, both stamens and carpels ranging from numerous to 4 or 5.
Spiral or irregular	Groups nearer the Ranales with the stamens and carpels spiral or irregular, the others cyclic.
Hypogynous	Flower perigynous or epigynous.

Group 4. ***OVARIFLORAE.*** OVARY FLOWERS

(Epigynous; sympetalous)

DEPARTURE FROM THE FLORAL CHARACTERS OF THE RANALES

Characters	***Departure from the Characters***
Free	Sepals, petals, and carpels each coalescent; stamens adnate with the petals.
Equal	Corollas mostly radially but in some groups bilaterally symmetrical.
Numerous	Stamens 5 or fewer; carpels 5 or fewer, often only 2.
Spiral or irregular	Stamens and carpels cyclic.
Hypogynous	Flower epigynous.

The following analyses indicate the trend of evolution of the more common groups of Monocotyledoneae (according to position on the chart.)

MOST ORDERS

DEPARTURE FROM THE FLORAL CHARACTERS OF THE ALISMALES

Characters	*Departure from the Characters*
Free	Perianth parts of some Liliales adnate edge to edge, i.e., sepal-with-petal-with-sepal, etc.; carpels coalescent except in some Arales, some Palmales, and some Juncaginales.
Equal	
Numerous	Stamens from numerous to 6 or rarely fewer; carpels numerous to 1.
Spiral or irregular	Stamens and carpels cyclic.
Hypogynous	

MUSALES, BURMANNIALES, ORCHIDALES, SOME LILIALES

DEPARTURE FROM THE FLORAL CHARACTERS OF THE ALISMALES

Characters	*Departure from the Characters*
Free	Perianth parts adnate edge to edge, i.e., sepal-with-petal-with-sepal, etc., and adnate with the contained inferior ovary; stamens adnate with the perianth tube; stamen and style adnate in the Orchidales.
Equal	Perianth sometimes bilaterally symmetrical, one petal far different from the others in the Orchidales; staminodia of the Musales varying in shape and size.
Numerous	Stamens 6 to mostly 3 to ½; carpels 3.
Spiral or irregular	Stamens and carpels cyclic.
Hypogynous	Flowers epigynous.

An understanding of this system of flowering plant classification is achieved most rapidly by filling out for each plant to be identified a mimeographed blank table similar to the following one made out for the garden snapdragon. This should be done before identifying the plant by means of the series of keys in order to assure a good preliminary floral analysis. **After the table has been filled out, the results should be compared with the large chart of relationships of the flowering plant orders (inside front cover).** In general, the distance of this order from the Ranales or Alismales on the chart will be roughly proportional to the number of departures from the floral characters occurring in the Ranales or Alismales. As more and more plants have been identified, it will become possible to make a fairly accurate estimate of the chart position of each new plant as soon as the table is filled out and before the keys have been used. It should be remembered that the characters of the plant will place it immediately in the dicot or monocot section of the chart and under the dicots in the Thalamiflorae, Corolliflorae, Calyciflorae, Ovariflorae. **The characters of these groups are important, and they should be memorized at once.**

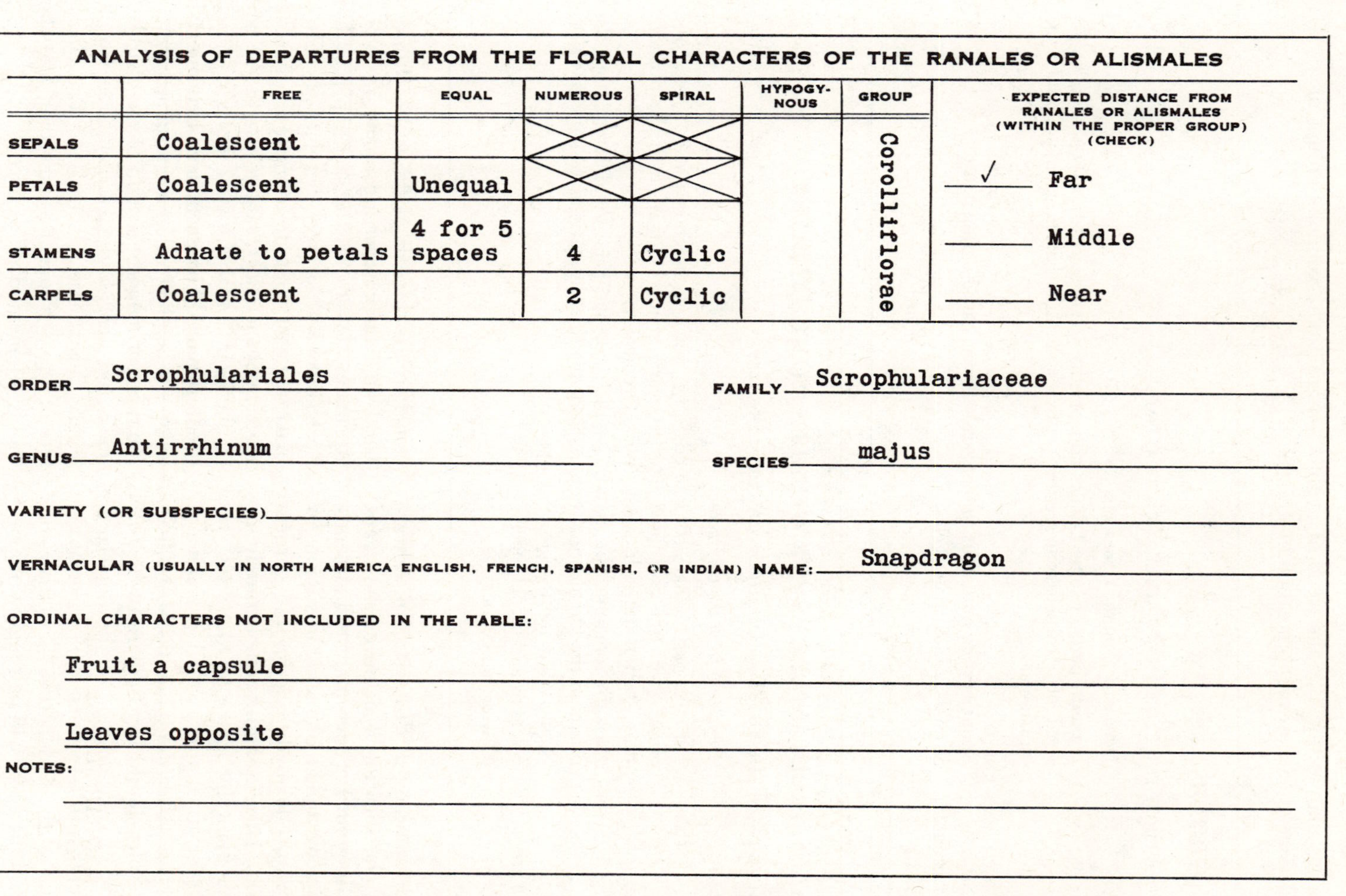

ANALYSIS OF DEPARTURES FROM THE FLORAL CHARACTERS OF THE RANALES OR ALISMALES

	FREE	EQUAL	NUMEROUS	SPIRAL	HYPOGYNOUS	GROUP	EXPECTED DISTANCE FROM RANALES OR ALISMALES (WITHIN THE PROPER GROUP) (CHECK)
SEPALS	Coalescent		X	X		Corolliflorae	✓ Far
PETALS	Coalescent	Unequal	X	X			___ Middle
STAMENS	Adnate to petals	4 for 5 spaces	4	Cyclic			___ Near
CARPELS	Coalescent		2	Cyclic			

ORDER Scrophulariales

FAMILY Scrophulariaceae

GENUS Antirrhinum

SPECIES majus

VARIETY (OR SUBSPECIES) ______

VERNACULAR (USUALLY IN NORTH AMERICA ENGLISH, FRENCH, SPANISH, OR INDIAN) NAME: Snapdragon

ORDINAL CHARACTERS NOT INCLUDED IN THE TABLE:

Fruit a capsule

Leaves opposite

NOTES:

If the plant fits the character in the table, the space is left blank. Answers of "yes" or "no" should be avoided because they may be misinterpreted. Use of the right-hand column of the table will become more effective with experience.

10

Manual of the Orders and Families of Dicotyledons

Subclass DICOTYLEDONEAE

TABLE OF THE MAJOR GROUPS OF DICOTYLEDONEAE

	Sympetalous	***Choripetalous or Apetalous***	
Hypogynous	Group 2 ***COROLLIFLORAE*** COROLLA FLOWERS (Page 248) Hypogynous; sympetalous	Group 1 ***THALAMIFLORAE*** RECEPTACLE FLOWERS (Page 128) Hypogynous; choripetalous or apetalous	***Hypogynous***
Perigynous or epigynous	Group 4 ***OVARIFLORAE*** OVARY FLOWERS (Page 330) Epigynous; sympetalous	Group 3 ***CALYCIFLORAE*** CUP FLOWERS (Page 284) Perigynous or epigynous; Choripetalous or apetalous	***Perigynous or epigynous***

KEY TO THE GROUPS

1. Flower hypogynous, sometimes with a hypogynous disc.
   2. Flower apetalous or choripetalous (petals *separate* basally or at least some separate from the others, rarely coalescent apically). Group 1. **Thalamiflorae** (receptacle flowers), page 128.
   2. Flowers sympetalous (petals *all coalescent* at the bases, though sometimes for only a short distance). Group 2. **Corolliflorae** (corolla flowers), page 248.
1. Flower *not* hypogynous, with a floral cup (or floral tube), therefore perigynous or epigynous, sometimes also with a perigynous or epignyous disc.
   2'. Flower apetalous or choripetalous (petals *separate* basally or at least some separate from the others). Group 3. **Calyciflorae** (cup flowers), page 284.
   2'. Flower sympetalous (petals *all coalescent* basally, though sometimes for only a short distance). Group 4. **Ovariflorae** (ovary flowers), page 330.

LIST OF ORDERS AND FAMILIES

NOTE: A few geographically restricted families scarcely known in the United States are assigned, subject to opportunity for study, to the orders accepted generally by other authors. Other families upon which there is less agreement or information are listed as of uncertain position. Doubtless various less known families need to be amalgamated or divided.

NOTE: The asterisk is used in the list below and in the text following to indicate families represented by one or more species occurring in North America north of Mexico (native or introduced plants but not those occurring only in cultivation).

1. RANALES
   A. Suborder Magnoliineae
      1. Winteraceae (*Drimys*)
      *2. Schisandraceae (bay-star-vine)
      *3. Illiciaceae (anise-tree, star-anise)
      4. Degeneriaceae (*Degeneria*)
      5. Himantandraceae (*Himantandra*)
      *6. Magnoliaceae (magnolia, tulip-tree)
      *7. Anonaceae (pawpaw, cherimoya)
      *8. Myristicaceae (nutmeg)
      9. Eupomatiaceae (*Eupomatia*)
   B. Suborder Laurineae
      *10. Canellaceae (Winteranaceae, *not* Winteraceae) (cinnamon-bark)
      11. Amborellaceae
      12. Austrobaileyaceae
      13. Trimeniaceae
      14. Chloranthaceae (*Chloranthus*)
      15. Lactoridaceae (*Lactoris*)
      16. Gomortegaceae (*Gomortega*)
      *17. Calycanthaceae (spicebush)
      18. Monimiaceae (boldo)
      *19. Lauraceae (laurel, avocado, bay, sassafras)
      20. Hernandiaceae (*Hernandia*)
   C. Suborder Ranunculineae
      *21. Ranunculaceae (Buttercup, *Clematis*, Anemone)
      22. Lardizabalaceae (*Lardizabala*)
      *23. Berberidaceae (barberry, Oregon-grape)
      *24. Menispermaceae (moonseed)
2. DILLENIALES
   1. Dilleniaceae (*Dillenia, Hibbertia*)
   *2. Paeoniaceae (peony)
3. NYMPHAEALES
   *1. Nymphaeceae (water-lily)
   *2. Ceratophyllaceae (hornwort)
4. PIPERALES
   *1. Saururaceae (lizard's-tail)
   *2. Piperaceae (pepper)
5. ARISTOLOCHIALES
   *1. Aristolochiaceae (Dutchman's-pipe, wild ginger, birthwort)
6. PAPAVERALES
   A. Suborder Papaverineae
      *1. Papaveraceae (poppy)
      *2. Fumariaceae (bleeding heart)
   B. Suborder Capparidineae
      *3. Resedaceae (mignonette)
      *4. Moringaceae (horseradish-tree)
      *5. Koeberliniaceae (desert crucifixion-thorn)
      6. Tovariaceae (*Tovaria*)
      *7. Capparidaceae (Rocky Mountain bee-plant, caper, spider-flower)
      *8. Cruciferae (mustard, radish, cabbage)

7. VIOLALES
   - *1. Cistaceae (rockrose)
   - 2. Bixaceae (annato)
   - *3. Cochlospermaceae (*Cochlospermum, Amoreuxia*)
   - 4. Flacourtiaceae (*Flacourtia, Xylosma*)
   - 5. Peridiscaceae
   - 6. Scyphostegiaceae
   - *7. Violaceae (violet, pansy)
   - *8. Turneraceae (*Turnera*)
   - 9. Malesherbiaceae (*Malesherbia*)
   - *10. Passifloraceae (passion-flower)
   - 11. Achariaceae (*Acharia*)
   - *12. Caricaceae (papaya)

8. LOASALES
   - *1. Loasaceae (*Loasa*, evening-star, *Mentzelia*)
   - *2. Datiscaceae (*Datisca* or durango root)
   - *3. Begoniaceae (*Begonia*)
   - *4. Cucurbitaceae (melon, pumpkin, cucumber, gourd, squash)

9. TAMARICALES
   - *1. Frankeniaceae (alkali-heath)
   - *2. Tamaricaceae (tamarisk, *not* tamarack)

10. SALICALES
   - *1. Salicaceae (willow, cottonwood, poplar, aspen)

11. THEALES
   - 1. Actinidiaceae (tara)
   - 2. Stachyuraceae (*Stachyurus*)
   - 3. Medusagynaceae (*Medusagyne*)
   - 4. Ochnaceae (*Ochna*)
   - 5. Strasburgeriaceae (*Strasburgeria*)
   - 6. Caryocaraceae (*Caryocar*)
   - 7. Marcgraviaceae (*Marcgravia*)
   - 8. Quiinaceae (*Quiina*)
   - *9. Theaceae (tea, *Camellia*)
   - *10. Guttiferae (*Clusia*, mangosteen, St.-John's-wort, St.-Peter's-wort)
   - 11. Dipterocarpaceae (*Dipterocarpus*)
   - 12. Pentaphylacaceae (*Pantaphylax*)
   - *13. Elatinaceae (waterwort)
   - 14. Nepenthaceae (pitcher-plant)

12. SARRACENIALES
   - *1. Sarraceniaceae (pitcher-plant)

13. MALVALES
   - 1. Elaeocarpaceae
   - *2. Malvaceae (mallow, hollyhock, *Hibiscus*)
   - 3. Bombacaceae (cotton-tree or kapok-tree, baobob)
   - *4. Tiliaceae (basswood, linden, jute)
   - *5. Sterculiaceae (flame-tree, bottle-tree, fremontia, cacao)
   - 6. Scytopetalaceae (*Scytopetalum*)
   - 7. Sarcolaenaceae (*Schizochlaena, Rhodolaena*)

14. EUPHORBIALES
   - *1. Euphorbiaceae (spurge, poinsettia, castor-bean, tung, rubber-tree, *Manihot* [source of cassava and tapioca], *Croton*)
   - *2. Buxaceae (box, jojoba or goatnut)
   - 3. Aextoxicaceae (*Aextoxicon*)
   - 4. Pandaceae (*Panda*)

15. URTICALES
   - *1. Ulmaceae (elm, hackberry)
   - *2. Moraceae (fig, mulberry, breadfruit)
   - *3. Cannabinaceae (hop, hemp [source of fiber, hashish, and marijuana])
   - *4. Urticaceae (nettle)

16. GERANIALES
   - *1. Limnanthaceae (meadowfoam)
   - *2. Linaceae (flax)
   - *3. Oxalidaceae (*Oxalis*)
   - *4. Tropaeolaceae (nasturtium)
   - *5. Balsaminaceae (touch-me-not)
   - *6. Geraniaceae (geranium, filaree)
   - 7. Erythroxylaceae (coca [source of cocaine])
   - *8. Zygophyllaceae (creosote-bush, puncture-vine or caltrop, lignum vitae)
   - *9. Malpighiaceae (*Janusia*, Barbados-cherry)

17. POLYGALALES
   - *1. Polygalaceae (milkwort)
   - 2. Trigoniaceae (*Trigonia*)
   - 3. Vochysiaceae (*Vochysia*)
   - *4. Krameriaceae (ratany)

18. SAPINDALES
   - A. Suborder Rutineae
     - *1. Simaroubaceae (desert crucifixion-thorn, tree-of-Heaven)
     - 2. Cneoraceae (*Cneorum*)
     - *3. Rutaceae (orange, lemon, grapefruit)
     - *4. Burseraceae (Boswellia [source of frankincense], *Commiphora* [source of myrrh], elephant-tree, gumbo limbo)
     - *5. Meliaceae (mahogany, Chinaberry or umbrella-tree)
     - *6. Anacardiaceae (sumac, poison ivy, poison-oak, cashew, pistachio, mango)
     - 7. Julianaceae (*Juliana*)
   - B. Suborder Celastrineae
     - *8. Aquifoliaceae (holly)
     - *9. Celastraceae (bittersweet, burning-bush)
     - *10. Hippocrateaceae (*Hippocratea*)
     - 11. Stackhousiaceae (*Stackhousia*)
     - 12. Icacinaceae (*Pennantia, Villarsia*)
     - 13. Salvadoraceae (tooth-brush tree)
   - C. Suborder Sapindineae
     - *14. Sapindaceae (soapberry, balloon-vine, white ironwood)
     - 15. Bretschneideraceae (*Bretschneidera*)
     - 16. Sabiaceae (*Meliosma*)

17. Melianthaceae (honey-bush)
18. Akaniaceae (*Akania*)
*19. Aceraceae (maple, box-elder)
*20. Hippocastanaceae (horse-chestnut, buckeye)
*21. Staphyleaceae (bladdernut)

19. JUGLANDALES
1. Rhoipteleaceae (*Rhoiptelea*)
*2. Juglandaceae (walnut, hickory, pecan)

20. MYRICALES
*1. Myricaceae (wax-myrtle, sweet-gale)

21. LEITNERIALES
*1. Leitneriaceae (corkwood)

22. RHAMNALES
*1. Rhamnaceae (coffeeberry, mountain-lilac)
*2. Vitaceae (grape, Virginia creeper)

23. CARYOPHYLLALES
*1. Phytolaccaceae (poke or pokeweed)
2. Gyrostemonaceae (*Gyrostemon*)
*3. Aizoaceae (ice-plant, *Lithops,* carpet-weed)
*4. Nyctaginaceae (four-o'-clock, sand-verbena)
*5. Polygonaceae (buckwheat, smartweed, *Eriogonum*)
*6. Chenopodiaceae (goosefoot, lamb's-quarter, beet, saltbush)
*7. Amaranthaceae (amaranth, pigweed, tumbleweed, bloodleaf)
*8. Portulacaceae (*Portulaca,* miner's-lettuce, spring-beauty)
*9. Basellaceae (Madeira-vine)
*10. Caryophyllaceae (chickweed, pink, campion, sweet-William)
11. Didiereaceae (*Didierea*)

24. CACTALES
*1. Cactaceae (cactus, prickly pear, cholla)

25. BATIDALES
*1. Batidaceae (*Batis*)

26. HAMAMELIDALES
A. Suborder Trochodendrineae
1. Trochodendraceae (*Trochodendron*)
2. Tetracentraceae (*Tetracentron*)
3. Eupteleaceae (*Euptelea*)
4. Cercidiphyllaceae (*Cercidiphyllum*)
B. Suborder Eucommineae
5. Eucommiaceae (*Eucommia*)
C. Suborder Hamamelidineae
*6. Hamamelidaceae (witch-hazel, *Liquidambar*)
*7. Platanaceae (sycamore or plane-tree)

27. FAGALES
*1. Betulaceae
*a. Betuloideae (birch, hazel)
*b. Coryloideae (hazel, hornbeam)
*2. Fagaceae (oak, beech, chinquapin, chestnut)

28. CASUARINALES
*1. Casuarinaceae (beefwood)

Group 2. *COROLLIFLORAE.* COROLLA FLOWERS

29. PRIMULALES
*1. Primulaceae (primrose, shooting-star)
*2. Plumbaginaceae (leadwort, marsh-rosemary, thrift)
*3. Myrsinaceae (*Myrsine,* marlberry)
*4. Theophastaceae (Joe-wood)

30. EBENALES
*1. Styraceae (*Styrax,* silverbell tree)
*2. Symplocaceae (sweetleaf, horse-sugar or yellowwood)
3. Lissocarpaceae (*Lissocarpa*)
*4. Ebenaceae (ebony, persimmon)
*5. Sapotaceae (sapodilla)

31. ERICALES
A. Suborder Ericineae
*1. Cyrillaceae (leatherwood, black ti-ti)
*2. Ericaceae
*a. Clethroideae (white-alder)
*b. Pyroloideae (wintergreen, pipsissewa)
*c. Monotropoideae (Indian-pipe, snow-plant, pine-drops)
*d. Ericoideae (heath, manzanita, *Rhododendron*)
*e. Vaccinioideae (blueberry, cranberry, huckleberry, bilberry)
3. Epacridaceae (*Epacris*)
B. Suborder Diapensineae
*4. Diapensiaceae (*Diapensia*)
C. Suborder Empetrineae
*5. Empetraceae (crowberry)

32. GENTIANALES
A. Suborder Gentianineae
*1. Gentianaceae (gentian)
*2. Loganiaceae (butterfly-bush, *Gelsemium, Strychnos* [source of strychnine and curare])
B. Suborder Apocynineae
*3. Apocynaceae (dogbane, periwinkle, Indian "hemp")
*4. Asclepiadaceae (milkweed)
C. Suborder Oleineae
*5. Oleaceae (olive, ash, lilac, jasmine)

33. FOUQUIERIALES
*1. Fouquieriaceae (ocotillo, boogum-tree or cirio)

34. POLEMONIALES
  A. Suborder Polemoniineae
   *1. Polemoniaceae (*Gilia, Phlox,* Jacob's-ladder)
   *2. Solanaceae (nightshade, tomato, potato, tobacco, ground-cherry)
   3. Nolanaceae (*Nolana*)
   *4. Convolvulaceae (morning-glory, bindweed, dodder)
   *5. Lennoaceae (sand-food, *Pholisma*)
  B. Suborder Boragineae
   *6. Hydrophyllaceae (*Phacelia,* baby blue-eyes, yerba santa, waterleaf)
   *7. Boraginaceae (heliotrope, fiddleneck, borage, forget-me-not)

35. LAMIALES
   *1. Verbenaceae (*Verbena, Lantana*)
   *2. Labiatae (mint, sage, hyssop)
   *3. Callitrichaceae (water-starwort)
   *4. Phrymaceae (lopseed)

36. SCROPHULARIALES
   *1. Scrophulariaceae (snapdragon, *Penstemon,* monkey-flower, foxglove, *Gerardia*)
   *2. Bignoniaceae (catalpa, desert-willow)
   *3. Pedaliaceae (benne [source of sesame])
   *4. Martyniaceae (devil's claw)
   *5. Orobanchaceae (broom-rape)
   6. Gesneriaceae (African-violet, gloxinia)
   *7. Lentibulariaceae (bladderwort, *Pinguicula*)
   8. Globulariaceae (*Globularia*)
   *9. Acanthaceae (water-willow, *Ruellia, Acanthus, Justicia*)
   10. Myoporaceae (*Myoporum*)

37. PLANTAGINALES
   *1. Plantaginaceae (plantain, Indian-wheat, psyllium)

Group 3. ***CALYCIFLORAE.*** CUP FLOWERS

38. ROSALES
   *1. Crassulaceae (stonecrop, hen-and-chickens, *Echeveria, Dudleya*)
   2. Connaraceae (*Connarus*)
   *3. Leguminosae
    *a. Mimosoideae (*Acacia,* sensitive-plant, mesquite)
    *b. Caesalpinoideae (senna, palo verde, redbud or Judas-tree, honey-locust)
    *c. Papilionoidaee (Lotoideae) (pea, bean, vetch)
   *4. Crossosomataceae (*Crossosoma, Apacheria*)
   *5. Rosaceae
    *a. Spiraeoideae (*Spiraea,* bridal-wreath)
    *b. Rosoideae (rose, *Potentilla,* strawberry, blackberry)
    *c. Prunoideae (plum, cherry, peach almond)
    d. Chrysobalanoideae (*Chrysobalanus*)
    e. Neuradoideae (*Neurada, Grielum*)
    *f. Pomoideae (apple, pear, quince)
   6. Myrothamnaceae (*Myrothamnus*)
   7. Bruniaceae (*Brunia*)
   8. Brunelliaceae (*Brunellia*)
   9. Cunoniaceae (*Cunonia*)
   10. Eucryphiaceae (*Eucryphia*)
   11. Byblidaceae (*Byblis, Roridula*)
   12. Pittosporaceae (*Pittosporum*)
   13. Cephalotaceae (*Cephalotus*)
   *14. Saxifragaceae
    *a. Penthoroidae (*Penthorum*)
    *b. Saxifragoideae (saxifrage, woodland star, alumroot)
    *c. Parnassioideae (*Parnassia*)
    d. Francoideae (*Francoa*)
    *e. Hydrangeoideae (*Hydrangea,* mock-orange)
    f. Pterostemonoideae (*Pterostemon*)
    g. Escallonioideae (*Escallonia*)
    *h. Ribesoideae (currant, gooseberry)
    i. Baueroideae (*Bauera*)
   *15. Droseraceae (sundew, Venus flytrap)

39. PODOSTEMALES
   *1. Podostemaceae (riverweed)

40. PROTEALES
   1. Proteaceae (*Protea,* silk-oak, *Macadamia*)

41. ELAEAGNALES
   1. Elaeagnaceae (buffaloberry, silverberry, oleaster)

42. MYRTALES
   *1. Lythraceae (*Lythrum, Rotala, Ammania,* crepe myrtle)
   2. Crypteroniaceae (*Crypteronia*)
   3. Sonneratiaceae (*Sonneratia*)
   4. Oliniaceae (*Olinia*)
   5. Peneaceae (*Penaea*)
   6. Geissolomataceae (*Geissoloma*)
   *7. Thymelaeaceae (mezereum, leatherwood)
   *8. Melastomataceae (meadow-beauty)
   9. Lecythidaceae (Brazil nut, paradise nut)
   *10. Rhizophoraceae (mangrove)
   *11. Combretaceae (white mangrove)
   *12. Myrtaceae (myrtle, *Eucalyptus, Eugenia,* guava)
   *13. Punicaceae (pomegranate)

*14. Onagraceae (evening-primrose, fireweed, *Clarkia*)
*15. Hydrocharyaceae (water-chestnut)
*16. Haloragidaceae (water-milfoil or *Myriophyllum*, mermaid-weed)
*17. Hippuridaceae (mare's-tail)

43. SANTALALES
*1. Olacaceae (whitewood, tallowwood)
2. Opiliaceae (*Opilia*)
*3. Loranthaceae (mistletoe)
*4. Santalaceae (sandalwood, *Commandra*, buffalo-nut)
5. Grubbiaceae (*Grubbia*)
6. Myzodendraceae (*Myzodendron*)
7. Balanophoraceae (*Balanophora*)
8. Cynomoriaceae (*Cynomorium*)

44. CORNALES
*1. Cornaceae (dogwood, bunchberry)
*2. Nyssaceae (sour gum, tupelo)
3. Alangiaceae (*Alangium*)
*4. Garryaceae (silktassel-bush)

45. UMBELLALES
*1. Araliaceae (ginseng, angelica-tree, devil's-club, ivy)
*2. Umbelliferae (carrot, dill, caraway, coriander, fennel, anise, parsley, parsnip, poison hemlock, celery)

46. RAFFLESIALES
*1. Rafflesiaceae (*Rafflesia, Pilostyles*)

Group 4. *OVARIFLORAE.* OVARY FLOWERS

47. RUBIALES
*1. Rubiaceae (madder, bedstraw, coffee, Cinchona [source of quinine], *Gardenia*, button-willow or button-bush)
• *2. Caprifoliaceae (honeysuckle, elderberry, twinflower, *Viburnum*)
*3. Adoxaceae (moschatel)
*4. Valerianaceae (valerian, lamb's-lettuce)
*5. Dipsacaceae (teasel, devil's-bit)

48. CAMPANULALES
*1. Campanulaceae (bellflower, Canterbury bells, *Lobelia*)
*2. Goodeniaceae (*Scaevola*)
3. Brunoniaceae (blue-pincushion)
*4. Calyceraceae (*Acicarpha*)
5. Stylidiaceae (*Stylidium*)

49. ASTERALES
• *1. Compositae
*a. Asteroideae
*1. Heliantheae (Sunflower Tribe)
*2. Madieae (Tarweed Tribe)
*3. Ambrosieae (Ragweed Tribe)
*4. Helenieae (Sneezeweed Tribe)
*5. Senecioneae (Groundsel Tribe)
*6. Astereae (Aster Tribe)
*7. Inuleae (Everlasting Tribe)
*8. Anthemideae (Mugwort or Sagebrush Tribe)
9. Calenduleae (Marigold Tribe)
*b. Cichorioideae
*10. Cichorieae (Chicory Tribe)
*11. Mutisieae (*Mutisia* Tribe)
*12. Vernonieae (Ironweed Tribe)
*13. Eupatorieae (Eupatory Tribe)
*14. Cynareae (Thistle Tribe)
15. Arctotideae (*Gazania* Tribe)

Families of Uncertain Position

Ancistrocladaceae
Balanopsidaceae
Cardiopteridaceae
Columelliaceae
Coriariaceae
Corynocarpaceae
Daphniphyllaceae
Dichapetalaceae
Didymelaceae
Dioncophyllaceae
Dipentodontaceae
Greyiaceae
Hoplestigmataceae
Huaceae
Hydrostachyaceae
Medusandraceae
Rhopalocarpaceae
Thelygonaceae
Tremandraceae

Group 1. *THALAMIFLORAE.* RECEPTACLE FLOWERS

(***Hypogynous; Choripetalous*** or ***Apetalous***)

NOTE: After determination of the order and family, proceed to the local flora or manual to identify the genus and species.

PRELIMINARY LIST FOR ELIMINATION OF EXCEPTIONAL CHARACTER COMBINATIONS

Before using the key to the orders, page 133, it is necessary to eliminate (1) exceptional plants belonging in other groups but duplicating the key characters of the Thalamiflorae, (2) plants not having the key characters marking most of their relatives, and (3) plants marked by obscure characteristics or those inconsistent if taken singly.

The plant must correspond with every character in the combination. If the specimen does not coincide with one character, even the first listed, the combination may be eliminated from further consideration. The characters in **boldface type** should be checked first to determine whether the combination requires further inspection.

1. **Aquatic herbs** with the characters listed in a or b, below.
   a. **Slender herbs; leaves whorled, dissected, submerged,** with small, weak, hornlike spines; fruit an achene with at least two almost basal spines; flowers minute, each with an 8–12-cleft involucre; plants monoecious; sepals and petals none. *Ceratophyllaceae* (hornwort family).
   b. **Perennial herbs with rhizomes and large peltate to cordate floating** (or sometimes emersed) **leaves;** carpels 8–30 (or rarely as few as 2), coalescent or usually separate; sepals 3 or 4–6; petals numerous or sometimes only 3 or 4, intergrading with the sepals. *Nymphaeaceae* (water-lily family).

   Order 1. **Ranales,** page 136.

2. **Desert shrubs** of the Southwest with the characters listed in a or b, below.
   a. **Longer branches wandlike below; leaf glutinous, composed of 2 sessile, opposite, coalescent, leaflets;** petals yellow, arranged like the blades of an electric fan; each filament with a conspicuous attached scale; flower with a hypogynous disc; fruit 5-chambered, forming a densely hairy ball; California to southwestern Utah and western Texas. *Larrea* (creosote-bush), *Zygophyllaceae* (caltrop family).

   Order 16. **Geraniales,** page 193.

   b. **Densely and intricately branched bushy shrubs** 0.5–2 m. high; leaves opposite but fascicled, up to about 1 cm. long, entire, leathery; flowers solitary; calyx 4-parted; petals none; stamens produced on the outside of the base of an elongated tubular hypogynous disc; fruit a solitary achene formed from a single carpel; southeastern California to southwestern Colorado and northern Arizona. *Coleogyne, Rosaceae* (rose family).

   Order 38. **Rosales** (Calyciflorae), page 290.

3. **Shrubs or trees** occurring in sand along the **southern Florida** coast and on the Keys; petals present; stamens 8 or 10, the series opposite the petals reduced; carpels 4 or 5, separate, each forming a separate pistil, opposite the petals, 4–5 mm. long in fruit; leaves fleshy, appressed-pubescent. *Suriana, Simaroubaceae* (quassia family).

   Order 18. **Sapindales,** page 201.

4. **Herbs; leaves opposite, usually in numerous pairs, usually dotted with dark or clear glands;** stipules none; flowers bisexual; sepals 5; petals 5, yellow, orange, flesh-color, or purplish; stamens (of the species in question) 5–15, often coalescent into groups; fruit a capsule, the styles 3–5, persistent in fruit; flowers in cymes. Species of *Hypericum, Guttiferae* (St.-John's-wort family).

   Order 11. **Theales,** page 172.

5. **Perennial herbs or suffrutescent plants; leaves alternate or falsely whorled; sepals 5, the 2 outer narrow and smaller, the 3 inner scarious; petals 3, reddish, persisting after withering;** stamens usually 5–15; ovary with a short stalk above the receptacle; carpels 3, the style obsolete, the stigmas 3, the ovary 1-chambered; placentae parietal; ovules 2 per carpel; fruit a capsule. *Lechea, Cistaceae* (rockrose family).

   Order 11. **Theales,** page 172.

6. **Trees or large shrubs** with the characters in a or b, below.
   a. **Leaves opposite; petals none; sepals very small;** stamens 2 or sometimes 4 or rarely 1 or 3, not adnate with the calyx; flowers *not* in spikes or elongated racemes; fruit *not* enclosed in a pair of longer bracts; hypogynous disc none. *Fraxinus* and *Forestiera, Oleaceae* (olive family).
   b. **Stamens 2** (or rarely 3 or 4); **petals 5–6, slightly and inconspicuously coalescent, 2–2.5 cm. long, narrowly linear,** the stamens adnate with their bases; fruit drupelike, globular; eastern North America. *Chionanthus, Oleaceae* (olive family).

   Order 32. **Gentianales** (Corolliflorae), page 261.

7. **Fruit a legume** (pea pod), normally splitting open along both margins (the suture and the midrib of the carpel), with 2 or more seeds; petals 5, sometimes with 2 coalescent, the corolla usually but not always somewhat to markedly bilaterally symmetrical; stipules usually present, but sometimes either lacking or obscure, or specialized as spines or glands. *Leguminosae* (pea family).

   Order 38. **Rosales** (Calyciflorae), page 290.

8. Corolla papilionaceous, that is, with the upper petal forming a banner, the two lateral petals wings, and the two lower (coalescent) petals a keel enclosing the stamens and style when they are young; stipules usually present. *Leguminosae* (pea family).

   Order 38. **Rosales** (Calyciflorae), page 290.

9. **Fruit a 1-seeded bur with barbed spines,** indehiscent, the ovules several at flowering time; small shrubs or herbs; **California to Georgia and Florida**; sepals 5, purple, the calyx bilaterally symmetrical; petals smaller than the sepals, the 2 lower ones scalelike; stamens 4, in 2 unequal pairs on the upper side of the flower; anthers opening by terminal pores; leaves small, simple, entire, alternate. *Krameriaceae* (ratany family).

   Order 17. **Polygalales,** page 199.

10. **Spiny trees of deserts and grasslands from California to Kansas and Louisiana; leaves bipinnate,** with 1 or 2 pairs of primary leaflets and numerous linear to linear-oblong secondary leaflets; flowers small, in dense elongate spikes, 5-merous; fruit a legumelike elongated pod, but indehiscent. *Prosopis* (mesquite), *Leguminosae* (pea family).

    Order 38. **Rosales** (Calyciflorae), page 290.

11. **Leaves adapted to capturing insects, flat and sticky** *or* **folding shut like steel traps;** stems scapose, the leaves all basal; plants of bogs or wet places; petals present; carpels 3–5, coalescent into a single pistil. Droseraceae (sundew family).

    Order 38. **Rosales** (Calyciflorae), page 290.

12. **Leaves markedly thick, succulent and fleshy,** without spines, green or green with red (having chlorophyll, not being parasitic or saprophytic); sepals, petals, stamens, and carpels all the same number (or the stamens twice the number), each in 1 series (or the stamens in 2 series), usually 5, but sometimes 3–30; fruit consisting of 4 or 5 follicles or technically a capsule by slight coalescence of the bases of the carpels, each carpel splitting along the midrib. *Crassulaceae* (stonecrop family).

    Order 38. **Rosales** (Calyciflorae), page 290.

13. **Herbs or low shrubs; calyx tubular, white or colored** but not green, **resembling a sympetalous corolla** (but petals none), **the base of the tube of the calyx constricted** just above the ovary, persistent, hardening and enclosing the small hard fruit; involucre green, enclosing a head of 1 to several flowers; carpel 1; leaves opposite. *Nyctaginaceae* (four-o'-clock family).

    Order 23. **Caryophyllales,** page 223.

14. **Petals none; placenta basal or forming a central column in the single chamber of the ovary; stamens as many as the sepals and alternate with them** (this point to be checked with care!); sepals coalescent, white or purplish; somewhat succulent perennial herbs; leaves opposite; fruit a capsule, opening along 5 lines. *Glaux, Primulaceae* (primrose family).

    Order 29. **Primulales** (Corolliflorae), page 251.

15. **Herbs; petals none; carpels coalescent only basally;** plants with the characters listed in a or b, below.
    a. Sepals 5 or sometimes 6 or 7; fruit a capsule with 5 or sometimes 6 or 7 carpels, the upper portions of the carpels falling away and exposing the seeds. *Penthorum, Saxifragaceae* (saxifrage family) (or in some books classified in the *Crassulaceae* [stonecrop family]).
    b. **Sepals 4;** plants small, slender; flowers minute, in leafy cymes; stamens 4–10, on the edge of a hypogynous disc; carpels 2, the ovary 1-chambered; placentae parietal, 2; seeds numerous. *Chrysosplenium, Saxifragaceae* (saxifrage family).

    Order 38. **Rosales** (Calyciflorae), page 290.

16. **Large trees; inflorescence a pendent series of ball-like clusters of staminate or pistillate flowers; leaves large, palmately lobed or cleft;** stems commonly 15–25 m. high, the bark usually white or greenish, exfoliating in broad sheets except sometimes on the lower portion of the trunk; petals none or minute and obscure; fruits dry, elongated, finally separating from the clusters (balls). *Platanaceae* (sycamore family) and seemingly applicable to some of the *Hamamelidaceae* (witch-hazel family in the same order).

    Order 26. **Hamamelidales,** page 238.

17. **Herbs; ovule (and seed) solitary, borne on a long thread arising from the base of the ovary,** the placenta basal; stamens as many as the sepals and *alternate* with them; sepals coalescent; petals barely coalescent; seacoasts, including the Arctic. *Staticeae, Plumbaginaceae* (leadwort family).

    Order 29. **Primulales** (Corolliflorae), page 251.

18. **Stipules none;** fruits not spiny; herbs, shrubs, or trees with the characters listed in a, b, c, or d, below.
    a. **Anthers opening by terminal, basal, or oblique pores or slits and not by full-length slits.** (Ordinarily anthers are plump until the flower bud opens, but as the anther splits open and the pollen is shed the remainder collapses and shrivels. Hence, if the anthers open by full-length slits, they are much larger in the buds than in the flowers. If the anthers have smaller openings, they do not collapse, but remain large.)
    b. **Anther with a special appendage** (a bristle, a cusp, a feathery structure, etc.)
    c. **Plants saprophytic** (deriving their food from leafmold), not green (without chlorophyll), **usually white, brown, or red,** sometimes turning black in drying; fleshy or succulent; flowers 4–6-merous; fruit a 4–6-chambered loculicidal capsule.
    d. **Shrubs** occurring from **New Jersey to Tennessee and Florida;** pollen grains remaining coalescent in groups of 4.

    Order 31. **Ericales** (Corolliflorae), page 257.

19. **Pistil elevated on a long stalk, 3-carpellate, the 3 stigmas usually forked; flower clusters simulating small perfect flowers,** each surrounded by a cup (cyathium) bearing as many as 4 or 5 marginal glands, each gland often having an appendage appearing to be a petal; each stamen an entire flower (as shown by the joint between the filament and the similar pedicel); the single central pistil (described above) composing an entire flower, its elongated stalk the pedicel. *Euphorbia, Euphorbiaceae* (spurge family).

    Order 14. **Euphorbiales,** page 186.

20. **Stamens numerous; plants either (1) monoecious or (2) dioecious and with the staminate and pistillate inflorescenses or flowers differing in other characters as well as in presence of stamens or pistils** (e.g., in the arrangement of the flowers or in the number or form of the sepals or petals); ovary 3-chambered, ovules 2 in each chamber, usually only 1 maturing into a seed. A few genera, *Euphorbiaceae* (spurge family).

    Order 14. **Euphorbiales,** page 186.

21. **Trees; leaves simple, narrow; southern Florida;** carpel 1; receptacle swollen beneath the ovary; style attached apparently on the side of the ovary; **fruit a drupe 5–10 cm. long,** ovoid, the stone fibrous-coated. *Mangifera* (mango), *Anacardiaceae* (cashew family).

    Order 18. **Sapindales,** page 201.

22. **Low, herbaceous or woody** (sometimes trailing or climbing) **plants; Arizona to Florida; flowers in racemes or racemelike panicles;** carpel 1; **fruit a flattened red or dark-colored berry;** seed 1, arising from the base of the ovary; flowers bisexual; petals none; sepals 4; stamens 4 or 8–16. *Rivineae, Phytolaccaceae* (poke family).

    Order 23. **Caryophyllales,** page 223.

23. **Herbs; fruit a 1-chambered capsule, opening at the apex and exposing the ovules before they mature;** leaves simple, alternate; sepals 4–8; petals 4–7; stamens 3–40, on one side of the flower; stigmas 2–6, sessile, minute; placentae parietal. *Resedaceae* (mignonette family).

    Order 6. **Papaverales,** page 152.

24. **Trees or shrubs; sepals 3 or 6, in one or two cycles; petals 6, in two cycles;** stamens in cycles of 3 or arranged spirally and numerous; **stipules present.** *Magnoliaceae* (*Magnolia* family) and *Berberidaceae* (barberry family); *Lauraceae* (laurel family, **stipules none.**

    Order 1. **Ranales,** page 136.

25. **Fruit with an elongated stipe of ovary tissue above the receptacle; flower with a hypogynous disc; sepals 2 or 4; petals 4 or 6;** leaves palmate or simple; **stamens long and protruding, 6 or 9,** *Capparidaceae* (caper family).

    Order 6. **Papaverales,** page 152.

26. **Stamens 5, the filaments coalescent into a tube; petals none, but the five large, yellow, coalescent sepals appearing to be a sympetalous corolla 2 or more cm. in diameter** and the 3–5 slender subtending involucral bractlets appearing to be sepals; leaves palmately lobed; carpels 5; placentae axile in the five chambers of the ovary; fruit a capsule; large shrubs or small trees occurring in California and Arizona. *Fremontodendron, Sterculiaceae* (chocolate family).

    Order 13. **Malvales** (Thalamiflorae), page 181.

27. **Flowers in catkins; juice of the plant milky;** fruit an achene, but this enclosed in the fleshy fruiting calyx, the entire spike or **head of fruits** forming a fleshy "aggregate fruit" or "berry" **appearing like a blackberry.** *Moraceae* (mulberry family).

    Order 15. **Urticales,** page 189.

28. Desert shrubs; branches all rigid, strongly divergent, thorn-pointed, dark green; leaves small and scalelike; flowers in small clusters, about 4 mm. long; stamens 5, the filaments broad; fruit a small berry, black, shining. *Koeberliniaceae* (desert crucifixion-thorn family).

    Order 6. **Papaverales,** page 152.

29. **Flowers in catkins; shrubs of deserts, alkaline areas, seacoasts, or dry grasslands;** herbage commonly scurfy (i.e., with dandrufflike white scales on the leaves and young stems, the scales originating from cells standing out from the surfaces); plant corresponding with *every character* listed in *either* a, b, *or* c, below.
    a. Pistillate flowers enclosed in a pair of bracts which enlarge at fruiting time, the plants dioecious. *Atriplex* (salt-bush).
    b. Pistillate flower enclosed basally by the calyx, which spreads into a large circular platform above the base of the fruit. *Sarcobatus* (black greasewood).
    c. Branches composed of fleshy, succulent joints, the leaves reduced to minute scales. *Allenrolfea* (iodine bush). *Chenopodiaceae* (goosefoot family).

    Order 23. **Caryophyllales,** page 223.

30. **Leaves adapted to capturing insects, forming pitcherlike hollow structures;** stems scapose, the leaves all basal; plants of bogs or wet places; petals present; carpels 3–5, coalescent into a single pistil.

    Order 12. **Sarraceniales,** page 180.

31. **Trees; Florida; leaves alternate, simple, entire, with dotlike glands;** stipules none; **stamens 12–15 or 20, coalescent the full length of the filaments;** carpels 2–5, fully coalescent; fruit a berry, 1-chambered, with 2-many seeds. *Canellaceae* (wild cinnamon family).

    Order 1. **Ranales,** page 136.

32. **Trees with milky sap and very large, alternate, palmately 7-9-lobed leaves; fruits melonlike, 2–18 cm. long;** stamens 10, the filaments usually adnate with the corolla; **Florida.** *Caricaceae* (papaya family).

    Order 7. **Violales,** page 159.

33. **Small aquatic annual herbs; leaves simple, opposite or whorled;** flowers very small, bisexual; carpels 3–5, coalescent, the ovary 3–5-chambered, the placentae axile, the ovules numerous; fruit a capsule. *Elatinaceae* (waterwort family).

    Order 11. **Theales,** page 172.

34. **Trees or** (nonparasitic) **shrubs; flowers small, unisexual** (very rarely a few bisexual), **all in catkins;** pistillate catkins sometimes globose; ovary shown to be superior through **absence** of small scalelike sepals arising from just below the style; the plants determined from the following brief key to groups long known as the Amentiferae:

1. Plant *not* corresponding with *every character* in lead 1 below.
    2. Ovary superior, the sepals (if any) not attached at the top of the ovary; flowers all in catkins.
        3. Leaves alternate.
            4. Fruits *not* with the surface obscured by waxy granules.
                5. Fruit not obscured by the scales of the catkin, clearly visible without dissection; plants dioecious.
                    6. Fruit a capsule less than 1 cm. long, 1–3 mm. in diameter, the seeds numerous, each with a tuft of hairs on one end; sepals none; placenta basal; carpels 2–4.

                    Order 10. **Salicales,** page 171.

                    6. Fruit a leathery drupe, 1–2 cm. long, 5–7 mm. in diameter, the seed solitary, glabrous; sepals coalescent, very small, present in only the pistillate flowers; placenta on the suture; carpel 1.

                    Order 21. **Leitneriales,** page 216.

                5. Fruit obscured by the scales of the catkin or by other bracts, visible only after dissection or when shed at maturity; plants monoecious. *Betulaceae* (birch family).

                Order 27. **Fagales,** page 243.

            4. Fruits covered with waxy granules, nutlets or sometimes drupelike, small, aggregated densely in the spikes; seed solitary, glabrous; carpels 2, the ovary 1-chambered.

            Order 20. **Myricales,** page 216.

        3. Leaves opposite.
            4'. Catkins erect; ovary 2-chambered, becoming 4-chambered by presence of false partitions, with one seed in each chamber; style none, the stigma sessile, capitate.

            Order 25. **Batidales,** page 238.

            4'. Catkins drooping; ovary 1-chambered, with 2 or sometimes 3 ovules; styles 2 or sometimes 3, elongate. *Garryaceae* (silktassel-bush family).

            Order 44. **Cornales,** page 325.

    2. Ovary inferior, the sepals minute, appearing

to grow from the top of the ovary; pistillate flowers often not in catkins.

See **CALYCIFLORAE,** p. 284.

1. Leaves scalelike, in whorls of 6 to 14 at the nodes of the branchlets, which resemble those of a horsetail (*Equisetum*); staminate flowers without a perianth, each consisting of a single stamen subtended by 2 scalelike bracts, arranged in long linear or clublike catkins; pistillate flowers in subglobose or ovoid eventually conelike heads; trees appearing like pines.

Order 28. **Casuarinales,** page 244.

KEY TO THE ORDERS

(First consult the Preliminary List for Elimination of Exceptional Character Combinations, page 129)

NOTE: After determination of the order and family, proceed to the local flora or manual to identify the genus and species.

1. Pistils more than one in each flower (sometimes in trees or shrubs numerous and packed tightly or even slightly coalescent into a large conelike structure).
   2. Plant *not* corresponding with *every character* in either combination in lead 2 below.
      3. Sepals usually neither persistent nor becoming leathery; stamens attached directly to the main receptacle, the primordia developing first toward the margin of the flower, then toward the center (centripetally); flowers of many sizes; fruit rarely both leathery and follicular; leaves various.

         Order 1. **Ranales,** page 136.

      3. Sepals persistent at fruiting time and becoming thick and leathery; flowers large and showy; fruits leathery follicles; leaves pinnately to ternately dissected. *Paeoniaceae* (peony family).

         Order 2. **Dilleniales,** page 148.

   2. Plant corresponding with *every character* in *one* of the combinations below.
      (1) Plants aquatic (usually in deep water); perennial herbs with rhizomes and large peltate to cordate floating (or sometimes emersed) leaves; carpels 8–30 (or rarely as few as 2), coalescent or usually separate; **sepals 3 or 4–6; petals numerous or sometimes only 3–4** (then dull purple or white with a basal yellow spot). Family *Nymphaeaceae* (water-lily family).

      (2) Plants aquatic; leaves whorled, dissected, with small, weak, hornlike spines; fruit a spiny or tuberculate achene; flowers minute, each with an 8–12-cleft involucre; plants monoecious; staminate flower composed of many sessile anthers, the pistillate of a single pistil. Family *Ceratophyllaceae* (hornwort family).

      Order 3. **Nymphaeales,** page 148.

1. Pistil clearly 1.
   2'. Carpel 1.
      3'. Stipules none (or rarely present in woody plants); stamens *not* curved, bent, or coiled inward, *not* springing outward as the flower opens; stems and leaves *not* with stinging hairs; petals usually present or the sepals petaloid.

         Order 1. **Ranales,** page 136.

      3'. Stipules present, often early deciduous but then detected by the attachment scars; plants herbaceous; stamens curved, bent, or coiled inward in the bud, springing outward as the flower opens and throwing the pollen into the air; stems and leaves usually with stinging hairs; petals none. *Urticaceae* (nettle family).

         Order 15. **Urticales,** page 189.

2'. Carpels 2 or more, *coalescent* (as seen from the presence of 2 or more stigmas or, failing this, by presence of the 2 or more double rows of seeds in the ovary or by its division into 2 or more seed chambers).

3''. Stamens numerous, more than 15, arranged spirally (at least a cyclic arrangement not detectable).

4. Sepals falling as the flower opens, 2 or 3, sometimes 2 coalescent into a deciduous cap (calyptra). *Papaveraceae* (poppy family).

Order 6. **Papaverales,** page 152.

4. Sepals persistent at least through flowering and often fruiting (when not persistent, 5).

5. Sepals in the bud with their edges at least slightly overlapping; stipules usually none.

6. Placentae parietal, the ovary 1-chambered.

Order 7. **Violales,** page 159.

6. Placentae axile or seemingly so, the ovary more than 1-chambered.

Order 11. **Theales,** page 172.

5. Sepals in the bud with their edges meeting but not overlapping; stipules present.

Order 13. **Malvales,** page 181.

3''. Stamens 3–15 or rarely more, in 1–3 cycles.

4'. Plant *not* corresponding with *every character* in *either* combination relating to the herbs described in lead 4' below (p. 136).

5'. Plant *not* corresponding with *every character* in lower lead 5', page 136.

6'. Flowers when bisexual with at least sepals and commonly with petals. (For lower lead 6, cf. p. 136).

7. Plants with clearly distinguished stems and leaves, terrestrial or sometimes aquatic but rarely attached to rocks in streams. (Cf. lower lead 7, p. 136).

8. Plant *not* corresponding with *every character* in the combination in lead 8, page 135.

9. Placentae basal or nearly so; ovules or seeds 2-many; ovary 1-chambered.

10. Placenta parietal but almost basal; ovule anatropous; embryo straight.

Order 9. **Tamaricales,** page 169.

10. Placenta on a central column; ovule campylotropous; embryo curved.

Order 23. **Caryophyllales,** page 223.

9. Placenta *not* basal, the ovary 1–several-chambered.

10'. Flowers *both* (1) all bisexual *and* (2) *without* hypogynous discs.

11. Placentae parietal; ovary 1-chambered or sometimes 2-chambered.

12. Petals 4 or rarely 6 or none; sepals usually 2 or 4, rarely 3 in tiny herbs of the Pacific Coast.

Order 6. **Papaverales,** page 152.

12. Petals 5; sepals commonly 5, rarely 3 in large trees of southern Florida.

Order 7. **Violales,** page 159.

11. Placentae axile; ovary 2–5- or rarely as many as 12-chambered.

12'. Flowers not in elongate simple racemes.

13. Sepals in the bud with their edges meeting but not overlapping (valvate); plants commonly with branched or stellate hairs. *Sterculiaceae* (chocolate family).

Order 13. **Malvales,** page 181.

13. Sepals in the bud with their edges overlapping at least slightly (imbricated) (except in one family of small annual herbs with dissected alternate leaves and no stipules).

Order 16. **Geraniales,** page 193.

12'. Flowers in elongate simple racemes, small; shrubs or small trees of the southern states; leaves entire, alternate, without stipules. *Cyrillaceae* (leatherwood family).

Order 31. **Ericales** (Corolliflorae), page 257.

10'. Flowers *either* (1) all or some of those on each plant unisexual (the plants being monoecious, dioecious, or polygamous) *or* (2) each with a hypogynous disc.

11'. Plant *not* corresponding with every character in the combination in lead 11' below.

12''. Plants either (1) monoecious or (2) dioecious and with the staminate and pistillate inflorescences or flowers differing in other characteristics as well as in the presence of stamens or pistils (e.g., in the number or arrangement of the flowers or in the number or form of the sepals or petals); ovary 3- (rarely 1–4-) chambered; ovules 2 in each chamber, usually only 1 maturing into a seed; capsule usually opening elastically through the septa between the carpels, each carpel then opening along the ventral angle (suture).

Order 14. **Euphorbiales,** page 186.

12''. Flowers bisexual or unisexual and the plants dioecious or polygamous or polygamodioecious, when unisexual the staminate and pistillate flowers (or inflorescences) differing in only the presence of either stamens or pistils; fruits not opening as described in lead 12'' above.

13'. Ovary with more than 1 chamber.

14. Stamens in 1 or 2 series with the outer or the single series opposite the sepals (i.e., alternate with the petals).

Order 18. **Sapindales,** page 201.

14. Stamens in a single series, alternate with the sepals (i.e., opposite the petals, if any).

Order 22. **Rhamnales,** page 217.

13'. Ovary 1-chambered, bearing 3 separate styles. *Anacardiaceae* (cashew family).

Order 18. **Sapindales,** page 201.

11'. Low shrubby plants with evergreen, alternate, simple, entire, linear, numerous, crowded, firm leaves; flowers unisexual or some on the plant bisexual; sepals 0 to 3 (scalelike bracts also present below the flower); petals none; stamens 2 or 3; carpels 2–9; fruit drupelike, with 2–9 seedlike internal stones. *Empetraceae* (crowberry family).

Order 31. **Ericales,** page 257.

8. Delicate annual herbs occurring in water or on mud or wet soil; leaves opposite or the upper in dense whorllike clusters, entire, without stipules; plants

monoecious, the staminate flower consisting of a single stamen, the pistillate of a single pistil with a 4-chambered deeply divided ovary and 2 slender styles; fruit separating into 4 nutlets. *Callitrichaceae* (water-starwort family).
Order 35. **Lamiales, page 274.**

7. Plant body thalluslike, not obviously differentiated into stems and leaves, resembling a seaweed, liverwort, or moss; plants attached to rocks in running streams; flowers bisexual, minute; petals none; stamens 1–4, all on one side of the flower; ovary 2–3-chambered, the styles 2–3; fruit a capsule.

Order 38. **Podostemales, page 313.**

6'. Flowers bisexual *and* lacking both sepals and petals, arranged in spikes or racemes (in one species the spike subtended by petallike white or reddish-tinged involucral bracts and therefore the inflorescence and bracts appearing like a large flower); herbs (as represented north of Mexico).

Order 4. **Piperales, page 151.**

5'. Ovary with only 1 ovule, the fruit indehiscent (not splitting open); petals none; flowers small, individually inconspicuous; plant, when shrubby, with no hypogynous disc in the flower.

6''. Placenta apical, the single (anatropous) ovule or seed suspended; stipules present at least on the young leaves, usually deciduous, *neither* (1) scarious (dry and parchmentlike) *nor* (2) persistent and sheathing the stem.

Order 15. **Urticales, page 189.**

6''. Placenta basal, the single (orthotropous, campylotropous, or amphitropous) ovule or seed erect; stipules none or, if present, *either* (1) scarious *or* (2) persistent and sheathing the stem. *Polygonaceae* (buckwheat family), *Chenopodiaceae* (goosefoot family), *Amaranthaceae* (amaranth family), and *Caryophyllaceae* (pink family).
Order 23. **Caryophyllales, page 223.**

4'. Herbs (with campylotropous or amphitropous ovules) corresponding with *every character* in *either* of the following combinations:

(1) Petals none; carpels and seed-chambers 3–12, with 1 seed in each chamber; fruit a capsule or a berry; leaves entire; flowers bisexual; stamens separate; seeds with curved or coiled embryos. *Phytolacca, Phytolaccaceae* (pokeweed family), and *Mollugo* and *Glinus, Aizoaceae* (carpetweed family).

(2) Petals present (or seemingly so); placenta on a column arising from the base of the single chamber of the ovary; ovules or seeds several to many.

Order 23. **Caryophyllales, page 223.**

Order 1. *RANALES.* BUTTERCUP ORDER

DEPARTURES AMONG THE RANALES FROM THE TYPICAL FLORAL CHARACTERS OF THE ORDER

Free	Stamens coalescent in all the *Cannellaceae*, the *Myristicaceae*, some *Lardizabalaceae*, and some *Menispermaceae;* carpels coalescent in the *Illiciaceae*, some *Magnoliaceae*, some *Annonaceae*, the *Gomortegaceae*, (perhaps) the *Berberidaceae*.
Equal	Calyx rarely strongly bilaterally symmetrical as in *Delphinium* (larkspur) and *Aconitum* (monkshood); corolla in these instances also bilaterally symmetrical but less obviously so.
Numerous	Stamens as few as 3–5 in some families, commonly in 2–several series of 3 in some families; carpels sometimes few, in some families or genera regularly 1.
Spiral	Stamens sometimes in 1–several cycles usually of 3; carpels sometimes several in a cycle or often

1 (and therefore with no arrangement).

Hypogynous Flowers perigynous in the *Calycanthaceae*, the *Eupomatiaceae*, the *Monimiaceae*, and some *Lauraceae*, epigynous in the *Gomortegaceae*, and the *Hernandiaceae*.

Herbs, shrubs, trees, or vines. Leaves commonly alternate, sometimes opposite; stipules usually none, but present in the *Magnoliaceae*, the *Cercidophyllaceae*, the *Lactoridaceae*, some *Berberidaceae*, and a few species of *Thalictrum* in the *Ranunculaceae*. Flowers corresponding ordinarily with all the primitive floral characters typical of the order but with rare departures as indicated above, these occurring mostly in members of relatively highly specialized small families, usually complete, but in some families unisexual, sometimes lacking petals; sepals and petals in some families 5-merous but in most families 3-merous (a character typical of the monocotyledons); stamens usually numerous but sometimes as few as 3–6 or 10; anthers usually with 2 pollen chambers; carpels numerous but sometimes few and in some families 1; ovary usually formed from a single carpel, with a single internal cavity, with 1 to many seeds; ovules anatropous; endosperm usually abundant. Fruit usually an achene, a utricle, a follicle, or a drupelike structure, or rarely a drupe or a berry, the fruits often occurring in dense heads on the enlarged conelike receptacle, rarely obviously coalescent, sometimes packed tightly together and slightly coalescent and thus forming a dense woody cone.

The probable primitive characters of the Ranales are discussed in full in Chapter 17. It is assumed tentatively that most or all other orders of dicotyledons have been derived directly to very indirectly from extinct ancestral groups related to the Ranales.

This evidently ancient order is composed of twenty-four families, more than in any other order. Many are small remnants of groups probably (in some cases certainly) more widespread in the geologic past. The number of species in the order is large (about 5,500) but less than in a number of other orders. The various departures from the typical floral characters of the Ranales reflect much evolutionary development along many lines within the order. According to a liberal policy the order could be divided.

Key to the Families

(North America north of Mexico)

1. Stamens coalescent through the full length of the filaments, 12–15 or 20; trees; Florida.

   Family 10. **Cannellaceae** (wild cinnamon family).

1. Stamens not coalescent, commonly numerous.
   2. Carpel not completely closed along the ventral margins, each margin bearing a ciliate stigmatic crest; true style and stigma none; filament of each stamen and the area between the 2 pollen chambers very broad, the pollen chambers thus spread far apart; bases of the filaments coalescent; woody vines with simple alternate leaves; Louisiana to South Carolina and Florida.

      Family 2. **Schisandraceae** (*Schisandra* family).

   2. Carpel margins closed along their entire length; a stigma and usually a style present; connectives narrow; filaments usually separate.
      3. Pistils and carpels more than 1 (the rare exceptions having the character given next in this lead); anthers splitting open along their entire length (as seen directly and as indicated by their collapse after the flower opens and by their consequent much smaller size than in the bud). (Cf. lower lead 3, p. 138.)
         4. Trees or shrubs.
            5. Flowers hypogynous; sepals and petals tending to be in cycles of 3 or sometimes 4, the flower usually with 1–3 cycles of each; leaves alternate.
               6. Petals rolled lengthwise or overlapping each other like shingles (i.e., imbricate).
                  7. Carpels in a single whorl of usually 7–15, coalescent basally; stipules none.

                     Family 3. **Illiciaceae** (*Illicium* family).

                  7. Carpels arranged spirally in a conelike head; stipules present, deciduous, leaving a ring around the node.

                     Family 6. **Magnoliaceae** (*Magnolia* family).

               6. Sepals and petals in the bud meeting only at the margins, not overlapping (i.e., valvate).

                  Family 7. **Annonaceae** (custard-apple family).

5. Flowers perigynous; bracts, sepals, and petals in an intergrading series, adnate with the floral cup (hypanthium); all flower parts numerous; leaves opposite.

Family 17. **Calycanthaceae** (spicebush family).

4. Herbs or woody vines.

5'. Herbs.

6'. Stamens indefinite in number, almost always numerous, arranged spirally, therefore *not* having definite positions in relationship to the petals; leaves *not* peltate.

Family 21. **Ranunculaceae** (buttercup family).

6'. Stamens twice as many as the petals, the petals 6 or 9; leaves peltate, palmately 7–9-lobed, large. *Podophyllum.*

Family 23. **Berberidaceae** (barberry family).

5'. Woody vines.

6''. Leaves opposite, compound; petals none or small and transitional to stamens; sepals large and petaloid, usually 4; stamens numerous, arranged spirally. Species of *Clematis.*

Family 21. **Ranunculaceae** (buttercup family).

6''. Leaves alternate, simple; petals usually present; sepals *not* very large, *not* markedly petaloid, 4–8, usually 6; stamens 6, 12, or up to 24, usually cyclic.

Family 24. **Menispermaceae** (moonseed family).

3. Pistil 1, composed of 1 carpel, forming a berry or a drupelike fruit; anthers not splitting open, opening by 2 or 4 uplifted trapdoorlike lids; sepals commonly 6 or sometimes 4 or 9, usually in cycles of 3; petals, if present, 6 or 9; stamens 6 or sometimes 8 or 9.

4'. Corolla none; fruit drupelike, with a single very large seed and a fleshy or leathery ovary wall (e.g., an avocado).

Family 19. **Lauraceae** (laurel family).

4'. Corolla present or sometimes none, in one genus the petals glandlike and much shorter than the sepals; fruit a berry, a pod opening horizontally, a follicle, or a peculiar structure with a single large exposed seed.

Family 23. **Berberidaceae** (barberry family).

Suborder A. **Magnoliineae**

Family 1. **Winteraceae.** DRIMYS FAMILY

Woody plants, the xylem of the primitive vesselless type (cf. p. 564). Leaves simple, entire, with clear glandular dots, alternate. Flowers bisexual; sepals 2–6, usually coalescent, in *Drimys* coalescent and forming a cap (calyptra) which falls as the flower opens; petals of various numbers; stamens numerous, variable in shape and in attachment of the sporangia, mostly not differentiated into anther and filament but more or less so in *Drimys,* section *Tasmannia* (South Pacific); carpels variable, open and with no style or stigma in *Drimys,* section *Tasmannia,* the stigmatic area being along the margins and the ovules being ventral on the middle leaf surface and attached in series between the veins, in *Drimys,* section *Wintera* (Mexico to South America), similar but the stigmatic area restricted to marginal crests near the apex of the carpel and the margins coalescent except near the crests, in other genera the same trend repeated or somewhat modified in detail.

The family is composed of six genera and 88 or more species occurring chiefly in the South Pacific and especially in Australia and New Guinea. *Drimys,* section *Wintera,* includes four species restricted to Latin America. *Drimys* is rare in cultivation in the United States. Fig. 17-3.

The primitive characters of the carpel, the series of carpel developmental stages, and the vesselless xylem of this family are of particular interest and significance for their bearing upon the origin of angiosperms (pp. 562–567). Although *Drimys,* section *Tasmannia,* has primitive carpels, its stamens and sepals are specialized and far from the primitive type in the family or the order.

*Family 2. **Schisandraceae.** SCHISANDRA FAMILY

(Sometimes included in *Magnoliaceae*)

Woody vines. Leaves simple, alternate, deciduous. Plants monoecious or dioecious; sepals and petals not clearly differentiated, in 2 or more series; stamens 4 to numerous, in the North American species 5 and the filaments and connectives very broad, the 2 pollen chambers therefore widely separated; filaments coalescent basally; carpels 12 to numerous, arranged spirally, the ventral margins not completely fused and the carpel partly open when young, each margin with a ciliate stigmatic crest somewhat resembling a style. Fruits fleshy, the receptacle often much enlarged at fruiting time.

The family is composed of two genera, one of which (*Schisandra*) is represented in North America north of Mexico by a single species (*Schisandra coccinea*, bay-star-vine) occurring from Louisiana to South Carolina and Florida. The remainder of the family is Asian or Malaysian).

The pollen shows primitive characters combined with the typical ones of the dicotyledons (cf. p. 565).

*Family 3. **Illiciaceae.** ILLICIUM FAMILY

Shrubs or moderate-sized trees. Leaves simple, entire, alternate. Flowers bisexual; sepals and petals not well differentiated, usually in several series commonly of 3, the innermost petals transitional to stamens; stamens usually numerous, carpels relatively few, usually 7–15, in a single cycle, coalescent basally. Fruit a follicle.

The single genus is represented from Louisiana to Georgia and Florida. The North American species are the star-anise and the purple-anise or stinkbush, the flowers of which are said to smell like decaying fish.

The family is related much more closely to the Schisandraceae than to any other.

Family 4. **Degeneriaceae.**

DEGENERIA FAMILY

Trees. Leaves simple, entire, alternate. Flowers bisexual, solitary; sepals and petals differentiated in size and form but not in texture; sepals 3; petals about 12, in 3–4 series; stamens about 20, not differentiated into anthers and filaments, each consisting of a broad structure, perhaps a specialized leaf, (microsporophyll) with four slender pollen chambers (microsporangia) imbedded under the dorsal surface (back) of the sporophyll; inner stamens sterile; pollen monocolpate (with a single furrow); carpel 1, folded (conduplicate), open on the ventral side, the margins and portions of the upper surfaces stigmatic, the stigmatic areas with numerous, loosely interlocking glandular hairs; style and stigma none; ovules attached to the upper surface of the carpel, distinctly not marginal. Fruit indehiscent, the sides of unequal length, more or less oblong-ellipsoid; seeds in 2 rows.

The single species, *Degeneria vitiensis*, has been collected on Viti Levu and Vanua Levu in the Fiji Islands. Its recent discovery (in 1934 and 1941) and subsequent study (published in 1942) stimulated new research on the Ranales displaying apparently primitive characters and a review of the related families, many of them previously little studied, especially from the standpoint of anatomy and reproductive structures.

Family 5. **Himantandraceae.**

HIMANTANDRA FAMILY

Resembling the Degeneriaceae. Branchlets, lower surfaces of the leaves, and inflorescences with numerous peltate scales. Calyx and corolla each forming a deciduous cap (calyptra) of fused parts; inner and outer stamens sterile; carpels several, closed along the ventral sutures; ovules 1 or 2. Carpels coalescent at fruiting time.

About 4 species; Australia, New Guinea and the adjacent South Pacific region.

*Family 6. **Magnoliaceae.**

MAGNOLIA FAMILY

Trees or shrubs. Leaves simple, usually entire, alternate; stipules present, enclosing the young bud, early deciduous, leaving a circular scar around the node. Flowers of the North American genera bisexual, usually large and solitary; sepals and petals usually in series of 3 but the petals sometimes numerous; stamens numerous, arranged spirally; carpels numerous, arranged spirally on a usually elongated axis, closed; stigma and style present; fruit a follicle or samaralike, the fruits of the head sometimes somewhat coalescent and forming a woody conelike structure.

The family includes several genera occurring in the temperate areas of the Northern Hemisphere. *Magnolia* is the most common representative of this family in eastern North America. Several species occur from southern Illinois and New York

Figure 10-1. Ranales; Magnoliaceae; *Magnolia grandiflora* I—flower just after anthesis, the stamens having fallen to the cupped upper surfaces of the petals.

(escaped in Massachusetts) southward, being more abundant in the Southern States. *Liriodendron* (tulip-tree) grows from Wisconsin to Ontario and Massachusetts and southward to the Southern States. The family is not represented in western North America.

The parenchyma (soft tissue) of the plant includes special oil-containing cells, as do plants of several other ranalian families of woody plants (*Annonaceae, Myristicaceae, Calycanthaceae, Monimiaceae,* and *Lauraceae*).

*Family 7. **Annonaceae.**

CHERIMOYA FAMILY

Aromatic trees, shrubs, or sometimes vines. Leaves simple, entire, alternate. Flowers bisexual or rarely unisexual; sepals usually 6; petals usually 6 in 2 series of 3, sometimes resembling the sepals; stamens numerous, arranged spirally; anther with 4 pollen chambers, the connective enlarged; carpels several to many, either separate or in *Annona* (cherimoya and custard-apple) forming a fleshy aggregate including the receptacle. Fruit a berry; seeds large.

This largely tropical family is represented in eastern North America by the pawpaws or custard-apples (*Asimina*). Two species of *Annona* known by various names, including custard-apple and sugar-apple, occur in south peninsular Florida and on the Keys, one of them being introduced from tropical America. These are the same genus as the cherimoya, one of the most important fruits of South America, sometimes cultivated along the southern edge of the United States.

Family 8. **Myristicaceae.**

NUTMEG FAMILY

Usually aromatic trees or shrubs. Leaves simple, entire, evergreen, alternate. Flowers unisexual, small, in clusters of various types; calyx rotate, usually 3-lobed; petals none; stamens 2–30, the filaments coalescent; anthers splitting lengthwise; carpel 1, the stigma and usually the style present; ovule 1. Fruit drupelike, but splitting open; seed with a colored fleshy aril and partly or wholly enveloped by it.

The family is distributed widely in the Tropics, some species occurring in southern Mexico.

Nutmeg is derived from the seed of *Myristica fragrans,* mace from the aril. The nutmeg is cultivated sparingly in southern Florida.

Family 9. **Eupomatiaceae.**

EUPOMATIA FAMILY

Shrubs. Leaves simple, alternate. Flowers bisexual, solitary, perigynous, the carpels being enclosed in the enlarged floral cup (a hypanthium) formed from the receptacle; sepals and petals indistinguishable, forming a deciduous cap on the rim of the floral cup; stamens numerous, the inner sterile and petallike; carpels numerous, the ovaries separate, but submerged in the tissue of the receptacle (floral cup); styles coalescent into a mass; ovules 1–2. Floral cup and fruits fleshy at maturity.

The family consists of a single genus restricted to New Guinea and Australia.

Suborder B. **Laurineae**

*Family 10. **Cannellaceae.**

WILD CINNAMON FAMILY

(Synonym: *Winteranaceae,* not *Winteraceae*)

Aromatic trees. Leaves simple, entire, coriaceous, alternate, with dotlike glands; stipules none. Flowers bisexual, radially symmetrical; sepals 5 or sometimes 4, thick; petals 5 or sometimes 4 or none, sometimes coalescent basally; stamens 12 or 15 or as many as 20, the filaments completely coalescent; anthers splitting lengthwise; carpels 2–5, the style 1, the ovary with one chamber; placentae parietal; ovules 2–many, barely anatropous. Fruit a berry.

Tropical America, tropical Africa, and Madagascar, the species and genera disjunct and localized in distribution. *Canella Winteriana,* wild cinnamon or cinnamon-bark, occurs in Florida and on the Keys.

The bark of *Canella* is both a condiment and of medicinal use.

Family 11. **Amborellaceae.**

AMBORELLA FAMILY

Shrubs. Leaves simple, alternate, entire, leathery; stipules none. Flowers unisexual, hypogynous, radially symmetrical, solitary; perianth of

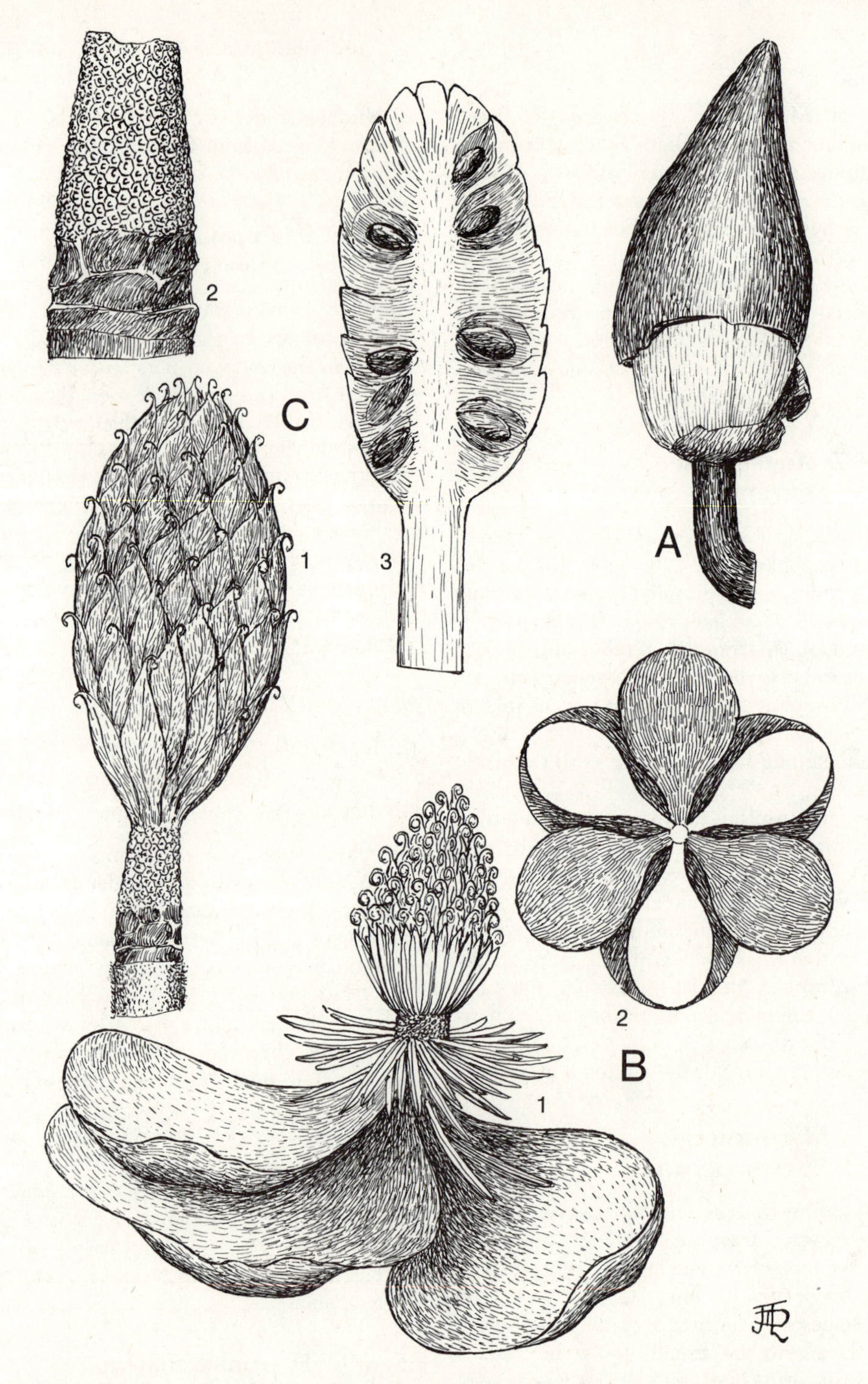

Figure 10-2. Ranales; Magnoliaceae; *Magnolia grandiflora* II—flower and fruit: **A** flower bud opening, the sepals being ruptured; **B** flower, **1** anthesis, the upper stamens having fallen, **2** arrangement of sepals and petals in cycles of three; **C** head of fruits (cf. **Figure 6-15**), each formed from a single carpel, these obscurely coalescent at the bases, **1** side view, **2** enlargement of a portion of the receptacle showing the scars at the points of attachment of the three sepals, the six petals (in two series), and the numerous spirally arranged stamens (cf. also **Figures 6-15** and **8-6 E**), **3** longitudinal section.

numerous spirally arranged members, with no sharp line between calyx and corolla; stamens numerous, the filaments very broad; carpel 1; ovule 1, orthotropous. Fruit a drupe or drupelike.

A single genus, *Amborella,* occurring in New Caledonia.

Family 12. **Austrobaileyaceae.**

AUSTROBAILEYA FAMILY

Shrubs, climbing lianas. Leaves simple, entire, opposite or nearly so; stipules small and deciduous. Flowers bisexual, hypogynous, choripetalous, radially symmetrical, solitary in the leaf axils, pedicelled; sepals and petals undifferentiated, but the inner ones becoming gradually larger and more petaloid, about 12 altogether, pale green; stamens about 12–25, the inner smaller and sterile, the anthers with 2 pollen chambers, the blade a scale leaf with the anther sacs embedded in the middle of the ventral side; carpels about 7–9, separate, on a swollen part of the receptacle; style with 2 branches; ovules 8–14.

A single genus, *Austrobaileya,* on the Atherton Tableland, Queensland, Australia.

Family 13. **Trimeniaceae.**

TRIMENIA FAMILY

Shrubs, trees, or lianas. Leaves simple, entire to toothed, alternate or opposite; stipules none. Flowers bisexual or unisexual, hypogynous, apetalous, radially symmetrical, in racemes, thyrses, or panicles; sepals 4 or 6 in 1 or 2 series, small; stamens 6–numerous, opening by longitudinal slits; carpels 1–2 and the ovary 1–2-chambered, with 1 ovule pendulous from the apex. Fruit berrylike, subglobose to ellipsoid.

A family of a few genera occurring in central and southern Africa, New Guinea, New Caledonia, New South Wales, and Fiji.

Family 14. **Chloranthaceae.**

CHLORANTHUS FAMILY

Usually aromatic herbs, shrubs, or trees. Leaves simple, opposite; stipules minute. Flowers bisexual or unisexual, often epigynous, minute, in clusters; sepals and petals none or the pistillate flower with a minute calyx; stamen 1 or stamens 3 and coalescent and adnate with one side of the base of the ovary; anthers with 2 (or sometimes 1) pollen chambers; carpel 1; ovule 1. Fruit a drupe.

The family is tropical and subtropical in the Eastern Hemisphere and in America from Jalapa, Mexico, southward.

The status of the proposed family *Lacistemaceae* has not been evaluated.

Family 15. **Lactoridaceae.**

LACTORIS FAMILY

Shrubs. Leaves simple, entire, small, covered with numerous clear glandular dots; stipules present, large and membranous. Flowers unisexual or bisexual (the plants polygamomonoecious, i.e., polygamous but largely monoecious), solitary or 2 or 3 in an axillary cluster; sepals 3; petals none; stamens 6 in 2 series; carpels 3; ovules 6. Fruit a follicle.

The family consists of a single genus restricted to the Juan Fernandez Islands (400 miles west of Chile).

Family 16. **Gomortegaceae.**

GOMORTEGA FAMILY

Aromatic trees with strong wood making excellent lumber. Leaves simple, opposite, evergreen; stipules none. Flowers bisexual, epigynous, apetalous (the perianth appearing to be of sepals), bilaterally symmetrical, in axillary and terminal racemes; sepals reported variously as 6–10, arranged spirally; stamens 2–3 or reported as up to 11 and with some sterile; anthers opening by 2 uplifted flaps covering pores; carpels 2–3, coalescent, the style 2–3-lobed, the ovary inferior but the carpels reported to be attached spirally. Fruit a drupe, 2–3-chambered but developing only 1 seed.

The family consists of a single species of *Gomortega* occurring in Chile. According to Stern (American Journal of Botany 42: 874–885, 1955) the closest relationship is to the *Monimiaceae.* Its anatomical features are prevailingly primitive.

*Family 17. **Calycanthaceae.**

SPICEBUSH FAMILY

Aromatic shrubs. Leaves simple, entire, opposite, with short petioles. Flowers bisexual, perigynous, the floral cup (a hypanthium) bearing numerous intergrading sepals and petals, in fruit suggesting the "hip" of a rose; stamens usually numerous, the inner ones sterile; carpels several to many; ovules 1–2. Achenes arranged spirally on the floral cup.

One species of spicebush occurs in California and three or four from Pennsylvania to Louisiana and Florida (mostly in the Southern States).

Being with perigynous flowers, the members of the family are unique among the North American Ranales, and some authors have placed them in the Rosales, where they may belong. The perigyny is of a peculiar type because the bracts, the sepals, and the petals shade off one into the other, and they are adnate basally to an enlarged hollow, bowllike receptacle.

Family 18. **Monimiaceae.**

MONIMIA FAMILY

Aromatic trees or shrubs or rarely vines. Leaves simple, evergreen, usually opposite or whorled but sometimes alternate. Flowers bisexual or sometimes unisexual, perigynous; sepals usually present, 4–8; petals present or absent (or these perhaps an inner petaloid series of sepals); stamens numerous or sometimes few and in 1 or 2 series; anthers splitting lengthwise; filaments with or without glands; carpels numerous; ovule 1. Fruits (achenes or drupes) enclosed by the floral cup.

The family occurs in the Tropics and in the Southern Hemisphere; one genus is native in Mexico.

A few species are cultivated along the southern margin of the United States. Boldo wood from Chile is used in cabinet making.

*Family 19. **Lauraceae.** LAUREL FAMILY

(Segregate: *Cassytheaceae*)

Usually aromatic trees or shrubs (or in one genus parasitic vines). Leaves simple, usually entire, usually leathery, usually evergreen. Flowers usually bisexual, relatively small, hypogynous or perigynous, commonly in panicles; sepals usually 4 or 6, greenish yellow, yellow, or white, coalescent basally; petals none; stamens commonly 12 in 4 cycles of 3; anthers with 2 or 4 pollen chambers, with 4 pores each covered by a flap, the 2 outer cycles opening inward, the third outward, the fourth usually sterile; carpel 1; ovule 1. Fruit a drupe or drupelike, the seed large, with no endosperm.

The family is largely tropical and most abundant in southeastern Asia and tropical America, but a few species occur in warm temperate regions. The only representative of the family on the Pacific Coast is *Umbellularia californica* (the bay-tree or pepperwood), of southwestern Oregon (known there as myrtle) and of California. A few representatives occur in the Middle West and the East, the most widespread being *Sassafras* and wild allspice (*Lindera*). The family is more common in the Southern States. The oriental-camphor-tree (*Cinnamomum Camphora*) is introduced in tidelands from Texas to Florida and in the Florida hammocks. The avocado (*Persea*) is grown from California to Florida, but mostly in these two states. It is a native of tropical America, but it has escaped from cultivation to the pinelands and hammocks in south peninsular Florida and on the Keys. *Persea* (red-bay and swamp-bay), *Glabraria* (pondspice), *Nectandra* or *Ocotea* (lancewood), and *Misanteca* occur near the coast.

Family 20. **Hernandiaceae.**

HERNANDIA FAMILY

Trees or shrubs (or rarely woody vines). Leaves simple or palmate. Flowers bisexual or unisexual (and the plants monoecious or polygamous), corymbose or thyrsoid, epigynous; sepals (or sepals and petals) 3–5 in each of 2 cycles or sometimes in 1 cycle of 4–8; stamens commonly 4 (3–5); anthers splitting open lengthwise; staminodia glandlike, often in 1–2 cycles outside the fertile stamens; carpel 1, embedded deeply in the tissue of the receptacle. Fruit fundamentally an achene or a samara but modified by attachment to the receptacle.

The family occurs in the Tropics—Central America; West Indies; northern edge of South America; west central Africa; Madagascar; southern Asia; South Pacific. *Hernandia* is cultivated in southern Florida.

Suborder C. **Ranunculineae**

*Family 21. **Ranunculaceae.**

BUTTERCUP FAMILY

Commonly herbs, but sometimes woody vines or bushes (*Clematis*) or annuals or perennials. Leaves exceedingly variable (even within a single genus (e.g., *Ranunculus*), mostly simple but sometimes

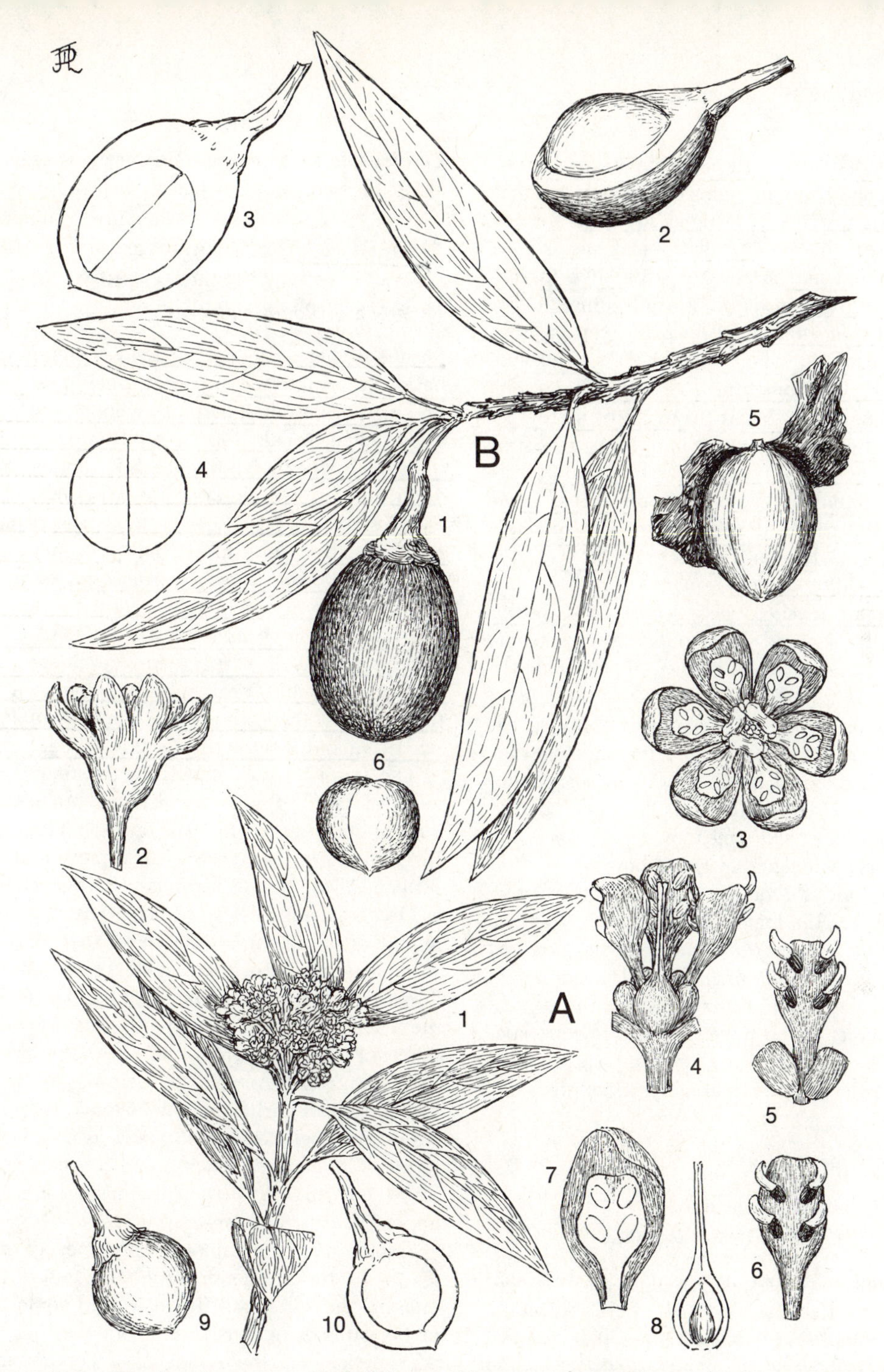

Figure 10-3. Ranales; Lauraceae; *Umbellularia californica,* mountain laurel, bay-tree, pepperwood, or (in Oregon) "myrtle": **A** flowers and young fruits, **1** flowering branch, **2** side view of a flower, **3** top view of a flower, **4** flower showing only the stamens, glands, and pistil, **5** stamen (showing the uplifted valves which cover the pores) and the glands basal to the stamen, **6** stamen, **7** sepal and stamen (in outline), **8** pistil, longitudinal section, **9** young fruit, **10** longitudinal section in outline; **B** fruit and seed, **1** branch with mature fruits, **2** fruit in longitudinal section, revealing the seed, **3** diagrammatic longitudinal section of the fruit and seed, showing the cotyledons, **4** cross section through the cotyledons, **5** old fruit with the seed exposed, **6** seed. (Cf. also **Figure 8-3 D** avocado, fruit.)

pinnate or palmate, entire to finely dissected, mostly alternate, but in some groups opposite or whorled; stipules rare (vestigial in some species of *Thalictrum*, the margins of the leaf base sometimes expanded and appearing stipulelike in *Ranunculus*). Flowers usually bisexual (unisexual in species of *Thalictrum* and *Clematis*); sepals present, often petaloid when the petals are absent, commonly 5, but sometimes 3–6; petals present or absent, commonly 5 but up to 23 or as few as 3 (calyx and corolla 3-merous in a number of species of *Ranunculus*, as in most of the woody families of the Ranales); stamens 5-numerous, commonly 20–50, arranged spirally; anthers splitting lengthwise, carpels usually numerous (rarely 1–4), commonly 5–300; ovules 1–many. Fruit usually a follicle or an achene, rarely a berry or capsule.

The family is world-wide, but it is most common in the northern Hemisphere and particularly northward. The largest genus is *Ranunculus*, which includes 98 species in North America including Mexico and Central America together with an almost equal number of varieties. Some other relatively common and important native genera are *Thalictrum* (meadow-rue), *Anemone* (windflower), *Clematis* (virgin's-bower), *Caltha* (marsh-marigold), *Trollius* (globeflower), *Coptis* (gold-thread), *Aquilegia* (columbine), *Delphinium* (larkspur), *Aconitum* (monkshood), *Actaea* (baneberry), and *Xanthorrhiza* (yellowroot).

Most members of the family match the typical floral characters of the Ranales, but a few are specialized. The flowers of the larkspurs and monkshoods in particular are remarkably bilaterally symmetrical.

Family 22. **Lardizabalaceae.**

LARDIZABALA FAMILY

Woody vines or sometimes shrubs. Leaves palmate or rarely pinnate, alternate. Flowers usually unisexual; sepals 3 or 6 in series of 3; petals, when present, 6 in series of 3, smaller than the sepals; stamens 6, sometimes coalescent basally; carpels 3–15; ovules usually many. Fruit berrylike, dehiscent lengthwise.

The family occurs in central and eastern Asia and in Chile. Some genera are cultivated in the United States (*Akebia* more generally; *Decaisnea*, *Lardizabala*, *Stauntonia*, and rarely *Sargentadoxa* along the southern margin of the United States).

*Family 23. **Berberidaceae.**

BARBERRY FAMILY

(Segregate: *Podophyllaceae*)

Shrubs or perennial herbs. Leaves simple or pinnately or ternately compound, deciduous or evergreen, alternate or sometimes wholly basal. Flowers bisexual; sepals 4–6, or rarely none, often like the petals; petals 4–6 or rarely none; stamens 4–18, the outer cycle opposite the petals, often in 2 series and commonly as many as the petals; anthers usually (and in all native genera except *Podophyllum* and rarely *Berberis*) opening by 2 pores each with a flaplike covering; carpel 1 (but according to some authors the result of previous coalescence of 3 carpels and reduction of 2); ovules 1–several. Fruit usually a berry or sometimes a follicle or the ovary (in *Caulophyllum*) withering and leaving the 2 seeds naked.

The family occurs mostly in the temperate regions of the Northern Hemisphere but it appears to a limited degree in the Southern Hemisphere. In western North America it is represented by the subgenus *Mahonia* of *Berberis* (barberry), known as Oregon-grape. These plants are used frequently in cultivation, being attractive for their dark green spiny leaves and for their clusters of bright yellow flowers as well as their bluish grapelike fruits. On the Pacific Coast there are two genera of herbs differing markedly from the shrubby barberries. In eastern North America the family is represented by the typical barberries and several other genera. An introduced species of *Berberis* is now being exterminated because it serves as a secondary host for wheat rust. Superficially these plants do not appear like the Oregon-grapes. However, the only character not in common is presence of pinnate leaves in the subgenus *Mahonia* and of simple ones in the subgenus *Berberis*, and some species of the latter subgenus show evidence that part of the "petiole" is the rachis of a compound leaf, which has retained only the terminal leaflet. (Cf. Ernst, Wallace R. Journal of the Arnold Arboretum 45: 1–35. 1964.) The May-apple (*Podophyllum*), the umbrella-leaf (*Diphylleia*), the twinleaf (*Jeffersonia*), and the blue-cohosh (*Caulophyllum*) are well-known eastern North American herbs.

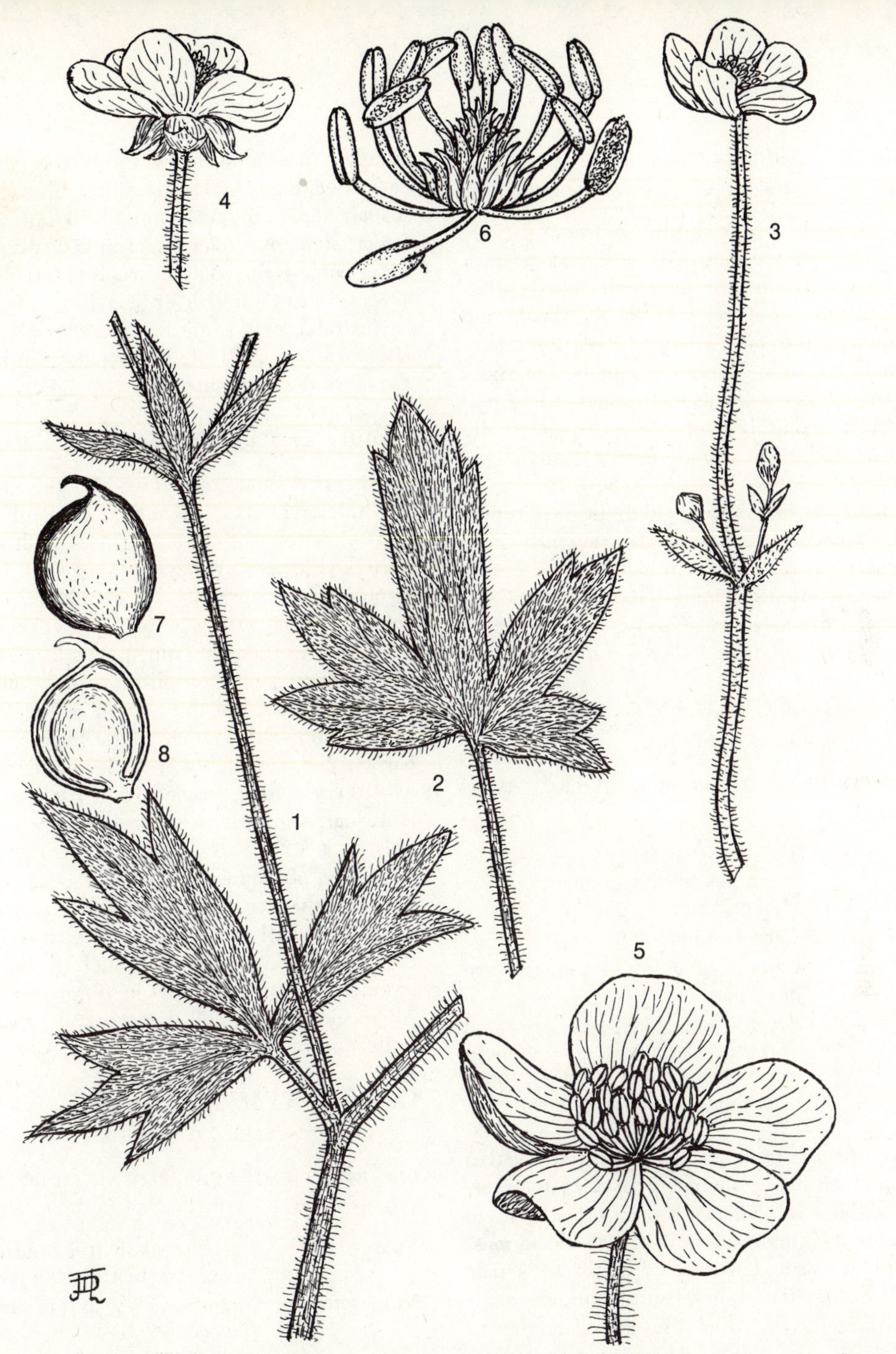

Figure 10-4. Ranales; Ranunculaceae; *Ranunculus occidentalis* var. *Eisenii,* a buttercup, flowers and fruit: **1** cauline leaf and bract, **2** basal leaf blade, **3** flowering branch, **4** flower, side view, **5** flower tipped to show petals (cf. also **Figure 9-7**) and stamens, **6** stamens and pistils, **7** single fruit (achene), **8** longitudinal section, showing the solitary seed. (Cf. also *Ranunculus* in **Figures 2-3 A** species with stolons, **8-4 B** fruits and seeds, and **9-6** to **9-12,** other species; *Delphinium* or larkspur, in **Figures 6-17 E** flower, **6-7 A** carpels, **8-6 D** fruit, and **9-14 A** flowers and fruits; *Aquilegia* or columbine in **Figure 9-14 B** flowers and fruits; *Aconitum* or monkshood in **Figure 9-14 C** flowers and fruits; *Caltha* or marsh-marigold, **Figure 6-1 C** flower.)

*Family 24. **Menispermaceae.** MOONSEED FAMILY

Woody vines or rarely shrubs or trees. Leaves simple or sometimes composed of 3 leaflets, usually simple and entire or palmately lobed, alternate. Flowers unisexual, very small; sepals and petals similar, in usually 3 series of 3 or sometimes of 2, small, greenish; stamens usually 6, sometimes coalescent; anthers with at least apparently 4 pollen chambers, splitting lengthwise; carpels usually 3–6; ovules 2 but one abortive. Fruit an achene or a drupe.

The family is distributed widely in the Tropics, and it extends into temparate regions. Four genera occur in eastern North America—*Cocculus* (coral-bead), *Menispermum* (moonseed), *Calycocarpum* (cupseed), and *Cissampelos* (on hammocks of Florida and the Everglade Keys).

Order 2. ***DILLENIALES.*** DILLENIA ORDER

DEPARTURES FROM THE FLORAL CHARACTERS OF THE RANALES

Free	Stamens sometimes coalescent at the bases of the filaments in the *Dilleniaceae*.
Numerous	Carpels 5 or 3–8.
Spiral	Carpels cyclic or in a single turn of a spiral.
Hypogynous	*Paeoniaceae* sometimes described as perigynous, but the stamens merely borne on a hypogynous disc, the flower hypogynous.

Herbs, shrubs, trees, or woody vines. Leaves simple and entire to pinnately or ternately dissected, without or sometimes with stipules, these when present forming wings and early deciduous. Flowers hypogynous, bisexual or unisexual; sepals 5; petals 5 (or 10); stamens numerous, developing centrifugally (i.e., the outermost first, the innermost last to be formed from the primordia); hypogynous stamen-bearing disc present or absent; carpels 5–several; ovules 1–many. Fruit a follicle or a berry.

Family 1. **Dilleniaceae.** DILLENIA FAMILY

Trees or shrubs or sometimes woody vines or herbs. Leaves simple, alternate, with or without stipules, these forming wings on the petioles and early deciduous. Flowers either bisexual or unisexual; sepals 5; petals usually 5 but sometimes fewer; stamens numerous, arranged spirally, sometimes coalescent basally; carpels several; ovules 1–several. Fruit a follicle or berrylike.

Australia and tropical America. *Dillenia* and *Hibbertia* are cultivated along the southern margin of the United States.

Family 2. **Paeoniaceae.** PEONY FAMILY

Herbs, perennial, rarely woody, the stems usually thick. Leaves dissected, commonly more or less pinnately or ternately twice dissected to parted. Flowers hypogynous, bisexual, bilaterally symmetrical, conspicuous; sepals 5, greenish or colored, irregular in size or shape, persistent and becoming leathery in fruit; petals 5 or sometimes 10, white to red or purplish or rarely yellow; stamens numerous, on the margin of the irregularly shaped and lobed rather fleshy hypogynous disc; carpels 2–8, commonly 5; ovules numerous. Fruit a leathery follicle formed from each of the separate carpels, with many seeds, these often with arils.

Mostly of Eurasia; two taxa occurring along and near the Pacific Coast of North America.

Until recently this genus has been included in the Ranunculaceae, from which it differs in a number of ways, mostly in obscure characters. Alliance with the Dilleniaceae and separation of both from the Ranales are open to debate.

Order 3. ***NYMPHAEALES.*** WATER-LILY ORDER

DEPARTURE FROM THE FLORAL CHARACTERS OF THE RANALES

Free	Carpels usually free but sometimes coalescent in the *Nymphaeaceae*.
Numerous	Stamens as few as 3 in some *Nymphaeaceae*, 10–20 in the *Ceratophyllaceae;* carpels as few as 3 in the *Nymphaeaceae*, only 1 in the *Ceratophyllaceae*.
Spiral	Stamens or carpels cyclic when the numbers are small (the carpel, of course, without arrangement when there is only 1).
Hypogynous	Flower epigynous in one section of *Victoria, Nymphaeaceae*.

Aquatic herbs, commonly perennial from stout rhizomes or free-floating and without rhizomes or roots; juice sometimes milky. Leaves simple, alternate and either peltate or appearing so (or sometimes dissected when submersed) or whorled and twice divided. Flowers hypogynous or rarely epigynous, bisexual or unisexual; sepals (or sepaloids), petals (or petaloids), and stamens intergrading in some genera, usually spiral; sepals 3, 4–5, 10–15, or numerous; petals 3–many or none; stamens 3–many, spiral or sometimes cyclic; anthers splitting lengthwise; carpel(s) 1 or 3–many, usually separate; ovules numerous or 1; fruit a follicle, nutlet (or achene), or leathery berry.

Key to the Families

(North America north of Mexico)

1. Perennial herbs with stout rhizomes; leaves alternate, large, cordate to peltate, floating (or sometimes large, submersed and dissected); sepals 3 or 4–6 or sometimes numerous; flowers large, bisexual; petals numerous or sometimes only 3–4 (then dull purple or white with a basal yellow spot); carpels 8–30; ovules numerous; fruit a follicle, nutlet, or leathery berry; ovules numerous (or rarely as few as 2); flower not subtended by an involucre.

   Family 1. **Nymphaeaceae** (water-lily family).

1. Stems free-floating, without rhizomes or even roots, slender; leaves submersed, whorled, small, once or twice dichotomously divided, the margins serrulate; flowers small, unisexual; sepals 10–15; petals none; carpel 1; ovule 1; fruit a spiny or tuberculate nutlet (or achene); flower subtended by an 8–12-cleft involucre.

   Family 2. **Ceratophyllaceae** (hornwort family).

*Family 1. Nymphaeaceae.

WATER-LILY FAMILY

(Segregates: *Cabombaceae, Nelumbaceae*)

Aquatic herbs, usually perennial from stout rhizomes; juice sometimes milky. Leaves simple, peltate or appearing so, floating or standing above the water or in *Cabomba* most leaves submersed and dissected, alternate. Flowers bisexual, solitary, floating, the pedicels long, arising to the surface of the water, the parts trimerous or spiral; sepals, petals, and stamens in some genera intergrading; sepals 3, 4–5, or numerous; petals 3–many; stamens 3–many, cyclic or arranged spirally (the stamens of some genera similar to those of the primitive woody Ranales [see pp. 562–564]); anthers splitting lengthwise; carpels 3–many, usually free but sometimes coalescent; ovules many; ovary rarely inferior (one section of *Victoria*). Fruit a follicle, a nutlet, or a leathery berry; the fruits sometimes in pockets in the top of the broad receptacle.

The family is distributed over most of the land masses of the Earth. Several genera occur in the eastern and northern portions of North America, and two species extend down the Pacific Coast to California. The large yellow pond-lilies or water-lilies are of the genus *Nuphar* and the small white water-lilies are *Nymphaea*, but in the older books the name *Nymphaea* is applied to the yellow water-lilies instead, while *Nymphaea* appears as *Castalia*. This point should be watched with care.

Figure 10-5. Nymphaeles; Nymphaeaceae; *Nymphaea* (hybrids), water-lily leaves and flowers. Photographs by Ladislaus Cutak.

Figure 10-6. Nymphaeales; Nymphaeaceae; *Nymphaea* (hybrid), water-lily: **1** leaf, from beneath, **2** opening flower bud, longitudinal section, **3** flower bud, cross section, **4** young flower, from above, **5** flower, from the side, showing the intergrading sepals and petals, **6** pistil, cut longitudinally, **7** ovary in cross section, **8** the same, one carpel enlarged. (Cf. also **Figures 5-3 C** *Nuphar polysepalum,* transition of sepals to petals to stamens, and **8-1 B** *Nelumbo lutea,* fruit.)

Other genera are *Nelumbo* (including the Oriental sacred-lotus) which is common in cultivation, *Brasenia* (water-shield), *Cabomba* (fanwort), and *Victoria* (noteworthy for its enormous floating leaves with rimmed margins).

Segregations and exclusions from the family are not fully evaluated. Mosely, a thorough student of the group, excludes *Nelumbo* from the order and segregates the *Barclayaceae* (*Barclaya*, 3 species in the Indo-Maylan region) and the *Cabombaceae* (*Cabomba* and *Brasenia*) as separate families within the Nymphaeales. Though probably this is justified, the problem has not been studied sufficiently for an evaluation to be made here.

*Family 2. **Ceratophyllaceae.** HORNWORT FAMILY

Aquatic herbs. Roots none, the plants free-floating. Leaves in whorls, divided once or twice dichotomously, the margins serrulate. Flowers unisexual, very small, solitary; sepals 10–15; petals none; stamens 10–20, arranged spirally; anther splitting lengthwise; carpel 1; the style elongate; ovule 1. Fruit a nutlet, the style persistent.

The family is composed of a single genus, *Ceratophyllum*, which occurs in lakes, ponds, and slow streams throughout the world.

Order 4. ***PIPERALES.*** PEPPER ORDER

DEPARTURES FROM THE FLORAL CHARACTERS OF THE RANALES

Free	Carpels sometimes coalescent.
Numerous	Stamens 1–10; carpels 2–4 or rarely 5.
Spiral	Stamens and carpels cyclic.
Hypogynous	Flowers epigynous in some *Sauruaceae*.

Herbs, shrubs, or rarely trees. Leaves alternate, opposite, or sometimes whorled; stipules present or absent; the leaves simple and entire. Flowers bisexual or sometimes unisexual, in spikes, small, individually inconspicuous, each in the axile of a small bract, with no perianth; anthers splitting lengthwise; carpels 1–4 or sometimes 5, free or (in North America) coalescent; ovule most frequently solitary, with 1–2 integuments.

Engler and Prantl considered the Piperales to be among the most primitive flowering plants, and they placed them at the beginning of the series. Rendle believed them to be related to the Polygonales (segregated from the Caryophyllales) because he thought the flower to be fundamentally 3-merous. It is to be noted that in both the *Polygonaceae* and the Piperales the ovule is characteristically solitary (not always so in Piperales) and orthotropous.

Key to the Families

(North America north of Mexico)

1. Flowers *not* each covered by a peltate, scalelike bract; vascular bundles of the stem in a single cylinder.

    Family 1. **Saururaceae** (lizard's-tail family).

1. Flowers each covered by a peltate, scalelike bract (in the North American species); vascular bundles of the stem "scattered" as in the Monocotyledoneae.

    Family 2. **Piperaceae** (pepper family).

*Family 1. **Saururaceae.** LIZARD'S-TAIL FAMILY

Perennial herbs of moist places. Leaves simple, usually ovate; stipules adnate with the petioles. Flowers bisexual, in dense spikes or racemes, the lower bracts sometimes large and white or colored; sepals and petals none, each flower subtended by a scalelike bract, the bract not peltate; stamens 2–8; anther with 2 pollen chambers; carpels 3–4, separate or (in North America) coalescent, when coalescent the placentae parietal or (in *Saururus*) axile; ovules 1–10 per carpel. Fruit a fleshy follicle or capsule, dehiscent apically. (Cf. Figure 6-1 **B** lizard's-tail, flower.)

The family consists of only three genera and about four species occurring in Southeastern Asia and in the United States and Mexico. It is represented in the United States by *Anemopsis*, the yerba mansa, which ranges from the warmer parts of California to Texas and southward into Mexico, and *Saururus*, the lizard's-tail, which occurs from southern Michigan and southeastern Kansas to southwestern Quebec, Texas, and Florida.

*Family 2. **Piperaceae.** PEPPER FAMILY

Herbs, shrubs, vines, or rarely trees; herbaceous plants sometimes succulent. Leaves simple, entire, sometimes succulent, alternate or rarely opposite or in cycles; stipules absent or present and adnate with the petiole. Flowers usually bisexual, minute, with bracts (the bracts in the North American genus peltate), usually in dense succulent spikes; sepals and petals none; stamens 1–10; anthers with 2 pollen chambers or these fused into 1; carpels 2–5, coalescent into a single pistil, the stigmas 1–5. Fruit a drupe.

The family occurs in the Tropics of both hemispheres. Three or more species of *Peperomia* occur in south peninsular Florida or on the Keys. These are relatively small herbaceous plants, some of them growing on the ground and others in trees (as epiphytes). Terrestrial species are succulent. The minute flowers are more or less obscured by flattened (peltate) bracts.

Pepper is produced from the berries of *Piper nigrum,* black pepper from the dried berries, white pepper from the dried seeds alone. The species is native in the East Indies, but it is cultivated widely through the Tropics. The plant is a climbing shrub with ovate leaves. The red berries are produced in spikes.

Order 5. *ARISTOLOCHIALES.*

BIRTHWORT ORDER

DEPARTURE FROM THE FLORAL CHARACTERS OF THE RANALES

Free	Sepals sometimes coalescent; stamens sometimes adnate with the style and forming a central column; carpels coalescent.
Equal	Calyx sometimes bilaterally symmetrical.
Numerous	Stamens as few as 6, but up to 36 or more; carpels 4–6.
Hypogynous	Flowers epigynous.

Decumbent or creeping herbs or woody vines. Leaves simple, entire, sympetalous, alternate. Flowers bisexual, radially or bilaterally symmetrical, epigynous; sepals present, coalescent, usually lurid or otherwise unusually colored, sometimes malodorous; calyx sometimes resembling the bowl of an old-fashioned Dutch pipe; petals none or in *Asarum* perhaps vestigial; stamens 6–26, sometimes adnate with the style, the filaments short and thick or elastic; carpels usually 4–6, the style 1, short, the ovary 4–6-chambered; placentae axile; ovules numerous in each carpel. Fruit a capsule.

The relationships of the Aristolochiales are in doubt. Bessey placed them in the Myrtales, and relationship to the Myrtales may be possible. Some have considered the order related to the Ranales, in view of the 3-merous (6-merous?) flowers, the anther type, and the presence of secreting cells in the leaves (these resembling somewhat those in the *Magnoliaceae* and other families of the Ranales).

*Family 1. **Aristolochiaceae.**

BIRTHWORT FAMILY

(Synonym: *Asaraceae*)

The family occurs throughout the Tropics, and limited numbers of species occur in temperate regions. The best-developed genus in North America is *Asarum,* the so-called wild-ginger. Species occur in most of the moist parts of the Continent; and their characteristic, attractive, heart-shaped leaves are one of the features of moist forest floors or stream banks. *Aristolochia,* the birthwort or Dutchman's-pipe, is represented by species in eastern North America and by one in California and another in Arizona. The plant is interesting for its markedly bilaterally symmetrical flowers which resemble an old-fashioned Dutch pipe.

Order 6. *PAPAVERALES.* POPPY ORDER

DEPARTURE FROM THE FLORAL CHARACTERS OF THE RANALES

Free	Carpels coalescent, the other parts rarely so.
Equal	Flowers bilaterally symmetrical in *Corydalis, Fumariaceae* (a spur on one petal) and slightly so in the *Cruciferae,* (deletion of 2 stamens from the outer series of 4).
Numerous	Stamens ranging from numerous to 6 or rarely fewer; carpels from numerous to 2.
Spiral	In groups with few stamens or carpels, these parts cyclic.
Hypogynous	Rarely perigynous (e.g., *Eschscholtzia* in the *Papaveraceae* and the *Moringaceae*); hypogynous disc present in the *Moringaceae, Resedaceae,* and *Bretschneideraceae.*

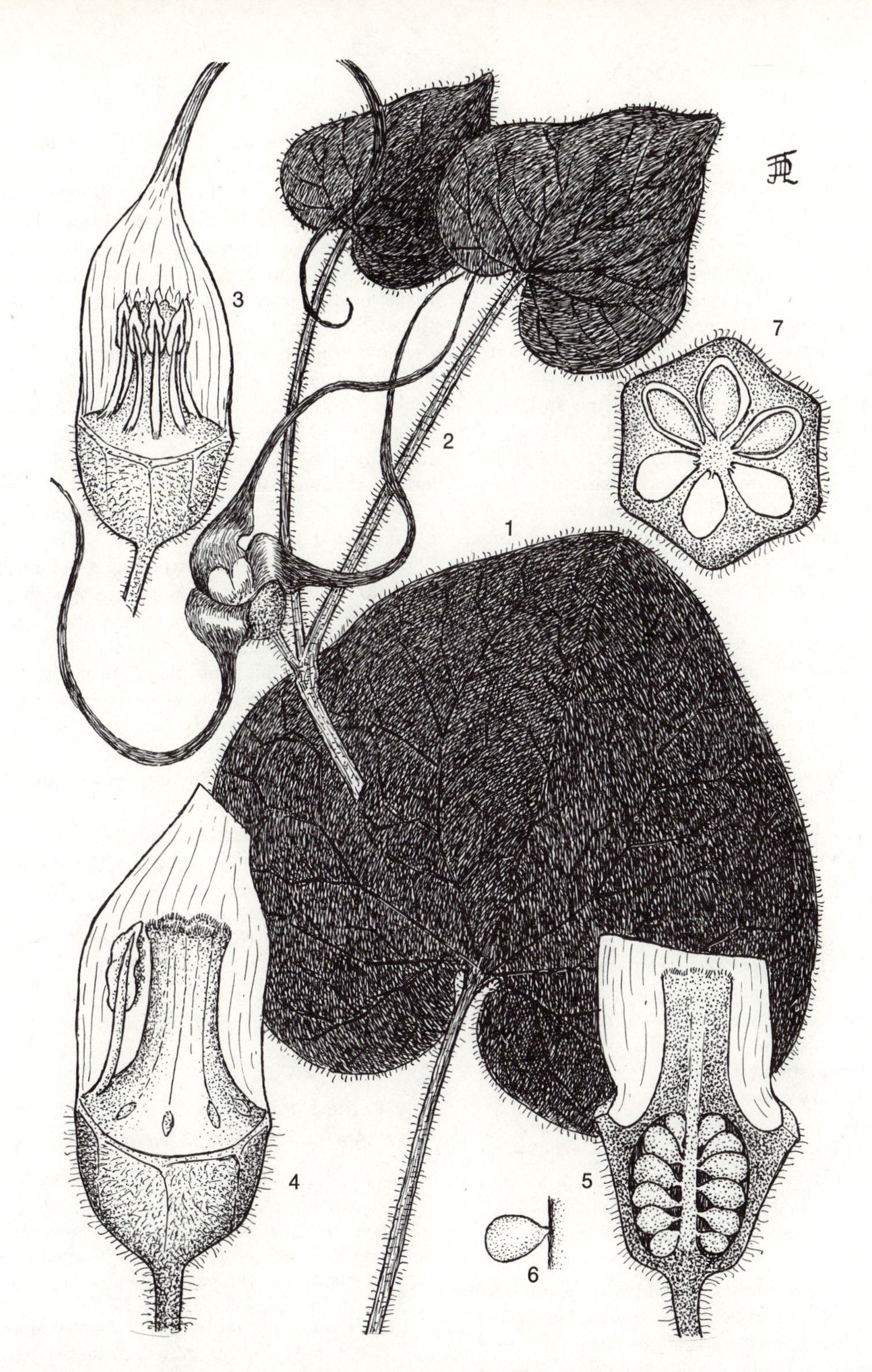

Figure 10-7. Aristolochiales; Aristolochiaceae; *Asarum caudatum,* a wild ginger or birthwort: **1** basal leaf, **2** branch with leaves and a flower, showing the long-attenuate sepals, **3** flower, two sepals removed, showing the inferior ovary and the adnate floral cup, the stamens, and the style and stigma, **4** the same, all but one stamen removed, showing the stigma more clearly, **5** longitudinal section of the ovary, **6** attachment of an ovule, **7** cross section of the ovary.

Usually herbs but sometimes trees or shrubs. Leaves usually alternate; stipules none. Flowers bisexual; sepals usually 2, 3, or 4; petals usually 4 or 6, rarely 8–12; anthers splitting lengthwise; placentae parietal; ovules anatropous or, in the *Capparidaceae,* campylotropous. Fruit usually a capsule.

One of the most primitive members of the order Papaverales is the cream-cup (*Platystemon*) of California. This plant resembles many of the Ranales in having the sepals and petals in cycles of 3. Furthermore, although the carpels are lightly coalescent at flowering time, at fruiting time they separate readily and promptly. They are relatively numerous (6–17 or rarely as many as 20). If the carpels were not coalescent, this genus would fulfill all the typical floral characters of the Ranales, and it could not be excluded from that order.

Most Papaverales differ from the Ranales in both the coalescent carpels with parietal placentae and the 2- or 4-merous flowers (instead of 3- or 5-merous). The latter character tends strongly to distinguish the Papaverales from the Malvales, Geraniales, Theales, and most other orders of Thalamiflorae, as well as the Ranales.

Proceeding from primitive types, like the cream-cup, there are various minor developmental series in the poppy family. A generalized and a specialized floral type are as follows.

Generalized. Papaver includes perhaps fifty species, most of them native in the Eastern Hemisphere, but a few in western North America. The carpels are coalescent by both the ovaries and the styles, but the stigmas are free. At maturity the compound ovary opens by pores at the apex rather than by splitting as in most genera. The seeds are arranged along the sides of the ovary on parietal placentae. Except for the coalescence of carpels and their cyclic arrangement, this plant satisfies all the floral characters of the Ranales.

Specialized. Eschscholtzia, which is known best for its most highly specialized species, the California poppy. The receptacle is elaborated by a vaselike basal portion and a distal platform (torus) spreading at right angles to the axis of the stem. Thus the flower is perigynous. The two sepals are coalescent into a fool's cap covering the bud. This becomes detached from the receptacle at flowering time (is caducous as are the sepals of all the members of the poppy family), and it slides off the petals as the flower opens. The California poppy is common not only in California but in certain areas of Arizona and Oregon, and it has been known long in cultivation throughout the world.

A much more highly specialized family departing considerably from the floral characters of the Ranales is the Cruciferae or the mustard family. Essentially the flower is 4-merous, but there are six stamens in two cycles, the inner series of four longer stamens, the outer of two shorter ones. Although the placentae are parietal, the ovary is 2-chambered by development of a "false partition" of complex origin extending from one placenta to the other. In a single rare species (*Tropidocarpum capparideum*) occurring in the San Francisco Bay Region and the Mt. Diablo Range and on adjacent alkaline plains in northern California, there are four developed carpels and no false partitions in the ovary.

Departure of the typical *Cruciferae* from the floral characters of the Ranales are as follows:

Free	Carpels coalescent.
Equal	Stamens 4 in the inner cycle, but only 2 in the outer.
Numerous	Stamens 6 (or rarely fewer); carpels 2 functional and 2 rudimentary or rarely all four functional.
Spiral	Stamens and carpels cyclic.
Hypogynous	Rarely perigynous.

Key to the Families

(North America north of Mexico)

1. Plant *not* corresponding with *every character* in the combination in lead 1 below.
   2. Fruit *not* with a *stipe* of ovary or receptacle tissue above the pedicel; leaves *not* palmate.
      3. Plant *not* corresponding with every character in the combination in lead 3 below.
         4. Fruit *not* opening at the apex before maturity and full growth of the seeds.
            5. Petals equal and similar, separate; stamens numerous or rarely 4–12, separate.

               Family 1. **Papaveraceae** (poppy family).

5. Petals in 2 series of 2, the series dissimilar, often coalescent; stamens 6, coalescent.

Family 2. **Fumariaceae** (fumitory family).

4. Fruit opening at the apex before maturity and full growth of the seeds; corolla or the stamens bilaterally symmetrical, being turned to one side of the flower.

Family 3. **Resedaceae** (mignonette family).

3. Desert shrubs; branches all rigid and thorn-pointed; leaves small, scalelike; stamens 5; fruit a small, black, shining berry.

Family 5. **Koeberliniaceae** (desert crucifixion-thorn family).

2. Plants with *one or both* of the following characters: (1) fruit with a stipe of ovary or receptacle tissue above the pedicel *or* (2) leaves palmate.

3'. Carpels 3; leaves bipinnate or tripinnate.

Family 4. **Moringaceae** (horseradish-tree family).

3'. Carpels 2; leaves palmate or occasionally simple.

Family 7. **Cappardidaceae** (caper family).

1. Capsule 2-chambered by development of a false partition in the ovary from the middle of one of the 2 parietal placental areas to the middle of the other; sepals 4; petals 4 (rarely none); stamens 6 (or rarely fewer), the inner series of 4 long ones and the outer of 2 almost always shorter than the inner but sometimes equal to them.

Family 8. **Cruciferae** (mustard family).

Suborder A. **Papaverineae**

*Family 1. **Papaveraceae.** POPPY FAMILY

Annual or perennial herbs or rarely trees or shrubs, usually with milky or colored juice. Leaves alternate or rarely whorled, entire to pinnately or palmately parted or finely ternately dissected. Flowers usually solitary, usually large and conspicuous; sepals 2–3, usually separate but coalescent and forming a cap (calyptra) in *Eschscholtzia,* falling away as the flower opens (caducous); petals 4–6 or sometimes 8–12, rarely none, equal and similar whether in 1 series or 2; stamens usually numerous and arranged spirally, sometimes few and cyclic; carpels 2–several, sometimes as many as 20, the ovary usually 1-chambered and with parietal placentae, rarely with several chambers through intrusion of the placentae; ovules usually numerous.

The family is common in temperate and subtropical regions of the Northern Hemisphere (particularly in eastern Asia and California). In California there are nine native genera, the chief one *Eschscholtzia* (California poppy) with about seven species. Others are *Platystemon* (cream-cup), *Argemone* (prickly poppy), *Papaver* (wind poppy), *Romneya* (matilija poppy), and *Dendromecon* (bush poppy). The last two genera are shrubs; in the Eastern Hemisphere the family includes a few trees. In eastern North America the poppy family is represented by the bloodroot (*Sanguinaria*), the wood poppy (*Stylophorum*), and a number of species introduced from the Eastern Hemisphere.

The opium poppy, *Papaver somniferum,* and species of *Argemone* yield opium and morphine.

*Family 2. **Fumariaceae.**

FUMITORY FAMILY

(Often included in the *Papaveraceae*)

Herbs or rarely vines. Leaves usually dissected into narrow segments, alternate or rarely opposite. Flowers bilaterally symmetrical, commonly in racemes; sepals 2, minute, falling as the flower opens (caducous); petals 4 in 2 series, all 4 or the 2 inner often coalescent or slightly so, one or both in the outer series usually with a basal sac or spur, the inner crested; stamens 6, 4 having anthers with 1 pollen chamber (perhaps an inner series) and 2 with 2 pollen chambers (outer series), seemingly in groups of 3; filaments winged, coalescent; carpels 2, the ovary with 1 chamber, the placentae parietal; ovules 2–many. Fruit a capsule or in *Fumaria* a nutlet.

The family is largely of the Eastern Hemisphere. The most widespread genera in North America are *Dicentra,* bleeding-heart, and *Corydalis,* both of which occur across the continent. The fumitory (*Fumaria*) is introduced sparingly. The climbing fumitory or mountain-fringe (*Adlumia*) occurs through southeastern Canada and the adjacent United States and southward into the mountains of Tennessee and North Carolina.

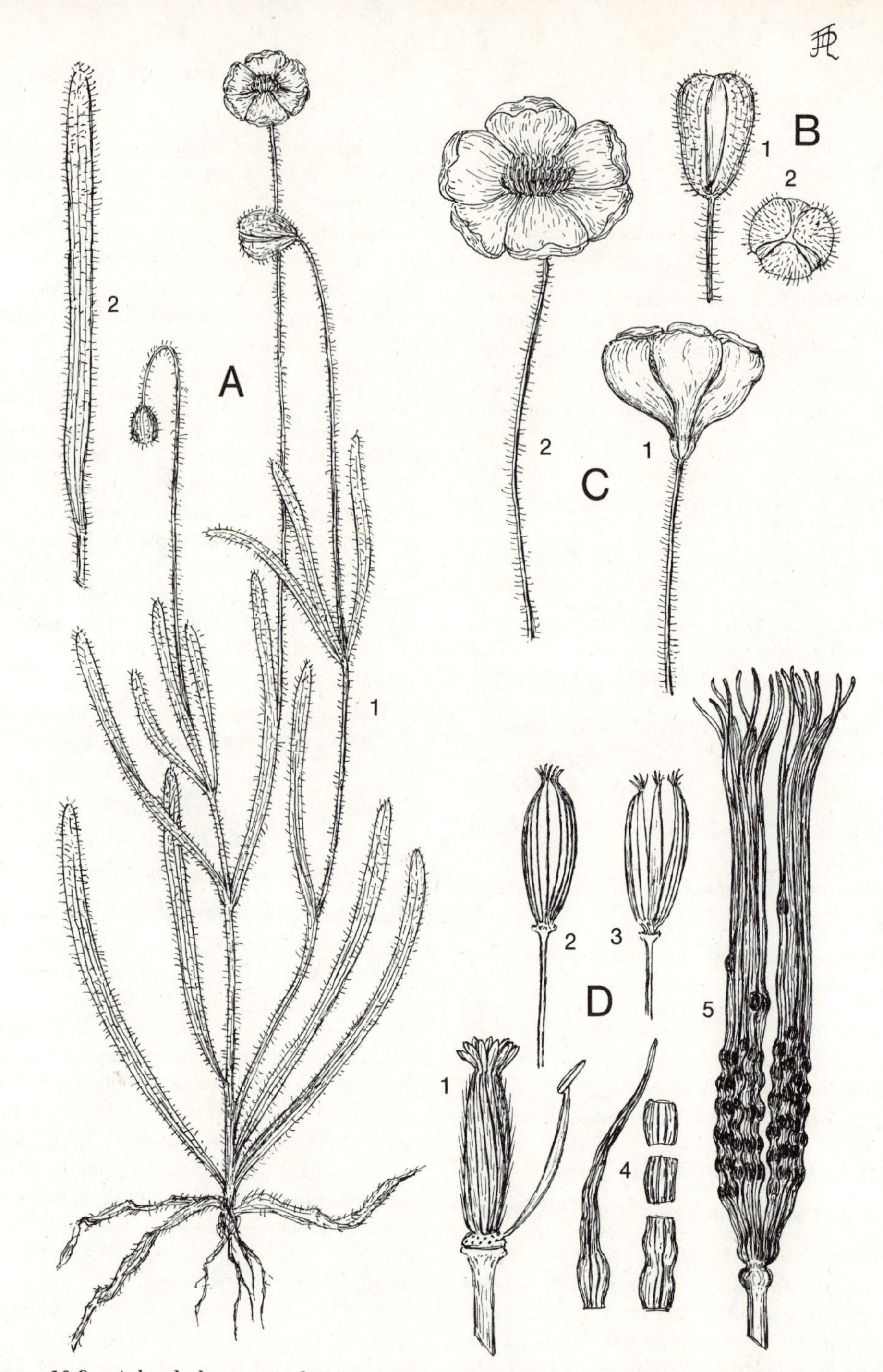

Figure 10-8. A borderline genus between the Papaverales and the Ranales—Papaverales, Papaveraceae, *Platystemon californicus,* creamcup, the sepals and petals 3-merous as in numerous Ranales, and a few Papaveraceae, the sepals caducous (falling at the opening of the flower, as in the Papaveraceae), the carpels rather numerous (usually ten to twenty) and lightly coalescent in flower but separating at fruiting time, the stamens numerous and arranged spirally as in nearly all the Ranales and most of the more primitive Papaverales (Papaveraceae): **A, 1–2** habit; **B, 1–2** buds, protected by the sepals; **C, 1–2** flowers, the sepals having fallen; **D** fruits, **1–3** separation of the carpels, **4** carpel and its segments enlarged, **5** old carpels bulging conspicuously in the areas of the mature seeds and no longer clearly coalescent.

Figure 10-9. Papaverales; Papaveraceae; *Romneya Coulteri,* matilija (mȧ-tĭl′ē-hä) poppy: **A** habit, including a bud, a flower, and a young fruit; **B** buds emphasizing the calyx, which is caducous, i.e., each sepal (**3**) falling away as the flower opens; **C** attachment of flower parts, **1** flower past anthesis with the parts falling away, **2** enlargement of the stamen scars showing the spiral arrangement; **D** fruits, **1–2** upper portions of fruits showing the sessile stigmas, **3** basal portion of a fruit and the receptacle, showing the scars left at the attachment points of fallen flower parts, **4** longitudinal section of a fruit, **5** cross section, showing the intruded parietal placentae, which may form partitions.

This beautiful and delicate herbaceous vine (biennial) scrambles over bushes, holding on by means of its slender petioles. It is cultivated, as well.

Suborder B. **Capparidineae**

*Family 3. **Resedaceae.** MIGNONETTE FAMILY

Annual or perennial herbs; rarely somewhat shrubby. Leaves simple or pinnately divided, alternate; stipules represented by glands. Flowers usually bisexual but sometimes unisexual (and then the plants monoecious), bilaterally symmetrical, racemose or spicate; sepals usually 4–7 or 8; petals 2–7 or 8 or sometimes wanting, 2 usually much larger than the others and sometimes lobed; stamens 3–40, on an asymmetrical hypogynous disc; carpels 2–6, the stigmas as many as the carpels; ovary open at the top before complete enlargement and maturity of the seeds; ovules many. Fruit a capsule or a berry.

Most of the species in the family are of the genus *Reseda* (mignonette) of the Mediterranean Region; some of these are introduced in various parts of North America. *Oligomeris linifolia,* a small weedy annual widespread in warm regions, is native in North America from the drier parts of California to Texas and Mexico; its relatives are African.

*Family 4. **Moringaceae.** HORSERADISH-TREE FAMILY

Trees. Leaves bipinnate or tripinnate, alternate; stipules, when present, reduced to glands. Flowers bisexual, bilaterally symmetrical, in cymose panicles, perigynous, the small floral cup partly a hypanthium and partly a "calyx tube," the flower with a perigynous disc lining the floral cup; sepals 5, unequal; petals 5, 2 small and reflexed, the one opposite them largest; stamens 5, unequal; anthers with 1 pollen chamber; staminodia 3–5, filiform; carpels 3, the style and stigma 1, the ovary on a stipe above the receptacle, 1-chambered, the placentae parietal. Fruit an elongated 3-angled capsule; seeds many, sometimes winged.

The family consists of a single genus, *Moringa* (horseradish-tree), with several species. One of these, a tree about 30 feet high introduced from southern Asia, has escaped from cultivation in south peninsular Florida and on the Florida Keys. The seeds produce oil; the roots have a horseradish odor.

*Family 5. **Koeberliniaceae.** DESERT CRUCIFIXION-THORN FAMILY

Shrubs or trees; plants not glandular or aromatic, the branches either hairy or with lines of wax flakes. Leaves simple, very small, scalelike, ephemeral, alternate; stipules none. Flower bisexual, radially symmetrical, with a hypogynous disc; sepals 5; petals 5; stamens 5 or 10; carpels 2 or 5; the ovaries 5- or 10-chambered. Fruit a capsule, dehiscent along 5 or 10 lines, or the fruit a berry (then composed of 2 carpels).

The family includes only one genus, *Koeberlinia,* restricted to the Mexican desert flora. Three unrelated desert genera are known in the United States as desert crucifixion-thorns, and the plants of each are composed almost wholly of sharp-pointed thorns, the leaves being small, simple, and inconspicuous. The other genera are *Castela* and *Canotia.*

Family 6. **Tovariaceae.** TOVARIA FAMILY

Odorous annual herbs. Leaves composed of 3 leaflets, membranous in texture, alternate; stipules none. Flowers bisexual, in terminal racemes; sepals 8, early deciduous; petals 8; stamens 8; carpels 6–8, the ovary with 6–8 chambers, on a short stipe above the receptacle; the placentae axile, ovules numerous. Fruit a berry.

The family consists of a single genus, *Tovaria,* of Tropical America including the West Indies.

*Family 7. **Capparidaceae.** CAPER FAMILY

Herbs, shrubs, trees, or sometimes woody vines. Leaves simple or palmate (sometimes with only 1 leaflet developed), alternate; stipules very small. Flowers bisexual or unisexual (and then the plants monoecious), either radially or bilaterally symmetrical, commonly racemose; sepals usually 4 but up to 8; petals 4 or sometimes none or many, 2 sometimes larger than the others; stamens 4 or sometimes more through division of the basic 4 primordia; anthers with 2 or 4 pollen chambers;

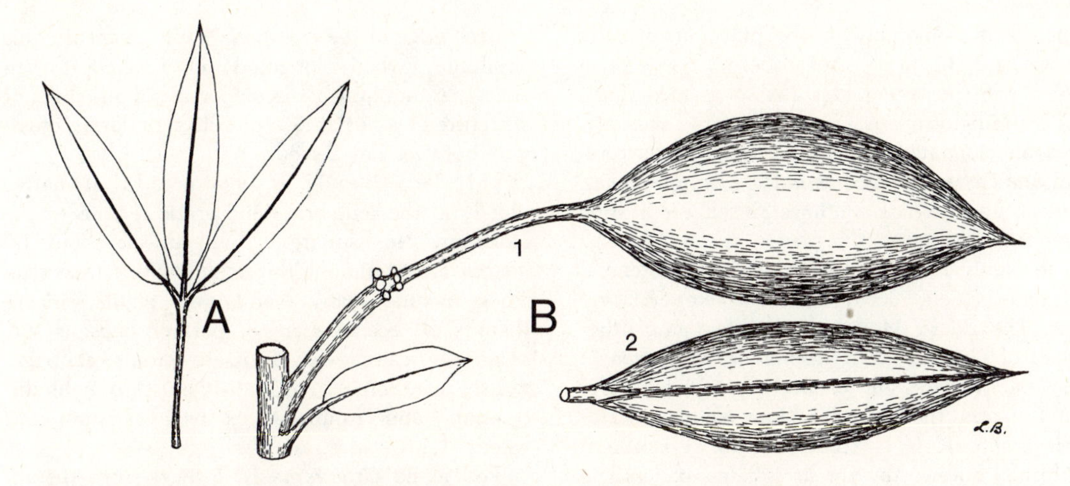

Figure 10-10. Papaverales; Capparidaceae; *Cleome Isomeris* (*Isomeris arborea*), burrofat, emphasizing the long stipe formed from the basal portion of the fruit, this *above* the receptacle: **A** leaf; **B, 1–2** fruit in two views.

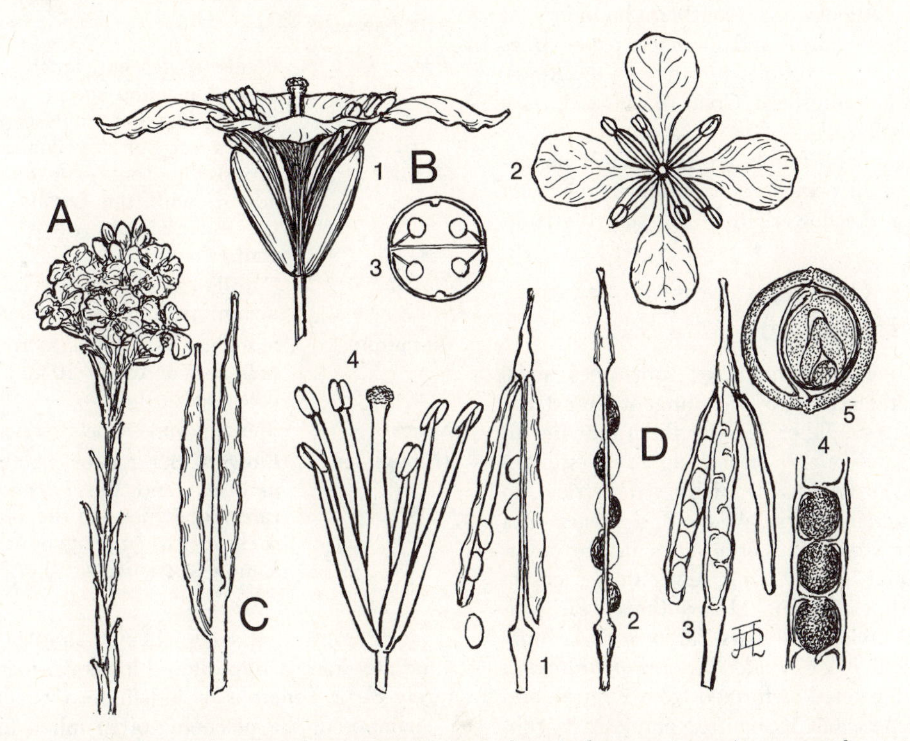

Figure 10-11. Papaverales; Cruciferae; *Brassica geniculata* (*adpressa, incana*), perennial mustard: **A** top of an inflorescence; **B** flower, **1** side view, **2** top view of the flower, **3** diagrammatic cross section of the ovary, drawn as though the ovules representing four rows were at the same level, **4** the six stamens (the four inner long, the two outer short) and the pistil; **C** young fruits; **D** mature fruits, **1–3** dehiscence, **4** longitudinal section, **5** cross section. (Cf. also **Figure 6-3 B** *Brassica Rapa*, flower.)

carpels 2 or sometimes 4, the placentae usually parietal and the ovary 1-chambered, borne on a stipe above the receptacle, this sometimes short; ovules campylotropous. Fruit a capsule or sometimes an elongated berry or a nut. Seeds reniform, the embryo curved as in the Caryophyllales. Stamens and pistil sometimes raised on a stipe above the receptacle.

The family includes numerous tropical genera, and some species occur in warm temperate regions. The Rocky Mountain bee-plant and other species of *Cleome* occur through a wide area in the Western States and on the Great Plains. *Cleome Isomeris,* the burro-fat (long known as *Isomeris arborea*), is a striking shrub of southern California, noteworthy for its inflated pods. The spider-plant, a spiny species, is introduced from cultivation in the eastern half of the Continent. Other members of the family are the jackass-clover (*Wislizenia, Southwest*), the clammyweed (*Polanisia,* widely distributed except on the Pacific Slope), *Atamisquea* (southern boundary of Arizona), the bayleaf and Jamaica caper trees (*Capparis,* south peninsular Florida and the Keys), and *Cleomella* (southern Great Basin and from California to Texas).

The pistils of this family are remarkable for development of an elongated stipe above the main receptacle of the flower, thus elevating the fruit.

*Family 8. **Cruciferae.** MUSTARD FAMILY
(Synonym: *Brassicaceae*)

Annual to perennial herbs or sometimes small shrubs; characterized by a pungent "mustard" taste. Leaves simple or some of them pinnate, alternate to rarely more or less opposite; stipules none. Flowers bisexual, radially symmetrical, in racemes; stamens nearly always 6, 4 longer ones in the inner series, 2 shorter ones in the outer, rarely of equal length, rarely fewer than 6 or (in an extraterritorial genus) 16; anthers nearly always with 2 pollen chambers, rarely with 1; functional carpels 2 (4 in *Tropidocarpum capparideum*), the pistil interpreted as formed from 2 fertile and 2 sterile carpels, but its nature complex and subject to debate, the ovary with a "false" partition apparently from the middle of one parietal placenta to the middle of the other, the 2 chambers each with 2 widely separated rows of ovules, one at each edge of the partition. Fruit a capsule, this a silique (when elongated) or a silicle (when short), sometimes a discoid or ovoid indehiscent structure or a nutlet or a pod disarticulating crosswise between any 2 seeds.

This large family is widespread but native chiefly in the temperate and arctic regions of the Northern Hemisphere. Members of about 50 genera are abundant throughout North America. They include many well-known plants, among them such economically important ones as turnips, kale, cauliflower, Brussels sprouts, cabbage, radish, water cress, horseradish, mustard, kohlrabi, broccoli, and rutabaga, as well as numerous weeds.

The name *Cruciferae* is derived from the resemblance of the 4 petals to a cross.

Order 7. ***VIOLALES.*** VIOLET ORDER

DEPARTURE FROM THE FLORAL CHARACTERS OF THE RANALES

Free	Petals coalescent in the *Achariaceae,* usually so in the *Caricaceae;* filaments coalescent in the *Caricaceae,* basally coalescent in groups in some *Flacourtiaceae,* adnate with the corolla tube in the *Achariaceae;* carpels coalescent.
Equal	Corolla and stamens bilaterally symmetrical in the *Violaceae.*
Numerous	Stamens 3–10 or numerous; carpels 2–5 or up to 10 in the *Flacourtiaceae.*
Spiral	Stamens and carpels cyclic.
Hypogynous	Flowers perigynous in the *Turneraceae* and the *Passifloraceae,* rarely epigynous in the *Flacourtiaceae,* with a hypogynous disc in some *Flacourtiaceae.*

Herbs, shrubs, or trees. Leaves simple, usually alternate, rarely opposite. Flowers 5-merous or rarely 3- or 4-merous or indefinite; ovary usually 1-chambered, the placentae often much intruded and the ovary sometimes therefore divided into chambers and with apparently axile placentae, in the *Passifloraceae* the intruded placentae parietal but fundamentally axile; ovule anatropous.

This order is related to the Theales. It has few and cyclic stamens and usually parietal placentae (sometimes also parietal in the Theales). Its departure from Ranalean floral characters is generally greater than in the Guttiferales. Some members of the *Flacourtiaceae* appear to be nearest the probable ancestral type of the order.

Key to the Families

(North America north of Mexico)

1. Plants *not* small trees with unbranched trunks or with milky juice, or with very large palmately lobed leaves; fruit *not* a very large, melonlike berry; petals separate.
    2. Stamens numerous, more than 10, arranged spirally (or at least a cyclic arrangement not detectable).
        3. Stipules none; anthers splitting lengthwise, collapsing as the pollen is shed; exocarp and endocarp not obviously distinguished.

            Family 1. **Cistaceae** (rockrose family).

        3. Stipules present; anthers opening by terminal pores, not splitting, not collapsing as the pollen is shed; fruit with a transparent endocarp through which the seeds may be seen after the falling or removal of the exocarp.

            Family 3. **Cochlospermaceae** (*Cochlospermum* family)

    2. Stamens 3–10, in 1 or 2 cycles.
        3'. Flowers hypogynous, bilaterally symmetrical, the lower petal with a spur or sac.

            Family 7. **Violaceae** (violet family).

        3'. Flowers perigynous, radially symmetrical, the petals without spurs or sacs.
            4. Flower not with a fringed crown arising from the floral cup; herbs or shrubs; tendrils none; stigmas brushlike.
            Family 8. **Turneraceae** (*Turnera* family).

            4. Flower with a conspicuous fringed crown (corona) between the petals and the stamens, this developed from the floral cup and bearing filiform appendages; perennial vines; tendrils present; stigmas not brushlike.

                Family 10. **Passifloraceae** (passion-flower family).

1. Plants small trees with unbranched trunks, with milky juice, and with very large palmately lobed leaves; fruit a very large, melonlike berry, edible; petals separate; stamens 5 or 10.

    Family 12. **Caricaceae** (papaya family).

*Family 1. Cistaceae. ROCKROSE FAMILY

Herbs or shrubs. Leaves simple, usually entire, opposite or alternate or the lower opposite and the upper alternate; stipules present or wanting. Flowers bisexual, radially symmetrical, solitary or in cymes; sepals typically 5, the outer 2 much smaller or sometimes wanting, thus leaving only 3; petals 5 or rarely 3 or none, separate, falling early; stamens numerous, separate; anthers splitting lengthwise; carpels 3–5, the style 1, the number of chambers of the ovary sometimes increased by intrusion of the basically parietal placentae; ovules 2–many per carpel, orthotropous or rarely anatropous. Fruit a capsule, more or less woody or leathery.

The family is tropical and subtropical except for some species groups extending into warm temperate regions. In California there are two species of *Helianthemum,* an herb known as rockrose; four species of a different section of the same genus occur in the Middle West and the East; others are in the Southern States. Species of *Hudsonia* (golden-heather) and *Lechea* (pinweed) occur in eastern North America.

The rockroses, *Cistus,* are cultivated from California to the Southern States; *Helianthemum* is cultivated more widely.

Family 2. Bixaceae. ANNATTO FAMILY

Shrubs or small trees; sap reddish. Leaves simple, alternate; stipules present. Flowers bisexual, radially symmetrical, in panicles; sepals 5 or rarely 4, early deciduous; petals 5, large, twisted in the bud; stamens numerous, separate; anthers splitting lengthwise or sometimes with apical pores; carpels 2, the style 1, the ovary 1- or 2-chambered; placentae parietal, but sometimes meeting in the center of the ovary. Fruit a capsule, often prickly.

The family consists of a single genus (*Bixa*) of one species widely distributed in the Tropics.

The single species, known as annato or arnotto, is cultivated in the Tropics and in subtropical areas for its fruit and for red or yellow dyes made from the fleshy red outer coat of the seed.

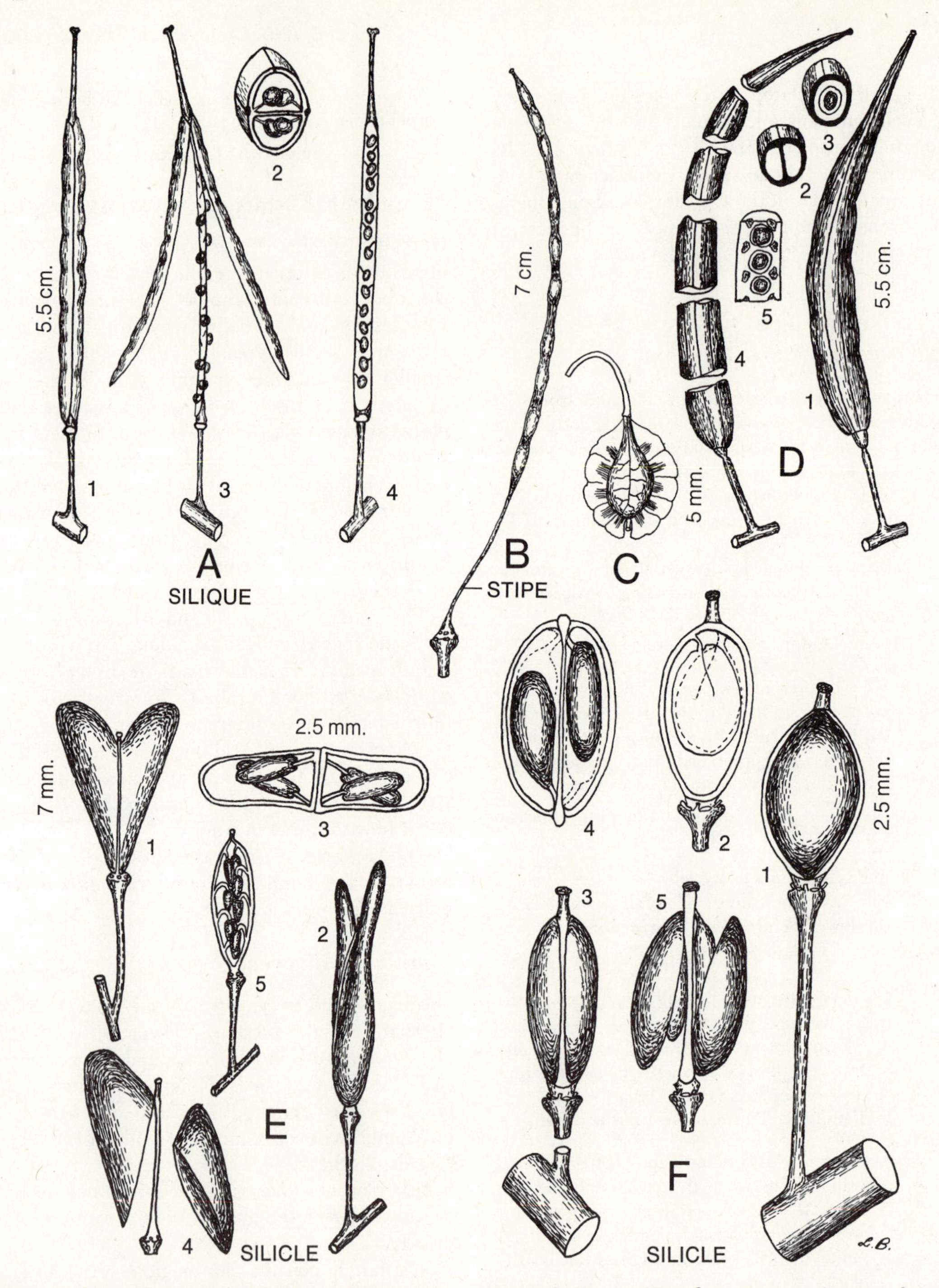

Figure 10-12. Papaverales; Cruciferae—fruits of the mustard family: **A** silique of mustard (or turnip) (*Brassica Rapa* [*campestris*]), **1** side view showing the valves (portions which separate away), **2** cross section, **3** dehiscence, **4** partition persistent after anthesis and (as sometimes) still bearing the seeds; **B** stiped silique of *Stanleya pinnata,* the stipe formed from ovary tissue and above the receptacle; **C** indehiscent short fruit of a fringepod (*Thysanocarpus curvipes*); **D, 1–5** indehiscent elongated fruit of a radish (*Raphanus sativus*), this disjointing crosswise into one-seeded segments (**3–4**) but not splitting

*Family 3. **Cochlospermaceae.**

COCHLOSPERMUM FAMILY

(Sometimes included in the *Bixaceae*)

Perennial herbs, shrubs, or trees; juice red to orange. Leaves simple, palmately lobed; stipules present. Flowers bisexual, radially (or sometimes barely bilaterally) symmetrical; sepals 4 or 5; petals 4–5; stamens numerous, separate; anthers with terminal pores; carpels 3–5, the style 1, the ovary 1- or 3–5-chambered; placentae parietal, but often intruded to the center and dividing the ovary into chambers. Fruit a capsule, the exocarp thick, the endocarp thin, often transparent and revealing the seeds after separation of the exocarp.

The family is largely tropical, but two species of *Amoreuxia* occur in the grasslands above the desert from southern Arizona to southern Texas.

Cochlospermum is cultivated in southern California.

Family 4. **Flacourtiaceae.**

FLACOURTIA FAMILY

Trees, shrubs, or rarely vines. Leaves simple, evergreen, coriaceous, usually alternate, rarely opposite; stipules present, falling while the leaf is young. Flower bisexual or sometimes unisexual (and then the plants either monoecious or dioecious), radially symmetrical, rarely epigynous, often with a hypogynous disc and sometimes a conspicuous appendaged corona which varies in form, mostly cymose; sepals 2–15, sometimes merging with the petals; petals usually the same number as the sepals, sometimes wanting or more numerous; stamens usually numerous, rarely few, sometimes coalescent into bundles; anthers splitting lengthwise; carpels 2–10, the style 1 or the styles often as numerous as the carpels, the ovary essentially 1-chambered, but the parietal placentae sometimes much intruded; ovules numerous. Fruit a capsule or a berry.

The family is tropical and subtropical. It is abundant in the warmer parts of Mexico.

Several genera, including *Xylosma, Azara, Berberidopsis, Idesia, Carrierea, Dovyalis* (Kei apple and Ceylon gooseberry), and *Flacourtia,* are cultivated especially in warmer areas. *Xylosma* is particularly attractive for its bright green foliage.

Lacistema is placed tentatively under this family. It is unknown to the author. It has been considered by some as a separate family, *Lacistemaceae*.

Family 5. **Peridiscaceae.**

PERIDISCUS FAMILY

Trees. Leaves simple, entire, alternate, elliptic but somewhat oblique; stipules present but attached to the petioles. Flowers bisexual, hypogynous, each with a large hypogynous disc, apetalous, in axillary clusters of short racemes; stamens numerous, attached to the disc and partly enclosed by it, broad and not well distinguished as anther and filament, the basal portions somewhat coalescent below; carpels 3–4; styles 3–4; ovary half enclosed by the disc, with 1 seed chamber, with several pendulous ovules at the apex. Fruit ovoid, with only 1 seed, this without endosperm.

A single genus, *Peridiscus,* composed of a single forest species occurring in Brazil.

Family 6. **Scyphostegiaceae.**

SCYPHOSTEGIA FAMILY

Trees. Leaves simple, acuminate, serrate to crenate; stipules small, early deciduous. Flowers unisexual, the plants dioecious, hypogynous, with large, fleshy hypogynous discs, apetalous or at least with no distinction of sepals and petals, the staminate flower with a tubular calyx with 6 lobes in 2 series, the pistillate with 6 perianth parts and these not coalescent, radially symmetrical; stamens 3, coalescent into a column projecting beyond the

---

lengthwise, **5** longitudinal section; **E** silicle of a shepherd's purse (*Capsella bursa-pastoris*), **1–2** two views, **3** cross section, **4** dehiscence, **5** partition and seeds after dehiscence when (as sometimes) the seeds are persistent (cf. also **Figure 8-7 C** embryo); **F** two-seeded silicle of sweet alyssum (*Lobularia maritima* [*Alyssum maritimum*]), **1** side view, **2** longitudinal section from the same side, **3** edge view, **4** longitudinal section from the same view, **5** dehiscence.

anthers; carpels numerous, separate; ovule 1 in each carpel or at least only 1 developing. Fruits achenes, enclosed in the much enlarged disc; seed with little endosperm.

A single genus, *Scyphostegia*, with one species occurring in Borneo.

*Family 7. **Violaceae.** VIOLET FAMILY

Herbs or tropical shrubs or rarely vines. Leaves simple, alternate; stipules present. Flowers bisexual, bilaterally symmetrical; sepals 5, usually persistent in fruit; petals 5, in the North American genera the lower one with a spur or sac; stamens 5, standing close around the pistil (connivent); anthers splitting lengthwise; carpels 3–5 (3 in the North American genera), the style 1, the ovary 1-chambered; ovules 1–numerous in each placenta. Fruit a capsule (North America) or a berry.

The family is world-wide but chiefly tropical in distribution. About 60 species of violets and pansies (*Viola*) and 2 of green violets (*Hybanthus*) occur in Canada and the United States.

The violets are advanced members of the family. The more primitive ones are tropical, particularly South American, some of them forest trees or shrubs with radially symmetrical flowers.

Some violet flowers may be produced on short pedicels or on stolons concealed beneath the leaves. These remain closed and do not develop petals; they are self-fertilized in the bud and are much more fruitful than the ordinary flowers. Flowers of this type are known as cleistogamous.

*Family 8. **Turneraceae.** TURNERA FAMILY

Herbs, shrubs, or trees. Leaves simple, alternate; stipules wanting or minute. Flowers bisexual, radially symmetrical, perigynous; sepals 5, early deciduous; petals 5; stamens 5; anthers splitting lengthwise; carpels 3; the styles 3, apically fringed and more or less brushlike, the ovary 1-chambered; placentae parietal but intruded into the ovary; ovules numerous. Fruit a capsule.

The family is tropical and subtropical in the Western Hemisphere. One species of *Turnera* is

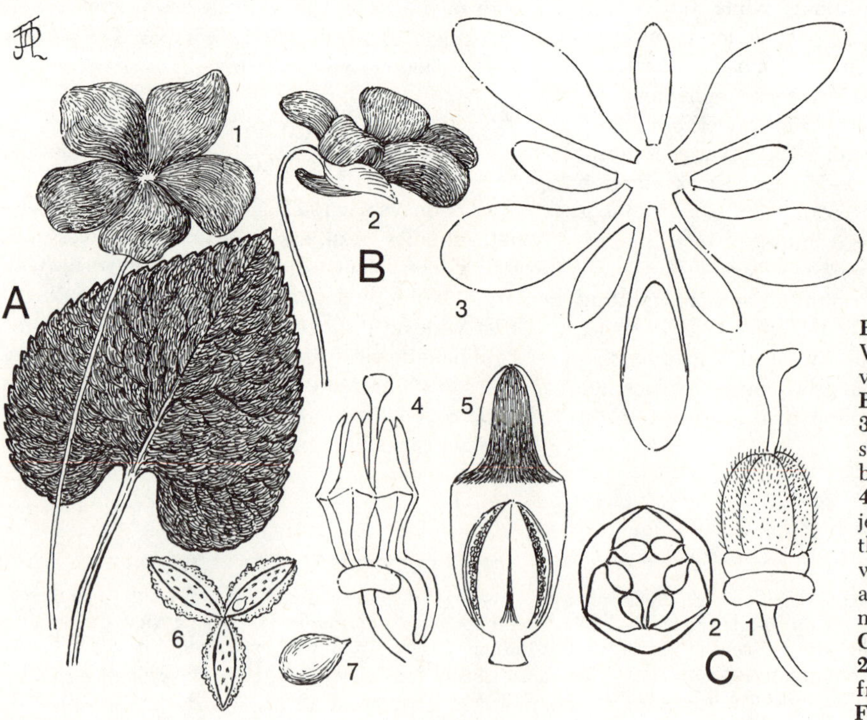

Figure 10-13. Violales; Violaceae; *Viola odorata*, violet: **A** basal leaf; **B** flower, **1–2** two views, **3** diagram of the removed sepals and petals, the spur being on one petal, **4** stamens, two with projections extending into the spur of the corolla, all with flat apical appendages and very short filaments, **5** single stamen; **C** pistil, **1** side view, **2** cross section, **6** open fruit, **7** seed. (Cf. also **Figures 6-7 B** and **6-17 D.**)

native in Texas; another is introduced from Louisiana to Florida; at least two species of *Piriqueta* are native on pinelands from North Carolina to Florida.

Family 9. **Malesherbiaceae.** MALESHERBIA FAMILY

Herbs or small shrubs. Leaves simple, alternate; stipules none. Flowers bisexual, radially symmetrical; sepals 5, the calyx long and tubular, persistent in fruit; petals 5; corona present, membranous, denticulate; stamens 5, produced on a villous lobed stalk (gynophore) bearing the pistil; anthers splitting lengthwise; carpels 3–4, the styles as many, the ovary 1-chambered; placentae parietal; ovules numerous. Fruit a capsule, supported by a stalk (stipe) above the receptacle.

The family is composed of two genera occurring in western South America.

*Family 10. **Passifloraceae.** PASSION-FLOWER FAMILY

Usually more or less woody vines, with tendrils opposite the leaves, rarely herbs or shrubs. Leaves simple or compound, alternate; stipules present. Flowers bisexual or rarely unisexual (and then the plants either monoecious or dioecious), radially symmetrical, perigynous, usually in pairs in the leaf axils; sepals usually 5 but sometimes 4, often petal-like or thickened, persistent; petals 5 or sometimes 4 or wanting, separate or practically so, sometimes smaller than the sepals; floral cup bearing a fleshy corona between the petals and the stamens, this with a fringe of filamentous structures and often a membrane as well; stamens 5 or more, usually opposite the petals; anthers splitting lengthwise; stamens and the pistil sometimes on a common raised stalk (androgynophore); carpels 3–5, the styles 3–5 or 1, the ovary 1-chambered; placentae parietal, fundamentally axile, much intruded; ovules numerous. Fruit a berry or a capsule.

The family occurs in the Tropics and in subtropical regions, especially in the Western Hemisphere. Several species of passion-flowers (*Passiflora*) occur across the southern edge of the United States from southeastern Arizona to the Southern States, two extending as far northward as Missouri, Illinois, the Ohio Valley, and southeastern Pennsylvania. The family does not occur (except in cultivation) on the Pacific Coast or in the Rocky Mountain System.

Family 11. **Achariaceae.** ACHARIA FAMILY

Herbs or small shrubs. Leaves simple, palmately lobed, alternate; stipules none. Flowers unisexual (and the plants monoecious), radially symmetrical; sepals 3–5, separate; petals 3–5, coalescent; stamens 3–5, adnate with the corolla tube; anthers splitting lengthwise; carpels 3–5, the stigma 2-lobed, the ovary with 1 chamber; placentae parietal; ovules several to many. Fruit a capsule, with a stalk (stipe) supporting it above the receptacle.

The family consists of two genera occurring in South Africa.

*Family 12. **Caricaceae.** PAPAYA FAMILY

(Synonym: *Papayaceae*)

Small trees with usually unbranched trunks, thin bark, and soft wood; juice milky. Leaves on only the upper portion of the trunk, simple, usually palmately lobed but rarely entire, very large, alternate; stipules none. Flowers unisexual or bisexual, radially symmetrical; sepals 5; petals 5, in most flowers coalescent but in some separate; stamens 5 or 10 (in 1 series or 2 according to the type of flower); anthers splitting lengthwise; carpels usually 5, the styles 5, the ovary with 1 chamber; placentae parietal: ovules numerous. Fruit a berry, large and melonlike.

The flowers are of various types, according to whether the plant is staminate, pistillate, or polygamous (with bisexual and unisexual flowers on the same individual), as follows.

Staminate plants. Flowers in long dangling racemes; stamens 10, in 2 series, sessile.

Pistillate plants (type A). Flowers 1–several in each leaf axil; fruit globose to ovoid or pear-shaped; (type B), flowers in corymbs; fruits elongate, cylindroidal.

Polygamous plants (type A). Corolla tube long, the 10 stamens sessile, attached in the corolla throat; (type B), corolla tube very short, the 5 stamens with long filaments.

The nearest relatives of the order are the Violales.

The family includes three genera, *Carica* being restricted to Tropical America and the other genera to Tropical Africa. *Carica Papaya*, the papaya, is grown throughout the tropics for its melonlike fruits. It has escaped from cultivation, as in peninsular Florida and on the Keys. The stems are up to about twenty feet tall, and the leaves may be up to two feet broad.

Order 8. *LOASALES.* LOASA ORDER

DEPARTURE FROM THE FLORAL CHARACTERS OF THE RANALES

Free	Petals (or rarely the whole perianth) coalescent; filaments sometimes coalescent or in some cases the filaments and the anthers coalescent into a column or sometimes coalescent in pairs with one left over.
Equal	Perianth sometimes bilaterally symmetrical in the *Begoniaceae*.
Numerous	Stamens 5–numerous; carpels usually 3, but sometimes 4–8.
Spiral	Stamens, when few, cyclic; carpels cyclic.
Hypogynous	Flowers epigynous.

Herbs (sometimes somewhat succulent), shrubs, trees, woody vines, or herbaceous semisucculent vines with tendrils; leaves alternate or opposite, simple to pinnate, entire or pinnately or palmately divided; stipules absent or sometimes present. Flowers bisexual or unisexual, choripetalous or sympetalous; sepals 2–9; petals 2–8 or 10; petals 2–8 or 0; stamens rarely 3, commonly 5–numerous; carpels 3–8, the styles separate or coalescent, the ovary 1-chambered or with as many chambers as carpels; placentae usually parietal but sometimes axile or seemingly so; ovules numerous or rarely only 1. Fruit a capsule or a pepo (usually a large one like a gourd or melon).

Key to the Families

(North America north of Mexico)

1. Petals separate or none; stamens numerous or 8–12, separate or rarely basally coalescent; herbs, shrubs, or woody vines, not with tendrils; fruit a capsule or rarely a berry.
   2. The two sides of base of the leaf blade equal; stipules none; sepals 3–9, not petaloid; ovary *not* winged.
      3. Petals present; flowers bisexual or unisexual (and then the plants monoecious); stems and leaves with *either* rough *or* hooked *or* stinging hairs.

         Family 1. **Loasaceae** (*Loasa* family).

      3. Petals none; flowers unisexual (and the plants dioecious) or some stamens present in the pistillate flowers; (North American) plants glabrous.

         Family 2. **Datiscaceae** (durango root family).

   2. The two sides of the base of the leaf blade unequal; stipules present, but early deciduous; sepals 2 and petaloid and petals 2 in the staminate flower, the perianth undifferentiated in the pistillate flower and consisting of 2–numerous petaloid members; ovary usually with a wing on each carpel.

      Family 3. **Begoniaceae** (*Begonia* family).

1. Petals coalescent (sometimes undeveloped); stamens 5 (or 3 or appearing to be 3), nearly always coalescent by the filaments and often the anthers or sometimes in 2 coalescent pairs with 1 left over; somewhat succulent herbaceous vines with tendrils; fruit a pepo (usually a gourd or melon).

   Family 4. **Curcurbitaceae** (melon family).

*Family 1. Loasaceae. LOASA FAMILY

Herbs or sometimes shrubs or woody vines; stems and leaves with either rough or hooked or stinging hairs. Leaves simple, often indented, alternate or opposite. Flowers usually bisexual; sepals 4 or usually 5; petals 4 or usually 5, sometimes coalescent basally; stamens numerous, sometimes the outer 5 with broad filaments, otherwise spirally arranged; carpels usually 3 or sometimes 4–7 or 8, style 1, the ovary 1-chambered; ovules numerous (or only 1 in the sandpaper-plant, a rough-hairy desert subshrub (*Petalonyx*). Fruit a capsule, dehiscent on the flat area at the top of the ovary or opening along spirally twisted lines.

The family occurs chiefly in South America and the warmer parts of North America. A single species occurs from South West Africa to Arabia. Various species of *Mentzelia* are endemic in western North America and on the Great Plains, two extending eastward barely to the Middle West. The sandpaper-plant (*Petalonyx*) and the rocknettle (*Eucnide*) occur in the Southwestern Deserts in California, Nevada, Utah, and Arizona. *Cevallia* occurs from Arizona to Texas.

Among the North American members of this largely tropical family is the evening-star (or blazing-star), *Mentzelia laevicaulis,* a night-flowering species common through the West especially in washes, creek beds, or other gravelly soil. The clear yellow flowers are 3–4 inches in diameter, and they open at twilight, pollination occurring at night. Transition between petals and stamens is shown clearly by many of the flowers. The stamens are numerous, but the outermost five alternate with the petals and they have somewhat broadened filaments. In occasional plants, these filaments are so broad that they are similar to the petals and the anthers may be missing, so that, in effect, the flower has ten petals. Sometimes others of the outer stamens are broadened as well, and they may or may not have anthers.

Some specimens of this species collected in the San Gabriel mountains in southern California are abnormal because the sepals and petals resemble vegetative leaves, as do the styles protruding from above the inferior ovary. In some flowers the stamens have green filaments and anthers, but in others they resemble vegetative leaves.

Individuals of the *Ranunculus occidentalis* complex exhibit similar leaflike floral parts. This occurs also in the carpels of roses.

*Family 2. **Datiscaceae.**

DURANGO ROOT FAMILY

Perennial herbs or sometimes small trees; extraterritorial species sometimes lepidote. Leaves simple to pinnate, alternate. Flowers unisexual or in extraterritorial plants sometimes bisexual, in North American plants the pistillate flowers sometimes with a few stamens, in spikes or racemes; sepals 3–9; petals 8 or (in the North American species) none; stamens of the North American species 8–12, in the family 4 to about 25; carpels 3, the styles 3 each with 2 branches; ovules numerous on each placenta. Fruit a capsule, dehiscent apically between the styles, which are persistent.

The family is composed of 3 genera with a total of 4 species. A single species of *Datisca* occurs in California, ranging into Mexico. This plant is known as durango root. In California it occurs along drying vernal creek beds or near the margins of streams. The other members of the family occur in eastern Asia and on the adjacent Pacific islands.

*Family 3. **Begoniaceae.**

BEGONIA FAMILY

Somewhat succulent herbs. Leaves simple, usually palmately veined and often palmately lobed, with the sides unequal at the bases, alternate and in 2 ranks; stipules present, early deciduous. Flowers unisexual, radially or bilaterally symmetrical; sepals 2, petaloid; petals 2, in the pistillate flowers the perianth undifferentiated, consisting of 2–numerous petaloid members; stamens numerous, arranged spirally, sometimes coalescent basally; carpels 3 or rarely 5 or 2, the styles 2–5, sometimes coalescent basally, the ovary with as many chambers as carpels and usually winged, the wings usually 1–3; placentae usually axile; ovules numerous, anatropous. Fruit a somewhat bony capsule or rarely a berry.

The family occurs through much of the tropical region of the Earth, but it is most abundant in northern South America. Many species and hybrids have been brought into cultivation. A single type known as perpetual *Begonia* is naturalized in the swamps and low grounds of peninsular Florida.

*Family 4. **Cucurbitaceae.**

GOURD FAMILY

Herbs, usually annual or occasionally perennial, tendril-bearing and somewhat succulent prostrate or climbing vines, rarely (extraterritorial genera) shrubs or small trees. Leaves simple, usually palmately or pinnately lobed, alternate; stipules none. Flowers nearly always unisexual (in one extraterritorial genus bisexual), usually 5-merous except the carpels and sometimes the stamens (choripetalous in 1 extraterritorial genus); anthers often twisted and the style 1 (or rarely 3), the ovary 1-chambered (rarely 3-chambered by intrusion of the placentae or sometimes completely filled with placental tissue); placentae at least fundamentally parietal; ovules usually numerous, rarely few or solitary. Fruit commonly a large pepo, that is, a berrylike fruit with a leathery external layer and a fleshy interior (including the

Figure 10-14. Loasales; Loasaceae; *Mentzelia laevicaulis*, evening (or blazing) star (cf. **Figure 5-3 A**): **1** branch with a leaf and with a smaller leaf from the axillary bud, **2** leaf and two smaller leaves from the axillary bud, **3** flower buds, **4** flower, showing the inferior ovary (covered by the adnate floral cup), the sepals, the petals, and the stamens (the five outer ones with broad filaments), **5** flower, longitudinal section, **6** enlargement of the area near the top of the ovary, **7** cross section of the ovary, showing the parietal placentae.

enlarged and often edible placentae), sometimes small and leathery, sometimes dehiscent at least apically or dehiscent almost explosively, rarely spiny.

The gourd family includes many genera, occurring mostly in the Tropics but also in the warmer temperate regions. *Cucurbita* (*Marah*) (gourd) is represented only sparingly in North America north of Mexico. Three species are native in southern California and the desert regions of Arizona. One of these ranges eastward to New Mexico and another to Missouri and Texas. Another species, the bottle gourd or calabash, is naturalized from the Tropics of the Eastern Hemisphere and, having escaped from cultivation, it occurs in thickets and waste places on the coastal plain from Texas to Florida. The genus *Echinocystis* includes about 5 species of wild cucumber occurring in California, one in Arizona, and one from Saskatchewan to New Brunswick and south to Texas and Florida. These plants are remarkable for the size of their underground tubers, which may be nearly as large as a man's body and which may weigh over 50 pounds. Because of the tuber, sometimes the plant is called man-in-the-ground. Goats browsing in the California Chaparral may live upon water from the tubers of this plant. Although most members of the family have gourdlike or melonlike fruits, some have small spiny fruits with a single seed and only a single ovule at flowering time. These include the desert species of *Brandegea, Sicyos,* and *Sicyosperma,* and a species of *Sicyos* occurring in eastern North America. Several other genera occur in the Southwestern Deserts from Arizona to Texas. In eastern North America there are several genera not occurring in the West. These are concentrated largely in the Southern States. They include *Cayaponia,* the melonette (*Melothria*), the balsam-apple (*Momordica*), and the dishcloth gourd or vegetable sponge (*Luffa*). In the South the watermelon, cantaloupe, squash, calabash, and pumpkin have escaped from cultivation, especially in Florida.

Usually the plants are monoecious or dioecious. On monoecious plants the staminate flowers may appear first on the younger parts of the stem, the pistillate later on the older parts of the same branch. The pistillate flowers are identified by the large inferior ovaries.

In some tropical climbing melons the fruit is supported by a tendril on the stem at the base of the pedicel. When the fruit ripens the tendril dries, releasing the melon, which bursts when it hits the ground, thus throwing seeds in all directions. The "curl" or tendril at the base of a watermelon dries similarly when the fruit matures.

Order 9. *TAMARICALES.*

TAMARISK ORDER

DEPARTURE FROM THE FLORAL CHARACTERS OF THE RANALES

Free	Stamens sometimes coalescent; carpels coalescent.
Numerous	Stamens 4–10; carpels 2–5.
Spiral	Stamens and carpels cyclic.
Hypogynous	Hypogynous disc present in the *Tamaricaceae*.

Herbs, shrubs, or trees occurring usually in saline or alkaline places. Leaves alternate or opposite; stipules none. Flowers bisexual; perianth 4–7-merous; calyx persistent in fruit, in 2 series. Ovary 1-chambered, the placentae basal or nearly so, fundamentally parietal. Ovules numerous, anatropous; embryo straight; endosperm present or absent.

Key to the Families

(North America north of Mexico)

1. Leaves opposite; herbs or small shrubby plants; seeds not hairy; stamens arising directly from the receptacle, there being no hypogynous disc; calyx tubular, only the tips of the sepals separate; funiculi of the ovules elongated.

   Family 1. **Frankeniaceae** (alkali-heath family).

1. Leaves alternate, small, scalelike or somewhat elongated, appressed against the branchlets; large shrubs or small trees; stamens arising from the glandular hypogynous disc; sepals separate; funiculi short.

   Family 2. **Tamaricaceae** (tamarisk family).

*Family 1. **Frankeniaceae.**

ALKALI-HEATH FAMILY

Perennial herbs or shrubby plants. Leaves simple, entire, opposite, the alternate pairs at right

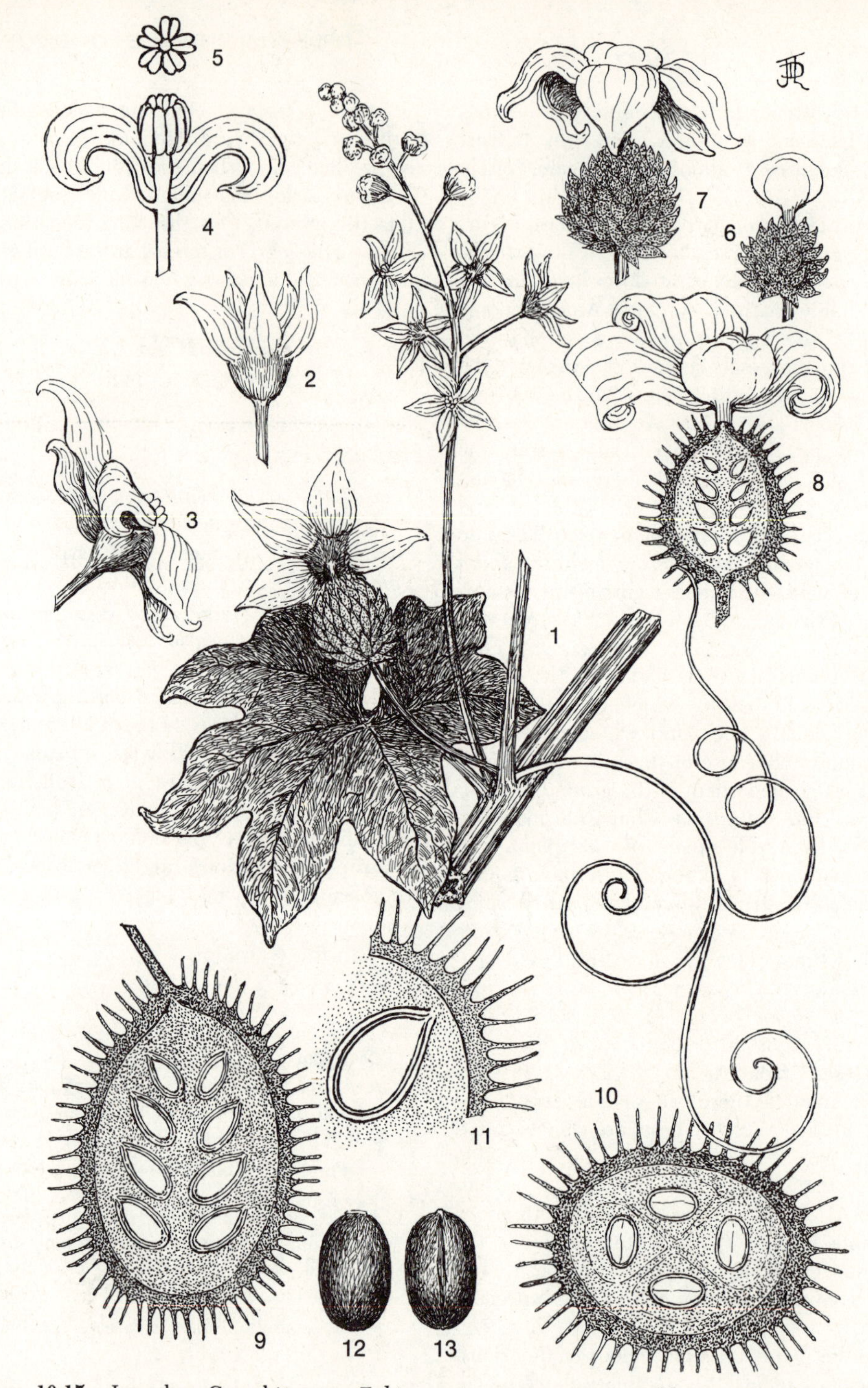

Figure 10-15. Loasales; Cucurbitaceae; *Echinocystis macrocarpa,* a wild cucumber (cf. **Figure 2-10 A**): **1** branch with a leaf, a tendril, a raceme of staminate flowers, and a pistillate flower, **2–3** staminate flower, showing the sympetalous corolla (the sepals minute or none) and the coalescent stamens, **4** longitudinal section, **5** anthers from above, **6** pistillate flower bud, **7** pistillate flower, showing the inferior ovary (covered by the adnate floral cup) and the sympetalous corolla, **8** longitudinal section, **9** fruit, longitudinal section, **10** cross section, **11** portion enlarged, showing the attachment of the seed, **12–13** seed, two views. (Cf. also **Figure 8-2 C** *Cucumis Melo,* fruit.)

angles, the leaves thus forming 4 rows (decussate); stipules none, but the bases of the leaves of the pair connected on each side by a ciliate line. Flowers radially symmetrical, solitary or cymose; sepals 4–7, the tube of the calyx more or less prismatic, persistent in fruit; petals 4–7; stamens 4–7; carpels 2–4, the style 1 but branched, the ovary with 1 seed chamber; ovules 2–many, the funiculi elongated. Fruit a capsule.

The family consists of four genera occurring sporadically throughout the world but most abundant in the Mediterranean Region. The genus *Frankenia* includes two herbaceous species in California and a shrubby one on the plains from Colorado to Texas.

*Family 2. **Tamaricaceae.**
TAMARISK FAMILY

Large shrubs or small trees of desert or saline habitats. Leaves simple, entire, alternate, very small, scalelike or relatively elongated, appressed against the twigs; stipules none. Flowers radially symmetrical, with hypogynous discs, small, in dense spikes or racemes or rarely solitary; sepals 4–5; petals 4–5, persistent at fruiting time; stamens 4–10; carpels usually 3–4, the styles the same number, the ovary 1-chambered; ovules usually numerous, the funiculi short. Fruit a capsule; each seed with an apical tuft of hairs.

The family includes 4 genera occurring largely about the Mediterranean Sea and in the drier regions of Western and Central Asia. The Asian salt-cedar, *Tamarix ramosisima,* and *Tamarix chinensis* and *parviflorus* have become established along water courses and irrigation ditches in the interior valleys of California, the Southwestern Deserts, and eastward to South Carolina.

Several species occur in cultivation, the most remarkable being *Tamarix aphylla.* This cypresslike plant is ubiquitous in cultivation in the Southwestern Deserts and the drier regions of California and the Southwest. It is used particularly as a windbreak and as a shade tree around desert houses. The spikes of pink flowers are attractive. Most of the individuals in cultivation are derivatives of half a dozen cuttings secured at the beginning of the twentieth century by Professor J. J. Thornber of the University of Arizona from a correspondent in Algeria.

The English name tamarisk should not be confused with tamarac or tamarack, the name of a gymnosperm of the pine family known also as larch.

The tuft of hair on the seed and the more or less basal placentae have suggested to several authors an affinity of this family with the *Salicaceae* (willow family). Presumably the willows were derived from an ancestral group with tamarisklike flowers by reduction and final loss of the perianth and change to unisexual flowers. However, the evidence for this theory is inconclusive.

An apical tuft of hairs occurs on each seed in the unrelated genus *Epilobium* (*Onagraceae,* Myrtales).

Order 10. ***SALICALES.*** WILLOW ORDER

DEPARTURE FROM THE FLORAL CHARACTERS OF THE RANALES

Free	Stamens coalescent in some species of *Salix;* carpels coalescent.
Numerous	Stamens in *Salix* 2–9 or rarely 1 or 12, in *Populus* usually numerous but sometimes only 4; carpels 2 or 4.
Spiral	Stamens perhaps cyclic when the numbers are small; carpels cyclic.
Hypogynous	Flower of *Populus* apparently perigynous.

Shrubs or trees. Leaves simple, alternate; stipules present. Flowers unisexual and the plants dioecious or abnormally otherwise, both types in catkins; staminate flower consisting of only stamens or in *Populus* these subtended by a disc (cf. below); pistillate flower consisting of a single pistil, subtended by a disc which may be either a floral cup, a vestigial perianth, or a gland of perhaps the same origin; developed perianth none, the flower usually subtended by a bractlet; ovary with a single chamber, but formed from 2 or 4 carpels, the placentae fundamentally parietal but often more or less basal. Fruit dehiscent along two lines; seeds very small, numerous, each with an apical tuft of fine hairs which aid in dissemination by wind, the seed coat thin, the embryo straight, with little endosperm.

The origin of the Salicales is obscure. Most authors agree that *Populus* (cottonwood, poplar, aspen) is more primitive than Salix (willow). This is in harmony with departure from the floral characters of the Ranales, but these criteria may not be applicable in all special cases, for lack of information. There is evidence from the vascular bundles that the nectar gland at the base of the flower of the willow and the more elaborate cupule or possible floral cup of the cottonwoods and their relatives are greatly reduced sepals or petals. Both genera are specialized highly in their catkin type of inflorescence. *Populus* is wind-pollinated, but *Salix* is insect-pollinated.

*Family 1. **Salicaceae.** WILLOW FAMILY

The family and the order consist of the two genera mentioned above. Both are most abundant in the Northern Hemisphere, *Populus* in temperate regions but *Salix* in temperate and Arctic regions. The family is almost world-wide but it is absent from much of the South Pacific. Dwarf willows with only the catkins projecting above the tundra occur far to the northward on the Arctic plains and islands. The willows are developed best in moist regions relatively far to the northward. The chief distributional areas are the following: (1) central Europe; (2) the mountains of central Asia; (3) the adjacent regions of Asia and North America centering about the Bering Sea; (4) the Pacific Northwestern Region of North America.

The species of willows may be difficult to classify because of their tendency to hybridize. Hybrids are relatively common in nature, and they may hybridize with other hybrids.

The species of both genera reproduce readily from cuttings. Often green fence posts may start to grow and form trees within a short time. In the dry areas of the Southwest, cottonwoods are used extensively as shade trees because of their rapid growth from relatively large pieces or even small logs stuck into the soil in wet or irrigated places.

Fossil remains of both willows and cottonwoods are common. This is due to antiquity of the family and also to the tendency of all species to occur in moist soil of areas where the remains of plants are preserved readily and to the relatively firm texture of the leaves, which are favorable to preservation.

Order 11. ***THEALES.*** TEA ORDER

DEPARTURE FROM THE FLORAL CHARACTERS OF THE RANALES

Free — Petals coalescent in some *Theaceae* and some *Dipterocarpaceae;* stamens sometimes coalescent, es-

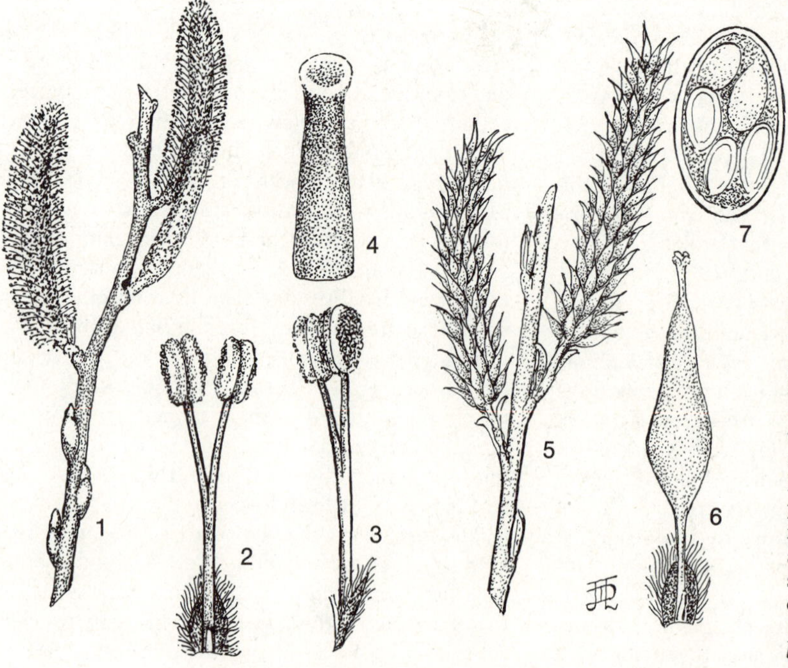

Figure 10-16. Salicales; Salicaceae; *Salix lasiolepsis,* arroyo willow: **1** branch of a staminate plant with catkins, **2–3** flower, consisting of two coalescent stamens and a gland (nectary), subtended by a hairy scalelike bract, **4** gland, enlarged, **5** branch of a pistillate plant with catkins, **6** pistillate flower, consisting of a pistil and a gland, subtended by a hairy scalelike bract, **7** cross section of the ovary, showing the seeds. (Cf. also **Figure 6-1 A** *Salix laevigata.*)

	pecially into groups; carpels coalescent.
Numerous	Stamens rarely reduced to 3–15 in some species of *Hypericum* (*Guttiferae*) and of *Lechea* (*Cistaceae*); carpels usually 2–5 but sometimes up to 20.
Spiral	Stamens rarely cyclic (when few); carpels cyclic.
Hypogynous	Flowers perigynous in some *Caryocaraceae,* perigynous or epigynous in the *Dipterocarpaceae,* with a hypogynous disc in the *Eucryphiaceae* and *Strasburgeriaceae.*

Herbs, shrubs, or trees. Leaves alternate or opposite; stipules present or absent. Perianth 5-merous; stamens numerous; carpels fewer than 5; ovary usually several-chambered, the placentae axile or sometimes parietal; ovules anatropous, or in the *Cistaceae* usually orthotropous, with usually 2 integuments.

The order is represented poorly in the flora of temperate North America, and, except in the Southern States, the native genera and species are few. Some families and numerous genera are in cultivation.

The stamens (with few exceptions) and sometimes the carpels are numerous as they are in the Ranales. In the camellias and the tea plant (*Camellia sinensis*) this primitive feature is combined with more advanced characters such as both coalescence and adnation of the petals and stamens (thus, the bulk of the flower, that is, the corolla and the stamens, falling to the ground without shattering).

Key to the Families

(North America north of Mexico)

1. Plants terrestrial, herbs, shrubs, or trees; stipules none; stamens nearly always numerous.
   2. Petals (of the North American species) white; trees or large shrubs (1–5 dm. or more in height); leaves alternate.

      Family 9. **Theaceae** (tea family).

   2. Petals, when present (rarely absent), yellow or rarely pink, purplish, brownish, or greenish; herbs or low shrubs usually under 1 m. in height (rarely as much as 2.5 m. high, the plants then with opposite leaves); leaves usually but not necessarily opposite.

      Family 10. **Guttiferae** (St.-John's-wort family).

1. Plants aquatic, very small herbs; stipules present, usually membranous; stamens 6–10 in the minute flower.

   Family 13. **Elatinaceae** (waterwort family).

Family 1. Actinidiaceae.

ACTINIDIA FAMILY

Trees, shrubs, or woody vines. Leaves simple, alternate; stipules none. Flowers bisexual or unisexual (and the plants dioecious); sepals 5; petals 5; stamens numerous or as few as 10, separate or adnate with the corolla; anthers with apical pores or splitting lengthwise; carpels 3–5 or sometimes more; styles usually 3–5; placentae axile. Fruit a berry or leathery.

The family is composed of four genera confined largely to the Tropics. The tara-vine (*Actinidia*) is an ornamental with edible fruits.

Family 2. Stachyuraceae.

STACHYURUS FAMILY

Shrubs or small trees. Leaves simple, deciduous, alternate; stipules minute or none. Flowers usually bisexual, radially symmetrical, racemose, appearing before the leaves; sepals 4; petals 4; stamens 8, separate, the anthers splitting lengthwise; carpels 4, the style 1, the ovary 4-chambered; placentae axile, but said to be fundamentally parietal. Fruit a berry.

The family is composed of a single small genus occurring in central and eastern Asia.

One species is cultivated in the United States.

Family 3. Medusagynaceae.

MEDUSAGYNE FAMILY

Shrubs. Leaves simple, opposite; stipules none. Flowers bisexual, radially symmetrical, in terminal panicles; sepals 5, early deciduous; petals 5, red; stamens numerous, the anthers splitting lengthwise; carpels 20–25, nearly separate; styles as many as the carpels; placentae axile; ovules 2 per carpel. Fruit a capsule, the carpels diverging after dehiscence.

The family consists of a single genus restricted to the Seychelles Islands just south of the equator in the Indian Ocean.

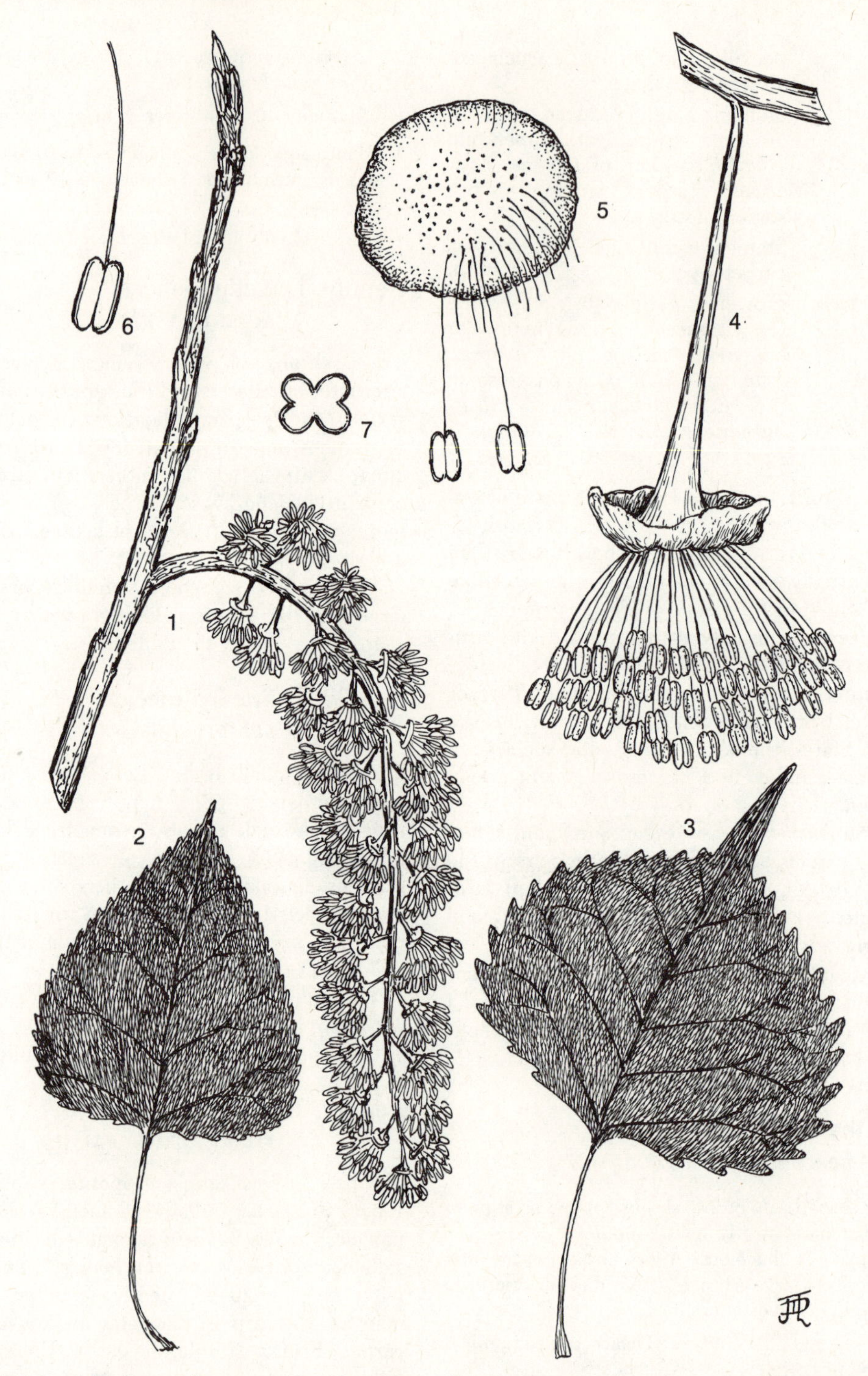

Figure 10-17. Salicales; Salicaceae; *Populus Fremontii,* Fremont cottonwood: **1** twig of a staminate plant with a catkin, **2–3** leaf forms, **4** staminate flower, showing the pedicel, the disc, and the numerous stamens, **5** disc, top view, with nearly all the stamens removed, **6** stamen, **7** diagram of the shape

of the anther in cross section, **8** twig of a pistillate plant with a catkin, **9** pistillate flower, showing the pedicel, the disc (cuplike), and the pistil with four styles and large stigmas, **10** stigmas, from above, **11** catkin with fruits, **12** fruits, **13** catkin with dehiscent fruits, **14** dehiscent fruit, **15** seed, showing the numerous hairs attached at one end, **16** hairs. (Cf. also **Figure 2-16**, *Populus nigra.*)

Family 4. **Ochnaceae.** OCHNA FAMILY

Trees, shrubs, or sometimes herbs. Leaves simple or very rarely pinnate, coriaceous, alternate; stipules present. Flowers bisexual, radially symmetrical, cymose, racemose, or paniculate; sepals 4–5 or rarely 10; petals the same number; stamens numerous or sometimes 5 or 10, separate; anthers with terminal pores or splitting lengthwise; carpels 2–5 or rarely 10–15; style usually 1, with separate stigmas; placentae usually axile but sometimes parietal and projecting inward; ovules 1–many per carpel. Fruit a berry or rarely a capsule.

Tropics, but centering in Brazil and adjacent tropical countries. *Ouratea* occurs in Mexico.

Ochna multiflora is cultivated from southern California to Florida.

Family 5. **Strasburgeriaceae.** STRASBURGERIA FAMILY

Trees. Leaves simple, coriaceous; stipules present, joined. Flowers bisexual, each with a hypogynous disc, bilaterally symmetrical, solitary in the leaf axils; sepals several; petals 5, unequal; stamens 10, the outer series opposite the petals; carpels 5; placentae axile; ovule 1 per carpel. Fruit woody, indehiscent.

The family consists of a single species, *Strasburgeria callicantha,* growing in New Caledonia. It has been included in the *Ochnaceae,* and it may belong there.

Family 6. **Caryocaraceae.** CARYOCAR FAMILY

Trees or shrubs. Leaves palmate, with 3–5 leaflets, alternate or opposite; stipules present. Flowers bisexual, radially symmetrical, sometimes perigynous; sepals 5–6; petals 5–6, sometimes coalescent at the apices; stamens numerous, coalescent into a tube or into groups; carpels 4–20 or rarely 1–3, the styles separate; placentae axile; ovule 1 per carpel. Fruit a drupe, but the carpels separating at maturity.

The family consists of two genera occurring in tropical America.

Souari nuts are the edible seeds of either of two species of *Caryocar.*

Family 7. **Marcgraviaceae.** MARCGRAVIA FAMILY

Climbing shrubs, usually living upon other trees or shrubs (epiphytic) but not parasitic; sometimes trees. Leaves simple, alternate, sometimes of 2 forms; stipules wanting. Flowers bisexual, usually racemose; sterile flowers forming pitcher- or saclike structures; sepals 4–5, coalescent, falling as the flower opens; petals 4–5; stamens 3–many, free or slightly coalescent; anthers splitting lengthwise; carpels 2 or more, commonly 5; placentae axile; ovules numerous. Fruit a berry.

The family consists of 5 genera occurring in Tropical America. It is related to the Theaceae.

Family 8. **Quiinaceae.** QUIINA FAMILY

Trees or shrubs; leaves simple to pinnate, opposite or whorled; stipules rigid and leaflike. Flowers bisexual or unisexual, in racemes or panicles; sepals 4–5, small, of various sizes; petals 4–8; stamens 15 or more, usually numerous; anthers splitting lengthwise; carpels 2–3; styles 2–3; placentae axile; ovary with 2–11 chambers; ovules 2 per carpel. Fruit a berry with 1–4 seeds; seeds tomentose.

The family is composed of 2 genera occurring in Tropical America. It has been included under the *Guttiferae.*

*Family 9. **Theaceae.** TEA FAMILY

(Synonym: *Ternstroemiaceae*)

Trees or usually shrubs. Leaves simple, usually toothed, alternate, usually evergreen; stipules none. Flowers bisexual or rarely unisexual (and then the plants dioecious), bilaterally symmetrical, solitary or clustered; sepals usually 5, sometimes more; petals 4–several but commonly 5, separate or coalescent basally; stamens numerous, rarely 10–15, separate or basally coalescent into a short tube or into 5 fascicles; anthers nearly always splitting lengthwise; carpels 3–5; styles as many as the carpels or 1; ovary 3–5-chambered, the placentae axile. Fruit a capsule.

The family is distributed widely in the Tropics and in warm temperate regions. The family is represented in the Southern States by four species of *Stewartia* and *Gordonia,* sometimes referred to as camellias. Like most of the other families of the order, the Theaceae do not occur in western North America.

The genus *Camellia* is best-known from its frequent cultivation for colorful flowers and beautiful dark green leathery leaves. The tea plant (*Camellia sinensis*) is segregated sometimes into another genus, *Thea.* Various kinds of tea come from the leaves of the same branch of the plant. They are numbered according to age as indicated by nearness to the bud at the end of the branch. Pekoe, orange pekoe, and various other types of tea are determined in this way. The three smallest leaves are pekoe; the last tiny one and the terminal bud are orange pekoe.

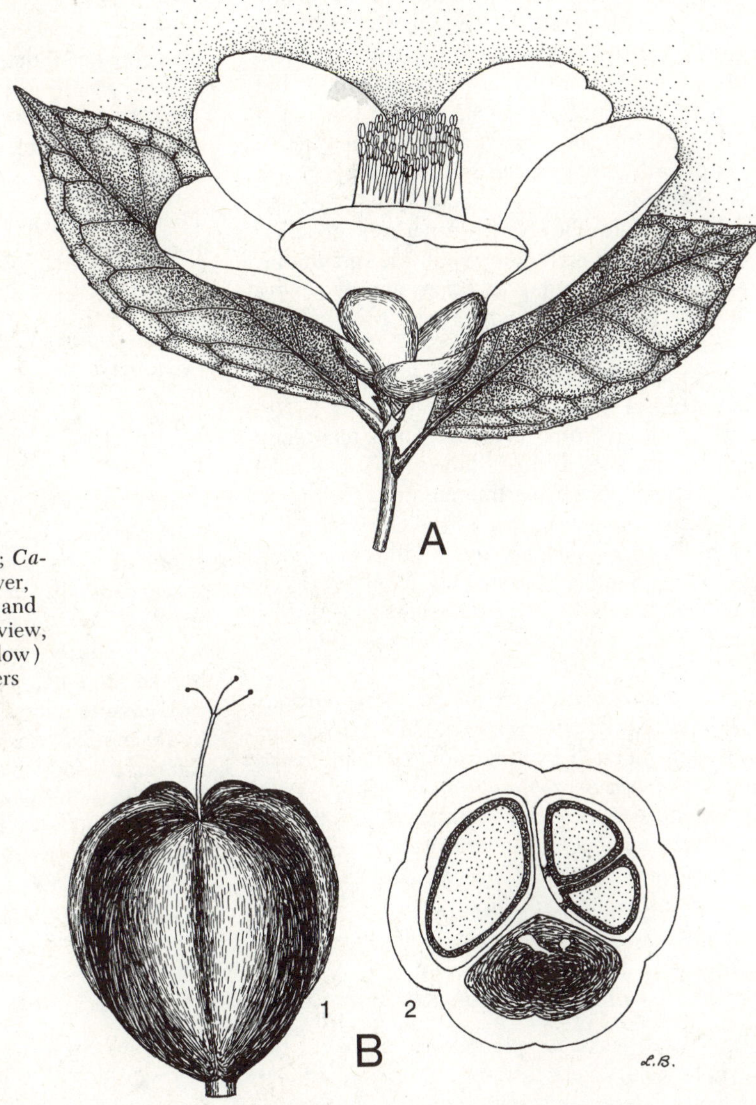

Figure 10-8. Theales; Theaceae; *Camellia japonica,* camellia: **A** flower, the petals and stamens coalescent and basally adnate; **B** fruit, **1** side view, **2** cut crosswise, one chamber (below) with two abortive ovules, the others with seeds.

*Family 10. **Guttiferae.**

ST.-JOHN'S-WORT FAMILY

(Synonym: *Clusiaceae.* Segregate: *Hypericaceae*)

Herbs, shrubs, or sometimes trees or woody vines. Leaves simple, at least usually entire, dotted with glands, opposite; stipules none. Flowers bisexual (North American genera) or unisexual, radially symmetrical; sepals 2–10 or more, in the North American genera *either* 4 in unequal pairs, the outer being much larger, *or* 5 (or 4) and unequal (with the outer ones smaller) to essentially equal; petals in the North American genera as many as the sepals, in other genera 4–12, separate; stamens usually numerous, but sometimes few in *Hypericum,* usually coalescent basally into 3–5 (or 6–8) fascicles, sometimes in 1 group; anthers splitting lengthwise; carpels 3–5 or 7, the styles 3–5 or 7, sometimes coalescent (only basally in the North American genera); placentae parietal or fundamentally so but the intruded placentae forming partitions and becoming apparently axile; ovules numerous. Fruit a capsule or a berry.

The family is typically tropical, but the subfamily Hypericoideae includes two genera, *Hypericum* and *Ascyrum,* extending into temperate areas, including the United States. The St.-John's-worts (*Hypericum*) are attractive plants with yellow flowers. They occur across the Continent, but they are most common in the Middle West and East and the Southern States. The St.-Peter's-wort (*Ascyrum*) is represented by Eastern and Middle Western species and by others in the Southern States.

Garcinia is cultivated in the southern portions of the United States, as are several species of *Hypericum.* The "Klamath-weed" of the Pacific Coast is *Hypericum perforatum,* introduced from Europe into eastern North America as well. It is exceedingly difficult to eradicate. Two edible fruits, the mangosteen (*Garcinia*) and the mammee-apple (*Mammea*) are important in the Tropics.

Family 11. **Dipterocarpaceae.**

DIPTEROCARPUS FAMILY

Trees; wood resinous; plant with stellate hairs or peltate scales. Leaves simple, alternate; stipules deciduous. Flowers bisexual, radially symmetrical, in axillary panicles; perigynous or epigynous; sepals 5, enlarged and winglike at fruiting time; petals 5, twisted, sometimes slightly coalescent; stamens numerous; anthers splitting lengthwise; carpels 3, the style entire or 3-lobed; placentae axile; ovules 2 in each chamber of the ovary. Fruit 1-seeded, indehiscent, the persistent sepals conspicuous.

Tropics of the Eastern Hemisphere, but rare in Africa.

The family includes the Sal- or Saul-tree, the camphor-tree of Sumatra, and the garjan of southeastern Asia.

Products include eng oil, piney-resin, and Indian copal of India, as well as a varnish.

Family 12. **Pentaphylacaceae.**

PENTAPHYLAX FAMILY

Shrubs. Leaves simple, coriaceous, alternate. Flowers bisexual, radially symmetrical, small; sepals, petals, stamens, and carpels 5; ovary 5-chambered; ovules 2 per carpel, pendulous. Fruit a capsule.

The family includes a single species restricted to Hongkong.

*Family 13. **Elatinaceae.**

WATERWORT FAMILY

Annual or perennial aquatic herbs or suffrutescent plants. Leaves simple, opposite or whorled; stipules present, usually membranous. Flowers bisexual, radially or bilaterally symmetrical, very small, solitary or cymose; sepals 3–5; petals 3–5; stamens 6–10; carpels 3–5, the styles and chambers of the ovary the same number; placentae axile; ovules numerous. Fruit a capsule.

The family consists of two genera, *Elatine* and *Bergia,* the former occurring in fresh water almost throughout the world, the latter restricted to the Tropics and usually warm temperate regions. Four species of *Elatine* are widespread in North America; one of *Bergia* occurs from Washington to southern Illinois, southern California, and Texas.

Family 14. **Nepenthaceae.**

NEPENTHES FAMILY

Herbs to shrubs, often climbing as described below. Leaves simple (cf. description above); stipules none; flowers unisexual and the plants dioecious, radially symmetrical, small, greenish,

Figure 10-19. Theales; Guttiferae; *Hypericum,* St.-John's-wort: **A** habit; **B** flower bud; **C** flower, **1** from above, **2** from below, **3** longitudinal section; **D** pistil, **1** pistil with adjacent petal, **2** pistil alone (extreme right), **3** cross section of the ovary, showing the partitions (intruded placentae); **E** enlargement of leaf surface showing dotlike glands. (Cf. also **Figure 8-6 B** *Hypericum,* fruit.)

in racemes or panicles; sepals 3–4, in 2 series; petals none; stamens 4–24, the filaments coalescent into a tube; carpels 3–4, the style 1 or obsolete, the ovary 3–4-chambered; placentae axile; ovules numerous in each carpel. Fruit a leathery capsule, elongate.

Asian pitcher-plants. The family occurs from southern China to Australia, being most abundant in Borneo.

The plants are cultivated as novelties in greenhouses and homes in the United States.

In the *Nepenthaceae,* the leaf consists of three parts: (1) a flat ordinary leaflike basal portion manufacturing food as in other leaves; (2) a twining middle portion which serves as a tendril, supporting the plant as it scrambles over other vegetation; and (3) a pitcherlike terminal portion with a "hinged" lid tipping like the top of a syrup pitcher. Nectar secreted about the rim of the pitcher, together with the brilliant coloring, attracts insects, some of which fall into the fluid at the bottom. The fluid contains digestive juices.

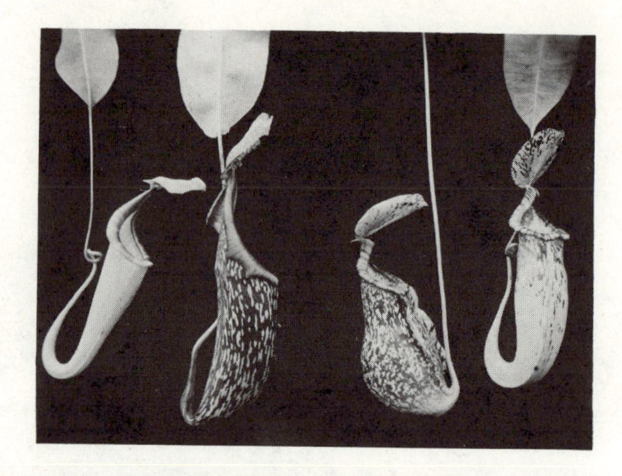

Figure 10-20. Theales; Nepenthaceae; *Nepenthes,* pitcher-plant I—leaves, each with (1) a flat basal portion, (2) a tendril-like twining middle portion, and (3) a pitcherlike apical portion, which captures insects. Photograph by Ladislaus Cutak.

Figure 10-21. Theales; Nepenthaceae; *Nepenthes,* pitcher-plant II—inflorescence. Photograph by Ladislaus Cutak.

Order 12. *SARRACENIALES.*

PITCHER-PLANT ORDER

DEPARTURE FROM THE FLORAL CHARACTERS OF THE RANALES

Free	Stamens sometimes coalescent; carpels coalescent.
Numerous	Carpels 3–5.
Spiral	Carpels cyclic.

Herbs growing in marshes or peat moss or sometimes in moist soil elsewhere. Leaves alternate but in North America largely or usually wholly basal. Flowers 5-merous except that the stamens are numerous and the carpels 3–5. Sepals overlapping in the bud. Ovary 3–5-chambered; placentae either axile or parietal; ovules numerous. Fruit a capsule; seeds with endosperm.

*Family 1. **Sarraceniaceae.**

PITCHER-PLANT FAMILY

Perennial herbs. Leaves basal, simple, tubular, usually with the terminal portion nontubular, the tube sometimes winged, often with internal downward-directed hairs; stipules none. Flower bisexual, radially symmetrical, solitary at the apex of a scape or rarely the flowers in racemes; sepals 4–5, separate, persistent in fruit, commonly colored; petals 5 or none; stamens numerous, irregularly arranged; carpels 3–5, the style 1, much enlarged and peltate apically, the stigmas as if on the ends of the ribs of an umbrella, the ovary 3–5-chambered; placentae axile. Fruit a capsule.

The family is restricted to North and South America, chiefly Eastern North America, where there are about 9 native species of *Sarracenia.* The single species of *Darlingtonia* with more elaborate pitcherlike leaves occurs in the mountains of extreme southwest Oregon and far northern California.

In the Sarraceniaceae, or pitcher-plant family, the entire leaf forms a pitcher in which insects are trapped. In some instances brilliant coloring of the upper part as well as the presence of nectar attract insects. In *Darlingtonia californica,* the only pitcher-plant occurring in Western North America, the upper part of the pitcher forms a hood which insects enter. Windowlike areas on the upper side of the hood admit light, attracting the insect, which beats against them using up its energy. Eventually it falls to the bottom of the elongated tube on the lower portion of the leaf, where it is retained by downward-directed hairs. Ultimately the insects are digested by bacterial action. The results are not pleasing to the olfactory sense. In the Eastern pitcher-plants (*Sarracenia*) digestion is by enzymes (digestive juices).

Order 13. ***MALVALES.*** MALLOW ORDER

DEPARTURE FROM THE FLORAL CHARACTERS OF THE RANALES

Free	Petals rarely slightly coalescent but often adnate with the base of the filament tube; stamens usually coalescent; carpels coalescent, though sometimes only lightly so until maturity.
Equal	Corolla rarely somewhat bilaterally symmetrical in the *Bombacaceae*.
Numerous	Stamens usually numerous but sometimes 5–10, the outermost 5 sometimes sterile; carpels many or usually 6–15, sometimes as few as 5 or 3 or reportedly 1.
Spiral	Stamens cyclic when they are few; carpels cyclic.

Herbs, shrubs, or trees. Stellate (star-shaped) or other branched hairs common. Leaves alternate, simple, often palmately lobed or veined; stipules present. Soft tissues (cortex and pith) commonly with special mucilage cells or ducts. Flowers bisexual, 5-merous except for the stamens and carpels; sepals in the bud not overlapping, thus being valvate; placentae axile but these and the partitions of the ovary reported to be derived from intrusion of parietal placentae. Ovules anatropous, with 2 integuments.

The most characteristic and best known members of this order are the family *Malvaceae* (mallow family), which includes such plants as the common garden hollyhock and the mallow or cheeseweed.

There is little departure from the typical flower structure of the Ranales, and the most significant difference is in the light coalescence of carpels which remain coalescent at least until fruiting time when they may separate. Usually the stamens are coalescent, too. The form of the leaf is characteristic, though modified in some genera, and it is reminiscent of many Ranunculaceae.

The order is related closely to the Geraniales, Theales, Papaverales, Urticales and Euphorbiales.

Key to the Families

(North America north of Mexico)

1. Stamens numerous, many more than the sepals.
   2. Stamens coalescent into a single group (monadelphous), the group often adnate with the petals and the bases of the petals sometimes coalescent into a very short tube.

      Family 2. **Malvaceae** (mallow family).

   2. Stamens separate or coalescent into several groups (polyadelphous), free from the petals; petals separate.

      Family 4. **Tiliaceae** (linden family).

1. Stamens as many or twice as many as the sepals, therefore few.

   Family 5. **Sterculiaceae** (chocolate family).

Family 1. Elaeocarpaceae.

ELAEOCARPUS FAMILY

Shrubs or trees. Leaves simple, entire, alternate or opposite; stipules present. Flowers bisexual,

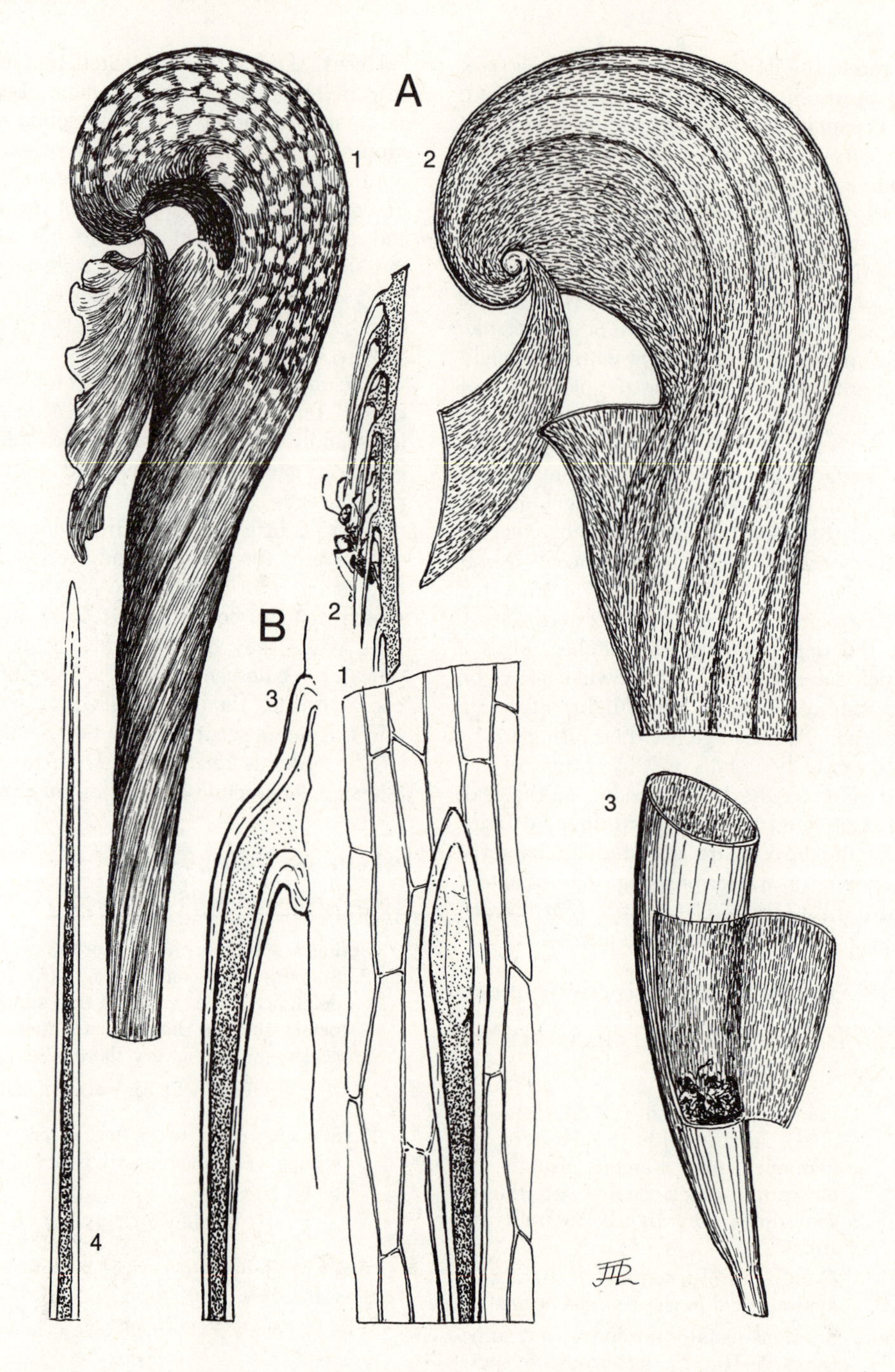

Figure 10-22. Sarraceniales; Sarraceniaceae; *Darlingtonia californica,* California pitcher-plant: **A** basal leaf, **1** upper portion showing "windows," **2** longitudinal section of the uppermost portion, showing the numerous downward-directed internal hairs, **3** basal portion cut open and revealing trapped insects; **B** enlargements of hairs, **1** cellular detail of the inner epidermis of the leaf and of a hair, **2** ant attempting to crawl upward among the hairs, **3–4** enlargement of a single hair.

radially symmetrical, with hypogynous discs, in cymes, racemes, or panicles; sepals 4–5; petals 4–5 or wanting, sometimes coalescent basally; stamens many, separate; anthers opening by terminal pores; carpels 2–many, the style 1, sometimes lobed; ovules 2–numerous in each chamber of the ovary. Fruit a capsule or a drupe.

The family is distributed widely in the Tropics; two genera occur in Mexico.

Elaeocarpus, *Muntingia*, *Crinodendron*, and *Aristotelia* are cultivated in the southern portions of the United States.

*Family 2. **Malvaceae.** MALLOW FAMILY

Herbs or less frequently shrubs or trees; the leaves, young branches, or reproductive parts often with stellate hairs. Leaves simple, entire to lobed or parted, most frequently palmately lobed, palmately veined; stipules present. Flowers bisexual, rarely unisexual, radially symmetrical, solitary or cymose; sometimes with an involucre of a few small calyxlike bracts; sepals 5; petals 5, separate, but sometimes slightly coalescent or frequently adnate with the base of the filament tube and thus indirectly with each other; stamens numerous, coalescent into a tube (monadelphous); anther with 1 pollen chamber, splitting lengthwise; carpels typically 6–many, uncommonly 3–5, usually in 1 series, the style 1, branched near the apex. Fruit a capsule, or with the carpels separating at maturity, or rarely a berry.

The family is large and world-wide, and it includes many attractive native and cultivated genera and species. Perhaps the best known plant is the common garden hollyhock (*Althaea*); other well-known plants are the mallows or cheeseweeds (*Malva*), so named because the fruits separate into sections resembling pieces of the circular cheeses of the old-fashioned grocery stores. *Hibiscus* is cultivated widely along the southern edge of the United States, and it is represented by native species occurring from the Southwestern Deserts eastward to the Southern States. The cotton plant (*Gossypium*) is a member of the *Malvaceae;* species occur in Arizona and perhaps Florida. The one in Arizona has been known as *Thurberia,* but the segregation is unwarranted. The plant is a shrub. Cotton fibers are the hairs on the seeds; oil and pulp meal are obtained also from the seeds. Okra is the fruit of a species of *Hibiscus.*

Family 3. **Bombacaceae.**
COTTON-TREE FAMILY

Trees; leaves and young branches with stellate hairs or peltate scales. Leaves simple or palmate, alternate, deciduous; stipules falling while the leaf is young. Flowers bisexual, large, often present while the tree is leafless, usually radially symmetrical, rarely slightly bilaterally symmetrical; sepals 5; petals 5 or sometimes none; stamens 5 to numerous, separate or coalescent into a tube; anthers with 1 or 2 pollen chambers; carpels 2–5, the style 1. Fruit a capsule or sometimes indehiscent or a berry.

Tropical, occurring especially in Tropical America. *Ceiba* (fruits the source of kapok), *Adansonia* (including the baobob, a tree of enormous trunk diameter), *Ochroma* (source of balsa wood), *Chorisia* (Brazilian silkfloss-tree), and *Bombax* (cotton-tree) are important genera.

Some genera are cultivated in the Southern States, especially in Florida.

*Family 4. **Tiliaceae.** LINDEN FAMILY

Trees or shrubs or rarely herbs; hairs mostly branched. Leaves simple, usually alternate, commonly deciduous; stipules present. Flowers usually bisexual, sometimes unisexual (and then the plants monoecious), radially symmetrical, in cymes; sepals usually 5, rarely fewer; petals usually 5, rarely fewer; stamens usually numerous, sometimes as few as 10, separate or coalescent into fascicles; anthers with 2 pollen chambers, with lengthwise slits or apical pores; carpels 2–10; placentae usually axile in position but fundamentally parietal. Fruit a capsule or indehiscent or a berry.

The family is large and mostly tropical. A few species of *Tilia* (basswood or linden) are native in eastern North America. Two Tropical American species of *Triumfetta* (burweed) are introduced in Florida; one yields a fiber similar to jute.

Three species of *Corchorus* (source of jute fiber) either occur along the southern edge of the United States, from perhaps though uncertainly Arizona eastward, or grow in Florida. Several other genera are cultivated in the warmer parts of the United States.

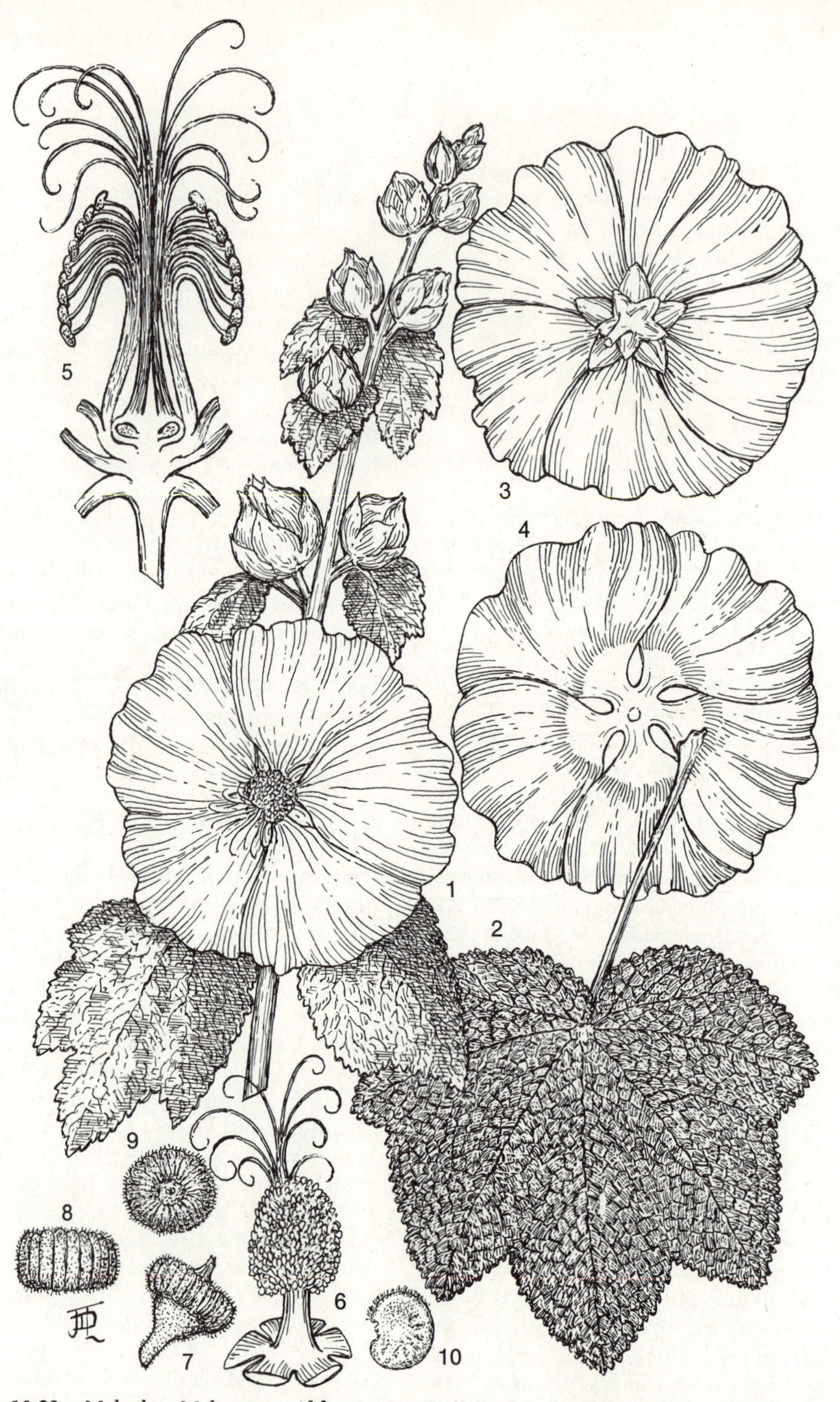

Figure 10-23. Malvales; Malvaceae; *Althaea rosea,* hollyhock: **1** habit, **2** leaf, **3** flower from beneath, showing one series each of bracts, sepals, and petals, **4** corolla from above, the petals basally coalescent (also adnate with the filament tube), **5** longitudinal section of the central portion of the flower, showing the coalescent stamens and coalescent carpels, **6** external view, showing the stamens and the stigmas, **7–8** young fruit, side view, **9** top view showing the coalescent carpels, **10** single carpel in side view.

Figure 10-24. Malvales; Tiliaceae; *Tilia,* linden or basswood: **1** habit, showing the peduncle of the inflorescence fused with the basal portion of the bract, **2–3** views of a flower, **4** one sepal, one petal, stamens, and the pistil, **5** stamens, these coalescent only basally, **6** pistil and the calyx, **7** diagrammatic cross section of the ovary, **8** side view of the fruit, **9** longitudinal section.

*Family 5. **Sterculiaceae.**

STERCULIA FAMILY

(Synonym: *Buettneriaceae*)

Herbs, shrubs, trees, or sometimes woody vines; leaves and young parts with stellate hairs. Leaves simple or compound, alternate; stipules falling away early. Flowers bisexual or sometimes unisexual (and then the plants monoecious), usually radially but sometimes bilaterally symmetrical; sepals 3–5, coalescent basally; petals 3–5 or none, often small or minute; stamens several, in usually 1 or 2 series, 1 series often sterile (of staminodia), coalescent or sometimes separate; anthers splitting lengthwise; carpels 4–5 or rarely up to 12, the styles as many as the carpels or more or less coalescent into 1. Fruit fleshy, woody, or coriaceous. (Cf. Figure 6-2 B, *Sterculia*, flower.)

The family is distributed throughout the Tropics and Subtropics, and sparingly in warm temperate regions. Five genera occur as native plants in the southern portions of the United States. These include *Waltheria* (Arizona and Florida), *Hermannia* (mountains near Tucson, Arizona), and *Fremontodendron*,* known as fremontia.

The two species (or the varieties of a single variable species) of *Fremontodendron* are attractive small trees or large shrubs much used in cultivation. The flowers, which lack petals, have a bright yellow calyx, and they are an inch or two in diameter. The most remarkable tree in this family is *Theobroma Cacao*, the cocoa tree, native in tropical America but cultivated widely in the Tropics of the Eastern Hemisphere. This plant has large reddish yellow relatively fleshy fruits. The seeds are the source of chocolate and cocoa. *Cola acuminata*, the cola nut native in tropical western Africa, is the source of the cola drinks. A magnificent plant cultivated in the southern portions of the United States is the flame-tree (*Sterculia*). This tree, like many other tropical trees, loses its leaves early in the summer, then bursts into a display of (red) flowers. The bottle-tree (*Sterculia*), native in eastern Asia, is cultivated from California eastward across the southern portions of the United States.

* *Fremontodendron* is accepted instead of the better-known name *Fremontia*, which was used first for a plant in the Chenopodiaceae. John Torrey, who a century ago applied *Fremontia* in both cases, did not anticipate the present rules, which do not permit withdrawing a name and reapplying it.

Family 6. **Scytopetalaceae.**

SCYTOPETALUM FAMILY

Trees. Leaves simple, alternate; stipules none, Flowers bisexual, radially symmetrical, perigynous, in fascicles, racemes, or panicles; sepals not distinguishable; petals 3–10; stamens numerous, sometimes coalescent basally; anthers opening by pores or slits; carpels 3–6, the style 1. Fruit woody.

The family consists of about ten species in West Africa.

Family 7. **Sarcolaenaceae.**

SARCOLAENA FAMILY

Shrubs or trees. Leaves simple, entire, alternate; stipules falling while the leaves are young. Flowers bisexual, radially symmetrical, sometimes with involucres, in cymes or panicles; sepals 3; petals 5–6; stamens 10 or more, surrounded by a cup formed from staminodia; anthers splitting lengthwise; carpels 3, the style 1. Fruit a capsule or sometimes with only 1 developed seed.

The family consists of seven small genera occurring in Madagascar.

Order 14. ***EUPHORBIALES.***

SPURGE ORDER

DEPARTURE FROM THE FLORAL CHARACTERS OF THE RANALES

Free	Stamens sometimes coalescent; carpels coalescent.
Numerous	Stamens from numerous to 1, mostly 5 or 10; carpels usually 3 but sometimes 2 or 4.
Spiral	Cyclic, except sometimes the stamens.

Herbs, shrubs, or sometimes trees. Leaves usually simple, sometimes deeply parted, or sometimes palmate; stipules none. Flowers unisexual (and the plants monoecious), usually with sepals but rarely with petals; stamens mostly the same number as the petals and opposite them or twice as many or numerous or 1; pistil 1, the ovary with usually as many chambers as carpels (2–4, usually 3), rarely with 1 chamber; each carpel producing on its inner angle 1–2 anatropous pendulous ovules. Fruit nearly always a capsule which opens elastically, otherwise indehiscent or berry-like.

Key to the Families
(North America north of Mexico)

1. Fruit splitting open (dehiscent) along the line of joining of the carpels, that is, through the partitions (septicidal); raphe (part of the stalk or funiculus) adhering to the body of the seed and forming a ridge on the front (ventral side) of the seed.

   Family 1. **Euphorbiaceae** (spurge family).

1. Fruit splitting open along the midrib of each carpel (loculicidal); raphe on the back of the seed (dorsal). (In North America only an opposite-leaved evergreen desert shrub and a procumbent, glabrous, alternate-leaved, rhizomatous perennial herb of the mountains of the Southern States.)

   Family 2. **Buxaceae** (box family).

*Family 1. Euphorbiaceae.
SPURGE FAMILY

Herbs, shrubs, or trees; juice often milky or acrid. Leaves simple or compound, usually alternate, but sometimes opposite or whorled; stipules present or none. Flowers radially symmetrical, sometimes with a hypogynous disc; sepals most commonly 5, sometimes none; petals most commonly 5 or usually none; stamens frequently the same number as the sepals and alternate with them, sometimes only 1 (*Euphorbia*), in 2 series, rarely numerous as in *Ricinus* and some species of *Croton*, separate or coalescent; carpels 3 or rarely 2 or 4, the styles 3, separate or coalescent basally, sometimes divided above, the ovary with as many chambers as carpels or rarely 1-chambered; ovules 1 or 2 in each carpel, pendulous, the raphe ventral. Fruit usually a capsule (rarely indehiscent or a berry), usually splitting elastically at maturity into 3 1-seeded segments, each of these opening on the ventral angle.

This is one of the four or five largest flowering plant families, ranking after the *Compositae, Leguminosae*, and *Orchidaceae* and about equal to the *Gramineae*. The family is world-wide but chiefly tropical in distribution. Most of the North American genera and species occur in the warm southern parts of the United States, and they are still more abundant in Mexico and South America. Northward the family is better represented in the Eastern and Middle Western states than on the Pacific Coast.

Among the more noteworthy plants of the family are the poinsettias (*Euphorbia*), striking for their large red bracts and for flowering at Christmas time and the castor-bean (*Ricinus*), the seeds of which yield castor oil. The whole castor-bean plant (especially the seeds) contains the acrid poison ricin. Many succulent Eastern Hemisphere species of *Euphorbia* superficially resemble the cacti of the Western Hemisphere, and they are cultivated as novelties. Other genera are *Croton*, the source of croton oil; *Mercurialis* (mercury); *Manihot* (cassava), the roots of which yield tapioca although they contain a poison which must be dissolved out; *Sapium* (native from Arizona southward) and *Sebastiana* (the Mexican jumping beans), the seeds of which "jump" because of the rapid movement of the contained larva of a moth. Several trees have rubber in their latex, and *Hevea braziliensis* is the chief rubber tree of commerce.

Flower structure varies greatly from genus to genus. The most highly specialized type is that of *Euphorbia*, which has a single pistillate flower aggregated with a number of staminate flowers and enclosed in a cuplike structure with glands on its margins. Appendages of the glands often are petal-like, and the whole structure then appears to be a flower. The pistil is elevated upon a pedicel, and the pedicels of the staminate flowers have a joint between them and a similar filament.

*Family 2. Buxaceae. BOX FAMILY

Herbs, shrubs, or trees. Leaves simple, leathery, evergreen, opposite or sometimes alternate; stipules none. Flowers radially symmetrical, usually small; sepals 4 or sometimes none, coalescent basally; petals none; stamens usually 4 or 10–12 or (in extraterritorial genera) numerous; carpels 3 or rarely 2 or 4, the styles and chambers of the ovary as many as the carpels; ovules usually 2 per carpel, pendulous, the raphe dorsal. Fruit a capsule, opening elastically along the midrib of each carpel or sometimes indehiscent or berrylike.

The family occurs chiefly in the tropical and subtropical portions of the Eastern Hemisphere.

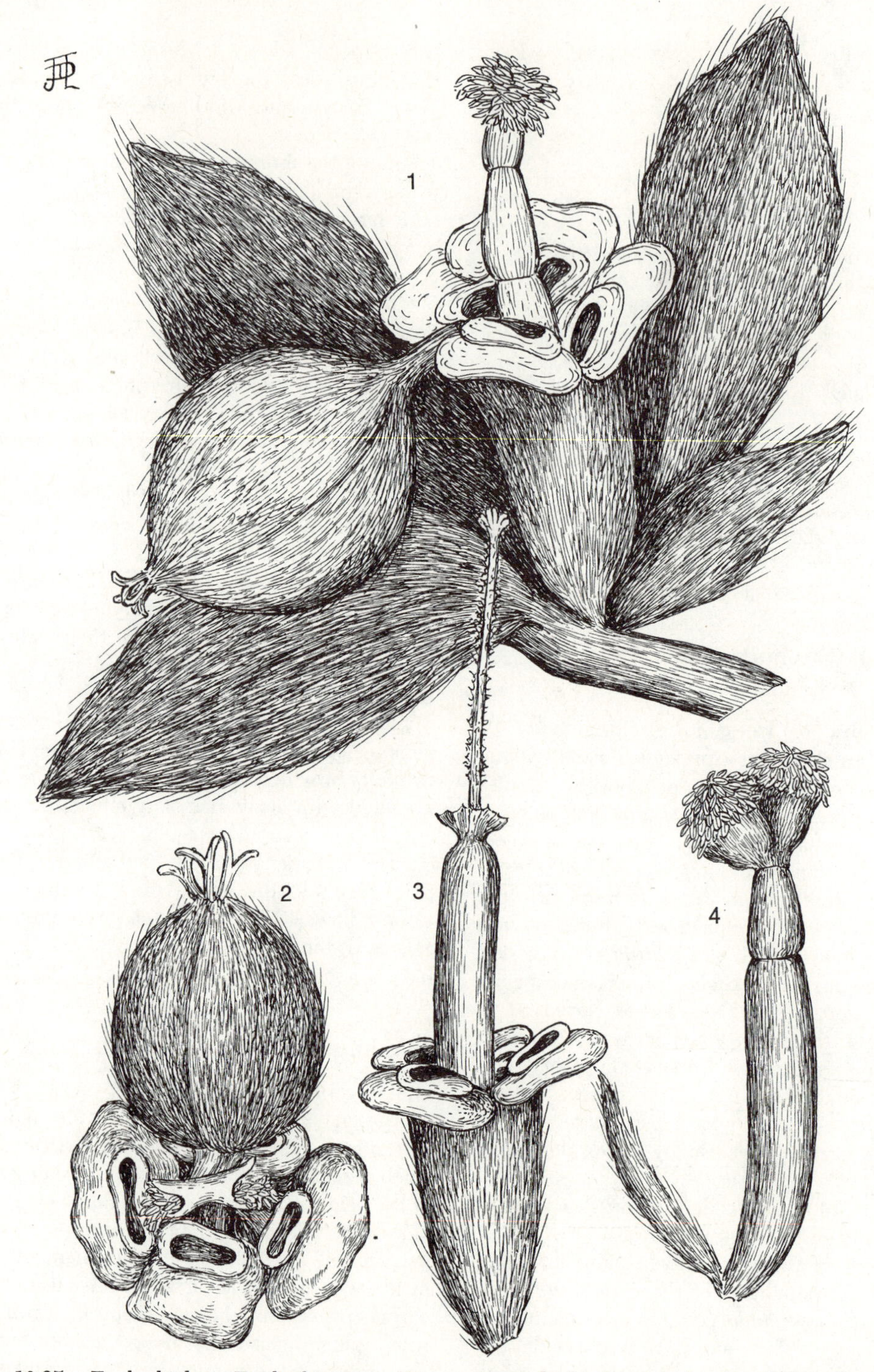

Figure 10-25. Euphorbiales; Euphorbiaceae; *Euphorbia,* a spurge: 1 upper portion of a branch with leaves and a flower cluster, one flower pistillate and composed of only a pistil on an elongated pedicel, the others (only one showing) staminate and each composed of a single stamen, the cluster surrounded by a

Only two species occur in North America. The Allegheny spurge (*Pachysandra*) is native from Kentucky to Louisiana and western Florida. Sometimes it is cultivated. A remarkable plant, the jojoba or goatnut (*Simmondsia*) is one of the most interesting in the Southwestern Deserts, particularly in the Colorado Desert of California and Arizona and the Arizona Desert in Arizona. The seeds yield a wax now under intensive study for its usefulness. The fruits resemble acorns, being surrounded basally by a cup formed from the persistent large green sepals. The shrubs are blue-green and rather attractive. *Simmondsia* may belong in a family by itself. Relationships are not clear.

The box (*Buxus*) is common in cultivation especially for borders and low hedges. An Oriental species of *Pachysandra* is cultivated also. The boxwood (*Buxus*) yields a fine-grained hard wood used for tool handles, carved objects, rules, and musical instruments.

Family 3. *Aextoxicaceae.* AEXTOXICON FAMILY

Shrubs. Leaves simple, lanceolate, alternate. Flowers unisexual, radially symmetrical, in racemes. Sepals 4–6, coalescent; petals none; stamens few; carpels 2, the style 2-parted, the ovary fundamentally 2-chambered, but with only 1 developing; ovules 2 per carpel, pendulous. Fruit a drupe.

The family includes a single species restricted to Chile.

Family 4. **Pandaceae.** PANDA FAMILY (Not *Pandanaceae*)

Small trees. Leaves simple, alternate; stipules present. Flowers unisexual (and the plant dioecious), radially symmetrical, with hypogynous discs; sepals 5, the calyx forming a cup; petals 5, stamens 10, in 2 series of differing lengths; carpels 3–4, the style with 3–4 branches, the ovary 3–4-chambered; ovule 1 in each carpel, pendulous from near the apex. Fruit a drupe or drupelike.

The family consists of the single genus *Panda*, occurring in tropical west Africa.

Order 15. *URTICALES.* NETTLE ORDER

DEPARTURE FROM THE FLORAL CHARACTERS OF THE RANALES

Free	Carpels coalescent when there are 2 or more.
Numerous	Stamens usually 4 or 5; carpels 1 or 2.
Spiral	Stamens usually cyclic; carpels cyclic.

Herbs, shrubs, or trees. Leaves alternate, simple; stipules present. Flowers unisexual or often bisexual in the *Ulmaceae*, usually 4- or 5-merous, except for the 1 or 2 carpels; ovary nearly always 1-chambered; ovule solitary. Fruit an achene (or sometimes each achene enclosed in a fleshy covering formed from the calyx and the many fruits aggregated together as for example in the mulberry, or sometimes the many achenes in a hollow, closed, compound receptacle as in the fig), a samara, a drupe, or an irregular warty nut; endosperm none, or, if present, fleshy or oily.

Some authors consider the four major families to be subfamilies of a single family (*Urticaceae*).

Key to the Families

(North America north of Mexico)

1. Fruit a samara (winged), or drupelike, *or* an irregular, warty nut.

   Family 1. **Ulmaceae** (elm family).

1. Fruit an achene, *sometimes* each achene enclosed in a fleshy covering (the calyx) and many fruits aggregated together (mulberry) *or sometimes* the many achenes in a hollow, closed compound receptacle (fig).
   2. Flowers numerous on or inside a large compound receptacle; *either* (1) each achene en-

---

cup (cyathium) bearing four glands each with a petallike appendage, **2** flower cluster from another view, **3** remains of the flower cluster after disappearance of the staminate flowers and dehiscence of the capsular fruit, leaving the axis of the ovary, **4** staminate flower; the joint in the "stalk" is between the filament and the pedicel (and receptacle). Note the peculiar anther and its pollen chambers.

closed in a fleshy calyx and all the achenes aggregated into an elongate or spherical mass (mulberry) *or* (2) all the achenes enclosed in the fleshy or somewhat leathery compound receptacle (fig).

Family 2. **Moraceae** (mulberry family).

2. Flowers not on or inside a compound receptacle; sepals not fleshy in fruit and not enveloping the achene.
   3. Stigmas 2; filaments erect in the bud; ovule hanging downward.

      Family 3. **Cannabinaceae** (hemp family).

   3. Stigma 1; filaments incurved or bent or coiled inward in the bud, springing outward as the flower opens and throwing the pollen into the air; herbs usually but not always with stinging hairs; ovule standing upward.

      Family 4. **Urticaceae** (nettle family).

*Family 1. Ulmaceae. ELM FAMILY

(Sometimes included in *Urticaceae*)

Shrubs or trees. Leaves simple, often asymmetrical basally (i.e., with one side developed differently from the other), alternate; stipules early deciduous. Flowers bisexual or all or some unisexual (and then the plant polygamous or polygamodioecious), radially symmetrical, sometimes perigynous, solitary or in fascicles; sepals 4–8; stamens usually as many as the sepals and opposite them; carpels 2, the styles separate, the ovary with 1 chamber; placentae apical; ovule 1, pendulous. Fruit a samara or a drupe.

The family occurs in much of the Northern Hemisphere, and it is abundant in tropical and subtropical regions. Elms (*Ulmus*) occur in Europe and Asia, and there are about seven species in eastern North America, but none in western North America. The trees flower very early, and the inconspicuous flowers are likely to be overlooked. The fruits are more conspicuous, and they catch attention as they fall from the tree when the leaves start to grow. The water elm (*Planera*) occurs in the Southern States and northward as far as southeastern Missouri and southern Illinois. About three species of hackberry (*Celtis*) occur in eastern North America. Only one of the treelike hackberries (*Celtis laevigata* var. *reticulata*) occurs in western North America. This plant ranges from Washington and south central and southern California eastward to New Mexico and northwestern Mexico. It is abundant in the Southwestern Deserts, as is another type of hackberry, the shrub (*Celtis Tala* var. *pallida*).

*Family 2. Moraceae. MULBERRY FAMILY

(Sometimes included in *Urticaceae*.
Synonym: *Artocarpaceae*)

Shrubs, trees, or rarely herbs; juice milky; leaves simple, sometimes palmately lobed, alternate or usually so; stipules early deciduous, leaving scars. Flowers unisexual, usually in heads or more or less catkinlike but sometimes (*Ficus*) the inflorescences lining the inside of the hollow end of a branch (fig), radially symmetrical; sepals usually 4, coalescent; stamens 4 or in some extraterritorial genera 1 or 2; carpels 2, the styles 2, the ovaries usually 1-chambered through the development of only 1 carpel; placenta apical, the ovule 1. Fruit an achene or a drupe, the fruits frequently massed in dense heads, each fruit often adnate with the perianth and the receptacles of other flowers and the whole forming an aggregate fruit, in *Ficus* the achenes enclosed in an enlarged fleshy branch enclosing the inflorescence.

This large family occurs throughout the Tropics and Subtropics, and some genera range northward in the warm temperate regions. Three species of mulberry (*Morus*) occur in eastern North America. Two, the white and black mulberry, are introduced from Eurasia; the other, the red mulberry, is native. The paper mulberry (*Broussonetia*) is introduced from Asia, and it has gone wild in some places. A shrubby mulberry occurs from Arizona to West Texas. The Osage-orange (the only species of *Maclura*) is native in Arkansas and Texas, and it has become naturalized through cultivation in many parts of eastern North America. It is cultivated also here and there westward to California. The strangler fig and one other native species of *Ficus* occur in peninsular Florida and on the Keys. The seedlings of the strangler fig grow on trees, their roots surrounding the trunks. The other species, known as the wild fig or wild banyan, does not enclose so tightly the trunk of the tree upon which it grows. Both the India-rubber-plant and the common fig have escaped from cultivation in the Southern States, the former in south peninsular Florida.

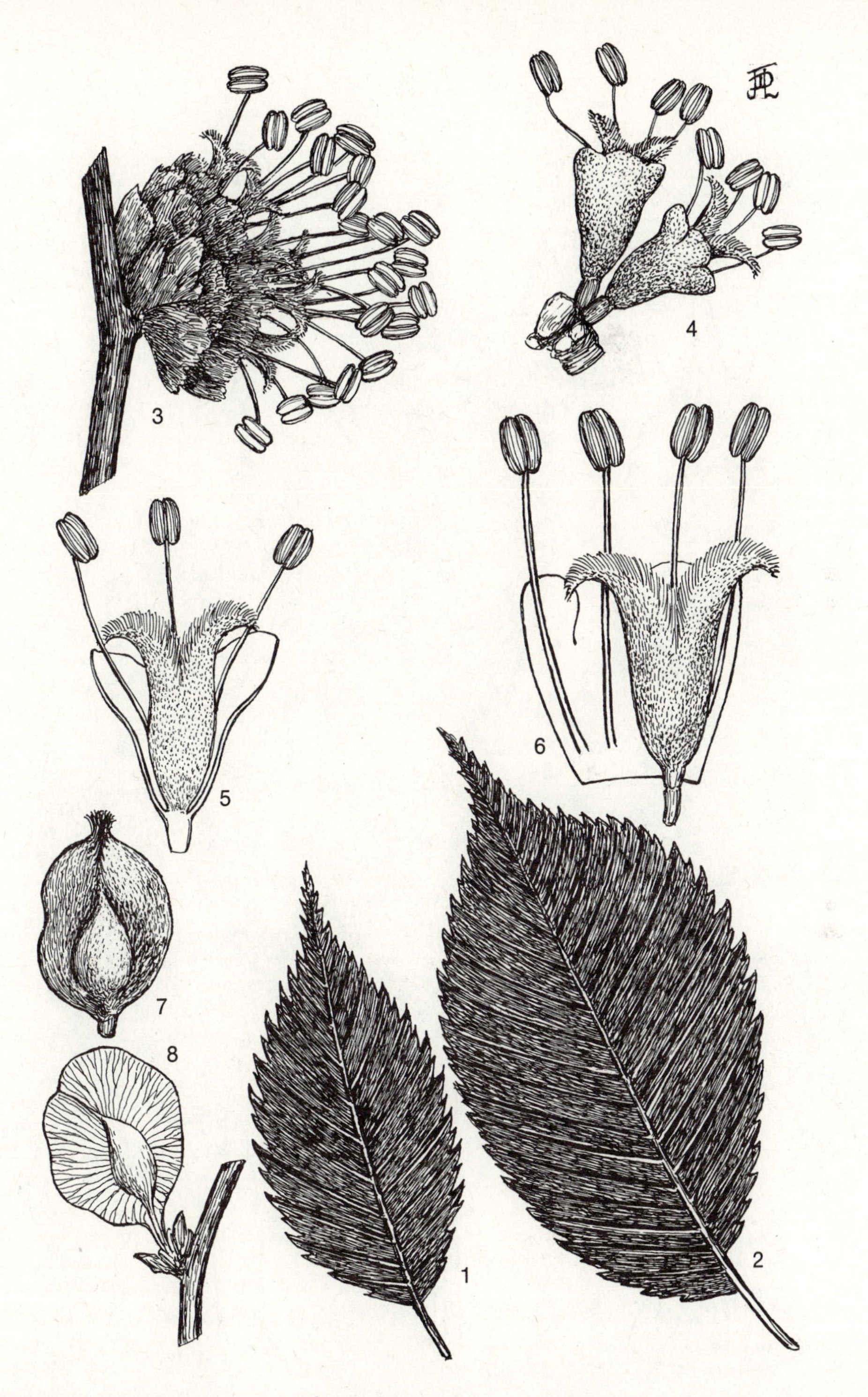

Figure 10-26. Urticales: Ulmaceae, *Ulmus,* elm: **1–2** leaves, the larger one with the unequal sides characteristic of the family, **3** flower cluster (appearing before the leaves), **4** flowers, **5** flower, longitudinal section, showing the sepals, stamens, and pistil, **6** flower laid open, **7** young fruit, **8** mature samara.

Figure 10-27. Urticales; Moraceae; *Ficus Carica,* fig: **1** leaf, **2–3** "fruit," formed from an enlarged hollow branch, the flowers and later the achenes inside, **4–7** pistillate flowers each composed of sepals and a pistil, **8–10** exterior development of flowers in an abnormal fig.

Other important economic plants in the family, in addition to the figs and mulberries, are the breadfruit and jackfruit of the genus *Artocarpus*.

*Family 3. **Cannabinaceae.**

HEMP FAMILY

(Sometimes included in *Urticaceae*)

Herbs, juice not milky. Leaves simple, palmately veined and usually palmately lobed to divided, alternate; stipules persistent. Flowers unisexual (and the plant dioecious), radially symmetrical, the staminate in loose racemes or panicles, the pistillate in dense clusters; sepals 5, in the pistillate flowers completely joined and forming a cup, which encloses the ovary; stamens 5; carpels 2, the styles 2, the ovary with only 1 chamber because of the abortion of 1 carpel; ovule 1. Fruit an achene, usually glandular.

The family is composed of two genera, both occurring in the Northern Hemisphere, including North America, at least as introduced plants. The hemp family includes only one native species, the hop, *Humulus Lupulus*.

Hops are used in brewing malt beverages and in medicine. They are native in eastern North America and as far westward as the mountains of eastern Arizona, and they have escaped from fields in many places, as well. The Japanese hop is naturalized also in eastern North America. A plant which has attracted much attention of late is the hemp, source of marijuana. This plant is occasional in waste places across the whole southern part of the United States from California to Florida and rarely as far northward as Quebec. For the most part it is sporadic in occurrence. At one time the seeds were a principal ingredient of pigeon feed and a great favorite of these birds. They contain a narcotic which may produce hallucinations, as do other parts of the plant, including the leaves. For many centuries the narcotic of the tender parts has been known as *hashish* or Indian Hemp. The drug known as marijuana is obtained from the staminate flowers. Marijuana or "reefer" cigarettes have been smuggled into the United States from areas to the southward where the southern narcotic-rich strain is more common. The hemp fiber of commerce comes from the stem of a northern strain of the same species with better fiber and less of the narcotic.

*Family 4. **Urticaceae.** NETTLE FAMILY

Herbs or rarely small trees, shrubs, or vines. Juice not milky. Leaves simple, alternate or opposite, stipules present or rarely none (*Parietaria*). Flowers unisexual or rarely more or less bisexual, radially symmetrical, small, green or yellowish; sepals 4 or 5; stamens usually 4, or sometimes 3 or 5, the filaments folded inward in the bud but springing outward as the flower opens and releasing the pollen; carpel 1, the ovule solitary, more or less basal, appearing to be orthotropous but actually anatropous. Fruit an achene or sometimes a drupe, sometimes enclosed by the calyx.

The family is distributed widely, but most abundant in the Tropics. Several species occur in the United States, one or a few in each part of the Continent. The nettles are chiefly of the widespread genus *Urtica*. In eastern North America and especially the Southern States there are species of wood nettle (*Laportea*), *Pilea*, pellitory (*Parietaria*, two species of which range westward to Arizona), and *Boehmeria*. The ramie of commerce (*Boehmeria nivea*) has escaped in the Southern States, where it is cultivated for its coarse fiber. A single species of *Hesperocnide* occurs in California. In general the family is represented poorly in the West.

The nettles are known for their stinging hairs. The sharp-pointed terminal cells have hard glass-like cell walls which penetrate human skin, forming effective hypodermic needles. Usually nettles may be picked up by using a quick firm pressure with the tough skin of the thumb and forefinger, but touching them casually with soft skin brings about disastrous results, the extent of which depends upon the sensitivity of the person to these plants.

Order 16. ***GERANIALES.***

GERANIUM ORDER

DEPARTURE FROM THE FLORAL CHARACTERS OF THE RANALES

Free	Carpels coalescent, sometimes only lightly so at flowering time and separating at fruiting time, in the *Limnanthaceae* only the bases of the styles coalescent and the ovaries free.
Equal	Flowers sometimes bilaterally symmetrical as in some garden gerani-

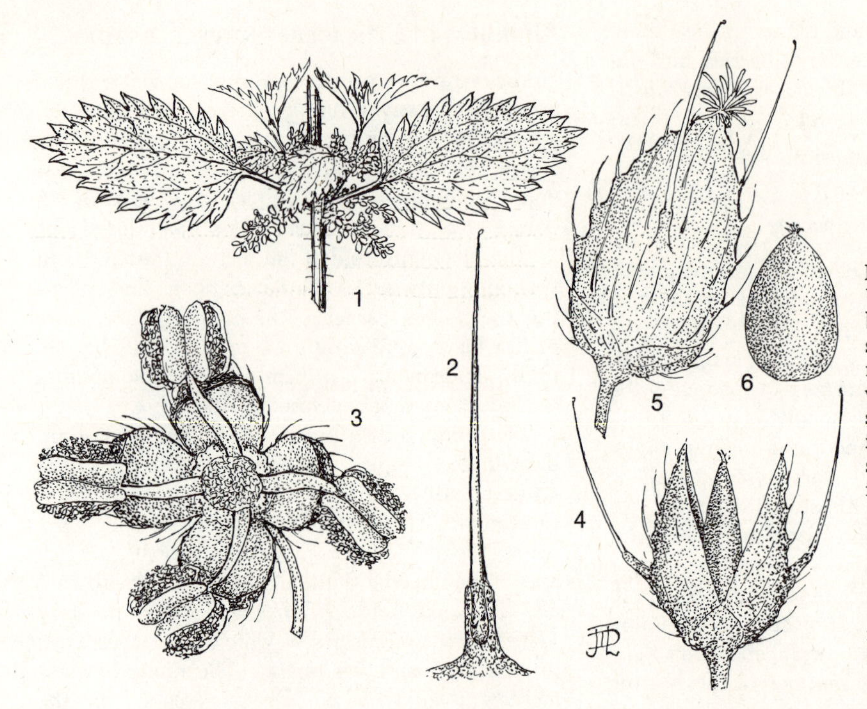

Figure 10-28. Urticales; Urticaceae; ***Urtica urens***, burning nettle, a common, small, inconspicuous, annual species: **1** branch with inflorescences, the staminate and pistillate flowers intermingled, **2** stinging hair, **3** staminate flower, composed of four sepals and four stamens, **4** pistillate flower, composed of four sepals in two unlike series, the outer pair minute, the inner much larger, hispid, **5** inner sepal enlarged, **6** achene.

ums (*Pelargonium*), the *Tropaeolaceae*, and the *Balsaminaceae*.

Numerous Stamens 5–15, usually 10 or 5; carpels commonly 5 or 3, rarely 2 or 4.

Spiral Stamens and carpels cyclic.

Herbs or sometimes woody vines, shrubs, or small trees. Flowers bisexual, 5-merous, or rarely 3-merous (*Floerkia*) or in a few extraterritorial plants 4-merous; sepals in the bud overlapping (imbricated); *sometimes* with 2 series of 5 stamens and the outer opposite (rather than as usual alternate with) the petals (**obdiplostemonous**); ovary usually 3- or 5-chambered; placentae axile or rarely basal but fundamentally parietal; ovules 1–2 or more per carpel, anatropous.

Union of the few carpels and the small number of the stamens distinguish the Geraniales from the Ranales. Often the coalescence of carpels is light, enduring only through flowering time, then disappearing. The obdiplostemonous arrangement of the stamens represents a tendency but not a key character of the order. Usually the stamens are separate; in the closely related Malvales usually they are coalescent into a tube or into groups. As the order is defined here, the ovules are usually but not necessarily pendulous and with the micropyle often turned away from the central placenta.

Key to the Families

(North America north of Mexico)

1. Calyx *not* with 2 large conspicuous external glands nearly covering each sepal.
   2. Stipules none or represented by mere glands; leaves alternate or rarely (in some *Linaceae* and *Balsaminaceae*) opposite.
      3. Calyx *not* with a spur.
         4. Leaves *not* palmate; petals separate.
            5. Fruit of 5 slightly coalescent carpels, these 1-seeded and separating at maturity; stamens 10, separate; leaves dissected; petals withering, persistent after flowering.

               Family 1. **Limnanthaceae** (meadowfoam family).

            5. Fruit a 4–10-chambered capsule, splitting (dehiscent) along 4–10 lines; stamens 5, slightly coalescent basally; leaves entire, small; petals quickly deciduous after flowering.

               Family 2. **Linaceae** (flax family).

4. Leaves palmate, with 3 (or rarely up to 10) obcordate leaflets; petals sometimes coalescent basally; juice watery, sour.

Family 3. **Oxalidaceae** (wood-sorrel family).

3. Calyx with a spur.

4'. Leaves (of the species introduced in North America) peltate; fruit 3-lobed, with 3 carpels each with 1 seed; anthers separate.

Family 4. **Tropaeolaceae** (nasturtium family).

4'. Leaves not peltate; fruit formed from 5 carpels each with several seeds; anthers coalescent.

Family 5. **Balsaminaceae** (touch-me-not family).

2. Stipules present; leaves opposite or sometimes whorled.

3'. Receptacle greatly elongated and tapering upward into a point, the 5 1-seeded carpels coalescent around it, separating elastically at maturity, the corresponding segment of the style remaining as a long twisted tail on each segment of the ovary; leaves simple or, if pinnate, the leaflets not entire.

Family 7. **Geraniaceae** (geranium family).

3'. Receptacle not greatly elongated, the ovary tipped by a single style; leaves pinnate (in one genus of shrubs with only 2 leaflets and these basally coalescent across the rachis), the leaflets entire.

Family 8. **Zygophyllaceae** (caltrop family).

1. Calyx with 2 large, conspicuous external glands nearly covering each sepal; stipules usually present; shrubs, trees, or somewhat woody vines; leaves opposite, simple.

Family 9. **Malpighiaceae** (*Malpighia* family).

*Family 1. Limnanthaceae. MEADOWFOAM FAMILY

Annuals. Leaves pinnate, alternate; stipules none. Flowers axillary, the pedicels elongated; sepals 3 or 5, persistent in fruit; petals 3 or 5, remaining after withering (marcescent-persistent), stamens in 2 series, the outer alternate with the petals (**diplostemonous**); carpels 3 or 5, the styles coalescent basally, the ovaries separate; placentae basal but fundamentally parietal; ovule 1 per carpel. Fruits achenes formed from the separate ovaries.

The family is composed of 2 North American genera, *Limnanthes* and *Floerkia*. *Limnanthes* (meadowfoam) is a Pacific Coast (predominantly Californian) genus of several species. The attractive low-growing plants often color pastures or meadowland in the lower portions of California. *Floerkia* includes a single inconspicuous species extending across the Continent.

*Family 2. Linaceae. FLAX FAMILY

Leaves simple, entire, alternate, opposite, or rarely whorled; stipules present or absent. Flowers radially symmetrical, 5-merous or in *Millegrana* 4-merous; calyx persistent in fruit; stamens 5 or 10; staminodia sometimes present (5 or 10); carpels 5 or rarely 3–4, the styles separate, the ovary sometimes with 10 chambers by intrusion of the midribs of the carpels; ovules usually 2 but sometimes 4 per carpel. Fruit a capsule or rarely a drupe.

The family is world-wide but most common in temperate regions. Only two genera occur in the United States. The chief of these is *Linum* (flax), which grows throughout the Continent except in northern regions. One of the most beautiful and widespread native species is *Linum Lewisii*, discovered by Captain Merriweather Lewis of the Lewis and Clark expedition. This plant occurs throughout the West and as far eastward as James Bay and Wisconsin. It is attractive for its blue flowers. The genus *Millegrana* (allseed), a native of Europe, is naturalized about roadsides and in waste places along the coast in Nova Scotia.

Linum usitatissimum, the flax of commerce, has escaped from cultivation in the Middle West and East and the Southern States. The plant is native in the area between the Caspian and Black Seas and the Persian Gulf. The fibers, the source of linen (strands of the pericycle at the outer edges of the vascular bundles), are separated from the softer tissues of the stem by a fermentation process brought about by exposure of the plant to the weather or allowing it to stand in water. After a time the soft tissues decay and mechanical manipulation is sufficient to remove the fibers. The process is called retting. The seeds of the flax plant yield linseed oil.

*Family 3. **Oxalidaceae.**

WOOD SORREL FAMILY

Extraterritorial plants; rarely woody. Leaves palmate, the leaflets 3 or rarely 1 or more numerous (3 in the North American species), alternate; stipules none. Petals separate or commonly coalescent; stamens in 2 series, the outer opposite the petals (obdiplostemonous); carpels 5, the styles separate; ovules 1–several per carpel. Fruit a capsule or rarely (in extraterritorial plants) a berry.

The family is tropical, and it includes perhaps 1,000 species, mostly of the genus *Oxalis* (wood-sorrel), which is also widespread in North America. The 25 native species are attractive for their palmate leaves with 3 obcordate leaflets. The sour juice is characteristic.

*Family 4. **Tropaeolaceae.**

NASTURTIUM FAMILY

Leaves simple (or sometimes pinnate), commonly peltate, sometimes lobed to divided. Flowers bilaterally symmetrical; upper sepal with a spur; petals 5, the upper 2 differentiated in shape and attached in the base of the spur of the calyx; stamens 8, in 2 series; carpels 3, the style 1; ovule 1 per carpel. Fruit separating into the 3 1-seeded carpels.

The family consists of a single genus, *Tropaeolum*, which occurs largely along the mountain axes from Mexico to the Andes. The garden nasturtium, a native of Peru, is an escape in moist waste places about gardens and particularly so in the Atlantic portions of the southern states. Another species, the canary-bird flower, is cultivated also.

*Family 5. **Balsaminaceae.**

TOUCH-ME-NOT FAMILY

Some plants shrubby. Leaves simple, alternate, opposite, or in threes; stipules none (in North American plants). Flower bilaterally symmetrical; the upper sepal with a spur; petals 5, separate or some coalescent and the corolla appearing 3-merous; stamens 5, sometimes coalescent by the anthers but not the filaments, the flattened filaments surrounding the ovary; carpels 5, the style 1 or practically obsolete; ovules 3–many per carpel. Fruit usually a capsule, this opening elastically, as if exploding, and throwing the seeds into the air (hence the name "touch-me-not"), sometimes a berry.

Although the family is a large one, it consists of only two genera, common in the tropical and subtropical areas of the Eastern Hemisphere. Two native species of *Impatiens* (touch-me-not) occur in the eastern portions of the United States and Canada, as well as northwestward to Alaska and Washington. Otherwise the family is not native in the Western Hemisphere. *Impatiens Balsamina*, the garden balsam, escapes from cultivation.

*Family 6. **Geraniaceae.**

GERANIUM FAMILY

Leaves simple or compound, the veins of simple leaves commonly palmate and the lobing tending to be palmate, alternate or (in the North American genera) opposite; stipules present. Flowers rarely bilaterally symmetrical in extraterritorial plants, usually in cymes or umbels; stamens 5–15, usually in 2 series (one sometimes sterile) with the outer opposite the petals (obdiplostemonous); carpels usually 5, sometimes 3, the styles separate or lightly coalescent around an axis; ovules 1 or 2 per carpel. Fruit separating into segments formed from the individual carpels, the styles usually persistent.

The family is a large one abundant in temperate and subtropical regions. The best-known members of the group are the large cultivated geraniums, belonging to the genus *Pelargonium*. Native species of the genus *Geranium* are much less conspicuous though some are attractive. They occur throughout the moist areas of the Continent except in arctic regions. The genus *Erodium* (filaree or alfilaria) is of special interest for the high development of its seed-planting mechanism. The five carpels are coalescent lightly around the elongated receptacle. At maturity the ovary segments separate, as in the family in general, and each remains attached to its own segment of the separated style. The style twists as the carpel dries, and it straightens out again whenever it becomes wet. With the alternate moisture and dryness of day and night or from one spring rain to the next the squirming of the style pushes and twists the sharp-pointed portion of the fruit into the ground. Backward-directed hairs serve as barbs. A similar arrangement exists in *Pelargonium*. The filarees

Figure 10-29. Geraniales; Geraniaceae; *Erodium moschatum,* white-stemmed filaree or alfilaria: **1** branch, pinnate leaf, and its stipules, **2–4** flowers, **5** flower after fall of the petals, showing the sepals, glands, staminodia, stamens, and pistil, **6** umbel of fruits, **7** fruit, **8–9** individual carpels separating from each other and from the elongate receptacle (cf. **Figure 8-1 A**), **10** receptacle, **11** single segment of the fruit, **12** cross section of the style.

are among the dominant plants of California, where they were introduced in the wool of sheep brought in by the Franciscan Fathers who founded the missions. Recently they have become abundant in Arizona and other states. Especially the red-stemmed filaree (*Erodium cicutarium*) is prized highly as a range forage plant. The most common species are not native in North America, but some are, as for example *Erodium texanum* of the southwestern deserts.

Family 7. **Erythroxylaceae.** COCA FAMILY

Shrubs or trees. Leaves simple, entire or toothed, alternate; stipules present. Flowers radially symmetrical, relatively small; petals 5, each with a 2-lobed appendage or projections of callus tissue on the upper surface; stamens 10, in 2 series, somewhat coalescent basally; carpels 3, the styles separate or coalescent basally. Fruit a berry or drupelike, with only 1 developed carpel.

The family consists of three genera, two small and the other (*Erythroxylum*) large. Species occur from Mexico and Cuba to the American Tropics and sparingly in Africa.

To man the most important plant is *Erythroxylum Coca* (coca), a large shrub the leaves of which are the source of cocaine.

*Family 8. **Zygophyllaceae.** CALTROP FAMILY

Herbs, shrubs, or rarely trees. Leaves mostly pinnate, usually opposite; stipules present, usually persistent, sometimes highly modified, e.g., spinescent. Flowers usually radially symmetrical, in *Larrea* each with a hypogynous disc; petals 5 or rarely 4 or none, separate or rarely coalescent basally; stamens 5–15, in 1–3 series, the outer series usually opposite the petals; carpels 4 or 5, the style usually 1; ovules 2–many per carpel. Fruit a capsule or rarely drupelike.

The family is widespread in tropical, subtropical, and warm temperate regions. The most widely introduced species is the caltrop (*Tribulus terrestris*). In recent times this weedy plant has come to be known as puncture-vine because the spines on the backs of the carpels are sharp and particularly effective against bicycle tires. In southern areas of the United States where puncture-vines are common the spines frequently break off in automobile tires, working their way gradually inward as the tires wear smooth. In older tire types they punctured the inner tube. However, they rarely caused any difficulty while the tires showed the tread. A related genus occurring from the Southwestern Deserts to the Southern States is *Kallstroemia*, sometimes called summer-poppy. Some of the species are exceptionally attractive. The creosote-bush (*Larrea tridentata*) is the characteristic shrub of the Southwestern Deserts. It covers many square miles from California to Texas, and its distributional range coincides almost exactly with that of the Mexican Desert Region. Lignum vitae (*Guaiacum*) is a shrub or tree sometimes as much as thirty feet tall. The wood is exceptionally hard and heavy and consequently strong. It is used for making many small objects requiring these features. A species of this genus is native on the Florida Keys. *Fagonia* occurs in the California and Arizona deserts.

*Family 9. **Malpighiaceae.** MALPIGHIA FAMILY

Usually woody vines but sometimes trees or shrubs. Leaves simple, usually opposite, sometimes whorled or alternate; petioles often glandular or jointed; stipules usually present. Flowers usually radially symmetrical; sepals (or some of them) with (usually 2) very large conspicuous external glands; petals 5, separate, sometimes fringed; stamens usually 10, in 2 series (1 series sometimes of staminodia), often coalescent basally; carpels 3 or sometimes 2–5, the styles usually separate; ovule 1 per carpel. Fruit various—a capsule, berry, drupe, or samara or dividing into the individual carpels.

A large family mostly of Mexico and Latin America. In the southern portions of the United States the family is represented by *Janusia* (a small twining desert vine, Arizona to Texas), *Byrsonima* (1 species occurring on the low pinelands and hammocks of southern Florida), *Thryallis* (1 species in Texas), *Aspicarpa* (Arizona and Texas), and *Malpighia* (1 species in Texas).

The family is transitional between the Geraniales and the Polygalales.

Order 17. *POLYGALALES.*

MILKWORT ORDER

DEPARTURE FROM THE FLORAL CHARACTERS OF THE RANALES

Free	Petals often coalescent and adnate with the stamens; the stamens often coalescent by their filaments into a split sheath about the pistil or into 2 sets; carpels coalescent.
Equal	Calyx and sometimes the corolla bilaterally symmetrical; stamens bilaterally symmetrical.
Numerous	Stamens 3–8 or 12; carpels 2 or 3.
Spiral	Stamens and carpels cyclic.
Hypogynous	Flower sometimes with a hypogynous disc.

Herbs, shrubs, small trees, or vines (these sometimes twining and saprophytic). Leaves simple, nearly always alternate; stipules absent or represented by glands. Flowers bisexual, hypogynous, radially symmetrical, sometimes with hypogynous discs, in spikes or panicles or commonly solitary; sepals 5 or 4–7, often petaloid; calyx bilaterally symmetrical; stamens 3–8; filaments free or often coalescent; anthers with 2 pollen chambers or these coalescing into one, opening by an apical pore; carpels fundamentally 2, but in the *Krameriaceae* one rudimentary, coalescent. Fruit a nut, drupelet, samara, or spiny and indehiscent bur.

Key to the Families
(North America north of Mexico)

1. Fruits *not* burlike, spineless; stamens 6 or commonly 8; petals coalescent and adnate with the stamens; filaments coalescent into 1 or 2 sheets.

   1. **Polygalaceae** (milkwort family).

1. Fruits each a spiny bur, indehiscent, the spines barbed; stamens 3–5, only those on the upper side of the flower developed; petals *not* coalescent or adnate with the stamens; filaments separate.

   4. **Krameriaceae** (ratany family).

*Family 1. **Polygalaceae.**

MILKWORT FAMILY

Herbs, shrubs, small trees or rarely twining saprophytes. Leaves most commonly alternate; stipules none or represented by glands. Flowers sometimes with hypogynous discs, solitary or in spikes or panicles; sepals usually 5, sometimes 4, 6, or 7, the 2 inner sepals larger than the others, often petaloid; petals usually 3, basically 5 but 2 usually lacking, the dorsal petal concave and sometimes crested with a fringe, the petals coalescent and adnate with the stamens; stamens 3–8, usually 8, coalescent but the filaments forming a sheet rather than a tube or sometimes in 2 sets; anthers with 1 pollen chamber (by coalescence of 2) with an apical pore; carpels usually 2, the ovary usually 2-chambered; ovule usually 1 per carpel. Fruit a capsule or sometimes a samara, a nut, or a drupelet.

The family is distributed widely in tropical to temperate regions. *Polygala* is abundant and with numerous species in the southern and eastern portions of North America (chiefly Arizona to the Southern States). Only 2 species occur on the Pacific Coast, these in California west of the Sierra Nevada summit. *Monnina Wrightii* occurs in Arizona and New Mexico.

Family 2. **Trigoniaceae.** TRIGONIA FAMILY

Trees or woody vines. Leaves alternate or opposite; stipules small, falling away early. Flowers bisexual, bilaterally symmetrical, sometimes with slight hypogynous discs, in racemes or panicles; sepals 5, in the bud overlapping, unequal; petals 5 or 3, the outer (dorsal) petal largest and basally saccate; stamens 3–12, including often some staminodia, all opposite the upper (ventral) petal, the filaments joined but not forming a tube; carpels 3, the ovary 3-chambered, woolly externally; ovules 2–many per carpel. Fruit a capsule.

The family is composed of 2 genera occurring in the American Tropics. It has been included (perhaps correctly) in the *Vochysiaceae.* The status of the family cannot be evaluated properly for lack of information.

Family 3. **Vochysiaceae.**

VOCHYSIA FAMILY

Shrubs, trees, or woody vines; juice resinous. Leaves simple, opposite, whorled, or rarely alternate; stipules none or small, sometimes glandular.

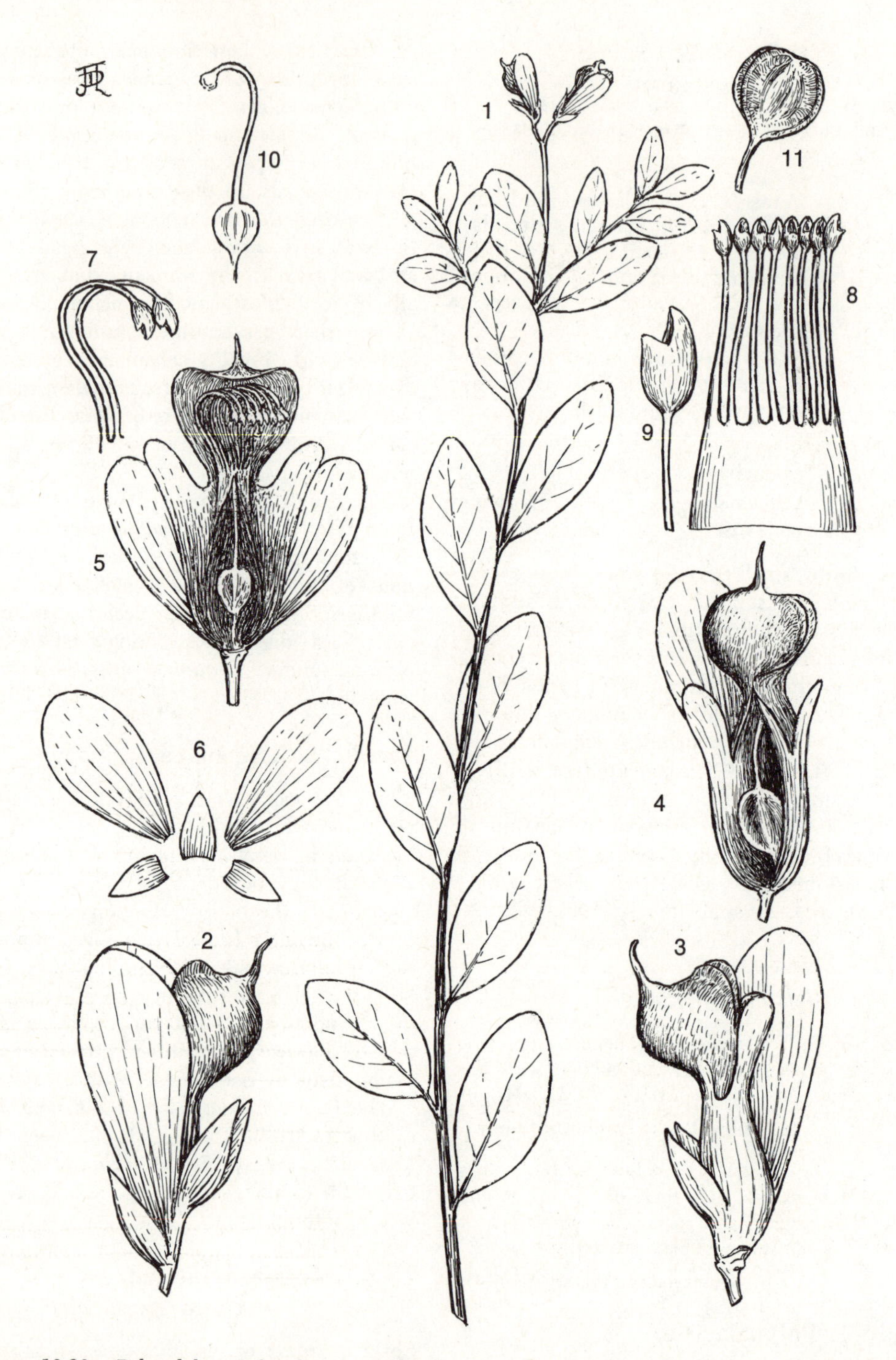

Figure 10-30. Polygalales: Polygalaceae; *Polygala cornuta,* a milkwort (*wort,* an old Anglo-Saxon word for plant, particularly an herb): **1** habit, **2–3** views of the flower, **4** interior showing the pistil, **5** the same from another view, showing both the stamens and the pistil, **6** the sepals, **7–9** stamens, the filaments basally coalescent, **10** the pistil, **11** a fruit.

Flowers bisexual, bilaterally symmetrical, sometimes with hypogynous discs, in racemes or panicles; sepals 5, in the bud overlapping, the lower (dorsal) one largest and with a basal sac or spur; petals 1–5; stamen 1, accompanied by staminodia; carpels 3, the ovary with 1 or 3 chambers; ovules 2 to many per carpel. Fruit a capsule or a samara.

The family is concentrated in tropical America but one species occurs in westernmost tropical Africa.

*Family 4. **Krameriaceae.**

RATANY FAMILY

Shrubs or herbs, usually less than 0.5 m. high; leaves simple, entire, linear, densely appressed-pubescent, as are the branchlets; stipules absent. Flowers bilaterally symmetrical, the sepals and petals and often the stamens purplish red, solitary; calyx much larger than the corolla, bilaterally symmetrical; corolla sometimes partly yellow, bilaterally symmetrical, the petals 3–5; stamens 3–5, those potentially on the lower side of the flower undeveloped. Fruit a dry, indehiscent burlike structure, with many barbed spines, the barbs following several patterns in different species.

The family is composed of a single genus, *Krameria,* long thought to be an anomalous one in the *Leguminosae* or pea family. Three shrubby species are important browse plants in the Southwestern Deserts of the United States.

T. H. Milby (American Journal of Botany 58: 569–576. 1971) showed the pistil to be formed from two carpels, instead of one as in the *Leguminosae.*

Order 18. ***SAPINDALES.***

SOAPBERRY ORDER

DEPARTURE FROM THE FLORAL CHARACTERS OF THE RANALES

Free	Petals sometimes coalescent basally; carpels coalescent.
Equal	Corolla sometimes bilaterally symmetrical in especially the *Sapindaceae* and *Hippocastanaceae.*
Numerous	Stamens 10 or less frequently 8, 5, or 4; carpels 2–5.
Spiral	Stamens and carpels cyclic.
Hypogynous	Flower nearly always with a hypogynous disc.

Herbs or commonly trees or shrubs. Leaves simple or compound, alternate or opposite; stipules none. Flowers bisexual or all or some unisexual, 5- or 4-merous, but the sepals 3–8, usually in 1–2 series and the carpels 2–8 or sometimes more, as many as the chambers or the ovary unless there is only 1; placentae axile. Ovary divided fundamentally into chambers, but often appearing 1-chambered through abortion of all but 1 carpel; ovules anatropous, or in the *Aceraceae* of several types arranged variously, often only 1 developing. Fruit a capsule, an indehiscent dry vessel, a berry, a drupe, a leathery structure with a single enormous seed, or a double (or sometimes triple) samara.

Key to the Families

(North America north of Mexico)

1. Plants with often obvious external or internal glands secreting volatile aromatic oils, balsams, or resins, usually either the crushed foliage (*poison ivy* should *not* be tested), the rinds of the fruits, the bark, or the wood strongly aromatic, the doubtful cases recognized as shrubs or trees with the following supplementary family or generic characters.
   (1) Leaves 3- (rarely 5-) foliolate; fruits in conspicuous clusters, each surrounded by a broad circular or elliptic wing. *Ptelea, Rutaceae* (citrus family).
   (2) *Either* the fruits (of the North American genera) (a) 2–5 from each flower, completely or practically separate or (b) 1 from a flower, flattened, about 3 times as long as broad, acute at both ends, the sides curved *or otherwise,* the bark bitter (*poison ivy* should *not* be tested). *Simaroubaceae* (*Quassia* family).
   (3) Filaments of the 10 stamens coalescent into a long tube. *Meliaceae* (mahogany family).
   2. Filaments of the stamens separate or coalescent only basally.
      3. Leaves and often other parts (e.g., fruits) covered with dotlike glands (or in *Ptelea* not gland-dotted, this genus being characterized by the 3- or rarely 5-foliolate leaves and conspicuous clusters of fruits with broad circular or elliptic wings).

         Family 3. **Rutaceae** (citrus family).

      3. Leaves and other parts *not* covered with dotlike glands; fruits not with broad circular or elliptic wings.
         4. Plants not strongly aromatic; fruits(s) of

Figure 10-31. Polygalales, Krameriaceae, ratany (recently separated from the Leguminosae, Rosales). *Krameria Grayi*: **1** branches with thorns, leaves, and flowers, **2** flower, showing the large sepals, small petals, 3–5 stamens, and the pistil, **3** fruit, with spines, these each with apical branches forming hooks or barbs arranged, in this species, like the ribs of an umbrella, **4** enlargement of the tip of a spine, showing the hooks or barbs. By permission from *Trees and Shrubs of the Southwestern Deserts,* LYMAN BENSON and ROBERT A. DARROW, Tucson: University of Arizona Press, copyright, 1979.

the North American genera *either* (1) 2–5 from each flower and completely or practically separate *or* (2) flattened, about 3 times as long as broad, acute at both ends, and the sides curved *or* (3) a 2–3-chambered berry, indehiscent, and circular in cross section.

Family 1. **Simaroubaceae** (*Quassia* family).

4. Plants strongly aromatic; fruit of the North American genus a berry, the exocarp splitting along 3 lines, more or less triangular in cross section, with a single large bony seed.

Family 4. **Burseraceae** (incense-tree family).

2. Filaments of the stamens coalescent through their full length, forming a tube or cup.

Family 5. **Meliaceae** (mahogany family).

1. Plants *not* with strongly aromatic foliage, fruit rinds, wood, or bark, *not* with the supplementary characters listed in lead 1 above.

2'. Plant *not* corresponding with *every character* in lead 2' below.

3'. Fruit *not* with 2 or 3 well-developed thin, leathery or membranous wings, sometimes inflated and with apical horns. (Cf. lower lead 2, below.)

4'. Flower *not* with a hypogynous disc developed from the receptacle below the ovary; stamens and petals attached to the main receptacle.

Family 8. **Aquifoliaceae** (holly family).

4'. Flower *with* a hypogynous disc developed from the receptacle below the ovary; stamens and sometimes the petals attached to the disc or sometimes under it.

5. Fruit not an inflated, thin-walled capsule.

6. Ovary 1-chambered (by abortion of all but 1 carpel).

Family 6. **Anacardiaceae** (cashew family).

6. Ovary with more than 1 chamber.

7. Leaves simple.

8. Anthers turned inward on the filaments (introrse).

Family 9. **Celastraceae** (staff-tree family).

8. Anthers turned outward on the filaments (extrorse).

Family 10. **Hippocrateaceae** (*Hippocratea* family).

7. Leaves compound.

Family 14. **Sapindaceae** (soapberry family).

5. Fruit an inflated, thin-walled capsule with 2 or 3 apical horns, dehiscent apically.

Family 21. **Staphyleaceae** (bladdernut family).

3'. Fruit with 2 or 3 well-developed thin, leathery or membranous wings.

4''. Fruit with 2 or rarely (in a few fruits) 3 elongated subapical wings (a double or rarely a triple samara), the wings much longer than the fruit.

Family 19. **Aceraceae** (maple family).

4''. Fruit with 3 wings, one on the back of each carpel. **Dodonaea.**

Family 14. **Sapindaceae** (soapberry family).

2'. Fruit leathery, enclosing a single seed or 2–3 seeds, 1–5 cm. in diameter; leaves palmate, with 5–7 leaflets; trees or large shrubs.

Family 20. **Hippocastanaceae** (horse-chestnut family).

Suborder A. **Rutineae**

*Family 1. **Simaroubaceae.**

QUASSIA FAMILY

(Segregate: *Surianaceae*)

Shrubs or trees (or rarely herbs). Leaves pinnate, rarely simple (*Suriana*) or reduced to scales (*Castela*), alternate, not with glandular dots. Flowers usually unisexual (and then the plants dioecious) but sometimes bisexual, radially symmetrical; sepals 3–8, coalescent at least basally; petals 3–5 or none; stamens 3–10, in 1 or 2 series; carpels 2–8, free (e.g., *Suriana*) or usually coalescent in various degrees, including sometimes even the styles. Fruit a capsule or samara (those in the cluster nearly separate) or splitting into the individual carpels, rarely a berry or drupe or achenelike.

The family occurs throughout the Tropics and it ranges to some extent into temperate regions. One species is *Castela Emoryi,* one of the three desert crucifixion-thorns of the Southwest. This practically leafless shrub with rigid branches all ending in thorns is confined to the low country near the Colorado River in California, the Gila River in Arizona, and adjacent Mexico. Other species occur in southwestern Texas. Genera occurring in Florida and on the adjacent Keys include those of the paradise-tree or bitterwood (*Simarouba*) and the bitter-bush (*Picramnia*) and the genus *Alvaradoa.* The tree-of-Heaven or copal-tree (*Ailanthus*), a native of China cultivated widely in the United States, has escaped here and there across the Continent. The fruits are produced in dense clusters, each flower yielding up to five samaras. The leaves are pinnate.

Family 2. **Cneoraceae.** CNEORUM FAMILY

Small shrubs. Leaves simple, alternate. Flower bisexual, radially symmetrical, attached to the subtending bract; sepals 3–4; petals 3–4, elongate; stamens 3–4; carpels 3–4, the styles 3–4; ovules pendulous. Fruit a drupe, composed usually of 3–4 separable segments.

The family includes a single genus restricted to the Mediterranean region.

*Family 3. **Rutaceae.** CITRUS FAMILY

Herbs, shrubs, or trees; glands yielding essential oils, abundant, small, most obvious on the fruits or leaves (except in *Ptelea*). Leaves simple, palmate or pinnate, alternate or opposite. Flowers bisexual (rarely unisexual and then the plants dioecious), nearly always radially symmetrical; sepals 3–5; petals 3–5 or rarely none, rarely coalescent; stamens 3–10 or sometimes more, usually in 2 series, the outer series usually opposite the petals (obdiplostemonous); carpels 4–5 or sometimes more, the styles separate or coalescent but each with a separate duct, the chambers of the ovary as many as the carpels, or in an extraterritorial genus 1 (and the placentae then parietal); ovules 1–several per carpel. Fruit usually a berry with a leathery rind or separating into segments or a capsule but sometimes a samara or a drupe.

The family is large; it occurs chiefly in tropical and subtropical regions, but it extends into warm temperate areas. Most citrus fruits are native in China and Indochina, but several species of *Citrus* occur in peninsular Florida and on the Keys, these being descendants of plants dispersed by early Spanish expeditions. Consequently they occur in some remote areas. The rue (*Ruta*), a strong-smelling herb with bipinnate leaves and yellow flowers, is cultivated, and it has escaped in some parts of the eastern and southern United States. The genus *Xanthoxylum* (*Zanthoxylum*) includes the wild lime of Florida, the yellowwood of the lower Florida Keys, and the prickly-ash. The hop-trees (*Ptelea*) occur through most of the southern portions of the Continent, and they range as far north as Minnesota and Ontario. The torchwoods (*Amyris*) are in Texas and Florida; the turpentine-broom (*Thamnosma*) is a low shrub of the deserts from California to Utah and Arizona; *Cneoridium* occurs from San Diego County, California, to Baja California. Other genera have become established in Florida. Species of Mexican orange, *choisya,* occur from Arizona to western Texas.

The cultivated members of the family include oranges, lemons, limes, grapefruits, tangerines, citrons, kumquats, mandarins, and many others. Cultivation is restricted by frost to essentially the southern tier of states from California to the Atlantic Coast.

The fruit of an orange is divided readily into its several carpels. The juice sacs within each carpel are the individual cells of hairlike structures projecting inward from the ventral surface of the carpel.

*Family 4. **Burseraceae.** INCENSE-TREE FAMILY

Shrubs to large trees; leaves, bark, and wood with essential oils and resins, the bark with resin ducts or chambers. Leaves pinnate, sometimes more than once so, usually alternate, deciduous. Flowers bisexual (North America) sometimes some unisexual and the plants polygamodioecious, radially symmetrical, small; sepals 3–5; petals 3–5 or rarely none, separate (except in some extraterritorial plants); stamens 6–10, in 1 or 2 series; carpels 2–5, the style 1 or obsolete; ovules 2 or rarely 1 per carpel. Fruit fleshy and essentially a berry (but splitting open) or sometimes a capsule, with 1–5 seeds.

The family occurs in the Tropics and the deserts in and near the Horse Latitudes of America, Africa, and (to a lesser extent) Asia. The elephant-tree (*Bursera microphylla*) is exceedingly

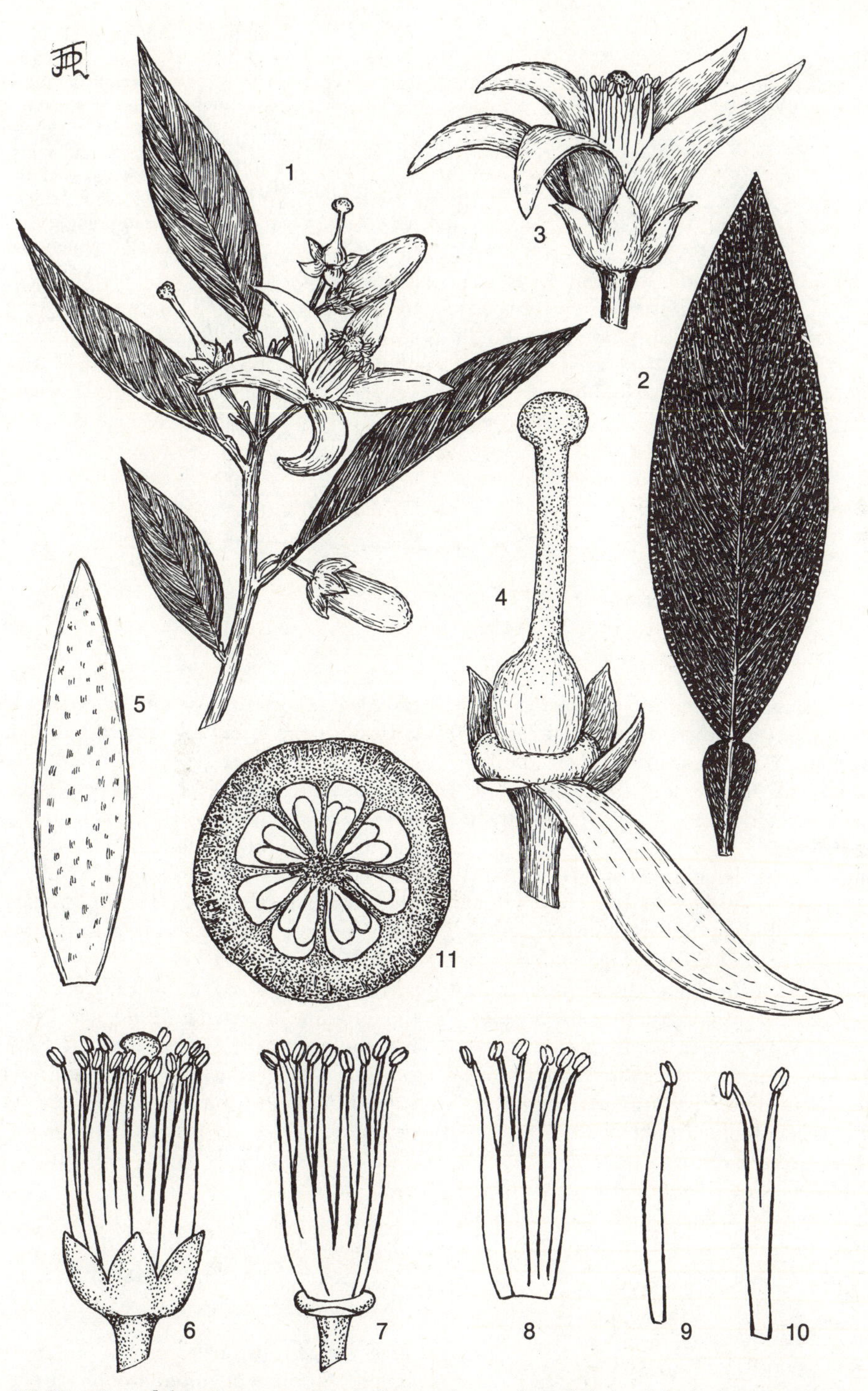

Figure 10-32. Sapindales; Rutaceae; *Citrus Limon,* lemon I—1 habit, **2** leaf, showing the wingéd petiole, **3** flower, **4** flower with two sepals, four petals, and the stamens removed, showing the pistil and the hypogynous disc, **5** petal, **6–10** stamens, these being coalescent in varying degrees, **11** cross section of the young ovary, showing the attachments of ovules.

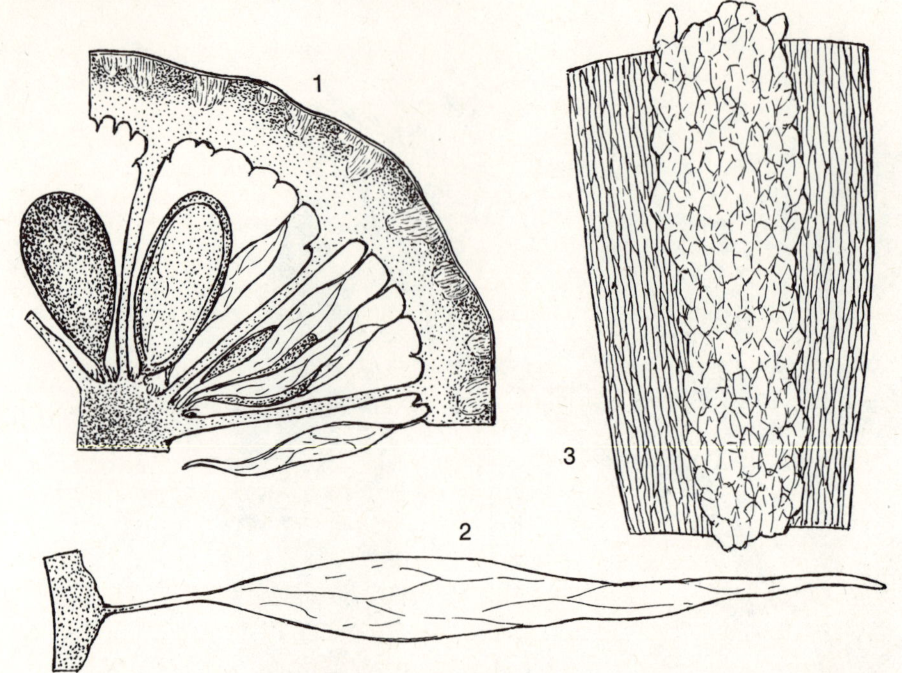

Figure 10-33. Sapindales; Rutaceae; *Citrus Limon*, lemon II—1 portion of a cross section of the fruit, showing the seeds and some of the juice sacs characteristic of citrus fruits, 2 juice sac enlarged, 3 further enlargement, showing the cellular structure. (Cf. also **Figure 8-2 D** kumquat, fruit.)

rare in the Colorado Desert of California and common in the same desert in extreme southwestern Arizona. The trees have short stubby trunks only a few feet long, and their odd shape, combined with their white bark, makes them particularly striking.

The genus *Bursera* is common in Mexico, and the resin (copal) is burned in the churches as incense. It was used in the same way by the Mayas and Aztecs. *Bursera fagarioides* var. *elongata* occurs from the Baboquivari Mountains, Arizona, to Puebla, Mexico. Another species *Bursera Simaruba* (West Indian birch, gumbo limbo, or gum elemi) occurs in south peninsular Florida and on the Keys. *Boswellia*, an east African genus, is the chief source of frankincense (pure incense) derived from a resinous gum containing a volatile oil. Species of *Commiphora*, particularly *Commiphora abyssinica*, yield myrrh. The myrrh of the Bible included myrrh mixed with labdanum, derived from a rockrose (*Cistus*).

*Family 5. **Meliaceae.**

MAHOGANY FAMILY

Shrubs or trees; wood hard, scented or often so. Leaves pinnate or more than once pinnate, rarely simple, usually alternate. Flowers bisexual or rarely some unisexual (and then the plants polygamodioecious), radially symmetrical, in clusters of cymes; sepals 4–5; petals usually 4–5, rarely 3 or up to 8, (in North America) separate; stamens usually 8–10, the filaments coalescent through their entire length; carpels 2–5, rarely many, the style 1 or obsolete; ovules usually 2 per carpel, rarely up to 12 (*Swietenia*). Fruit a berry or a capsule or rarely a drupe; seeds often with wings.

The family is large, and it occurs throughout the Tropics. *Swietenia Mahogani*, the tree yielding the mahogany much used in cabinet work, furniture, and paneling, occurs in south peninsular Florida and on the Keys. The Chinaberry, or umbrella-tree, as it is called in California, has escaped occasionally from cultivation, especially in the South.

*Family 6. **Anacardiaceae.**

CASHEW FAMILY

(Synonym: *Spondiaceae*)

Trees or shrubs; bark usually resinous. Leaves simple to pinnate, alternate or rarely opposite; stipules none or minute. Flowers of the North American species usually (at least some of them) unisexual (the plants being either polygamous, monoecious, or dioecious), usually radially symmetrical, with hypogynous discs, in elaborate panicles; sepals 3 or usually 5; petals 3 or usually 5 or sometimes none; stamens usually 5 or 10, in two series; carpels usually 3, the styles 3 in the North American genera, the developed ovaries 1-chambered, only one carpel being functional. Fruit a resinous drupe.

The family is chiefly tropical in distribution, but it extends into the warmer regions and some strictly temperate areas of the Northern Hemisphere. The principal representative in North America is *Rhus*, which includes the sumacs, poison-oaks, and poison-ivy. On the Pacific Coast a local species of poison-oak occurs from the Columbia River southward to California. The poison-oak of the Middle West, East, and South is a different species. Poison ivy occurs chiefly through the Middle West, the East, and the South, and the species, *Rhus radicans*, is represented sparingly in the Rocky Mountain region by *Rhus radicans* var. *Rydbergii*. It is absent from the Pacific Coast, the Canadian Rocky Mountains, and the Great Basin. Poison-sumac, poison-elder, or poison-dogwood of the East is also a species of *Rhus*. These plants are troublesome to those who are allergic to them. Other species of *Rhus* as, for example, the sumacs and lemonade berries, are nontoxic. The common other native genera are *Cotinus*, the southern smoke-tree, occurring in the Southern States and as far northward as Tennessee, and *Metopium* (the poison tree), in south peninsular Florida. *Spondias*, the tropical American hog plum, is naturalized on disturbed sites and shell mounds in southern Florida. A pistachio, *Pistacia*, occurs in southern Texas.

The mango (*Mangifera indica*) is cultivated in the South, especially in Florida, and it has escaped in south peninsular Florida and on the Keys. Members of this family yield pistachio and cashew nuts; resins, lacquers, and oils (from the varnish-tree); and mastic. The Peruvian pepper-tree (*Schinus*) is cultivated widely through the southern portions of the United States.

Family 7. **Julianaceae.** JULIANA FAMILY

Shrubs or small trees. Leaves alternate, bearing resin glands; stipules none. Plants dioecious, the staminate flowers in catkins, the pistillate in

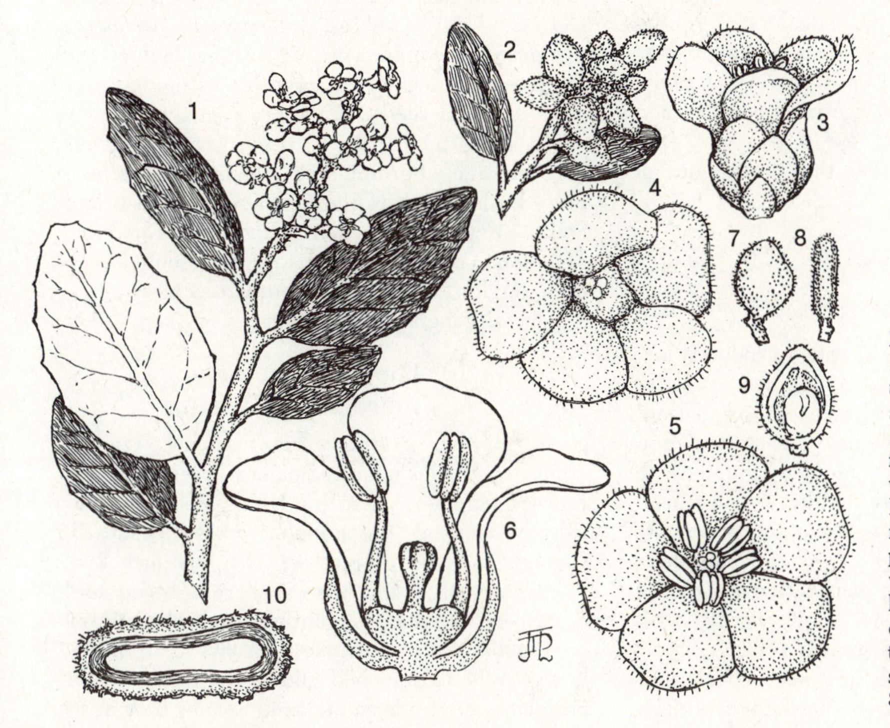

Figure 10-34. Sapindales; Anacardiaceae; *Rhus integrifolia*, a species of sumac with simple leaves: **1** flowering branch, **2** fruiting branch, **3–5** views of the flower, **6** longitudinal section of the flower, showing the hypogynous disc, **7–8** two views of the fruit, **9** longitudinal section, **10** cross section, the detail of the solitary seed not shown. (Cf. also **Figure 3-3 C** poison-oak.)

groups of 3 or 4 enclosed by an involucre; staminate flower with 3–9 sepals and 3–9 stamens, without a rudiment of a pistil; pistillate flower consisting of only a pistil; ovary with a single chamber and a solitary ovule; ovule half anatropous, with one integument partly covered by a large fleshy structure developed from the funiculus. Fruit a samara, nutlike, the fruits three or four in a cluster, enclosed in and attached to the persistent involucre, the stalk of the segment of the inflorescence flattened and winglike; endosperm none.

The two genera are *Juliana,* with about four species in Mexico, and *Orthopterygium,* with a single species confined to Peru. According to Stern probably the two genera should be united.

Various relationships have been suggested for this family. It resembles the following: (1) the Juglandales—leaves alternate, pinnate, without stipules; staminate and pistillate flowers dissimilar; stigmas broad and elongated; ovule with a single integument; (2) the Fagales—pollen grains small and spheroidal; involucres of the pistillate flowers like those of the beeches and chestnuts (however, each ovary clearly 1-chambered [and with only one ovule] instead of 3-chambered and the endosperm and the cotyledons different from those of the Fagales); (3) the *Anacardiaceae* (Sapindales)—leaves pinnate (as in many *Anacardiaceae*); seed solitary, without endosperm; embryo with large planoconvex cotyledons, wood anatomy of a similar type (but not similar to that in either the Juglandales or Fagales).

Suborder B. **Celastrineae**

*Family 8. **Aquifoliaceae.** HOLLY FAMILY

Shrubs or trees. Leaves simple, usually spiny, toothed or lobed, alternate; stipules none or minute. Flowers or some of them unisexual (and the plants usually polygamous or dioecious, commonly polygamodioecious), radially symmetrical, with no hypogynous disc, white or greenish; sepals, petals, and stamens 4–9, most commonly 4–5; carpels 3–many, the style 1 or obsolete, ovary 3–many-chambered; placentae axile; ovules 1–2 per carpel, pendulous. Fruit fleshy, with usually 4 stony internal segments.

The family includes only 3 genera, the largest being *Ilex* (holly), which is distributed widely. Species of holly are abundant in the eastern half of North America, and the single species of catberry (*Nemopanthus*) is endemic in the same area.

*Family 9. **Celastraceae.**

STAFF-TREE FAMILY

Shrubs, trees, or woody vines. Leaves simple, alternate or opposite; stipules none or small and falling early. Flowers bisexual or sometimes some unisexual (and then the plants polygamodioecious), radially symmetrical, with a hypogynous disc, this sometimes adhering to the ovary and the flower appearing epigynous; sepals 4–5; petals 4–5 or rarely none; stamens 4–5 or rarely 10; the anthers turned inward on the filaments (introrse); carpels 2–5, the style 1, the ovary 2–5-chambered; placentae axile; ovules 2 per carpel, usually erect. Fruit a capsule, samara, drupe, or berry; seed with a conspicuous, colored, fleshy aril.

The family is large and widely distributed. *Pachystima* (mountain-lover) and a single species of *Euonymus* (burning-bush) occur on the Pacific Coast, the former extending eastward to the Rocky Mountains. *Glossopetalon* or *Forsellesia* (Crossosomataceae?) is in the Great Basin Region and *Mortonia* chiefly in the Southwestern Deserts. The climbing bittersweet (*Celastrus scandens*) is common in the eastern half of the Continent, and it is attractive for its orange capsules and red seeds. Another species of *Pachystima* and three native species of *Euonymus* (spindle-tree, burning-bush, wahoo, or strawberry-bush) occur in the Middle West, East, and South. Four other genera (*Rhacoma, Gyminda, Maytenus,* and *Schaefferia*) are represented each by a single species in the Southern States.

Several genera are cultivated.

*Family 10. **Hippocrateaceae.**

HIPPOCRATEA FAMILY

Trees, shrubs, or woody vines. Leaves simple, opposite; stipules none or small. Flowers bisexual, radially symmetrical, with a hypogynous disc outside the stamens, this minute; sepals 5; petals 5; stamens 3 (North America) or sometimes 2–5, the filaments (North America) very broad and coalescent basally, the anthers turned outward on the filaments (extrorse); carpels 3 or rarely 5, the style 1, branched above, the ovary 3- (or 5-) chambered; placentae axile; ovules 2 or more

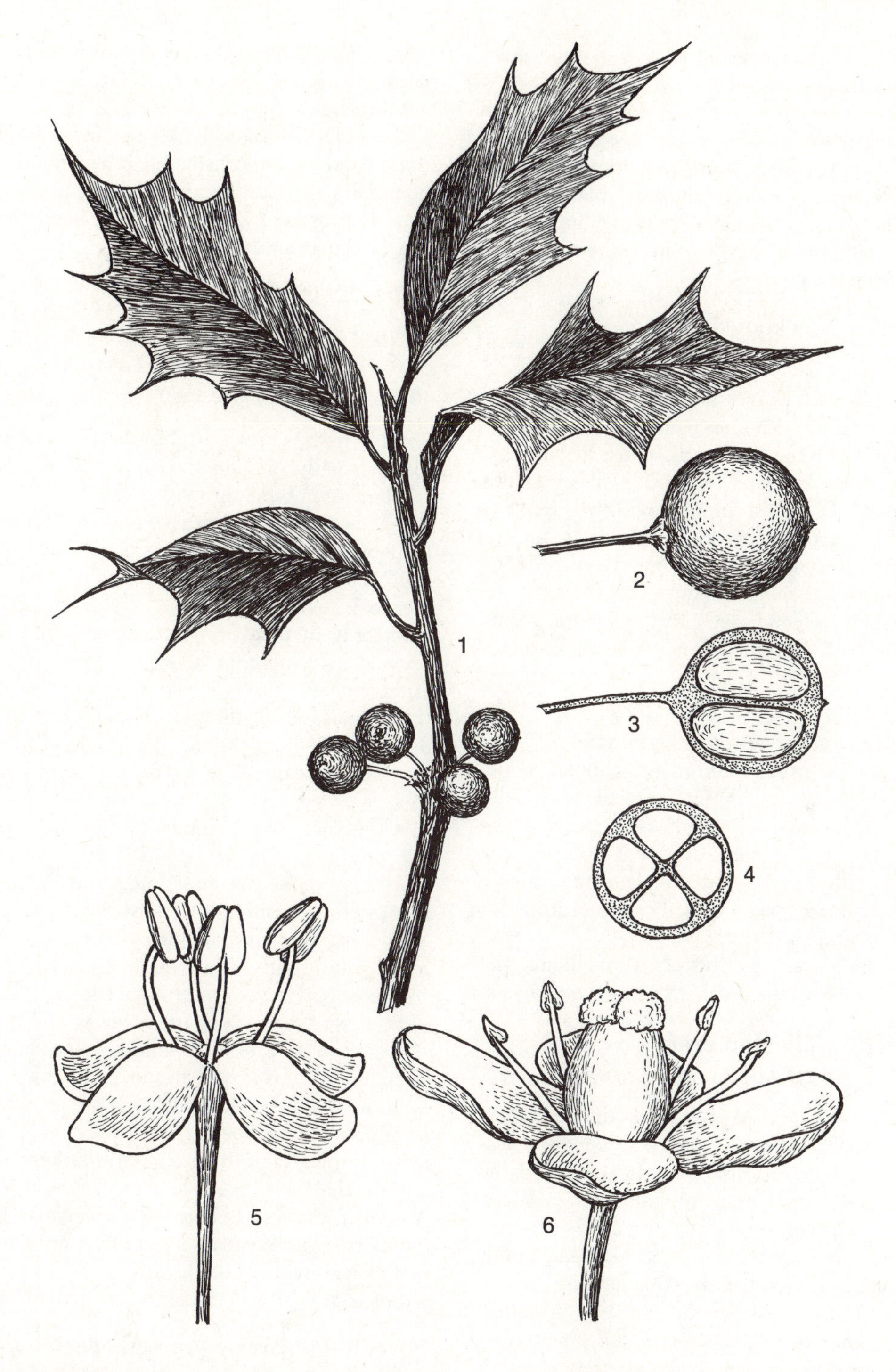

Figure 10-35. Sapindales; Aquifoliaceae; *Ilex,* holly: **1** branch with leaves and fruits, **2** fruit, side view, **3** longitudinal section, **4** diagrammatic cross section (cf. **Figure 8-3 B**) **5** staminate flower, the sepals covered by petals, **6** complete flower. (Cf. also **Figure 2-15 E** holly, leaf.)

per carpel. Fruit a capsule or a berry or separating into the carpels.

The family occurs throughout the Tropics and in subtropical areas. The single species of *Hippocratea* is restricted in North America to stream beds and hammocks in south peninsular Florida and on the Keys. The hooking and coiling of the branches and branchlets of this vine result in impenetrable thickets.

Family 11. **Stackhousiaceae.**

STACKHOUSIA FAMILY

Perennial herb. Leaves simple, coriaceous or slightly succulent; stipules none. Flowers bisexual, radially symmetrical, perigynous, in racemes, spikes, or dense clusters; sepals 5, coalescent into a tube; petals 5, coalescent above but free basally; stamens 5; carpels 2–5, the styles separate or coalescent, the ovaries 2–5-chambered. Fruits separating into 1-seeded segments.

The family is composed of two genera occurring from the Philippine Islands to New Zealand.

Family 12. **Icacinaceae.** ICACINA FAMILY

Trees, shrubs, or woody vines; rarely herbs. Leaves simple, usually alternate; stipules none. Flowers bisexual or rarely unisexual, radially symmetrical; sepals 4–5; petals 4–5, usually separate; stamens 4–5; carpels usually 3 or 5, but the ovary of only 1 developed; placentae apical; ovules usually 2, pendulous. Fruit a drupe or rarely a samara; seed 1.

The family is distributed throughout the Tropics, including the southern portion of Mexico.

Family 13. **Salvadoraceae.**

SALVADORA FAMILY

Shrubs or trees; thorns present in the leaf axils. Leaves simple, opposite; stipules none or rudimentary. Flowers bisexual or unisexual (and the plants then dioecious), radially symmetrical, sometimes with hypogynous discs, in dense clusters or panicles; sepals 3 or 4, coalescent; petals 4, free or partly coalescent, overlapping in the bud; stamens 4, the filaments sometimes coalescent or adnate with the bases of the petals; carpels 2, the style 1, the ovary 1–2-chambered; ovules 1–2, erect. Fruit a drupe or a berry.

The family is composed of 3 genera occurring in tropical Africa, Madagascar, and western Asia.

Salvadora persica is the toothbrush tree, the twigs (bound in clusters) having been used for toothbrushes.

Rendle included this family in the Oleales; Engler and Prantl considered it a member of the Sapindales. It may be a connecting link between these two orders. The family has been classified also in the Primulales.

Suborder C. **Sapindineae**

*Family 14. **Sapindaceae.**

SOAPBERRY FAMILY

(Segregate: *Dodonaeaceae*)

Shrubs, trees, vines, or rarely herbs. Leaves simple or usually pinnate, alternate; stipules none or in some climbing species present. Flowers bisexual or usually unisexual (and the plants polygamodioecious), the bisexual flowers usually functionally unisexual, radially to bilaterally symmetrical, with hypogynous discs, small; sepals commonly 5; petals 5 (sometimes with one suppressed or sometimes none); stamens usually 4–10, commonly in two series; carpels usually 3, the style usually 1, the ovary usually 3-chambered; ovules 1–2 per carpel. Fruit usually (North America) a capsule, a berry, or a winged structure.

Altogether the Sapindaceae include perhaps 125 or 150 genera with about 1,000 species. The family is one of the most important in the Tropics of both hemispheres. The soapberry (*Sapindus Saponaria* var. *Drummondii*) occurs from Arizona to Missouri and Louisiana and southward into Mexico. The large attractive berries contain saponins, or poisonous soapy substances. The hop-bush (*Dodonaea*) occurs in southern Arizona. The balloon-vine (*Cardiospermum*) is restricted to eastern North America. Four other genera (*Talisia, Exothea, Hypelate,* and *Cupania*) of this large tropical family occur in southern Florida and on the Keys. Such genera as *Paullinia* of the American tropics are noteworthy as lianas (woody climbers) with tendrils resembling watch springs and usually with winged fruits.

Some genera are valuable as timber trees.

Family 15. **Bretschneideraceae.**

BRETSCHNEIDERA FAMILY

Trees. Leaves pinnate. Flowers bisexual, bilaterally symmetrical, with a hypogynous disc; sepals

5; petals 5, bilaterally symmetrical, these and the stamens on the hypogynous disc; stamens 8; carpels 3, the ovary not on a stipe above the receptacle; ovules 2 in each chamber of the ovary.

The family consists of a single species growing in Yünnan, China.

Family 16. **Sabiaceae.** SABIA FAMILY

Shrubs, trees, or woody vines. Leaves simple or pinnate, alternate; stipules present. Flowers bisexual or some of them unisexual (and the plants polygamous), bilaterally symmetrical, with hypogynous discs; sepals 3–5; petals 4–5, sometimes coalescent basally, the inner two very small; stamens 3–5, separate, opposite the petals, or at least opposite the outer petals, sometimes only two fertile; carpels 2, the styles separate or coalescent, the ovaries 2-chambered; placentae axile; ovules usually 2 per carpel. Fruit a berry, or sometimes dry and coriaceous.

The family is tropical, the four genera being native chiefly in eastern Asia. The largest genus, *Meliosma,* is of scattering distribution in the Tropics of both hemispheres, and it occurs in southern Mexico.

Family 17. **Melianthaceae.** HONEY-BUSH FAMILY

Shrubs or trees or slightly woody bushes. Leaves pinnate or sometimes simple, alternate; stipules present. Flowers bisexual or some of them unisexual (and the plant then polygamodioecious), bilaterally symmetrical, with hypogynous discs, twisting halfway around on the pedicels after flowering, in racemes; sepals 5 or sometimes 2, coalescent and indistinguishable; petals 4 or 5, separate; stamens 4, 5, or 10; carpels 4 or 5, the style 1, the ovary 4–5-chambered; placentae axile; ovules 1 or several per carpel. Fruit a capsule, sometimes dehiscent only at the apex.

The family is small and restricted to Africa.

A few species of honey-bush (*Melianthus*) are cultivated in the warmer parts of the United States.

Family 18. **Akaniaceae.** AKANIA FAMILY

Trees. Leaves odd-pinnate, alternate. Flowers bisexual, radially symmetrical, with *no* hypogynous discs, in panicles; sepals 5; petals 5; stamens usually 8, the outer 5 alternate with the petals, the others small and near the lower part of the ovary, separate; carpels 3, the style 1, the ovary 3-chambered; placentae axile; ovules 2 per carpel, pendulous. Fruit a capsule.

The family consists of a single species occurring in eastern Australia. It has been classified in the Geraniales (Engler and Prantl) and in the Sapindales (Hutchinson, Gundersen, and others).

*Family 19. **Aceraceae.** MAPLE FAMILY

Shrubs or trees; juice often milky. Leaves simple or sometimes pinnate; stipules none. Flowers some bisexual and some unisexual (and then the plants polygamous) or the flowers all unisexual, radially symmetrical, with hypogynous discs, in racemes, corymbs, or panicles; sepals 4 or commonly 5; petals 4 or commonly 5 or sometimes none; stamens 4–10, or usually 8; carpels 2 (or occasionally 3 on the same plant), the styles 2, the ovaries ordinarily 2-chambered; placentae axile; ovules 2 per carpel, orthotropous, campylotropous, amphitropous, or anatropous. Fruit a double or occasionally triple samara, splitting into sections at maturity.

The family consists of two genera occurring in the Northern Hemisphere. The common genus is *Acer,* which includes the maples and box elders. Species of maple occur in all the moist regions of North America north of Mexico. The sugar maple, the most useful species, occurs only in the eastern portion of the Continent. It not only yields sweet sap boiled down into maple sugar and syrup but also it is of importance as a source of hardwood, as is the black maple of the same general region.

*Family 20. **Hippocastanaceae.** HORSE-CHESTNUT FAMILY

(Synonym: *Aesculaceae*)

Shrubs or trees. Leaves palmate, the leaflets 3–9, opposite; stipules none. Flowers bisexual or unisexual (and the plants usually polygamous, having numerous staminate flowers and a few complete ones), bilaterally symmetrical, with hypogynous discs, the inflorescence cymose and terminal, appearing racemose; sepals 4–5; coalescent at least basally; petals 4 or 5, separate; stamens 4–9, in two series, separate; carpels 3, the style 1, the ovary 3-chambered. Fruit with only one carpel

Figure 10-36. Sapindales; Aceraceae; *Acer macrophyllum*, big-leaf-maple: **1** branch with a young leaf (cf. **Figure 2-17**) and flowers, **2** flowers, enlarged, the middle one complete, the other two staminate, **3** flower, **4** sepals and petals removed, showing the hypogynous disc; **5** pistil, showing the young wings, **6** fruit (double samara, cf. also **Figure 8-5 A**), the sepals and stamens persistent but dry, **7** single carpel.

developing, this usually very large and containing only a single seed of great size (or a few large seeds present) the wall of the capsule leathery and husklike, dehiscent along three lines.

The family includes only two genera, these native largely in the Western Hemisphere. The horse-chestnut and the buckeyes (*Aesculus*) include two species occurring on the Pacific slope, one in California, the other in Baja California, Mexico. The Californian species is a common plant of the oak woodland and sometimes of the chaparral in the foothills of northern California. Several species known as buckeyes occur in the eastern portion of the Continent.

The horse-chestnut of the Old World is cultivated in eastern North America. It was brought to northern and western Europe from Constantinople in the Sixteenth Century.

*Family 21. Staphyleaceae.

BLADDERNUT FAMILY

Shrubs or trees. Leaves pinnate, usually opposite; stipules present. Flowers usually bisexual, radially symmetrical, in drooping racemes or panicles, with hypogynous discs; petals 5; stamens 5, separate; carpels 2–3, the styles usually separate, ovaries 2–3-chambered; placentae axile; ovules numerous in each carpel. Fruit an inflated capsule (dehiscent at the apex) or sometimes a berry; leaves not numerous.

The family consists of several genera occurring in the temperate portions of the Northern Hemisphere and in northern South America. Two species of *Staphylea* occur in the United States, one in California and one in the eastern portion of the Continent.

Order 19. *JUGLANDALES.*

WALNUT ORDER

DEPARTURE FROM THE FLORAL CHARACTERS OF THE RANALES

Free	Carpels coalescent.
Numerous	Stamens usually numerous, sometimes as few as 3 but usually as many as 20–40 or even 100; carpels 2 or rarely 3.
Spiral	Stamens probably usually spiral; carpels cyclic.
Hypogynous	Flowers epigynous in the North American genera, but not necessarily so in other groups.

Trees or rarely shrubs. Leaves pinnate, large, with small glands producing an aromatic substance. Plants monoecious, or in the *Rhoipteleaceae* polygamomonoecious, the staminate flowers commonly (in North America) in catkins, the pistillate solitary or a few in a cluster; staminate flower commonly in the axil of a bract and a pair of smaller bractlets, consisting of usually 4 sepals and the stamens, sometimes with a rudimentary pistil; pistillate flower subtended by bractlets, these sometimes adnate with the floral cup, the flower consisting of 4 sepals and a pistil, the sepals commonly appearing at the top of the floral cup just below the style, the stigmas elongated and conspicuous but not colored; ovary usually 1-chambered. Fruit a large nut (as a walnut, hickory nut, or pecan), the outer layer fleshy or ultimately leathery or fibrous, the inner stony with 2–4 usually incomplete partitions near the margin of the ovary, these hard like the inner layer but usually thin; embryo usually very large, the cotyledons usually leafy or complexly corrugated and lobed.

The relationships of the Juglandales are obscure, as with most other orders of the old Amentiferae. Engler and Prantl considered the order to be primitive, but it displays a high degree of specialization. The origin of the order is most likely from the forerunners of the Thalamiflorae or the Calyciflorae, but these, of course, are unknown. Often the order has been defined to include the Myricales. An affinity to the *Anacardiaceae* in the Sapindales has been suggested. This is not supported by study of the wood anatomy. Possibly the natural position for the Juglandales is somewhere in the Calyciflorae, but, as with the other "Amentiferae," this is open to debate.

Family 1. Rhoipteleaceae.

RHOIPTELEA FAMILY

Leaves with stipules. Catkin with pistillate and perfect flowers, these in clusters of 3 axillary to each bract, the two lateral pistillate, the middle perfect and with 4 sepals, 6 stamens, and 1 pistil; ovary 2-chambered but with only 1 ovule, superior. Fruit a nutlet, 2-winged.

Figure 10-37. Sapindales; Hippocastanaceae; *Aesculus californica,* California buckeye I—**1** branch (portions of leaves being detached, cf. **Figure 3-3 A**) and a flowering raceme, the terminal flowers bisexual, the lower staminate, **2** young fruits developing from the bisexual flowers at the terminus of the inflorescence, **3** fruit, swelling open and rupturing along the sutures after wetting, **4** dry seed, showing the hilum.

Figure 10-38. Sapindales; Hippocastanaceae; *Aesculus californica,* California buckeye II—**1** bisexual flower from the terminal portion of the inflorescence, **2** the same with sepals and petals removed, showing the stamens, the pistil, and the irregular hypogynous disc, **3** cross section of the ovary, **4–5** staminate flowers, **6–7** petals.

The family consists of a single species occurring in southwestern China.

*Family 2. **Juglandaceae.**

WALNUT FAMILY

Leaves without stipules. Staminate flowers in catkins, the pistillate solitary or in small clusters or (in genera not occurring in Noth America) in catkins; staminate flower usually with 4 or fewer sepals, 3–100 stamens, and sometimes a rudimentary pistil; pistillate flower epigynous, with 4 sepals appearing at the top of the ovary above the attached floral cup; stigmas 2 or rarely 3; ovary 1-chambered, with 1 ovule, inferior. Fruit a nut, covered by a fleshy, leathery, or fibrous husk formed from the floral cup and the attached bracts, sometimes (in genera not occurring in North America) with 2–3 wings. Embryo large, the cotyledons of a complex form. (Pp. 218–219.)

The family includes about six genera and approximately 60 species. The best known genus is *Juglans,* which includes the walnuts. There are four species of walnuts in the United States: one in California, one in the Southwest along the Mexican boundary, and two (the black walnut and the butternut) in the East, Middle West, and the Southern States. Ten species of *Carya,* which includes the hickories, pignuts and pecans, occur in the Middle West, the East, and the Southern States.

Several species are important for their edible seeds. The commonly cultivated walnut is *Juglans regia* of southeastern Europe and Asia. Often this is grafted on the root stock of one of the native black walnuts (for example, in California *Juglans californica* var. *Hindsii,* for resistance to oak root fungus). Various species are of outstanding importance for their wood, which is prized highly for cabinet work and furniture. It is used now mostly as a thin veneer.

Order 20. ***MYRICALES.***

WAX-MYRTLE ORDER

DEPARTURE FROM THE FLORAL CHARACTERS OF THE RANALES

Free	Carpels coalescent.
Numerous	Stamens 2–20; carpels 2.
Spiral	Stamens probably cyclic when in small numbers; carpels cyclic.

Shrubs or small trees. Leaves simple, rarely pinnately lobed, leathery, with yellow resin glands, the plant aromatic. Plants usually dioecious (but the individual or its branches or even the flowers variable as to sex); both kinds of flowers in catkins, each flower subtended by a scaly bract; staminate flower consisting of 2–20 stamens; pistillate flower composed of a single pistil; stigmas 2; ovary with a single chamber producing a single basal orthotropous ovule, usually with one integument, the pollen tube often entering through the base (chalaza) of the ovule. Fruit a small drupe, the surface covered with wax-secreting glands; endosperm containing oil and proteins.

Rendle and other authors have considered the family Myricaceae, the only one in the order, to be a member of the small order Juglandales, which includes the walnuts, hickories, and pecans. A character in common is a single one-celled ovary with a large seed formed from an orthotropous ovule. Both the wax-myrtle family and the walnut family have aromatic compounds in their leaves. In the *Juglandaceae* there are various degrees of adnation of the calyx (or floral cup) with the ovary. It is possible that the order is of common origin with the *Hamamelidaceae* in the Rosales.

*Family 1. **Myricaceae.**

WAX-MYRTLE FAMILY

The family is composed of only two genera, *Myrica* and *Comptonia.* These occur in cool areas in the North Temperate Zone and in South Africa.

Some species, as, for example, *Myrica Gale,* are not necessarily dioecious, and there is a small percentage of monoecious individuals. The sex of an individual plant or of one of its branches may vary from year to year.

Occasionally there are extra bractlets in the pistillate flower. These have been interpreted as sepals.

Order 21. ***LEITNERIALES.***

CORKWOOD ORDER

DEPARTURE FROM THE FLORAL CHARACTERS OF THE RANALES

Numerous	Stamens 3–12; carpel 1.

Spiral	Stamens probably cyclic except perhaps when the numbers are larger; carpel solitary, therefore not either spiral or cyclic.

Large shrubs. Leaves alternate, stipules none or possibly rudimentary. Flowers unisexual (and the plants dioecious), both types in catkins, the catkins resembling those of willows, but the bracts larger and more conspicuous, the staminate catkins drooping, the pistillate erect; staminate "flower" consisting of 3–12 stamens (these actually a cymule); pistillate flower composed of a single carpel and a very small perianth, with a series of bracts at the base, these usually four, coalescent; ovary with a single internal chamber containing 1 parietal but subapical ovule; style thickened, elongated, with a groove along the side; ovule amphitropous, with two integuments. Fruit a leathery drupe 1–2 cm. long, 5–7 mm. in diameter, somewhat flattened; endosperm thin and fleshy.

The relationships of this order are unknown. Study of the vascular bundles leading to the structures of the pistillate flower suggests former presence of several carpels and several ovules and indicates that the bractlike structures in the pistillate flower are 3–8 (usually 4) sepals. The relationship of the order has been suggested as being possibly with the willows or the Myricales but more likely the *Hamamelidaceae* in the order Rosales or some group in the Geraniales or Ranales. According to Engler and Prantl, the order is primitive, but this is unlikely.

*Family 1. **Leitneriaceae.**

CORKWOOD FAMILY

The entire order consists of a single species, the corkwood, *Leitneria floridana,* growing in the southeastern United States from southern Missouri to northern Georgia and southward to Texas and Florida.

Order 22. *RHAMNALES.*

COFFEEBERRY ORDER

DEPARTURE FROM THE FLORAL CHARACTERS OF THE RANALES

Free	Petals rarely coalescent basally; carpels coalescent.
Numerous	Stamens 4–5; carpels 2–5.
Spiral	Stamens and carpels cyclic.
Hypogynous	The *Rhamnaceae* perigynous or sometimes epigynous, in either case also with a disc bearing both petals and stamens; the *Vitaceae* with hypogynous discs.

Shrubs, trees, or woody vines. Leaves alternate or opposite, simple or palmately compound; stipules none. Flowers bisexual or unisexual; stamens in a single series, alternate with the sepals, i.e., opposite the petals, except when these are absent, each petal often forming a hood over a stamen, but the petals sometimes wanting; ovary with 2–5 chambers, according to the number of carpels, each containing 1–2 ovules; ovules anatropous, ascending, with 2 integuments; endosperm surrounding the embryo. Fruit a capsule or a berry.

This order is related closely to the Sapindales. Its primary differences are (1) arrangement of the single series of stamens *opposite the petals* (the outer series *opposite the sepals* having disappeared) and (2) the often perigynous flower. In the *Rhamnaceae* the disc spreads upward inside the floral cup, and sometimes it is attached also to the at least partly inferior ovary of an epigynous flower.

Key to the Families

(North America north of Mexico)

1. Sepals well developed; fruit either a capsule or berrylike or drupelike, usually not markedly juicy; plants mostly trees or shrubs but sometimes woody vines.

   Family 1. **Rhamnaceae** (coffeeberry family).

1. Sepals minute or wanting; fruit a usually juicy berry (grape); woody or fleshy vines usually with tendrils.

   Family 2. **Vitaceae** (grape family).

*Family 1. **Rhamnaceae.**

COFFEEBERRY FAMILY

(Synonym: *Frangulaceae*)

Shrubs or trees or sometimes climbing vines or rarely herbs. Leaves simple, usually alternate; stipules present. Flowers bisexual or rarely unisexual (and then the plants usually monoecious), radially symmetrical, perigynous and with a disc

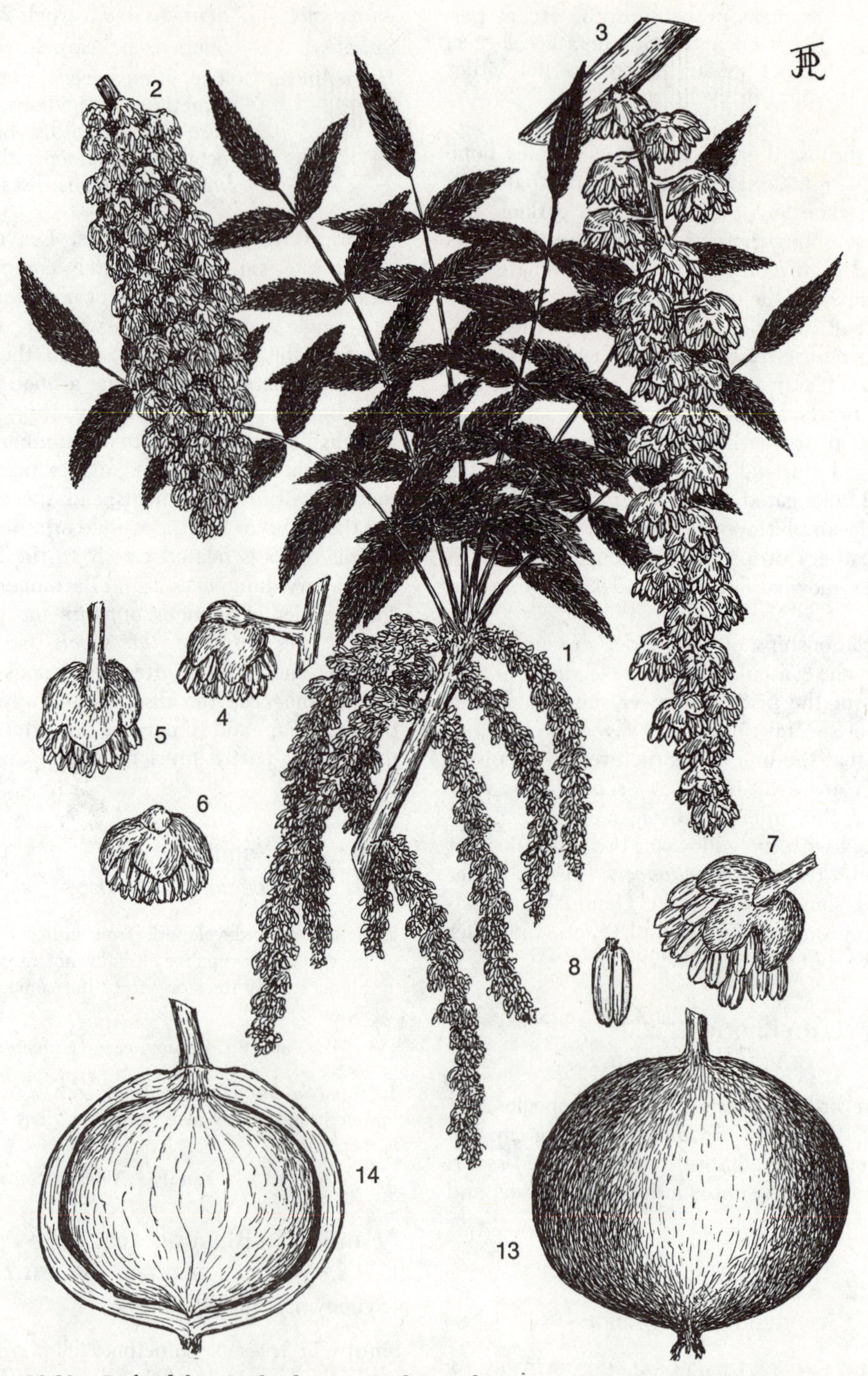

Figure 10-39. Juglandales; Juglandaceae; *Juglans californica,* California black walnut: 1 branch with young leaves and staminate catkins, 2–3 staminate catkins of differing age, 4–7 views of staminate flowers, each consisting of sepals and numerous stamens, 8 anther, 9 twig with a ma-

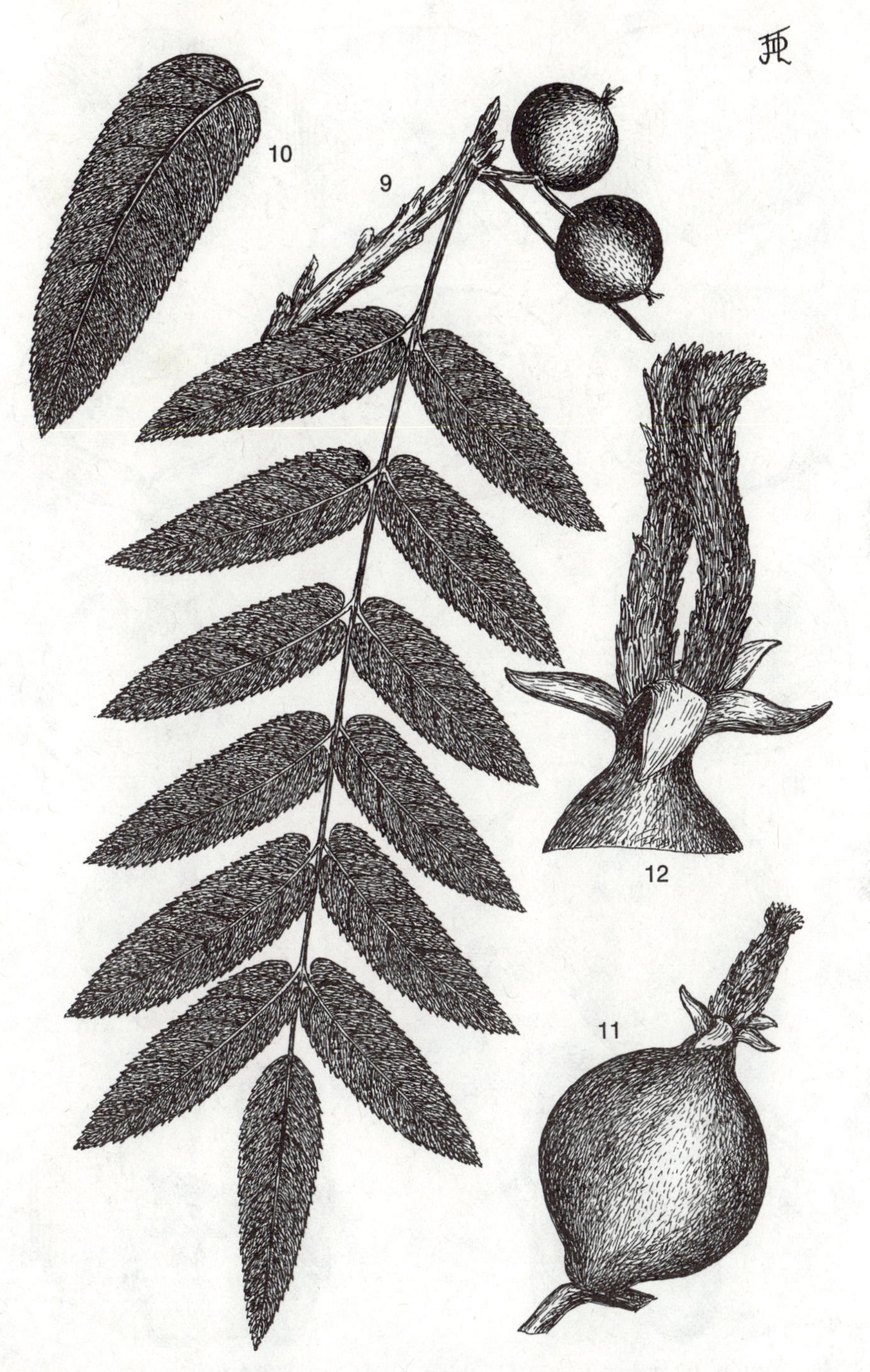

ture leaf and nearly mature fruits, **10** leaflet, **11–12** young fruits, showing the structures essential to the pistillate flower, i.e., the floral cup (adnate with the inferior ovary), the sepals, and the two stigmas, **13** mature fruit, **14** half of the husk removed, showing the hard interior layer. (Cf. also **Figures 6-9 D** pistillate flower and **8-4 G** fruit.)

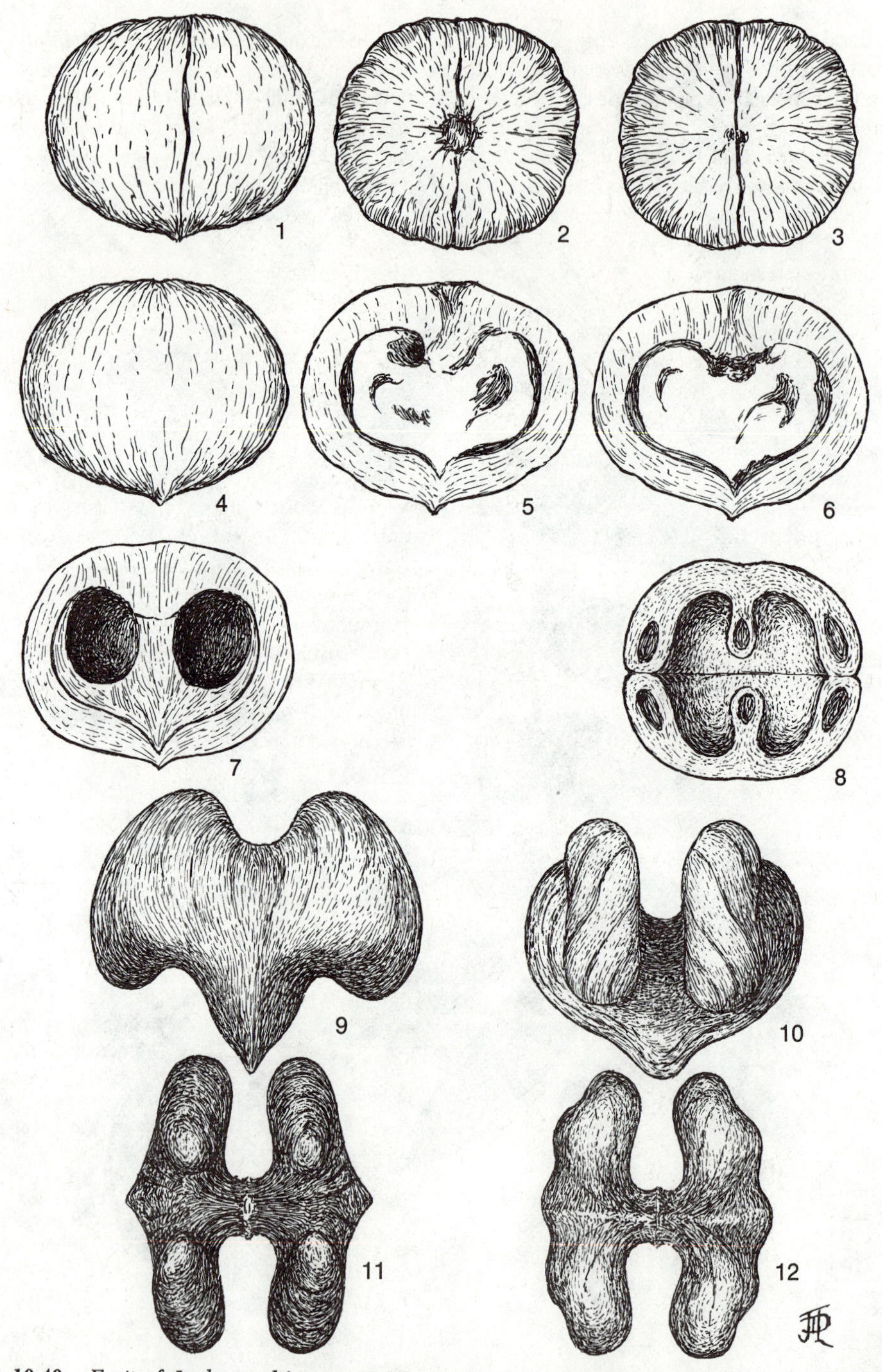

Figure 10-40. Fruit of *Juglans californica,* California black walnut, the husk (floral cup) removed, the pericarp (shell) remaining: **1–4** external views, some showing the suture of the two carpels, **5–6** ovary, longitudinal section, divided in the plane of the suture, **7** ovary, longitudinal section, the embryo removed, divided crosswise to the suture, **8** cross section, **9–12** cotyledons of the four-lobed embryo, four views.

lining the floral cup or sometimes epigynous, i.e., with at least the base of the ovary adnate with the floral cup; sepals 5 or sometimes 4; petals 5 or sometimes 4 or rarely none; stamens 5 or sometimes 4; carpels usually 3, sometimes 2 or 4, the style 1 but commonly cleft, the ovary with as many chambers as carpels; placentae basal; ovules 1 (or rarely 2) per carpel. Fruit a berry, a dupe, a capsule, or rarely a samara.

The family is almost world-wide in distribution. In North America north of Mexico it is developed best in California. *Ceanothus* (mountain-lilac) includes about 30 complex species in or near California (40–50 or more by more recent counts applying a very liberal policy), and these shrubs are one of the principal ingredients of the chaparral. They are highly attractive, and many are cultivated. The genus occurs sparingly elsewhere on the Pacific Coast and in the eastern portion of the Continent. *Rhamnus* (buckthorn) includes the cascaras used in medicine. These are brewed from the bark and used as a very bitter purgative. Species occur here and there across the Continent, with the chief concentration in California. *Condalia, Colubrina,* and *Ziziphus* occur from California to Texas. *Colubrina* and *Ziziphus* are also in the Southeast. Other genera occur along the southern edge of the United States, particularly in south peninsular Florida and on the Keys. These are *Sageretia, Gouania* (chew-stick), *Berchemia* (supplejack or rattan-vine), *Reynosia* (red ironwood), and *Krugiodendron*. The jujube (*Zizyphus*) has been introduced from the Eastern Hemisphere into cultivation, and in some areas of the southeastern coastal plain from Louisiana to Alabama it has escaped, forming thickets along roadsides. The drupelike fruit is only an inch long and red to nearly black.

*Family 2. **Vitaceae.** GRAPE FAMILY

Woody vines with tendrils. Leaves simple, pinnate or palmate, alternate; stipules present or absent. Flowers bisexual or unisexual (and the plant then monoecious), radially symmetrical, with hypogynous discs, in elaborate clusters of cymes; sepals usually 4 or 5; petals commonly 4 or 5 and minute or sometimes none, often coalescent at the apex but separate below for by far the greater part of their length, the corolla then early deciduous;

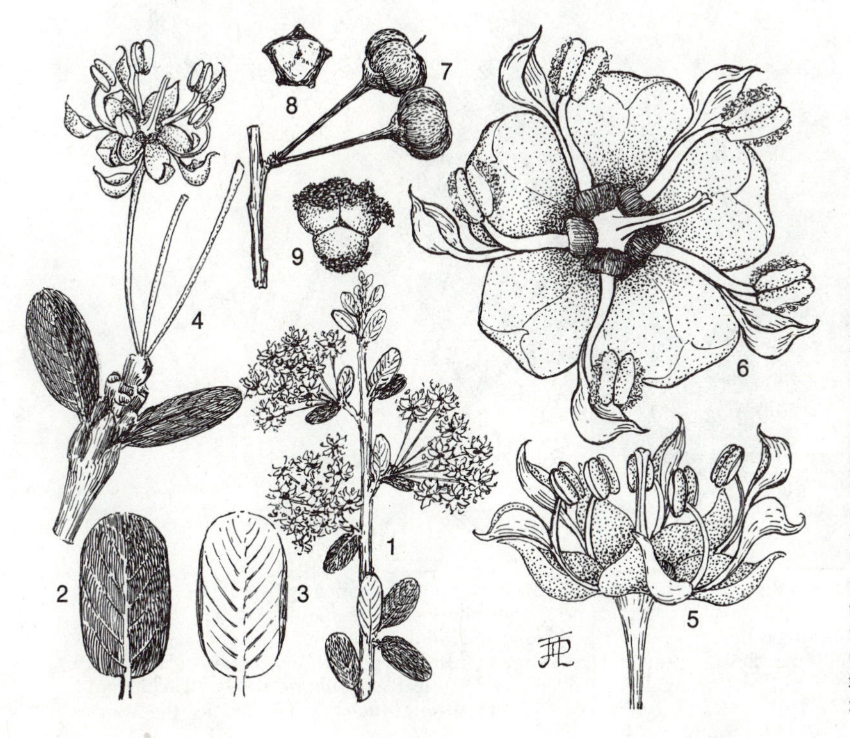

Figure 10-41. Rhamnales; Rhamnaceae, *Ceanothus crassifolius,* a mountain-lilac: **1** branch with flowers, **2** leaf, upper surface, **3** lower surface, **4** portion of a branch and a flower, enlarged, **5** flower, side view, **6** flower, top view, with the floral cup and perigynous disc area, the small petals each forming a hood subtending the opposite stamen, **7** fruits, **8** fruit and calyx, top view, **9** old fruit with the soft outer tissues sloughing away. (Cf. also **Figure 6-14 A** jujube, flower and fruit, and **B** coffee berry, flower and fruit.)

Figure 10-42. Rhamnales; Vitaceae; *Vitis,* grape: **1** (background) branch with leaves and tendrils, **2** inflorescence with buds and flowers, **3** portion of the inflorescence enlarged, **4** flower bud, longitudinal section, **5–7** flower bud opening, the petals remaining coalescent apically, separate basally, **8–9** views of the open flower, showing the stamens, pistil, hypogynous disc, and glands, **10** young fruit, **11** young fruit, longitudinal section (above with leaf as background), **12** fruits, **13** longitudinal section of a fruit. (Cf. also **Figures 2–10 D** *Vitis girdiana,* desert grape, the species illustrated here, and **2–10 E** Boston ivy.)

stamens the same number as the sepals and petals; carpels usually 2, rarely 3–6, the style 1, the ovary with as many chambers as carpels; placentae axile; ovules 1–2 per carpel. Fruit a berry.

The family occurs chiefly in the Tropics, Subtropics, and warm temperate regions, some species ranging into cooler areas. Two species of wild grape (*Vitis*) occur in California, another in Arizona, and numerous species in the Middle West, East, and South. Some American species have been brought into cultivation, and many hybrids used as commercial grapes are derived from them. The genus *Parthenocissus* includes the Virginia creepers or woodbines and Boston ivy. One species of Virginia creeper ranges as far westward as Arizona. The 'possum grape of the Southern States belongs to the genus *Cissus*. *Ampelopsis* occurs, as do the preceding two genera, in the eastern portion of the Continent.

The Boston ivy and the Virginia creeper are used as covers for walls and buildings. Several species of *Cissus* are cultivated outdoors in the southern part of the United States and used as houseplants throughout the country.

Order 23. *CARYOPHYLLALES.*

PINK ORDER

DEPARTURE FROM THE FLORAL CHARACTERS OF THE RANALES

Free	Sepals sometimes coalescent, as in the tribe *Sileneae, Caryophyllaceae;* petals rarely coalescent (*Sileneae*); carpels coalescent, except in some *Phytolaccaceae,* some *Aizoaceae,* and some *Nyctaginaceae.*
Numerous	Stamens 3–10 or rarely more by division of the 5 primary initial areas; carpels 1–5 or rarely 12.
Spiral	Stamens and carpels cyclic.
Hypogynous	Flowers rarely perigynous or epigynous (in several families).

Usually herbs but sometimes shrubs or vines or even trees. Leaves alternate or opposite; stipules rarely present. Sepals 2 or 4–8 (usually 5) in one series or sometimes 6 in 2 series of 3 (or rarely 4 in 2 series of 2); petals none except in most *Caryophyllaceae* and *Portulacaceae* (in this family actually sepals); stamens in 1 or 2 series or sometimes in 3 series of 3 or numerous; ovule orthotropous (*Polygonaceae*) or commonly campylotropous or sometimes amphitropous or rarely anatropous; embryo folded or straight (*Polygonaceae*) or commonly curved or coiled in seeds developed from campylotropous ovules; placenta basal (and with solitary ovules) or axile (and with numerous ovules), the axile placentae sometimes at the inner angles of seed chambers, but usually on a central raised structure, this sometimes post-like. Fruit an achene, a utricle, a nutlet, a capsule, or rarely a berry, usually with only 1 seed chamber, but rarely with 3–12.

Under the classification system of Engler and Prantl, this order was considered to be transitional from the supposedly more primitive group of flowering plants lacking petals to the more advanced group with petals. Under the Bessey System, the flowering plants considered most primtive (Ranales) had petals, and consequently it was necessary to consider the apetalous Caryophyllales as specialized types derived from ancestors with petals. This discrepancy emphasizes the difficulty of determination of even the beginning point in a supposed evolutionary series.

The name Centrospermae used for the order by Engler and Prantl refers to the arrangement of the seeds in the ovary. The fruit has a placenta often resembling a hitching post standing in the middle of the ovary, its seeds being attached by relatively long stalks (funiculi). The solitary seeds of many members of the order, e.g., the goosefoot and wild buckwheat (*Eriogonum*), are attached to the base of the ovary. The central (axile) or basal placenta is characteristic of nearly all the order.

The ovule structure of nearly all the Caryophyllales (except the *Polygonaceae*) differs from that general through any other order (though occasional in families of several orders) because the ovules are campylotropous (i.e., the ovule is bent back upon itself but without fusion of the curved ovule to its own basal portion or to the funiculus). The curve of the ovule results in a curved embryo largely characteristic of the whole order (except the *Polygonaceae*). An old name for the order is Curvembryonae.

Key to the Families

(North America north of Mexico)

1. Fruit a berry, hypogynous (or in *Agdestis*, occurring in Texas and Florida, an achene formed from an inferior ovary and with the 4 or rarely 5 sepals

persistent as wings), indehiscent.

Family 1. **Phytolaccaceae** (poke family).

1. Fruit not a berry, if fleshy, at least partly inferior.
2. Ovary with more than 1 seed chamber; petals none, but the flower of *Mesembryanthemum* (ice-plant) with numerous petal-like staminodia; calyx *not* globose or tubular, sometimes cuplike.

Family 3. **Aizoaceae** (carpetweed family).

2. Ovary with 1 seed chamber (except in plants with both petals and a globose or tubular calyx).
3. Petals none, the calyx tubular, white or colored but not green, resembling a sympetalous corolla, the base of the tube constricted above the ovary, persistent, hardening and enclosing the fruit; flower(s) sometimes with a green calyx-like involucre, this (when only 1 flower is enclosed) readily mistaken for the calyx.

Family 4. **Nyctaginaceae** (four-o'clock family).

3. Plant *not* corresponding with *every character* in the combination in lead 3 above.
4. Seed or ovule solitary in the ovary, attached at the base of the chamber, the fruit an achene or a utricle.
5. Plant *not* with scarious stipules *or,* if so, the stipules sheathing the stem.
6. Plant *not* corresponding with *every character* in the combination in lead 6 below.
7. Flowers *not* obscured among many dry, scarious, persistent bracts, sometimes subtended by an involucre of coalescent bracts.
8. Plants *not* scurfy; characterized by *one or more* of the following features.
(1) Stipules present, sheathing the stem.
(2) Flowers produced in an involucre of coalescent bracts *or* rarely each flower with a single bract which finally becomes large and membranous, encloses the achene, and develops 2 saclike structures on its back.
(3) Sepals 6, *or,* if only 5, each with a hooked bristle.

Family 5. **Polgonaceae** (buckwheat family).

8. Plants usually scurfy (with external protruding cells which dry into scales resembling dandruff); stipules none; involucres and large membranous bracts none; sepals 5 or fewer.

Family 6. **Chenopodiaceae** (goosefoot family).

7. Flowers obscured among many dry, scarious, persistent, separate bracts, these often colored, the sepals also often scarious; involucre none.

Family 7. **Amaranthaceae** (amaranth family).

6. Somewhat succulent vines; rhizomes with tubers; sepals apparently 2 (actually bracts), 1–1.5 cm. long; petals apparently 5 (actually the sepals); tendrils or spines none.

Family 9. **Basellaceae** (Madeira-vine family).

5. Plant with scarious (thin and membranous) stipules, these *not* sheathing the stem. *Illecebroideae.*

Family 10. **Caryophyllaceae** (pink family).

4. Seeds or ovules more than 1 in the ovary, usually numerous, attached to a raised basal placenta, usually forming a central column; fruit a capsule; petals usually present or seemingly so, not numerous.
5'. Sepals seemingly 2 (actually bracts) (or in some species of *Lewisia* 4–8, these plants having the leaves all basal and the flowers solitary on scapes).

Family 8. **Portulacaceae** (*Portulaca* family).

5'. Sepals 4 or 5; plant *not* with the leaves all basal and *not* with solitary flowers on scapes.

Family 10. **Caryophyllaceae** (pink family).

*Family 1. **Phytolaccaceae.** POKE FAMILY

(Segregate: *Petiveriaceae*)

Herbs, shrubs, or trees, or rarely woody vines. Leaves simple, entire, alternate; stipules none or very small. Flowers bisexual or rarely unisexual (and then the plants monoecious), radially or rarely bilaterally symmetrical, often with a hypogynous disc, epigynous in *Agdestis*, in cymes or racemes; sepals 4–5, persistent; petals none; stamens 3–10 or more, usually in 1 or 2 series; carpels 1 to about 15, usually 5–12, the style 1, the ovary with as many chambers as the carpels, the carpels solitary in the *Rivineae;* placentae axile when there is more than 1 carpel; ovule 1 per carpel, campylotropous or amphitropous. Fruit a berry, a drupe, an achene, or a utricle or dividing into the carpels at maturity.

The family occurs chiefly in tropical America. The common poke or pokeweed (*Phytolacca*) is widespread. Species of *Rivina* occur from Arizona eastward to Florida. *Petiveria* and *Agdestis* are native in the Southern States, especially on the Coastal Plain and in Florida.

The genus *Achatocarpus* and a related genus have been considered as a separate family *Achatacarpaceae.* This group of plants is not known to the author, and its status cannot be evaluated.

Family 2. **Gyrostemaceae.**

GYROSTEMON FAMILY

Shrubs. Leaves somewhat succulent; stipules none. Flowers unisexual; sepals present; stamens 6–numerous; filaments none; carpels usually numerous, rarely 1 or 2; placentae axile; ovule 1 per carpel.

The family is restricted to Australia.

*Family 3. **Aizoaceae.**

CARPETWEED FAMILY

(Synonym: *Tetragoniaceae*)

Herbs or rarely low shrubs. Leaves simple, often either small or scalelike or succulent, commonly opposite or appearing to be in whorls or sometimes alternate; stipules none or sometimes present. Flowers bisexual, radially symmetrical, hypogynous, perigynous, or epigynous; sepals 5–8, coalescent; petals none, apparently numerous in *Mesembryanthemum*, but these actually staminodia; stamens in large groups from 5 primordia carpels 3–20, styles usually several; ovary 1–10-chambered or sometimes up to 20-chambered; placentae axile or rarely parietal or basal; ovules usually many, campylotropous or anatropous. Fruit a capsule or sometimes coriaceous or berrylike.

The family is distributed widely, but it is centered most strongly in South Africa where there are numerous genera and species, many related to *Mesembryanthemum,* these being remarkably succulent and sometimes of bizarre appearance. *Sesuvium, Trianthema, Mollugo* (the Indian-chickweed or carpetweed), *Glinus,* and *Tetragonia* (New Zealand spinach) are representatives of the family native or introduced in warm regions of the United States. The most striking plants are in the genus *Mesembryanthemum,* native on the seacoast in California and in South Africa. Sometimes they are called fig-marigolds, but in America the most used name is ice-plant. They are known also as *Carpobrotus.*

Many species of *Mesembryanthemum* and *Lithops* are cultivated as succulent novelties, but often under other generic names.

The family is noteworthy for diversity of flower structure.

*Family 4. **Nyctaginaceae.**

FOUR-O'CLOCK FAMILY

(Synonym: *Allioniaceae*. Segregate: *Pisoniaceae*)

Herbs or (in an extraterritorial group) shrubs or trees. Leaves simple, entire, opposite or usually so; stipules none. Flowers bisexual or rarely unisexual, radially symmetrical, rarely perigynous (*Abronia*), the clusters of flowers or the solitary flowers often surrounded by 2–5 green and leaflike or highly colored bracts, these forming an involucre and appearing to be a calyx when the flower is solitary; sepals 5, in the North American species coalescent into an elongated tube and often appearing (especially because of the calyx-like involucre) to be a sympetalous corolla, the calyx tube constricted just above the ovary and deciduous at that level, the base persisting around the fruit; petals none; stamens usually 5 but varying from 1 to 3; carpel 1, with 1 ovule, this campylotropous or anatropous. Fruit an achene, usually enveloped by the hardened persistent lower portion of the calyx, in many species this highly specialized and facilitating dispersal of the seeds.

The family occurs throughout the Tropics and Subtropics, and northward into the warmer regions of the Northern Hemisphere. The four-o'clocks (*Mirabilis*) are common in cultivation,

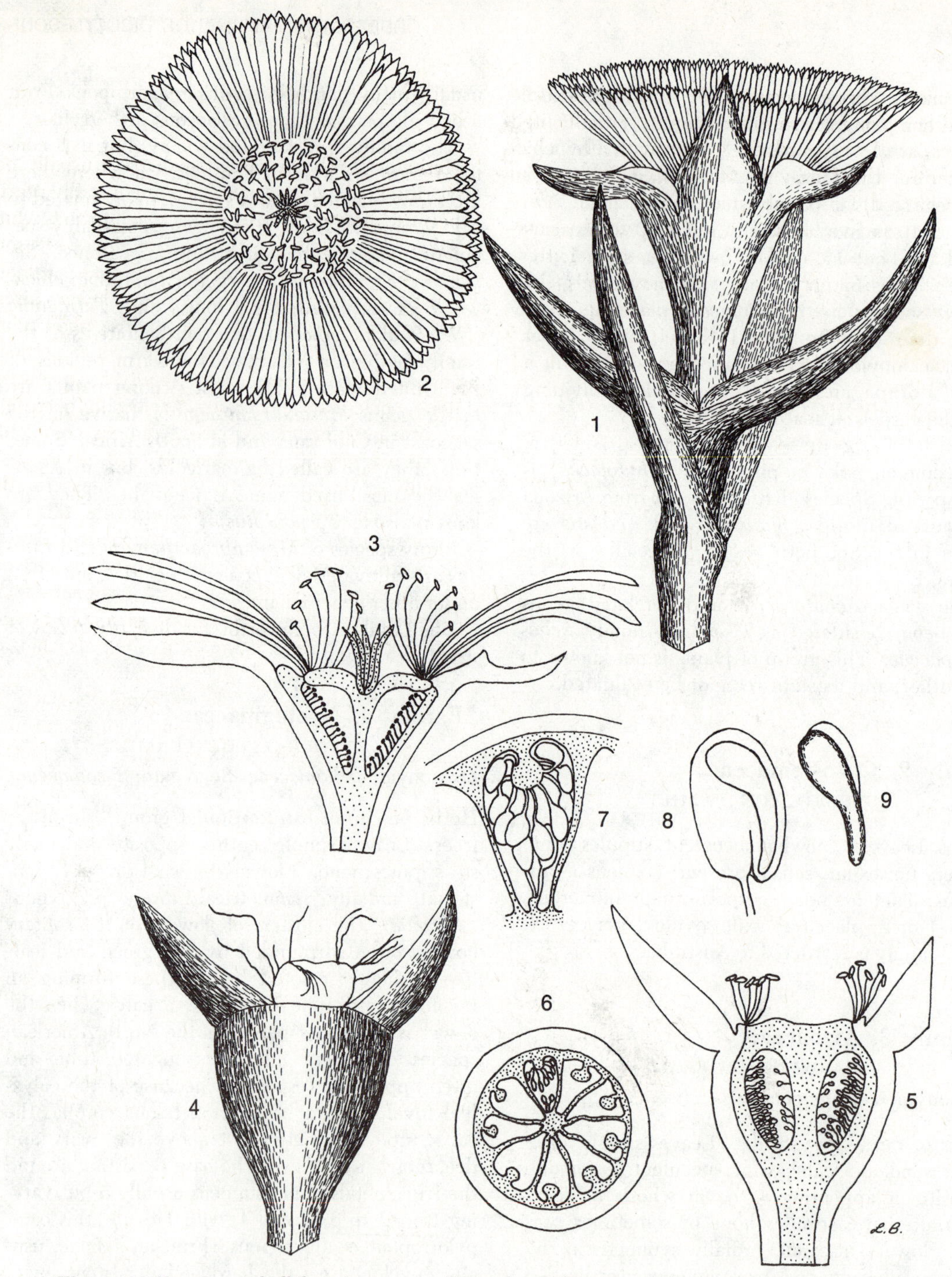

Figure 10-43. Caryophyllales; Aizoaceae; *Mesembryanthemum chilense,* ice-plant or sea fig: **1** branch and flower, showing the fleshy opposite leaves, the fleshy sepals, and the numerous petallike staminodia, **2** top view of the flower, showing the staminodia, stamens, and stigmas, **3** longitudinal section of the flower, showing the sepals, staminodia, stamens, and stigmas and the inferior ovary including the ovules, **4** fruit, the sepals persistent, **5** fruit, longitudinal section, the sepals and some stamens persistent, **6** cross section of the ovary, showing the seed chambers, **7** enlargement of one chamber showing the ovules and placentae, **8** longitudinal section of the seed, **9** embryo. (The plant is also known as *Carpobrotus aequilaterus.*)

and native species occur in the East and Middle West and along the southern margin of the United States southwestward to California. The chief center for the family in the United States is in Arizona and Texas (about eleven genera). *Abronia,* the sand-verbena, is a striking genus occurring on dunes along the Pacific Coast and in the deserts of Southern California and Arizona. It covers many acres of sandy areas, adding on wet years a brilliant springtime touch of color. Another colorful genus is *Allionia.*

*Family 5. **Polygonaceae.**

BUCKWHEAT FAMILY

Herbs or rarely shrubs or trees or sometimes vines. Leaves simple, alternate; stipules present (except in the *Eriogoneae*), sheathing the stem (known as ochreae). Flowers bisexual or rarely unisexual, radially symmetrical, cymose or in panicles, spikes, or heads; sepals 3–6, commonly in 2 series of 3, sometimes 5 as a result of fusing of a sepal from the inner series with one from the outer; stamens usually 6 or 9 in 2 or 3 series; carpels usually 3, sometimes 2 or 4, the style 1, the ovary 1-chambered, with 1 basal orthotropous ovule. Fruit an achene, usually 3-angled.

The family occurs chiefly through the temperate regions of the Northern Hemisphere. In North America it is largely Californian and to a lesser degree of the other Western States. It includes such well-known plants as *Polygonum* (knotweed, smartweed, and wire-grass), *Rumex* (docks, including sour dock), *Oxyria* (mountain sorrel, a plant characteristic of the areas above timberline), and *Fagopyrum* (buckwheat). In California and in the West there is an enormous development of the Eriogoneae and especially *Chorizanthe* and *Eriogonum.* By conservative count in California alone there are about 70 species of *Eriogonum* (known as wild buckwheat), and they are one of the characteristic features not only of the California landscape but also of vast areas throughout the Western States. Two species of the evergreen shrub or tree genus *Coccoloba* occurring in south Peninsular Florida and on the Keys are known as pigeon-plum and sea-grape.

Buckwheat and rhubarb (*Rheum*) are the only important food plants of the family grown in North America.

The sourness of the juice is characteristic of the family.

*Family 6. **Chenopodiaceae.**

GOOSEFOOT FAMILY

Herbs or shrubs or rarely small trees usually occurring in alkaline situations; plants usually scurfy because of external cells which dry into thin white flakes. Leaves simple, sometimes more or less succulent or reduced to small scales, usually alternate but rarely opposite; stipules none. Flowers bisexual or rarely unisexual (and then the plant monoecious), radially or rarely bilaterally symmetrical, commonly numerous; sepals usually 5, sometimes none, coalescent at least basally, usually persistent at fruiting time; petals none; stamens commonly 5; carpels usually 2 or sometimes 3; styles 1–3, usually 2 or at least the stigmas separate, the ovary 1-chambered, with 1 basal ovule, this campylotropous. Fruit a nutlet or an achene or said to be a utricle; the embryo coiled or strongly curved.

The family is world-wide, but it is centered in alkaline areas. Some species are restricted to wet, strongly salty or alkaline soil such as that of coastal salt marshes or alkaline plains and playas in the deserts. As a whole, the family is made up of weedy plants, some of the more important genera being as follows: *Chenopodium* (goosefoot or lamb's-quarter or sometimes called pigweed); *Kochia* (red sage), and *Salsola* (the so-called Russian thistle, which is an abundant plant in the arid parts of the West). Genera occurring in markedly alkaline situations include *Sarcobatus* (the black greasewood of the desert alkali flats of the West), *Suaeda* (a desert or coastal genus), *Salicornia* (the pickleweed or samphire, common also along seacoasts); *Allenrolfea* (a succulent shrub of desert alkaline places, resembling *Salicornia* in its remarkably swollen stem joints); *Atriplex* (the saltbushes, abundant not only along the salt marshes of the ocean but also on the alkali flats of the deserts); *Grayia* (the hop sage of the Sagebrush Desert); *Ceratoides,* formerly known as *Eurotia* (the winter-fat of the Sagebrush Desert and the Mojave Desert, an important plant for winter forage of cattle).

Cultivated plants in the family are the beet and chard (*Beta*) and spinach (*Spinacia*).

*Family 7. **Amaranthaceae.**

AMARANTH FAMILY

Herbs, or rarely shrubs, trees or vines. Leaves simple, usually entire, alternate or opposite; stipules

Figure 10-44. Caryophyllales; Nyctaginaceae; *Mirabilis Jalapa,* garden four-o'clock: **1** branch with leaves and flowers, **2** outline of a leaf, **3** opposite leaf bases and axillary buds, **4** flowers, showing the spreading upper portion of the colored calyx, the lower portion tubular (**1**), **5** longitudinal view of the basal portion of the flower, showing the involucre (green and calyxlike), the enlarged basal portion of the calyx

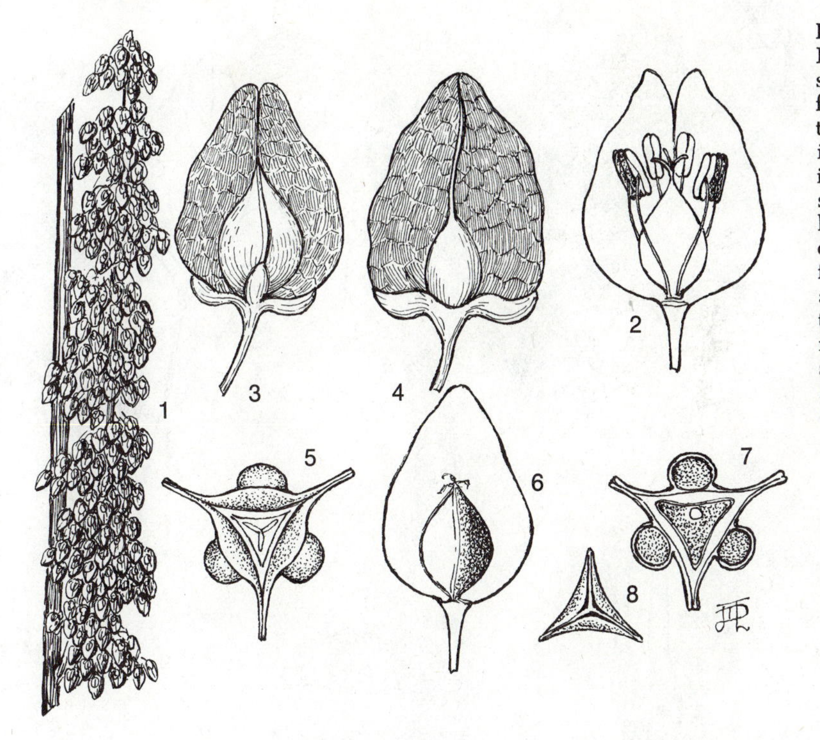

Figure 10-45. Caryophyllales; Polygonaceae; *Rumex crispus*, sour dock or curly dock: **1** inflorescence, **2** diagram showing the stamens and pistil enclosed in the calyx, **3–4** calyx at fruiting time, showing the two dissimilar series of three sepals, the large inner ones each with a basal callus grain, **5** top view of the flower, showing the inner sepals and the stigmas, **6** fruit within the calyx, **7** cross section of the inner sepals and the fruit, **8** shape of the fruit from the top. (Cf. also **Figures 8-4 D** and **8-7 A** *Eriogonum*, fruit and seed.)

none. Flowers bisexual or sometimes unisexual (and then the plant dioecious or polygamodioecious), radially symmetrical, with membranous or scarious bracts (these persistent in fruit), usually in dense inflorescences; sepals 3 to usually 5, usually dry and membranous; petals none; stamens usually 5, with either 2 or 4 pollen chambers; carpels usually 2 or sometimes 3–5; the style(s) 1 or usually 2 or sometimes more, the ovary 1-chambered, with a single basal ovule, this campylotropous. Fruit usually a utricle or a nutlet, rarely a drupe, a berry, or a capsule.

The family is abundant in the Tropics and the adjacent warmer regions, and its greatest development is in tropical America and Africa. It is a family of weeds, but some species of *Amaranthus* and *Iresine* are cultivated for their red pigmentation. Among the worst weeds are the ubiquitous pigweeds and tumbleweeds (both *Amaranthus*). The tumbleweeds have curving branches forming a sort of basket or ball several feet in diameter. They are rolled along by the wind, and finally they collect along fences or among bushes. Most of the seeds are shed along the way. *Iresine*, the bloodleaf, occurs from Arizona to Texas, with another species in the Middle West, East, and South. The family is more common southward toward the Tropics, and numerous species and several genera occur from Arizona to the Southern States.

*Family 8. **Portulacaceae.**

PORTULACA FAMILY

Herbs or sometimes shrubby plants; leaves simple, usually somewhat succulent, alternate or opposite, often in basal rosettes; stipules hairlike or scarious

---

tube (colored and petallike) and its constriction above the ovary, filaments of two stamens, and the pistil, **6** style and stigma, **7** persistent involucre enclosing the fruit (achene), within the persistent basal enlargement of the calyx, **8** achene, **9** achene, longitudinal section, **10** curved embryo surrounding the endosperm, **11** embryo in three dimensions, showing the outer cotyledon, **12** another view (the inner cotyledon smaller than the outer). (Cf. also **Figure 6-9 C** *Abronia villosa*, flower.)

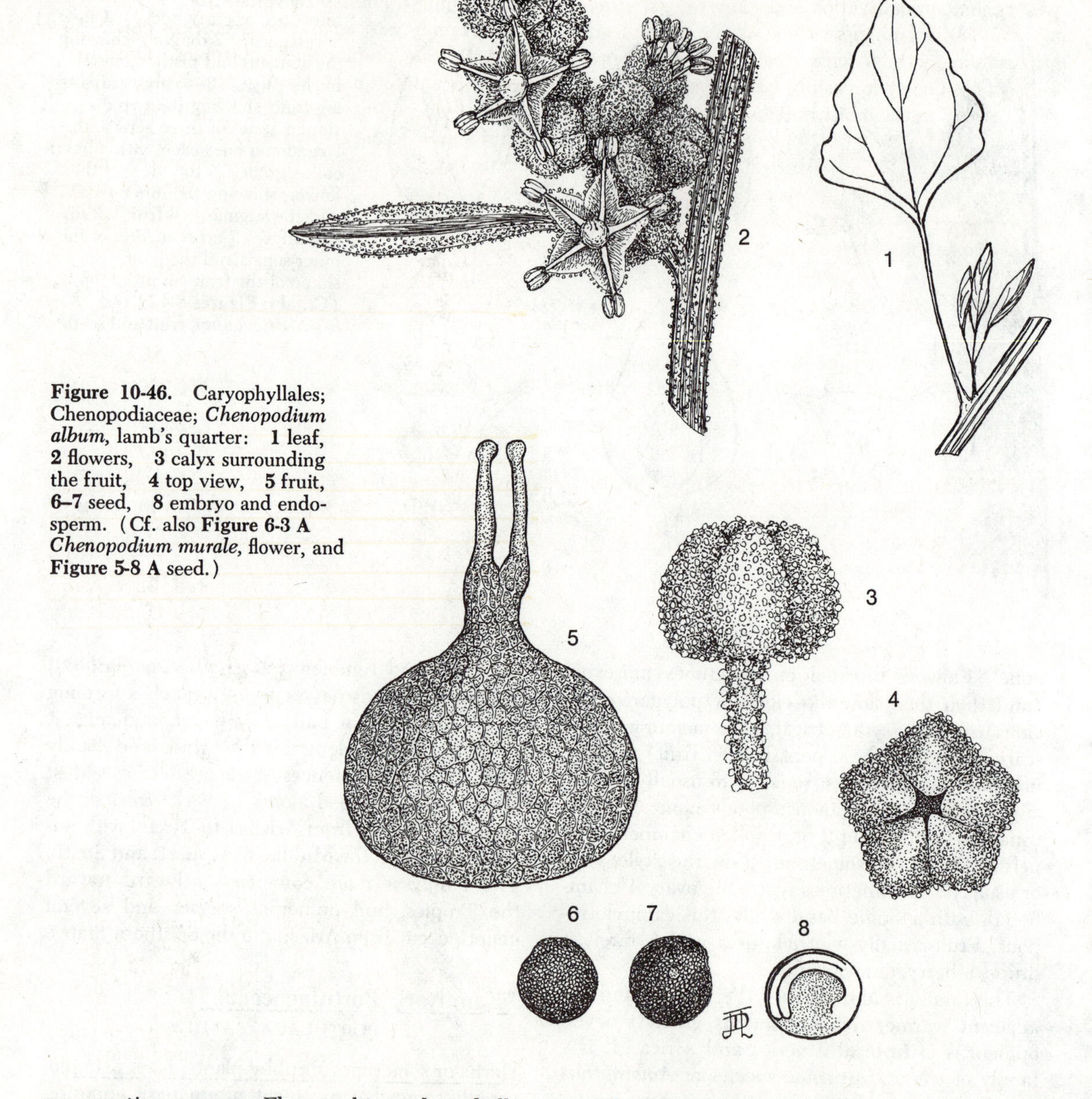

Figure 10-46. Caryophyllales; Chenopodiaceae; *Chenopodium album,* lamb's quarter: **1** leaf, **2** flowers, **3** calyx surrounding the fruit, **4** top view, **5** fruit, **6–7** seed, **8** embryo and endosperm. (Cf. also **Figure 6-3 A** *Chenopodium murale,* flower, and **Figure 5-8 A** seed.)

or sometimes none. Flowers bisexual, radially symmetrical; sepals apparently 2 (these being bracts commonly interpreted as sepals); petals actually none but the calyx simulating a corolla of usually 5 or sometimes 2, 4, or 6 or rarely 3 petals (the common interpretation in books); stamens commonly 5, opposite the petallike sepals, sometimes fewer than the apparent petals; carpels usually 2 or sometimes 3, the styles separate, the ovary 1-chambered, the ovules several to numerous, on a raised basal placenta, campylotropous. Fruit a capsule, circumscissile or more commonly splitting lengthwise, the fruit rarely indehiscent.

The family is distributed widely, but it is centered in Pacific North America and in the southern part of South America. In North America north of Mexico there are numerous native species as well as some in cultivation and some weeds. *Claytonia,* including miner's-lettuce, is one of the leading genera on the Pacific Coast, but it is represented only poorly in eastern North America. The genus occurs in the moist regions of the Continent, and it is known best for the spring-beauty of the Middle West, East, and South. The red-maids (*Calandrinia*) of California and Arizona, the pussypaws of the Sierra Nevada (*Calyptridium*), and the bitterroot (*Lewisia*) are particularly noteworthy plants in the West. Species of *Talinum* occur from Arizona to the Southern States.

*Family 9. **Basellaceae.**

MADEIRA-VINE FAMILY

Perennial, herbaceous or frutescent vines; the plants sometimes rather succulent. Leaves simple, sometimes fleshy, sometimes succulent, alternate; stipules none. Flowers bisexual, radially symmetrical, in racemes, cymes, or panicles; sepallike bracts 2; petaloid sepals usually 5, often colored, persistent in fruit; petals none; stamens 5; carpels 3, the style 1, the ovary 1-chambered, with a single basal ovule, this campylotropous. Fruit a drupe; embryo circular or spiral.

The family is a small one occurring chiefly in the American Tropics and Subtropics. Only one species is native in the Eastern Hemisphere. The Madeira-vine (*Boussingaultia*) is cultivated for its bright green foilage and fragrant flowers. It occurs in pinelands, along roadsides, and in hammocks in Florida. Some species of *Anredera* occur in southern Texas. The malabar (*Basella*) is cooked in the same way as spinach.

*Family 10. **Caryophyllaceae.**

PINK FAMILY

(Synonym: *Silenaceae.* Segregates: *Alsinaceae, Illecebraceae, Scleranthaceae, Corrigiolaceae*)

Herbs or rarely the stems slightly woody basally. Leaves simple, usually elongated, opposite or rarely alternate; stipules none. Flowers bisexual, radially symmetrical, the inflorescence usually complex; sepals 5 or rarely 4; petals 5 or rarely 4, separate; stamens 3 to 10; carpels 2–5 or rarely 1, the ovary nearly always 1-chambered, rarely 2–5-chambered (*Sileneae*); placenta on a central col-

Figure 10-47. Caryophyllales; Caryophyllaceae; *Stellaria media,* chickweed (cf. also **Figure Int.-1**): **1** branch with fruits, **2** flower, top view, showing the sepals, deeply parted petals, stamens, and pistil (with three styles and stigmas), **3** stamens and pistil alone, **4** fruit surrounded by the persistent sepals, **5** the same, the sepals pulled back, **6** mature fruit enlarged, showing the styles and the sutures, **7** capsule opening and exposing the seeds, **8** longitudinal section of the fruit, showing the central placenta arising from the base of the single chamber of the ovary, **9** cross section, showing two of the seeds and the central placenta, **10** seed, **11** longitudinal section of the seed showing the curved embryo and the endosperm (cf. **Figure 8-7 B** embryo).

Figure 10-48. Caryophyllales; Caryophyllaceae; *Dianthus Caryophyllus,* clove-pink or carnation (applied mostly to the "double" forms)—the name of the species, used in pre-Linnaean times as a generic name, being the basis for the names of the family and the order: 1 branch, flower, and fruits, 2 top view of the flower, emphasizing the corolla, 3 flower dissected, showing the bracts (involucre), calyx (coalescent se-

umn arising from the base of the single chamber or rarely (*Sileneae*) an axile placenta present within each of the chambers; ovules usually numerous. Fruit a capsule, dehiscent at the apex or on a horizontal line (circumscissile).

This large family occurs in temperate areas mostly in the Northern Hemisphere, and it is most abundant in the Mediterranean Region. About 20 genera occur in North America north of Mexico. Some of the best known plants of the order are the carnation (*Dianthus*), the chickweeds (*Stellaria* and *Cerastium*), the spurry and sand spurry (*Spergularia* and *Spergula*), the sandworts (*Arenaria*), the campions (*Lychnis*), the Indian pink of California and the catchflies (*Silene*), and the bouncing-bet (*Saponaria*). The family is one of the largest in North America, and it is represented well throughout the Continent.

The subfamily *Illecebroideae* often is considered to be a distinct family, *Illecebraceae*. It has not been included in the description of the *Caryophyllaceae* in order that those who wish may segregate it readily. The following are distinguishing characters: stipules present, thin, membranous; flowers often perigynous; petals none or represented by threadlike structures; style 1, 2–3-cleft or parted or sometimes the styles 2; fruit a utricle or an achene with a solitary seed attached basally. The plants are small, inconspicuous, often prostrate weeds such as the whitlowwort (*Paronychia*) and the knawel (*Scleranthus*).

Family 11. **Didiereaceae.**

DIDIEREA FAMILY

Trees; wood soft; plant spiny. Leaves simple, small, alternate; stipules none. Flowers unisexual (and the plants dioecious), radially symmetrical, with hypogynous discs; sepals 2, petaloid; petals 2, in 2 series; stamens 8–10, slightly coalescent basally; carpels 3, the style 1, the ovary fundamentally 3-chambered but with only 1 developing; ovule 1, erect. Fruit indehiscent, triangular in cross section.

The family is composed of 3 genera growing on Madagascar.

The family has been proposed as related to the Cactaceae, but this seems improbable. (Cf. Lyman Benson, *The Cacti of the United States and Canada*, Stanford University Press, in press.)

Order 24. ***CACTALES.*** CACTUS ORDER

DEPARTURE FROM THE FLORAL CHARACTERS OF THE RANALES

Free	Sepaloids and petaloids coalescent and adnate basally, forming an often deciduous part of the floral tube above or continuous with the floral cup or tube; carpels (including the styles but not the stigmas) coalescent.
Numerous	Carpels 3–20 or more commonly 5–12.
Spiral	Carpels cyclic.
Hypogynous	Flower epigynous, the floral cup or tube bearing areoles like those on the stem, vegetative branches or the flowers of the next season rarely growing from them; the floral cup or tube a hypanthium.

See pp. 236, 237, 240–242 for figures.

Cacti are characterized by large fleshy usually leafless stems, and by spines (large or diminutive) produced always in clusters within definite circumscribed areas (*areoles*). Each areole is directly above a leaf (or the potential position of a leaf) on the stem, and the structures in the areole are developed from the axillary bud of the stem. Many other succulent plants have spines, but the spines are not arranged in areoles. The cactus family is characterized by the following diagnostic features of flower structure: (1) The sepaloids and petaloids are numerous, and they intergrade with one another; the sepaloids, in turn, may intergrade with scalelike leaves borne on the floral cup or tube enclosing the inferior ovary. (2) The pistil consists of an inferior ovary with 1 chamber

pals), petals, stamens, and pistil, **4–6** views of the bracts and calyx at fruiting time, **7** calyx laid open, showing the fruit, **8** fruit, **9** longitudinal section of the fruit, showing the central placenta, **10** ovary, cross section, **11–12** seeds.

and several parietal placentae bearing numerous usually campylotropous ovules and of a single style bearing several stigmas.

The plants commonly confused with cacti include such desert types as the ocotillo, the yuccas (one of which is the Joshua-tree) the sotol or desert-spoon, and various century-plants or agaves. These plants are members of several families, all mentioned above except the ocotillo being monocotyledons. The spiny structures and the considerable degree of succulence of certain parts of the plants are the only characters in common with the cacti.

The cactus plant body is similar fundamentally to that of any other flowering plant, but it is unusual in the slight development or usually the complete elimination of leaves and the remarkable relative size of the stem. Small fleshy ephemeral leaves appear on the new stem joints of chollas and prickly pears, and well-developed persistent leaves similar to those of many other plants are found on the stems of the primitive tropical cacti of the genus *Pereskia.* Most cacti do not have leaves at all or the leaves are developed for a short time on only the growing part of the stem. Food manufacture is by the green cells of the stem. A large part of the plant is occupied by storage cells containing mucilaginous materials retaining water, and the stem surface is covered by a waxlike layer preventing or at least retarding evaporation. The water-retaining powers of the surface layer may be demonstrated by cutting a detached joint of a "spineless" prickly pear in two and peeling one half and leaving the other unpeeled. After a few hours the peeled portion is shriveled and the unpeeled part unchanged.

The root systems of a few cacti, such as the night-blooming cereus (*Cereus Greggii*) of the Southwestern Deserts, include tuberous structures; but the majority of cacti have shallow systems of elongated slender but fleshy roots, well-adapted to absorption of large quantities of water during the brief period in which it is available after a rain. The detailed structure of a mature cactus is as follows.

The stem may be a simple unbranched columnar axis as it is in the barrel cactus, but in most cacti it is branched either at or near the base, as in the organ-pipe cactus, or much higher as in the saguaro or giant cactus. In the chollas and prickly pears the stem consists of a branching series of joints. The joints of chollas are cylindroidal, and those of prickly pears flat. Nearly all prickly pears have a smooth stem surface, but most other cactus stems have prominent ribs (e.g., saguaro and barrel cactus) or tubercles (e.g., chollas and *Mammillaria*). The areoles are arranged spirally on the stem. Tuberculate stems have one areole at the apex of each tubercle; ribbed stems have areoles along the summits of the ridges.

In many cacti, some spines in each areole may be distinguished as *central spines,* others as marginal or *radial spines.* The distinction is often arbitrary and a matter of individual interpretation. Chollas and prickly pears have numerous small or even minute barbed bristles as well as spines. These are the *glochids* ("ch" pronounced as "k"). To humans prickly-pear glochids are more troublesome than the spines.

The cactus flower structure is similar to that in the Myrtales except for the intergradation of sepals and petals mentioned above, production of leaves and areoles (which may serve as growing points) on the floral tube or cup, and the always parietal placentae.

Some features of the Cactaceae are similar to, though not necessarily homologous with, certain characters of the Ficoideae (subfamily) of the Aizoaceae in the Caryophyllales, and many authors have accepted the idea that the cacti are members of that order. The following is a resumé of similarities and differences of the Cactales and the Caryophyllales as exemplified especially by the Ficoideae.

Similarities

1. *Succulence.*

2. *Presence of betacyanins* and beta-xanthins (betalains) in the place of anthocyanins and anthroxanthins. These pigments change in the same manner as litmus paper, from near blue in a basic medium to near red in an acid one, and they are responsible for many plant colors, especially those of flowers and fruits. The pigments are not related chemically, but they perform the same function. The betalains are known to occur only in *most* of the Caryophyllales and in the Cactales. Almost certainly this indicates relationship in some degree.

3. *Similarity in structure of the pollen grains.* The surface characters resemble each other, and there are three nuclei in each grain.

4. *Campylotropous or amphitropous ovules.* However, occurrence of this type of ovule is not universal in either order.

5. *Similar development of the megagametophyte.*
6. *Storage of the reserve food of the seed commonly in the perisperm,* i.e., the megasporangium (nucellus) rather than the endosperm.
7. *Type of sieve tube plastid.* Fine structure studies indicate the phloem to be of the starch-accumulating type, not the type with or without starch grains but with protein inclusions. This was based on study of only 33 species in 11 families of Caryophyllales and two species of the cactus *Pereskia.* Thus the difficulty is the same as in nearly all studies of obscure characters; the known examples are too few for evaluation.

If one considers only the similarities, inclusion of the Cactaceae in the Caryophyllales seems justified — if the differences have been ignored. They include the following:

Differences

1. *Succulence.* The cacti are the ultimate in stem succulence, and only a few of them produce leaves; the Ficoideae are the ultimate in leaf succulence, and in some the stems are virtually obsolete, except as an attachment point for the leaves. Thus succulence probably is an analogous feature and not homologous.
2. *Arrangement of leaves.* The leaves of Cactaceae are strictly alternate, and this includes the leaf position microscopically when there are no leaves. Those of the Ficoideae are opposite.
3. *Presence or absence of spines.* Almost all cacti have spines in the juvenile or adult stage or both. The Caryophyllales are spineless, except for the Didiereaceae, which probably belong to the order.
4. *Presence of areoles, bearing spines.* This feature, universal in the Cactaceae, probably is not duplicated in other flowering plants. In 1931 Chorinsky reported a structure presumed to be homologous in *Acampseros* (Portulacaceae, Caryophyllales), but, if the interpretation is correct, it applies to only 1 species in 8,000 of the Caryophyllales, and it is not a strong basis for merging orders, whatever its significance in indicating remote relationship. The Didieraceae have a structure in some ways similar to the cactus areole, though it could be interpreted otherwise and the plants have virtually no other indication of relationship to the cacti. In the Caryophyllales the Didiereaceae are anomalous.
5. *Perianth type.* The transition of sepaloids to petaloids to stamens occurring in the cacti is not duplicated in the Ficoideae. The sepals of the Ficoideae are few and not transitional to the petals, which have arisen through sterilization of stamens produced in bundles from a fundamental group of 5 primordia (Rendle, 1952, 2: 112).
6. *Origin of stamens.* The stamens of the Cactaceae occur in great numbers (up to more than 3,000 in a flower), but so far as known they arise from separate primordia, as in *Pereskia pititache,* Boke, 1963.*
7. *Nodes, leaves, and spines on the floral cup.* The floral tube or cup is partly a stem, and commonly it bears all the usual structures (scale leaves, spines in areoles) for cactus stems; nothing similar appears in the Caryophyllales.
8. *Style(s).* In the Cactaceae there is always one compound style; in the Aizoaceae the styles are separate.
9. *Numbers of chambers in the ovary.* Cactaceae one chamber; Aizoaceae as many as the carpels.
10. *Placentation.* In the Cactaceae the placentae begin as axile, but ultimately, as development proceeds, they become parietal. The placentae of the Ficoideae are axile. The development in the Cactaceae is unique; each carpel is like a pea pod slit open lengthwise midway between the suture and the midrib. It is opened with the dorsal segment projecting upward and the ventral (to which, of course, the ovules are attached) downward. Commonly the external wall of each carpel is adnate with the floral tube (hypanthium). So far as known, this is not matched in the Caryophyllales or any other order. See Boke, 1964.*
11. *Type of inferior ovary.* The ovary of the Cactaceae is inferior in appearance, but, according to Boke, actually superior. The ovary is sunken in the receptacle (stem) tissues in a manner not matched in more than a few plant families, if any, and certainly not in the Ficoideae or other Caryophyllales.

Thus placing the Cactaceae in the Caryophyllales is based on scanty information. However, there are enough characters in common to indicate, despite the overwhelming differences, that there is a remote relationship through descent from a far distant common ancestor. The Cactales stand alone, but they do show this one plausible link with another order. (See Lyman Benson, *The Cacti of the United States and Canada,* Stanford University Press, in press.)

*Family 1. **Cactaceae.** CACTUS FAMILY

(Synonym: *Opuntiaceae*)

Special references: Benson, Lyman. *The Cacti of Arizona.* Editions 1–3. 1940–1969; and Britton, N. L. and J. N. Rose. *The Cactaceae.* Carnegie

* Boke, Norman H. American Journal of Botany 50: 848–858. 1963; 51: 598–610. 1964.

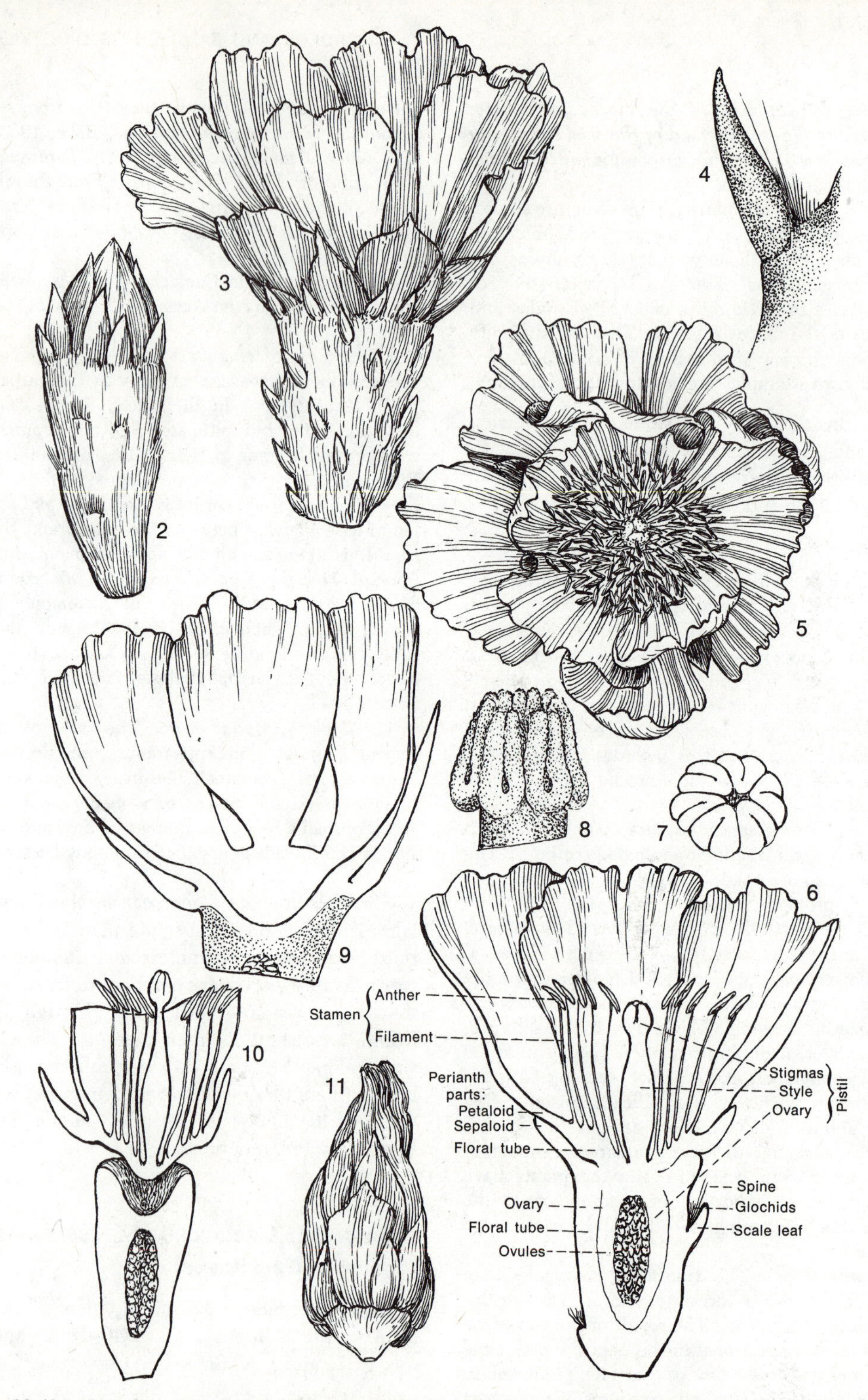

Figure 10-49. Cactales; Cactaceae; *Opuntia "occidentalis,"* a prickly pear: **1** young stem joint with a leaf subtending each areole, the older joint below leafless, **2** flower bud, **3** flower, side view, showing the inferior ovary with the adnate basal portion of the floral cup, this a hypanthium (as shown by the leaves and areoles on the surface), the sepaloids grading into petaloids, **4** leaf and axillary areole on the hypan-

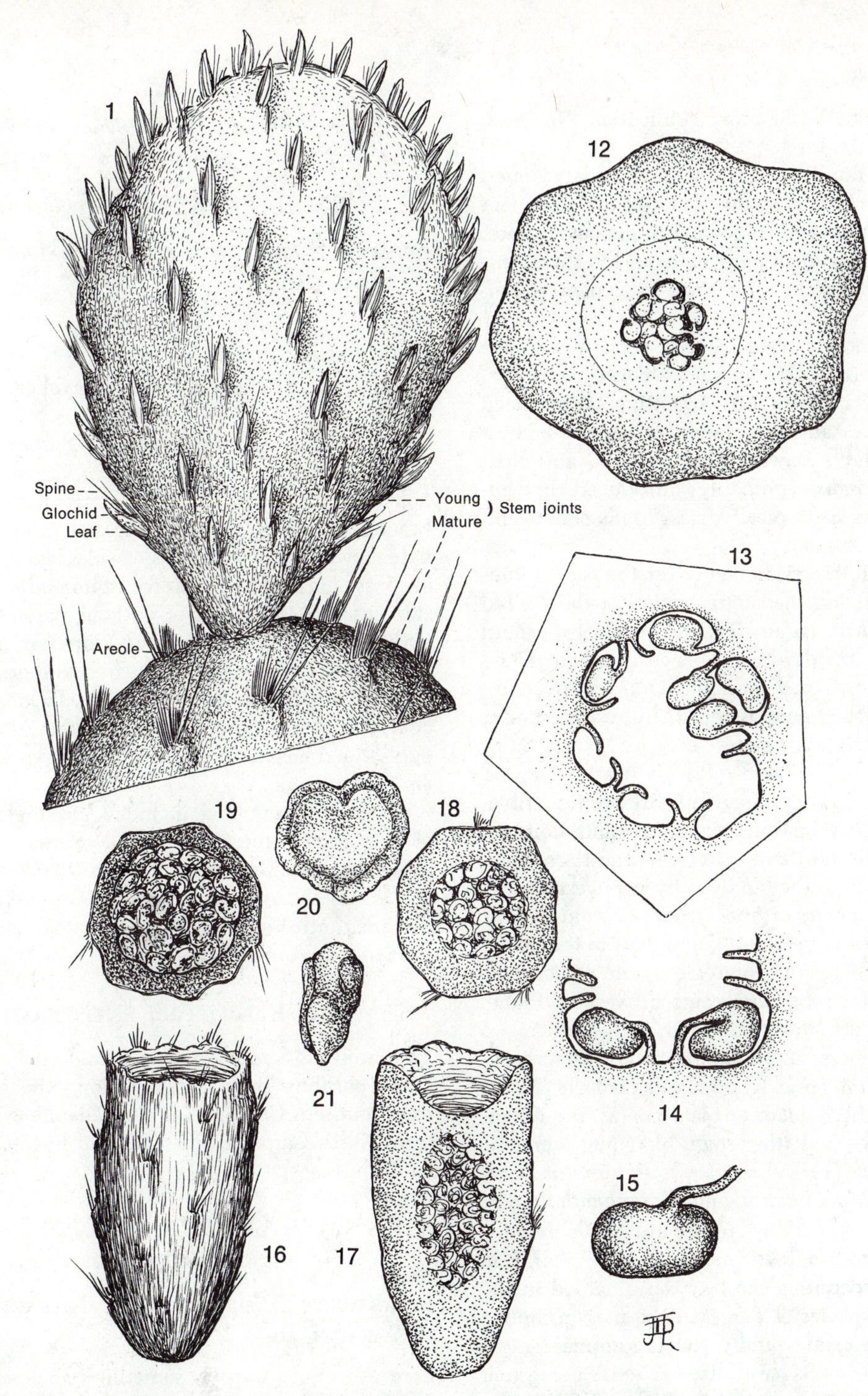

thium, **5** flower, top view, showing the petalloids, stamens, and stigmas, **6** flower, longitudinal section, **7–8** enlargements of the stigma, **9** edge of the upper portion of the floral cup (perianth tube), showing sepaloids and petaloids on its rim and interior, **10** upper portion of the floral cup and the flower parts becoming detached by abscission from the ovary and the lower portion of the floral cup, **11** external view, **12** cross section of the ovary at flowering time, **13–14** enlargements, showing the ovules and placentae, **15** ovule, **16** fruit, **17** longitudinal section, **18–19** cross sections, **20–21** seed, two views. (Cf. cacti in **Figures 24-12** and **25-41**.)

Institution of Washington Publication No. 248. 4 volumes. 1923.

The only family, consisting of an undetermined number (perhaps as many as 30 to 40) of genera with uncounted and undefined numerous species, but with only a small percentage of the thousands of described species or numerous genera is tenable.

North America from Puget Sound and Peace River in British Columbia and Alberta to Ontario, Massachusetts, Chile, and Argentina. Introduced in other warm and especially dry regions such as Palestine, Italy, Hawaii, South Africa, and Australia. *Rhipsalis* (probably introduced by humans) occurs in tropical Africa, Madagascar, and Sri Lanka (Ceylon).

The North American center for the cactus family is Mexico and the cactus capital of the United States is in Arizona and Texas. North of southern California, Arizona, New Mexico, southern Colorado, western Texas, and the Gulf Coast representation of the family is sparse, but cacti occur in all the contiguous states except Maine, New Hampshire, and Vermont. Prickly pears are found in the Canadian provinces of British Columbia, Alberta, Saskatchewan, Manitoba, and Ontario. Through the northern area the most common representation of the family is by two western and one eastern species of prickly pear (*Opuntia*); but two species of *Coryphantha* range far to the northward in the Great Plains region, and *Pediocactus Simpsonii* var. *robustior* occurs in western Idaho, the center of Washington, and the center of Oregon. Genera in the Southwestern Deserts include *Opuntia* (prickly pear and cholla), *Cereus* (including the saguaro or giant cactus, the organ-pipe cactus, and the night-blooming cereus), *Echinocereus* (hedgehog cacti), *Ferocactus* (barrel cacti), *Mammillaria* and *Coryphantha* (pincushion cactus). Many species are native in western Texas, and a few (mostly prickly pears) in Florida (discounting most of those named in the books). A species of *Pereskia,* the most primitive genus in the cactus family (having normal leaves, as stated above), is naturalized in south peninsular Florida, where it is known as West Indian gooseberry or lemon-vine. A species of *Rhipsalis* (pencil or mistletoe cactus) is rare in peninsular Florida and reported from the Everglade Keys. In Florida there are also several other cacti described under generic names for *Cereus,* e.g., *Cephalocereus, Selenicereus, Hylocereus, Acanthocereus,* and *Harrisia.*

Order 25. *BATIDALES.* BATIS ORDER

DEPARTURE FROM THE FLORAL CHARACTERS OF THE RANALES

Free	Carpels coalescent.
Numerous	Stamens 4; carpels 4.
Spiral	Stamens and carpels cyclic.

Low odoriferous bushes or small shrubs. Leaves opposite, fleshy, simple, sessile, entire; stipules none. Plants dioecious, both kinds of flowers in bracteate catkins or catkinlike structures; staminate flower with two partly coalescent sepals, 4 or 5 stamens, and petal-like staminodia (sterile stamens); pistillate flower without perianth, consisting of a single pistil; ovary superior, 4-chambered, with one ovule in each chamber; stigma capitate, sessile, the style practically lacking; ovule anatropous. Fruits coalescent and forming a fleshy egg-shaped mass; seed oblong; embryo straight; endosperm none.

Several authors have included the single family of this order in the Caryophyllales, and the plants show considerable resemblance to the *Chenopodiaceae* and the *Amaranthaceae,* the goosefoot and the amaranth families. However, relationships of the order are obscure.

*Family 1. **Batidaceae.** BATIS FAMILY

The family is composed of the single genus *Batis,* a low shrubby bush growing along the seashore from southern California to the Galapagos Islands, from North Carolina to Brazil, and in the Hawaiian Islands.

Order 26. *HAMAMELIDALES.* WITCH-HAZEL ORDER

DEPARTURE FROM THE FLORAL CHARACTERS OF THE RANALES

Free	Carpels sometimes coalescent.
Numerous	Stamens 2-numerous; carpels 2–18.
Spiral	Stamens spiral or in 1–4 cycles; carpels cyclic, at least when few.
Hypogynous	Flowers epigynous in the *Hamamelidaceae.*

Shrubs or small trees. Leaves simple, entire to toothed or cleft, alternate, sometimes appearing to be in apical whorls or 4 or 6–12; stipules present or absent. Flowers in loose clusters or racemelike cymes or ball-like clusters, bisexual, radially symmetrical; sepals 0–5 or 7, mostly none; petals 0–5, commonly none; carpels 1–18, separate or coalescent, sometimes unsealed along the bases of the styles or the apices of the ovaries; ovules 1–several or numerous. Fruit a follicle, achene, samara, or capsule.

The Hamamelidaceae together with the *Platanaceae* (sycamore family) and the *Myrothamnaceae* were segregated long ago by Wettstein as an order Hamamelidales. According to Tippo this order was derived from the Ranales of the *Magnolia* group and it has given rise to the Casuarinales, Fagales, and the Urticales.

Key to the Families
(North America north of Mexico)

1. Petals 4–5; flowers epigynous, in spikes or heads, these sometimes ball-like; carpels 2, coalescent, developing into a 2-chambered capsule; stipules separate, developing on the young growth, but soon deciduous; shrubs or trees, the leaves simple to lobed.

   Family 6. **Hamamelidaceae** (witch-hazel family).

1. Petals none or minute; flowers (very simple ones) hypogynous, compacted into remote balls along an axis or terminal to it, carpels 3–5, each developing into an achene; each pair of stipules large and coalescent into a very large circular shelf around the twig; large trees with exfoliating bark and with palmately lobed to cleft leaves.

   Family 7. **Platanaceae** (sycamore family).

Suborder A. **Trochodendrineae**

Family 1. **Trochodendraceae.**

TROCHODENDRON FAMILY

Trees. Leaves simple, toothed, alternate but appearing to be in whorls of 4 or commonly 6–12 at the apices of the branchlets. Flowers in racemelike cymes, bisexual; sepals and petals none (or perhaps represented by 2–5 inconspicuous bractlike scales just below the receptacle); stamens numerous, in 3–4 series, composed of anthers and filaments, the anthers with 4 pollen chambers; carpels several, in a single series, adnate basally with the receptacle and coalescent; ovules several or rather numerous; style present, folded (conduplicate). Fruit a cluster of coalescent follicles.

The family includes a single species, *Trochodendron aralioides,* occurring from Formosa to Japan. It is cultivated occasionally in the Southern States.

Family 2. **Tetracentraceae.**

TETRACENTRON FAMILY

Trees. Leaves simple, deciduous, alternate, the leaf bases broad and sheathing the bud. Flowers numerous, bisexual, in clusters of 4; sepals 4 in 2 series, persistent in fruit; petals none; stamens 4, opposite the sepals; composed of anthers and filaments, the anthers with 4 pollen chambers; carpels 4, alternate with the stamens and sepals, coalescent; style folded (conduplicate), stigmatic along most of the ventral suture; ovules several. Fruit composed of the 4 coalescent follicles.

The family includes a single species, *Tetracentron sinense,* occurring in south central China and northern Burma. It is cultivated occasionally.

The *Tetracentraceae* and *Trochodendraceae* are related fairly closely. Both have primitive vesselless xylem.

Family 3. **Eupteleaceae.**

EUPTELEA FAMILY

Trees or shrubs. Leaves simple, toothed, alternate. Flowers appearing before the leaves, bisexual; sepals and petals none; stamens 6–18, composed of anthers and filaments, produced in a single cycle; carpels 6–18, separate, in a single series; stigmas none, the ventral margin of the carpel or the apical portion of it stigmatic, the carpel closed; ovules 1–3 or perhaps 4. Fruit a stalked samara, the stalk expanded gradually into the wing, the wing thin.

The family includes only two species, *Euptelea polyandra* of Japan and *Euptelea pleiosperma* of China and northeastern India, both species being cultivated occasionally.

Family 4. **Cercidiphyllaceae.**

CERCIDIPHYLLUM FAMILY

Trees. Leaves simple, toothed, those on the terminal twigs opposite, those on the short lateral

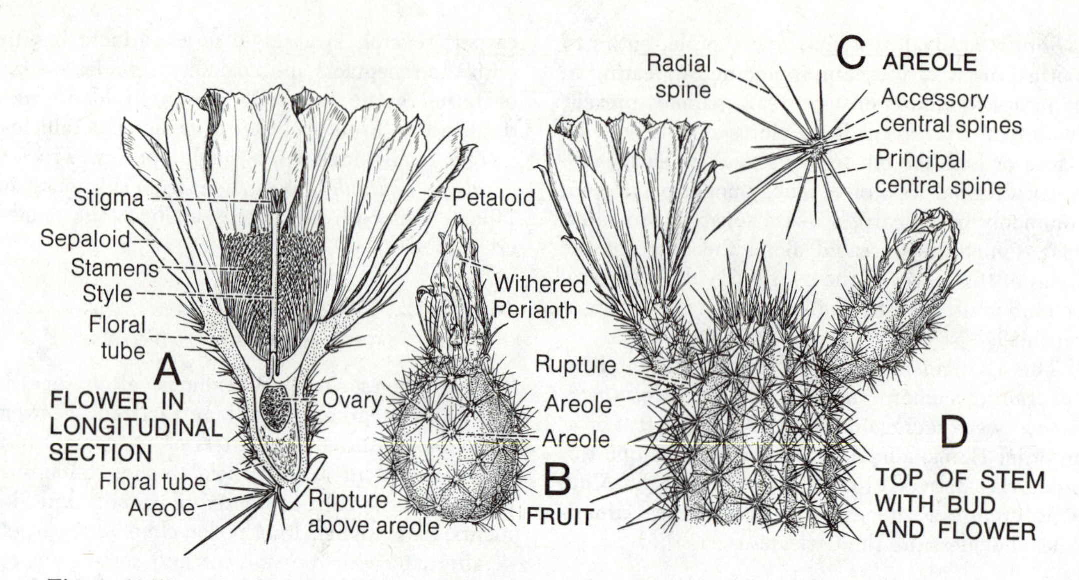

Figure 10-50. Cactales; Cactaceae; *Echinocereus fasciculatus* var. *Boyce-Thompsonii,* a hedgehog cactus: **A–D** as labelled. In this variety the principal central spine turns downward, and there are two shorter accessory central spines above it in the areole; radial spines surround these. The fruits are red, and the interlocking clusters of spines fall away at maturity. (Cf. also **Figure 6-9 E** *Cereus giganteus,* flower.) By Lucretia Breazeale Hamilton. From LYMAN BENSON. *The Cacti of Arizona.* Ed. 2. 1950. Courtesy of the University of Arizona and University of New Mexico presses.

shoots alternate; stipules present, early deciduous. Flowers unisexual; sepals and petals none; staminate flower (or cluster of flowers) composed of 8–13 stamens subtended by 2 bractlets; stamens composed of anthers and filaments; pistillate flower composed of a single carpel, but the flowers in flowerlike clusters of 2–6; style with two stigmatic lines (ridges) extending along the margins of the carpel. Fruit a follicle; seed winged.

The family includes a single species, *Cercidiphyllum japonicum,* occurring in China and Japan. The species is cultivated.

Suborder B. **Eucommineae**

Family 5. **Eucommiaceae.**

EUCOMMIA FAMILY

Trees; juice milky, containing rubber. Leaves simple, serrate, alternate; stipules present. Flowers unisexual (and the plant dioecious), radially symmetrical, solitary; sepals none; stamens 4–10, the filaments short, the anthers long; carpels 2, the styles 2, the ovary 1-chambered by abortion of 1 carpel; placenta apical, the ovule solitary, pendulous. Fruit a samara.

The family consists of a single Asiatic species native in China, but cultivated in other continents.

Suborder C. **Hamamelidineae**

*Family 6. **Hamamelidaceae.**

WITCH-HAZEL FAMILY

(Segregate: *Altingiaceae*)

Shrubs or trees. Leaves simple, toothed to lobed, alternate; stipules present. Flowers bisexual or unisexual (and the plants monoecious or polygamous), radially symmetrical, epigynous (the ovary inferior through at least part of its length) or sometimes hypogynous (extraterritorial genera), in spikes or heads; sepals 4–5; petals 4–5, separate; stamens 2–8 or numerous, in 1 series; anthers splitting lengthwise or opening by pores with lids; carpels 2, coalescent, the styles separate, the ovary 2-chambered; placentae axile; ovule(s) 1 or few per carpel. Fruit a capsule, often coriaceous or woody, with only 1 or 2 bony seeds in each chamber, the apices of the carpels free and spreading, dehiscent along the sutures; endosperm present, thin, fleshy.

The family is largely of East Asia, but 2 genera occur in Africa, 1 in Australia, and 3 in eastern North America. *Hamamelis* (witch-hazel) is represented by two species in eastern North America, *Fothergilla* by perhaps three species, and *Liquidambar* by one. The family does not occur in western North America.

The sweet gum (*Liquidambar*) is a large tree with beautiful starlike palmately lobed leaves, which turn red and yellow in the autumn. The witch-hazel yields the liniment of the same name, which is prepared from the bark. Various plants of the family are cultivated, and *Liquidambar* is one of the finest of ornamental trees. It is considered the best street tree in southern California.

*Family 7. **Platanaceae.**

SYCAMORE FAMILY

Large trees; under bark white or greenish when exposed by exfoliating of the outer layers, outer bark sometimes very dark and roughly fissured on old stems of some species. Leaves simple, palmately lobed to cleft, the lobes 3–9, mostly 5–7; the base enclosing the axillary bud; stipules very large, coalescent and encircling the node, often remaining loose around the twig after abscission. Flowers unisexual (and the plants monoecious), radially symmetrical, hypogynous, the inflorescence unisexual, a pendulous series of dense ball-like clusters of flowers (or a single cluster in *Platanus occidentalis*), these either sessile and surrounding the main axis (e.g., *Platanus racemosa*) or peduncled and one from each node (var. *Wrightii*); sepals 3–5 or 7, very small; petals usu-

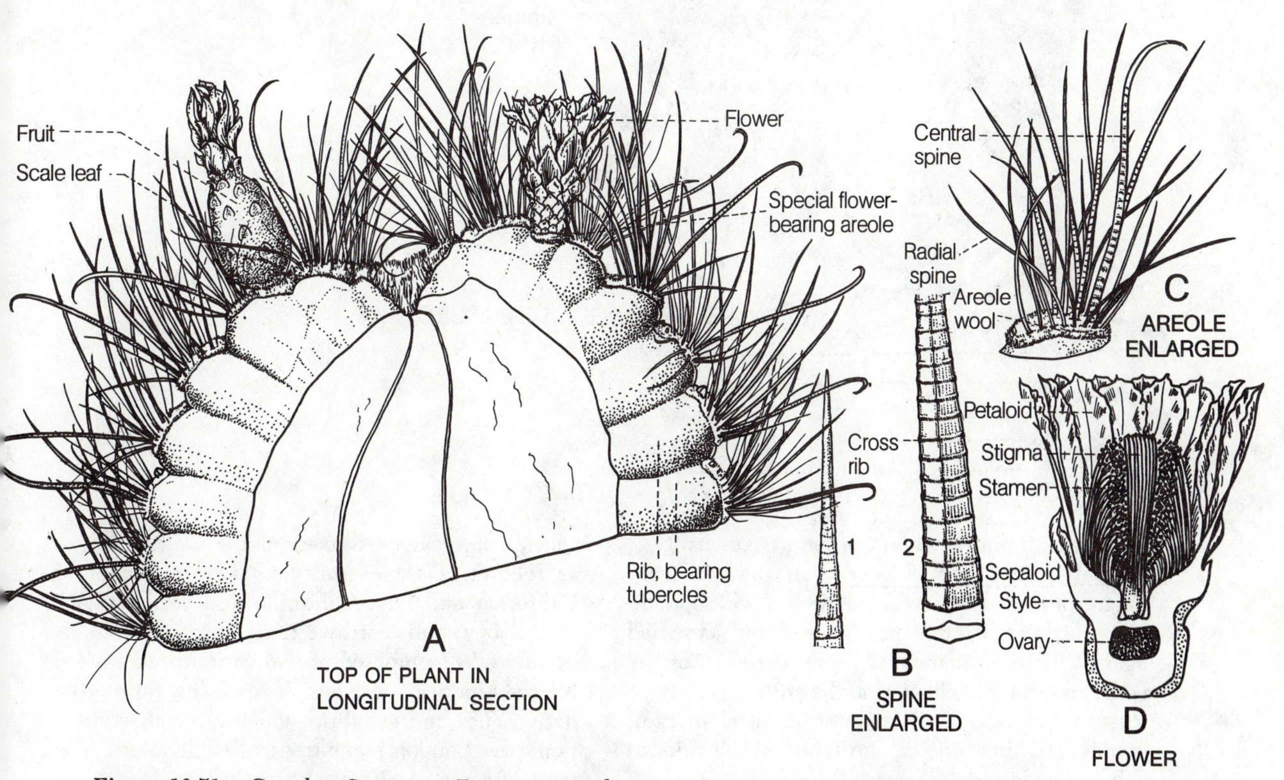

Figure 10-51. Cactales; Cactaceae; *Ferocactus Wislizenii,* a barrel cactus: **A–D** as labelled. The fleshy area in the center of the stem may yield a considerable amount of flavorless water. By Lucretia Breazeale Hamilton. From Lyman Benson. *The Cacti of Arizona.* Ed. 2. 1950. Courtesy of the University of Arizona and University of New Mexico presses.

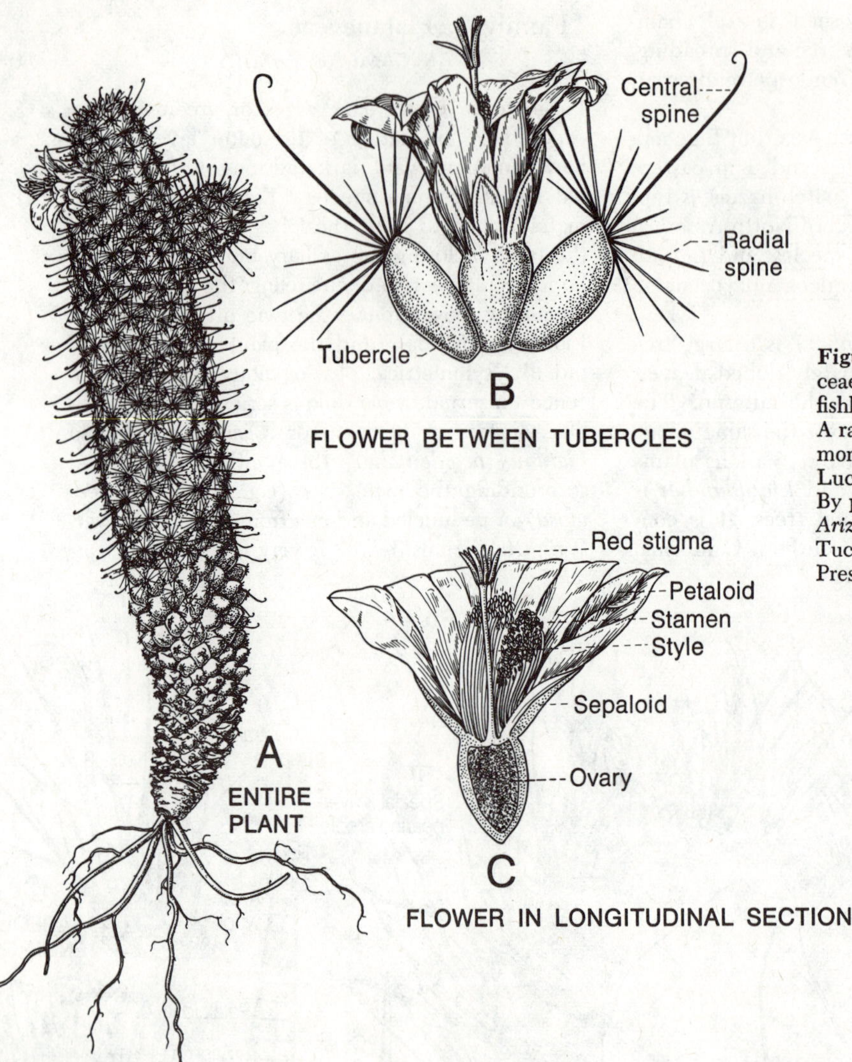

Figure 10-52. Cactales; Cactaceae; *Mammillaria Thornberi*, a fishhook cactus: **A–C** as labelled. A rare species similar to the common *Mammillaria microcarpa*. By Lucretia Breazeale Hamilton. By permission from *Cacti of Arizona* by LYMAN BENSON, Tucson: University of Arizona Press, copyright 1969.

ally none, minute when present; stamens 3–7, separate, the filaments very short; anthers very long, splitting lengthwise; carpels 3–5, separate or sometimes 9, the pistils or fruits crowded against those of adjacent flowers, the apex of the carpel not completely closed, the stigma (or style) stigmatic along almost its entire ventral margin; ovules 1–2 per carpel, orthotropous. Fruit an achene; endosperm present, thin, fleshy.

The family is composed of a single genus (*Platanus*, sycamore and plane-tree) of about seven species occurring in the Northern Hemisphere. Two species of sycamores occur in the United States, as follows: *Platanus racemosa* var. *racemosa* of California and Baja California and var. *Wrightii* of Arizona and northwestern Mexico; *Platanus occidentalis* (composed of two varieties) in eastern North America. Various species, including the native ones and a hybrid which arose in cultivation (the London plane-tree) are cultivated. The trees of some species are attractive for their white bark (e.g., *Platanus racemosa*) as well as their large sometimes basally decumbent trunks and large leaves.

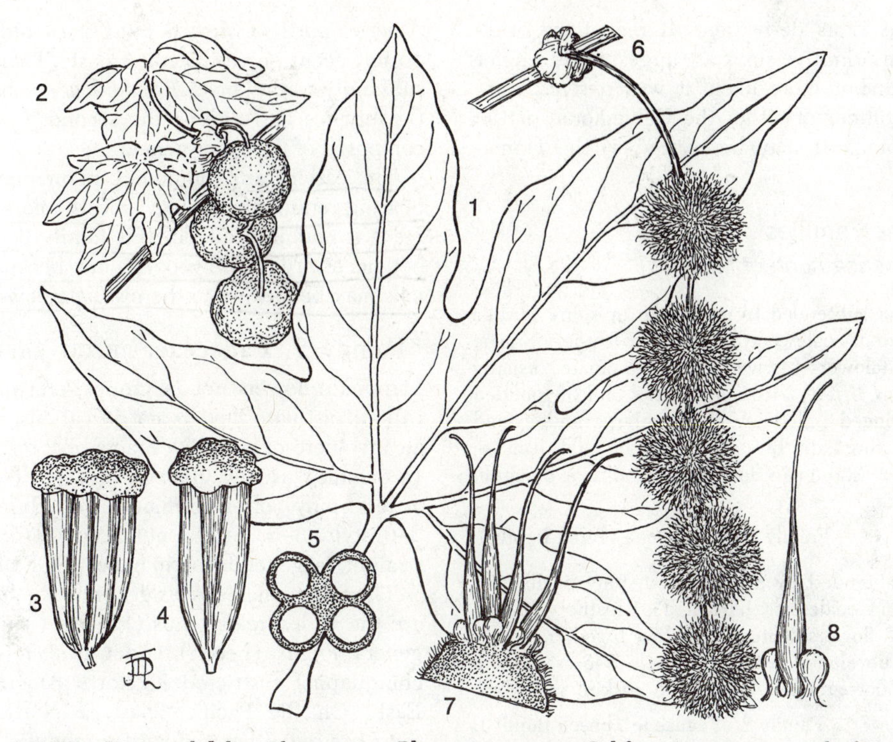

Figure 10-53. Hamamelidales; Platanaceae; *Platanus racemosa,* California sycamore: **1** leaf outline, **2** branch with a staminate inflorescence, **3–4** stamen, two views, **5** anther, diagrammatic cross section, **6** pistillate inflorescence, **7** pistillate flower composed of sepals and carpels, **8** carpel and the two adjacent sepals. (Cf. also stipules, **Figure 3-2 F.**)

Order 27. *FAGALES.* BEECH ORDER

DEPARTURE FROM THE FLORAL CHARACTERS OF THE RANALES

Free	Carpels coalescent.
Numerous	Stamens mostly 2–12 but sometimes as many as 40; carpels 2–3 or rarely as many as 6.
Spiral	Stamens apparently cyclic when the numbers are small; carpels cyclic.
Hypogynous	Flower epigynous but sometimes not visibly so in the *Betulaceae.*

Trees or shrubs. Leaves alternate, simple, sometimes deeply lobed or divided; stipules commonly deciduous. Plants monoecious, the staminate flowers in catkins, the pistillate solitary or in clusters or catkins; staminate flower consisting of several free or coalescent sepals and 2–12 or sometimes as many as 40 stamens; pistillate flower consisting of a single pistil and usually 3–4 minute sepals appearing to be adnate with the top of the inferior ovary (these sometimes obsolete and the ovary appearing inferior); ovary 2–3- or rarely 6-chambered with only 2 ovules (or only 1 by abortion) per chamber and with only one chamber developing, after flowering time the others remaining small and rudimentary; pollen tube entering the ovule through its base (chalaza) rather than the micropyle. Fruit a nut or a nutlet, associated with persistent bractlets (as in the catkins of birches or the "cones" of alders) or an involucre (as in the cup of an acorn or the spiny bur of a chinquapin or a chestnut) or a green husk of bracts (as in the hazelnut); seed 1, the embryo straight; endosperm none.

The relationships of this specialized order are

obscure, as is its derivation. It may have originated from primitive stock as suggested by Engler and Prantl, but more likely it was derived from some forerunner of either the Calyciflorae or the Thalamiflorae. It may be related to the Hamamelidales.

Key to the Families

(North America north of Mexico)

1. Fruit *not* subtended by a cuplike or spiny burlike involucre of coalescent hardened or leathery bracts; pistillate flowers (as well as the staminate) usually in catkins, *if few,* 2 to each bractlet and surrounded by 2 fringed bractlets, which enlarge and extend beyond the fruit but remain green and leaflike; carpels 2; staminate flowers in groups of 3 within the catkin.

   Family 1. **Betulaceae** (birch family).

1. Fruit subtended by a cuplike or spiny burlike involucre of coalescent hardened or leathery bracts; pistillate flowers solitary or a few together, not in clearly developed catkins; carpels 3–6; staminate flowers not grouped within the catkin, racemose.

   Family 2. **Fagaceae** (beech family).

*Family 1. **Betulaceae.** BIRCH FAMILY

(Segregate: *Corylaceae*)

Staminate flowers in 3's (reduced cymules) along the axes of the catkins, but the structure obscured in some genera by absence of bracts or sepals (consequently the stamens of the 3 flowers of the cymule apparently one flower); pistillate flowers in 2's or (in *Betula,* birch) 3's in cymules along the axis of the catkin or in a cluster, typically with sepals upon the ovary near the styles and the ovary inferior, but in *Betula* (birch) and *Alnus* (alder) the sepals obsolete and the ovary therefore appearing to be superior; carpels 2, stigmas 2; ovary 2-chambered only basally or 1-chambered. Fruit a small samara or sometimes a nut, 1-chambered; seed 1; embryo straight; endosperm none.

The family is common in North Temperate regions, but species of *Alnus* (alder) extend as far southward as Argentina and Bengal. There are 6 genera and about 80 species. The genera are *Betula* (birch), *Alnus* (alder), *Corylus* (hazel), *Ostrya* (hop hornbean), *Carpinus* (hornbean), and *Ostryopsis.* The first five occur in North America north of Mexico, but *Ostrya* and *Carpinus* do not occur on the Pacific Coast. There are two subfamilies, the first, *Betuloideae,* consisting of the birches and alders, the second, *Coryloideae,* composed of the other genera.

The catkins are compound inflorescences (axes bearing reduced cymules), and the floral arrangement is complicated. In this family the flower is not necessarily visibly epigynous because the sepals may be lacking in the pistillate flower.

*Family 2. **Fagaceae.** BEECH FAMILY

Staminate flowers not in groups, racemose in the catkin; pistillate flowers not in catkins, solitary in an involucre of scalelike bracts or 2 or 3 together (a cymule) usually in an involucre which is more or less spiny (often forming a spiny bur); carpels 3–6, styles 3–6. Fruit a nut, seed 1, large; embryo straight, large; endosperm none. (cf. Figure 12–1.)

The family includes six genera. The best known are the widespread genus *Quercus* (oak) and the genera *Fagus* (beech) and *Castanea* (chestnut, chinquapin) restricted in North America to the East. On the Pacific Coast of North America there are two other genera, *Castanopsis* (chinquapin) and *Lithocarpus* (tan oak).

Many species of *Quercus* are used for lumber, and particularly for cabinet wood and flooring. Acorns are used as feed for hogs. An oak occurring along the shores of the western portion of the Mediterranean Sea is the source of cork. The bark of a large cork oak is peeled away in a sheet, cured, and sold as cork. Cork oaks are cultivated occasionally in the southernmost states. The bark of the tan oaks of two varieties are known as tan bark, and it is used extensively in tanning leather.

Order 28. *CASUARINALES* (Verticillatae). BEEFWOOD ORDER

DEPARTURE FROM THE FLORAL CHARACTERS OF THE RANALES

Free	Carpels coalescent.
Numerous	Stamen 1; carpels 2.
Spiral	Stamens and carpels cyclic.

Trees with (from a distance) the appearance of pines. Branchlets green, several-grooved, with obvious nodes. Leaves scalelike, minute, in whorls

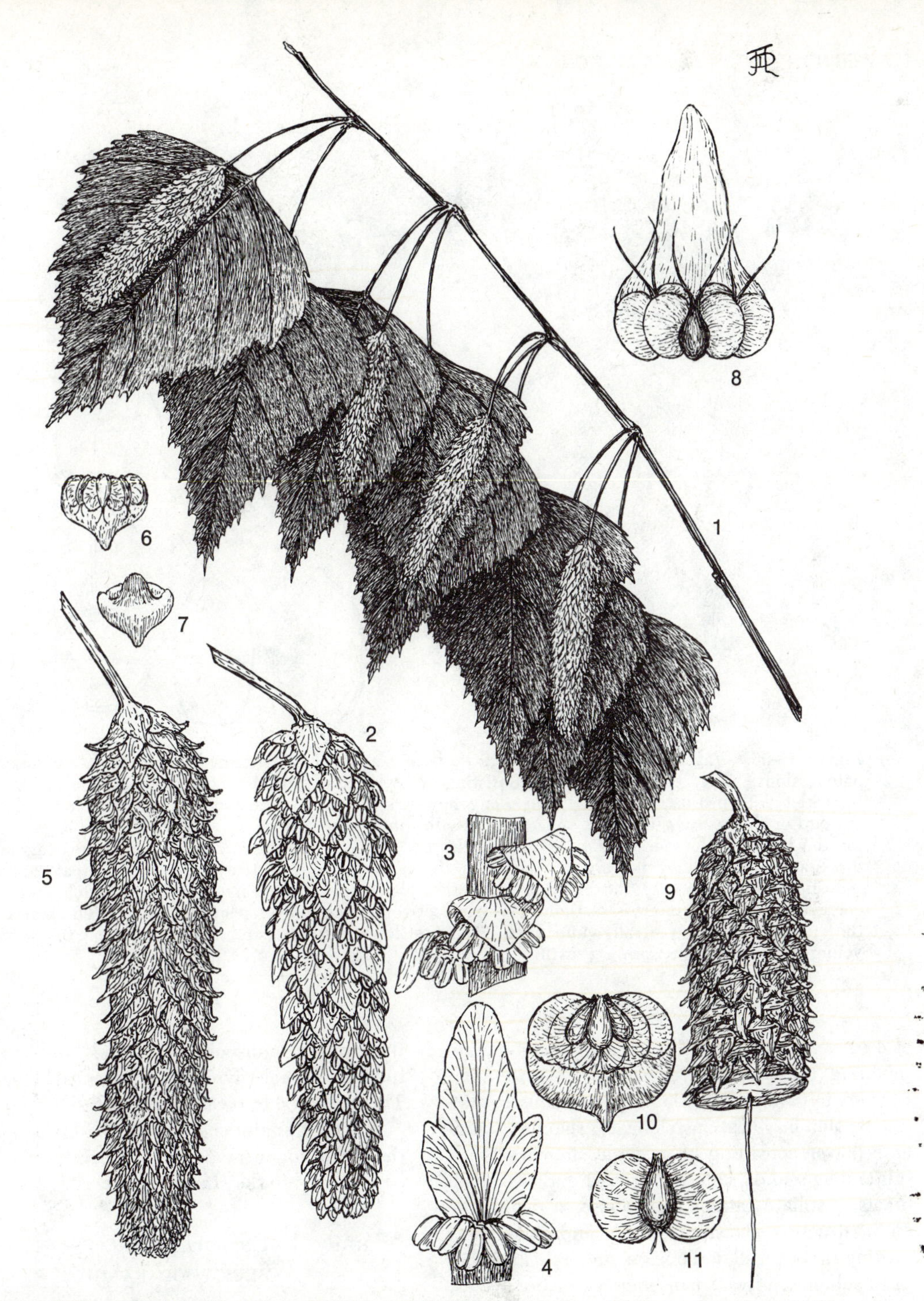

Figure 10-54. Fagales; Corylaceae; *Betula pendula,* European white birch: **1** branch with leaves and catkins, **2** staminate catkin, **3** staminate flowers and the bracts next to them, **4** stamens (the scalelike calyx not shown) and the bracts associated with the flower, **5** pistillate catkin, **6–8** pistillate flowers (each consisting of a pistil) and the associated bracts, **9** catkin in fruit, the fruits and bracts disarticulating from the axis, **10** three fruits of a cymule (as shown at flowering time in **8**) and the large bract subtending them, **11** the wingéd fruit.

Figure 10-55. Fagales; Fagaceae; *Quercus dumosa,* scrub oak: **1** branchlet with a leaf and staminate catkins, **2–3** staminate flower, consisting of calyx and stamens, two views, **4** stamen, **5** branchlet with pistillate flowers, **6** pistillate flower, showing the basal scales of the involucre (later the cup of the acorn), the floral cup (adnate with the inferior ovary), the scalelike sepals (surrounding the bases of the styles), and the three styles and stigmas, **7** longitudinal section of the ovary and the bracts, **8** pistillate flower, **9** fruit with the basal cup (involucre), **10** longitudinal section, showing the embryo, **11** the embryo, in another plane, showing the hypocotyl and radicle and one cotyledon, **12** fruit in cross section. The fruits of this species develop in one season. The scales of the cup are warty, as in all white oaks. Pistillate flowers of oaks are to be sought on the tender young stems of the new spring growth.

of 4 or commonly 6 to 14, the leafy branchlets resembling the stems of horsetails (*Equisetum*). Flowers in catkins, the plants monoecious or dioecious; staminate catkins elongate, slender; staminate flower consisting of a single stamen with 2 subtending scales (sepals or bracts) and two small bracts; pistillate catkins subglobose to ovoid, the entire structure becoming woody and conelike at fruiting time; pistillate flower a single 2-carpellate pistil subtended by a small bract and 2 bracteoles; styles 2, filiform; ovules 2–4, erect, orthotropous. Fruit 1-chambered through abortion of 1 carpel, woody, indehiscent.

The development of the sporogenous tissue in the ovule is unusual, for several megaspores are functional, each giving rise to a megagametophyte. This is argued to be a primitive feature.

A recent anatomical study by Mosely indicates the order to be moderately specialized and perhaps related to the *Hamamelidaceae*.

*Family 1. **Casuarinaceae.**

BEEFWOOD FAMILY

The family is composed of a single genus, *Casuarina,* occurring from the Indo-Malayan Region to Australia. It is cultivated from California to the Southern States, and it has escaped in Florida.

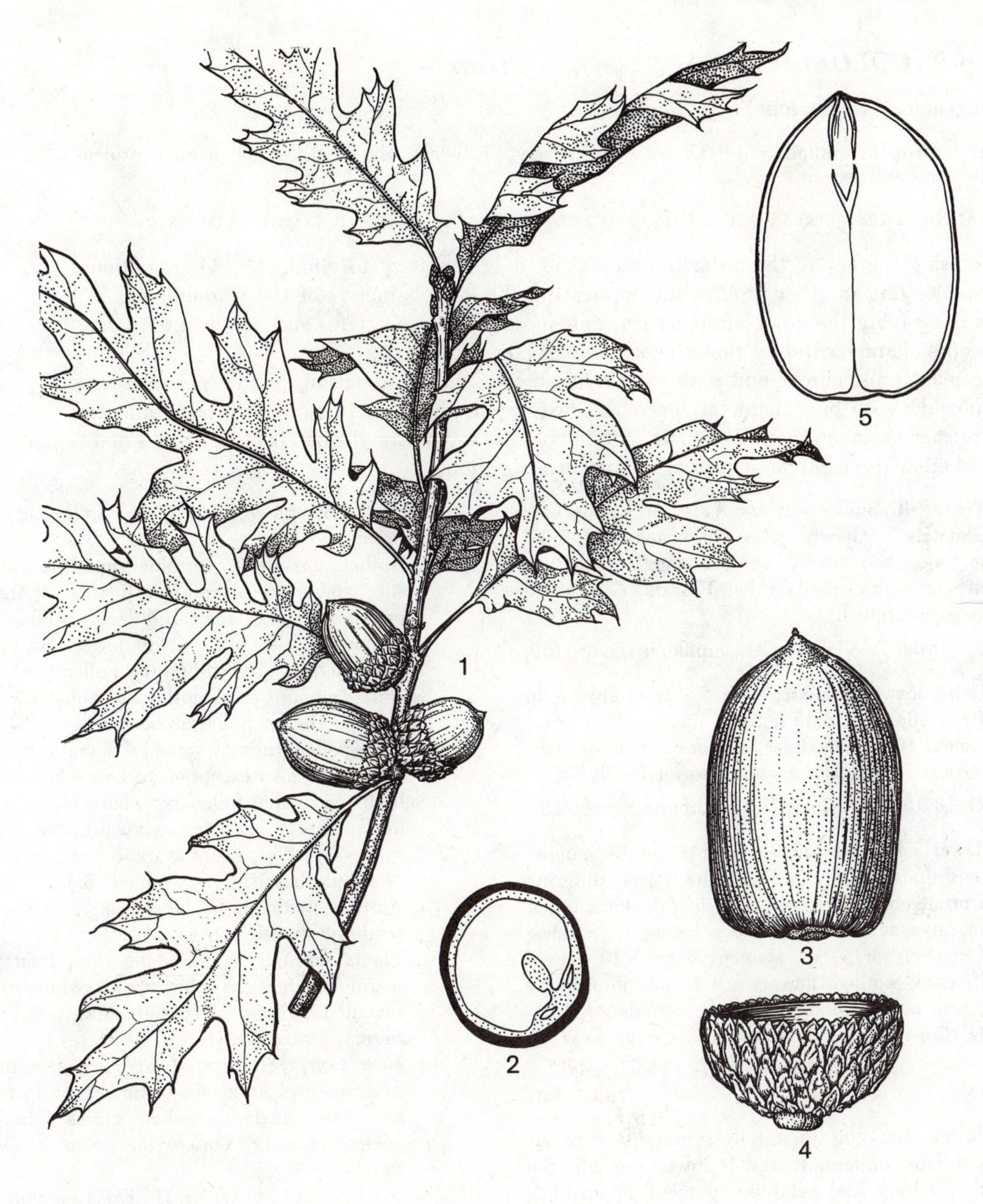

Figure 10-56. Fagales; Fagaceae; *Quercus Kelloggii,* California black oak (similar to the red oak of the East and Middle West): 1 branch with leaves and acorns (from the flowers of two seasons earlier) and young fruits (budlike, at the upper end of the branch, from one season earlier), 2 cross section of the ovary of the young fruit, showing the two abortive chambers, and the single developing ovule, 3 acorn, 4 cup (involucre), the scales thin as in the black oaks, 5 embryo, showing the hypocotyl and radicle (above), the epicotyl (below), and the cotyledons. By Lucretia Breazeale Hamilton.

Group 2. *COROLLIFLORAE.* COROLLA FLOWERS

(Hypogynous; *Sympetalous*)

NOTE: After determination of the order and family, proceed to the local flora or manual to identify the genus and species.

PRELIMINARY LIST FOR ELIMINATION OF EXCEPTIONAL CHARACTER COMBINATIONS

Before using the key to the orders on page 250, it is necessary to eliminate (1) exceptional plants belonging to other groups but duplicating the key characters of the Corolliflorae, (2) plants not having the key characters marking most of their relatives, and (3) plants marked by obscure characteristics or those inconsistent if taken singly.

The plant must correspond with every character in the combination. If the specimen does not coincide with one character, even the first listed, the combination may be eliminated from further consideration. The characters in **boldface type** should be checked first to determine whether the combination requires further inspection.

1. **Trees with milky sap and very large, alternate, palmately 7–9-lobed leaves; fruits melonlike, 2–18 cm. long;** stamens 10, the filaments usually adnate with the corolla tube; **Florida.** *Caricaceae* (papaya family).

   Order 7. **Violales** (Thalamiflorae), page 159.

2. **Herbs; leaves palmate, with 3** (or rarely up to 10) **sessile obcordate leaflets;** juice watery, sour; stamens 10 or sometimes 15; corolla radially symmetrical. *Oxalidaceae* (wood-sorrel family).

   Order 16. **Geraniales** (Thalamiflorae), page 193.

3. **Flowers all unisexual,** the plants usually monoecious, the staminate and pistillate flowers differing in arrangement or position or in the character of the calyx as well as in the presence or absence of stamens or pistils; **stamens 6 or 8–10;** leaves alternate, toothed; flowers green, racemose; ovary 1, fruit a 3-chambered capsule; everglades, **Florida.** *Caperonia, Euphorbiaceae* (spurge family).

   Order 14. **Euphorbiales** (Thalamiflorae), page 186.

4. **Flowers markedly bilaterally symmetrical; sepals 5, 3 (the uppermost and 2 lowest) small, the other 2 large and petal-like; petals 3 or up to 5, coalescent below and adnate with the coalescent filaments, the middle petal keeled** and often crested; **stamens 6–8,** the filaments coalescent below except on one side or coalescent into 2 groups; fruit a 2-chambered pod with one seed in each chamber.

   Order 17. **Polygalales** *Polygalaceae* (milkwort family), page 199.

5. Fruit *not* a 2-chambered, many-seeded, juicy berry *and* the plants corresponding with *every character* in *one or more* of the following combinations:

   a. **Anthers opening by terminal, basal, or oblique slits and not by full-length slits.** (Anthers opening through their full length collapse as the pollen is shed when the flower bud opens. This method of shedding the pollen is obvious from comparison of anthers of the buds with those of the open flowers because anthers shedding pollen through smaller openings do not collapse.) (An exception to be excluded is an herb with the following characters: corolla lobes turned strongly backward, the corolla tube very short; stems scapose above a rosette of simple basal leaves; stamens 5.)

   b. **Anthers with appendages** (e.g., bristles or feathery structures).

   c. **Plants saprophytic** (deriving their food from leafmold), not green but **brown, white, or red,** succulent. *Pterospora, Sarcodes,* and *Newberrya, Ericaceae* (heath family).

   d. **Low, evergreen shrub of arctic regions ranging southward to areas above timberline in northern New England;** pollen grains remaining coalescent in 4's. *Loiseleuria, Ericaceae* (heath family).

   Order 31. **Ericales,** page 257.

6. **Stamens 5, the filaments coalescent into a tube; petals none, but the five large, yellow coalescent sepals appearing to be a sympetalous corolla 2 or more cm. in diameter** and the 3–5 slender subtending involucral bractlets appearing to be sepals; leaves palmately lobed; carpels 5; fruit a capsule; large shrubs or small trees occurring in California and Arizona. *Fremontodendron, Sterculiaceae* (chocolate family).

   Order 14. **Malvales** (Thalamiflorae), page 181.

7. **Stamens 10 or rarely 7–9; fruit a legume; corolla markedly bilaterally symmetrical,** the petals 5, specialized as a banner (above) 2 wings (on the sides), and a keel (below and formed from 2 coalescent petals); leaves usually palmate or pinnate. *Papilionoideae, Leguminosae* (pea family).

   Order 38. **Rosales** (Calyciflorae), page 290.

8. **Large desert shrubs; fruits coiled tightly into cylindroidal bodies, occurring in clusters,** indehiscent; flowers small, in dense spikes. *Prosopis* (screw bean), *Leguminosae* (pea family).

   Order 38. **Rosales** (Calyciflorae), page 290.

9. **Stamens numerous, the filaments coalescent into an elongated tube or into groups,** as indicated in a or b, below; bases of the 5 petals adnate with the filaments and sometimes for a very short distance coalescent; corolla radially symmetrical; flowers perfect.

   a. Filaments coalescent into groups.

   Order 11. **Theales** (Thalamiflorae), page 172.

   b. Filaments coalescent into a tube.

   Order 13. **Malvales** (Thalamiflorae), page 181.

10. **Herbs; petals 5, equal, slightly coalescent far above the bases,** with elongated claws (stalks) below; sepals forming a tube; stamens 10. *Silene, Caryophyllaceae.*

    Order 23. **Caryophyllales** (Thalamiflorae), page 223.

11. **Herbs; sepals seemingly 2** (actually bracts); **placenta a central column arising from the base of the 1-chambered ovary;** corolla radially symmetrical, the petals slightly coalescent basally; stamens adnate with the petals for a considerable distance above the minute corolla tube, styles 3. *Montia, Portulacaceae* (*Portulaca* family).

    Order 23. **Caryophyllales** (Thalamiflorae), page 223.

12. **Shrubs or trees; leaves alternate, simple; flowers white, yellowish, or greenish,** usually some unisexual and some bisexual on the same individual; **fruit a drupe with 4–9 small hardened seed chambers;** flowers 4–9-merous; sepals minute; **corolla radially symmetrical,** the petals slightly coalescent basally; stamens adnate with the bases of the petals, as many as the petals; ovary 1. *Ilex. Aquifoliaceae* (holly family).

    Order 18. **Sapindales** (Thalamiflorae), page 201.

13. **Herbs; eastern North America; corolla radially symmetrical, rotate; carpels 2; style slender, 2-cleft or -parted;** anthers soon recurved and becoming arcuate or revolute; sepals, petals, and stamens 4–13-merous. *Sabatia, Gentianaceae* (gentian family).

    Order 32. **Gentianales**, page 261.

14. **Perennial herbs with slender rhizomes; leaves opposite, ovate, entire, 2–4 or 5 cm. long; Sierra Nevada and extreme northern California; flowers in a terminal small cluster of helicoid (1-sided) cymes;** carpels 2, the capsule with 2 seed-chambers and 2 seeds in each chamber; style filiform, cleft near the apex. *Draperia, Hydrophyllaceae* (waterleaf family).

    Order 35. **Lamiales**, page 274.

15. **Shrubs or small trees; leaves simple, usually coriaceous; stipules none; flowers bisexual, with hypogynous discs; petals barely coalescent basally; stamens 10, in 2 series;** carpels 2–4, the ovary 2–4-chambered; fruit a capsule or a leathery or fleshy drupe or a berry. *Cyrillaceae* (*Cyrilla* family).

    Order 31. **Ericales** (Thalamiflorae), page 257.

16. **Trees, shrubs, or rarely herbs; flowers small, in dense heads or spikes, the filaments of the stamens the most conspicuous feature of the inflorescence; leaves bipinnate, the leaflets usually small;** sepals coalescent, the calyx usually campanulate or turbinate; fruit a legume, rarely indehiscent and coiled into an elongate spiral structure; mostly of the deserts and other southern portions of the United States. *Mimosoideae, Leguminosae* (pea family).

    Order 38. **Rosales** (Calyciflorae), page 290.

17. **Herbs or low shrubs; calyx tubular, white or colored** but not green, **resembling a sympetalous corolla** (but petals none), **the base of the tube of the calyx constricted just above the ovary,** persistent, hardening and enclosing the small hard fruit; involucre green, enclosing a head of 1 or several flowers; carpel 1; **leaves opposite.** *Nyctaginaceae* (four-o'clock family).

    Order 23. **Caryophyllales**, page 223.

18. **Herbs; leaves markedly succulent, simple, usually entire;** stems somewhat succulent; sepals and petals 5, stamens 5 or 10; **carpels 5, separate or sometimes coalescent basally;** fruit a follicle or, when the carpels are coalescent basally, the follicles barely coalesced into a capsule. *Crassulaceae* (stone-crop family).

    Order 38. **Rosales**, page 290.

19. **Large shrubs** or sometimes small trees **occurring in swamps along the coast from Texas to Florida; leaves elliptic, mostly 4–7 cm. long, leathery; corolla white, bilaterally symmetrical;** petals 5; stamens 4; **fruit a flat elliptical capsule, apically acute, 3–5 cm. long.** *Avicennia* (black mangrove), *Verbenaceae* (*Verbena* family).

Order 35. **Lamiales, page 274.**

20. **Low shrubs or woody-based perennials of arctic and cool temperate regions; leaves simple, entire, alternate,** *either* numerous and narrow or few and reniform; corolla twisted in the bud; **each stamen attached at the sinus between two petals** (not lower in the corolla tube); fruit a 3-chambered capsule. *Diapensiaceae* (*Diapensia* family).

Order 31. **Ericales, page 257.**

21. **Aquatic or marsh herbs;** stamens 5, alternate with the petals, all functional, the filaments not woolly; corolla of 5 equal petals, **each lobe** *either* (1) **white-bearded,** (2) **densely hairy on the upper surface,** *or* (3) **glandular-appendaged on the upper side near the base;** carpels 2, the fruit a 1-chambered capsule. *Menyanthoideae, Gentianaceae* (gentian family).

Order 32. **Gentianales, page 261.**

22. **Herbs** or rarely woody plants; stamens 5, alternate with the petals, all functional, the filaments not woolly; **corolla twisted in the bud, of 5 equal petals; fruit a 3-chambered capsule.** *Polemoniaceae* (phlox family).

Order 34. **Polemoniales, page 269.**

KEY TO THE ORDERS

(First consult the Preliminary List for Elimination of Exceptional Character Combinations, page 248)

NOTE: After determination of the order and family, proceed to the local flora or manual to identify the genus and species.

1. Functional stamens twice as many *or* more than twice as many as the petals (corolla lobes) *or,* if *otherwise,* the stamens and usually the petals 8 or more.
   2. Stamens more than 10, usually 3–4 times as many as the petals.
      3. Plant *not* corresponding with *every character* in the combination in lead 2 above.

Order 30. **Ebenales, page 254.**

      3. Desert shrubs or trees; branches spiny, the spines developed from the first leaves (petioles and midribs), 1 at each node; leaves produced in dense clusters axillary to the spines, appearing only in the rainy seasons; fruit an incompletely 3-chambered capsule.

Order 33. **Fouquieriales, page 269.**

   2. Stamens 10, twice as many as the petals.

Order 31. **Ericales, page 257.**

1. Functional stamens not exceeding the petals in number, not more than 7.
   2'. Pollen-producing stamens as many as the petals, usually 5; corolla nearly always radially symmetrical. (Cf. lower lead 2' below, p. 251)
      3'. Stamens opposite the petals, sometimes accompanied by additional sterile stamens (staminodia) alternating with the petals.
         4. Juice of the plant not milky; usually herbs or rarely somewhat woody plants (these being trees or large shrubs only in southern Florida and on the Keys).

Order 29. **Primulales, page 251.**

         4. Juice of the plant milky; trees or large shrubs. *Sapotaceae* (sapodilla family).

Order 30. **Ebenales, page 254.**

      3'. Stamens alternate with the petals (if this character is obscure, almost always the plant falls under this key lead); sterile stamens none.
         4'. Plant *not* corresponding with *every character* in the combination in lead 4' (p. 251).

5. Ovaries 2, each 1-chambered, the pair with a single swollen common style or stigma above them; juice of the plant milky, usually obviously so; seed usually with a tuft of hairs on one or both ends. *Apocynaceae* (dogbane family) and *Asclepiadaceae* (milkweed family).

Order 32. **Gentianales, page 261.**

5. Ovary 1, the styles and stigmas 1 or more (or rarely the 2 carpels nearly separate and appearing to be separate pistils); juice of the plant rarely milky.

6. Leaves cauline, *all* opposite or whorled, simple, and entire, *either* (1) sessile *or* (2) with stipules. *Gentianaceae* (gentian family) and *Loganiaceae* (*Logania* family).

Order 32. **Gentianales, page 261.**

6. Leaves *not* corresponding with *every character* in the combination in lead 6 above.

Order 34. **Polemoniales, page 269.**

4'. Filaments (at least 3 of the 5) woolly; corolla barely bilaterally symmetrical, almost always yellow, but sometimes white, saucer-shaped; flowers in racemes. *Verbascum, Scrophulariaceae* (snapdragon family).

Order 36. **Scrophulariales, page 277.**

2'. Pollen-producing stamens, 2 or 4, usually fewer than the petals; the number of petals sometimes obscured by union into 2 groups, these being the upper and lower lips of a bilaterally symmetrical corolla, with 2 petals above and 3 below, the upper 2 sometimes appearing as 1, then usually forming a hood (galea).

3'. Corolla *not* scarious.

4''. Corolla radially symmetrical. *Oleaceae* (olive family).

Order 32. **Gentianales, page 261.**

4''. Corolla bilaterally symmetrical.

5'. Fruit composed commonly of 2 or usually 4 nutlets, these usually separating at maturity, or the fruit a drupe or in 1 family an achene formed from 2 carpels.

Order 35. **Lamiales, page 274.**

5'. Fruit a capsule, 1- or usually (in the North American species) 2-chambered.

Order 36. **Scrophulariales, page 277.**

3'. Corolla scarious (thin, dry, transparent, and parchmentlike).

Order 37. **Plantaginales, page 283.**

Order 29. *PRIMULALES.*

PRIMROSE ORDER

DEPARTURE FROM THE FLORAL CHARACTERS OF THE RANALES

Free	Sepals, petals, and carpels coalescent; stamens almost always adnate with the petals.
Numerous	Stamens 5 or sometimes 4; carpels 5 or fewer (rarely 10 in the *Theophrastaceae*).
Spiral	Stamens and carpels cyclic.
Hypogynous	Flowers rarely epigynous (tribe *Samoleae, Primulaceae,* and tribe *Maesoideae, Myrsinaceae*).

Herbs, shrubs, or trees. Leaves alternate or opposite, simple, entire or toothed; stipules none except in some members of the *Plumbaginaceae.* Flowers usually 5-merous but sometimes 4-merous and the carpels sometimes fewer than 5 or 4; stamens in a single series, *opposite the petals* (an outer series of staminodia alternate with the petals being present in the *Theophrastaceae*); ovary 1-chambered, the placenta basal and usually produced into a central column or fleshy structure rising from the base of the ovary (or in the *Plumbaginaceae* the solitary ovule attached by a long funiculus to the base of the ovary); ovules anatropous, with two integuments. Seeds small;

embryo straight; endosperm fleshy or horny or (in the *Plumbaginaceae*) mealy.

There is a close resemblance between the *Primulaceae*, or primrose family, and the *Caryophyllaceae* (Caryophyllales). The 5-merous flower, the 1-chambered ovary, and the basal placenta usually produced into a column or into spongy tissue are similar in the two families, and the ovules of nearly all Caryophyllales and the *Primulaceae* are not anatropous. It is probable that these two orders are related distantly and derived at least from a similar primitive ancestral stock.

The most noteworthy feature of the Primulales is the arrangement of stamens. The Ericales are characterized by 10 stamens and the Ebenales by more than 10, but nearly all the Primulales have only 5. In this respect they agree with the Polemoniales and the Gentianales, but in those orders the five stamens are alternate with the petals, whereas in the Primulales they are opposite. In evolution of most Primulales from ancestors having ten stamens in two series of five the outer series has been eliminated, leaving the five inner stamens, which are in positions opposite the petals; in evolution of the Polemoniales and Gentianales an inner set of five stamens has been eliminated, leaving the five remaining stamens alternate with the petals. In the family *Theophrastaceae* (Primulales) neither set has been eliminated completely, but the outer (alternate with the petals) is sterile. In some *Primulaceae* there are vestiges of the outer set of stamens.

Key to the Families

(North America north of Mexico)

1. Fruit not fleshy; herbs or large bushy shrubs or reclining or twining (scandent) plants.
   2. Style and stigma 1; fruit a capsule with a basally-attached central placenta; ovules several to numerous.

      Family 1. **Primulaceae** (primrose family).

   2. Styles or stigmas 4 or 5; fruit an achene or a utricle; ovule 1, attached to a long stalk from the base of the ovary; plants of seacoasts or of the southern parts of Arizona, Texas, and Florida.

      Family 2. **Plumbaginaceae** (leadwort family).

1. Fruit fleshy, a drupelike berry; large shrubs or trees; peninsular Florida and the Keys.
   2'. Staminodia absent; fruit with only 1 seed.

      Family 3. **Myrsinaceae** (*Myrsine* family).

   2'. Staminodia present; fruit with more than 1 seed.

      Family 4. **Theophrastaceae** (Joe-wood family).

*Family 1. **Primulaceae.**

PRIMROSE FAMILY

Herbs or rarely slightly woody plants. Leaves usually simple (sometimes pinnate in aquatic plants), usually opposite or whorled. Flowers usually hypogynous, but in the Samoleae epigynous; sepals usually 5 but sometimes 4–9; petals usually 5, coalescent at least basally or rarely absent (*Glaux*) or separate (*Pelletiera*); stamens usually 5, as many as the petals and opposite them, the vestiges of an outer whorl of stamens sometimes present and alternate with the petals; carpels usually 5, the style 1, the ovary 1-chambered; placenta a central stalk or mound of tissue arising from the base of the ovary; ovules usually several to numerous, semianatropous. Fruit a capsule or a pyxis (circumscissile); embryo straight.

This family is essentially world-wide in distribution. On the Pacific Coast the primrose family is represented by about 8 or 9 species of the genus *Dodecatheon* (shooting-star) and by one or two species of each of about seven other genera. The family is about equally represented through the Rocky Mountain region, but the emphasis is not quite so strongly upon *Dodecatheon*. There are three species of native primroses (*Primula*). The family is more abundant in the Middle West and East, where there are two species of shooting-stars, four of primroses, thirteen of *Lysimachia* (loosestrife), and representatives of other genera. The same groups occur in the Southern States, but there are fewer species.

Species of *Primula* are common and attractive in cultivation. Other cultivated genera include *Cyclamen, Soldanella,* and *Cortusa.*

*Family 2. **Plumbaginaceae.**

LEADWORT FAMILY

(Synonym: Armeriaceae)

Perennial herbs or shrubs or sometimes woody vines. Leaves simple, entire, alternate, often in rosettes. Sepals 5, coalescent, folded lengthwise like a Japanese fan, sometimes scarious, commonly

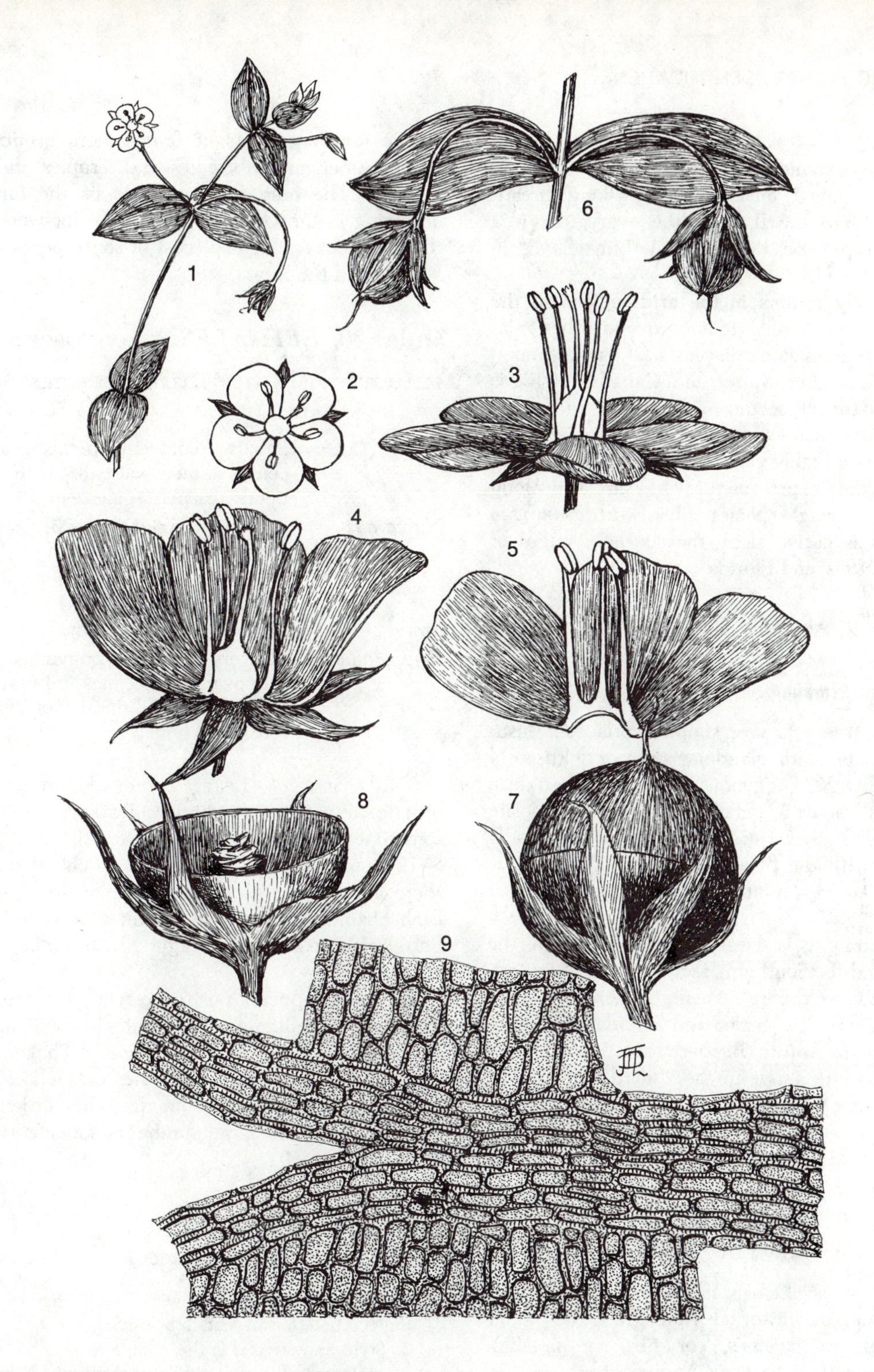

Figure 10-57. Primulales; Primulaceae; *Anagallis arvensis,* scarlet pimpernel (cf. also **Figure 8-6 C**): **1** branch with leaves, flowers, and fruits, **2** flower, top view, showing the sepals, the slightly coalescent petals, the stamens *opposite* the petals, and the pistil, **3–4** views of the flower, **5** three petals and the stamens opposite them, **6** fruiting branch, **7–8** circumscissile dehiscence of the pyxis, the central placenta appearing in **8**, **9** cellular detail along the line of dehiscence of the pyxis.

persistent; petals 5, sometimes only slightly coalescent basally; stamens 5; carpels 5, the styles or stigmas 5; ovule 1, anatropous, on a long threadlike stalk from the base of the ovary. Fruit a utricle or sometimes circumscissile but delayed in opening.

The family centers in the arid regions of the Eastern Hemisphere, but it occurs in other relatively dry regions along seacoasts. It has few members in the United States and Canada. *Armeria maritima* (thrift) occurs in the north and along the Pacific Coast where it is common near the ocean. A few highly variable species of *Limonium* (sea-lavender, marsh-rosemary) occur also along the coasts. A single species of leadwort (*Plumbago scandens*) is native along the southern edges of Arizona, Texas, and Florida.

*Family 3. **Myrsinaceae.**

MYRSINE FAMILY

(Synonym: *Myrsinaceae*)

Shrubs or trees. Leaves simple, coriaceous, usually alternate, with glandular dots or with resin ducts. Flowers hypogynous or in the Maesoideae epigynous; sepals 5; petals 5; stamens 5, opposite the petals, but an outer series of 5 staminodia alternate with the petals; carpels 5, the style 1; ovules numerous, anatropous. Fruit ordinarily a berry.

The family is a large one occurring in the Tropics and the Southern Hemisphere and extending northward through Pacific regions to Japan and in America to Mexico and Florida. The largest genera are *Ardisia, Rapanea,* and *Maesa.* Only two species are native in the United States. One is a species of *Rapanea* and the other the marlberry (*Ardisia*). These occur in peninsular Florida and on the Keys.

*Family 4. **Theophrastaceae.**

JOE-WOOD FAMILY

Shrubs or trees. Leaves simple, alternate. Flowers bisexual or unisexual (and then the plants dioecious), in racemes, corymbs, or panicles; sepals 5; petals 5; stamens 5, opposite the petals, accompanied by an outer series of 5 staminodia alternate with the petals; carpels 5, the style 1; ovules numerous, anatropous. Fruit a berry or rarely a drupe.

The family consists of four genera occurring in the American Subtropics and Tropics and in Hawaii. The only representative of the family occurring in the United States is the Joe-wood of the hammocks along the coast of south peninsular Florida and the Keys.

Order 30. ***EBENALES.*** EBONY ORDER

DEPARTURE FROM THE FLORAL CHARACTERS OF THE RANALES

Free	Sepals coalescent; petals coalescent; stamens adnate with the petals; carpels coalescent.
Numerous	Stamens about 12–25, rarely fewer, 3–4 times as many as the sepals and petals; carpels 3 to several.
Spiral	Stamens and carpels cyclic.
Hypogynous	Flowers usually hypogynous, but the ovary sometimes inferior or half inferior and the flower, therefore, epigynous.

Shrubs or trees. Leaves commonly simple, alternate, opposite, or whorled, leathery; stipules sometimes present. Flowers bisexual, complete, 3–12-merous; ovary with as many chambers as there are carpels; placentae axile; ovules 1–2 in each chamber, anatropous. Fruit a capsule or a berry; embryo straight or slightly bent; endosperm present.

The order appears not to be related closely to the other Corolliflorae, and quite likely it arose independently from some group in the Thalamiflorae, perhaps near the Guttiferales or Violales. It is distinguished readily from the other orders of Corolliflorae by the large number of stamens (usually more than 10).

Key to the Families

(North America north of Mexico)

1. Plant *not* corresponding with *every character* in the combination in lead 1 below.
   2. Style and stigma 1, the components completely coalescent (or the stigma slightly lobed).
      3. Flowers all bisexual; *either* (1) the flower epigynous *or* (2) the stamens in a single series of 8–16.

         Family 1. **Styracaceae** (styrax family).

3. Some flowers unisexual; flowers hypogynous; stamens in several series.

Family 2. **Symplocaceae** (sweetleaf family).

2. Styles (or their upper portions) and the stigmas separate; fruit of the single North American species (persimmon) a berry 2.5–4 cm. in diameter.

Family 4. **Ebenaceae** (ebony family).

1. Sap of the plant milky; functional stamens as many as the petals and opposite them, staminodia often present, filaments not forming an elongated tube free from the corolla tube; flower hypogynous.

Family 5. **Sapotaceae** (sapote family).

*Family 1. **Styracaceae.** STYRAX FAMILY

Shrubs or small trees; hairs of the young parts of the plant either stellate or scalelike (lepidote). Leaves simple, alternate. Flowers usually hypogynous but sometimes epigynous; sepals 4 or 5, coalescent; petals mostly 4–8; stamens 8 or usually 10–16; carpels 2–5, the style 1, the ovary more or less 2–5-chambered, but the partition restricted to the lower half and therefore not forming strictly separate chambers; placentae axile below and parietal above; ovules 2–8 per carpel. Fruit a drupe or sometimes the pericarp papery and the fruit then a capsule.

The family occurs in the warmer parts of the Americas, in the Mediterranean Region, and in eastern Asia. One species of *Styrax* occurs in California and two others in eastern North America, particularly in the Southern States. The snowdrop-tree or silverbell-tree (*Halesia*) occurs also in the South and northward to the southern edges of the states of the latitude of Illinois. This tree is cultivated, and occasionally it escapes from cultivation outside its native range.

*Family 2. **Symplocaceae.** SWEETLEAF FAMILY

Shrubs or trees. Leaves simple, leathery, alternate. Flowers ordinarily bisexual but rarely unisexual (and then the plant polygamodioecious), in racemes or panicles; sepals 5, coalescent, persistent; petals 5–11, coalescent only basally; stamens usually numerous, a cluster at the base of each petal; carpels 2–5, the style 1, the ovary 2–5-chambered; placentae axile; ovules usually 2 per carpel. Fruit a berry or a drupe.

The family consists of a single large genus occurring in the Tropics of Asia and America and extending northward through subtropical regions to Mexico and the southeastern United States.

The sweetleaf, horse-sugar, or yellowwood (*Symplocos*) occurs in swamps, bottomlands, and moist woods through the southern states and as far northward as Delaware. The sweet leaves are eaten by cattle and other livestock.

Family 3. **Lissocarpaceae.** LISSOCARPA FAMILY

Small trees. Leaves entire, alternate. Flowers epigynous, in cymes; sepals 4; petals 4; stamens 8, the filaments coalescent; carpels 4, the style 1, the ovary 4-chambered; placentae axile; ovules 2 per carpel. Fruit indehiscent, containing 1 or 2 seeds; seed with 3 ribs.

The single genus, *Lissocarpa*, occurs in tropical South America.

*Family 4. **Ebenaceae.** EBONY FAMILY

Shrubs or trees; juice milky. Leaves simple, entire, coriaceous, alternate. Flowers usually unisexual and the plant ordinarily monoecious; sepals 3–7; petals 3–7, leathery; stamens usually 2–3 times as many as the petals but sometimes the same number and in a single series, coalescent in pairs or separate; carpels 2–16, the styles separate or coalescent basally, the ovary 2–16-chambered; placentae axile; ovules usually 2 per carpel, pendulous. Fruit a berry.

The family occurs in many parts of the world, but only one species is native to the United States. The persimmon (*Diospyros*) is distributed widely in eastern North America south of the Canadian boundary.

Ebony (*Diospyros Ebenum*), a tropical tree of India and the East Indies, is a member of the same genus. The heartwood of most genera is black, red, or brown and the family is one of the most important sources of cabinet wood. Persimmons are cultivated in the warmer parts of the United States, being a commercial crop in several areas.

*Family 5. **Sapotaceae.** SAPOTE FAMILY

Shrubs or trees; juice milky. Leaves simple, usually entire, coriaceous, alternate or rarely oppo-

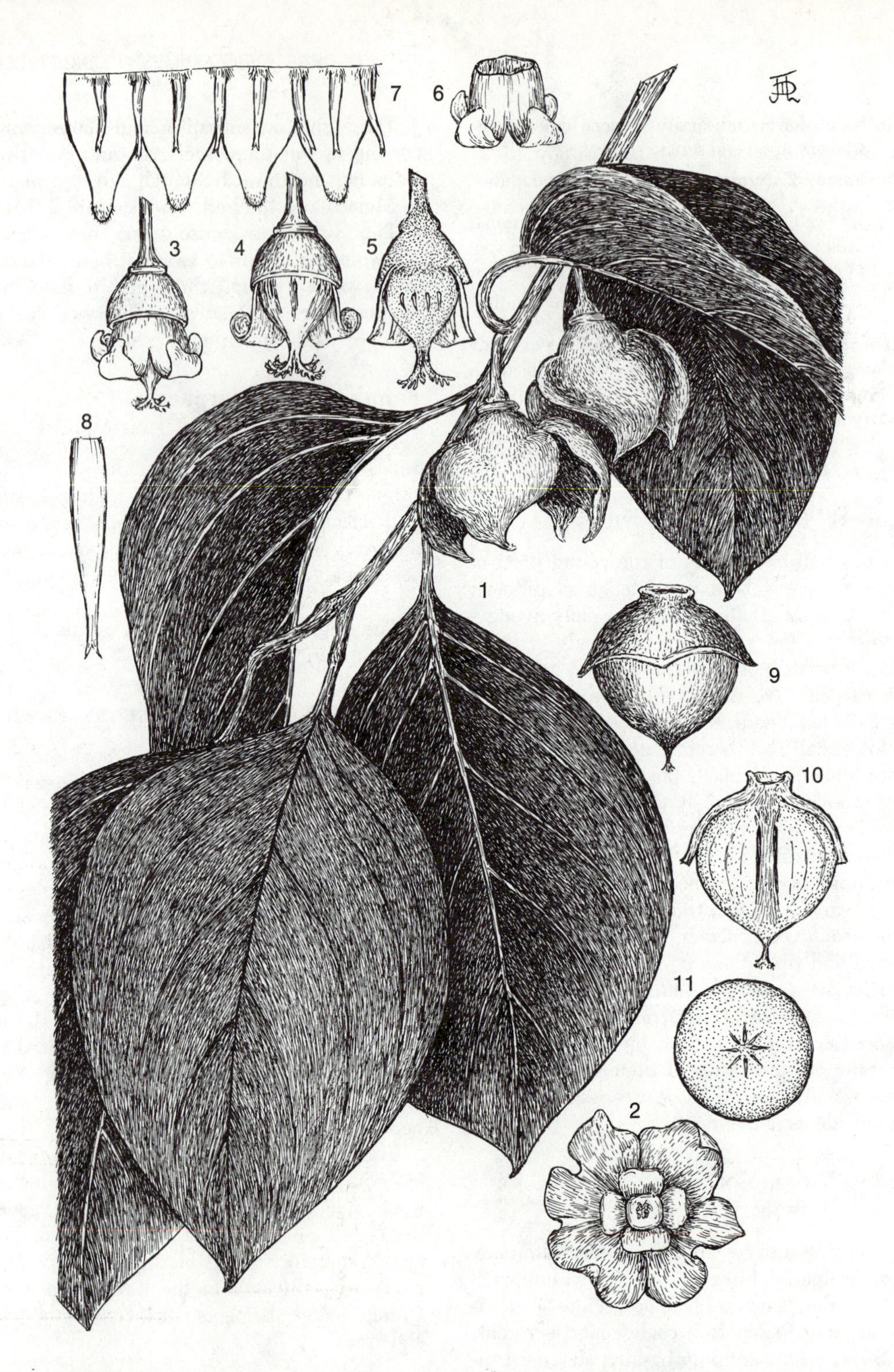

Figure 10-58. Ebenales; Ebenaceae; *Diospyros Kaki,* Japanese persimmon: **1** branch with leaves and flowers, **2** flower, top view, showing the four sepals (coalescent), the four petals, the stamens, and the stigmas, **3** flower, the top of the calyx cut away, showing the coalescent petals and the pistil, **4** another view, **5** longitudinal section, **6** corolla, **7** corolla laid open, showing the filaments of the stamens, **8** filament enlarged, **9** young fruit, **10** longitudinal section, **11** cross section.

site. Flowers solitary or usually cymose, occurring on the older stems; sepals 4–12, usually in 2 series or arranged spirally; petals usually the same number as the sepals; stamens about 8–15, commonly in 2 or 3 series, but the outer 1 or 2 series usually sterile; carpels usually 4 or 5, sometimes 1–14, the style 1, the ovary with as many chambers as there are carpels; placentae axile; ovule 1 per carpel. Fruit a berry and the outer layer usually thin and horny or leathery.

The family is distributed widely through tropical and warm areas. The noted member is the sapodilla (*Manilkara* [*Achrys*] *Sapota*), a triply useful plant. The fruit is edible, the skin being brownish and the meat brown also; the wood is hard, reddish, and durable; the latex of the trunk yields chicle, which, after addition of flavoring matters, becomes chewing gum. The sapodilla is cultivated in Florida and it has gone wild in old fields and hammocks. Other Florida representatives of the family are the star-apple or satinleaf (*Chrysophyllum*), the mastic or wild-olive (*Mastichondendron*), the egg-fruit (*Pouteria campechiana*), the bustic or cassada (*Dipholis*), and the wild dilly or wild sapodilla (*Mimusops*). A number of species of *Bumelia* occurs from the southeastern corner of Arizona to Florida. These go by a variety of names, common among them buckthorn, chittamwood, and gum elastic.

Gutta-percha is derived from several Eastern Hemisphere representatives of the family. Fruits includes the sapote and eggfruit.

Order 31. *ERICALES.* HEATH ORDER

DEPARTURE FROM THE FLORAL CHARACTERS OF THE RANALES

Free	Petals often coalescent; stamens usually (except in the *Diapensiaceae* and some *Epacridaceae*) either free or barely adnate with the petals; carpels coalescent.
Numerous	Stamens 2–3 or usually 8 or 10 or sometimes as few as 5; carpels 2–9, usually 4–5.
Spiral	Stamens and carpels cyclic.
Hypogynous	Flower in some instances with a hypogynous disc, but this often only slightly developed; flower epigynous in the *Vaccinioideae, Ericaceae.*

Herbs (then often saprophytic and not green but commonly red, white, or brown) or usually shrubs or sometimes trees. Leaves simple, alternate, entire or toothed; stipules none. Flowers bisexual or sometimes unisexual; sepals 2–3 or usually 4–5; petals 4–5 or rarely none; stamens usually 10 but sometimes 8 or in some cases 5, inserted on the margin of the hypogynous disc, the disc nectar-secreting; carpels usually 4–5, usually opposite the petals; ovary divided into as many chambers as there are carpels; placentae axile, with numerous ovules; style 1, the stigma lobes usually as many as the carpels; ovule anatropous, with 1 integument.

The order is difficult to key out according to usual lines of classification, for it includes apetalous, choripetalous, and sympetalous plants as well as hypogynous and epigynous ones and some which could be interpreted as perigynous. All these types are found in the *Ericaceae.*

Key to the Families

(North America north of Mexico)

1. Petals 4–5 or rarely none; stamens usually 8–10 in 2 series; sepals 4–5; flowers nearly always bisexual.
   2. Ovules 1–3 per carpel; seed coat none; anthers dehiscent lengthwise by full-length slits; petals slightly coalescent basally or separate; fruit a 2-valved capsule; trees or shrubs with long racemes.

      Family 1. **Cyrillaceae** (leatherwood family).

   2. Ovules numerous in each carpel; seed coat present; anthers dehiscent by pores, slits, or chinks, these more or less terminal; petals ranging from separate to mostly markedly coalescent; fruit, if dehiscent, opening along 3 or commonly 5–10 lines.
      3. Stamens *either* free from the corolla *or* barely adnate with its base *or* in special pouches of the corolla.

         Family 2. **Ericaceae** (heath family).

      3. Stamens clearly adnate with the corolla, not in special pouches.

         Family 4. **Diapensiaceae** (*Diapensia* family).

1. Petals 0; stamens 2–3; sepals 3 (or 2–6) or 0; flowers unisexual or on the same plant some bisexual; low shrubs with evergreen, alternate, simple, entire, linear, numerous, crowded, firm leaves.

   Family 5. **Empetraceae** (crowberry family).

Suborder A. **Ericineae**

*Family 1. **Cyrillaceae.**

LEATHERWOOD FAMILY

Shrubs or small trees. Leaves simple, entire, alternate, rather thick, coriaceous; stipules none. Flowers bisexual, radially symmetrical, with no hypogynous discs, in racemes; sepals 5, persistent at fruiting time; petals, sometimes slightly coalescent basally, 5; stamens 10, in 2 series, the inner sometimes sterile or small, the filaments dilated below; carpels 2–4, the style 1; the ovary 2–4-chambered; placentae axile; ovules usually 1 or 2, pendulous. Fruit a capsule or leathery to fleshy and indehiscent.

The family is composed of 3 small genera, 2 including solitary species and restricted to the Southern States and the third with 3 species in tropical America. The north American species are the leatherwood or black ti-ti (*Cyrilla racemiflora*) and another plant known as black ti-ti or buckwheat tree (*Cliftonia monophylla*).

Both species are attractive, and they are cultivated.

*Family 2. **Ericaceae.** HEATH FAMILY

(Each subfamily in the key below has appeared in family rank.)

Perennial herbs, shrubs, or trees. Leaves simple, usually evergreen and coriaceous, usually alternate but sometimes opposite. Flowers hypogynous and with hypogynous discs or sometimes epigynous, radially symmetrical or bilaterally symmetrical in *Rhododendron;* sepals 4 or 5; petals 4 or 5 or rarely none, sometimes separate but usually coalescent, white or red; stamens usually 8 or 10, rarely as few as 5, free from the corolla or essentially so, the anthers opening by terminal pores or chinks, or rarely opening along their full length, often with feathery or bristlelike terminal appendages or points; carpels usually 4–10 or rarely 2–3, the style 1, the ovary with as many chambers as carpels; placentae axile; ovules numerous in each carpel.

The family occurs throughout the temperate parts of the world and in the mountains of the Tropics, and it ranges northward to middle arctic regions. It includes a great variety of plants. Among them are the white-alder of the eastern part of the Continent and the winter-green, pipsissewa, Indian-pipe, pine-drops, pine-sap, Labrador tea, rhododendrons (including azaleas), American laurels, various heaths, trailing arbutus, bearberries, manzanitas, Scotch heather, cranberries, huckleberries, blueberries, and bilberries. The genus *Arctostaphylos* has a remarkable distribution. Two species of low matted plants are circumpolar and another nearly circumpolar. These are known as bearberries. The major development of the genus is in California where there are perhaps 25 species, these being complex and difficult to segregate. A few of these species or closely related ones occur to the north and southeast of California in Washington, Oregon, Arizona, and northern Mexico. The *Vaccinioideae* (blueberry subfamily) is developed extensively in the eastern half of the Continent, and a relatively small number of species occurs westward to the Rocky Mountains and the Pacific Coast. Most of these grow near timberline or in moist forests. Among the more attractive plants are the species of *Rhododendron,* represented on the Pacific Coast but much more common in the Appalachian Region. The genus includes the azaleas as well as the typical rhododendrons. Particularly attractive small plants are the various heaths of several genera occurring near timberline in the northern mountains. In many ways the most striking plants in this family are the colorful saprophytic herbs of many forests and especially the snow-plant of the California mountains. The snow-plant is reputed to appear just after the snow melts, though usually this is not true. It is about a foot high, an inch in diameter, and a brilliant red. The winter-greens (*Pyrola*) are abundant in northern and other moist woods. A common northern and eastern plant is the Indian-pipe, which is white while it is alive but which turns black in drying.

The *Ericaceae* consist of several subfamilies proposed by many authors as separate families because the extreme in each group is striking. These are separated by the following key.

Key to the Subfamilies of the Ericaceae

1. Ovary superior.
   2. Carpels 3; petals separate; trees or shrubs; pubescence stellate; pollen grains single.

Subfamily A. **Clethroideae** (white-alder subfamily).

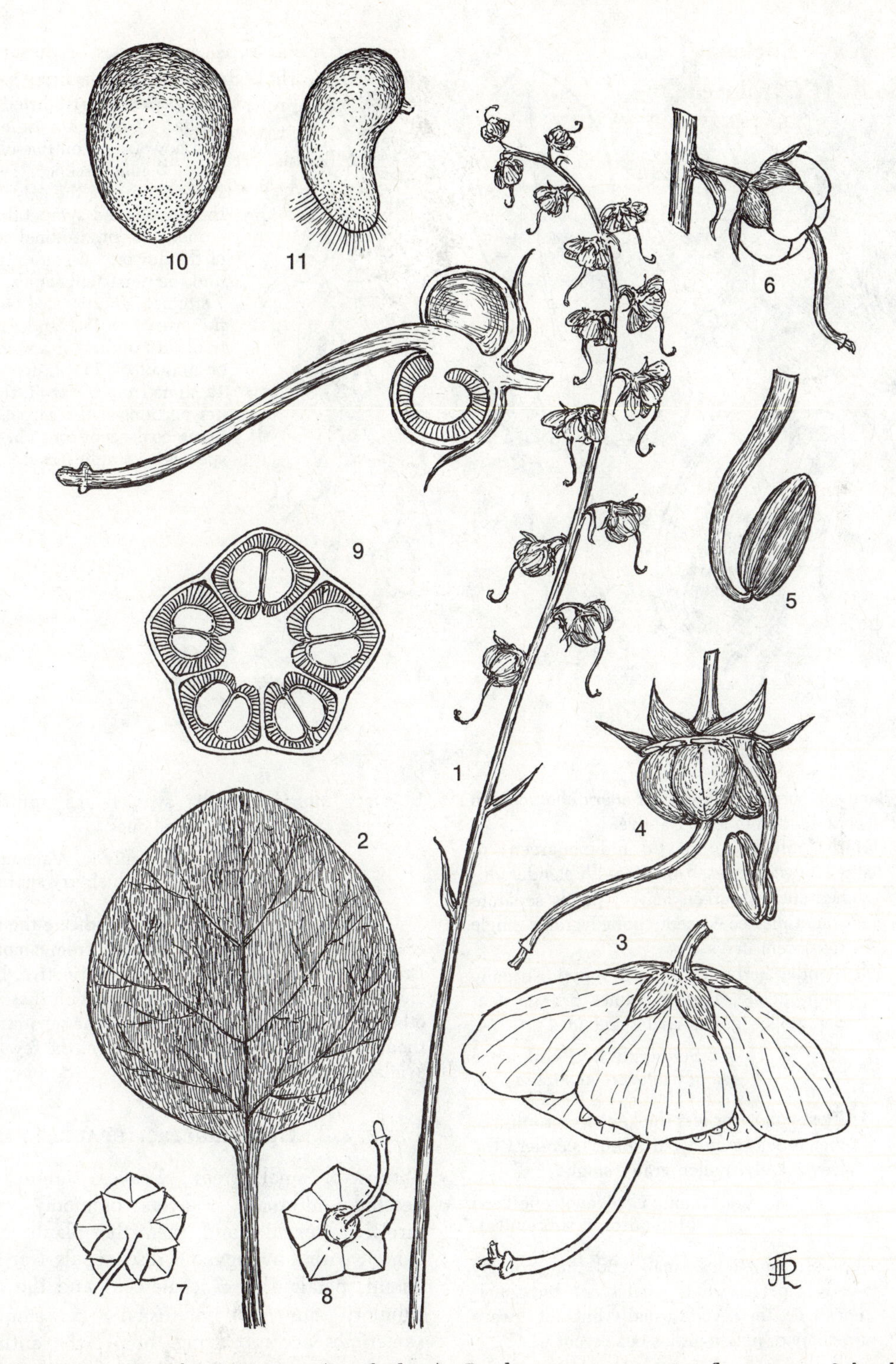

Figure 10-59. Ericales; Ericaceae (Pyroloideae); *Pyrola,* wintergreen: **1** inflorescence, **2** basal leaf, **3** flower, the petals separate, **4** flower with the petals and all but one stamen removed, **5** stamen, **6–8** fruit with the persistent calyx, **9** cross section of the fruit, **10–11** views of the seed.

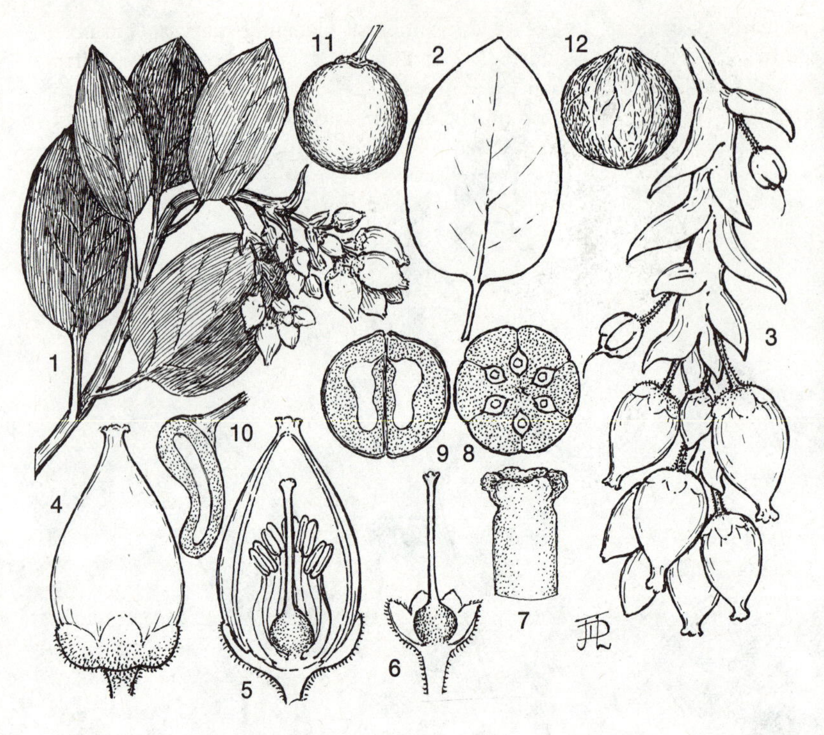

Figure 10-60. Ericales; Ericaceae (Ericoideae); *Arctostaphylos glauca*, great-berried manzanita: **1** branch with leaves and flowers, **2** outline of a leaf, **3** inflorescence, **4** flower, showing the calyx and the urn-shaped sympetalous corolla, **5** longitudinal section of the flower, **6** young fruit and the persistent sepals, **7** stigma, **8** cross section of the ovary, **9** longitudinal section, **10** outline of a section of an ovule, **11** mature fruit, **12** stone from the fruit, the inner portions of the carpels being coalescent into (in this species) a single stone.

2. Plant *not* corresponding with *every character* in the combination in lead 1 above.
   3. Herbs (often saprophytic and nongreen) or slightly woody-based plants with slender rhizomes and evergreen leaves; petals separate or sometimes coalescent; pollen grains single or coalescent in 4's.
      4. Slender herbs, rarely saprophytic, usually with green leaves; rhizomes prsent, slender, pollen grains coalescent in 4's.

         Subfamily B. **Pyroloideae** (wintergreen subfamily).

      4. Fleshy more or less thickened saprophytes, the underground parts not slender, the leaves scaly; pollen grains single.

         Subfamily C. **Monotropoideae** (Indian-pipe subfamily).

   3. Shrubs (sometimes small and not strongly woody); petals usually coalescent, but sometimes free, the corolla usually but not always urn-shaped; pollen grains coalescent in 4's.

      Subfamily D. **Ericoideae** (heath subfamily).

1. Ovary inferior; corolla sympetalous, urn-shaped; shrubs, but sometimes small ones.

   Subfamily E. **Vaccinioideae** (blueberry subfamily).

The characters in the key above indicate the futility of attempting to segregate families from among the Ericaceae. The extreme forms are distinctive, but the characters of one subfamily interlock with those of the other, and there is no place for a clear separation, although it would seem so on the basis of the few genera available in most localities.

Family 3. **Epacridaceae.** EPACRIS FAMILY

Shrubs or small trees. Leaves simple, small, heathlike, alternate. Flowers commonly bisexual, rarely unisexual (and then the plants monoecious), with a hypogynous disc; sepals 4 or 5, persistent; petals 4 or 5, coalescent and the corolla tubular; stamens 4 or usually 5 (staminodia sometimes accompanying them), the anthers 1-chambered, splitting lengthwise; carpels 4–5, the style 1, the ovary 1–10-chambered; placentae usually axile; ovules 1–many per carpel. Fruit a capsule or a drupe with 1–5 seeds.

The family is a fairly large one, centering in Australia but extending to Malaya and Hawaii.

Suborder B. **Diapensineae**

*Family 4. **Diapensiaceae.**

DIAPENSIA FAMILY

(Segregate: *Galacaceae*)

Low shrubs or sometimes slightly woody-based herbs (*Galax*). Leaves simple, either numerous, narrow, and imbricated or sometimes fewer and then reniform, alternate; stipules present but not noticeable. Flowers hypogynous, *not* with hypogynous discs; sepals 5, persistent; petals 5, coalescent; stamens 5, adnate with the petals, sometimes accompanied by staminodia, the anthers splitting lengthwise or in one genus crosswise; carpels 3, the style 1, the ovary 3-chambered; placentae axile; ovules several to numerous in each carpel. Fruit a capsule.

The family is composed of six genera with a few species occurring in the Arctic and cool temperate regions of the Northern Hemisphere. Single species of the genera *Diapensia, Shortia, Galax,* and *Pyxidanthera* occur in eastern North America. *Diapensia* is circumpolar.

Order 32. ***GENTIANALES.***

GENTIAN ORDER

DEPARTURE FROM THE FLORAL CHARACTERS OF THE RANALES

Free	Sepals coalescent; petals coalescent or 0; stamens adnate with the petals; carpels coalescent, sometimes by only the stigmas and styles, these sometimes adnate with the stamens.
Numerous	Stamens 2–5 or rarely up to 13; carpels 2.
Spiral	Stamens and carpels cyclic.

Herbs, shrubs, or trees. Stem often with internal phloem (i.e., internal to the xylem of the young stem). Leaves simple, trifoliolate, or pinnate, opposite, with or without stipules. Flowers bisexual or unisexual, complete or sometimes lacking petals, 4- or 5-merous, rarely up to 13-merous in some species of *Sabatia;* ovary with 1–2 seed chambers; placentae parietal or axile; ovules 1–2 or numerous, anatropous or amphitropous, each with 1–2 integuments; endosperm present, the amount variable. Fruit a capsule, follicle, samara, berry, or drupe.

In many works the more inclusive order Gentianales has been segregated from other orders and especially the Polemoniales on the basis of the twisting (contortion) of the corolla in the bud. This character is present in some *Gentianaceae,* some *Loganiaceae,* and some *Apocynaceae,* but it is inconsistent in occurrence and, like the obdiplostemonous character of the Geraniales, it has been overemphasized. The same character occurs in the Polemoniales (e.g., among the *Polemoniaceae* and *Hydrophyllaceae*). Segregation of the Gentianales from the Polemoniales is difficult and perhaps not warranted. The most consistent difference is opposite or whorled leaves as opposed to alternate leaves, but this is inconsistent in both orders. The other characters used to segregate the orders are more consistent in occurrence in books than in plants. Cf. discussion in Chapter 14.

Key to the Families

(North America north of Mexico)

1. Stamens 4 or 5 or rarely up to 13, the same number as the sepals and petals; fruit a capsule or a follicle; herbs or uncommonly woody plants.
   2. Ovary 1, the styles and stigmas 1 or more; juice of the plant not milky; seeds not with tufts of hairs.
      3. Ovary with 1 chamber; stipules none.

         Family 1. **Gentianaceae** (gentian family).

      3. Ovary with 2 chambers; stipules present at least on young leaves or represented by lines on the adjacent surface of the stem.

         Family 2. **Loganiaceae** (Logania family).

   2. Ovaries 2, each 1-chambered, each becoming a follicle in fruit, the pair with a single swollen common style or stigma above them; juice of the plant milky, usually obviously so; seed with a tuft of hairs on one or both ends.
      3'. Anthers *not* adnate with the compound stigma; style and stigma 1; hoods none; pollen not in waxy masses; corolla twisted in the bud.

         Family 3. **Apocynaceae** (dogbane family).

3'. Anthers adnate with the compound stigma; styles 2, the stigma 1; each petal bearing a hood, which may have a horn as an internal appendage; pollen in pear-shaped waxy masses; corolla usually valvate (with the petals edge to edge) in the bud.

Family 4. **Asclepiadaceae** (milkweed family).

1. Stamens 2, fewer than the sepals and petals when (usually) these are present; fruit a berry, drupe, samara, capsule, or pyxis; trees or shrubs.

Family 5. **Oleaceae** (olive family).

Suborder A. **Gentianineae**

*Family 1. Gentianaceae.

GENTIAN FAMILY

(Segregate: *Menyanthaceae*)

Herbs or rarely woody plants. Leaves simple or sometimes trifoliolate, sessile except in the Menyanthoideae; stipules none. Flowers ordinarily bisexual or rarely some unisexual and the plant polygamous; sepals 4 or 5; petals 4 or 5; stamens 4 or 5; carpels 2, the style 1, the ovary 1-chambered; placentae parietal, usually intruded; ovules numerous in each carpel. Fruit a capsule.

The family is large, and it occurs throughout the world, being particularly common in temperate areas. In the drier lowlands of the Pacific Slope and particularly in California the gentian family is represented by *Centaurium*, a genus of colorful small annuals with pink or red flowers. Several species of the genus are native in eastern North America. *Gentiana* (gentian) occurs through Canada and from coast to coast and border to border of the United States. *Swertia* (deer's-tongue) includes about 10 species in the West, and it occurs in the Middle West and East from Michigan to New York, Louisiana, and Georgia. *Sabatia* includes about 10 species in the eastern portions of the Continent, and several other genera (including *Lomatogonium, Halenia* or spurred gentian, *Bartonia*, and *Obolaria* or pennywort) occur in eastern North America. Three genera of the specialized subfamily Menyanthoideae occur in North America. The buckbean (*Menyanthes trifoliata*) occurs in bogs around the world in the northern plains and mountains. The typical variety (Eurasia and Pacific North America) is replaced by a local variety in eastern North America. *Nymphoides* (floating-heart), an

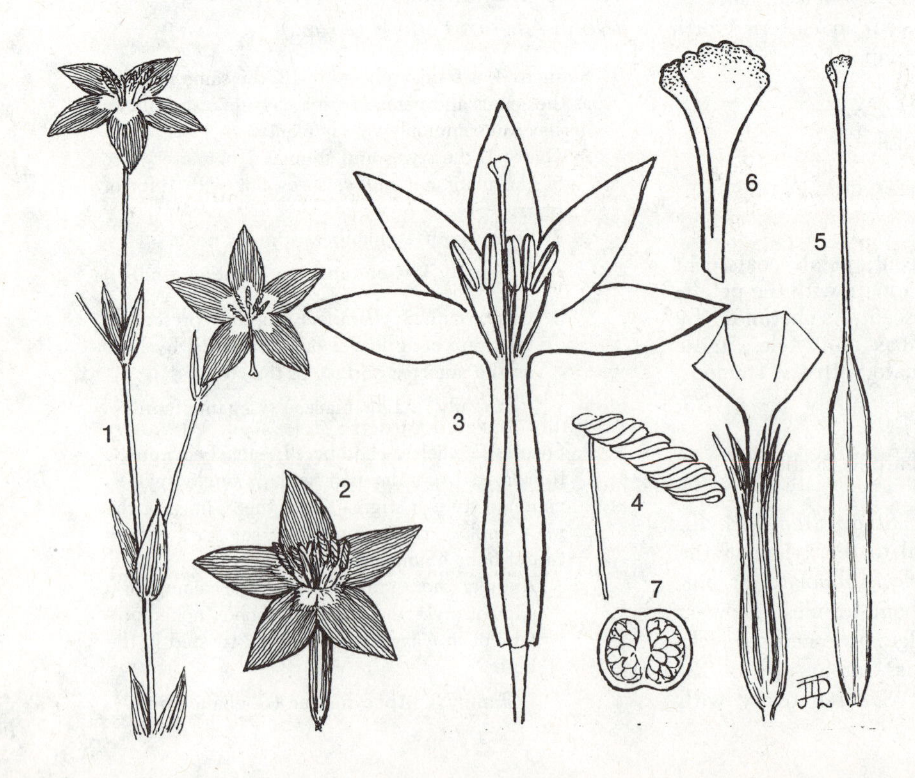

Figure 10-61. Gentianales; Gentianaceae; *Centaurium venustum*, centaury or canchalagua: **1** branch with opposite leaves and flowers, **2** flower, **3** flower, the corolla laid open, **4** anther after dehiscence, **5** pistil, **6** stigma, **7** cross section of the ovary showing the axile placentae. Usually, the ovary is 1-chambered with parietal placentae.

aquatic genus, occurs in eastern North America especially on the coastal plain in the Southern States. A species of *Nephrophyllidium* occurs from Alaska to Washington and in Japan.

*Family 2. **Loganiaceae.**

LOGANIA FAMILY

(Segregate: *Spigeliaceae*)

Herbs, shrubs, trees, or woody vines. Leaves simple, usually opposite, but alternate in some species of *Buddleia;* stipules present, at least on the young leaves. Flowers hypogynous or in *Mitreola* epigynous, in cymes or thyrses; sepals 4 or 5; petals 4 or 5, but sometimes each petal 2-lobed; stamens 4 or 5 or rarely 1; carpels usually 2, the style 1, the ovary 2-chambered; placentae axile or in the rare instances in which there is only 1 chamber, parietal; ovules several to many in each carpel. Fruit a capsule or rarely a berry or a drupe.

The family occurs throughout the warmer and tropical regions of the Earth. *Buddleia* (butterfly-bush) is common in cultivation, and it includes about 100 species native mostly in the Tropics. A local species occurs in Utah, and a wider-ranging Mexican species occurs in the upper Arizona Desert in southern Arizona. A third species grows on the coastal plain from Texas to Florida and Georgia, but it is a native of China escaped from cultivation. The family is absent from the Pacific Coast, but it appears in eastern North America, where the yellow jessamine (*Gelsemium sempervirens*), also known as evening-trumpet-flower, occurs in the Southern States. Other genera in that region are *Spigelia,* known as Indian-pink or worm-grass (occurring as far north as Ohio), *Polypremum,* and *Cynoctonum* (miterwort). *Coelostylis* occurs in Florida.

The tropical genus *Strychnos* yields the poison strychnine.

The genus *Desfontainea* has been separated as a family *Desfontainaceae.* The plant is not known to the author and its status cannot be evaluated.

Suborder B. **Apocynineae**

Herbs, or two species of *Asclepias* shrubby; plants with milky juice carried in latex vessels. Leaves usually opposite, simple, entire; stipules none. Flowers complete, 5-merous, but the carpels 2; ovaries 2, a common style and stigma above them; ovules amphitropous or anatropous. Fruit a follicle, 2 fruits being produced by each flower; embryo large, straight, the seed usually bearing a long tuft of soft hairs on one or both ends at least at maturity; endosperm large.

The two families of this suborder have been included by most authors in the Gentianales (*Contortae*). Whether they are to be included in the Gentianales or not, they form a natural unit. As stated in the description, in both families the styles and stigmas are coalescent but the ovaries of the two carpels are separate. This peculiar feature is almost unique among the flowering plants.

Recently authors have proposed union of the families, as Apocynaceae.

*Family 3. **Apocynaceae.**

DOGBANE FAMILY

Herbs, or extraterritorial plants sometimes shrubs or trees. Leaves simple, entire, ordinarily opposite or whorled but sometimes alternate; stipules usually none. Flowers solitary or in racemes or cymes; sepals 5 or rarely 4; petals 5, twisted in the bud (contorted), funnelform or salverform; anthers often adherent to viscid materials on the stigma. Fruits follicular or sometimes fleshy; seed often with a tuft of hairs at each end.

In some extraterritorial genera the ovaries and the 2 carpels are coalescent and with axile or parietal placentae. In one primitive group there are 5 to 8 carpels. Sometimes the ovary is inferior. (Cf. Figure 12-14.)

The family is large and almost world-wide in distribution. *Vinca major* is the common periwinkle of gardens; it may escape frequently in moist places usually in bottomlands along streams. A smaller-flowered garden species is *Vinca minor.* Frequently these plants are referred to as myrtle, but that name belongs to *Myrtus* (Myrtales). Other important genera are *Amsonia* and *Apocynum,* the dogbane or Indian-"hemp." As (American) Indian-"hemp" this is not to be confused with the true hemp or Indian Hemp of India, which is the source of hashish and marijuana. The dogbane has entire opposite leaves; the plant yielding marijuana has palmatifid leaves. Most of the species of Apocynaceae are acrid poisonous

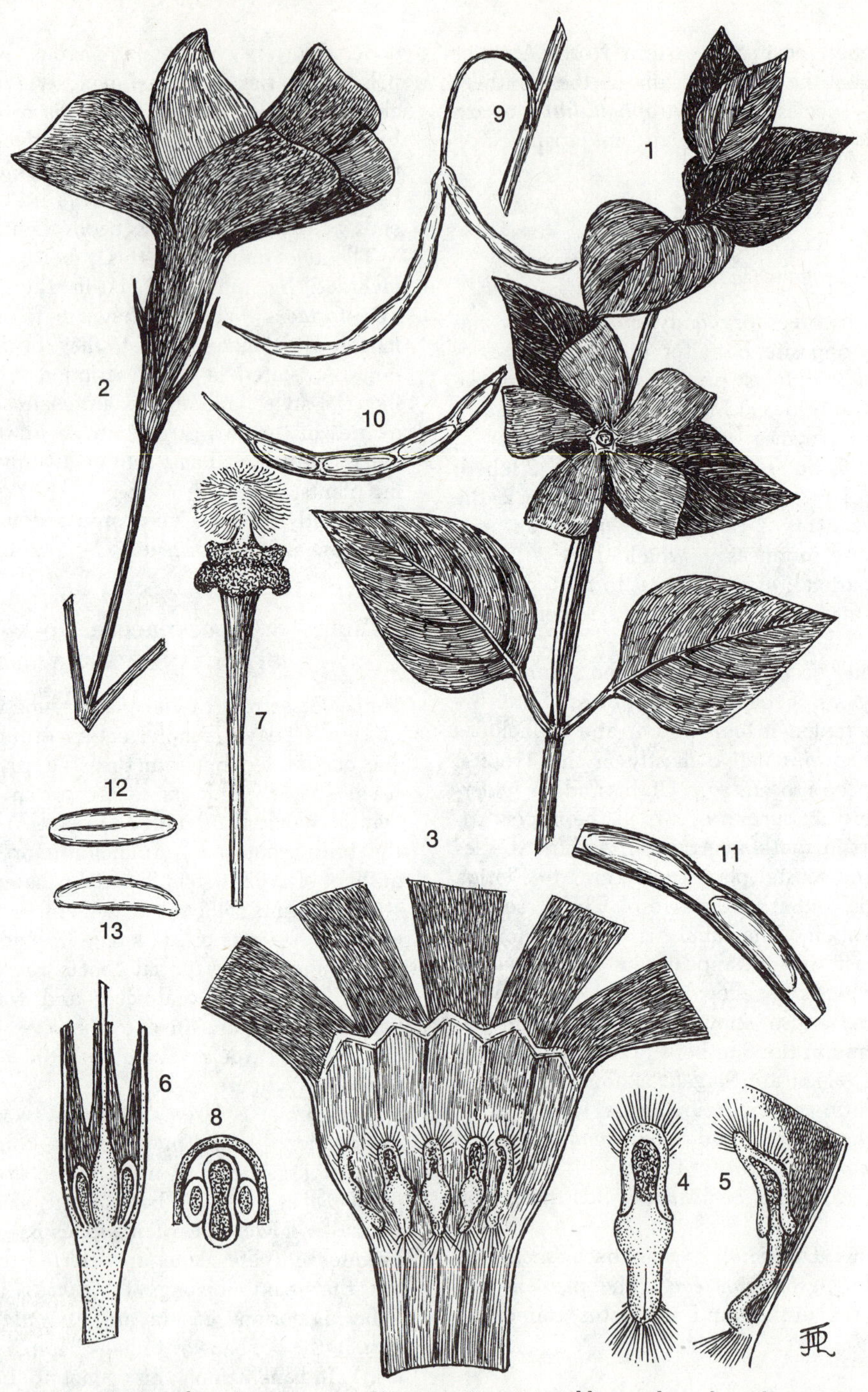

Figure 10-62. Gentianales; Apocynaceae; *Vinca major*, periwinkle: **1** branch with opposite leaves and a flower, **2** flower showing the coalescent sepals and coalescent petals, **3** corolla laid open, showing the adnate stamens, **4–5** stamens, **6** longitudinal section of the base of the flower, showing the two separate ovaries beside the coalescent styles of the carpels, **7** upper portion of the style and the stigma, **8** portion of a cross section of flower at ovary level, **9** the two ovaries developing as independent fruits (follicles), **10–11** seeds in the fruit, **12–13** seeds.

tropical plants. Four tropical genera (*Echites, Urechites, Vallesia,* and *Angadenia*) are represented each by 1 or 2 species in Florida. *Cycladenia* includes only one species native in the mountains of California.

*Family 4. **Asclepiadaceae.**
MILKWEED FAMILY

Perennial herbs (sometimes herbaceous vines) or rarely shrubs (or extraterritorial plants small trees or woody vines). Leaves simple, usually entire, opposite or whorled or rarely alternate; stipules minute. Corolla twisted or valvate in the bud, the corolla tube with a terminal internal appendage, this commonly called a corona, but consisting of (1) an outgrowth of the corolla tube or (2) of the filaments or (3) of the anthers (sometimes 2 or all 3 types present); stamens nearly always adnate with the stigma, the pollen of each anther sac a waxy mass connecting with a similar mass in the adjacent sac of the neighboring anther, the double mass of pollen termed a pollinium; carpels 2, separate except for the common stigma, the stigma 5-angled, with an anther adnate on each side, the stigmatic surface of the lower side of the stigma exposed between each pair of anthers. Fruits 2 from each flower (or 1 abortive), each fruit a follicle; each seed almost always with a tuft of long silky hairs arising near the micropyle. (Cf. Figure 12-12, *below.*) (Cf. pollinating mechanism, below.)

The family occurs throughout the Tropics, but its greatest development is in South America. The outstanding North American genus is *Asclepias* (milkweed), which includes many native species. The other North American genera include *Cynanchum, Matelea, Gonolobus,* and *Sarcostemma,* but many other names appear in the literature because the family has received a thorough study only recently. There are many other genera, including the silk-vine (*Periploca*), a twining vine.

In the milkweed family the stamens are attached to the stigma so that the spaces between the anthers form grooves down the sides of the compound structure composed of stigma and anthers. The apex of the stigma is polished and slippery, and when an insect alights its leg is likely to slide down into the groove between the two portions of adjacent anthers and to be brought back up with waxy masses of pollen hanging from it. These pollen masses or **pollinia** are rare among flowering plants. They occur in the orchid family in the Monocotyledoneae.

In the milkweed family a crown or corona develops from the upper surface of the petal. In some genera this resembles somewhat the corona in the flowers of the monocotyledon genus *Narcissus* which includes the jonquils, China-lilies, and daffodils. However, in the milkweeds the crowns from the individual petals are hoodlike and separate, and in some species each may bear an inward-protruding horn. The nature of the corona is discussed in the family description.

The latex of various genera contains rubber and in some the percentage of rubber is relatively high, this being true for not only the *Asclepiadaceae* but also the *Apocynaceae.* According to bromination tests, the pure latex of some species of *Sarcostemma* (the twining vine species, known also as *Funastrum*) has a rubber content approaching that of rubber cement. However, the latex is too small a percentage of the plant as a whole to make recovery practical. A number of species of both families have been under investigation as possible commercial sources of rubber. The silk-vine, desert species of milkweed, and *Amsonia* (*Apocynaceae*), mentioned above, are among them.

Suborder C. **Oleineae**

*Family 5. **Oleaceae.** OLIVE FAMILY

Trees or shrubs. Leaves simple or pinnate, opposite; stipules absent or sometimes present and small. Flowers bisexual or unisexual; sepals 4 (or 5–6) or sometimes none; petals 4 or sometimes none or sometimes 5–6 or even more; carpels 2, the style 1 or obsolete; ovary with 2 chambers; ovules 1–2 per carpel, anatropous, pendulous, with 1–2 integuments. Fruit a berry, drupe, samara, capsule or pyxis.

Here this family is included in the Gentianales (*Contortae*), but Wettstein has pointed out a number of differences or tendencies separating the Oleales from the Gentianales. These include the position of the ovules, absence of internal phloem, the two stamens, and the frequent presence of compound leaves (*Oleaceae*). He suggests that the group may have originated from the Sapindales (perhaps from the *Staphyleaceae*), but the origin of the order is obscure. (Note the Salvadoraceae, page 210.)

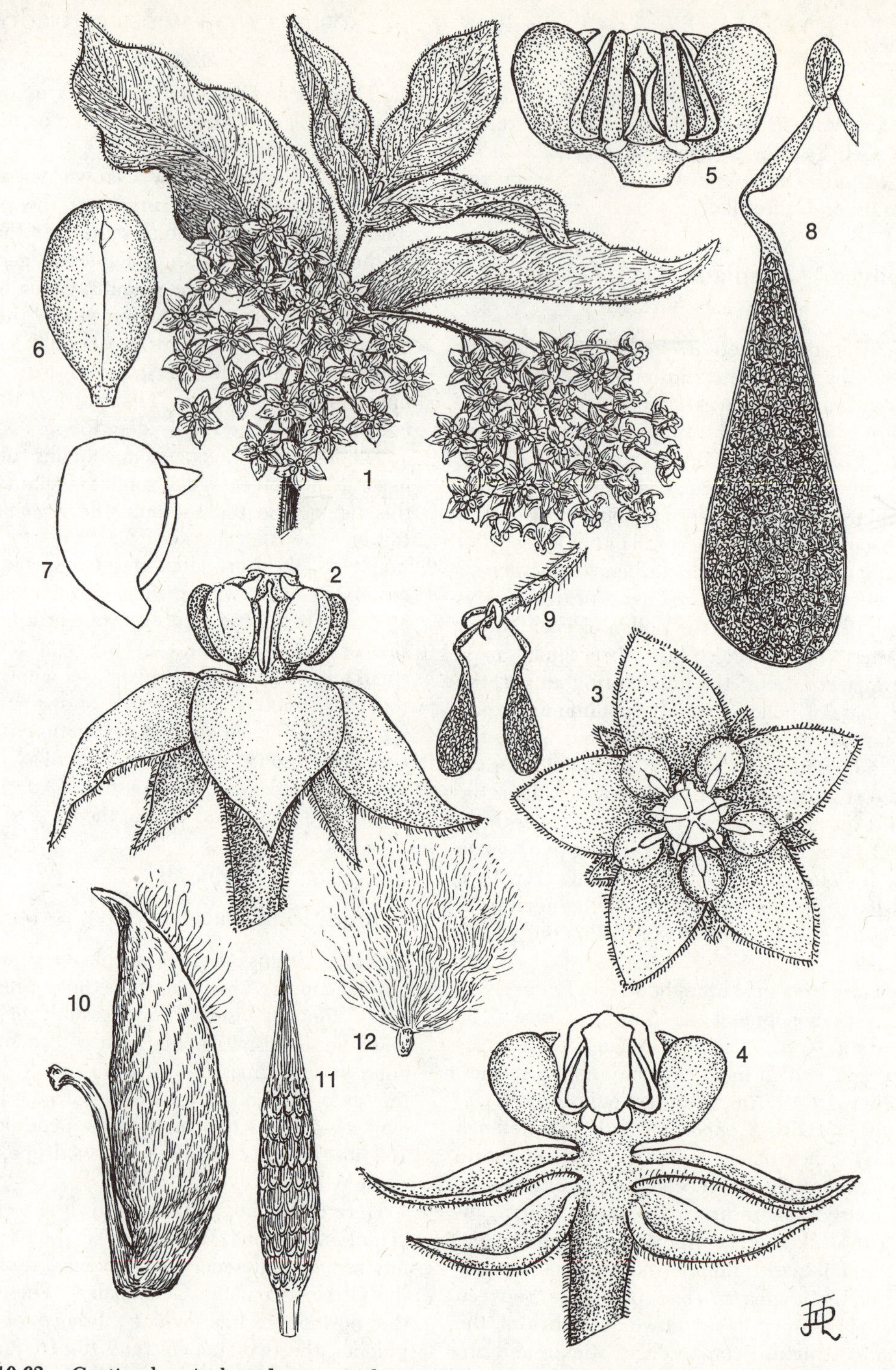

Figure 10-63. Gentianales; Asclepiadaceae; *Asclepias eriocarpa*, a milkweed: **1** branch with opposite leaves and with flowers, **2** flower, side view, showing the petals (coalescent), the hoods arising from them, the horns of the hoods (projecting inward), and portions of the stamens and stigma, **3** flower, top view, showing the tips of the sepals, the corolla, the hoods and their horns, and the five stamens adnate with the stigma, the stigma with five apical grooves continuing into deeper grooves between the anthers, **4** flower, longitudinal section. **5** two anthers flanked by hoods, between them the groove across which the waxy pol-

Figure 10-64. Gentianales; Oleaceae; *Jasminum officinale* (*grandiflorum*), jasmine: **1** branch with opposite pinnatifid leaves, flowers, and young fruits, **2** longitudinal section of the flower, **3** young fruits, **4** young fruit, longitudinal section, showing the merging of the persistent calyx with the pedicel, **5** cross section of the ovary.

len masses (pollinia) of the adjacent half-anthers are joined, **6** hood, ventral view, the horn protruding, **7** outline of another view, **8** pollinium, **9** pair of pollina attached to an insect's leg, which slipped into the groove between anthers, **10** follicle, **11** mass of seeds within the follicle (smaller scale), **12** seed after drying of the attached hairs.

Figure 10-65. Gentianales; Oleaceae; *Fraxinus velutina,* ash: **1** branch with staminate flowers, the flowers appearing before the leaves, **2** staminate inflorescence, **3** pistillate inflorescence, **4** pistillate flower, consisting of calyx and pistil (with two stigmas), **5** branch with leaves and fruits (samaras), **6–7** samaras. (Cf. also **Figure 8-5 B** ash, fruit.)

The family occurs chiefly in temperate and warm regions and the Tropics of the Eastern Hemisphere. Among the common plants of the family are *Fraxinus* (ash), various species of which occur across the Continent; *Chionanthus* (fringe-tree); *Osmanthus* (devilwood); *Ligustrum* (privet), which is naturalized in the Southern States; *Forestiera* (swamp privet, which occurs in the East and in the Southern States, and desert olive, which grows chiefly in the Southwestern Deserts); and *Menodora*, of the Southwest.

The olive is the outstanding plant of the family. The wood of ash trees is used for cabinet work and baseball bats. Lilacs (*Syringa*), privets (*Ligustrum*), *Osmanthus*, ashes (*Fraxinus*), and other groups are cultivated widely.

Order 33. *FOUQUIERIALES.*

OCOTILLO ORDER

DEPARTURE FROM THE FLORAL CHARACTERS OF THE RANALES

Free	Petals coalescent; stamens adnate with the base of the corolla tube; carpels coalescent.
Numerous	Stamens 10–19; carpels 3–4.
Spiral	Carpels cyclic.

Shrubs or fantastic trees. Leaves simple, entire, alternate, the original one at each node persistent as a spine formed from the petiole and the midrib. Flowers in terminal panicles; sepals 5; petals 5, the corolla tubular; stamens 10–19; carpels 3 or 4, the styles 3 or 4, the ovary 1-chambered; placentae parietal, intruded; ovules 4–6 per carpel. Fruit a capsule; each seed with wings or hairs along the margins. See Figure 2–15 **C.**

The genus *Cantua* is a possible relative, and perhaps the family should be classified in the Polemoniales near the *Polemoniaceae*.

*Family 1. Fouquieriaceae.

OCOTILLO FAMILY

The family includes two genera with about nine species; California to Texas and Mexico. The only species occurring in the United States is *Fouquieria splendens*, the ocotillo or coach-whip of the Southwestern Deserts from California to western Texas. The plant has a trunk only about one foot long and large wandlike branches as much as 20 feet long and 1 to 2½ inches in diameter. At every node there is a spine (formed from the petiole and midrib of a leaf), and in the rainy season leaf clusters appear in the axiles of the spines. After the cloudburst which initiated the summer rainy season at Tucson, Arizona, in 1938 the ocotillo leaves had grown to full size, that is up to two inches long, within a week. In the spring or rarely in the late summer large tassels of red flowers appear at the ends of the long branches.

The boogum-tree or cirio of Baja California and Sonora, Mexico, is cultivated sometimes as a curiosity. This weird tree is like a giant carrot root turned upside down, and only small twigs appear at the top of the trunk. These bear white flowers.

Order 34. *POLEMONIALES.*

PHLOX ORDER

DEPARTURE FROM THE FLORAL CHARACTERS OF THE RANALES

Free	Sepals coalescent; petals coalescent; stamens adnate with the petals; carpels coalescent, except in some *Convolvulaceae*.
Equal	Corolla rarely bilaterally symmetrical, as in *Petunia* or *Langloisia*.
Numerous	Stamens 5, alternate with the petals; carpels usually 2, but 3 in the *Polemoniaceae*, 2–3 or rarely 4 in the *Convolvulaceae*, and 3–7 in the *Lennoaceae*.
Spiral	Stamens and carpels cyclic.

Herbs, or rarely shrubs or small trees. Leaves alternate or sometimes opposite (some *Hydrophyllaceae*) or in a few instances in the *Polemoniaceae;* stipules none. Flowers 5-merous except for commonly 2 or rarely 3 or more carpels; ovaries 1–several-chambered, most commonly 2-chambered but in the *Boraginaceae* commonly 4-chambered and each chamber forming a nutlet in fruit, the placentae axile except when the pistil is 1-chambered (some *Hydrophyllaceae*); ovules anatropous, with 1 integument. Fruit a capsule or sometimes a berry or 4 nutlets formed from 2 carpels.

The order Polemoniales includes numerous species, but its members are relatively homogeneous in flower structure. They are distinguished from most other Corolliflorae by either the radially symmetrical corolla or presence of five stamens alternate with the petals. The corolla is radially symmetrical in all Corolliflorae except the Scrophulariales and Lamiales. The order, combined with the Lamiales and Scrophulariales, has been known as the Tubiflorae or Bicarpellatae; cf. discussion in Chapter 14.

Key to the Families

(North America north of Mexico)

1. Ovary 1–4-chambered, the carpels 2–4; plants not parasitic on the *roots* of other plants; coalescent petals usually 5.
   2. Fruit a capsule or a berry with 1, 2, or 3 chambers (rarely the 2 almost separate carpels with separate styles and each giving rise to a utricular 1- or 2-seeded fruit).
      3. Carpels 3, coalescent except for the stigmas; plants not twining.

         Family 1. **Polemoniaceae** (*Phlox* family).

      3. Carpels 2 or 4–5 (if rarely 3, *then either* the 3 styles wholly separate *or* the plant twining).
         4. Sepals coalescent into a cup or a tube, the sepal tips appearing as teeth or lobes; style 1.

            Family 3. **Solanaceae** (nightshade family).

         4. Sepals separate or coalescent only below, scarcely forming a cup or a tube, the separate portions of the sepals usually at least as long as the coalescent portions; style(s) more than 1 or else 1 but cleft or branched.
            5. Twining or trailing vinelike or rarely (in North America north of Mexico) erect plants; corolla plaited in the bud; flowers *never* in coiled, 1-sided helicoid cymes ("scorpioid spikes or racemes").

               Family 4. **Convolvulaceae** (morning-glory family).

            5. Erect or more or less spreading plants, but not twining and only rarely trailing irregularly; corolla not plaited in the bud; flowers usually in coiled, 1-sided helicoid cymes ("scorpioid spikes or racemes").

               Family 6. **Hydrophyllaceae** (waterleaf family).

   2. Fruit *either* (1) of 2 or 4 nutlets (these sometimes remaining attached together) *or* (2) (in plants of southern Texas and Florida) the fruit drupelike; style 1; flowers usually in coiled, 1-sided (helicoid) cymes.

      Family 7. **Boraginaceae** (borage family).

1. Ovary 6–17-chambered, the carpels 3–7; plants fleshy parasitic herbs attached to the *roots* of other plants; coalescent petals 5–7; California, Arizona, and Mexico.

   Family 5. **Lennoaceae** (*Lennoa* family).

Suborder A. **Polemoniineae**

*Family 1. **Polemoniaceae.**

PHLOX FAMILY

Herbs or shrubs or rarely small trees or vines. Leaves entire, lobed, dissected, or compound. Flower radially symmetrical or rarely bilaterally symmetrical, with a hypogynous disc; corolla twisted (contorted) in the bud, saucerlike to tubular; stamens separating from the corolla tube at various levels; carpels 3 (rarely 2 or 5), the style 1, sometimes 3-branched above, the ovary usually 3-chambered; placentae axile; ovules 1–many per carpel. Fruit a capsule or rarely indehiscent.

The family is restricted to the Western Hemisphere except for a few species occurring in northern Eurasia. Its center is in California and to some extent the adjacent states. The common genus of the East and Middle West is *Phlox*, which includes about 14 species in that region as well as many more in the West. A number of species of *Polemonium* (Jacob's-ladder) occur in the mountains of the West and across the northern part of eastern North America. The largest genus is *Gilia* which, interpreted broadly, includes many species in western North America but only one in the eastern half of the Continent. Several authors have segregated *Gilia* into a number of smaller genera, a few of which appear to be justified. The better-segregated genera include *Eriastrum* (*Hugelia*), *Navarretia, Linanthus, Leptodactylon,* and *Langloisia.* These groups are typically Californian but fairly common in Arizona. Several species of *Collomia* occur in the West and one as far eastward as Ontaria and Quebec as well as Nebraska.

*Family 2. **Solanaceae.**

NIGHTSHADE FAMILY

Herbs, shrubs, or sometimes trees or woody

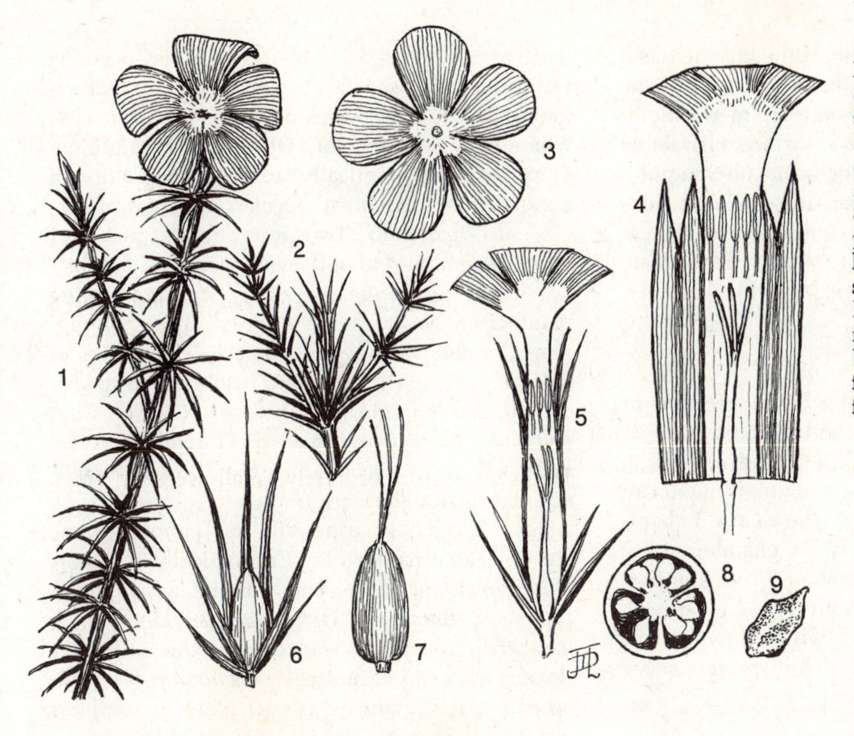

Figure 10-66. Polemoniales; Polemoniaceae; *Leptodactylon californicum* (*Gilia californica*), prickly phlox: **1** branch with leaves (in fascicles from the axillary buds) and a flower, **2** branch tips, **3** flower, top view of the sympetalous corolla, **4** corolla laid open, showing the adnate stamens and the pistil, **5** longitudinal section of the flower, **6** fruit surrounded by the persistent calyx, **7** fruit, **8** cross section of the fruit, **9** seed.

vines. Leaves simple or sometimes pinnate, alternate except sometimes on the upper part of the plant. Flowers radially or rarely bilaterally symmetrical, with hypogynous discs, sepals usually coalescent for almost their full length, the calyx persistent and often enlarging in fruit; corolla rotate to tubular; carpels 2, the style 1, the ovary usually 2-chambered or sometimes 3–5-chambered through irregularities of the placentae; placentae axile, ovules numerous in each carpel, antropous or tending to be amphitropous. Fruit a berry or sometimes a capsule. (Cf. Figures 6–2 **A**, tobacco, flower, and 6–10 *Datura*, flower).

This large family of about 2,000 species occurs chiefly in Latin America, but various genera are native to the United States. The largest genus, *Solanum*, includes nearly three-quarters of the species. The family is abundant along the southern edge of the United States from Arizona to Florida. In Arizona the genera *Datura* (Jimsonweed), *Nicotiana* (tobacco), *Petunia, Lycium* (wolfberry and desert-thorn), *Solanum* (nightshade), *Capsicum* (red pepper), *Saracha, Marganthus, Chamaesaracha*, and *Physalis* (groundcherry) are represented by varying numbers of species. Other species of the same genera are native in Florida, and a few tropical genera are represented there by introduced members. Some species of *Datura, Petunia, Nicotiana, Lycium, Salanum*, and *Physalis* occur westward into California. There are a few endemic species in the Middle West and East, including about five native species of *Solanum*, one of *Chamaesaracha*, and fifteen of *Physalis*. A few other genera are introduced there.

The nightshade family is one of the leading sources of foods, drugs, and ornamental plants. The underground stems of a series of species of *Solanum* form tubers or potatoes; one of these is the white potato taken from the Western Hemisphere for cultivation throughout the world. Tomatoes (*Lycopersicon*), groundcherries (*Physalis*), eggplant (*Solanum*), and red peppers (*Capsicum*) are other important food plants. Drugs include the nicotine of tobacco (from *Nicotiana*), belladonna and atropine (from *Atropa*), henbane (from *Hyoscyamus*), and stramonium (from *Datura*). Many of the genera are cultivated for their showy flowers. Among the best known are *Petunia, Solanum* (nightshade), *Streptosolen, Solandra* (cup-of-gold), *Cestrum* (nightblooming jessamine), and *Datura* (thornapple).

The Solanaceae are a connecting link between the Polemoniales and the Scrophulariales. The closest relationship of the family is to the Scrophulariaceae. Segregation by surface characters of some extraterritorial groups from the Scrophulariales is arbitrary. However, the vascular bundles of the Solanaceae are unusual in arrangement of xylem and phloem, strands of phloem being on both sides of the xylem (bicollateral).

Family 3. **Nolanaceae.** NOLANA FAMILY

Herbs or small shrubs. Leaves simple, more or less succulent, alternate or the upper ones sometimes opposite. Flower with a hypogynous disc, solitary, axillary; corolla campanulate, funnelform; carpels usually 5, the style 1, the ovary 5-chambered or appearing to be 10–15-chambered by extrusion of the axile placentae. Fruit splitting into segments, these often stony, with 1–7 seeds.

The family is composed of two or three genera native to Chile and Peru. The species occur mostly along the ocean shore.

Family 4. **Convolvulaceae.** MORNING-GLORY FAMILY

(Segregates: *Dichondraceae, Cuscutaceae*)

Herbs, shrubs, or small trees, usually twining herbs but sometimes erect or nearly prostrate; juice commonly milky. Leaves simple. Flowers usually with a hypogynous disc or sometimes epigynous, usually subtended by pairs of bracts, these sometimes forming an involucre; sepals imbricate, persistent; corolla commonly funnelform, or salverform, plaited or twisted in the bud; carpels usually 2 but sometimes 3–5, the styles sometimes compound, but usually separate, the ovary with as many chambers as carpels; placentae axile; ovules 1 or 2 per carpel. Fruit usually a capsule, but sometimes indehiscent or fleshy. (Cf. Figures 2-10 **C**, and 6-3 **C**, *Convolvulus.*)

The family includes numerous species occurring chiefly in the Tropics and Subtropics, but some genera extend into warm temperate regions. The morning-glories (*Ipomoea*) and most other members of the family are characteristic of the southern edge of the United States from Arizona to Florida. The morning-glories of California are of the genus *Convolvulus,* known elsewhere mostly as bindweeds. About 14 species occur in California, and several are restricted to that region. Only a few species of the genus occur elsewhere in North America. Otherwise the morning-glory family is practically lacking from California except for a widespread species of *Cressa,* a species of *Dichondra,* two introduced species of other genera, and about seven species of *Cuscuta* (dodder), an orange or yellow parasitic twining plant often covering large areas. The morning-glory family is more abundant in the East and Middle West where there are half a dozen species of true morning-glory (*Ipomoea*), four of *Convolvulus,* and about fifteen of *Cuscuta.* Several additional genera occur along the Mexican boundary and in the Southern States.

Several species, especially of morning-glory, are used as ornamentals. Economically the most important plant of the family is the sweet potato (*Ipomoea Batatas*). Often the dodders, which parasitize crop plants including alfalfa and other clovers, are a significant pest. *Dichondra repens* is becoming important as a lawn plant in southern regions. It makes a dense, low, ground cover requiring no mowing and making an effective dark green lawn if competing plants can be kept out. Some species of *Convolvulus* are among the most difficult weeds to eradicate from fields because the rhizomes penetrate to a considerable depth in the soil and they start up readily after cutting. Several other genera are cultivated.

*Family 5. **Lennoaceae.** LENNOA FAMILY

Parasitic herbs; plants succulent and lacking chlorophyll, parasitic upon the roots of other plants. Leaves short and scalelike. Flowers sometimes bilaterally symmetrical, in a thyrse or head; sepals 5 or 10; petals usually 5 but sometimes 6–8, the corolla tubular or nearly so; stamens 5 (or in an extraterritorial genus 10); carpels 6–14, the style 1, the ovary with as many chambers as there are carpels, but often appearing to have twice as many; placentae axile; ovules 2 per carpel. Fruit capsular but somewhat fleshy, ultimately irregularly circumscissile.

The family consists of three genera, made up of four species restricted to California, Arizona, Mexico, and Colombia. The plants occur in sandy soil along the seacoast and in the deserts. *Pholisma* is confined to California and Baja California. The plants are about five to ten inches high and the branching stems may be up to an inch in diameter at the bases. The purple and white flowers occur in great numbers on the upper portions of the stem or its branches. Sandfood (*Ammobroma*) is an exceedingly fleshy plant usually buried in the sand, except for the saucer-shaped upper portion covered thickly with violet flowers. The large subterranean portion of the plant (4–16 inches long) is edible, and the Papago Indians of western Sonora roasted them, ground them into meal, or ate them raw.

Because of the similarity of the plant bodies to saprophytic members of the *Ericaceae,* the family has been classified in the order *Ericales,* but both the obvious features of flower structure and recent microscopic studies indicate affinity to the other families of the Polemoniales.

Suborder B. **Boragineae**

The two families of this suborder form a connecting link between the Polemoniales and the Lamiales. See discussion under the Lamiales.

*Family 6. **Hydrophyllaceae.**

WATERLEAF FAMILY

(Synonym: *Hydroleaceae*)

Herbs or shrubs (*Eriodictyon* and *Turricula*). Leaves simple, entire to pinnate or rarely palmately divided. Flowers usually in coiled, 1-sided, helicoid cymes ("scorpioid spikes or racemes"), hypogynous or in some species of *Nama* epigynous (the ovary being half inferior); corolla with the lobes overlapping or sometimes twisted (contorted) in the bud; stamens sometimes with scalelike or hairy appendages at the bases of the filaments or the appendages attached to the corolla; carpels 2, the style 1 but divided into 2 branches or the styles rarely 2, the ovary usually 1-chambered or divided into 2 by intrusion of the parietal placentae; ovules 2–many per carpel, anatropous or amphitropous. Fruit a capsule or sometimes indehiscent.

The family is a moderately large one, occurring throughout the world except in Australia. It is most abundant in western North America, and it is centered in California and to a lesser

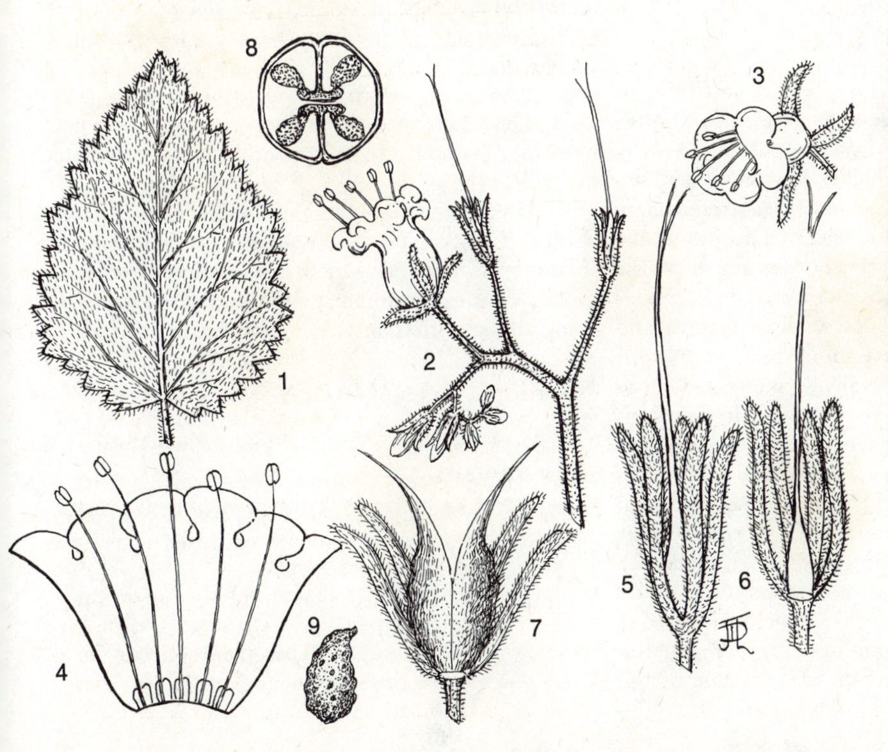

Figure 10-67. Polemoniales; Hydrophyllaceae; *Phacelia minor,* a phacelia: **1** leaf, **2** inflorescence, a helicoid cyme ("scorpioid raceme"), **3** flower, another view, **4** corolla laid open, showing the adnate stamens and the scales growing from the base of the corolla, adnate in some species with the filaments, **5–6** young fruit surrounded by the calyx, **7** fruit, **8** fruit, cross section, **9** seed.

extent in the adjacent states. A few species of waterleaf (*Hydrophyllum*) occur across mostly the northern portions of the continent and in the more moist areas southward. In California there are about sixty species of *Phacelia,* thirteen of *Nemophila,* ten of *Nama,* and smaller numbers of several other genera including the nearly endemic shrub genus *Eriodictyon* (yerba santa). About twenty-five or thirty species of *Phacelia* and half a dozen of *Nama* occur in Arizona. Three species of *Hydrophyllum,* one of *Nemophila,* and a few of *Phacelia* and *Nama* occur in the Southern States and some of these range northward into the southern portions of the Middle West and to the Middle Atlantic States.

The waterleaf family is of little economic importance. Species of *Hydrophyllum* were used by Indians and American pioneers as pot herbs. Some species of *Phacelia,* particularly the Southwestern desert species *Phacelia crenulata,* are poisonous to those who are allergic to them. Many of the species of *Phacelia* and *Nemophila* are colorful, but they are just coming into cultivation, except as their seeds have been included in packets of California wild flower seeds.

*Family 7. **Boraginaceae.**

BORAGE FAMILY

(Segregates: *Ehretaceae, Heliotropaceae*)

Herbs; extraterritorial genera sometimes shrubs, trees, or vines. Leaves simple, usually entire, alternate. Flowers in helicoid (sometimes described as "scorpioid") cymes (these referred to frequently as "scorpioid spikes or racemes"), the structure spikelike with the flowers on one (the "upper") side of the axis, which uncoils gradually at the apex as the flower buds mature and open, the inflorescences sometimes in clusters, the flowers rarely bilaterally symmetrical (*Lycopsis* and *Echium*), sometimes with hypogynous discs; corolla rotate to salverform or campanulate; carpels 2, the style 1, the ovary deeply divided into 4 externally visible segments which become 4 nutlets in fruit, with 4 internal chambers, each carpel being deeply divided; placentae fundamentally axile, but basal in position; ovule 1 per chamber of the ovary. Fruit usually of 4 nutlets, but sometimes only 1–3 developing, the nutlets varying in the texture and coloring of the surface, the keys to genera and species being based largely upon the characters of the nutlet.

The family occurs over most of the Earth. In North America it centers in California and to a lesser extent in the adjacent states. Some of the more widely known genera are *Heliotropium* (heliotrope), *Borago* (borage), *Symphytum* (comfrey), *Lycopsis* (bugloss), *Echium* (viper's-bugloss), *Lithospermum* (puccoon), *Cynoglossum* (hound's-tongue), *Myosotis* (forget-me-not), *Amsinckia* (fiddleneck), *Mertensia* (blue-bells or lungwort), *Lappula* (stickseed), and *Hackellia* (stickseed or beggar's-lice). These genera are represented in the Middle West and the East and more sparingly in Southern States, and several occur in the West. In California there are many local species of *Amsinckia, Cryptantha, Plagiobothrys* (popcorn-flower), and several smaller genera. Many of the desert species occur also in Arizona, southern Nevada, and southwestern Utah, and local species occur especially in Arizona.

The fruits are at the bottom of the inflorescence, often far below the flowers and buds. The upper end of the inflorescence curves like a fiddle neck, and the open flowers are at the top of the curve, the buds being at the extreme apex.

Specimens of the borage family are difficult to identify unless the fruit is mature, and often it is impossible to determine the name of the species if the nutlets are not present. In some instances plants without fruits are difficult to determine even to the genus.

A similar inflorescence is in the *Phacelia* family. The two families may be distinguished by the type of fruit, which is a capsule (with one or two seed chambers) in the Hydrophyllaceae and four nutlets in the Boraginaceae.

Order 35. ***LAMIALES.*** MINT ORDER

DEPARTURE FROM THE FLORAL CHARACTERS OF THE RANALES

Free	Sepals coalescent; petals coalescent; stamens adnate with the petals; carpels coalescent.
Equal	Corolla markedly bilaterally symmetrical; stamens 4 or 2, with 1 or 3 of the 5 places vacant, usually in 2 pairs of unequal length.

Figure 10-68. Polemoniales; Boraginaceae; *Amsinckia intermedia,* yellow fiddleneck: **1** inflorescence, a helicoid cyme ("scorpioid spike"), **2** flower, showing the sepals and the sympetalous corolla, **3** flower, top view, showing the corolla and the adnate stamens, **4** corolla laid open, showing the adnate stamens and the pistil, the ovary deeply divided, **5** stamen adnate with the corolla, **6** calyx enclosing the fruit, **7** fruit composed of four nutlets (two from each carpel), these separating at maturity.

Numerous Stamens 4 or 2 or rarely 1; carpels 2 (often appearing as 4 because of the common division of each carpel into 2 nutlets in fruit, but the stigmas usually 2).

Spiral Stamens and carpels cyclic.

Herbs or rarely shrubs or trees; young stems usually nearly square in cross section. Leaves opposite or whorled, usually simple, rarely palmate or pinnate; stipules none. Flowers complete; ovary consisting of 2 carpels; ovules anatropous, the micropyles directed downward. Fruit usually of 2–4 nutlets or segments, these hard and seedlike, each containing a single seed, the fruit rarely an achene; fruits of the *Verbenaceae* often fleshy (small drupes).

The Lamiales often are included in a larger more comprehensive order, the Tubiflorae of Engler and Prantl and of Rendle (cf. Chapter 14). Division is somewhat arbitrary, and this is a borderline case in classification. The more inclusive order is comprised of not only the Lamiales but the Scrophulariales and the Polemoniales. The fruit type of the mint order is duplicated in the *Boraginaceae* of the Polemoniales. However, the inflorescence type of the *Boraginaceae* is duplicated in combination with capsular fruit in the *Hydrophyllaceae* (Polemoniales), and these families are definitely related closely to each other. This indicates derivation of the Lamiales from some phase of the forerunners of the Polemoniales, probably from the same earlier phase as the *Boraginaceae* and the *Hydrophyllaceae*. Despite the closeness of relationship of the Lamiales to some elements of the Polemoniales, separation is relatively simple and clear on the basis of the bilaterally symmetrical corolla, the reduction of stamens to 4 or 2 instead of 5, and the opposite leaves. The order is distinguished from all but one species (forming a monotypic family) of the Scrophulariales by the formation of nutlets or small drupes (e.g., *Lantana*) as opposed to a capsular fruit.

Key to the Families

(North America north of Mexico)

1. Fruit *not* an achene, with more than 1 chamber and 1 seed; the persistent calyx *not* strongly reflexed at fruiting time, or rarely so.

  2. Styles coalescent into one or separate at the apices; flowers bisexual; fruits nearly always of 4 nutlets, sometimes small drupes.

    3. Ovary at flowering time not lobed, 2–4-chambered, in fruit splitting into 2–4 nutlets, or fruit a drupe; style 1, entire.

      Family 1. **Verbenaceae** (verbena family).

    3. Ovary at flowering time lobed, 2–4-chambered, in fruit splitting into 2–4 nutlets; style 2-cleft at the apex; stems usually square in cross section; herbage usually with a mint odor due to an essential (volatile) oil.

      Family 2. **Labiatae** (mint family).

  2. Styles separate; flowers unisexual, the staminate of 1 stamen, the pistillate of 1 pistil; fruits ultimately splitting into 2 segments, each with 1 seed; small, usually aquatic, annuals.

    Family 3. **Callitrichaceae** (water starwort family).

1. Fruit an achene (or interpreted as a nutlet), the ovary with only 1 chamber and 1 ovule; calyx strongly reflexed at fruiting time.

  Family 4. **Phrymaceae** (lopseed family).

Family 1. Verbenaceae. VERBENA FAMILY

(Segregate: *Avicenniaceae*)

Herbs or shrubs or in other areas trees. Leaves usually simple, but sometimes pinnate or palmate; stamens rarely 5 in extraterritorial genera, the fifth stamen represented by a staminodium; style 1, entire; fruit of 2 or 4 nutlets or a drupe with the hard segments internal to the fleshy parts (the most common type of fruit in the family outside the United States).

The family is largely tropical and subtropical, but a few species (and especially the large genus *Verbena*), occur in temperate regions of the Western Hemisphere. Species of *Verbena* are native throughout especially the southern portions of North America. *Lippia* is a matted plant sometimes used for lawns and often carpeting large areas of moist low land. The French mulberry (*Callicarpa*) occurs in rich woods and thickets in the Southern States. A single species of *Lantana* (*Lantana horrida*) occurs in the bank of a stream near Sells in south central Arizona. The same species is native and also introduced on the Coastal Plain from Texas to Georgia and Florida. Other native and introduced species occur in

Florida, and one occurs in southern Texas. Various other genera are native or introduced in Florida, among them the fiddlewood (*Citharexylum*), the golden-dewdrop (*Duranta*), the glory-bowers (*Clerodendron,* introduced from China), the Turk's-turban (*Siphonanthus,* naturalized from the East Indies), and the monk's-pepper (*Vitex,* introduced from the Eastern Hemisphere). Another plant, the black mangrove (*Avicennia*), occurs on the coasts of southern Florida. Sometimes it is classified in a separate family, *Avicenniaceae.*

*Family 2. **Labiatae.** MINT FAMILY
(Synonym: *Lamiaceae*)

Herbs or sometimes shrubs or rarely trees or woody vines; plants with aromatic (essential) oils. Leaves simple or sometimes pinnate. Flowers sometimes with hypogynous discs, in dense clusters in the leaf axils and commonly appearing to be in dense whorls; fifth stamen rarely present, vestigial; style branched apically; fruit composed of 4 nutlets, the calyx persisting and enclosing them. (Cf. pp. 116–117.)

The family includes about 2,500 species distributed throughout the world. It is particularly abundant in the Mediterranean region. It is one of the most common families in North America, and it includes large numbers of well-known plants, some of them used in medicines or for various other purposes. Among them are the pennyroyals, blue-curls, bugle-weed, germander, skull-cap, horehound, hyssop, catnip, dragonhead, self-heal, hedge-nettle, sage (the true sages, *Salvia,* as opposed to sagebrush), savory, thyme, stoneroot, henbit, peppermint, tule mint, and desert-lavender.

*Family 3. **Callitrichaceae.**
WATER-STARWORT FAMILY

Delicate annual herbs; aquatic or sometimes terrestrial. Leaves entire or in terminal rosettes, opposite; stipules none. Flowers unisexual, each staminate flower consisting of a single stamen and growing immediately beside a pistillate flower composed of a single pistil, the two appearing to be a single flower; style 2; ovary with 2 chambers and sometimes each of these divided and thus forming 4 chambers, placentae axile; ovules 1 per carpel or 1 in each of the 4 chambers. Fruits splitting into 2 or 4 segments each with 1 seed.

The small family consists of a solitary genus, *Callitriche,* which occurs almost throughout the world. Several species occur in various parts of North America, including areas from coast to coast.

*Family 4. **Phrymaceae.** LOPSEED FAMILY

Perennial herbs. Leaves simple, opposite. Flowers in slender spikes, the persistent calyx strongly reflexed at fruiting time; ovary 1-chambered, with 1 ovule. Fruit an achene or interpreted as a nutlet.

The lopseed (*Phryma Leptostachya*) is the only species in the family. It occurs in eastern North America and in northeastern Asia.

The relationships of this family are in doubt. It differs from the *Verbenaceae* chiefly in the presence of a single ovule in a 1-chambered ovary, and there is as much to be said for classification in either the Lamiales or the Scrophulariales.

Order 36. ***SCROPHULARIALES.***
SNAPDRAGON ORDER

DEPARTURE FROM THE FLORAL CHARACTERS OF THE RANALES

Free	Sepals coalescent; petals coalescent; stamens adnate with the petals; carpels coalescent.
Equal	Corolla nearly always bilaterally symmetrical, two petals forming an upper lip, the lower three forming a spreading lower lip; stamens bilaterally symmetrical by reduction in number from the original pattern of 5 (or in *Verbascum,* which has 5 stamens, the stamens unlike).
Numerous	Stamens 4 or 2 (or rarely 5); carpels 2.
Spiral	Stamens and carpels cyclic.

Herbs, shrubs, or trees. Leaves opposite or whorled or sometimes alternate. Flowers 5-merous (except the stamens and the carpels); ovary 1- or 2-chambered, the placentae parietal in 1-chambered ovaries and axile on the partition in 2-chambered ovaries, or in the *Lentibularia-*

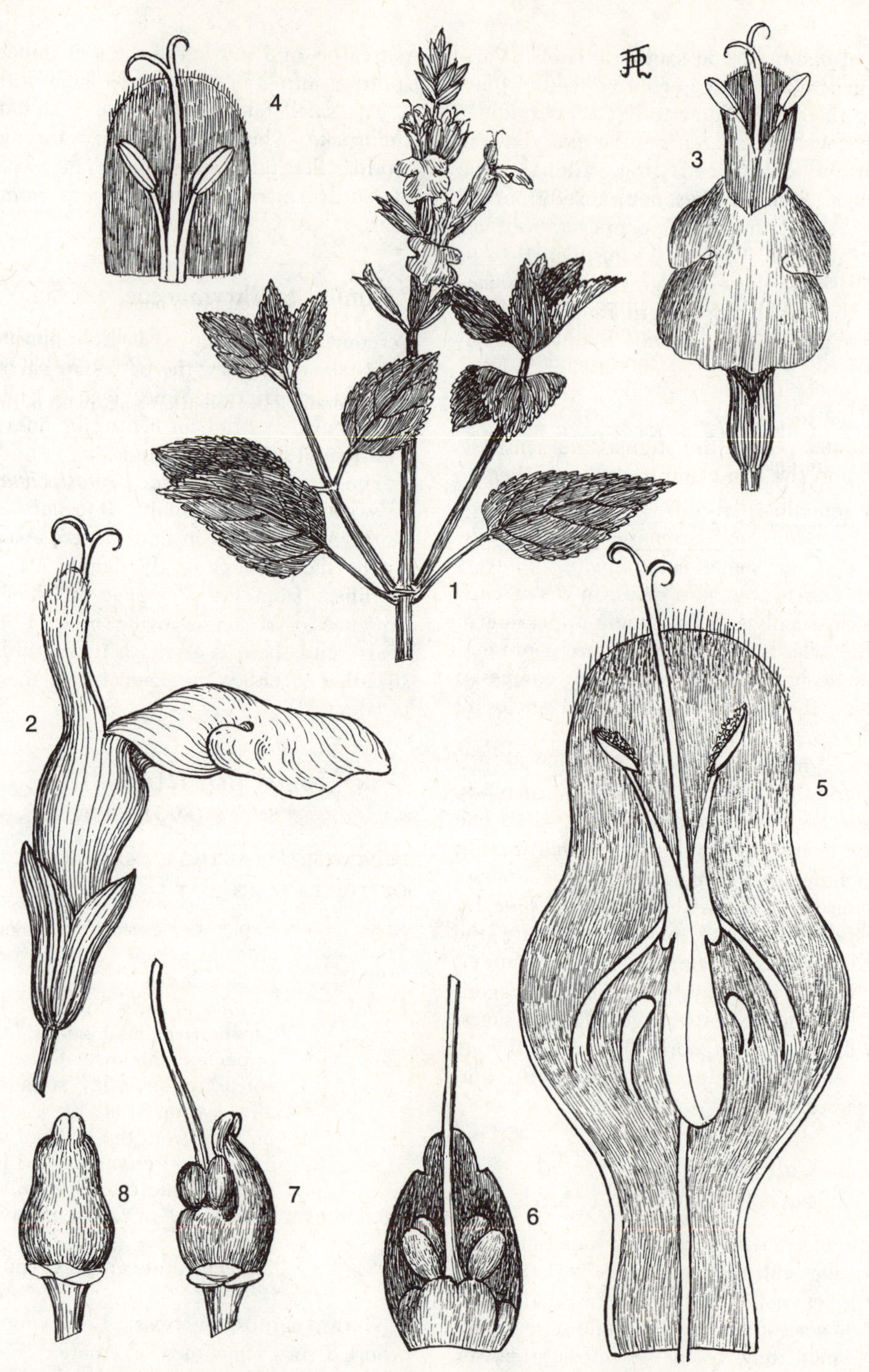

Figure 10-69. Lamiales; Labiatae; *Salvia microphylla*, a sage: **1** branch with opposite leaves and an inflorescence, **2** flower, side view, showing the sepals, the bilaterally symmetrical sympetalous corolla, and the stigmas, **3** flower, ventral view, with the stigmas and the functional portions of the stamens protruding from the upper lip of the corolla, **4** another view, **5** upper lip of the corolla, showing the style and stigma and the stamens, each stamen consisting of a filament (outside) and an elongated connective between the

ceae 1-chambered but with axile placentae; ovules anatropous. Fruit a capsule.

This is one of the largest and most important of plant orders. Many plants of several families are cultivated as ornamentals.

The flower structure of the Scrophulariales is illustrated by the monkey-flower, *Mimulus*. This flower is markedly bilaterally symmetrical as in nearly all members of the order, and it has the characteristic development of the upper two petals into an erect upper lip of the corolla which attracts insects and of the lower three petals into a lower lip functioning as a landing platform from which the insect crawls into the tubular portion of the corolla. Often the stigmas are sensitive to touch, and the two clamp tightly together when an insect brushes them as it crawls into the corolla tube. Pollen brought from other flowers may be deposited upon the stigma. By the time the insect has entered the corolla tube and become dusted with additional pollen the stigmas are closed and therefore not receptive to pollen from the same flower. A fair degree of precision in pollination is attained, and the four stamens characteristic of the order are sufficient to produce an adequate supply of pollen. In a few genera there are only two stamens.

Bilateral flowers and few stamens mark the order. Distinction from the Lamiales, or mint order, is by the fruit which is practically always a capsule with one or two seed chambers. The fruit of the Lamiales is composed of either two or four nutlets or is a drupe, as described under that order. In *Avicennia* (black mangrove), however, it is a capsule. The opposite or whorled leaves distinguish nearly all Scrophulariales and Lamiales from all but a few Polemoniales (some *Hydrophyllaceae*).

Key to the Families

(North America north of Mexico)

1. Plants green (with chlorophyll), not wholly parasitic upon other plants, usually not at all so.

2. Ovary 2-chambered.

3. Corolla lobes not rolled up lengthwise in the bud; capsule not springing open elastically, not stalked.

4. Seeds not wingéd; mostly herbs but sometimes shrubs or vines; endosperm present.

Family 1. **Scrophulariaceae** (snapdragon family).

4. Seeds wingéd; shrubs, trees, or woody vines; endosperm none.

Family 2. **Bignoniaceae** (trumpet-vine family).

3. Corolla lobes rolled up lengthwise in the bud; capsule springing open elastically, often on a special stalk of ovary tissue.

Family 9. **Acanthaceae** (*Acanthus* family).

2. Ovary 1-chambered.

3'. Fertile stamens 2 or 4; corolla *not* with a spur or sac, sometimes bulging on one side (gibbous); terrestrial plants; leaves simple.

4'. Capsule not with an elongate beak; stamens 4, accompanied by 1 staminodium; pollen sacs of the anther parallel.

Family 3. **Pedaliaceae** (benne family).

4'. Capsule with a conspicuous elongate beak, at maturity this separating into 2 divergent, long, curved horns or claws; stamens 4 or 2 of them staminodia; pollen sacs of the anther divergent.

Family 4. **Martyniaceae** (devil's-claw family).

3'. Fertile stamens 2; corolla tube with a spur or sac extending from one side; capsule not spiny; plants aquatic or of wet places. Leaves sometimes highly specialized, with the divisions producing bladders serving as traps capturing water insects.

Family 7. **Lentibulariaceae** (bladderwort family).

1. Plants wholly parasitic, not green (without chlorophyll), attaching themselves underground to the roots of other plants; ovary 1-chambered.

Family 5. **Orobanchaceae** (broom-rape family).

---

functional half anther and the position of the other half anther (which is absent) (cf. pages 116–117), this being a special feature of the genus *Salvia*, **6** gynobase (receptacular structure), ovary (four sections, two from each carpel), and base of the style, **7–8** other views (the bilaterally symmetrical gynobase overarching the nutlets as shown in **7**). (Cf. also **Figures 8-4 H** bee-sage, fruit, and **9-13 A** black sage, flower, **B** *Salvia microphylla*, flower in longitudinal section.

*Family 1. **Scrophulariaceae.**

SNAPDRAGON FAMILY

(Synonym: *Rhinanthaceae*)

Usually herbs but sometimes shrubs or vines. Leaves simple to pinnately parted. Flowers usually with hypogynous discs, rarely radially symmetrical (*Verbascum*), stamens rarely 5 and differentiated into 2 groups (*Verbascum*); rudiment of a fifth stamen often present; ovary 2-chambered; placentae axile; ovules numerous. Fruit a capsule.

The family includes about 2,500 species, and it occurs throughout the world. It is common through all of North America. Some of the members of the family are as follows: mullein (*Verbascum*), toad-flax (*Linaria*), snapdragon (*Antirrhinum*), figwort (*Scrophularia*), Chinese-houses (*Collinsia*), turtle-head (*Chelone*), beard-tongue (*Penstemon*), monkey-flower (*Mimulus*), hedge-hyssop (*Gratiola*), water-hyssop (*Bacopa*), foxglove (*Digitalis*), speedwell (*Veronica*), *Gerardia* (a genus occurring in eastern but not western North America), Indian-paintbrush (*Castilleia*, a genus abundant in the West and especially California), owl's-clover (*Orthocarpus*), cow-wheat (*Melampyrum*), eye-bright (*Euphrasia*), yellow-rattle (*Rhinanthus*), and lousewort and elephant-snout (*Pedicularis*). A beautiful localized genus occurring in the deserts of California and Arizona is *Mohavea*. Other genera occur north of Mexico only in Arizona or the Southern States.

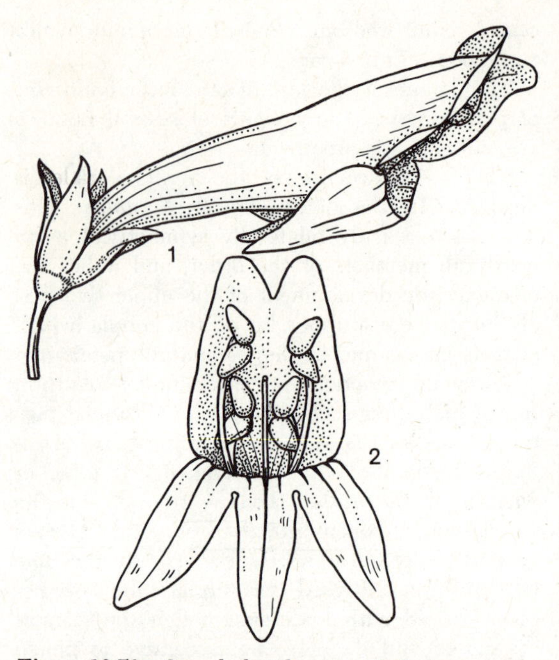

Figure 10-70. Scrophulariales; Scrophulariaceae; *Penstemon*, a penstemon; flower, **1** side view, **2** top view, showing the upper and lower lips of the corolla, the fertile stamens, and the style and stigma. By Lucretia Breazeale Hamilton.

*Family 2. **Bignoniaceae.**

TRUMPET-VINE FAMILY

Shrubs, woody vines, or rarely herbs. Leaf simple or pinnate, the terminal leaflet sometimes forming a tendril; leaves rarely alternate. Flower with a hypogynous disc; fifth stamen sometimes present, vestigial; ovary usually 2-chambered; placentae usually parietal; ovules numerous. Fruit a capsule or sometimes fleshy and indehiscent; seeds wingéd, without endosperm (this present in the related families).

The trumpet-vine family is a large one, though smaller than the snapdragon family. It is composed chiefly of tropical woody vines, and it is especially abundant in South America. The best known plants include the catalpas (*Catalpa*) and the vines of the genus *Bignonia* (in horticultural works often divided into several proposed genera). The conspicuous flowers of *Tecoma* are a familiar sight from Arizona to the Southern States. The calabash-tree (*Crescentia*) occurs in southeastern North America. The desert-willow (*Chilopsis*) occurs in the Southwestern Deserts from California to New Mexico. The family is absent from the Pacific Coast and the Rocky Mountains.

Family 3. **Pedaliaceae.** BENNE FAMILY

Herbs or rarely shrubs. Leaves simple, entire or indented, the uppermost sometimes alternate; fifth stamen a small staminodium; pollen chambers parallel; ovary 2-chambered or sometimes 4-chambered by extrusion of the axile placentae; ovules 1–numerous on each placenta. Fruit a capsule or a nut, often spiny or wingéd.

The family occurs along seacoasts and in the deserts of the warm parts of the Eastern Hemisphere. Two species of this family are introduced in the Southern States. Benne (*Sesamum indicum*, native in the East Indies) occurs in waste

Figure 10-71. Scrophulariales; Scrophulariaceae; *Penstemon spectabilis,* a penstemon: **1** inflorescence, **2** pair of opposite leaves, **3** flower, top view, showing the four fertile stamens and the style and stigma, **4** corolla laid open, showing the adnate fertile stamens and sterile stamen (staminodium), this in some species of *Penstemon* elaborate and with hairs on one side like a toothbrush, in other Scrophulariaceae rudimentary, glandular, or wanting, **5** sterile stamen, **6** young fruit enclosed in the calyx, **7** longitudinal section, **8** cross section, showing the axile placentae on the partition. (Cf. also **Figures 6-17 A** snapdragon, flower, and **9-12,** bush monkey flower, pollinating mechanism.)

places and cultivated ground on the Coastal Plain from Texas to Florida. It is the source of sesame oil and seeds. *Ceratotheca triloba* (native in Africa) has escaped on the higher pine lands and along roadsides in peninsular Florida.

*Family 4. **Martyniaceae.**

DEVIL'S-CLAW FAMILY

Spreading herbs; surfaces with viscid hairs. Leaves simple, sometimes lobed, the upper ones alternate. Flower with a hypogynous disc; pollen chambers divergent; ovary 1-chambered; placentae parietal, sometimes coalescent and forming a false partition in the ovary; ovules several to many. Fruit a capsule, the halves of the apical portion widely divergent after dehiscence, the exocarp deciduous, leaving a woody endocarp and a crest along the line of coalescence of the carpels.

The family includes about 5 genera restricted to the Tropics of the Western Hemisphere and 4 species of *Proboscidea* native from California to Louisiana. These unicorn-plants, known also as *Martynia,* have spread northward sparingly into especially the southern portions of the Middle West.

The family is remarkable for the elongated beak of the fruit. At dehiscence the beak is divided into two parts or sometimes into three, the chief portions diverging in more or less the manner of the horns of Longhorn cattle. The desert Indians split long fibrous segments from the edges of the pods to be used as the black portions of the designs in their baskets. The plants are spreading annual herbs resembling small squash or pumpkin vines.

*Family 5. **Orobanchaceae.**

BROOM-RAPE FAMILY

Succulent root-parasitic herbs with no evidence of chlorophyll. Leaves scalelike, simple, alternate. Flowers each in the axil of a scale-like bract, along a condensed spikelike axis; ovary 1-chambered; placentae parietal; ovules numerous. Fruit a capsule, often enclosed in a persistent calyx, somewhat coriaceous.

The family is common through the North Temperate regions and the warm temperate areas of the Eastern Hemisphere. The largest genus and the most widespread in North America is *Orobanche*, the broom-rape. Some species are known also as cancerroot. The same name is applied to two species of *Conopholis* (known also as squawroot), one occurring in eastern North America, the other in Arizona, New Mexico, and Mexico. The eastern North American beech-drops (*Epifagus*) is known also as cancerroot. A single species of the genus *Boschniakia* occurs on the Pacific Coast.

Family 6. **Gesneriaceae**

Herbs, shrubs, or sometimes trees or woody vines. Leaves simple, often in basal rosettes. Flowers with hypogynous discs, sometimes epigynous (the ovary then being wholly or partly inferior), solitary or in cymes; ovary 1-chambered; placentae parietal; ovules numerous. Fruit a capsule or rarely berrylike.

The family is large and is composed of about 1,000 species, most of which occur in the Tropics and Subtropics, but 2 genera occur in temperate Europe. Several members of the family occur in Mexico. The African-violet (*Saintpaulia*) and the gloxinias (*Sinningia*) are popular as house plants. The Cape-primrose and plants of various other genera are in cultivation as ornamentals.

*Family 7. **Lentibulariaceae.**

BLADDERWORT FAMILY

(Synonym: *Pinguiculaceae*)

Herbs of moist places or water; often insectivorous. Leaves alternate and in basal rosettes, often highly specialized and the leaflets forming insect traps. Lower portion of the corolla with a sac or spur; stamens 2, accompanied by 2 staminodia; anther with 1 pollen chamber; ovary 1-chambered; placenta a globular mass of ovules, these sometimes "sunken" into it. Fruit a capsule or a pyxis or opening irregularly.

The family is world-wide. The genus with the most numerous species is *Utricularia,* which grows in lakes and ponds across the Continent but most abundantly in the glaciated regions of the North and in the Southern States. The leaflets serve as insect traps. These are bladderlike, and each one opens by a trap door through which a water insect or other aquatic animal may enter. Some tropical terrestrial genera also have some leaves specialized as insect traps. The genus *Pinguicula* is composed of small, attractive herbs with markedly bilaterally symmetrical flowers. The spur developed from the base of the corolla is conspicuous.

Family 8. **Globulariaceae.**

GLOBULARIA FAMILY

Herbs or shrubs. Leaves simple, alternate. Flowers very small, often with hypogynous discs, lower limb of the corolla often very small, 2-lobed, the petals therefore appearing to be 4; the 2 anther cells becoming joined into 1 before flowering, each opening by a lengthwise slit; ovary 1-chambered; ovule 1, pendulous on an apical placenta. Fruit a nutlet enclosed within the persistent calyx.

This small family occurs chiefly in the Mediterranean region. *Globularia* is cultivated widely.

*Family 9. **Acanthaceae.**

ACANTHUS FAMILY

Herbs, shrubs, or rarely trees or woody vines. Leaves simple. Flower with a hypogynous disc, subtended by bracts and bracteoles; petals rolled up lengthwise in the bud; fifth stamen sometimes present; ovary 2-chambered; placentae axile; ovules 2 to several per carpel. Fruit a capsule or in a few genera a drupe, commonly elastically dehiscent, that is, the carpels springing apart at maturity and leaving a persistent central column.

The family occurs throughout the Tropics, and it includes nearly 2,000 species. In the Western Hemisphere it is particularly abundant in Central America and Mexico, but it occurs only sparingly in the United States. It does not occur on the Pacific Slope, and only a nearly endemic species of *Justicia* grows in the deserts of southeastern California (and eastward into Arizona). The genus best represented in the United States is *Ruellia,* which occurs from Arizona to the eastern half of the Continent. Several genera are represented each by one or two species occurring along the international boundary between the United States and Mexico. Among these are *Elytraria, Carlowrightia, Tetramerium, Dicliptera, Siphonoglossa, Anisacanthus,* and *Justicia.* Other genera as well as some of those mentioned above occur in the Southern States and particularly in Florida. *Justicia, Dicliptera,* and *Ruellia* occur in the Middle West and East.

Family 10. **Myoporaceae.**

MYOPORUM FAMILY

Shrubs or sometimes trees. Leaves simple, entire, alternate or rarely opposite. Fifth stamen sometimes present as a staminodium or rarely a fertile stamen; ovary 2-chambered or sometimes 3–10-chambered by development of false partitions from the axile placentae; ovules 2–8 per carpel. Fruit a berry or a drupe.

The family is composed of 5 genera occurring in the Eastern Hemisphere.

Order 37. ***PLANTAGINALES.***

PLANTAIN ORDER

DEPARTURE FROM THE FLORAL CHARACTERS OF THE RANALES

Free	Sepals coalescent; petals coalescent; stamens adnate with the petals; carpels coalescent.
Numerous	Stamens 4 or rarely 2; carpels 2.
Spiral	Stamens and carpels cyclic.

Herbs with very short caudices and with scapes. Leaves basal. Flowers complete, 4-merous; corolla salverform or rotate, translucent like parchment (dry-scarious), after withering persistent on the apex of the fruit; stamens exserted on elongated filaments far above the top of the corolla; carpels 2; ovary with 2 chambers, with 2 to several seeds; ovule anatropous, produced on a central placenta. Fruit a pyxis, opening along a horizontal ring (circumscissile), the top falling away like a lid and the partition falling away too (in one genus the fruit an achene with 1 chamber and 1 seed); embryo straight; endosperm fleshy.

The relationships of the order are obscure. By some it has been classified near the Primulales. Others have suggested a similarity of the flower to that of *Veronica* in the *Scrophulariaceae* or a relationship to the Polemoniales. Any of these interpretations may be correct.

*Family 1. **Plantaginaceae.**

PLANTAIN FAMILY

The only important genus is *Plantago* (plantain). A number of species grows in California and the Southwestern Deserts, and a greater number occurs in the eastern portion of the Continent. One of the more important species is *Plantago Psyllium,* the source of psyllium seed, important for the swelling of its mucilaginous material when wet. After desert rainstorms, the numerous seeds of native species may swell up into colloidal masses spreading over considerable areas. A single species of the genus *Littorella* occurs in the Middle West and the East. This is an aquatic plant.

Group 3. *CALYCIFLORAE.* CUP FLOWERS

(*Perigynous or Epigynous; Choripetalous or Apetalous*)

NOTE: After determination of the order and family, proceed to the local flora or manual to identify the genus and species.

PRELIMINARY LIST FOR ELIMINATION OF EXCEPTIONAL CHARACTER COMBINATIONS

Before using the key to the orders on page 287, it is necessary to eliminate (1) exceptional plants belonging to other groups but duplicating the key characters of the Calyciflorae, (2) plants not having the key characters marking most of their relatives, and (3) plants marked by obscure characters or those inconsistent if taken singly.

The plant must correspond with every character in the combination. If the specimen does not coincide with one character, even the first listed, the combination may be eliminated from further consideration. The characters in **boldface type** should be checked first to determine whether the combination requires further inspection.

1. Plant corresponding with *every character* in combination a, b, c, or d, below.
   a. **Herbs with *either* pinnate *or* palmate or palmately lobed leaves; flowers perigynous; petals none; carpel solitary;** achene enclosed in the floral cup; sepals persistent in fruit. *Alchemilla* and *Sanguisorba, Rosaceae* (rose family).
   b. **Western large shrubs** or small trees; **flowers perigynous; petals none; carpel solitary, the hairy style forming an elongated feathery persistent tail on the achene;** body of the fruit enclosed in the basal portion of the elongate floral tube; leaves simple. *Cercocarpus, Rosaceae* (rose family).
   c. **Shrubs or trailing or twining woody vines; leaves opposite, simple, without stipules; ovary inferior; fruit a capsule,** with 3–10 seed chambers (*or* with 2 chambers and then opening through a cleft between the 2 styles); **styles or style branches as many as the seed chambers** or rarely wholly or partly coalescent (completely so in only a twining vine of the Southern States and a rare Californian shrub occurring between the San Joaquin and Kings rivers, these plants with 20–200 stamens). *Hydrangeoideae, Saxifragaceae* (saxifrage family).

   Order 38. **Rosales,** page 290.

2. **Large trees with palmately lobed to palmately parted leaves and with the fruits numerous and dry and occurring in pendulous, ball-like clusters;** petals none. *Hamamelidaceae* (witch-hazel family) (apparently applicable to the *Platanaceae,* sycamore family).

   Order 26. **Hamamelidales** (Thalamiflorae), page 238.

3. **Aromatic shrubs with perigynous flowers; leaves opposite, usually entire; stipules none; flowers produced singly, red to brownish maroon, 3–5 cm. in diameter, terminating the branchlets; bracts, sepals, and petals forming a gradually intergrading series,** numerous, all attached to a large, **urnlike floral cup;** stamens numerous, the inner ones sterile; carpels numerous, separate, enclosed in the floral cup. *Calycanthaceae* (sweet-shrub family).

   Order 1. **Ranales** (Thalamiflorae), page 136.

4. Plant corresponding with *every character* in combination a or b, below.
   a. **Herbs; Pacific Slope to southwestern Utah and Arizona, rarely escaping elsewhere; receptacle developed into a deep cup with a spreading disclike margin protruding outside the attachment points of the flower parts; sepals 2, in bud united into a "fool's cap," which slides off as the flower opens; petals usually 4, orange to yellow** or rarely cream or white or partly purple; stamens numerous; carpels about 4 or 6; fruit a capsule with parietal placentae, dehiscent upward along two lines; leaves finely dissected. *Eschscholtzia, Papaveraceae* (poppy family).

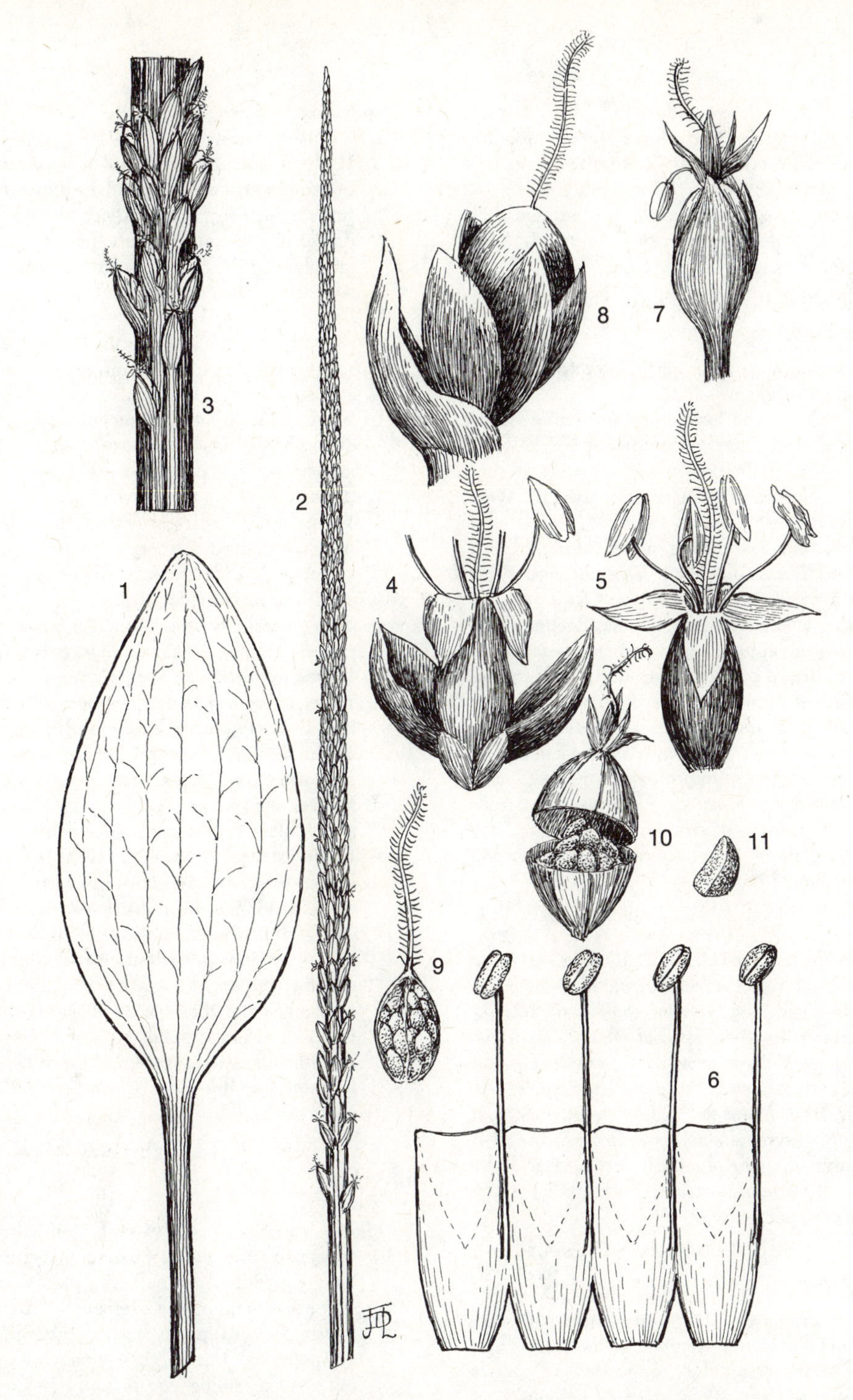

Figure 10-72. Plantaginales; Plantaginaceae; *Plantago* spp., common plantains: **1** basal leaf, **2** spike, **3** portion of the spike, enlarged, **4** flower, showing the calyx, the sympetalous scarious corolla, the stamens (three with anthers omitted), and the stigma, **5** another view (calyx omitted), **6** corolla laid open, showing the adnate stamens, **7** old flower, **8** fruit and the persistent calyx, **9** fruit, longitudinal section, **10** circumscissile dehiscence of the pyxis, **11** seed, **1–3** *Plantago Rugelii,* **4–11** *Plantago major.* In **6** note the alternation of petals and stamens.

b. **Trees; Florida; leaves alternate, bi- or tripinnate; ovary on a stipe above the receptacle;** flowers perigynous, barely bilaterally symmetrical, in panicles; sepals 5, unequal; petals 5, similar to the sepals; stamens 5, staminodia 5; carpels 3, coalescent; placentae parietal. *Moringaceae* (horseradish-tree family).

Order 6. **Papaverales** (Thalamiflorae), page 152.

5. Plant corresponding with *every character* in combination a or b, below.
   a. **Herbs or shrubs; Florida to Louisiana;** leaves alternate, entire, toothed, or pinnatifid; flowers perigynous; sepals, petals, and stamens 5; carpels 3, coalescent, the placentae parietal; styles 3. *Turneraceae* (*Turnera* family).
   b. **Perennial vines occurring in southern Arizona and from Texas to Kansas, Virginia, and Florida; corona present between the petals and the stamens, developed from the floral cup, bearing filiform appendages and a membrane;** leaves entire to parted; alternate; stipules present; tendrils borne opposite the leaves; flowers bisexual, perigynous; sepals and petals (when these are present) 5 (or sometimes 4); stamens 5 or more, usually opposite the petals; pistil often raised on a stalk; ovary 1-chambered, with 3–5 parietal placentae, the stigmas 3–5; fruit a capsule or a berry. *Passifloraceae* (passion-flower family).

Order 7. **Violales** (Thalamiflorae), page 159.

6. **Small trees or large shrubs of California, Arizona,** and Baja California; **leaves palmately lobed; sepals 5, 1.5–3 cm. long, yellow, coalescent, basally and adnate with the tube of the 5 stamens;** pubescence of stellate (starlike) branched hairs; floral cup very narrow and saucerlike; carpels 4–5, coalescent; fruit a capsule. (The calyx is mistaken readily for a sympetalous corolla and the 3–5 slender subtending involucral bracts appear to be the calyx.) *Fremontodendron* (fremontia), *Sterculiaceae* (chocolate family).

Order 13. **Malvales** (Thalamiflorae), page 181.

7. **Either (1) shrubs or (2) woody vines with tendrils; stamens the same number as the sepals and alternate with them** (i.e., opposite the petals when these are present); carpels coalescent.

Order 22. **Rhamnales** (Thalamiflorae), page 217.

8. **Stipules none; carpels coalescent;** plant corresponding with every character in combination a, b, c, d, or e, below.
   a. **Herbs or sometimes vines** (or very rarely small **Southwestern shrubs** with the flowers in involucrate heads); **fruit an achene or a utricle; flower perigynous; petals none;** the stamens attached to the basal portion of the calyx (this sometimes but not necessarily composed of 2 series of 3 sepals each); stamens 3–10. *Polygonaceae* (buckwheat family), *Amaranthaceae* (amaranth family), and *Illecebroideae, Caryophyllaceae* (pink family).
   b. **Fleshy (i.e., somewhat succulent) herbs; stems decumbent; leaves opposite; flowers solitary, axillary, perigynous, petals none; sepals 5, pinkish or lavender on the upper sides; stamens** 5–60; styles 2–5, the ovary 2–5-chambered; fruit a membranous capsule, circumscissile, **the upper part a deciduous lid.** *Sesuvium, Aizoaceae* (carpetweed family).
   c. **Petals** (actually staminodia) **numerous; plants, especially the leaves, markedly succulent; leaves opposite; flower epigynous;** sepals 5 or more, unequal; **stamens numerous;** fruit fleshy, but dehiscing apically by radiating openings in the flattened top, 5–12-chambered. *Mesembryanthemum, Aizoaceae* (carpetweed family).
   d. **Flower epigynous; sepals 2; petals numerous; plant often somewhat succulent.** *Portulaca, Portulacaceae* (*Portulaca* family).
   e. **Prostrate herbs of sandy places; floral tube colored, elongated, constricted just above the ovary and disarticulating at that level, the basal portion persistent, hardening and enclosing the dry 1-seeded fruit,** the ovary appearing (from the outside of the flower) to be inferior; carpel 1; petals none; flowers in involucrate heads; leaves opposite, those of each pair more or less unequal. *Abronia, Nyctaginaceae* (four-o'clock family).

Order 23. **Caryophyllales** (Thalamiflorae), page 223.

9. **Plants vinelike, usually with tendrils; flowers epigynous, the plants monoecious; fruits melonlike, gourdlike, or leathery, sometimes spiny;** stamens commonly coalescent. (Actually the flower is with petals rather than sepals and sympetalous, but rarely when the sepals are minute the single conspicuous perianth series may be taken for the calyx.) *Cucurbitaceae* (gourd family).

Order 7. **Loasales** (Thalamiflorae), page 166.

10. **Slender herbs or rarely small shrubs, the plants usually delicate; stems 4-angled; leaves appearing to be in whorls of 4–6** (actually all but 2 of these stipules); calyx obsolete, the petals (sympetalous) appearing to be the calyx; stamens 4 or rarely 3; styles 2; the carpels separating at maturity of the fruit into 2 indehiscent, 1-seeded structures. *Galium, Rubiaceae* (madder family).

Order 47. **Rubiales** (Ovariflorae), page 332.

11. **Aromatic shrubs or trees; leaves entire or lobed; anthers not splitting open, opening by 2 or 4 uplifted valves, these resembling trap doors but swinging upward,** the stamens in 2–4 series of usually 3; carpel 1; **fruit a drupe.** *Lauraceae* (laurel family).

Order 1. **Ranales** (Thalamiflorae), page 136.

12. **Trees with a smooth ash-gray bark; leaves simple,** toothed, with straight veins; the flowers unisexual, the plant monoecious; **staminate flowers in rounded, pendulous, pedunculate heads; pistillate flowers usually in pairs,** the pair with a short peduncle; **fruit a sharply triangular nut, the pair of fruits enclosed in a soft, prickly, leathery involucre, dividing part way into 4 segments.** *Fagus* (beech), *Fagaceae* (beech family).

Order 27. **Fagales** (Thalamiflorae), page 243.

13. Herbs or bushes; calyx synsepalous, corollalike, tubular, basally constricted above the ovary, enclosing the small, hard fruit; involucre green, calyxlike; carpel 1; leaves opposite. *Nyctaginaceae* (four-o'clock family).

Order 23. **Caryophyllales** (Thalamiflorae), page 223.

14. Trees or nonparasitic shrubs; flowers small, unisexual (very rarely a few bisexual), at least the staminate in catkins, the pistillate usually solitary or 2 or 3 together, but sometimes in catkins; the ovary inferior, as shown by presence of sepals on the adnate floral tube at the apex of the ovary—plants (long known as among the Amentiferae) determined by the following brief key:

1. Leaves opposite, simple, entire; plants dioecious; both the staminate and pistillate flowers in drooping catkins; large shrubs. *Garryaceae* (silk-tassel bush family).

Order 44. **Cornales,** page 325.

1. Leaves alternate; plants monoecious; pistillate flowers usually not in catkins; mostly trees.
   2. Leaves pinnate, large; fruit with a hard shell as an inner layer and with a fleshy outer layer, this becoming a leathery or fibrous husk in age (the fruit similar to a walnut, pecan, or hickory nut).

   Order 19. **Juglandales** (Thalamiflorae), page 213.

   2. Leaves simple; fruit *not* with the combination of (1) a hard shell as an inner layer and (2) a fleshy, leathery, or fibrous husk as an outer layer, the enveloping structures, if any, completely free from the fruit (the fruit often but not always an acorn). *Fagaceae* (beech family).

   Order 27. **Fagales** (Thalamiflorae), page 243.

KEY TO THE ORDERS

(First consult the Preliminary List for Elimination of Exceptional Character Combinations, page 284).

NOTE: After determination of the order and family, proceed to the local flora or manual to identify the genus and species.

1. Flowers perigynous.
   2. Calyx and floral cup not colored externally (except by chlorophyll and its associated pigments, the green being sometimes pale or yellowish; reddish in *Cercis*), not covered with silvery or rusty scales.
      3. Carpels *either* (1) solitary *or* (2) separate *or* (3) incompletely coalescent, as shown by presence of more than 1 style (two styles sometimes lightly coherent but their individuality evident); stipules usually present, of various forms (large and green, scalelike and not green, or reduced to glands), sometimes deciduous when the leaf matures, therefore often to be found only on the leaves of young shoots.

      Order 38. **Rosales,** page 290.

3. Carpels 2–several, completely coalescent, except sometimes the stigmas, the style therefore 1; stipules none.

Order 42. **Myrtales,** page 314.

2. Calyx and floral cup conspicuous for their external color or whiteness or for their silvery or rusty covering of scales; petals none.
   3'. Fruit a follicle, sometimes 1-seeded; leaves (of the tree escaped in Florida) pinnate or pinnatifid; inflorescence a conspicuous raceme, spike, or head of orange to yellow flowers.

Order 40. **Proteales,** page 313.

3'. Fruit *either* (1) a berrylike drupe *or* (2) appearing drupelike because the single achene is enclosed in the persistent basal portion of the floral cup which may become fleshy at fruiting time.
   4. Leaves with silvery or rusty scales or starlike (stellate) hairs; floral cup and sepals not colored or corollalike at anthesis; ovary appearing inferior, being invested closely by the floral cup; fruit an achene, but appearing berrylike or drupelike because of the fleshy, colored floral cup surrounding it.

Order 41. **Elaeagnales,** page 314.

4. Leaves not with silvery or rusty scales or starlike (stellate) hairs; floral cup and sepals colored and corollalike at anthesis; ovary obviously superior, not closely invested by the floral cup and not appearing inferior; fruit berrylike; floral cup not fleshy. *Thymelaeaceae* (mezereum family).

Order 42. **Myrtales,** page 314.

1. Flowers epigynous.
   2'. Plants *not* parasitic on the branches of other plants. (Cf. lower lead 2', p. 290.)
   3''. Stems bearing spines in areoles (small clearly defined areas) at the nodes (these commonly not appearing as nodes, however, but arranged spirally on flat surfaces or on ridges or tubercules); segments of the stem markedly succulent; petals intergrading with the sepals, both numerous, all the perianth parts adnate at the bases and forming a perianth tube; style 1, the stigmas several; ovary 1-chambered, the placentae parietal.

Order 24. **Cactales** (Thalamiflorae), page 233.

3''. Plant *not* corresponding with *every character* in the combination in lead 3'' above.
   4'. Corolla, when present, usually *not* bilaterally symmetrical; plant *not* corresponding with *every character* in lead 4' below (p. 290).
   5. Carpels coalescent by the bases of the ovaries, but the upper portions of the ovaries (together with the styles) free; fruit *not* 2–4-horned; plants terrestrial.

Order 38. **Rosales,** page 290.

5. Carpels coalescent through the full length of the ovaries, the styles often separate.
   6. Stipules present, often deciduous and therefore found on only the young leaves of new growth, sometimes specialized as glands or spines.
   7. Plants *not* trees with prop roots growing from the trunks and branches.

Order 38. **Rosales,** page 290.

7. Plants trees with prop roots growing from the trunks and branches; Florida. *Rhizophoraceae* (mangrove family).

Order 42. **Myrtales,** page 314.

6. Stipules none.
   7'. Ovary incompletely inferior, that is, with the floral cup attached to only the lower half or less of the ovary, leaving the upper portion free; sepals and petals present and both well developed.

Order 38. **Rosales,** page 290.

7'. Ovary (in the North American species) completely inferior or at least more than half so.

8. Ovary *not* with a single chamber containing numerous ovules or seeds, *either* (a) with several chambers *or* (b) with 1 chamber containing only 2–6 ovules and usually only one of these maturing into a seed.

9. Leaves or some of those on each plant basally cordate (some sometimes ovate); prostrate or rhizomatous herbs or twining shrubs; petals none or vestigial; calyx and floral tube conspicuous, together 1–7 cm. long, dingy brown to purplish; stamens 5–12; ovary with 6 chambers, forming a capsule or a berry in fruit.

Order 5. **Aristolochiales** (Thalamiflorae), page 152.

9. Plant *not* corresponding with *every character* in the combination in lead 9 above.

10. Style 1.

11. Stamens borne on the margin of the floral cup or tube.

Order 42. **Myrtales,** page 314.

11. Stamens borne on an epigynous disc or its margin, the disc covering the top of the inferior ovary where it is free from the floral cup.

12. Trees or shrubs occurring in the eastern half of the United States (mostly Southern States); flowers small, greenish; plants dioecious; fruit a drupe, ovoid or obovoid to ellipsoid, 1–2.5 cm. long, resembling an olive. *Nyssaceae* (sour gum family).

Order 44. **Cornales,** page 325.

12. Plant *not* corresponding with *every character* in the combination in lead 12 above.

13. Stamens more numerous than the sepals.

Order 42. **Myrtales,** page 314.

13. Stamens the same number as the sepals.

14. Petals present.

Order 44. **Cornales,** page 325.

14. Petals none.

Order 43. **Santalales,** page 322.

10. Styles 2–5, separate.

11'. Flowers sessile in the leaf axils or the axils of bracts, solitary or whorled on the stem; plants aquatic, except in Florida and southward. *Haloragaceae* (water-milfoil family).

Order 42. **Myrtales,** page 314.

11'. Flowers in umbels, compound umbels, or sometimes heads, numerous; plants usually but not necessarily terrestrial.

Order 45. **Umbellales,** page 328.

8. Ovary with a single seed chamber containing numerous ovules or seeds (or only 1 in a rough-hairy desert bush), the placentae parietal, 2–5.

9'. Shrubs; fruit a berry; sepals usually larger than the petals; sepals, petals, and stamens 5 (or rarely 4); withered flower parts remaining attached to the fruit; leaves palmately veined and nearly always palmately lobed or parted. *Ribes, Saxifragaceae* (saxifrage family).

Order 38. **Rosales,** page 290.

9'. Herbs or bushy plants; fruit a capsule; sepals smaller than the petals *or* the petals none; only the sepals remaining upon the fruit; leaves *not* palmately veined, lobed or parted.

Order 7. **Loasales** (Thalamiflorae), page 166.

4'. Terrestrial (land) rather succulent herbs, monoecious, with bilaterally symmetrical flowers; leaf blade not symmetrical, one side curving more widely from the base than the other; fruit a wingéd capsule, 2–several-chambered. *Begoniaceae* (Begonia family).

Order 7. **Loasales** (Thalamiflorae), page 166.

2'. Plants parasitic on the branches of trees or shrubs; flowers unisexual, the plants at least usually dioecious; petals none.

3''. Plants minute, less than 1 cm. long, only the flowers and bracts occurring outside the host plant; fruit a capsule, many-seeded; Southwestern Deserts. *Pilostyles, Rafflesiaceae* (*Rafflesia* family).

Order 46. **Rafflesiales,** page 330.

3''. Plants 2 cm. or more in length; the branches and often the leaves well developed; fruit berry-like, 1-seeded. *Loranthaceae* (mistletoe family).

Order 43. **Santalales,** page 322.

Order 38. *ROSALES.* ROSE ORDER

DEPARTURE FROM THE FLORAL CHARACTERS OF THE RANALES

Free	In some extraterritorial families the petals sometimes coalescent basally, in the *Mimosoideae* and a few *Papilionoideae* (*Leguminosae*) the petals coalescent basally, in all *Papilionoideae* the 2 lower petals and in some the tips of the 2 lateral petals coalescent; stamens sometimes coalescent (as in many *Leguminosae*); carpels free or coalescent, but the styles and stigmas usually free (rarely both coalescent as in a few *Saxifragaceae*).
Equal	Corolla rarely bilaterally symmetrical (except in most *Leguminosae*).
Numerous	Stamens numerous to 10 or fewer; carpels numerous to 1.
Spiral	Stamens often cyclic, the carpels usually so.
Hypogynous	Most species perigynous but most families hypogynous; some groups epigynous.

Herbs, shrubs, or trees. Leaves simple or compound; half the families with stipules. Flowers commonly 5-merous; but the stamens often numerous and the carpels 1-numerous; ovary, if formed by coalescence of carpels, usually divided into chambers; ovules usually anatropous, but orthotropous or campylotropous in some families, the integuments usually 2. Fruit a legume, follicle, achene, drupe, pyxis, capsule, pome, or berry.

The order Rosales is the second largest among the flowering plants. It includes about 18,000 species, 13,000 of them in the pea family (*Leguminosae*).

The Rosales are placed in the Calyciflorae, but several of the smaller families in the order are hypogynous, and the *Leguminosae* include some hypogynous members and many barely perigynous. The more primitive hypogynous families are related closely to the Ranales and not strongly differentiated from them. Although several characters occur among nearly all Rosales but infrequently in the Ranales and vice versa, each individual character has many exceptions in both groups.

Whether the Rosales are the terminus of a relatively short line of evolution from near ancestors of the Ranales or they are nearer the ancestral type of several other orders of the Calyciflorae is debatable. At least some other orders of Calyciflorae have arisen independently, and this may be true of all.

The Myrtales resemble the Rosales in general flower structure. However, the carpels of the

Myrtales are (almost without exception) coalescent through not only the ovaries but also the styles so that the pistil has but 1 compound style. In the Rosales when the carpels are coalescent, only the ovaries are coherent, and the pistil has nearly always more than 1 style or a branched style. Exceptions are the extraterritorial *Byblidaceae* and *Pittosporaceae* and (rarely) *Ribes* in the *Saxifragaceae*.

Various authors have attempted to subdivide the Rosales into smaller more coherent orders, but the forms connecting the extremes are numerous and there are no clearly reasonable dividing lines. The most common proposal is segregation of one group centered about the *Rosaceae* and of another about the *Saxifragaceae*. The cornerstone of any such segregation is a clear distinction between the *Rosaceae* and the *Saxifragaceae*. The two families display close relationship, and they are not keyed apart simply. The tabulated characters distinguish the North American representatives and with fair accuracy the entire families as they occur in the remainder of the world.

From the parallel lists below, it is clear that the *Rosaceae* and *Saxifragaceae* can be distinguished with some degree of clarity despite the great diversity in both. Application of the same and other characters to the segregation of the group of families related to the *Rosaceae* and *Saxifragaceae* reveals great inconsistencies, as shown in the table. The most nearly consistent single character which might serve as a basis for an ordinal segregation is presence or absence of endosperm in the seeds. Even this, as recorded in various books, is somewhat inconsistent. Furthermore, obscure characters of this sort and those requiring special techniques for investigation must be considered cautiously. Usually the records are based upon relatively few cases—a few individuals of a few species of a few genera. There is no reason to believe that the quantity of endosperm may not vary greatly within each family. From monographing large genera it is evident that nearly every character may vary from species to species or from individual to individual, and that constancy may not be assumed from a few cases. Although intensive study of the Rosales may reveal natural lines of segregation into smaller orders, these are not apparent from the data at hand. Even a division into suborders would be dubious.

Key to the Families

(North America north of Mexico)

1. Sepals, petals, stamens, and carpels the same number, usually 5, sometimes 3 or 4, rarely up to 30, each in a single series or the stamens in 2 series and double the number; Leaves (and to a lesser degree the stems) markedly succulent, thick and fleshy; follicles separate or barely coalescent basally; plants *not* insectivorous.

   Family 1. **Crassulaceae** (stonecrop family).

Rosaceae	*Saxifragaceae*
Fruit indehiscent, except when there are 3 or more follicles per flower (these sometimes barely coalescent basally but splitting lengthwise at maturity). (Note: when the ovary is inferior, in fruit it becomes a pome with 2–5 chambers and axile placentae.)	Fruit dehiscent, a capsule (the carpels rarely 5–7 and nearly separate, then dehiscent horizontally and forming a pyxis) or in *Ribes* a berry (which is inferior, 1-chambered, and with parietal placentae) and **indehiscent.**
Stipules present, except in *Spiraea* and *Holodiscus*, but often early deciduous.	Stipules none, except sometimes in *Heuchera* and sometimes in *Ribes*.
Stamens usually numerous, more than 10, except (1) in herbs with pinnate or 3-foliolate leaves (*Alchemilla, Acaena,* and some species of *Agrimonia, Potentilla,* and *Sanguisorba*) and (2) in shrubs or trees with alternate leaves which are not palmately veined or lobed (10 in *Photinia* and 5–10 in some species of *Crataegus* and perhaps *Spiraea* and *Adenostoma*); leaves alternate.	Stamens 3–10, except in shrubs with opposite leaves (*Carpenteria, Philadelphus, Decumaria,* and sometimes *Whipplea* in the *Hydrageoideae*).
Carpels sometimes 1 or rarely 2–4, commonly 5–many.	Carpels usually 2, rarely 3–10.
Seeds with little or no endosperm.	Seeds with endosperm.

OCCURRENCE OF SOME DISTINGUISHING CHARACTERS AMONG FAMILIES OF THE ROSALES

Group	Family	*Hypogynous, perigynous, or epigynous*	*Stipules*	*Endosperm*	*Carpels free or coalescent*	*Carpel number*	*Fruit type*	*Stamen number*
GROUP 1 (Based upon the *Rosaceae* as typical)	2. *Connaraceae*	Hypogynous	None	Present or absent	Free	5	Follicle	10 or sometimes 5
	3. *Leguminosae*	Mostly perigynous; some hypogynous	Present	None or scanty	Free	1	Legume or loment (usually)	10 or rarely fewer
	4. *Crossosomataceae*	Perigynous	None	Thin	Free	2–9	Follicle	15–50
	5. *Rosaceae*	Perigynous or epigynous	Present (except rarely)	None (usually)	Free or sometimes coalescent	5–many (or rarely 1–4)	Indehiscent or follicle or pome	Numerous (rarely 10 or fewer)
GROUP 3 (Based upon the *Saxifragaceae* as typical)	1. *Crassulaceae*	Hypogynous	None	Scanty or none	Free or barely coalescent	5 (or rarely 3–30)	Follicle or barely a capsule	5 (or [illegible]–30, same as other parts)
	6. *Myrothamnaceae*	Hypogynous	Present	Present	Coalescent	3	Capsule	5 (sometimes 4–8)
	7. *Bruniaceae*	Epigynous	Present	Present	Coalescent	4–5	Capsule	8–10
	8. *Brunelliaceae*	Hypogynous	Present	Present	Free	4–5	Follicles	8–10
	9. *Cunoniaceae*	Hypogynous	Present	Present	Coalescent	2–5	Capsule or fleshy	4–5 or numerous
	10. *Eucryphiaceae*	Hypogynous	Present	Present	Coalescent	5–12 or 18	Capsule	Numerous
	11. *Byblidaceae*	Hypogynous	None	Present	Coalescent	2–3	Capsule	5
	12. *Pittosporaceae*	Hypogynous	None	Present	Coalescent	2–5	Berry or capsule	5
	13. *Cephalotaceae*	Perigynous	None	Present	Free	6	Follicular (1-seeded)	12
	14. *Saxifragaceae*	Perigynous or epigynous or rarely hypogynous	Rare	Present	Coalescent or rarely free	2 or rarely 3–12	Capsule or berry	3 or commonly 5–10 or sometimes numerous
	15. *Droseraceae*	Hypogynous	None	Present	Coalescent	3–5	Capsule	5–20

1. Sepals, petals, stamens, and carpels *not* all of the same number (or the same number except that the stamens are double the number); leaves *not* markedly succulent, thick, or fleshy.
   2. Fruit *not* clearly a capsule, *not* a pyxis or a berry, one of the following: a follicle (or a cluster of 3 or usually more follicles slightly coalescent basally and therefore barely technically a capsule), an achene, a legume, a drupe, a drupelet (then the fruits numerous in a compact head), or a pome.
      3. Carpel solitary, usually a legume in fruit, sometimes specialized but the fruit not an achene or a drupe (if possibly interpreted as an achene the next character receives precedence); corolla usually but not necessarily papilionaceous, i.e., markedly bilaterally symmetrical, with a (ventral) banner, 2 (lateral) wings, and a dorsal and enclosed keel of 2 coalescent petals enclosing the 10 stamens and the pistil.

         Family 3. **Leguminosae** (pea family).

      3. Carpels 1–numerous, the fruit never a legume, when formed from a solitary carpel either an achene or a drupe.
         4. Stipules none or minute; follicles usually 1–7 (rarely 2–9); glabrous shrubs; leaves bluish, simple, entire; petals white; flower perigynous; southern California to Arizona and northwestern Mexico.

            Family 4. **Crossosomataceae** (*Crossosoma* family).

         4. Stipules nearly always present on the young growth, often early deciduous but leaving scars, separate or adnate with the petiole; fruit an achene, a drupe, a drupelet, a pome, or sometimes a follicle (then not corresponding with the rest of the description in lead 4 above).

            Family 5. **Rosaceae** (rose family).

   2. Fruit a capsule (the carpels being coalescent in the North American species) or rarely a berry (then with 2 carpels, 1 chamber, and parietal placentae); flowers (of the North American species) perigynous or epigynous; stipules none.
      3'. Plants not insectivorous, the leaves not adapted to capturing insects and not bearing abundant stalked glands; stamens 5–10 or rarely 4 or 3, in 1 or 2 series, if in 2, the outer opposite the petals (obdiplostemonous); carpels 2 or rarely 3–10; flowers nearly always perigynous or epigynous.

         Family 14. **Saxifragaceae** (saxifrage family).

      3'. Plants insectivorous, the leaves with glandular hairs that secrete sticky material holding the insect, the leaf lobes sometimes snapping shut over the insect; stamens 5–20, in series of 5; carpels 3–5; flowers hypogynous.

         Family 15. **Droseraceae** (sundew family).

*Family 1. Crassulaceae.

STONECROP FAMILY

(Synonym: *Sedaceae*)

Annual or chiefly perennial herbs; stems somewhat succulent. Leaves simple, rarely compound, usually entire, almost always markedly succulent; stipules none. Flowers nearly always bisexual, radially symmetrical, hypogynous; sepals, petals, (each series of) stamens, and carpels the same number, commonly 5, sometimes 3 or 4, rarely up to 30; petals usually separate (coalescent in some genera); stamens usually in 2 series, separate; carpels usually separate or sometimes coalescent basally. Fruit a follicle or sometimes the follicles somewhat coalescent basally and barely forming a capsule; seeds with some endosperm. (Cf. Fig. 12-2, *right.*)

The family is large and widely distributed except in the South Pacific. Members of the group include such well-known succulents as the garden hen-and-chickens (*Sempervivum*), *Crassula* (with some species resembling miniature trees), numerous species of stonecrops (*Sedum*), bryophyllums (*Kalanchoë*) *Echeveria Graptopetalum,* and *Dudleya.* The genus *Sedum* occurs through most of North America, being represented by many regional species. *Tillaea,* a genus of minute herbs, occurs in both western North America and the coastal region of eastern North America. A widespread species occurs on mud; another is abundant on soil in Oregon, California, and Chile. *Echeveria* occurs in Mexico. *Dudleya* is represented by species in California, Arizona, and Baja California. The family is much more highly developed in Mexico than in the United States. In relatively recent years, the genera and species have been subdivided beyond recognition, and the family is much in need of a thorough and comprehensive study.

The *Crassulaceae* are distinguished from all cacti (*Cactaceae*) by lack of spines and from most cacti by presence of leaves.

Several characters of the family are similar to

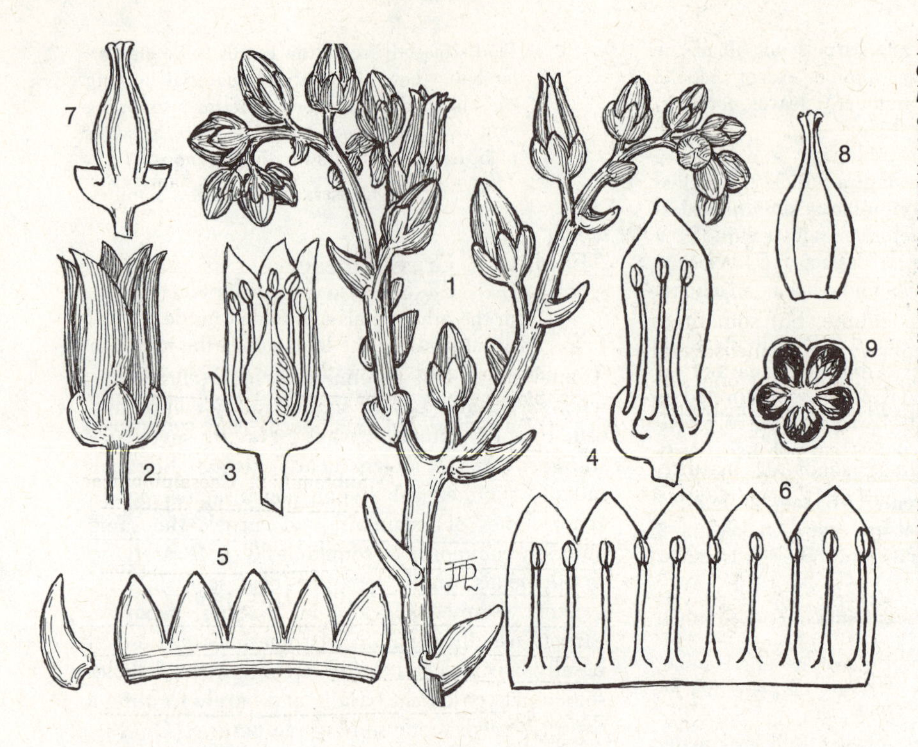

Figure 10-73. Rosales; Crassulaceae; *Dudleya lanceolata,* live-forever or dudleya: **1** inflorescence, **2** flower, side view, showing the sepals and the basally coalescent petals, **3** flower, longitudinal section, **4** petal and the adjacent stamens and sepals, **5** calyx, showing the coalescent sepals, one sepal detached and redrawn from another view, **6** corolla laid open, showing the adnate stamens and the alternate and opposite positions of the two series, **7–8** pistil formed from five incompletely coalescent carpels, **9** cross section of the ovary.

those of the order Ranales: mostly the carpels are separate; the flowers are hypogynous; there are no stipules. Nevertheless, the nearest affinity of the family is distinctly with the *Saxifragaceae* or saxifrage family, primitive members of which differ only in lack of succulence and the lack of equal numbers of sepals, petals, and stamens (within in each series). The position of the *Saxifragaceae* at the end of the list with the *Crassulaceae* at the beginning does not indicate remoteness of relationship. Most *Saxifragaceae* are specialized in one way or another. The relatively primitive flower characters of the *Crassulaceae* are similar to those of the primitive members of each of several family groups, including the saxifrage line.

Family 2. **Connaraceae.**

CONNARUS FAMILY

Shrubs, trees, or usually woody vines. Leaves mostly odd-pinnate, sometimes simple, alternate; stipules none. Flowers nearly always unisexual, radially symmetrical or tending to bilateral symmetry, hypogynous, in racemes or panicles; sepals 5; petals 5, sometimes slightly coalescent; stamens 10 or sometimes 5, usually in 2 series; filaments usually coalescent basally; carpels 5, separate; ovules 2 per carpel. Fruit a follicle, solitary, only one of the carpels developing, it with only 1 seed; endosperm present or absent.

The family occurs in the Tropics, and it includes about 20 genera.

The *Connaraceae* are related most nearly to the *Leguminosae,* and, according to Wettstein, also to the *Chrysobalanoideae* in the *Rosaceae.*

*Family 3. **Leguminosae.** PEA FAMILY

(Synonym: *Fabaceae.* Segregates: *Mimosaceae, Caesalpinaceae* or *Cassiaceae*)

Predominantly herbs but frequently shrubs and less commonly trees. Leaves usually pinnate to tripinnate, sometimes palmate or simple; stipules present or sometimes none, sometimes reduced to glands, often falling away as the leaf matures. Flowers bisexual, bilaterally symmetrical except in the *Mimosoideae,* usually barely perigynous; sepals 5, coalescent, often the lobes of the calyx fewer than 5; petals 5, rarely none or 1 (*Amorpha*), separate or commonly 2 coalescent (*Papilionoideae*) or rarely all coalescent (some *Mimosoideae* and species of *Dalea*); stamens 10, rarely fewer, in some *Mimosoideae* (as in *Acacia, Lysiloma,* and *Calliandra*) 20 or more, separate, coalescent into a tube (monadelphous), or 9 coalescent into a sheet and 1 free (diadelphous); carpel 1; ovules 2 to numerous, ranging from (rarely) campylotropous to anatropous (anatropous in the *Caesalpinoideae,* and amphitropous in the *Papilionoideae*). Fruit typically a legume, but sometimes indehiscent, as in the genus *Prosopis* and several genera of *Papilionoideae,* also coiled in one section of *Prosopis* and in *Medicago*) or constricted and breaking crosswise between the seeds, i.e., a loment (e.g., *Desmodium*); seeds at maturity without endosperm (e.g., pea or bean) or with a small quantity of hard endosperm.

The pea family, which includes about 13,000 species, is the second largest of the three hundred families of flowering plants, and it is exceeded in numbers of species by only the sunflower family (perhaps 15,000 to 20,000 species). Distribution is world-wide. Through a large part of the United States only one of the three subfamilies of the *Leguminosae* is represented or at least is well known. This is the typical or bean subfamily (*Papilionoideae* or *Lotoideae*) which includes most of the common legumes in the world as a whole. In the Southwest (especially in Arizona and Texas) and in the South members of the other subfamilies are fairly abundant either as native plants or in cultivation. The acacias, cassias, bird-of-paradise, palos verdes, and their relatives are fairly common in cultivation. These trees and shrubs together with various herbaceous legumes are the most outstanding element in the flora of the Southwestern Deserts. Although they are not bizarre in appearance as are the cacti, yuccas, and century-plants, they form the real backbone of the desert vegetation.

Key to the Subfamilies

(North America north of Mexico)

1. Corolla radially symmetrical or usually tending slightly to bilateral symmetry, not papilionaceous (like a butterfly or actually like a sweet pea, cf. lead 1 below).
    2. Flowers crowded into dense spikes or heads; corolla radially symmetrical, inconspicuous, less than 5 mm. long; stamens the conspicuous part of the flower, exceeding the petals in length, not more than 7 mm. long.

        Subfamily 1. **Mimosoideae** (*Acacia* subfamily).

    2. Flowers not crowded into very dense spikes or heads, usually 1 cm. or more in diameter or length; corolla bilaterally symmetrical but usually not conspicuously so except in *Cercis* (which is distinguished by the simple cordate-reniform leaves), with the banner somewhat differentiated from the other petals and folded inside the two adjacent petals in the bud; corolla the most conspicuous part of the flower except in an occasional genus, as for example, *Caesalpinia* (bird-of-paradise), but not distinctly more conspicuous than the petals.

        Subfamily 2. **Caesalpinoideae** (bird-of-paradise subfamily).

1. Corolla markedly bilaterally symmetrical, papilionaceous (like a butterfly or actually like a sweet pea), that is, the five petals differentiated into a banner, two wings, and a keel formed from two petals coalescent by their adjacent margins and enclosing the stamens and the pistil.

    Subfamily 3. **Papilionoideae** (bean subfamily).

1. **Mimosioideae.** *Acacia* includes five species occurring mostly in the Southwestern Deserts and the adjacent mountains and others occurring in Florida. Other members of the family include the sensitive-briars (*Schrankia*), fairy-dusters (*Calliandra*), *Lysiloma* (a single species in one locality in Arizona; another in Florida), sensitive-plants (*Mimosa*), mesquites and screw-beans (*Prosopis*), cat-claws (*Pithecolobium* of Florida or *Acacia Greggii* of the Southwestern Deserts), the silk-tree and woman's-tongue-tree (*Albizzia*) and the lead-tree (*Leucaena*).

2. **Caesalpinoideae.** Members of the subfamily include the bird-of-paradise-tree (*Caesalpinia*) (introduced from the West Indies in Arizona and eastward to Florida), redbud (*Cercis*), senna (*Cassia*), Kentucky coffee-tree (*Gymnocladus*), honey-locust (*Gleditsia*), palos verdes (*Parkinsonia* and *Cercidium*) (occurring along the southern edge of the United States), wait-a-bit vines (*Guilandina*), and tamarind (*Tamarindus*).

3. **Papilionoideae.** This enormous subfamily includes many well-known plants, among them beans, peas, vetches, clovers, lupines, brooms, gorse, *Lotus,* indigo, locusts, *Wisteria,* locoweeds, licorice, tick trefoils, peanuts, desert ironwood, rattle-boxes, and coral-trees.

To man the legumes are particularly important because many establish a symbiotic (mutually

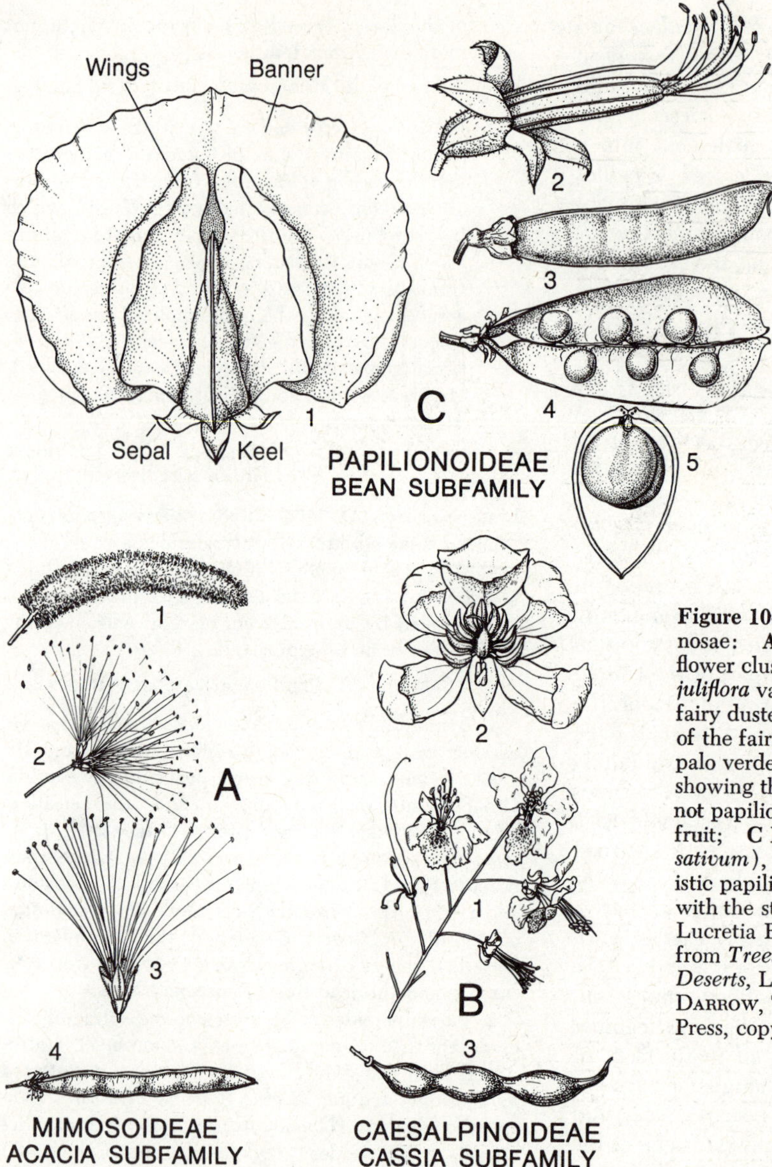

Figure 10-74. Subfamilies of the Leguminosae: **A** Mimosoideae, **1** typical dense flower cluster (velvet mesquite, *Prosopis juliflora* var. *velutina*), **2–3** flowers of the fairy duster (*Calliandra eriophylla*), **4** fruit of the fairy duster; **B** Caesalpinoideae, a palo verde (*Parkinsonia aculeata*), **1** flower, showing the bilaterally symmetrical (though not papilionaceous) corolla, **2** raceme, **3** fruit; **C** Papilionoideae, garden pea (*Pisum sativum*), **1** flower, showing the characteristic papilionaceous corolla, **2** young fruit with the stamens persistent, **3–5** fruit. By Lucretia Breazeale Hamilton. By permission from *Trees and Shrubs of the Southwestern Deserts*, LYMAN BENSON and ROBERT A. DARROW, Tucson: University of Arizona Press, copyright 1979.

beneficial) relationship with certain bacteria, which penetrate the tissues of the roots where their secretion stimulates formation of a nodule of parenchymatous (spongy or pithy) tissue. Although these bacteria are partly parasitic upon the leguminous plant, they fix nitrogen from the air, the excess being used within the leguminous plant in the formation of proteins. The family includes the various clovers, alfalfa, bur clovers, peas, beans, vetches, lentils, soybeans, and innumerable other plants used either for purposes of soil building by addition of nitrogenous compounds or for food (starch and proteins) for man or his livestock. From an economic point of view the pea family is second in importance to only the grass family.

Recognition of the pea family is relatively easy because the familiar pea pod is an outstanding and unique characteristic of the group, and it occurs in nearly all the members. This pod, known

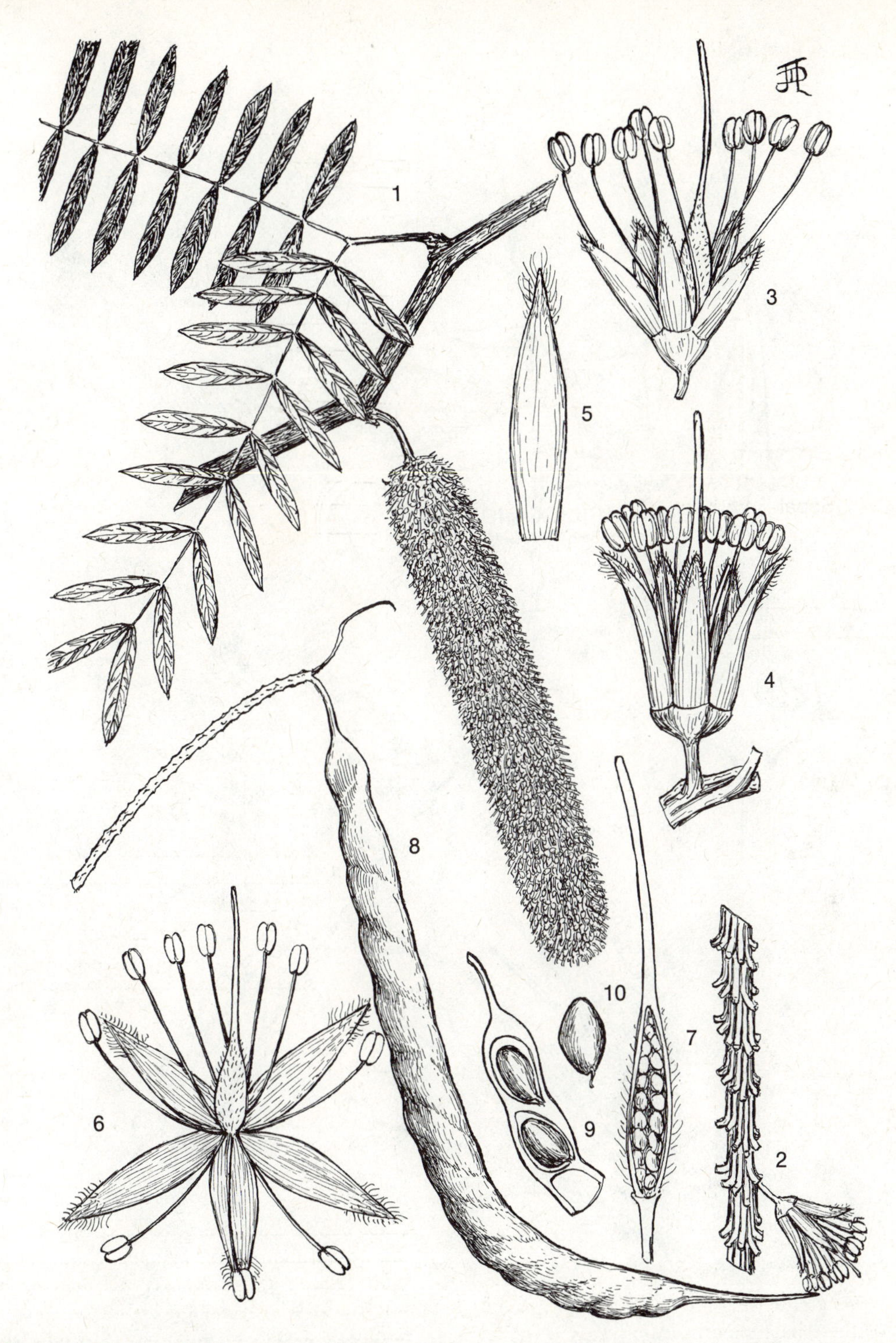

Figure 10-75. Rosales; Leguminosae (Mimosoideae): *Prosopis juliflora* var. *Torreyana,* western honey mesquite: **1** branch with a bipinnate leaf and a dense raceme, **2** enlargement of a portion of the axis of the raceme and one flower, **3–4** side views of the flower, showing the shallowly lobed calyx, the separate petals, the stamens, and the pistil, **5** petal, **6** flower from above (the petals spread out), **7** longitudinal section of the pistil, showing the double row of ovules along the suture, **8** fruit (leguminous type but indehiscent), **9** seeds in the fruit, **10** seed. (Cf. also **Figure 1-1** *Acacia.*)

Figure 10-76. Rosales: Leguminosae (Caesalpinoideae): *Parkinsonia aculeata,* a palo verde: **1** branch with a bipinnate leaf and an inflorescence, **2** flower, top view, showing the bilaterally symmetrical (but not papilionaceous) corolla and the stamens, **3** view of the flower from above (the banner, the only markedly differentiated petal), **4** flower with all petals but the banner removed, **5** pistil in longitudinal section, **6** legume. Fundamentally the flowers of the Leguminosae are generally perigynous, cf. floral cup in **5**.

Figure 10-77. Rosales; Leguminosae (Papilionoideae): *Pisum sativum,* garden pea: **1** branch bearing leaves (the upper leaflets specialized as tendrils) and a flower, **2** flower opening, **3** open flower, the banner being reflexed, **4** the petals, i.e. the banner, two wings, and the keel (formed by coalescence of the two lower petals), **5** keel, **6** flower with only the stamens and the pistil, the stamens diadelphous, only nine of the ten being coalescent, **7** pistil, the stigmatic surface being hairy, **8** legume, **9** longitudinal section, **10** seed and placenta. (Cf. also various details in **Figures 5-4** to **5-7** and **6-17, B** Spanish broom and garden sweet pea, and **Figure 8-6, F** *Lathyrus splendens,* dehiscence of the fruit, **2–10 B,** sweet pea, *Laphyris odoratus,* tendrils.)

as a legume, may be distinguished from other similar appearing dry fruits by (a) presence of only one double row of seeds, the fruit formed from a single carpel, and (b) dehiscence along both edges instead of along only one (as for example in *Delphinium,* the larkspur, which opens along only the suture). The name of the family is derived from this fruiting structure (legume). However, as indicated in the description above, the pods of a few leguminous plants are not strictly legumes.

The *Papilionoideae* represent a high stage of specialization among the *Leguminosae.* The term *Papilionoideae* refers to the butterflylike appearance of the flowers (a typical family of butterflies being the *Papilionidae*). In many of the more highly specialized members of the subfamily, such as the sweet pea and the Spanish broom, the flower is reminiscent of a butterfly because of the great development of the banner. (*Lotoideae* may replace *Papilionoideae.*)

DEPARTURE OF THE PAPILIONOIDEAE FROM THE FLORAL CHARACTERS OF THE RANALES

Free	Sepals usually coalescent; lower 2 petals coalescent and forming a keel; stamens usually coalescent into a single group of ten or into one group of nine with the tenth free.
Equal	Corolla markedly bilaterally symmetrical, differentiated into a banner, two wings, and a keel.
Numerous	Stamens ten, carpel one.
Spiral	Stamens cyclic; (carpel solitary).
Hypogynous	Flowers usually barely and obscurely perigynous.

*Family 4. **Crossosomataceae.** CROSSOSOMA FAMILY

Shrubs. Leaves simple, entire, coriaceous, glaucous, alternate or opposite; plant glabrous; stipules none or minute. Flowers bisexual, radially symmetrical, perigynous and with a hypogynous disc lining the floral cup; sepals 4–5; petals 4–5, white or sometimes with a rose tinge; stamens 8 or 15–50, arranged spirally; carpels 1–9, separate. Fruit a follicle, green at maturity, drying after dehiscence; seeds several; endosperm thin.

The family consists of three species in two genera, as follows: *Crossosoma californicum,* restricted to Santa Catalina, San Clemente, and Guadalupe islands off the coast of California and Baja California; and *Crossosoma Bigelovii,* ranging through the Mexican Desert Region from southeastern California to Arizona, Baja California and Sonora. These shrubs are among the most attractive in the deserts and among the earliest plants to flower in the spring. *Apacheria,* a newly discovered genus consists of one species in the Chiricahua Mountains, Arizona.

Glossopetalon (*Forsellesia*), a small genus of Southwestern Desert shrubs, commonly classified in the Celastraceae, may be of this family, instead. It is similar in some ways to the Crossosomataceae, as pointed out by Thorne. The separate carpel(s) have caused doubt of classification in the Celastraceae.

*Family 5. **Rosaceae.** ROSE FAMILY

(Segregates: *Amygdalaceae* [the *Prunoideae*], *Malaceae* [the *Pomoideae*])

NOTE: Exceptions to certain characters are given in the discussion of differentiation of the *Rosaceae* and *Saxifragaceae*

Herbs, shrubs, or trees or sometimes vines. Leaves simple, pinnate, practically always alternate; stipules present (except as noted) at least on the young growth, often falling before the leaf matures but leaving a scar, sometimes coalescent with the edges of the petiole. Flowers bisexual or rarely some or all unisexual, radially symmetrical, perigynous or epigynous (or in *Coleogyne* hypogynous); sepals and petals 5 (but often a few individual flowers on the plant 6- or 7-merous); stamens numerous, more than 10 (except as noted), usually more than 20, separate; carpels sometimes solitary, but usually 3–many, usually separate (coalescent basally in some *Spiraeoideae* [perigynous] or more fully in the Pomoideae [epigynous]), the styles when joined remaining separate, the ovary with as many chambers as there are carpels, the placentae axile. Fruit indehiscent (except as noted), an achene, a follicle, a drupe or drupelet, or a pome; seeds usually without endosperm.

The family is large, including more than 100 genera and probably 2,000 to 3,000 species. It is

Figure 10-78. Rosales; Rosaceae (Rosoideae); *Fragaria,* hybrid of the commonly cultivated strawberry, a hybrid of *Fragaria vesca,* and the beach strawberry, *Fragaria chiloensis:* **1** 3-foliolate leaf, showing the stipules, **2** inflorescence with flower buds and flowers, **3** flower in longitudinal section, showing the floral cup bearing the sepals, petals, and stamens, and the pistils, **4** the same, enlarged, **5** fruiting head with the "cap" composed of floral cup, bracts, and calyx, **6** a head of fruits in longitudinal section, showing the large edible receptacle and, in pits on its surface, the small seedlike achenes, **7** pistil in a depression in the receptacle, **8** fruit in a recess in the receptacle. (Cf. also **Figures 2-11 A, 8-4 A,** and **8-3 C.**) (Cynthia Smith.)

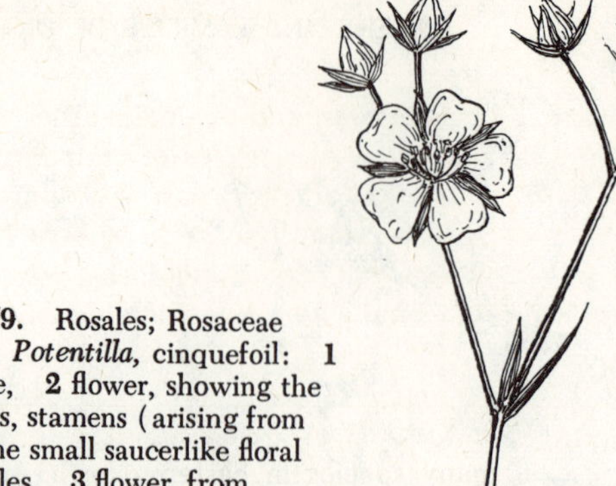

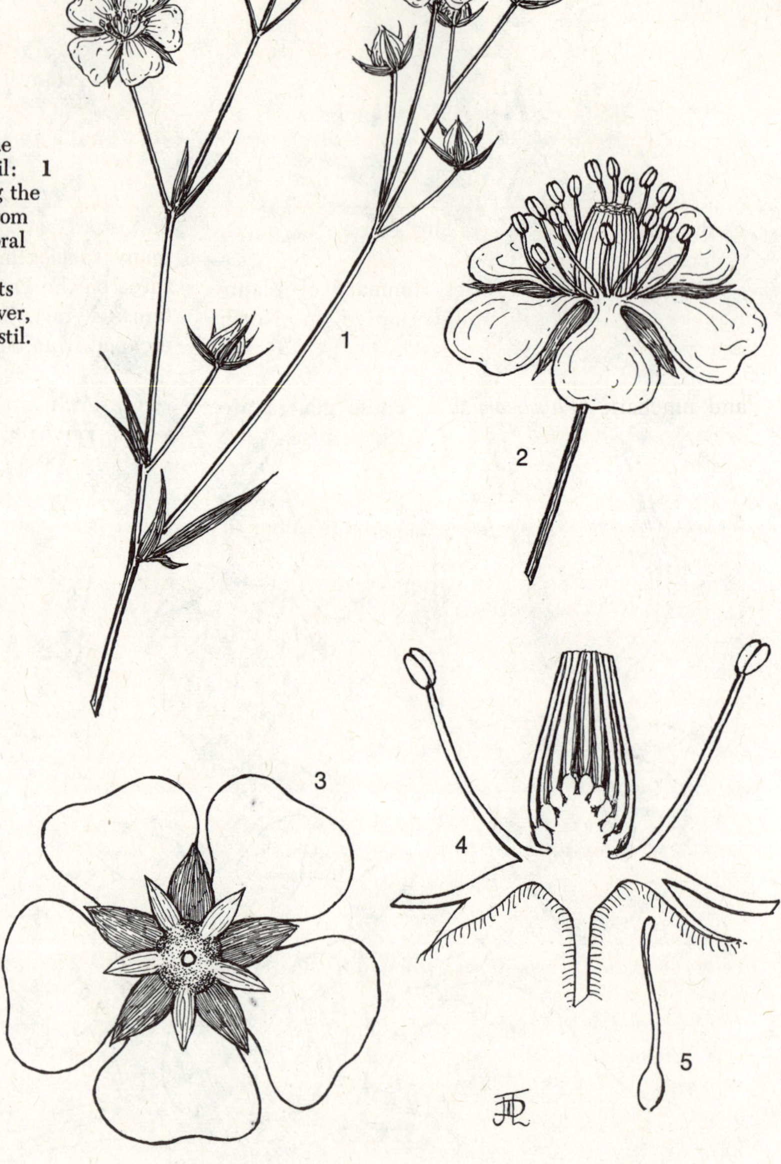

Figure 10-79. Rosales; Rosaceae (Rosoideae); *Potentilla,* cinquefoil: **1** inflorescence, **2** flower, showing the sepals, petals, stamens (arising from the rim of the small saucerlike floral cup, and styles, **3** flower, from beneath, showing a series of bracts and the sepals and petals, **4** flower, longitudinal section, **5** single pistil.

distributed nearly everywhere, being prominent in North Temperate regions. Almost half the genera occur in Canada and the United States.

Key to the Subfamilies

(North America north of Mexico)

1. Ovary superior, the flower perigynous. (Cf. p. 304.)
   2. Fruit a follicle or 3–5 more of these slightly coalescent and forming a capsule, dehiscent.

      Subfamily 1. **Spiraeoideae.** (*Spireaea* subfamily).

   2. Fruit an achene, a drupe, or a drupelet.
      3. Fruit an achene (the achene or achenes sometimes enclosed by the floral cup) or a drupelet (several to many drupelets being packed closely together); pistils usually several to many, rarely few or even one.

         Subfamily 2. **Rosoideae.** (rose subfamily).

      3. Fruit usually a solitary drupe; pistil usually 1, but the pistils rarely 2–5.

         Subfamily 3. **Prunoideae.** (plum subfamily).

1. Ovary inferior; the flower epigynous; fruit a pome; trees or shrubs.

Subfamily 4. **Pomoideae.** (apple subfamily).

Rendle lists the following subfamilies not occurring in North America: *Neuradoideae,* consisting of two genera occurring in Africa and Southern Asia; *Chrysobalanoideae,* 13 genera including 200 species occurring chiefly in South America.

The following is a brief summary of plants representing the subfamilies native in North America.

1. **Spiraeoideae.** *Spiraea,* goatsbeard (*Aruncus*), and ninebark (*Physocarpus*). These genera are widespread, especially in the northern part of the Continent. The following are restricted to the West Coast: cream-bush (*Holodiscus*), Catalina ironwood (*Lyonothamnus,* a remarkable genus of a single species composed of two varieties restricted to the islands off the coast of southern California). Arizona rosewood (*Vauquelinia*), desert-sweet (*Chamaebatiaria*), and another species of creambush (*Holodiscus*) occur in the deserts and the adjacent mountains of the Southwest.

2. **Rosoideae.** A few North American examples of this enormous subfamily are as follows: strawberry (*Fragaria*), cinquefoil (*Potentilla,* a very large genus), *Dryas,* avens (*Geum*), bramble or blackberry (*Rubus*), agrimony (*Agrimonia*), rose (*Rosa*). The following are restricted to the Southwest: cliff rose (*Cowania*), mountain-misery (*Chamaebatia*), mountain-mahogany (*Cerocarpus*), chamisebrush (*Adenostoma*), black-brush (*Coleogyne*), and antelope-brush (*Purshia*).

3. **Prunoideae.** The chief representative of the subfamily is the genus *Prunus* which includes the prunes, plums, cherries, apricots, almonds, peaches, and nectarines, that is, the pitted, or stone, fruits. Many species of the following subgenera are represented in North America: *Prunophora* (plum), *Amygdalus* (peach, almond, and apricot), *Cerasus* (cherry), *Padus* (chokecherry). A single species of *Osmaronia* (osoberry) occurs on the Pacific Coast from Washington to California. It has 1 to 5 carpels in the flower.

4. **Pomoideae.** About 15 species of apples and pears and their immediate relatives (*Pyrus*) are native or introduced from cultivation in North America. The native species are crab-apples, chokeberries, and mountain-ashes (known also as *Sorbus*). One species of crab-apple ranges from Alaska to the north coast of California; three other crab-apples are native in eastern North America. Various species of serviceberries (*Amelanchier*) are abundant in the East and Middle West, and some occur on the Pacific Coast and in the Great Basin and the Rocky Mountain Region. The haws or hawthorns (*Crataegus*) are represented by many species in eastern North America and by three on the Pacific Coast or in the Rocky Mountains. Pyracantha (*Cotoneaster Pyracantha*) has escaped from cultivation in the East. The Christmasberry of California (*Photinia*) is an attractive large shrub brought successfully into cultivation.

The flower of a wild rose corresponds exactly with the floral characteristics of the Ranales except for one feature—it is perigynous. The floral cup is the rose "hip," which later surrounds the numerous small, hard, dry fruits. The hips of many species turn red or yellow at maturity, and they become fleshy, but because they are floral cups they are not fruits. If roses did not have floral cups and stipules, they could be included in the order Ranales.

The fundamental plan of a strawberry flower is almost identical with that of a rose. The floral cup is smaller, and it is shallow and saucerlike. Ultimately it and the sepals together form the green cap or husk which is removed from the "berry" before it is eaten. The edible part is not a fruit, but the greatly enlarged red receptacle, bearing the fruits in pits on its surface (Figure 8-4 A). These troublesome so-called "seeds" are achenes. Each has all the parts of a typical pistil, and the style and stigma may be seen with the naked eye.

The flower of a blackberry, loganberry, raspberry, or any other typical member of the genus *Rubus* is identical fundamentally with that of the strawberry, but the receptacle is elongated and not colored at fruiting time (Figure 8-3 **C**). It is covered densely by small fruits (drupelets) each structurally identical with drupes such as a plum or cherry, even to the small one-seeded pit. Close inspection reveals the style and stigma, which, despite drying, are persistent even in the mature fruits. A plum, prune, cherry, apricot (Figure 8-3 **A**), peach, nectarine, or almond flower differs only in the presence of a single large pistil instead of many small ones. At flowering time the single

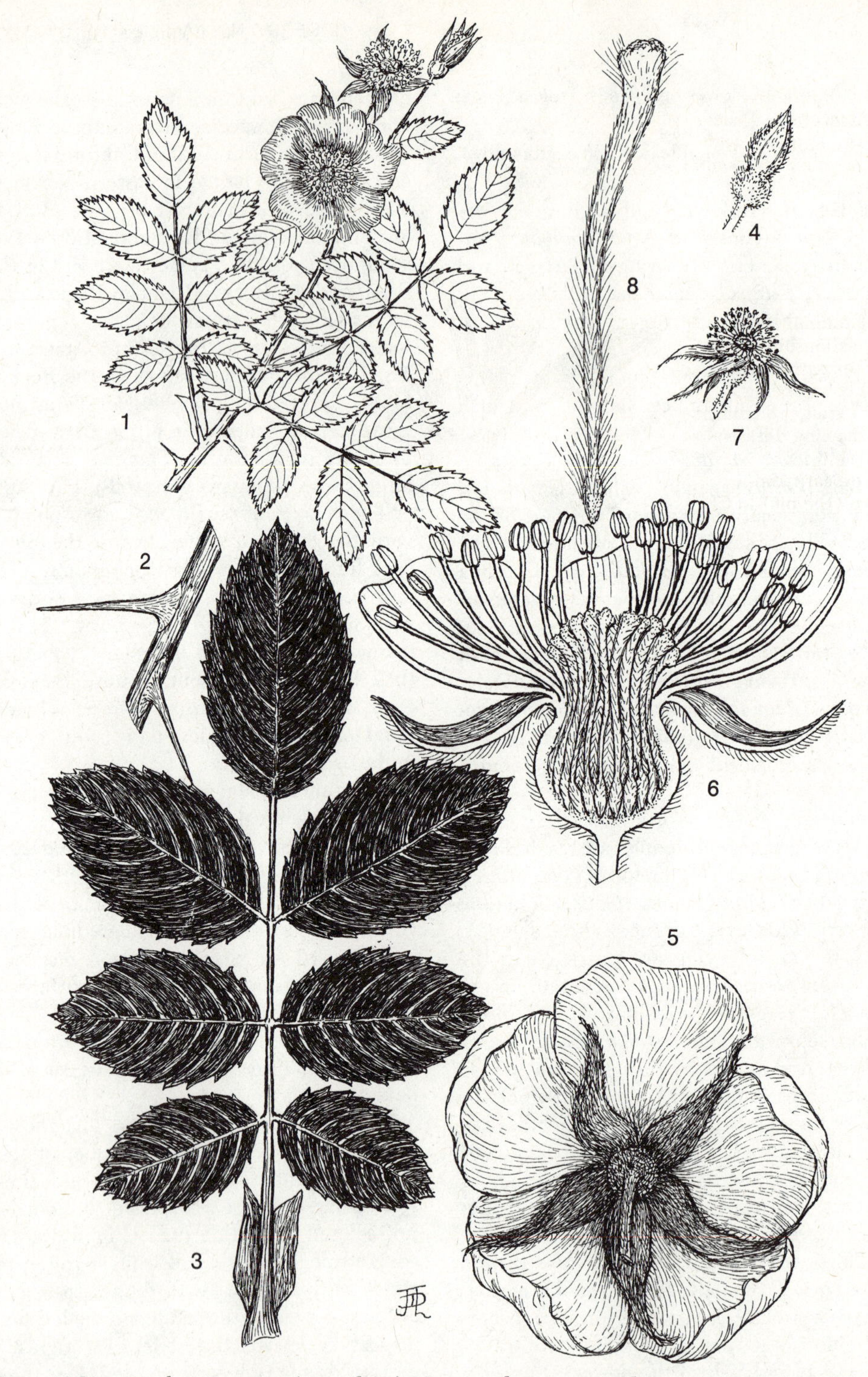

Figure 10-80. Rosales; Rosaceae (Rosoideae); *Rosa spithamea,* ground rose: **1** branch with leaves, prickles, and flowers, **2** prickles (in this species straight and elongate, in others curved and short, cf. **Figure 2-15 B**), **3** leaf with stipules, **4** flower bud, showing the calyx and the floral cup, **5** flower, from beneath, showing the floral cup, sepals, and petals, **6** flower, longitudinal section, showing the floral cup (hypanthium), sepals, petals, stamens, and pistils, **7** beginning of the fruiting stage, **8** pistil. (Cf. also **Figure 5-2,** *Rosa,* transition of stamens to petals.)

Figure 10-81. Rosales; Rosaceae (Prunoideae); *Prunus Persica*, peach: **1** branch with young leaves, a young fruit, and the last flower, **2** flower, showing the floral cup, sepals, petals, and stamens, **3** flower, from above, showing the petals and stamens, **4** old flower after fall of the petals, **5** top view, **6** longitudinal section showing the relationship of the remaining flower parts to the floral cup, the flower being perigynous, **7** fruit, **8** fruit in cross section showing differentiation of the pericarp into a fleshy, edible exocarp and a stony endocarp (the pit). (Cf. also **Figure 8-3 A** apricot, fruit.)

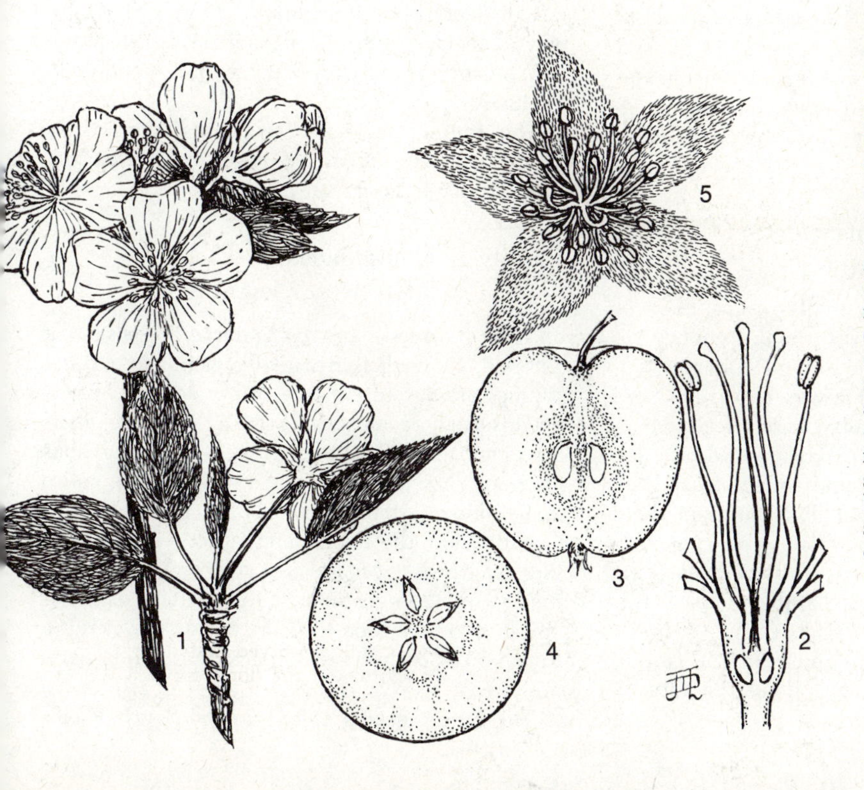

Figure 10-82. Rosales; Rosaceae (Pomoideae); *Pyrus Malus,* apple: **1** two flowering branches, **2** longitudinal section of the flower, showing the floral cup attached to the inferior ovary and the sepals, petals and stamens attached to the rim of the floral cup, **3** longitudinal section of the fruit, showing the pedicel (above) and the "blossom end" which retains the sepals and the dried stamens and styles (the core is the ovary, the fleshy part largely floral cup), **4** cross section showing the five seed chambers, **5** the "blossom end," showing the sepals and the dried stamens and styles. (Cf. also **Figure 8-2 B.**)

pistil is surrounded by the floral cup. The fruit enlarges greatly, and it consists ultimately of a single seed surrounded by two layers of ovary wall tissue—the fleshy and usually edible exocarp, and the hard and stony endocarp. At first thought the almond seems not to belong with the rest of this group, because the outer fleshy part of the fruit becomes dry and leathery at maturity, but this is identical in origin with the exocarp of a peach, plum, or cherry. The stony endocarp forms the shell of the almond, and the edible seed is enclosed within. Usually there is only one seed, but occasionally there are two.

The flowers of apples, pears, and quinces are epigynous, and the ovaries of the carpels are coalescent and (as in all epigynous flowers) adnate externally with the floral cup. The true ovary of the apple is the core plus a small amount of the surrounding fleshy tissue. The edible portion is largely the floral cup, which becomes exceedingly thick and fleshy. In common language, one end of the apple or pear is known as the stem end, the other as the blossom end. Usually at the blossom end all the part of the flower except the petals may be seen attached to mature fruits—the sepals become slightly enlarged and fleshy; the dried stamens usually are persistent; the styles and stigmas are dry but persistent. As in the rest of the family, the floral characters of the Ranales are present except that the flowers are not hypogynous. In this case, though, the carpels are coalescent.

Family 6. **Myrothamnaceae.** MYROTHAMNUS FAMILY

Small shrubs; twigs quadrangular, rigid, more or less spiny; resinous. Leaves simple, cuneate to fan-shaped, plicately folded along the nerves, opposite; stipules present. Flowers unisexual (and the plants dioecious), radially symmetrical, hypogynous, in erect terminal catkinlike spikes, each flower subtended by a single bract; sepals and petals none; stamens commonly 5 but sometimes 4–8, the filaments coherent except above; anthers splitting lengthwise; carpels 3, coalescent, the styles separate, the ovary 3-chambered; placentae axile; ovules numerous in each carpel. Fruit a small leathery capsule, the apices of the carpels free and spreading, dehiscent along the sutures; endosperm abundant.

The only genus is *Myothamnus* with 2-species occurring in South Africa.

Family 7. **Bruniaceae.** BRUNIA FAMILY

Suffrutescent to fruticose plants. Leaves simple, entire, crowded, alternate; stipules present. Flowers generally bisexual, radially symmetrical, epigynous, in heads or spikes; sepals 4–5, coalescent, persistent in fruit, subtended by a cycle of 5 bracts; petals 4–5, usually separate but sometimes coalescent basally; stamens 4–5, in 1 series, the filaments separate or sometimes adnate with the bases of the coalescent petals; anthers splitting lengthwise; carpels 2–3, coalescent, the styles separate or partly coalescent, the ovary 1–3-chambered; placentae apical; ovules 1–2 per carpel. Fruit a capsule, dehiscent along the suture or dry but indehiscent; endosperm abundant.

The family consists of several genera occurring in South Africa.

Family 8. **Brunelliaceae.** BRUNELLIA FAMILY

Similar to the Cunoniaceae (below) and perhaps not distinct from that family. Trees. Plants always dioecious; petals none; stamens 8–10; carpels 4–5, *separate; ovules 1–2* per carpel, pendulous. Fruit a follicle.

The family, whatever its status, consists of about 10 species of the genus *Brunellia* occurring from Mexico to Peru.

Family 9. **Cunoniaceae.** LIGHTWOOD FAMILY

Shrubs or trees. Leaves pinnate, opposite or whorled, the leaflets frequently glandular-serrate; stipules present, often large and coalescent. Flowers bisexual or unisexual (and then the plants dioecious), radially symmetrical, with hypogynous discs, small; sepals 4–5; petals 4–5 or none, entire to 2–3-lobed; stamens 4 or 5 to numerous, attached below the hypogynous disc; anthers splitting lengthwise; carpels 2–5, the styles separate, the ovary 2–5-chambered, placentae axile or apical; ovules several to many per carpel. Fruit a capsule or fleshy and indehiscent; endosperm present.

The family occurs chiefly in the South Pacific

and to a limited extent from Mexico to South America; one species is native in South Africa.

Family 10. **Eucryphiaceae.**

EUCRYPHIA FAMILY

Trees or shrubs. Leaves simple or pinnate, opposite, evergreen; stipules very small. Flowers bisexual, usually radially symmetrical, each with a hypogynous disc, solitary; sepals 4, coalescent apically and falling as the flower opens; petals 4, white; stamens numerous; anthers splitting lengthwise; carpels 5–12 or 18, the styles the same number; placentae axile; ovules a few in each carpel. Fruit woody or leathery, capsular.

The family consists of a single small genus occurring in Australia and Chile, one of many connecting links in the floras of the two sides of the southern portion of the Pacific Ocean.

Eucryphia is cultivated in some part of the United States.

Family 11. **Byblidaceae.** BYBLIS FAMILY

Herbs or small shrubs. Leaves simple, linear, numerous, pilose-glandular, alternate; stipules none. Flowers bisexual, radially symmetrical, hypogynous, solitary or in short racemes; sepals 5; petals 5, slightly coalescent basally; stamens 5, sometimes slightly adnate with the base of the corolla; anthers opening by pores or short slits; carpels 2–3, coalescent, the style 1, the ovary 2–3-chambered; placentae axile or apical (then with only 1–2 ovules). Fruit a capsule; seeds with endosperm.

The family consists of two species of the genus *Byblis* occurring in northern and western Australia and of perhaps the genus *Roridula* occurring in the mountains of southeast Africa (also interpreted as a separate family, *Roridulaceae*). The leaves of *Roridula* have long-stalked viscid (sticky) glands.

Family 12. **Pittosporaceae.**

PITTOSPORUM FAMILY

Shrubs or trees or sometimes woody vines. Leaves simple, leathery, alternate or sometimes whorled; stipules none. Flowers bisexual, radially symmetrical, hypogynous, in cymes or panicles; sepals 5; petals 5, usually separate, sometimes coalescent basally; stamens 5, separate; anthers opening usually by pores; carpels 2–5, often 3, coalescent, the style 1, the ovary either 1-chambered and with parietal placentae or 2–5-chambered and with axile placentae. Fruit a berry or a capsule; seeds with endosperm.

The family occurs through the warm temperate to tropical regions of the Eastern Hemisphere, and it is particularly abundant in the South Pacific. Several species of *Pittosporum* are common in cultivation from southern California to the Southern States.

Family 13. **Cephalotaceae.**

CEPHALOTUS FAMILY

Herbs; perennial from rhizomes. Leaves basal; foliage leaves entire, flat, elliptic or more or less elongate and acute at both ends, with no evident veins; other leaves with the blades specialized as pitchers resembling somewhat those of the *Nepenthaceae* (Sarraceniales, page 180), these with lids. Flowers bisexual, radially symmetrical, perigynous and with a disc; sepals 6, coalescent, colored, the tips forming hoods; petals none; stamens 12, in 2 series, separate; anthers splitting lengthwise; carpels 6, separate, in 1 series. Fruits follicular but 1-seeded; endosperm present.

The family includes a single species restricted to the southwestern corner of Australia. *Cephalotus* occurs in bogs, as do the pitcher-plants of the Sarraceniales.

According to Diels the development of the pitcher in *Cephalotus* is not the same as in the ultimately similar pitchers of *Nepenthes.* Whether the similarity is due to genetic relationship or parallel evolution can be determined only by thorough study.

*Family 14. **Saxifragaceae.**

SAXIFRAGE FAMILY

(Segregates: Each subfamily in family rank; also *Iteaceae*)

Herbs or shrubs or sometimes small trees; leaves simple or sometimes compound, commonly deciduous, alternate or in the *Hydrangeoideae* opposite; stipules none. Flowers bisexual or rarely unisexual, radially symmetrical (bilaterally symmetrical in *Tolmiea* and sometimes *Saxifraga*), perigynous or epigynous or (in *Penthorum, Chrysosplenium,* and extraterritorial groups) sometimes hypogynous; sepals and petals usually 5, sometimes 4, the petals sometimes reduced or none; stamens usually 5–10, rarely 4 or 3, when in 2 series the outer opposite the petals (obdiplostemonous); carpels usually 2, rarely 3–10, the styles and usually the upper portions of the ovaries free or the style rarely solitary because of coalescence, the compound ovary 1-chambered (placentae parietal or almost basal) or 2- (or rarely 3–10-) chambered (placentae axile); ovules few to many per carpel. Fruit a capsule or in *Penthorum* a pyxis (with 5–7 carpels slightly coalescent basally and circumscissile) or in *Ribes* a berry (ovary inferior, 1-chambered, with axile placentae); seeds with endosperm, this usually abundant.

The family occurs over most of the earth, and about one third of the 70 or 80 genera occur in North America. The *Saxifragaceae* center particularly on the Pacific Coast, where there are numerous endemic genera, but they are common in all moist regions of the Continent from the Arctic Coast southward.

Key to the Subfamilies

(North America north of Mexico)

1. Herbs.
    2. Staminodia none.
        3. Carpels 5–7; fruit opening along a horizontal line and the top coming off like a lid.

            Subfamily 1. **Penthoroideae.** (Penthorum subfamily).

        3. Carpels 1–2 or rarely 3; fruit not opening along a horizontal line.

            Subfamily 2. **Saxifragoideae.** (saxifrage subfamily).

    2. Staminodia present, palmately cleft into elongated threadlike structures terminated by glands; stem a caudex producing simple and entire basal leaves and 1-leaved scapes.

        Subfamily 3. **Parnassioideae.** (*Parnassia* subfamily).

1. Shrubs.
    2'. Fruit a capsule.
        3'. Leaves opposite; ovary at least partly inferior (i.e., at least the base of the ovary attached to the floral cup); carpels 3–10.

            Subfamily 4. **Hydrangeoideae.** (*Hydrangea* subfamily).

        3'. Leaves alternate; ovary superior, carpels (of the North American genus) 2.

            Subfamily 5. **Escallonioideae.** (*Escallonia* subfamily).

    2'. Fruit a berry; ovary inferior; superior floral cup, calyx, and stamens remaining attached to the upper part of the fruit; ovary 1-chambered, with 2 parietal placentae; leaves alternate.

        Subfamily 6. **Ribesioideae.** (currant subfamily).

Engler describes three other subfamilies not occurring in the United States. These are the *Francoöideae,* including two genera occurring in the mountains of Chile; the *Pterostemonoideae,* consisting of a single genus of two species occurring in Mexico; and the *Baueroideae,* composed of a single genus of three species occurring in New South Wales and Tasmania.

1. **Penthoroideae.** *Penthorum sedoides* occurs in eastern North America. It is transitional to the *Crassulaceae,* where it has been classified.

2. **Saxifragoideae.** By far the largest subfamily in North America is the *Saxifragoideae,* including the saxifrages, miterworts, alumroots, woodland-stars, and coral-bells.

3. **Parnassioideae.** Various species of *Parnassia* (grass of Parnassus) occur in moist regions across the Continent (including the major mountain ranges of the West).

4. **Hydrangeoideae.** The most widespread members of the subfamily are the mock-oranges (sometimes known as syringas). These shrubs (*Philadelphus*) occur mostly in moist areas from coast to coast, but the individual species are restricted to particular regions. The climbing hydrangea (*Decumaria*) occurs in the Southern States and as far north as Tennessee and Virginia. A highly variable species of *Hydrangea* occurs in eastern North America. *Jamesia* is abundant through especially the southern portion of the Rocky Mountain Region. The remarkable genus *Carpenteria* occurs in a limited area in the foothills of the Sierra Nevada of California in the vicinity of the San Joaquin River and the Kings River. Its foliage and its large white flowers make it a favorite in cultivation. An obscure genus (*Whipplea*) occurs in the forest belt of the Pacific Coast.

5. **Escallonioideae.** The subfamily is represented in North America by only the Virginia

Figure 10-83. Rosales; Saxifragaceae (Saxifragoideae); *Heuchera* (hybrid), alum root: **1** basal leaf, **2** inflorescence, **3** flower, side view, showing the floral cup and sepals and the tips of the petals, stamens, and styles, **4** floral cup laid open, showing the sepals, petals, and stamens attached to it and the pistil composed of two incompletely coalescent carpels, **5** the same with a section of the ovary wall removed, **6** cross section of the ovary showing the ovules and the placentae. Cf. also **Figure 2-14 B** *Saxifraga flagellaris.*)

Figure 10-84. Rosales; Saxifragaceae (Hydrangeoideae); *Philadelphus*, mock orange or "syringa": **1** branch with leaves and flowers, **2** flower with one petal removed, showing the floral cup (adherent with the inferior ovary), sepals, petals, stamens, and free portion of the pistil, **3** flower, from beneath, **4–5** relationship of petals and stamens, **6** young fruit, **7** fruit, longitudinal section, showing the elongate seeds and a partition, **8–9** seeds, **10** half of a cross section of the ovary.

willow (*Itea virginica*), which occurs in the Southern States and as far northward as Pennsylvania.

6. **Ribesioideae.** The subfamily includes only the genus *Ribes* (currant and gooseberry), consisting of about 125 species occurring through the temperate regions of the Northern Hemisphere and southward along the Andes in South America. Some authors segregate the gooseberries from *Ribes* as a separate genus (*Grossularia*), but there are transitional species in which the characters of the two groups are scrambled. Usually the gooseberries can be distinguished by the presence of prickles.

The family is highly variable, and it includes hypogynous, perigynous, and epigynous flowers even within the single genus *Saxifraga*. The degree of cohesion of the carpels varies from a slight basal coalescence to complete coalescence including even the stigmas. Most of the genera are herbs, but certain groups are large shrubs.

*Family 15. **Droseraceae.**

SUNDEW FAMILY

(Segregate: *Dionaeaceae*)

Annual or perennial herbs or sometimes shrubby plants; glands abundant, stalked. Leaves simple, basal or a few alternate, covered with stalked glands effective in capturing insects (cf. discussion of the order); stipules none. Flowers bisexual, radially symmetrical, in raceme- or panicle-like inflorescences; sepals 4–5, persistent in fruit; petals 5, separate; stamens 5–20, in series of 5, separate or rarely coalescent basally; carpels 3–5, the styles 3–5, the ovary 1-chambered; placentae parietal; ovules several to many per carpel. Fruit a capsule.

The largest genus is *Drosera* (sundew), which occurs almost throughout the world; the other 3 genera are composed of single species. Two species of sundew (*Drosera*) occur across the northern end of the Continent and southward to the

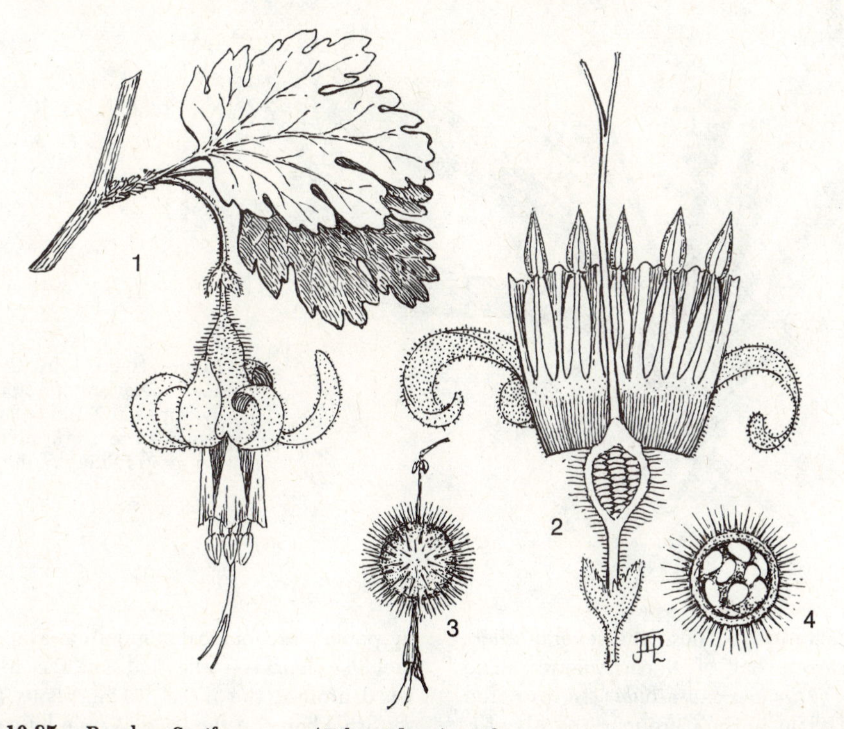

Figure 10-85. Rosales; Saxifragaceae (Ribesoideae); *Ribes amarum*, a gooseberry: **1** branch with leaves and a flower, this showing the floral cup adherent with the inferior ovary and the sepals, petals (smaller than the sepals), stamens, and style (branched), **2** flower, laid open, showing the pistil (in longitudinal section) and the sepals, petals, and stamens, **3** fruit, **4** fruit, cross section, the placentae parietal.

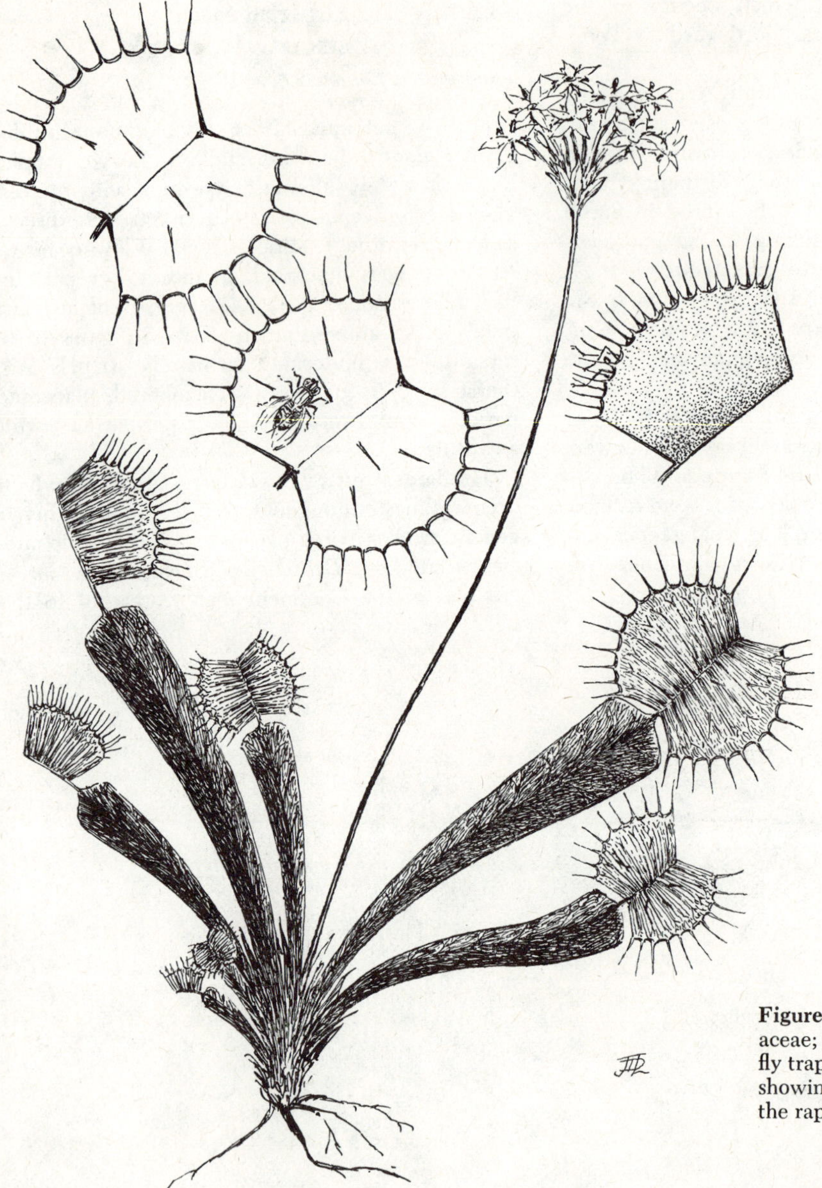

Figure 10-86. Rosales; Sarraceniaceae; *Dionaea muscipula,* Venus fly trap: habit and enlargements showing the capturing of insects by the rapid folding of the basal leaf.

mountains of California. These and several others occur in the eastern half of North America. The Venus flytrap (*Dionaea muscipula*) is restricted to the Coastal Plain in the Carolinas (cf. above).

In the *Droseraceae* leaf glands, for example the sundew, secrete a clear, sticky fluid in which small insects may become trapped as though on fly paper. Mechanical stimulation also causes tentaclelike hairs on the leaf surface to curve inward around the insect. The Venus flytrap has sensitive hairs on the base of each lobe of the leaf. When an insect touches these, the pair of leaf lobes snaps shut like a steel trap, capturing the insect. The elongated hairs of the leaf surface surround the victim until it is digested and absorbed.

Order 39. *PODOSTEMALES.*

RIVERWEED ORDER

DEPARTURE FROM THE FLORAL CHARACTERS OF THE RANALES

Free	Sepals (when present) sometimes coalescent; stamens sometimes coalescent; carpels coalescent.
Equal	Stamens at one side of the flower, the lateral ones sterile.
Numerous	Stamens 1–4 or rarely numerous; carpels 2–3.
Spiral	Stamens solitary or usually cyclic; carpels cyclic.

Aquatic herbs, usually submerged, commonly resembling algae, mosses, or liverworts. Flowers minute, hypogynous; petals none. Seeds without endosperm.

An order of peculiar plants possibly related to the *Crassulaceae* and *Saxifragaceae*. However, their floral parts are much reduced, and relationships are obscured. Tentatively the order is accepted as with affinity to the primitive Rosales but highly specialized in its own ways.

*Family 1. **Podostemaceae.**

RIVERWEED FAMILY

Aquatic herbs, the North American species submerged and growing attached by green, creeping roots of various forms to rocks in running streams, often about waterfalls; stem resembling a seaweed, a liverwort, or a moss, branching and forming a structure resembling a thallus, the branches ranging from threadlike to broad and flattened, the leaves, if any, being on special branches; juice of the plant often milky. Leaves, when present, simple, linear or broad, alternate. Flowers bisexual, bilaterally symmetrical, hypogynous, minute, solitary or in cymes, flower bud enclosed in a spathe (or coalescent bracts); sepals 2–3, often coalescent; petals none; stamens 1–4 or rarely numerous, coalescent at least basally on one side of the flower; anther usually with 4 pollen chambers; carpels 2–3, the styles separate, the ovary 2–3-chambered (or rarely 1-chambered); placentae axile (the placenta free and central when the ovary is 1-chambered); ovules numerous in each carpel. Fruit a capsule.

The family (100 to 150 species) occurs throughout the Tropics, and it extends here and there into temperate regions. One species of *Podostemum* occurs in eastern North America as far north as western Quebec.

Order 40. *PROTEALES.* SILK-OAK ORDER

DEPARTURE FROM THE FLORAL CHARACTERS OF THE RANALES

Free	Sepals coalescent.
Equal	Calyx sometimes bilaterally symmetrical.
Numerous	Stamens 4; carpels 1.
Spiral	Stamens cyclic; carpel solitary.

Commonly trees or shrubs. Leaves alternate, entire to bipinnate; stipules none. Flower perigynous, with a long, colored floral tube, 4-merous; sepals coalescent, petal-like, colored; stamens opposite the sepals; ovary 1-chambered, the placentae parietal, the ovules 1–many, each with 2 integuments. Fruit a follicle or sometimes indehiscent. Endosperm none.

*Family 1. **Proteaceae.** SILK-OAK FAMILY

The family includes about 50 genera and 1,000 species occurring chiefly in the arid regions of the Southern Hemisphere, i.e., South Africa and Australia. The silk-oak (*Grevillea robusta*) has become naturalized in waste places in Florida. A host of genera and species are endemic to Cape Province, South Africa, where they are a major part of the chaparral-type vegetation. *Protea,* a major genus there, is coming into cultivation through the warmer parts of the world.

The silk-oak is cultivated commonly in California and other states along the southern edge of the United States. The Queensland nut (*Macadamia*) is cultivated for its fruit. Other genera, including *Hakea,* are cultivated as ornamental plants.

In *Grevillea* the style elongates more rapidly than the floral tube while the stigma remains enclosed in the calyx; the tension splits the tube along one side, and the long curve of the style bursts through the side of the tube.

Order 41. *ELAEAGNALES.*

OLEASTER ORDER

DEPARTURE FROM THE FLORAL CHARACTERS OF THE RANALES

Numerous	Stamens 4 or 8; carpel 1.
Spiral	Stamens cyclic.
Hypogynous	Perigynous (appearing epigynous as described below).

Shrubs or sometimes trees; vegetative parts and fruits covered densely with rusty, silvery, or golden scalelike hairs (lepidote or stellate). Leaves simple, entire, either alternate, opposite, or whorled. Stipules none. Flowers bisexual or unisexual (and then the plant dioecious or polygamodioecious), radially symmetrical, perigynous, the upper enlarged portion of the floral tube often deciduous, leaving the lower portions investing the fruit, often giving the appearance of an inferior ovary; sepals 4; the single carpel with only 1 ovule, this basal, anatropous, the integuments 2. Fruit an achene, enveloped by the remains of the lower portion of the floral tube. (There are no petals.)

*Family 1. Elaegnaceae.

OLEASTER FAMILY

The family is composed of 3 genera ranging from North America to Europe and southern Asia. The oleaster (*Elaeagnus*) occurs from the Rocky Mountain Region to the Atlantic Coast. One species is known as silverberry. The genus *Shepherdia* includes two common species, the soapberry (Alaska to Newfoundland and southward to Western Washington, the Rocky Mountains, and the more northerly portions of the Middle West and the East) and the buffaloberry (eastern portion of the Rocky Mountain System and the Great Plains and sparingly westward to southeastern California. The buffaloberry is a somewhat thorny shrub as much as 20 feet high. The leaves are silvery on both sides and striking in appearance.

Order 42. *MYRTALES.* MYRTLE ORDER

DEPARTURE FROM THE FLORAL CHARACTERS OF THE RANALES

Free	Petals rarely coalescent, as in *Eucalyptus;* carpels, including (except in the *Haloragidaceae*) the styles and usually the stigmas, coalescent.
Numerous	Stamens usually numerous or sometimes 3, 4, 5, 6, 8, or up to 15, or frequently double these numbers; carpels sometimes numerous, but usually about 4 (as few as 2).
Spiral	Stamens sometimes cyclic; carpels cyclic.
Hypogynous	Flowers epigynous or in the *Lythraceae,* some *Melastomataceae, Thymelaeaceae,* and some extra-territorial families perigynous. Extreme epigyny is typical of the order.

Herbs or mostly shrubs or trees. Stems of a group of families with accessory phloem internal to the xylem. Leaves opposite or less often alternate; stipules none or rarely present and minute or early deciduous. Flowers bisexual, 3–15-merous, but commonly 4- or 5-merous (with the exception of the often numerous stamens), the stamens, when few, frequently in 2 series; ovary divided into chambers; placentae axile; ovule anatropous or sometimes campylotropous (some *Myrtaceae*).

Even though a few members of the Myrtales are perigynous, the prevailing flower type is epigynous (usually strikingly so) and the flower structure is more or less similar to that of an apple, pear, or quince. However, the carpels are completely coalescent or nearly so, the styles being coalesced into a single compound style and the stigmas usually coalescent. This is a major point of difference between the Myrtales and the Rosales. In the Rosales there may be one or more separate carpels each with its own style. If the ovary is inferior the carpels are coalescent, but nearly always they are coalescent by only the ovaries and the styles and stigmas remain free. This may be summarized as follows:

Rosales	***Myrtales***
Styles rarely coalescent, the carpels either separate or solitary or only the ovaries coalescent.	Carpels coalescent by both the ovaries and the styles and usually by the stigmas.

Another useful distinguishing feature of the Rosales and Myrtales is the almost complete absence of stipules in the latter order and their presence in the great majority of species (but only half the families) of the former. 4-merous flowers and presence of some phloem internal to the xylem are common features of the Myrtales but not of the Rosales.

Key to the Families

(North America north of Mexico)

1. Plant *not* corresponding with *every character* in the combination in lead 1 below.
   2. Anthers opening by apical pores, not splitting lengthwise.

      Family 5. **Melastomataceae** (meadow-beauty family).

   2. Anthers splitting lengthwise, opening through their entire length.
      3. Flowers perigynous.
         4. Petals usually present; carpels 2–6, the ovary 2–6-chambered; ovules 2–many per carpel; fruit a capsule.

            Family 1. **Lythraceae** (*Lythrum* family).

         4. Petals none, the sepals usually colored; carpels 2, the ovary 1-chambered through abortion of 1 carpel, sometimes 2-chambered by intrusion of the placenta; ovule 1 per carpel, but usually only 1 carpel developed; fruit a drupe or a small nut, or sometimes a berry or a capsule.

            Family 7. **Thymelaeaceae** (mezereum family).

      3. Flowers epigynous.
         4'. Stamens twice or rarely thrice or even four times as many as the sepals or petals, rarely the same number as the petals (or sepals, if petals are lacking), cyclic, in 1 or more series.
            5. Leaves leathery; trees, shrubs, or woody vines.
               6. Carpels 2–5, coalescent; seed germinating in the fruit and forming a protruding embryo 1–3 dm. long.

                  Family 10. **Rhizophoraceae** (mangrove family).

               6. Carpel 1; seed germinating on the ground.

                  Family 11. **Combretaceae** (white mangrove family).

            5. Leaves not leathery; herbs or sometimes suffrutescent plants or (extraterritorial) shrubs; sepals, petals, and carpels usually 4, very rarely 2, 5, or 6 (or very rarely the carpel 1); stamens usually 8, in any case twice as many as the sepals or petals.

               Family 14. **Onagraceae** (evening-primrose family).

         4'. Stamens numerous and of an indefinite number, arranged spirally.
            5'. Seeds not embedded in watery, edible pulp; petals not crinkled; floral cup or tube usually not leathery, but sometimes so.

               Family 12. **Myrtaceae** (myrtle family).

            5'. Seeds each embedded in a watery, edible pulp; petals crinkled; floral tube and calyx leathery.

               Family 13. **Punicaceae** (pomegranate family).

1. Fruit a single nut or nutlet or the ovary separating at maturity into 4 nutlets; aquatic or mud plants.
   2'. Fruit a large nut, with 2–4 sharp conspicuous horns.

      Family 15. **Hydrocaryaceae** (water-chestnut family).

   2'. Fruit a nutlet (not more than 2–3 mm. long), sometimes with small horns.
      3'. Submersed leaves from toothed to dissected into fine lobes or divisions; stamens 3, 4, or 8; stigmas 3–4; ovary 3–4-chambered.

         Family 16. **Haloragidaceae** (water-milfoil family).

      3'. Submersed leaves (and all others) entire, whorled; stamen 1; style and stigma 1; ovary 1-chambered.

         Family 17. **Hippuridaceae** (mare's-tail family).

*Family 1. Lythraceae. LYTHRUM FAMILY

Herbs, shrubs, or trees. Leaves simple, usually entire, opposite or whorled; stipules occasionally present, minute. Flowers bisexual, ordinarily radially symmetrical, perigynous; sepals 4–8 or sometimes none, crumpled in the bud; stamens in 2 series, commonly twice as many as the petals; carpels 2–6 or commonly so, the ovary ordinarily with as many chambers as carpels, rarely with 1 chamber; placentae ordinarily axile; ovules several to many per carpel. Fruit a capsule.

The family is distributed widely, and it is particularly abundant in the tropical regions of South America. The North American plants are herbaceous. Several genera of small marsh and ditch plants occur through most of the Continent except the extreme north. These are *Lythrum, Ammania, Rotala,* and *Peplis* (water-purslane). A larger plant ranging from 2 to 3 feet or more in height is the Northern and Eastern water-willow (*Decodon*). Species of *Cuphea* occur in Arizona and Texas, *Heimia* in Texas.

An Asiatic species, the crepe myrtle (*Lagerstroemia*), is an excellent small tree for avenue planting in the southern part of the United States. It bears a mass of flowers through the latter half of the summer. It has gone wild in parts of the South.

Family 2. **Crypteroniaceae.** CRYPTERONIA FAMILY

Trees. Leaves simple, entire, opposite. Flowers perigynous, in spikelike racemes or panicles, small; sepals 4–5, white or green; petals none; stamens 4–5, alternate with the sepals; carpels 2, the ovary 2-chambered; placentae axile; ovules many. Fruit a capsule, the style persistent.

The family consists of a single genus occurring in India and the Malay Archipelago.

Family 3. **Sonneratiaceae.** SONNERATIA FAMILY

Trees. Leaves simple, entire, opposite. Flowers perigynous or epigynous, the ovary being sometimes partly inferior; flowers solitary or in groups of 3; sepals 4–8; petals 4–8 or sometimes none, small; stamens numerous, arranged spirally; carpels numerous, about 15–25, the ovary many-chambered; placentae axile; ovules numerous, ascending. Fruit a berry.

The family consists of 2 genera associated with the mangrove along tropical shores from east Africa to Australia.

Family 4. **Oliniaceae.** OLINIA FAMILY

Shrubs or small trees; branches square in cross section. Leaves simple, opposite; stipules none. Flowers bisexual, radially symmetrical, epigynous; sepals 4–5; petals 5 or rarely 4, pilose basally; stamens 4–5, the filaments short; carpels 3–5, the style 1, the ovary 3–5-chambered; placentae axile; ovules 1–3 per carpel, pendulous. Fruit drupelike, with 1 seed in each carpel.

The family includes a single genus occurring on Saint Helena Island in the Atlantic Ocean and of others in southern and eastern Africa.

Family 5. **Penaeaceae.** PENAEA FAMILY

Small shrubs. Leaves simple, entire, usually sessile, opposite, decussate and often imbricate; stipules none or minute and glandular. Flowers bisexual, radially symmetrical, perigynous, subtended by bracteoles; sepals 4, persistent in fruit; petals none; stamens 4, alternate with the sepals, the filaments short; carpels 4, the style 1, the ovary 4-chambered; ovules 2–4 per carpel. Fruit a capsule; seed often 1 per chamber.

The family consists of about 5 genera occurring in South Africa.

Family 6. **Geissolomataceae.** GEISSOLOMA FAMILY

Similar to the *Penaeaceae;* floral cup short, the sepals being more elongate, petaloid, imbricate instead of valvate; stamens 8, inserted near the base of the floral cup, in 2 series, the filaments slender; the ovary with 4 narrow wings; styles 4, slightly coalescent at the *apices;* ovules 2 in each chamber of the ovary; placentae apical, the ovules pendulous.

The family consists of a single genus occurring in the mountains of South Africa. According to Carlquist, the wood anatomy is unlike that of the Myrtales, and doubtless the family will be shown to belong elsewhere.

*Family 7. **Thymelaeaceae.** MEZEREUM FAMILY

(Synonym: *Daphnaceae*)

Shrubs, small trees, or rarely herbs. Leaves simple, entire, alternate or opposite, stipules none. Flowers usually bisexual, when unisexual the plants dioecious, radially symmetrical except for the carpels; petals none, the petaloid scales appendages of the sepals; stamens in 1 or 2 series. Ovary usually 1-chambered through abortion of 1 carpel, sometimes 2-chambered by placental intrusion; ovule 1, anatropous, suspended from the top of the ovary. Endosperm little or none.

Fruit a drupe or a small nut, sometimes a berry or a capsule.

The family is represented in North America by two species of leatherwood (*Dirca*), one being known also as moosewood. One occurs in the San Francisco Bay Region of California and the other in eastern North America. Mezereum (*Daphne*) is cultivated widely, and here and there in eastern North America it has become naturalized.

A recent study by Heinig shows that the apparently solitary carpel is accompanied by a rudiment of another. The family appears possibly to be related also to the *Flacourtiaceae* (Violales) or the Malvales.

*Family 8. **Melastomataceae.**

MEADOW-BEAUTY FAMILY

Herbs, shrubs, trees, or woody vines. Leaves simple, commonly palmately veined, usually opposite, rarely alternate. Flowers bisexual, radially symmetrical, perigynous or epigynous; sepals 5 or rarely 3–6; petals usually 5; stamens usually 10 in 2 series, rarely in 1 series, each anther opening by a single apical pore; carpels 2 or 4–14, the ovary usually 4–14-chambered; placentae axile or more or less basal; ovules numerous in each carpel. Fruit a capsule or a berry.

This large tropical family includes between 3,000 and 4,000 species, these being chiefly in the warm areas of the Western Hemisphere. In eastern North America there are several species of meadow-beauty (*Rhexia*). A typical West Indian tree or large shrub (*Tetrazygia*) is an abundant native plant on hammocks and pinelands on the Everglade Keys in Florida. According to Small, this plant and other West Indian species are so numerous on the Everglade Keys as to afford evidence that these islands should be considered as a part of the West Indies. However, this point of view is open to question (cf. p. 833).

Family 9. **Lecythidaceae.**

BRAZIL NUT FAMILY

Shrubs or trees. Leaves simple, alternate. Flowers epigynous (the ovary being inferior or partly inferior), large; sepals 4–6; petals 4–6, sometimes coalescent, the corolla then being campanulate and ribbed; stamens numerous, arranged spirally, commonly coalescent on one side of the flower; carpels 2–several, the ovary with as many chambers as there are carpels; placentae axile; ovules 1–many. Fruit indehiscent or opening at the apex, woody, fibrous, or fleshy.

The family is tropical and it includes about 20 genera.

The edible products of the family include Brazil nuts, paradise nuts, and anchovy pears. Various species are used for their wood, and the genus *Couratari* yields the fiber of the same name.

*Family 10. **Rhizophoraceae.**

MANGROVE FAMILY

Shrubs or usually trees; branches swollen at the nodes. Leaves simple, coriaceous, usually opposite; stipules either none or present but early deciduous. Flowers sometimes unisexual (and the plant monoecious), perigynous or epigynous (the ovary then being partly or wholly inferior), also with a perigynous or epigynous disc, in cymes or solitary; sepals 3–14; petals 3–14, somewhat succulent or leathery, folded inward in the bud; stamens 2–4 times as numerous as the petals; anthers usually with 4 pollen chambers; carpels 2–4, the ovary usually 4-chambered; placentae axile; ovules usually 2 per carpel, pendulous. Fruit commonly a berry, rarely a capsule or a drupe.

The family is distributed throughout the Tropics, and mangrove vegetation, including not only the mangrove but other genera of this family, is characteristic of the beaches and tidelands in the Tropics of especially the Eastern Hemisphere. The mangrove, *Rhizophora Mangle,* is called red mangrove, and it occurs on the coast of peninsular Florida and the Florida Keys along the shores of creeks and rivers as far upward as the water is brackish. The plant grows best in brackish water though it is found most commonly in salt water. The heartwood is reddish brown and streaked. Since it is hard and heavy and close-grained, it is used for cabinet work. The tree is supported partly by prop roots which arise from the main trunk and from the large branches. These roots interlace about the base of the trunk. The embryo develops while the fruit is still on the tree, and the enlarged embryo turns downward when the fruit floats in the water. When the embryos are stranded on muddy places, they grow rapidly.

Figure 10-87. Myrtales; Rhizophoraceae; *Rhizophora Mangle,* mangrove (or red mangrove), growing on Big Pine Key, Florida: *above,* plant growing in salt water on the beach, showing the prop roots; *middle,* close up view of the prop roots; *below,* seedlings with conspicuous hypocotyls and tap roots, developed from the fruits.

*Family 11. **Combretaceae.** WHITE MANGROVE FAMILY

(Synonym: *Terminaliaceae*)

Shrubs, trees, or woody vines. Leaves simple, alternate or sometimes opposite. Flowers rarely unisexual, occasionally bilaterally symmetrical, epigynous, in racemes, spikes, or panicles; sepals usually 4–5 or sometimes 8, small; stamens either twice as many as the sepals, that is, in 2 series, or sometimes only 2–5; carpel 1; ovules 2–6; pendulous from an apical placenta. Fruit drupelike, coriaceous.

The family occurs throughout the Tropics. Four genera occur in south-peninsular Florida and on the Keys. One of these, the West Indian almond (*Terminalia*) is introduced; the others are native. They are as follows: *Conocarpus* (buttonwood), the bark of which is used medicinally and for tanning and the wood for making charcoal; *Bucida* (black-olive), which has a light yellow-brown, close-grained, very heavy, and exceedingly hard wood; *Languncularia* (white mangrove), which has fragrant flowers and a dark, yellow-brown, close-grained, hard, heavy wood.

*Family 12. **Myrtaceae.** MYRTLE FAMILY

Shrubs or trees. Leaves simple, usually entire, coriaceous, markedly aromatic, producing essential oils, usually opposite. Flowers bisexual, radially symmetrical, epigynous or in some extraterritorial groups perigynous; sepals usually 4–5 (in *Eucalyptus* small or wanting); petals usually 4–5 (in *Eucalyptus* forming a cap which falls away as the flower bud opens); stamens numerous; carpels 3 or more, often 4, the ovary 1-chambered; placentae parietal, but strongly intruded and sometimes coalescent and thus becoming axile and dividing the ovary into chambers; ovules 2–many per carpel, anatropous or campylotropous. Fruit a berry or a capsule, often dehiscent apically about the attachment of the floral cup or tube.

The family includes about 2,500 species distributed widely through the Tropics but particularly abundant in Australia and the American Tropics. Several species of *Eugenia* are native in south peninsular Florida and on the Keys. These are known as Spanish stopper, white stopper, red stopper, and ironwood. The spicewood and the

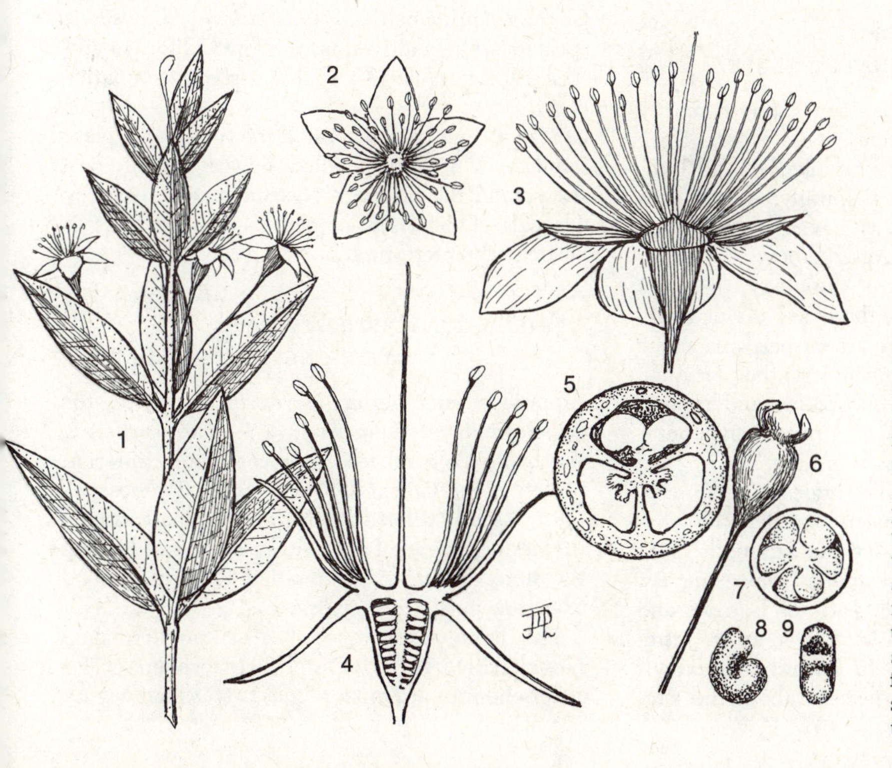

Figure 10-88. Myrtales; Myrtaceae; *Myrtus communis,* myrtle: **1** leafy branch with flowers **2** flower, top view, showing the petals, stamens and style, **3** flower with two petals removed, showing the floral cup (covering the adnate inferior ovary) and the sepals, petals, stamens, and style, **4** longitudinal section, **5** cross section of the ovary, **6** fruit with the persistent sepals, **7** fruit, cross section, **8–9** views of a seed.

myrtle-of-the-river (species of *Calyptranthes*) occur in the same area. The bottle-brush (*Melaleuca Leucadendra*) and the guava (*Psidium Guajava*) have escaped from cultivation in peninsular Florida and on the Keys. The guava is an edible fruit.

Numerous species of *Myrtaceae,* including *Feijoa* (pineapple guava), *Eugenia, Melaleuca* (bottle-brush), *Callistemon* (bottle-brush), *Leucadendron* (Australian tea-tree), and *Tristania* (Brisban-box), are cultivated from California to the Southern States; and the myrtle (*Myrtus communis*) is planted more widely. The genus *Eucalyptus* is native in Australia where there are great numbers of species, many having been introduced in America especially in California. Ordinarily it does not maintain itself outside of cultivation, but in one locality 2 or 3 species are reported to be established locally in California. Some products of economic importance are allspice (*Pimenta dioica,* immature fruit), bay rum (*Pimenta racemosa,* oil), and cloves (*Syzygium,* flower buds). A number of plants have edible fruits, the guavas being cultivated in the southern parts of the United States.

*Family 13. **Punicaceae.**

POMEGRANATE FAMILY

Large shrubs or small trees. Leaves simple, mostly opposite. Flowers epigynous; sepals 5–8, thick and leathery; petals 5–8, crumpled in the bud; stamens numerous, arranged spirally; carpels usually 8–12 but sometimes as few as 3, arranged in 2 series, the outer elevated above the inner during early development of the ovary, the cross section therefore complex, the ovary divided into as many chambers as there are carpels; placentae axile, protruding into the chambers, the placentae of the outer series of carpels displaced and appearing parietal; ovules numerous. Fruit a berry; outer coating of the seed fleshy and edible.

The family consists of the single genus *Punica,* which is composed of 2 Asiatic species, the common one being the cultivated pomegranate. This plant has been cultivated at least since ancient Greek times in the warmer parts of Europe, and more recently of the United States, as an introduction from the Orient. In favorable situations in the Southern States it persists about the sites of the old homesteads. It has become established in Florida.

*Family 14. **Onagraceae.**

EVENING-PRIMROSE FAMILY

(Synonym: *Epilobiaceae*)

Herbs or in extraterritorial genera rarely shrubs or trees. Leaves alternate or opposite; stipules rarely present and early deciduous in extraterritorial genera. Flowers epigynous, solitary or in racemes or spikes; sepals 4 or rarely 2–5; petals 4 or none; stamens ordinarily in 2 series, the total therefore usually 8; carpels usually 4, rarely 2 or sometimes 5, the ovary with as many chambers as there are carpels; ovule 1 to many per carpel. Fruit a capsule or in *Fuchsia* a berry and in *Circaea* and *Gaura* a nut or a nutlet.

The family is distributed throughout the world, but it is most abundant in the Western Hemisphere. The largest genera are *Oenothera,* the evening-primroses, and *Epilobium.* A well-known plant in northern and mountain forests is the fireweed (*Epilobium angustifolium*). Other plants are the water-primroses (*Ludwigia*) and the West-American *Clarkia* and (or better including) *Godetia* (principally of California). *Fuchsia* is spectacular in cultivation, and in California and Arizona, *Zauschneria* is equally striking as a native plant.

This is one of the most attractive of all plant families. The four-petalled flowers usually are showy and the primness of arrangement (usually 4 sepals, 4 petals, 8 stamens, and 4 carpels) is particularly interesting.

*Family 15. **Hydrocaryaceae.**

WATER-CHESTNUT FAMILY

Aquatic annual herbs; leaves in a rosette, the petioles inflated; stamens 4, in 1 series; carpels 2, the ovary 2-chambered, not completely inferior; placentae fundamentally axile; ovules 2 per carpel, but 1 usually abortive. Fruit a turbinate drupe, the fleshy pericarp falling away early and leaving the stony endocarp, this with 2–4-horns formed from the persistent sepals.

The family is composed of a single genus, *Trapa,* native to the Eastern Hemisphere. The water-chestnut (*Trapa natans*) is called some-

Figure 10-89. Myrtales; Onagraceae; *Oenothera Hookeri*, yellow evening primrose: **1** upper portion of a branch with flowers, **2** lower portion with fruits, **3** leaf, **4** flower, showing the inferior ovary enclosed in the adnate floral tube, the superior floral tube above the ovary, the four sepals still partly coalescent and turned to one side in anthesis, the petals, and the stigmas, **5** flower laid open, **6** anther with cobwebby masses of pollen, **7** stigma with pollen, **8** pollen grain, **9** ovary, cross section, **10** attachment of ovules to the placenta.

times the water-caltrop because of its horned edible "nut." The seed is an important article of diet in the Orient, and it appears in Chinese-American dishes. The species has become established in lakes and rivers from New York to Massachusetts and Maryland.

*Family 16. **Haloragidaceae.**

WATER-MILFOIL FAMILY

(Synonym: *Gunneraceae*)

Aquatic herbs or some extraterritorial genera terrestrial. Leaves simple or pectinately pinatifid. alternate, opposite, or whorled. Flowers usually unisexual (and then the plant monoecious or polygamomonoecious), epigynous; sepals 2–4 or none; petals usually none, when present 2–4, early deciduous; stamens 4 or 8, the outer series, when present, opposite the petals; carpels 4, the ovary 4-chambered; placentae axile; ovule 1 per carpel, pendulous or, when occasionally the ovary has only 1 chamber, the ovule 1. Fruit a nutlet or a drupe, sometimes wingéd.

The family occurs throughout the world but mostly in the Southern Hemisphere. The water-milfoil (*Myriophyllum*) is represented by a number of species in eastern North America and in the northern part of western North America as well as southward along some of the mountain chains. The parrot-feather, cultivated in fish ponds and pools, is of this genus. Another genus is *Proserpinaca,* the mermaid-weed.

*Family 17. **Hippuridaceae.**

MARE'S-TAIL FAMILY

Perennial herbs. Leaves simple, entire, narrow, whorled. Flowers minute, bisexual and some unisexual (and then the plants polygamous), epigynous, sessile in the leaf axils; sepals completely coalescent but not individually visible; petals none; stamen 1; carpel 1, with 1 ovule suspended from the top. Fruit nutlike.

The family consists of a single genus, *Hippuris.* Some consider the genus to consist of a single polymorphic species; others consider it to include about 3. The mare's-tail (*Hippuris vulgaris*) is a common plant in the main mountain chains of the West. The same or a segregated species occurs from Alaska to Labrador and Quebec.

Order 43. *SANTALALES.*

SANDALWOOD ORDER

DEPARTURE FROM THE FLORAL CHARACTERS OF THE RANALES

Free	Petals sometimes coalescent (*Olacaceae*), the stamens adnate with the petals; carpels coalescent (or rarely solitary).
Numerous	Stamens 1–5 or sometimes double the number, that is, in two series; carpel 1 or the carpels commonly 2–3.
Spiral	Stamens and carpels cyclic.
Hypogynous	Flowers epigynous.

Herbs or shrubs or sometimes small trees, the plants commonly but not always partly or wholly parasitic. Leaves simple, alternate or opposite, the plants rarely leafless; stipules none. Flowers radially symmetrical, 2–5-merous; petals 0–5, stamens usually in one series, but sometimes in two (as in many *Olacaceae*), style 1; ovary usually 1-chambered; placentae basal or axile and free from the walls of the carpels or subapical and the ovules hanging from the upper portion of the ovary or more or less fused with the placentae; ovules few or solitary. Fruit a drupe, a berry, or an achene; endosperm present, of considerable quantity, usually enveloping the embryo.

According to Rendle, the order is allied closely to the Proteales, the chief difference being the inferior ovary.

Key to the Families

(North America north of Mexico)

1. Petals coalescent at least basally; stamens usually but not always twice as many as the petals.

 Family 1. **Olacaceae** (tallowwood family).

1. Petals separate or none; stamens as many as the sepals.
 2. Plants parasitic (and growing) upon the branches of trees or shrubs; leaves opposite; fruit a berry.

 Family 2. **Loranthaceae** (mistletoe family).

 2. Plants free-living or parasitic upon the roots of other plants; leaves chiefly alternate; fruit a drupe or a nut.

 Family 3. **Santalaceae** (sandalwood family).

*Family 1. Olacaceae.

TALLOWWOOD FAMILY

Shrubs, trees, or vines. Leaves simple and usually entire, alternate; stipules present. Flowers usually bisexual but some unisexual (and the plant polygamodioecious), in cymes or thyrses; sepals 4–6; petals 4–6, separate, or in the North American genera coalescent at least basally; stamens 4–12, in 1 or 2 series, if in 1 series, opposite the petals, in North America usually in 2 series; carpels 3–4, the style 1, the ovary usually 3–4-chambered or the partition sometimes incomplete; placenta axile; ovule 1 per carpel, pendulous. Fruit a drupe or a berry.

The family occurs throughout the Tropics. In North America north of Mexico there are only two genera, each represented by a single species occurring on the hammocks and pinelands of peninsular Florida and the Keys. These are the whitewood (*Schoepfia*) and the tallowwood (*Ximenia*).

*Family 2. Loranthaceae.

MISTLETOE FAMILY

Small shrubs parasitic on the branches of trees or in extraterritorial genera the plants sometimes trees; stems brittle, forking; leaves simple, entire, thick and leathery, either green (with chlorophyll) or yellow, sometimes reduced to small yellowish scales, usually opposite or whorled. Flowers bisexual or (in North America) unisexual (and then the plants dioecious), solitary or in racemes or panicles, greenish or colored, small; sepals 2–5 or usually 3; petals none; stamens as many as the sepals; carpels 3 or sometimes 4; placenta basal, the style 1 or none, the ovary 1-chambered; ovules

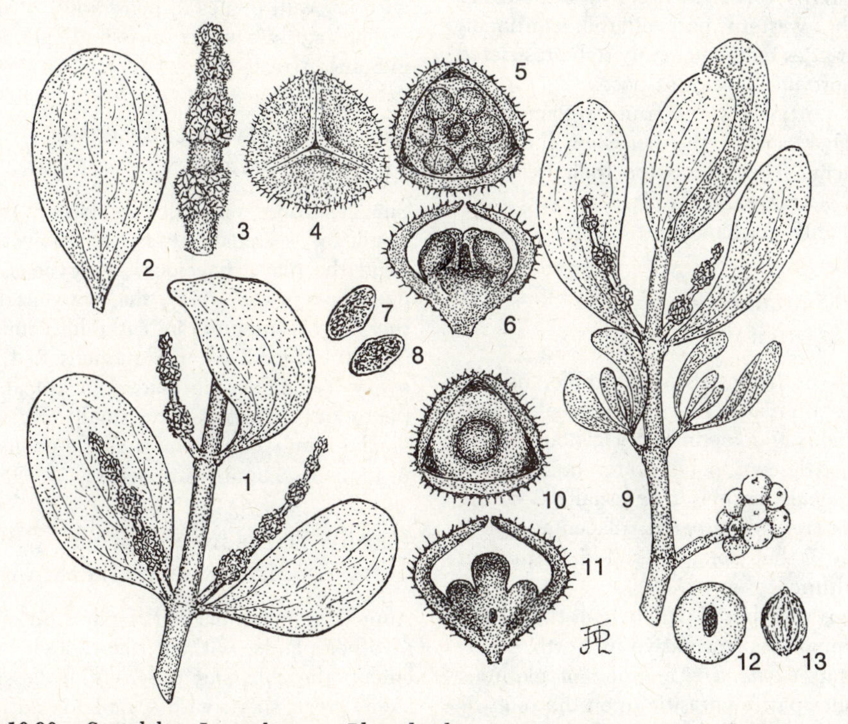

Figure 10-90. Santalales; Loranthaceae; *Phoradendron tomentosum* var. *macrophyllum,* a mistletoe parasitic on softwood trees of stream banks: **1** staminate plant with spikes of flowers, **2** leaf, **3** inflorescence segment, **4** flower, top view, showing the sepals, **5** cross section, showing the sepals and stamens, **6** longitudinal section, **7–8** pollen, **9** pistillate plant with flowers and fruits, **10** pistillate flower, cross section, showing the sepals and the style above the inferior ovary, **11** longitudinal section, **12** fruit, **13** seed.

as many as the carpels, the cavity of the ovary often very small. Fruit a berry or a drupe.

The family occurs throughout the Tropics, and some species extend into the Temperate Zones. There are two genera in North America. *Phoradendron* includes the common leafy green mistletoes partly parasitic upon particularly oaks and softwood trees occurring along streams (e.g., cottonwoods and alders) and even upon some conifers, as for example, junipers, cypresses, and firs. Some species in the Southwestern Deserts occur chiefly upon the leguminous trees and shrubs such as the mesquites and palos verdes. The chief center of the genus in North America is in California and to a lesser extent in the adjacent states. A single species occurs in eastern North America. In one desert species of *Phoradendron* the leaves are reduced to mere scales, but the stems are green. *Arceuthobium* includes the common yellow leafless mistletoes parasitic upon conifers. One species occurs in eastern North America, and there are several in the western part of the Continent. Species or varieties tend commonly to be restricted to special coniferous species as hosts.

The fleshy portions of the fruits are mucilaginous, and they stick to the beaks of the birds which eat them. The birds wipe their beaks on the branches of trees, and the hard-covered seeds adhere there until germination.

*Family 3. **Santalaceae.** SANDALWOOD FAMILY

Woody plants or herbs. Leaves simple, usually entire, usually alternate. Flowers bisexual or unisexual, epigynous or sometimes perigynous, small; sepals 4 or 5, coalescent; petals none; stamens 4–5, opposite the sepals; carpels 2 or usually 3–5, the style 1, the ovary 1-chambered; placentae basal; ovules usually 3 but sometimes 1–5. Fruit an achene or a drupe.

The family is distributed widely over the Earth. The most common representative in North America is the genus *Comandra,* a group of glabrous herbs sometimes partly parasitic upon the roots of other plants. The buffalo nut (*Pyrularia*), known also as oil nut, is a small shrub occurring in the Appalachian chain. Three other genera, *Nestronia, Buckleya,* and *Geocaulon* occur in the east or north.

Family 4. **Opilaceae.** OPILIA FAMILY

Shrubs, trees, or woody vines. Leaves simple, alternate; flowers usually bisexual, perigynous or epigynous, the ovaries being partly inferior; sepals 4–5, minute; petals 3 or 4–5, separate or sometimes coalescent; stamens 4–5, opposite the petals, sometimes adnate with the bases of the petals; carpels 3, the style 1, the ovary 1-chambered; ovule 1. Fruit a drupe.

The family occurs in the tropical regions of Africa, Asia, and Brazil.

Family 5. **Grubbiaceae.** GRUBBIA FAMILY

Shrubs. Leaves simple, linear, opposite. Flowers bisexual, epigynous, with epigynous discs, small; sepals 4; stamens 8, in 2 series; carpels 2, the style with 2 branches above, the ovary 1-chambered; placenta a central column arising from the base of the ovary or sometimes adnate with its wall; ovules 2, pendulous. Fruit a drupe.

The family consists of a single genus occurring in South Africa.

Family 6. **Myzodendraceae.** MYZODENDRON FAMILY

Small shrubs, parasitic on trees. Leaves very small and scalelike, alternate. Flowers unisexual (and the plant dioecious), epigynous, the staminate flowers with discs, the ovary partly inferior, the flowers minute, in catkinlike inflorescences; sepals and petals none; stamens 2–4; carpels 3, ovary 1-chambered; placentae apical; ovules 3, pendulous. Fruit nutlike, small.

The family consists of a single genus, occurring in temperate South America.

Family 7. **Balanophoraceae.** BALANOPHORA FAMILY

Annual or perennial herbs parasitic on the roots of other plants; without chlorophyll. Stems succulent, the rhizomes tuberous, leafless; flowering stems erect, short, with scale-leaves at least when young. Flowers unisexual or rarely bisexual (the plants ordinarily monoecious or dioecious); staminate flowers with a 3–4- or 8-lobed perianth (calyx ?); stamens as many as the perianth lobes; pistillate flowers epigynous, the perianth usually wanting, however; carpels 1–3, the ovary 1-chambered; ovules 1–3, sometimes consisting of only megagametophytes, frequently adnate with the pericarp; fruit a nut; seed 1.

About 15 genera occurring in the Tropics. The principal genus is *Balanophora*.

Family 8. **Cynomoriaceae.**

CYNOMORIUM FAMILY

Parasitic herbs growing attached to the roots of other plants; sap violet and astringent; stems succulent. Leaves scalelike, numerous. Flowers unisexual or bisexual, epigynous; sepals 1–5; petals none; stamen 1; carpel 1, with 1 ovule.

The family is composed of a single species occurring on alkaline steppes fom the Mediterranean Region to western Asia.

Order 44. ***CORNALES.*** DOGWOOD ORDER

DEPARTURE FROM THE FLORAL CHARACTERS OF THE RANALES

Free	Carpels coalescent.
Numerous	Stamens 4–10 or 12; carpels 2–10.
Spiral	Stamens and carpels cyclic.
Hypogynous	Flowers epigynous, with an epigynous disc, in the *Garryaceae* many flowers appearing to be hypogynous, showing no signs of sepals at the tops of the ovaries, but in some these appearing as minute scales just below the styles and the flower therefore epigynous.

Shrubs or trees or rarely herbs or woody vines. Leaves simple, deciduous or usually so, opposite or alternate; stipules none or rarely present (*Helwingia*). Flowers small, bisexual or unisexual, in clusters of various kinds; sepals usually 4 or 5; petals usually 4 or 5; style 1 (or rarely several). Fruit a drupe or rarely a berry; embryo small, endosperm abundant.

Key to the Families
(North America north of Mexico)

1. Flowers bisexual or unisexual, in clusters of various kinds but not in catkins; petals usually 4–5, rarely 0; fruit a drupe or rarely a berry; ovule 1 per carpel or 1 in the ovary.
   2. Leaves opposite or rarely alternate; carpels 2–4, usually 2; drupe less than 1 cm. long, usually about 5 mm. long.
      Family 1. **Cornaceae** (dogwood family).
   2. Leaves alternate; carpels and chambers of the ovary 6–10; fruit a 1-seeded drupe, the drupe 1–2.5 cm. long.
      Family 2. **Nyssaceae** (sour gum family).
1. Flowers unisexual, those of both sexes in catkins with opposite pairs of bracts, the plants dioecious; petals 0; fruit fleshy until maturity, then dry; ovules and seeds 1–2 per ovary.
   Family 4. **Garryaceae** (silk-tassel bush family).

*Family 1. **Cornaceae.** DOGWOOD FAMILY

Perennial herbs, shrubs, trees, or, in an extraterritorial genus, rarely woody vines. Leaves simple, usually deciduous, opposite, or rarely alternate; flowers bisexual or sometimes unisexual (and then the plants monoecious or polygamodioecious or in one genus dioecious), in flat-topped panicles or cymes; sepals 4–5, petals 4–5 or rarely none; stamens 4–5; carpels 2–4, usually 2, the ovaries with as many chambers as there are carpels or rarely with 1; placentae axile, or parietal if there is only 1 chamber; ovule 1 per carpel, pendulous.

The family is distributed widely in tropical and temperate regions. Although there are several genera in the family and between 75 and 100 species in temperate and tropical regions, only one genus (*Cornus*) occurs in North America north of Mexico. *Cornus* is remarkable for inclusion of herbs, trees, and shrubs, the trees being evidently more closely related to the herbs than either is to the shrubs. Perhaps 15 species occur in North America north of Mexico. Among the most attractive are *Cornus florida*, a beautiful tree of eastern North America, and *Cornus Nuttallii*, an attractive tree of the Pacific Coast.

*Family 2. **Nyssaceae.** SOUR GUM FAMILY

Shrubs or trees. Leaves simple, deciduous, alternate; stipules none. Flowers bisexual or usually some unisexual (and then the plant commonly polygamodioecious); sepals 5, reduced to small teeth, or sometimes none; petals 5 or sometimes

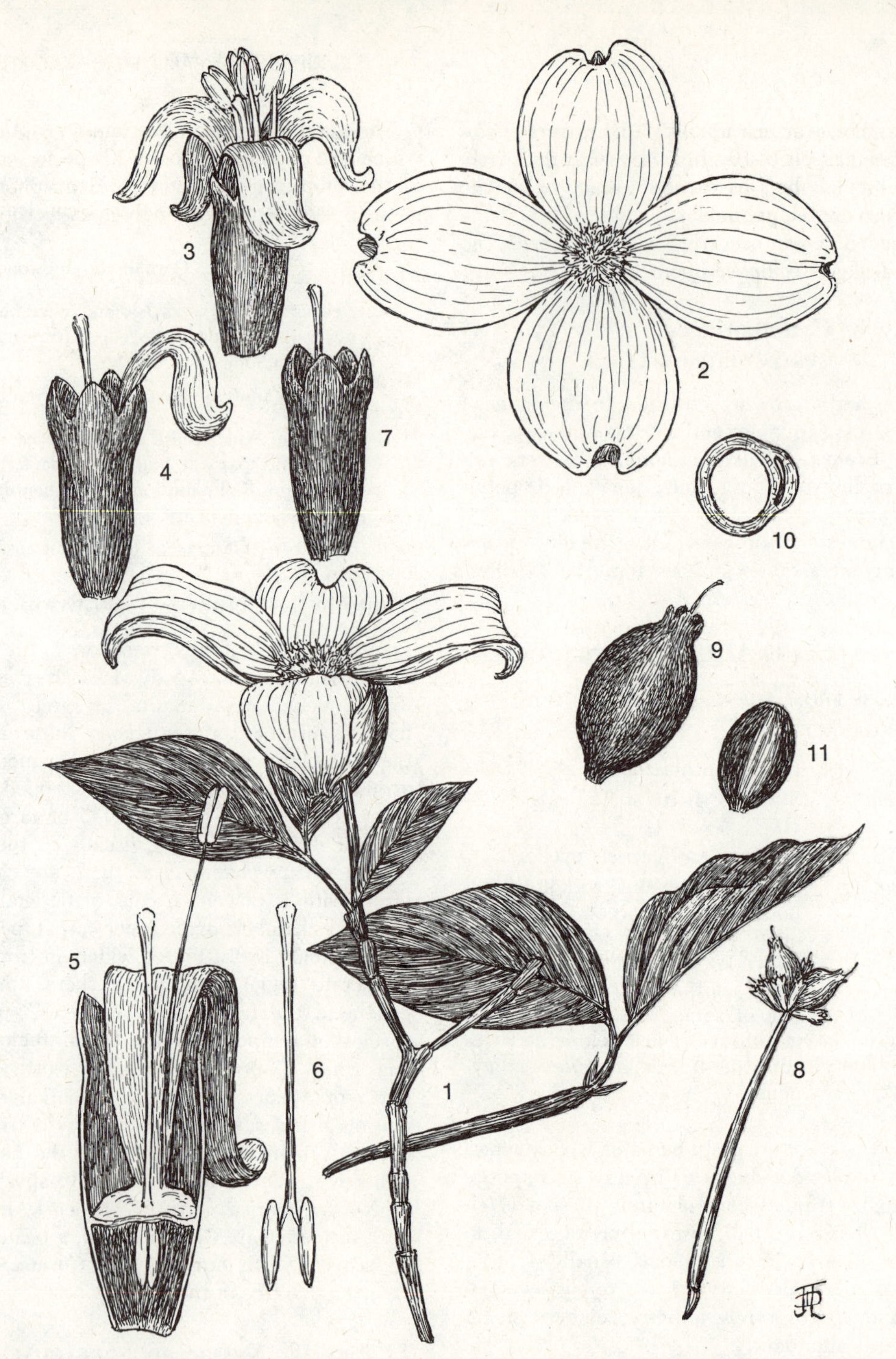

Figure 10-91. Cornales; Cornaceae; *Cornus florida,* flowering dogwood: **1** branches with leaves, one bearing an involucre and the included flowers, **2** white (or pink) involucre with a mass of flowers in the center, **3** flower, showing the floral cup (investing and adnate with the inferior ovary), the small sepals, the relatively large petals, and the stamens, **4** old flower with only one petal persistent, the style and stigma visible, **5** longitudinal section showing the ovary, the two chambers connected to the style, this enlarged in **6**, **7** young fruit, **8** mature and abortive fruits, **9** fruit, **10** cross section, one chamber abortive, **11** seed. In **5** the epigynous disc is visible.

more or none; stamens usually 5–10 or rarely 12, in 2 series; carpels 6–10, the ovary with the same number of chambers as carpels; placentae axile; ovule 1 per carpel, pendulous. Fruit a drupe with a 1–6-seeded stone.

The family is composed of 3 genera occurring in eastern North America and Asia.

In North America there is only one genus (*Nyssa*), the sour gum or tupelo. The two principal species are *Nyssa aquatica,* the cotton gum, and *Nyssa sylvatica,* the black gum. The first species occurs in swamps where it may grow in water; the other appears in low woods, in swamps, and on shores of lakes and marshes.

Family 3. **Alangiaceae.**
ALANGIUM FAMILY

Shrubs or trees. Leaves simple, entire or lobed, alternate. Flowers bisexual; sepals 4–10; petals 4–10, narrow, sometimes coalescent basally and the corolla tubular; stamens in series of the same number as the petals, the series 1–4; carpels 2–3, the ovary with 1 chamber; ovule 1, pendulous. Fruit a drupe.

The single genus of the family is *Alangium,* which is composed of about 20 species occurring in the Tropics of the Eastern Hemisphere. The genus is represented sparingly in cultivation.

*Family 4. **Garryaceae.**
SILKTASSEL-BUSH FAMILY

Medium-sized shrubs or sometimes small trees. Leaves opposite, entire, leathery, evergreen. Plants dioecious, both the staminate and pistillate flowers in drooping catkins, the catkins with opposite pairs of bracts; staminate flowers in 2's or 3's or solitary in the axils of the bracts; sepals (bracts ?) 4; stamens 4, separate; rudiment of a pistil present in the center of the flower; pistillate flowers in 2's or 3's, each consisting of a single pistil with 2 or 3 coalescent carpels and a single chamber, the ovules pendulous from near the top of the chamber, the 2 or 3 styles long and spreading, each with stigmatic papillae inside; ovary with 2 ovules, these anatropous, with 1 integument. Fruit rounded or obovoid, with 1–2 seeds, fleshy until maturity (then dry); seed membranous-coated; embryo straight; endosperm abundant.

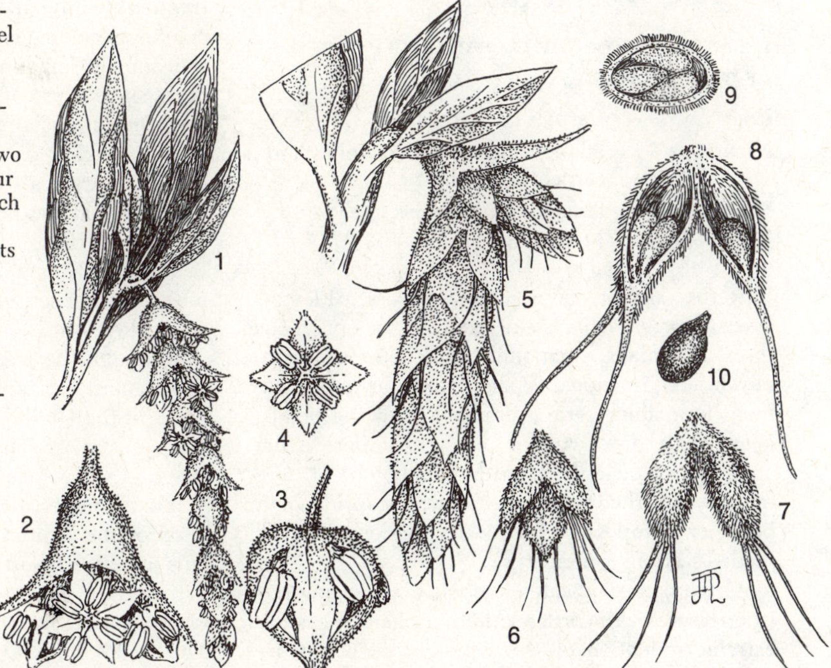

Figure 10-92. Garryales; Garryaceae; *Garrya Veatchii,* a silktassel bush: **1** branch of a staminate plant with opposite leaves and a catkin, **2** pair of opposite coalescent bracts subtending staminate flowers, **3–4** staminate flower, two views, the flower consisting of four sepals and four stamens, **5** branch of a pistillate plant with a catkin, **6** pair of opposite coalescent bracts subtending pistillate flowers, **7** two pistillate flowers, each consisting of a single pistil (most flowers with two styles but some with three), **8** pistillate flowers, longitudinal section, **9** cross section, **10** seed.

The family is made up of a single genus, *Garrya*, with about 10 or 15 species occurring in the southwestern United States, Mexico, and Jamaica. The interrelationships of the four Californian species and their classification are uncertain.

The shrubs are large and attractive, and *Garrya elliptica*, a California coastal species, is cultivated as an ornamental of interest for its pendulous silky catkins against a contrasting background of green leathery leaves.

Many authors have considered the Garryaceae to be related closely to the *Cornaceae* or dogwood family. *Garrya* and *Aucuba* of the *Cornaceae* have been grafted. It has been proposed also that *Garrya* may be related to the *Hamamelidaceae*. The affinities of the family are not clear enough for the time being to warrant union with any other group. According to Engler and Prantl, the family is primitive, being a representative of the Amentiferae. However, the inflorescence is highly specialized, and the simplicity of the flowers is more likely due to reduction than to primitiveness. The inferior ovary is by no means a primitive feature; neither is coalescence of carpels.

Order 45. *UMBELLALES.* PARSLEY ORDER

DEPARTURE FROM THE FLORAL CHARACTERS OF THE RANALES

Free	Carpels coalescent.
Numerous	Stamens usually 5, rarely more; carpels 2–5 or up to 15.
Spiral	Stamens and carpels cyclic.
Hypogynous	Flowers epigynous.

Herbs, shrubs, or sometimes trees. Flowers bisexual or sometimes unisexual, with epigynous discs, in umbels, compound umbels, or occasionally heads, 5-merous; sepals minute or none; stamens in a single series, alternate with the petals; ovary with one ovule in each chamber; ovule anatropous, pendulous, with 1 integument, the micropyle turned outward. Fruit a capsule or a berry or composed of two dry, indehiscent, 1-seeded sections (mericarps).

As the name indicates, umbels are a characteristic feature of the order. Another character is the extreme reduction of the sepals, which may be either minute scales or ridges or may be missing altogether.

According to Thorne, the two families should be united because of connecting links occurring extraterritorially.

Key to the Families
(North America north of Mexico)

1. Fruit a drupe or a berry; carpels commonly 2–5 (when the carpels are 2 the stigmas turned inward); stylopodium none; herbs, vines, shrubs, or trees.

   Family 1. **Araliaceae** (ginseng family).

1. Fruit dry, with the carpels separating at maturity; carpels 2, the stigmas not turned inward, each with a stylopodium at the base; herbs.

   Family 2. **Umbelliferae** (parsley family).

*Family 1. **Araliaceae.** GINSENG FAMILY
(Synonym: *Hederaceae*)

Herbs, shrubs, trees, or vines; stems markedly pithy; young parts often covered by stellate hairs. Leaves simple to palmate or pinnate, always alternate; stipules often present, sometimes forming a membranous border along the edge of the petiole or sometimes wanting. Flowers bisexual or more frequently unisexual (and the plant then dioecious or polygamodioecious), with hypogynous discs, the umbels sometimes reduced to heads, the inflorescence in some extraterritorial genera sometimes not umbellate; petals sometimes 10 or rarely 3; carpels 2–5 or sometimes up to 15. Fruit a berry or rarely a drupe or sometimes the carpels separating at maturity. (Cf. Figure 6-14 **C,** ivy, flower and fruit.)

The family is a large one, occurring primarily in the Tropics, especially in India and Malaya and tropical America.

Although the ginseng family includes about 60 genera and perhaps 750 species, only 4 genera occur in the United States and Canada. The largest genus is *Aralia,* which includes the angelica-tree and some species known as wild sarsaparilla. A few characteristic local species occur in various parts of the United States. Ginseng (*Panax*) occurs in eastern North America. At one time it was gathered for its roots which were employed in Chinese medicine; now it is cultivated for sale. The devil's-club (*Oplopanax*) occurs from Kodiak Island, Alaska, to northern California and here and there eastward to northern Michigan and adjacent

Ontario. This is a striking species with very large leaves and an exceedingly spiny stem. In some areas the common English ivy (*Hedera*) has escaped.

Rice paper, used for cleaning lenses, is derived from the pith of the rice-paper plant (*Tetrapanax*) of the Orient. Several genera are cultivated as ornamentals.

*Family 2. Umbelliferae.

PARSLEY FAMILY

(Synonyms: *Apiaceae, Ammiaceae*)

Annual or most commonly perennial herbs or some extraterritorial genera shrubs. Leaves usually pinnate but sometimes palmate or rarely simple, the leaf bases broad, the leaves alternate; stipules none. Flowers bisexual or occasionally unisexual (and the plant then polygamomonoecious or sometimes monoecious); stamens bent inward in the bud; carpels 2; the 2 styles with swollen spreading bases, these forming a stylopodium; fruit separating at maturity into 2 dry segments, these opening on their internal faces, each with 5 primary ribs and 4 alternating secondary ribs, these sometimes winged or corky, oil tubes usually present in the intervals between the ribs (in cross section these visible with magnification), flattened (1) *dorsally* (that is, as if "pinched" inward against the plane of coalescence of the carpels, i.e., against the midribs of the carpels) or (2) *laterally* (as if "pinched" at right angles to plane of coalescence of the carpels, i.e., midrib against suture), each segment suspended on a wirelike structure after dehiscence of the fruit.

The family includes about 2,500 species occurring chiefly in the temperate regions of the Northern Hemisphere. About 75 of the genera and 350 species occur in North America north of Mexico. Only a few of the more important plants appear in the following list: water-pennywort (*Hydrocotyle*), snakeroot (*Sanicula*), *Eryngium*, sweet-cicely (*Osmorhiza*), coriander (*Coriandrum*), poison hemlock (*Conium*), parsley (*Petroselinum*), water hemlock (*Cicuta*), celery (*Apium*), caraway (*Carum*), water parsnip (*Sium* and *Berula*), fennel (*Foeniculum*), dill (*Anethum*), anise (*Pimpinella*), lovage (*Levisticum*), *Angelica*, hog-fennel (*Oxypolis*), parsnip (*Pastinaca*), cow parsnip (*Heracleum*), carrot (*Daucus*), and *Lomatium*.

Figure 10-93. Umbellales; Umbelliferae: *Velaea arguta*, an umbellifer: **1** pinnate leaf, **2** inflorescence, a compound umbel, **3** flower, side view, showing the inferior ovary enclosed in the adnate floral cup, the petals (the sepals being minute), the stamens, and the two drooping styles, with the petals turned in, **4** flower, top view, **5** fruit with the persistent minute sepals and the recurved styles, **6** cross section, showing the two chambers and the oil tubes in the pericarp, **7** dehiscence of the fruit into two segments each suspended on a wirelike structure, the two styles each arising from a stylopodium.

As shown by the list above this is one of the most significant families from an economic point of view because it yields vegetables, drugs, and flavoring materials. Poison hemlock (given to Socrates), water hemlock, and fool's-parsley (*Aethusa*) are poisonous.

Order 46. *RAFFLESIALES.*

RAFFLESIA FAMILY

DEPARTURE FROM THE FLORAL CHARACTERS OF THE RANALES

Free	Carpels coalescent.
Numerous	Carpels 4–8.
Spiral	Carpels probably cyclic.
Hypogynous	Flowers epigynous.

Parasitic herbs growing on the roots or stems of other plants. Leaves none or scalelike. Flowers unisexual or bisexual, radially symmetrical, epigynous, the ovary wholly or partly inferior; sepals 3–10; petals none; stamens numerous; carpels 4–8, the style 1 or obsolete; ovary 1-chambered; placentae parietal or apical; ovules numerous. Fruit fleshy, usually a berry.

Only the *Rafflesiaceae* occur in the United States; there are no representatives of the order in Canada.

*Family 1. Rafflesiaceae.

RAFFLESIA FAMILY

Parasitic herbs growing on the roots or branches of other plants, the plant body almost wholly within the host plant. Leaves ordinarily scalelike. Flowers unisexual (the plant being monoecious or dioecious), radially symmetrical, epigynous, the ovary inferior or partly so; sepals 4–10; petals none; stamens numerous, the pollen chambers opening by slits or apical pores; carpels 4–8, the style 1 or obsolete; the ovary 1-chambered; placentae parietal; ovules numerous. Fruit a berry.

The family occurs in subtropical regions mostly in the Eastern Hemisphere. The largest flowers in the world, reported to measure up to 3 feet across and to weigh 10–20 pounds, belong to *Rafflesia Arnoldii,* which occurs in the Malay Archipelago. In contrast, the only representative of the family in the United States is *Pilostyles Thurberi,* occurring near the mouth of the Colorado River in California and Arizona and probably in Baja California and Sonora. This plant is a parasite on the twigs of certain shrubs of the pea family, including species of *Dalea.* Its total length is only 5–6 mm., including the stem under the flower. The plant is brown and without leaves.

Family 2. Hydnoraceae.

HYDNORA FAMILY

Parasitic herbs occurring on the roots of trees and shrubs. Leaves none. Flowers bisexual, radially symmetrical, epigynous, solitary, nearly sessile; sepals 3–4, very thick; petals none; stamens numerous, the filaments obsolete; carpels several, the style obsolete, the ovary 1-chambered; placentae parietal or apical; ovules numerous. Fruit fleshy.

The family is distributed widely in the Tropics. It does not occur in temperate North America.

Group 4. *OVARIFLORAE.* OVARY FLOWERS

(Epigynous or rarely perigynous; Sympetalous)

NOTE: After determination of the order and family, proceed to the local flora or manual to identify the genus and species.

PRELIMINARY LIST FOR ELIMINATION OF EXCEPTIONAL CHARACTER COMBINATIONS

Before using the key to the orders on page 331, it is necessary to eliminate (1) exceptional plants belonging to other groups, but duplicating the key characters of the Ovariflorae, (2) plants not having the key characters marking most of their relatives, and (3) plants marked by obscure characters or those inconsistent if taken singly.

The plant must correspond with every character in the combination. If the specimen does not coincide with one character, even the first listed, the combination may be eliminated from further consideration. The characters in **boldface type** should be checked first to determine whether the combination requires further inspection.

1. **Stems bearing spines in areoles** (small clearly defined areas) at the nodes (these commonly not appearing as nodes, however, but arranged spirally on flat surfaces or on ridges or tubercles), the segments of the stem markedly succulent; **petaloids intergrading with sepaloids,** both numerous, all the perianth parts coalescent at the bases and forming a long or very short perianth tube above the hypanthium; style 1, the stigmas several; **ovary 1-chambered, inferior,** the placentae parietal.

   Order 24. **Cactales** (Thalamiflorae), page 233.

2. **Stamens 8–16, more numerous than the petals; flowers epigynous;** trees, shrubs, bushes, or sometimes small but woody creeping plants; determined by the brief key below.
   1. Plant *not* thorny; petals glabrous or at least *not* conspicuously and densely pubescent on the upper sides.
      2. Anthers splitting along their entire length (dehiscent). Some *Styracaceae* (styrax family).

         Order 30. **Ebenales** (Corolliflorae), page 254.

      2. Anthers *not* splitting lengthwise (indehiscent), opening by pores at the apices of terminal tubes. *Vaccinioideae, Ericaceae* (heath family).

         Order 31. **Ericales** (Corolliflorae), page 257.

   1. Plants thorny; petals conspicuously and densely pubescent on the upper sides. *Ximenia, Olacaceae* (tallowwood family).

      Order 43. **Santalales** (Calyciflorae), page 322.

3. **Flowers perigynous;** petals 5, coalescent only basally; stamens 5, opposite the petals; seed 1, arising from an elongated thread attached at the base of the ovary; carpels 5. *Staticeae, Plumbaginaceae* (primrose family).

   Order 29. **Primulales** (Corolliflorae), page 251.

4. **Stamens opposite the petals, not more than 5,** ***not*** **more numerous than the petals, the anthers not coalescent; flowers epigynous,** of a single kind, *not* in dense involucrate heads; sepals *not* forming a pappus of hairlike threads or membranous scales; plant with the character in a or b, below.
   a. Herbs. *Samoleae, Primulaceae* (primrose family).

      Order 29. **Primulales** (Corolliflorae), page 251.

   b. Trees or shrubs. *Schoepfia, Olacaceae* (tallowwood family).

      Order 43. **Santalales** (Calyciflorae), page 322.

5. **Perennial herbs occurring in marshes from Alaska to Washington; caudex bearing reniform basal leaves with palmate veins;** flowers epigynous; corolla campanulate; stamens 5; style thick, the stigmas 2; ovary 1-chambered; ovules numerous, on parietal placentae; capsule bursting irregularly at the apex. *Nephrophyllidium, Menyanthoideae, Gentianaceae* (gentian family).

   Order 32. **Gentianales** (Corolliflorae), page 261.

6. **Desert herbs or low bushes; plants with both barbed hairs and larger stinging hairs;** leaves alternate, ovate; **flowers yellow, 5–7 cm. in diameter;** petals 5, coalescent only basally; stamens numerous, adnate and deciduous with the petals; carpels 5, the styles 5, the ovary 1-chambered, with numerous ovules on parietal placentae; fruit a capsule. *Eucnide, Loasaceae* (*Loasa* family).

   Order 7. **Loasales** (Thalamiflorae), page 166.

KEY TO THE ORDERS

(First consult the Preliminary List for Elimination of Exceptional Character Combinations, page 330.)

NOTE: After determination of the order and family, proceed to the local flora or manual to identify the genus and species.

1. Plant *not* corresponding with *every character* in lead 1 below.
   2. Flowers unisexual; herbaceous vines, usually with tendrils; fruit a fleshy, leathery, or spiny pepo (gourd, melon, pumpkin, etc.) or sometimes small and leathery or spiny. *Cucurbitaceae* (gourd family).

      Order 7. **Loasales** (Thalamiflorae), page 166.

2. Flowers usually bisexual; herbs, shrubs, trees, or vines, but *not* with tendrils; fruit *not* a pepo.
   3. Leaves opposite or whorled.

      Order 44. **Rubiales,** page 332.

   3. Leaves alternate.

      Order 45. **Campanulales,** page 337.

1. Flowers in dense heads, each head being surrounded by a series of involucral bracts; corollas often of 2 kinds, those of the inner (disc) flowers radially symmetrical, those of marginal (ray) flowers with the corollas opened on one side and flattened out, the entire head resembling a single flower, the ray flowers appearing to be the petals and the disc flowers the other parts; fruit an achene formed from 2 carpels; sepals specialized as a pappus of scales or bristles (sometimes wanting) and usually aiding in dissemination of the fruits by wind; stamens usually coalescent by the anthers but *not* by the filaments; style 1; stigmas 2, each with a stigmatic line on each margin, the pair of stigmas serving with their external hairs as a brush which sweeps the pollen from the inside of the anther tube as the style elongates.

   Order 46. **Asterales,** page 339.

Order 47. *RUBIALES.*

HONEYSUCKLE ORDER

DEPARTURE FROM THE FLORAL CHARACTERS OF THE RANALES

Free	Petals and carpels coalescent; stamens adnate with the corolla.
Equal	Corolla often bilaterally symmetrical (in the advanced members of the order).
Numerous	Stamens 5 or 4, rarely fewer; carpels 2–5, usually 2.
Spiral	Stamens and carpels cyclic.
Hypogynous	Flowers epigynous.

Herbs, shrubs, or trees. Leaves opposite or (often falsely) whorled. Flowers usually bisexual, 5- or 4-merous; stamens alternate with the petals; ovary 1- or commonly 2–5-chambered. Fruit a capsule or a berry or sometimes dry and with only 1 seed.

As shown by the table above, the more highly specialized members of the order, as, for example, the honeysuckle, depart in at least some respects from every one of the floral characters of the primitive Ranales.

The chief characters distinguishing the Rubiales from the Umbellales are coalescence of petals and its usual corollary, adnation of stamens and petals. The flower structure of the elderberry (*Sambucus*) approaches closely that of the Umbellales. It is identical with that of *Aralia* and *Hedera* (ivy) except in the two associated features mentioned above.

On the other hand, many Rubiales are more highly specialized, as for example, honeysuckles and *Abelia,* a plant relatively common in cultivation through the southern parts of the United States. These flowers differ from the elderberry in both the greater coalescence of petals and the bilateral symmetry of the corolla.

Key to the Families

(North America north of Mexico)

1. Ovary 1–5-chambered; corolla radially or sometimes bilaterally symmetrical.
   2. Leaves *either* opposite and with stipules *or* whorled and without stipules.

      Family 1. **Rubiaceae** (madder family).

   2. Leaves *both* opposite and without stipules.
      3. Plant *not* corresponding with *every character* in the combination in lead 3 below.

         Family 2. **Caprifoliaceae** (honeysuckle family).

      3. Herbs with rhizomes; each stamen divided to the base of the filament, each segment bearing a half anther; leaves mostly basal, the cauline in a single pair; terminal flower 4-merous, the other four to six flowers 5-merous; each flower subtended by a 2- or 3-lobed involucre.

         Family 3. **Adoxaceae** (moschatel family).

1. Ovary 1-chambered; corolla bilaterally symmetrical.

2'. Flowers *not* in dense *involucrate* heads or spikes; stamens 4, as many as the petals, rarely 2.

Family 4. **Valerianaceae** (valerian family).

2'. Flowers in dense *involucrate* heads or spikes; stamens 1–3, fewer than the petals.

Family 5. **Dipsacaceae** (teasel family).

*Family 1. **Rubiaceae.** MADDER FAMILY

Herbs, shrubs, trees, or woody vines, the tropical genera usually woody. Leaves simple, usually entire or sometimes toothed; stipules present, sometimes (as in *Galium*) as large as the leaves, the 2 opposite leaves with their 4 stipules appearing to be a whorl of 6 leaves, or their numbers sometimes reduced (e.g., 4) by lack of development of some members. Flowers radially or sometimes bilaterally symmetrical; corolla usually salverform to rotate or funnel-form; carpels usually 2 or sometimes up to 4, the style 1, the ovary 2–4-chambered; placentae axile or apparently basal (in *Gardenia* the ovary 1-chambered with 1 parietal placenta); ovules usually numerous. Fruit a capsule or separating into 2 indehiscent, 1-seeded segments (*Galium*) or a berry.

The family occurs throughout the Tropics and Subtropics. It is one of the largest plant families, including perhaps 4,500 species. The largest North American genus is *Galium* (bedstraw) which includes about 30 species in eastern North America and many others in the West. *Cephalanthus*, the button-bush or button-willow, is distributed widely in North America. The genus *Houstonia* includes about 10 species in eastern North America. *Bouvardia* is represented in Arizona and Southern New Mexico by a small shrub with brilliant red flowers. Many other genera (the fever-tree, *Pinckneya; Hedyotis* or *Oldenlandia;* the princewood, *Exostema;* the seven-year apple, *Casasia* or *Genipa; Randia; Catesbaea; Hamelia;* the velvet-seeds, *Guettarda; Erithalis;* the snowberries, *Chiococca; Strumpfia;* the wild coffees, species of *Psychotria;* the partridgeberry, *Mitchella; Morinda; Richardia; Ernodea;* the buttonweeds, *Diodia; Borreria;* and *Spermacoce*) and additional species of the genera mentioned above occur in the Southern States. Several of these are restricted in the United States to Florida.

Economically the family is important as the source of quinine (*Cinchona*) and coffee (*Coffea*). Several genera are cultivated as ornamentals.

*Family 2. **Caprifoliaceae.**

HONEYSUCKLE FAMILY

Shrubs, woody vines, trees, or herbs. Leaves usually simple, sometimes pinnate, as in the elderberries, stipules usually none, but sometimes present and glandular. Flowers radially or bilaterally symmetrical, nearly always bisexual but sometimes unisexual; corolla rotate to salverform and then often markedly bilaterally symmetrical; stamens 5 (as also the petals); or reduced to 4 in some genera; carpels 3–5, the style 1, the ovary 2-, 3-, 5-, or 1-chambered; placentae usually axile, but becoming parietal when there is only 1 chamber; ovule usually 1 per carpel, pendulous. Fruit a berry or a drupe.

The family is distributed widely in the Northern Hemisphere, being most abundant in the deciduous forests of eastern North America and eastern Asia. It is represented more sparingly in South America and New Zealand. The chief groups occurring in North America are the honeysuckles and twinberries (*Lonicera*), the bush honeysuckles (*Diervilla*), the snowberries (*Symphoricarpos*), the twinflower (*Linnaea*), the feverwort (*Triosteum*), *Viburnum,* and the elderberries (*Sambucus*).

The elderberries are of some importance for their fruit. There are many ornamental shrubs and vines, especially the viburnums, honeysuckles, *Abelia, Weigelia,* and the bush honeysuckles.

*Family 3. **Adoxaceae.**

MOSCHATEL FAMILY

Small perennial herbs; main stem a rhizome. Leaves mostly basal, divided; cauline leaves 2, 3-foliolate, opposite. Flowers radially symmetrical, small, in heads of about 5, the terminal flower usually 4-merous, the lateral 5–6-merous; sepals 2–3; petals 4–6, the corolla rotate; stamens 4–6, the filaments divided and each segment bearing a half anther; carpels 3–5, the styles 3–5-cleft, the ovaries 3–5-chambered; ovule 1 per carpel, pendulous. Fruit a drupe or composed of 1–5 segments which separate at fruiting time.

The family occurs in subarctic and cold temperate regions of the Northern Hemisphere. It is composed of a single species. The moschatel (*Adoxa Moschatellina*) occurs in eastern North America. The name is a diminutive meaning musk, in reference to the odor of the plant.

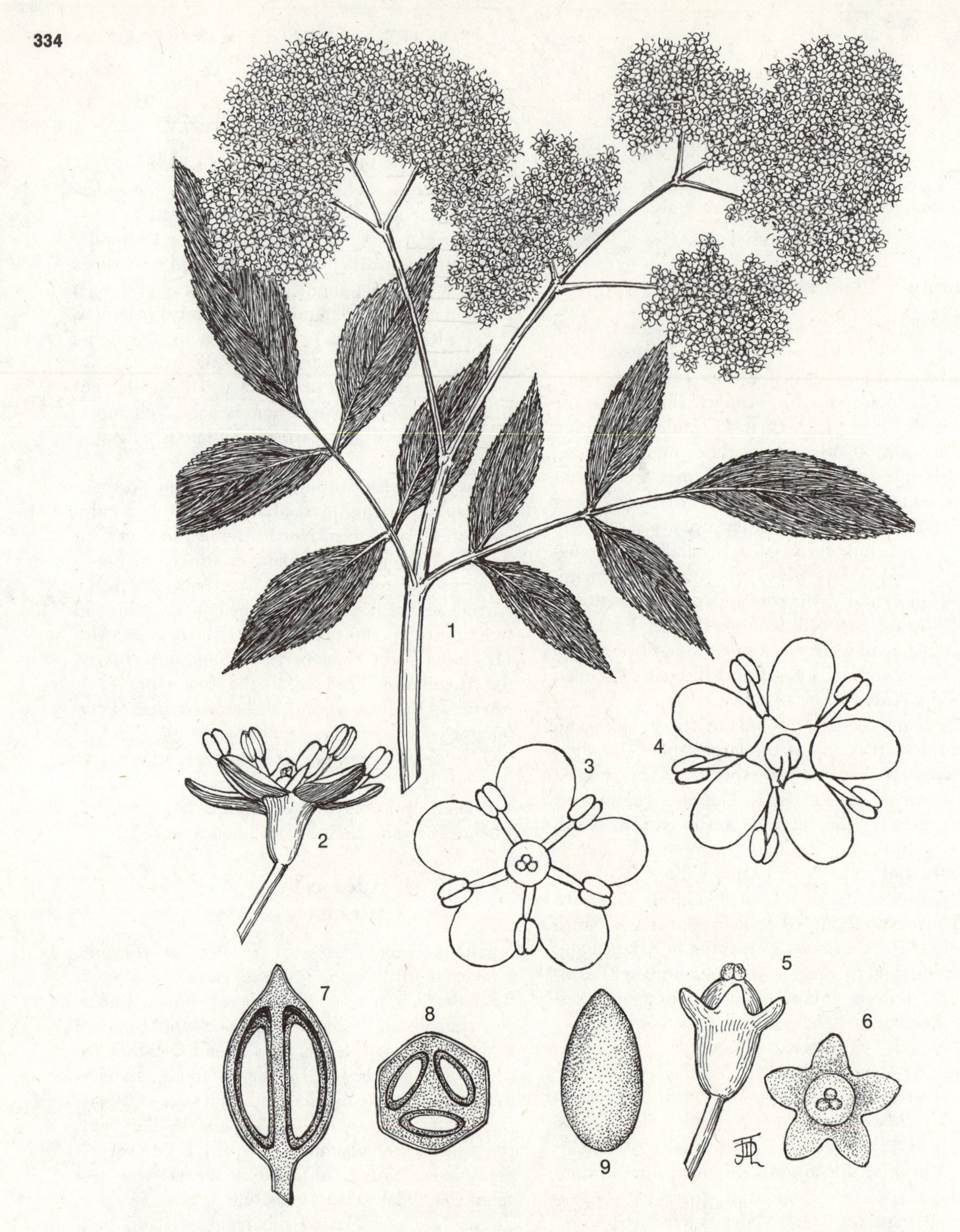

Figure 10-94. Rubiales; Caprifoliaceae; *Sambucus caerulea* var. *mexicana,* desert elderberry: **1** branch with opposite pinnate leaves and a complex compound inflorescence, **2** flower, with side view, showing the floral cup (adnate with the inferior ovary), the small sepals, the barely sympetalous corolla, the stamens, and the style and stigmas, **3–4** outline drawings of the flower from above and below, **5** young fruit, the sepals persistent, **6** from above, **7** diagrammatic longitudinal section, **8** diagrammatic cross section, **9** young seed.

Figure 10-95. Rubiales; Caprifoliaceae; *Lonicera japonica,* honeysuckle: **1** branch with opposite leaves and flowers, **2** pair of flowers, one just past anthesis, **3** longitudinal section, showing the inferior ovaries, the sepals, and the bases of the corolla and the style, **4** upper portion of the flower, showing the corolla, the adnate stamens, and the style and stigma, **5** calyx persistent around the young fruit, **6** fruit, **7** longitudinal section, **8** cross section, **9–10** seed, two views.

*Family 4. **Valerianaceae.**

VALERIAN FAMILY

Herbs or rarely woody plants or shrubs. Leaves simple or pinnate. Flowers usually unisexual (and the plant usually dioecious), asymmetrical, with dense inflorescences of the cymose type; sepals 4 or none, slow in developing; corolla composed of petals of irregular sizes and shapes, the basal portion tubular and often with a spur on the sac; carpels 3, but sometimes only 1 developed and the other sterile, the style 1, the ovary about half inferior, usually with only 1 chamber and a solitary pendulous ovule; placentae axile but apparently parietal. Fruit an achene, often surrounded by the elaborated calyx, this winged or plumose.

The family occurs in the North Temperate Zone and in the Andes in South America. It is represented in North America by a few species of the genera *Plectritis, Valerianella* (corn-salad or lamb's-lettuce), and *Valeriana* (valerian). *Plectritis* is restricted in the United States to California and Arizona; it occurs also in Chile. *Valerianella* occurs only in the East; *Valeriana* spans the Continent.

*Family 5. **Dipsacaceae.** TEASEL FAMILY

(Synonym: *Morinaceae*)

Herbs or rarely shrubs. Leaves various but usually simple or pinnatifid. Flowers bisexual, bilaterally symmetrical or rarely asymmetrical, in dense heads or dense, whorl-like subapical clusters, these densely bracteate; sepals forming a pappuslike structure of 5–10 segments; stamens usually 4; carpels 2, but only 1 developed, the style 1, the ovary with a solitary, pendulous ovule. Fruit an achene, enclosed in a sac.

The family is composed of about 9 genera occurring in the Eastern Hemisphere from the Mediterranean Region to the Russian steppes and India.

The teasel (*Dipsacus*), devil's-bit (*Succisa*), and various kinds of *Scabiosa* escape from cultivation.

The teasel has many bracts about the inflorescence, and these have strong recurving hooked tips. Because of the use of the bracts by fullers and in textile mills, the plant is known as fuller's teasel.

Order 48. *CAMPANULALES.*

BELLFLOWER ORDER

DEPARTURE FROM THE FLORAL CHARACTERS OF THE RANALES

Free	Petals coalescent; stamens adnate with the petals; carpels coalescent; anthers sometimes coalescent.
Equal	Corolla sometimes bilaterally symmetrical.
Numerous	Stamens usually 5; carpels 2–3.
Spiral	Stamens and carpels cyclic.
Hypogynous	Flower epigynous except in the *Brunoniaceae*.

Herbs or rarely woody plants. Leaves alternate (or rarely otherwise); stipules none. Flowers bisexual or rarely unisexual, 5-merous; anthers converging (connivent) and sometimes coalescent; ovary inferior, either with as many chambers as there are carpels and with one to many ovules in each carpel or sometimes 1-chambered and with a single ovule. Ovules anatropous.

One of the features of the order is the method of removal of the pollen from the anthers, which is similar to that in the Asterales. The anthers may form a tube or stand close together. Pollen is released inside the tube, and the elongation of the style bearing the stigmas brings the pollen out by an action similar to that of a bottle-brush.

In a few members of the order the ovary is not inferior. On the basis of this and the similarity of flower structure some authors have thought the Campanulales to have been derived from some group in the Corolliflorae, probably near the Polemoniales. On the other hand, the order shows marked similarity to the Rubiales, which in turn resemble the Umbellales (Calyciflorae) as well as some of the Gentianales (Corolliflorae). Derivation may have been from either of these stocks or from some ancestor in common with one or both. There is also similarity to the Cucurbitaceae.

Key to the Families

1. Flowers *not* in a dense head or headlike spike.
   2. Plants *not* succulent; stigma usually *not* surrounded by a cuplike structure.

      Family 1. **Campanulaceae** (bellflower family).

2. Plants succulent; stigma surrounded by a cup-like structure.

Family 2. **Goodeniaceae** (*Goodenia* family).

1. Flowers in a dense head or headlike spike.

Family 4. **Calyceraceae** (*Calycera* family).

*Family 1. Campanulaceae.

BELLFLOWER FAMILY

Herbs or extraterritorial genera sometimes shrubs or trees. Leaves simple, alternate, or rarely opposite or whorled. Flowers bisexual, radially or bilaterally symmetrical; sepals sometimes 3–10 but commonly 5; corolla campanulate to tubular and strongly bilaterally symmetrical (one of the sutures between petals much shorter than the other); carpels usually 5 or 2, the style 1, sometimes branched above, the ovary with 5 or 2 or rarely with 1 (*Downingia*), 3, or 10 chambers, the larger number resulting from intrusion of the placentae; placentae usually axile but (when the ovary has only 1 chamber) parietal or sometimes apparently basal; ovules numerous. Fruit a capsule opening by apical or basal pores, or sometimes a berry.

The family is distributed widely, particularly in the temperate regions and subtropical areas. The bellflowers and Canterbury-bells (*Campanula*) are typical members of the family. About ten or more species occur in eastern North America and a few in the West. Other common plants are in the genus *Specularia.* The genera mentioned above belong to the subfamily *Campanuloideae* or bellflower subfamily. The subfamily *Lobelioideae* or *Lobelia* subfamily has strongly bilaterally symmetrical flowers and coalescent anthers. Numerous species of *Lobelia* are in eastern North America and southwestward, one as far as the mountains of southern California. A colorful and interesting genus in California is *Downingia,* a plant with flowers similar to *Lobelia.* Several species occur about drying vernal pools. Several other genera occur in the Western States, most of them being endemic or nearly so to California. One of the most attractive is *Nemacladus,* which includes half a dozen species of minute desert annuals. The West Indian genus *Hipprobroma* occurs in southern Florida.

*Family 2. Goodeniaceae.

GOODENIA FAMILY

Herbs or sometimes small shrubs. Leaves simple, alternate, or rarely in extraterritorial genera opposite. Flowers bisexual, usually radially symmetrical, in cymes, racemes, or sometimes heads; corolla strongly lipped, the lips 2 or rarely 1; stamens 5, free or rarely adnate with the base of the corolla, the anthers sometimes coalescent; carpels 2, the style 1, the ovary sometimes only partly inferior or in 1 genus superior, 1–2-chambered; placentae axile or basal; ovules 1–2–many. Fruit a capsule or a berry, a drupe, or a nut.

The family includes about 10 genera occurring in the South Pacific region. A few species of *Scaevola* occur along the ocean throughout the Tropics, one in sand dunes in peninsular Florida and on the Keys. It is a succulent herb.

Family 3. Brunoniaceae.

BRUNONIA FAMILY

Herbs. Leaves in a basal rosette. Flowers bisexual, radially symmetrical, hypogynous, in a dense head terminating each scape, small; anthers coalescent; carpel apparently 1; ovule 1, attached at the bottom of the ovary. Fruit a small nut.

The family is composed of a single species occurring in Australia.

*Family 4. Calyceraceae.

CALYCERA FAMILY

Herbs or suffrutescent plants. Leaves simple, entire or pinnately lobed, alternate or in basal rosettes. Flowers usually bisexual, radially or bilaterally symmetrical, 4–6-merous; anthers often coalescent; carpel apparently 1; placenta apical; ovule 1. Fruit an achene, the calyx persistent.

The family is native to tropical America, and it is composed of genera. The only representative in North America is a species of *Acicarpha* naturalized in fields and along roadsides in northern Florida. The achene appears to be a spiny bur because of the spinelike sepals.

The family is related to both the Campanulales and the Rubiales.

Family 5. Stylidiaceae.

STYLIDIUM FAMILY

Leaves simple, narrow, alternate or in basal rosettes or in fascicles on the stem. Flowers bisexual

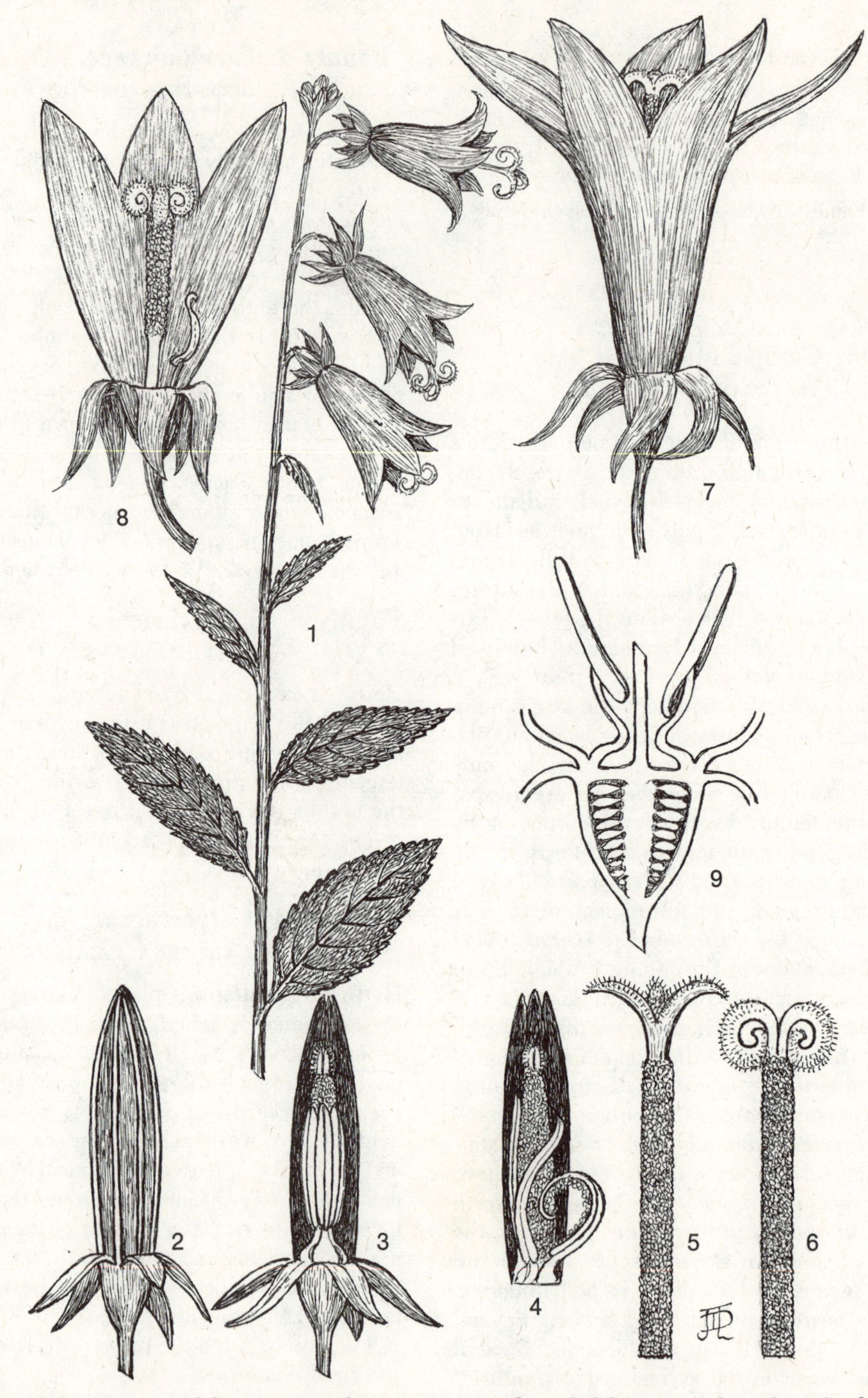

Figure 10-96. Campanulales; Campanulaceae; *Campanula*, a bellflower: **1** branch with alternate leaves and flowers, **2** flower bud, showing the floral cup (adnate with the inferior ovary), the sepals, and the unopened corolla, **3** internal view, showing the style and stigmas pushing upward among the connivent (closely associated) anthers, **4** later, as the corolla begins to open, **5–6** stages in expansion of the stigmas, **7** open flower, **8** part of the corolla and all but one stamen removed, **9** longitudinal section.

or unisexual (and then the plant monoecious), usually bilaterally symmetrical, 1 petal being much larger than the others; stamens 2–3, adnate with the style, the tube forming a column (gynadrium) or the stamens and styles sometimes separate; carpels 2, but sometimes only 1 developing, the style 1, the ovary 1-chambered, with 1 parietal placenta or with the placenta on a column in the center of the ovary; ovules numerous. Fruit a capsule or sometimes indehiscent.

The family consists of several genera occurring in Australia and New Zealand.

Order 49. *ASTERALES.*

ASTER OR SUNFLOWER ORDER

DEPARTURE FROM THE FLORAL CHARACTERS OF THE RANALES

Free	The petals, (usually) the anthers, and the carpels coalescent; stamens adnate with the corolla.
Equal	Ray corollas bilaterally symmetrical through failure of coalescence (except basally) between two of the five petals.
Numerous	Stamens 5; carpels 2.
Spiral	Stamens and carpels cyclic.
Hypogynous	Flowers epigynous.

The order consists of a single family, the *Compositae* or sunflower family, the largest of all plant families, including perhaps 20,000 species.

The flowers are small, but they are arranged in heads, each of which seems at first glance to be an individual flower, although actually it may consist of several to one hundred or more. The head is surrounded by a series of involucral bracts (phyllaries). The flowers are produced on a compound receptacle formed by coalescence of the individual receptacles, and the entire head is supported by a peduncle. The flowers in the center of the head are disc flowers; usually those of the outermost single series are specialized as ray flowers.

The Disc Flower. The ovary is inferior. The pappus usually present at the top probably corresponds to the calyx; it may consist of either scales or bristles. The corolla above and partly enclosed by the pappus consists of five coalescent petals. The anthers protrude from the corolla; they are coalescent into a tube, but the filaments are not. The style, bearing two stigmas (or style branches), protrudes from the summit of the anther tube. Each style branch is stigmatic along two marginal lines. These **stigmatic lines** may be seen under a dissecting microscope or a powerful hand lens or by removal of the stigma and examination with low power of a compound microscope. The stigmatic line is differentiated partly by color and partly by texture from the rest of the stigma. The pollen grains adhering to it may make it conspicuous.

The Ray Flower. Typically, ray flowers are similar to disc flowers, except for the flat (**ligulate** or strap-shaped) corollas. The ray corolla forms a tube for only a short distance above the base; farther up the five petals are coalescent side-by-side except between two petals and are laid out in a flat structure (**ligule**).

There is great variation among members of the family with respect to sterility of either the ray or disc flowers. In some genera the anthers may be lacking from one or the other or in other genera the ovary may not develop in one type of flower.

In addition to the characters in the table above the following are outstanding characteristics of the order: (1) flowers in heads, each head subtended by an involucre; (2) calyx not green or leaflike, consisting of a pappus of bristles or scales, or sometimes wanting altogether; (3) corollas differentiated into two types, ray and disc, the ray corollas simulating petals and the disc flowers more or less giving the impression of being the stamens and pistils, the entire head therefore appearing superfically like a single flower; (4) style elongating rapidly in anthesis (flowering time) and elevating the stigmas through the anther tube, the folded style branches usually hairy on the backs and therefore acting as a "bottle-brush" pushing the pollen from the inside of the anther tube; (5) fruit an achene formed from an inferior ovary with two carpels. Nearly all the Compositae are herbs, but a few are large shrubs.

The Asterales depart from the floral characteristics of the Ranales in practically every respect, and several specialized characters of the order are concerned with other features. Such factors as the pappus, which may carry the achene through the air as if by parachute, are responsible for rapid dissemination of this family over the Earth. For intangible physiological and structural reasons the group seems to be well-adapted to many habitats

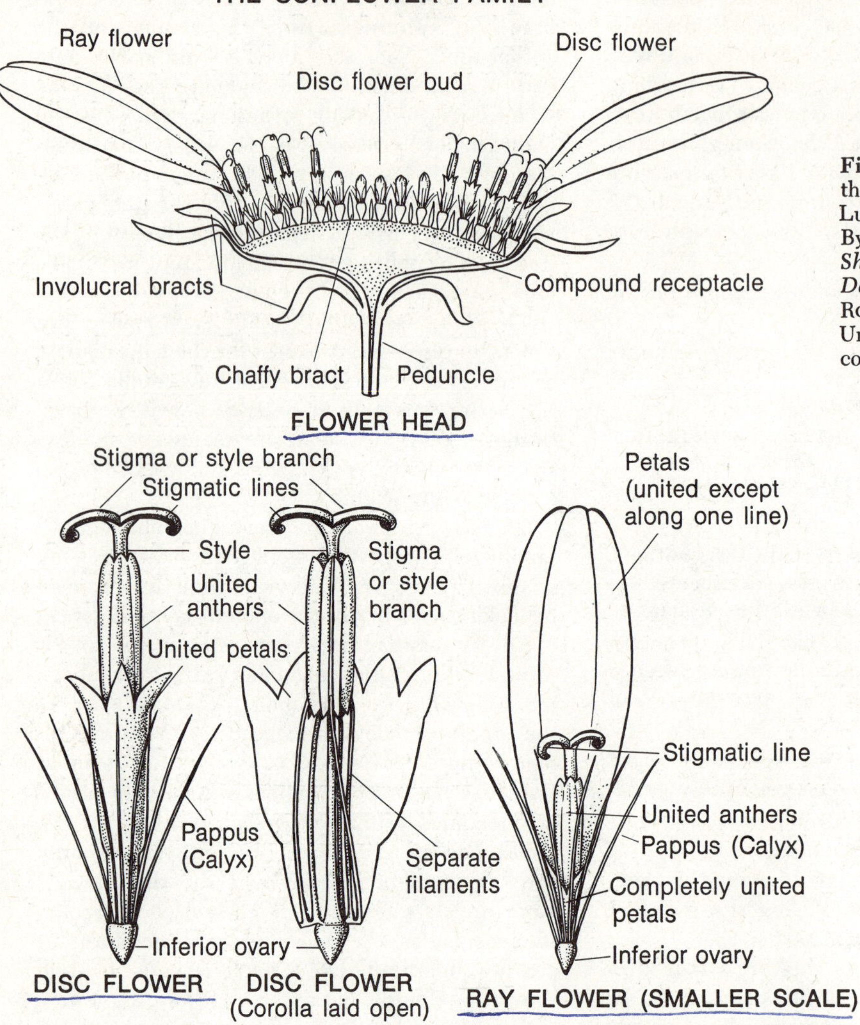

Figure 10-97. The flowers of the Asterales: as labelled. By Lucretia Breazeale Hamilton. By permission from *Trees and Shrubs of the Southwestern Deserts,* LYMAN BENSON and ROBERT A. DARROW, Tucson: University of Arizona Press, copyright 1979.

such as deserts which are difficult for other plants. Many *Compositae* thrive also in the more common types of habitats of temperate regions or in the Tropics. As a result, the *Compositae* include innumerable successful species.

Efficiency in pollination is a matter of speculation. A butterfly visiting a head of *Zinnia* flowers inserts its proboscis rapidly into first one and then another of the flowers in the turn or two of the spiral of open disc flowers, thus pollinating sometimes 15 or 25 flowers at one sitting. Since the pathway of the insect's proboscis within each flower is guided, the pollination methods are effective and precise. Perhaps this represents an introduction of assembly line methods into pollination, but it is to be noted that flowers arranged in this way and pollinated one after the other are not adapted to prevention of self-pollination, at least in the sense that the pollen from one flower of the plant is carried to another of the same individual.

*Family 1. **Compositae.**

SUNFLOWER FAMILY

(Synonyms: *Asteraceae, Carduaceae.* Segregates: *Chichoriaceae, Ambrosiaceae*)

Tribes of the Compositae. The *Compositae* are not only a large family abundant everywhere but, as may be expected, a diverse one. Evolution has proceeded in many directions and the principal developmental lines are summarized by grouping of related genera into tribes.

There have been proposals for raising each tribe listed below to family rank. This would require a radical point of view, for various genera are difficult to place in the tribes, and no key is completely satisfactory. Although the tribes are not wholly distinct, they represent partly differentiated natural groups of genera, and knowledge of them is indispensable to understanding the family. A less radical though dubious proposal is segregation of the *Ambrosieae* as *Ambrosiaceae* and the *Cichorieae* as *Chichoriaceae,* retaining the other tribes as *Compositae* (a change of name for the residue being unnecessary).

Some tribes (e.g., the *Heliantheae* or sunflower tribe and the *Astereae* or *Aster* tribe) are relatively generalized, but others (e.g., the *Madieae* or tarweed tribe; the *Ambrosieae* or ragweed tribe, which includes the cockleburs; and the *Inuleae* or everlasting tribe) are highly specialized, and with much variation in the amount and type of specialization in individual genera. The lines of development are so complex and so complicated by parallel evolution in some characters accompanied by divergent evolution in others that construction of a reliable key is difficult. As in all segregations of plant or animal groups, a certain complex or combination of characters marks each tribe but no one character is found always in every member and never in related tribes. On page 342 is a list of the tribes. For those occurring in North America north of Mexico the name is accompanied by one or two (or a few) characters which best set it off from at least its nearest relatives. Commonly these characters are present in at least 95 to 99 per cent of the members of the tribe.

The *Compositae* are considered by many to be a "difficult" group. The reason for this impression is inadequate study. The family is recognizable at a glance, and following the keys to the orders becomes unnecessary once the characters of the Asterales are mastered. The most effective method of study is collection of as many *Compositae* as possible. When 50 or 100 are available, they should be sorted out as a group step-by-step in the key. Some will fit the upper lead 1, some the lower 1 (*Cichorieae*). Some of the upper group will fit the upper lead 2, some may fit the lower 2 (*Mutisieae*). As the process of division continues, nearly every lead will be illustrated by a plant. When the tribes are mastered, all the plants of one tribe, then another, should be keyed to genera, then *all* the species of each genus keyed out. Identification of a single species at a time is a tedious and ineffective procedure to be avoided as soon as sufficient material becomes available. Fortunately the *Compositae* are studied almost as readily pressed as fresh once the fundamental pattern is learned.

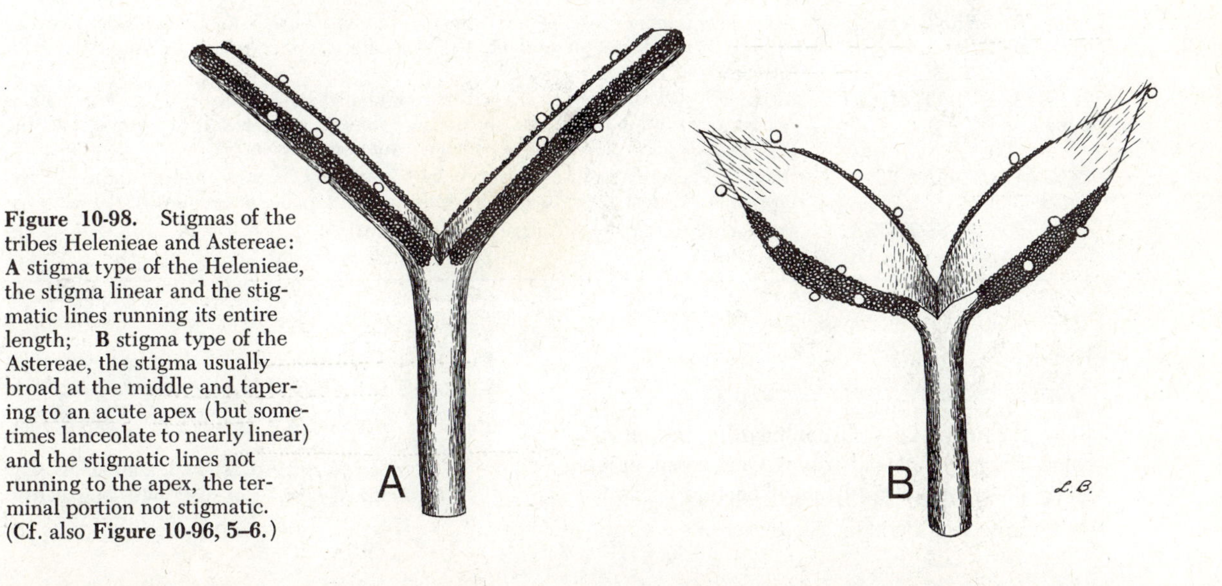

Figure 10-98. Stigmas of the tribes Helenieae and Astereae: **A** stigma type of the Helenieae, the stigma linear and the stigmatic lines running its entire length; **B** stigma type of the Astereae, the stigma usually broad at the middle and tapering to an acute apex (but sometimes lanceolate to nearly linear) and the stigmatic lines not running to the apex, the terminal portion not stigmatic. (Cf. also **Figure 10-96, 5–6.**)

SUBFAMILIES AND TRIBES OF THE COMPOSITAE

Subfamily a. **Asteroideae**

Tribe	Characters
Tribe 1. *HELIANTHEAE* Sunflower Tribe	Receptacle with a chafflike bract (scale) subtending each disc flower.
Tribe 2. *MADIEAE* Tarweed Tribe	Involucral bracts each enfolding or half-enfolding a ray achene; chafflike bracts in a single series between the ray flowers and disc flowers.
Tribe 3. *AMBROSIEAE* Ragweed Tribe	Anthers not coalescent or coalescent only slightly at the bases; some or all the flowers unisexual, the plants nearly always monoecious, the staminate heads in terminal racemes or spikes and the pistillate heads (usually burlike) produced lower on the same branch or on an adjacent one; pappus none.
Tribe 4. *HELENIEAE* Sneezeweed Tribe	Style branch with the stigmatic lines running to the apex, linear, each branch commonly with an apical tuft of hair, the apex "chopped off" abruptly (truncate) or rarely with a short bristle; otherwise similar to the *Aster* tribe (Tribe 6, *Astereae*).
Tribe 5. *SENECIONEAE* Groundsel Tribe	Pappus remarkably soft, fluffy, copious, usually white, the bristles capillary in the extreme sense.
Tribe 6. *ASTEREAE* *Aster* Tribe	Style branch with a stigmatic line along each margin of the basal half, the terminal portion not stigmatic and usually short-hairy, usually broad at the middle and tapering to an acute apex.
Tribe 7. *INULEAE* Everlasting Tribe	Involucral bracts scarious (thin, white or colorless, and translucent); ray flowers none; pappus of bristles; anthers basally caudate (with "tails" other than the filaments).
Tribe 8. *ANTHEMIDEAE* Mugwort or Sagebrush Tribe	Involucral bracts scarious over most of their surface; pappus none or consisting of scales; anthers not caudate, i.e., not with tails; involucral bracts usually imbricated.
Tribe 9. *CALENDULEAE* Marigold Tribe	Not occurring in North America north of Mexico.

Subfamily b. **Cichorioideae**

Tribe	Characters
Tribe 10. *CICHORIEAE* Chicory Tribe	Disc flowers none, only "ray" (ligulate) flowers present; juice of the plants nearly always milky; cauline leaves alternate.
Tribe 11. *MUTISIEAE* *Mutisia* Tribe	Ray flowers none; disc flowers with two-lipped corollas, that is, with the petals in two strongly differentiated groups, the upper lip consisting of two petals and the lower of three, the lower lips of the marginal flowers often more elongated than those of the inner.
Tribe 12. *VERNONIEAE* Ironweed Tribe	Ray flowers none; disc flowers bisexual, rarely yellow; style branches long, cylindroidal, threadlike, puberulent, the stigmatic lines restricted to the basal half, obscure.
Tribe 13. *EUPATORIEAE* Eupatory Tribe	Stigmas (style branches) shaped like baseball bats or clubs, that is, *either* elongated and gradually enlarged upward *or* sometimes relatively short and with the tips knoblike; ray flowers none; flowers white or greenish.
Tribe 14. *CYNAREAE* Thistle Tribe	Ray flowers none; disc flowers cleft into long narrow lobes; anthers long-appendaged; leaf margins usually spiny and the plants therefore thistlelike.
Tribe 15. *ARCTOTIDEAE* *Gazania* Tribe	Not occurring in North America north of Mexico.

Key to the Tribes

(North America north of Mexico)

(The following key is simplified to set off the chief characters of the tribes. Occasional exceptional genera have not been taken into account.)

1. Disc flowers present; juice of the plant not milky.
   2. Corolla *not 2-lipped in all of the flowers of the head,* or rarely so in the outer disc flowers when ray flowers are lacking.
      3. Corolla (except in *Lessingia,* a spineless slender annual herb) not cleft into long narrow

lobes; anthers not long-appendaged; plants usually not spiny on the leaf margins, not thistlelike.

4. Pappus never soft and fluffy (i.e., capillary in the extreme sense).
   5. Stamens separate from each other or practically so; flowers or some of them unisexual, the plants practically always monoecious, the staminate heads in terminal racemes or spikes and the pistillate (usually burlike heads) produced lower on the same branch or an adjacent one; pappus none.

      Tribe *3.* **Ambrosieae** (ragweed tribe).

   5. Stamens coalescent by the anthers, forming a tube around the style; flowers usually bisexual but sometimes all or some of them unisexual; heads not burlike.
      6. Involucral bracts *not* markedly scarious (i.e., the bract *not* thin, white or colorless, and *not* translucent) or rarely so, sometimes the margins slightly scarious.
         7. Receptacle with chafflike bracts (scales), one subtending each disc flower (or at least each of the outermost disc flowers, then forming a series between ray and disc flowers).
            8. Involucral bracts *not* enclosing the bases of the ray achenes; each disc flower subtended by a chaffy bract.

               Tribe *1.* **Heliantheae** (sunflower tribe).

            8. Involucral bracts each folded around or halfway around a ray achene; chaffy bracts in a single series between the ray and disc flowers.

               Tribe *2.* **Madieae** (tarweed tribe).

         7. Receptacle almost always naked, that is, with no elongated bristles and with no chaffy bracts subtending or surrounding the disc flowers.
            8'. Stigmas or style branches *not* shaped like clubs or baseball bats, usually *not* greatly elongated, linear or broad at the middles, flat, *not* enlarged markedly at the ends; involucral bracts *not* striate.
               9. Stigma or style branch with the stigmatic lines running to the apex, linear, commonly with an apical tuft of hairs, the apex truncate or rarely with a short bristle; involucral bracts usually in one series.

                  Tribe *4.* **Helenieae** (sneezeweed tribe).

               9. Stigma or style branch with a stigmatic line along each margin of the lower half, the terminal portion not stigmatic; involucral bracts usually in several series of graduated lengths and overlapping like shingles.
                  10. Flowers of various colors or sometimes white; ray flowers usually present, usually yellow; each style branch flattened, broadest at the middle and tapering to an acute apex, the apical sterile portion usually hairy.

                      Tribe *6.* **Astereae** (*Aster* tribe).

                  10. Flowers purple to rarely white; ray flowers none; each style branch threadlike, cylindroidal, minutely hairy over its entire surface.

                      Tribe *12.* **Vernonieae** (ironweed tribe).

            8'. Stigmas or style branches shaped like clubs or baseball bats (sometimes greatly elongated but gradually enlarged upward); involucral bracts of the semishrubby species commonly striate; ray flowers none; disc flowers white or greenish.

               Tribe *13.* **Eupatorieae** (eupatory tribe).

6. Involucral bracts (inner ones) scarious over most of their surface.
   7'. Pappus of bristles, anthers basally caudate (with "tails").

   Tribe 7. **Inuleae** (everlasting tribe).

   7'. Pappus not of scales; anthers not caudate.

   Tribe 8. **Anthemideae** (mugwort or sagebrush tribe).

4. Pappus of soft, fluffy, copious, usually white bristles, these capillary in the extreme sense or in *Lepidospartum* (a Southwestern shrub with scalelike leaves) coarser.

   Tribe 5. **Senecioneae** (groundsel tribe).

3. Corolla cleft into long narrow lobes; ray flowers none; anthers basally long-appendaged (caudate, i.e., with "tails"); plants usually spiny on the leaf margins, thistlelike.

   Tribe *14.* **Cynareae** (thistle tribe).

2. Corolla in *all* the flowers 2-lipped, that is, with the petals in 2 strongly differentiated groups with 2 petals in the upper lip and 3 in the lower, the lower lips of the marginal flowers often elongated; ray flowers none; anthers basally caudate (with long tails).

   Tribe *11.* **Mutisieae** (*Mutisia* tribe).

1. Disc flowers none, only ray flowers present; juice of the plant milky.

   Tribe *10.* **Cichorieae** (chicory tribe).

Figures 10-99 to 10-107 illustrate representatives of the tribes of the Compositae, showing detailed diagnostic characters.

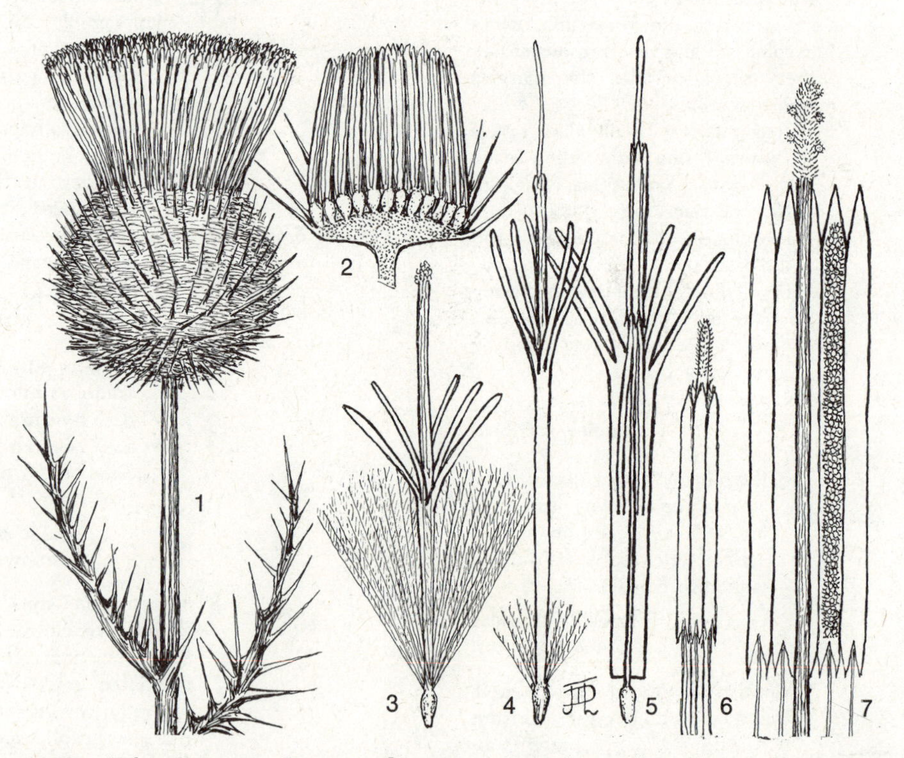

Figure 10-99. Tribe Cynareae; *Cirsium californicum,* a thistle: **1** branch with the upper spiny leaves and a flower head, **2** flower head, longitudinal section, showing the spiny involucral bracts, the compound receptacle, and the flowers (all disc), **3–4** young and older flower, showing the floral cup (adnate with the inferior ovary), the plumose pappus, the corolla (with the individual petals free for one third their length), the elongate anther tube, and the appressed stigmas, **5** flower dissected, showing the filaments, **6–7** anthers, filaments, and appressed stigmas.

Figure 10-100. Tribe Heliantheae; *Helianthus annuus* (or a hybrid), sunflower: **1** leaf and flower bud, **2** head of flowers, showing the areas of ray and disc flowers, **3** head of flowers, showing the involucral bracts and the corollas of the ray flowers, **4** head of flowers, longitudinal section, showing the ray flowers (marginal) and disc flowers above the involucral bracts and compound receptacle, **5** ray flower, showing the floral cup (adnate with the inferior ovary), the corolla (even the tips of the petals coalescent), and no stamens or style (the ovary being sterile), **6** disc flower subtended by a chaffy bract, **7** disc flower, showing the floral cup (adnate with the inferior ovary), the pappus (of scales), the tubular corolla, the anther tube, and the style and stigmas, **8** disc flower, dissected, showing the separate filaments.

Figure 10-101. Tribe Madieae; *Calycadenia ciliosa,* a tarweed: **1** branch with the upper leaves (each ending in a stalked, "tack-shaped" gland), a head of flowers developed from each axillary bud and the terminal bud, **2** terminal flower head and the subtending leaves, **3** head of flowers with the involucral bracts and ray flowers pulled downward, revealing the single series of chaffy bracts between the ray and disc flowers, **4–5** ray flower, two views, showing the floral cup (adnate with the inferior ovary) and the corolla, **6** ray flower dissected, showing the style and stigmas, **7–9** incomplete enfolding of the ray flower (especially the ovary) by the adjacent involucral bract, **10** disc flower, showing the floral cup (adnate with the inferior ovary), the narrow scales of the pappus, and the sympetalous corolla, **11** corolla laid open, showing the filaments (the anthers separated), **12** achene half-enfolded in the involucral bract.

Figure 10-102. Tribe Ambrosieae; *Xanthium strumarium* var. *canadense,* cocklebur: **1** branch with an inflorescence, **2** inflorescence of staminate and pistillate flowers, **3** staminate flower, showing the corolla and the stamens (anthers practically separate, the filaments coalescent), **4** staminate flower, laid open, **5** two pistillate heads, showing the hooked involucral bracts and two protruding pistillate flowers, **6** involucral bract, **7–8** styles protruding from the enclosing terminal involucral bracts, **9** longitudinal section of the head, showing the two enclosed flowers (each consisting of a pistil), **10** cross section, **11** achene.

Figure 10-103. Tribe Senecionieae; *Senecio Douglasii,* a shrubby groundsel: **1** branch with pinnate leaves and with flower heads, **2** head, showing leaves, the involucre, the ray corollas, and the tops of the disc flowers, **3** ray flower, showing the floral cup (adnate with the inferior ovary), the pappus, the corolla, and the stigmas, **4** disc flower, **5** stigmas, **6** enlargement of one stigma showing one stigmatic line with pollen grains clinging to it.

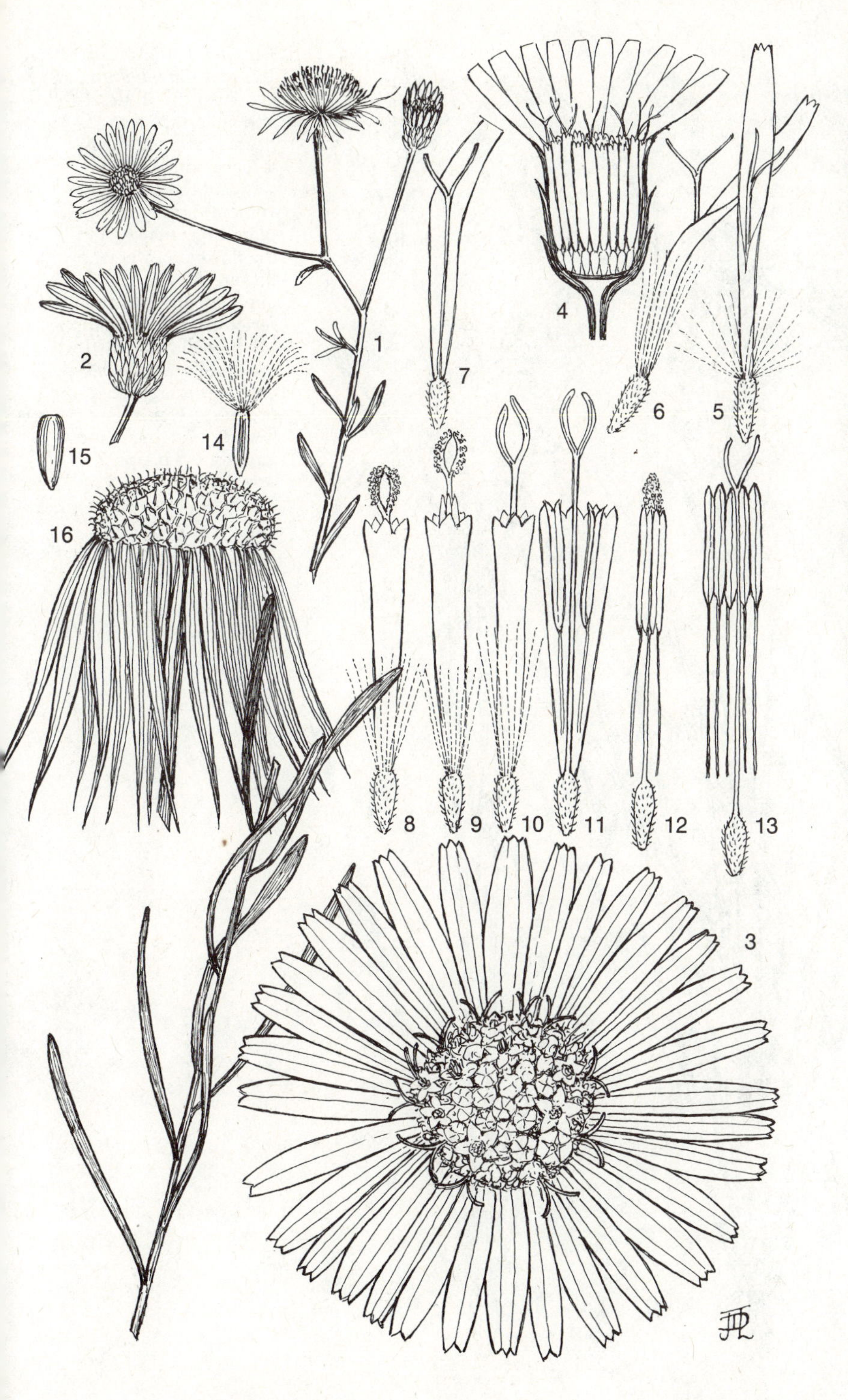

Figure 10-104. Tribe Astereae; *Aster exilis,* an aster: **1** branch with leaves and flower heads, **2** flower head, side view, showing the involucral bracts and the ray corollas, **3** top view, showing the ray and disc flowers, **4** longitudinal section, showing also the involucral bracts and the compound receptacle, **5–6** ray flower, showing the floral cup (adnate with the inferior ovary), the pappus of slender bristles, the coalescent petals, and the style and stigmas, **7** corolla dissected and the pappus removed, showing the complete style, **8–10** disc flower series, stigmas emerging from the anther tube, each covered dorsally with pollen, **11–13** disc flower, dissections showing the stamens (the anthers coalescent, the filaments separate), **14–15** achene, with and without pappus, **16** receptacle and reflexed involucral bracts after flowering.

Figure 10-105. Tribe Inuleae; *Gnaphalium*, an everlasting: **1** bracts and inflorescence, **2–3** heads, external views, **4** longitudinal section, showing the scarious involucral bracts, the compound receptacle, and the flowers, **5** perfect flower, showing the floral cup (adnate with the inferior ovary), the pappus (of barbellulate bristles), the corolla, the coalescent anthers, and the stigmas, **6** pistillate flower, **7** achene and pappus, **8** pappus bristles.

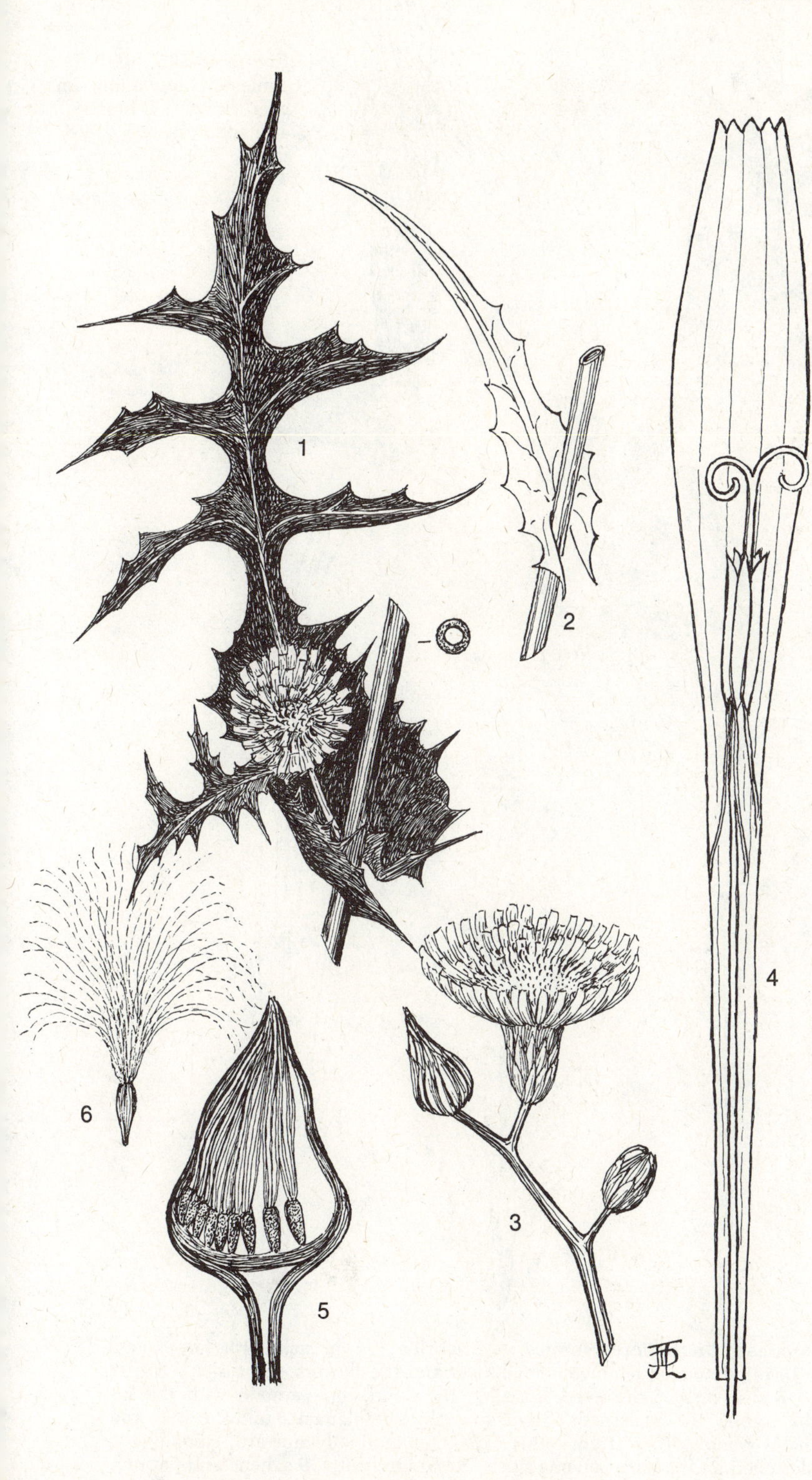

Figure 10-106. Tribe Cichorieae; *Sonchus oleraceus*, sow-thistle: **1** branch with a leaf and a flower head, **2** leaf outline, upper leaf, **3** inflorescence, the flower head showing the involucral bracts and the flowers (all ray), **4** flower, dissected, showing the corolla, the stamens, and the style and stigma, **5** head of achenes, longitudinal section, **6** achene. (Cf. also **Figures 12-5,** sow-thistle, and **8-4 C** dandelion, fruit.)

Figure 10-107. Tribe Eupatorieae; *Brickellia californica*, a brickellia: **1** branch with leaves and heads of flowers, **2** flower head, showing the involucral bracts and the flowers (all disc), **3** disc flower bud, **4** disc flower, **5** disc flower, dissected, showing the floral cup (adnate with the inferior ovary), the pappus, the corolla, the coalescent anthers and separate filaments, and the style and stigmas, **6** corolla and stamens removed, **7** stigma, this expanded gradually upward, similar to a baseball bat but somewhat flattened and grooved on one side, **8** head in fruit, **9** achene and pappus.

FAMILIES OF UNCERTAIN POSITION

The following extraterritorial families are of uncertain relationships, and they are not assigned to orders, pending the results of further investigation.

Family **Ancistrocladaceae.**

ANCISTROCLADUS FAMILY

Climbing shrubs or vines; with short, hooked tendrils. Leaves simple, alternate; stipules present but small; flowers bisexual, the petals slightly coalescent, i.e., barely sympetalous, the inflorescence a panicle, the flower barely epigynous; sepals 5, overlapping, enlarged and wingéd in fruit; petals 5, contorted; stamens 5 or 10, anthers with 2 pollen chambers, splitting lengthwise; carpels 3; stigmas 3; ovary 1-chambered, with 1 ovule. Fruit a nut, enclosed in the wingéd calyx; seeds with the cotyledons strongly folded and the seed coat and the root tip folded within them.

A single genus occurring in tropical West Africa and tropical Asia.

Family **Balanopsidaceae.**

BALANOPS FAMILY

Shrubs or trees. Leaves alternate, stipules none, simple. Plants dioecious, the staminate flowers in catkins, the pistillate solitary and in involucres; flowers hypogynous; sepals and petals none; staminate flowers sometimes with 2–12 stamens but most commonly with 5–6; pistillate flower consisting of a single pistil of two carpels; ovary partly divided into two chambers, but the partition incomplete; ovules 4; erect, the embryo straight. Fruit a drupe, surrounded by the persistent involucre, ovoid, sometimes divisible into 2 sections; endosperm fleshy.

The family occurs in New Caledonia and tropical Australia.

Family **Cardiopteridaceae.**

PERIPTERYGIUM FAMILY

Climbing shrubs; glabrous; juice milky. Leaves simple, entire to palmately lobed or cleft, basally cordate; stipules none. Flowers bisexual, sympetalous; hypogynous disc none; inflorescence of axillary cymes, these branching; sepals 5, coalescent, overlapping; petals 5, the corolla lobes spreading; stamens 5, adnate with the corolla tube, alternate with the petals; carpels 2, the ovary 1-chambered, the styles and stigmas strongly dissimilar, one columnar, elongate, and persistent, the other short and with a columnar style and a knoblike stigma; ovules 2, apical and pendulous. Fruit indehiscent, oblong-obovate, indented apically, with two broad longitudinal wings; seeds grooved, elongate, the embryos very small, with fleshy granular endosperm.

A single genus, *Peripterygium* (*Cardiopteris*), in the Indo-Malay region.

Family **Columelliaceae.**

COLUMELLIA FAMILY

Shrubs or trees. Leaves simple, opposite, estipulate. Flowers bisexual, epigynous, barely bilaterally symmetrical, sympetalous; petals 5, the corolla tube very short, more or less campanulate; stamens 2; pollen chambers twisted and plaited; ovary incompletely 2-chambered; placentae parietal, almost meeting in the middle of the ovary. Fruit a capsule.

The family is composed of a single genus occurring in Columbia and Peru.

Family **Coriariaceae.** CORIARIA FAMILY

Suffrutescent to fruticose plants; twigs angled. Leaves simple, entire, opposite or whorled; stipules present. Flowers bisexual or sometimes unisexual, hypogynous, choripetalous, radially symmetrical, solitary or in racemes; sepals 5, in the bud overlapping, larger than the petals; petals 5, with keels on the ventral surfaces; stamens 10, in 2 series, separate, but each stamen of the inner series adnate with the keel of a petal; carpels 5–10, separate, each with a single pendulous ovule. Fruit an achene, but surrounded by the persistent fleshy petals and appearing drupelike.

The single genus is distributed widely. One species occurs in Mexico.

The fruits, leaves, and roots are said to be strongly poisonous.

The family is thought by various authors to be related to the *Empetraceae* (both families being in the Sapindales according to Engler and Prantl) or to constitute a distinct order (Hutchinson).

Family **Corynocarpaceae.**

CORYNOCARPUS FAMILY

Shrubs. Leaves simple, entire, alternate; stipules none. Flowers bisexual, radially symmetrical, hypogynous, sympetalous, in racemes; sepals 5, separate; petals 5; stamens 5, adnate with the corolla; petaloid staminodia 5; carpels 2, the styles 1–2, the ovary 1–2-chambered; ovule 1 per carpel, pendulous. Fruit a drupe.

The family is composed of a single genus occurring in the South Pacific.

Family **Daphniphyllaceae.**

DAPHNIPHYLLUM FAMILY

Shrubs or trees. Leaves evergreen, alternate or more or less verticillate, with long petioles, smooth, leathery, glaucous beneath; stipules none. Flowers unisexual, the plants dioecious, in terminal umbels or axillary racemes; sepals 3–6 or none; petals none; stamens 6–12; carpels 2, the style(s) short, the stigmas 2, the ovary incompletely 2-chambered; ovules 2 in each chamber. Fruit a drupe; seed 1.

A single species occurring in China and Japan.

Family **Dichapetalaceae.**

DICHAPETALUM FAMILY

Shubs, trees, or woody vines; juice poisonous. Leaves simple, alternate; stipules present. Flowers bisexual or rarely unisexual, choripetalous or sympetalous, radially (or slightly bilaterally) symmetrical, with glandular hypogynous discs, sometimes perigynous or epigynous; sepals 5, in the bud overlapping; petals 5, usually 2-lobed to -parted, sometimes adnate with the stamen tube; stamens 5, sometimes coalescent; anthers splitting lengthwise; carpels 2–3, the style 1, sometimes lobed, the ovary 2–3-chambered; placentae axile; ovules 2 per carpel, pendulous. Fruit a capsule or drupelike or berrylike, sometimes with only the exocarp splitting open.

The family consists of three genera occurring in the Tropics and most abundant in Africa. It has been classified in the Euphorbiales (Gundersen), in the Rosales, and in the Geraniales (Engler and Prantl).

Family **Didymelaceae.**

DIDYMELES FAMILY

Trees. Flowers unisexual (the plants dioecious), hypogynous, apetalous, the staminate flower consisting of 2 stamens and 1–2 scales below the sessile anthers, the pistillate flower consisting of 1 carpel and 4 scales, the stigmatic surface oblique. Fruit a drupe, with a groove on one side; ovule and seed 1.

A single genus, *Didymeles,* composed of 1 species occurring in Madagascar.

Family **Dipentodontaceae.**

DIPENTODON FAMILY

Shrubs or small trees. Leaves simple, alternate, elliptic-ovate or elliptic, acuminate, denticulate; stipules present, deciduous. Flowers bisexual, hypogynous, choripetalous, radially symmetrical, small and numerous in axillary long-pedunculate dense umbels, these at first involucrate, the bracts 4 or 5; sepals 5–7, not coalescent, not overlapping; petals 5–7, similar to the sepals; stamens 5–7, separate, alternate with the petals; carpels 3, the ovary partly (basally) 3-chambered; ovules 2 in each chamber. Fruit capsular, but with only 1 seed, late in dehiscence.

A single genus, *Dipentodon,* with 1 species in southeastern Asia.

Family **Dioncophyllaceae.**

DIONCOPHYLLUM FAMILY

Shrubs, but the stems scrambling over other vegetation. Leaves simple, alternate, some apically modified into hooks or an elongate midrib bearing capitate glands; stipules none. Flowers bisexual, hypogynous, choripetalous, radially symmetrical; hypogynous disc none; sepals 5, sometimes basally coalescent; petals 5; stamens 10, not coalescent or adnate; carpels 2–5, the ovary 1-chambered; placentae parietal, with numerous ovules. Fruit a capsule, the seeds ultimately projecting on long funiculi after opening of the fruit.

The family is thought to be composed of 3 genera, *Dioncophyllum, Habropealum,* and *Triphyophyllum,* these being considered by other authors as a single genus of 3 species. Tropical West Africa.

Family **Greyiaceae.** GREYIA FAMILY

Small trees. Leaves simple, alternate, with long petioles; stipules none. Flowers conspicuous, in dense racemes from the leaf axils; sepals 5, persistent; petals 5, not coalescent; stamens 10, these and the petals on a hypogynous disc, alternating with marginal glands; filaments slender; carpels 5, the ovary deeply grooved, 5-chambered, the style long. Fruit a leathery capsule, septicidally dehiscent.

A single genus of 3 species; South Africa.

Family **Huaceae.** HUA FAMILY

Trees. Leaves simple, alternate, elliptic-oblong, acuminate, entire; stipules present, small and deciduous; flowers bisexual, hypogynous, choripetalous, radially symmetrical, small, a few in each axillary cluster, long-pedicelled; sepals 3–5, not overlapping; petals 4–5, the blades villous; stamens 10, not coalescent or adnate; carpels 5, coalescent; style 1; ovule 1 on a basal placenta, erect, anatropous. Fruit capsular but with only 1 seed.

A single genus, *Hua,* with one species; Gabon and Zaire (Belgian Congo), tropical West Africa.

Family **Hoplestigmataceae.**
HOPLESTIGMA FAMILY

Large trees. Leaves large, alternate. Sepals 4; petals about 12–16 in 3 or 4 irregular series; stamens numerous, arranged irregularly; carpels 2, the ovary 1-chambered; placentae parietal; the ovules 2 per carpel, pendulous. Fruit a drupe.

The family is composed of a single genus, *Hoplestigma,* occurring in the tropical parts of West Africa.

Family **Hydrostachyaceae.**
HYDROSTACHYS FAMILY

Submerged aquatic herbs; stems tuberous. Leaves simple to bi- or tripinnate. Flowers unisexual (and the plants dioecious), hypogynous, in dense spikes borne singly in the leaf axils, sessile; sepals and petals none; stamen 1; anther splitting lengthwise; carpels 2, the styles separate, the ovary 1-chambered; placentae parietal; ovules numerous. Fruit a capsule.

The family is restricted to South Africa and Madagascar.

Family **Medusandraceae.**
MEDUSANDRA FAMILY

Trees. Leaves simple, alternate, minutely crenulate, the apex of the petiole swollen; stipules small and early deciduous. Flowers bisexual, hypogynous, choripetalous, radially symmetrical, in axillary drooping catkinlike racemes; sepals 5, separate or almost so, overlapping; petals 5; stamens 5, opposite the petals, barely adnate with the petal bases; staminodia 5, opposite the sepals, elongate; carpels 3–4, coalescent but with 1 seed chamber; placenta central and free from the ovary wall, the ovules 6–8, apical, pendulous, anatropous. Fruit a capsule; seed 1, pendulous.

A single genus, *Medusandra,* occurring in the Camaroons, tropical West Africa.

Family **Rhopalocarpaceae.**
RHOPALOCARPUS FAMILY

Shrubs or trees. Leaves simple, alternate, ovate to elliptic; stipules present, but deciduous. Flowers bisexual, hypogynous, choripetalous, bilaterally symmetrical, in terminal cymes with opposite branching; sepals 4, separate, unequal; petals 4, unequal; stamens numerous, on a hypogynous disc, which envelopes the base of the fruit; carpels 2, the ovary 2-chambered; placentae basal; ovules commonly 3 in each chamber, anatropous. Fruit globose to reniform; seed 1, large.

A single genus in Madagascar.

Family **Thelygonaceae.**

THELYGONUM FAMILY

Herbs, somewhat succulent; lower leaves opposite, the others alternate; stipules present. Flowers unisexual and the plant monoecious; stamens about 10–20, the filaments slender; sepals 5, the calyx 2-parted but with 5 veins or in pistillate flowers pushed to one side and not parted; ovary enlarging irregularly on one side, the style becoming lateral at fruiting time; ovule 1, erect from the base of the placenta. Fruit a nut, partly enclosed in the membranous calyx.

The single genus occurs in the Mediterranean Region and in China and Japan.

Family **Tremandraceae.**

TETRATHECA FAMILY

Suffrutescent or fruticose plants. Leaves simple, alternate, opposite, or whorled; stipules none. Flowers bisexual, radially symmetrical, with no hypogynous discs, solitary, axillary; sepals 4–5 or rarely 3, valvate in the bud; petals 4–5 or rarely 3; stamens twice the number of the petals, in 2 series; anther with 2–4 pollen chambers, opening by a single terminal pore; carpels 2, the style 1, the ovary 2-chambered; placentae axile; ovules 1–2 per carpel, pendulous. Fruit a flattened capsule; seeds often hairy.

The family consists of three genera occurring in temperate (except eastern) Australia. *Tetratheca* is the principal genus; one species has been introduced into cultivation in the southern areas of the United States.

The family has been classified in the Geraniales (Engler and Prantl) and in the Sapindales (Gundersen), as well as in other orders.

11

Manual of the Orders and Families of Monocotyledons

Subclass MONOCOTYLEDONEAE

KEY TO THE ORDERS

NOTE: After determination of the order and family, proceed to the local flora or manual for determination of the genus and species.

1. Plant *not* corresponding with *every character* in the combination in lead 1 below (p. 359).
   2. Pistils commonly 10 to numerous, rarely 6–9, separate, each formed from a single carpel; stamens usually 7–many or rarely 6 in pairs, i.e., 2 opposite each petal.

      Order 1. **Alismales,** page 360.

   2. Pistil(s) 1 or sometimes as many as 4 or very rarely 5 or 6; stamens 1–6 or very rarely more, if 6 then 1 opposite each petal and sepal.
      3. Flowers not hidden in the axils of dry chaffy bracts. (Cf. lower lead 3, p. 359.)
         4. Herbs (or rarely trees or shrubs with simple, entire, usually linear leaves). (Cf. lower lead 4, p. 359.)
            5. Sepals and petals not scalelike, at least one series petal-like, the perianth conspicuous, usually white or conspicuously colored. (Cf. lower lead 5, p. 358.)
               6. Ovary superior. (Cf. lower lead 6, p. 358.)
                  7. Inflorescence *not* a fleshy spike (spadix), *not* enclosed by a spathe; sepals and petals well-developed, at least the petals and often the sepals petaloid.

                     Order 2. **Liliales,** page 361.

                  7. Inflorescence a fleshy spike (spadix) with a spathe formed by a single white or colored but nongreen bract enclosing the entire inflorescence or flower cluster at least when it is young; sepals and petals, if any, small and scalelike.

8. Spike with flowers on only one side, flattened; plants aquatic in salt water, growing on either the rocks of the ocean or in brackish ocean inlets. *Zosteraceae* (pondweed family).

Order 10. **Naiadales, page 407.**

8. Spike with flowers on all sides of a cylindroid, ovoid, or ellipsoid axis or rarely (*Pistia*) the pistillate flower solitary and the staminate above it; plants growing on land or in fresh-water marshes or bogs, but usually not strictly aquatic.

Order 14. **Arales, page 413.**

6. Ovary inferior.

7'. Plants *not* corresponding with *every character* in lead 7' below.

8'. Flower radially symmetrical or nearly so, the perianth sometimes somewhat bilaterally symmetrical, but, if so, no single petal markedly dissimilar from the other two.

9. Seeds *not* minute and powderlike; embryo differentiated in the seed into a cotyledon, a hypocotyl, and an epicotyl; capsule usually not wingéd or angled; sepals and petals usually but not necessarily more or less similar.

Order 2. **Liliales, page 361.**

9. Seeds minute, powderlike; embryo *not* differentiated; capsule wingéd or angled; sepals much smaller than the petals, scalelike; Louisiana to Virginia and Florida, perhaps still rare in occurrence near Chicago, Illinois.

Order 4. **Burmanniales, page 378.**

8'. Flower markedly bilaterally symmetrical; fertile stamen(s) 1 or 2 or rarely more.

9'. Corolla radially symmetrical; staminodia petallike, usually conspicuous, unequal and therefore the flower bilaterally symmetrical; seeds *not* small and powderlike; embryo differentiated into a cotyledon, a hypocotyl, and an epicotyl; pollen not in waxy masses; style and stamen(s) separate.

Order 3. **Musales, page 374.**

9'. Corolla strongly bilaterally symmetrical, the apparently lower petal (the flower being twisted 180°) markedly dissimilar to the others; staminodia none; seeds small; embryo undifferentiated; pollen in waxy masses (pollinia); style and stamen(s) adnate (forming a gynandrium).

Order 5. **Orchidales, page 379.**

7'. Plants aquatic; fruit a capsule or a berry with 2 or commonly 3–15 coalescent carpels; flowers radially symmetrical, unisexual or rarely bisexual, the plants usually dioecious, the staminate flowers becoming detached and floating to the pistillate, the pistillate with elongate submerged perianth tubes; sepals and usually petals present, the series clearly differentiated; fruit berrylike, submerged.

Order 8. **Hydrocharitales, page 405.**

5. Sepals and petals (if any) scalelike or bristlelike, none petal-like, the perianth inconspicuous; herbs (in North America).

6'. Fruit a capsule, the carpels coalescent at least basally.

7''. Carpels clearly and fully coalescent through the entire length of the compound ovary, *not* adnate with a central axis. *Juncaceae* (rush family) and *Eriocaulaceae* (pipewort family).

Order 2. **Liliales, page 361.**

7''. Carpels *either* coalescent only basally *or* adnate with a central axis formed by prolongation of the receptacle (in that case each separating at maturity and then hanging from the tip of the axis).

Order 11. **Juncaginales,** page 408.

6'. Fruit 1-seeded.

7'''. Flowers *either* relatively few (i.e., solitary or a few together) *or sometimes* up to about 25 or perhaps 35 in an elongated spike, usually *not* packed very tightly together except in the bud stage; perianth calyxlike or fleshy.

Order 10. **Naiadales,** page 407.

7'''. Flowers usually very numerous, packed tightly together in cylindroidal spikes or spheroidal clusters, rarely as few as about 25 in a spheroidal cluster.

8''. Inflorescence not hooded by a green spathe; the staminate spikes or spheroidal flower clusters above, the pistillate below.

Order 12. **Typhales,** page 410.

8''. Inflorescence hooded by a green spathe *or* the single spike with bisexual flowers. *Acorus* (sweet-flag), *Orontium* (golden-club), and *Peltandra* (arrow arum), *Araceae* (arum family).

Order 24. **Arales,** page 413.

4. Trees or shrubs (or sometimes with no stem above ground); leaves very broad, plaited, lobed to compound; inflorescence (or each segment of it) enclosed while it is young in a spathe (large bract).

Order 15. **Palmales,** page 416.

3. Flowers hidden in the axils of chaffy bracts, these small bracts dry at least in age, the flowers and subtending bracts arranged in spikelets *or* the flower, *when solitary* (in a reduced spikelet) enclosed in 4 bracts (2 glumes, a lemma, and a palea); perianth parts minute, scalelike or threadlike; fruit an achene or a caryopsis.

Order 7. **Graminales,** page 384.

1. Plant floating, *without* well-differentiated and apparent stems and leaves, the plant body thalluslike, small, circular or elongated, budding asexually; individuals sometimes remaining attached to each other; flower seldom seen, minute, with a spathe, the plant monoecious (but staminate and pistillate flowers sometimes appearing to be one perfect flower). *Lemnaceae* (duckweed family).

Order 14. **Arales,** page 413.

LIST OF ORDERS AND FAMILIES

NOTE: Small geographically restricted families scarcely known in the United States are assigned, subject to opportunity for study, to the orders accepted generally by other authors. Other families upon which there is less agreement or information are listed as of uncertain position. Doubtless various less known families need to be amalgamated or divided.

NOTE: The asterisk is used in the list below and in the text following to indicate families represented by one or more species occurring in North America north of Mexico (native or introduced plants but not those occurring only in cultivation).

1. ALISMALES
   - *1. Alismaceae (arrowhead, water-plantain, burhead)
   - *2. Butomaceae (water-poppy, flowering-rush)

2. LILIALES
   - A. Suborder Liliineae
     - *1. Liliaceae (lily, onion, Solomon's-seal)
     - *2. Stemonaceae (*Croomia*)

*3. Pontederiaceae (water-hyacinth, pickerel-weed)
4. Cyanastraceae (*Cyanastrum*)
5. Philydraceae (*Philydrum*)
*6. Dioscoreaceae (yam)
*7. Haemodoraceae (bloodwort)
*8. Iridaceae (*Iris* or flag)
9. Velloziaceae (*Vellozia, Barbacenia*)
10. Taccaceae (*Tacca*)

B. Suborder Commelinineae
*11. Commelinaceae (dayflower, spiderwort)
*12. Xyridaceae (yellow-eyed grass)
*13. Mayacaceae (*Mayaca* or bogmoss)
*14. Eriocaulaceae (pipewort)
15. Rapateaceae (*Rapatea*)
16. Bromeliaceae (*Bromelia,* pineapple, Spanish-moss)

C. Suborder Juncinieae
*17. Juncaceae (rush)
*18. Thurniaceae (*Thurnia*)

3. MUSALES
*1. Musaceae (banana)
*2. Zingiberaceae (ginger)
*3. Cannaceae (canna)
*4. Marantaceae (arrowroot)

4. BURMANNIALES
*1. Burmanniaceae (*Burmannia*)

5. ORCHIDALES
*1. Orchidaceae (orchid)

6. RESTIONALES
1. Restionaceae (*Restio*)
2. Centrolepidaceae (*Centrolepis*)
3. Flagellariaceae (*Flagellaria*)

7. GRAMINALES
*1. Cyperaceae (sedge)
*2. Gramineae (grass)

8. HYDROCHARITALES
*1. Hydrocharitaceae (elodea or *Anacharis, Vallisneria,* frog's-bit)

9. TRIURIDALES
1. Triuridaceae (*Triuris*)

10. NAIADALES
1. Aponogetonaceae (*Aponogeton*)
*2. Zosteraceae (pondweed)
*3. Naiadaceae (naiad)

11. JUNCAGINALES
*1. Juncaginaceae (arrow grass)
*2. Lilaeaceae (*Lilaea*)

12. TYPHALES
*1. Sparganiaceae (bur-reed)
*2. Typhaceae (cat-tail)

13. PANDANALES
1. Pandanaceae (screw-pine)

14. ARALES
1. Cyclanthaceae (Panama-hat-palm)
*2. Araceae (*Calla,* jack-in-the-pulpit, skunk-cabbage, taro)
*3. Lemnaceae (duckweed)

15. PALMALES
*1. Palmae (palm)

Order 1. *ALISMALES.*

WATER-PLANTAIN ORDER

DEPARTURE AMONG THE ALISMALES FROM THE TYPICAL FLORAL CHARACTERS OF THE ORDER

Numerous	Stamens rarely as few as 6; carpels rarely as few as 6–9.
Spiral	Stamens and carpels rarely in cycles of 3.

Marsh or aquatic plants; stems scapelike. Leaf bases sheathing the stem. Flowers bisexual or unisexual, radially symmetrical; sepals and petals 3 each, the calyx and corolla strongly differentiated, the petals usually white, sometimes pink with lavender or rose; stamens and carpels usually numerous, separate; ovules campylotropous or anatropous. Fruit an achene or a follicle; seeds erect; endosperm none.

The fundamental pattern of flower structure in the Alismales is identical with that in the Ranales, and the closest approach of monocotyledons and dicotyledons to each other is in these two orders. This is evident not only from the identical flower structure but in addition from frequent occurrence in the Ranales of several typical characters of the monocotyledons. These are as follows: (1) the fibrous root system, which is common among the herbaceous members of the Ranales, as for example in the larkspurs (*Delphinium*) or the buttercups (*Ranunculus*); (2) the arrangement of sepals, petals, and stamens in threes in many Ranales, as for example, in the magnolia family, the barberry family, and the laurel family; (3) the conspicuous differentiation (for example in *Ranunculus*) of the sieve tubes and companion cells of the phloem, these resembling the same tissues in the vascular bundles of corn and other monocotyledons; (4) perhaps rarely the single cotyledon. An essentially solitary cotyledon occurs in *Ranunculus Ficaria,* one being well-developed while the other is abortive. However, this does not necessarily show definite relationship of *Ranunculus Ficaria* to the monocotyledons because the abortion of one cotyledon may have arisen through separate development parallel to that in the monocotyledons.

Differentiation of the orders Alismales and Ranales is dependent upon four of the six typical opposed external characters of the monocotyledons and the dicotyledons, as follows: (1) leaves parallel-veined; (2) vascular bundles apparently irregularly scattered as they appear in a cross section of the stem; (3) cambium none; (4) cotyledon 1.

Key to the Families

(North America north of Mexico)

1. Fruit an achene.

   Family 1. **Alismaceae** (water-plantain family).

1. Fruit a follicle.

   Family 2. **Butomaceae** (flowering-rush family).

*Family 1. Alismaceae.

WATER-PLANTAIN FAMILY

Aquatic or marshland herbs. Leaves basal or mostly so, simple, linear to usually ovate or sagittate, only the main veins parallel, the smaller veins pinnate or running crosswise. Flowers bisexual or in *Sagittaria* unisexual (and the plants then monoecious), in racemes or panicles, often in whorls at the nodes, radially symmetrical; sepals green, separate; petals 3, white, early deciduous; stamens numerous and rarely in 2 or 3 series; carpels ordinarily numerous and arranged spirally, but sometimes as few as 6 and in a single series, the ovary superior, 1-chambered, the ovule usually 1 or occasionally several per carpel, basal. Fruit an achene or rarely a follicle, opening at the base, the style persistent in fruit.

The family is distributed over most of the Earth, but the chief concentration is in the temperate regions of the Northern Hemisphere. Four genera are common in the United States and Canada. The water-plantain (*Alisma*) is represented by species occurring commonly northward and eastward but less commonly in the drier portions of the West because of the restricted water supply (as with the species of the other aquatic and palustrine genera). *Echinodorus* and *Lophotocarpus* occur mostly through the southern part of the Continent. The largest of the four common genera is *Sagittaria*, the arrowhead, which includes sixteen or more species in eastern North America and several in western North America. One species of *Damasonium* occurs in and adjacent to California.

*Family 2. Butomaceae.

FLOWERING-RUSH FAMILY

Aquatic or marsh plants; juice often milky. Leaves basal or some cauline, the blades orbicular to elliptic or swordlike, the main veins parallel but the secondary veins at right angles to them. Flowers solitary or several in an umbel; stamens 6–9; carpels 6–numerous and sometimes slightly coalescent basally, the ovary superior; ovules numerous in each carpel, anatropous. Fruit a follicle, the group of follices sometimes slightly coalescent basally.

The family is distributed widely through the Eastern Hemisphere and in tropical America. The flowering-rush (*Botomus*) is grown in water gardens, and it is naturalized from the St. Lawrence River in Quebec to the Great Lakes and Vermont. The species is native in Europe.

Order 2. *LILIALES.* LILY ORDER

DEPARTURE FROM THE FLORAL CHARACTERS OF THE ALISMALES

Free	In some genera the bases of the sepals and petals adnate edge to edge as if in a single series, thus forming a perianth tube with which the stamens also are adnate; carpels coalescent.
Equal	Flowers very rarely bilaterally symmetrical.
Numerous	Stamens 6, in two series, or only 3; carpels 3. (Very rarely the flower 2- or 4-merous.)
Spiral	Stamens and carpels cyclic.
Hypogynous	A few genera of the *Liliaceae* perigynous (e.g., *Brodiaea*) or epigynous (e.g., *Sansevieria, Liriope,* and *Aletris*); several families epigynous.

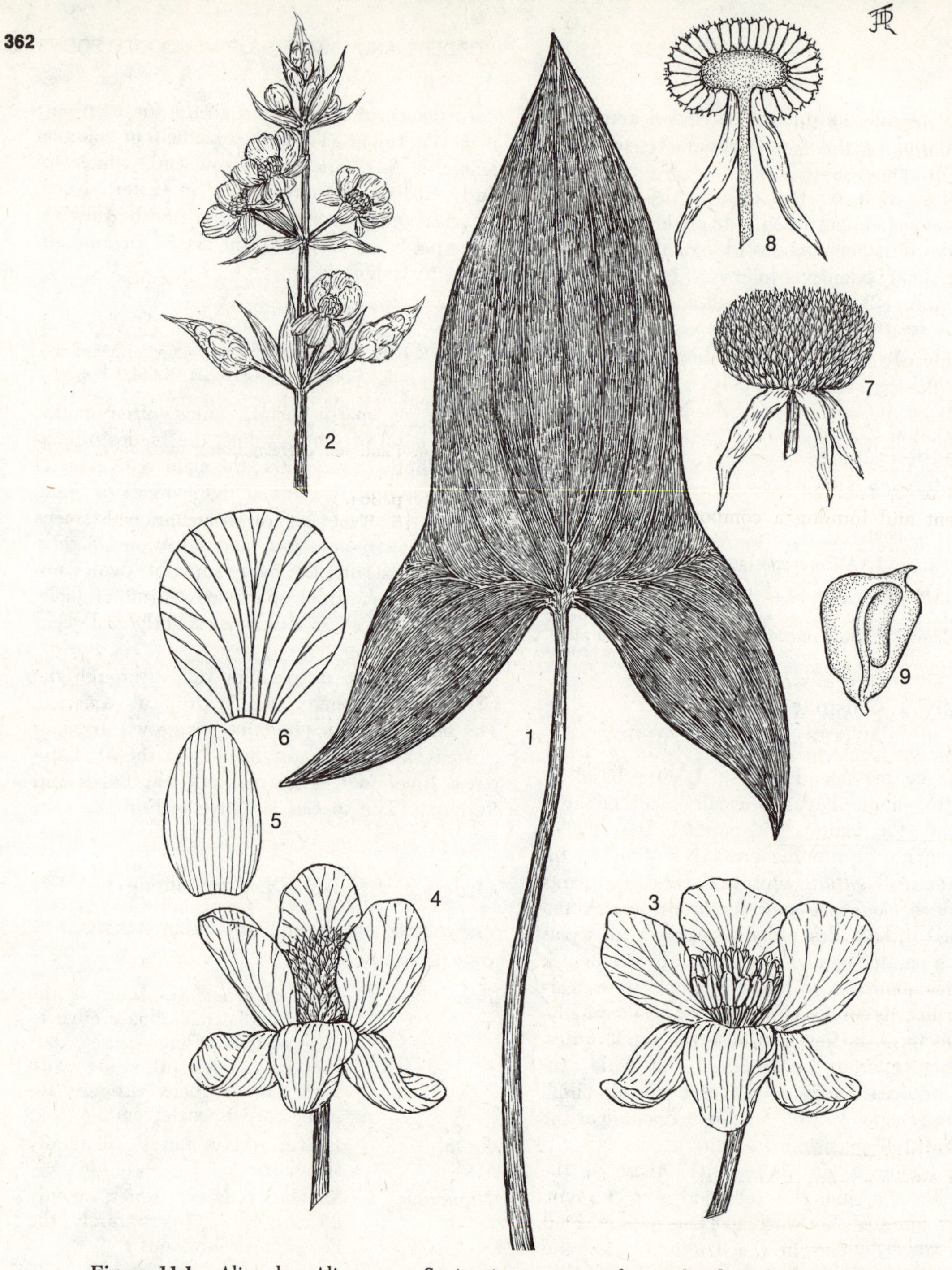

Figure 11-1. Alismales; Alismaceae; *Sagittaria,* a species of arrowhead or duck potato or tule potato: **1** basal leaf, **2** inflorescence, upper portion staminate, lower pistillate, **3** staminate flower, composed of three sepals, three petals and numerous spirally arranged stamens, **4** pistillate flower, composed of three sepals, three petals and numerous spirally arranged pistils (each from one carpel), **5** sepal, **6** petal, **7** head of young fruits, **8** head of fruits, longitudinal section, the receptacle enlarged, the sepals persistent, **9** fruit, showing the solitary seed attached at the base of the single chamber.

The characters of this order are more or less representative of the monocotyledon series as a whole. Most orders appear to be offshoots from the developmental line ancestral also to the *Liliaceae*, and they are characterized readily by contrast with the Liliales, that is, by picking out the characters which differ from those of this central order. Within the Liliales there is remarkably little variation in flower structure, and nearly any plant of the group is characterized as follows: flower 3-merous; sepals 3; petals 3, the segments of the two series of the perianth being commonly but not always nearly indistinguishable except by their position on the receptacle; stamens 6 or 3, in 1 or 2 series of 3; carpels 3, coalescent and forming a compound pistil, usually with a single style, the ovary 3-chambered; ovules anatropous or (in the *Commelinaceae, Xyridaceae, Mayacaceae,* and *Eriocaulaceae*) orthotropous. Fruit nearly always a capsule or a berry.

The leaves are simple, as in nearly all monocotyledons.

Distinction by some authors (e.g., Bessey) of the "Iridales" from the typical Liliales is based upon relatively few characters, the chief ones being as follows: (1) flowers epigynous and the ovary therefore inferior, (2) stamens 3 instead of 6.

In this edition the *Amaryllydaceae* are placed in the *Liliaceae,* but the group appears in the key to facilitate use of manuals in which the family is separated.

Engler and Prantl's segregation of the *Farinosae* was based upon strong emphasis of the presence of mealy endosperm in the seeds of certain families.

Key to the Families

(North America north of Mexico)

1. Plant corresponding with *every character* in *one* of the combinations below.
   (1) Ovary inferior; flowers in a leafy-bracted spike terminating a scape, at fruiting time the lower part of the spike, including both the ovaries and the bracts, forming a juicy conelike mass of edible material. *Ananas* (pineapple).
   (2) Plant epiphytic (growing upon but not parasitic on trees or shrubs or sometimes on rocks, fences, or other objects), commonly with scurfy leaves (with dandrufflike particles or scales on the surfaces); corolla radially or barely bilaterally symmetrical; stamens 6.

   Family 13. **Bromeliaceae** (*Bromelia* family).

1. Plant *not* corresponding with *every character* in *either* combination in lead 1 above.
  2. Ovary superior. (Cf. lower lead 2, p. 364.)
    3. Sepals and petals *not* scalelike, at least the petals petaloid and conspicuous, usually either white or conspicuously colored. (Cf. lower lead 3, p. 364.)
      4. Plant *not* corresponding with *every character* in the combination in lead 4 below, p. 364.
        5. Plant *not* corresponding with *every character* in the combination in lead 5 below, p. 364.
          6. Plant *not* corresponding with *every character* in the combination in lead 6 below, p. 364.
            7. Calyx and corolla similar, *not* or only slightly differing, except in the following 3 genera: (1) *Trillium* (with the scape bearing a whorl of 3 large leaves below a single flower); (2) *Scoliopus* (Pacific Coast from western Oregon to northwestern California; leaves 2, basal, spotted or mottled); (3) *Calochortus* (western North America and Mexico; stem a scape, this and the basal leaves arising from a corm, i.e., a "bulb" formed from a thickening of the stem-base); endosperm of the seeds of the whole family fleshy, horny, or cartilaginous.
              8. Flowers 3-merous or rarely 2-merous (in *Maianthemum,* which has a 2-chambered ovary).

                 Family 1. **Liliaceae** (lily family).

              8. Flowers 2-merous (sepals 2, petals 2, stamens 4, carpels 2); ovary **1-chambered.**

                 Family 2. **Stemonaceae** (*Stemona* family).

            7. Calyx and corolla strongly differentiated; endosperm of the seeds

mealy (farinose); eastern Arizona to the area east of the Rocky Mountains.

Family 11. **Commelinaceae** (spiderwort family).

6. Plants aquatic or stranded on muddy shores in the East, Middle West, and South (one species growing in wet pinelands in peninsular Florida); perianth petal-like, tubular below and spreading above (salverform) or bilaterally symmetrical; flowers with a spathe; stamens 3–6, mostly unequal or dissimilar, attached to the throat of the perianth tube.

Family 3. **Pontederiaceae** (pickerelweed family).

5. Sepals 3, 1 larger than the others and deciduous, the 2 smaller ones firm, keeled, sometimes with special appendages; rushlike herbs with linear leaves sheathing the base of the scape; flowers in a terminal head, this conelike, having numerous scaly bracts; fruit a 1-chambered capsule; flowers perfect, the petals yellow or rarely white.

Family 12. **Xyridaceae** (yellow-eyed grass family).

4. Plants aquatic, mosslike, the stems covered with narrowly linear, sessile, small, translucent leaves; flowers axillary, pedicelled, bisexual, with green sepals and white petals; stamens 3; fruit a 1-chambered capsule; Ohio to Virginia, Texas, and Florida.

Family 13. **Mayacaceae** (bogmoss family).

3. Sepals and petals scalelike, none of them petal-like, the perianth inconspicuous, not white or brightly colored, usually greenish or brown.

4'. Flowers bisexual.

Family 17. **Juncaceae** (rush family).

4'. Flowers unisexual; Newfoundland to Florida and west to Texas.

Family 14. **Eriocaulaceae** (pipewort family).

2. Ovary wholly or partly inferior.

3'. Stamens 6; leaves *not* folded and *not* with the margins grown together, *not* equitant (see lower 3').

4''. Flowers bisexual; plant *not* a climbing vine; leaves parallel-veined. "Amaryllidaceae" and "Agavaceae."

Family 1. **Liliaceae** (lily family).

4''. Flowers unisexual, the plants dioecious; plant a climbing vine; leaves netted-veined (an exception in the Monocotyledoneae).

Family 6. **Dioscoreaceae** (yam family).

3'. Stamens 3; leaves folded and the margins grown together, equitant (astride the stem like a rider astride a horse).

4'''. Perianth woolly; ovary incompletely inferior, adnate with the perianth tube through only part of its length.

Family 7. **Haemodoraceae** (bloodwort family).

4'''. Perianth not woolly; ovary completely inferior, adnate with the perianth tube through its entire length.

Family 8. **Iridaceae** (*Iris* family).

Suborder A. **Liliineae**

*Family 1. **Liliaceae.** LILY FAMILY

(Segregates: *Amaryllidaceae, Nartheciaceae, Alliaceae, Convallariaceae, Aloaceae, Dracaenaceae, Trilliaceae, Roxburghiaceae, Smilacaceae, Melanthaceae, Agavaceae, Xanthorrhoeaceae*)

Herbs or occasionally woody plants; rhizomes, corms, bulbs, or other fleshy structures often present; stems of various types. Leaves alternate or whorled or very rarely opposite, sometimes all basal. Flowers bisexual or rarely unisexual (and then the plants usually dioecious), radially symmetrical or rarely tending to be bilaterally symmetrical, hypogynous or rarely perigynous or epigynous, rarely other than 3-merous; perianth nearly always large and conspicuous, both sepals and petals corollalike (differentiated in North America in only the three genera indicated in the key to the families), usually white or conspicuously colored; stamens rarely of other numbers than 6 (3–12); styles rarely separate. Fruit a capsule or sometimes a berry.

The lily family includes over 200 genera and perhaps 3,000 species. It is world-wide, and about 50 genera occur in North America. Several subfamilies have been distinguished and some authors have separated these as families. Some of the principal representatives of the family are the following: bog-asphodel (*Narthecium*), false-asphodel (*Tofieldia*), Zygadene (*Zygadenus*), false-hellebore (*Veratrum*), onion, garlic, and leek (*Allium*), *Brodiaea,* day lily (*Hemerocallis*), lily (*Lilium*), dog-tooth-violet, avalanche lily, and chamise lily (*Erythronium*), camas (*Camassia*), Spanish-bayonet and Joshua-tree (*Yucca*), *Asparagus, Clintonia,* Solomon's-seal (*Polygonatum*), false Solomon's-seal (*Smilacina*), fairy-lantern (*Disporum*), twisted stalk (*Streptopus*), lily-of-the-valley (*Convallaria*), wake-robin (*Trillium*), *Smilax,* mariposa lily and cat's-ear (*Calochortus*), bear-grass (*Xerophyllum*), desert lily (*Hesperocallis*), soaproot (*Chlorogalum*), fritillary (*Fritillaria*), *Nolina,* sotol (*Dasylirion*). Among the largest genera on the Pacific Coast, and especially in California, are *Brodiaea* and *Calochortus.*

Possibly the family needs redefinition, but this will require much additional study. The proposed union of the yuccas and their relatives and the century-plants (*Agave, Amaryllidaceae*) forming a family, *Agavaceae,* may or may not be justified.

*Family 2. **Stemonaceae.**

STEMONA FAMILY

Flowering stems arising from a rhizome or from storage roots. Leaves simple, alternate, opposite, or whorled. Flowers bisexual, radially symmetrical, hypogynous or epigynous (the ovary being sometimes partly inferior), solitary in the leaf axils, 2-merous, the stamens in 2 series; carpels 2, the ovary in the North American species sometimes superior; ovules numerous or sometimes as few as 2; placentae basal or apical. Fruit a capsule.

The family consists of 3 genera, two occurring in southeastern Asia and the South Pacific, and the other (*Croomia*) composed of 2 species, one occurring in Japan and the other in the Southern States.

The peculiar distribution of the family and particularly the genus *Croomia* is similar to that of some of the ancient families of the *Ranales,* particularly the *Schisandraceae. Croomia pauciflora* occurs from Alabama to Georgia and northern Florida.

*Family 3. **Pontederiaceae.**

PICKERELWEED FAMILY

Aquatic, usually perennial herbs; plants either floating or growing on muddy shores; rhizomes present, the flowering stems short. Leaves ovate, elliptic, or lanceolate or reduced to flattened petioles, usually opposite or 3–4 in a whorl; flowers bisexual, usually radially but sometimes bilaterally symmetrical, hypogynous; sepals 3, usually not distinguished clearly from the petals, sometimes adnate with them and forming a perianth tube; stamens usually 6 but sometimes 3 or 1, usually unequal or otherwise dissimilar; carpels 3, the style 1, the ovary either 3-chambered and with axile placentae or 1-chambered and with parietal placentae; ovule(s) 1-numerous in each carpel, anatropous. Fruit a capsule or sometimes a utricle; endosperm abundant, mealy.

The family includes half a dozen genera occurring throughout the Tropics, and a few species extend northward to warm temperate regions. The pickerelweed (*Pontederia*) includes two species in the eastern part of the Continent. Three or four species of mud-plantain (*Heteranthera*) occur chiefly in the Southern States. The water-hyacinth (*Eichornia*) has been introduced from South America into some of the streams and lakes of central and southern California, and it is abundant in the Southern States. Near the Gulf of Mexico and the Mississippi Delta it is a pest because it clogs waterways.

Family 4. **Cyanastraceae.**

CYANASTRUM FAMILY

Flowers bisexual, radially symmetrical, epigynous; perianth tube very short; stamens 6, coalescent basally; carpels 3, the style 1, the ovary 3-chambered, partly inferior; ovules 2 per chamber. Fruit deeply 3-parted, with only 1 seed.

The family consists of a single genus, *Cyanastrum,* composed of four species occurring in tropical Africa.

Family 5. **Philydraceae.**

PHILYDRUM FAMILY

Herbs; rhizomes present. Leaves all basal or near the base of the stem, linear. Flowers bisexual, bilaterally symmetrical, hypogynous, soli-

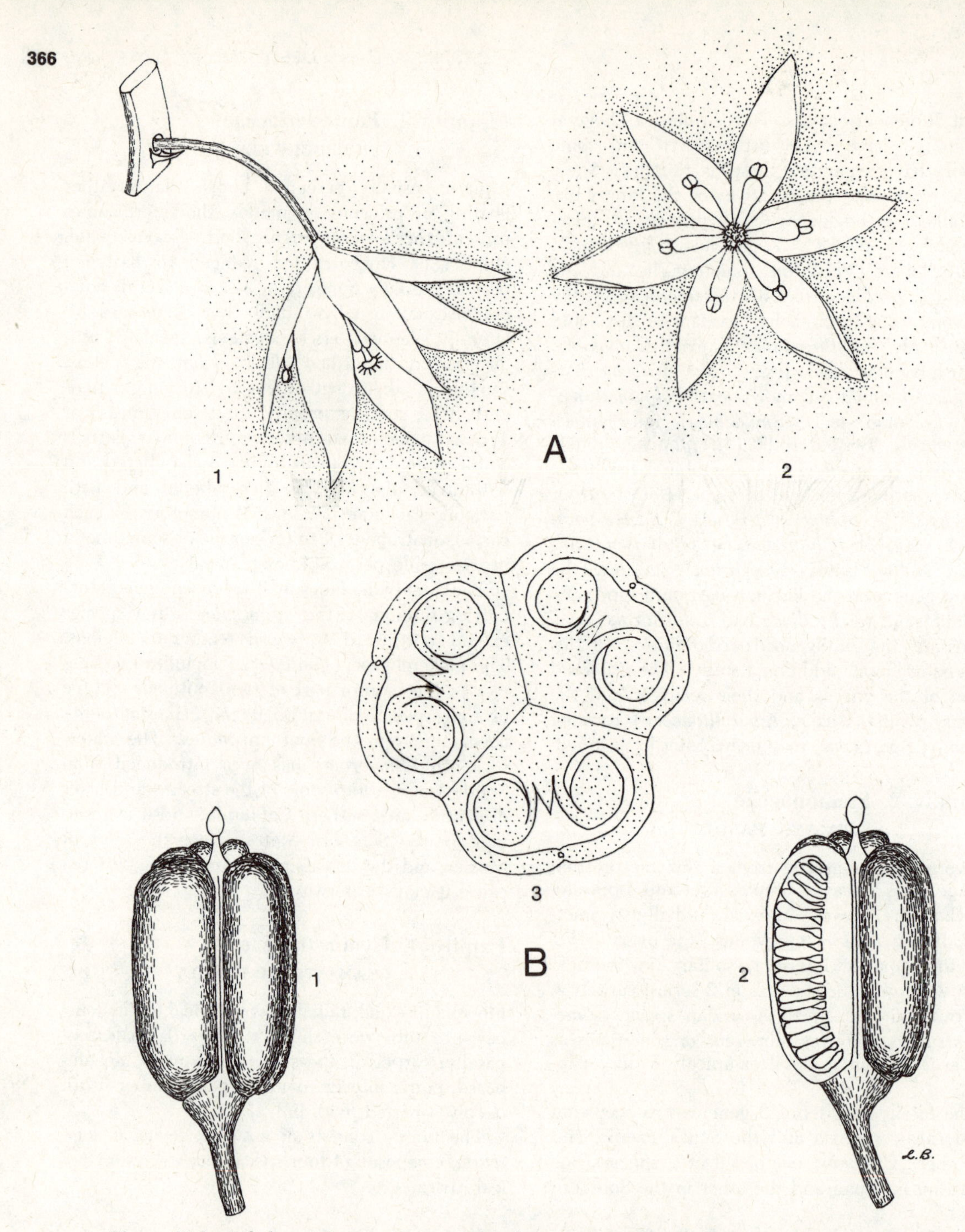

Figure 11-2. Liliales; Lilaceae; *Yucca Whipplei,* yucca or Spanish bayonet: **A** flower, **1** side view, showing the sepals, petals, a stamen, and the pistil, **2** top view, showing the three sepals, three petals, six stamens (in two cycles of three), and compound pistil (of three carpels) characteristic of the Liliaceae; **B, 1** young fruit, **2** external view, half the external portion of one carpel removed, showing the ovules, **3** ovary, cross section, showing the axile placentae and the double row of young seeds in each chamber. (Cf. also **Figures 2-12** *Lilium Humboldtii,* bulb, **2-13 B** onion, bulb, **C** *Brodiaea pulchella,* corm, **2-14 D** *Lilium tigrinum,* bulbils, **6-7 C** *Agapanthus africanus,* fruit, **6-9 A** *Agapanthus africanus,* flower.)

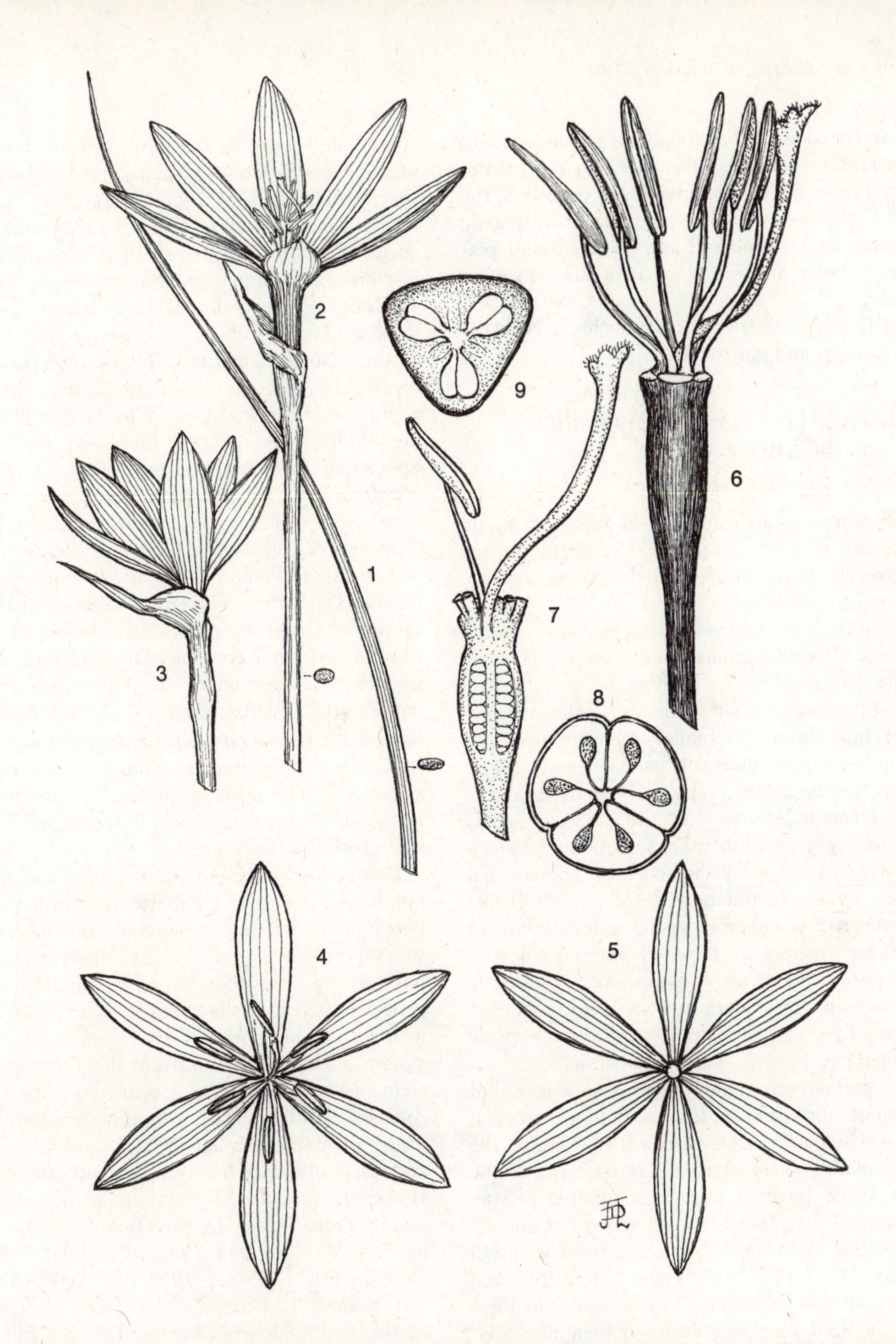

Figure 11-3. Liliales; Liliaceae; *Zephyranthes candida,* zephyr-lily: **1** basal leaf, **2–3** flower, two views, showing the subtending bracts and the three sepals, three petals, six stamens, and the style and stigma, **4–5** flower, from above and below, **6** flower with the bracts and the perianth removed, perianth tube (adnate with inferior ovary) below, **7** longitudinal section, **8–9** ovary in flower and in fruit, cross sections, showing the ovules and the seeds. (Cf. also **Figures 2-14 C** century plant, bulbils, **6-9 B** *Amaryllis vittata,* flower.) These plants are of the *Amaryllis* group of genera in the Liliaceae or lily family.

tary, in the axils of sheathing bracts, the perianth 2-merous; sepals petal-like, not readily distinguished from the corolla; stamen 1; carpels 3, the style 1, the ovary 3-chambered and with axile placentae or 1-chambered and with parietal placentae; ovules numerous, anatropous. Fruit a capsule.

The family is restricted to Indochina, Malaya, New Guinea, and Australia.

"AMARYLIDACEAE" or *Amaryllis* group in the Liliaceae

(Synonym: *Leucojaceae*)

This group of genera appears in many books as a separate family. This is not warranted, but the group is keyed above and described here for student convenience.

Perennial herbs; stems forming rhizomes, bulbs, or corms; flowering stems usually scapes. Leaves usually basal or rarely cauline as well, linear, often stiff, sometimes succulent (as in the century-plant) but fibrous internally. Flowers bisexual, radially or rarely bilaterally symmetrical, epigynous or rarely not so; sepals 3; petals 3, sometimes adnate edgewise with the adjacent sepals and forming a perianth tube, sometimes (*Narcissus*) with an accessory cup or tube known as a corona; stamens 6, adnate with the perianth tube, the anthers 2-chambered, splitting lengthwise or sometimes opening by terminal pores; carpels 3, the style 1, the ovary 3-chambered; placentae axile or sometimes more or less basal; ovules 2 or rarely 1 per carpel, anatropous. Fruit a capsule or sometimes a berry; endosperm fleshy.

The group occurs throughout the world, but it is most abundant in tropical and subtropical regions. In North America north of Mexico the largest plants of the *Amaryllis* group are of the genus *Agave* (century-plant). A number of species occurs in (or largely at elevations just above) the Southwestern Deserts, ranging from southern California to Texas. Other species are introduced in the Southern States, and one is native in Florida. The sisal plant, source of a hemplike fiber obtained from the leaves, has become established in Florida, and *Agave americana*, a common large cultivated century-plant, has escaped in Texas along the Rio Grande and in southern Florida. A species of *Zephyranthes*, the zephyr lily or atamosco, occurs in southern Arizona and eastward to southeastern Virginia and to Florida. A species of golden-eyed grass (*Hypoxis*) is native in the mountains of southern Arizona, and others occur in eastern North America. Certain species of *Narcissus* common in cultivation have escaped, and one of these, the daffodil, has became naturalized locally in fields and open woods in northeastern North America. The snowflake (*Leucojum*) has become abundant in some localities along the Atlantic Coast. The spider lily (*Hymenocallis*) has escaped here and there, also. Species of *Crinum* occur from Texas to Florida. *Manfreda, Aletris,* and *Cooperia* occur in eastern North America, largely in the Southern States. A species of *Habranthus* occurs in Texas.

The *Amaryllidaceae* are of economic importance. The leaves of certain century-plants are sources of coarse fibers (sisal and henequen) used in rope and other cordage. Other genera are also sources of coarse fibers. In Latin America and particularly in Mexico species of *Agave* are cultivated for the sugary sap of the budding flower stalk, which resembles a gigantic young stem of asparagus. The juice is allowed to ferment, then it is distilled, forming mescal and pulque. Tequila is of greater potency.

Distinction of the *Amaryllis* family and the lily family is a subject of debate. Separation of the two families has been commonly on the basis of the superior ovary of the lily family versus the inferior ovary of the *Amaryllis* family, but the generally accepted lines of division are scarcely tenable. Some authors have combined several genera, including *Yucca,* from the *Liliaceae* with elements such as *Agave* (century-plant) of the *Amaryllis* family into a new family, the *Agavaceae*. These plants are similar in habit and appearance and their chromosome complements are similar. However, probably the group is of less than family rank. (See Lyman Benson, *Plant Taxonomy, Methods and Principles,* John Wiley & Sons [ex Ronald Press], 1962, and Lyman Benson and Robert A. Darrow, *The Trees and Shrubs of the Southwestern Deserts,* Ed. 3, University of Arizona Press, in press.)

In some members of the *Amaryllis* group the structure of the flower appears to be complicated. In the genus *Narcissus* (which includes the common garden narcissus, the China-lily, the jonquils, and the daffodils) a special structure known as the corona, or crown, is developed from the inner edges of the perianth parts; this forms the cup of the China-lily or the greatly elongated trumpet or tube of the daffodil flower; however, the corolla is radially symmetrical, and the corona is the only unusual feature.

*Family 6. **Dioscoreaceae.** YAM FAMILY

(Synonym: *Tamacaceae*)

Herbaceous or woody vines; stem below ground a rhizome or a caudex; roots often tuberous. Leaves alternate or sometimes opposite, entire, palmatifid. Flowers of the North American species unisexual (and the plants commonly dioecious), radially symmetrical, epigynous, small and inconspicuous, in spikes, racemes, or panicles; petals 3, small, usually adnate with the sepals and the two series forming a perianth tube; stamens 6, the inner series sometimes sterile, the anthers 2-chambered, splitting lengthwise; carpels 3, the style(s) 1 or sometimes 3, the ovary 3-chambered; placentae axile; ovules 2 or more per carpel. Fruit a capsule or a berry; seeds with endosperm.

The family is abundant in the Tropics and Subtropics and rare in the North Temperate zone. Several species of yams occur in eastern North America and particularly on the Coastal Plain in the Southern States. The Chinese yam of cultivation (*Dioscorea Batatas*) is naturalized in waste places.

*Family 7. **Haemodoraceae.** BLOODWORT FAMILY

Perennial herbs; juice red or orange. Leaves mostly basal, those on the stem alternate and reduced, linear. Flowers bisexual, radially or rarely bilaterally symmetrical, epigynous, in racemes, cymes, or panicles; sepals and petals not clearly distinguished, the perianth segments 6, sometimes coalescent basally; stamens usually 3 but sometimes 6, the anthers 2-chambered, splitting lengthwise; carpels 3, the style 1, the ovary 3-chambered; placentae axile; ovules 1–many. Fruit a capsule.

The family centers in the Southern Hemisphere, but one genus of a single species is restricted to eastern North America. The red-root (*Lachnanthes*) occurs along the ocean and in swamps in Queens County, Nova Scotia, and from southeastern Massachusetts to Florida and thence westward along the Gulf Coast to Louisiana. It is also a weed in artificial cranberry bogs.

*Family 8. **Iridaceae.** IRIS FAMILY

(Synonym: *Ixiaceae*)

Perennial herbs (or extraterritorial genera more or less shrubby); underground stem a bulb, a corm, or a rhizome, producing leaflets or leafy flowering stems. Leaves basal or alternate, commonly numerous, usually equitant (folded around the stem like a rider astride a horse), usually linear or curving like the blade of a sword (ensiform). Flowers bisexual, usually radially but sometimes bilaterally symmetrical (in extraterritorial genera), usually solitary or in racemes or panicles, each flower usually subtended by a pair of bracts; sepals 3, petaloid; petals 3, these and the sepals often similar or at least both colored, the perianth forming a long or short tube, the adjacent sepals and petals being adnate edge to edge; stamens 3, the anthers 2-chambered, splitting lengthwise; carpels 3, the style 1, the ovary ordinarily 3-chambered; placentae axile or rarely the ovary with one chamber and the placentae parietal; ovules several per carpel, anatropous. Fruit a capsule; seed with abundant endosperm.

The family includes more than 1,000 species, and it occurs throughout the world, most abundantly in Africa. The principal native genus in North America is *Iris,* which includes several species on the Pacific Coast, a single widespread one in the Great Basin and the Rocky Mountain Region, and about a dozen in eastern North America. Some of the Eastern species hybridize freely, and in the Southern States plants with special combinations of hereditary characters resulting from hybridization have been shown to have become established in different agricultural fields or pasture areas in accordance with the slightly differing ecological conditions maintained on neighboring farms through long periods. Many of these local colonies have been named as "species." Several species of *Sisyrinchium* (blue-eyed grass) occur in the Western States, and perhaps ten occur in eastern North America. The genus of the celestial lily (*Nemastylis*) is represented by

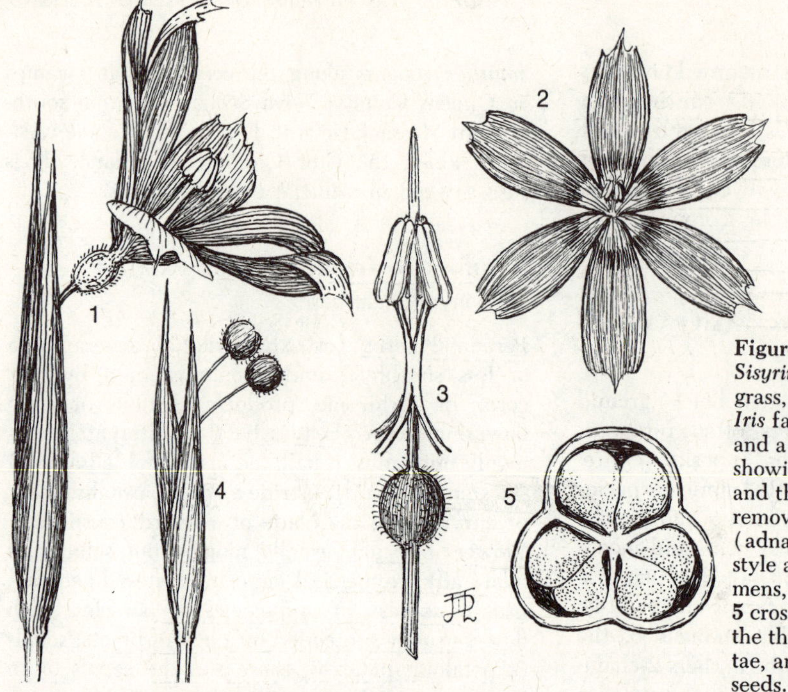

Figure 11-4. Liliales; Iridaceae; *Sisyrinchium bellum,* a blue-eyed grass, a less-specialized member of the *Iris* family (cf. **Figure 11-5**): **1** bracts and a flower, **2** flower, top view, showing the similar sepals and petals and the stamens, **3** flower, perianth removed, showing the perianth tube (adnate with the inferior ovary), the style and stigmas, and the three stamens, **4** bracts and two fruits, **5** cross section of the fruit, showing the three chambers, the axile placentae, and the three double rows of seeds.

a single species in Arizona and Texas and by others occurring near the Gulf of Mexico and northward to the southern edge of the Middle West. The blackberry lily (*Belamcanda*) is introduced in eastern North America. Three species of *Alophia* occur in Texas. *Eustylis* is endemic in Texas and Louisiana.

The flowers of *Iris* are characterized by unusual development of the segments of the perianth and by broad colored styles and stigmas. Despite its elaborate appearance, the flower is radially symmetrical and fundamentally simple. The ovary is inferior and just above it there are three relatively large sepals. The upper surface of each sepal may be bearded and often highly colored. The petals stand more or less erect above and between the sepals. The styles are petal-like, and each one overarches the basal and middle (usually bearded portion) of the opposite sepal. The stigmatic surface is under a little flap near the distal end of the style. A stamen stands under the style, and so each stamen is more or less enclosed between a style and the basal portion of a sepal. The unit of sepal and style forms a passageway through which an insect must travel if it is to reach the nectar. It encounters first the stigma, then the stamen (which is shorter than the style). Each of the three similar units in the flower functions as if it were a single flower. The pollination apparatus is unique, and the *Iris* flower may be looked upon as highly specialized.

DEPARTURE OF THE GENUS IRIS FROM THE FLORAL CHARACTERS OF THE ALISMALES

Free	Carpels coalescent.
Numerous	Stamens 3; carpels 3.
Spiral	Stamens and carpels cyclic.
Hypogynous	Flowers epigynous.

In other members of the *Iris* family, flower structure is essentially the same as in *Iris,* but because the individual sepals, petals, and styles are less elaborate, the flower appears simpler.

Figure 11-5. Liliales; Iridaceae; *Iris* (hybrid), an iris or flag, a specialized member of the *Iris* family (cf. **Figure 11-5**): **1** flower, showing the bracts sheathing the inferior ovary, two of the three sepals (at the sides), the three petals, and the tips of two (forked) styles, **2** flower, the near petal removed, showing the three sepals (drooping), the other two petals (above), and the three styles, **3** flower, from above, showing the sepals (one next to the number), the petals (alternating with the sepals), and the styles (one above

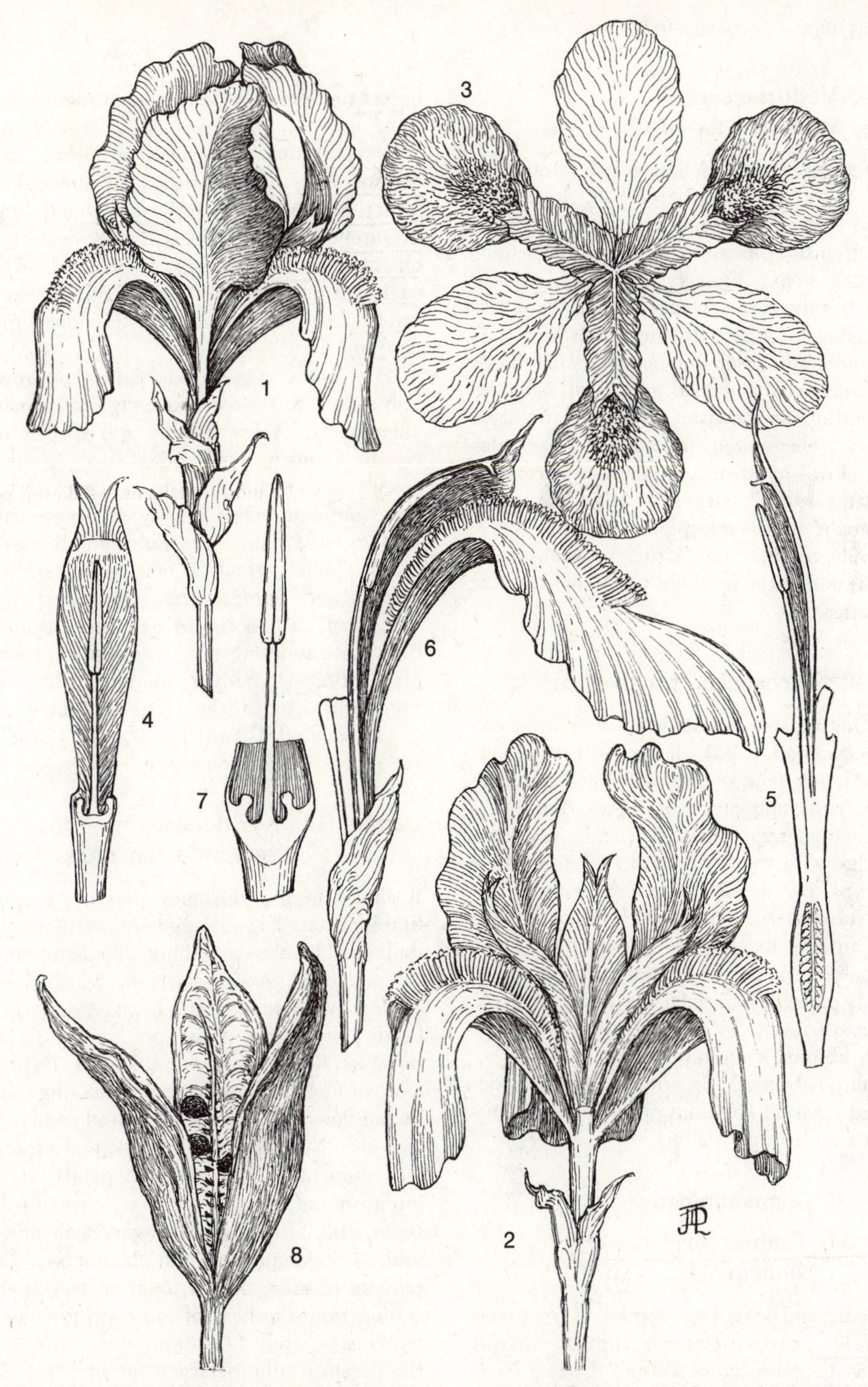

each sepal), **4** style and stamen, the stigma being covered by the flap just below the forked apex of the style, **5** style and stamen and a longitudinal section of the inferior ovary, **6** relationship of style, stamen, and the bearded portion of a sepal, **7** attachment of a stamen and of two glands, **8** dehiscent fruit. Each unit like that in **6** functions in pollination as though it were a separate flower. (Cf. also **Figures 2-13 D, 8-6 A,** and **9-5 B.**)

Family 9. **Velloziaceae.**
VELLOZIA FAMILY

Shrubs or trees; stems dichotomously branched (each fork with two equal branches). Leaves in tufts at the ends of the branches, linear or more or less so, often sharp-pointed, the bases persistent, covering the stem. Flowers bisexual, radially symmetrical, epigynous, solitary; sepals 3; petals 3, adnate edge to edge with the sepals and thus forming a perianth tube; stamens 6 or sometimes numerous and occurring in 6 groups of 2–6, the anthers splitting lengthwise; carpels 3, the style 1, the ovary 3-chambered; placentae axile; ovules numerous. Fruit a capsule, sometimes hard, frequently flattened or concave at the apex; seed with abundant hard endosperm.

The family occurs from Africa to Arabia and Madagascar and in the southern portion of tropical South America.

Family 10. **Taccaceae.** TACCA FAMILY

Perennial herbs; stem forming a tuber or a rhizome. Leaves all basal, large and relatively broad, entire or lobed. Flowers bisexual, radially symmetrical, epigynous, in umbels, the bracts forming an involucre; sepals 3; petals 3, adnate edge to edge with the sepals and thus forming a perianth tube, the sepals and petals more or less petaloid; stamens 6, adnate with the perianth tube, the anthers 2-chambered, splitting lengthwise; carpels 3, the style 1, short, the ovary 1-chambered; placentae parietal; ovules numerous, anatropous. Fruit a berry or rarely dehiscent; seeds with abundant endosperm.

The family is tropical. In addition to the widespread genus *Tacca,* Hutchinson includes a genus *Schizocapsa* occurring in China.

Suborder B. **Commelinineae**

*Family 11. **Commelinaceae.**
SPIDERWORT FAMILY

Herbs; stems and leaves composed of relatively soft tissues. Leaves relatively broad, sharply folded or v-shaped in cross section, the leaf bases sheathing the stem, usually relatively broad. Flowers bisexual, usually radially but sometimes bilaterally symmetrical, hypogynous; sepals and petals clearly differentiated; sepals green and more or less the texture of leaves; petals white or colored (not green), withering early or becoming deliquescent, that is, formed of liquid, sometimes the members of the series of 3 unequal, 1 being sometimes proportionately very small; stamens 6 or rarely 3 or 1; carpels 3, the ovary rarely 2-chambered; placentae axile; ovules 1 or a few in each carpel, orthotropous. Fruit a capsule or rarely fleshy and indehiscent; endosperm abundant, mealy.

The family occurs chiefly in the Tropics and Subtropics and more sparingly in warm temperate regions. A few genera and species occur in North America north of Mexico from Arizona eastward, most abundantly in the eastern part of the Continent. The family is absent from the Pacific Coast and the northern Rocky Mountains. The important genera are *Commelina* (dayflower) and *Tradescantia* (spiderwort). Several species of each occur in eastern North America. Both the wandering-Jew (*Zebrina*) and the oysterplant (*Rhoeo*), natives of Mexico, have escaped from cultivation in the Southern States. *Callissia* (*Tradescantella*) and *Tripogandra* (*Cuthbertia*) are native in the Southeast.

Family 12. **Xyridaceae.**
YELLOW-EYED GRASS FAMILY

Rushlike herbs; rhizomes present, short, sometimes forming large storage organs, then somewhat bulbous. Leaves sheathing, elongate, sometimes threadlike, basal or nearly so. Flowers bisexual, bilaterally symmetrical, hypogynous, in dense heads terminating the scapes; sepals 3 (or sometimes 2), 2 boat-shaped and keeled, 1 (the inner) more or less membranous and enclosing the corolla before flowering; petals equal and similar, usually coalescent, usually yellow, persistent after withering; stamens 3, opposite the petals; staminodia often present; carpels 3, the style usually 1, sometimes with 3 branches, the ovary 1-chambered (or with 3 incomplete basal chambers); placentae parietal or more or less basal or central and then arising from the base of the ovary; ovules several to many or rarely 1, orthotropous. Fruit a capsule, the perianth tube persistent around it.

The family is composed of two genera occurring in the tropical and subtropical areas of Africa, Australia, and the Western Hemisphere. A dozen species of *Xyris,* the yellow-eyed grass, occur in eastern North America. They are small herbs growing in wet places or sometimes in dry habitats.

*Family 13. **Mayacaceae.**

BOGMOSS FAMILY

Aquatic herbs occurring in fresh water. Leaves alternate, very narrow, either filiform or linear, sessile, with a single vein, with two apical teeth. Flowers bisexual, radially symmetrical, hypogynous, solitary in the axils of the leaves, with slender pedicels, often abundant near the apex of a branch; sepals 3, petaloid; petals 3, white with pink or violet; stamens 3, the pollen chambers opening by apical pores or slits; carpels 3, style 1, the ovary 1-chambered; placentae parietal; ovules orthotropous. Fruit a capsule, triangular in cross section; seeds reticulate; endosperm present.

The family consists of a single genus (*Mayaca*) occurring chiefly in tropical and subtropical America, but one species grows in tropical west Africa. Two species of bogmoss, small mosslike aquatic herbs, occur in pools, springs, and streams on the Coastal Plain of the Southern States.

*Family 14. **Eriocaulaceae.**

PIPEWORT FAMILY

Small perennial or sometimes annual herbs; caudices very short, the flowers borne on naked scapes. Leaves basal or essentially so, linear, resembling those of grasses, sometimes membranous. Flowers unisexual (and the plants usually monoecious but sometimes dioecious), radially symmetrical, hypogynous, in heads or umbels of heads, each head with an involucre, each flower subtended by a chafflike bract, this sometimes colored, the heads including both staminate and pistillate flowers except when the plants are dioecious; perianth not clearly differentiated, but the sepals 2–3 and the petals 2–3; each perianth segment scarious or membranous, sometimes very thin and translucent; stamens 2–3, in one series or twice this number and in two series, the anthers 1–2-chambered, turned toward the center of the flower, splitting lengthwise; carpels 2–3, the style 1 but branched above, the ovary 2–3-chambered; placentae axile; ovule 1 per carpel, orthotropous. Fruit a capsule, membranous; embryo minute.

The family includes about one thousand species occurring in wet places in the Tropics, but a few species grow in temperate regions. The chief center of distribution is in South America. About eight species of *Eriocaulon* (pipewort) occur in the eastern portion of the Continent and more especially on the southeastern Coastal Plain. The shoe-button (*Syngonanthus*) grows near the coast from Alabama to North Carolina and Florida. Species of *Lachnocaulon* occur in the Southern States and as far northward as southeastern Virginia.

Family 15. **Rapateaceae.**

RAPATEA FAMILY

Perennial herbs; rhizomes present, the flowers on scapes. Leaves basal, linear-elongate. Flowers bisexual, radially symmetrical, hypogynous, in heads or one-sided spikes; sepals 3, thin and translucent, rigid; petals hyaline, 3, usually coalescent below; stamens 6, adnate with the corolla tube, the anthers 4-chambered, opening by terminal pores or slits; carpels 3, the style 1, the ovary 3-chambered; placentae axile or more or less basal; ovule(s) 1 or several per carpel, anatropous. Fruit a capsule; seeds with abundant mealy endosperm.

The family occurs almost exclusively in tropical South America, but one genus occurs in Liberia.

*Family 16. **Bromeliaceae.**

BROMELIA FAMILY

Herbs; usually epiphytic (growing on trees or other large plants). Leaves usually all basal, alternate, usually stiff, narrow, sometimes more or less grasslike, lepidote (with scalelike hairs), therefore more or less scurfy, the bases forming sheaths. Flowers bisexual or rarely unisexual, radially or sometimes bilaterally symmetrical, hypogynous or sometimes (as in the pineapple) epigynous, in heads or panicles terminating the scapes, the bracts sometimes colored; sepals 3, greenish; petals large and colored or white, sometimes coalescent; stamens 6, often adnate with the petals, sometimes coalescent basally, each anther with 2 pollen chambers, splitting lengthwise; carpels 3, the style 1, the ovary 3-chambered; placentae axile; ovules numerous in each carpel. Fruit a berry or a capsule, commonly enclosed in the perianth, in the pineapple (*Ananas*) the fruits of adjacent flowers and the branches between them fused into a large fruiting structure; seed with abundant mealy endosperm.

This large family occurs in the warmer parts of North and South America. A single species occurs in French West Africa. The most widely occurring genus of the family in North America north of Mexico is *Tillandsia,* which includes the well-known Spanish-moss and a dozen other species of the Southern States. The plants are epiphytic, that is, they grow upon other plants or sometimes upon sticks or rocks. One species occurs from Southern Arizona to Florida, another, the Spanish-moss, occurs as far northward as central Texas and southeastern Virginia. Other species occur in southern (tropical) Florida. Three species of *Hechtia* occur in Texas and others in and near Florida; two species of *Catopsis* and one of *Guzmania* occur in Florida. The pineapple (*Ananas*) occurs in the Southern States but the specimens are said to have originated from slips thrown away in the area where the plant is cultivated. In the South it does not reproduce by seeds. The pistils are coalescent and adnate with the bracts of the inflorescence into an aggregate of juicy material forming the ovoid fruiting structure or pineapple. The axis of the inflorescence continues to grow upward, and it is leafy above the flowers or fruits.

Many plants of the *Bromeliaceae* are cultivated, especially in greenhouses, in the United States. Various plants in the family are known as air plants. These include some of the species of *Tillandsia* and other genera.

Suborder C. **Juncineae**

*Family 17. **Juncaceae.** RUSH FAMILY

Rushes, sometimes grasslike in appearance. Leaves linear or sometimes threadlike, often resembling those of grasses, either flattened or circular in cross section (terete), when flat the leaf blades often folded and through part of their length enfolding a portion of the stem (equitant), when terete a bract often appearing to be a continuation of the stem above the inflorescence. Flowers bisexual or sometimes unisexual (and then the plants dioecious), radially symmetrical, hypogynous, small; sepals and petals not white or highly colored, greenish, brownish, or yellowish, resembling chaff; stamens 6 in 2 series or 3 in 1 series; carpels 3, the style 1, the ovary 1- or 3-chambered; placentae axile or in 1-chambered ovaries parietal; ovule(s) 1 or numerous in each carpel (according to the genus). Fruit a capsule.

The family is composed of 8 genera occurring throughout the world, but half a dozen others are restricted to the Southern Hemisphere. These are small genera of few species. The common genus is *Juncus* (rush), which includes many species in all parts of North America. The other North American genus, *Luzula* (wood rush), is not as well known because it includes relatively few species.

The flowers of the rush family are almost identical with those of the lily family except that they are very small and that the perianth is composed of chafflike scales instead of petal-like structures.

The family is old geologically, being known from Cretaceous time. This fact probably reflects occurrence in wet places where commonly fossils are formed, together with the toughness of the stems, leaves, floral parts, and fruits as well as the fact that the genus *Juncus* is relatively easily recognized from an impression. It is not unlikely that many other living families are just as old and that their members have left no trace.

Family 18. **Thurniaceae.** THURNIA FAMILY

Herbs; stems scapose. Leaves elongated, somewhat sedgelike, sheathing basally. Flowers bisexual, radially symmetrical, hypogynous, in a dense head terminating an elongated leafy branch; perianth parts 6, arranged irregularly, persistent in fruit; stamens 6, much longer than the perianth segments, conspicuous; anthers splitting lengthwise; carpels 3, the style 1, the ovary 3-chambered; placentae axile; ovule(s) 1 or a few per carpel, anatropous. Fruit a capsule, triangular in cross section, with 3 seeds, elongated and pointed; endosperm present.

The family consists of a single genus composed of two species restricted to British Guiana.

Order 3. ***MUSALES.*** BANANA ORDER

DEPARTURE FROM THE FLORAL CHARACTERS OF THE ALISMALES

Free Carpels coalescent.

Equal	Flowers bilaterally symmetrical through development of elaborate staminodia.
Numerous	Stamen 1 (or sometimes up to 5 or rarely 6), the staminodia usually 5 but sometimes fewer; carpels 3.
Spiral	Stamens and carpels cyclic.
Hypogynous	Flowers epigynous.

Large, sometimes treelike, perennial herbs; rhizomes present and the aërial stems developed from them. Leaves usually broad, glabrous, the lower part of each one sheathlike; blades large, asymmetrical, simple, pinnately veined, with the main veins parallel to each other but diverging from the midrib. Flowers bisexual; sepals 3; petals 3, sometimes dissimilar, differing somewhat from the sepals; stamens fundamentally in 2 series but usually all but one developed into elaborate staminodia, these often highly colored and petal-like; ovary 3-chambered, inferior; ovules 1 to many in each chamber of the ovary. Seeds with hard or soft endosperm; perisperm abundant.

Key to the Families

(North America north of Mexico)

1\. Giant treelike herbs; fertile stamens 5 or 6 per flower; south peninsular Florida and the Keys.

Family 1. **Musaceae** (banana family).

1\. Herbs at most a few feet in height; fertile stamen 1 per flower.

2\. Ovary 3-chambered; seeds numerous in each chamber; fruit a capsule.

3\. Sepals partly coalescent, anthers with 2 pollen chambers; south peninsular Florida.

Family 2. **Zingiberaceae** (ginger family).

3\. Sepals separate; anthers with 1 pollen chamber; Texas to South Carolina and Florida.

Family 3. **Cannaceae** (*Canna* family).

2\. Ovary with 1 seed-bearing chamber, producing only 1 seed, the other 2 carpels abortive, the fruit utricular; Missouri to Texas and South Carolina.

Family 4. **Marantaceae** (arrowroot family).

*Family 1. **Musaceae.** BANANA FAMILY

Large herbs, some so large as to appear to be trees, reaching several meters in height. Leaves alternate, large, broad, entire. Flowers bisexual or unisexual (and then the plants monoecious, with pistillate flowers below and staminate above), with spathelike bracts, in spikes, panicles, or heads; sepals 3; petals 3, both the sepals and the petals sometimes unequal within their own series; stamens usually 5, accompanied by 1 staminodium, or in *Ravenala* the stamens 6; carpels 3, the style 1, the ovary 3-chambered; placentae axile, but appearing to be basal; ovule 1 per carpel, anatropous. Fruit a capsule or an elongate berry (e.g., a banana); endosperm present.

The family consists of five genera occurring in the Tropics. Two species of banana (*Musa*) have escaped from cultivation in southern Louisiana or in south peninsular Florida. One of these is the common banana (*Musa sapientum*), the other the dwarf banana (*Musa Cavendishii*). The common banana is a native of the East Indies, the dwarf banana of China. Both are cultivated widely in the Subtropics and Tropics.

Some banana plants are sources of food and fiber. Many species are in cultivation for fruit, only a few types of which ordinarily reach the North American market. One species is grown in the Orient and the western islands of the Pacific Ocean, the fiber being woven into Abaca cloth. The same fiber is used for cordage and Manila hemp.

Strelitzia is common in cultivation in warmer areas of the United States, and the flowers are a commercial crop in southern California. The traveller's-palm (*Ravenala*) is grown in the warmer parts of southern California and in Florida. The petioles of this tree contain a clear watery fluid which serves as a drink.

*Family 2. **Zingiberaceae.**

GINGER FAMILY

(Synonym: *Alpiniaceae*)

Perennial herbs; stems usually forming rhizomes or tubers, the flowering stems either scapose or leafy. Leaves alternate or basal or both, the basal portion of the leaf forming a sheath, the blade linear to elliptic. Flowers bisexual, in spikes or racemes or solitary, usually with conspicuous bracts; sepals 3; petals 3, unequal; stamen 1, staminodia 2, one much larger than the other; carpels 3, the style 1, the ovary 1- or 3-chambered; placentae axile or, when there is only one chamber

Figure 11-6. Liliales; Juncaceae; *Juncus*, rush: **A** flowering plant of *Juncus phaeocephalus* var. *paniculatus,* **1** bract and inflorescence, **2** head of flowers, **3** flower, enlarged, showing the three sepals, three petals, six (in other species three) stamens, and pistil (formed from three carpels), **4** young fruit; **B** fruiting plant of *Juncus ensifolius* var. *montanus,* **1** bract and inflorescence,

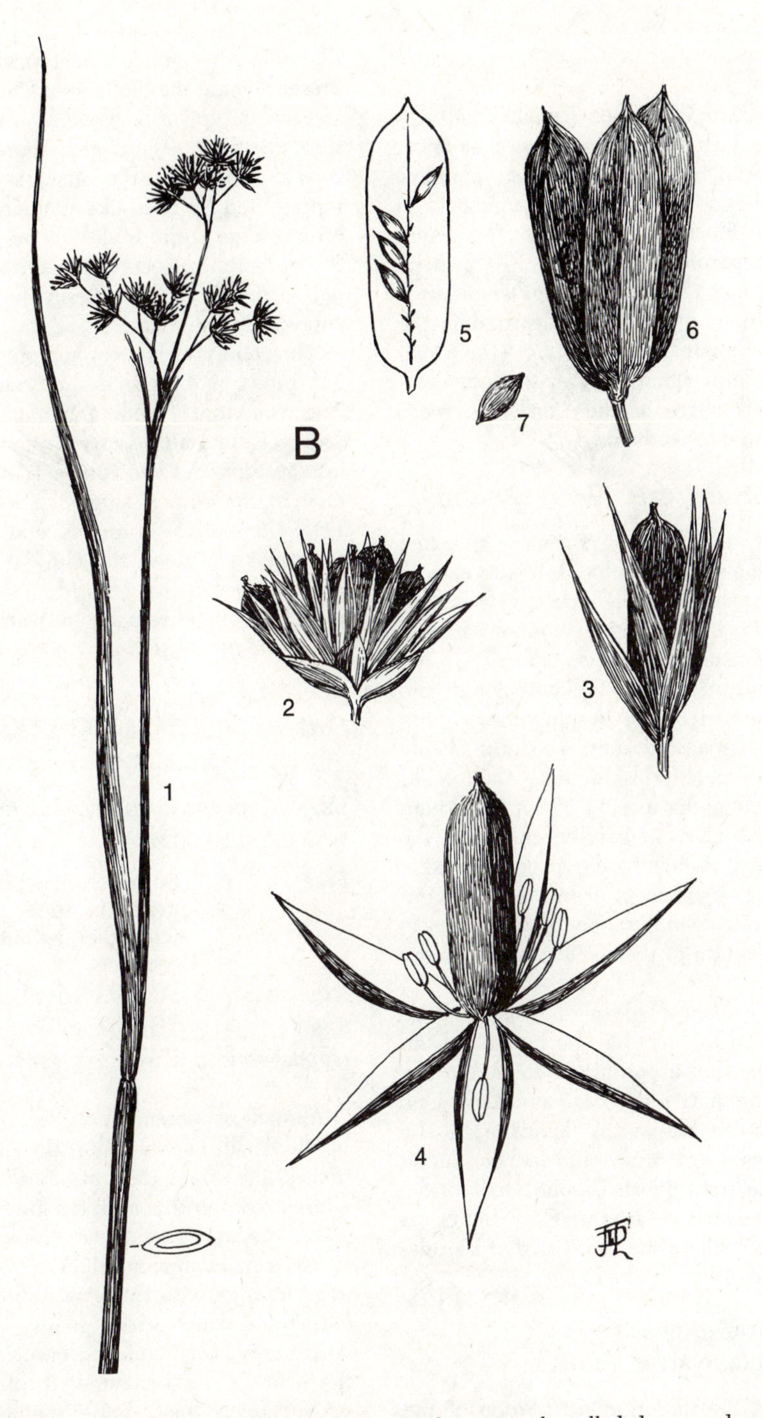

2 head of fruits, 3 fruit surrounded by persistent sepals, 4 sepals pulled downward, revealing the dried persistent stamens and the fruit, 5 fruit, longitudinal section, showing seeds on the partition, 6 capsule in dehiscence, 7 seed.

in the ovary, parietal; ovules numerous in each carpel. Fruit a berry or a capsule; endosperm mealy, abundant.

The family occurs throughout tropical and subtropical regions. The shell ginger or shellflower (*Alpinia*) referred to also as *Languas,* a native of the East Indies, is naturalized in hammocks in south peninsular Florida. A species of *Hedychium* occurs in marshes from the Mississippi River delta to Georgia. Ginger root (from *Zingiber*) is used as a spice. Various genera are cultivated in the United States, among them *Alpinia* (the shellflower), *Hedychium* (ginger-lily or torch-flower), and *Cardamon* (source of the *Cardamon* seeds used as a spice and as medicine).

*Family 3. **Cannaceae.** CANNA FAMILY

Perennial herbs; underground portion of the stem a tuberous rhizome. Leaves basal or alternate or both, broad, the base of each petiole sheathing the stem. Flowers bisexual, in racemes or panicles, with usually conspicuous bracts; sepals 3, not strongly petaloid; petals 3, highly colored and conspicuous; stamen(s) 1 or sometimes 2 and coalescent, the remaining stamens sterile, being highly elaborated petaloid staminodia, the total of stamens and staminodia usually 6, but sometimes only 4 or 5, all coalescent basally; carpels 3, the style 1, the ovary 3-chambered; placentae axile; ovules numerous. Fruit more or less capsular, but dehiscent through disintegration of the pericarp rather than by splitting, the surface warty; endosperm hard.

The family is composed of the single large genus *Canna.* This genus may be native only in Tropical America, but according to some authors, certain species are native in Africa and Asia. One species, the golden canna or "bandana of the everglades," occurs in swamps and marshes along the coastal plain from South Carolina to Florida. Another, perhaps native in the East Indies, is naturalized widely along the Gulf Coast of Florida.

Several species are cultivated.

*Family 4. **Marantaceae.** ARROWROOT FAMILY

Perennial herbs; the underground portion of the stem a rhizome. Leaves alternate or mostly basal, the blades linear or broad, sometimes asymmetrical, the basal portion of the petiole sheathing the stem, the petiole sometimes wingéd, a ligule present where the blade joins the petiole. Flowers bisexual, in spikes or panicles, usually with bracts; sepals 3; petals 3, unequal; stamens 1 or 2, when 2 often coalescent, the remaining 5 or 6 stamens represented by petal-like staminodia; carpels 3 (2 often sterile), the style 1, the ovaries basically 3-chambered; placentae axile; ovule 1, in one carpel. Fruit a capsule, fleshy but splitting open; endosperm abundant.

The family occurs throughout the Tropics and Subtropics and in some adjacent warm regions. The arrowroot (*Maranta*), a native of the East Indies, is naturalized in wet ground in south peninsular Florida. The tubers (rhizomes) yield arrowroot starch. A single species of the genus *Thalia* is native in ponds and swamps of the Coastal Plain along the Gulf of Mexico and the Atlantic Ocean from Florida to Southeastern Virginia. Several genera are cultivated, e.g., *Calathea, Maranta,* and *Thalia.*

Order 4. ***BURMANNIALES.*** BURMANNIA ORDER

DEPARTURE FROM THE FLORAL CHARACTERS OF THE ALISMALES

Free	Sepals and petals adnate into a perianth tube; stamens adnate with the perianth tube; carpels coalescent.
Numerous	Stamens 3 or 6; carpels 3.
Spiral	Stamens and carpels cyclic.
Hypogynous	Flowers epigynous.

Annual or perennial herbs, sometimes saprophytic, small and delicate; the underground portions of the stem either rhizomes or tubers. Leaves alternate or mostly basal, the lower linear to lanceolate, the upper scalelike. Flowers bisexual, in racemes or cymes or solitary; sepals 3; petals 3, edge to edge with the sepals and forming a perianth tube, which varies greatly in form; stamens with very short filaments; carpels 3, the style 1, the ovary 1- or 3-chambered, placentae axile, or when there is one chamber the placentae parietal; ovules numerous, minute. Fruit a capsule; seeds minute, numerous; embryo undifferentiated.

*Family 1. **Burmanniaceae.**

BURMANNIA FAMILY

This largely tropical family is represented in the United States by three genera. Three species of *Burmannia* occur in the Southern States; a single species of *Apteria* is native on the Coastal Plain in the Southern States; a species of the genus *Thismia* is known from a single collection on low prairies near Chicago, Illinois.

Order 5. ***ORCHIDALES.*** ORCHID ORDER

DEPARTURE FROM THE FLORAL CHARACTERS OF THE ALISMALES

Free	Carpels coalescent; style and stamens adnate, forming a **gynandrium** or **column.**
Equal	"Lower" (actually upper, the flower being twisted) petal differing greatly from the others; the gynandrium deflected to the "upper" side of the flower.
Numerous	Stamens 1 or rarely 2, or reduced to even half an anther; carpels 3.
Spiral	Stamens and carpels cyclic.
Hypogynous	Flowers epigynous.

Perennial herbs. Leaves alternate or rarely opposite or whorled, the basal sheath usually closed around the stem. Flowers bisexual or rarely unisexual, with bracts; sepals 3; petals 3, usually all different from the sepals but 2 not necessarily so, the third petal highly differentiated from the others; style 1, adnate with the stamen or stamens, the ovary 3-chambered; placentae parietal or sometimes axile; ovules minute, numerous, anatropous. Fruit a capsule; seeds minute, numerous, without endosperm, the embryo not differentiated into a cotyledon or a hypocotyl or an epicotyl.

*Family 1. **Orchidaceae.** ORCHID FAMILY

The *Orchidaceae* are the third or possibly the second largest plant family. Estimates of species numbers vary from 5,000 to 17,000, but about 10,000 to 12,000 may be correct. Because the family centers in the Tropics, many species are still to be discovered, but many named species are duplicates.

Common orchids occurring in the northern portions of the Continent include the lady's-slippers or moccasin-flowers (*Cypripedium*), *Orchis*, rein-orchis (*Habenaria*), beard-flowers (*Pogonia, Cleistes, Isotria,* and *Triphora*), the grass-pink (*Calopogon*), the brook orchid (*Epipactis*), the lady's-tresses (*Spiranthes*), the rattlesnake-plantain (*Goodyera*), the twayblades (*Listera* and *Liparis*), the coralroot (*Corallorhiza*), the adder's-mouth (*Malaxis*), the puttyroot (*Aplectrum*), and *Calypso.* Other genera and species occur in the Southern States, and various tropical and subtropical genera and species range into southern Florida.

Orchids attain their greatest development in the tropical rain forests. Much of the jungle has relatively little undergrowth because of the dense shade. The trees standing high above the bare forest floor are covered with scrambling lianas or twining woody vines. Innumerable small plants are perched on the upper branches of the trees and in the mass of vines overhead. These epiphytes (plants living upon other plants) include many species of orchids. Water is abundant in the humid jungle, and some of the epiphytic orchids and other plants have aërial roots which catch and absorb water from the frequent rain.

Most of the orchids occurring to the northward in temperate or even relatively cold regions, live in the moist habitats either of mountain meadows or of the relatively moist forest floor. Some of those inhabiting the forest floor are saprophytic, that is, they live upon the decaying material of other plants (i.e., on leaf mold).

A few of the native orchids of North America live in dry situations. Particularly noteworthy among these is *Habenaria unalaschensis,* which occurs not only in more moist situations but in the hot, dry chaparral belt of southern California.

The seeds of orchids are numerous and exceedingly small, the embryo being undifferentiated and consisting of only a few unorganized cells. There is only a small amount of reserve food and stored water. This is one of the factors making growing of orchids relatively difficult. Formation of the typical structures of the seed, including the single cotyledon characteristic of the Monocotyledoneae, does not occur until the seed germinates. Since the small seeds are a characteristic feature of the order Orchidales, another name for this order is Microspermae.

Figure 11-7. Musales; Cannaceae; *Canna* (hybrid), garden canna: **1** leaf, **2** stem, enfolded by a bract, **3** section through the stem and the bract, **4** inflorescence, **5** flower with the bracts removed, revealing the perianth tube (adnate with the inferior ovary), the short sepals, the slender petals, the much larger petal-like staminodia, and (centrally) the stamen and the style and stigma,

6 style and the apical stigmatic surface, **7** mostly longitudinal sectional view of the central parts of the flower, showing the ovary, and the stamen (with the elaborate development of the filament), and the style (sectional and edge view), **8** ovary, cross section showing the adnate perianth tube, the three chambers, the axile placentae, and the ovules. (Cf. also **Figure 6-17 F.**)

Figure 11-8. Orchidales; Orchidaceae; *Habenaria unalaschensis,* a rein-orchis: **1** inflorescence, **2** hair and epidermis cells, **3** flower, showing the perianth tube (adnate with the inferior ovary), this twisted 180° (resupinate), the three sepals, the three petals (the lower [really upper] one with a spur), and the two anther sacs, **4** longitudinal section through the spur. (Cf. also **Figures 2-5** *Cattleya* and **6-17 C** *Cymbidium,* flower.)

Figure 11-9. Orchidales; Orchidaceae; *Peristeria elata,* dove-orchid: portion of the inflorescence, showing the perianth tube (adnate with the inferior ovary) in the buds (above) and especially the lowest flower, the flower at left-center showing the three relatively large sepals and the three petals, the lower petal highly specialized, its marginal wings forming the wings of the "dove," the gynandrium forming the body and head, the anther the bill. Photograph by Ladislaus Cutak.

The pollen within each of the anther chambers is imbedded in a structure known as a **pollinium.** This is a waxy mass of pollen, and the entire mass is transported as a unit by the visiting insect. This feature is not quite unique among the flowering plants for it is simulated in the *Asclepiadaceae* or milkweed family (Dicotyledoneae).

The petal spoken of as the "lower" one may be termed more accurately as the "upper" because in the flowers of the Orchidales there is a characteristic twisting of the ovary which (being inferior) supports the rest of the flower. This twisting of the ovary and also of the perianth tube around it results in an inversion of the flower parts (the flower being termed **resupinate**). The specialized petal sometimes functions as a passageway into which insects are guided. An insect (or its proboscis) entering an orchid flower in which the lower petal is spurred passes by the stigma, then the anther or anthers, and on down into the spur where the nectar is produced. In a slipper type of flower (e.g., lady's-slipper) the insect likewise passes first by the stigma and then by the anther or anthers, acquiring a mass of pollen after having had an opportunity to deposit pollen from other flowers on the stigma. The guiding of the insect is elaborate. Since most orchids have only one or two stamens, or in some cases only half an anther, this method must be efficient.

The gynandrium (from gynoecium, a female organ or pistil, and androecium, a male organ) is one of the characteristic features of the family, and it is a distinguishing feature from the other orders of monocotyledons. Oddly enough, as with the waxy masses of pollen, simulation of this feature occurs in the *Asclepiadaceae* or milkweed family and in almost no other group. From the standpoint of relationship, these two groups, *Orchidaceae* and *Asclepiadaceae,* are about as far removed from each other as any flowering plants, and this is a striking example of parallel evolution or the production of analogous structures independently by two only distantly related groups of living organisms.

The following is a summary of the characters of the order Orchidales which serve to distinguish the group from the epigynous Liliales: (1) corolla bilaterally symmetrical; (2) stamens mostly one or two, rarely three to six; (3) flowers resupinate (twisted 180°); (4) stamen or stamens and the style adnate, forming a gynandrium; (5) pollen imbedded in wax and forming pollinia; (6) seed small, the embryo not differentiated into a cotyledon or into an epicotyl or a hypocotyl.

Order 6. ***RESTIONALES.*** RESTIO ORDER

DEPARTURE FROM THE FLORAL CHARACTERS OF THE ALISMALES

Numerous	Stamens 1–6; carpels 3.
Spiral	Stamens cyclic; carpels cyclic.

Herbs or sometimes vines. Leaves simple and entire, linear or filiform. Flowers bisexual or unisexual, hypogynous, choripetalous or apetalous, radially symmetrical; perianth of 3–6 scales or none or of 3 sepals and 3 petals; stamens 1–6; carpels 3, the ovary 1–3-chambered, the styles 1–3 or 1 and divided to entire; ovule 1 per carpel, pendulous or sometimes horizontal, orthotropous. Fruit a capsule, nut, drupe, or berry; seed with abundant endosperm.

None of the families occurs in North America north of Mexico.

Family 1. **Restionaceae.** RESTIO FAMILY

Perennial herbs; underground stem a rhizome, the flowering stems quadrangular in cross section or flattened. Leaves alternate, the blades reduced, linear. Flowers unisexual (and the plants usually dioecious), radially symmetrical, hypogynous, in spikelets each with 1–many flowers, the spikelet usually subtended by a spathe; perianth composed of 3–6 chafflike scales (or sometimes wanting), these scarious or hyaline, in 2 series; stamens 3, the anthers 2-chambered, splitting lengthwise; carpels 3, the styles 1–3, the stigmas sometimes plumose, the ovary 1–3-chambered; ovule 1 in each carpel, pendulous, orthotropous. Fruit a nut or a capsule, triangular in cross section; endosperm abundant.

The family occurs in southern and eastern Africa, Indochina, New Zealand, and Chile. The South African species are used for thatches and brooms.

The plants are abundant in Cape Province, South Africa. Their distribution, and that of other families like the *Proteaceae,* across large bodies of water in the Southern Hemisphere was one of the early bits of evidence of continent drift.

Segregates proposed as families include the *Anarthriaceae* and *Ecdeiocoleaceae*.

The *Restionaceae* show some similarity to the *Juncaceae,* the Graminales, and the Arales. They have been classified in various orders.

Family 2. **Centrolepidaceae.** CENTROLEPIS FAMILY

Annual or perennial herbs resemblings grasses, sedges, or mosses in appearance. Leaves alternate, linear or filiform, usually densely crowded. Flowers bisexual or unisexual, radially symmetrical, hypogynous, very small, usually in spikes or heads or sometimes solitary, usually subtended by 1–3 scalelike bracts; perianth none; stamen 1 (or rarely 2), anthers 1–2-chambered, splitting lengthwise; carpels 3, the styles separate or coalescent, the ovary 1–3-chambered, or sometimes the carpels more numerous and in 1 or 2 series; ovule 1 per carpel, pendulous, orthotropous. Fruit membranous, splitting lengthwise; endosperm abundant.

The family occurs chiefly in Australia and New Zealand, but it ranges northward to the Philippine Islands, and it is represented in South America.

Family 3. **Flagellariaceae.** FLAGELLARIA FAMILY

Herbs or sometimes vines. Leaves elongate, sometimes each one with a terminal tendril, the sheath enclosing the stem closed. Flowers bisexual or unisexual (and then the plants dioecious), radially symmetrical, hypogynous, in panicles terminating the stem; sepals 3; petals 3, not coalescent, small and dry but somewhat petaloid as are the sepals; stamens 6, sometimes slightly adnate with the perianth, the anthers 2-chambered, splitting lengthwise; carpels 3, the style 1 with 3 divisions, the ovary 3-chambered; ovule 1 per carpel, either pendulous or horizontal. Fruit a berry or a drupe; endosperm abundant.

The family occurs in the Tropics and Subtropics of the Eastern Hemisphere.

Order 7. ***GRAMINALES.*** GRASS ORDER

DEPARTURE FROM THE FLORAL CHARACTERS OF THE ALISMALES

Free	Carpels coalescent.
Equal	Perianth of the *Gramineae* consisting of two scales on one side of the flower; stigmas of the *Gramineae* 2, but the carpels 3.
Numerous	Stamens 3 or rarely 1–6 or more; carpels 2 or more frequently 3.
Spiral	Stamens and carpels cyclic.

Herbs or rarely large, woody, canelike plants, such as the bamboos and giant reeds. Leaves linear and commonly greatly elongated, alternate. Flowers bisexual or unisexual, radially or bilaterally symmetrical, hypogynous, arranged in spikelets, each subtended by one or two chafflike bracts;

perianth represented by minute scales or bristles or threadlike structures; style(s) 1 or 2, the stigmas 2 or 3; ovule 1. Fruit an achene or a caryopsis; endosperm abundant.

The characteristic features of the order are as follows: (1) reduction of the perianth to scales; (2) small, dry, scalelike bracts enclosing the flowers; (3) dry, indehiscent, one-seeded fruits; (4) arrangement of the flowers in **spikelets** (the diminutive of spike, in this instance the small spikes being so distributed that they form particularly conspicuous units of the floral arrangement). Because each of the flowers is subtended by and usually completely enclosed in a special scalelike bract, there appear to be no flowers at all.

In the orders as shown in the portion of "Bessey's Cactus" (chart of relationships of flowering plant orders, cf. p. 538) representing various monocotyledons, the direction of the development of flowering plant orders is generally toward a precise type of insect pollination (represented by an upward slant). In the Graminales the direction of development is toward adaptation to wind pollination, and Bessey gave the figure representing Graminales a downward slope from the top of the figure representing the Liliales. Apparently he intended to convey the impression that the Graminales are retrogressing from the general direction of development among the flowering plants. Because specialization in any direction represents a departure from a primitive type, Bessey's students (e.g., Clements) have constructed new charts in which the Graminales and other wind-pollinated plants are represented by ascending instead of descending figures.

Abundance of individuals, the lightness of the pollen grains, the flexibility of the filaments, and the feathery character of the stigmas serve as an excellent adaptation to wind-pollination; consequently the Graminales have a peculiar significance as hay fever plants. Hay fever is one of the class of maladies known as allergies. The cells of an individual may become sensitized to particular foreign proteins or other substances in the blood stream. The typical reaction to pollen of individual plants, to particular kinds of fur, to wool, to dandruff, to poison ivy or poison-oak, to house dust, or to particular foods may be the result of allergy. In many instances, immunization to the effects of the foreign substance may be produced by hypodermic injection, which may give temporary relief from the sneezing or choking or rash produced by the allergy or which eventually may give complete immunization which may be permanent. The degree of sensitivity of the individual may vary from time to time, and some persons may be completely without hay fever at certain stages of their careers but at other times be exceedingly sensitive to the pollen of grasses or other plants. Hay fever due to grass pollen reaches its peak when grasses flower in the early morning. Many species are in full anthesis before daylight and some reach their peak of flowering at two or three o'clock in the morning. Others flower at about dawn or later in the morning, but, unless the weather is cloudy, most grasses are through with flowering for the day not later than midmorning. One of the worst hay fever plants is Bermuda grass (*Cynodon dactylon*), sometimes known also as devil grass, the most common plant in lawns and fields and along streams throughout the southern portions of the United States. Persons so sensitive to this grass that they do not dare to walk across the lawn in the early morning may be able to mow the same law without sneezing at about five o'clock in the afternoon because the pollen is light and the great bulk of it is blown away during the day.

Alliance of the two families in a single order is open to question. Dr. G. Ledyard Stebbins points out their considerable difference and possible different origin. The writer has not been able to investigate the question with sufficient thoroughness.

Key to the Families

(North America north of Mexico)

1. Flower subtended by a single bractlet, the spikelet as a whole usually not subtended by extra bractlets; fruit an achene; stem solid or with pithy partitions in the internodes, often triangular but sometimes circular in cross section; leaf sheaths with the margins coalescent and enclosing the stem; perianth of bristles or scales, these usually 6, but sometimes wanting.

   Family 1. **Cyperaceae** (sedge family).

1. Flower subtended by a lemma (lower bract) which encloses both the flower and the palea (upper bract), the spikelet as a whole subtended by two bractlets (glumes) which do not enclose individual flowers; fruit a caryopsis (with the ovary wall

[pericarp] attached to the single seed); stem hollow in the internodes, circular in cross section; leaf sheaths not coalescent by the margins, enfolding the stem; perianth of 2 minute scales (lodicules) on one side of the flower.

Family 2. **Gramineae** (grass family).

*Family 1. **Cyperaceae.** SEDGE FAMILY

Annual or usually perennial more or less grasslike plants occurring usually in bogs, marshes, or meadows or along streams; main stem a rhizome, this sometimes short or tuberous, each flowering stem a usually solid culm, this often triangular in cross section, usually little branched below the area of flowering, sometimes scapose. Leaves alternate, the basal ones commonly abundant, each blade linear and elongated, the basal portion forming a sheath, this usually closed around the stem. Flowers bisexual or unisexual (and then the plants usually monoecious but sometimes dioecious), radially symmetrical, each flower subtended by a chaffy bract; sepals and petals represented by bristles or scales or sometimes absent; stamens usually 3 but sometimes 1–6; carpels 2–3, the style 1, with 2 or 3 branches above, the ovary 1-chambered; placenta basal; ovule solitary, erect, anatropous. Fruit an achene, when formed from 2 carpels lenticular (lens-shaped, that is, more or less like a biconvex lens), when formed from 3 carpels triangular in cross section, in the genus *Carex* enclosed in a special sac (**perigynium**).

The family is cosmopolitan, but most abundant in the cooler regions of both hemispheres. The largest genus is *Carex*. The number of species runs into the hundreds, but it is not nearly as large as most books would indicate, because the "species" have been divided in the extreme. The entire genus is in need of careful study and reorganization. In the present state of knowledge the described species are difficult to classify partly because they are complex and poorly evaluated. *Scirpus* is a large genus which includes the bulrushes, tules, and many smaller plants in marshes and other wet places. *Eleocharis* is common in wet places also. The cotton grass (*Eriophorum*) is abundant in Arctic regions, and it occurs in the mountains and wet places farther southward. *Cyperus*, the galingale or umbrella sedge, includes many species in all parts of the Continent. Various other genera of sedges occur in wet places throughout North America, some being restricted to special regions.

The pith of the papyrus plant (*Cyperus Papyrus*) is the classical material used by the Egyptians for making paper. The plant is grown frequently in gardens, particularly about pools. Certain species of *Cyperus* known as nut grass are weeds particularly difficult to eradicate because of the tuberous underground portions of the rhizomes.

The flowers of the sedge family resemble those of the Liliales. The principal difference is the extreme reduction of the perianth, which typically consists of six threadlike (filiform) scales.

DEPARTURE FROM THE FLORAL CHARACTERS OF THE ALISMALES

Free	Carpels coalescent.
Numerous	Stamens 3 or rarely 1–6 or more; carpels 2–3.
Spiral	Stamens and carpels cyclic.

The leaf sheaths (corresponding to petioles) are closed so that each one completely encircles the stem; in the grass family the sheaths are open on one side. The two or three coalescent carpels have corresponding numbers of stigmas, and none lack stigmas as in the grasses.

Family 2. **Gramineae.** GRASS FAMILY

(Synonym: *Poaceae*)

Annual or perennial herbs, or sometimes, as the bamboos or giant reeds, woody plants of considerable size; stems often composed of rhizomes below ground and always of **culms** above, the culms hollow except at the nodes or infrequently solid, that is with a central pithy tissue. Leaves basal or alternate or both, linear and greatly elongated or rarely broader, alternate and fundamentally in two ranks on the stem, the basal portion forming a sheath enclosing the stem but the margins not coalescent, a membranous ligule usually present or represented by a row of hairs at the point of joining of the sheath and the blade. Flowers usually bisexual, but sometimes unisexual, bilaterally symmetrical but not obviously so, each flower of the spikelet enclosed in a lower bract (**lemma**) and an upper (**palea** or palet), the lemma enclosing the palea as well as the flower, the combina-

Figure 11-10. Graminales; Cyperaceae; *Cyperus rotundus*, a nut-grass or (more generally) a sedge (cf. **Figure 2-14 A**): **1** inflorescence, **2** spikelet, **3** bractlet (scale) and the axillary flower, **4** flower, enlarged, showing the three stamens and the tricarpellate pistil, the perianth none in this genus. (Cf. also **Figure 2-11 C** *Eleocharis* or spike rush.)

Figure 11-11. Graminales; Cyperaceae; Carex (perhaps *Carex amplifolia*), a sedge: **1** inflorescence and bract, **2** terminal staminate spikelet, **3** staminate flower consisting of three stamens (the perianth none in this genus), in the axil of a bractlet (scale), **4** pistillate spikelet, **5–6** pistillate flowers in the axils of bractlets, each flower (a single pistil) enclosed in a sac (perigynium) typical of the genus *Carex*, **7** perigynium cut lengthwise, exposing the pistil with its style and three (in some species two) stigmas.

tion of the lemma, palea, and flower known as a **floret,** the spikelet as a whole subtended by a pair of **glumes,** one glume (usually the lower) sometimes missing, the spikelet sometimes with extra (sterile) lemmas or paleas; in one inflorescence type the sterile lemma adjacent to the glumes and similar to them in texture; perianth represented by two hyaline scales on one side of the flower, these being possibly a sepal and a petal; stamens 3 or rarely 6 or more; carpels 3, the styles and stigmas 2, the stigmas being feathery, the third style and stigma not developing, the ovary 1-chambered; ovule 1, attached to the pericarp. Fruit a **caryopsis** (a special type of fruit occurring only in this family and characterized by the adherence of the seed to the pericarp) or rarely a nut or a berry (in the *Bambuseae*) or a utricle; endosperm abundant.

DEPARTURE FROM THE FLORAL CHARACTERS OF THE ALISMALES

Free	Carpels coalescent.
Equal	Perianth parts (lodicules) 2, on one side of the flower; carpels 3, but the stigmas only 2.
Numerous	Stamens 3, carpels 3.
Spiral	Stamens and carpels cyclic.

The *Gramineae* are one of the larger families of plants. The number of species is difficult to estimate, but it is thought to be about 4,000 or more. The family includes undoubtedly more individual plants than any other. The genera occurring in North America north of Mexico are too numerous to mention, there being about 160 of them and over 1,000 species in the United States, as well as great numbers in Canada and the Arctic portions of North America.

As a special reference for study of the family, the following is of the greatest importance: Hitchcock, A. S., *Manual of the Grasses of the United States,* U. S. Dept. of Agriculture Miscellaneous Publication 200. Washington, D. C., 1935. Edition 2, revised by Agnes Chase. 1950. Reprinted 1971, Dover Books, New York.

The following features of the organization of the inflorescence and of the flowers are unique.

1. The three types of bracts, i.e., glumes, lemmas, and paleas. The palea is smaller than the lemma and, except during anthesis (the period during which the flower is shedding pollen and the stigma is receptive), the palea and the flower are both enclosed within the margins of the lemma.

2. Reduction of the perianth to 2 scales (**lodicules**) occurring side-by-side at the base of the flower and representing possibly 1 petal and 1 sepal, or perhaps 2 sepals or 2 petals.

3. The 3 carpels with only 2 developed stigmas, these being large and feathery and well adapted to capturing windblown pollen.

4. The caryopsis (dry, one-seeded, and indehiscent, the wall of the fruit and the single seed being adnate, e.g., a grain of wheat or corn).

To man the *Gramineae* constitute the most important plant family. They include such staple food-producing plants as wheat, oats, barley, rye, corn, sorghum, rice, the various millets, sugar cane, and bamboo. As wheat is to the Western World, rice is to the Orient. Bamboo is nearly as important because it is put to many uses, ranging from innumerable types of building materials to food derived from the young shoots. The family includes not only the cereals but also the far more numerous pasture and range grasses. The value of these forage grasses together with corn (maize) and other grass seeds used for feeding animals is beyond calculation.

Systems of Grass Classification

Although the classification of grasses has been revised during the period since publication of Edition 1 in 1957, the system of Hitchcock, still available in reprint form, is used in the great majority of local or regional floras and manuals. (A. S. HITCHCOCK, revised by AGNES CHASE, *Manual of the Grasses of the United States.* Edition 2, 1950. Reprinted by Dover Books, New York. 1971.) A version of this system presented in Edition 1 of this book appears here with slight modification, as **Track 1,** below.

As a result of many recent investigations, classification of the grasses has been revised. The chief basis for revision is study of obscure, minute, or microscopic characters including the following: the internal structure of the stems, including the arrangement of the vascular bundles and other tissues; the microscopic epidermal cells of the roothair region; the epidermal cells of the leaf, including those of the stomata and those forming the basis for surface characters; the internal anatomy of the leaf; the structure of the components of the flower or its bracts; the cellular structure of the embryo; chromosome numbers and morphology; the nature of the food reserve in the seed; the germination of the seed.

Investigation of these characters is the basis for a new system of classification of the grasses approached first by Avdulov in 1931 and elaborated in various versions by Prat in 1936 and 1960; Parodi in 1961; Stebbins and Crampton in 1961; and Gould in 1968. The subject is presented in full by Frank W. Gould (*Grass Systematics.* McGraw-Hill, New York, 1968.)

Despite the probable soundness of the new arrangement, systematic study is incomplete. A key along natural lines has not been worked out, and the known basic characters are obscure and difficult to detect, except by time-consuming techniques unknown or unavailable to undergraduate students in classification courses except advanced ones. The problem of developing a workable natural key may not be solved soon, because it is complex and numerous taxa are involved. It is a subject for many years of research, and the key presented here (**Track 2**) is only a beginning toward solution of the problem. There will be many exceptions not taken into account.

Gould presents no key to the subfamilies or tribes, though they are described in detail, and the characters of all types, including the obscure ones, are discussed. The genera are keyed out directly, regardless of relationships, by an artificial key on pages 134–145. Local floras or manuals employing this system are likely to have keys to the subfamilies, tribes, and genera appearing in the areas they cover, but these, too, are likely to be artificial keys, which are practical sometimes for identification, but which contribute nothing to learning the nature of the taxa.

The system used in the local or regional flora or manual may be determined by the number of subfamilies. For **Track 1,** there are 2, for **Track 2,** 5 or 6.

TRACK 1

(Bentham, as revised by A. S. Hitchcock) See **Track 2,** Avdulov, Prat, Parodi, Stebbins and Crampton, and Gould, p. 394.

The genera fall apparently into two subfamilies, the *Festucoideae* and the *Panicoideae.*

In the *Festucoideae* the first (lowest) lemma is fertile and any sterile lemmas are at the apex of the spikelet; usually there are several fertile lemmas. The lemmas usually are relatively soft but dry at fruiting time.

In the spikelet of *Panicum,* a member of the *Panicoideae,* the first glume is relatively short; the second glume is considerably longer, and it is matched by a third structure of the same texture. This is a sterile lemma, that is, a lemma which subtends no flower (or sometimes it encloses stamens alone). In some members of the *Panicoideae* there is a sterile palea matching the sterile lemma. Enclosed in the first glume, second glume, and the sterile lemma is a solitary fertile lemma. This and the enclosed palea become tough and leathery or hard and bony at fruiting time. In the *Panicoideae* another flower type is illustrated by *Andropogon* (beard grass). The glumes are firm in texture and of about equal length; the sterile lemma directly above and enclosed by them is thin and membranous as are also the fertile lemma and its palea. However the basic arrangement is the same as in *Panicum* because the spikelet consists of two glumes, a sterile lemma directly above them, and then a fertile lemma with its palea and flower. The table on page 392 summarizes the differences of the *Festucoideae* and the *Panicoideae.*

No single character listed in the table (page 392) is completely reliable because there are exceptions to all, but ordinarily the preponderance of characters of any particular grass will place the plant within the proper subfamily. Students wishing to learn well the application of the characters of the subfamilies, tribes, and genera should follow the procedure outlined below.

As many grasses as possible should be collected and pressed, and the specimens should include the roots, stems, and leaves and the spikelets in both the early and the mature condition. Preferably fifty or one hundred or more species should be brought together. After checking the preliminary list for elimination of exceptional character combinations (page 392), the grasses should be sorted out into the two subfamilies according to the characters outlined above, each plant being checked for all four characters. Fortunately the characters of grasses are clear in dried specimens provided the general pattern of flower structure for the family is understood. Nevertheless, the

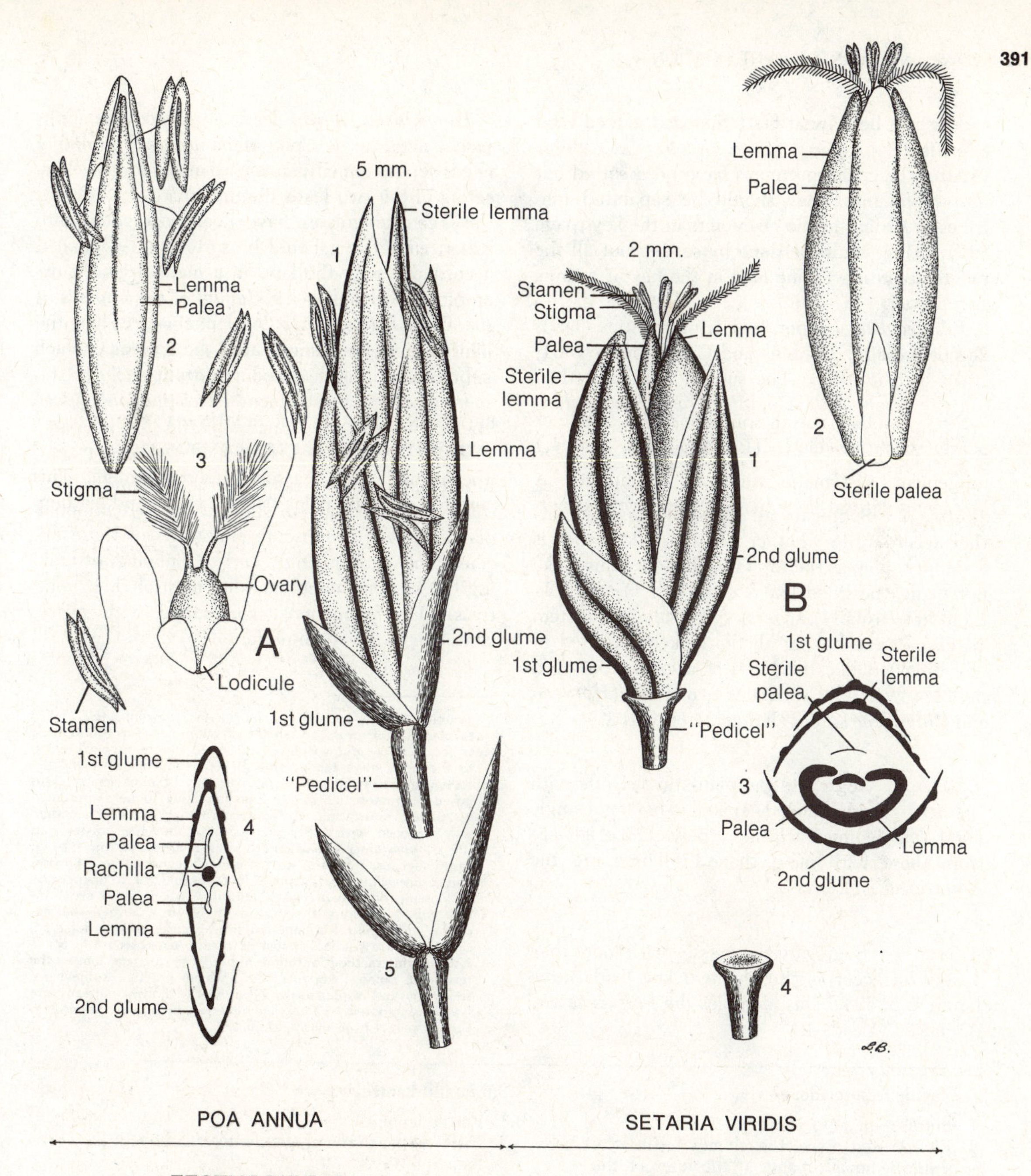

Figure 11-12. The grass flower and the characters of the subfamilies Festucoideae and Panicoideae: **A** Festucoideae, represented by *Poa annua*, the common weedy annual bluegrass. **1** spikelet, **2** floret, ventral view (toward the rachilla of the spikelet), **3** flower, the enveloping bracts (lemma and palea) removed, **4** diagram showing the bracts in a cross section of the spikelet, **5** persistent glumes after the fall of the lemmas and the enclosed fruits (caryopses); **B** Panicoideae, represented by *Setaria viridis*, green foxtail-millet or bristly foxtail, **1** spikelet, **2** the solitary fertile floret ventral view (toward the rachis) (the palea of the sterile floret shown also), **3** diagram showing the bracts in a cross section of the spikelet, **4** "pedicel" (peduncle) after the fall of the spikelet including the glumes. (Cf. also **Figure 8-4 F** wheat, detail of the fruit and seed.)

sorting will be slow at first. Speed is gained after a few hours.

After the grass specimens have been sorted out into subfamilies, they should be separated into tribes according to the characters in the key (page 392). This should be done by sorting out all the plants according to one lead in the key at a time, as follows.

Festucoideae (upper lead 1; note various leads 2–4 below it).

STEP 1. Pick out all specimens with sessile or nearly sessile spikelets. (Lower lead 2, p. 394.)

STEP 2. Segregate all plants in this pile further according to whether (1) the infloresence is a solitary spike at the top of each stem (this spike not being one-sided like a comb) *or instead* (2) there are (usually) several spikes along the stem (each of these having all the spikelets turned in one direction like the teeth of a comb, but in a double row). This separates two tribes (*Hordeae* and *Chlorideae*) from all others. (Leads 3′.)

STEP 3. Segregate the plants in the pile with "pedicelled" (peduncled) spikelets having a single floret (one lemma enclosing a palea and a flower) from those with more than 1. These are the *Agrostideae*. (Leads 3.)

STEP 4. Segregate the grasses with more than 1 spikelet according to the two key leads numbered 4, p. 394. This identifies the *Festuceae* and *Aveneae*. (Leads 4.)

Panicoideae (lower lead 1). Since the *Tripsaceae* are rare, the pile of *Panicoideae* probably needs separation only in accordance with the characters in the two leads numbered 3″.

When the grasses have been segregated into tribes, each tribe should be sorted into genera in accordance with the keys in a manual, as for example A. S. Hitchcock's *Manual of the Grasses of the United States* or a local manual. When the genera have been segregated the species of each genus should be identified as a group.*

PRELIMINARY LIST FOR ELIMINATION OF EXCEPTIONAL CHARACTER COMBINATIONS

Before using the key below to the subfamilies and tribes, it is necessary to eliminate (1) exceptional plants marked by obscure characters or those inconsistent if taken singly and (2) plants *not* having the key characters marking most of their relatives. **The plant must correspond with every character in the combination.**

* Practice in this special use of the key develops the ability to classify any grass. Each set of characters for subfamilies, tribes, genera, and species is mastered before proceeding on to the next lower taxon, and special study of tribes by dissection and drawing representatives may be unnecessary. Opposed characteristics in the key are likely to be represented by plants illustrating both alternatives. The same method may be used with other plant groups, but the grasses and the *Compositae* are particularly well-adapted to this type of study because species are abundant everywhere. Grasses and *Compositae* are particularly favorable for a first course in classification given in the autumn (where there are summer rains), because they are available in quantity and because, once their fundamental flower pattern is understood, the plants are studied almost as readily dried as fresh. Studied by the method outlined above they are no longer the terror of botany students but a good practice subject for freshmen and sophomores, because the flower patterns are few as compared to those encountered in random identification of plants from many families.

Subfamily **Festucoideae**	Subfamily **Panicoideae**
1. Fertile lemma(s) 1 or more.	1. Fertile lemma 1.
2. First lemma above the glumes fertile, the sterile lemmas, if any, at the apex of the spikelet (rarely the whole spikelet staminate, pistillate, or sterile).	2. First lemma sterile or staminate, the second lemma fertile.
3. Fully matured and ripened spikelets usually disarticulating above the glumes and leaving the glumes on the plant.	3. Fully matured spikelets disarticulating below the glumes, the entire spikelet falling away from the inflorescence.
4. Glumes and lemmas *usually* compressed from side to side, i.e., laterally, often creased or tending to be creased along the midribs.	4. Glumes and lemmas flattened from back to front, i.e., dorsoventrally, not creased.

NOTE: The tribes below are all in the subfamily ***FESTUCOIDEAE.***

1. Usually woody grasses several meters tall, the culms 1–several cm. in diameter; lodicules usually 3; stamens 6 or numerous or sometimes only 3.

   Tribe A. **Bambuseae** (bamboo tribe).

2. Glumes of each group of 3 adjacent spikelets forming an involucre, the cluster deciduous as a unit, the 2 lateral spikelets of the cluster staminate and usually 2-flowered, the middle spikelet perfect and 1-flowered; inflorescence a single terminal spikelike panicle.

   Tribe G. **Zoyzieae** (tobosa grass tribe).

3. Spikelet 1-flowered, flattened; glumes both vestigial or *commonly none;* stamens 6; lemma and palea folded and keeled (like boats); plant aquatic or subaquatic.

   Tribe H. **Oryzeae** (rice tribe)

4. Spikelets unisexual, 1-flowered; one or both glumes vestigial or none; plant aquatic or subaquatic.

   Tribe I. **Zizaneae** (Indian rice tribe).

5. Lowest fertile lemma (the only one) subtended by 2 small lemmas, these reduced to tiny scales or subulate structures much shorter than the fertile lemma.

   Tribe J. **Phalarideae** (Canary grass tribe).

NOTE: The taxa below are miscellaneous exceptional plants not having some of the usual key characters for their tribes.

1. Perennials with creeping rhizomes, low-growing, the leaf blades pungent; spikelets 1-flowered, laterally compressed, disarticulating below the glumes; first glume absent, the second enfolding the floret; lemma (and palea when present) very thin. *Zoysia.*

   Tribe G. **Zoysieae** (tobosa grass tribe).

2. Small arctic or sometimes alpine plants less than 1 dm. high; spikelets 1-flowered; glumes minute, unequal, the first sometimes not present; lemma 3-nerved, slightly keeled, abruptly acute; spikelets small, with few flowers. *Phippsia.*

   Tribe B. **Festuceae** (fescue tribe).

3. Stems forming mats, the branches less than 5 cm. long; on mud flats in Eurasia and (native or introduced) along the lower Columbia River in Washington and Oregon; glumes none; lemma very thin and membranous, ovate, with a short terminal awn; palea with 2 apical teeth, the keel with a terminal awn; panicles very small. *Coleanthus.*

   Tribe B. **Festuceae** (fescue tribe).

4. Annual, 1.5–3 dm. high; nodes pubescent; spikelets nearly sessile, with several to many flowers, the lemmas long-awned (i.e., 1–2 cm. long) or at least mucronate; the plant, especially the inflorescence, with the aspect of the foxtail (*Hordeum, Hordeae*) but the spikelets not in sessile clusters of 3 and the rachilla disarticulating between each pair of spikelets rather than between clusters of spikelets; palea as long as the lemma, the keels with cilia like the teeth of a comb; rachis joints alternately concave. *Brachypodium.*

   Tribe B. **Festuceae** (fescue tribe).

5. Spikelets 1-flowered, compressed laterally; glumes equal, thin and hyaline, terminating abruptly, with terminal awns (to sometimes acute); lemma much shorter than the glumes, thin and membranous; florets in dense cylindroid panicles; spikelets sessile or nearly so, the inflorescence a dense, elongate panicle appearing like a spike; leaf blades broad. *Phleum.*

   Tribe C. **Aveneae** (oat tribe).

Key to the Subfamilies and Tribes

(North America north of Mexico)

Refer first to the Preliminary list above. (The following key is simplified slightly to set off the chief characters of the tribes. **Occasional exceptional genera or species have not been taken into account.**)

1. Lowest lemma of the spikelet fertile, that is, enclosing a complete flower, the sterile lemmas, if any, above it; fertile lemma(s) 1 or more, except in completely sterile or staminate spikelets; glumes and lemmas *usually* laterally compressed; *ripe* spikelet *usually* disarticulating above the glumes.

   Subfamily 1. **Festucoideae.**

   2. Spikelets or some of them "pedicelled" (except in *Lycurus,* in which one of the glumes is parted into two awns and in which there is only 1 floret per spikelet).

3. Spikelet 2–many-flowered (sometimes only 1 floret per spikelet).
    4. First glume *not* extending as far as the tip of the first lemma exclusive of the awn (except in *Pappophorum*, which has several awns per lemma, and *Tridens*, which has hairy-nerved lemmas); lemma not awned from the middle or base of the back, the awn an apical or nearly apical continuation of the midrib.

    Tribe B. **Festuceae** (fescue tribe).

    4. First glume extending at least as far as the first lemma (except sometimes in *Trisetum spicatum*); lemma usually awned from the middle or base of the back.

    Tribe C. **Aveneae** (oat tribe).

Figure 11-13. Subfamily Festucoideae; Tribe Zoyzieae; *Hilaria Belangieri*, curly mesquite: **1** panicle of spikelets, **2** involucre of the glumes surrounding three adjacent associated spikelets, **3** two dissimilar glumes. In **Track 2,** Hilaria is in the Subfamily Eragrostoideae, Chlorideae.

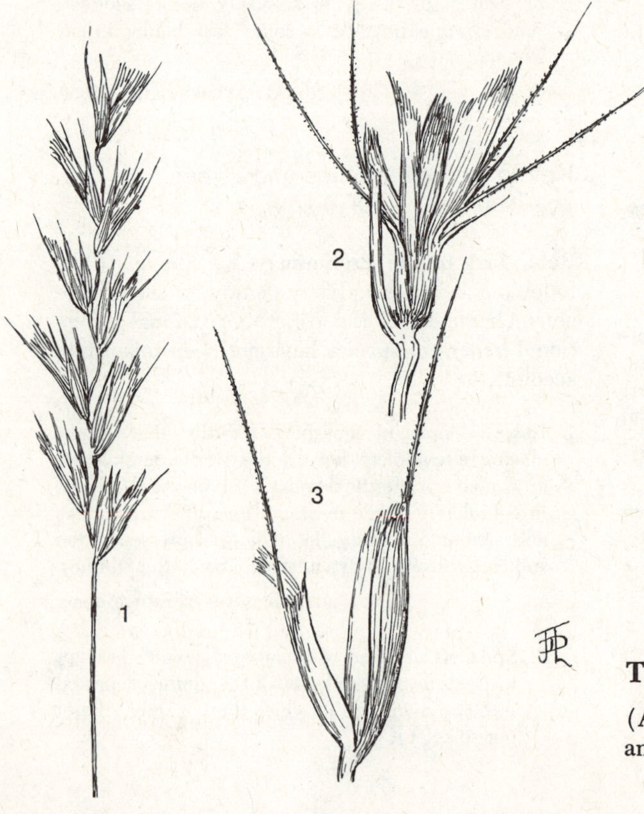

3. Spikelet with 1 perfect floret and *no* sterile florets (sometimes 2-flowered in *Muhlenbergia asperifolia* and *M. arenacea*).

    Tribe D. **Agrostideae** (bent tribe).

2. Spikelets sessile (or subsessile) *or* in sessile or subsessile clusters of 3.
    3'. Spike *not* one-sided, solitary and terminating the culm.

    Tribe E. **Hordeae** (barley tribe).

    3'. Spikes one-sided, pectinate (comblike in appearance), usually more than one per culm, lateral to the main stem, the spikelets usually in two rows in each spike.

    Tribe F. **Chlorideae** (grama tribe).

1. Lowest lemma of each bisexual spikelet sterile, that is, *not* enclosing a perfect flower, the second lemma fertile; glumes and lemmas dosally compressed; ripe spikelet disarticulating below the glumes.

    Subfamily 2. **Panicoideae.**

2'. Flowers or most of them usually perfect *or*, if otherwise, the plant not with the character combination in lead 2' below.
    3''. Each rachis joint *not* producing one pedicellate and one sessile spikelet; fertile lemma never awned; sterile lemma usually similar to the glumes, the fertile lemma and palea hard or tough.

    Tribe K. **Paniceae** (Panicum tribe).

    3''. Each rachis joint producing one pedicellate and one sessile spikelet; fertile lemma almost always long-awned; sterile lemma thin and membranous like the fertile lemma and palea, the glumes indurated (leathery or tough).

    Tribe L. **Andropogoneae** (Sorghum tribe).

2'. Flowers unisexual; staminate spikelets borne in pairs in spicate racemes or in terminal panicles; pistillate spikelets embedded in the rachis below the staminate spikelets or on thick spikes sheathed in bracts and borne in the leaf axils on the middle portion of the stem; glumes of the pistillate flowers either markedly hardened or chafflike or membranous.

    Tribe M. **Tripsaceae** (Indian corn tribe).

TRACK 2

(Avdulov, Prat, Parodi, Stebbins and Crampton, and Gould).

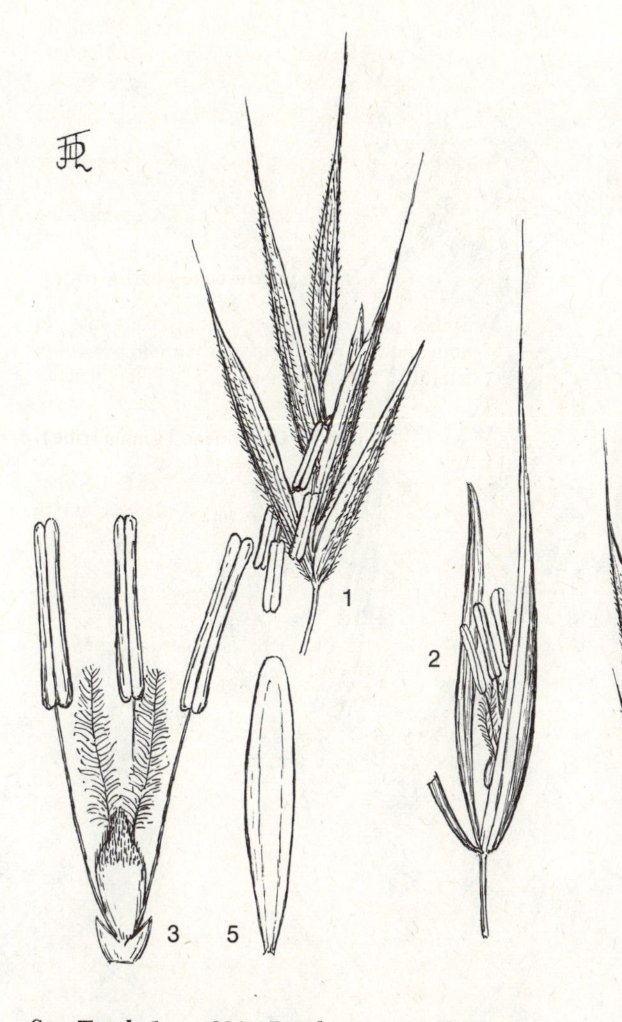

Figure 11-14. Subfamily Festucoideae; Tribe Festuceae; *Bromus carinatus,* a brome: **1** flowering spikelet, **2** floret, the lemma, palea, stamens, and a stigma visible, **3** flower, **4** fruiting spikelet, **5** caryopsis.

See **Track 1**, p. 390; Bentham, revised by A. S. Hitchcock.

As presented here, the system of **Track 2** is modified slightly by elimination of the tribe *Meliceae,* as well as the three appearing in the proposed subfamily *Arundinoideae.* These four tribes do not seem to have any substantial macroscopically visible basis, and their validity for segregation according to microscopic characters is in need of clarification. They are retained, though with no certainty, in their earlier positions in tribes of the *Festucoideae* pending better distinction but not with the thought that they necessarily will remain there. Further clarification may require their removal. See footnotes following the key to the tribes of the *Festucoideae.*

The proposed subfamily *Arundinoideae* seems to have no firm status, and Gould characterizes it as a "somewhat heterogeneous assemblage," even after some modification of the group. The characters binding the proposed taxon together are microscopic, and none is usable for a key to separate the subfamily from its peers. From the standpoint of outward appearance, the group has the aspect of a three-animal team composed of a jackrabbit, a mule, and a kangaroo. Some of the microscopic characters are variable from one proposed tribe to another, and they are similarly variable among the grasses as a family. At this point no evaluation is attemped because of the inadequacy of information concerning the occurrence of the microscopic characters.

Figure 11-15. Subfamily Festucoideae; Tribe Aveneae; *Avena fatua,* a wild oat: **1** panicle of spikelets, **2** one glume (left) and the adjacent lemma, this with an awn from the middle of the back, **3** floret, **4** floret, another view, **5** pistil, **6** persistent glumes after the lemmas have fallen. In this tribe (cf. **2**) the first lemma is shorter than the adjacent glume.

Figure 11-16. Subfamily Festucoideae; Tribe Agrostideae; *Aristida hamulosa*, three-awn: **1** panicle and the tip of a leaf, **2** spikelet, showing the two glumes and the solitary three-awned hard elongate lemma, **3** disarticulated lemma, **4** stigmas and anthers protruding from the lemma at anthesis, **5–6** the sharp tip of the lemma after natural disarticulation due to a diagonal abscission layer. In **Track 2**, *Aristida* composes the Tribe Aristideae of the subfamily Eragrostoideae.

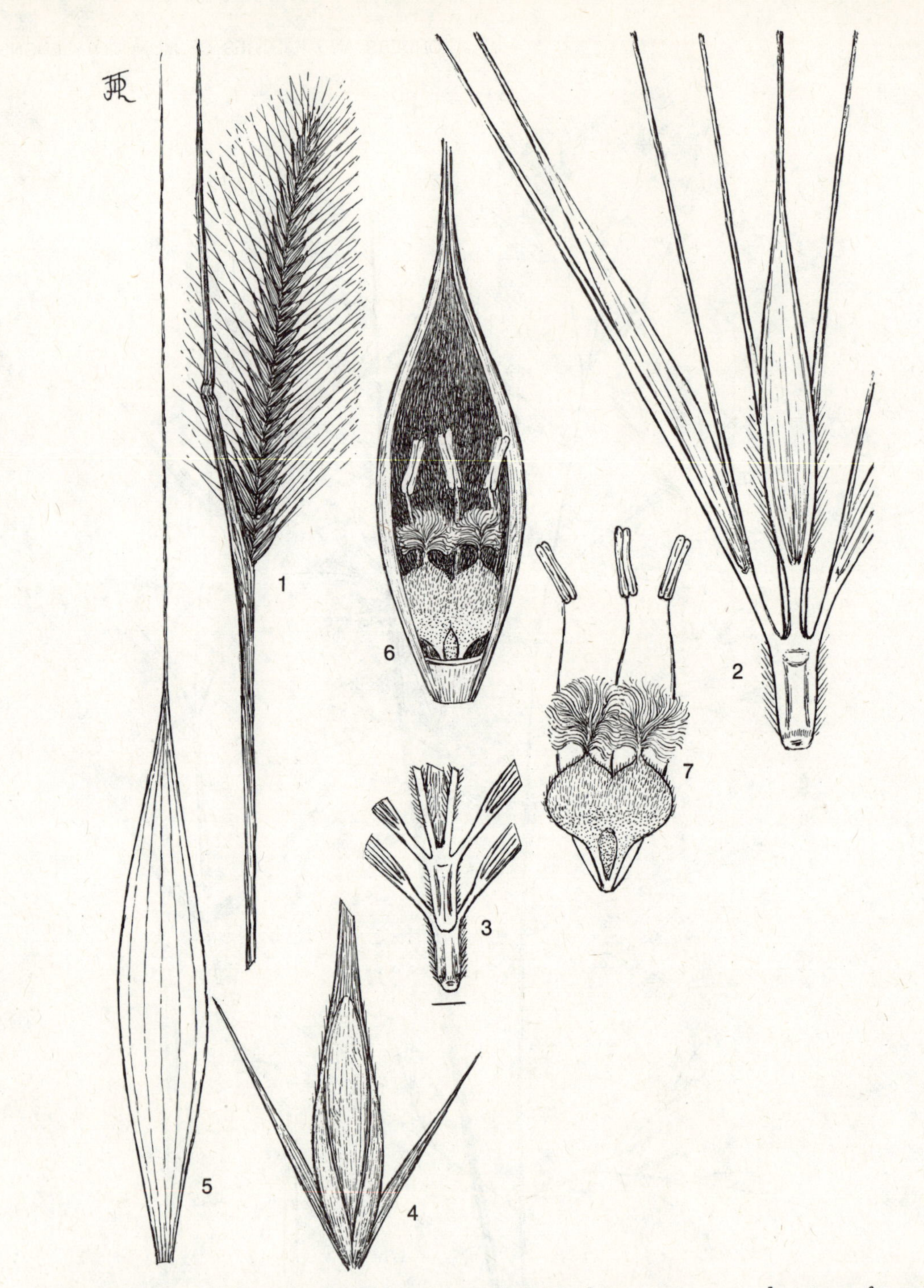

Figure 11-17. Subfamily Festucoideae; Tribe Hordeae; *Hordeum murinum* var. *leporinum,* farmer's foxtail (a barley), a common weed from California to the Southern States: **1** the spikelike terminal panicle, **2** group of three spikelets, the glumes awnlike, the lemmas broader, **3** view of the rachis of the panicle, **4** bases of the glumes and the lemma and palea, **5** lemma, **6** floret, the palea removed, revealing the stamens and the pistil, **7** flower, showing the basally incomplete coalescence of the carpels.

Figure 11-18. Subfamily Festucoideae; Tribe Chlorideae; *Bouteloua,* grama grass: **A** *Bouteloua gracilis,* blue grama, **1** inflorescence, **2** aspect of spikelet, when closed, **3** spikelets, one opened and showing the glumes (outside), fertile lemma (left-center), and sterile lemma (right-center) (about half developed in size in this species), **4** a glume and the fertile lemma (left), **5–6** fertile and sterile lemma, two views; **B** *Bouteloua eriopoda,* black grama, a species with greater reduction of the sterile lemma, **1** inflorescence, **2** fertile lemma and palea and (left) the three-awned sterile lemma terminating the rachilla, **3** lemma (more deeply lobed than usual) and palea. In **Track 2,** the tribe is in the subfamily Eragrostoideae.

Figure 11-19. Subfamily Panicoideae; Tribe Paniceae; *Setaria viridis*, green foxtail-millet or bristly foxtail: **1** leaves and inflorescence, **2** inflorescence, enlarged, **3–6** spikelets and reduced panicle branches (bristles), **7** spikelet (for detail, cf. also **Figure 11-12 B**), **8** fertile (near) and sterile (far) lemma, **9** fertile lemma and the enclosed palea, the sterile lemma pulled aside.

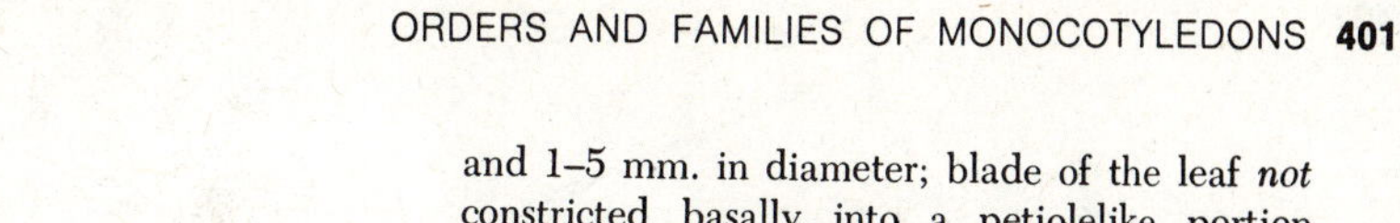

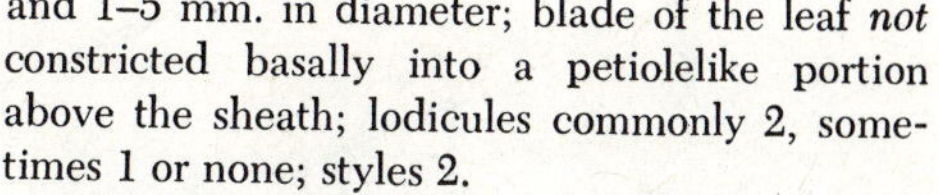

Figure 11-20. Subfamily Panicoideae; Tribe Paniceae; *Panicum bulbosum*, a panicum: **1** inflorescence, **2** young spikelet, showing (right to left) the first glume, sterile lemma, palea, fertile lemma, and second glume, **3** spikelet showing the base of the first glume, the second glume and the tip of the sterile lemma **4** (fertile) lemma and palea spread apart, **5** back (dorsal) side of the lemma, **6** lemma and palea from the other (ventral) side.

Key to the Subfamilies

1. Lowest lemma of the spikelet fertile, that is, enclosing a complete flower, the sterile lemmas, if any, above it; fertile lemma(s) 1 or more, except in completely sterile or staminate spikelets; glumes and lemmas *usually* laterally compressed; *ripe* spikelet *usually* disarticulating above the glumes. (Emphasis on whole combination.)
   2. Usually large woody grasses several meters tall, the culms 1–several cm. in diameter; blade of the leaf constricted basally into a well-defined petiolelike portion above the sheath; lodicules usually 3; stamens usually 6 but sometimes 3 or up to 120; styles usually 3.

      Subfamily 1. **Bambusioideae.**

   2. Annual or perennial herbs or rarely slightly woody, the culms usually less than 1–2 m. high and 1–5 mm. in diameter; blade of the leaf *not* constricted basally into a petiolelike portion above the sheath; lodicules commonly 2, sometimes 1 or none; styles 2.
      3. Glumes probably none (1 or 2 scales in their position now interpreted as vestigial florets); aquatic or marsh plants; spikelet with 1 flower, this unisexual or bisexual; stamens 6; lemma folded and keeled like a boat.

         Subfamily 2. **Oryzoideae.**

      3. Plant *not* corresponding with *every character* in the combination in lead 3 above.
         4. Lemma with 5–several nerves (vascular bundles); culms hollow; a swelling (pulvinus) present at the base of each leaf sheath; lodicules mostly elongate, acute, basally thick and apically membranous.

            Subfamily 3. **Festucoideae.**

         4. Lemma with 3 nerves or sometimes only 1; culms usually solid or nearly so; a swelling (pulvinus) present at the base of each internode of the stem, only a minor one at the base of the leaf sheath; lodicules small, cuneate to obcuneate or sometimes oblong.

            Subfamily 4. **Eragrostoideae.**

1. Lowest lemma of each bisexual spikelet sterile, that is, *not* enclosing a perfect flower, the second lemma (the only other one) fertile; glumes and lemmas dorsally compressed; ripe spikelet disarticulating below the glumes. (Emphasis on whole combination.)

   Subfamily 5. **Panicoideae.**

Subfamily 1. BAMBUSOIDEAE

Key to the Tribes

1. Large woody grasses several meters tall, the culms 1–several cm. in diameter; flowers, so far as known, bisexual; fruits not burlike, not with hooked projections from the surfaces.

   Tribe 1. **Bambuseae.**

1. Perennials 3–5 dm. high, the culms only a few mm. in diameter; flowers unisexual, in pairs along the panicle branches, one pistillate and nearly sessile, the other staminate, much smaller, and pedicelled; fruit with hooked projections and burlike.

   Tribe 2. **Phareae.**

Subfamily 2. ORYZOIDEAE

A single tribe. Tribe 1. **Oryzeae.**

Figure 11-21. Subfamily Panicoideae; Tribe Andropogoneae; *Andropogon cirratus*, Texas bluestem: **1** portion of a panicle of spikelets, **2** unit associated with one node of the inflorescence, including the fertile spikelet (left), the pedicel of the sterile spikelet (middle), and the rachilla internode (right), **3** fertile spikelet (right) and sterile spikelet (left), **4–5** fertile spikelet, two views, **6** the two glumes spread apart revealing the fertile lemma (with the long awn) and behind it the palea and the sterile lemma, these three structures being thin and scarious in the Andropogoneae, **7** fertile lemma, enlarged, **8** the structures behind the fertile lemma in **6**, including one glume, **9–10** palea.

Subfamily 3. FESTUCOIDEAE

PRELIMINARY LIST FOR ELIMINATION OF SOME EXCEPTIONAL CHARACTER COMBINATIONS

Before using the key to the tribes, this page, it is necessary to eliminate some tribes or genera not having visible diagnostic characters making them an obvious unit or an obvious part of the tribe, the distinguishing features being very obscure or microscopic.

The plant must correspond with every character in the combination. If the specimen does not coincide with one character, even the first listed, the combination may be eliminated from further consideration.

1. Plant corresponding with all the characters in one of the combinations below:
   a. Spikelets sessile, occurring singly edgewise to the rachis; first glume none, except in the terminal spikelet. *Lolium.*
   b. Small arctic or sometimes alpine plants less than 1 dm. high; spikelets 1-flowered; glumes minute, unequal, the first sometimes not present; lemma 3-nerved, slightly keeled, abruptly acute; spikelets small, with few flowers. *Phippsia.*
   c. Stems forming mats, the branches less than 5 cm. long; on mud flats in Eurasia and (native or introduced) along the lower Columbia River in Washington and Oregon; glumes none; lemma very thin and membranous, ovate, with a short terminal awn; palea with 2 apical teeth, the keel with a terminal awn; panicles very small. *Coleanthus.*
   d. Annual 1.5–3 dm. high; nodes pubescent; spikelets nearly sessile, with several to many flowers, the lemmeas long-awned (1–2 cm. long) or at least mucronate, the spikes of the inflorescence (and the entire plant) with the aspect of the foxtail (*Hordeum, Hordeae*) but the spikelets not in sessile clusters of 3 and the rachilla disarticulating in between all spikelets rather than between clusters of spikelets; palea as long as the lemma, the keels with cilia like the teeth of a comb; rachis joints alternately concave. *Brachypodium.*

Tribe 1. **Festuceae.**

1. Plant corresponding with *all* characters in one combination below:
   a. Inconspicuous small annuals; Washington to California; spikelets 1-flowered, solitary along the stem, sessile and sunken in hollows in the thickened rachis, laterally compressed, turned flatwise to the rachis; lemma shorter than the glumes, membranous; stamen 1; leaves almost filiform. *Scribneria.*
   b. Spikelets sessile and overlapping like shingles in 2 rows along one side of the axis of the spike; glumes equal, boat-shaped to circular; spikelet disarticulating below the glumes, with 1 fertile and 1 sterile floret, the flower bisexual. *Beckmannia.*

Tribe 4. **Aveneae.**

Key to the Tribes

1. Spikelets or some of them peduncled ("pedicelled").
   2. Longer (usually the second) glume *not* extending as far as the first lemma, usually as long as or longer than the rest of the spikelet.
      3. Florets 2 or more per spikelet; both glumes usually well-developed.
         4. Lemmas membranous (thin, soft, and pliable); caryopsis *not* with a large turgid beak, the body usually not spreading the lemma and palea apart.

Tribe 1. **Festuceae.***

---

* The tribes *Festuceae* and *Meliceae* are combined for lack of distinguishing features, either visible or minute. The *Festuceae* have, so far as known, 7 pairs of chromosomes, the *Meliceae* 8, 9, or 10. There may be other distinguishing characters, but, if so, none of them have come to light. All the other character pairs investigated so far are inconsistent in occurrence in the two proposed tribes. The *Meliceae* form an obviously related unit, but they may or may not be a group worthy of tribal status. The question is whether the distinction of chromosome numbers has been allowed to override other considerations. The number varies as much within the *Meliceae* as between the two groups, and an increase of chromosome numbers is a character that can arise more than once, as obviously it has in producing 8, 9, or 10 from the presumably basic 7 of the *Festucoideae.* Like numbers represent only the packages in which genes are carried, not necessarily a like selection of genes. Seven may be increased to 8 by more than one method, as, for example, most simply by separation into two of any pair of the 7. Even this yields a possibility of 7 differing sets of 8 chromosomes and even more possibilities when 8, 9, and 10 are involved. Eight also may be reduced to 7, if, for example, chromosomes of different pairs join end-to-end. There is a wide variation in the numbers of chromosomes in the other grasses, and the original number can be inferred but not proved.

The proposed tribe *Arundineae,* under the subfamily *Arundinoideae,* tentatively not accepted here, is left under the *Festuceae,* though not with firm conviction. These plants are the large reeds of the genera *Arundo, Phragmites,* and *Cortaderia* and the smaller plants of *Molinia.* The status of the group needs to be worked out more clearly.

The status of the proposed tribe of the *Arundinoideae,* the *Centotheceae,* based for the United States upon species segregated from *Uniola* as *Chasmanthium,* cannot be determined at present, and no opinion concerning validity of the proposed tribe is presented here. Separation of *Chasmanthium* as a genus doubtless is correct.

4. Lemma chartaceous (the texture of writing paper); caryopsis with a large turgid beak, the distended body and beak spreading the lemma and palea apart at maturity.

Tribe 2. **Diarrheneae.**

3. Floret 1 per spikelet; glumes minute, the first often missing, the second sometimes awned.

Tribe 3. **Brachyeletreae.**

2. Longer (usually the second) glume extending as far as or farther than the first lemma, usually longer than the rest of the spikelet.

3'. Lemma *not* becoming hard or leathery at fruiting time, sometimes with a terminal awn but this usually relatively short and deciduous, the callus point of detachment not very strongly developed.

Tribe 4. **Aveneae.***

3'. Lemma becoming very hard or leathery at fruiting time, tightly enclosing the caryopsis, circular in cross section, the surface smooth and shining, terminated by an awn 3 or 4 to commonly many times its length, this commonly bent and twisted, deciduous by a basal diagonal abscission layer leaving a sharp to blunt callus point at the base of the fruiting floret.

Tribe 5. **Stipeae.**

1. Spikelets sessile or subsessile or in sessile or subsessile clusters of 3.

2'. Spikelets 2–several-flowered, 1 or more at each node of the rachis; both glumes usually well-developed; styles and stigmas 2.

Tribe 6. **Hordeae** (Triticeae).

2'. Spikelets 1-flowered.

3''. Spikelets *not* embedded in a hard, cyclindroid, jointed rachis; style 1, elongate; first glume none, the second minute; lemma awned.

Tribe 7. **Nardeae.**

3''. Spikelets embedded in the hard, cyclindroid, jointed rachis; styles 2; first glume present or none, the second or both hard or leathery and enclosing the floret and covering its cavity; lemma *not* awned.

Tribe 8. **Monermeae.**

* The proposed tribe *Danthoneae*, under the subfamily *Arundinoideae*, tentatively not accepted here, is left under the *Aveneae*, but with no certainty that this is correct. The status of the group as a tribe may be of greater validity than any assumption concerning the position of the group in classification. Morphological, anatomical, and cytological evidence for separation seem to be adequate, but a visible distinction is difficult to find. This raises questions applicable to all minute and difficultly studied characters. Have they been investigated in a truly representative sample of the members of each genus and species? Can visible characters be ignored if they do not correlate with the newly discovered minute ones? Somehow the entire complex of characters must be evaluated without more emphasis upon one than another. Any new system of classification goes through a wave of great enthusiasm for a new approach, resulting in overemphasis for a time, but ultimately all items, new and old, are considered in perspective.

Subfamily 4. **ERAGROSTOIDEAE**

PRELIMINARY LIST FOR ELIMINATION OF EXCEPTIONAL CHARACTER COMBINATIONS

1. Plants with *all* the characters in one of the following combinations:

a. Low, spreading annual; spikelets sessile, 1-flowered, disarticulating below the glumes; lemma broad, thin, 3-nerved, the palea similar, splitting; caryopsis separating readily from the lemma and palea, the pericarp thin and the seed free from it; inflorescence capitate, above a pair of proportionately broad bracts. *Crypsis.*

b. Spikelets with peduncles ("pedicels"), without sterile florets, with only 1 flower, this bisexual; lemma not with 3–9 awns or awnlike lobes. Several genera.

c. Spikelets in dense spikelike panicles, clusters, or heads, in pairs or threes on the short rachis, the group really a reduced spike, enclosed in the broadened sheaths of short bracts; rachilla disarticulating above the glumes and between the florets; lemma with a tuft of hairs about the middle of the margin; glumes of the lower spikelets equal, of the upper unequal; mat-forming grasses of the Great Plains and adjacent grasslands. *Munroa.*

d. Flowers unisexual, the plants dioecious or sometimes monoecious; lemmas of the pistillate flowers with 3 very long, twisted, divergent awns, the lemmas of the staminate flowers entire or mucronate. *Scleropogon.*

Tribe 1. **Eragrostideae.**

Key to the Tribes

1. Spikelets or some of them peduncled ("pedicelled").

2. Lemma *not* divided into 3–9 awns or awnlike lobes.

3. Lower florets of the spikelet fertile, or at least the upper and lower florets *not* both sterile.

4. Flowers usually bisexual, the plants *not* dioecious; stems *not* widely spreading rhi-

zomes or stolons; plants not of alkali flats or markedly saline soil.

5. Spikelets 2–several-flowered, usually disarticulating above the glumes and sometimes between florets along the rachis; both glumes usually present, neither glume completely enfolding the lemma and palea of a floret.

Tribe 1. **Eragrostideae.**

5. Spikelet 1-flowered, disarticulating below the solitary second glume, which is leathery and which completely enfolds the thin lemma and (if present) palea.

Tribe 2. **Zizaneae.**

4. Flowers all unisexual, the plants dioecious; stems widely spreading rhizomes or stolons; plants of alkali flats or markedly saline soil.

Tribe 3. **Aeluropodeae.**

3. Lower 4–6 lemmas of the spikelet sterile, the terminal lemmas also sterile; plants of coastal sand dunes.

Tribe 4. **Unioleae.**

2. Lemma divided into 3–9 awns or awnlike lobes.

3'. Spikelets 3–several-flowered; lemma divided into 9-numerous awns or awnlike teeth.

Tribe 5. **Pappophoreae.**

3'. Spikelets 1-flowered, the lemma divided into 3 apical awns (or branches of an awn).

Tribe 6. **Aristideae.**

1. Spikelets sessile or subsessile or in sessile or subsessile clusters.

2'. Spikelets on 1 side of the rachis, the spikes solitary, racemose, or digitate like the fingers of a hand.

Tribe 7. **Chlorideae.**

2'. Spikelets arranged spirally around the rachis, the spikes crowded or widely spaced.

Tribe 8. **Orcutteae.**

Subfamily 5. PANICOIDEAE

Key to the Tribes

1. Spikelets *not* in pairs of 1 sessile and 1 pedicelled at each joint of the rachis; fertile lemma never awned; sterile lemma usually similar to the glumes, the fertile lemma hard and tough, the palea similar.

Tribe 1. **Paniceae.**

1. Spikelets in pairs of 1 sessile and 1 pedicelled at each joint of the rachis, the pedicelled spikelet often reduced in size and development; fertile lemma awned, often consisting largely of an elongate awn; sterile lemma thin and hyaline like the fertile lemma and palea.

Tribe 2. **Andropogoneae.**

Order 8. *HYDROCHARITALES.*

ELODEA ORDER

DEPARTURE FROM THE FLORAL CHARACTERS OF THE ALISMALES

Free	Carpels coalescent.
Equal	Flowers sometimes bilaterally symmetrical in *Vallisneria.*
Numerous	Stamens 3–12 or 15 (in 1–5 series of 3); carpels 2–15, but in the North American species usually 3 or 6.
Spiral	Stamens and carpels cyclic.
Hypogynous	Flowers epigynous.

Herbs, the plants always aquatic (some floating, others completely submerged, still others with only the flowers coming to the surface). Leaves either basal or cauline and opposite or whorled. Flowers (or some of them on each plant) unisexual; sepals 3, clearly distinguished from the petals; petals 3, white; style 1; ovary 1–3-chambered; placentae parietal but often almost meeting in the center of the ovary; ovules numerous. Fruit ripening under water, indehiscent; seeds without endosperm.

In *Vallisneria* the plants are dioecious. The staminate flowers are produced in clusters under water, but they soon become detached, rise to the surface, and float. The sepals expand and form a float above which the more or less erect stamens project. The pistillate flowers are solitary. They rise to the surface through the straightening of a long, spiral stalk. After pollen is received from the floating staminate flowers which lodge against the pistillate, the stalk of the pistillate flower becomes spirally coiled, pulling the fruit to the bottom of the pond where it matures and ripens.

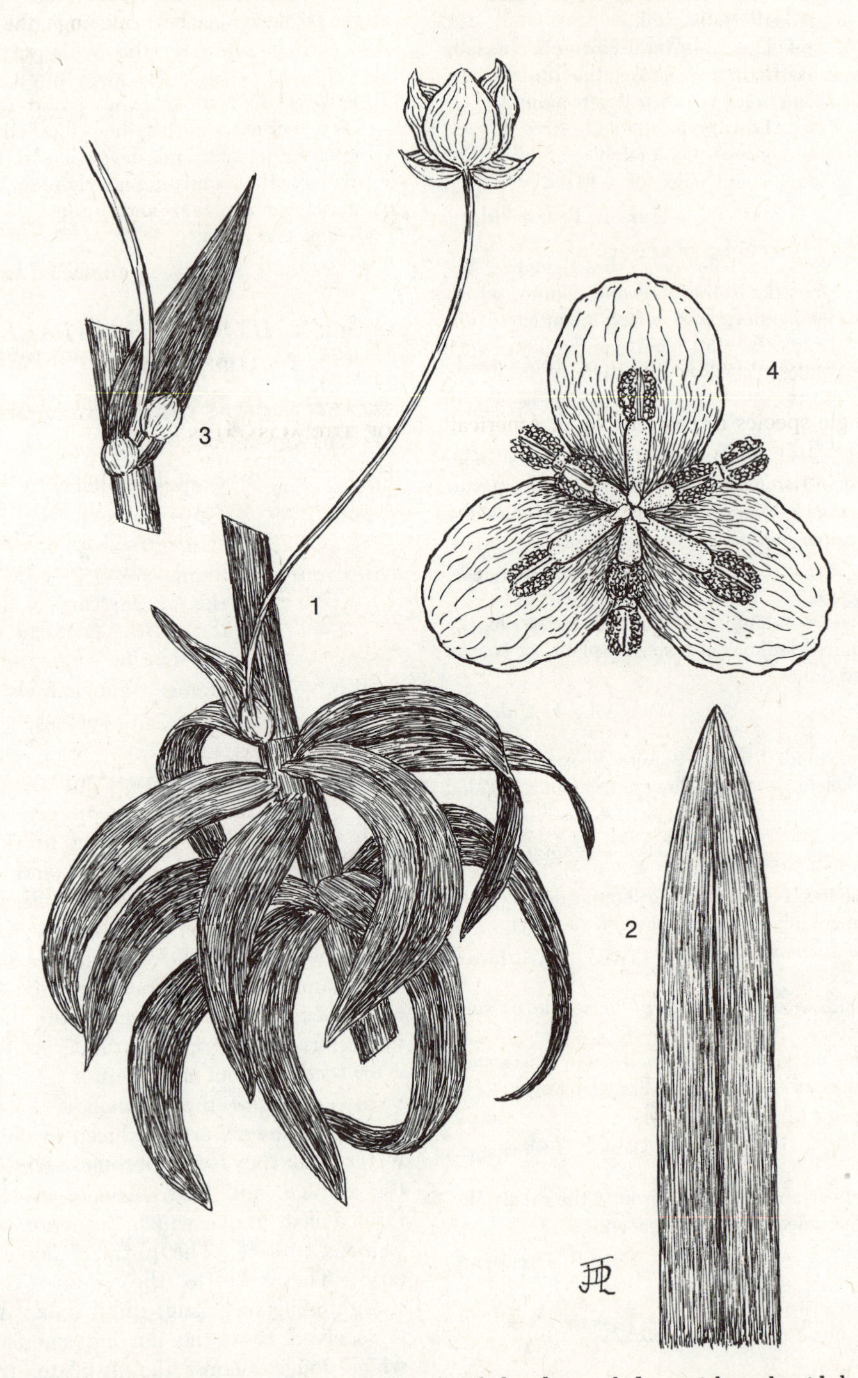

Figure 11-22. Hydrocharitales; Hydrocharitaceae; *Elodea densa,* elodea: **1** branch with leaves and staminate flowers, these showing the sepals and petals, **2** leaf tip, enlarged, **3** flower buds, **4** top view of the staminate flower, showing the petals and stamens.

In *Elodea* (*Anacharis*) the flowers may be unisexual or polygamous (some unisexual and some bisexual). The staminate flowers become detached and both they and their pollen float on the surface of the water. Pistillate flowers rise to the surface of the water and receive the floating pollen.

*Family 1. **Hydrocharitaceae.** ELODEA FAMILY

(Synonyms: *Vallisneriaceae, Elodeaceae*)

The family occurs through much of the world. *Vallisneria,* tape-grass or eel-grass, is represented by but a single species in eastern North America. The American frog's-bit (*Limnobium*) occurs in the Southern States, and it is rare in the southern portions of the Middle West and East. *Elodea canadensis* occurs in the mountains of the West as well as in the eastern portion of the Continent. Another species is native in eastern North America. The common elodea of fish bowls and ornamental ponds is a larger plant than either native species. In southern areas from California to Florida and here and there farther northward in eastern North America, staminate plants have become established, spreading by vegetative propagation. Some genera of the family occur in salt water. One of these (*Thalassia*), known as turtle-grass, or sometimes as "seaweed," occurs in enormous submarine fields off the Florida coast. Great quantities of the leaves washed ashore are gathered for fertilizer.

Order 9. ***TRIURIDALES.*** TRIURIS ORDER

DEPARTURE FROM THE FLORAL CHARACTERS OF THE ALISMALES

Equal	Perianth sometimes bilaterally symmetrical.
Numerous	Stamens 2–6; carpels several.
Spiral	Stamens and carpels cyclic.

Herbs; saprophytes; stems simple, not green. Leaves scalelike, colorless. Flowers unisexual, radially or bilaterally symmetrical, hypogynous, in racemes or corymbs, very small, with scalelike bracts; perianth segments 3–8, in a single series, not distinguished as sepals and petals, reflexed after anthesis; anther opening by a transverse slit; carpels separate, the ovule solitary in each carpel, arising from the inner angle of the base, with a single integument. Fruit opening by a vertical slit.

Family 1. **Triuridaceae.** TRIURIS FAMILY

The family is a small one restricted to the Tropics.

Order 10. ***NAIDALES.*** PONDWEED ORDER

DEPARTURE FROM THE FLORAL CHARACTERS OF THE ALISMALES

Free	Carpels sometimes coalescent.
Numerous	Stamens 1, 2, or 4; carpels 1–4 or rarely 6.
Spiral	Stamens and carpels cyclic.

Aquatic herbs. Leaves either all immersed or the upper ones floating and often of a different type, that is, usually broader. Leaves simple, either with stipules or with basal sheaths which clasp the stem. Flowers bisexual or unisexual, radially symmetrical, hypogynous, often in spikes or clusters or solitary, the inflorescence often enclosed in a small green spathe; perianth segments 0–6, when present sepal-like; pistils 2–4, each formed from a single carpel or the pistil 1 and formed from 2 or 3 carpels, in either case containing only 1 seed chamber and 1 ovule. Fruit a nutlet or somewhat fleshy and drupletlike. Endosperm none.

Key to the Families
(North America north of Mexico)

1. Carpels 1–4, separate; flowers in clusters or spikes.
   Family 2. **Zosteraceae** (pondweed family).

1. Carpels 2 or 3, coalescent, distinguished by the 2 or 3 stigmas; flowers solitary in the axils of the leaves or sheathing leaf bases (the staminate flower of a single stamen, the perianth forming a sac around it and this in turn enclosed in a spathe resembling a bottle, the pistillate flower consisting of a single naked pistil).
   Family 3. **Naiadaceae** (naiad family).

Family 1. **Aponogetonaceae.**

APONOGETON FAMILY

Aquatic herbs occurring in fresh water; the main stem a tuberous rhizome. Leaves linear to oblong; petioles elongate. Flowers bisexual or rarely unisexual, radially or bilaterally symmetrical, in spikes or spicate panicles; perianth segments 1–3 or none, sometimes petaloid, usually persistent in fruit; stamens 6 or more numerous, persistent in fruit; carpels 3–6, separate; placentae basal, ovules 2 or more per carpel, anatropous. Fruit splitting open on the dorsal side; endosperm present.

The family consists of a single genus occurring in or near the Tropics or Subtropics of the Eastern Hemisphere. Sporadic in California.

*Family 2. **Zosteraceae.**

PONDWEED FAMILY

(Synonym: *Potamogetonaceae.*
Segregates: *Zanichelliaceae, Cymodoceaceae*)

Perennial herbs; plants aquatic. Leaves often but not necessarily of two types, the submersed ones then broad and the floating ones narrow. Flowers bisexual or unisexual, in spikes arranged in whorls; perianth none, in *Potamogeton* the connectives of the stamens producing 4 sepal-like outgrowths; stamens 1–4; carpel 1; ovule 1, attached apically or parietally, pendulous. Fruit a nutlet or a drupelet; endosperm present.

The family occurs almost throughout the world. Most members occur in fresh water but a few grow in tidal brackish waters or in the surf on the rocks along the ocean.

By far the largest, most complicated genus of the family in North America is *Potamogeton* (pondweed), which includes more than 30 species. It occurs throughout North America, but it is particularly abundant in the northern lakes of the glaciated region. The ditch-grass (*Ruppia maritima*), a highly variable species, occurs from coast to coast. The widgeon-grass (*Ruppia occidentalis*) occurs from Alaska to British Columbia and Nebraska. The horned pondweed (*Zannichellia*) grows practically throughout the world, including most of North America. The eel-grass (*Zostera*) inhabits shallow waters of bays along the ocean. On the Pacific Coast the genus *Phyllospadix* is represented by two species known as surf-grass. These occur in the ocean near and below low tide level, and they are the only flowering plants growing on the intertidal rocks of the Pacific Coast. Manatee-grass (*Cymodocea*) grows in creeks and bays along the coast from Louisiana to Florida. It is washed ashore during storms or sometimes it is dredged from the bottom. A species of *Halodule* occurs in bays and creeks and on reefs off the coast of southern Florida.

*Family 3. **Naiadaceae.** NAIAD FAMILY

Aquatic annuals, plants occurring in fresh or brackish water, slender, delicate. Leaves almost opposite or appearing to be whorled, linear-lanceolate, sometimes with prominent stipular appendages. Flowers unisexual, solitary or in clusters in the leaf axils, very small; perianth none; the staminate flower consisting of a single stamen enclosed in a tiny membranous bract; pistillate flower consisting of a single pistil, this either naked or with a hyaline bract adhering to the ovary; carpel 1, the style 1 but the stigmas 2–4 (usually 3), ovule 1, basal, anatropous. Fruit an achene, usually enclosed by a green sheath; endosperm present.

The family is composed of the genus *Naias*, which includes about 30 or 40 widely distributed species. Half a dozen occur in eastern North America, and two of these appear in lakes and ponds westward to the Pacific Coast.

Order 11. ***JUNCAGINALES.***

ARROW GRASS ORDER

DEPARTURE FROM THE FLORAL CHARACTERS OF THE ALISMALES

Free	Carpels coalescent at flowering time; separating at maturity.
Numerous	Stamens 1–6; carpels 3–6.
Spiral	Stamens and carpels cyclic.

Herbs of marshes, the plants sometimes semiaquatic (the Lilaeaceae occurring in vernal pools or rills). Leaves terete, all or mostly basal, rushlike or grasslike, with no distinction of petioles and blades. Flowers perfect or unisexual, radially symmetrical, hypogynous, arranged in spikes or racemes; sepals 3 or none; petals 3, similar to

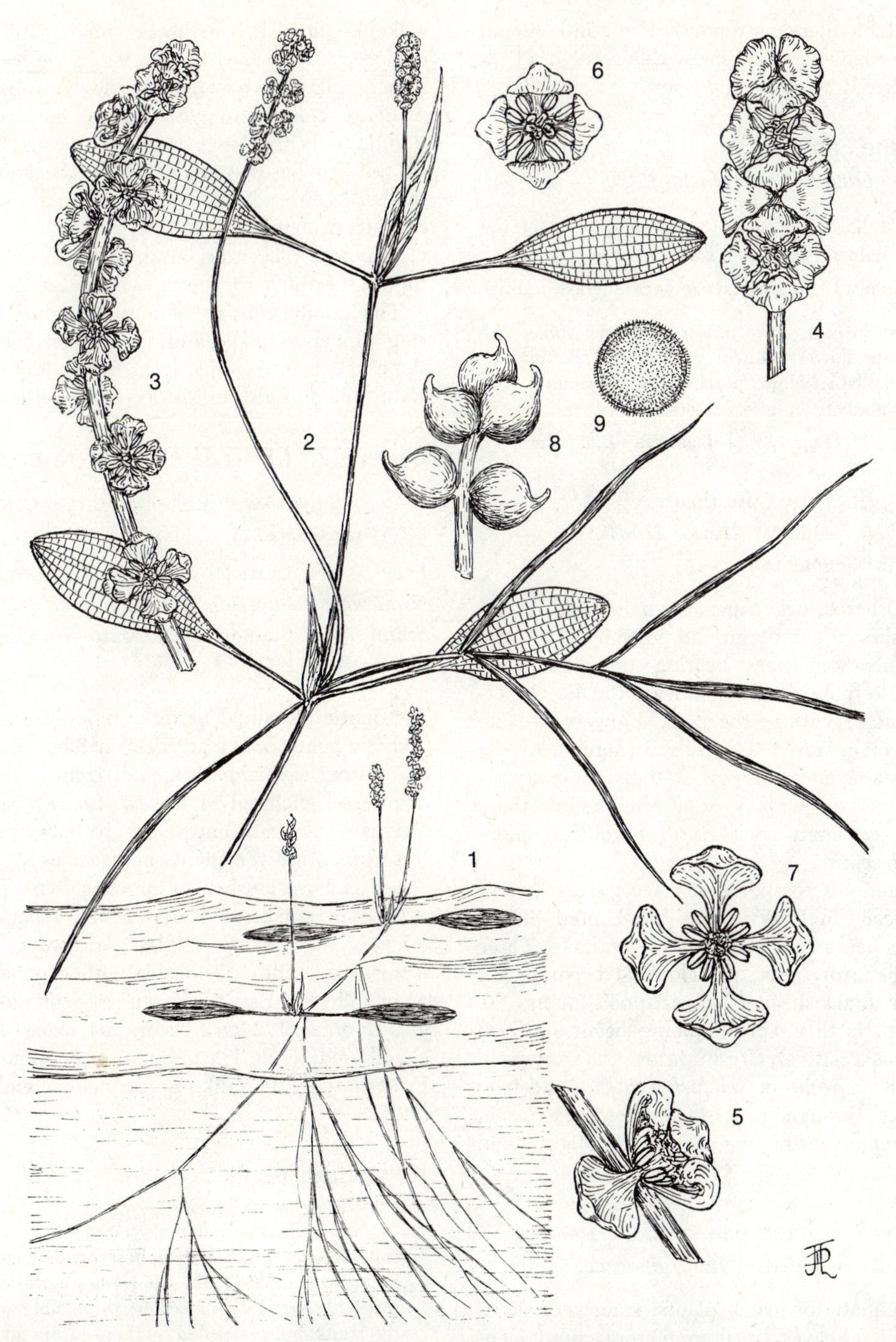

Figure 11-23. Naiadales; Zosteraceae; *Potomogeton diversifolius,* a pondweed: **1** habit, showing the floating leaves (dilated), the submersed leaves (linear), and the spikes of flowers, **2** plant, enlarged, **3–4** flowers on the spike, **5–7** flower, three views, showing the stamens, pistil, and sepal-like appendages of the connectives of the anthers, **8** young fruits, **9** seed.

the sepals, ovules anatropous. Fruit indehiscent and 1-seeded or sometimes dehiscent and 2-seeded; seeds without endosperm.

Key to the Families

(North America north of Mexico)

1. Flowers bisexual, racemose; perianth present but small; stamens usually 6, sometimes 3.

   Family 1. **Juncaginaceae** (arrow grass family).

1. Flowers in each spike unisexual or the spike with staminate flowers above, bisexual in the middle, and pistillate below; perianth none, stamen 1 in each staminate or bisexual flower.

   Family 2. **Lilaeaceae** (*Lilaea* family).

*Family 1. **Juncaginaceae.** ARROW GRASS FAMILY

(Synonym: *Scheuchzeriaceae*)

Perennial herbs; occurring in marshes and resembling rushes; the underground portion of the stem a rhizome, sometimes bearing tuberous roots. Leaves alternate or basal, linear, the basal portion of each sheathing the stem. Flowers bisexual or unisexual; sepals 3; petals 3, similar to the sepals; stamens 6; carpels 3–6, each ovary 1-chambered and with 1 or a few ovules, these basal, erect, anatropous. Fruit a follicle; endosperm present.

The family is composed of three genera, two of which occur in Canada and the United States, the other being restricted to the Antarctic. Three species of arrow grass (*Triglochin*) occur in sea inlets or brackish marshes or pools along the coasts of North America. Some occur in fresh or at least relatively fresh water. An American variety of a species of *Scheuchzeria* is found here and there throughout the Continent, except in the extreme north, the Southern States, and Mexico.

*Family 2. **Lilaeaceae.** LILAEA FAMILY

(Sometimes included in *Juncaginaceae*)

Annual aquatic or marsh plants; stem very short, a rhizome. Leaves basal, numerous, fundamentally alternate, terete, spongy, the basal portion of each sheathing the stem. Flowers unisexual, in a spikelike raceme, the lower spike pistillate and enclosed by the sheathing bases of the leaves, the upper spike, with staminate flowers above and pistillate below and with perfect flowers in the middle, only the lower pistillate flowers sheathed by the leaf bases; perianth none, the flower subtended by a bract or possibly a perianth segment; carpels probably 3, the style 1, the ovary 1-chambered; placentae basal, ovule 1, erect, anatropous. Fruit a caryopsis, winged or ribbed.

The family consists of a single species occurring in vernal pools and rivulets on the Pacific slope of North America from British Columbia to California and also in Mexico and South America.

Order 12. ***TYPHALES.*** CAT-TAIL ORDER

DEPARTURE FROM THE FLORAL CHARACTERS OF THE ALISMALES

Free	Carpel 1 or carpels 2 and coalescent.
Numerous	Stamens 3–5 or more.
Spiral	Stamens cyclic; carpels cyclic or solitary.

Aquatic perennial herbs commonly growing in lakes or ponds or the adjacent marshes or ditches; main stem a rhizome or erect from a short rhizome and enclosed in leaves. Leaves basal and alternate, elongate and narrowly linear. Flowers unisexual (and the plants monoecious), either in dense globose clusters or in very dense elongate cylindroid spikes, the upper inflorescences staminate, the lower pistillate; perianth of 3–6 elongate membranous scales with expanded tips or of filiform bristles; stamens 3–5 or more; carpels 1 or 2, if 2 coalescent, the ovary 1- or 2-chambered; ovule 1, pendulous or basal and erect. Fruit hard and nutlike or drupelike; endosperm mealy.

Key to the Families

(North America north of Mexico)

1. Inflorescence of several globose heads (each up to 2 cm. in diameter), the upper staminate, the lower pistillate; perianth of scales, these not concealing the fruits, but concealed by them (from an external view of the head).

   Family 1. **Sparganiaceae** (bur-reed family).

1. Inflorescence of two cylindroidal spikes one above the other, the upper staminate, the lower pistillate; perianth of bristles, these concealing the pistils and later the fruits.

   Family 2. **Typhaceae** (cat-tail family).

*Family 1. Sparganiaceae.
BUR-REED FAMILY

Aquatic perennial herbs; main stem a rhizome, the leafy flowering stem derived from it. Leaves alternate, linear, basally sheathing the stem, in two ranks. Flowers unisexual (the plants being monoecious), in panicles composed of dense globose clusters of sessile flowers, the staminate in the upper clusters, the pistillate in the lower; perianth composed of 3–6 elongated membranous scales, these dilated at the tips; stamens 3 or more; carpels 2, the style branched above, the ovary 1–2-chambered; ovule 1 in each chamber of the ovary, either pendulous or basal. Fruit hard and nutlike or drupelike (the outer layer spongy and the inner bony), indehiscent; endosperm abundant, mealy.

The family includes a single genus, *Sparganium,* occurring in cool and temperate regions of the Northern and Southern hemispheres. About 10 species occur in North America, most of them in the eastern part of the continent. About five grow as far westward as the Rocky Mountain System; four reach Washington and Oregon and three California.

*Family 2. Typhaceae. CAT-TAIL FAMILY

Perennial herbs; plants growing in marshes and ditches; main stem a rhizome, the leafy flowering stem arising from it. Leaves alternate and basal, linear, elongated, without petioles, standing erect. Flowers unisexual (and the plants monoecious), in 2 elongate, cylindroidal spikes, the upper spike staminate and the lower pistillate; perianth composed of filiform bristles; stamens 2–5, the filaments more or less coalescent, with long, slender hairs, carpel 1, the style filiform, the ovary on a stipe bearing slender hairs; ovule 1, pendulous. Fruit a nutlet, minute, the style persistent; endosperm mealy.

The family includes a single genus, *Typha,* the occurrence of which is world-wide except for the colder regions. Four taxa occur in North America. Recently these have been accorded specific rank, but this may be a matter of individual opinion. All are wide-ranging, but one extends through only about the southern half of the United States.

Order 13. *PANDANALES.*
SCREW-PINE ORDER

DEPARTURE FROM THE FLORAL CHARACTERS OF THE ALISMALES

Free	Stamens sometimes coalescent; carpels coalescent or confluent in fascicles and coalescent by at least the stigmas.
Numerous	Carpels several to numerous.

Trees, shrubs, or sometimes vines; plants sometimes with aërial prop roots supporting the trunk, these produced from the branches as well as the main trunk. Leaves alternate, in dense tufts at the ends of the branches or of the main stem, linear, sessile, the bases sheathing, the blades tough and fibrous and with strong keel-like midribs, these commonly spiny. Flowers unisexual (the plant dioecious), in dense panicles; perianth none or rudimentary; staminate flowers in a dense mass, but the individual flowers difficult to distinguish, borne on an axis resembling a spadix, the filaments coalescent or not; carpels several to numerous, or confluent in fascicles and coalescent by at least the stigmas. Fruit a syncarp (that is, formed from the coalescence of numerous sessile ovaries of adjacent flowers), each segment a berry, a drupe, or a woody structure with pulp inside; endosperm fleshy.

Family 1. Pandanaceae.
SCREW-PINE FAMILY

The family is composed of three genera occurring through the Tropics of the Eastern Hemisphere, including the Pacific islands as far eastward as Hawaii. The principal genus is *Pandanus* (screw-pine), species of which are cultivated out-of-doors in Southern Florida and similar warm areas and under glass northward.

The leaves are used for clothing, matting, thatches, and baskets. The fruits are edible.

Figure 11-24. Typhales; Sparganiaceae; *Sparganium eurycarpum,* a bur-reed: **1** portion of a leaf, **2** bract and fruiting portion of the inflorescence, **3** inflorescence, the upper flower clusters staminate, the lower pistillate, **4** stamen, **5** stamen releasing pollen, **6** pollen grains, **7** pistillate head, section, **8** pistil, **9** fruiting head, section, **10** fruit, external view, **11–12** longitudinal sections.

Order 14. *ARALES.* CALLA ORDER

DEPARTURE FROM THE FLORAL CHARACTERS OF THE ALISMALES

Numerous Stamens often 1–8; carpels usually few.

Spiral Stamens and carpels cyclic or solitary.

Herbs (commonly of marshes) or tropical shrubs or large treelike plants. Leaves commonly simple but sometimes compound; in the Lemnaceae the main plant body *not* obviously differentiated into a stem and leaves. Flower subtended by a spathe (that is, a bract which is commonly large, conspicuous, and either white or colored but sometimes green), the bract enclosing the entire inflorescence or the solitary flower, at least in the early stages of development. Flowers or some of them unisexual, small or minute, sometimes composed of a single stamen or a single carpel; perianth usually none, but sometimes composed of 6 segments (sepals and petals); carpels separate and solitary, sometimes coalescent in the *Araceae,* the ovaries sometimes many-chambered; ovules few to numerous, of varying types and attached in various ways. Fruit a berry, a follicle, or a utricle; seeds with or without endosperm.

The most outstanding feature of the order is the spathe which usually is large, conspicuous, and either colored or white.

Key to the Families

(North America north of Mexico)

1. Plants with well-differentiated stems and leaves, terrestrial or palustrine (in bogs or marshes) or rarely aquatic and floating; inflorescence a spadix (fleshy spike), almost always with a spathe formed by a single white or colored but usually nongreen bract enclosing the entire inflorescence at least when it is young.

Family 2. **Araceae** (*Calla* family).

1. Plants without well-differentiated and apparent stems and leaves, the plant body thalluslike, small, circular or elongated, aquatic and floating except when stranded on mud; plants monoecious, the flowers produced singly or a few together in a sac-like spathe on the margin or upper surface of the thallus.

Family 3. **Lemnaceae** (duckweed family).

Family 1. **Cyclanthaceae.** PANAMA-HAT-PALM FAMILY

Perennial herbs or somewhat woody climbing vines. Leaves simple and entire or palmate, fan-like and resembling palm leaves, sometimes with two primary divisions and these often again much divided. Flowers unisexual (and the plants monoecious), radially symmetrical, hypogynous, the spadix enclosed at first by two relatively conspicuous spathes, these deciduous; pistillate flowers either surrounded by staminate or the two in alternate spirals; perianth in the staminate flower either none or forming a tube or a cup, in the pistillate flower either none or of 4 segments, these either coalescent or separate, sometimes leathery in fruit, those of an adjacent flower sometimes adnate; stamens numerous; carpels 2–4, the style short or wanting, the ovary 1-chambered, inferior or appearing so, often imbedded in the spadix; placentae parietal; ovules numerous, anatropous. Fruit a berry, the adjacent berries often adnate each other and with the spadix; seed coat fleshy at maturity; endosperm abundant.

The family occurs in tropical America. The genus *Carludovica* includes the plant which yields the fiber of Panama hats, and it is known as the Panama-hat-palm. Sometimes this plant is cultivated.

*Family 2. **Araceae.** ARUM FAMILY

Terrestrial or rarely aquatic herbs; stems either rhizomes or tubers, producing either leafy flowering stems or scapes; juice sometimes pungent or milky. Leaves simple or compound, all basal, or the cauline alternate, petioled, the bases sheathing. Flowers bisexual or unisexual (and then the plants monoecious or rarely dioecious), radially symmetrical, hypogynous, small; perianth present in bisexual flowers, usually absent in unisexual ones, consisting of 4–6 scalelike members; stamens 1, 2, 4, or 8; carpels 1–many, the style 1, the ovary 1–many-chambered; placentae axile, parietal, apical, or basal; ovules 1–many. Fruit a berry; endosperm usually present.

Figure 11-25. Arales; Araceae; *Zantedeschia aethiopica,* the garden white calla: **1** basal leaf, **2** spathe with the tip of the spadix protruding, **3** one half of the white spathe removed, showing the spadix, which bears staminate flowers above and perfect below, **4** spadix enlarged, **5** crowded perfect flowers from the lower part of the spadix, stamens surrounding each pistil, **6** pollen grains, **7** diagrammatic longitudinal section of the spadix, **8** perfect flower, stamens removed, showing the pistil and the perianth of scales,

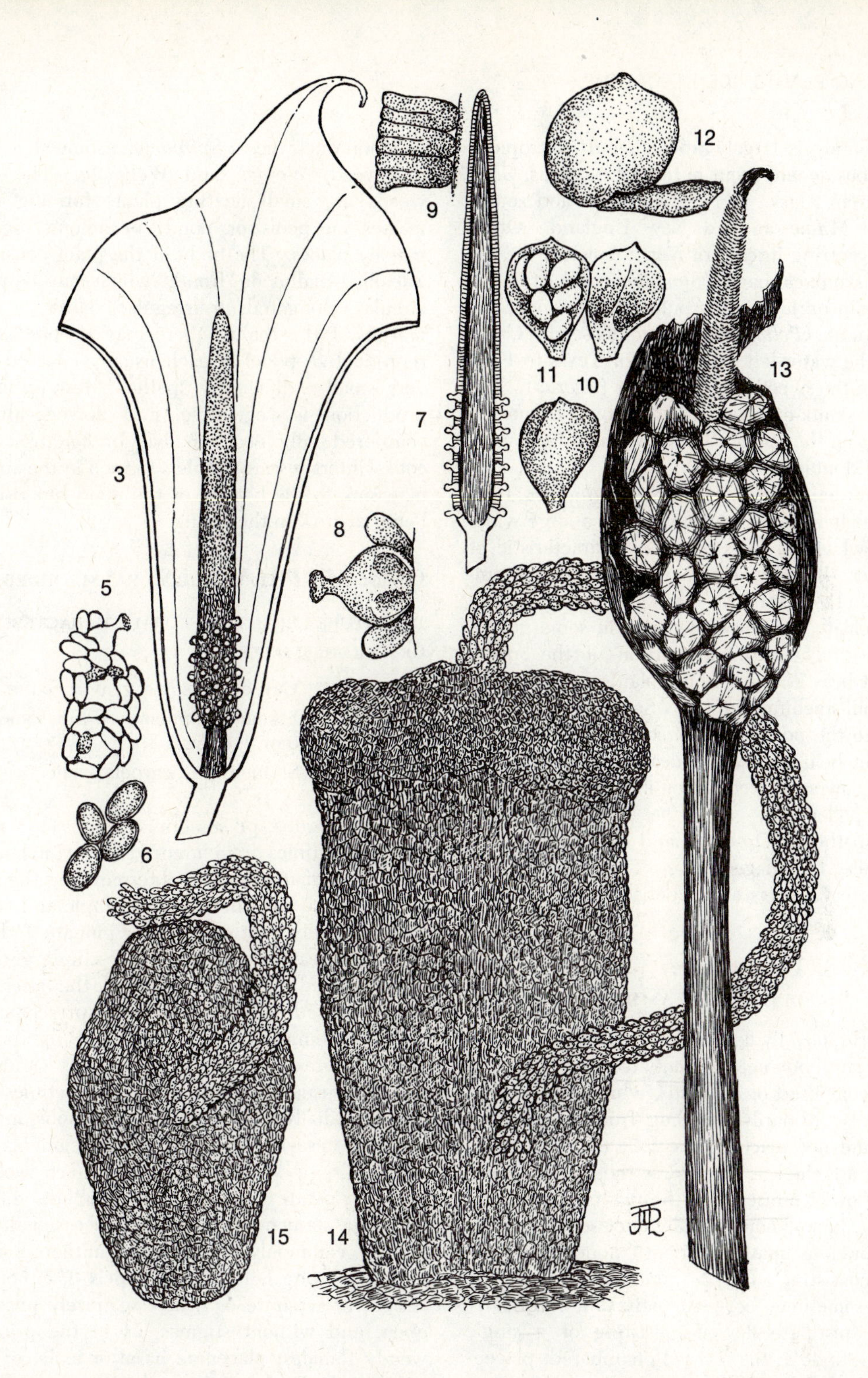

9 stamens, upper portion of the spadix, the staminate flowers with no perianth and crowded together, **10** young fruit, **11** this laid open, showing the developing seeds, **12** fruit approaching maturity, **13** spadix at fruiting time, **14–15** views of the anther (dorsal and top) showing the extrusion of the long filament of sticky pollen grains.

The family is largely tropical and subtropical, but various genera occur in temperate areas. *Calla* occurs from Alaska to Newfoundland and southward to Minnesota and New England. Other aroids occurring in eastern North America are the eastern skunk-cabbage (*Symplocarpus*), the Indian-turnip or Jack-in-the-pulpit (*Arisaema*), the arrow arums, (*Peltandra*), the golden-club (*Orontium*), the water-lettuce (*Pistia*, of Texas to Florida), and the sweet-flag or calamus (*Acorus*). The western skunk-cabbage (*Lysichiton*) occurs in lowlands on the Pacific Coast and inland to northwestern Montana.

The garden calla, *Zantedeschia aethiopica*, typical of the family, is characterized by a large white spathe, which encloses a spadix characteristic of the order. The flowers may be somewhat embedded in the axis of the spadix; the upper usually are staminate, the lower perfect (in some genera pistillate). The terminal portion of the spadix does not bear flowers. Many plants in the family have foul-smelling flowers. Some bear resemblance to the odor of a skunk, others to carrion, one plant being known as dead-horse-lily. Many of these flowers attract carrion insects which transport the pollen.

The spathes in *Orontium* and *Acorus* are green and similar to foliage leaves. These genera also have perfect flowers and a perianth of six sepals.

*Family 3. **Lemnaceae.** DUCKWEED FAMILY

Tiny herbs, usually floating on the surface of quiet water; plant body not differentiated into stems and leaves, composed of a thallus, which buds asexually; roots ordinarily dangling from the floating plants and not reaching the soil; thalli often remaining attached in "chains" as the result of budding. Flowers unisexual, (the plants being monoecious), hypogynous, usually enclosed at first by a membranous spathe; perianth none; staminate flower consisting of 1 or rarely 2 stamens, the flowers sometimes borne in pairs but frequently solitary; pistillate flower consisting of a single pistil, the style 1, the ovary 1-chambered; placentae basal; ovules 1–7. Fruit a utricle; seed without endosperm.

The family is nearly cosmopolitan. Four genera occur in North America. These are *Lemna*, the common duckweeds; *Spirodela*, a somewhat larger duckweed; *Wolffia;* and *Wolffiella*. The duckweeds are small floating plants forming dense masses on pools or ponds or among reeds in marshy places. The body of the plant consists of a leaflike thallus or "frond," which may be nearly circular, elongated, or irregular. Near the basal margin of the thallus there may be one or two reproductive pouches each usually covered by a very small rudimentary spathe. Most of the reproduction is vegetative, and flowers are encountered only occasionally. In countries with cold winters resting bulblets formed in the autumn may sink to the bottom of the pond but rise and begin to grow in the spring.

Order 15. *PALMALES.* PALM ORDER

DEPARTURE FROM THE FLORAL CHARACTERS OF THE ALISMALES

Free	Carpels of some genera coalescent.
Numerous	Stamens commonly 6 but sometimes 3 or 9; carpels 3.
Spiral	Stamens and carpels cyclic.

Trees, shrubs, or sometimes vines (the rattan palms sometimes up to hundreds of feet in length). Leaves alternate, in clusters terminating the stems of the arboreous genera, *either* simple and fanlike and palmately cleft or lobed *or* pinnate and with numerous leaflets (*or* sometimes simple and pinnately veined); petioles elongated, the bases commonly dilated, a ligule sometimes present at the point of joining of the blade and the petiole. Flowers bisexual or usually unisexual (and then the plant usually monoecious but sometimes dioecious), radially symmetrical, hypogynous, in panicles, inflorescence or each of its segments subtended by a spathe (this sometimes woody); sepals 3; petals 3; the perianth segments small or scalelike; stamens usually 6 but occasionally numerous, commonly in two series, anthers 2-chambered, splitting lengthwise; carpels 3, either coalescent or separate or nearly so, rarely unsealed, open, and without stigmas, as in the primitive woody Ranales; placentae basal or axile; ovule 1 per carpel. Fruit usually a drupe but sometimes a nut, the exocarp usually fleshy or leathery but sometimes fibrous (as in the fruit of the coconut); endosperm present.

Figure 11-26. Arales; Lemnaceae; **1–2** *Spirodela polyrhiza;* **3–7** *Lemna minor*, the body consisting of a thallus and roots, the root caps evident, **4** plant with flowers (the ones enlarged in **5**) right, staminate (a single stamen), left, pistillate (a single pistil), the flowers in a saclike spathe, **6–7** stamen opening and releasing the pollen. The entire plant is less than one-quarter inch long.

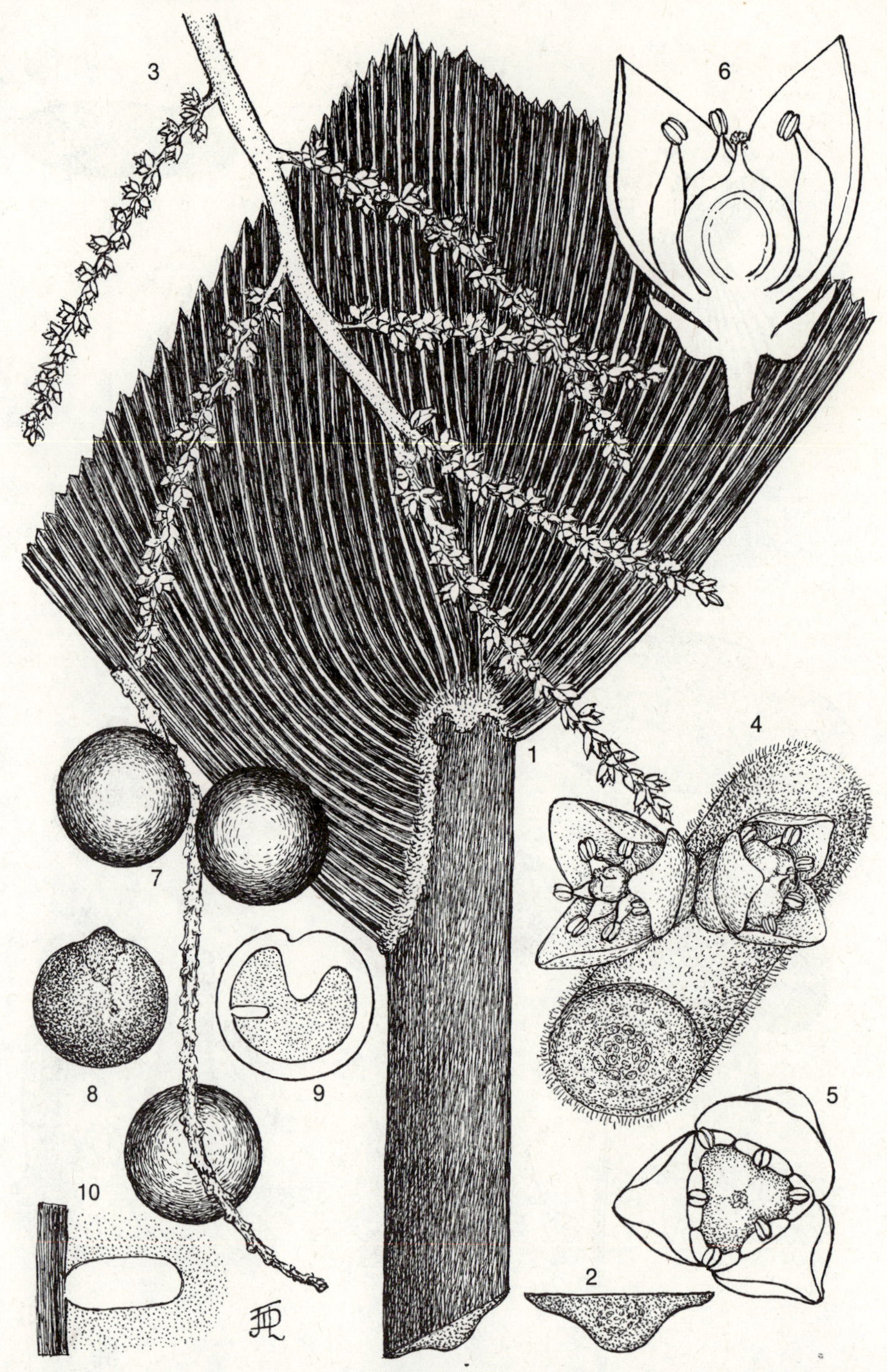

Figure 11-27. Palmales; Palmae; *Erythea edulis,* Guadalupe palm: **1** attachment of the petiole and blade of the leaf, **2** cross section of the adjacent petiole, **3** a few branchlets of the inflorescence, **4** flowers, **5** flower from above, showing the petals, stamens, and pistil (sepals shown in **6**), **6** flower, longitudinal section, **7** fruits, **8** fruit from below, **9** cross section, showing the position of the embryo (at the left), **10** this area enlarged, the endosperm surrounding the embryo large, hard (in a date forming the stony "pit"). Cf. **Figure 25-43,** *above.*

The woody trunks of the trees are a diagnostic character of the order—a feature duplicated in few other monocotyledons. Woody plants of other orders include the arborescent to arboreous yuccas (including the Joshua-tree), the dragon-tree (*Dracaena*), the tropical families *Velloziaceae* and *Pandanaceae*, and a few of the larger grasses such as the bamboos.

The structure of most palm flowers does not depart markedly from that in the hypogynous Liliales, and the distinguishing features of the group are based upon vegetative, inflorescence, and fruiting characters. In contrast to the Liliales, (1) the flowers are unisexual, the plants being either monoecious or dioecious; (2) the perianth parts are small and scalelike as opposed to usually though not always large and showy; (3) the 3 carpels may be separate (often only 1 or 2 developing as pistils, the rest abortive). As in the Arales, the inflorescence (or in this case each of its principal segments) is subtended by a spathe, that is, a large bract which at first encloses the entire unit. Usually the spathe is not green like ordinary leaves but colored—in the palms usually yellow or yellowish-brown. In many palms the inflorescence is compound and greatly elongated with a spathe below each portion of the panicle.

The endosperm of the seed is abundant and either oily or bony. Sometimes it is exceedingly hard as in the common date or in the palm which is the source of "vegetable ivory." The embryo is imbedded in one side of the much larger endosperm of the solitary seed.

A rough popular classification of palms is the following: "fan palms" have either palmate or more frequently palmately lobed, parted, or cleft leaves; "feather palms" have pinnate leaves. This mode of classification is nontechnical, and it cannot be carried very far.

Family 1. **Palmae.** PALM FAMILY

(Synonyms: *Arecaceae*, *Phoenicaceae*)

The family is one of the largest among flowering plants, but an accurate estimate of the number of species is difficult. It is distributed through and near the Tropics and Subtropics. A single species of native fan palm (*Washingtonia filifera*) occurs in isolated canyons on the edge of the Colorado desert in southeastern California and in the Kofa Mountains of western Arizona. A similar palm in Baja California and Sonora is the tall slender form often cultivated. Other genera and species of palms occur along the Gulf Coast, on the southern Atlantic Coastal Plain, and especially in southern Florida. The coconut palm (*Cocos*), a native of the Malayan and South Pacific regions, is naturalized widely in the Tropics and Subtropics, including peninsular Florida and the Keys. The date palm (*Phoenix*) is naturalized in some areas of southern Florida. It is native in North Africa and Arabia. The royal palm (*Roystonea*) occurs in Florida and the West Indies. Species of palmetto (*Sabal*) occur from southeastern Texas to Florida and the Carolinas. The best known is the cabbage palmetto (Florida to the islands off the coast of North Carolina). The Seminole Indians built their villages with the trunks and leaves of these trees, and the young part of the bud is edible. The saw palmetto (*Serenoa*) occurs from Louisiana to South Carolina and Florida. This plant has edible fruits. The petioles have particularly strong spines. The saw-cabbage palm or Cuban palm (*Paurotis*) and the needle palm or blue palmetto (*Rhapidophyllum*) occur in Florida, the needle palm having a broader range from Mississippi to South Carolina as well. *Thrinax*, *Pseudothrinax*, and *Pseudophoenix* occur in Florida.

Flowering Plants

Part Three

PREPARATION AND PRESERVATION OF PLANT SPECIMENS

12

Preparation and Preservation of Plant Specimens

The Value of a Student Plant Collection

For the beginner living plants are easier to identify than pressed specimens because the flower is not flattened. Pressing may distort the structural pattern and obscure the chief characters differentiating orders and families.

With experience the student becomes so familiar with the floral characters of the orders and families that he can detect their presence or absence readily even in dried specimens. Often the characters used in classifying genera and species are seen as easily in dried as in living material because they rarely include the fundamental pattern of the flower. For the advanced student it is efficient to accumulate large numbers of pressed specimens, to sort them into orders, families, and genera, and then to identify all the available species of one genus at a time.

Accuracy in identification is increased by study of genera one by one because comparison of specimens reduces errors. The key is easier to use because one plant may correspond to one lead while another represents the opposed lead. If four of the five species of a particular genus described in a manual are represented in the collection, the chances of error in identification are much less than in determination of a single species.

Speed in identification is acquired chiefly by making "flock shots." With the specimens organized into related groups, a student may identify a dozen or more species in a genus during an afternoon laboratory period instead of carrying two to four unrelated plants individually through the keys to the classes, orders, families, genera, and species.

A well-kept collection is a fundamental aid to learning the best method of identification. As the beginner's collection grows and as he accumulates experience, he may change over gradually to a comparative method of study, working on one genus at a time and integrating new specimens with those collected earlier.

Selecting Material for Specimens

Few situations can be more frustrating to a botanical student than to have only a fragment of a plant consisting of perhaps a single flower and a leaf from the top of the stem. If the flower of such a "specimen" is dissected, nothing is left for a permanent record or later study; if it is not dissected, the key cannot be followed. The roots, stems, lower leaves, and fruits cannot be studied.

If possible, a specimen should include the entire plant.* Obtaining the roots and any associated tubers or rhizomes is of the utmost importance. The root system of an herb is useful for its own characters, and almost inevitably a specimen without roots lacks also the basal leaves, which commonly differ from those on the stem. The entire stem is significant because often the upper and lower parts differ in degree or type of hairiness or other surface characters, and the upper cauline leaves and bracts may be unlike the lower leaves. Although the leaves and flowers usually are more conspicuous, the fruits are indispensable to the use of the keys, and care should be taken to find them. Fruits are not necessarily edible fleshy structures, and they are likely to be overlooked unless the development of the flower after pollination is followed through.

Few would attempt to press a pine tree, but its essential characters may be represented in specimens by leafy branches, seed-producing and pollen-producing cones, and bark. Specimens of a typical woody flowering plant, such as an oak tree, should include branches with representives of all types of leaves, together with acorns (when collected in the fall) or both kinds of flowers (when collected in the spring), and bark showing the distinctive pattern for the species.

Specimens of large succulent plants, such as century-plants, should include at least one of the largest basal leaves, showing every part from the broad base to the spine at the tip, even though this may require several herbarium sheets or several sections mounted side by side on the same sheet. If possible, specimens should include both flowers and fruits together with representative sections of the flowering stem. The smaller species of century-plants may be split lengthwise so that the specimen will include half the basal leaf rosette as well as the reproductive parts.

Specimens of small cacti may be made from all or half of each plant, while those from larger ones should include representative portions of the stem. For example, several joints of a prickly pear or a cholla are necessary, and these should be in various developmental stages. For a large cactus a stem section showing several ribs is needed. This should exhibit all types of spines, taking into account the variability at different levels on the plant. If possible, the angles of divergence of the spines from the stem should be preserved. When it is practical, the root system of the plant should be shown also. The flowers and fruits, as always, are of great importance.

If specimens are to be used ultimately for research purposes, particular care should be taken to select individuals representative of all the phases of the natural population of which they are a part. A species or variety may be either a somewhat homogeneous or a relatively variable population or system of populations of genetically related individuals. In order to define such populations a research student needs specimens of as many types of individuals as can be made available. This gives him clues to follow up in the field as he examines the plants of many hillsides and valleys to determine whether a segregated natural species or variety actually exists or whether, instead, individuals on the herbarium sheets represent only one of many associations of hereditary factors in a highly variable population, just as

* Notes may form a part of the "specimen" also. These should cover data not included in a pressed specimen or characters not persisting after pressing (cf. p. 427). The object is to record complete data concerning the plant.

Figure 12-1. Tree specimens: an oak (*left*) and a conifer (*right*). Larger fruits or cones and thick bark often are filed separately. Photograph by Frank B. Salisbury.

people with red curly hair and blue eyes represent only one human genetic combination among many.

Plant-Collecting Tools, Equipment, and Methods

Various tools may be useful in collecting specimens. The most important are digging and cutting instruments. For digging a shovel or pick is convenient to use but difficult to carry. A geology pick, a heavy screwdriver, a sharpened tire iron, an unusually strong trowel, or almost any other sharp instrument may suffice. An ordinary box- or crate-opener is recommended because it may be used for prying or breaking rocks as well as digging. It has the advantage of being flat, small, and easy to carry. A pair of pruning shears or a pocket or hunting knife is useful for collecting shrub and tree specimens. For some plants a machete is helpful, and in the Tropics it is indispensable. One collector recommends a large corn knife which may be used for either cutting or digging. For collecting cactus specimens a supply of paper bags, cardboard cartons, and a large knife and fork or a pair of tongs are recommended. For specialized purposes, such

Figure 12-2. Herbarium specimens of succulent plants. The prickly pear stems (*Opuntia phaeacantha* var. *major*, a large prickly pear) (*left*) were split, scraped on the cut surfaces, salted, and pressed; the flowers and fruits were split, salted, and pressed separately, and they were mounted some with the cut surface and some with the outside up. The specimen includes a full representation of the visible genetic characters of the plant. It may be compared with hundreds of other genetic combinations of other plants, and this indicates the presence of problems of genetics, classification, evolution, and ecology to be studied in the field. The overall pattern of characters can be plotted in the herbarium but further correlations must be sought in the living natural populations. The plant with succulent leaves, *Dudleya* (*right*), was split lengthwise through the caudex and dried by salting. Leaves (and flowering stems) of this type may be slashed through the epidermis at various points and salted before pressing. Many changes of newspapers, cardboards, and blotters may be necessary in drying the specimens. (Photograph at right by Frank B. Salisbury.)

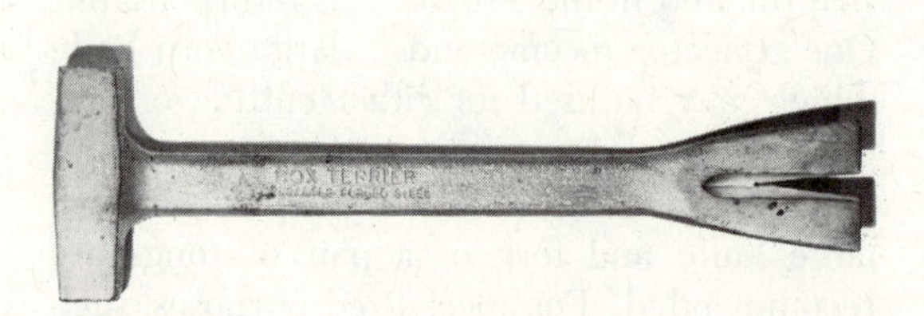

Figure 12-3. A digging and prying tool with many uses in the field. Photograph by Frank B. Salisbury.

as securing the cones from the top of a fir tree, a pair of pole-type pruning shears may be handy to use though difficult to carry.

It is best to press specimens in the field as soon as they are collected. However, if time is short, as on many class field trips, the plants must be brought indoors for pressing. The traditional specimen carrier is a special metal box called a vasculum. Most vascula are not large enough to hold adequate specimens, but good results may be obtained with large, specially built vascula. A medium-sized cardboard carton with newspaper dividers separating *thin layers* of plants is effective, especially if each layer is sprinkled lightly with water to maintain the humidity of the air surrounding the specimens. Very large paper sacks, preferably shopping bags, are carried more easily than cartons, and they are effective, provided not too many plants are put into each one. A plastic bag may be used instead of a paper bag or be carried inside a shopping bag. A light sprinkling with water keeps the specimens in good condition.

Plants which cannot be pressed in the field should be pressed immediately after being taken to the laboratory. Many species will remain in fairly good condition enclosed in water-soaked paper overnight, but some kinds of flowers may lose their petals or other parts if this procedure is followed. On the other hand, if the plants are arranged like a bouquet, some buds may open during the next day or two, replacing those that wither. However, not all flowers opening indoors are exactly like those developed under normal conditions outside. For example, the species of buttercups normally having the sepals **reflexed** (bent downward) may have spreading sepals if the flowers open in the laboratory. Color may differ, too.

Pressing and Drying Specimens

The chief factor limiting the quality of plant specimens is their preparation. Usually there is relatively little further change after plants are dried thoroughly, although some species, including the California poppy, lose their flower color in the course of several years, and all specimens lose their color in strong light. When the average plant is pressed, if the job is done well, the original color of both vegetative and floral parts is likely to be preserved or at least to follow a standard pattern of change. The plastid pigments, including the green (chlorophyll) in the stems and leaves of most plants or the many special colorings (particularly some reds or yellows) of flowers, are altered little by proper drying. These pigments are insoluble in water, and they are fixed in solid bodies (**plastids**) within the living cells. Ordinarily, extraction of water from the cell does not affect their color. On the other hand, dissolved pigments, such as the anthocyanins responsible for most coloring in the magenta series (ranging from lavender to purple), are altered by removal of the water or change of the pH (alkalinity or acidity) of the cell. Many of these pigments behave as does litmus paper, changing from red in an acid solution to blue in a basic (alkaline) medium. In herbarium specimens anthocyanins are usually nondescript blue even after the best pressing methods. The "secret" of retaining the original colors of the plant is rapid drying. Ideally, the average plant specimen should be dry within one or two days.

Drying of specimens is accomplished with a plant press and the associated equipment and materials listed below.

1. The press itself may be constructed in any of several ways, and it may range from two pieces of 1″ x 12″ lumber to a cabinetmaker's dream. Ordinarily, the two sections of the press should be from the size of standard herbarium mounting sheets (11½″ x 16½″) up to 12″ x 18″. Many presses are made as a latticework of hardwood, but the simplest and cheapest kind of press, and yet one which is effective, consists of two sheets of half-inch plywood. These and the enclosed materials are tied together into a neat bundle by

either straps or ropes. Window-sash cord is particularly good because it may be tied or untied quickly. A six- or seven-foot length of sash cord with a loop (preferably a bowline knot) at one end is used to tie each end of the press. The cords may be fastened to the outside (bottom) of the lower half of the press by long tacks, staples, or screws, with the loops extending slightly beyond one edge. When the press is to be tied, the ropes are slipped through the loops, cinched up, and tied. Proper pressure upon the plant specimens is secured by kneeling on top of the press as it is tied.

2. Driers, for example, cardboards and blotters, are cut to the dimensions of the press (e.g., 12″ x 18″). The cardboards may be of the corrugated type used for cardboard cartons. The corrugations should be coarse enough for ready passage of an air current, and they should run crosswise to make the air passages as short as possible. Some collectors prefer a cardboard with deep corrugations on only one side, that is, a corrugated board with one of the smooth outer surfaces left off. Probably the cheapest and best kind of effective blotter is cut from the rolls of deadening felt used in the construction of houses, but this material is becoming difficult to find because house construction procedures have changed.

Instead of blotters, foam sheets about ⅜ to ½ inch thick may be used. These are superior to the felts obtainable, and they both fit well around the specimen and permit transfer of moisture through their porous interiors. Any foam to be used should be tested cautiously to be certain it is not inflammable.

3. Old newspapers may be used to enclose the specimens. Standard-sized newspapers should be torn or cut into single page-sized sheets to make folders approximately the size of the press. The usual order of arrangement for driers and specimens in the press is as follows: cardboard, specimen in a newspaper folder, blotter, specimen in a newspaper folder, cardboard, etc. The cardboards and blotters alternate; the specimens in newspaper folders fill in the spaces between. Each specimen in newspaper, then, has a blotter on one side as an absorbent layer and a cardboard on the other as a layer presenting a smooth, flat surface and making aëration possible through its corrugations. Thick succulent specimens, like cacti, may require extra driers.

Although it may seem wasteful of space, **specimens of only one species should be placed on each newspaper sheet.** This practice works against the tendency to press small, inadequate specimens, and, as shown farther on, it is essential for recording data by specimen number.

Arrangement of the specimens for pressing will vary with the taste, judgment, and artistic sense of the individual. Usually it takes little more work to make attractive specimens than to make unattractive ones. Unnecessary overlapping of leaves or stems is unattractive; it should be avoided also because superimposed parts dry slowly and stick together. Often arrangement of parts is easier after the plants have been in the press for a short time and have lost a little of their initial crispness. At this time some leaves should be turned to expose the lower surfaces, others turned to expose the upper. Full use should be made of each specimen sheet, but the plants should not be crowded upon it. Although slender plants may be folded to resemble a letter "V", "N", or "M" and thus fitted on a single specimen sheet, tall broad ones should be cut into sections and the segments placed on different sheets. If there are open places in the newspaper folder it is well to fill these in with extra flowers and fruits to be dissected when the plant is being identified or compared with other specimens.

After the specimens are in the press, drying should be completed quickly. The most effective method, at least in the western United States where humidity is low, is to place the press in a cabinet through which an air current is drawn by an electric fan. The same results may be achieved by simply placing the press in front of a fan in such a way that the air current passes through the corrugations of the cardboards. Within limits, satisfactory results may be obtained in dry climates by placing the press out-of-doors where the prevailing winds will blow through the corrugations and the sun will warm the press. With this method it is necessary to change the driers every day or to dry them in the sun for brief periods. In

Figure 12-4. A plant press with driers (cardboards and blotters) shown (*above*) open with a plant about to be pressed in a newspaper folder and (*below*) closed. Photograph by Frank B. Salisbury.

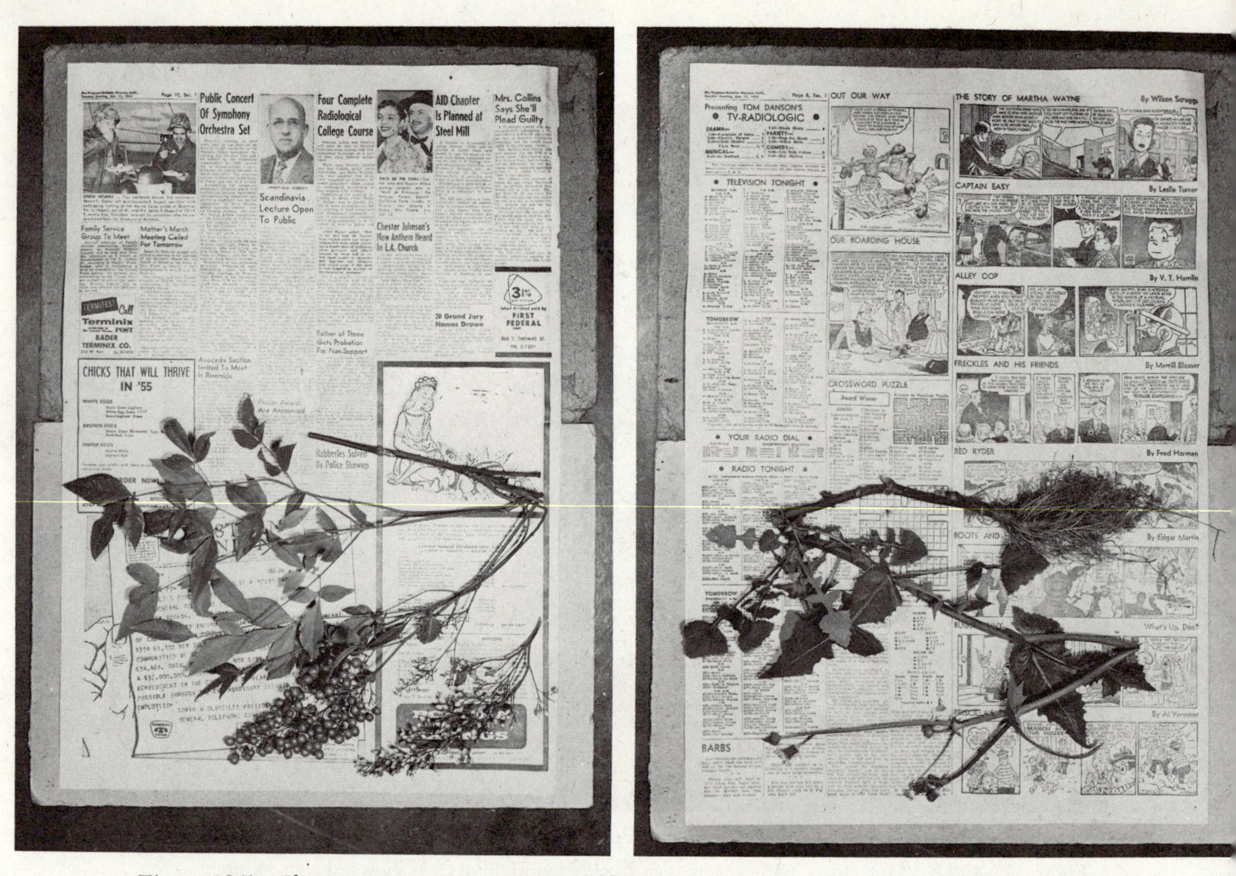

Figure 12-5. Plant specimens in newspaper folders ready for pressing between cardboards and blotters. The plant at the right is a sow-thistle, *Sonchus oleraceus*, that at the left is a Japanese shrub, *Nandina domestica*. Photographs by Frank B. Salisbury.

dry country, carrying the press in an exposed position on an automobile results in rapid drying and good specimens. Application of artificial heat to the specimen, particularly if the heat is accompanied by some air movement, is effective also, and in humid areas like the East and Middle West it is necessary. Drying plants on a steam radiator is fairly satisfactory, and there are many types of cabinets designed for drying by heat from small stoves or electric light bulbs. Perhaps the most efficient source of both heat and air movement is one of the now common type of small electric heaters with built-in fans. In the field, drying may be hastened by placing the presses on a rack of poles above lanterns or a small kerosene stove and surrounding them by a canvas skirt. Artificial heat from any source should be used only with caution because if the specimens are too hot they will became brittle and they may turn brown. Some plants have special beads of wax forming a "bloom" on the leaves, stems, or fruits. This feature may be lost if the specimens are heated above about 100° F.

Plants should not be removed from the press until they are thoroughly dry, but if artificial heat is used in drying, they should not be left in the press longer than necessary. Plants almost but not quite dry are slightly cool and moist to the touch. This coolness is recognized after a little experience. Plants still wet upon removal from the press may wrinkle, curl, or even mildew. Upon being taken from the press, the specimens are kept in the newspaper sheets until they can be mounted.

Succulent Plants. Drying succulent plants is a special process. Various methods are used by individual collectors. Unless the specimen is very bulky, preparation for an ordinary herbarium sheet (or a series of them) is more desirable than preparation for storage in a box or jar. An herbarium sheet may be filed with the others of its kind, and it is not so likely to be lost to sight on an obscure shelf apart from the main collection. Nevertheless, often it is necessary to file bulky specimens in boxes or jars because they cannot be adapted to herbarium sheets.

The less bulky cacti may be split with a knife and fork and hollowed out like dugout canoes. Flowers should be split lengthwise but not hollowed. After the cut surfaces have been covered heavily with salt and the material has

Figure 12-6. A plant-drying cabinet with an electric exhaust fan at the top and with a slit for entrance of air at the bottom. Electric light sockets allow for warming the air as it enters. This cabinet is designed for cardboards with the corrugations running lengthwise but it is preferable for them to run crosswise so that the air passages are shorter. In any case, they should parallel the flow of air. At the left, the same cabinet is shown closed. From LYMAN BENSON, *Torreya* 39: 74. 1939. Courtesy of the Torrey Botanical Club.

stood for several hours or perhaps overnight, large quantities of water may be poured or wiped off and so eliminated before pressing. Just as the well-salted specimen starts to curl it should be placed in the press and dried in the ordinary manner except for adding extra blotters and changing them each day. An alternative method is drying the specimens under the sun with an iron grating for a weight. This method is quicker and often more satisfactory. If the salt will be visible on the herbarium sheet, it may be removed from the dried specimen by brushing or by washing with water and redrying. Salt not only aids in removal of water from the specimen but also discourages growth of molds. It is effective only in dry climates because in regions of high humidity it may absorb moisture from the atmosphere. Instead of using salt, many collectors prefer to kill the specimens with formaldehyde, 25% alcohol, or quick freezing before drying. Another method is to immerse the plant in hot water (with some salt) for a minute or less.

Some cacti, or parts of them, cannot be dried by the usual method of pressing. Large segments, as, for example, several ribs from the stem of a barrel cactus, may be salted and then dried in the open air for preservation in boxes. Parts also may be preserved in 5 to 12% formaldehyde or various formalin-alcohol solutions (usually 30% to 50% alcohol and sometimes a small amount of glycerine). Some small- or medium-sized cacti may be hollowed out like a jack-o'-lantern (but through an off-center hole at the base) and then dried with the aid of salt.

Specimens of most other succulent plants are prepared more simply than those of the cacti. Members of the stonecrop family (Crassulaceae) are pressed effectively by the usual methods, provided either the plants are killed by freeze drying or application of alcohol or the epidermis of the thick portions of the stems and leaves is slashed and then salted heavily. This can be done most effectively by turning the slashed side of the specimen downward so that it rests on a layer of salt. The salt kills the plant by removing most of its water. Without use of salt or of some other method of killing the cells, many somewhat succulent plants and the bulbs of onions may continue to grow in the press or even on herbarium sheets. The bitterroot, *Lewisia rediviva,* was named because of this tendency. Usually it is desirable to change the driers each day, although, if a current of warm air passes through the corrugations of the cardboards, it may be necessary only to check up now and then. Usually a few extra blotters are necessary. If a specimen is particularly bulky, it should be split lengthwise from the base upward to form two specimens or split a little off-center so that only one is formed. The cut surfaces should be salted heavily.

Water Plants. Many aquatic flowering plants require particular care in pressing. They may be collected in a large, flat, white, dissecting pan and kept in water until they are lifted carefully onto the pressing papers or placed between two sheets of muslin, from which they separate readily after drying. Paper or (better) cloth may be slipped under the smaller, more delicate species while they are floating. This material, carrying the plants, is placed in the newspaper folder used for pressing. After a time, the driers must be changed to remove the water carried from the pan with the paper and specimens. Some aquatics must be loosened from paper occasionally to prevent permanent adhesion, or if the paper is clean and without printing, some species may be allowed to adhere to it.

Tropical Plants. In the Tropics high humidity makes the drying of specimens much more difficult than in temperate regions. Artificial heat is necessary to dry the plant and to prevent molding, and constant changing of driers is required. Corrugated cardboard may soften and flatten out, becoming ineffective for

Figure 12-7. A method of drying plant specimens in the field. The presses are placed above lanterns (for heat) and then surrounded by a canvas skirt. From C. L. LUNDELL, Wrightia 1: 161–162. fig. 4–5. 1946. Plate supplied through the courtesy of Southern Methodist University and of Dr. Lundell, Texas Research Foundation.

ventilation. Woven frames of split bamboo or corrugates of metal (usually aluminum) are substituted for cardboards.

Spruces and Hemlocks. These plants lose their leaves while in the press because an abscission layer at the base of each leaf proceeds rapidly with its development while the specimen is in the press. According to Little, placing the specimens in boiling water for one minute or more before pressing is sufficient to kill the cells and prevent dropping of the leaves. Because the glaucousness of the leaves may be removed in the boiling process a twig should be pressed also in the usual way, be specially labelled, and be placed in a fragment packet or envelope.

Recording Data

The data for specimen labels are recorded first in a field notebook or "field book" when the plants are pressed. This eliminates the need for writing any more than a serial number on the margin of each newspaper folder. The numbering system will be referred to again below with the other data required for the label. Figure 12-8 shows a sample of three pages from a field book.

The data for the specimen labels need be entered in the field book only once for each locality where specimens are gathered on a given date. The serial numbers applying to the plants collected at a particular time and place are listed beneath the data, and the blank line opposite each number is a convenient place to enter the name and author reference(s) for the plant as it is identified. Thus all the data needed for writing or typing the labels* are recorded in the field book, and the labels may be prepared without opening the

* Since ordinary fountain-pen ink may fade out in the course of less than a century and since specimens should last indefinitely, all labels are prepared either with typewriter or with a carbon suspension ink, such as India or eternal ink. The labels should be of all-rag (linen) paper because other papers discolor or disintegrate in time. Ball-point pen marks seem to endure.

Colorado, 1950

Montrose County. South
rim of the Black Canyon
of the Gunnison River.
8,000 feet. Colorado
River Drainage. North
slope. Oct. 29.
14744 Mammillaria
vivipara Haw.
Variety or form known
as "M. radiosa".
14745 Opuntia
fragilis (Nutt.) Haw.
Delta County. Delta,
North fork of the
Gunnison River. 5000 feet.
Colorado River Drainage
North slope. Oct 29.
14746 Echinocactus.
Topotype of "Sclerocactus
Franklinii".

Colorado, 1950

San Juan County. South
of the second summit
on the road from Ouray
to Silverton, U.S. 550.
10,000 feet. Animas
River and Colorado
River Drainage. Rocky
Mountain Subalpine
Forest. South slope. Oct. 30.
14747 Ranunculus
inamoenus Greene
Montezuma County.
McElmo Canyon 3
miles west of Cortez
on State Road No. 146.
5,700 feet. Colorado River
Drainage. Light tan
sandstone. Juniper-
Pinyon Woodland.

Colorado, 1950

South slope. Oct. 31.
14748 Opuntia erinacea
Engelm. & Bigel. var.
hystricina (Engelm. &
Bigel.) L. Benson.
14749-50 Same (other forms)
Data above but 7 miles
west of Cortez; red
sandstone.
14751 Echinocereus tri-
glochidiatus Engelm.
var. melanacanthus
(Engelm.) L. Benson.
14752 Opuntia erinacea
Engelm. & Bigel. var. hystricina
(Engelm. & Bigel.) L. Benson
14753-4 Same (other forms)
14755 Echinocactus Whipplei
Engelm. & Bigel.

Figure 12-8. Examples of pages from a field book, illustrating the recording of data for use on the labels of numbered plant specimens. The names applied to plants do not necessarily remain constant; they may be altered by research changing the limits or definition of particular genera, species, or varieties. Since 1950 several changes have occurred in preparation of *The Cacti of the United States and Canada* (Stanford University Press, in press), and names appearing on the field book pages are altered as follows: *Mammillaria vivipara* to *Coryphantha vivipara* var. *arizonica; Echinocactus* (as applied to no. *14746*) to *Sclerocactus glaucus; Opuntia erinacea* var. *hystricina,* indicated for the variable forms *14748–14750* and *14752–14754,* is reinterpreted as a hybrid swarm of *Opuntia erinacea* var. *utahensis* and *Opuntia phaeacantha* var. *phaeacantha; Echinocactus Whipplei* to *Sclerocactus parviflorus* var. *intermedius.* The field book and the label on the plant specimens still reflect the basic data, and only the identification needs to be changed, usually by adding a special small identification label.

bundle of plants. Full data should be recorded except that those supplied readily from a map may be omitted from the book provided the key information needed in reference to the map is recorded.

The selection of data to be recorded depends largely upon the intended use for the specimen. An herbarium may fulfill any of several functions. It may be the hobby of an individual, a reference collection for checking the identity of plants, a selection of material for teaching, or a research collection. The type of label varies accordingly. For *his own immediate purposes* an amateur botanist may need no more than the name of a familiar town, hill, lake, or ditch. A reference collection for use of forest rangers in checking the identity of local plants may require special information adapted to forest or range management. A collection designed for teaching and student reference may differ in its data requirements from one intended primarily for research. However, because many small collections prepared for limited use ultimately become parts of large research herbaria, it is advisable to use labels covering the principal points necessary for scientific purposes. To be of value for research each specimen should include in one form or another the data required by the upper label in Figure 12-9, or at least it should include sufficient data so that missing points may be figured out from a map. This label was constructed after collecting specimens in the field and using these and many others in research. The form of the label may vary, and many prefer essentially the same data written up as in the lower label.

The top blank line of the first sample label is intended for the name of the plant, that is, the genus and species. This should be followed by the full or abbreviated author's name found in technical books. Although this name is not important to the beginner, it is of greatest significance for advanced study in plant classification, as shown in the following paragraph.

Author's names are not included to honor anyone, but as a tool for reference. Author references are necessary because inadvertently two authors may use duplicate combinations of generic and specific names for different plants, as, for example, *Ranunculus tenellus* Viviani (published in 1831) and *Ranunculus tenellus* Nuttall (1838). Although only one of such duplicating names (the earlier one) is valid, both may appear in the botanical literature for a time until the duplication is discovered. Consequently, to avoid confusion, it is important to cite the author reference. If the genus, species, and author's name or its abbreviation (e.g., *Cercidium floridum* Benth.) are known, the rest of the reference to the original place of publication may be found in the Index Kewensis or the Gray Herbarium Card Index. Consultation of the original publication of the name of the species yields the information

HERBARIUM OF POMONA COLLEGE

Echinocactus Whipplei Engelm. & Bigel.

Var.

State Colorado County Montezuma

Locality McElmo Canyon 7 miles west of Cortez on State Road No. 146.

Mt. Range Alt. 5,700 ft.

Drainage Area San Juan River Soil red sandstone

Veg. Type Juniper-Pinyon Woodl. Slope south

Lyman Benson No. 14755 Date Oct. 31 19 50

(Ultimately Colorado River Drainage.)

HERBARIUM OF POMONA COLLEGE

Echinocactus Whipplei Engelm. & Bigel.

McElmo Canyon 7 miles west of Cortez on State Road No. 146, Montezuma County, Colorado. Red sandstone; south slope; Juniper-Pinyon Woodland; San Juan River Drainage (Colorado Drainage).

Date October 31, 1950 Alt. 5,700 feet

Coll. Lyman Benson No. 14755

Figure 12-9. Labels for plant specimens, the same data being recorded on two for illustration. (Cf. specimen number 14755 and the data above it in **Figure 12-8.**) Research by Dorothea Woodruff shows reclassification to be necessary, and the name of the plant becomes *Sclerocactus parviflorus* (Clover and Jotter) Woodruff and Benson var. *intermedius* (Peebles) Woodruff and Benson. Ordinarily this is added on a special small label for reidentification.

available concerning the specimen upon which the original description was based; finding the specimen shows exactly what plant was described; visiting the place of its collection reveals the nature of the population from which the original plants were taken. Comparison of these plants with each other and with neighbors determines whether or not they constitute a separate natural population or system of populations worthy of a name as a species or variety. When two authors' names or abbreviations are given, the first in parenthesis and the second not, there are two references, e.g., *Prosopis juliflora* (Swartz) DeCandolle. The name in parenthesis refers to the original description of the plants upon which the species was based. The author's name following the close of parenthesis refers to publication of the currently recognized combination of plant names. In this instance, the specific (species) epithet (*juliflora*) was not published in combination with *Prosopis*, but under another genus, that is, as *Mimosa juliflora* Swartz. Consequently, there are two authors to be considered, one (Swartz) having supplied the specific epithet (*juliflora*) and the other (DeCandolle, usually abbreviated to DC.)* having supplied the classification of the species with respect to a genus (*Prosopis* instead of *Mimosa*) accepted by the present author. Similar use of authors' names may indicate transfer of a variety from one species to another. Although for reference or teaching purposes inclusion of the author's name has no importance, it is necessary for precision in research.

The second line of the label, beginning with "Var." (the abbreviation of *varietas* or "variety") is intended for the varietal epithet, if any, and its author references.

The next three items deal with location of the place where the specimen was collected. The locality should be specific, and it should be associated with a permanent geographic feature. A crossroads post office may become a ghost office in forty years, but a well-established farming community is more likely to persist. Usually, but not always, the name of a valley or a mountain is permanent. Judgment must be used in any statement of locality. Usually a place of collection should be indicated in terms of mileage in a particular direction from the nearest clearly permanent and readily located point of reference.

A mountain range may serve both as a definite location and as important data because the flora of a mountain range often differs from that of its neighbors. Knowing the mountain range inhabited by a plant may contribute to recognition of populations characteristic of only one or a few ranges. It may be helpful also in working out the general area inhabited by a more widely distributed population or in determining the range of each of its local phases.

In many instances it is important to know on which side of a mountain range a specimen was collected. For example, on the western slope of the Cascade Mountains in Oregon the average rainfall is about forty to seventy inches, but on the eastern edge of the same mountain range at a similar elevation the average may be only about ten inches. Consequently, there is a profound difference in the associated plants on the two sides of the Cascades. These differences are emphasized in the item called "Drainage Area." Western Oregon is on the moist coastal forested drainage area of the Pacific Ocean; eastern Oregon is on the semidesert inland drainage area of the Columbia River. The Cascade Range acts as a barrier to the Prevailing Westerly Winds bearing moisture from the Pacific Ocean.

Vegetation zonation according to altitude in the mountains is equally important. Only ten airline miles from certain orange groves in

* Many of the more common author abbreviations may be found in an alphabetical list in the Eighth Edition of *Gray's Manual of Botany* by M. L. Fernald, pages liii to lviii, 1950, and in Philip A. Munz and David D. Keck, *A California Flora* 1551–1576, University of California Press, 1959.

Southern California is a mountain summit of 10,000 feet elevation. In this local 9,000-foot differentiation of altitude are three well-marked vegetation zones, each harboring many species different from those in the others. On a winter or spring day in the orange belt the temperature may be seventy or eighty degrees while at the summit of the mountain there are five or six feet of snow. Study of species involves their correlation with altitudes as well as with localties and general ranges of occurrence.

The data to be recorded in the space marked "Soil" depend largely upon the individual and his background. A soil chemist may designate a particular soil as "Aiken Clay Loam"; another person may describe the same soil as red clay; another may call it "red," "fine," or "red volcanic" soil. It is important to record as much as one can and particularly to indicate derivation of soil from some recognizable kind of rock, such as limestone or serpentine, because some plants are associated with specific soil types although others are not. In some instances this item or others must be left blank for lack of information.

Recording data concerning vegetation types depends also upon the background of the individual. The more specific the designation for any vegetation type the better, but there are inevitable variations according to differing systems of vegetation classification. Even if no system is known, a description in ordinary terms is helpful. The words "brush," "among oak trees," "yellow pine forest," "redwood forest," "hardwood forest," "deciduous forest," "prairie," "piney woods," or "hammock" may tell the professional botanist at least part of what he needs to know.

In botany the significant slopes are north-facing and south-facing, simplified to "north" and "south." The difference in the vegetation growing on the north and south sides of the same hill may be equal to that produced by a variation of 1,000 to 3,000 feet of altitude because in the Northern Hemisphere the south side is exposed to the sun, while the north is not. On the north the temperature is lower and more moisture is retained. East and west slopes are usually of much less importance.

The last item in the list is a specimen or "field" number. Many systems are possible, but a specific one has been adopted almost universally in North America, and there are technical reasons for using it rather than a more original one. This system consists of giving each collection a serial number and recording it, as indicated above, both in the field book and on the newspaper sheet used in pressing the specimen. All plants collected by the same person are numbered consecutively in a simple series continued unbroken year after year. If several duplicate sheets of a specimen are collected, each is given the same number. Specimens received from other individuals may be included in the collection but should not be included in the numbering system; they should bear their collector's name and whatever data he has supplied with the specimens.

In technical botanical literature the collector's number serves another important purpose, for it facilitates reference to the collections of an individual. Since the number is a part of a series and is not duplicated, a reference to *Frazier 7,451* is precise, though it may refer to any one of a set of duplicates. In the literature, addition of a standard symbol (e.g., *US* for United States National Herbarium, Smithsonian Institution, Washington, D.C.) to the field number indicates the exact collection in which the specimen examined is housed. Use of several symbols indicates that duplicate specimens occur in different herbaria.

As stated above, in some instances certain items called for on the labels may be omitted, and good specimens already collected should not be thrown away for lack of full data if they include certain essentials. Botanists who have travelled in the area from which a specimen came may be able to fill in much of the information required even though the label

Figure 12-10. The effect of altitude upon climate and vegetation. The ripe lemons are in a grove at 1,500 feet elevation near Claremont, California; the mountain peak ten miles away (Mt. San Antonio or Baldy) is 10,000 feet high. The snow on the summit when the picture was taken was six to eight feet deep.

carries no more data than "Steamboat Springs, Colorado" or "Asheville, North Carolina."

Since many points concerning a plant are not shown by an herbarium specimen, these should be recorded in notes on the label. The height of a tree and the diameter of its trunk together with the prevailing height and diameter of other individuals in the region may be noted. This is true, also, for the color, odor, taste, and other characters of flowers, fruits, and bark — provided, of course, these matters can be recorded accurately so that they may be understood by another person using the specimens. On the other hand, recording of such matters should not be carried to extremes, and specimen labels should not be cluttered with insignificant data. The selection of what is to be recorded should be based upon what may be needed in classification or in ecological or other studies. A little practice in use of the keys will make the selection of significant data relatively easy.

Often a good photograph of a plant and especially of a tree records data which supplement the herbarium specimen. The photograph should be placed in a paper pocket on the specimen sheet. Mounting with rubber cement is unsatisfactory because the cement adheres for only a short time. Photographs are not satisfactory substitutes for specimens because they show only incomplete surface data and because they cannot be dissected for study. Few plants, if any, could be keyed completely and accurately from photographs.

Placing Specimens in a Collection

There are many ways of keeping specimens for study. In some institutions students preserve their collections in large books. These are satisfactory if the principal objective of the course is learning names of plants, but for learning systematic identification methods the books have obvious drawbacks. Even if they are of adequate size for herbarium specimens, they are designed for only small, temporary collections and are soon outgrown. Unless the books are of the loose-leaf type it is not possible to bring together the related plants as the collection grows, and the main object of making collections — that is, to emphasize relationships and to facilitate comparison of related species — is not achieved. Even if the books are of the loose-leaf type, shifting of specimens from one place to another and inserting new ones is a relatively slow procedure. A more practical method is to place the specimens on inexpensive herbarium sheets made of unprinted newspaper stock, cut either to standard herbarium-sheet size (11½″ x 16½″) or to a slightly larger size (up to 12″ x 18″); the specimens may be fastened down with strips of adhesive tape or left loose on the sheet. The sheets are placed in folders made of wrapping paper. If the sheet is 11½″ x 16½″ the paper folders may be about 30″ x 17″ to allow for two flaps (one 12 and one 6 inches wide) to lap over each other above the specimens. If the specimens are handled with care by only one or two individuals, this system is satisfactory. Collections of over 20,000 specimens have been put up in this way and maintained without any particular difficulty. The specimens require relatively less space, and they are less difficult to transport from one location to another than are those mounted by the more elaborate system to be described in the second paragraph below.

A variant on the system mentioned above is use of newspaper sheets cut to double size (16½″ x 23″ up to 18″ x 24″) and used as folders. In this case each specimen has an individual folder and consequently somewhat better protection, but both the weight and the bulk of the collection are increased and the cost is somewhat higher.

Use of newspaper sheets and wrapping paper folders is satisfactory only for individual collections. Collections to be used more extensively should be mounted on herbarium paper. As stated above, the standard herbarium sheet is 11½″ x 16½″. The paper should be of a type which will last indefinitely, that is, heavy all-rag (linen) paper because once a specimen is glued to the sheet it cannot be removed satisfactorily, and any paper but the linen type will either disintegrate or discolor in the course of time. In some of the larger herbaria there are old specimen sheets which are so brittle that they tear with ordinary handling; other sheets have turned black in the course of about forty years. Unfortunately, all rag paper, and especially linen paper, has almost disappeared, and the best substitute available must be used. Usually cotton is substituted for linen.

The majority of the herbaria use one of many variations on the "glass plate" method of mounting. A high quality glue or tin-can label paste is brushed out into a thin layer on a smooth plate of glass; the specimen is picked up with tweezers, laid in the paste, and then transferred to the herbarium sheet. An unprinted newspaper sheet is placed above the specimen and a pressing blotter on top of that. Then light pressure is applied to the stack of newly mounted specimens until the paste is dry. Next, gummed linen strips are used to fasten down the bulkier parts of the plant. These should be placed across solid portions, such as the stems or the petioles of leaves

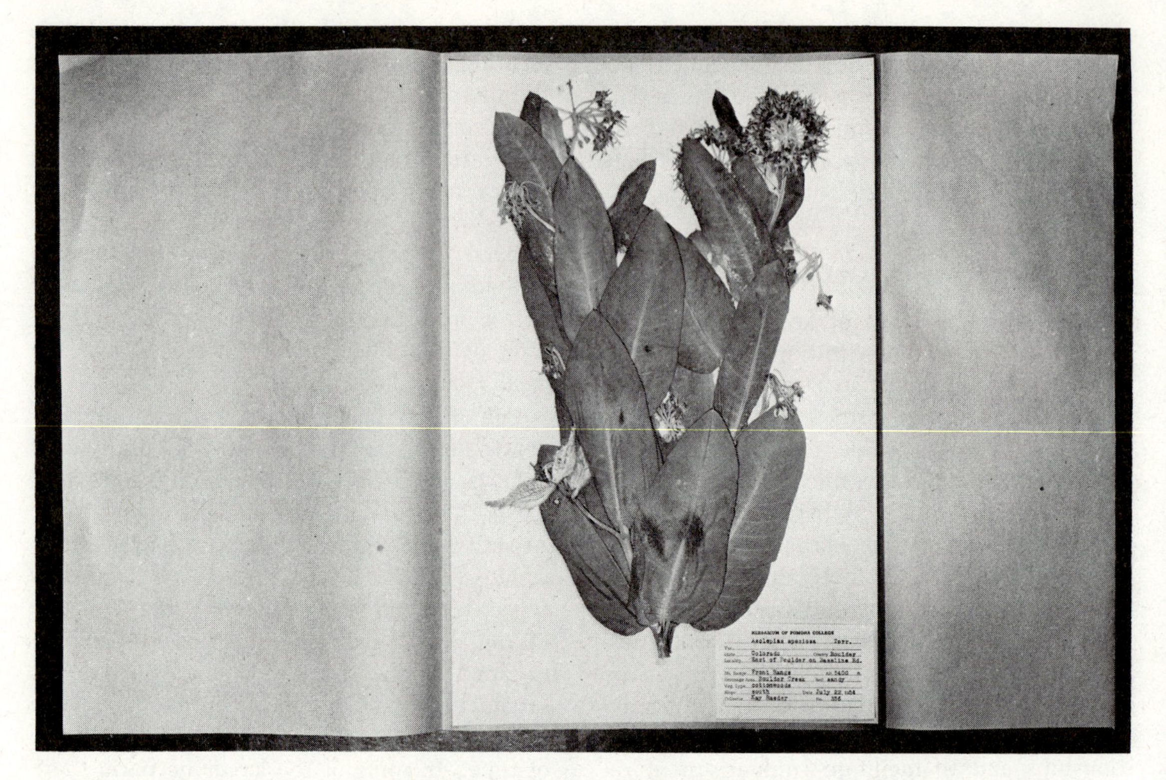

Figure 12-11. Specimen of a student plant collection. The specimen is on a sheet of newsprint (12 x 18 inches); the folder is of either Manila or kraft paper (18 to 19 x 30 inches); the label is fastened lightly with paste or mucilage at the two upper corners, facilitating removal if later the collection may be mounted on standard herbarium sheets. Photograph by Frank B. Salisbury.

(rarely across leaf blades) and pressed tightly around the plant with a pair of forceps, the ends being pushed against the herbarium sheet until they stick. Some large leaves may be fastened down by inserting the ends of gummed linen strips through punctures on both sides of the midvein and sticking them to the specimen sheet beneath the leaf. At least one leaf should be mounted with the lower surface exposed.

Some institutional herbaria do not glue the specimens to the sheet, but, instead, fasten them down only by means of gummed linen strips accompanied by stitching with thread if the plants are thick. This has the *advantage* of permitting removal of the specimen in case it is necessary to examine the underside, but the *disadvantage* of possible breakage of the specimen.

Other methods involve use of the best available type of glue or other adhesive to fasten bulky specimens down directly. The method of application of the glue (adhesive) depends on the type of specimens, as developed in the next paragraph. Unfortunately, a special plastic designed for plant mounting over 20 years ago cannot be used because it contains toluene. Hopefully another satisfactory plastic may be developed.

Exceedingly Woody Plants. All parts touching the herbarium sheet should be covered with a thin layer of glue before the specimen is placed on the sheet. Parts not lying flat should be weighted with metal (strap iron, heavy washers, etc.). After the specimen is weighted, strings of adhesive should be placed across the main stems at points selected for the greatest effect. The best method is placing a drop of glue on top of the stem and allowing it to divide and run down the sides. The adhesive should form a band across the stem.

Semi-woody Specimens. In some cases the glue alone may be used, provided it is applied under the specimen where this is necessary and over the stem where most effective.

Objects such as pine cones or acorns may be cemented to the sheet with a liberal application of glue. Acorns also may be cemented to their cups

Herbaceous Plants. Usually these should be pasted to the sheet in the standard herbarium manner. After pasting, bands of glue applied across stems or other firm parts of the specimen are more effective and applied more quickly than tape.

After application of the adhesive the sheet should be laid aside to dry. This may require about two hours. During the drying period anything touching the glue will adhere to it. The specimen should be left untouched even after the glue appears to be dry because the surface hardens while the interior is still fluid. Under even slight pressure an apparently hard surface may break and allow the liquid adhesive to flow out. It is convenient to place each herbarium sheet upon a sheet of cardboard for drying. The cardboards may be stacked, being separated by wooden slats thicker than the thickest specimen. After drying, the sheets may be piled up without danger of sticking.

Finally the specimen label (typed or written in ink) is glued usually in the lower right-hand corner of the sheet (about one-quarter inch from the margin to prevent soiling when the specimens are handled). A collection mounted on herbarium sheets requires heavier paper folders than does one kept in newspaper. Usually a heavy manila paper is used.

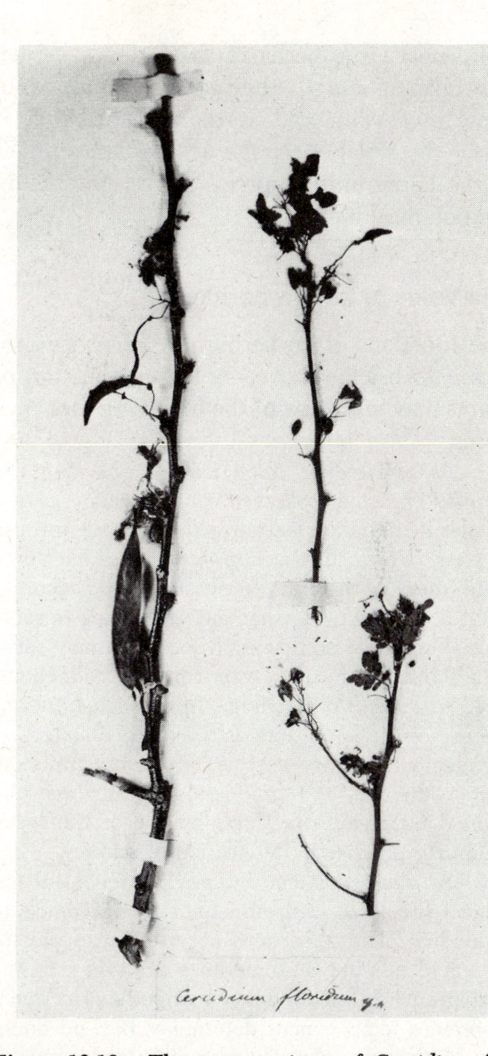

Figure 12-12. The type specimen of *Cercidium floridum* Benth., the blue palo verde of the Southwestern Deserts of California, Arizona, and northwestern Mexico. This specimen, collected near Hermosillo, Sonora, northwestern Mexico, by Thomas Coulter in 1830, is labelled, "Cercidium floridum, sp. n." It was deposited in the herbarium of Trinity College, Dublin, Ireland, and American botanists have not seen it. This photograph determined the correct use of the name *Cercidium floridum,* which for many years was applied incorrectly to a species of palo verde (now known as *Cercidium macrum*) occurring in southern Texas and adjacent northeastern Mexico. The type specimen shows clearly which plant was named *Cercidium floridum,* despite the confusion which appears in the botanical literature. From LYMAN BENSON, American Journal of Botany 27: 187. 1940. Used by permission.

In most large herbaria, after the specimen has been mounted, the sheet is stamped to show ownership and at the same time it is given the serial specimen acquisition number of the herbarium for precision in reference in the botanical literature.

The Value of the Herbarium

The functions of an herbarium have been referred to briefly above. A large collection of plants may serve any of the following four purposes.

1. *As a reference collection for checking the identity of newly collected plants.* Many of the smaller herbaria are started with only this purpose in view. For an herbarium with reference as its main function, the chief emphasis is upon accurate identification of every specimen according to some special authority so that each specimen may serve as a standard of comparison for new collections.

2. *As an aid to teaching.* In many institutions the use of the herbarium as a reference collection for elementary class work is its primary function. Upper division and graduate students may be trained in research methods by use of the herbarium for practice in delineation of species.

3. *As a historical collection.* The technical botanical literature is documented by reference to individual plant specimens according to collector and field number or sometimes according to the herbarium accession number stamped on the sheet. Reference to specimens determines beyond doubt which plant was described or discussed. According to modern practice, each species or variety is based upon a designated "type specimen" in a particular herbarium. Certain herbaria are rich in type specimens, and some, as, for example, the Herbarium of Linnaeus at the headquarters of the Linnaean Society in London or the DeCandolle Herbarium at Geneva, Switzerland, are made up very largely of type specimens. It would not be possible to delimit species or varieties by study at these herbaria, but often it is possible to find the exact plant upon which a particular species "name" (epithet) was based. Nearly any large herbarium is the basis for at least some of the botanical literature, and one of the greatest functions of an herbarium is to serve as a record of plants described in technical books and papers.

Figure 12-13. A type specimen in the herbarium of Linnaeus, Linnaean Society of London—the Indian "hemp" (not the true Indian hemp, source of marijuana) native through much of North America. This is the basis for the name *Apocynum androsaemifolium* L., i.e. the specimen determines which population system of plants was given this name by Linnaeus. Photograph, courtesy of Missouri Botanical Garden.

Nearly every "manual" of plants or "flora" of an area is based upon specimens preserved in one or more herbaria, and these are available for study by properly qualified persons.

The herbarium should serve also as a depository for samples of the plants used in every experimental research project. The infinite variability

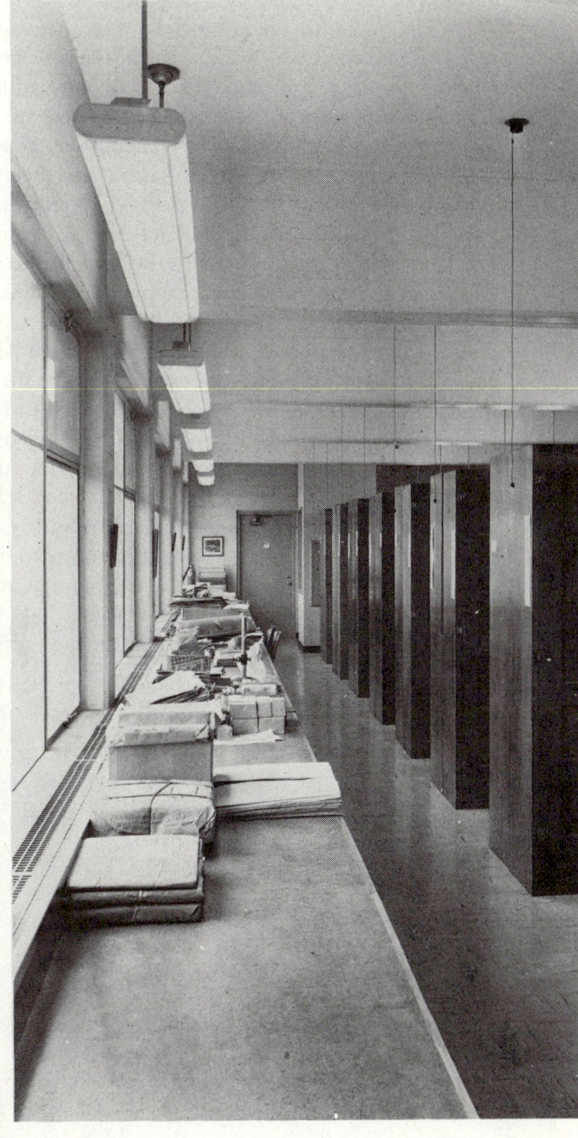

Figure 12-14. An example of an herbarium (Pomona College Herbarium). The open cabinet shows the specimens mounted on standard herbarium sheets filed in folders according to genus and species. Photographs by Frank B. Salisbury.

of natural populations and the inaccuracies in identification of the plants used in experimental research make many otherwise sound data or conclusions invalid because no one can be sure which organism was actually the subject of the experiment. As an example, had one chemist not been aware of the possibility of error and had he not made a last-minute check he would have published a paper on chemical analysis of the "common *Malva borealis*" though actually he had studied a local species of *Sphaeralcea* occurring on the vacant lots in Tucson, Arizona. In another case, data on the reaction of a slime mold to a current of water were based upon experiments with a lichen. Many chromosome counts recorded in the best technical journals are not worth the paper they are printed on because no one can be sure of the species or varieties they represent. An herbarium specimen of the individual plant used in the experiment or furnishing the material for the chromosome count serves as a permanent record available at any time for consultation.

4. *As a body of data for research* on the nature and delimitation of plant groups, such as families, genera, species, and varieties, and for mapping and analyzing their distribution. As an herbarium becomes larger it becomes more and more valuable for research but less valuable as a reference collection for checking the identity of newly collected plants because the herbarium staff is not likely to be adequate for keeping up with identifying each specimen according to some particular outside authority. On the other hand, the staff is likely to include members who are themselves "authorities" actively revising the limits of genera, species, and varieties and writing their findings on annotation labels appended to the specimen sheets.

A large herbarium is looked upon as a body of raw data which may be used in working out the classification of any particular plant group chosen for intensive study. These intensively studied groups form the basis for much of the literature of plant classification, and, as time goes on, the herbarium takes on the character not only of a body of raw data for research but also the values of an historical collection.

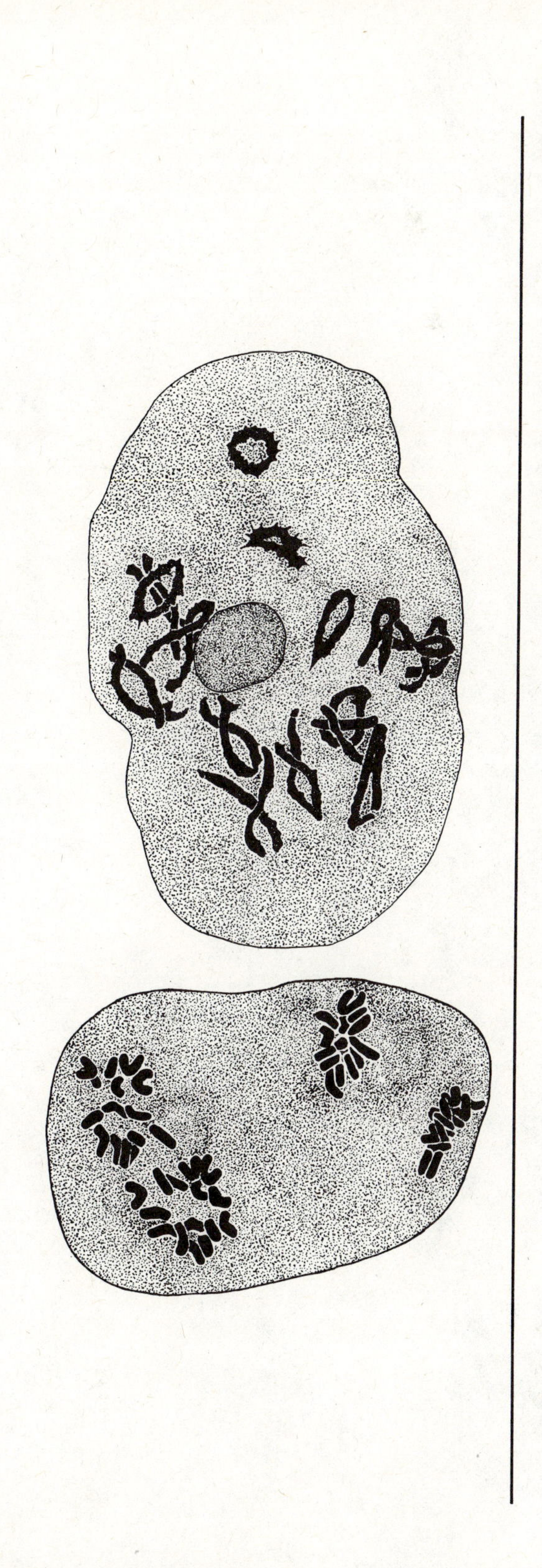

Flowering Plants

Part Four

THE BASIS FOR CLASSIFICATION

13

Evolution

The Concept of Evolution

Organic evolution deals with the *origin of the kinds of living organisms.* As with automobiles, "old models" of plants and animals have been supplanted in the course of time by "new models" better adapted to success in particular habitats or environments. The theory of evolution has been proposed * to explain the succession of living organisms occurring from period to period in the history of the Earth. According to these explanations the plants and animals of any particular time are descendants of those of the preceding eras, but some are not wholly like their ancestors.

* The first significant explanation of evolution was by Lamarck (1802). Part of its basis was the notion that modifications of living plants and animals are produced in response to the environment and that these acquired characters are inherited. As an example he used the giraffe, which he supposed to have acquired its high shoulders and long neck by straining to reach higher and higher into trees for leaves. Lamarck's explanation of the origins of new characters was false, but he focussed attention upon evolution. Charles Darwin (1859) and Alfred Russel Wallace proposed a more reasonable basis for interpretation of evolution, and the concept of evolution became accepted by scientists throughout the world. Darwin did not produce a significant explanation of the origin of new inheritable characteristics; he was concerned with natural selection and preservation of characters after they had arisen and the development of elements within a species into new species.

Evidences of Evolution

The sources of evidence of evolution are many, but they may be summed up and illustrated under the fields listed below: (1) classification of living organisms; (2) paleontology; (3) embryology and comparative anatomy; (4) geographical distribution of living organisms; (5) artificial selection of plants and animals; (6) genetics and cytology; (7) biochemistry.

CLASSIFICATION OF LIVING ORGANISMS

Until the close of the Eighteen Century few realized that there were problems in classification beyond bringing together brief descriptions and illustrations of the various kinds of plants or animals and giving each a name, just

Figure 13-1. Charles Darwin (1809–1882). Photograph at the Arnold Arboretum, supplied through the courtesy of the Arnold Arboretum and the Gray Herbarium, Harvard University.

as one might accumulate and arrange the various types of postage stamps. Those inclined toward philosophy applied to classification the beliefs of the times, hoping to discover the plan of creation and to arrange plants and animals according to it. In the Nineteenth Century it became clear to many that the problem was not as simple as had been supposed. As the complexities of classification emerged, the simple explanation of origin of precise unchangeable species according to a prearranged plan was seen not to fit the facts. Consequently, a more reasonable explanation was sought, and this led to the original proposal of the theory of evolution.

Evidence of evolution from classification is based upon (a) internal variability of taxa and shading off of one into another (b) series of plants and animals from simple to complex or from less to more specialized types.

Internal Variability of Taxa and Shading Off of One into Another. Although some living organisms fit neatly into their assigned classification pigeonholes, nearly all plants and most animals do not. As pointed out in the discussion of the plant and animal kingdoms, any category such as a genus or species is not necessarily wholly distinct from its nearest relatives; characteristically each shades off gradually into related types. Most species are variable internally; they include a wide range of individuals resembling each other but differing in individual characteristics. The combinations of characters of some individuals of any species may approximate those of some members of closely related species. No two botanists or zoölogists agree completely upon where to "draw the line" between related species or other groups. If species are variable internally and they cannot be distinguished clearly, evidently they are not fixed, and there is the possibility of change from time to time when certain specialized members survive and others die out as the environments of the Earth are altered.

When we assume a genetic (hereditary) relationship, we assign a common ancestry to the groups involved. We accept this principle in, for example, the classification of human beings. Few would argue that such nationalities as the Germans and the Spanish, despite their considerable average differences, are not descended from the same ancestral stock because the differences are inconsistent in occurrence and the similarities far outweigh them. On a broader basis, few would not interpret similarly the Caucasian, Oriental, or African races of mankind. If we admit this principle for minor groups such as the races and varieties (or subspecies) within a species, then we must realize that the differences between species or higher groups may be just as arbitrary and inconsistent as those between races. Therefore, if we assume a common ancestry for the races within a species there is no reason for not carrying over the same line of reasoning to groups of all higher ranks. As pointed out by S. J. Holmes,

> Linnaeus, later in life, came to hold that the species of a genus descended from a common created form [in other words, good churchman though he was, Linnaeus, as a

result of his vast experience in classification, finally admitted changing of species but thought that genera and higher groups were fixed], but if we assume a common descent for the species of a genus, we can with as much reason make the same assumption for the genera of a family, or the families of an order. The position of Linnaeus is much like that of a geologist who would admit that the natural processes of sedimentation might produce strata ten feet thick, but that they could not produce strata a hundred feet thick.*

Many species, as those of oak trees, interbreed freely, and through large areas there may be so many recombinations and intergradations of characters that no two individuals are alike and none may be assigned without question to any species. This is true, for example, in California where through an area 300 miles long the blue oak (a deciduous tree) and the desert scrub oak (an evergreen shrub) have interbred at least in the past (cf. page 508 and Figures 14-1 to 14-3, **6**). Interbreeding of oak species occurs throughout the Continent. It is noticed in these trees because they are large and conspicuous and present all year, but similar lack of sharp differentiation of species occurs in nearly all other genera.

Series of Plants and Animals from Simple to Complex. The individuals of some major plant and animal groups are simple and their cells or organs are relatively unspecialized; the members of others are complex and their cells or organs are specialized. For example, in the Volvocales, an order of fresh-water green algae, the simplest member is *Chlamydomonas.* The other members of the order retain the essential features of *Chlamydomonas,* in which each individual has only one cell, but some of the essential functions are carried out by coördination of the entire cell group or by specialized cells. In *Eudorina* sixteen, thirty-two, or sixty-four cells are organized into a colony, which moves by the coördinated lashing of all the pairs of flagella. Under the proper conditions each cell may divide internally and produce a new colony (asexual reproduction). Under others some cells may form groups of (male) gametes which escape and swim about in the water. Ultimately each swimming gamete may join a (female) gamete similar to a nonsexual cell in another colony. Each united pair of gametes forms a single cell which later gives rise to a new colony. This is sexual reproduction. Although in most species of *Chlamydomonas* all the gametes are alike, gradual differentiation of two kinds like those of *Eudorina* may be traced within the genus. In the primitive species the only difference is in the chemical substances (hormones) secreted by gametes of opposite sex; in others the female gamete is motile but larger and more sluggish than the male; in a few the female gamete is nonmotile. In *Volvox* there are usually about one or more thousand (500–40,000) cells arranged in a hollow sphere. Nearly all participate in movement of the colony, but only a few form new colonies by asexual reproduction or form male gametes like those of *Eudorina* or specialized nonflagellate and nonmotile female gametes. In the series from *Chlamydomonas* to *Eudorina* and *Volvox* the sequence of gradual specialization and division of labor among cells results in restriction of both types of reproduction to special cells, leaving the other functions such as food manufacture and locomotion to the remainder. In more advanced types of plants all functions tend to become restricted to specialized cells, some of which are adapted to manufacturing food, exchanging gases with the air, secreting substances preventing loss of water to the air, conducting food downward, conducting water upward, strengthening and supporting, storing food, or absorbing water from the soil. If a plant is to be large, it must have many specialized cells. Therefore it must be complex. The more complex plants may be traced back to simple beginnings. The possibility of development of the more complex members of the series from the simpler is obvious; the reverse is unlikely in this case. (Fig.

---

* From *General Biology,* Ed. 2. 356. 1937, by S. J. Holmes. Copyright by Harcourt, Brace and Co. Used by permission.

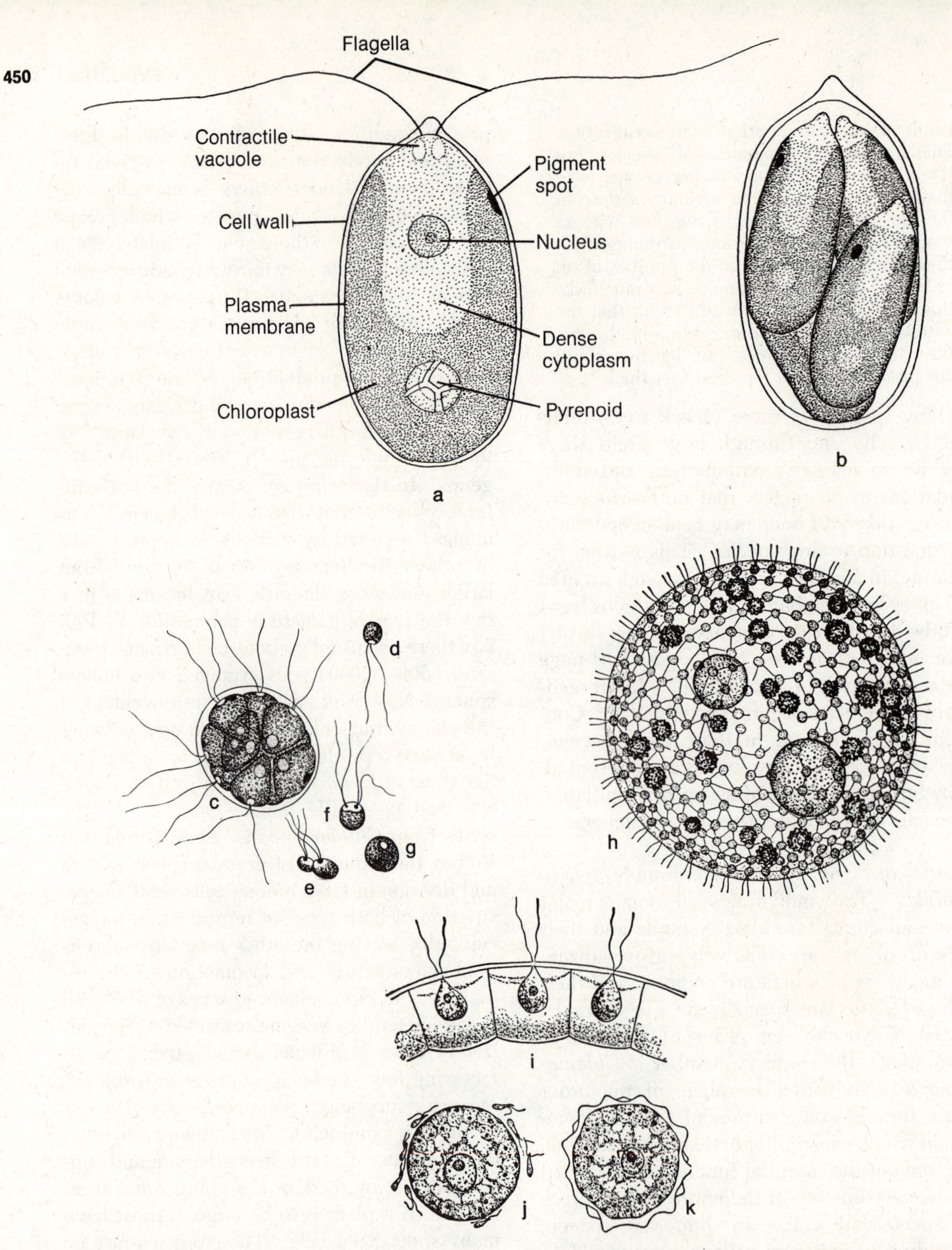

Figure 13-2. Series of plants from simple to complex. **a** *Chlamydomonas,* a one-celled green alga, the cell of the type fundamental in the entire series within the Volvocales to follow, **b** cell of *Chlamydomonas* with the flagella retracted and with internal division having formed four daughter cells, which will escape from the old wall and form independent new plants, this by asexual reproduction, the single cell carrying out all physiological functions and either asexual reproduction or sexual reproduction (then by combining with another), **c** colony of cells of *Pandorina,* each similar to one of *Chlamydomonas,* each cell of the colony either forming a new colony inside and this escaping or forming one kind of gamete, either smaller

13-2. Note that *Pandorina* is similar to *Eudorina* but with fewer cells and less difference between gametes.)

However, simplicity is not necessarily always an indicator of primitiveness. For example, many parasites have simpler structures than their probable forebears, adaptation to a specialized mode of nutrition having rendered some features of the ancestral plants unnecessary or undesirable.

PALEONTOLOGY

The possibility of occurrence of evolution requires evidence of (1) *time*—the availability of an enormous period for gradual change of living organisms from the simplest ancestors to the complex organisms of today, and (2) *environmental change*—evidence of change of climates and other features of the Earth from period to period, requiring constant selection from among the members of each species on the basis of ability to survive under new conditions.

Time. Everywhere on the land surface of the Earth there is evidence of erosion. Gradually the surface is being washed down into the sea, and mud, sand, and gravel are being deposited layer upon layer in the shallower water. But in many parts of the world, layers of this sort have been covered by later layers. For example, in the high desert country on the Utah-Arizona boundary just south of Kanab there are beautiful colored beach pebbles worn smooth by wave action and embedded in loosely consolidated beach sand. Similar sand and wave-worn pebbles forming a conglomerate occur from Zion National Park, Utah, southeastward to the Petrified Forest National Monument near Holbrook, Arizona (more than 200 miles), and thence eastward through the Colorado Plateau. This conglomerate contains no fossils of sea or ocean organisms, but it includes remains of some ancient animals of stream and lake types and great petrified driftwood logs of an ancient forest gymnosperm (*Araucarioxylon*) now extinct. Near Kanab the same conglomerate formation is overlain by at least 3,000 feet of more recent rocks laid down partly by water and partly by wind (as sand dunes). Disappearance of the lake and laying down of the great thicknesses of later rocks as well as the mile of earlier rocks beneath (exposed at the Grand Canyon) must have required millions of years. This, together with similar evidence from other parts of the world, indicates that the time of development of the Earth as we know it today must be traced back through billions of years.

Evidence of the length of geologic time comes from three sources—sedimentation, saltiness of the ocean, and the disintegration rate of uranium and other radioactive elements or isotopes, such as thorium and radioactive potassium.

Sedimentation. About 450 B.C. the Greek historian Herodotus concluded that, judging by the amount of the sediment deposited each year, building of the delta of the Nile River must have required many thousands of years.

(fundamentally male) or larger (fundamentally female), the colonies of cells thus becoming specialized somewhat for reproduction, **c–f** fusion of gametes, **g** zygote, which becomes a resting spore from which after reduction divisions (meiosis) a new colony emerges, **h** colony of *Volvox,* made up of as many as 1,000 or even 50,000 cells, mostly unspecialized as in *Chlamydomonas* but a few specialized as reproductive cells, the asexual ones entering the center of the hollow sphere of cells and dividing and forming, here, two new colonies that will escape when the parent plant dies, the sexual ones, specialized as male and female gametes (antherozoid is preferable to sperm in plants); numerous resting spores, each formed after the fertilization of an egg by an antherozoid appear in the interior of the colony—each will form a new colony later, **i** enlargement of three vegetative cells in the wall of the colony, **j** egg cell surrounded by antherozoids, one of which will enter it and join with it in fertilization, **k** fertilized egg forming a heavy wall and becoming a resting spore. Sources: **a, b,** G. M. SMITH ET AL., *A Textbook of General Botany,* Macmillan Co., New York, Ed. 5. 264. fig. 154. 1953; fifth edition copyright, 1953, by the Macmillan Company. **c–k,** J. M. COULTER, C. R. BARNES, and H. C. COWLES, *A Textbook of Botany,* 17, 19. figs. 23–27, 29–32. 1910, 1930. Used by permission.

Figure 13-3. Smooth water-worn beach pebbles and beach sand in conglomerate and the fossilized log of an extinct conifer (*Araucarioxylon*) in the same formation. *Left,* Arizona west of Holbrook; *right,* Arizona near Kanab, Utah.

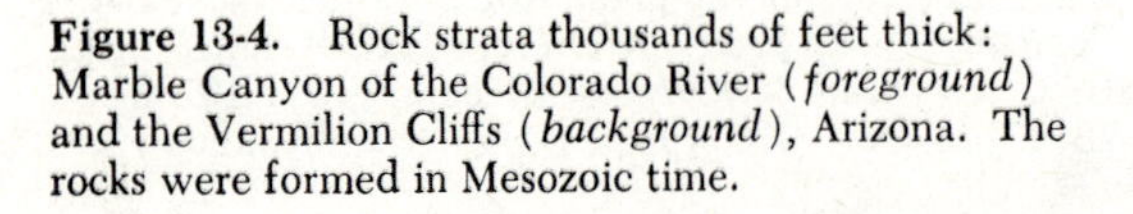

Figure 13-4. Rock strata thousands of feet thick: Marble Canyon of the Colorado River (*foreground*) and the Vermilion Cliffs (*background*), Arizona. The rocks were formed in Mesozoic time.

Figure 13-5. Horizontal and tilted rock strata. Both types were formed under water, therefore on the level; the tilting required time. *Left,* Salt River Canyon, Arizona; *right,* Piru Gorge, southern California.

It is obvious that deposition of layers of sediment thousands of feet thick requires a great deal of time. However, since rates of deposition of silt in the ocean vary greatly from place to place and from time to time, only approximations of geologic time can be based upon sedimentation. Nevertheless, the thickness of sedimentary rock indicates the antiquity of the Earth because, for example, a reasonable estimate of the time required to lay down only the Green River lake deposits in Colorado and Wyoming is approximately six and one-half million years.

Saltiness of the Ocean. As the sediments are washed out into the ocean, salts are carried along in solution in the water. The water returns to the atmosphere by evaporation but the salt is left behind in the sea. Consequently, the salinity of the ocean increases appreciably from geological epoch to epoch. The annual increase of salt carried into the sea by the streams of the world can be estimated with a fair degree of accuracy, estimates being based upon measurements of the volume and the chemical content of the more important rivers of the world. According to the present rate of annual addition, the tremendous accumulation of salt in the sea has required many millions of years. However, this method of calculation is limited in its accuracy by a number of factors, and only a rough estimate of the antiquity of the earth can be based upon it.

Disintegration Rates of Radioactive Elements. The modern, more precise, methods of calculating geologic time are based upon the disintegration rates of various radioactive elements, of which uranium is an example. Briefly, checking the age of a rock by use of uranium is based upon the following conditions:

1. An atom of uranium (atomic weight 238) disintegrates by emitting one helium atom (atomic weight 4) at a time until 8 helium atoms have been lost. The residue is nonradioactive lead of atomic weight 206 (238–[8x4]), which is slightly different in weight from ordinary lead.
2. This disintegration occurs at a known rate, and the ratio of lead to uranium indicates the age of the rocks.

In actual dating with uranium, three decay rates are used and checked against each other:

Uranium 238 to lead 206 (^{238}U to ^{206}Pb)
Uranium 235 to lead 207 (^{235}U to ^{207}Pb)
Thorium 232 to lead 208 (^{232}Th to ^{208}Pb)

If the three agree, the chances are very high that the dating is correct. This method is especially useful in dating very old formations, such as those of Precambrian time.

The ages of only igneous rocks, those formed by cooling of lava, may be determined directly by use of most radioactive elements. However, sedimentary rocks, those formed by cementing of layers of silt, often may be dated by their position in relation to igneous rocks of age determined by uranium decomposition. Uranium-bearing lava, which has intruded its way upward through sedimentary rock layers and flowed over them, is younger and therefore determines the minimum age of the penetrated rocks. Such igneous rock is older than overlying sedimentary strata, and so it determines the maximum age of the higher rocks. In other words, the position of a dated igneous formation sets one limit (either maximum or minimum) of the age of each adjacent sedimentary stratum.

Radioactive potassium 40 (^{40}K), decaying to argon (^{40}Ar), provides a particularly useful dating method, because potassium-bearing minerals are abundant and widespread, and they occur even in some sedimentary rocks. This makes possible absolute dating of some strata. New volcanic rocks do not contain argon, but they do include small amounts of potassium, some of which is radioactive. Argon, a heavy gas, is retained in the rock as it is formed, and it can be differentiated from the argon in the atmosphere.

Decay of rubidium to strontium may be used also in dating rocks.

Dating rocks with the elements discussed so far cannot be applied to more recent strata, but radioactive carbon 14 is used for objects containing carbon (for example, fossils) and (with usual methods) no more than about 40,000 years old. Some of the normal carbon 12 (^{12}C) in the atmosphere is changed to carbon 14 (^{14}C) by cosmic radiation, producing a balanced mixture of the two isotopes. Because both forms of carbon are incorporated into the carbohyu ates produced in photosynthesis, both the plants and the animals eating them or their parts include the mixture of carbon atoms in the same percentages as the atmosphere. However, during the course of up to 40,000 years, the proportions change as carbon 14 decays to carbon 12. Thus, the age of the dead organism can be determined accurately, i.e., with only a small probable error. This method of age determination is of great importance in many branches of biology, including archeology and tracing the history of vegetation. An ingenious method of determining past vegetation developed by Wells and Berger is dependent upon the habits of wood-

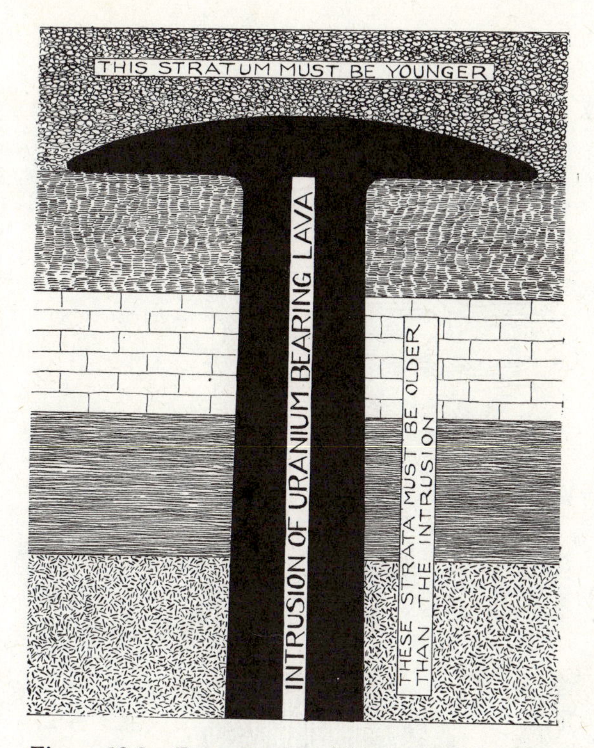

Figure 13-6. Determination of the age of sedimentary rock strata from an intrusion of uranium-bearing lava. The lava was forced upward through a crack in the strata, and it spread out over them. Subsequently later (younger) strata were laid down upon the lava. By J. D. Laudermilk.

rats, or packrats, which bring plant materials of all kinds to their nests from a radius of a hundred feet or so. Occasionally nests are built under rock ledges or in shallow caves, and these have been preserved for periods known to be up to 40,000 years. Thus the natural vegetation changes of deserts, woodlands, and forests during the ice ages and since have been determined according to the plants carried into woodrat nests at different times.

The estimate of the minimum age of the Earth based upon deposition of sediments was confirmed by study of rocks in Karelia, near the Finnish border of Russia. According to the lead-uranium ratio, these igneous rocks are 1,800,000,000 years old, but even they are intruded into still older formations. Recent studies indicate the age of the Earth to be about 4.6 billion years—that is, since formation of the crust. A football field is 100 yards or 90.68 m. long, in its American version. If the length of the field represents geologic time or the age of the Earth, the length of the Christian era, represented by 0.045 mm., is equal to about half the thickness of a blade of grass growing on the field. The 10,000 years since the last ice age is less than the width of a narrow grass leaf. The 120 million years during which flowering plants, insects, birds, and mammals have moved toward dominating the earth are represented by 2.6 yards, about the height of the drum major and his cap.

Environmental change. The evidence from paleontology indicates great fluctuations in the climate, topography, and other features of the Earth and of its individual regions. From time to time these features have been favorable to different types of plants and animals. For example, the great coal age was a period of relative warmth and abundant moisture favorable to lush plant growth. On the other hand, the coal age was followed by a period when much of the land was dry and when there was glaciation in several areas. Under these severe conditions many plants and animals became extinct, and better-adapted types slowly replaced them. Great fluctuations of climate have occurred within the last million years, and on four occasions great ice sheets have formed. These glaciations, extending in North America as far south at various times as Washington (Olympia), western Montana, Wisconsin, Kansas, southern Ohio, and Long Island were sufficient to produce profound changes in the vegetation and the fauna. In view of climatic fluctuations like these, it has not been surprising to find striking variation in the plants and animals from one geological period to another, or even within much shorter intervals such as the ice ages.

Evidence that through geologic time organisms have changed in company with climatic change is derived from **fossils.** Obvious remains of plants and animals occur in the rocks

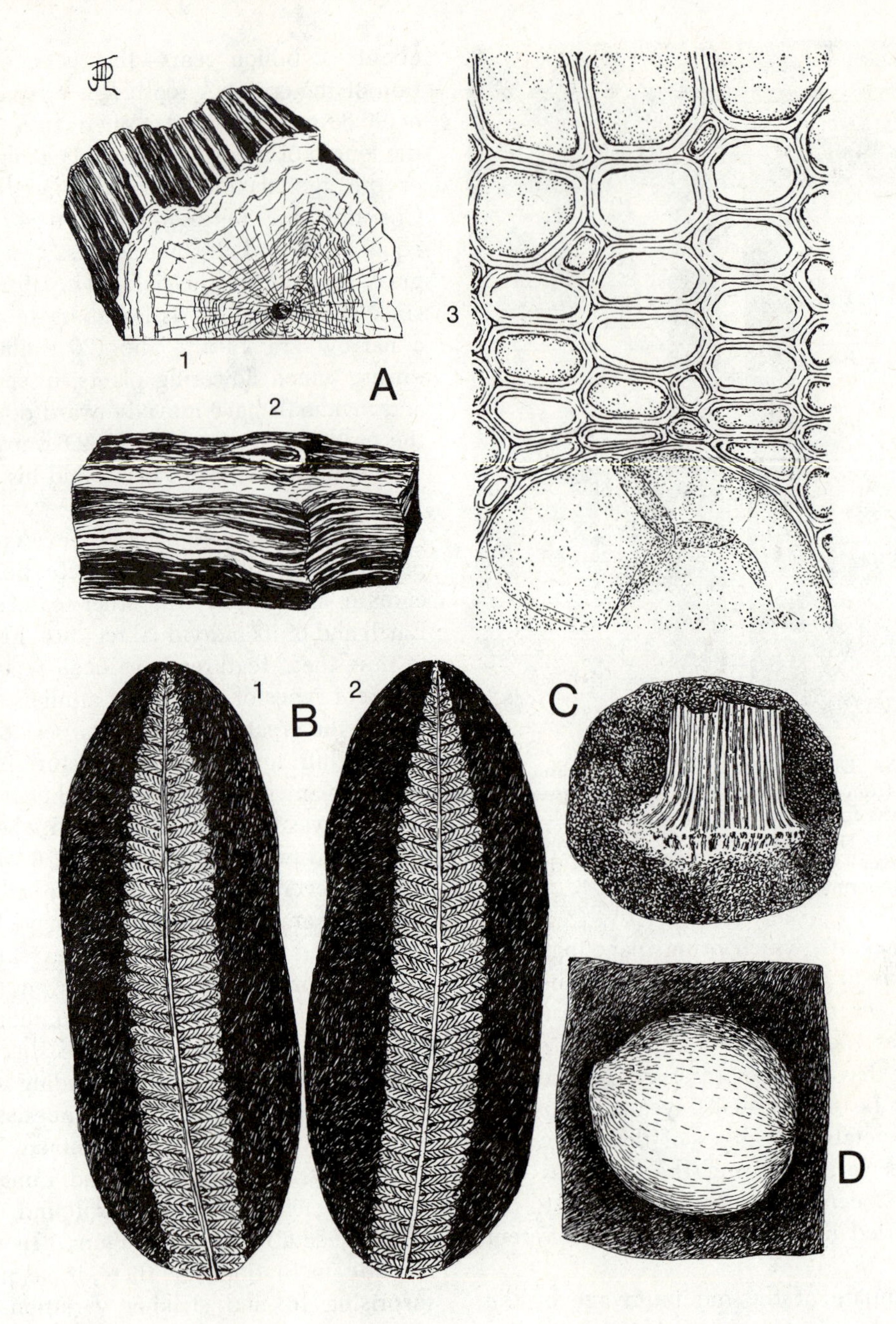

Figure 13-7. Plant fossils: **A**, **1–2** fossilized wood from beneath 200 feet of gravel at Parker Dam, Arizona, **3** section showing the cellular structure; **B** the two parts of a concretion found in coal, separated along the fern or seed-fern leaf around which the material of the rock accumulated; **C** portion of the stem of a giant horsetail (Calamitales, pteridophytes) in coal; **D** seed of a cycad (Cycadales, gymnosperms) from a Mesozoic shale.

of only about the last 570 million years. The apparent absence or rarity of fossils in older strata has been thought to be due to the lack or scarcity of living organisms, but now it is known to stem from the fact that simple organisms lacking hard parts do not leave easily interpreted records. However, new combined methods of chemical, paleontological, and biological analysis have made possible determination of the molecular character of very old fossils. The evidence consists of degradation products from the decay of once-living organisms. Undoubted chemico-fossils were found by Tyler and Barghoorn in a dark chert more or less coating stromatolites (bunlike masses of colonial algae with calcareous material and silicates) in the Gunflint Iron Formation in Ontario. These are 1,900,000,000 years old. The fossils are interpreted as thread bacteria, bluegreen algae, and other thallophytes. Possible fossil chemical material in the Soudan Iron Formation in Minnesota, 2,800,000,000 years old, may represent degradation of living materials, but the substances in question can be formed inorganically. Amino acids and porphyrins (basis for hemoglobin) have been found in sediments over 3 billion years old, as have pristane and phytane, which are typical products of the disintegration of chlorophyll.

Macrofossils usually represent the firm portions of plants and animals, and they fall into the following types.

1. Some of the actual substance of the original organism. This may be a piece of carbonized wood, that is, coal, or some teeth or bones or, as in Rampart Cave, near the Colorado River, the excrement of an animal (giant sloth).

2. Petrified material, as the wood of the ancient trees in the Petrified Forest in Arizona. In this case, mineral matter has been deposited molecule by molecule in the wood of the tree, preserving perfectly the original structural form of each cell wall. The original materials of the cell walls may be preserved at least in part within the mineral covering. The petrified wood may be cut with a diamond saw into sections, and these may be ground with abrasives until they are thin enough to see through under a microscope. Even the minute details of thickening of the cell walls are preserved. In many cases just as much can be determined concerning the structure of hard parts of the extinct plants of many millions of years ago as can be seen of the corresponding features of living plants. Less detailed structure may be studied by dissolving the mineral matter with acids, thus leaving the original walls exposed in relief. Cellulose nitrate added to the exposed surface of a section through petrified wood hardens into a thin transparent layer, which may be peeled off and placed under a microscope, revealing the general arrangement of cells.

3. Casts and impressions of organisms. Examples: a cast of mineral matter filling in exactly the area in which a dead organism lay; the tracks of a dinosaur left in mud now cemented into rock; an impression of a leaf with only a carbonized film remaining from the original tissue.

Examination of sedimentary rocks reveal a gradation from relatively simple organisms in the lower strata to a greater abundance of more complex or specialized ones in the higher. In the region around Kanab, Utah, rocks laid down through the entire period of macrofossil history may be found. They include rocks from the following eras.

1. *Paleozoic,* the lower level of strata exposed at the Grand Canyon. These alone are about one mile in thickness.

2. *Mesozoic,* the mid-level rocks exposed in Zion Canyon.

3. *Cenozoic,* the highest level rocks now being eroded away into the beautiful and fantastic formations of Bryce Canyon.

The *Paleozoic Era* began about 570,000,000 years ago. In the lower strata there is no evidence of any of the higher plants but only of various algae. In the later Paleozoic there were many pteridophytes, including apparent forerunners of the modern ferns and their relatives. Other prominent members of the flora included the so-called seed ferns which were really gymnosperms of a type far different from the present day conifers (e.g., pines and firs). Despite a wealth of plants characteristic

Figure 13-8. A track of the gigantic herbivorous dinosaur, *Brontosaurus,* made in mud, cemented to shale, in Texas. One track from the trails left by a herd of the animals, at least 23 having gone by. The track is about 2 feet long and 20 inches across. Walter G. Moore, Turtox News 38: 49. 1950. Used with permission.

of the time, there were included probably (except at the very end of the Era) few, if any, species living today and probably not a single representative of any of the prominent groups of plants most abundant in the living flora. For example, there were no flowering plants and no trees with cones like those of the living conifers. Likewise the remains of animals show no evidence of birds or mammals. The largest animals of the time were Stegocephalians, "giant" relatives of the modern salamanders.

The *Mesozoic Era* began about 225,000,000 years ago. A new flora replaced the plants which had dominated the landscapes of the great coal age toward the close of Paleozoic time. There were many plants related to the modern cycads (a family largely of small tropical trees resembling dwarf palms but with cones) and to groups similar to the modern cone-bearing trees. Until toward the middle of the Mesozoic Era no known fossils of flowering plants were formed. With respect to the plant kingdom, the Mesozoic is called the Age of Cycads. From the point of view of the animal kingdom, however, it is referred to as the Age of Reptiles because it was dominated for long periods by the giant dinosaurs. During the latter part of the Mesozoic Era flowering plants, mammals, birds, and insects appeared and started the trend toward their dominance.

The *Cenozoic Era* began about 65,000,000 years ago. The predominant fossil records of plants are from flowering plants and cone-bearing trees, together with modern types of ferns. Along with these plants are the remains of mammals, birds, and a tremendous variety of insects. The great development of these groups of animals was correlated with the rise of the flowering plants upon which they depend for food.

Fossil remains in successive strata of any Era present outstanding series showing gradual development of highly specialized modern plants or animals from much less differentiated earlier types. Two examples are as follows.

Horses. The oldest known horse, eohippus (*Hyracotherium*), was a small animal adapted to forested regions of about 60 million years ago. Its four spreading toes were excellent for walking on soft ground or mud, and its teeth were suitable for browsing on shrubs or eating soft plants. As grassy plains developed later, horses adapted to living in the open, securing safety by running and cutting off and eating tough range grasses. There were few trees and no places to hide, hence a better running mechanism was imperative. The modern horse runs or walks upon a single toe of each foot, the hoof being a highly developed toenail. Two small bones beside the remaining toe are remnants of two other toes which were far better developed in the ancestral horses, as, for example, those of about 15 million years ago. The single functional toenail has

SCALE OF GEOLOGIC TIME

Era	*Period* *	*Epoch or Notes*	*Plants*	*Animals*
CENOZOIC	Quaternary 2.5	Recent Pleistocene	Dominance of flowering plants	Passing of large mammals
Began about 65 million years ago	Tertiary 65	Pliocene Miocene Oligocene Eocene Paleocene		Dominance of mammals, birds, and insects, the insects with more species than all other organisms
MESOZOIC	Cretaceous 135		Ascendency and dominance of flowering plants	Rise of insects; end of dinosaurs and pterodactyls
Began about 225 million years ago	Jurassic 190 Triassic 225		Climax of gymnosperms, including cycads	Age of reptiles including dinosaurs and pterodactyls
PALEOZOIC	Permian 280	Cold, dry	Reduction of swamp vascular spore plants	Rise of reptiles
Began about 570 million years ago	Carboniferous, i.e., Pennsylvanian and Mississippian 345	Warm, moist	Abundance of pteridophytes and gymnosperms	Dominance of amphibians; in Mississippian rise of sharks
	Devonian 395		Rise of vascular spore plants, i.e., of land plants	Rise and dominance of fishes
	Silurian 430		Earliest vascular spore plants (cf. pp. 584–585)	Abundance of corals
	Ordovician 500		Algae; fossil record meager	First chordates
	Cambrian 570		Algae	Dominance of invertebrates
PRECAMBRIAN 4 billion years ago	Living organisms minute and simple, but there has been amazing progress in analysis of chemical remains (chemico-fossils) of bacteria, bluegreen, algae, etc.			

FORMATION OF THE EARTH'S CRUST. 4.6 billion years ago

* and millions of years since beginning
** In this table, *pteridophytes* in the broad sense—all vascular plants reproducing by spores.

Figure 13-9. Sedimentary rocks accumulated and eroded through geologic time: Paleozoic strata about one mile thick, Grand Canyon, Arizona; Mesozoic strata exposed for 800 feet, Canyon de Chelly, Arizona; Cenozoic strata, exposed through thousands of feet, Bryce Canyon, Utah.

Figure 13-10. Early stages in the history of the Earth: *above,* Archeozoic time—probably before the beginning of life; *below,* Proterozoic time—masses of algae appearing in the water, other organisms leaving macroscopic fossil remains having been those with some kind of mineral skeleton or shell.

Figure 13-11. Simple vascular plants of early Devonian time (about 375–395 million years ago): taller plants, *Psilophyton* (left) and *Asteroxylon* (right); shorter plants, *Rhynia* (left) and *Horneophyton* (*Hornea*) (right). (See also the Introduction to Vascular Plants.)

Figure 13-13. The great coal age or Carboniferous (Mississippian) time: *left,* giant horsetail, *Calamites; background,* a relative of the club-mosses, *Sigillaria; foreground and right,* seed-ferns, *Lyginopteris* (Cycadofilicales); *middle above and right* (large, entire elongate leaves), Cordaitales; *right* (trunk), *Lepidodendron.* (Cf. Chapters 20 and 23). The salamanderlike animals were stegocephalians, giant relatives of the living salamanders.

'igure 13-12. A forest in Devonian time (including ome plants known from middle, others from late De- onian): *left, Aneurophyton* (*Eospermatopteris*), a ern relative; *middle* (small plant *center*), *Calamophy- on,* a relative of the horsetails, and (tall plant *cen- er*), *Archaesigillaria,* a relative of the club-mosses; *ıpper right,* leaves of *Archaeopteris,* a fern-relative, hese also at the upper left.

Figure 13-14. Permian time, a colder, drier period than Carboniferous: *upper left, Ulmannia,* an early conifer; *lower left-center,* leaf whorls of a species of the Calamitales (giant horsetails); seed-ferns. The reptiles were (*left to right*) *Diasparactus, Dimetrodon,* and *Limnoscelis.*

Figure 13-15. Triassic time: *upper left,* leaves in the immediate foreground, *Baieria,* a relative of the maidenhair tree (*Ginkgo,* a gymnosperm); *left margin,* a cycad with pollen-bearing cones; *upper right,* an early conifer; *foreground,* ferns including *Phebopteris* (with several pinnate primary leaflets arising palmately) and *Neuropteris* just to the right of it (this name being used for certain forms of seed-fern and fern leaves); *Macrotaeniopteris* (*Nilssoniales*) a relative of the cycads (gymnosperms). The large reptile was *Mystriosuchus.*

Figure 13-16. Dinosaurs; Mesozoic time: *above,* pterodactyl; *next,* the three-horned dinosaur, *Triceratops; middle level, Stegosaurus* and *Struthiomimus; swimming, Tylosaurus; lower right,* duckbill dinosaur, *Trachodon.*

Figure 13-17. The four-toed horse, *Hyracotherium* (*Eohippus*); Eocene time, Wyoming. Photograph by courtesy of the American Museum of Natural History, New York.

become a hard hoof pounded against the ground in running, and the toe joints form a springlike mechanism. The increased size of the animal enables it to see for long distances across the plains and to find areas with good grass or to locate approaching enemies. The capacity to cut off and chew tough grasses is associated with development of high-crowned teeth, growing continuously from beneath and replacing the worn-off upper parts. These teeth were adapted to cropping the rough, silicate-impregnated plains grasses, the stems and leaves of which provided a much more formidable diet than that of the domesticated horse. The age of a domestic horse is determined by the length of the teeth, which, with a relatively bland diet, continue to elongate for many years—a point well known to horse traders and dealers and sometimes not understood by buyers.

Elephants. Cenozoic fossils show gradual differentiation of the characteristic features of elephants, including the long stabbing teeth (tusks), the long prehensile (grasping) nose (trunk), and the complex grinding teeth used for chewing grasses. One extinct species of elephant is known from one of the most remarkable kinds of fossils, individuals having been trapped in the ice in Siberia and preserved perfectly as if in a "deep freeze" locker. Specimens of muscles, blood, hair, and other parts of these animals are at the Smithsonian Institution in Washington, D.C. A fascinating and thorough review of this subject appears in *Smithsonian* 8: 61–68. 1977. Finding of an intact baby mammoth is reported.

Fossils often fill in the gaps in series of related living organisms or connect widely differentiated groups. One of the most significant "missing links" is *Archaeopteryx,* the first known gliding bird, known from four collections of fossils and linking reptiles similar to lizards with birds. Through a fortunate circumstance this creature was preserved in the very fine calcareous ooze of an atoll in the ocean of Jurassic time. In constrast to the toothless bills of modern birds, the jaws bore teeth similar to those of lizards. Furthermore, it had many long joints in its tail, each vertebra bearing a pair of feathers. This tail was similar to the tails of reptiles but unlike the short stubby tails of modern birds. In addition, there were claws on the wings, suggesting the loss of this reptilian feature by degrees. On the feet one claw was opposed to the others. Among modern birds only the young of the hoatzin occurring in Brazil has similar wing claws. The feathers of birds are thought to be derived from the scales of reptiles. It is to be noted that bantam chickens having feathers instead of scales on their shanks differ only slightly in hereditary factors from those having scales, though these scales differ from those of reptiles.

EMBRYOLOGY AND COMPARATIVE ANATOMY

At some stage every living organism consists of a single cell. Many plants and animals

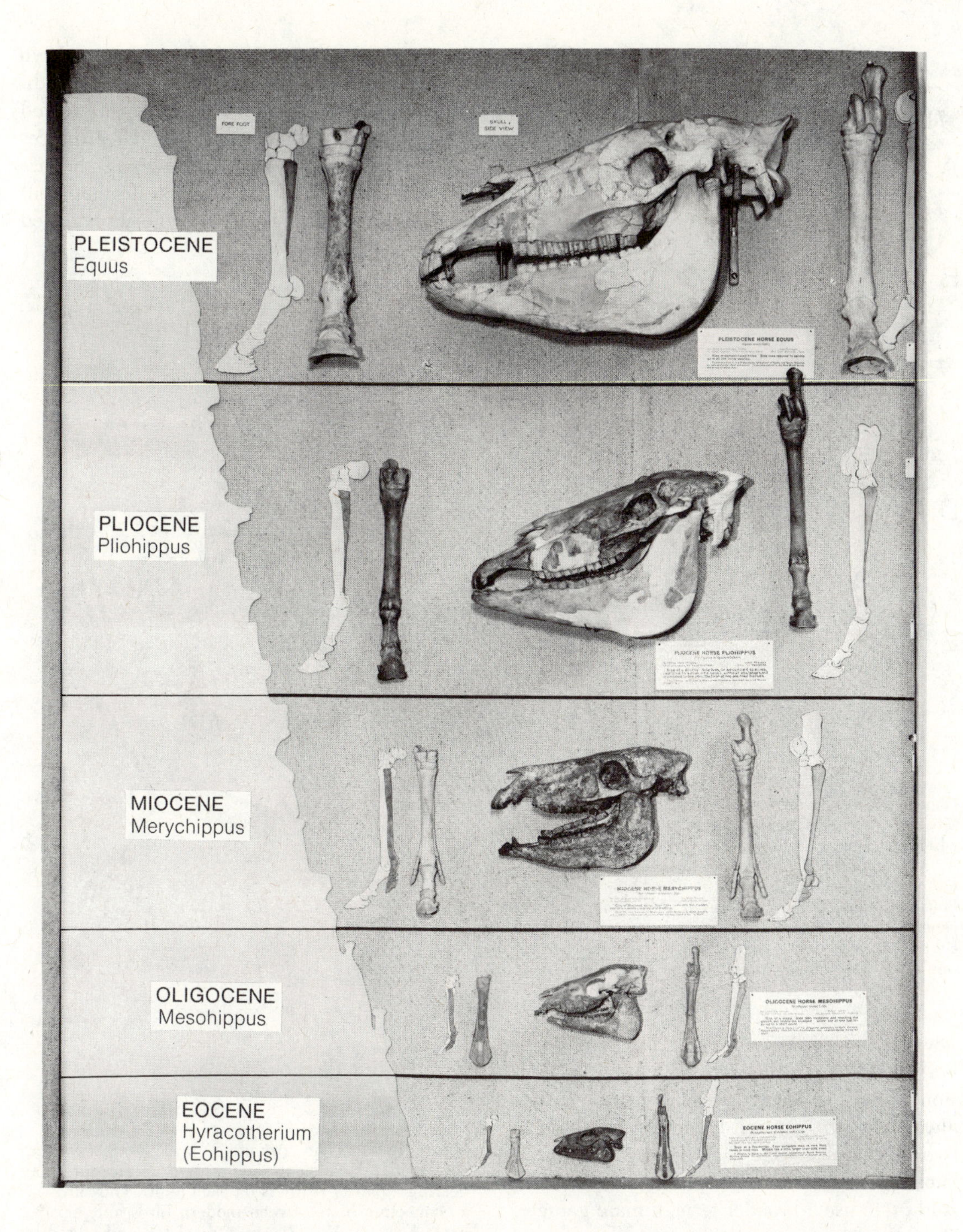

Figure 13-18. Evolution of horses: representing the gradual change from a small four-toed animal to a large single-toed animal. Adaptation to living on plains, running for security, and biting through the tough stems of grasses. Photograph by courtesy of the American Museum of Natural History, New York.

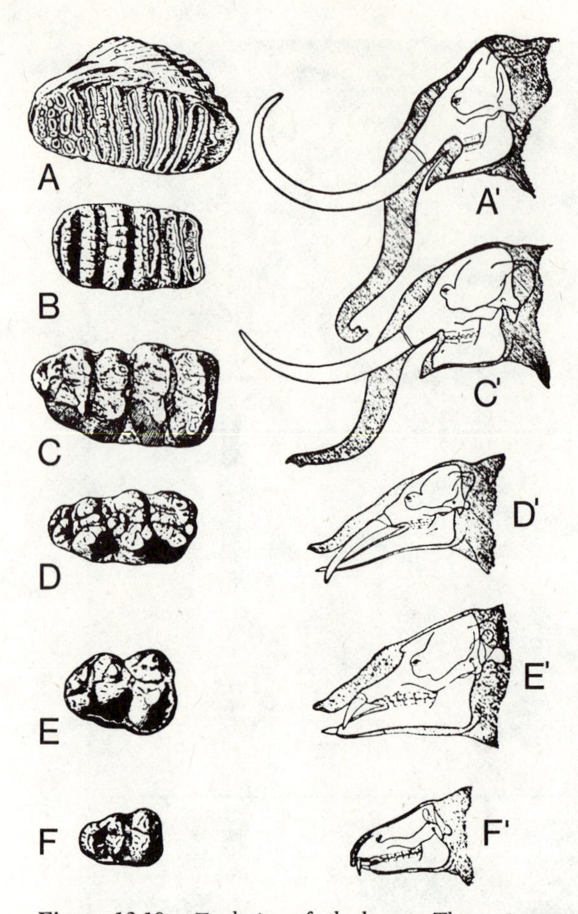

Figure 13-19. Evolution of elephants. The series at the left represents the transition of the molar teeth, which became grinding organs with several ridges. **A, A'** *Elephas;* **B** *Stegodon,* Pliocene; **C, C'** *Mastodon,* Pliocene; **D, D'** *Trilophodon,* Miocene; **E, E'** *Paleomastodon,* Oligocene; **F, F'** *Moeritherium,* Eocene. From R. S. LULL, *Organic Evolution,* Revised Edition of 1947 and earlier editions. Used by permission of the Macmillan Company.

never have more than one cell, but great numbers of others develop into adulthood by division of this solitary cell into many. Nearly always the primary cell is formed by union of two other cells. One of these is a male cell known as a **sperm** or (in plants) an **antherozoid** or, to use a broader term, a **male gamete.** The other is a female cell known as an **egg,** or an **ovum,** or a **female gamete.** The single cell (fertilized egg or **zygote**) resulting from the union of these two develops into an **embryo** which continues further into an adult individual. In portions of its sex organs called testes and ovaries the adult animal produces sperms or eggs or both, completing the life cycle. The study of the stages of development of the fertilized egg into an embryo is called embryology.

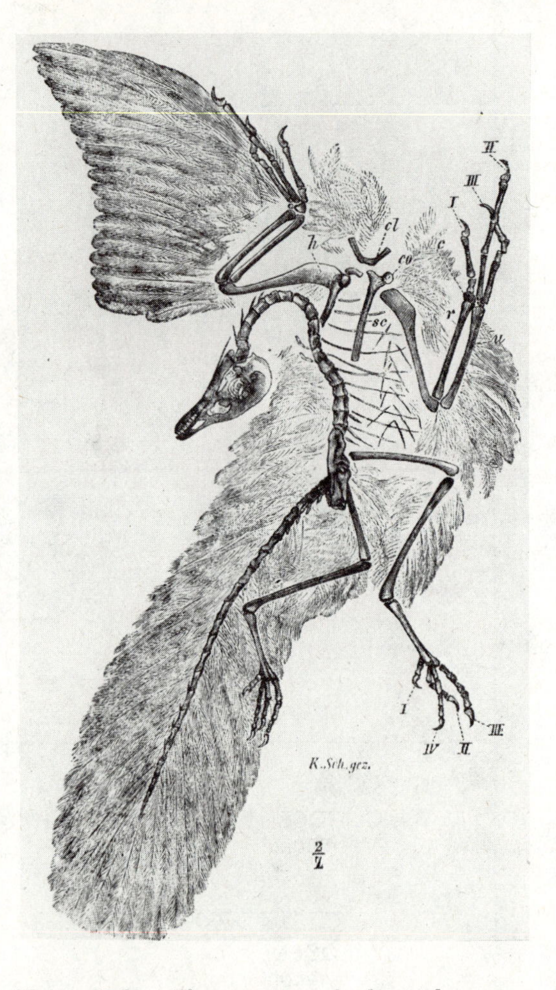

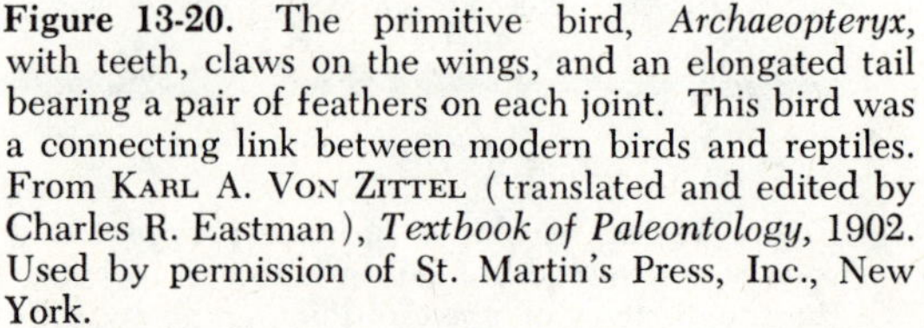

Figure 13-20. The primitive bird, *Archaeopteryx,* with teeth, claws on the wings, and an elongated tail bearing a pair of feathers on each joint. This bird was a connecting link between modern birds and reptiles. From KARL A. VON ZITTEL (translated and edited by Charles R. Eastman), *Textbook of Paleontology,* 1902. Used by permission of St. Martin's Press, Inc., New York.

In nearly all animals having more than one cell the first stages of embryonic development are somewhat alike. Some rather simple animals go through only the very beginning stages and, even after some specialization of their own, remain relatively primitive through their adult life. Animals of various other phyla pass through similar early stages and develop a little or much beyond them. The most highly developed group, the vertebrates (animals having backbones), continues through a very long series in embryonic development. Up to varying points the vertebrate embryos are seemingly alike, and differentiation occurs at later and later stages partly in accordance with the degree of relationship between any two animals. The embryos of a man and a rabbit, for example, become distinguishable only in relatively late stages; those of a man and an ape even later. On the other hand, the early human embryo is differentiated visibly from that of a fish.

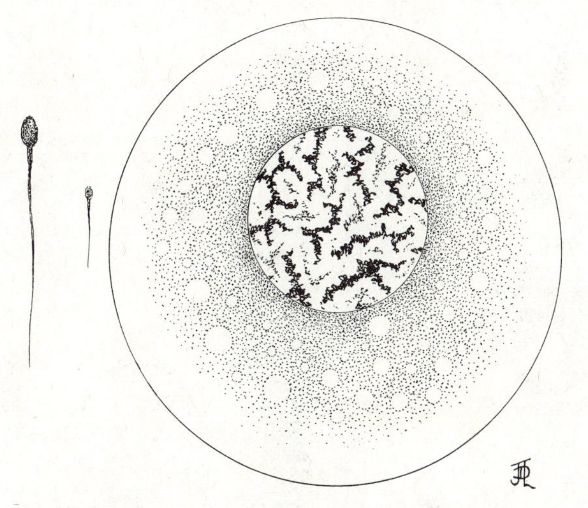

Figure 13-21. Human sperm and egg; the reduced sperm is proportional in size to the egg.

The adults of primitive vertebrate animals often have characters found only in the embryos of the more advanced types. In many cases these characters disappear during the course of the development of the embryo, though they may persist either as vestigial organs of no use or as specialized organs of specific value to the adult animal. Vertebrates in early stages of development have structures similar to those of the embryo of a shark. For example, there are relics of a circulatory system correlated with securing oxygen from gills instead of lungs. Functional gills are developed in the embryos of aquatic vertebrates but not in those of such groups as the birds, mammals, and reptiles that are adapted to life on land. Even in man there are remnants not only in the embryo but also during the adult life of the individual of features of a system correlated with securing oxygen and releasing carbon dioxide through gills. The eustachian tubes between the throat and the middle ears are remnants of one pair of gill clefts. They are separated by only the thin tympanic membranes from the corresponding passagways of the outer ears. In sharks and fish there is no membrane across the passageway, and gills along it absorb oxygen from water currents from the "throat" to the outside. During the course of human and other higher vertebrate embryonic development the blood vessels supplying gills are developed, but most of them disappear. In man the arch of the aorta (the great artery from the heart) and some other arteries are relics of the blood supply to the gill system. Various vertebrate groups retain different relics of the blood supply system of the embryo. In its early development there are only two chambers in the hearts of humans as in the adult shark, and later this heart develops into the typical four-chambered heart of a mammal.

The evidence of evolution derived from embryology was summed up a century ago under the statement of Haeckel, "Ontogeny recapitulates phylogeny". Ontogeny is the development of the individual; phylogeny is the development of the species or race. In other words, then, according to the recapitulation theory individual development repeats race

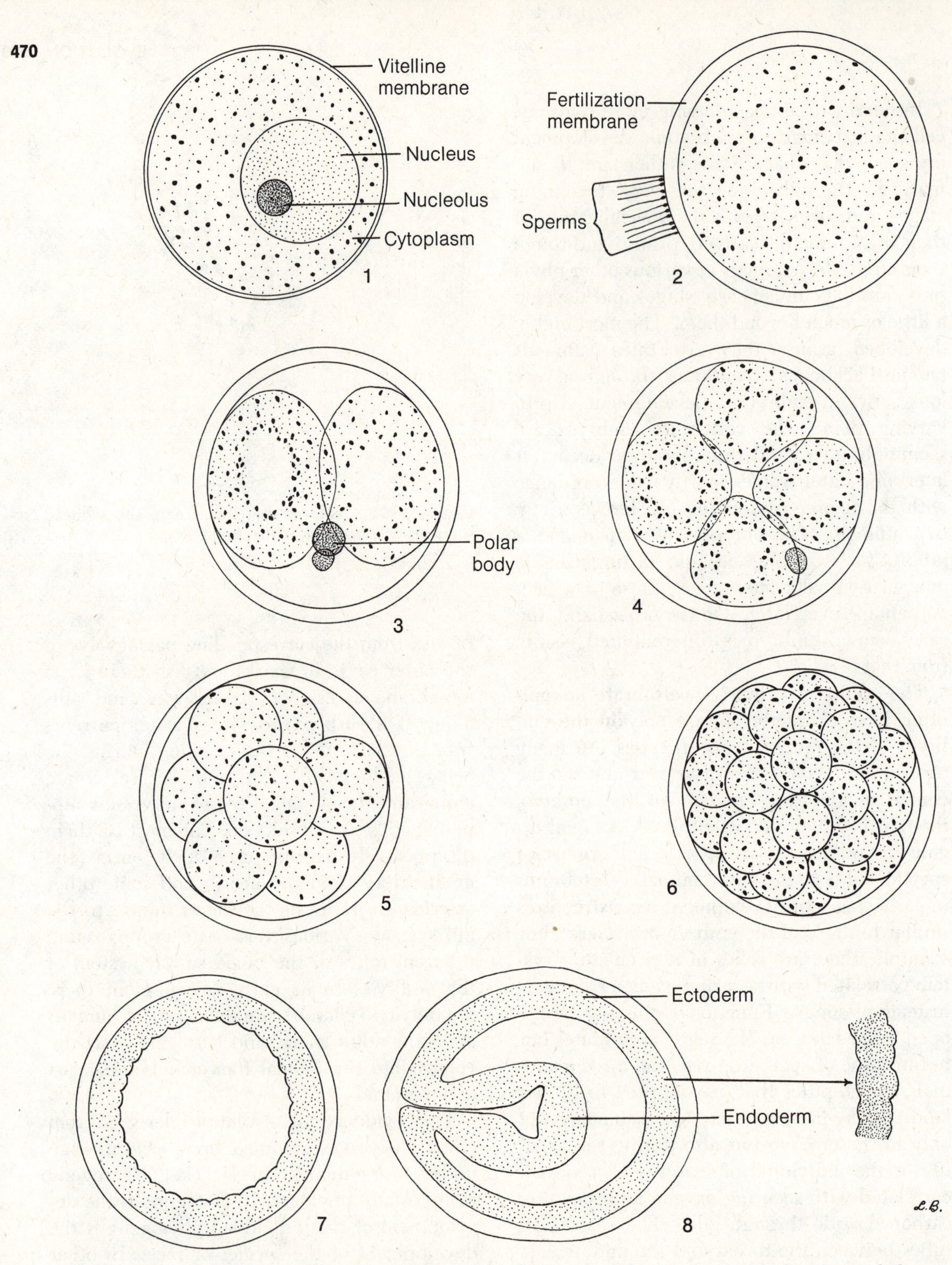

Figure 13-22. Fertilization and early stages in the development of the embryo of a tube worm (*Eurechis*) of salf-water flats along the Pacific Ocean: **1** egg, **2** fertilized egg and sperms, the fertilization membrane, in combination with complex chemical changes following entry of the sperm triggering its formation, excluding entry of other sperms after the first one, **3–6** division of the fertilized egg and passage through 2-, 4-,

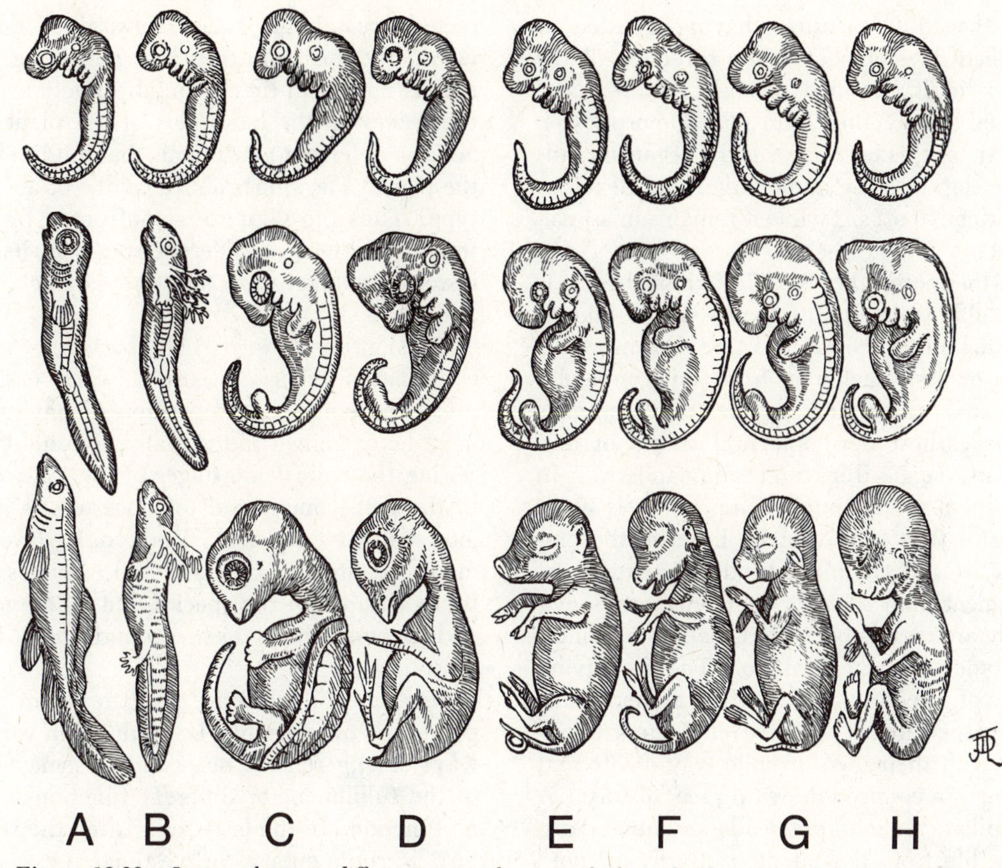

Figure 13-23. Late embryonic differentiation: the animals less closely related becoming differentiated at an earlier stage; **A** fish, **B** salamander, **C** turtle, **D** bird, **E** pig, **F** calf, **G** rabbit, **H** man. After Haeckel.

history. The embryonic history of related groups of some animals, and possibly in some cases plants, is similar except in the final stages. Presumably it parallels the history of the groups, which probably became separate only in the latter part of their historical development.

The recapitulation theory as a whole is discarded today, because it fits only some of the facts. The great similarity of the embryos of related organisms and even of the early embryonic stages of more remotely related organisms must be evidence of their relationship, as are all characters in common, except those superficially similar ones that have arisen by parallel evolution. However, this similarity is not evidence of recapitulation in the embryo of the whole course of evolution from primitive to advanced. Evolutionary changes may have been added either toward the end of the series of development leading to the present adult form or at an earlier point in the developmental series. In many cases the adults resemble the young of their ancestors rather

---

8-, 16- (not shown), and 32-celled stages, 7 blastula, a hollow-sphere stage, 8 gastrula, differentiation of an inner layer of cells (endoderm) by invagination (as if a tennis ball were pushed in by thumb-pressure) of cells at one point, the endoderm giving rise ultimately to the digestive tract and its appendages, the ectoderm in human beings to the skin, hair, nails, and nervous systems. Drawings from living cells.

than the adult ancestors, having arrested or modified development in the direction of the earlier adult, retained juvenile features, and evolved a new adult form (paedomorphosis). Sometimes this may occur if the organism fails to develop into the adult stages and develops reproductive organs while it remains in a juvenile state.

Furthermore, the recapitulation theory is applicable only doubtfully to plants and *at most* to animal groups in which the embryo is more or less sheltered from environmental fluctuations, as in birds, reptiles, and mammals or the earliest developmental stages of amphibians (e.g., frogs and salamanders). In even these groups much of ontogeny is not a recapitulation of phylogeny because the embryos are not completely independent of environment, and some of their characters evidently are correlated with it, being hereditary adaptations of the embryo. The free-living larvae of insects and many other animals have highly specialized adaptive characters correlated with their own juvenile way of life and having no connection with race history. A caterpillar, for example, leads an entirely different life from that of the butterfly or moth into which it matures. Natural selection operates upon populations at both stages in development. This is true in plants also. For example, the seedling of a saguaro or giant cactus (*Cereus giganteus*) must establish immediate contact with seasonally and briefly moist soil, and at the same time it must not provide too juicy a morsel for animals to eat. The young saguaro soon develops relatively large, stiff, downward-directed spines helpful in warding off rodents and other animals that might eat the young stem. The moisture of even a light rain is concentrated on the sharp points of these spines, forming drops which fall to the ground, moistening the soil much more than would a misty rain. When the plant starts to flower and it reaches a height of about four feet it ceases to produce the original type of spine, and from there on up the stem the spines are more slender and not directed prevailingly either upward or downward. By this time the chief absorbing root area is far out in the soil in all directions, but the spines on the basal part of the plant still act as a deterrent for animals that might climb the trunk. The spines above can be of another type. Thus the saguaro is confronted by different problems at different stages in its life history, and there are adaptive characters at both times.

Thus, although in part the basis for the recapitulation theory is correct, each case of presumed embryonic repetition of phylogeny must be examined individually. Resemblance during the embryonic stages of the individual to the embryonic stages of ancestors is common but not universal. Embryonic development is controlled by genes and enzymes, not by the history of the species, but many genes and enzymes have been around for a long time.

Fundamental patterns of anatomy in both plants and animals may be modified in various ways among related taxa, and this may lead to the fulfillment of different functions or of no function. In the latter case often the structure is rudimentary, or vestigial, i.e., much reduced as compared with the more elaborate homologous structures in its relatives.

A simple case of divergence of structure and function occurs in the well-known members of the pea family, the garden pea and bean. In the pea the cotyledons never emerge from the seed (Figures 5-6 and 5-7) but remain inside where they constitute a storage place or food reservoir upon which the developing embryonic plant draws even after it emerges from the seed coat. On the other hand, in the related bean the cotyledons enlarge, emerge from the seed coat, develop chlorophyll, become photosynthetic, and supply food for the young plant both from the reserve stored in them by the female parent plant and from their own manufacture as the seedling grows. The cotyledons of the pea and bean are homologous, but their development and methods of functioning are not the same.

Figure 13-24. Young saguaro, or giant cactus, in the Arizona Desert. Plant about 18 inches high and a number of years old, the principal spines in the areoles turned downward, stouter and much longer than the others, forming both (1) a defense against thirsty or hungry rodents and rabbits and (2) drip tips. The moisture of even a light rain condenses on the spines and runs down to the sharp points, where on each spine a drop is formed. This soon falls off, and the soil beneath receives moisture that otherwise would pass on with the air currents. Later, when the saguaro is about 4 feet high, a different type of spreading spine is produced, but the lower spines still inhibit climbing the stem. The water-absorbing area has moved outward from the base of the plant as the roots have grown, and the drip tips of the spines are no longer important for this purpose.

All leaves of plants are of the same fundamental origin, but they may have different functions. Most are flat and photosynthetic, but they are highly variable in form and sometimes structure (Figures 3-1 to 3-8). They may be further specialized as spines (Figure 2-15 **C**), tendrils (Figure 2-10 **A** and **B**), underground storage organs (Figures 2-12 and 2-13 **B**, onion), glands, flower parts, or fruits or parts of them. Similarly, portions of leaves, such as the teeth or lobes, may be modified into special useful or useless structures. Stipules may be green and photosynthetic, nongreen and either glandular or scalelike and functionless, or sharp and forming spines (Figure 3-2).

The monkey flowers (*Mimulus*, Scrophulariaceae) have four pollen-producing (fertile) stamens, and this is true of nearly all the members of the snapdragon, or figwort, family, except *Verbascum*, the mulleins, which have five stamens, and the genus *Veronica* and some relatives, which have two. In most instances the flowers of this family are specialized in having two upper petals differing from the lower three, and the forms of the tubular corolla and the free tips of the petals (together with size, color, and guideline markers) determine which insects are likely to enter the flower in search of nectar or pollen. Although in *Verbascum* there are five petals with five stamens in alternating positions, in most of the genera of the family one stamen is either missing or rudimentary or specialized. In the genus *Penstemon* the fifth stamen is present but sterile and specialized, and the name of the genus is derived from its presence as a fifth stamen. This stamen varies remarkably in form, and in some species it resembles a spatula, in others a toothbrush, and in still others a bottle brush. Although it bears no anthers or pollen, it may bear a close relationship to pollination, as, for example, in some species it lies in the right position to enable small bees to reach the nectar, stamens, and stigma. In other species within the same family the fifth stamen is specialized as a nectar gland, attracting insects. In others it is a useless rudiment of no particular value to the plant but nevertheless a marker of a line of development from ancestors with five stamens. Similar flowers with five stamens, differing only in having all the petals equal and similar to each other, are characteristic of the Solanaceae or nightshade family, which, as indicated by numerous characters in common, bears a close relationship to the Scrophulariaceae, the flower being less specialized and presumably more primitive. A transition toward unequal petals may be seen in the garden petunia (Solanaceae).

Rudimentary leaflets of the apricot. The apricot (Rosaceae) has a simple leaf blade perhaps similar to the terminal leaflet of a rose leaf. The pairs of lateral glands may possibly represent vestiges of pairs of lateral leaflets corresponding to those of roses. Each gland is produced at the apex of a vascular bundle, showing that it is not a superficial growth like many glands. Occasionally "abnormal" apricot leaves have some glands replaced by small green leaflets. Regardless of what may be the primitive condition, evidently apricot leaflets and those of other members of the rose family may be specialized in different ways and they may serve different functions or no function, even though they represent the same fundamental pattern. Some functionless types must be considered vestigial.

Comparative anatomy of various adult animals shows that a fundamental pattern may be modified in many ways and adapted to various uses or none. For example, the mouth parts of a mosquito are developed in such an unusual way that they may be used for piercing and sucking. However, they are not a special organ designed strictly for this purpose, but merely a late embryonic modification of ordinary mandibles developed in other insects into a form used for chewing. The forearms of the vertebrates are used for many purposes, but they all have the same fundamental pattern and their development is the same up to a very late stage. The human hand has five digits or fingers; so, also, has the foreleg of a cat, the wing of a bat, or the flipper of a whale or seal, even though each is used differently. The foreleg of a horse is essentially the same except that (as shown previously) it includes only one finger and vestigial bones of two others. The wing of a bird has also only three of the five fingers, but a fourth may be made out in the embryo. The wing of a bat includes all five fingers, one of them bearing its claw as in many other mammals, the other four forming the framework of an elaborate web used in flying. In view of the similar early embryonic development of all these organs, a logical conclusion is that the common fundamental pattern indicates a genetic relationship and a common extinct ancestor with some kind of generalized foreleg bearing five digits. Similarly, the salivary glands may secrete an ordinary digestive enzyme, but the specialized glands of some snakes secrete venoms of several kinds, and these are used to kill prey and for defense. The teeth of these snakes are specialized in various ways. The injecting fangs may be short and hollow or elongate and tubular or grooved. The injection may be by a sudden stab or by holding and chewing. In all cases both the glands and their poison and the teeth are specially adapted. *Bothrops insularis* of Queimada Island, Brazil, would lose the birds upon which it preys if the poison did not kill them very quickly. Other snakes can be more leisurely.

Presence of vestigial, nonfunctional organs is common among animals. The (vermiform) appendix of man is known as a troublesome organ, because it is noticed only when it becomes infected and usually must be removed. In many mammals this is a highly useful organ; in rabbits it forms a large part of the intestinal tract, being an area in which grasses and other food materials are held while they are being digested partly by bacteria. In man the corresponding structure is a relatively small pouch called the caecum, and the appendix is a smaller pouch associated with it. The latter is considered to be a remnant of little use in man.

Other vestigial organs among vertebrates are homologous with useful structures having

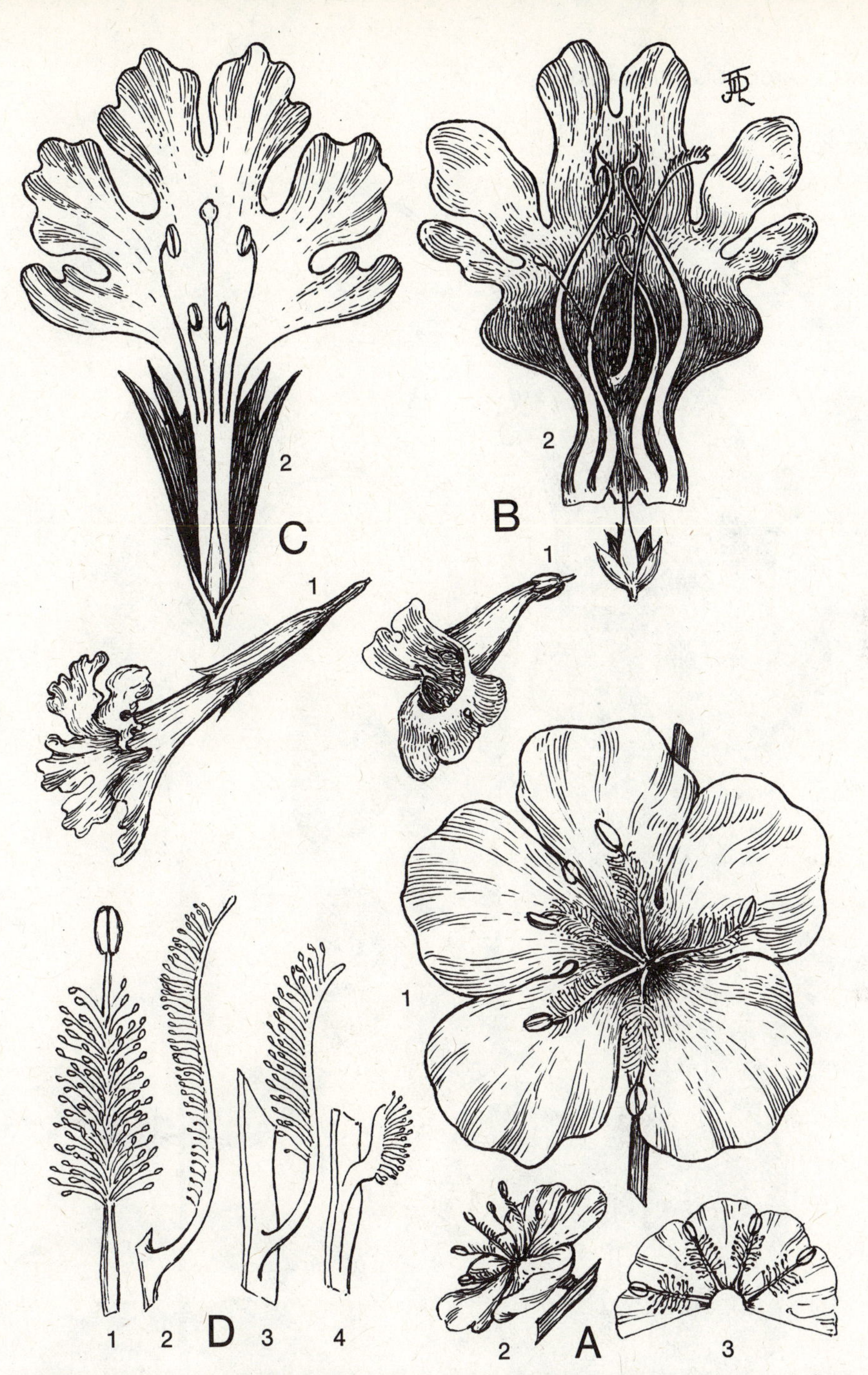

Figure 13-25. Vestigial or specialized organs in plants: the fifth stamen in the Scrophulariaceae: **A** mullein (*Verbascum*) with usually five stamens, but one or two sometimes reduced or different from the others, **1, 2** views of the flower, **3** corolla with attached stamens, laid open; **B** penstemon (*Penstemon* undetermined) with one stamen differentiated into a structure resembling a toothbrush, this sometimes of importance in pollination because the insect climbs upon it to reach the nectar; **C** bush monkey flower (*Mimulus* [*Diplacus*] *longiflorus*) with only four stamens and with no visible vestige of the fifth; **D, 1** stamen of *Verbascum*, **2–4** sterile stamens of species of *Penstemon*. In other members of the family the fifth stamen may be represented by all or part of a filament or by a gland.

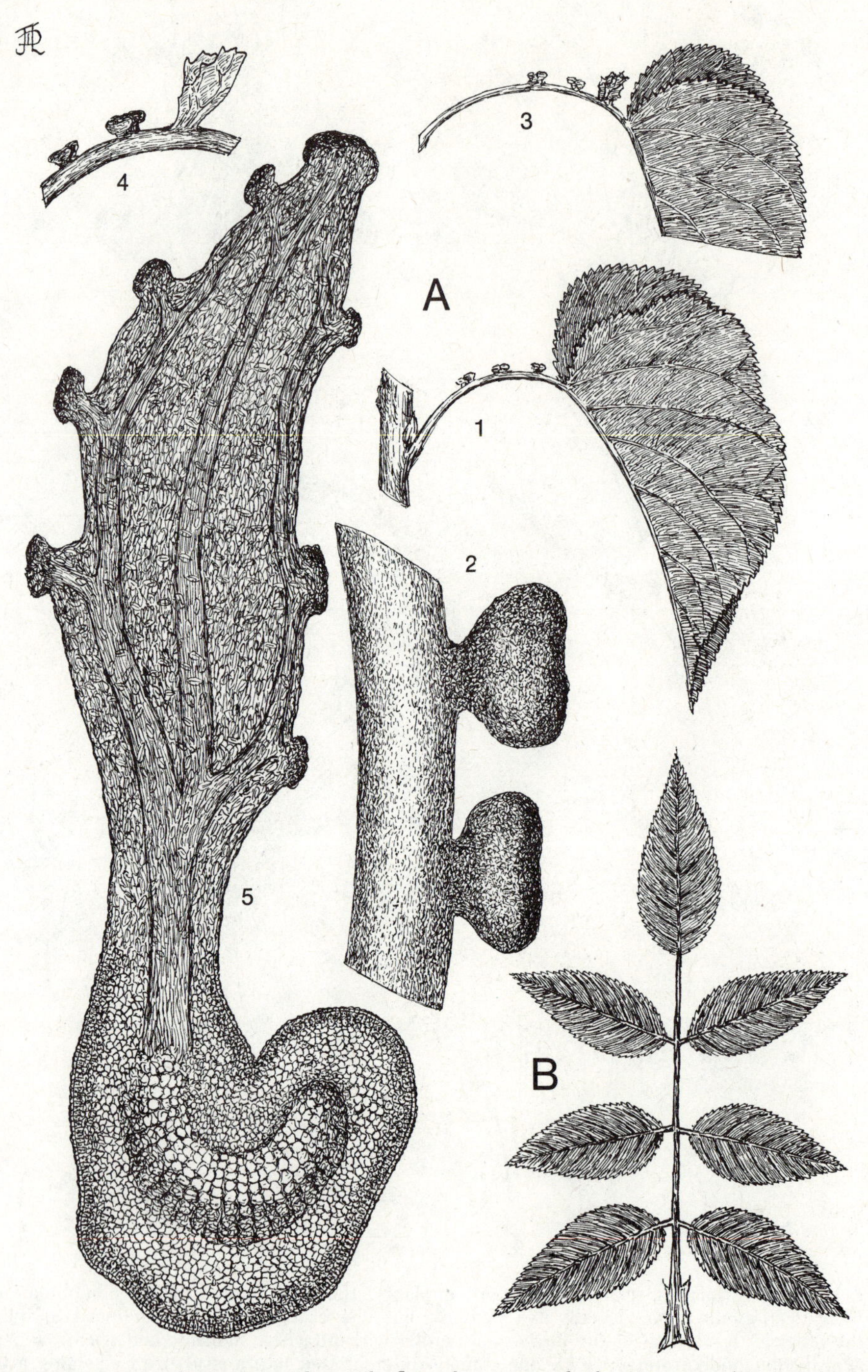

Figure 13-26. Vestigial organs in plants: leaflets of an apricot leaf: **A** apricot (*Prunus Armeniaca, Rosaceae*), **1** normal "simple" leaf with paired glands (vestigial leaflets) along the "petiole," **2** glands enlarged, **3** unusual leaf with one pair of glands developed into lateral leaflets, **4** glands and lateral leaflet enlarged, **5** cross section through the rachis of the pinnate leaf, the attached lateral leaflet cleared and showing the vascular bundles and lateral glands; **B** rose (*Rosa*) leaf approximating the pinnate type from which the apricot leaf may (or may not) have been derived.

the same embryonic origin in other vertebrate animals. If one watches a chicken or a captive eagle or hawk at a zoo, it is evident that the bird does not wink its eye. Instead a transparent membrane glides occasionally across the eyeball, moistening its surface. This is known as the third eyelid or nictitating membrane, and it is a common feature of birds and reptiles. A trace of it is to be discerned in the *plica semilunaris* at the inner angle of the human eye. The presence of such a useless structure in man and other mammals would indicate a "blood" or genetic relationship, in other words, a common descent of birds, reptiles, and mammals.

Useless muscles. These occur in various parts of the human body and in other animals. They include three which extend from the cartilage of the human ear to the bones of the skull. In many mammals they are used to move or flick the ears. Most people cannot use them, but some are amazingly proficient. It is to be noted that certain muscles occurring in some men and not others are developed regularly among the apes.

A python's hind legs. On either side of the anus or opening of the large intestine of a python is a single claw which, together with a few bones embedded in the body, is the only vestige of a hind leg corresponding to the well-developed legs of lizards.

The structure of water plants. Aquatic plants of many families have poorly developed internal water-conducting tissue (**xylem**) in contrast to the better-developed corresponding tissue of their terrestrial relatives. Plants living immersed in water do not need a conducting and supporting tissue like xylem (wood), and their stems are better adapted with only a little xylem and this flexible like a cord in the current.

GEOGRAPHICAL DISTRIBUTION OF LIVING ORGANISMS

Despite occurrence of similar habitats in various parts of the world, the individual continents and large islands do not have the same flora or fauna. For example, the hundreds of species of the cactus family (with the exception of a few introduced as in the Mediterranean Region and Australia and of four species of *Rhipsalis* probably introduced in the Tropics of the Eastern Hemisphere) are confined to North and South America. The three hundred species of hummingbirds also are native to only the Western Hemisphere. On the other hand, aloes, banyan-trees, lions, tigers, elephants, and many other organisms are restricted to Africa and Asia. The Eastern and Western Hemisphere monkeys are entirely different types, only those of the Americas having prehensile tails as adults, though the young of some Old World species make great use of their tails.

Study of the floras and faunas of the Earth shows that regions isolated for only a short period of geological time have, in general, similar though not identical plants and animals. The following are examples.

Northern North America and Northern Eurasia. The forests of the northern portions of both North America and Eurasia have many plant and animal genera in common, but relatively few of the species are the same. For example, the coniferous tree genera *Picea* (spruce), *Abies* (fir), and *Pinus* (pine) occur on both land masses, but no single species of any of them is found in both hemispheres. The continents appear to have had island stepping stones within about the last 5–10 million years. According to Hultén the last far northern connections occurred even within 1,000,000 or 2,000,000 years during the time of the great ice sheets. Formation of the ice sheets required sufficient water to be withdrawn from the ocean to expose the bottom of a large portion of the northern Bering Sea. However, it is not likely that the glacial period provided a land connection suitable for migration of any species but those of treeless Arctic regions, which are largely the same around the world. Almost certainly the forest plants and animals have been separated longer.

Western California and the Highlands of Arizona and Northwestern Mexico. About 20,000,000 years ago, during Miocene, a woodland dominated by oak trees covered the plain extending from the Pacific Coast of North America to the present states of Colorado and New Mexico. Uplift of the California mountain axis and cessation of the last glacial period and the development of a desert region east of the mountains has separated western California and southeastern Arizona and adjacent Mexico as effectively as though by an ocean. In both areas there is oak woodland, but, although there are many plant genera common to both woodlands, nearly all the species are different.

The Galapagos Islands and South America. The flora and fauna of the volcanic Galapagos Islands in the Pacific Ocean are largely different from those of the adjacent South American mainland. Not only are many of the native species **endemic** (found nowhere else in the world), but each island has its own peculiar plant varieties or animal subspecies. This is true, for example, of the cacti and giant tortoises. Some common groups such as the mammals (except bats) are missing. Observation during a visit to these islands was one of the chief factors resulting in the proposal of Darwin's theory.

In the cases discussed in the preceding paragraphs, segregation is mostly at the species level; in instances of longer separation it may be at the levels of genera, families, or orders. In the examples cited in the next paragraphs a prolonged period of geographical isolation is correlated with occurrence of plants and animals belonging to different higher taxa.

Warm Regions of the Eastern and Western Hemispheres. The cacti and the hummingbirds mentioned above are instances of restriction of entire families of organisms to the warm regions of one hemisphere and not the other. Of approximately 300 families of flowering plants, 56 are confined to the warm and tropical parts of the Eastern Hemisphere and 24 to similar areas of the Western Hemisphere. These regions have been separated by oceans for a long time, and the supposed land connections far to the northward could have been significant in plant migration only at a time when the climate of the Earth or its northern areas was relatively warm. According to recent evidence, areas of tropical climates connected eastern Asia and North America at the close of Cretaceous time and until Oligocene.

The 56 families restricted to the Eastern Hemisphere are characteristic of several areas also long isolated from each other by oceans or climatic belts. They fall into the following geographical groups.

Warm Regions of Africa and Asia (and in some cases Mediterranean Europe and the islands of the Pacific Ocean)	30	families
Tropical Africa	5	"
South Africa	6	"
Madagascar	2	"
Pacific Islands (in some cases including Australia and New Zealand)	7	"
Australia (alone)	6	"

The land mass of the Western Hemisphere is continuous, and the 24 families restricted to its warmer regions are not segregated by oceans as are those of the Eastern Hemisphere. Most of these families occur in the Tropics or through the warm regions in general; a few are restricted to Chile or to Chile and Peru at the south. Thus the chief segregation is according to ability to span the Tropics.

Australia and Other Land Masses. Australia was separated from Asia at or just before the beginning of the Cenozoic Era (about 60,000,000 or 70,000,000 years ago) just when the most primitive mammals (the monotremes and the marsupials) were in the process of development. Although in the main the marsupials were crowded out elsewhere, the group has continued to develop in Australia, and it is represented by many forms, some of these paralleling closely the chief types of the higher mammals occurring in other areas. Examples among marsupials are the wolflike "Tasmanian tiger," a molelike burrowing animal, and animals resembling rabbits and squirrels. Other forms, such as the kangaroo, are unique. The higher mammals developed elsewhere, and, having had no natural access to Australia, they have not occurred there as native organisms even though there have been habitats favorable to them. This is illustrated by the introduction of rabbits, which have been so successful as to become a pest. A similar catastrophe resulted from introducing cacti (prickly pears) from the Western Hemisphere because the plants spread rapidly as weeds, excluding sheep from forage.

New Zealand and Other Land Masses. New Zealand, likewise, has been separated from the rest of the world for a very long time, and most of its plants and animals are unique. Although the flora includes some genera and species in common with both Australia and Chile, many are not duplicated elsewhere. Some higher taxa common in other countries are missing evidently because following their development in other areas there has been no access to New Zealand. For example, before the coming of explorers from Europe there were no carnivores or other mammals; consequently, birds (some of them wingless) were safe on the ground until the introduction of cats wrought havoc with them.

Madagascar and Africa. The island of Madagascar is separated by only a relatively short sea distance from Africa. Nevertheless, there is a profound difference in the plants and animals of the two areas. For example, despite their abundance on the mainland, many plants and such well-known animals as elephants, lions, zebras, apes, and giraffes do not occur on Madagascar. This is correlated with long geological separation of the island.

The only reasonable explanation for this correlation of the degree of differentiation between isolated floras and faunas and their relative length of separation is evolution. Presumably, ever since the members of each pair of land masses were separated, evolution has continued in both places but along different lines. Each has a characteristic flora and fauna resulting from independent development from many of the same original groups of plants and animals.

In some cases current geographical distribution of particular plant or animal groups may be correlated directly with a known paleontological family history. The family of camels, for example, originated in North America, where it no longer occurs. One group is known to have migrated to South America, and it survives in the llamas and alpacas. Another migrated to the Eastern Hemisphere, and it survives in the modern camels and dromedaries.

ARTIFICIAL SELECTION OF PLANTS AND ANIMALS

Since prehistoric days man has exerted conscious or unconscious control over the pathway of evolution of the plants he used. Some of the seeds he threw on kitchen middens and other dump heaps grew in the disturbed habitats about the cave or other habitation, and there was variation among the plants on the waste ground. Some types were better than others, and he propagated them, commonly choosing those slightly better for his use. During the course of many centuries of chance variation and of selection according to the wishes of man, new types of plants appeared and became stabilized genetically. These included grains, fruits, and fiber plants much improved for his purposes over the original wild ones. Most of these could survive only in cultivation, often with extra water or at least with removal of unwanted competitors known as weeds.

Thus, in many instances man has controlled the pathway of evolution. By means of long continued selection he has derived new plant or animal types differing greatly from the originals and from each other.

In some instances the original wild plant is known as, for example, the single ancestor of cabbage, brussels sprouts, broccoli, kale, kohlrabi, and cauliflower. All these were developed from a single wild species, *Brassica oleracea,* the rock cabbage native on the sea bluffs of Western Europe. The edible part of the cultivated cabbage is a solitary, tremendous, terminal bud; that of brussels sprouts consists of numerous small lateral buds along the stem. Kale is used for leaf vegetables or green; kohlrabi for swollen leaf bases and edible stems; cauliflower for the thickened edible succulent branches of the flower-bearing portion of the plant, these being grown together (**fasciated**).

As man was choosing plants to be used, he also was selecting animals and controlling their evolution as well. At some point, wolflike species came to his kitchen middens, and, being fed, came ultimately to the campfire. Some young ones were domesticated, but doubtless many were rejected because they became vicious as they grew older. The first great selection must have been on the basis of controllability. The present-day poodle, dachshund, malamute, greyhound, or Chihuahua is a far cry from the original. If these breeds were encountered in nature they would be considered members of different species or perhaps of different genera. These and many other races have come into being in relatively recent times, and all have arisen under the direction of man, who has exerted careful control over the breeding of some of the animals, keeping certain characters in the stock and eliminating others.

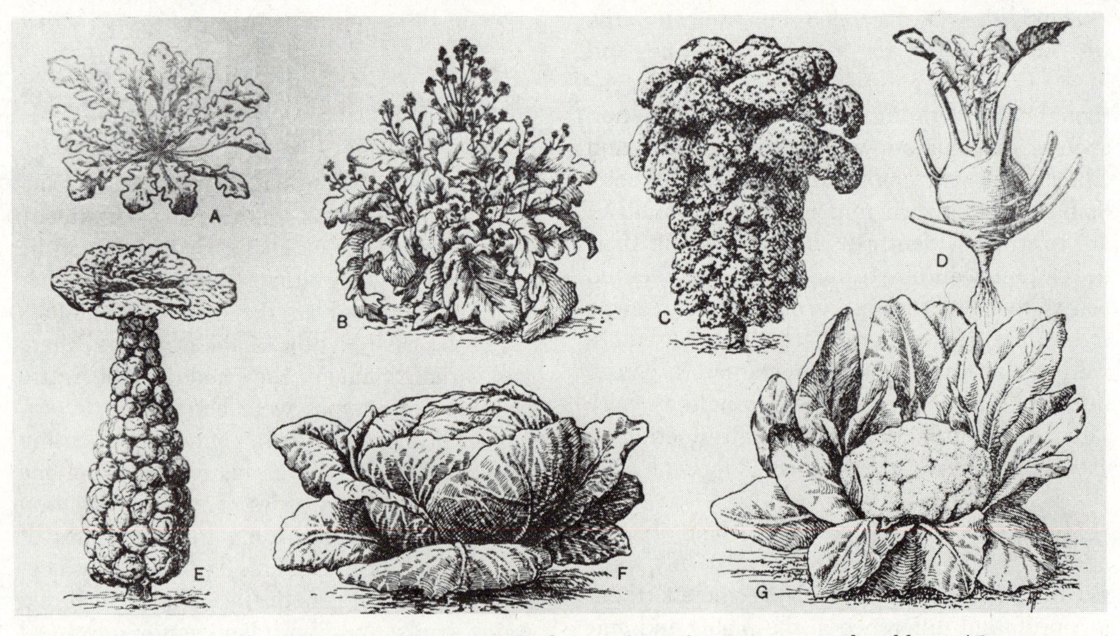

Figure 13-27. Vegetables derived by artificial selection from the European rock cabbage (*Brassica oleracea*), **A** rosette of basal leaves of the wild plant, a mustard, **B** broccoli, the inflorescence thickened, **C** kale, **D** kohlrabi, the leaf bases thickened, **E** Brussels sprouts, the large lateral buds edible, **F** cabbage, the single exceedingly large terminal bud edible, **G** cauliflower, the thickened, fasciated inflorescence edible. Each vegetable includes various cultivated types, as well. By permission from *Fundamentals of Botany* by G. Stuart Gager, copyright 1916. McGraw-Hill Book Company, Inc.

Figure 13-28. Artificial selection of animals: some breeds of dogs derived from wolflike ancestors domesticated by prehistoric man. Courtesy of the Kellogg Company, Battle Creek, Michigan, and of the United States Printing and Lithograph Company and Divisions (copyrighted by the Eastern Division), Brooklyn.

Similar care in selection has been devoted to other animals, including such household pets as cats and useful farm and range animals. These include many breeds of cattle selected either to produce meat or to supply the maximum of milk. Rich creamy milk from light brown Jerseys and Guernseys long was preferred because of nutritional value, but it has been replaced by thin milk from black-and-white Holsteins, now favored to aid in control of the waistlines of an affluent society. In the Iberian Peninsula and Latin America, some cattle are bred for male ferocity in the bull ring rather than for usefulness. The numerous kinds of horses are adapted, uncommonly now, to pulling loads or, more commonly, to carrying a rider across the range or around a track or for recreation. Chickens are bred for producing meat or for laying eggs, especially so in winter, which was once the off season, or for fighting (game cocks). The American wild turkey, a slim rangy bird, has been modified into a broad-breasted one with a preponderance of white meat, including a lesser model for small families which do not like to eat turkey for weeks. About 150 or more recognized varieties of domestic pigeons differ strongly from each other, but the ancestry of the entire group has been traced back to the wild rock dove native in Europe and North America. This point was established by Charles Darwin.

As will be shown farther on, during the current century man has learned the laws and controlling mechanisms of inheritance. Consequently, he now directs evolution according to the known principles of genetics. Plant and animal breeding have passed from the long stage of dependence upon chance genetic recombinations into the recently acquired status of a more or less exact science. Instead of working blindly by trial and error, man now controls the course of evolution for plants and animals according to definite objectives, and he proceeds systematically by precise methods based upon knowledge of underlying principles.

In recent years man has gone even farther in the breeding of both plants and animals. In development of grains, for example, he has gone to the ends of the Earth to secure strains which happen to have inherited factors producing rust- and smut-resistance. By cross breeding and selection of desirable offspring, he has combined these with factors producing resistance to cold or drought, flowering or fruiting at a particular time of the year, color or size of grain, higher starch or protein content, one flour texture or another, relatively small chaffy structures (glumes, lemmas, or paleas) around the fruit (grain), making threshing easier, or tall or short stems according to the conditions under which the plant might grow or be harvested. By precise manipulation of breeding he has produced races which combine all the characters he wishes for a particular locality. When corn (maize) picking machinery was introduced, it was desirable also to have the ears always at the same height above the ground, and new races with this characteristic were developed. For many years the beach strawberry, *Fragaria chiloensis*, occurring near the Pacific Ocean from Alaska to Patagonia, has been planted as a ground cover. For this it has not been wholly satisfactory, because in cultivation often it is not a vigorous grower, and a bed of it is spotty, reflecting minor differences of watering or the underlying soil. Lenz obtained hybrids of the beach strawberry and the commercial strawberry, derived from other species, then bred selected lines for cultivation. These plants had the vigor of hybrids and the robust size often associated with polyploids (being allotetraploids, see p. 486), and they did not reflect the slight variations of watering or substrate. They produced a uniform covering of the ground, and, furthermore, they were selected for delicious large fruits, a feature not found in the beach strawberries used for ground covers. Thus, by careful planning and controlling of breeding man now directs evolution in accordance with known laws of heredity.

What has occurred in the man-directed evolution of grains has been paralleled with countless other crop plants, resulting in the numerous varieties of apples, pears, peaches, plums, apricots, blackberries, melons, lettuce, peas, beans, citrus fruits, sugarcane, avocados, bananas, cherimoyas, mangoes, pineapples, rice, onions, potatoes, sweet potatoes, yams, artichokes, and other fruits and vegetables. In each case one, two, or sometimes a few ancestral types have given rise to many.

GENETICS AND CYTOLOGY

Background. The foundation of the science of **genetics** (heredity) as well as the basis of the related science of **cytology** (study of the cell) is continuity of the germ plasm (proposed as a theory by August Weismann). The hereditary characteristics of an individual are carried in the living part or germ plasm of the individual cell. This is passed on from generation to generation through the sperms and eggs. These unite forming the fertilized eggs, each of which develops into an embryo, later to become a mature individual, which again produces sperms or eggs (or antherozoids and eggs).

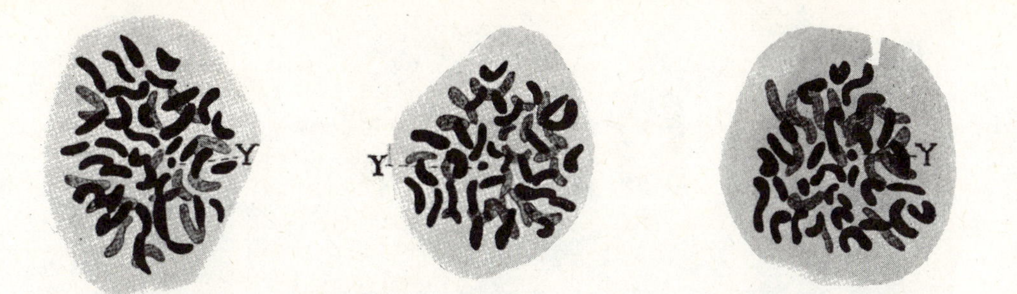

Figure 13-29. The chromosomes of man; *above,* three views of the chromosomes in the cells which form sperms in the testis of the male; *below,* the pairs of chromosomes lined up in order of size, those of the male in the upper row, those of the female in the lower. The chromosomes of the sex pair are at the right in each row, those of the male unlike, those of the female alike. From EVANS and SWEZY in the *Memoirs of the University of California* and from L. H. SNYDER, *Principles of Heredity,* D. C. Heath and Company. (Refined techniques show only 23 pairs of chromosomes.)

Experiments show that the factors governing inherited characters are carried in the nucleus of the cell, in special structures called **chromosomes.** The chromosomes are seen usually only when the nucleus is dividing. Ordinarily in any given species they are of a constant number. The number in man, for example, is twenty-three in the sperm and twenty-three in the egg, making a total of forty-six in the fertilized egg and in every derived cell in the direct line of development within the individual leading again to development of sperms or eggs. At this time the number is reduced once more to twenty-three by special "reduction divisions" (meiosis). Chromosome numbers vary widely among organisms. The typical number in the sperms and eggs of the common vinegar fly (i.e., *Drosophila melanogaster,* which occurs about overripe fruit) is four, and the body cells of the adult individual contain eight chromosomes, two of each kind. In the garden onion the number of chromosomes in the gamete cells is eight; in the mature plant there are twice as many, that is, sixteen. The smaller number, as, for example, twenty-three in the sperms or eggs of man, is called the "reduced number" and it is referred to, also, as **haploid (n).** The larger number, forty-six in the body cells of man, is known as **diploid (2n).**

By numerous ingenious experiments, some of them involving the study of the giant chromosomes occurring in the salivary gland cells of flies, it has been proved that the **genes** or factors determining inheritance of particular characteristics are arranged in a line along each chromosome. These occur at random or in "blocks" (series) insofar as their effect upon individual parts of the plant or animal are concerned. For example, in one chromosome of a plant a gene or block of genes affecting the color of the flower may be next to one affecting the structure of the stem or the shape of the leaves. However, in a single species the genes are arranged commonly in identical order in the chromosomes of every individual. The chromosomes, then, are the "packages" in which the genes affecting individual characteristics are contained.

In a given organism the chromosomes occur in pairs. For example, in the onion there are two long chromosomes each nearly straight except for a double bend in the middle. There are, also, two large similar ones resembling the

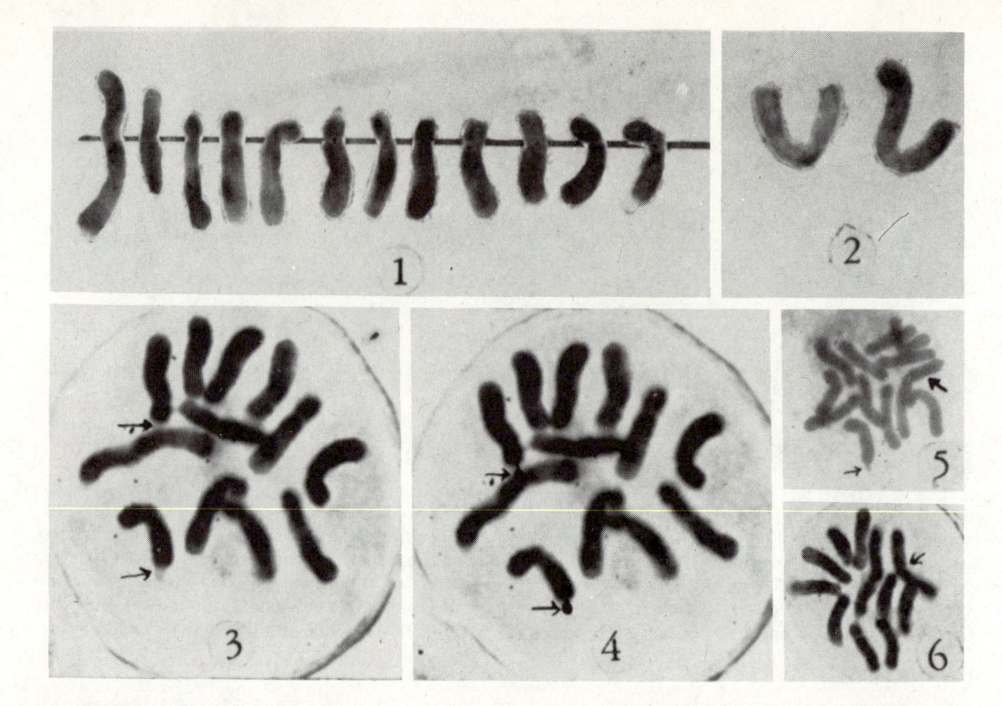

Figure 13-30. The chromosomes of the maidenhair tree, *Ginkgo biloba,* a dioecious plant: **1** one chromosome from each of the twelve pairs (taken from **3**), **2** the unlike members of the pair thought to be associated with sex determination (taken from **3**), **3** the chromosomes in their normal position in the cell, the cell with the haploid (i.e. reduced) number of chromosomes—one from each pair, **4** the same cell as in **3**, but the satellites of two chromosomes inked in for emphasis, **5** arrow at the right pointing to one member of the supposed sex pair of chromosomes, **6** arrow at the right pointing to the other member of the pair of supposed sex chromosomes. From Earl H. Newcomer, American Journal of Botany 41: 543. fig. 1–6. 1954. Used by permission.

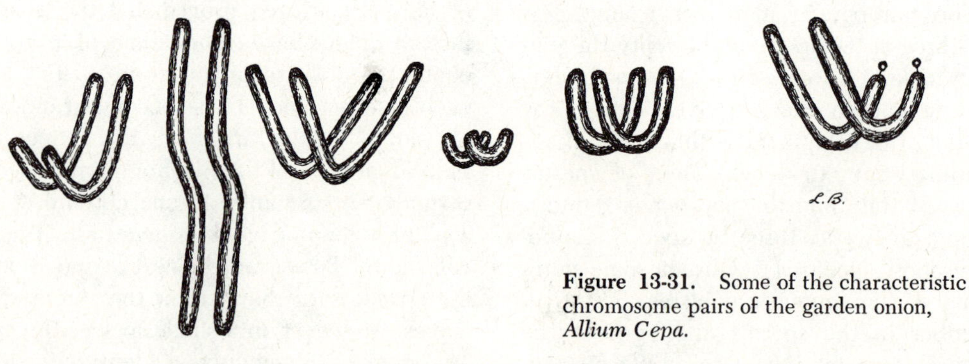

Figure 13-31. Some of the characteristic chromosome pairs of the garden onion, *Allium Cepa.*

letter "J" and another large pair also resembling the letter "J" but differing slightly from the first pair of "J's." There are five other pairs of chromosomes of "U" and "V" shape, members of which can be associated clearly with each other.

Genes each resulting in presence or absence of a particular characteristic or group of characteristics are located in regular order along each chromosome. These genes may be alike in the two chromosomes of the pair. For example, both may be of the type tending to produce a red flower or they may be different, one tending to produce red and the other white, but both are concerned with a single characteristic—flower color. A plant with two genes of the type yielding red flowers will have red flowers; an individual with two genes of the type yielding white flowers will have white; but ordinarily an individual with one of each, will have the flower color produced by the dominant gene. A dominant gene produces its standard effect regardless of the presence of another like itself or of one ordinarily yielding a different effect. If the gene resulting by itself in red is dominant, white is recessive. That is, when both are present, red takes precedence over white. In other plants white may be dominant to red. If neither is dominant, the two opposed genes produce pink flowers, as they do in the four-o'clock plant.

Recent evidence indicates the gene to be an alterable point in a very long double-stranded, spiral molecule of DNA (deoxyribonucleic acid). The DNA molecule is duplicated when the two spiral strands held together by hydrogen bonds are separated. Each strand serves as a template in synthesis of a similar strand. Thus two double strands identical at all points are formed. These carry the hereditary pattern of the set of genes governing the makeup of the organism. DNA controls the synthesis of RNA (ribonucleic acid) in the cell nucleus. The RNA is stored in the nucleus and later moved into the endoplasmic reticulum radiating into the cytoplasm from the nuclear membrane. RNA serves as a template, governing the formation of enzymes in the cell, and this controls the synthesis of enzymes, then proteins, then the development of the organism. This subject will be developed further below under evidence from biochemistry.

New Hereditary Characters. Evolution would be impossible if there were no means of producing new hereditary characters. They arise primarily by *gene mutation* and "reshuffling" (i.e., recombination) of existing genes (i.e., those which have arisen by earlier mutations, some of which may have been unexpressed or modified because of occurrence in combination with other genes inhibiting or changing their effect or the effect of a group of genes acting in combination).

The earlier workers, e.g., Lamarck and Darwin, who proposed the theory of evolution, knew nothing about chromosomes. This was true, also, of even Gregor Mendel, an Austrian monk, who (in 1866) determined by experimental crosses the regularity of inheritance of seven pairs of characters in the common garden pea. About 1900 Hugo DeVries proposed the Mutation Theory, which postulates sudden changes in hereditary characters. He based his conclusions upon observations of a species of evening-primrose (*Oenothera Lamarckiana*) in which sudden and formerly unaccountable changes in hereditary characteristics occur. Although the changes in this evening-primrose are not due to changes in individual genes but to certain cytological peculiarities of the plant, DeVries' theory of mutation now has been modified into the principle of gene mutation, which is supported by abundant evidence. Gene mutation, as it is understood today, is a chemical change in the gene itself resulting in an effect upon the production of an enzymatic action within the plant or animal and the production of new and unexpected characters. Mutations of this sort have been demonstrated over and over in the vinegar fly, Indian corn, barley, and many other organisms. Their rate of occurrence has been affected by irradiation of the germ cells with X-rays and other rays and by application of heat and cold and certain chemicals. Under normal circumstances mutations occur relatively frequently in some genes and only very rarely in others.

Changes in the evident characteristics of the developed plant may result from addition or deletion of chromosomes through peculiarities in nuclear division. Addition of a single chromosome makes a trisomic, a plant having three chromosomes instead of a pair of one type, but all the rest in pairs (a total of $2n + 1$ chromosomes). Such plants may have a marked change in hereditary characters. In the Jimsonweed (*Datura*) there are twelve pairs of chromosomes, and, since a chromosome may be added in any pair, there are twelve possible kinds of trisomics. Each trisomic has different

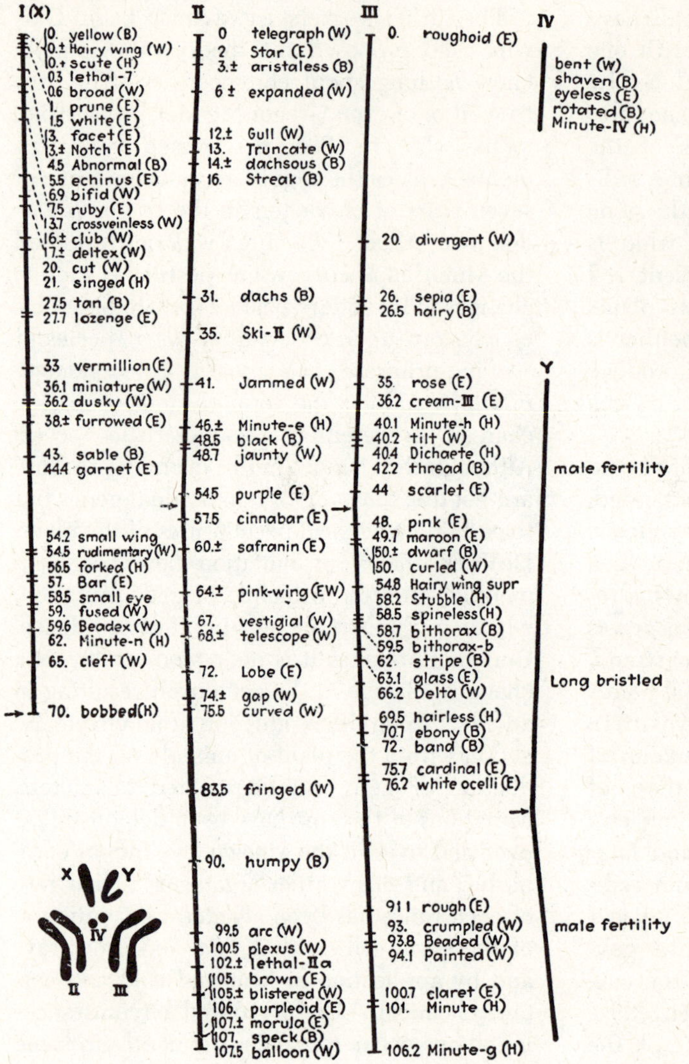

I (X)
0. yellow (B)
0.± Hairy wing (W)
0.+ scute (H)
0.3 lethal-7
0.6 broad (W)
1. prune (E)
1.5 white (E)
13. facet (E)
3.± Notch (E)
4.5 Abnormal (B)
5.5 echinus (E)
6.9 bifid (W)
7.5 ruby (E)
13.7 crossveinless (W)
16.± club (W)
17.± deltex (W)
20. cut (W)
21. singed (H)
27.5 tan (B)
27.7 lozenge (E)
33 vermillion (E)
36.1 miniature (W)
36.2 dusky (W)
38.± furrowed (E)
43. sable (B)
44.4 garnet (E)
54.2 small wing
54.5 rudimentary (W)
56.5 forked (H)
57. Bar (E)
58.5 small eye
59. fused (W)
59.6 Beadex (W)
62. Minute-n (H)
65. cleft (W)
70. bobbed (H)

II
0 telegraph (W)
2 Star (E)
3.± aristaless (B)
6 ± expanded (W)
12.± Gull (W)
13. Truncate (W)
14.± dachsous (B)
16. Streak (B)
31. dachs (B)
35. Ski-II (W)
41. Jammed (W)
46.± Minute-e (H)
48.5 black (B)
48.7 jaunty (W)
54.5 purple (E)
57.5 cinnabar (E)
60.± safranin (E)
64.± pink-wing (EW)
67. vestigial (W)
68.± telescope (W)
72. Lobe (E)
74.± gap (W)
75.5 curved (W)
83.5 fringed (W)
90. humpy (B)
99.5 arc (W)
100.5 plexus (W)
102.± lethal-IIa
105. brown (E)
105.± blistered (W)
106. purpleoid (E)
107.± morula (E)
107. speck (B)
107.5 balloon (W)

III
0. roughoid (E)
20. divergent (W)
26. sepia (E)
26.5 hairy (B)
35. rose (E)
36.2 cream-III (E)
40.1 Minute-h (H)
40.2 tilt (W)
40.4 Dichaete (H)
42.2 thread (B)
44 scarlet (E)
48. pink (E)
49.7 maroon (E)
50.± dwarf (B)
50. curled (W)
54.8 Hairy wing supr
58.2 Stubble (H)
58.5 spineless (H)
58.7 bithorax (B)
59.5 bithorax-b
62. stripe (B)
63.1 glass (E)
66.2 Delta (W)
69.5 hairless (H)
70.7 ebony (B)
72. band (B)
75.7 cardinal (E)
76.2 white ocelli (E)
91.1 rough (E)
93. crumpled (W)
93.8 Beaded (W)
94.1 Painted (W)
100.7 claret (E)
101. Minute (H)
106.2 Minute-g (H)

IV
bent (W)
shaven (B)
eyeless (E)
rotated (B)
Minute-IV (H)

Y
male fertility
Long bristled
male fertility

X Y IV II III

Figure 13-32. Map showing the chromosomes (*lower left*) and the relative positions of known genes on each chromosome as determined by experiments involving the hereditary phenomena known as linkage and crossing over. The letters in parentheses indicate the part of the insect in which the characters appear: **B** body, **E** eye, **H** hairs, **W** wing. The arrows of the diagrams shown the positions of the bends in the chromosomes. Adapted by L. H. SNYDER, *Principles of Heredity*, D. C. Heath and Company, from MORGAN, STURTEVANT and BRIDGES, and STERN; from SHARP, *Introduction to Cytology*, McGraw-Hill Book Company, New York.

fruit characters as a result of the presence of an extra chromosome of a particular kind. Sometimes two chromosomes, that is, a full extra pair, may be added or a single chromosome may be added in each of two or more pairs. Through deletion one pair of chromosomes may be represented by a single member. A portion of a chromosome may break off and be lost during one of the divisions of a nucleus. Sometimes what is deleted from one cell is added to another.

Polyploidy. Occasionally there are abnormalities in the reduction divisions or in the preceding ordinary nuclear divisions, and the chromosome numbers fail to be reduced normally from diploid (2n) to haploid (n) before sperms or eggs are produced. This may happen without human intervention or may be induced in plants by application of an alkaloid, colchicine, to dividing cells. When meiosis fails to occur, the sperms and eggs are diploid (2n). If a diploid sperm and a diploid egg are

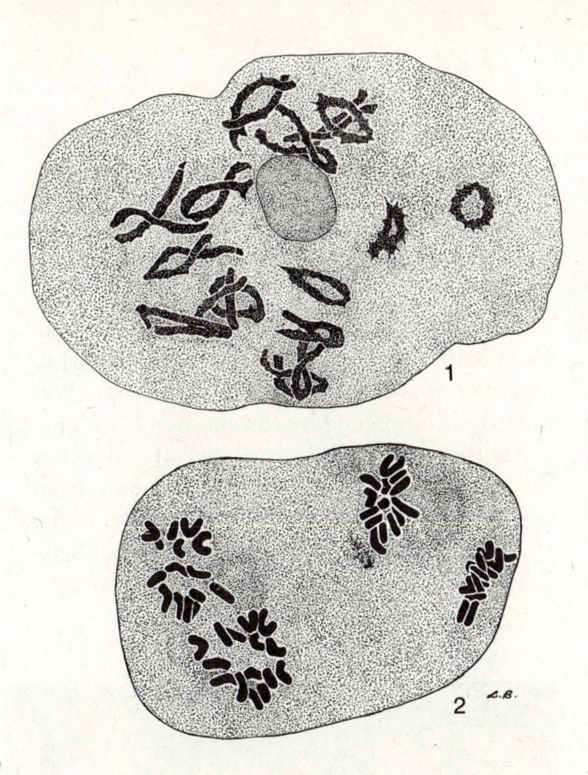

Figure 13-33. Chromosomes of a plant during reduction divisions (meiosis): **1** the 16 pairs of chromosomes of *Ranunculus orthorhynchus* var. *Bloomeri* when they first shorten and become obvious in an early stage (diaphase or diakinesis) of meiosis, each pair representing actually four strands one of which is distributed ultimately to each of the four cells resulting from the reduction divisions, **2** four sets of 15 (or possibly 16) chromosomes of *Ranunculus occidentalis* var. *Eisenii* near the end of meiosis, just before organization of four nuclei and division of the original cell into four, each of which will develop into a pollen grain. (*R. orthorhynchus* var. *Bloomeri*, Kelseyville, Cailfornia, *L. Benson 6311; R. occidentalis* var. *Eisenii*, Kelseyville, California, *L. Benson 6375.* Specimens and slides in the Pomona College Herbarium.)

united, the result is a tetraploid plant (4n). Tetraploidy may be induced in other ways. For example, very severe pruning of some members of the nightshade family has produced tetraploid shoots. Such tetraploids usually differ from ordinary diploids in that the cells and in special cases the organs of the plant are larger. The additional chromosomes may change the balance within the cell and within the plant as a whole, but the effects are predictable in only a few well-known cases.

Polyploidy is common in the plant kingdom, though rare among sexually reproducing animals, where the accompanying physiological changes are more likely to be disastrous. Many plant genera include species with varying chromosome numbers. In *Ranunculus* (buttercups), for example, there are two basic numbers. In one of these, n equals seven and in the other n equals eight, and there are various multiples of both these numbers, but particularly the latter, among the two hundred and fifty species. Among the two hundred species of *Crepis* (hawk's-beard, in the sunflower family) there are four basic chromosome numbers (3, 4, 5, and 6) and four others (7, 8, 11, and 20) are derived from these.

Synthesis of Species and Genera. One of the most convincing sources of evidence of evolution is the artificial synthesis of a well-known species from two others, proving the possibility of development of new species. A species of mint, *Galeopsis Tetrahit,* has been known since before the time of Linnaeus (1753), and its status as a species is clear. Two other well-known and well-defined species are *Galeopsis pubescens* and *Galeopsis speciosa.* The reduced number of chromosomes in *Galeopsis pubescens* and *Galeopsis speciosa* is eight (n equals 8); in *Galeopsis Tetrahit* the reduced number is 16. A particular series of crosses of *Galeopsis pubescens* with *Galeopsis speciosa* yielded a tetraploid race, i.e., individuals with thirty-two (4n) chromosomes (not reduced). Sixteen were derived from *Galeopsis pubescens* and sixteen from *Galeopsis speciosa.* This tetraploid race matched in every respect the well-known *Galeopsis Tetrahit,* and in general it could not be

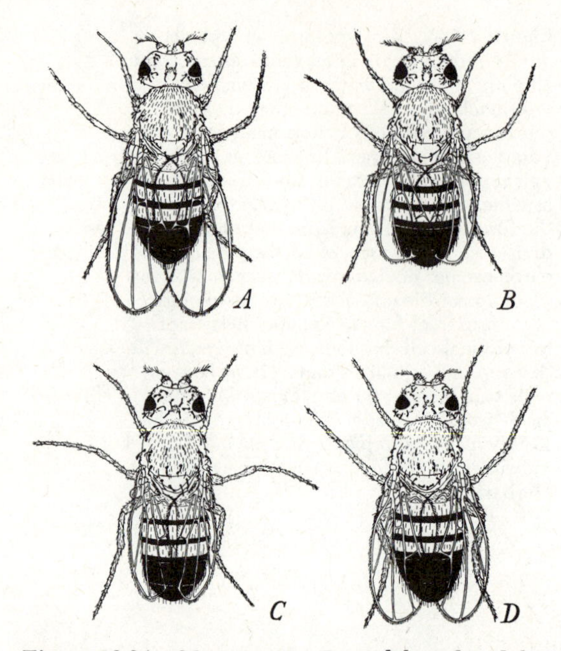

Figure 13-34. Mutations in *Drosophila* induced by irradiation from radium: **A** normal vinegar fly, **B** rudimentary wing, **C** miniature wing, **D** small wing. These mutations are inherited in the normal fashion as are other characters. From L. H. SNYDER, *Principles of Heredity*, D. C. Heath and Company; from HANSON in the *Journal of Heredity*.

distinguished from it either structurally or genetically. It crossed with natural *Galeopsis Tetrahit*, yielding some fertile hybrids with regular chromosome pairing at meiosis. It was *Galeopsis Tetrahit*, despite its peculiar method of synthesis. This experiment has been duplicated in other groups.

In 1870 a large grass soon to be known as *Spartina Townsendii*, or (according to author) *Spartina anglica*, was discovered growing in a single locality in southern England. Because of its intermediate character, it was thought to be a possible hybrid of two more common species, *Spartina maritima*, native in Europe, and *Spartina alterniflora*, introduced into Europe from the north Atlantic Coast of North America. *Spartina Townsendii* proved to be an aggressive plant, and it spread rapidly, covering large areas in England by 1902 and reaching France by 1906. *Spartina maritima* has 60 chromosomes (diploid or 2n, unreduced) and *Spartina alterniflora* has 62 (2n). *Spartina Townsendii* has 120, 122, or 124 chromosomes (unreduced; essentially tetraploid). Sterile diploids have 62 chromosomes (2n). This species could well have been derived by hybridization of *Spartina maritima* and *Spartina alterniflora*, accompanied by chromosome doubling.

Another classical case of special interest is the synthesis of a potential new genus from two well-known ones in the mustard family. The reduced chromosome number of cabbage (*Brassica*) is nine; the same is true of the

Figure 13-35. Mutation in barley (*Hordeum vulgare*); induced by irradiation. The plant at the left is the mutant known as vine; that at the right is normal. From L. H. SNYDER, *Principles of Heredity*, D. C. Heath and Company, from STADLER in *Scientific Agriculture.*

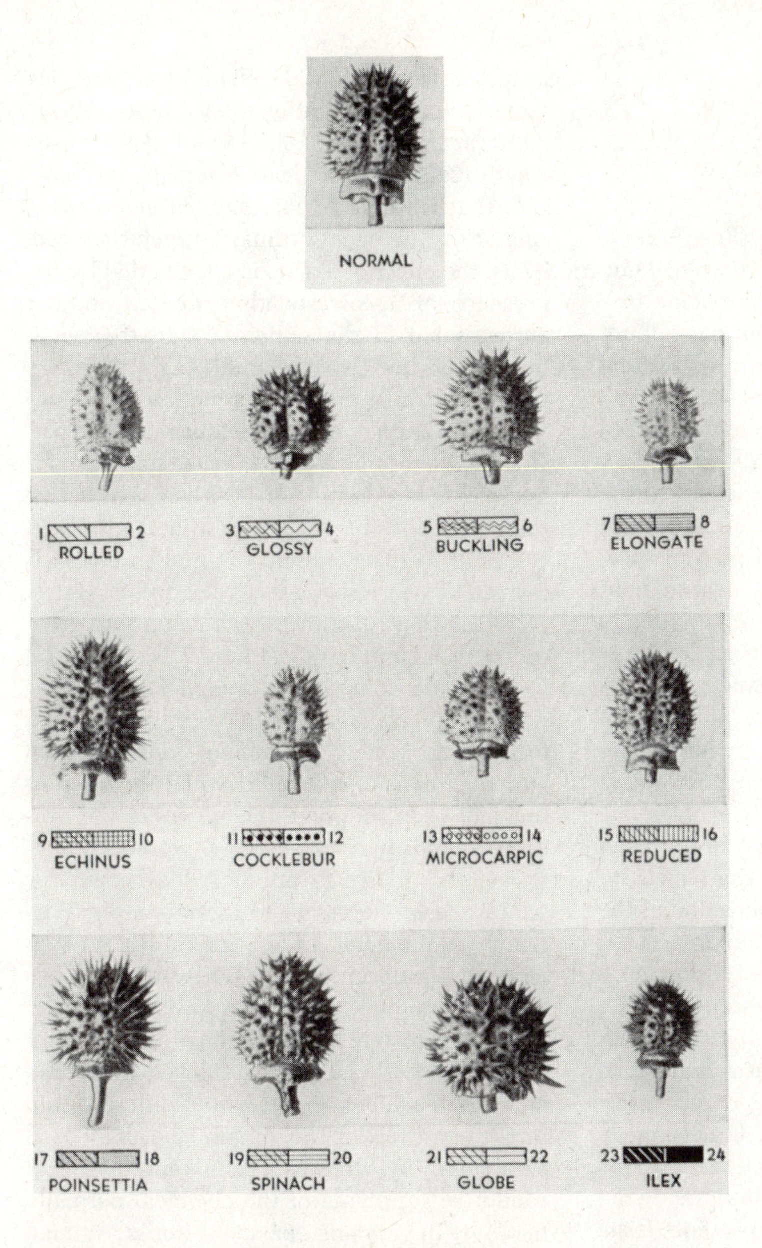

Figure 13-36. Variation in the fruits (capsules) of Jimson weeds (*Datura*) according to addition of a chromosome in any one pair. Each capsule is from a different kind of trisomic plant, there being twelve possible types because there are twelve pairs of chromosomes. A diagram of the extra chromosome is shown in each case. The laboratory name assigned to each variant is given beneath the illustration. The characters of the whole plant are affected by addition of an extra chromosome. From L. H. SNYDER, *Principles of Heredity*, D. C. Heath and Company; from BLAKESLEE in the *Journal of Heredity*.

radish (*Raphanus*). Crossing these plants produces sterile hybrids. However, in one instance polyploid plants were derived from a cross because in the highly sterile F_1 hybrid the usual reduction in chromosome numbers did not occur before production of pollen and ovules. The resulting plants had 36 (4n) chromosomes altogether, two full sets of cabbage chromosomes (eighteen) and two full sets of radish chromosomes (eighteen). Thus, in the reduction divisions every chromosome might pair with a homologue. Consequently, the plant could reproduce normally by sexual methods. This hybrid has been given the generic name of *Raphanobrassica*. Unfortunately from an agricultural point of view, this peculiar plant had the root of a cabbage and the top of a radish.

Biochemistry*

THE BIOCHEMICAL NATURE OF THE GENE

The gene is a segment of a very long organic chemical molecule called DNA (deoxyribonucleic acid). Alteration of a gene results in a change in a protein synthesized in the cell under the indirect but precise control of that particular gene. Every cell has a large number of genes, not all of which are actively functioning at all times. Some genes are active only during early development of the organism, influencing ultimately either the formation of structures or the development of physiological characteristics. Others are active throughout the life of the cell; still others are active only under certain conditions.

The gene is a sequence of chemical subunits called nucleotides occurring in a molecule of DNA. The DNA molecule is a very long chain of the basic nucleotide units, carrying coded information of many genes. The elongate double helical molecules of DNA are located in the chromosomes, and genes exert their influence on the cell through transcription of the information recorded in them upon related substances known as messenger ribonucleic acids, or messenger RNA. Messenger RNA moves from the nucleus through the nuclear membrane and into the cytoplasm, where, in the presence of the ribosomes, genetic information is translated (decoded) into complex proteins. Knowledge of the action of DNA and RNA has been obtained as follows.

In 1941 Beadle and Tatum presented the first clear experimental evidence indicating the manner in which genes act by controlling the synthesis of proteins in the cells. Through a wise choice of an organism for experiment, they were able to formulate the *one gene, one enzyme hypothesis,* ultimately modified to substitute "one polypeptide chain" for "one enzyme." For investigation they chose *Neurospora crassa,* the red bread mold. Like many other fungi and most algae, *Neurospora* is haploid, that is, there is *only one chromosome of each type* in each ordinary vegetative cell. Thus, the effect of a gene is not clouded by the presence of a corresponding one in another chromosome in the same cell and the possibility that the two are unlike. Furthermore the mold can be grown on a medium made up of known and controlled chemical substances. It must be supplied with salts, a simple sugar, and a vitamin, biotin. It is able to synthesize the other substances it needs. If an amino acid is added to the medium, the mold can use it instead of synthesizing its own, and this is true of some other substances. The ability to grow on the usual minimum medium was discovered to be lost by some X-ray-irradiated molds and not by others. Some would grow only in a medium to which one amino acid had been added, others in media to which a different one had been supplied. The genes of the irradiated molds had undergone mutations, eliminating the ability to produce one enzyme or another, each necessary to formation of a particular amino acid. Identification of a missing enzyme became possible. For example, three separate mutant strains of *Neurospora* lacked the ability to synthesize arginine. Breeding of the strains with each other showed that three enzymes, each concerned with a different step in the synthesis of this amino acid, had been eliminated by different gene mutations. In each case supplying of the chemical normally made by the missing enzyme restored arginine synthesis. Consequently, a single gene was shown to control the synthesis of a single enzyme.

While the additional experimental evidence supporting the *one gene, one polypeptide chain* hypothesis was being obtained, chemists were studying the nature of nucleic acids with special emphasis upon DNA. DNA was found to be responsible for the transmission of hered-

* This section is written by Robert L. Benson, Department of Biochemistry and Biophysics, University of California, Davis.

itary characters from generation to generation and indirectly for their expression in the organism.

DNA was isolated first in 1869 by Friedrich Miescher in Switzerland from a pepsin enzymatic digest of cells, although this material did not receive the name DNA until much later. Interestingly, this work was published about the same time as the genetic work of the Austrian Monk, Gregor Mendel, in 1866. As has often been the case, the relationship of two seemingly unrelated results was not appreciated fully for many years. Eighty-four years passed before Watson and Crick published their model of DNA structure in 1953, and only then the precise relationship between these two early experiments could be established.

About 1924, a German chemist named Robert Feulgen applied a selective strain for DNA to cells and showed that DNA was contained almost wholly within the nucleus (it is known now that mitochondria and chloroplasts have DNA and are partly genetically autonomous). This information, coupled with the growing body of cytogenetic evidence that chromosomes are involved in heredity, suggested to some that DNA might be the genetic material. However, since the nuclei also contained large amounts of protein, better evidence was required in order to determine conclusively whether genetic information was stored in protein or in DNA. In 1944, Avery, McCarty, and MacLeod in New York discovered that nonvirulent strains *Pneumococci* could be "transformed" into virulent strains by the addition of highly purified DNA isolated from virulent strains of the bacteria. While these experiments strongly indicated that DNA was the hereditary material, skeptics suggested that the DNA preparations used in these experiments were contaminated with protein which was the "true" hereditary material.

The interest in DNA was heightened by "transformation" experiments with bacteria, and in the late 1940's and early 1950's Chargaff's laboratory published a series of meticulous chemical analyses of DNA which showed that DNA was the same in all developmental stages and tissues of the same species, but differed among species. Moreover, Chargaff determined a relationship among the four "bases" in DNA: the content of the base adenine (A) was equal to thymine (T), and that of guanine (G) was equal to cytosine (C), or $A = T$, and $G = C$. Since adenine and guanine are purine bases, and thymine and cytosine are pyrimidine bases, then there is always one purine for each pyrimidine.

In 1952, Hershey and Chase published a paper on their experiments with bacteriophage replication. They incorporated two radioisotopes into the phage, radioactive phosphorus (^{32}P) specifically into the DNA and radioactive sulfur (^{35}S) specifically into the proteins. They then incubated the doubly labeled phage with bacteria and checked to see whether just ^{32}P, just ^{35}S, or both isotopes were injected into the hosts. The results were unambiguous; only ^{32}P, the specific marker for DNA, was injected into the bacteria, very strongly indicating that DNA is the hereditary material.

In 1953 Watson and Crick presented a model showing their interpretation of the DNA molecule. This consisted of a double helix (spiral ladder), the two antiparallel sides being composed of alternating units of the sugar deoxyribose and of a phosphate group. In other words, these two alternating groups are polymerized into a long chain. From each of the sugar elements of the chain, there is a chemical side chain known as a "base." Each base specifically pairs with a specific complementary base on the other side of the molecule. Thus, paired bases form each rung of the ladder in double-stranded DNA. However, the rungs are not all alike, because the pairs of side chains which compose them differ. Each rung is composed of one *purine* (adenine or guanine), one *pyrimidine* (thymine or cytosine). The pairing is always specifically either *adenine with thymine* or *guanine with cytosine*, and this explains why Chargaff observed that A = T and C = G. The members of the pair forming a rung may be in reverse order with respect to the two sides of the ladder, and the rungs may occur in any order. Thus, there are four types of rungs: A-T, T-A, G-C, and C-G.

The fundamental importance of this interpretation, now universally accepted for the structure of DNA, is based upon the ability of the hydrogen bonds (represented in Figure 13-37 **A** by short cross lines) between the two members forming each rung of the ladder to be broken and thus for the two sides of the ladder to be separated (Figure 13-37 **B**). According to the sequence of bases attached to it, each half acts as a template used for polymerization of each new half from deoxyribonucleotide units present in the nucleus. Consequently, the newly formed single strands are identical with the original single strands they replace, and therefore, each of the two new double-stranded DNA molecules is identical with the original double-stranded structure from which both were derived (Figure 13-37). Thus, identical DNA molecules are made available to be passed on to each cell in the organism or to its offspring. Each DNA molecule is composed of one parental DNA strand and one newly synthesized strand, as shown by Meselson and Stahl in 1957, and this is referred to as "semiconservative" replication of DNA (Figure 13-38). Each new double-stranded DNA molecule continues to control indirectly the process of synthesis of proteins in the two daughter cells. Consequently, the power of replication by each of the two parts of the DNA molecule of the portion which has been lost to it by lengthwise division is fundamental to transmission of hereditary characters.

The sequence of the four types of rungs of the ladder forming the DNA molecule is capable of enormous variation in the pattern of the arrangement of units, and this variation constitutes the hereditary information. According to the sequence theory, a gene is a segment of the DNA molecule which contains information in the sequence of bases on one of the two strands (the "sense strand") to specify the number and types of amino acids contained in one particular protein. Three bases (a triplet) are needed to specify one amino acid, which is the primary subunit of which proteins are made. There are twenty types of amino acids, each specified by one or a few specific triplets.

DNA does not control directly the synthesis of proteins, and this synthesis does not occur in the nucleus but in the cytoplasm on small particles called ribosomes. The information in the DNA gene is passed on through messenger RNA (m-RNA) as follows.

Messenger RNA is formed in the nucleus according to the sequence of bases specified by a particular gene segment on the sense strand of the DNA molecule. RNA has the same basic structure as DNA, except that the sugar ribose replaces deoxyribose, uracil replaces thymine, and the molecule is single-stranded. The sequence of bases in m-RNA is complementary to that of the "sense strand" of DNA which serves as a template for m-RNA synthesis, and therefore the m-RNA has the sequence of bases in the "non-sense strand" of DNA (except that uracil replaces thymine). The molecule of messenger RNA is a single long strand corresponding with one or more genes. When it moves into the cytoplasm, it serves as a message through codons controlling the specific sequence of the amino acids in the protein(s) to be formed. The messenger RNA molecule becomes attached by one specific end to a ribosome in the cytoplasm.

Transfer RNA (t-RNA) is a shorter molecule of RNA with a looped configuration that in a two-dimensional figure looks like a four-leaf clover. An amino acid is attached to the t-RNA, and each specific type of amino acid has one or a few corresponding specific types of transfer RNA. The amino acid-transfer RNA's contain *"anticodons,"* and they line up along the messenger RNA in a sequence determined by the *"codons"* of the messenger RNA.

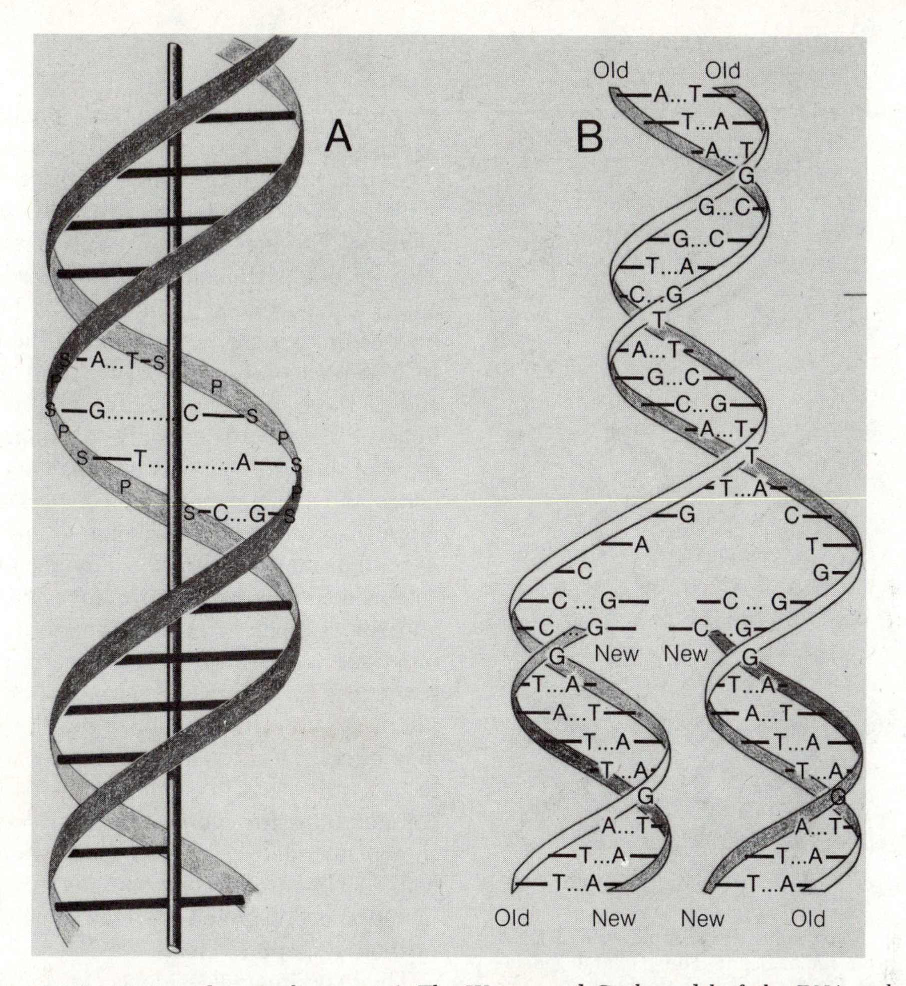

Figure 13-37. DNA and its replication. **A** The Watson and Crick model of the DNA molecule, which consists of two elongate spiral polynucleotide chains. Their adjacent bases (indicated by letters: S for sugar; P for phosphates; A or G for purines, A for adenine and G for guanine; T or C for pyrimidines, T for thymine and C for cytosine) are held together by hydrogen bonds. One purine (A or G) and one pyrimidine (T or C) occur in each of the "rungs" connecting the long spiral columns of the "ladder." The ladder, or double-chained structure, is a double helix. The central rod is hypothetical, and it marks the linear axis. **B** Replication of the DNA molecule, shown on the lower part of the drawing. The strands separate after formation of complementary units on the new strand of DNA, dictated by the occurrence of units in the original strand. **A**, modified as it appears in *Biological Science* by William T. Keeton, 2nd ed. Illustrated by Paula DiSanto Bensadoun. With permission of W. W. Norton and Company, Inc. © 1972, 1967 by W. W. Norton and Company, Inc. **B**, from James D. Watson, *Molecular Biology*. Ed. 3. W. A. Benjamin, Inc.

(Note that anticodons make sense!) Under the influence of the ribosome, the amino acids are combined in sequence, forming the protein. This is accomplished with the aid of ribosomal enzymes and several other proteins.

The ribosomes contain a third type of RNA called *ribosomal RNA* (r-RNA) which is essential to the structure of the ribosome, but the precise role of r-RNA in protein synthesis remains to be determined.

The preceding information on replication of DNA, and the production of specific proteins according to information contained in DNA, is summarized in Figure 13-39. The nature of the coding system is described more fully below.

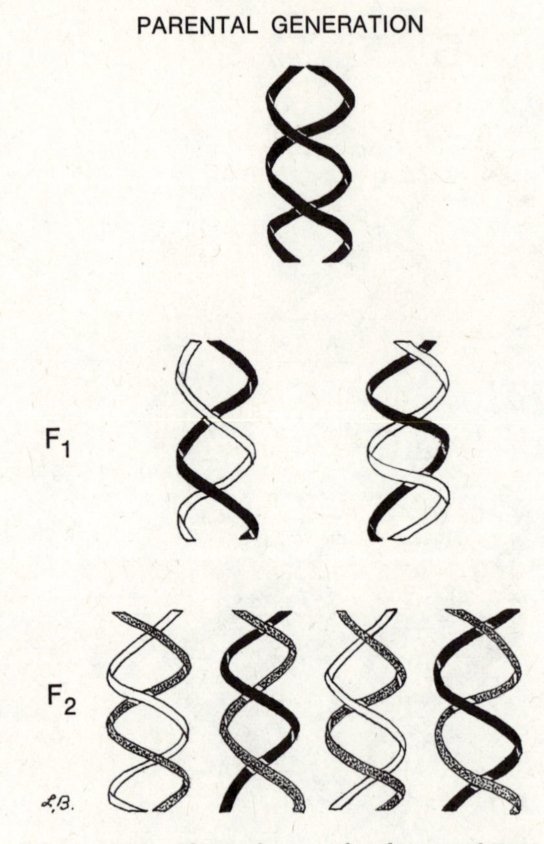

Figure 13-38. The mechanism of replication of DNA (semiconservative replication). The parental generation includes one DNA molecule, a double helix. The F_1 generation includes two double-stranded DNA molecules, each with one parental strand (shown in black) and one new strand (white). In the F_2 generation, each double-stranded DNA molecule consists of one of those occurring in its F_1 parent (black or white) and one newly formed strand (stippled).

The DNA in the nucleus precisely controls the formation of the proper combinations of amino acids in the synthesis of proteins through the genetic code. The genetic code for arrangement of amino acids in proteins, imparted through RNA, is based upon groups of three nucleotides in a linear sequence in the DNA and thus the messenger RNA molecule. Each of the m-RNA triplets is a codon, and the codons are read linearly from one end to the other. The codons are made up of a combination of any three of the following: *guanine*, *cytosine*, *adenine*, and *uracil*, any of which may or may not be repeated once, twice, or thrice in the group of three.

One codon may be, for example, *adenine*, *guanine*, *cytosine*—or AGC. This combination calls for one particular amino acid (serine) to enter at a particular point into the linear sequence of amino acids making up the protein. In a sequence of three drawn from A, G, C, and U there are 64 distinguishable combinations. Since there are only 20 amino acids, some of them must have more than one codon, hence the genetic code is degenerate (with more than one codon for some amino acids). A protein may consist of a number of amino acids linked together in any of many sequences, and the number of possible amino acid combinations to be derived from many different sequences of codons is enormous. A change in any group of three nucleotides in the DNA may affect the ultimate structure of a protein to be synthesized in a cell. Of the 64 possible combinations for codons, all but 3 have been shown to code the incorporation of an amino acid. These three combinations are the termination codons, each of which signals the end of the polypeptide chain.

The experimental evidence for DNA as the genetic material which controls the synthesis of proteins and, therefore, morphological and physiological characteristics of all living organisms is now conclusive. Synthetic messenger RNA's have been prepared which code for specific polypeptides. The entire genetic code is known, and specific amino acid changes in proteins (mutations) have been traced to specific known changes in DNA. Natural messenger RNA's for specific proteins have been isolated, and even specific genes have been obtained in essentially pure form.

Variation in types of living organisms is possible because genes may change and, therefore, the protein specified by a gene is capable of modification. Gene mutation (or change) is due to a change in the sequence of bases in a segment of the giant DNA molecule, the

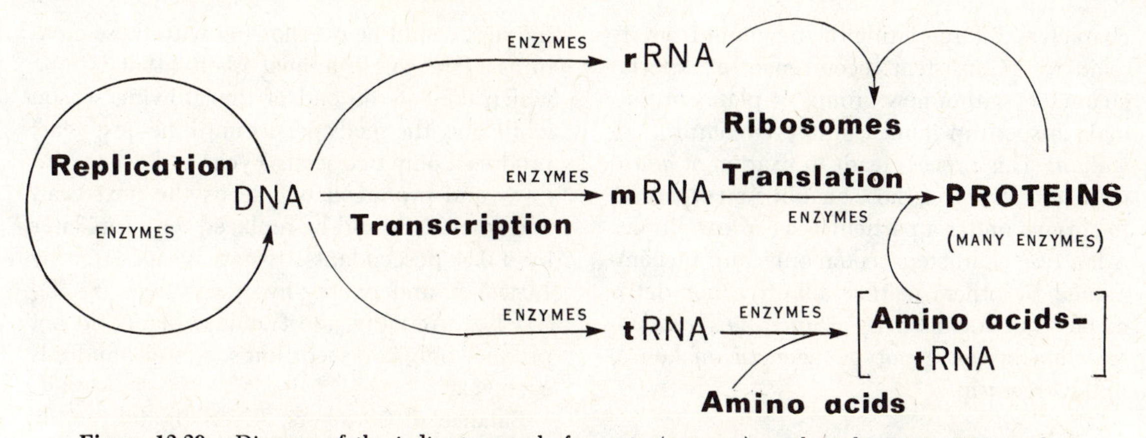

Figure 13-39. Diagram of the indirect control of protein (enzyme) synthesis by DNA. DNA replicates and produces more DNA in the presence of specific enzymes. The enzymes for all cellular processes are produced as follows. Genetic information in DNA is transcribed into messenger RNA which is then translated into specific proteins in the presence of ribosomes which contain ribosomal RNA and amino acid-transfer RNA complexes. Note that each step in the scheme is catalyzed by specific enzymes (proteins). "Genes" for ribosomal RNA and transfer RNA are in the DNA. The whole system is interlocking in that each component is necessary for the production of every other component shown, and for the many other parts of the cell which are not shown in this diagram.

segment constituting an individual gene. The sequence of bases is capable of an enormous degree of variation without changing the central biochemical organization of the DNA molecule, and this permits the tremendous diversity of proteins and, therefore, the diversity of living organisms.

The presence of DNA in all types of living organisms from bacteria to elephants and oak trees indicates their common heritage and genetic relationship. DNA molecules are found in all living things except some viruses having only RNA. Although other coding mechanisms would have been possible, none are known to exist and all known organisms have the same genetic code. *Thus, presence of DNA is the unifying force throughout the world of life, and its universal occurrence indicates a common origin and the divergence of all known living things.*

The then seemingly separate works of Darwin, Mendel, and Miescher on evolution, genetics, and biochemistry which began in the horse and buggy age have coalesced and entered the space age. Man is now learning to manipulate the gene through biochemistry and molecular genetics and may ultimately learn to control the genome and, therefore, manipulate evolution. Already genetic engineers are at work changing the genome of domesticated bacteria. Other genetic engineers are attempting to introduce the genes for nitrogen fixation from microbes into higher plants. The direction of future work in this area is not totally predictable but is certain to have dramatic effects upon the future of man and other animals, plants, and microbes.

The Development of New Taxa

Evolutionary development of new taxa requires two interlocking largely simultaneous processes—*differentiation* and *isolation.*

DIFFERENTIATION

Formation of a new population or of systems of closely interrelated populations composing a new taxon is attended by accumulation of characters differentiating the new unit from its relatives. Consistent occurrence of special characters within new groups of plants or animals is sped up immeasurably by *natural selection.* This process leads to fixation of *adaptive characters,* i.e., those enabling an organism to survive under a particular set of conditions. Adaptive characters commonly are accompanied by others neither adaptive nor detrimental but included by chance, i.e., *random selection* in the genotype (genetic make-up) of the organism.

Natural Selection. Once a new characteristic has arisen by mutation or reshuffling of genes, it still may or may not become established in nature. In his presentation of the theory of evolution, Charles Darwin was influenced somewhat by Thomas Malthus's well-known *Essay on the Principle of Population* (1798). According to this work, the human population of any area is limited in numbers by the supply of food and by other factors necessary for existence. The full title of Darwin's outstanding work on evolution reflects the influence of Malthus. This title is *On the Origin of Species by Means of Natural Selection, or the Preservation of Favoured Races in the Struggle for Life* (1859). It is based upon the principle of natural selection, that is, that under natural conditions of keen competition organisms having characters well adapted to a particular environment tend to crowd out and replace their less well adapted competitors, and this idea was Darwin's own, his greatest contribution. It involves the following points.

Overreproduction. Most natural species have the capacity for producing far greater numbers than actually exist. A single conger eel, for example, produces about fifteen million eggs per year, the codfish about nine million, the Virginia oyster about sixty million. Should all the eggs of conger eels, codfish, and Virginia oysters produced in a single year be fertilized and matured into adult individuals, the seas would be overflowing with these creatures. If a single annual plant (that is, one which dies at the end of the growing season while only the seeds persist until the next year) produced only two seeds a year and every seed grew and produced two seeds the next year, and so on, it would be replaced ten years later by 1,024 descendants, twenty years later by 1,048,576, and twenty-five years later by 33,554,432. Actually, most annual plants do not produce only two seeds but scores or hundreds each year.

Balance of Numbers. Even though nearly all species have great powers of overreproduction and potentially any one might cover the Earth in a short time, this does not occur because of the many casualties. Instead, the numbers of any particular species are apt to remain nearly stationary or to fluctuate within relatively narrow limits from year to year. Therefore a struggle for existence may be inferred to exist.

The Struggle for Existence. Among natural species the struggle for existence is intense. Each is in competition with many others for food, space, water, exposure to light, and other necessities. The reality of the struggle for existence in nature is evident from watching any wild bird as it hops across a lawn looking for insects, worms, or seeds. It is constantly on the alert—there is never a time in its life when it is wholly safe from its enemies. Furthermore, its quest for food is interminable. Food is plentiful during only a short period of the year, and whenever it becomes abundant competition also increases. No particular area will support more than a certain number of any given species of plant or animal, and overreproduction makes competition for a place in nature keen.

In a classical experiment Darwin demonstrated the competition among plants by clearing a small plot of ground and recording each of the many individuals which germinated there from seeds. Later, after the plants had matured, he took another census and found that very few had reached maturity, most having been crowded out, particularly by being shaded out in the seedling stage.

Survival of the Best Adapted. In the struggle for existence, ordinarily the survivors are those best adapted to the prevailing conditions of the time and place. The working of natural selection may be illustrated by the following example: Protected by a gray-brown coat which matches its surroundings, the common pocket gopher of California searches for food usually within a few feet of the mouth of its burrow. Periods spent above ground are short, and the slightest disturbance causes the gopher to back rapidly into its hole. Very rarely there is a mutant albino gopher, but it does not survive long because it is in sharp contrast to the surrounding soil and vegetation and therefore highly vulnerable to its enemies, especially so at night when owls are about. Ingles and Biglioni have shown that small degrees of color occur in the races of pocket gophers living in dark, humus-filled soils and light desert-type soils in the San Joaquin Valley of California, the races having been sorted according to these fine differences in their color backgrounds.

Although having white fur is a disadvantage under the conditions described above, it may be advantageous under other conditions. In the far north white winter fur or plumage is common among mammals and birds. Under the conditions of the northern winter it is a distinct advantage both to species attempting to escape enemies and to those attempting to catch other animals for food. Its drawbacks in summer usually are obviated by a spring molt and seasonal replacement with fur or feathers of another color.

The advantages of color patterns may be greater than they seem, as with colored snakes. Sedentary species tend to have complex camouflage patterns of rectangles, triangles, bands, or blotches, the figures being sharply angular and tending to run crosswise and to confuse the form of the coil. Active snakes have either solid color or stripes, reducing the sensation of motion. Convergence of stripes near the tip of the tail creates the illusion of sudden disappearance.

Thus, organisms with special advantages arising from gene mutation or (much more frequently) reshuffling or chromosome rearrangement tend to persist and their competitors to lose out. These advantages are with respect to the environment of the time and place where the organism and its competitors live.

Special adaptations among plants include the following: By having natural lenses in their cells, some specialized mosses are able to live in the very dim light of caves where other plants cannot endure. These lenses concentrate the light on the chlorophyll, thus enabling photosynthesis to occur. The plants of the fantastic cactus family survive in deserts because they have (1) extensive shallow root systems which absorb water rapidly during the infrequent rains, (2) much fleshy water storage tissue in the stems, (3) effective surface coats of waxy material which prevent rapid evaporation except through the stomata (pores) which are few and sometimes in protected locations, and (4) a special type of metabolism in which carbon dioxide is absorbed from the air at night, when the pores (stomata) can be open because it is cool and water loss is low. The carbon dioxide is stored in organic acids until daylight, then released and used in photosynthesis. The stomata then can be closed during the day, reducing water loss while it is hot.

Another group of desert plants, the mesquites, can survive in competition because the roots reach far into the earth and absorb water not available to other plants. However, mesquites absorb water through roots at higher levels as well, and on cattle ranges they are a serious pest because they take up much of the water otherwise available to grasses. According to Went, where there is almost no rainfall in the Atacama Desert in north coastal Chile, a mesquite absorbs water from the sea fog through its leaves, then secretes it into the soil surrounding the roots, storing it there until it is needed when it is reabsorbed through the roots.

Other special adaptations include those aiding in defense, as, for example: thorns, spines, or prickles (century-plants, cacti, mesquites, locusts, or roses); hypodermic injections (nettle); fire- and insect-resistant bark (redwoods).

Among animals, in general the individuals of a species with the greatest speed, fighting ability, general vigor, cunning, or the keenest senses tend to survive. However, the possessors of any one of these characters or any combination of them may lose out in competition to those with some unusual adaptation suited to the special conditions of the period and location. The extinct giraffe camels and the modern giraffe are two kinds of browsing mammals with remarkably long necks which have been able to survive for long periods by eating leaves far above the reach of their competitors. Elephants living in the woodland savannah of equatorial Africa do not have long necks or very long trunks (noses), but they reach the leaves high above through having great weight and strength. They simply lean against the trees and push them over. Other special abilities or features facilitating escape, defense, or attack of prey are as follows: fleetness (antelopes, horses, rabbits); climbing trees (monkeys, tree squirrels, cats); climbing rocky ledges (mountain sheep and goats); burrowing (moles, gophers, rabbits, earthworms, ground squirrels, marmots); hiding in mud or sand (frogs, turtles, clams); spines (porcupines); armor (armadillos, turtles, clams, snails); "borrowed" armor (hermit crabs); poison (rattlesnakes, copperheads, water moccasins, spiders, scorpions, centipedes, and some insects); electricity (electric eels and rays); clouds of black or colored fluid (squids, sea hares); portable shelters of cemented sticks or gravel (caddice fly larvae).

Some organisms secure an advantage over their competitors by resembling other plants or animals in superficial ways. Some cuckoos resemble in form and coloring small hawks occurring in the same region, and they scare other birds from their nests. This permits laying eggs in the nests, and the care and feeding of a young cuckoo is assured. Some female cuckoos can even lay eggs matching those of one host bird, some those of another. Certain butterflies are distasteful to birds; others are palatable. Sometimes the palatable butterflies are able to avoid being eaten, because birds have become conditioned to the bad taste of the other species. Thus, resemblance brings salvation. Some orchids develop a portion of the flower into a structure appearing like the female of a particular species of bee or fly. The male insects try to copulate with the segment of the flower and effect pollination of the orchid in the process. Other orchids duplicate the odors female bees and flies use to attract the males. The orchid mantis of Malaysia bears the color of the orchid flowers it inhabits, and it lies in wait for visiting insects. Thus, the combinations of genes producing resemblance to another organism may be highly beneficial.

The balance in nature may be so delicate that more than one form of the same species may be adapted to different local environmental niches. Although this phenomenon is to be expected in the light of the great variation in genetic combinations within species, its demonstration is not easy. In a common species of European snails, there may be yellow (but green while the snail is alive), pink, or brown. Green (yellow) shells predominate in grassy areas, the brown on the dead leaf surfaces of the floors of beech forests. The song thrush carries the snails to rocks and opens the shells. The shells in the middens about the rocks indicate that the thrush brings in the snails for which the pattern of color provides the poorest local camouflage, those in a particular midden being similar. The middens vary with the habitat, and natural selection according to the microhabitat proceeds rapidly. There is also a seasonal selection. The shells are not only of different colors but also plain or with 1 to 5 bands, and this affects their visibility against different backgrounds. For example, the song thrushes eat more of the green snails during the winter when the background is primarily of brown leaves, but during the spring and summer they eat mostly pink-and-brown-banded snails, which show up against the green and yellow background of the same spot. Thus, the snails have predator insurance against various circumstances.

About 100 species of moths occurring in England and others in Europe and the United States have been found to vary in color forms according to presence of industrial pollution of the atmosphere. The moths have genes determining whether they produce melanins (dark pigments) and others determining the quantity of the pigment. The original members of the species, which are nocturnal, rested on tree trunks, fences, buildings, and the like during the day time. Their lightly mottled more or less pale gray patterns were adapted particularly to the lichen-covered tree trunks. However, the smog killed the lichens and blackened the tree trunks. The habits of the moths did not change, but birds nearly eliminated the light-colored moths, and the populations became in some places 90% to 97% black. In Britain, control of atmospheric pollution has reversed the trend and the moth populations are becoming lighter.

The theory of natural selection presented above is essentially the proposal of Darwin. The chief difference is inclusion of recent knowledge of variation and heredity, which supersedes Darwin's idea that all minor variations are inheritable and Lamarck's earlier assumption that characters acquired during the lifetime of the individual are passed on to the offspring. Although modern knowledge of genetics and cytology replaces these theories of the origin of hereditary differences, the natural selection theory of Darwin, limited somewhat by more recent study, stands as the most reasonable explanation of the method by means of which many but not all new characters become established.

An unfortunate expression is "survival of the fittest," which gives the wrong impression of the operation of natural selection. This has been improved by substituting "survival of the best adapted." However, those surviving are simply the individuals able to survive in the conditions of the environment (including the effects of other living organisms) of the time and place. Not all the characters surviving in a natural population are necessarily advantageous, or they may be so in one way but disadvantageous in another, so long as the taxon or a phase of it survives. The following are two examples of disadvantageous features surviving in human populations.

Sickle cell anemia in man is caused by a gene producing distortion of the shape of the red cells in the blood and the synthesis of an abnormal hemoglobin. The elongate, sickle-shaped cells are prone to breaking down, and the result is anemia, thrombosis (clotting in the minute blood vessels and lymph channels), or death. If the gene pair is heterozygous, the effects are limited, and difficulty occurs only under stress from oxygen deficiency. However, the unusual hemoglobin prevents the malarial parasite from entering the red cells and continuing its life cycle there. Thus, the homozygous conditions have perils. If the two genes produce sickle cell anemia, death may result; if there are no genes for sickle cell, death may result from malaria. If the individual is heterozygous, he is likely to survive, but some of his offspring are likely to die of either malady. Sickle cell genes are rare, except in black human stocks from equatorial west Africa, where malaria determined their presence. The survival value of being heterozygous results in about 20% of the population having sickle cell anemia.

Species of *Rhus*, including poison ivy and the various poison oaks, contain a substance, urushiol, to which some people become sensitized. Contact with the plant at any season produces a rash, if there has been a sensitizing earlier encounter and if the person is capable of being sensitized to urushiol. The blood stream contains an almost infinite number of T-lymphocytes, each with a particular single chemical "receptor" on the cell surface, and there are millions of kinds of receptors. These constitute essentially a "shot-gun prescription" for any of the innumerable foreign substances that may enter the body. The reaction of the blood stream to any invasion is specific for the invading substance, whether or not the body has encountered it before, but it is much quicker and more effective if there has been a previous invasion and immunity has been built up earlier. When the urushiol of poison ivy or poison oak encounters the skin cells, it binds to their surfaces. T-lymphocytes with the receptor fitting the substance bind to the invading substance in turn, and then they both reproduce rapidly and release "lymphokine factors," which bind to other molecules of the urushiol. "Killer" lymphocytes are attracted to the lymphokine factors bound to the urushiol, and they bind to these factors, forming a chain of three. The killer lymphocytes release enzymes that digest the skin cell membranes and kill the cells, thus producing the rash. Thus, a mechanism combatting the invasion of foreign substances and a part of the body's defense against infection and other maladies in this instance produces discomfort and harm to the person normally protected. If the reaction did not occur, the urushiol would be harmless. However, this is a by-product of the hit-or-miss course of evolution. Development of lymphocytes provides protection from almost any invader, but some invaders are not harmful and damage can be done in combatting them. Without the immunization mechanism, man and other animals would not have evolved, but its presence is not necessarily all beneficial. The characteristic reaction to urushiol, by all but the few not having T-lymphocytes opposing it, is unpleasant but it does not kill people. Consequently, there has been no selection against it. (For an interesting and readable account, see Gary Reynolds, *Fremontia* 6: 19–23, 1978.)

Integration of New Genes into the System. When there is ecological change, there must be also a change in the gene balance of each organism, because only individuals with the necessary expression of genes can survive in the new environment. Those with other gene combinations die out. However, although mutations occur only rarely, new ones are not needed often. Over millions of years of evolution each organism (or its forebears) has built up such a great reservoir of genes accumulated from past mutations that it is likely to be able to produce some individuals able to meet any environmental shift but a very severe one. Adding a new (mutant) gene to the balanced system is more likely to cause harm through upsetting the balance than to be beneficial.

Most of the genes expressed in naturally occurring organisms are dominant. In part, this is the work of natural selection. Dominant genes tend to be selected or eliminated rapidly, because they cannot be covered up; they are expressed in even heterozygous* individuals. On the other hand, many recessive genes, including innocuous to deleterious ones, persist indefinitely without being expressed frequently. Homozygous* recessives are subject to selective pressures; heterozygous ones are not completely so. As the undesirable homozygous types are eliminated, ultimately the reservoir of recessive genes is almost wholly in heterozygous individuals. This is because homozygous types occur only rarely, the heterozygous individuals having become so rare in the populations that they seldom mate with each other. Thus, the recessive gene is not likely to be completely eliminated but only to

* Heterozygous, with unlike genes in the pair or pairs under consideration: homozygous, with like genes. Heterozygous organisms do not breed true.

become rare or uncommon. Numerous recessive genes tend to be held over as a gene bank of factors expressed only rarely but forming a ready reserve of characters that may become useful when the environment changes.

When a new gene arises by mutation, it does not revolutionize the system of genes of the individual possessing it through any automatic process. The gene system of the taxon has resulted from millions of years of gradual evolution resulting from stirring up the combination of genes during reproduction and the formation of each new generation. The genes of an individual are in balance with each other, and the addition of a mutant gene to the system may produce a continuing major effect only through changing the balance of the entire system, especially so if it is undesirable. If a gene deleterious to the plant or animal arises, the gene complex tends to reduce its effect, and the nature of the complex tending to neutralize the deleterious gene automatically becomes a feature having survival value for the individual. According to de Beer, the obviously harmful dominant eyeless gene in *Drosophila* appears in a few individuals capable of breeding though producing few surviving offspring. The offspring have the eyeless trait, but after a number of generations there are descendants with normal eyes. During the period of breeding of the eyeless flies the greatest number of individuals has died, but in the survivors there has been much reshuffling of the gene complex, and the complexes of some individuals have obliterated the effect of the eyeless mutation. Thus, there are flies with normal eyes, despite presence of the originally dominant gene for eyelessness, now in effect recessive. Ford's experiments on the currant moth have carried the progression of dominance in opposite directions, making the same gene dominant in one line of descent and recessive in another.

Thus, if a gene harmful to the organism arises, it tends either to be eliminated by natural selection or to be covered up or nullified by the gene complex. According to Fisher, a gene may become recessive because it is deleterious, its status being altered by the action of the gene complex as a whole. Most mutations are harmful to the organism under its current relationship to the environment, and most are or become recessive. The status of a gene as dominant or recessive merely reflects its position in the gene balance with regard to another. If one is dominant, the other is per se recessive, but the effect of either gene is subject to the action of all the genes in the complex.

Thus, evolution does not occur as the immediate effect of adding mutant genes, except in rare instances in which the mutant gene may be highly desirable. Evolutionary change is the result of selection from among the vast number of genes already present in the organism as an accumulation from the past, which is recombined as each generation is formed and which is forever under the jurisdiction of the natural selection process. The effect of mutation is addition to the bank of stored genes of factors likely to be useless and rarely expressed but with the possibility of being important in the course of time and change.

Random Selection of Characters. The pathway of evolution may be wholly at random or partly so as in the following examples. When certain members of a taxon become differentiated from the others because they have adaptive characters permitting their survival under differing conditions from those suitable to their relatives, they may carry along with them characters having little or nothing to do with selection according to the environment. These may be neither advantageous nor disadvantageous, or they may be even somewhat detrimental, provided the adaptive characters of the organism secure an advantage which overweighs their effect.

Nonadaptive allelomorphic genes (alternative contrasting members of the same pair) may exist through long periods in the evolution of species and higher taxa without either type becoming established and the other eliminated. Some humans can taste a bitter organic chemical substance, phenyl-thio-carbamide; others cannot. The (dominant) genes producing the ability to taste this synthetic laboratory substance, nonexistent in nature, have had no effect upon survival, yet they or their recessive alleles must have arisen far back in the common ancestry of man and apes, for some individual apes can taste it and others cannot, just as in man. The blood groups (AB, A, B, and O) occur similarly in man and apes. These may have had no adaptive significance until blood transfusions were attempted. Insect pests may be destroyed rapidly by new insecticides when they are first applied, but often resistant members of the population of insects survive, multiply, and replace the original members of the species. The factors producing resistance could have had no survival value in the earlier evolution of the insect species, but application of a new insecticide may make resistance an adaptive character.

Well-established cases of random selection and fixation of characters among natural plant populations are few, but the following is an example. *Hutchinsia* is a genus in the mustard family. Two populations in the Swiss Alps have been referred to as species, although probably they should be considered as varieties or races of the same species. The petals of *Hutchinsia alpina* have claws (narrow stalk-like bases), and the plants grow on talus slopes of limestone fragments; the petals of the plants known as "*Hutchinsia brevicaulis*" are spathulate (the bases not being narrowed into claws), and the plants grow on talus slopes of igneous rocks such as granite. Plants of *Hutchinsia alpina* grown in water culture survive the presence of far more calcium than do those of *Hutchinsia brevicaulis,* and evidently this physiological characteristic is primarily responsible for the differentiation of the two varieties or races. The petal types are reported to have no adaptive significance, and as a matter of fact, in one instance the combination of characters is reversed. In the Dolomite Mountains southeastward a form restricted to limestone and with a high calcium requirement in water culture has petals of the *brevicaulis* type.

In localized small populations of plants or animals with relatively few breeding individuals, random fixation of genes tends to be significant. Some genes are lost by chance; others become fixed, i.e., homozygous in the population; and thus there is a genetic drift in one direction or another unless natural selection overshadows it as adaptive characters are preserved and the trend of the drift is nullified or reversed.

Irrelevant characters are perpetuated in the same way among domesticated species. Innumerable breeds of plants and animals useful to man include unimportant characters which accompanied by chance the desirable ones which led to human selection and preservation or the strain. Obvious characters of this sort are used as "markers" of horticultural varieties of animal breeds. No matter how an apple might taste a commercial buyer would not pay a high price for it as representative of a particular "variety" if the color or shape were unusual. In cattle black-and-white coloring marks the Holstein breed; black, Aberdeen Angus; red-and-white, Hereford; fawn-and-white, Guernsey. Coat color has nothing to do with the important characters of the breed, e.g., beef production for Aberdeen Angus or milk production and quality for Guernsey, but a purchaser would be suspicious of the ancestry of a fawn-and-white "Aberdeen Angus" or a black "Guernsey." Any individual not having all the characters associated commonly with a domestic breed or horticultural variety is likely to be *heterozygous* not only in the unimportant visible characters but also in the critical ones which may be inconspicuous.

Random preservation of particular character combinations when plants or animals migrate or are carried from one region to another may be illustrated by the following example. *Ra-*

nunculus acris is a common and conspicuous buttercup occurring abundantly in southern Canada and across the adjacent northernmost tier of states from Washington to New England. It exists in two forms differing most obviously in the degree of dissection of their leaves. A third form with very finely dissected leaves and possibly other characters once existed at Gansevoort, New York, but it has not been collected since 1919. Some North American botanists have considered these to be botanical varieties of the species. In Northern Europe, where the plants are native (or at least where they have occurred since they migrated into areas burned over by man in the Stone Ages) there are numerous character combinations including many degrees of leaf dissection occurring often side by side in the same meadow or old field. The three strains introduced in North America are a random selection from the assemblage in Europe. Other gene combinations might have been transported across the ocean just as readily.

Similarly, early migration of man into North America was selective in that the strain which crossed from Siberia to Alaska included char-

Figure 13-40. Forms of *Ranunculus acris,* tall buttercup, introduced into North America from Europe. The two forms illustrated here, especially the one at the left, have spread over southern Canada, southern Coastal Alaska, and the northern two tiers of the contiguous 48 states, their northward occurrence being limited to the vicinities of villages where there are houseflies. Pollination occurs during rainy weather, when the flower turns downward and forms an umbrella under which flies congregate, effecting pollination. Various intergrading forms occur in Eurasia, but only two (and probably for a short time a third) have been introduced into Canada and the United States.

acters of the North American Indian. Had the later exploration by the Vikings been followed up by permanent settlers, the second invasion would have introduced only a Scandinavian strain instead of the still later influx primarily of French, English, and Spanish peoples.

ISOLATION

New characters may occur without isolation of individuals from the rest of a general natural population of plants or animals, but so long as all members of a species are free to breed with each other and the offspring are fertile, new characteristics may tend to yield only a more variable or gradually altered single general population. When segments of the species are isolated, evolution may proceed independently among the derivatives of the original population because there can be no further gene interchange. Thus isolation of some populations from their relatives is an important factor in development of varieties, species, and higher taxa.

Isolation of taxa may be either *reproductive* or *geographical*. *Ecological isolation* may be under either heading or treated separately.

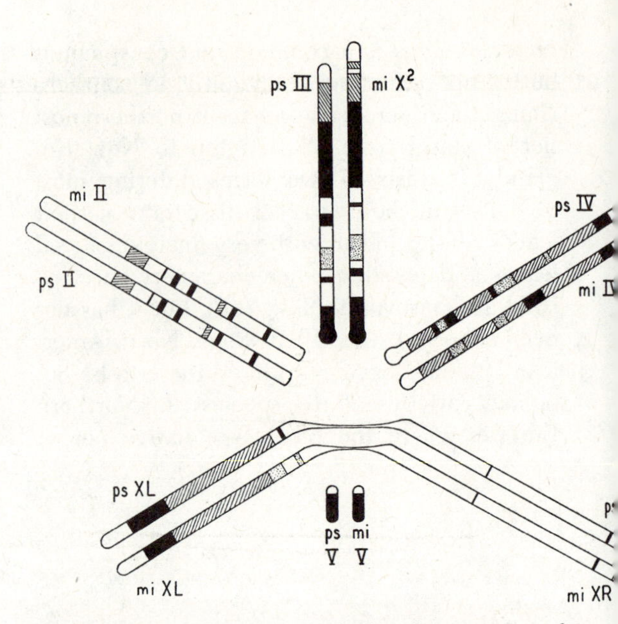

Figure 13-41. Comparison of the chromosomes of two related species of vinegar flies, *Drosophila pseudoöbscura* and *Drosophila miranda.* Sections with the same gene arrangements are shown in white; inverted sections are cross-hatched; translocated portions are stippled; non-homologous sections are in black. From L. H. SNYDER, *Principles of Heredity,* D. C. Heath and Company; from DOBZHANSKY, *Genetics and the Origin of Species,* courtesy of the Columbia University Press.

Reproductive Isolation. In two species of *Drosophila* less widespread than *Drosophila melanogaster* the basic number of chromosomes is five (i.e., n equals 5). In these very closely related flies (known as *Drosophila pseudoöbscura* and *Drosophila miranda*) the arrangement of chromosome detail (crossbanding) has been studied thoroughly and mapped from the large chromosomes in the cells of the salivary glands. Some parts are homologous, containing corresponding genetic factors; others are inverted in position but otherwise homologous; some are translocated to different chromosome pairs. Presumably these flies originated from a common ancestor, two strains of the original type having become isolated from each other insofar as reproduction is concerned. Male individuals produced by crossing these two strains were sterile because sterility genes made them incapable of developing normal testes and sperms.

Reproductive isolation may result from failure of the chromosomes to pair (synapse) normally at the beginning of the reduction divisions. For example, in the ordinary crosses of *Galeopsis speciosa* and *Galeopsis pubescens* referred to above, the hybrid is diploid, i.e., with only 16 chromosomes, eight from each parental species. Although some of these are homologous or partly so, others are not. Consequently the reduction divisions do not proceed normally, and fertile pollen and ovules are not produced or the percentage is low. As a result, usually the hybrids do not reproduce. Most crosses of cabbages and radishes have similar results, the nine chromosomes from each parent lacking homologous mates and the hybrids being sterile. The syntheses of *Galeopsis Tetrahit*, (perhaps) *Spartina Townsendii*, and *Raphanobrassica* were possible because some of the gametes produced by the hybrid had two full sets of the chromosomes of each parent species.

Other factors restricting or eliminating interbreeding of natural populations or subsequent reproduction of the hybrids between them may include such factors as those of the following examples: (1) *Ecological factors*—restriction of two or more related populations to local environmental conditions either separating them and preventing interbreeding or resulting in death of the offspring except in instances of removal from the habitat of the female parent to one intermediate between those occupied by the two parents. (2) *Seasonal factors*—failure to interbreed because of differing reproductive periods. (3) *Internal factors*—restriction of interbreeding because of incompatibilities of the reproductive structures, as, for example, differing size of the pollen tubes, the progress of the larger ones through the passageway of the style of the flower being impeded; differing relative growth rates of pollen; arresting of the growth of foreign pollen by substances on the stigma; failure of the gametes to be attracted or to unite; death of hybrid embryos not physiologically adapted to development in the ovules or other structures of the female parent; sterility of the hybrids.

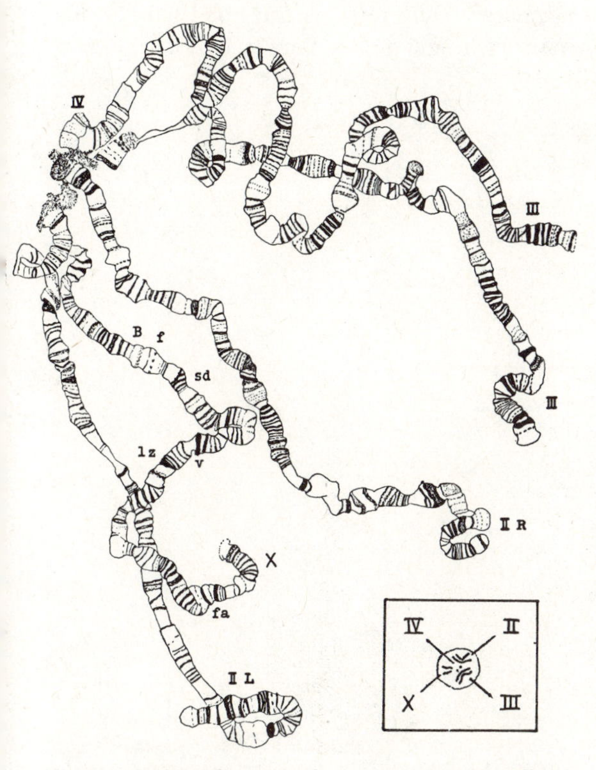

Figure 13-42. The giant chromosomes in the cells of the salivary glands of the vinegar fly (*Drosophila*). Experiment has made possible association of the bands on the chromosomes with the location of genes. The relative size of the chromosomes in the reproductive cells is shown by the inset. From L. H. SNYDER, *Principles of Heredity*, D. C. Heath and Company; from PAINTER in *Genetics*.

In the flowering plants methods of pollination (e.g., by hummingbirds or by insects) correlated with flower structure may tend to keep species largely though not necessarily strictly segregated.

Geographical Isolation. Just as some plant or animal populations cannot breed together even though they grow side by side, others cannot because they are isolated geographically, as discussed above under the evidences of evolution. Geographical isolation is just as effective as reproductive isolation—but only as long as geographical separation continues. If continents are reconnected, if islands are rejoined to the mainland or to each other, or if deserts or other barriers disappear, isolated units are brought into contact once more. Unless reproductive isolating mechanisms have developed along with or during geographical isolation, interbreeding may begin immediately. This has occurred between the North American strains of moose and also of grouse separated in Pleistocene time by the great ice sheets which covered the northern portion of the Continent. Some differentiation of these strains occurred during the Pleistocene Epoch, but it is being obliterated in the areas of the former ice fields where the races now come into contact with each other. The genus *Platanus* (sycamore, plane-tree) is represented by markedly differentiated species in the Eastern and Western Hemispheres. Where these species are cultivated side by side, they prove to be interfertile, and obviously, removal of their geographical barriers might lead to mixed populations.

If advantageous new characters appear within one isolated unit of a population the individuals possessing them may (1) crowd out and replace all others, (2) replace them in only the geographical or ecological habitat in which the new characters are most advantageous or in which the segregated population is isolated, or (3) occupy a new geographical or ecological niche not previously available to the larger population.

In the earlier stages of differentiation and isolation, often both processes are incomplete. Usually differentiation is gradual, and there are many degrees of isolation. As more or less consistent occurrence of associated characters differing from those of closely related organisms is accompanied by more and more effective isolation, the derivatives of a species may become two or more varieties. Later, as differentiation proceeds still farther and isolation becomes still more effective (though not necessarily complete), some or all the varieties may become species. Eventually any of them may give rise to other varieties or species, and ultimately each may be the potential starting point of a new genus, family, or higher taxon.

14

Some Fundamental Problems of Plant Classification

Before classifying plants it is necessary to determine (a) *the method of segregating taxa* and (b) *the relative stability of characters* which may be employed in the segregation.

The Method of Segregating Taxa

Natural populations of plants and animals are self-perpetuating. Their reproductive mechanisms transmit the same or a similar association of genes through countless generations, often with few or minor changes. Therefore, taxa are real entities, and the chief problem of classification is to distinguish them from each other.

An example of a genus of relatively stable species is *Lathyrus* (sweet pea). According to Senn there have been few or no successful experimental crosses between species. He obtained seeds by only 4 of 458 attempts. Only three lots germinated, and these resembled the female parent in all details, indicating pollen contamination or some form of self-fertilization. In this instance reproductive isolation is complete or nearly so, although more recently Hitchcock obtained a few interspecific crosses of West American species and reported evidence of natural hybridization of a few species.

In other genera usually reproductive isolation (in the most strict sense) between species is incomplete, but partly it may be compensated for by geographical isolation or ecological or other factors related to reproduction such as strong adaptation of the parents to different environmental niches. Even species which hybridize freely in gardens, cultivated fields, or areas disturbed by fire, floods, or overgrazing rarely produce many hybrids in localities of stable natural vegetation. This is because almost always the seeds bearing hybrid embryos must germinate in the environmental niche occupied by and favorable to the female parent. The hybrid seedling includes genes of the male parent, which is adapted to a different environment, and in competition with seedlings similar to the female parent it is not likely to survive. Its survival might be expected only if it grew in an intermediate habitat where it had an advantage over other seedlings or in a neutral environment such as a garden or other disturbed area. Thus genetic and environmental factors work together in areas of well-balanced local natural vegetation, tending to maintain the status quo and within the area the integrity of species.

On the other hand, according to Senn, although the boundary lines between some species of *Lathyrus* are clear cut and definite, others are vague. This is true in nearly every genus because the phases of a species present in the conditions prevailing in one locality may be replaced in other areas by differing populations. From place to place each species varies not only in visible characters but also in the often more significant but intangible physiological characters producing adaptation to slightly different environments. Thus, although in one locality or habitat two population systems may be markedly differentiated and reproductively isolated, in the numerous other areas inhabited by one or both, commonly there are gradual shifts of character combinations. Often these include the factors responsible for isolation. For example, as far as has been observed, two markedly different Californian oaks, the common type of scrub oak (*Quercus dumosa*) and the blue oak (*Quercus Douglasii*) are completely without hybridization where they occur side by side in the coastal mountains along abandoned United States Highway 99 at Oak Flats Fire Guard Station about 45 miles north of Los Angeles. Yet, within a few miles *Quercus dumosa* shades through intermediate populations into *Quercus turbinella* var. *californica*, and in a narrow area 300 miles long (Cajon Pass to Salinas Valley), var. *californica* passes gradually through many variable local populations into *Quercus Douglasii*. (Cf. p. 856.)

Taxa are tangible entities, but, although their existence is definite, the distinctions we are able to discover often are indefinite, and their segregation may be complex and difficult to interpret. Reproductive isolation may occur between closely related species growing side by side, but it may exist also between individuals of the same species, including those with little or no differentiation of other characters. Yet it may *not* be complete between even genera of the same family. Usually, however, species are isolated at least partly (more so than varieties); genera are isolated completely or interbreeding and production of fertile offspring occur usually only between the more nearly related species; families and higher taxa are isolated except that certain species of closely related genera may be interfertile. Taxa may be markedly different in some of their subordinate groups but closely similar and gradually intergrading in others. The flowering plant orders Polemoniales, Lamiales, and Scrophulariales, as defined here, include the families on pp. 269–283.

Advantages of the realignment of families shown in the box are as follows: (a) The fruit (4 nutlets) of the Boraginaceae is similar to that of the Labiatae, and this fruit type occurs in only these two families and (as 2 or 4 nutlets or sometimes a drupe) in the Verbenaceae. (b) Relationship of the Solanaceae (and the related Nolanaceae of Peru and Chile) to the Scrophulariaceae is very close. In general the Solanaceae have 5 stamens, radially symmetrical corollas, and alternate leaves; the Scrophulariacea 4 (or 2) stamens, bilaterally symmetrical corollas, and opposite leaves. However, several tropical genera of Solanaceae, including *Petunia,* may have bilaterally symmetrical corollas, and some plants such as *Verbascum* (mullein) in the Scrophulariales have 5 stamens and radially symmetrical co-

POLEMONIALES	LAMIALES	SCROPHULARIALES
Suborder Polemonineae Polemoniaceae (*Gila, Phlox*) Convolvulaceae (morning-glory, bindweed) Nolanaceae Solanaceae (tomato, potato, tobacco) Lennoaceae (sand-food, *Pholisma*) Suborder Boragineae Hydrophyllaceae (*Phacelia,* waterleaf) Boraginaceae (forget-me-not)	Verbenaceae (*Verbena, Lantana*) Labiatae (mint, sage) Phrymaceae (*Phryma*)	Scrophulariaceae (snapdragon, *Penstemon,* monkey-flower, foxglove) Bignoniaceae (*Catalpa,* desert-willow) Pedaliaceae Martyniaceae (devil's-claw) Orobanchaceae (broom-rape) Gesneriaceae Lentibulariaceae (bladder-wort, *Pinguicula*) Acanthaceae (water willow, *Ruellia*) Globulariaceae Myoporaceae
Corolla radially symmetrical; stamens 5; leaves nearly always alternate; fruit a capsule or a berry or of 4 nutlets.	Corolla bilaterally symmetrical; stamens 4 or 2; leaves opposite; fruit composed of 4 or 2 nutlets or sometimes a drupe.	Corolla bilaterally symmetrical; stamens 4 or 2; leaves opposite; fruit a capsule.

At least two other alignments are about equally satisfactory: (1) redistribution of the families within the orders, (2) union of the three orders into one. Redistribution might be as follows:

POLEMONIALES	LAMIALES	SCROPHULARIALES
Polemoniaceae	Hydrophyllaceae	Nolanaceae
Convolvulaceae	Boraginaceae	Solanaceae
Nolanaceae	Phrymaceae	All families included in the order in the list above.
Solanaceae	Verbenaceae	
Lennoaceae	Labiatae	

rollas. (c) The Phrymaceae resemble the Verbenaceae, and the fruit is dry and 1-seeded, but an achene instead of composed of nutlets.

Disadvantages are as follows: (a) If the Boraginaceae are transferred, the closely related Hydrophyllaceae (with a similar almost unique arrangement of flowers) should be also, but they have far more in common with the Polemoniales than with the Verbenaceae or Labiatae, and many of their characters are similar to those of the Polemoniaceae. (b) Although the Solanaceae (and Nolanaceae) are related closely to the Scrophulariaceae, they have much in common with the Polemoniaceae and Hydrophyllaceae, as well.

Union may be the best solution, but the question then becomes how far it should go. The distinctions of the Polemoniales and the Gentianales, themselves heterogeneous, are even more doubtful. Shall the ultimate enlarged order engulf the Gentianales, including the Oleales and Apocynales?

As shown in the preceding chapter, the living organisms of today have arisen by gradual evolution through more than 500,000,000 years. Their differentiation and isolation have proceeded as inheritable innovations have arisen by chance and either have succeeded (or at least persisted) or have failed in the environment of the time and place. Present-day plants

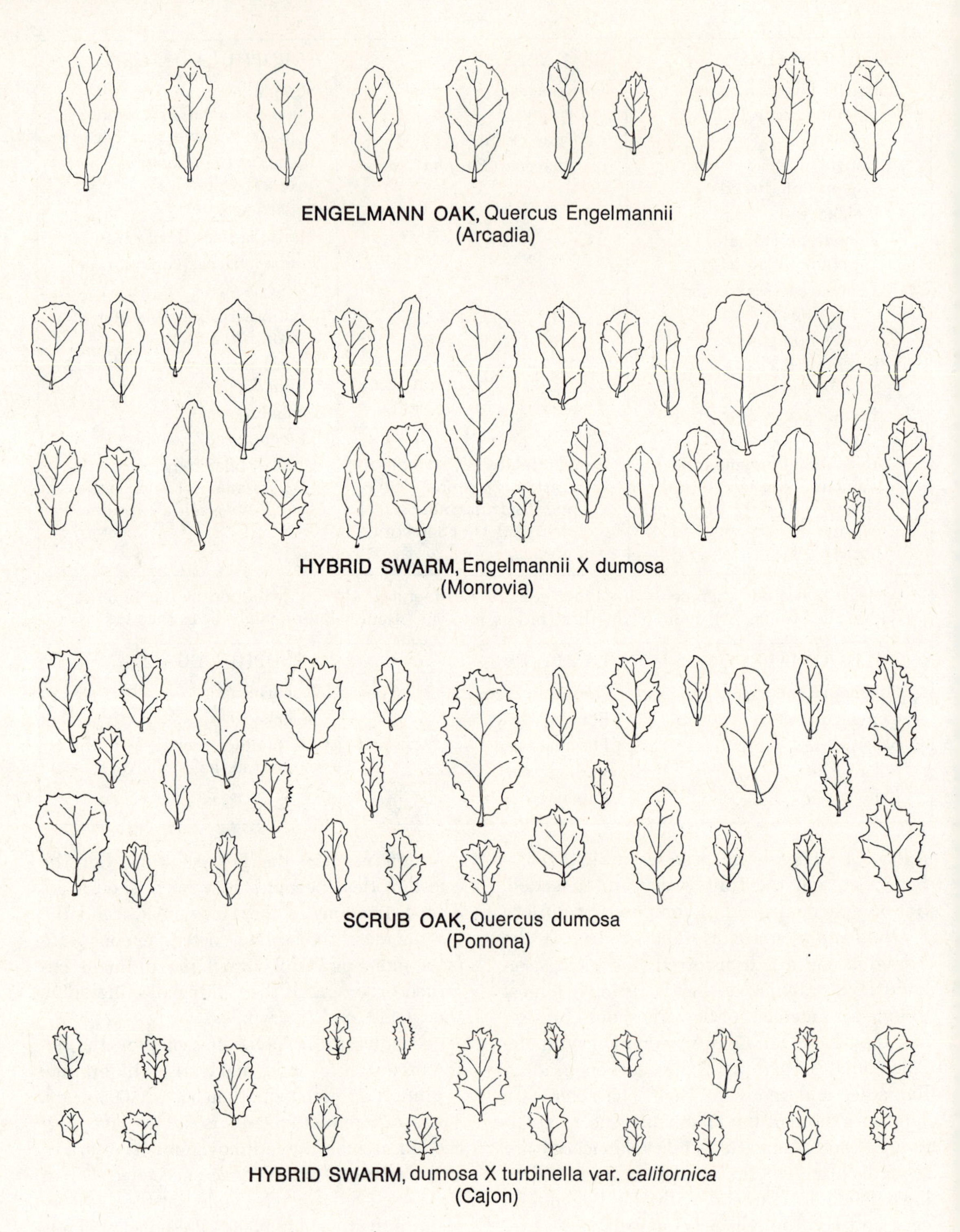
ENGELMANN OAK, Quercus Engelmannii
(Arcadia)
HYBRID SWARM, Engelmannii X dumosa
(Monrovia)
SCRUB OAK, Quercus dumosa
(Pomona)
HYBRID SWARM, dumosa X turbinella var. *californica*
(Cajon)

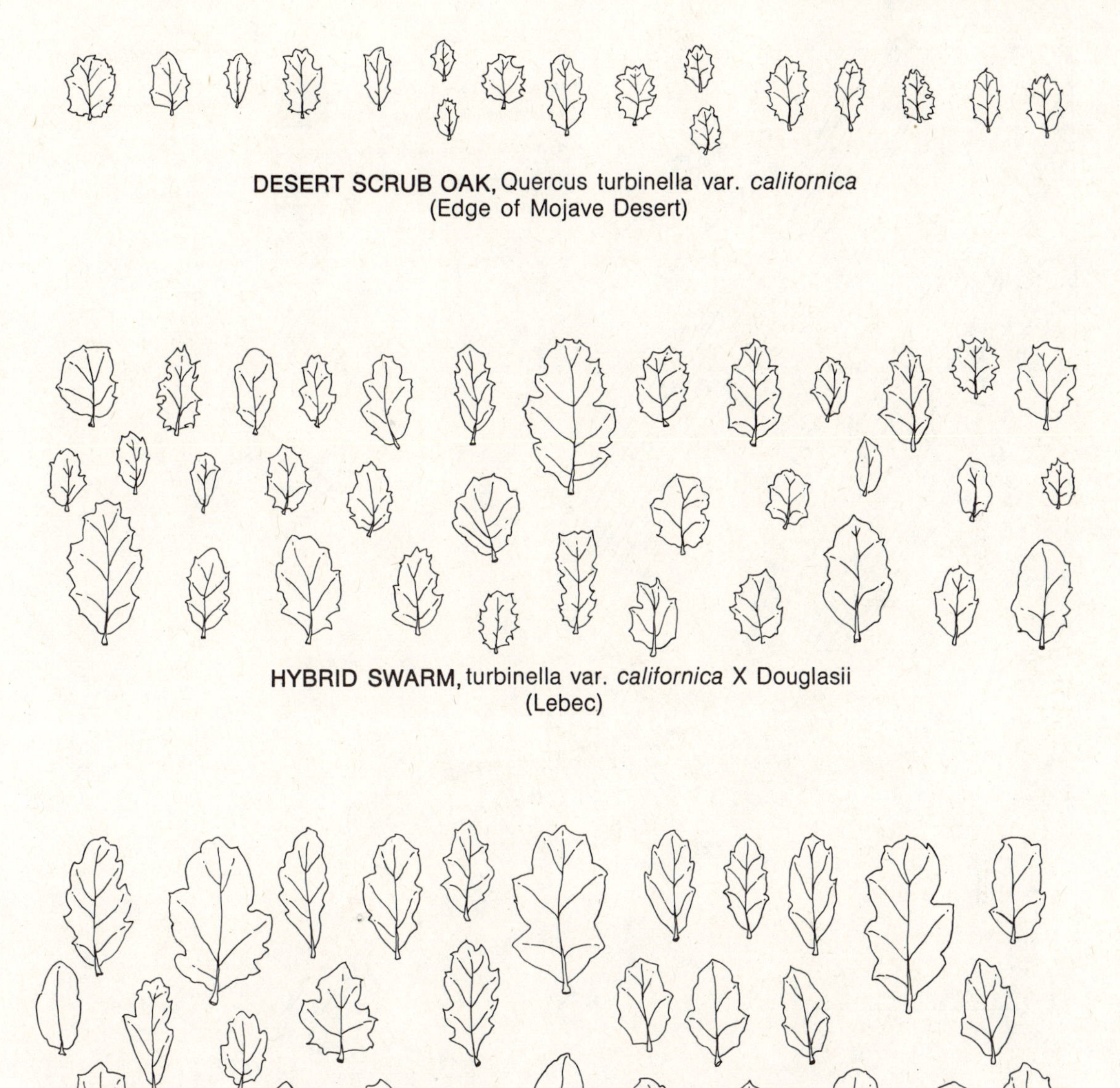

Figure 14-1. Intergradation of some species of Californian oaks: a representative leaf from every individual within a given area in populations of four species or varieties and in areas where the geographical ranges of species overlap. The other characters vary as do those represented by leaf diagrams.

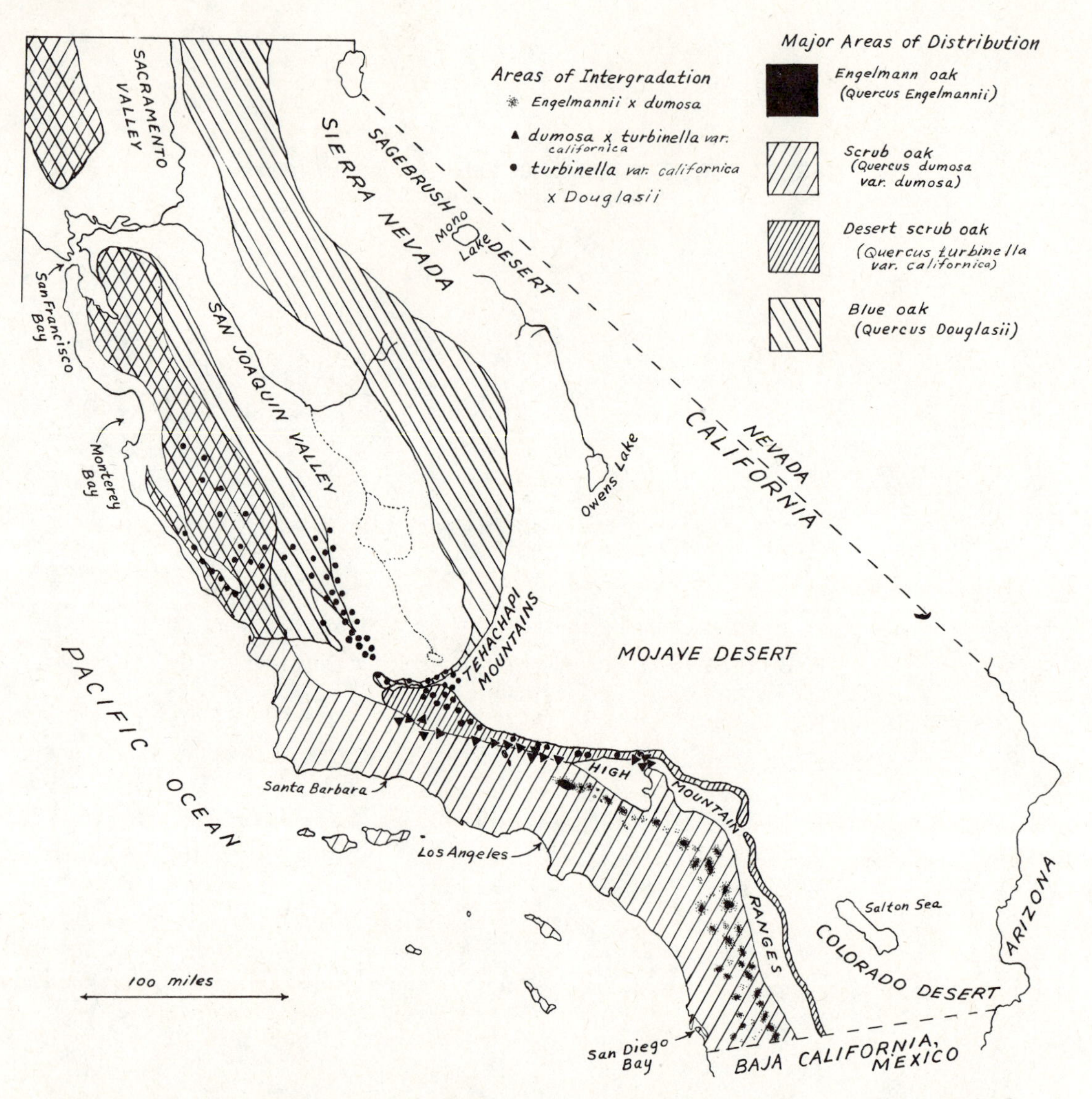

Figure 14-2. Map showing the geographical distribution and the areas of intergradation (hybrid swarms) of the four species or varieties of Californian oaks illustrated in **Figure 14-1.**

and animals often represent only the momentary results of evolution—a changing mass of individuals in innumerable stages of development in an almost infinite number of directions. Hard and fast distinctions of taxa are not necessarily to be expected because both differentiation and isolation of species and other taxa may exist in an infinite number of degrees. Taxa of all ranks may shade off gradually into their relatives as do the Plant Kingdom and the Animal Kingdom. The groupings described in books often stand for species and other taxa only partly differentiated, some well, some poorly, and the complexity of classifying many of them may be likened to distinction of the races of men or of dogs. **The**

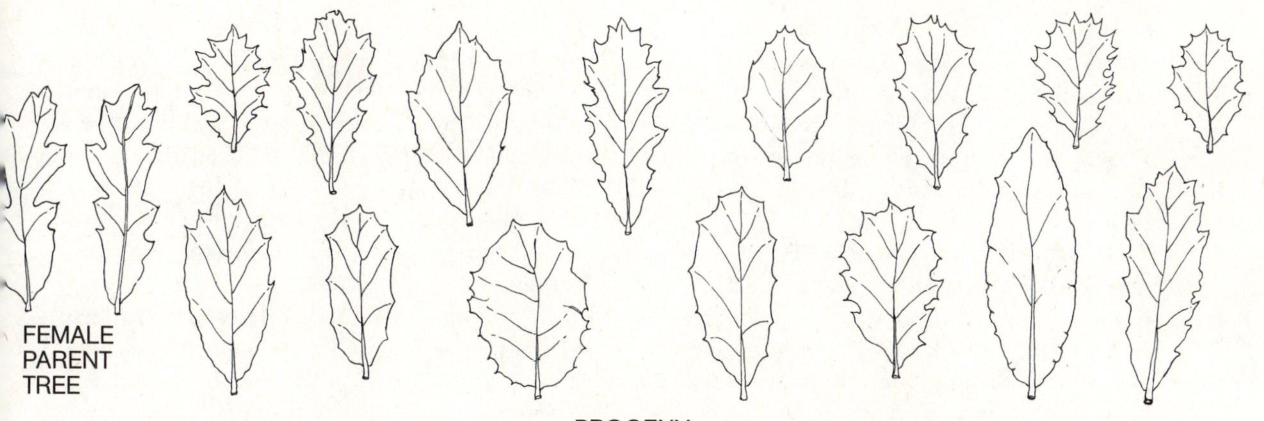

Figure 14-3. Results of a progeny test of the type tree (the one originally named) of *"Quercus gradidentata,"* which grew in the hybrid swarm of *Quercus Engelmannii* and *Quercus dumosa* in Monrovia Canyon, California. The leaves of the progeny varied from types identical with members of *Quercus Engelmannii* to others identical with members of *Quercus dumosa.* Although pollination was not controlled, this indicates a high degree of heterozygosity (dissimilarity of genes within each pair) in the plant named as *Quercus grandidentata.*

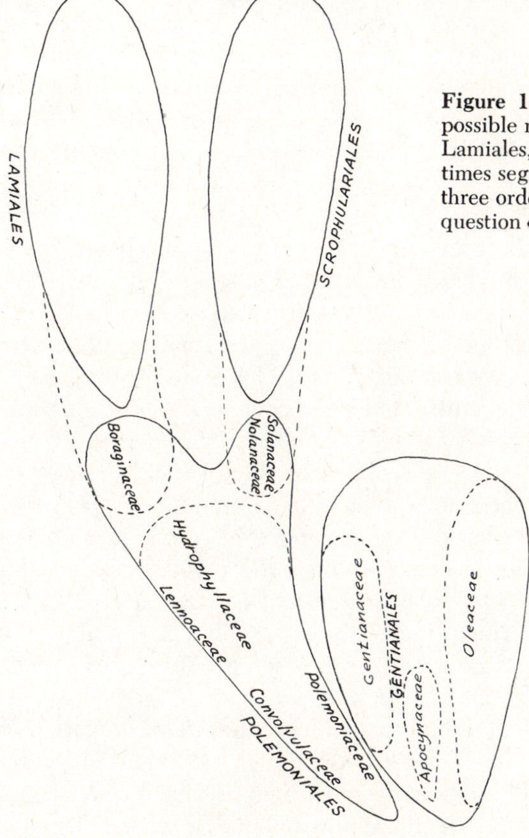

Figure 14-4. Chart illustrating the relationships and possible reclassification of the orders Polemoniales, Lamiales, Scrophulariales, Gentianales and the sometimes segregated Oleales and Apocynales. If the first three orders are merged, as seems reasonable, the question of inclusion of the last three is raised.

degrees of both differentiation and isolation of major taxa or of lesser natural populations from each other are infinite, but the ranks of the taxa into which the populations or systems of populations may be classified must be few. Consequently, assignment of natural populations to taxonomic ranks is a matter involving educated judgment.

The relative complexity of classification of plants and animals and of placing them in taxa may be illustrated by the following series of diagrams.

Two related taxa of the same rank are represented by a pair of circles. Degrees of differentiation in characters are indicated by stippling one circle, scattered dots representing slight separation of characters from those of the other taxon and crowded dots greater differentiation. More dense stippling in one part of the circle indicates that some individuals are more highly differentiated than others. Similar variation in the degree of differentiation of the components of the other taxon may be indicated by x's.

If the circles do not touch, they may represent natural populations completely isolated from each other. Varying lesser degrees of isolation are indicated by bringing them together and making them overlap more and more until ultimately one is superimposed exactly upon the other.

The species* of a genus or the components of a taxon of another rank may be represented diagrammatically by several or many circles with the differentiation of each from the others indicated by dots, crosses, o's, or other symbols and degrees of isolation indicated by varying amounts of overlapping.

Parallel Evolution. When all the members of two or more taxa have become completely isolated reproductively or geographically, they still may have no single character always present in one and never in the other. Often this is due to parallel evolution, that is, for example, characters arising in some families of one order may duplicate those occurring in at least some families or genera of the other. This may be due either to chance or to parallel adaptation to similar environmental conditions.

An example of a character which must have arisen by parallel evolution occurring in many plant groups is seen in the occurrence of willowlike leaves, that is, elongated more or less lanceolate leaf blades. Such leaves occur in some members of a very high percentage of the orders of Dicotyledoneae, regardless of closeness or distance of relationship or of the other leaf forms occurring in the same order. They occur also among the monocotyledons and even the gymnosperms and the ferns.

Another example is occurrence of (1) pollen embedded in waxy masses (pollinia) and (2) union of the stamens(s) with the stigma or style in the Asclepiadaceae or milkweed family (Dicotyledoneae) and in the Orchidaceae (Monocotyledoneae).

The Ranales and Rosales are two of the best-known orders of flowering plants (Dicotyledoneae). Each is of relatively clear status, and, although some authors have proposed dividing both into smaller units, few have favored either (1) transfer of fundamental units from one order to the other, or (2) union of the Ranales and Rosales or of elements selected from them. Careful study of the numerous families, genera, and species of either order does not reveal great gaps among its members, that is, the plants included, despite great differentiation of the extreme forms, are connected by essentially continuous series of intermediate species groups (cf. pp. 290, 292).

Despite the evidently natural segregation of the Ranales and Rosales, no single character has been demonstrated to occur throughout one order and not at all in the other. The most

* The definition of species accepted here is not exactly that of the evolutionary geneticist because the taxonomist or classifier of plants has a different objective. (The definition of species and other taxa is complex. It is discussed in an advanced book: Lyman Benson, *Plant Taxonomy, Methods and Principles*, John Wiley & Sons, New York. 1962, reprinted 1978. Chapter 9.)

An evolutionary geneticist may be concerned primarily with the mechanisms by means of which species arise, are perpetuated, develop internal variation, and become still further differentiated into other species. He may not be much concerned with their clear distinction from each other because he does not have to produce literature enabling other persons to recognize them. Consequently, for his own purposes he may accept as "species" plant population systems which would be scarcely intelligible to anyone who has not studied them experimentally. Often these may be reproductively isolated populations with relatively little differentiation from each other. If the author is interested primarily in genetics, he may tend to accept isolation as the basis for his definition of species because, as a criterion, isolation may serve his purpose well. He may come to accept a narrow definition of species and to accord this status to minor taxa, which the conservative taxonomist would consider to be varieties (or subspecies) of mere races within the species.

Some of the tasks of the taxonomist are the production of floras and manuals and of monographs of plant genera. His work must include preparation of workable keys and descriptions and the cataloging of plants in such a way that others may recognize them. Consequently, he insists that to qualify for specific rank, population systems must be both fairly well isolated and visibly differentiated. The occurrence of reproductive isolation (as between diploid and tetraploid entities) with no or little differentiation is not enough for his purpose in distinguishing species.

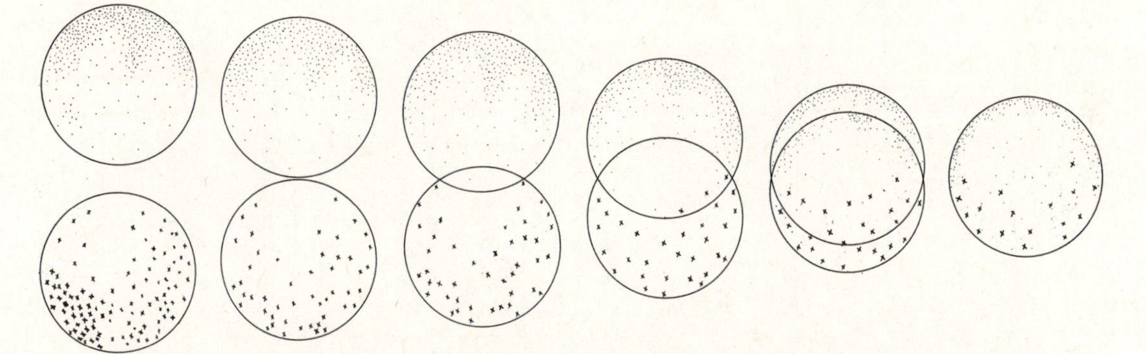

Figure 14-5. Degrees of isolation (represented by the merging circles) of taxa (e.g. species); degrees of differentiation of the members of the population systems included in each taxon (represented by the stippling within one circle and the x's within the other). As isolation disappears, so does differentiation or vice versa.

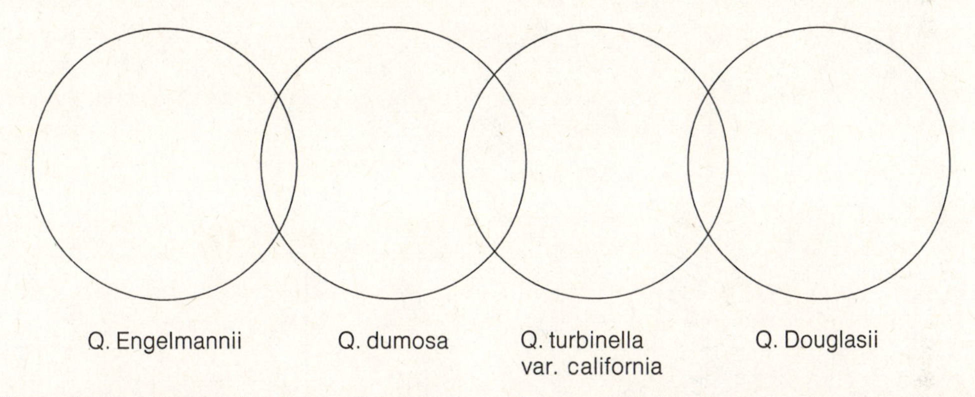

Figure 14-6. Degrees of isolation of the species and varieties of the oaks in **Figures 14-1** and **14-2.**

nearly consistent differentiating character has been thought to be presence of stipules in the Rosales and their absence in Ranales. However, although the great majority of the species of Rosales have stipules, various small groups, including several entire families, do not, and a few Ranales have either stipules (Magnoliaceae, Cercidiphyllaceae, Lactoridaceae, some Dilleniaceae, some Berberidaceae, and species of *Thalictrum* in the Ranunculaceae) or stipulelike broadened leaf bases (some species of *Ranunculus*). A group of differentiating characters marks off most members of the two orders, but no single factor in the complex is infallible by itself. Parallel evolution within the two groups apparently has broken down the effectiveness of every distinguishing feature.

The Rôle of Evolution in Classification. The taxa of plants and animals have arisen through evolution, and this mode of origin underlies their classification. Nevertheless, the pathway evolution has followed through large plant or animal groups is not proved unless a series of fossils records the history of development. Series of fossils are likely to represent only those organisms which occurred in or near areas where silt or similar material was being deposited, ordinarily in shallow water; and

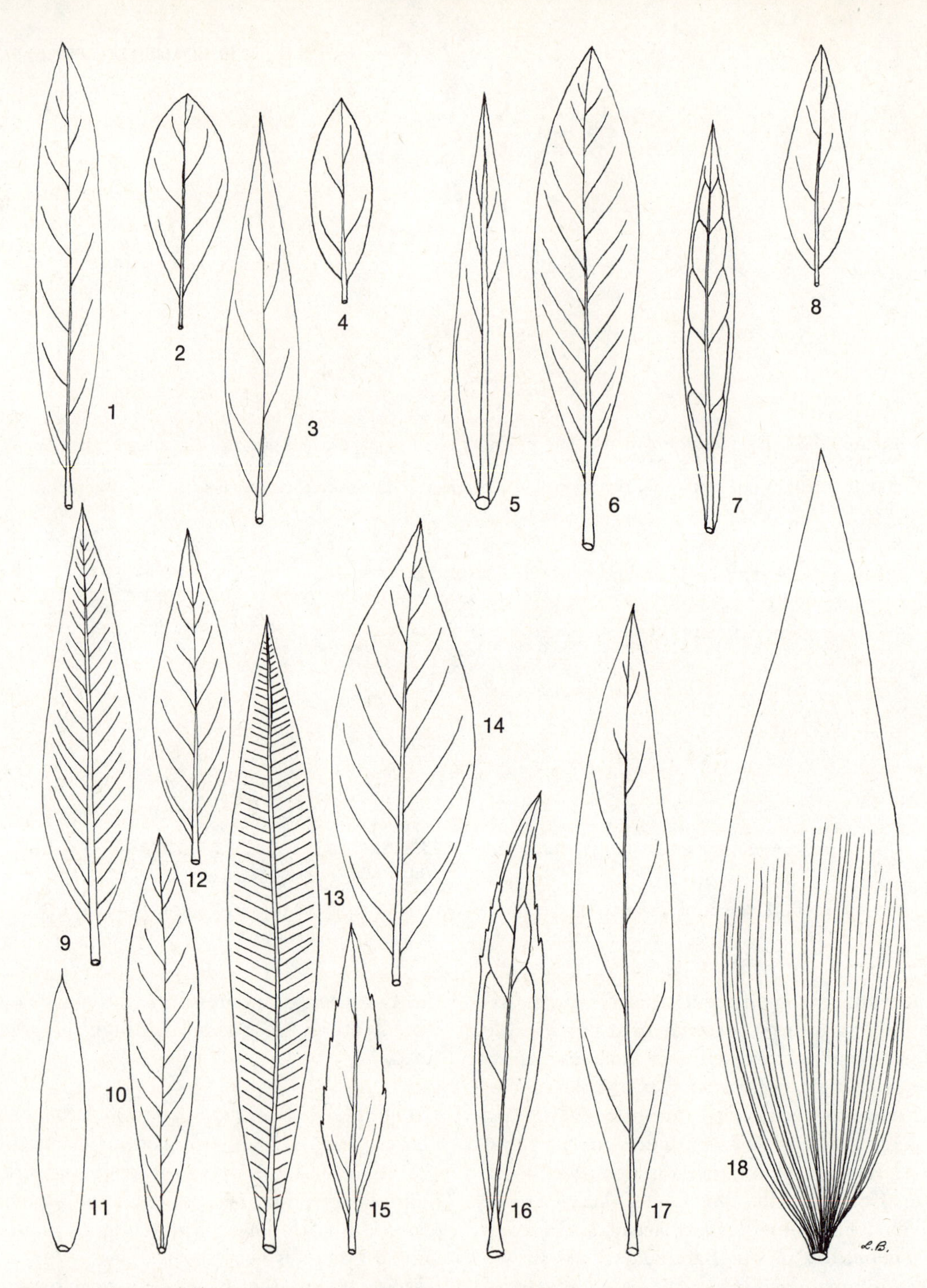

Figure 14-7. Parallel evolution: willowlike leaves in members of widely divergent major taxa or vascular plants: **1–4** four species of willows, *Salix*, Salicales, **5** *Penstemon*, Scrophulariales, **6** *Pittosporum*, Rosales, **7** goldenrod, *Solidago*, Asterales, **8** *Abelia*, Rubiales, **9** *Eucalyptus*, Myrtales, **10** olive, *Olea*, Gentianales, **11** *Araucaria*, Pinales (a gymnosperm order), **12** Carolina jessamine, *Gelsemium*, Gentianales, **13** *oleander, Nerium*, Gentianales, **14** cherry, *Prunus*, Rosales, **15** buttercup, *Ranunculus*, Ranales, **16** *Baccharis*, Asterales, **17** *Cestrum*, Polemoniales, **18** *Agathis*, Pinales. Similar leaves occur also among the ferns and the club mosses.

plants which occurred on plains and mountains far from sites of deposition usually have not been fossilized. Furthermore, ordinarily only the hard parts of plants and animals have been preserved, and plants whose essential structures were soft have left no decipherable fossil record. In some taxa, as, for example, the members of the order Pinales (pines, firs, etc.) the record is relatively clear. Differentiation of the present families can be traced through Mesozoic time with considerable accuracy. The same is true of the differentiation of major groups of vascular spore plants in Devonian time, but in most other instances the fossil record is either nonexistent or fragmentary and unreliable.

Usually knowledge of the pathway of evolution from one group of plants or animals to another is incomplete, and the most reliable basis for their classification is a composite of their degree of differentiation and their degree of isolation. Weighing these factors results in bringing into each recognized taxon individuals having much in common and relatively few differences. In general such groups are composed of truly related organisms because, despite parallel evolution of a few characters, coincidence of the great bulk of characters is likely to be due to a common heritage.

Classification of the Higher Taxa of Vascular Plants. The greatest difficulty in classifying the higher taxa of vascular plants is preparation of a reasonable alignment of the orders of angiosperms. Classification of the living gymnosperms and pteridophytes is relatively simple because evidently the few surviving groups are the well-differentiated remnants of once complex major taxa. The flowering plants, on the other hand, are the dominant plants of the present geological period, and among them the plants of many successful lines of evolutionary development are changing rapidly. Most groups are represented by a wealth of forms, and intermediate types between major groups are abundant. Consequently, classification is difficult.

The Relative Stability of Characters

Despite the importance of physiological characters usually their use in classification is not practical because they are intangible and difficult to work with. Consequently, for the distinction of taxa it is necessary to rely chiefly upon variations in the complex patterns of visible structural characters. These variations in character patterns may occur in any organ of the plant but in some organs the character combinations are relatively stable while in others they are less so because they are affected significantly by a greater range of environmental factors. Differences from taxon to taxon in the form and structure of an organ (or its parts) depend upon its relative complexity, its capacity to be modified, and its potentiality for fitting into different environmental niches.

VEGETATIVE ORGANS

Roots. Roots tend to retain ancient structures rather than to change over rapidly to new ones. It is obvious that conservatism is possible for the root because nearly all roots live and function underground in an environment with no light and with lesser and slower fluctuations of temperature and usually of moisture than in air. It is true that some roots are adapted to living above ground, as are those of many tropical orchids, that some live in swampy soil saturated with water, and that some grow in very dry soil, but most do not. The most common variation is not in physical factors or moisture but in the chemical composition of the soil. The variety and changeability of physical environments with which all but a few roots are in contact are not comparable to those affecting the form and structure of aërial organs, and this is reflected in the nearly constant make-up of roots. New root types produced genetically at random are not very likely to survive because only rarely do they provide an advantage to the plant, yielding success in a specialized environment.

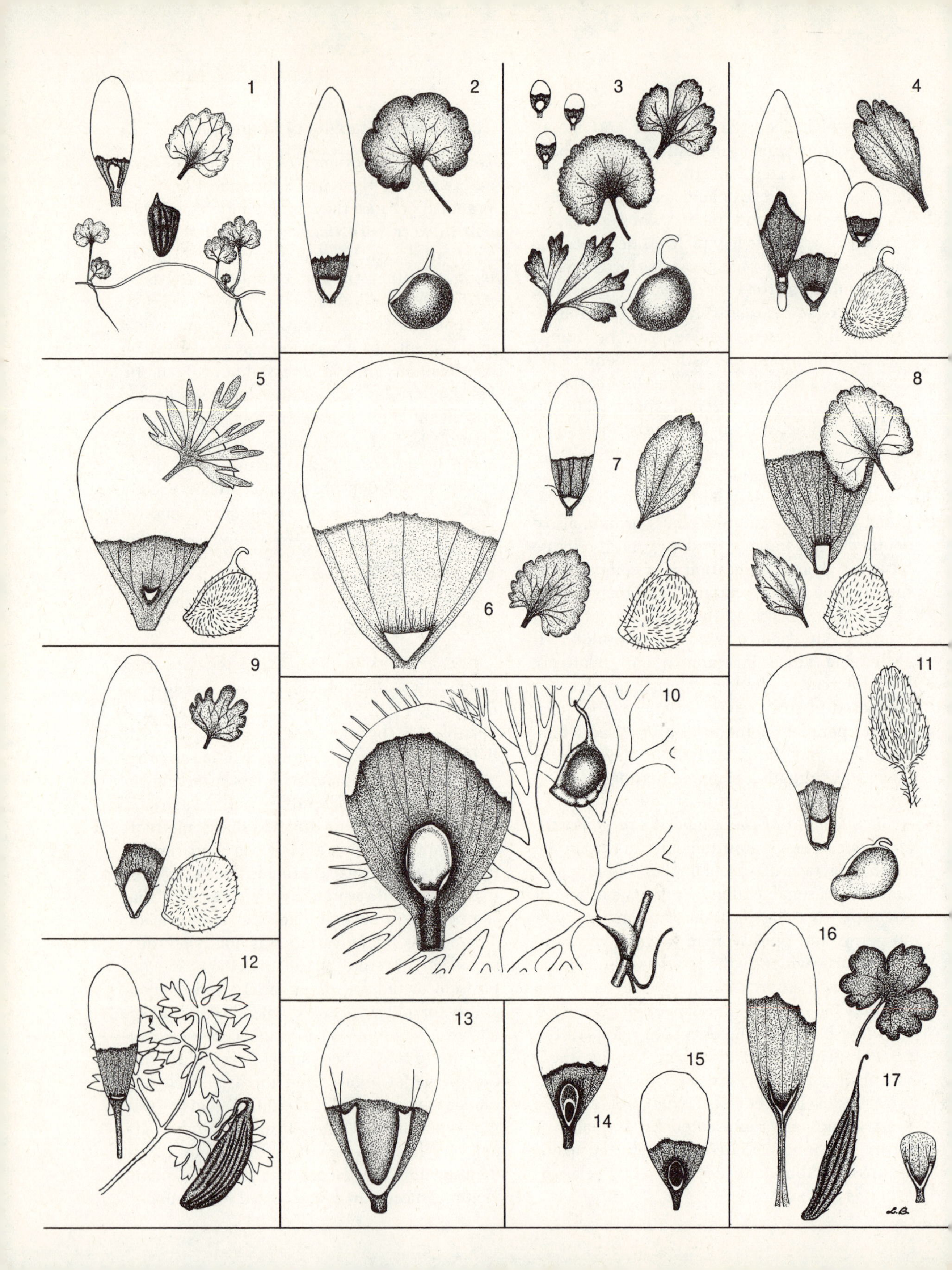
1
2
3
4
5
6
7
8
9
10
11
12
13
14
15
16
17
L.B.

The underground portion of the axis of *Rhynia*, which lived in mid-Paleozoic time perhaps 375 million years ago, differed relatively little in structure from the roots of modern plants. However, the similar aërial portion of the axis of *Rhynia* or its relatives evidently has evolved into the various types of stems of existing plants.

Stems. Stems, most of which live in air and support the leaves in positions favorable to receiving the light necessary to photosynthesis, include a wider range of forms than do roots. However, these supporting (and conducting) organs do not differ *externally* from one another nearly as much as do leaves, whose efficiency as food-manufacturing organs is affected more by environmental factors.

Leaves. A green plant is more likely to survive in competition if as many as possible of the leaf cells are exposed to light, and so the thickness of the leaf is limited in part by necessity for illumination of a high percentage of the cells. But in food production the leaf cells must be so arranged that exchange of gases with the air is possible, and, as a consequence of exposure of some cells, water is lost by evaporation. Since too rapid evaporation results in death, leaf structure may be limited in accordance with the type of environment peculiar to the plant. Herbs or shrubs growing on the floors of moist deciduous forests such as those east of the Mississippi River may have large, broad, thin leaves using as much as possible of the relatively dim light beneath the trees. With these plants moisture lost to the atmosphere is replaced readily, and protection from drying is relatively unimportant. On the other hand, vascular plants growing in the Southwestern Deserts can survive only if they grow in the brief rainy seasons and live through the dry seasons as seeds or dormant underground parts or if they have special features facilitating conservation of water. Among these characters are presence of varnishlike resin over the leaf surface, sunken *stomata* (pores), persistence of leaves only in the more moist seasons, development of thick leaves presenting little evaporation surface in proportion to volume (commonly correlated with presence of internal water storage tissue as in a century-plant), or absence or early disappearance of leaves (as in nearly all cacti). The leaf, then, in keeping with its functions and its relationship to the environment, has many adaptive characters of form and structure as well as physiology. Each leaf type may represent at least one characteristic which has enabled the plant to survive because of special ability to live with or benefit from some feature(s) of the environment.

Roots have too few patterns to appear more than occasionally as key characters in classification of vascular plants; stems have more external patterns than roots but fewer than leaves. The organs with more forms and struc-

Figure 14-8. Divergent evolution of petals, nectary scales, leaves, and fruits (achenes) within a single genus, *Ranunculus:* **1** *R. Cymbalaria,* nectary scale a flap, **2** *R. Harveyi,* nectary scale a pocket, **3** *R. allegheniensis,* nectary scale a pocket, **4** *R. inamoenus,* nectary scale a pocket, **5** *R. pedatifidus,* nectary scale a pocket, **6** *R. cardiophyllus,* nectary scale a pocket with marginal hairs, **7** *R. coloradensis* (a local species), **8** *R. cardiophyllus* var. *subsagittatus,* **9.** *R. arizonicus,* nectary scale a pocket with marginal hairs, **10** *R. flabellaris,* nectary scale a pocket on the ventral surface of the scale, **11** *R. rhomboideus,* nectary scale a pocket, **12** *R. ranunculinus,* nectary scale a thickened ridge, **13** *R. Eastwoodianus,* nectary scale forked by prolongation of the edges of the pocket, **14, 15** *R. Gmelinii* and *R. sceleratus,* nectary scale a border around the gland, **16, 17** *R. Cooleyae* and *R. hystriculus,* nectary scale forked. From Lyman Benson, American Journal of Botany 27: 800. plate 1. 1940. Used by permission.

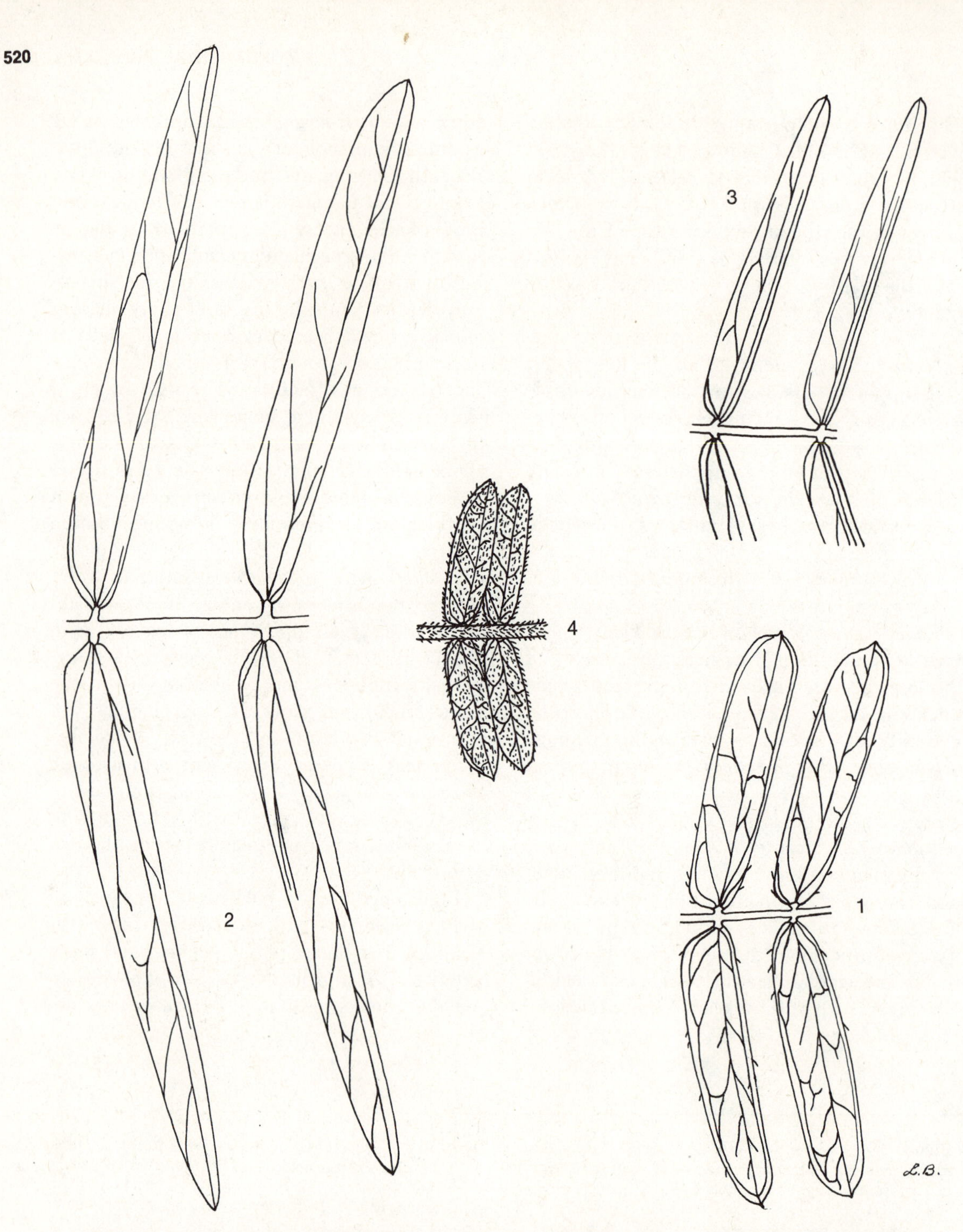

Figure 14-9. Divergent evolution of leaflets among the varieties of a single species, *Prosopis juliflora,* mesquite: **1** var. *juliflora* (West Indies and Mexico), **2** var. *glandulosa* (mostly Texas), **3** var. *Torreyana* (California to New Mexico and northwestern Mexico), **4** var. *velutina* (mostly southern Arizona and Sonora outside the range of var. *Torreyana*). From LYMAN BENSON, American Journal of Botany 28: 751. plate 1. 1941. Used with permission.

tures most frequently provide the key character segregating taxa within even a large class like the Angiospermae.

REPRODUCTIVE ORGANS

Reproductive organs become modified in form and structure more frequently than do roots or even stems, but most of them are considerably more stable in large groups of related plants than are leaves because their features are correlated less frequently with physical factors of the environment. Nevertheless, the parts of flowers and fruits are affected by other factors, and sometimes they may vary even more than leaves. Some changes of characters may promote pollination by different agents; others may make the transportation of fruits and seeds more effective. Seeds are affected by innumerable factors connected with their dispersal, and they are highly variable.

In the Angiosperms the fundamental patterns of the flowers and the structures of the fruits commonly distinguish orders, families, and in some cases genera. Within families or among the species of a genus the reproductive organs vary most commonly in minor (often quantitative) details such as shape, size, color, or tendency to produce glands or hairs. The characters of the most variable organs, the leaves, more commonly distinguish the lesser taxa, such as genera or more often species or varieties; they are constant only occasionally throughout a larger group. Since new characters or character combinations are more likely to be retained in leaves where they may determine the adaptability of the plant to one environmental niche or another, instances of parallel evolutionary development of leaf forms in groups not closely related but adapted to similar environments are much more common among the flowering plants than are similar duplications of flower organization or fruit structure. Consequently on the whole, leaves less frequently distinguish large groups. **Nevertheless, no caste system should be assigned rigidly to the characters of the various organs with regard to their "importance" in classification or to the ranks of the groups they may distinguish.** The relative values enumerated above for classification at different levels of rank represent only general tendencies, and **the characters to be selected as distinguishing any two groups are the ones which occur most consistently in combination, regardless of the organ in which they occur. The only importance of a character in classification is the degree of its consistency of occurrence in combination with others.**

The discussion of relative stability of characters of organs presented above is not to be confused with Jeffrey's Doctrine of Conservative Organs,* which deals with retention of primitive features of the vascular bundles (conducting tissue) in plant organs. In its vascular anatomy the root is also most conservative, but the stem is radical in the changeability of its *internal* pattern as a conducting organ carrying water, dissolved salts, and food between leaf and root, this being (aside from support of the leaves) its primary function. Often leaves and reproductive parts, which are not primarily conducting organs, retain the primitive vascular features occurring in evidently ancestral plants, despite occurrence of much more advanced types of conducting tissues in the stems of the same species. Thus, in vascular anatomy as well as external characters, inconstancy of structure is correlated with the possibility of differences in method of carrying out the primary function(s) of the organ.

* Jeffrey, Edward Charles. *The Anatomy of Woody Plants.* pp. 238–241. 1917.

15

Historical Development of Classification Systems

The largest and most conspicuous plants are trees, and through all history man has considered them as a group apart from the soft and lowly herbs.

Ancient Greece—Theophrastus

The layman's classification of plants as trees and herbs appears in the writings of Theophrastus, a pupil of Aristotle. Recognizing the differences between trees and shrubs and between herbaceous annuals, biennials, and perennials, Theophrastus classified plants as woody or herbaceous. Although he did not use the following characters to set off major divisions in classification, he knew the seed, stem, and leaf differences of monocotyledons and dicotyledons, the anatomy of stems including the formation of annual cylinders of xylem and phloem, the distinction between roots and rhizomes, the possible nature of sepals and petals as specialized leaves, the existence of apetalous flowers, and the distinctions between hypogynous, perigynous, and epigynous flowers. These features were to be brought gradually into the flowering plant classification system more than 2,000 years later.

The Period of the Roman Empire

With the conquest and decline of Greek civilization, botany, like many other fields of knowledge, became dormant. In Roman times Cato, Virgil, and others wrote on agriculture, and Dioscorides and Galen published works on medicine. Primarily medicine was lore concerning herbs, and Dioscorides (circa 64 A.D.) discussed many plants, grouping some of them into more or less natural families. Pliny is known widely as a naturalist, but his work was little more than a compilation from Theophrastus and others plus a mass of amazing fables.

The Middle Ages

From the time of the Roman Empire to the beginning of the Fifteenth Century almost nothing was accomplished in botany. Ancient manuscripts were copied, and their substance appeared occasionally in a new book. Copied illustrations went through gradual evolution. For example, the strawberry plant came to have new features, including five leaflets instead of three. During this scientifically dark period botany, like other subjects, was chiefly a matter of discovering what the ancients had said, and the books were considered to be sources of information preferable to the plants.

The Renaissance

Valerius Cordus (born 1515 A.D.) appears to have been the first to accept once more the plant as the true source of data and to describe species from living individuals. From Wittenberg he explored the German mountains and forests. Unfortunately, while visiting universities in Italy he contracted malaria and died when he was only twenty-nine years old, and his book, embellished with relatively poor plates from another author, was published after his death. Had Cordus lived longer, application of the scientific method to botany might have received a great impetus at that time.

Although other botanists of the Fifteenth and Sixteenth Centuries were not as outstanding as Cordus, several of them (including Fuchs, Brunfels, Obelius, and Bauhin) had sufficient originality to examine living plants and to make illustrations directly from nature. The typical publications of both the Fifteenth and Sixteenth Centuries and much of the Seventeenth Century were herbals—highly illustrated works covering plants thought to have medicinal virtue. This restriction did not leave out many plants, and the texts illustrate the flight of human fancy when information is needed but not available. Here and there were flickerings of perception of classification—for example, Bauhin's sequences of related plants, bringing together such natural groups as the members of the mustard and mint families.

Fifteenth- and Sixteenth-Century attempts at arrangements of plants into major groupings were few and unsuccessful. For example, Obelius devised a classification system based upon the width of the leaf. In 1583 Caesalpino, an Italian, published a book in which botany was studied for its own sake rather than as supporting material for the study of medicine, but unfortunately, the system of classification was based upon preconceived theoretical principles. Although Caesalpino studied plants directly, he did so with his mind made up in advance; consequently, he contributed little toward making botany a science.

The Seventeenth Century was marked by acceleration of original activity in botany. Jung (a German) cleared up several plant structural problems of significance in classification. Robert Hooke (an Englishman) discovered the cell, but he used the microscope more as a plaything than as an instrument for research, and the real study of microscopic anatomy of plants was begun simultaneously in England and Italy by Grew and Malpighi. Each of these men was a brilliant research worker—a person of great originality and wisdom. In the next generation another great and original man, Stephen Hales, conducted remarkable investigations in plant physiology. Thus, botany left the Dark Ages and entered upon a true renaissance, an era of investigation of the plant itself and of facts related to its processes. The simple scientific method—drawing conclusions only from the results of investigation—was applied to plants, and their study became a science.

Natural and Artificial Systems of Classification—Linnaeus

The greatest figure in systematic botany is Carolus Linnaeus (1707–1778), known in his later years as Carl von Linné. Many authors have pointed out the abilities of Linnaeus in analysis, description, and diagnosis, crying out in the same breath that he was not an experimenter, that is, not one who delves into the underlying reasons for the existence of natural phenomena. This is true, and the greatness of Linnaeus rests upon his broad outlook, his supreme ability as an organizer, and his usually good judgment in classification. Before the reasons for natural phenomena could be determined, the phenomena themselves had to be discovered, organized, and classified, and this was the legacy of Linnaeus to all later investigators, as to Darwin in the formulation of the theory of evolution through natural selection. Organizing ability and judgment are exceedingly important factors in construction of any kind of classification system, for without it no amount of data from any source or group of sources can be put to effective use.

The greatest botanical contributions of Linnaeus were as follows.

1. His publications, particularly *Genera Plantarum* (5th Ed. 1754) and *Species Plantarum* (1753), which in their various editions were a remarkable summary of the vegetation known during his lifetime.

2. The first extensive use of the binomial system of naming plants, which he applied also to animals and diseases. Linnaeus did not originate this system of naming organisms according to genera and species. It appears as far back as the work of Bauhin, and there are suggestions of it even in the writings of Theophrastus. However, Linnaeus was the first author to apply the principle consistently, seeming to abandon the prevailing method of naming plants by a generic name followed by a series of descriptive words, the series ranging from one to several Latin adjectives. The first consistent use of the binomial system was in the two books by Linnaeus mentioned above; consequently these are the cornerstone of the procedure of naming plants. They are the earliest works used as sources of scientific names, and earlier publications are invalid, except as Linnaeus's references to them clarify his meaning. According to W. M. Harlow, Linnaeus actually considered the species names (as they are interpreted) entered in the margins as trivial and intended them as an indexing device. At any rate they soon became the designations of the species.

3. Two systems of plant classification—an artificial system and the basis for a natural system.

The Artificial System. The better known classification system devised by Linnaeus was called the Sexual System. This was a practical device for pigeonholing plants for rapid identification. The Plant Kingdom was divided into twenty-four classes. The first ten of these were based on the number of stamens, the others mostly on various features of the stamens or their relationship to the carpels. For example, in classes sixteen to eighteen the stamens were coalescent by the filaments (monadelphous, diadelphous, or polyadelphous); in class nineteen the anthers but not the filaments were coalescent; in class twenty the stamens and styles were adnate; in classes twenty-one and twenty-two the flowers were unisexual and the plants monoecious or dioecious; in class twenty-four the plants had no obvious sexual reproduction, and they were said to be of "clandestine marriage."

Publication of the Sexual System made botany a practical and popular science. Applied to a relatively limited group of plants, the system provided a workable method of cataloging, and it dominated botanical publications for approximately a century.

The Sexual System was artificial because it was based upon predetermined grounds, that is, an arbitrary importance was attached pri-

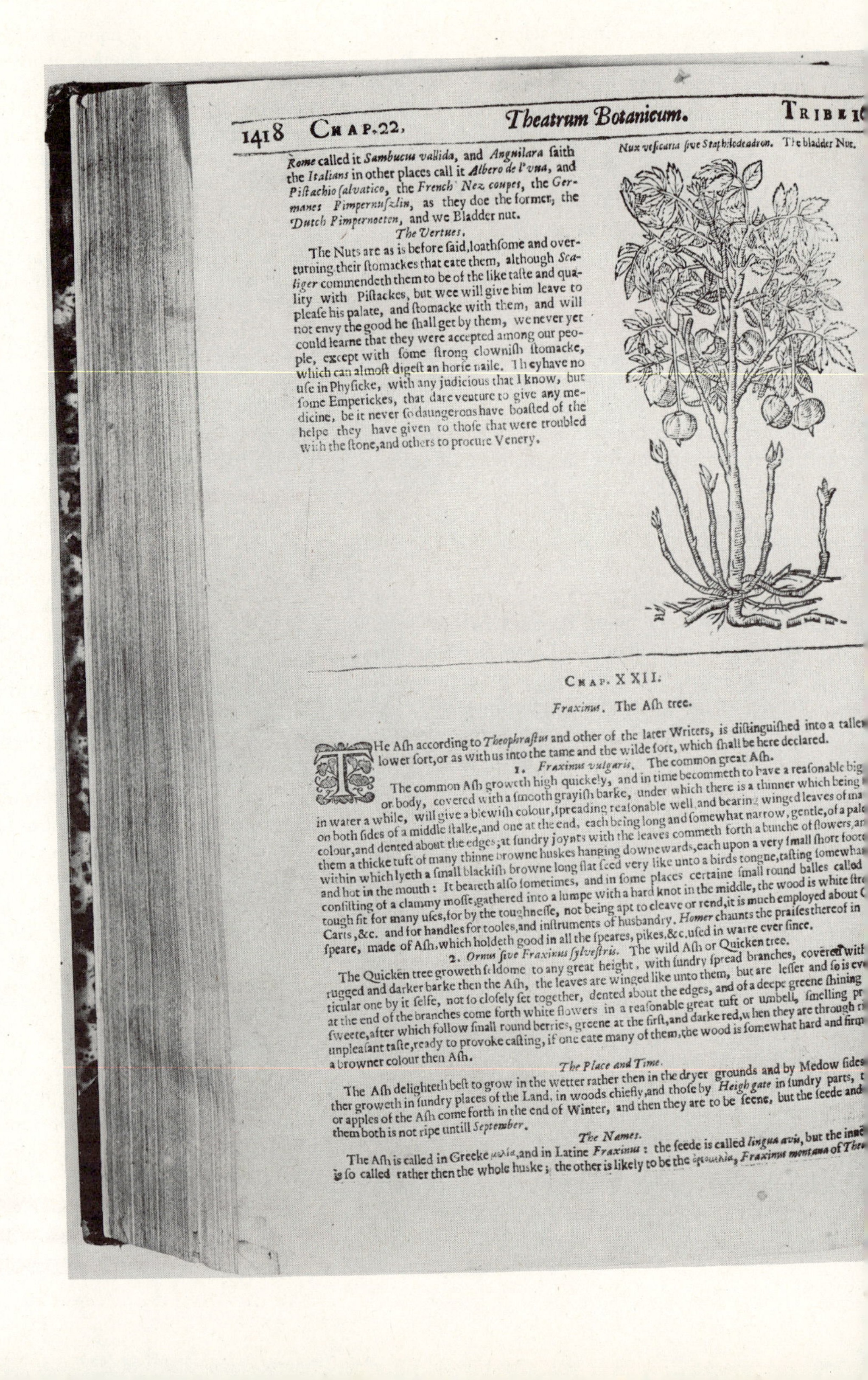

1418 CHAP. 22. *Theatrum Botanicum.* TRIBE 1[illegible]

Rome called it *Sambucus vallida*, and *Angnilara* ſaith the *Italians* in other places call it *Albero de l'vua*, and *Piſtachio ſalvatico*, the *French* *Nez coupes*, the *Germanes* *Pimpernuſzlin*, as they doe the former, the *Dutch Pimpernoeten*, and we Bladder nut.

The Vertues.

The Nuts are as is before ſaid, loathſome and overturning their ſtomackes that eate them, although *Scaliger* commendeth them to be of the like taſte and quality with Piſtackes, but wee will give him leave to pleaſe his palate, and ſtomacke with them, and will not envy the good he ſhall get by them, we never yet could learne that they were accepted among our people, except with ſome ſtrong clowniſh ſtomacke, which can almoſt digeſt an horſe naile. They have no uſe in Phyſicke, with any judicious that I know, but ſome Empericks, that dare venture to give any medicine, be it never ſo daungerous have boaſted of the helpe they have given to thoſe that were troubled with the ſtone, and others to procure Venery.

Nux veſicaria ſive Staphilodeadron. The bladder Nut.

CHAP. XXII.

Fraxinus. The Aſh tree.

THe Aſh according to *Theophraſtus* and other of the later Writers, is diſtinguiſhed into a talle[…]
lower ſort, or as with us into the tame and the wilde ſort, which ſhall be here declared.

1. *Fraxinus vulgaris.* The common great Aſh.

The common Aſh groweth high quickely, and in time becommeth to have a reaſonable big[…]
or body, covered with a ſmooth grayiſh barke, under which there is a thinner which being[…]
in water a while, will give a blewiſh colour, ſpreading reaſonable well, and bearing winged leaves of ma[…]
on both ſides of a middle ſtalke, and one at the end, each being long and ſomewhat narrow, gentle, of a pale[…]
colour, and dented about the edges; at ſundry joynts with the leaves commeth forth a bunche of flowers, an[…]
them a thicke tuft of many thinne browne huskes hanging downewards, each upon a very ſmall ſhort foot[…]
within which lyeth a ſmall blackiſh browne long flat ſeed very like unto a birds tongue, taſting ſomewha[…]
and hot in the mouth: It beareth alſo ſometimes, and in ſome places certaine ſmall round balles called[…]
conſiſting of a clammy moſſe, gathered into a lumpe with a hard knot in the middle, the wood is white ſtr[…]
tough fit for many uſes, for by the toughneſſe, not being apt to cleave or rend, it is much employed about C[…]
Carts, &c. and for handles for tooles, and inſtruments of husbandry. *Homer* chaunts the praiſes thereof in[…]
ſpeare, made of Aſh, which holdeth good in all the ſpeares, pikes, &c. uſed in warre ever ſince.

2. *Ornus ſive Fraxinus ſylveſtris.* The wild Aſh or Quicken tree.

The Quicken tree groweth ſeldome to any great height, with ſundry ſpread branches, covered wit[…]
rugged and darker barke then the Aſh, the leaves are winged like unto them, but are leſſer and ſo is ev[…]
ticular one by it ſelfe, not ſo cloſely ſet together, dented about the edges, and of a deepe greene ſhining[…]
at the end of the branches come forth white flowers in a reaſonable great tuft or umbell, ſmelling pr[…]
ſweete, after which follow ſmall round berries, greene at the firſt, and darke red, when they are through ri[…]
unpleaſant taſte, ready to provoke caſting, if one eate many of them, the wood is ſomewhat hard and firm[…]
a browner colour then Aſh.

The Place and Time.

The Aſh delighteth beſt to grow in the wetter rather then in the dryer grounds and by Medow ſides[…]
ther groweth in ſundry places of the Land, in woods chiefly, and thoſe by *Heighgate* in ſundry parts, t[…]
or apples of the Aſh come forth in the end of Winter, and then they are to be ſeene, but the ſeede and[…]
them both is not ripe untill *September*.

The Names.

The Aſh is called in Greeke μελία, and in Latine *Fraxinus*: the ſeede is called *lingua avis*, but the inne[…]
is ſo called rather then the whole huske; the other is likely to be the [illegible], *Fraxinus montana* of *The*[…]

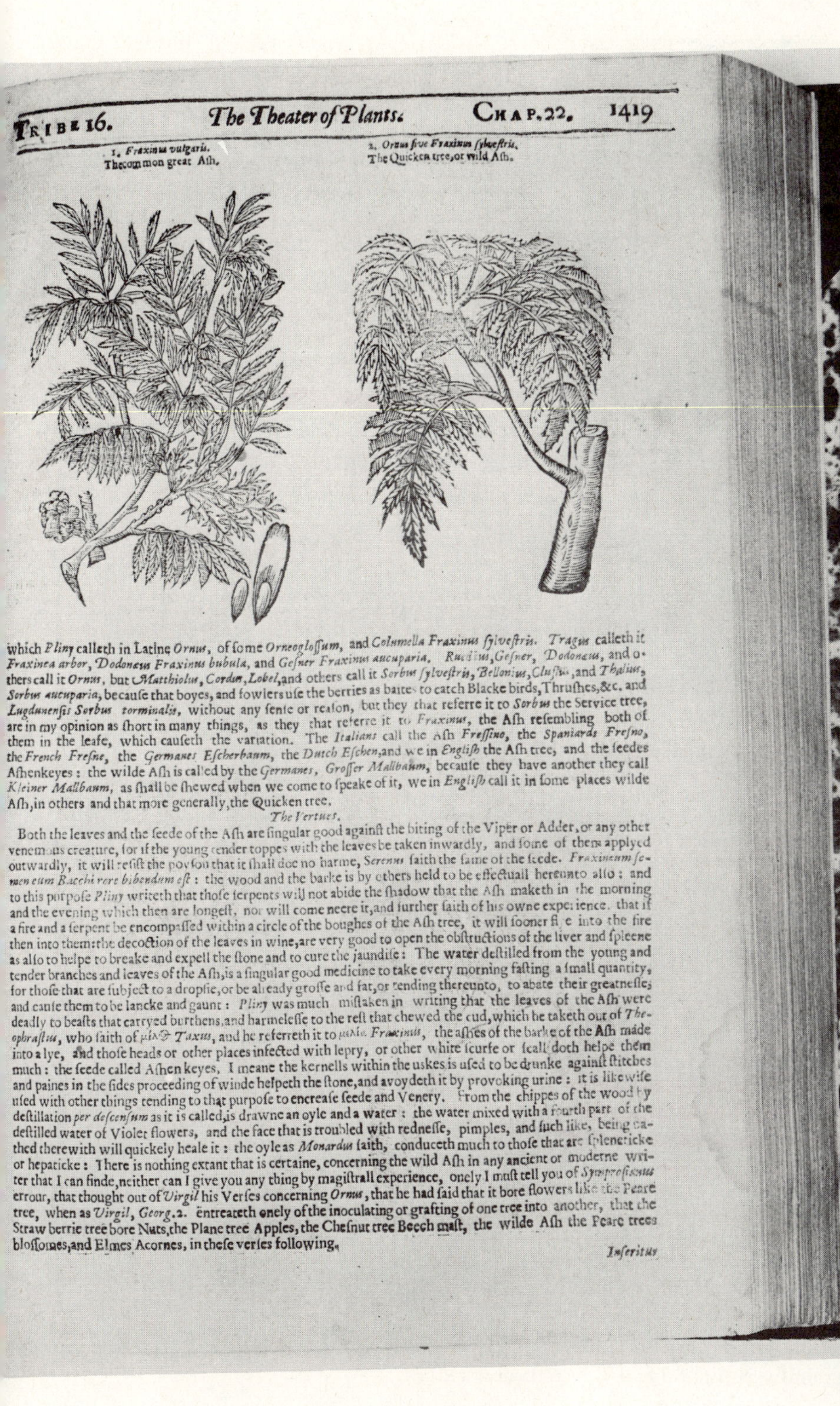

TRIBE 16. *The Theater of Plants.* CHAP.22. 1419

1. *Fraxinus vulgaris.* The common great Aſh.

2. *Ornus ſive Fraxinus ſylveſtris.* The Quicken tree, or wild Aſh.

which *Pliny* calleth in Latine *Ornus*, of ſome *Orneogloſſum*, and *Columella Fraxinus ſylveſtris*. *Tragus* calleth it *Fraxinea arbor*, *Dodoneus Fraxinus bubula*, and *Geſner Fraxinus aucuparia*. *Ruellius*, *Geſner*, *Dodoneus*, and others call it *Ornus*, but *Matthiolus*, *Cordus*, *Lobel*, and others call it *Sorbus ſylveſtris*, *Bellonius*, *Cluſius*, and *Thalius*, *Sorbus aucuparia*, becauſe that boyes, and fowlers uſe the berries as baites to catch Blacke birds, Thruſhes, &c. and *Lugdunenſis Sorbus torminalis*, without any ſenſe or reaſon, but they that referre it to *Sorbus* the Service tree, are in my opinion as ſhort in many things, as they that referre it to *Fraxinus*, the Aſh reſembling both of them in the leafe, which cauſeth the variation. The *Italians* call the Aſh *Freſſino*, the *Spaniards Freſno*, the *French Freſne*, the *Germanes Eſcherbaum*, the *Dutch Eſchen*, and we in *Engliſh* the Aſh tree, and the ſeedes Aſhenkeyes: the wilde Aſh is called by the *Germanes*, *Groſſer Mallbaum*, becauſe they have another they call *Kleiner Mallbaum*, as ſhall be ſhewed when we come to ſpeake of it, we in *Engliſh* call it in ſome places wilde Aſh, in others and that more generally, the Quicken tree.

The Vertues.

Both the leaves and the ſeede of the Aſh are ſingular good againſt the biting of the Viper or Adder, or any other venemous creature, for if the young tender toppes with the leaves be taken inwardly, and ſome of them applyed outwardly, it will reſiſt the poyſon that it ſhall doe no harme, *Serenus* ſaith the ſame of the ſeede. *Fraxineum ſemen cum Bacchi rore bibendum eſt*: the wood and the barke is by others held to be effectuall hereunto alſo: and to this purpoſe *Pliny* writeth that thoſe ſerpents will not abide the ſhadow that the Aſh maketh in the morning and the evening which then are longeſt, nor will come neere it, and further ſaith of his owne experience, that if a fire and a ſerpent be encompaſſed within a circle of the boughes of the Aſh tree, it will ſooner flie into the fire then into them: the decoction of the leaves in wine, are very good to open the obſtructions of the liver and ſpleene as alſo to helpe to breake and expell the ſtone and to cure the jaundiſe: The water deſtilled from the young and tender branches and leaves of the Aſh, is a ſingular good medicine to take every morning faſting a ſmall quantity, for thoſe that are ſubject to a dropſie, or be already groſſe and fat, or tending thereunto, to abate their greatneſſe, and cauſe them to be lancke and gaunt: *Pliny* was much miſtaken in writing that the leaves of the Aſh were deadly to beaſts that carryed burthens, and harmeleſſe to the reſt that chewed the cud, which he taketh out of *Theophraſtus*, who ſaith of μίλ⁊ *Taxus*, and he referreth it to μελία *Fraxinus*, the aſhes of the barke of the Aſh made into a lye, and thoſe heads or other places infected with lepry, or other white ſcurfe or ſcall doth helpe them much: the ſeede called Aſhen keyes, I meane the kernells within the uskes is uſed to be drunke againſt ſtitches and paines in the ſides proceeding of winde helpeth the ſtone, and avoydeth it by provoking urine: it is likewiſe uſed with other things tending to that purpoſe to encreaſe ſeede and Venery. From the chippes of the wood by deſtillation *per deſcenſum* as it is called, is drawne an oyle and a water: the water mixed with a fourth part of the deſtilled water of Violet flowers, and the face that is troubled with redneſſe, pimples, and ſuch like, being bathed therewith will quickely heale it: the oyle as *Monardus* ſaith, conduceth much to thoſe that are ſpleneticke or hepaticke: There is nothing extant that is certaine, concerning the wild Aſh in any ancient or moderne writer that I can finde, neither can I give you any thing by magiſtrall experience, onely I muſt tell you of *Sympreſianus* errour, that thought out of *Virgil* his Verſes concerning *Ornus*, that he had ſaid that it bore flowers like the Peare tree, when as *Virgil*, *Georg*. 2. entreateth onely of the inoculating or grafting of one tree into another, that the Straw berrie tree bore Nuts, the Plane tree Apples, the Cheſnut tree Beech maſt, the wilde Aſh the Peare trees bloſſomes, and Elmes Acornes, in theſe verſes following.

Inſeritur

Figure 15-1. An herbal of the Seventeenth Century (John Parkinson, *Theatrum Botanicum, The Theater of Plants.* 2: 1418–1419. 1640). The treatment of each species is typical of herbals, as are the fabulous "vertues" of each plant.

Figure 15-2. Carolus Linnaeus (Carl von Linné) (1707–1778). Founder of modern plant classification; first to use the binomial system consistently; classifier of plant genera and species in his *Genera Plantarum* and *Species Plantarum* and of animal genera and species and of genera and species of diseases; founder of a useful artificial system and a fragment of a natural system of plant classification. Professor at Uppsala, Sweden. Photograph by Frank B. Salisbury from an etching supplied through the courtesy of the Botaniska Avdelningen, Naturhistoriska Riksmuseet, from the K. Vetenskapsakademien, Stockholm.

marily to the number and arrangement of stamens and secondarily to the number and arrangement of carpels. Plants were fitted into the system regardless of relationship; for example, the class Octandria (stamens eight) included members of the heath family, the evening-primrose family, and the buckwheat family, none of these plants having any close genetic relationship to the others. In many instances monocotyledons and dicotyledons were thrown together indiscriminately.

Some students of botanical history have criticized Linnaeus because his Sexual System cut across natural relationships, and some have dismissed him with contempt, claiming the Sexual System set classification back for many years. Certainly it did not contribute anything directly to knowledge of plant relationships, but it did give a tremendous impetus to study of the Plant Kingdom. Accumulation of information was the first step necessary to constructing a system based upon relationships.

The Natural System. The Sexual System fooled most of Linnaeus' contemporaries and many of the later historians who have attempted to evaluate him, but it did not fool

CAROLI LINNÆI

S:Æ R:GIÆ M:TIS SVECIÆ ARCHIATRI; MEDIC. & BOTAN. PROFESS. UPSAL; EQUITIS AUR. DE STELLA POLARI; nec non ACAD. IMPER. MONSPEL. BEROL. TOLOS. UPSAL. STOCKH. SOC. & PARIS. CORESP.

SPECIES PLANTARUM,

EXHIBENTES PLANTAS RITE COGNITAS, AD GENERA RELATAS, CUM DIFFERENTIIS SPECIFICIS, NOMINIBUS TRIVIALIBUS, SYNONYMIS SELECTIS, LOCIS NATALIBUS, SECUNDUM SYSTEMA SEXUALE DIGESTAS.

TOMUS I.

Cum Privilegio S. R. M:tis Sueciæ & S. R. M:tis Polonicæ ac Electoris Saxon.

HOLMIÆ, IMPENSIS LAURENTII SALVII. 1753.

Figure 15-3. Title page of Linnaeus's *Species Plantarum*, First Edition, 1753. Photograph by Frank B. Salisbury.

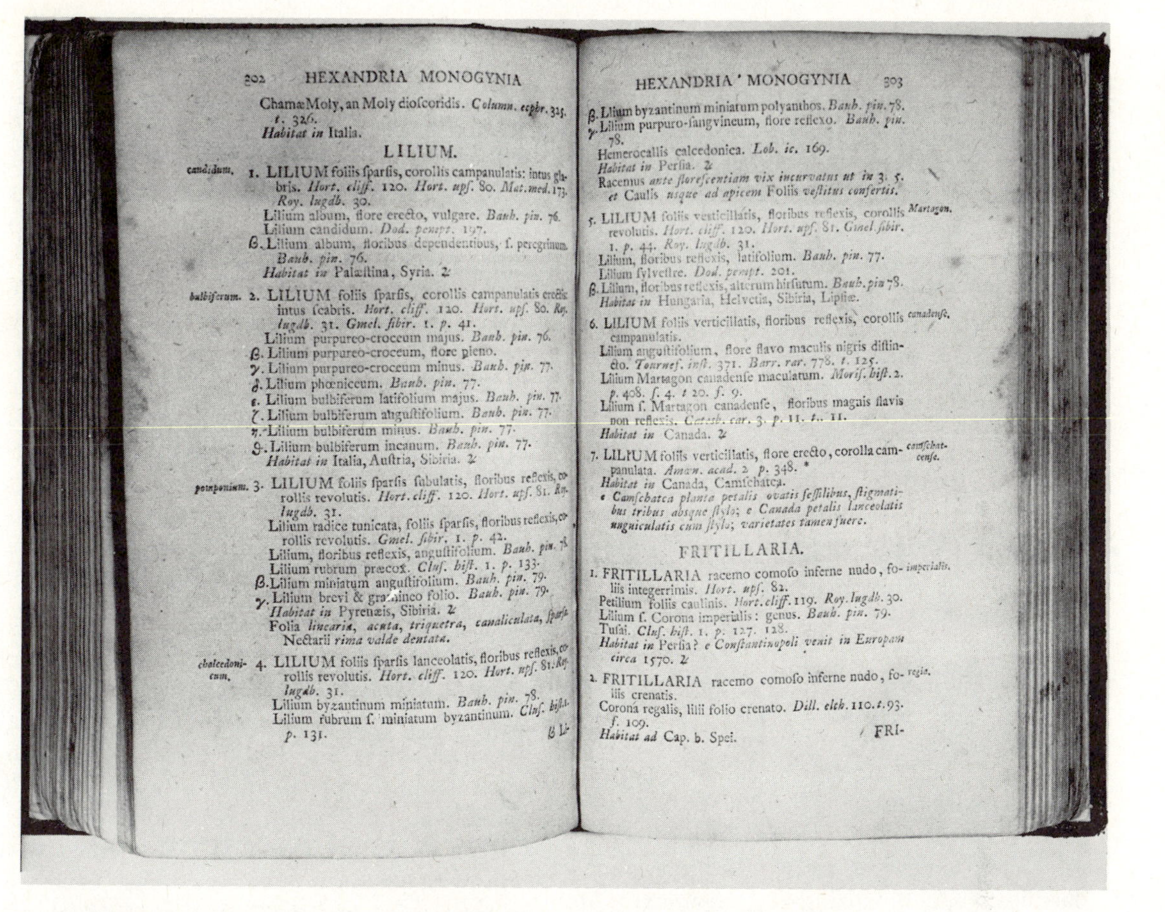

202 HEXANDRIA MONOGYNIA

ChamæMoly, an Moly dioſcoridis. *Column. ecphr.* 325. *t.* 326.
Habitat in Italia.

LILIUM.

candidum. 1. LILIUM foliis ſparſis, corollis campanulatis: intus glabris. *Hort. cliff.* 120. *Hort. upſ.* 80. *Mat. med.* 173. *Roy. lugdb.* 30.
Lilium album, flore erecto, vulgare. *Bauh. pin.* 76.
Lilium candidum. *Dod. pempt.* 197.
β. Lilium album, floribus dependentibus, ſ. peregrinum. *Bauh. pin.* 76.
Habitat in Palæſtina, Syria. ♃

bulbiferum. 2. LILIUM foliis ſparſis, corollis campanulatis erectis intus ſcabris. *Hort. cliff.* 120. *Hort. upſ.* 80. *Roy. lugdb.* 31. *Gmel. ſibir.* 1. *p.* 41.
Lilium purpureo-croceum majus. *Bauh. pin.* 76.
β. Lilium purpureo-croceum, flore pleno.
γ. Lilium purpureo-croceum minus. *Bauh. pin.* 77.
δ. Lilium phœniceum. *Bauh. pin.* 77.
ε. Lilium bulbiferum latifolium majus. *Bauh. pin.* 77.
ζ. Lilium bulbiferum anguſtifolium. *Bauh. pin.* 77.
η. Lilium bulbiferum minus. *Bauh. pin.* 77.
θ. Lilium bulbiferum incanum. *Bauh. pin.* 77.
Habitat in Italia, Auſtria, Sibiria. ♃

pomponium. 3. LILIUM foliis ſparſis ſubulatis, floribus reflexis, corollis revolutis. *Hort. cliff.* 120. *Hort. upſ.* 81. *Roy. lugdb.* 31.
Lilium radice tunicata, foliis ſparſis, floribus reflexis, corollis revolutis. *Gmel. ſibir.* 1. *p.* 42.
Lilium, floribus reflexis, anguſtifolium. *Bauh. pin.* 78.
Lilium rubrum præcox. *Cluſ. hiſt.* 1. *p.* 133.
β. Lilium miniatum anguſtifolium. *Bauh. pin.* 79.
γ. Lilium brevi & gramineo folio. *Bauh. pin.* 79.
Habitat in Pyrenæis, Sibiria. ♃
Folia *linearia, acuta, triquetra, canaliculata, ſparſa.*
Nectarii *rima valde dentata.*

chalcedonicum. 4. LILIUM foliis ſparſis lanceolatis, floribus reflexis, corollis revolutis. *Hort. cliff.* 120. *Hort. upſ.* 81. *Roy. lugdb.* 31.
Lilium byzantinum miniatum. *Bauh. pin.* 78.
Lilium rubrum ſ. miniatum byzantinum. *Cluſ. hiſt.* 1. *p.* 131.
β Li-

HEXANDRIA MONOGYNIA 203

β. Lilium byzantinum miniatum polyanthos. *Bauh. pin.* 78.
γ. Lilium purpuro-ſangvineum, flore reflexo. *Bauh. pin.* 78.
Hemerocallis calcedonica. *Lob. ic.* 169.
Habitat in Perſia. ♃
Racemus *ante floreſcentiam vix incurvatus ut in* 3. 5. *et* Caulis *usque ad apicem* Foliis *veſtitus confertis.*

5. LILIUM foliis verticillatis, floribus reflexis, corollis revolutis. *Hort. cliff.* 120. *Hort. upſ.* 81. *Gmel. ſibir.* 1. *p.* 44. *Roy. lugdb.* 31. *Martagon.*
Lilium, floribus reflexis, latifolium. *Bauh. pin.* 77.
Lilium ſylveſtre. *Dod. pempt.* 201.
β. Lilium, floribus reflexis, alterum hirſutum. *Bauh. pin.* 78.
Habitat in Hungaria, Helvetia, Sibiria, Lipſiæ.

6. LILIUM foliis verticillatis, floribus reflexis, corollis campanulatis. *canadenſe.*
Lilium anguſtifolium, flore flavo maculis nigris diſtincto. *Tournef. inſt.* 371. *Barr. rar.* 778. *t.* 125.
Lilium Martagon canadenſe maculatum. *Moriſ. hiſt.* 2. *p.* 408. *ſ.* 4. *t* 20. *f.* 9.
Lilium ſ. Martagon canadenſe, floribus magnis flavis non reflexis. *Catesb. car.* 3. *p.* 11. *t.* 11.
Habitat in Canada. ♃

7. LILIUM foliis verticillatis, flore erecto, corolla campanulata. *Amœn. acad.* 2 *p.* 348. * *camſchatcenſe.*
Habitat in Canada, Camſchatca.
* *Camſchatca planta petalis ovatis ſeſſilibus, ſtigmatibus tribus absque ſtylo; e Canada petalis lanceolatis unguiculatis cum ſtylo; varietates tamen fuere.*

FRITILLARIA.

1. FRITILLARIA racemo comoſo inferne nudo, foliis integerrimis. *Hort. upſ.* 82. *imperialis.*
Petilium foliis caulinis. *Hort. cliff.* 119. *Roy. lugdb.* 30.
Lilium ſ. Corona imperialis: genus. *Bauh. pin.* 79.
Tuſai. *Cluſ. hiſt.* 1. *p.* 127. 128.
Habitat in Perſia? *e Conſtantinopoli venit in Europam circa* 1570. ♃

2. FRITILLARIA racemo comoſo inferne nudo, foliis crenatis. *regia.*
Corona regalis, lilii folio crenato. *Dill. elth.* 110. *t.* 93. *f.* 109.
Habitat ad Cap. b. Spei.
FRI-

Figure 15-4. Pages from Vol. 1, Ed. 1, of *Species Plantarum* (1753) illustrating the use of the binomial system and the method of describing species. The epithets of species, as *imperialis* (*lower right*), i.e. *Fritillaria imperialis,* appear in the margins opposite the names (often descriptions consisting of several words) used in previous works (indicated by the italicized abbreviations and page references). Photograph by Frank B. Salisbury.

Linnaeus. He wrote to Haller, "I have never pretended that the method was natural." Linnaeus presented in his *Classes Plantarum* (1738) a fragment of a natural system, but he saw correctly that construction of a complete natural system was a matter for the future when more information might become available. In the "Fragmenta" he arranged together plants with characters in common, regardless of the nature of the characters. He enumerated sixty-eight natural "orders," each being about the equivalent of a family in modern classification. Linnaeus proclaimed development of a natural system of classification to be the great aim of botany. He went as far as he could with it, but he left the main job for the future.

The two systems of Linnaeus illustrate two approaches to the problem of classification. One of these is the *dividing method*—an attempt to separate the Plant Kingdom into large segments according to a plan devised after an overview of the whole problem. Inevitably the result is an artificial system, that is, one based on predetermined notions of which characters are important and which are not. The other method, the one used by Linnaeus in formation of his fragment of a natural system, is the *grouping method*—an attempt to build up from the "grass roots" by bringing together those plants with many characters in common and few differences. This approach is sound, for the occurrence in two or more plant groups of many common genetic characters is not likely to be the result of mere chance but rather of an underlying relationship. As more and more natural groups are discovered, the major outlines of a classification system begin to emerge.

The basis for natural relationships is derivation from a common ancestry. The first theories of evolution were not formulated until the Nineteenth Century, and the concept of evolution did not approach its present form until the publications of Darwin and particularly his *Origin of Species* (1859), which appeared a century after the principal writings of Linnaeus. The prevailing philosophy at the time of Linnaeus was based upon simultaneous creation of all plants and animals, and the natural system was thought to be the plan of Creation. Several historians have criticized Linnaeus for not giving an evolutionary interpretation to classification, not realizing that the impetus he gave to the study of plants and animals was necessary to bring about accumulation of sufficient information for formulation of even the earlier theories of evolution. Essentially, Darwin stood on the shoulders of Linnaeus. Darwin realized this, but others have not.

Gropings toward the Natural System

Attempts to develop the natural system of plant classification have employed both division and synthesis. The dividing method has attracted more attention, for it seems logical and the results are quick and spectacular. A system can be devised after a brief study, and its shortcomings are not apparent until someone makes extensive use of it. The grouping method requires many slow and laborious syntheses; it is not spectacular, and the results contributed by any one person are not likely to add up by themselves to a system of classification. They are the stones which someday will make a cathedral. Each of the greater authors has employed both methods, adding as many stones as he could and then drawing an artist's sketch of his conception of the cathedral. But to make the sketch, inevitably he has resorted to the dividing method, proposing artificial distinctions to develop the main outlines of classification. Consequently, even the best modern systems are queer mixtures of artificial and natural features, the main outlines being largely artificial and the groupings of the families and lesser categories mostly natural. Attempts to organize the superstructure of the system of classification have placed emphasis on the following organs or characters.

The Corolla—Caesalpino, Rivinus, and Tournefort. In 1583, Caesalpino founded a classification system based upon the embryo and seed but in part upon the "flower," and to Caesalpino this meant the corolla. Rivinus (1690) and Tournefort (1700) presented systems with emphasis upon the corolla. All three systems were artificial, and they did little to bring related plants together. Nonetheless, each of them contributed toward attracting attention to plants and to the fact that there is a problem in classification. Emphasis upon the corolla reappeared later in the writings of many authors, and particularly so in the systems constructed by Jussieu and by Engler and Prantl.

The Cotyledons—Ray. As mentioned earlier, Theophrastus correlated the presence of one or two cotyledons with occurrence of other characters. Nonetheless, he did not use these characters as a major basis for his classification system. It remained for John Ray to appreciate the distinctions between monocotyledons and dicotyledons, and it was he who introduced these groupings into the botanical literature in 1703. John Ray's system of classification of all plants was as follows:

I. Herbae.
  A. Imperfectae (flowerless plants).
  B. Perfectae (flowering plants).
    1. Dicotyledones.
    2. Monocotyledones.
II. Arbores.
  A. Monocotyledones.
  B. Dicotyledones.

Thus, although Ray preserved for his primary division the herbaceous and woody categories employed by Theophrastus, he gave strong emphasis to the cotyledons. At first glance appearance of monocotyledons and dicotyledons under both herbaceous and woody plants seems strange. However, there are few woody monocotyledons; and only the palms, bamboos, and dragon-trees were known to Ray. Probably the palms seemed more like the woody dicots than the herbaceous monocots.

Hypogyny, Perigyny, and Epigyny—Jussieu. Several Jussieus were botanists. The most outstanding were Bernard and Antoine, who refined and developed the Linnaean fragment of a natural system and consequently made one of the greatest contributions to botany, pointing the way for others. In 1789 Antoine published as a result of their studies an arrangement of plants containing most of the elements of the modern classification systems. In addition to cotyledon and corolla characters, Jussieu employed in third rank the distinction of hypogynous, perigynous, and epigynous flowers noted by Theophrastus but neglected in the intervening centuries. The outlines of Jussieu's system are as follows:

I. Acotyledones (plants not known to have seeds or flowers). Mostly thallophytes, Bryophyta, and vascular spore plants.
II. Monocotyledones.
  A. Hypogynous group.
  B. Perigynous group.
  C. Epigynous group.
III. Dicotyledones.
  A. Apetalae (apetalous group).
    1. Hypogynous group.
    2. Perigynous group.
    3. Epigynous group.
  B. Monopetalae (sympetalous group).
    1. Hypogynous group.
    2. Perigynous group.
    3. Epigynous group with free anthers.
    4. Epigynous group with coalescent anthers (Compositae or sunflower family).
  C. Polypetalae (choripetalous group).
    1. Hypogynous group.
    2. Perigynous group.
    3. Epigynous group.
  D. Diclines irregulars (unisexual plants with no corolla).

(The names of the groups of the third degree have been changed to English, and the groups have been rearranged.)

The Vascular System—DeCandolle. The DeCandolles, like the Jussieus, were a botanical family. The leader was A. P. DeCandolle, whose outstanding work included his two-volume "Systema" (1818) and his encyclopedic "Prodromus" (begun in 1824 and continued for half a century). DeCandolle reinforced segregation of monocotyledons and dicotyledons by bringing into the classification system recently developed knowledge of plant anatomy and particularly of the vascular systems. He arranged the plant kingdom as follows:

I. Vasculares.
  A. Exogenae (plants possessing exogenous growth in thickness, that is, seasonal additions of xylem and phloem by the cambium). These are the dicotyledons.
    1. Diplochlamydeae (with both calyx and corolla).
      a. Thalamiflorae (choripetalous and hypogynous).
      b. Calyciflorae (*either* choripetalous and perigynous or epigynous, *or* sympetalous and epigynous).
      c. Corolliflorae (sympetalous and hypogynous).
    2. Monochlamydeae (with only calyx and no corolla). (This group included the conifers, the true nature of which was shown soon afterward by Robert Brown.)

Figure 15-5. Sir Joseph Dalton Hooker (1817–1911). Author of many botanical works. With George Bentham, author of the most intensive study of the flora of the world to date. Photograph from the Arnold Arboretum; courtesy of the Arnold Arboretum and the Gray Herbarium, Harvard University.

B. Endogenae (plants of endogenous growth, that is, without cambial activity). These were primarily the monocotyledons, but they included also the cycads and vascular spore plants.

II. Cellulares (plants without a vascular system). Thallophytes and bryophytes.

General Organization—Bentham and Hooker. The classification system of the British botanists Bentham and Hooker (1862–1893) was based upon an unexcelled study of a great section of the world's flora by detailed research upon families, genera, and species. It is an ancestor (in some degree) of every recent system. The main outline is as follows:

I. Dicotyledones.
   A. Polypetalae (choripetalous).
      1. Thalamiflorae (hypogynous).
      2. Disciflorae (with a hypogynous disc).
      3. Calyciflorae (with a floral cup).
   B. Gamopetalae (sympetalous).
      1. Inferae (with an inferior ovary).
      2. Heteromerae (hypogynous and with more than two carpels).
      3. Bicarpellatae (hypogynous and with nearly always two carpels).
   C. Monochlamydeae (apetalous).

II. Gymnospermae. (Obviously the nature of the gymnosperms was not appreciated fully.)

III. Monocotyledones.

16

Comparison and Evaluation of Recent Systems of Classification

In the Twentieth Century systems of flowering plant classification still include mixtures of natural and artificial features. Although the families and lesser categories are largely natural, the orders and higher groupings within the subclasses tend to be at least partly artificial because differences of emphasis upon "important" characters are applied at the higher levels of organization. Progress has been slow since Jussieu formulated the broad outlines of angiosperm classification in 1789. The data for attacking this complex problem are inadequate, and it is no wonder that so many lances have been bent.

The Engler and Prantl System

The classification system of Jussieu was revised in Germany and Austria by Endlicher, Eichler, and finally Adolf Engler (University of Berlin) and his associate, K. Prantl. According to Engler and Prantl the most primitive flowering plants in both dicotyledons and monocotyledons are those without petals and with very small, simple, wind-pollinated flowers. In their opinion the two subclasses arose independently from an ancestral stock of unknown, hypothetical, extinct gymnosperms; and the living plants most nearly approaching the archetype (original type) are the "Amentiferae" (dicotyledons) and the Pandanales (monocotyledons).

The Amentiferae (willow, poplar, cottonwood, oak, beech, alder, birch, beefwood, waxmyrtle, walnut, and hickory) are characterized as follows:

1. Trees or shrubs, mostly of the Northern Hemisphere.
2. Flowers apetalous, small, unisexual, one or both kinds in catkins (aments).
3. Embryo straight, the ovules orthotropous to anatropous.

Figure 16-1. Adolf Engler (1844–1930), Berlin. Author (or joint author with K. Prantl), of *Das Pflanzenreich, Die Natürliche Pflanzenfamilien,* and *Syllabus der Pflanzenfamilien* based upon revision and elaboration of the classification system of Eichler. These works dealt with the flora of the Earth as a whole. Photograph at the Arnold Arboretum; courtesy of the Arnold Arboretum and the Gray Herbarium, Harvard University.

Engler and Prantl considered the Amentiferae primitive because the flowers are simple and the catkins seem to resemble strobili. In the willow, for example, each flower in the staminate catkin consists of two to nine stamens subtended by a scalelike bract, and each flower in the pistillate catkin consists of a single pistil with a bract. In other Amentiferae either the staminate or the pistillate flower or both may have a calyx. In the more advanced members of the group (birches, oaks, and walnuts) the flower is epigynous, and the minute sepals appear to be near the top of the ovary. In the genus *Populus* (cottonwoods, poplars, and aspens) the flower includes a cup or "disc" which may be a floral cup, and possibly the flower should be interpreted as perigynous. Except in the willows, the pollen of the Amentiferae is carried commonly from one catkin to another by wind as it is between the strobili (cones) of the gymnosperms. (Figs. 6-1 **A**; 10-16; 10-17.)

Engler and Prantl's classification of the monocotyledons begins also with plants having small apetalous flowers crowded into dense spikes or heads. The group selected as most primitive is an enlarged Pandanales that includes the cat-tails, bur-reeds, and screw-pines. In the cat-tail, the staminate flower consists of one to seven but usually three stamens associated according to no definite arrangement with a few hairs or scales. The pistillate flower consists of a single carpel on a long stalk bearing many slender hairs. In the bur-reed the staminate flower consists of a perianth of scales and three or more stamens sometimes coalescent at their bases. The pistillate flower consists of a perianth of scales and a single pistil formed from one or two carpels. The screw-pines (*Pandanus*) are trees restricted to the Tropics of the Eastern Hemisphere. The flowers are simple, as in the related herbaceous groups, consisting of only stamens or pistils but with the number of stamens or carpels varying from one to many. In rare instances the stamens surround a single abortive carpel. (Fig. 11-24)

Since Engler and Prantl did not believe clear continuous lines of relationship could be demonstrated in the flowering plants, they did not summarize supposed lines of phylogenetic development from a single order as an evolutionary starting point. They considered the existing Angiospermae to be composed of many fragmentary lines of evolution, some running through a single order, others through a series of orders. The trend of development was thought to begin with flowers with no perianth and to continue through the following series: sepals alone, sepals and separate petals, and sepals and coalescent petals. This was not interpreted as a single linear series, but as several parallel series in both the dicotyledons and the monocotyledons.

The emphasis in the classification by Engler and Prantl is upon apetaly, choripetaly, and sympetaly; hypogyny, perigyny, and epigyny are a secondary matter, and plants with each of these features are scattered through the major groupings. In the dicotyledons there are two chief groups—Archichlamydeae (with no perianth, only a calyx, or a calyx and separate petals) and Metachlamydeae (with a calyx and coalescent petals). In the monocotyledons no large group is characterized by a corolla tube, but there are some genera, families, and entire orders in which all six members of the perianth (three sepals and three petals) are coalescent edge to edge into a perianth tube.

The Revision by Wettstein

According to Richard von Wettstein (University of Vienna), the most primitive Amentiferae are the Casuarinales (Thalamiflorae, p. 244) (beefwoods of Australia, cultivated in the southern United States). Superficially, the beefwoods resemble either the horsetail (*Equisetum,* a spore plant, cf. pp. 689 to 695) or the Mormon or Mexican tea (*Ephedra,* a gymnosperm, cf. pp. 606, 629), and Wettstein has presented a theory of derivation of the *Casuarina* flower from a strobilus (cone) similar to that of *Ephedra.* (Figs. 18-7, 18-8.)

There is no evidence from fossils that the Angiospermae and *Ephedra* are close relatives. There is evidence for a remote common ancestry for the two groups, but the ancestor may have been extinct since either Paleozoic or early Mesozoic time. A positive relationship seems obvious, but probably it is not close.

Wettstein thought the dicotyledons started from several extinct groups, no contemporary flowering plant order being the sole primitive one.

Wettstein's system differs from Engler and Prantl's classification in the following points.

1. The monocotyledons are thought to have been derived from the Polycarpicae (another name for the Ranales).
2. The dicotyledons are thought to be of diverse origin, and several groups are considered primitive and independent.
3. The Dialypetalae (plants with choripetalous flowers) are thought to have been derived primarily from two orders, the equivalents of the Ranales (Polycarpicae) and the Euphorbiales (Tricoccae). These groups gave rise directly to the Sympetalae (plants with sympetalous flowers) except for two orders developed from the Caryophyllales (Centrospermae).
4. Many plants lacking corollas are thought to have been derived from choripetalous groups. Wettstein admitted many more cases of perianth reduction than did Engler and Prantl. For example, the Graminales (sedges and grasses) are thought to be derived from the Liliales (lilies and rushes).

The Wettstein System resembles the Bessey System (p. 537) in some of the points enumerated above, but it resembles the Engler and Prantl System in the following ways.

1. Many plants with unisexual apetalous flowers, including some orders of the Amentiferae, are thought to be primitive.
2. Primary emphasis in formulating the main divisions is upon apetaly, choripetaly, and sympetaly rather than hypogyny, perigyny, and epigyny.

Wettstein's summary of the supposed primitive characters of the Amentiferae is particularly thorough. It includes the following external features of the plants.

1. *The stems and roots are woody.* In many classification systems this character is described as primitive. Many authors have believed the angiosperms were derived from the gymnosperms, all of which are woody, and that, *per se,* woodiness must be primitive among flowering plants.
2. *The flowers are unisexual.* Interpretation of this character as primitive falls in the same category as number 1.

3. *The flowers usually are wind-pollinated.* This falls also in the same category as number 1.

4. *Petals are absent.* This point may be considered as either a manifestation of the philosophy of number 1 because gymnosperms have no petals or as a point in its own right because absence of petals is characteristic of simple flowers, which may be primitive.

The summary by Wettstein includes also the following internal features of the plant.

1. *Vascular bundles are present occasionally in the integument of the ovule* (later the seed coat). This is a gymnosperm feature.

2. *The gametophytes may have a larger number of cells than in most angiosperms.* In *Casuarina* the magagametophyte is multicellular. This is a feature possessed by both pteridophytes and gymnosperms.

The Revision by Rendle

Alfred Barton Rendle (former Keeper of the Department of Botany of the British Museum of Natural History) accepted the main outlines of the Engler and Prantl System. Since he thought classification of angiosperms too complex for supposed phylogenetic systems to be convincing, he did not claim his extensive revision of Engler and Prantl's work to be a strictly natural system, i.e., phylogenetic in its broader outlines. He believed that the first flowers were simple types like those of the Amentiferae or Pandanales and that the ranalean flower is a relatively advanced type developed by evolution from the Amentiferae. On the other hand, he did not suppose that all flowers having only a calyx and no corolla were necessarily representative of a more primitive condition than that in all having petals. He thought that the first angiosperms were wind-pollinated and that insect-pollinated plants were derived from them.

Rendle arranged the dicotyledons in three "grades" beginning with the most primitive types—Monochlamydeae (without petals but sometimes with a calyx); Dialypetalae (with separate petals); Sympetalae (with coalescent petals). These grades correspond to the Engler and Prantl arrangement except for Rendle's division of the Archichlamydeae into two groups according to whether or not petals are present.

Monochlamydeae. In this group there is a gradual transition from the orders of the Amentiferae to a middle group and an upper group. The middle group (Urticales, Proteales, Santalales, and Aristolochiales) has the following characters: plants woody or herbaceous; flowers unisexual, not in catkins; sepals present, embryo straight; ovules anatropous or amphitropous. It includes such common plants as the nettles, figs, hemp, mistletoes, Dutchman's-pipes, and birthworts or "wild-gingers" (*Asarum*). The upper group consists of only two orders—Polygonales (e.g., buckwheat, smartweed, sour dock) and Centrospermae (including the chickweed, carnation, pink, lamb's-quarter, pigweed, and poke), together the equivalent of the Caryophyllales. This group has the following characters: plants nearly always herbaceous; flowers bisexual, not in catkins; embryo usually curved but sometimes folded or straight; ovules orthotropous or campylotropous or rarely anatropous.

Dialypetalae. Rendle considered the Ranales to be the most primitive order of the Dialypetalae and the starting point in their evolution. The chief tendency of development within the Dialypetalae was essentially departure from one or more of the five floral characters of the Ranales about as enumerated earlier (*free, equal, numerous, spiral,* and *hypogynous,* cf. p. 114).

Sympetalae. These more advanced plants continued development along several lines from the Dialypetalae, with further departure from the five characters. The more primitive members of the group are pentacyclic (with five cycles of flower parts—sepals, petals, two series of stamens, and carpels); the higher groups are tetracyclic (with one series of stamens eliminated); the very highest group is tetracyclic and epigynous. According to Rendle this is the highest form of floral development.

The Bessey System

The classification system of Jussieu was revised by DeCandolle (Geneva), Bentham and Hooker (Kew, near London) and later Charles Bessey (University of Nebraska). In 1875 Alexander Braun pointed out the similarity of a flower to the vegetative shoot or to the cone of a gymnosperm, these being considered homologous in many respects. This was fundamental to the Bessey System of classification. According to Bessey the most primitive flowering plants are of the order Ranales (buttercups, anemones, peonies, *Clematis,* marsh-marigolds, magnolias, tulip-trees). He summarized the supposed primitive characters of flowering plants in twenty-eight more or less random *dicta,* but the essential points of primitiveness (with some additions) are derived from the floral characters of the primitive Ranales.

Figure 16-2. Charles E. Bessey (1845–1915), University of Nebraska. Outstanding teacher of many American botanists and author of a widely used system of classification of plants, the portion dealing with flowering plants being particularly well known. Photograph at the Arnold Arboretum; courtesy of the Arnold Arboretum and the Gray Herbarium, Harvard University.

The Origin of Angiosperms. Bessey (following Arber and Parkin) held that the flowering plants arose from a group similar to the Bennettitales in the Cycad Line of the gymnosperms (cf. pp. 650, 654, Figs. 20-9, 10). In the Bennettitales the strobilus (conelike structure) consisted of many spirally arranged bracts, microsporophylls (pollen-bearing leaves), and megasporophylls (ovule-bearing leaves) one above the other like the parts of a flower. Bessey thought the flowers of the Ranales originated from such a strobilus by transformation of the bracts into a perianth, the microsporophylls into stamens, and the megasporophylls into carpels. However, the stems were of a far different structure from those of flowering plants; the leaves were bipinnate and not at all similar to sepals and petals; the microsporophylls were bipinnate with innumerable microsporangia (pollen chambers) and far more elaborate than bladeless stamens with usually 2 microsporangia; the megasporophylls (unlike carpels) had lost their blades in earlier evolution; and the ovules were different from those of flowering plants.

Probably the Ranales did not develop from the Bennettitales. It is possible that flowering plants did originate from an ancestral group near the Cycad Line, this having had bracts, microsporophylls, and megasporophylls in a single strobilus, but, if so, probably it was a member or relative of the Cycadofilicales (p. 649) rather than the Bennettitales.

The Trend of Evolution Among Flowering Plants. Bessey summarized his ideas of the orders of flowering plants and of their evolutionary lines of development in a chart of relationships known familiarly as "Bessey's Cactus" (Figure 16-3). In this chart, there are three major series of figures, each figure representing an order and resembling in shape a joint of the stem of a prickly pear. The branch at the left represents the monocotyledons, the other two branches (middle and right) major groupings within the dicotyledons.

Bessey considered the monocotyledons to have been derived from the Ranales through the Alismales (Alismatales), and certainly the two orders are related more closely than any other dicotyledons and monocotyledons (cf. p. 569).

The Dicotyledoneae were divided into two major groupings: *Strobiloideae* (strobilus flowers, in reference to similarity of a hypogynous flower to a strobilus), represented by the right-hand branch in the chart, and *Cotyloideae* (cup flowers, in reference to the floral cup), the middle branch. Prevailingly, the Strobiloideae are hypogynous, and the Cotyloideae, except for a few primitive groups in the Rosales and some Sapindales (and the segregated Celastrales) are either perigynous or epigynous. The perigynous families (ignoring the catchall order Sapindales and the Celastrales) were confined to about the middle level in the Rosales and to the lower level of the Myrtales.

In each of the two lines of dicotyledons and in the monocotyledons there is gradual departure from one or all five floral characters of the Ranales or Alismales. According to the degree of departure from these characters, the plant may be placed at approximately the proper distance from the Ranales in the proper branch in Bessey's chart.

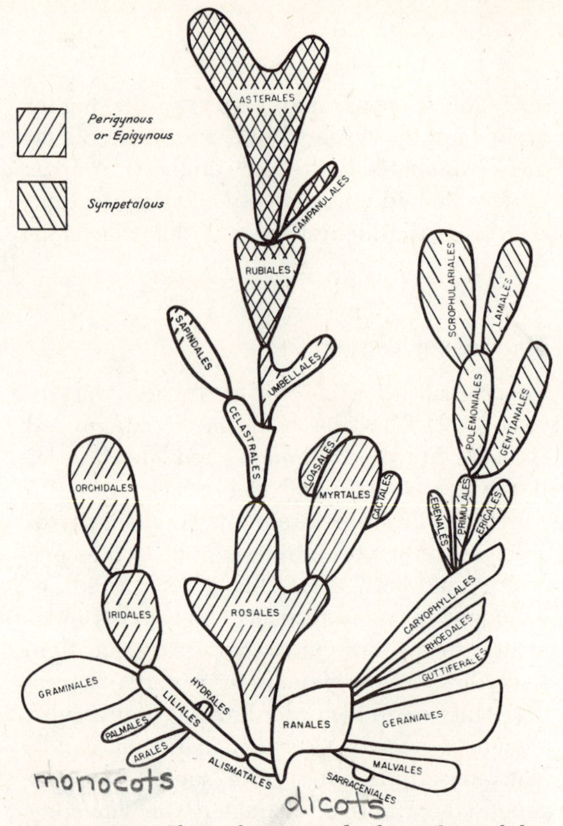

Figure 16-3. Chart of origin and relationships of the orders of flowering plants according to Charles E. Bessey (the chart known familiarly as "Bessey's Cactus" or "Opuntia Besseyi"). The left branch represents the monocotyledons; the middle branch includes (except basally in the lower Rosales) perigynous and epigynous dicotyledons (and some with hypogynous discs probably interpreted as perigynous); the right branch represents the hypogynous dicotyledons. Sympetaly is attained in the orders represented in the upper portion of each branch of dicotyledons. Relationship of orders is indicated by relative position, origin by attachment of the figure to the one below. The area of each figure is approximately proportional to the number of species occurring in the order. Adapted from CHARLES E. BESSEY, *Annals of the Missouri Botanical Garden*. 2: 118. fig. 1. 1915. Used by permission. (Elaborated.)

The Bessey System Versus the Engler and Prantl System

In North America the Bessey System has been popular; in the world as a whole the Engler and Prantl System was the leading one. The Bessey classification was common in American classrooms because the significance of the simple chart was easy to grasp and because the fairly reasonable phylogenetic arrangement was desirable for illustration of evolution. The Engler and Prantl classification was used almost universally for arrangement of families in regional, state, or local "floras" and "manuals," and most herbaria still are arranged according to it. During recent years both systems have been fading away.

The differences between the two systems are as follows.

1. *Derivation of the Angiospermae.* According to Bessey the angiosperms were derived from the Cycad Line—from gymnosperms with a bisexual strobilus similar to that of the Bennettitales; according to Engler and Prantl the flowering plants originated from unknown gymnosperms with a unisexual strobilus similar to that of the living conifers.

2. *Starting points within the Angiospermae.* Bessey considered all angiosperms to have been derived from the Ranales* by evolutionary departure from plants with a flower type fulfilling the floral characters of the Ranales; Engler and Prantl considered the dicotyledons and monocotyledons to have been derived from the Amentiferae and the Pandanales, respectively, in each case the simple flower gradually acquiring a perianth. According to Bessey the monocotyledons were derived from a starting point in the dicotyledons (the Ranales), and they continued their independent development; according to Engler and Prantl they arose independently, being in general more primitive than the dicotyledons.

3. *Characters Given Primary Emphasis.* According to Bessey the primary division of both subclasses is based upon hypogyny, perigyny, or epigyny; characters of the perianth are secondary. Plants lacking petals were derived from types having had petals. They constitute no major group, being distributed among the choripetalous orders according to nearest apparent relationship. Plants with coalescent petals are considered to be advanced types within the major lines of development (choripetaly being one of the floral characters of the Ranales). According to Engler and Prantl, on the other hand, the primary division is based upon apetaly, choripetaly, or sympetaly; characters of the floral cup are secondary.

Bessey and Engler and Prantl may be imagined as surveying the angiosperms from a mountain top. Bessey's glasses emphasize the dividing lines running one way, Engler and Prantl's those running at right angles to them. This is illustrated by a chart representing the dicotyledons with hypogynous flowers as placed on the right side of a vertical dividing line and those with perigynous or epigynous flowers on the left. According to Bessey the vertical line should be heavy, and horizontal lines separating the plants according to the perianth should be light. According to Engler and Prantl the vertical line should be dashed and the horizontal ones heavy.

Until the last decade or two neither faction has had the better of the argument because there have been points to be made on both sides and because no one had enough information for a definite answer.

The various evolutionary series built up chiefly according to gradual departure from the floral characters of the Ranales, as well as according to specialization in other ways, present, as phylogenetic systems go, a fairly plausible pattern of development from the Ranales to some other groups. This principle was accepted by Rendle, who considered the Ranales to be the most primitive choripetalous dicotyledons and the immediate source of the group.

Evolution of the Amentiferae from the Conifer Line would require loss of so many technical conifer features that a direct connection does not seem likely. Descent from a common ancestor is a more reasonable explanation, but the ancestor must have been remote. Separation of the lines of development leading to the angiosperms and the conifers probably occurred among the extinct Devonian plants which formed the border zone between the pteridophytes and the early spermatophytes.

As will be shown in the next chapter, evidence is accumulating that the Ranales retain various primitive features, but the proposition that the simple-flowered Amentiferae and Pandanales retain some primitive characters is not necessarily to be discarded. Possibly some of the small divergent orders of the Amentiferae

* The primitive character of the Ranales was proposed by Jussieu (1789) and accepted by Bentham and Hooker (1862).

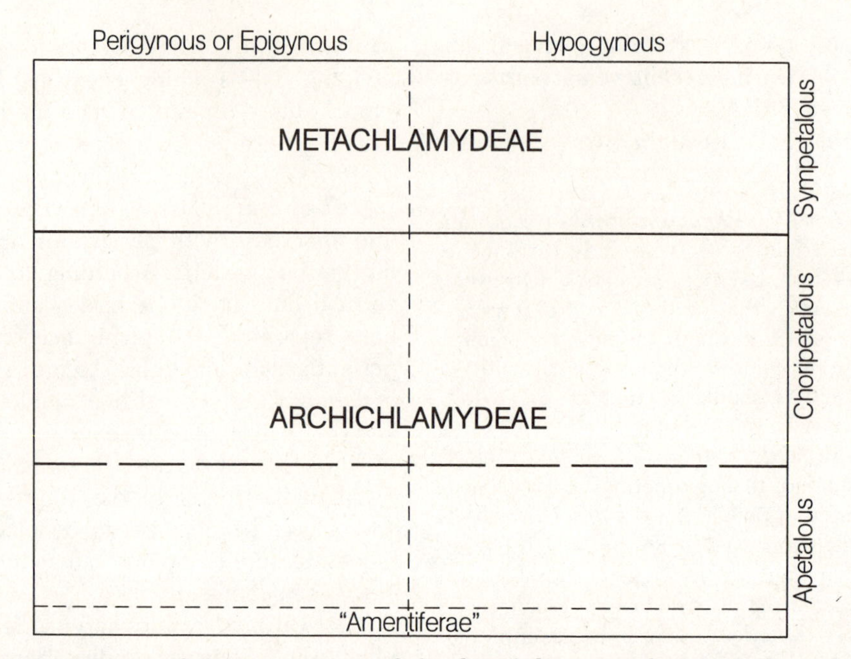

Figure 16-4. Diagram of the major groups of the dicotyledons, according to Engler and Prantl. Emphasis is upon the dividing lines shown horizontally, which represent characters of the perianth (apetaly, choripetaly, and sympetaly). Hypogyny, perigyny, and epigyny were considered to be of minor importance.

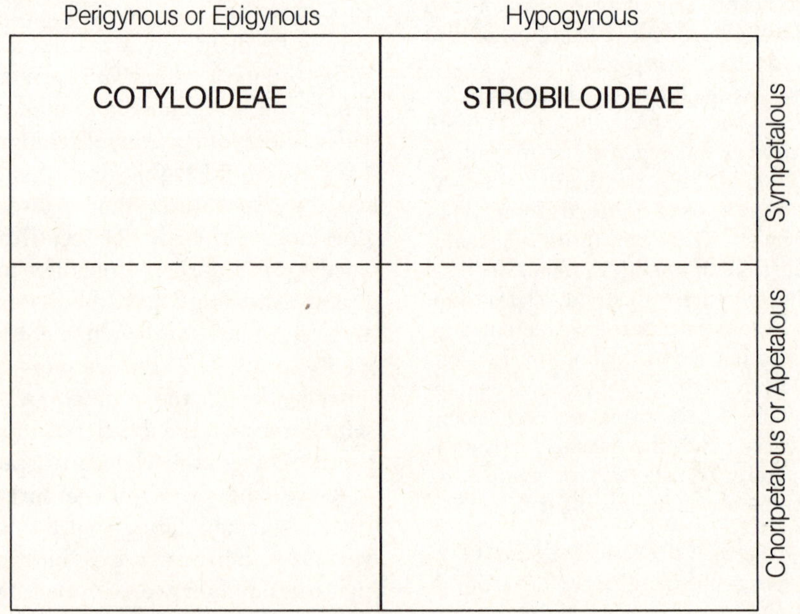

Figure 16-5. Diagram of the major groups of the dicotyledons, according to Bessey. Emphasis is upon the dividing line shown vertically, which represents the distinction of perigyny or epigyny (presence of a floral cup) as opposed to hypogyny. Choripetaly and sympetaly were assigned secondary importance, and apetaly minor significance.

may be survivors of nearly extinct lines of development and in some features more primitive than are most other living angiosperms. New evidence indicates these groups to have been derived from various choripetalous flowering plants through reduction in complexity of the flower. Recent morphological studies all support the latter view.

The Bessey System is stronger in its choice of a possible evolutionary starting point and its general outline. Its principal weaknesses are overemphasis of speculation about evolutionary series and oversimplification of a complex problem. The chart is too simple, and the large orders include diverse elements often not very closely related to each other.

The Engler and Prantl System is weak in its starting point and its main outline, but the treatment of orders and families reflects both a broad and a deep knowledge of plants, and in many instances the orders are more nearly natural than Bessey's. Wettstein's and Rendle's arrangements of the flowering plants are even stronger in the treatments of orders and families. All through the books by these authors there is evidence of deep insight into the floras of the world and full appreciation of the complexities of their classification.

Since 1789, the main outlines of each new flowering plant classification system have been developed by stirring around the principal characters used by Jussieu. Argument has centered upon whether the corolla (apetaly, choripetaly, and sympetaly) should be considered more important than the floral cup (hypogyny, perigyny, and epigyny) or vice versa. No decisive results have been produced during the last 190 years, and the sterility of such an argument is apparent because one character is not *per se* more important than another. Potentially any character of the plant is as important as any other. The characters actually segregating taxa in each particular case are the significant ones in classification (cf. especially p. 521); their predetermination on philosophical grounds leads to an artificial classification.

Other Classification Systems (to 1956)

The following systems of classification are of special importance.

Hutchinson. John Hutchinson (formerly Keeper of the Museum of Botany of the Royal Botanic Gardens of Kew, England) classifies the flowering plants according to a system resembling Bessey's. Essentially, the starting point for the dicotyledons is the Ranales. The primitive monocotyledons, the Alismales, are considered not to have been derived directly from the Ranales but from a common ancestor. The most evident feature of the system is division of the dicotyledons into two lines of development—one woody, arising from an order Magnoliales (segregated from the Ranales), the other herbaceous and arising from the herbaceous Ranales. Hutchinson's work is original, and often the treatment of individual families and especially of some groups of monocotyledons is good. Unfortunately, the primary division into woody and herbaceous plants results in an unnatural division of orders into fragments, as would any other character selected arbitrarily.

Mez. Serum diagnosis has been used extensively in studying interrelationships of animals, and it was applied to plants by Mez, who published an elaborate system of classification based upon it. Briefly, the method is applied as follows: A protein extract from a particular plant is injected into a laboratory animal (e.g., rabbit or Guinea pig). After a period some of the serum (liquid from the blood) of the animal and the protein extract of a second plant are placed in a test tube. The amount of the resulting precipitate indicates the degree of relationship. Recently this technique has been refined, particularly by Chester and by Boyden at Rutgers University and their successors.

Data from serodiagnosis used in conjunction with other characters are valuable for classification, but the protein content of the juice of a plant is no more an infallible index of relationship than is any other single feature.

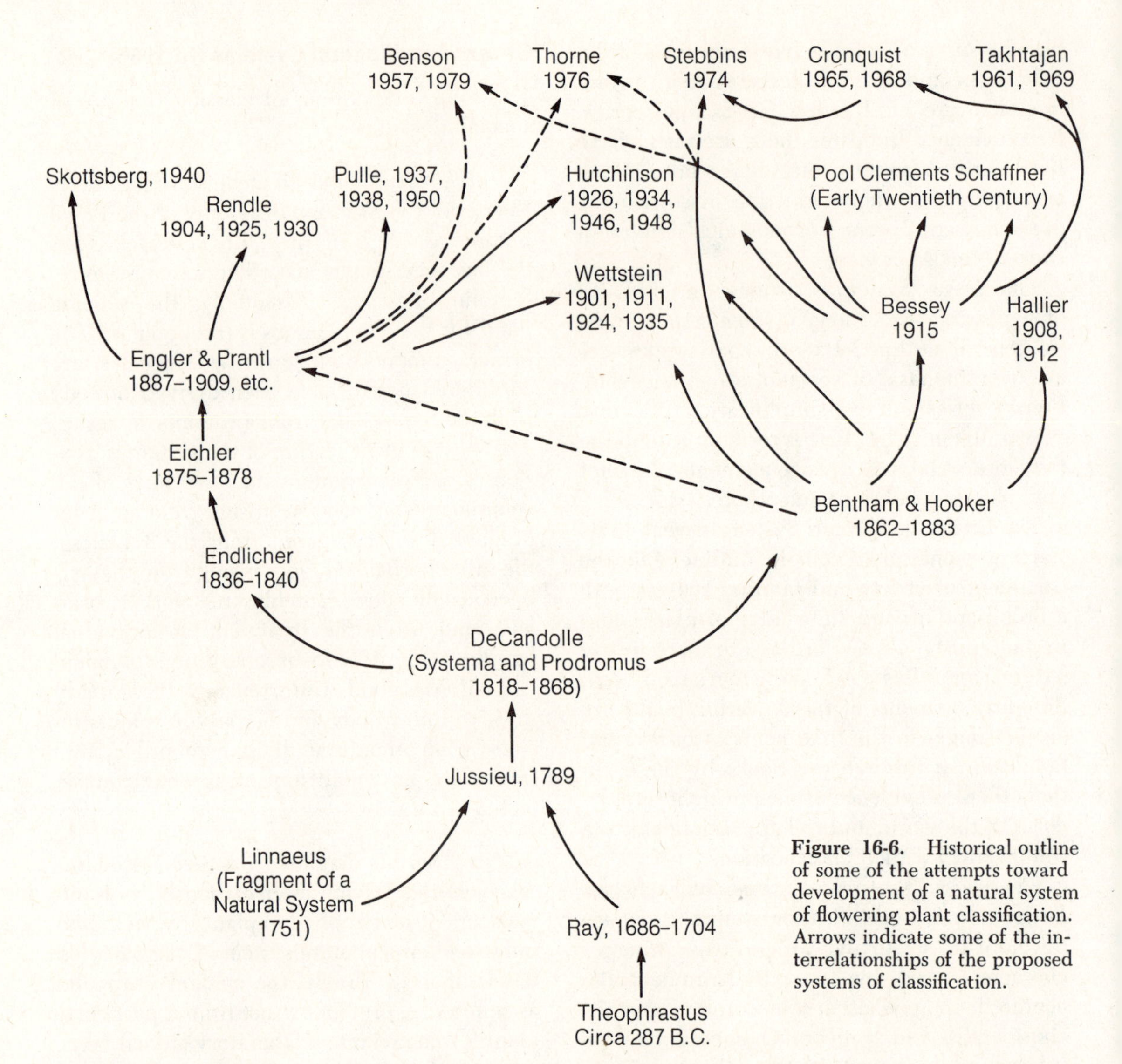

Figure 16-6. Historical outline of some of the attempts toward development of a natural system of flowering plant classification. Arrows indicate some of the interrelationships of the proposed systems of classification.

Thomas. According to H. Hamshaw Thomas, perigynous flowers with whorled stamens and small carpels predominated among the earliest angiosperms, which were derived from the seed ferns. Several seed ferns or Cycadofilicales of Carboniferous time (discussion, p. 171) had sporophylls arranged in structures much like the flowers of *Populus* (cottonwoods, aspens, and poplars). The staminate or pistillate flower of *Populus* is subtended by a broad disc of unknown origin, which Thomas interpreted as a floral cup.

Thomas' theory of the origin of flowers was based upon his study of the order Caytoniales, fossils of which he discovered in 1925 in rocks of the Jurassic Period on the coast of Yorkshire, England. In this order the microsporangia and the seeds were reported to have terminated branches, and there was no direct association of sporangia with leaves or their derivatives. According to Thomas, the flower originated as a sorus (cluster of sporangia) terminating a branch. Consequently, he argues, stamens are not microsporophylls (spore-bearing leaves) but structures developed from small branches, and similarly carpels are not megasporophylls. In the Caytoniales short-stalked "fruits" were produced in lateral more or less opposite pairs. Each of these structures was a rounded saclike body only about five millimeters in diameter. A narrow passageway opened to the interior where orthotropous ovules were arranged in a row. This passageway was open at the time of pollination, but it closed soon afterward, and the seeds were matured within the enclosure. At first it was supposed that the pollen grains germinated outside the opening to the ovule-containing cavity as in the angiosperms, but more recently they have been found in the micropyles of the ovules as in gymnosperms. Apparently the Caytoniales were not angiosperms but gymnosperms related to the Cycadofilicales.

The significance of the Caytoniales as possible ancestors of the angiosperms cannot be either accepted or rejected. Stebbins (1974) presents a possible mode of derivation of the flowering plant ovule from the gymnosperm type with a single integument (seed coat at maturity). First, the ovule was surrounded by a cupule, as in the Caytoniales, the cupule being on a short sporangiophore, or stalk. This may have been followed by adnation of the sporangiophore with the subtending bract and reduction of the ovule and seed number to one, the cupule becoming the outer integument of the ovule. This, of course, is presented as theory, not as fact, but as a theory worthy of further investigation. It is a promising theory. (See p. 560.)

Recent Systems of Classification

Stebbins. George Ledyard Stebbins has made an enormous contribution to understanding of the basis for classification of flowering plants through numerous papers and his book, *Flowering Plants: Evolution above the Species Level.* He presents no meticulous classification of the angiosperm orders and families, but a list with modifications of the system of Cronquist described below, and his book deals with the principles involved in the evolution of the flowering plants and their special structures and the fulfilment of their functions. He summarizes the application of principles of population genetics, microevolution, mutation, genetic recombination, natural selection, reproductive isolation, paleobotany, and paleoecology to problems concerning the origin and later development of flowering plants. He discusses the kinds of adaptive radiation* or diversification that may have given rise to the higher taxa and the kinds of environmental change to which the radiations may have been responses. Information from many fields is applied to development of a hypothesis of angiosperm origin. His many-faceted approach to the problems of evolution are in strong contrast to the dogmatic dicta of Bessey (see p. 537), and this underlines the enormous progress that has occurred since 1915. Nevertheless, despite the rashness of Bessey's statements, the overfirmness of his convictions, and the necessarily limited information available to him, such progress in understanding the origin and evolution of flowering plants as has been made so far would have been long delayed without this great teacher's overall insight. For all the highly sophisticated material, much of it original, appearing in Stebbins' writing, the most significant statement of all is the following: "This book should be regarded as a progress report rather than a final exposition of the nature of angiosperm evolution." This humble statement summarizes the essence of science and the scientific method—*facts first, and, as they accumulate, tentative conclusions from them, but these always subject to revision.* The entire book is presented with the attitude of the open-minded—that is, here

* Adaptive radiation is the filling of the ecological niches available to a group of organisms. For example, when the Hawaiian Islands began to arise one by one from the ocean as volcanic cones and the cones began to erode into soil, only a few plants had access to them. The Hawaiian Flora (Chapter 25) has arisen from whatever plants could get there, each group, with little competition, giving rise to derivatives able to occupy special niches.

is what we know so far, where do we go from here? The last chapter of the book is devoted to suggestions concerning the kinds of research needed to advance knowledge of angiosperm origin and the evolution of the flowering plants to their present stage of diversity and difficulty of classification.

The principles of taxonomy have been discussed mostly with respect to the taxa of lesser rank from which they are gathered by observation and experiment. Classification of the higher taxa has been discussed in *Plant Taxonomy, Methods and Principles* (Benson, 1962) and in *Principles of Angiosperm Taxonomy* (Davis and Heywood, 1963), though these books are not devoted primarily to the orders and families of flowering plants.

Takhtajan. The work of Armen Takhtajan became available to most botanists in western countries through republication (from the Russian, 1961) in English of *Flowering Plants, Origin and Dispersal,* 1969. The bulk of the book is devoted to the origin of angiosperms, the nature of primitive taxa, the origin of some of the larger extant groups, the question of where they arose (presumably uplands of Southeast Asia and moderate levels in the mountains), and their spread from this area. Unfortunately, publication occurred originally before general acceptance of continental drift. Unfortunately, also, the classification system of orders and families is given only as an appendix to the English version, where it is stated to include "only the bare bones of the system." In Russian, the system is explained in detail in a work with a title translated by the author as *A System and Phylogeny of the Flowering Plants* (1967). Takhtajan and Cronquist (see below) have been in close contact, and their systems have many similar, though other differing, features. The system of Takhtajan is basically a continuation of development from the Bessey System, but on firmer ground and with much new information concerning particular families and with recent ideas of evolution taken into account. It retains the arrangement of orders in a presumed evolutionary tree with much of Bessey's chart merely modified but with much of it rearranged. A major line of evolution is thought to begin with the Trochodendraceae and other Hamamelidales and to culminate in a number of the orders like the Fagales and Juglandales formerly included in the Amentiferae of Engler and Prantl. A major contribution is evaluation of many little known groups barely covered in earlier works.

Cronquist. Arthur Cronquist has presented a system of classification of the angiosperms based on his interpretation of phylogeny in *The Evolution and Classification of Flowering Plants,* 1968. His book includes an extended discussion of taxonomic principles and the origins of angiosperms and the evolution of their characters as the higher taxa arose. As with Takhtajan, this is a phylogenetic system with an evolutionary tree indicating the accepted derivation of the angiosperm orders from the primitive Ranales (these divided into six orders, the Magnoliales being the most primitive). The tree of phylogeny is divided into sections, these appearing with the treatment of each of six subclasses of the Magnoliatae (Dicotyledoneae) and four of the Liliatae (Monocotyledoneae). Although such a system is based on the proposition that the living orders were derived one from another, or there would be no phylogenetic tree, great caution is exercised throughout the book and the author takes care not to be dogmatic about the course of evolution or the correctness of his theories of classification, these matters being considered as problems. The book is a good one for discussion of the problems of phylogeny and classification, and this is its purpose. It is not designed as a manual for identification of orders and families, and it does not have keys or descriptions of the higher taxa. The "synoptical arrangement of families" under each order appears like a key, but it is not intended as one, and there are only brief characterizations of the orders and families and discussions of classification, relationships, and phylogeny.

Thorne. Robert F. Thorne has published papers on the phylogeny and classification of the higher taxa of angiosperms, the most recent and comprehensive being entitled, *A Phylogenetic Classification of the Angiospermae,* 1976. This includes a discussion of his philosophy of classification, the possible origin of angiosperms, the primitive features of the group and trends of specialization among its members, and a synopsis of his elaborate arrangement of subclasses (Dicotyledoneae and Monocotyledoneae), superorders, orders, suborders, families, and subfamilies. These are not keyed or described, the objective being construction of the outlines of a new classification system. To this project, Thorne brings a background of extensive travel for study of the families of flowering plants, and he has added immeasurably to the plant material available at the herbaria of Pomona College and the Rancho Santa Ana Botanic Garden in Claremont by his orientation of the specimen exchange program of the Garden toward securing material of the small and, from the viewpoint of Americans, obscure families restricted to other parts of the world. This has included the woody Ranales retaining primitive characters (as Annonales). His acute observations and his integration of data from various fields have contributed much to understanding the flowering plants and their origin and their higher taxa.

A new type of system based on relationships of orders and families but with classification of the taxa not according to "importance" of characters in a presumed phylogeny was introduced in the presidential address of the Western Society of Naturalists at Davis, California, in 1955 and in the first edition of *Plant Classification* in 1957.* This system was based on the premise that groups of characters remaining more or less together reflect an underlying natural relationship of the organisms having them and that they are the only practical basis for delimitation of taxa. The course of evolution involves both the arising of new characters or character combinations and the replacement of taxa, the "old models" of living organisms most frequently disappearing as they are replaced by one or more taxa of their offspring living under new conditions aiding their survival but not that of their forebears. As time goes on, the earlier derivative taxa either disappear or they are modified in other directions. Thus, in the absence of coherent series of fossils and of more than a few with reproductive parts, a phylogenetic series of the angiosperm orders is not likely to be worked out, and family trees are with little validity. They are based on the false assumption that the higher taxa of the present or any other thin slice of geologic time were derived one from the other, this being only rarely true and then for only a few taxa. The basic classification of the angiosperms in the chart on the front end leaves of *Plant Classification* and on page 550 reflect these principles. The figures representing the orders are not joined, and there is no indication that one order was derived from another. The orders are grouped according to probable relationships, and their arrangement reflects correspondence or lack of it to a presumed extinct ancestral group from which both dicotyledons and monocotyledons were descended. The living orders thought to retain the largest numbers of primitive or ancestral characters are the Ranales (especially the Magnoliineae) and the Alismales, and these are placed nearest the ancestral order or complex of orders. The orders farthest away in the chart, like the Asterales, are those most unlike the Ranales and Alismales and therefore presumably the ancestral group. The heavy lines across the chart segregating groups of orders of the dicotyledons are a practical expedient designed to make a little order of the maze of orders and families. As expected, this results in large, mostly natural but partly

* Simultaneously and independently a similar approach to classification was presented by Sporne (in 1956). It was used for grasses by Stebbins (in 1956).

artificial groups, the Thalamiflorae, Corolliflorae, and Calyciflorae of DeCandolle and Bentham and Hooker and the newly added Ovariflorae. This arbitrary division by predetermined characters does not add taxa to the classification system, but groups with no formal status. Only a brief summary is given here. Full discussion appears partly in Chapter 15 and primarily in Chapter 17.

In the recent work of Stebbins (1974) and of Thorne (1976) on the orders of angiosperms, there are no family trees of the flowering plants but instead charts fundamentally similar to those of both editions of *Plant Classification*. These indicate the closeness of relationships of orders and their general trends of divergence from a common extinct ancestor with affinities to the primitive Ranales (as Magnoliales or Annonales). This represents a fundamental difference from the family trees of other recent authors, as indicated by the following statement of Thorne (1976), paralleling those in *Plant Classification* (1957): "The reader should not infer that these beads [representations of taxa in the diagram] have arisen from the preceding bead. It is likely that no order or suborder has arisen from existing orders and suborders, and very likely no families from extant families or subfamilies. At best, these related higher categories have arisen from common generalized ancestors, probably long extinct."

Above, only the proposers of elaborate systems of classification of the flowering plants have been mentioned. To this list the names of hundreds of individuals should be added, and the extensive bibliographies of Stebbins and of Takhtajan are filled with them. These individuals have contributed the bulk of the basic progress in classification through their analyses of individual orders and families through all or at least several of the possible avenues of approach to learning their relationships and especially from data concerning neglected or obscure or minute characters of anatomy, reproduction, physiology, ecology, chemistry, cytology, or development.* Investigation of these features and their correlation with the macroscopic characters, themselves often heretofore poorly known or understood, have provided the real advances. These bits of progress are less spectacular than the systems of classification, but they are basic to all else. They are behind the bulk of the proposals of all authors, and often they confirm the judgments of workers of long ago, such as Bentham and Hooker, Engler and Prantl, or Wettstein.

It is to be hoped that during the course of time someone may be able to bring together the world's flora as did Bentham and Hooker and Engler and Prantl, despite the much greater amount of information to be sifted through now than a century to a century and a half ago. It is to be hoped, further, that this individual may have the insight and the judgment to develop not only a flora but also a sound and workable classification system for the innumerable taxa with which he must work, this accompanied by usable keys and descriptions for the orders and families of angiosperms.

In the higher taxa of flowering plants, as in any other group of plants or animals, often there are discrepancies in the nomenclature appearing in various classification systems. Often these reflect differences in underlying classification. The orders recognized in this book are interpreted conservatively, that is, the taxa are broad and they include a wide range of lesser taxa such as families. Other authors subdivide various orders into proposed orders of lesser groups of families. On theoretical grounds grouping and dividing are of equal justification. However, if subdivision

---

* These obscure characters contribute much to the devising of a classification system, but they are not ordinarily usable in the keys and descriptions for a book like this, designed as a working manual. The primary function of the book is for student and other use in the field and laboratory for determining the orders and families of vascular plants. The obscure features are essentially research characters in the background.

is carried too far, the broad outlines of classification are lost, or, once they are comprehended, they are difficult to remember unless one is working with them every day, as would a professional taxonomist. For this reason more inclusive orders are considered in this book to be more useful and practical units, provided, as with all taxa, they are natural (composed of truly related organisms) and of essentially equal status. Consequently, numerous proposed taxa in the hierarchies of recent authors are not accepted for formal use in this book. Their value is for discussion and consideration in technical papers and the advanced books in which they appear, if they are to be used at all. The number of names of higher taxa appearing in even this book, where their proliferation is restricted, is staggering for a student or a layman; adding more would be catastrophic.

SELECTED READING REFERENCES

BENSON, LYMAN. *Plant Taxonomy, Methods and Principles.* 1962. Ronald Press, New York. 1978. Reprinted by John Wiley & Sons, New York.

BENTHAM, GEORGE, and J. D. HOOKER. *Genera Plantarum.* 1–3: 1862–1883. London.

BESSEY, CHARLES E. *Phylogeny and Taxonomy of the Angiosperms.* Botanical Gazette 24: 145–178. 1897.

———. *The Phylogenetic Taxonomy of Vascular Plants.* Annals of the Missouri Botanical Garden 2: 109–164. 1915.

CRONQUIST, ARTHUR. *The Status of the General System of Classification of Flowering Plants.* Annals of the Missouri Botanical Garden 52: 281–303. 1965.

———. *The Evolution and Classification of Flowering Plants.* 1968. Houghton Mifflin Co., Boston.

DAVIS, P. H., and V. H. HEYWOOD. *Principles of Angiosperm Taxonomy.* 1963. D. Van Nostrand, Inc., Princeton, New Jersey.

EICHLER, A. W. *Blüthendiagramme construirt und erläutert.* 1–2: 1875–1878. Leipzig.

ENDLICHER, S. L. *Genera Plantarum.* 1836–1850. Vienna.

ENGLER, ADOLF. *Syllabus der Pflanzenfamilien.* Ed. 10, revised by Ernst Gilg, 1924. Ed. 11, revised by L. Diels, 1936. Berlin.

——— and K. PRANTL. *Die natürliche Pflanzenfamilien.* 1–20: 1897–1915; ed. 2. 1924– (incomplete). Leipzig.

GUNDERSEN, ALFRED. *Families of Dicotyledons.* 1950. Waltham, Mass.

HALLER, H. *Provisional Scheme of the Natural (Phylogenetic) System of Flowering Plants.* New Phytologist 4: 152–162. 1905.

———. *On the Origin of Angiosperms.* Botanical Gazette 45: 196–198. 1908.

———. *L'origine et le système phylétique des angiospermes.* Arch. Néerl. Sci. Exact. Nat. IIIB. 1: 146–234. 1912.

HUTCHINSON, J. *The Families of Flowering Plants.* 1–2: 1926, 1934; Ed. 2, 1959; ed. 3, 1973, Claredon Press, Oxford.

———. *A Botanist in South Africa.* 1946. London.

———. *British Flowering Plants.* 1948. London.

———. *The Genera of Flowering Plants.* 1967. Clarendon Press, Oxford.

———. *Evolution and Phylogeny of Flowering Plants: Dicotyledons; Facts and Theory.* 1969. Academic Press, London, New York.

LAWRENCE, GEORGE H. M. *Taxonomy of Vascular Plants.* 1951. New York.

MEZ, C. *Die Bedeutung der Serodiagnostik für die stammesgeschichtliche Forschung.* Bot. Arch. 16: 1–23. 1926.

PULLE, A. A. *Compendium van de Terminologie, Nomenclatuur en Systematiek der Zaadplanten.* 1938. Ed. 2. 1950. Utrecht.

RENDLE, ALFRED BARTON. *The Classification of Flowering Plants.* 1–2: 1925, 1930. Cambridge.

SKOTTSBERG, C. *Växternas Liv.* 1–5: 1940. Stockholm.

SPORNE, K. R. *The Phylogenetic Classification of Angiosperms.* Biological Review 31: 1–29. 1956.

STEBBINS, G. LEDYARD, JR. *Cytogenetics and Evolution of the Grass Family.* American Journal of Botany 43: 890–895. 1956.

———. *Flowering Plants, Evolution Above the Species Level.* 1974. Harvard University Press, Cambridge, Massachusetts.

TAKHTAJAN, ARMEN. *Flowering Plants, Origin and Dispersal.* Translation from Russian, 1969. Smithsonian Institution Press, Washington (distributed by Random House).

THOMAS, H. HAMSHAW. *The Early Evolution of Angiosperms.* Annals of Botany 45: 645–672. 1931.

———. *Paleobotany and the Origin of Angiosperms.* Botanical Review 2: 397–418. 1936.

THORNE, ROBERT F. *A Phylogenetic Classification of the Angiospermae.* In *Evolutionary Biology,* edited by HECHT, STEERE, and WALLACE. 1976. Plenum Press.

———. *When and Where Might the Tropical Angiospermous Flora Have Originated?* Gardener's Bulletin 29: 183–189. 1976.

WETTSTEIN, RICHARD VON. *Handbuch der systematischen Botanik.* Ed. 4. 1935. Leipzig and Vienna.

17

A System of Classification

The Limitations of a System

A reclassification of flowering plants is a major need in botany. But, although criticism and evaluation of existing work is not difficult, synthesizing a new system is most difficult. Its formulation is subject to the tremendous complexities of the subject matter, to the limitations of knowledge, and to the restricted experience possible for any one individual. He who attempts to organize orders of flowering plants and to prepare a chart of relationships finds himself committed at least tentatively to propositions for which evidence is insufficient.

The classification system for the flowering plants presented in Edition 1 in 1957 and here with revision is not the ultimate but one intended to be as nearly natural as possible in representing the orders and families but necessarily partly artificial insofar as the four great divisions of dicotyledons are concerned. Because the data at hand do not justify formulation of a wholly natural grouping of the orders of dicots, the system is frankly partly artificial, though no more so than other extant systems. All systems of flowering plant classification are handicapped by the same factors—the problem is complex, and the data concerning the plants are insufficient. The chart here and on the front end leaves must be interpreted as follows:

The Figures Representing the Orders. As pointed out briefly in Chapter 16, these are arranged according to relationships, as nearly as the affinities have been determined. The orders do not appear in phylogenetic series, because such series are unknown from fossils of the flowering plants and because one living order or family was not derived from another living one, or, if this phenomenon occurred, it was very rare. The gaps between figures for the orders represent separations that have occurred during geologic time as the present orders were derived from preexisting orders that have left no trace. The arrangement of

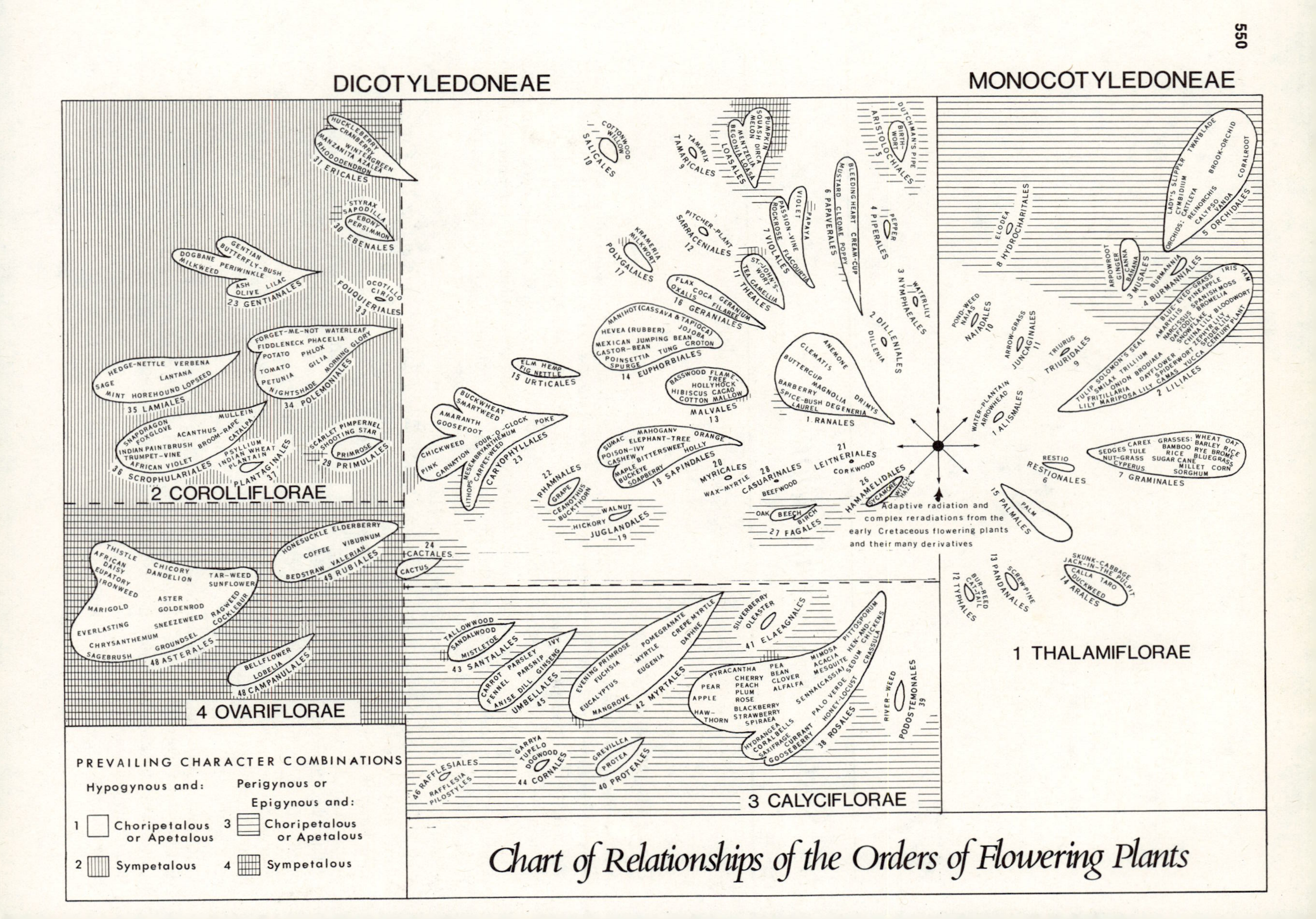

Chart of Relationships of the Orders of Flowering Plants

orders portrays to some extent possible groupings of related orders, those toward the area of the extinct forerunners of the angiosperms having retained a greater abundance of primitive characters. The orders farther away on the chart are more advanced, insofar as this can be determined to be likely. Thus, the basic outline of the chart is intended to be representative of a natural system of relationships, but it does not portray lines of supposed phylogeny, because these are little known.

The feature underlying natural groups of organisms is a common long phylogeny, or race history, which is responsible for all or most of the cluster of characters tending to remain together and marking the order, family, or other taxon. The characters in this cluster tend to be well represented in all members of the taxonomic group, though often no single feature is found in all members. The presence of these clusters is the basis for classification of the taxa, and for the flowering plant orders the association of characters is the chief existing evidence of common descent of groups of orders. However, the ordinary process of evolution involves shifting of the alliances of characters through geologic time, and with each passing epoch the characters marking each alliance must have gained and lost some members. Thus, phylogeny is a subject of absorbing interest, but, in this case, it is based on only fragmentary evidence.

The taxa of today are composed of and their limits are determined by the character combinations of the existing individuals and not by those of their relatives of the past. The present classification is the result of phylogenetic development in the light of natural

---

Figure 17-1. According to the evidence available, the first flowering plants appeared during early Cretaceous time, though there may have been more primitive or transitional types a little or even much earlier. During the Cretaceous period, the new group of plants underwent adaptive radiation, developing many new taxa capable of exploiting the environmental niches available or becoming available. As the environment changed from time to time, some groups died out and others prospered in each area, and there were new radiations in many directions within each ecosystem. Thus, the pathway of evolution of flowering plants is no simple route or routes that can be indicated effectively by a two-dimensional representation in the form of a tree, cactus, shrub, or any other figure. In the absence of more than a few intelligible fossils, only the evolutionarily disconnected living plant groups can be represented, and even the relationships of many orders are obscure. No living flowering plant order is thought to have been derived from any other living order, i.e., in its current definition. Related orders have evolved from preexisting groups, with some members similar at least in some ways to the species now composing the living taxa.

The origin of the angiosperms as a class is obscure. Some woody Ranales have retained more primitive features than any other group, and it is inferred that the ancestors of the flowering plants, or at least the early angiosperms, had these characters in combination with other primitive characters, some of which have disappeared and others of which have persisted in the other orders.

The chart represents the orders as the surviving groups after the appearance and disappearance of a host of others. No connections between them are shown because information is lacking and because most plant groups persist for only a limited geologic time before dying out and usually being replaced by their descendants, a few of which are likely to be better adapted than they to the changing conditions. Thus, in most instances, ideas of persistent observable connections between living orders are largely matters of speculation. In most cases only relationships can be indicated, the related orders tending to be placed closer to each other in the chart. Relationships are worked out to some extent from the presence of numerous characters in common and of relatively few differences. Ordinarily this indicates an underlying genetic relationship and thus derivation from a common heritage.

The chart distance of any order from the position for the Cretaceous plants at the center of origin and radiation shown in the chart is in rough proportion to the number of features of flower structure and other characters departing from the primitive ones retained in some orders of living angiosperms. These characters (Chapter 17) appear in the woody Ranales and in the Alismales, Palmales, and some other orders, as indicated by their position near the center of radiation. The size of the figure representing each order is in rough proportion to the number of species included. See discussion in Chapters 16 and 17.

selection, but the definition and scope of its taxa are not determined by phylogeny. Any classification system of living organisms must be for the present or for some other restricted part of geologic time, such as an epoch of the past. Most of the plants and animals of the present or past occur without great change during only their own epoch. Since life began, organisms have been always in flux, and the taxa of even one epoch, like Oligocene, may be similar to those of another, like Miocene, but they are not necessarily the same. With the shift of climatic conditions, in this case, in western North America for example, from more or less tropical to temperate, there has been moving of floras across land surfaces and up and down the levels of altitude. Taxa on the move and under changing stresses tend to lose the individuals with some characters and to retain those with others. At least the lesser or subordinate taxa are modified, and their differences constitute part of the modification of the higher taxa they compose. These, too, are forever changing. Presumably, if all the organisms of the past could be studied, each line of development would be a continuum or virtually so, and there would be few gaps between taxa, except some in the lowest ranks. In fact, there would be no taxa, because the concept of taxa is dependent upon presence of gaps between them. These discontinuities are present because some plants or animals once connecting the taxa are extinct, as are nearly always the ancestors of both groups. Thus, the taxa of today must be classified according to the consistency of occurrence of clusters of characters which tend to stay together rather than to disperse and not according to presence or absence of presumably ancestral characters. The orders and families of today are what they are, not what they were.

The Bold Mechanical Lines between Major Groups of Orders. Again, as stated briefly in Chapter 16, some major boundary lines indicate wholly natural separations into groups with internal affinity, as between the dicotyledons and monocotyledons. Others separate groups with natural affinity, but in part the groups are artificial, mostly around the edges. The classification of the dicotyledons into primary groups of orders has baffled all authors, and many approaches to the problem have been presented. Each solution to the basic question has some merit, but no proposal solves the problem, which is complex to a degree difficult to appreciate. The existing orders are composed of nearly 300,000 species not forming an evolutionary tree but a complex tangle of the survivors of many evolutionary lines of adaptive radiation in all directions. Nearly all these originally irregular and crisscrossing lines of development have disappeared without leaving a record.

Among the dicotyledons the lines on the chart separate the orders into choripetalous or apetalous versus sympetalous groups and into hypogynous versus perigynous or epigynous groups. The primary division could have been along lines representing other opposed character pairs instead, and the results would have been fundamentally the same, the groups partly natural and partly artificial, though probably in different degree. However, these do not present any definite advantage, and adopting any such groupings that come to mind involves some loss or at most slight gain. DeCandolle and Bentham and Hooker knew plants; they had monographed the known flora of the world in their time. Their adoption of the groups Thalamiflorae, Corolliflorae, and Calyciflorae did not represent either chance or an arbitrary choice. All these men were searching for a natural system of classification, and the adoption of the large groups represented a step in that direction. The old masters must have recognized the cohesiveness in these groups, though they were too wise to miss the inconsistencies. For the time being we are wise to retain these groupings as a means of bringing some degree of organization and understandability to the dicots but with the understanding that when truly natural lines of classification of the major groups become apparent, the bold lines in the chart will be removed or modified or they will just fade away, leaving the still valid or remodeled underlying arrangement of the orders.

Designation of any character as "important" results in at least a partly artificial system of classification. Its only true importance in classification is its occurrence in company with others in the combinations characteristic of taxa. Thus, neither the characters emphasized here for practical reasons nor those designated by others as of phylogenetic importance are likely to occur consistently. Nearly all the common characters of flowering plants have arisen more than once in evolution. Perigyny and epigyny certainly have arisen from hypogyny in several groups of dicotyledons and a number in the monocotyledons. According to Eyde and Tseng (Science 166: 506–508. 1969), *Tetraplasandra gymnocarpa* (Araliaceae, Dicotyledoneae) has hypogynous flowers that have arisen from epigynous ones. Thus, the characters are not stable, but neither are others. The characters chosen have the advantages of being *relatively* stable, easily detected, and well known because they are traditional.

The formulation of a truly accurate and natural system of classification is hampered further because each judgment of relationships must be based on only a small number of the nearly 300,000 species or the numerous families and genera of flowering plants. If study is to be more intensive, as in the preparation of a monograph of a genus or in experimental studies, only a limited number of species can be investigated or have their bearing on the classification of the higher taxa evaluated. Thus, reaching conclusions concerning the major groupings of dicotyledons (above the rank of order) must be postponed for many years, and truly justifiable wholly natural groups may never emerge. The necessary data for their formulation are not available now.

It is hoped that the classification presented here will facilitate teaching by giving students a way of cutting through the complexity in their path, by making the orders easier to comprehend for both the academic world and the layman, and by providing a beginning point for research. Good teaching does not leave the impression that the last word on any subject is known. Classification of living organisms is dynamic—like all science forever approaching the truth, yet never reaching it in all particulars. No system of flowering plant classification represents more than a mere stage in accumulation of knowledge through investigation. Research continues.

Evidence Concerning Evolutionary Starting Points

The soft parts of flowers do not lend themselves readily to formation of adequate fossils. If the intricate and delicate structure is not lost through decay, it is crushed and distorted as the plants are preserved in mud or sand later to be cemented into shale or sandstone. Preserved woody stems and tough leaves from the critical period when the angiosperms were evolving cannot be correlated with complete records of reproduction because flowers do not make good fossils. Hard or tough fruits tend to be preserved, but most fruits are soft and decay is likely to preclude their preservation. Furthermore, herbaceous plants—the kind most prevalent among modern angiosperms at least in temperate regions and, because of their short life cycles, the types evolving most rapidly in any period—may have existed by the thousands of species without leaving any clear trace. The apparently rapid rise of angiosperms could reflect derivation from either an herbaceous or a woody line that left no record. Most evidence favors a shrubby line, and this is even considered proved. However, since herbs leave no record, the point really proved is that existing herbaceous groups were derived from woody plants. (See footnote on p. 565.) Furthermore, the innumerable upland species growing away from swamps seldom are fossilized, and we may never know more than a little of their evolutionary history. There is little in the fossil record to indicate whether (1) the angiosperms evolved from the gymnosperms; (2) both were derived from some fairly near ancestor in common; or (3) both were derived from some primitive group of ferns or their forerunners, the Psilophytales (pp. 588 to 590), each having followed a long and independent course. Probably the most reasonable origin proposed so far is that suggested by Stebbins (Chapter 16) for derivation from the Caytoniales.

Angiosperms were seemingly suddenly abundant in Cretaceous time, but the fossil record does not show them clearly to have been present during the Triassic and Jurassic periods. Most of the geologically early groups were similar in general characters to modern plants.

Summaries of the relative abundance of the larger groups of dicotyledons in various periods and epochs of geological history were attempted by both Bessey and Wettstein. However, the fossil record is too fragmentary to yield reliable information. The data were drawn from leaves of a few species in a small number of deposits. Some identifications were inaccurate and others open to question, and the fossils represent chiefly the ecological associations near water where at least today the "Amentiferae" and other special groups are abundant. The probable error in any of the figures is high, in fact, so high as to discount the value of making an analysis.

The time of origin of the angiosperms is debated hotly by two major groups, (1) the botanists and the paleontologists, who emphasize the evidence from macrofossils, and (2) the palynologists, who are influenced by the occurrence of fossil pollen grains. The evidence of occurrence of flowering plants before Cretaceous time is scanty and uncertain, and, although origin of the class during Jurassic or even Paleozoic seems likely, rather than for it to have appeared full-fledged during Cretaceous, evidence from either type of fossil is uncommon or lacking. It may be argued that a few individuals are preserved as macrofossils, while pollen accumulates in great quantity in bodies of quiet water. According to some, if there were flowering plants before Cretaceous, almost surely there would be fossil pollen. There is some reason to accept this in part, but pollen is irregular in accumulation and that in lakes and ponds is from mostly wind-pollinated plants and from others growing along the watercourses and still water. Most flowering plants have insect pollination and heavy pollen not carried far otherwise. The pollen of some flowers is too delicate to be likely to be preserved. The great majority of species occur away from water, some far from it in grasslands, woodlands, chaparrals, or deserts, which also are areas in which macrofossils are formed only uncommonly or rarely. Thus, the microfossil record must be considered useful but to be evaluated carefully because its data are incomplete and out of proportion. If the angiosperms originated on seasonally dry uplands and mountains, where evolution proceeds rapidly, only a few of innumerable species may be preserved and their pollen may have had no likelihood of accumulating in lakes, ponds, or bogs. According to Axelrod, the Dakota Formation, of middle Cretaceous (p. 557), which is rich in angiosperms insofar as macrofossils are concerned, includes only about 5% of flowering plant microfossils, the accumulation of the two therefore not being necessarily proportional. Although origin of the flowering plants in restricted areas at moderate elevations in the mountains, where erosion is faster than deposition, seems likely, evidence is insufficient. The formation of abundant fossils of all kinds was likely to be delayed until the plants had radiated from their places of origin and early diversification into extensive habitats of the moist kinds in which fossil formation occurs more frequently. The original angiosperms may have occurred in the mountains and temperate uplands of Southeast Asia and the larger islands farther southeast, where the

Ranales retaining primitive features are most common today and where, in some places, conditions favorable to rapid evolution still exist, supporting not only angiosperms with primitive features but a wealth of others. Only time and accumulation of evidence can provide an answer to the problem.

Some evidence may come from fossils of flowering plant reproductive parts. For example, according to Crepet, Dilcher, and Potter (Science 185: 783. 1974) recent fossils from middle Eocene sediments of southeastern North America include an interesting and rather extensive array of well-preserved flowers of angiosperms. These represent insect and wind-pollinated types and a variety of flower structures. One now studied in detail is allied to the living genera of the Juglandaceae, or walnut family. Crepet and Dilcher (American Journal of Botany 64: 714–725. 1977) have found considerable detail in flowers of the Mimosoideae (Leguminosae) of Middle Eocene.

Since the fossil record is meager, it is necessary to depend chiefly upon living species for evidence concerning which group or groups of flowering plants may be most primitive and therefore near a possible starting point in their evolution. Determination of the origin of flowering plants and the starting point for any discernible evolutionary lines among the orders of angiosperms requires, as with the classification process, a synthesis and evaluation of data from all possible fields, including the fossil record, plant geography, gross structure, microscopic anatomy, embryology, chemistry, developmental morphology, and cytology. Significant evidence of primitive groups among angiosperms comes now from at least three sources representing segments of the fields above. These are flower structure, wood anatomy, and pollen grains.

Flower structure. The floral characters of the Ranales and Alismales have been summarized under the key words *free, equal, numerous, spiral or irregular,* and *hypogynous* (pp. 114 to 122), and the similarity of this floral structure to the arrangement and characters of vegetative structures has been discussed. Degrees of departure from the floral characters of the Ranales and Alismales have been correlated roughly with high efficiency in pollination, and orders with highly specialized methods of pollen transfer have been eliminated as possible starting points in the general evolution of angiosperms.

The carpels. The genus *Degeneria*, one of the most primitive known members of the order Ranales, was discovered recently in the Fiji Islands, and it constitutes a whole family (refer to p. 139). The flower includes only one pistil formed from a single carpel, this having the three vascular bundles characteristic of the leaves and sepals. The carpel is folded, and it encloses the ovules. These are not marginal but are on the ventral (upper) surface away from the margin. There is no style, and the stigmatic surface is along each margin of the carpel in the position of the suture in ordinary carpels. *However, since the margins are not fused, there is no suture.* When the carpels are young, there is a slight separation between the margins, though they are pressed together later. This is one of the most primitive known pistils.

Discovery of *Degeneria* growing in the Fiji Islands by A. C. Smith provided the stimulus for review of carpels of other Ranales by a Harvard group, lead by I. W. Bailey and including A. C. Smith, B. G. L. Swamy, Charlotte G. Nast, and their associates. In the order this group found developmental series of carpels leading from some even more primitive than those of *Degeneria* (e.g., *Drimys,* section *Tasmannia*) to various advanced types. They found evidence that the carpel is a folded (conduplicate) leaf with the ovules on the upper surface and fundamentally not marginal but, as in *Drimys piperita,* at a considerable distance from the margins. Stages in the development of the style and stigma are seen in other Ranales in which there is specialization of stigmatic crests on the margins near the

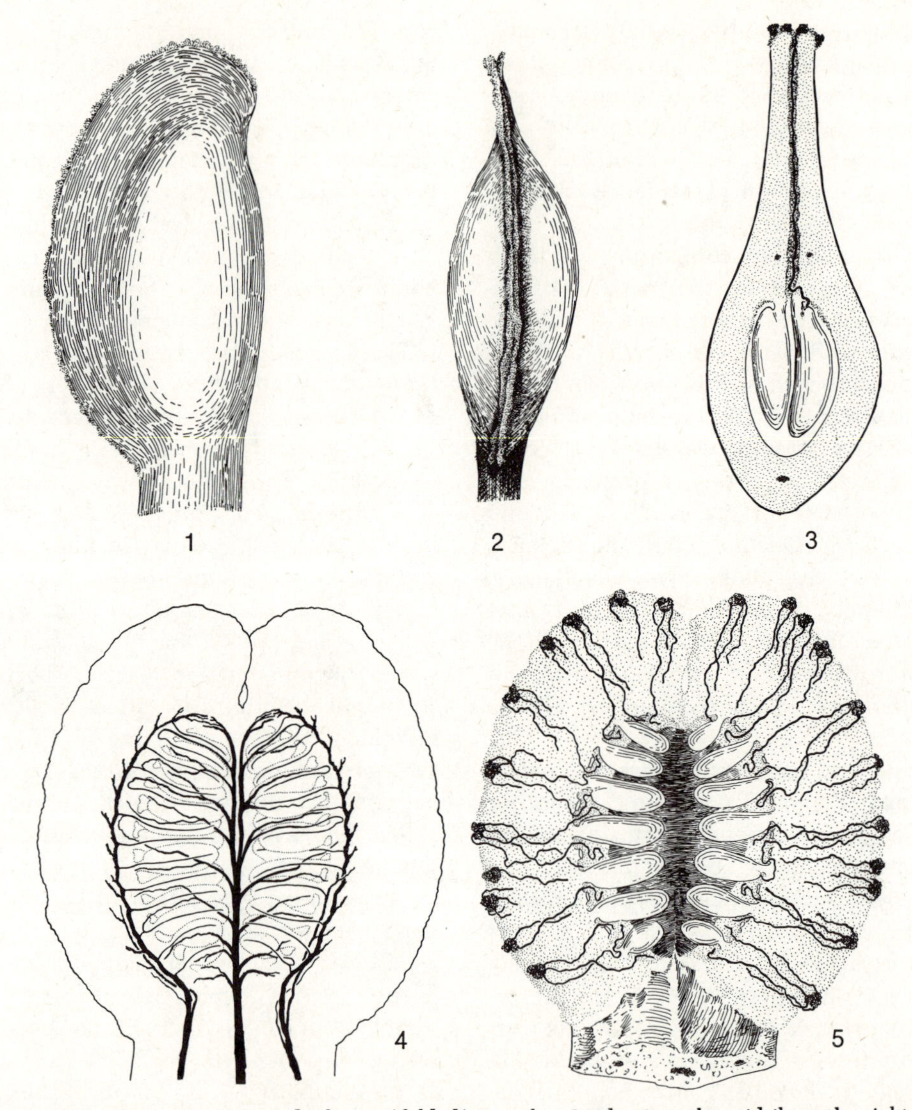

Figure 17-2. The primitive conduplicate (folded) carpel: **1** side view, the midrib on the right, the margins folded together on the left, **2** ventral view showing the margins folded together, these specialized as stigmatic crests and receptive to pollen, **3** cross section of the carpel, the midrib below, the margins above, a pollen tube penetrating among the hairs of the infolded ventral surface of the carpel and reaching the micropyle of an ovule, **4** carpel unfolded and cleared, showing the vascular bundles and (by dotted lines) the positions of the ovules, **5** carpel unfolded, showing the placentae, ovules, distribution of glandular hairs (stippling), and courses of pollen tubes. From IRVING W. BAILEY, American Journal of Botany 38: 374. figs. 1–5. 1951 and *Contributions to Plant Anatomy,* the Chronica Botanica Company. 1954. Drawings supplied through the courtesy of the author; used by permission.

apex of the carpel. Development of the stigma is not from the midrib. In the primitive unsealed carpel of *Drimys* there is no style; the pollen grains lodge among glandular hairs of the free margins, and the pollen tubes grow to the ovules through a mat of similar hairs on the inner (ventral) surface. In other Winteraceae and related families of the Ranales, the stigmatic marginal surfaces are restricted in various ways to small crests near the apex of the carpel, and a developmental series representing the origin of the stigma and style is preserved in the living members of the order.

More of the gloom surrounding failure of the angiosperm reproductive structures to be fossilized is disspelled by the study of *Liriophyllum,* a genus reported by Lesquereaux in in 1874, 1883, and 1892. New research by Dilcher, Crepet, and Beeker (Science 191: 854–856. 1976) reveals some of the structure of the carpels of the fruits of a plant, linked by identical yellow surface bodies (glands?) with the leaf genus *Liriophyllum* Lesquereaux occurring with them. The carpels of *Liriophyllum* (assuming identity) were arranged spirally on an elongate axis. Each was the shape of a pea pod, and the margins formed a suture bearing a stigmatic crest through its entire length, there being no stylar extension. The seeds were in two rows on the upper side of the carpel in the same position as in the woody, presumably primitive Ranales. Presence of the suture may require some explanation in view of the elongate stigmatic crest and lack of a style. There must have been an area in which the suture was incomplete or the pollen tubes would have had no way of reaching the ovules (unless closure of the suture occurred after pollination). Thus, carpels of the type in the woody Ranales in most features were present in the Dakota Formation of middle Cretaceous. *Liriophyllum* was accompanied by various other flowering plants not so far studied for morphological detail.

The unsealed carpel is not confined to the woody Ranales. It occurs in several genera of the Alismales (Kaul, American Journal of Botany 63: 175–182. 1976) and among greatly differing groups like the cacti and the palms. The relatively primitive cactus, *Pereskia pititache* (Boke, American Journal of Botany 50: 843–858. 1963) and often the palms (Uhl and Moore, American Journal of Botany 58: 945–992. 1971) have open carpels. This could indicate retention of this feature from an ancestral group from which all the flowering plants were derived and loss of the unsealed carpel in most surviving lineages but its retention in some taxa. Presence of primitive features among the flowers of cacti is striking, and in the palms they are perhaps as abundant as in the Ranales.

According to Stebbins, the closed carpel may have developed partly as an adaptation to drought occurring seasonally in the areas in which the angiosperms originated. The canal of the style forms an avenue favoring the rapid growth of the pollen tube to the micropyle of the ovule, and in a seasonally variable climate this is important. Provision of a ready-made passageway through the style, ovary, and integuments to the egg, together with protection of the internal structures, including the ovules, from drying, must have allowed considerable flexibility for the early angiosperms in coping with variations of seasonal climate. In most cases there could be special adaptation to a favorable season for pollination, quick fertilization, and fast development of the fruits and seeds. As postulated by Grant, the ovary wall also provided protection from the chewing mouth parts of beetles, which visited the flowers for nectar and pollen and commonly pollinated them but also ate ovules and later seeds. In contrast, the gymnosperms have become adapted to seasonal variation only through having the ovule enclosed in a very thick integument, and digestion of a passageway by the pollen tube requires waiting and active periods through several months or a year. In many gymnosperms protection of the seed is by a woody cone surrounding it during the stages after fertilization.

The cones of pines and other gymnosperms

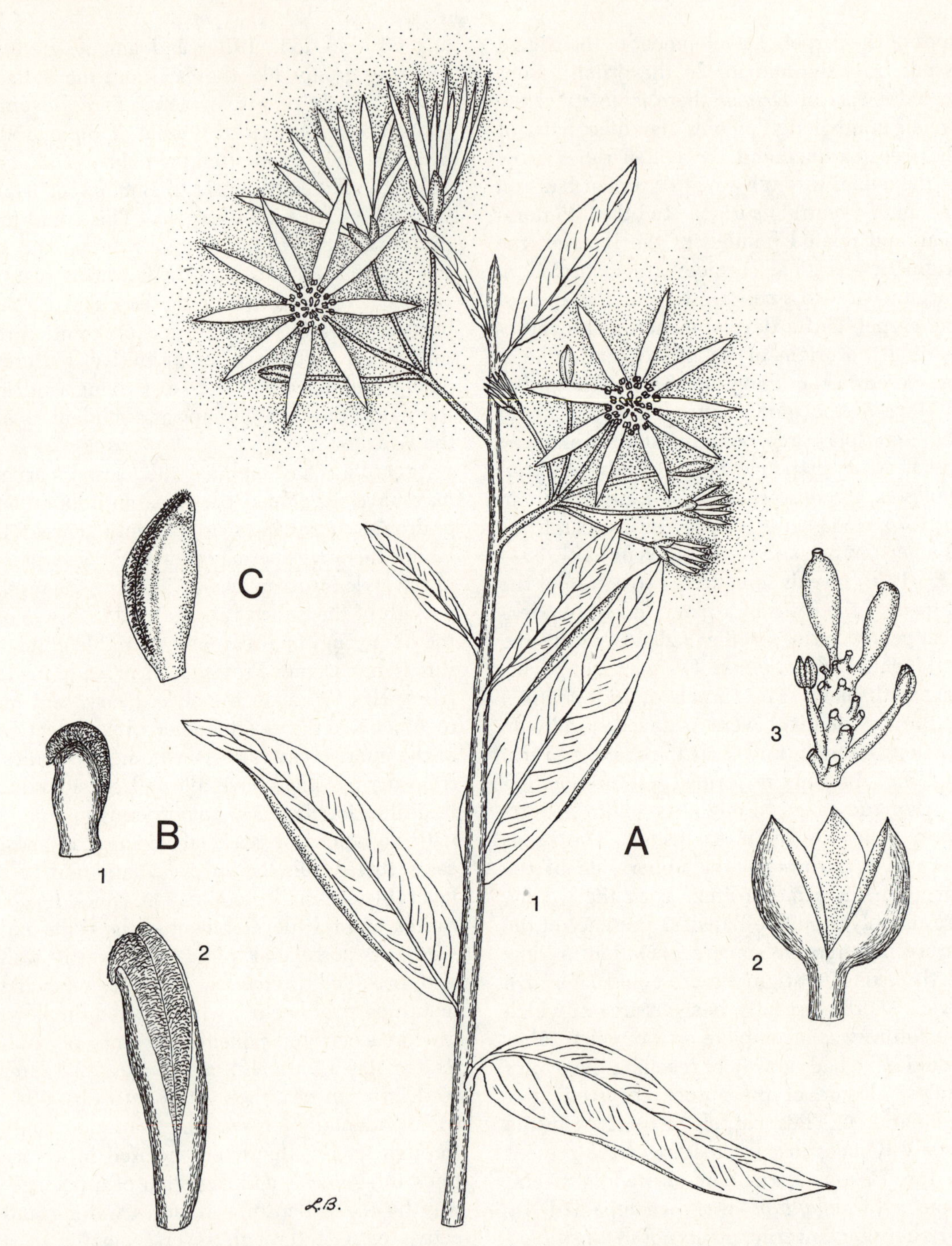

Figure 17-3. The primitive woody Ranales—*Drimys and Degeneria*: **A** *Drimys Winteri*, **1** branch with leaves, flower buds, and flowers in various views, showing the sepals, petals, stamens, and pistils, **2** lower part of the flower, the sepals, **3** upper part of the flower, the detached receptacle bearing stamens below and pistils above, some of each cut away, each pistil formed from a single carpel, with no stigma and

are not a choice item of food for most herbivorous animals, though the seeds of some are edible. Furthermore, the leaves tend to be thick and leathery or tough. The herbivorous reptiles and other animals of Jurassic and Cretaceous doubtless enjoyed having these plants to eat, but they did not provide the rich and varied array of foods becoming available during Cretaceous as the angiosperms spread out over the land of the dividing continents. A diet essentially of fern leaves and pine nuts* does not sound attractive to those used to a world dominated by flowering plants.

From the viewpoint of an omnivorous mammal, after the flowering plants arose during Cretaceous there was something good to eat in the world. The flower, and especially its great innovation, the carpel, which becomes the fruit, producing an enormous variety of seeds and taking a great range of external form and texture, offered innumerable possibilities for attracting animals and inducing them to carry pollen or seeds. Some orders of insects were adaptable to radiating into the enormous number of new niches associated with presence of nectar and pollen; other orders spread into the niches provided by the new plants with an amazing ability to produce differing kinds of tender young stems or leaves or even of wood which may be invaded and of diverse fruits and seeds. Mammals and birds carried pollen only rarely, but they throve on the great variety of fruits and seeds and on tender vegetative structures. Carnivores of all groups appeared, and soon there must have been at least the beginnings of new food chains based mostly on the flowering plants. The presence of angiosperms must have contributed toward revolutionizing the ecosystems. Certainly, the great shift from floristic dominance* of gymnosperms like cycads and conifers, with tough leaves and hard or leathery seed coats and no edible fruits, to the flowering plants must have been triggering an enormous ecological change favoring new groups of animals like the mammals and birds and some older ones like the insects (present since Devonian), but these with modification. Thus even during Cretaceous a potential new combination of organisms was being forged with the angiosperms, mammals, birds, and insects as the dominants. Nevertheless, the other groups of organisms were abundant until the end of Cretaceous. The angiosperms and their dependents had not yet conquered the world, but at least the seeds of conquest were present or already germinated and growing rapidly.

Most organisms of the time of the dinosaurs apparently were not on the wane during late Cretaceous, but perhaps they were being crowded into the remaining ecosystems favorable to them or those in which competition could be overcome. To whatever extent the decline and fall of the Saurian Empire may have been portended during Cretaceous, reptilian dominance was extinguished quickly at

* Oversimplified, of course.

* During Cenozoic, even in coniferous forests, gymnosperms have been dominant only in an ecological sense, that is, through providing shade for the lower levels of the forest, soil-building litter for the forest floor, and food for specialized animals like tree squirrels and some birds and insects. Floristically they are minor; their numbers of species and individuals being far overshadowed by those of the angiosperms forming the middle and lower stories of the forest. These, too, contribute to the litter and they provide enormous quantities of diverse foods required to sustain a great array of animal species. With only ferns and gymnosperms, the present complex of ecosystems would not be possible. (P. 771.)

an unsealed opening through which the pollen tube grows, **B** *Drimys piperita*, **1** side view of the unsealed carpel, showing the stigmatic margin, **2** ventral (edge) view showing the stigmatic crests along the unsealed margins (size double), **D** *Degeneria vitiensis*, view of the carpel, showing the stigmatic crests and the unsealed opening between the ventral margins of the carpel. After JOHN HUTCHINSON, Families of the Flowering Plants, ed. 2, 1: 127. fig. 3. 1959, Clarendon Press, Oxford.

the end of the period. Between then and the start of Cenozoic time a few million years later, one-quarter of the known animal families disappeared. Although this might have been expected to occur gradually in view of the changes of the foods available and the new food chains, competitors, and predators, the changes were sudden and perhaps catastrophic. Moreover, they involved not only land organisms but also marine taxa. Thus, there may have been other factors at work in addition to the great shift of land plants. Ecological changes due to the presence of flowering plants, alone, can account for the change, though probably not for its speed of occurrence.

A number of explanations have been advanced for the sudden disappearance of many of the Cretaceous organisms and especially the dinosaurs. Those under discussion now include flooding of much of the lowlands of continents by shallow seas, this having occurred during Cretaceous time when various epeiric seas formed and withdrew. Fluctuations of the shore lines may have been an important factor in extinction of Mesozoic taxa, because study of the numerous oscillations of the earth's crust indicates evolutionary change to have reached its maximum rate when the inland seas were most extensive and extinctions to have been most frequent when the land rose and the waters receded. The flowering plants may have become diverse and have spread over the continents during the great periods of flooding now and then during Cretaceous. By the end of the period they had become the dominant land plants. Many extinctions must have occurred because of the rising and falling of the water level, and there must have been repercussions in the regions beyond reach of the water as newly evolved species migrated into the surrounding areas, upsetting the ecological balance. During the last 20 million years before the sudden extinctions at the end of Cretaceous, temperatures were relatively low, but various lines of evidence indicate an increase of a few degrees at the end of the period. This could have meant death or impaired reproduction to various animals and especially the large ones like the dinosaurs (see Axelrod and Bailey, Evolution 22: 595–611. 1968; McLean, Science 201: 401–406. 1978). Experiments with living reptiles show temperatures only a few degrees above normal to be detrimental or lethal to the sperm cells. Some large reptiles may have had internal temperature-regulating mechanisms, but others doubtless did not, and no land animals weighing more than 25 kg., or 55 pounds, survived from Cretaceous into Cenozoic. This may reflect the lower ratio of surface to volume in larger organisms and the consequent difficulty in excluding excess heat. Elimination of the large reptiles gave a great advantage to the still relatively small mammals, which during Paleocene radiated into the niches becoming available and, having internal heat regulation mechanisms, gave rise to large species of their own. The realm* of *Tyrannosaurus rex* was replaced by a new empire.

The most likely candidates for ancestors of the angiosperms were the seed ferns, or Cycadofilicales and Caytoniales, abundant during Carboniferous and Permian (pp. 543, 649). These were fernlike in appearance, and their leaves were like those of ferns, but they bore seeds on the leaves, along with (on the same or different plants) microsporangia bearing small spores, which in the angiosperms are the precursors of pollen grains. As summarized by Andrews (Science 142: 925–931. 1963), single seeds or groups of seeds of seed ferns (Carboniferous) sometimes were enclosed in a cupule of leaf tissue, as, for example, in *Lyginopteris Oldhamia,* in which a solitary seed is surrounded by a cupule of segments coalescent on the lower quarter or third of their length. In *Eurystoma* the seed was invested by a series of terete branches of the primitive frond;

* Actually *Tyrannosaurus* is known from only late Cretaceous and from North America and eastern Asia; he did not conquer the whole world, at least so far as the evidence goes. (Neither did the Caesars or Alexander.) The animals were up to nearly 50 feet long, the skulls 4 feet.

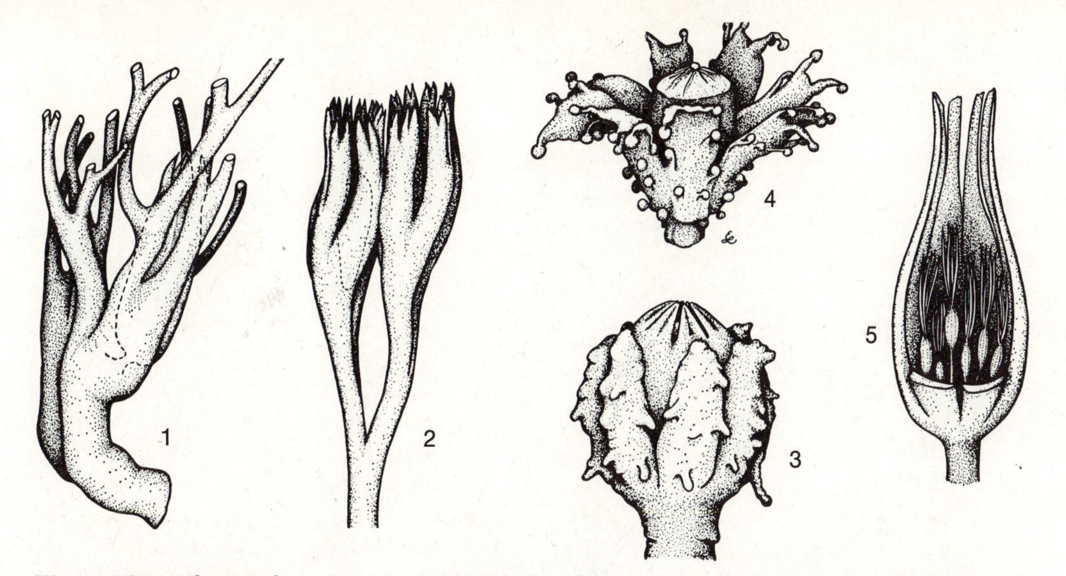

Figure 17-4. The cupule in the Cycadofilicales; the lobes of the cupule showing various degrees of fusion: **1** *Eurystoma angulare,* the ovule (shown by dotted lines) surrounded irregularly by essentially terete branches of the primitive frond or leaf system and only the beginning of a cupule, **2** *Stamnostoma huttonense,* more definite beginning of a cupule (seed indicated by dotted line), **3** *Tyliosperma orbiculatum,* with a definite cupule, **4** *Lyginopteris Oldhamia,* similar cupule, **5** *Calathospermum scoticum,* the large cupule enclosing several seeds. The cupule may be the forerunner of the carpel of a flower of an angiosperm. HENRY N. ANDREWS, Science 142: 929. 1963. Copyright 1963 by The American Association for the Advancement of Science (**1, 2** after A. G. Long, Transactions of the Royal Society of Edinburgh **64**: 29, 201, 261, 281, 401. 1960, 1961; **3,** after S. H. Mamay, U.S. Geol. Surv. Prof. Papers No. 254-D. 81–95. 1954; **4,** after F. W. Oliver and D. H. Scott, Philosophical Transitions of the Royal Society, 197B: 193. 1904; **5,** after J. Walton, *Introduction to the Study of Fossil Plants* 133. *f.110.* 1940. Used by permission of Adam and Charles Black, Publishers). Used by permission.

these continued to branch dichotomously. In *Stamnostoma* the seed was surrounded by lesser branchlets of the frond, these dichotomous only at the tips. Only one seed was found, but scars indicate the possibility of four surrounded by each cupule of four frond branches. In *Calathospermum* several stalked ovules were surrounded by a cupule of segments open only in its upper half or less. These plants were not necessarily ancestors of the flowering plants, but they do indicate a manner in which the carpel may have arisen. (See also, Delevoryas, T. 1962. *Morphology and Evolution of Fossil Plants.* Holt, Rinehart, Winston, New York.)

The theory suggested by Stebbins for the origin of the carpel is presented in part on p. 543. The carpel may be a derivative of evolution of structures occurring in the Caytoniales, an order of Jurassic gymnosperms related to the extinct Cycadofilicales (p. 649 in Chapter 20). According to Stebbins, evolution of the female reproductive structures may have proceeded by the following steps:

1. Presence of plants of the seed fern type with forking sporangiophores with terminal sporangia, with differentiation of micro- and megasporangia, the megasporangia with few spores and ultimately only one.
2. Formation of an integument (later the inner) around the megasporangium, paralleling seed formation in the Paelozoic gymnosperms, such as *Lyginopteris,* Cycadofilicales.
3. Formation of a bowlike cupule around the seeds as in *Calathosperimum* (Carboniferous), the cupules and ovules becoming small, and on short sporangiophores probably attached on the sides of the cupules, the cupules becoming adnate with the subtending bract, as in several genera of the *Glossopteris* group.
4. Change of the now solitary megasporangium into an ovule with two integuments, the outer being formed from the adnate cupule. In many families of angiosperms the outer integument and the inner do not match up very well around the

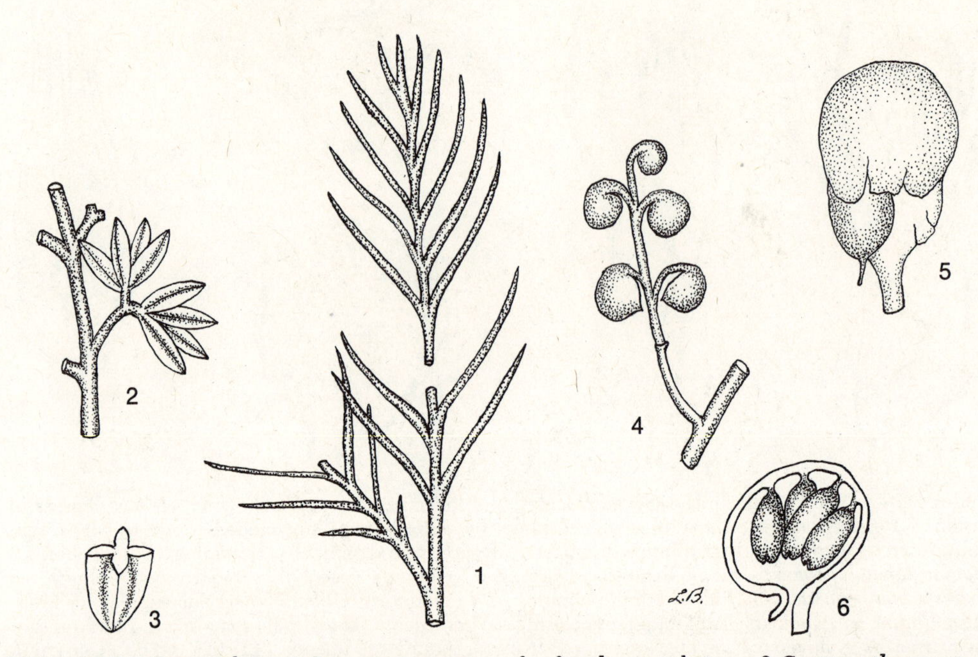

Figure 17-5. Caytoniales: **1** *Stenopteris,* portion of a frond, natural size, **2** *Caytonanthus,* part of a microsporophyll with clusters of microsporangia, **3** section through a single microsporangium, **4** *Umkomasia,* raceme of ovule-bearing cupules, **5** *Umkomasia,* opening cupule with a projecting seed and the elongate tube around the micropyle, **6** *Caytonia,* longitudinal section of the cupule and the enclosed seeds. Based upon H. HAMSHAW THOMAS, **1, 4, 5,** Philosophical Transactions of the Royal Society 222B: 193–265. 1933; **2, 3, 6,** *ibid.* 213B: 299–363. 1925. Used by permission.

micropyle, and this gives rise to a "zigzag" opening perhaps corresponding to the independent origins of the integuments. The microscopic, including cellular, structures of the two integuments are markedly different in the angiosperms, and Stebbins' theory provides a possible solution to their origins. (Andrews, above, pointed out the possibility that the cupule is the forerunner of the carpel, rather than the outer integument of the ovule.)

5. Infolding of the bract in forming the carpel, as in the carpels of primitive types in the woody Ranales.

6. Adnation of the vascular bundles of the sporangiophore and the ventral bundles of the bract or scale-leaf, providing the three bundles typical of carpels.

The stamens. The stamens of the Degeneriaceae (p. 563) and the Himantandraceae (Fig. 17-7) in the order Ranales exhibit putatively primitive characters, as shown by Bailey, Nast, and Smith. The stamen in these families is a broad 3-veined microsporophyll with two pairs of linear microsporangia (pollen chambers) embedded deeply nearer the midrib than the margins, i.e. between the midrib and the lateral veins. There is no division of the stamen into anther and filament, but only the broad blade. Canright described in 1952 and 1960 stamens in the tropical Magnoliaceae similar to those of the Degeneriaceae and Himantandraceae and series within the family leading to more or less conventional stamens. The series includes the following trends: (1) elongation of the apex of the stamen; (2) differentiation of a filament; (3) elimination of the two lateral veins; (4) increase in relative sporangium size and protuberance; (5) transition of the sporangia to a marginal position, and (6) development of fibrous tissue (in the "connective") between the anthers.

Stebbins (*Flowering Plants: Evolution Above the Species Level,* 1974) presents strong evi-

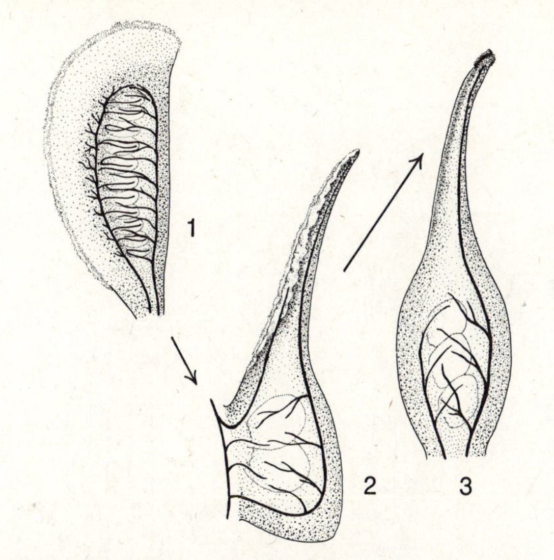

Figure 17-6. One type of modification of the primitive carpel (in each instance the midrib at the right, the margins at the left): **1** primitive conduplicate (folded) carpel (cf. **Figure 17-1**), **2** carpel sealed basally, the apical portion (style) folded but not sealed, bearing elongate stigmatic crests, **3** pistil formed from a single carpel with the ovary and style sealed and the stigmatic crest restricted to the apex, as in many Ranales. From IRVING W. BAILEY, American Journal of Botany 38: 374. figs. 6, 13–14. 1951 and *Contributions to Plant Anatomy,* the Chronica Botanica Company. 1954. Drawings supplied through the courtesy of the author; used by permission.

dence that stamens are not fundamentally individual structures formed from leaves but appendages of a branching structure originating from the axis of the flower, or receptacle. This structure does not emerge ordinarily from the receptacle, but its presence is indicated by stamen vascular bundles originating in groups rather than singly. Solitary stamens are produced by elimination of all but one of the fascicles. The origin of stamens in this manner has been known for a long time for certain families, but now it is to be regarded as the rule rather than the exception. The fascicle of stamens may represent a specialized leaf bearing sporangia (the stamen being a group of sporangia) or a branch bearing them, or an appendage of uncertain derivation. This point remains to be settled. Thus, in the woody Ranales retaining primitive characters, the stamens are either *spiral or irregular* in arrangement insofar as this can be determined readily with the naked eye, but actually they are in clusters, which themselves may be spiral in attachment. When the numbers of

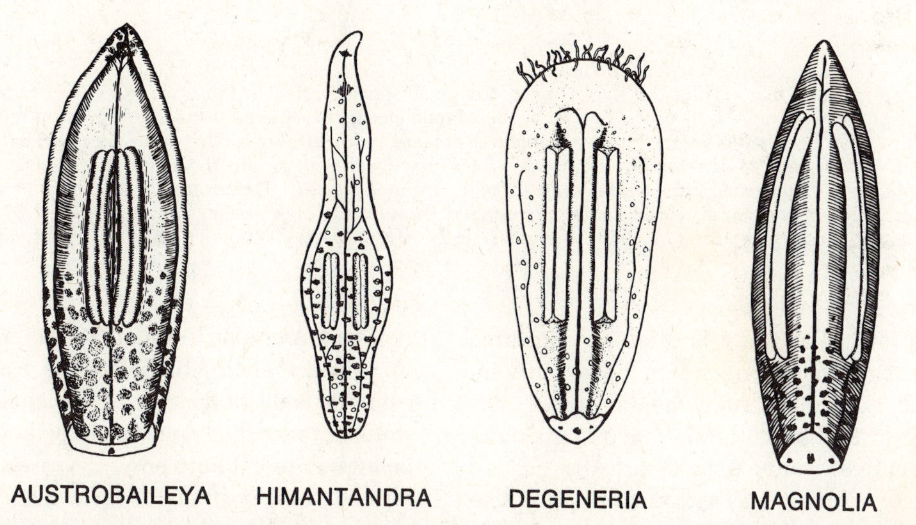

Figure 17-7. Stamens of the primitive woody Ranales, a diagrammatic comparison of microsporophylls (spore-bearing leaves or pollen-bearing leaves). The pollen chambers (microsporangia) are embedded in either the dorsal or the ventral surface of the specialized leaf (microsporophyll) often nearer the midrib than the margins. The blade is about 5–10 mm. long; there is no differentiation of a petiole or of anther and filament. From JAMES E. CANRIGHT, American Journal of Botany 39: 488. fig. 13. 1952. Drawings supplied through the courtesy of the author; used by permission.

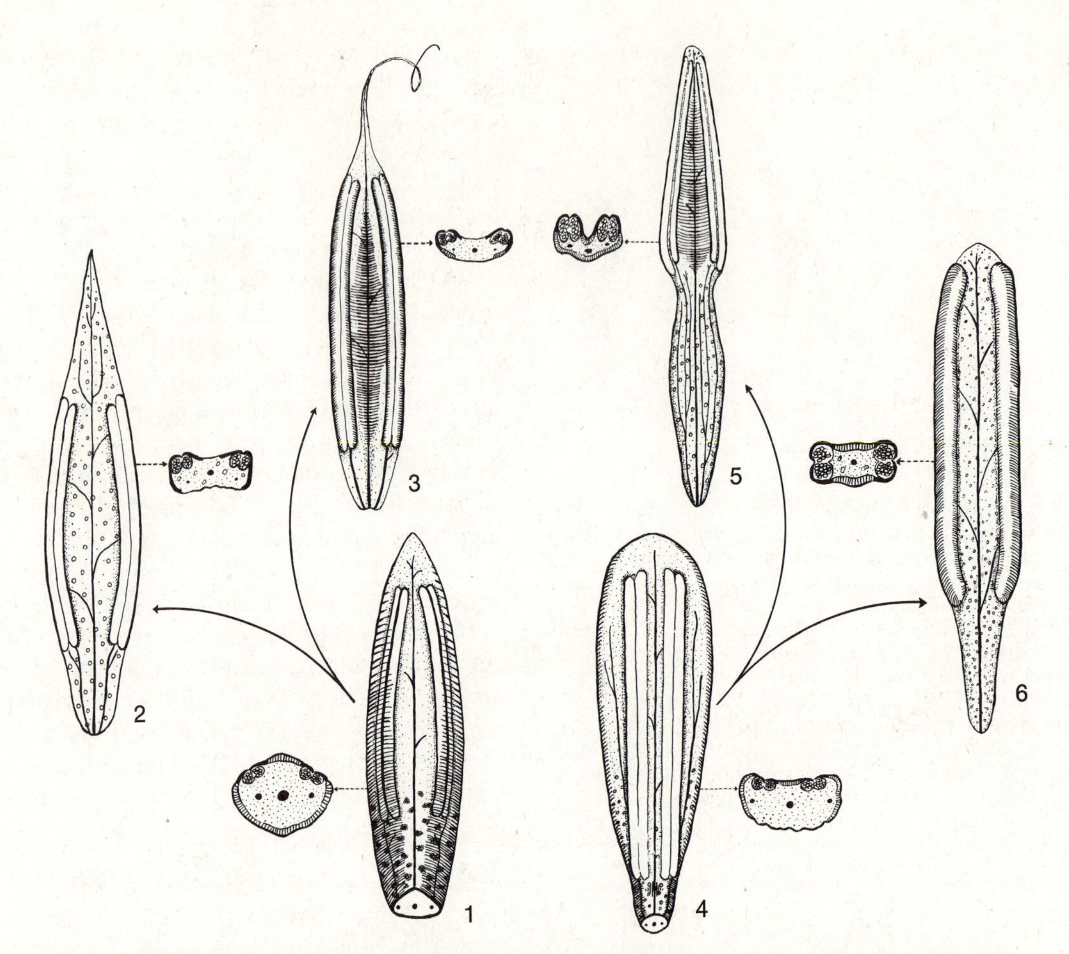

Figure 17-8. Diagram illustrating the chief trends of specialization in the stamens of the Magnoliaceae (Ranales) (stamens in ventral view, in the Magnoliaceae, except the tulip-tree, *Liriodendron,* the pollen chambers being ventral): 1 *Magnolia Maingayi,* with arrows to the more specialized types, 2 *M. nitida,* and 3 *M. Hamori;* 4 *Manglietia Forestii,* with arrows to the more specialized types, 5 *Magnolia hypoleuca,* and 6 *Magnolia fuscata.* From JAMES E. CANRIGHT, American Journal of Botany 39: 489. fig. 14. 1952. Drawing supplied through the courtesy of the author; used by permission. Moseley reports (in 1956) a similar series in the water-lilies (Nymphaeaceae), cf. **Figure 5-3 C.**

stamens are small, they become concentrated in apparent whorls, which may be actually individual turns of a fundamental spiral. According to Stebbins, the flattened stamens of the primitive woody Ranales are shaped and arranged in such a way as to favor pollination by beetles, the insects that were available when the angiosperms were evolving, and still of primary importance in pollination of the Ranales.

Wood Anatomy. Wood or xylem includes two kinds of conducting elements, each formed from the walls of dead cells. **Tracheids** are the walls of individual spindle-shaped cells (these being pointed at both ends). **Vessels** are tubes formed from portions of the walls of a series of several cylindroidal cells, the end walls having been dissolved away.

Vessels are characteristic of the angiosperms and not ordinarily of the pteridophytes or gymnosperms, except the Ephedropsida and Gnetopsida.* Only about 100 living species of dicotyledons have primitive vesselless xylem, and all these are Ranales. A nearly complete record of the origin of vessels and other trends of anatomical specialization in the flowering plants is preserved within this order.† Vessels probably have arisen several times among the angiosperms. Their nature in monocotyledons is different from that in dicotyledons. Thus the evidence from wood anatomy points strongly toward the woody Ranales as living angiosperms retaining primitive features.

Pollen Grains. Pollen grains have varied patterns of sculpture on their walls, and certain markings are characteristic of large groups of plants. Spores‡ of ferns and the extinct Cycadofilicales (primitive Cycadopsida) show three crests radiating from one point. This characteristic occurs in only one group of flowering plants—the woody Ranales. Monocolpate (one-furrowed) pollen grains are characteristic of most gymnosperms, the monocotyledons, and certain families in the order Ranales, probably also indicating a genetic tie among these plants because in the other dicotyledons the pollen is tricolpate (three-furrowed).

Another pollen character with a similar pattern of occurrence is called to attention by James W. Walker (Science 187: 445–448. 1975). The exine, or outer wall of the pollen grain of nearly all angiosperms includes columellae, or rodlike, internal upright structures. These do not occur in the gymnosperms. However, they are absent, also, from some of the woody Ranales, including the Magnoliaceae and Annonaceae (in which a transition series from one type to the other is present) and in the Degeneriaceae, the Eupomatiaceae, and the more primitive Annonaceae. This may be another feature retained in a few of the woody Ranales but not in the rest of the angiosperms.

According to Walker, the specialization of pollen types begins even among the primitive woody Ranales, and he describes and illustrates 10 types among them, as well as many transitional types (American Journal of Botany 61: 1112–1136. 1974). The origin of the tricolpate type of pollen prevalent in dicotyledons and its evolution into innumerable types

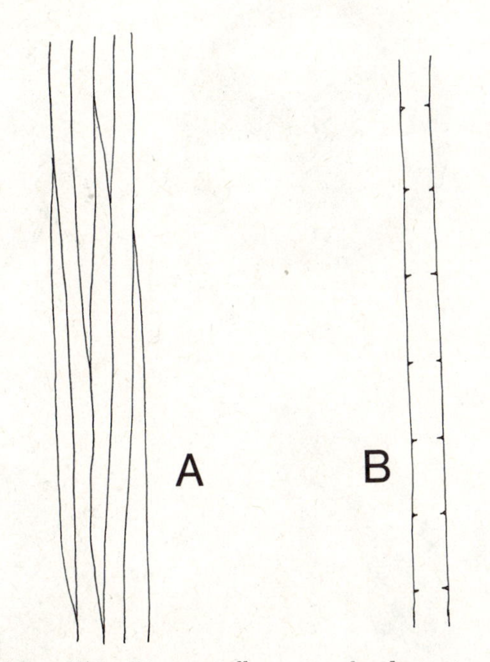

Figure 17-9. Diagrams illustrating the distinction of tracheids and vessels of the xylem: **A** tracheids; **B** vessels, the end walls of the cells dissolved away.

* The vessels of the Ephedropsida and Gnetopsida are of a different origin from those in the Angiospermae. Their sequence of wall thickening is different.

† Many authors have thought that the primitive angiosperms were woody, basing this supposition largely upon the acceptance that the angiosperms were derived from the gymnosperms. Bailey argues that the living dicotyledonous herbs have a high degree of specialization in their vascular tissues and that primitive vesselless xylem and the less modified forms of vessel-bearing xylem are found only in certain woody dicotyledons. This is a better-founded argument for the derivation of the flowering plants from woody immediate ancestors—not necessarily gymnosperms. Unfortunately, the herbs of Triassic and Jurassic times have left no clear record, and no one knows whether their anatomical characters were more primitive or advanced than those of their woody contemporaries.

‡ Pollen grains develop from single cells, which fundamentally are spores, and their walls are developed from those of these original cells. However, the maturing of the walls occurs after the internal cell of the spore has divided and thus passed on to another stage (the microgametophyte).

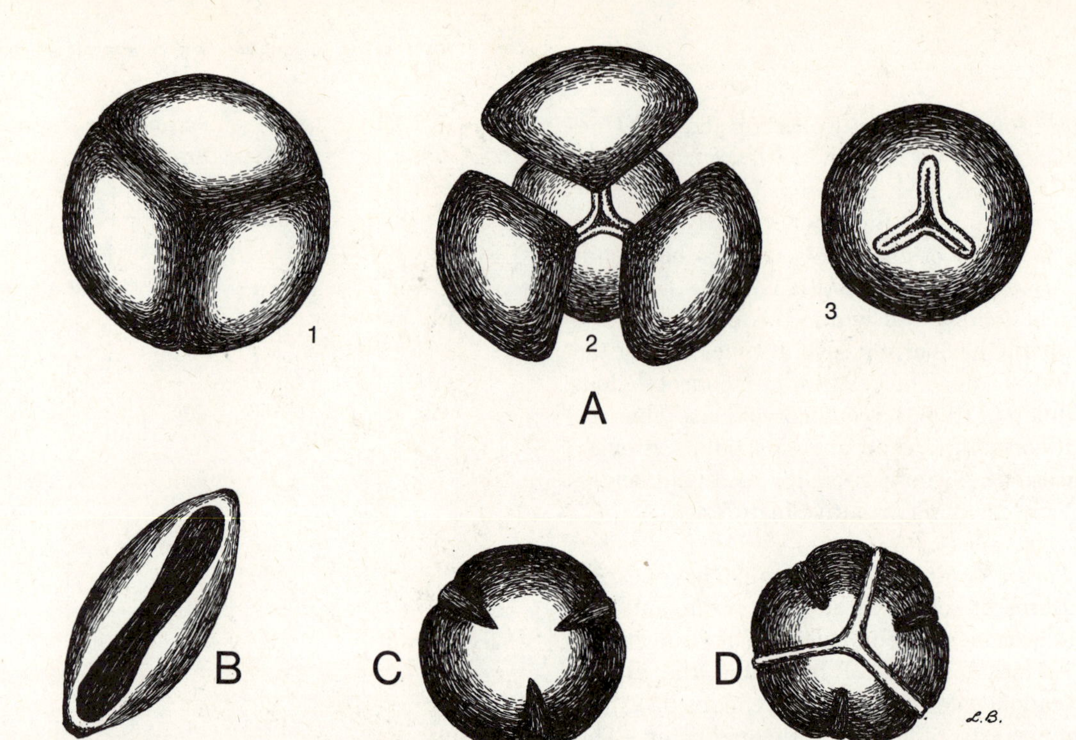

Figure 17-10. Evolution of pollen grains: **A** the triradiate crests of fern spores, **1** group of four spores just formed by the reduction divisions (meiosis) (only three visible in this view), **2** spores separating, **3** single spore showing the triradiate crest, the crests formed along the margins where the four spores were tightly against each other; **B** the single furrow of monocolpate pollen of a gymnosperm; **C** the three furrows of the pollen of all dicotyledons, except some of the woody Ranales, **D** the pollen grain of *Schisandra* (Schisandraceae, woody Ranales), showing both the triradiate crest of the ferns and the three furrows of the dicotyledons, an area at the opposite end (not shown) being homologous with the single furrow of monocolpate pollen. Redrawn from ROGER P. WODEHOUSE, Botanical Review 2: 67–84. figs. 1; 2a, b, c; 8a, b, c. 1936. Used by permission.

can be traced back to the Ranales. The scanning electron microscope has revealed much that was obscure.

SUMMARY OF POSSIBLY PRIMITIVE CHARACTERS IN THE RANALES

1. *Flower Structure.* The flower parts are *free* from each other, essentially *equal* within a series, *numerous* and not definite numbers, arranged *spirally* or the stamens seemingly so but actually in clusters or appearing *irregular,* and *hypogynous,* thus approximating the arrangement of vegetative leaves or branches. Departure from these floral characters commonly is associated with specialized methods of pollination and presumably an advanced flower type.

However, the amount and direction of specialization for particular types of pollination is difficult to judge, and high degrees of adaptation to particular methods may exist where none is apparent. According to Thien (American Journal of Botany 61: 1037–1045. 1974) probably the flowers of *Magnolia,* though they are with the parts corresponding with all the Ranalian criteria, are highly specialized for pollination by beetles, and other insects are excluded or ineffective. Beetles can enter or leave the flower even when it is closed at night. At this time the flower has large quantities of nectar, pollen, and soft stigma tissue, which become available only to beetles, these insects spending long periods (several hours) in the closed flowers. During their stay the beetles crawl through much of the flower, carrying pollen to the stigmas as well as other parts. When the flower opens in the morning the supply of food is much reduced, and the stigmas have turned brown and unreceptive to pollen. Thus, specialization of the *Magnolia* flower is great, though it does not result in attraction of the pollinating insects like bees and butterflies, for which many modern flowering plants are well prepared. It favors beetles, the insects abundant when the angiosperms were evolving (as they still are).

Carpels. *Degeneria* and *Drimys* have unsealed carpels without stigmas, and the stages of development of a sealed carpel with a stigma can be traced among living Ranales.

Stamens. The stamens of some Ranales are similar to scale leaves, having the pollen chambers (sporangia) sunken near the middle of the blade. In the Magnoliaceae there are gradual transitions to the stamens (with anther and filament) characteristic of angiosperms.

2. *Wood Anatomy.* Vessels of the angiosperm type occur in no other plants. However, about 100 species of flowering plants have primitive vesselless xylem. All are Ranales.

3. *Pollen grains.* The spores of ferns have a triradiate crest resulting from compression of the four formed from one microspore mother cell against each other. These occurred also in the extinct Cycadofilicales. Among flowering plants this character appears in only the woody Ranales. The pollen grains of gymnosperms, monocotyledons, and some families of Ranales have one furrow (are monocolpate). Pollen of the other dicotyledons has three furrows (is tricolpate). Thus, evidently the Ranales are a transitional group or connecting link in pollen grain features.

Brown has published an extensive study of the types of nectar glands occurring in angiosperms. The Magnoliaceae (Ranales) have no glands. *Illicium* (Illiciaceae, Ranales) has glands among the carpels on the rounded receptacle, and this is true also in some genera of the Connaraceae (a family in the Rosales possibly transitional to the Leguminosae from the primitive members of the order). In *Connarus* the glands are in a ring and transitional to the gland type in the Leguminosae. According to Brown there are several other series in gland development, and he uses this character to trace supposed evolutionary series to various orders, the woody Ranales serving more or less as a starting point. Other nectary types are reported as follows:

1. Modified stamens (Ranunculaceae, Geraniales).
2. A disc around the base of the ovary (Ericales, Polemoniales, Lamiales, Scrophulariales).
3. A ring beneath the stamens (Caryophyllales).
4. A cushion of multicellular hairs (Malvales).
5. The floral cup (Rosales and lines of development leading to
    a. Ultimately the Rubiales, Campanulales, and Asterales.
    b. Ultimately the Euphorbiales, Sapindales, Rhamnales, Rutales, and Santalales.
    c. The Myrtales.
6. A disc attached to the receptacle (Flacourtiaceae [Violales] and lines leading to
    a. The Papaverales.
    b. The Datiscaceae, Begoniales, and Cucurbitales) [Loasales].

Although the supposed lines of development are debatable as evolutionary series from one living order to another, the suggested conservation in living plants of various degrees of departure from the primitive condition apparently preserved in the Magnoliaceae may be correct. Departure from the type of flower common in the Ranales is accompanied commonly by specialization—leading to control of the pathway of the pollinating insects.

Bailey emphasizes the occurrence of many primitive Ranales in the living floras of northern Australia, New Guinea, New Caledonia, the Fiji Islands, and adjacent regions ranging northward to southern China. There are ten genera of dicotyledons with primitive vesselless xylem; five occur on New Caledonia, three being endemic to the island. The living floras as well as the fossil flora of the Mesozoic and

Tertiary rocks of these islands are relatively little explored, and they may be the future source of data solving the problem of origin of angiosperms.

According to the available evidence, the Ranales include the most primitive living dicotyledons, and there is little evidence of primitiveness among the Amentiferae. Investigation of floral anatomy and other features of first one order of the Amentiferae and then another indicates a derivation in common with choripetalous orders. The flowers of the Salicales and the Juglandales have rudimentary vascular bundles evidently once supplying a perianth. The fairly common occurrence of rudiments of pistils in the staminate flowers suggests derivation from the groups with bisexual flowers. The presence among the Amentiferae of catkins derived from reduction of cymules along an axis as well as of the typical spicate and racemose types suggests derivation from more than one ancestry. According to Tippo, the wood anatomy of the Amentiferae shows little evidence that the group as a whole is primitive. Abbe has published a thorough account and analysis of all the "Amentiferae" (Botanical Review 40: 159–261. 1974).

In addition to conducting water and dissolved materials, xylem serves as a mechanical support, adding strength and rigidity to the organs in which it occurs. The first-formed xylem elements of an immature stem or root are strong but capable of elongation. The wall thickening material is added in either of two ways—in rings at intervals along the cell wall (**annular thickening**) or in spirals (**spiral thickening**). These two types of thickening allow stretching of the thinner part of the cell wall between thickened areas as growth continues. Later-formed xylem cells have cross-connections between the coils of the spiral, making a reticulate (meshwork) pattern (**scalariform thickening**) which allows no stretching. This is the typical sequence of thickening in the pteridophytes. In the Cycad Line of the gymnosperms (p. 649), the most primitive extinct members of the Conifer Line of the gymnosperms (the Cordaitales, p. 645), and the angiosperms the last xylem to develop shows transitions between scalariform thickening surrounding large thin areas in the meshwork pattern and the more extreme **pitted thickening** with small, nearly circular thin areas (**pits**). This "normal" development sequence of xylem elements is **annular, spiral, scalariform, pitted.**

In the more highly developed living gymnosperms (the Conopsida, Ephedropsida, and Gnetopsida) the normal sequence is different, for typical scalariform elements do not occur and circular bordered pits are formed in early stages of development. The sequence is **annular, spiral, pitted.** For this reason, Bailey argues that the angiosperms could not have been derived from these groups of gymnosperms. Their sequence of development of the primary vascular tissues is indicative of an ancient and irrevocable trend of anatomical specialization. Thus the propositions of Engler and Prantl that the Amentiferae were derived from the conifer group and of Wettstein that they were derived from the *Ephedra* group are eliminated from consideration. The supposed similarity of the Amentiferae or some of them to these groups of gymnosperms was the cornerstone of the theory that the Amentiferae are primitive.

The Ranales are accepted here as the existing dicot order among whose members the greatest number of primitive features has been retained, and probably the order is near the starting point for at least some short evolutionary series. The Amentiferae appear to be mostly an artificial group, and in edition 1 they were retained as a unit because their relationships were not settled. Some or all the included orders probably were derived from the same stocks as existing choripetalous orders, though some may have originated from developmental lines segregated far back in the history of angiosperms.

Despite the expression of the thoughts above in edition 1, the Amentiferae were retained because students encounter the group in many

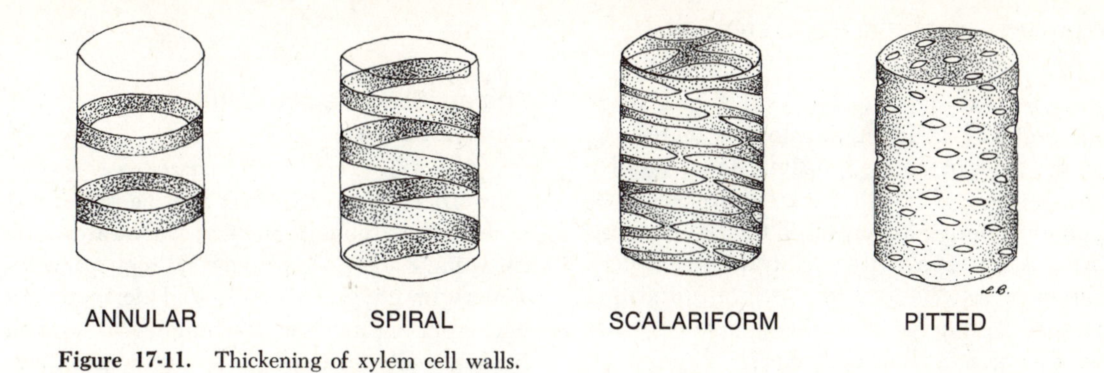

Figure 17-11. Thickening of xylem cell walls.

of the older floras and manuals. Now the evidence against the group is even stronger than then because of much new research, and there are new books on regional and local flora. The Amentiferae, therefore, are abandoned.

According to Bailey, the dicotyledons and the monocotyledons must have diverged very early, because vessels have arisen independently in the two groups. Nonetheless, as pointed out earlier, the Ranales and Alismales have much in common, for there are no important differences in flower structure and the orders differ in only about two thirds of the characters segregating most dicotyledons and monocotyledons. According to Hutchinson, the genus *Ranalisma,* with species in the Mayay Peninsula and West Africa, is placed in the Alismaceae rather than the Ranunculaceae partly because its members have one cotyledon and no endosperm. Obviously, there are three possibilities as follows: the Alismales may have been derived from the Ranales; the Ranales may have been derived from the Alismales; both may have been derived from a common ancestry. The hypothetical ancestors must have composed a great Mesozoic complex long extinct. Apparently the earliest angiosperms may be sought somewhere in or among the near relatives of this vast group of related plants perhaps including many orders. According to the fragmentary available fossil record, both dicotyledons and monocotyledons were present during Cretaceous time.

No single living plant or group of plants may be considered to be the ancestor of all or with certainty any other major angiosperm group. This principle is illustrated in Figure 17-12. The primitive features discussed above occur in various members of the Ranales and Alismales and in lower concentration in other orders. However, plants which retain one primitive feature may be advanced in others. For example, in the section *Tasmannia* of *Drimys,* the most primitive known open carpels are associated with a relatively advanced type of ranalian stamen, and in the Himantandraceae primitive stamens are associated with closed carpels. The more primitive living Ranales include remnants of many lines of evolution from an ancestral type which combined the primitive features now retained individually or in associations of two, three, or a few among some of its derivatives. Primitive or specialized features, such as orthotropous or campylotropous ovules, appear here and there among many of the orders of dicotyledons and monocotyledons* in combination with characters pre-

* The following are examples of families in which nonanatropous ovules have been reported.

Dicotyledoneae. *Orthotropous.* THEALES, Cistaceae, or rockrose family (usually). CARYOPHYLLALES, Polygonaceae, or buckwheat family. HAMAMELIDALES, Platanaceae, or sycamore family. MYRICALES, Myricaceae, or wax-myrtle family. CASUARINALES, Casuarinaceae, or beefwood family. *Campylotropous.* PAPAVERALES, Capparidaceae, or caper family. CARYOPHYLLALES, almost the entire order. MYRTALES, some of the Myrtaceae, or myrtle family. The ovules of the Aceraceae (maple family, SAPINDALES) range from orthotropous to anatropous and those of the Leguminosae (pea family, ROSALES) from campylotropous to anatropous (with the Papilionoideae amphitropous). The ovules of the Canellaceae (wild cinnamon family, RANALES), Leitneriaceae (corkwood family, LEITNERIALES), and Julianaceae (*Juliana* family, SAPINDALES) are amphitropous to nearly anatropous.

Monocotyledoneae. *Orthotropous.* LILIALES, Commelinaceae, or spiderwort family; Xyridaceae, or yellow-eyed grass family; Mayacaceae, or bogmoss family; Eriocaulaceae, or pipewort family. *Campylotropous.* ALISMALES, Alismaceae, or water-plantain family.

sumed to be more (or less) advanced, while the ovules of the living Ranales are anatropous (now thought to be possibly the primitive type). Thus, although the greatest concentration of primitive features is in the Ranales, this order is not necessarily ancestral to the others. Either it has diverged less from a preëxisting primitive group of flowering plants or it has not diverged in all its characters.

Evolutionary Series

It is not possible to discern an entire family tree among either the dicotyledons or the monocotyledons. Here and there a twig or a fragment of a large branch may be made out, but these fragments are inadequate to form the framework of a classification system. The chart of relationships (front end leaves) gives no implication of derivation of one order from another existing order. In some instances the members of one order may have been derived actually from an unmodified contemporary member of another, but this is unlikely and impossible to prove. Commonly both were derived from preexisting plants not exactly like any of the living species. In many cases two or more orders of the present constituted a single order in Tertiary or Cretaceous time. Divergent evolution or sometimes convergent evolution in segments of this order have resulted in the present separate orders.

The orders of flowering plants may be likened to coral "islands" in a rising sea of time. Both the coral and the "water" level are rising slowly. The portions of the islands barely protruding above the surface of the sea and visible to the eye are the living orders. Just as it is likely that two or more adjacent similar islands are connected beneath the surface, it is probable that two or more similar flowering plant orders were connected in past geologic time. Just as the connections between islands may be proved by soundings, past links between orders may be established sometimes by fossils. Fossils are soundings of the past; they come from various depths in the sea of time.

Primary Subdivisions of the Dicotyledoneae

The orders, the real basis for the system of classification, may be made largely natural by thrusting aside preconceived notions of evolutionary or phylogenetic lines of development and grouping together the families with much in common, relatively few differences, and often incomplete isolation of their related genera. In the present state of knowledge, this does not permit bringing together large groups of families but only those with apparently clear relationships. The result, for the time being, is small orders. This is not due to a policy of subdividing the large orders of the older systems but to inability at present to carry a synthesis farther.

Since the orders of dicotyledons are too numerous for immediate comprehension by the beginner, larger groups are needed, but currently these must be partly artificial because there is insufficient information to make them natural.

Thus, it is necessary to adopt in milder form the expedient chosen by Linnaeus for his Sexual System. As, in the Greek legend, Theseus used a ball of thread given to him by Ariadne to find his way back through the labyrinth, Linnaeus considered the sexual system to be a ball of thread to guide him through the labyrinth of facts. It is to be hoped that this later resort to some features of an artificial system may serve in the same way.

In Chapter 16 (p. 540) Bessey's emphasis upon the floral cup and Engler and Prantl's upon the perianth were compared by two diagrams representing the dicotyledons. These differed primarily in whether the vertical or the horizontal lines were stronger. In this chapter the two diagrams are superimposed upon each other, the vertical and horizontal lines being given equal weight. If the Amentiferae are slipped out of position and placed to one side, the residue is represented by a rectangle divided into quarters (cf. p. 123). The resulting figure has been rotated one-

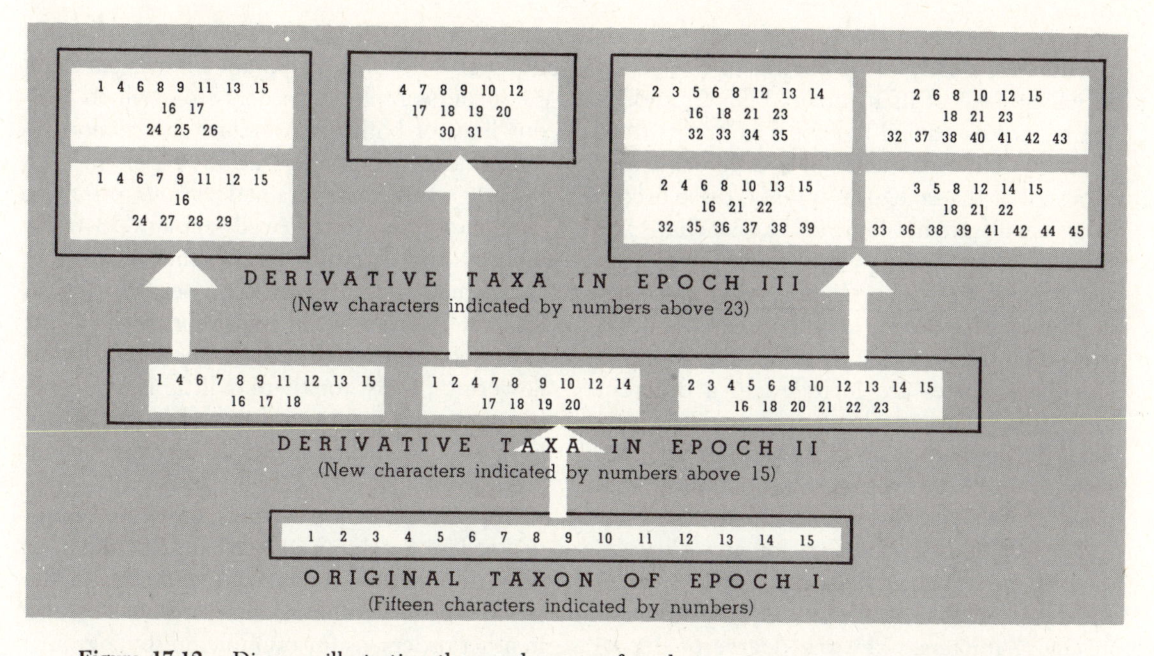

Figure 17-12. Diagram illustrating the usual course of evolution. Sometimes, as with formation of an allotetraploid, a potential new taxon arises suddenly, but ordinarily the taxa of one geological epoch gradually gain and lose characters in succeeding epochs. Some characters (as 1–15 above) are retained by some derivative taxa but not others, and some may appear through several epochs. The taxa of any one epoch are rarely derived from each other; they are descendants from a common ancestor of an earlier epoch. Construction of evolutionary trees featuring the taxa of a single epoch as derived from each other is not in harmony with this concept of evolution. Relationship of one taxon to another may be indicated by numerous characters in common as opposed to few differences. However the origin of existing higher taxa is lost in the past, except in the few instances in which well-preserved series of fossils are available.

quarter turn counterclockwise, and for the following discussion it must be viewed with the left margin of the page taken as the bottom. The quadrant at the upper right is reserved for hypogynous plants with separate petals (or none); the upper left for hypogynous plants with coalescent petals; the lower right for perigynous and epigynous plants with separate petals (or none); the lower left for epigynous plants with coalescent petals.

The names adopted for three of the groups of dicotyledons are classical ones used by DeCandolle and Bentham and Hooker, as follows: upper right, Thalamiflorae (receptacle flowers); upper left, Corolliflorae (corolla flowers); lower right, (Calyciflorae (cup flowers). A new name, Ovariflorae (ovary flowers), is necessary for the group at the upper left.* These four constitute the major groups of Dicotyledoneae.

The Amentiferae were an almost wholly artificial group, but they have been eliminated. The four remaining are more natural than artificial. The Amentiferae have been absorbed into the Thalamiflorae.

In some instances portions of an order could be represented as extending across the arbitrary line separating two or more of the major groups of dicotyledons. For example, the Rosales include some hypogynous members, but the bulk of the order is composed of perigy-

* These convenient but partly artificial "groups" are not taxa and their names are not proposed for any official status.

nous or epigynous plants. Such an order is considered usually as belonging to the group in which most of its members fall. In numerous cases exceptional members of orders duplicate the chief character of a different major group. For the sake of simplicity, these exceptions usually are not indicated in the chart, but they are noted in the text in the discussion of the order and in the preliminary lists for elimination of exceptional character combinations preceding the keys to the orders. Some orders like the Cactales are included in a different group than their relationship to presence of a floral cup and adnation of the petals would indicate. This is to avoid placing them far from their nearest relatives.

The position in the chart for any dicotyledonous or monocotyledonous order corresponds roughly to its degree of specialization in any characters, often the degree of departure from the five floral features characteristic of the primitive Ranales and Alismales. This is expressed in rough proportion by the chart distance of each order from the Ranales or Alismales. It must be emphasized that (1) specialization in other characters is potentially and often actually at least as significant as departure from the "five characters" and that (2) there may be many exceptions to the relative primitiveness of any two opposed characters, including correspondence with or departure from any of the five. These usually are rough indicators, but they must not be taken as the only basis for classification of flowering plants. The result would be an artificial classification system.

The Monocotyledoneae

The broad outlines of classification of the monocotyledons proposed by Bessey are in need of considerable modification. His proposed lines of evolution do not gain much support from further study, and the development of the orders has been far more complex than he supposed. The monocotyledonous orders probably are surviving end points of many lines of evolution from an extinct ancestral Mesozoic complex of orders from which various lines of both dicotyledons and monocotyledons arose. Probably various orders in both major groups have retained various primitive characters that fitted well enough in some niche or niches either to give their possessors an advantage or at least not to cause them to lose out. Certainly the separation of monocots and dicots occurred early in the evolution of flowering plants, for there is little if any overlap of the groups, even though few of the distinguishing features occur exclusively in one group.

The Alismales are considered to be one of the orders retaining the greatest number of primitive features but not as ancestral to the other orders. Primitive characters occur also in other orders, as for example, in some groups of the inclusive Liliales and especially in the Palmales, which according to Moore include about as many such features as the woody Ranales. The Alismales of Bessey were a catch-all, and in the present work the Pandanales, Naiadales, Juncaginales, and Triuridales are considered to be separate orders.

The evolution of the monocotyledons has been complex, at least as varied, irregular, and difficult to work out as that of the dicotyledons. No clear lines of possible segregation into major groupings have emerged, unless a partly arbitrary method like that for the dicots is used. In the monocots there is little value in this because a number of small groups, some of a single small order, would be necessary, and they would not contribute much toward clarity.

According to Brown (see discussion on p. 567) the development of the monocotyledon type of nectary, consisting of septal glands (between the members of the cycle of coalescent carpels), may be traced in the Alismales. In a species of arrowhead (*Sagittaria sagittaefolia*), a plant of the Alismaceae, glands are produced on the receptacle among the petals, stamens, and numerous separate carpels. In the waterplantain (*Alisma Plantago*) the glands are restricted to positions between the members of the single ring (cycle) of separate carpels.

During the course of years Hutchinson and others have suggested several lines of development from the hypogynous Liliales to epigynous plants; and consequently, the bulk of the Iridales are joined with the Liliales, forming a comprehensive order including many lines of internal development. Doubtless epigyny has arisen a number of times in hypogynous members of the order.

Relationships of the System to the Engler and Prantl and Bessey Systems

Data have been drawn from many authors. The features derived from the two leading systems appear with modification as follows.

THE ENGLER AND PRANTL SYSTEM

1. The Amentiferae are distributed into the Thalamiflorae (but not Calyciflorae). They are considered to be not primitive but advanced derivatives of the stocks which gave rise to various Thalamiflorae and primitive Calyciflorae, some or most of the orders being of uncertain derivation and relationship and not necessarily closely related to each other.
2. Division of the dicotyledons into choripetalous (or apetalous) vs. sympetalous groups. This is retained but deëmphasized and considered partly artificial. Distinction of major groups according to choripetaly and apetaly is considered almost wholly artificial.
3. Definition of many orders of dicotyledons and some of monocotyledons, these being often though not necessarily well-defined by Engler and Prantl, whose orders were more often natural than Bessey's.

THE BESSEY SYSTEM

1. Consideration of the Ranales and Alismales as including members with relatively primitive features.
2. Recognition of gradual departures from ranalean and alismalean floral characters. These, however, are organized and somewhat augmented, and departure from them is correlated often with specialization resulting in efficient insect or other pollination.
3. Division of the dicotyledons into hypogynous vs. perigynous or epigynous groups. This is retained but deëmphasized and considered partly artificial.
4. Proposal of a chart of relationships of angiosperm orders, but *not* of a chart depicting evolution from order to order. *This is not a chart of purported phylogeny.*
5. The general classification of the monocotyledons is considerably modified and with less emphasis upon superior and inferior ovaries and with realignment of several orders.

Relationship of the System to Recent Ones

This subject has appeared in Chapter 16. Much has been brought to attention by the recent systems of Takhtajan, Cronquist, Stebbins, and Thorne; and various ideas, interpretations, and facts have been drawn or modified from these authors. This has not yielded any great modification of the system of classification proposed in edition 1, but it has modified the interpretations of various orders and families. The greatest impact on the system is from the background material of Stebbins and from the contributions of a host of authors, including those mentioned, to knowledge of the characters, especially minute or obscure ones, of various orders and families.

Introduction to Vascular Plants in General

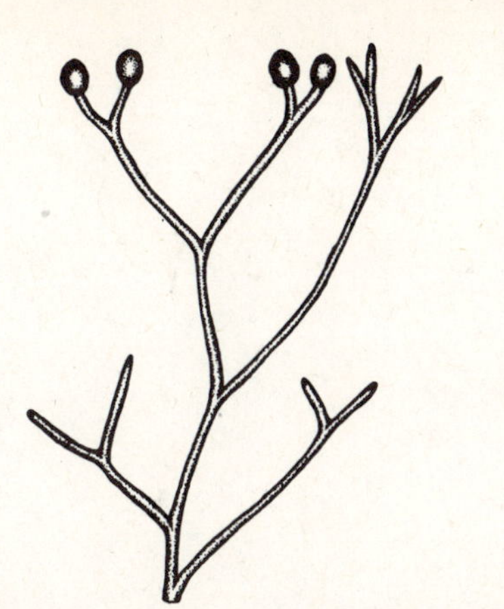

The organization of this book is designed to facilitate its use as an elementary textbook on either the Angiospermae or flowering plants alone or as one on all the vascular plants. Its use as an elementary or advanced reference book has been common, as well. The first seventeen chapters are devoted to the angiosperms, the next seven relate to the other vascular plants (gymnosperms, ferns, horsetails, *Psilotum,* and the club mosses). Some of the material between Chapters 17 and 18 is basic to understanding what little is known of the origin of the flowering plants. This has been discussed in Chapter 17 with references to later chapters on the gymnosperms. However, so little is known about the subject that no direct connection can be made out as more than a vague possible derivation from one group of extinct gymnosperms, and otherwise the flowering plants barely enter into a discussion of the vascular plants in general and their origin. The material in the following chapter deals with the earliest known (late Silurian and up to middle Devonian) Division Rhyniophyta, but the connections between the flowering plants and the early vascular plants are nebulous. On the other hand, most other vascular plants are tied in fairly well, as the result of new recent research, to the Rhyniophyta. For this reason the following material appears at this point, instead of as part of Chapter 1, where it would be remote from the other vascular plants and with little bearing on the seventeen chapters on flowering plants immediately following. This facilitates study of the flowering plants and the other vascular plants as two units, if this is desired. The necessary introduction to evolution and (briefly) to heredity has been covered in Chapter 13.

The First Vascular Plants

As shown in Chapters 13 to 17, plants and animals always are changing, and an ultimate classification of organisms of both past and present would have to be divided into many systems, each representing a thin section of geologic time. The plants or animals of a segment of time include some similar to those of the preceding unit and the next, but nearly all are in flux, and their antecedents and descendants differ from slightly to greatly from them. If all the organisms of geologic time were known, there would be few breaks in the continua from taxon to taxon along lines running in innumerable directions or changing course frequently. Thus, combining all the organisms of the past and present into one system is not an attainable goal. They could not be assigned to taxa, because there would be no lines of demarcation.

Sufficient information for construction of a system of classification exists for only one point in time—the present. Relatively few taxa from the past are known, and large numbers, so far, are represented by only fragments of individuals. Often these are not connected, and the parts cannot be assembled accurately to represent the whole organism. Thus, because the known record is sketchy, plant classification tends to be divided into only two temporal units—*plants of the present,* which are fairly well known, and *plants of the past,* which are not, though enough is known to make out some of the outlines of probable origin and relationships of the major groups, like divisions and classes.

Even though a separate classification system cannot be set up for each unit of geologic time, every effort is made to associate each organism represented by a fossil with its place in the time series. Although often the sample is meager and perhaps not wholly representative, something of the major aspects of the flora of each major segment of time can be made out. For the purpose of classification of plants, the primary interest in fossils is for the evidence they provide for the past evolutionary development of which each extinct plant is a part. Classification depends upon knowledge of the relationships of plants, those of common origin being placed together if this relationship can be determined. If fossils are found, they may provide the key to origin and relationships of living and extinct plants.

The Trend of Evolution in Vascular Plants

The earliest vascular plants were also the earliest true land plants. Some algae, fungi, mosses, and liverworts are able to live in very dry places, but this is accomplished by these small plants through cellular structural or physiological features enabling them to withstand drying. For example some species of *Riccia*, a liverwort most species of which live on mud or in water, occur on the floor of the chaparral in southern California, being dry and brown all summer but returning to the normal green state within minutes to an hour or two if the plants are placed in water. A 1- to few-celled green alga, *Protococcus*, lives in a semidry state on rocks, fences, trees, or buildings during the dry season in almost any region, but the cells absorb water and grow rapidly with the first rains. Lichens (combinations of algae and fungi) live in the same way, often becoming dry and brown to black during the dry season, but resuming their normal colors (greens, reds, yellows, etc.) after a rain. This method of coping with the dryness of land has limitations, and a plant using it never could attain much size or rise above its competitors into the light. Furthermore, the reproduction of all these plants requires water; the antherozoid, or male gamete cell, must swim to the egg. Thus, at times the plants must be bathed in moisture. Only the vascular plants have overcome the inherent difficulties of living on land that stood in the way of attaining stature. This has been accomplished by deep absorbing and anchoring *roots* growing into the moist parts of the soil beneath the surface; elaboration of *stems* able to stand erect and support the rest of the plant and to conduct water to the parts above and manufactured food to the nongreen parts; development of a photosynthetic area, usually the *leaves*, exposing much surface to the light but restricted in water loss to the atmosphere; and alteration of the mechanisms of *reproduction* in ways eliminating the necessity for an antherozoid to swim through water, at least outside the plant.

Roots were lacking in the first vascular plants, which were only a few inches high. Probably they grew in marshes, and this is a reason for their preservation as fossils. Absorption of water and salts was through underground parts of the branching axis, or stem, which performed all functions, being doubtless photosynthetic above, there being no leaves. The need for anchorage was minimal. Later vascular plants had roots extending deep into the soil and providing (1) specialized absorbing and conducting cells and (2) effective anchorage.

Stems hold the leaves and reproductive parts aloft, where a maximum of photosynthetic surface and suitable contact with the carbon dioxide of the air may be developed. This requires strong and largely rigid construction, with xylem (wood) and other cells specialized for support. The water from the soil must be transported from the roots to all parts of the plant and particularly to its place of use in photosynthesis. The manufactured food must be carried to places of storage or use. Commonly the excess is stored in the soft tissues of the stem and roots until it is needed, often during an unfavorable season, according to the climate of the region. Xylem facilitates transportation of water and salts upward, and phloem conducts food downward.

Leaves are primarily photosynthetic organs,

and their raw materials are supplied by (1) the xylem of the veins bringing water (as well as salts) and (2) the stomata or openings in the protective outermost layer of cells bringing carbon dioxide from the air into an aërated tissue next to the primary photosynthetic tissue. The manufactured food is transported away by the phloem of the veins. The leaves occur along the smaller branches, and they are elevated by the stem into a position where they are exposed to the air. There they provide a much greater amount of surface than would be available in a sheet of cells on the ground, and there is access to both light and carbon dioxide. Elevation also provides protection from some herbivores and from shading by competitors.

Leaves are derivatives of the stem and its branches through (1) elaboration, including flattening, of single terminal branchlets, though this occurred in only the most primitive vascular plants, (2) development by fusion of the branchlets of a seemingly lateral system, "overtopped" by the continuation of another axis as the main one, the lesser system then becoming flattened (planation) and the spaces between branchlets filled in with soft tissues (webbing), and (3) development of small leaves from enations (outgrowths) along the sides of the stems, a method of leaf formation occurring in only one line of development, that leading to the club mosses.

Reproduction in the vascular plants involves the alternate occurrence of two unlike generations of the plant. One, the **sporophyte** or *spore-producing generation* is well known; it is the large plant. The other, the **gametophyte** or *gamete-producing generation,* is known to few, because it is inconspicuous and simple in its structure. This point is fundamental, because only the sporophyte could have developed into a plant capable of coping with the many problems inherent in living with the complexities of the environment occurring in air and on land as opposed to the simplicity of adapting to the ecology of living in water or in contact with it in wet places.

The direction of evolution has been dictated by the nature of the plants of the two generations, and these are parts of the following life cycle, illustrated by a fern, a plant in which both generations are independent. The fern sporophyte is dominant but combined with a self-sustaining gametophyte generation, much like that dominant in liverworts and hornworts (bryophytes) and corresponding to the usually dominant generation of freshwater green algae. The fern life cycle is given in detail and illustrated in Chapter 21, pp. 657–659, Figures 21-1 through 21-4, the gametophyte (from beneath) in Figure 21-4-4. As in the liverworts, the fern gametophyte is only a few cells thick, and it forms a flat sheet on the ground from which it absorbs water and salts directly. It must be not only in contact with moist soil (at least seasonally) but also covered, at least during rains or mists or when there is dew, by a film of water through which the antherozoids (male gametes) must swim to the eggs (female gametes). Pairs of gametes join each other, effecting fertilization. All cells of the gametophyte generation are haploid, i.e., with **n** chromosomes, and the union of the gametes marks the change to the diploid (**2n**) sporophyte generation. Thus, the gametophyte, or gamete-producing, generation is haploid, and the sporophyte, or spore-producing generation, is diploid. The first diploid cell, the zygote, or fertilized egg, of the fern divides, forming the first two cells of the embryo, then divides again into four. One develops into a *foot* absorbing water and food from the female parental plant for a short time until independence is possible. The others form a *root,* a *stem,* and a *leaf.* The stem, the last to develop, then grows into a rhizome which produces other roots, stems, and leaves. At maturity, sporangia on the backs of the leaves include a tissue which undergoes the reduction divisions or meiosis in each cell, producing four

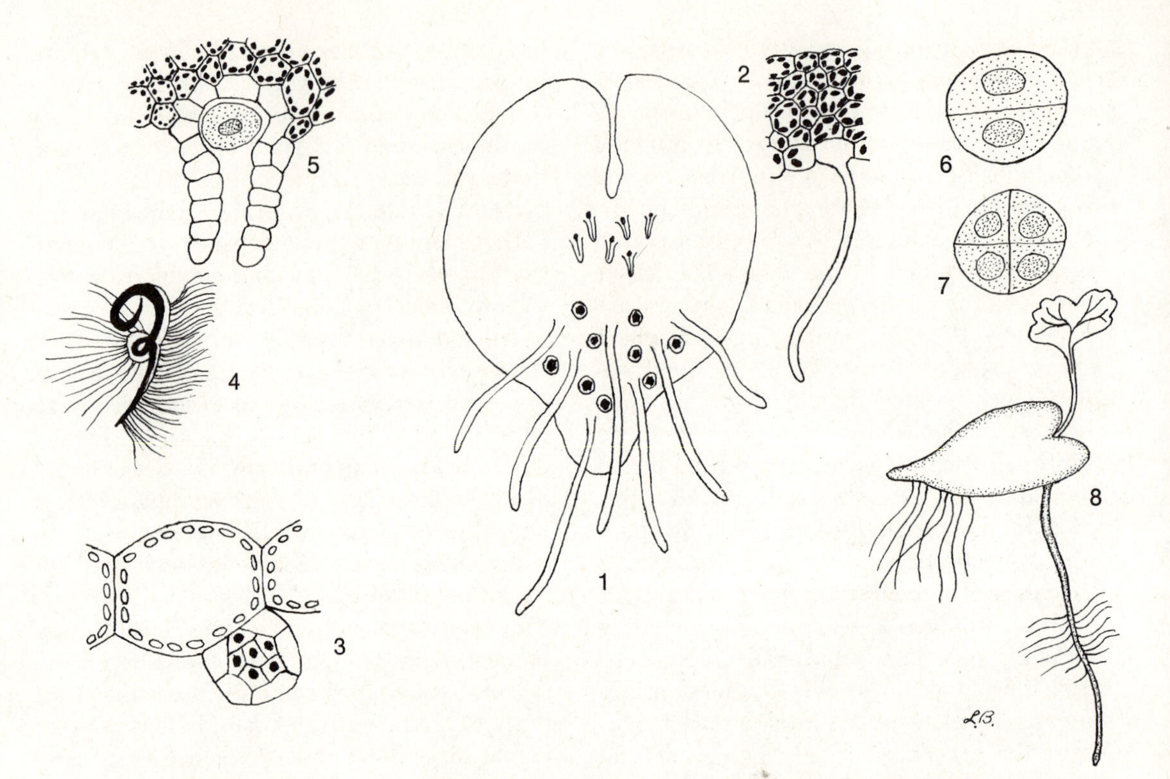

The Life History of a Fern. The sporophyte (**2n** chromosomes), or large fern plant, produces spores by reduction divisions (meiosis), and these are the first cells of the gametophyte generation (**n** chromosomes), alternating with the sporophyte. Each spore grows into a gametophyte plant, or prothallium, consisting of a flat sheet a few cells thick. The prothallium produces sex organs on its under surface, and these develop male and female gamete cells (each with **n** chromosomes, as in all the cells of the gametophyte generation). Joining of the gametes produces the embryo, the first cell of the sporophyte generation. Here, only some stages are shown; the life history of a fern is presented fully in Chapter 21. **1** Prothallium from beneath (characteristically cordate and the notch apical) showing the antheridia, or male sex organs (posterior) and the archegonia, or female organs (anterior) and rhizoids, or absorbing organs, **2** portion of a cross section of the prothallium, showing the simplicity of vegetative structure, the cells being little differentiated and nearly all or all photosynthetic, some cells of the lower surface layer bearing rhizoids (extensions increasing the absorbing surface in contact with the soil; these similar to the root hairs of flowering plants), **3** an antheridium on the lower surface of the prothallium, consisting of a jacket and a few enclosed cells, each of which becomes an antherozoid or male gamete cell, **4** an antherozoid, formed largely from a nucleus (elongate) and many flagella, which lash back and forth propelling the male gamete through the film of water on the plant to the female organ or archegonium, **5** archegonium, a flask-shaped body of cells in which the egg, or female gamete cell, is contained, **6** embryo after union of the antherozoid and egg, doubling the chromosome number, this becoming **2n**, and the embryo being of the sporophyte generation, **7** four-celled embryo, the cells giving rise to a foot (absorbing organ bringing food, salts, and water from the parental gametophyte), a primary root, a primary leaf, and a stem (developing much later than the other three organs and giving rise to the ultimate mature sporophyte or large fern plant), **8** prothallium with the young sporophyte attached, the foot embedded in the gametophyte, the primary root and leaf visible, and the stem undeveloped.

haploid (**n**) spores. The spore develops into a gametophyte, completing the cycle, summarized as follows:

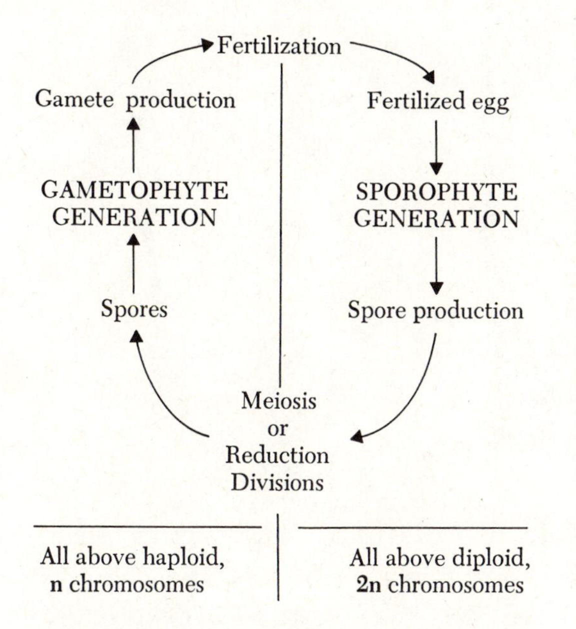

Under water the environment is relatively stable; fluctuation of any factor is within narrow limits, and some factors do not change. Water is always present, and the temperature fluctuates slowly and within relatively few degrees. There is no need for supporting structures or tissues. Chemical substances vary, but only slightly in the same body of water. Reproduction and development of the new generation present few problems. For example, the freshwater (as well as the marine) algae live, in almost all cases, under water, and the antherozoid can reach the egg through swimming by lashing slender threads of cytoplasm (flagella) or by drifting if a method of locomotion is lacking. Out of water this mechanism is preserved only with difficulty, and in the more advanced plants it has been eliminated.

In the aquatic environment, either the gametophyte or the sporophyte is satisfactory as the major generation, and both relationships occur. However, in fresh water the green algae, the nearest simpler relatives of both the bryophytes and the vascular plants, perhaps by chance the gametophyte is the more elaborate generation in all but a few genera. Commonly the sporophyte consists of a single cell, the fertilized egg or zygote, and the reduction divisions or meiosis occur in it before the development of derivative cells into haploid swimming spores that form new gametophytes. On the other hand, diploids have the capacity of storing innumerable unexpressed or seldom expressed genes, these being hidden as recessives, many of which become rare because of elimination of the few individuals in which they are homozygous. Thus, the diploid generation may be adaptable to a variety of irregularities or changes of the environment in which it lives and to many markedly different environments.

In wet places on land, the haploid generation of bryophytes serves well as the dominant, and its chlorophyll contributes the food for the simple sporophyte as well. The gametophyte absorbs the water and salts for both. The bryophyte gametophyte is the most elaborate known to have occurred in the history of the world. The haploid plant of a liverwort is a sheet of cells flat on the ground but sometimes with differentiation of internal photosynthetic, aërated, and storage tissues (as in *Marchantia*). It is more complex than the gametophyte of a fern which is only a few undifferentiated cells in thickness. On the lower surface, rhizoids (extensions like roothairs) and sometimes scales of cells reach minute distances into the wet soil, thus increasing the absorbing surface but not tapping the soil below the surface. In mosses there is a slender stemlike axis bearing leaflike scales, but there is no effective conducting tissue, and size is limited.

Invasion of the land, except for moist spots, required a reversal of the dominant plant generation, from gametophyte to sporophyte. Only plants with a dominant sporophyte generation could have invaded the land, because there a haploid plant is always at a disadvan-

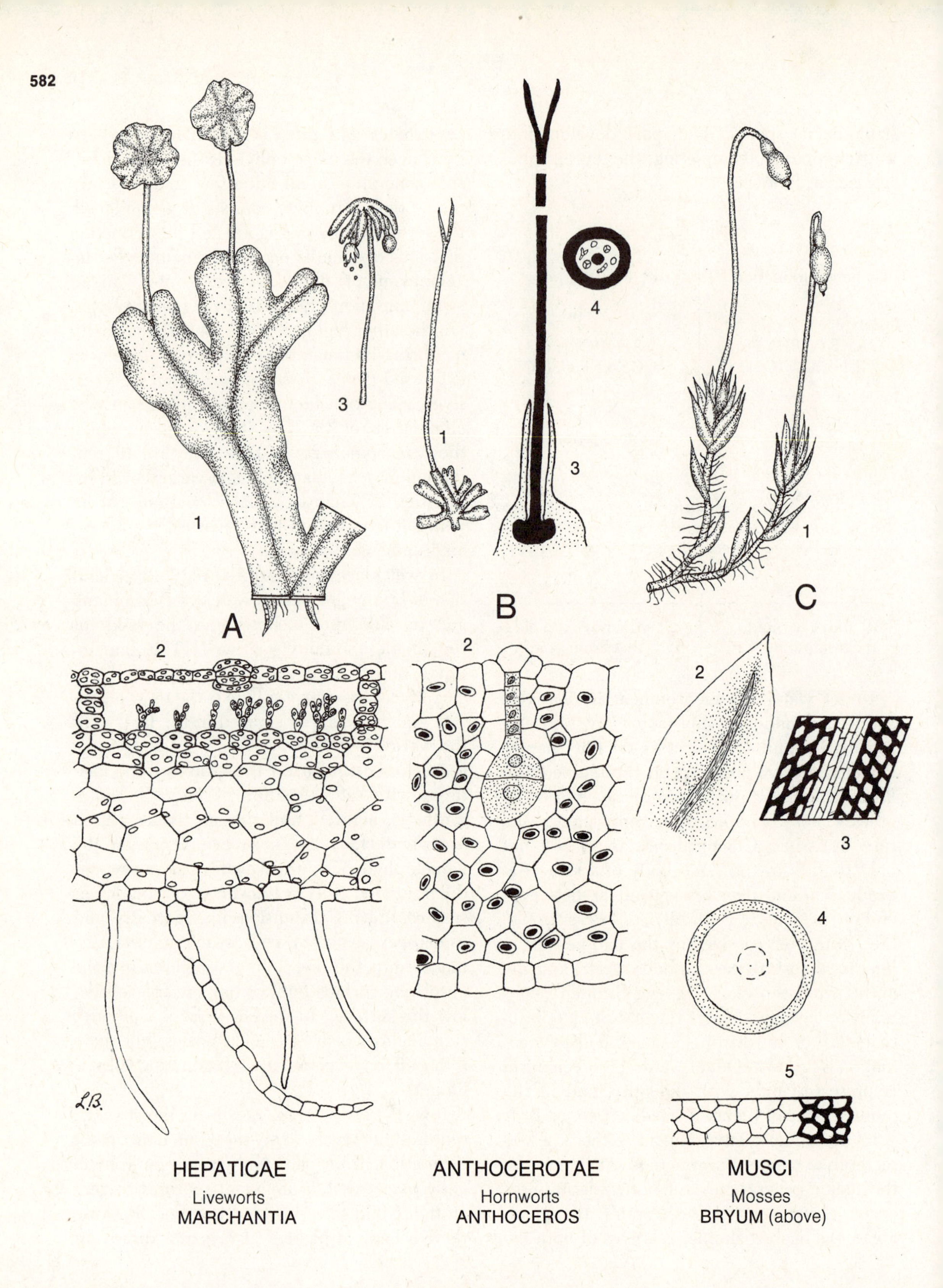
1
2
3
4
5
A
B
C
L.B.
HEPATICAE
Liveworts
MARCHANTIA
ANTHOCEROTAE
Hornworts
ANTHOCEROS
MUSCI
Mosses
BRYUM (above)

tage, except in limited specially favorable areas. The limitation is due to the chromosome number (n), which inhibits variability and therefore the adaptability of the gametophyte generation. Haploids have only one gene governing each feature, and it is always expressed. In the gametophyte there are no dominants or recessives; all genes are, in effect, dominant, and the plant lives or dies according to the single set of genetic factors. There is no mechanism for preservation of genes not immediately beneficial but potentially desirable under changed conditions of the environment. The gametophyte is vulnerable to any alteration of the ecological balance, and an essentially stable set of conditions is necessary.

The sporophyte has made the adjustments necessary to its vegetative success on land, but it could not reproduce and live there without the gametophyte link in the chain of the life cycle. The fate of the invasion hinged on (1) emancipation of the gametophyte from living in a moist place; (2) a new method of transportation of the male gamete to the egg; (3) formation of the egg in a place protected from herbivores and from drying; and (4) continuation of the development and safety of the embryo in the same place with accommodation to its increase in size and its need to wait for favorable conditions before growth as an independent sporophyte plant. These needs have been met by modifications of the stages leading to spore production in the sporophyte and by major changes in the gametophyte generation, which has been reduced in complexity and transformed into minute male plants (microgametophytes) and small female plants (megagametophytes). (See pp. 601–606, 46–54.)

In the most highly developed vascular plants, the seed plants, neither gametophyte has chlorophyll or contact with the soil, and

---

The bryophytes, or Bryophyta, the three classes of the division. All are characterized by a simple gametophyte generation, that of *Marchantia* (**A**) being nevertheless the most complex living gametophyte on Earth. In the bryophytes the sporophyte is simpler than the gametophyte, and it is dependent on the gametophyte for water, salts, and food (or it may manufacture some of its own sugar). The most complex sporophyte of the three is that of *Anthoceros*. **A** *Marchantia*, **1** dichotomously forking thallus with discoid antheridial and umbrellalike archegonia branches, **2** cross section of part of the thallus, showing an air chamber near the surface and synthetic cactuslike branching filaments of cells arising from the floor of the chamber, a (primarily) storage layer of cells beneath, and rhizoids and flat scales of cells from the lower surface layer of cells, these increasing the absorbing surface next to the soil, **3** an archegonial branch with two sporophytes dangling from the old archegonia on the under sides of the rays, one open and shedding spores; **B** *Anthoceros*, **1** a branching thallus with a sporophyte attached, its base surrounded by the sheath developed from the old archegonium, the apex of the sporophyte splitting and releasing spores, the base with a meristematic tissue which adds cells of an undetermined number through the period of favorable growth, causing the sporophyte to elongate for an indefinite period, **2** cross section of a portion of the thallus, showing its simple structure and the single chloroplast in each cell and a pyrenoid (food storage body) in each chloroplast (as in the Ulotrichaceae, a family of green algae), a single archegonium, this embedded in the gametophyte, **3** longitudinal section of a sporophyte and the adjacent gametophyte tissue, the sporophyte black for emphasis, **4** cross section, showing the spores (some seemingly in threes but actually fours as formed by the reduction divisions, others separate) and the peculiar multicellular elaters which change shape with moisture changes and aid in ejecting the spores from the sporophyte; **C** *Bryum* (above) and an unidentified moss (detail below), **1** gametophyte with a branching stemlike axis and leaflike appendages (true stems and leaves are sporophyte structures) and with sporophytes arising from the old archegonia in each cluster of leaves, **2** leaflike appendage, showing the "midrib" of more elongate cells, **3** elongate cells of the midrib and the beautifully arranged adjacent cells, these with thickened walls providing support, **4** diagrammatic cross section of the "stem" indicating the area of thicker-walled supporting cells on the periphery, dotted lines indicating a central conducting tissue occurring in the stems of some mosses, though not this one (this shown by radioactive tracers to conduct water and salts upward, though movement occurs also through the film of water on the surface of the plant), **5** outer tissues (right) and inner tissues of the stem.

both are dependent upon the sporophyte for food, water, and salts. Thus, removal of the need for living in wet places has been accomplished by entering a state of virtual symbiosis with the independent sporophyte, for which, in turn, sexual reproduction is provided.

The microgametophyte, known as a pollen grain, has become so small as to be transportable by wind, insects, some birds, bats, or other agencies. It develops within the coat of the spore that began the gametophyte generation, and there it is protected from drying. In the gymnosperms (Figure 18-5, p. 608) ordinarily it consists of four or five cells, in the angiosperms (Figure 5-4, p. 62) of only two. The pollen grains, carried to the ovules of the gymnosperms or to the stigmas of the flowering plants, germinate, and each one develops an elongate pollen tube. In the gymnosperms the tube digests its way to the egg; in the angiosperms it grows through the stigma, style, and ovary to the egg. The male gamete has no power of locomotion; it is carried by the growth of the pollen tube. Flagella and a film of water are not needed. A transitional stage occurs in the living cycads and in *Ginkgo;* the antherozoids bear small flagella or cilia, developed in the same position as the longer flagella of a fern antherozoid. However, swimming is confined to the fluid in the pollen tube.

For its part, the sporophyte generation has become modified, and sporangia do not occur on the vegetative leaves. The sporangia ultimately giving rise to the microgametophytes through microspores are microsporangia, or pollen sacs. They are produced in the gymnosperms in pollen cones (Figure 18-3, p. 605), and in the flowering plants they are the pollen sacs of the stamens. They are parts of the microsporophylls (stamens in the flowering plants).

The megagametophyte is enclosed in an ovule, which includes the surrounding megasporangium (nucellus), one or two integuments or investing sheaths also of sporophyte tissue, and in the flowering plants the ovary. Thus, the sporophyte provides ample protection from injury, herbivores, and drying, through a sheltering megasporophyll (part of the cone in the gymnosperms; the carpel in the flowering plants) and its outgrowths [the integument(s)] and the megasporangium. There is a passageway for pollen or the pollen tube to the tip of the megasporangium and a means of access to the egg.

The structures protecting the egg continue to enclose the seed, as it develops from the ovule after fertilization. The integument(s) ordinarily enlarge and harden, forming a still more effective seed coat, and either the sporangium (nucellus) or in the flowering plants a special tissue, the endosperm, or both may become a great food reservoir around or near the developing, then resting, embryo. Thus, the embryo may go through a waiting stage until conditions are right for germination and beginning independent growth in the soil.

In summary, the history of vascular plants and their invasion of the land is an elaboration of the sporophyte as the dominant vegetative generation and diminution of the gametophyte and its ultimate restriction to its basic function of producing gametes. This has required elimination of the relationship of the antherozoid to water for transportation to the egg and substitution of a means of carrying the microgametophyte through air to the vicinity of the egg. Further movement of the gamete results from growth of the pollen tube containing it, and flagella and the power of locomotion have been lost. Special development of the structures of the sporophyte, the megasporophyll and megasporangium, protect the megagametophyte, and later the embryo from drying and injury and enable the new embryonic plant to develop during a favorable season.

The Earliest Vascular Plants

Great strides in understanding the late Silurian and early Devonian vascular plants and their connections with later taxa have been in large part due to the excellent research of Harlan P. Banks, Henry N. Andrews, Charles B. Beck, Andrew E. Kasper, and their associates and former students. The interpretation of these plants presented in this book rests primarily upon their studies. The slight differences are in the rank accorded some taxa and in nomenclature. As stated by Banks (Taxon 24: 411. 1975), "Whether these groups [classified as Rhyniophytina, Trimerophytina, and Zosterophyllophytina] be accepted as Divisions, as Subdivisions of Tracheophyta, as orders, or as families depends on the opinion of the observer." Obviously the known fossils are but a tiny sample of Devonian plants, and classification is difficult.

Division RHYNIOPHYTA*

Roots none; leaves none; stems naked to papillate or prickly; branching dichotomous or partly trichotomous, spiral, or unequal and the branches simulating a continuous main stem; sporangia terminal and ovoid to ellipsoid or narrowly ellipsoid, dehiscent usually lengthwise, and erect or drooping *or* lateral and somewhat stalked, more or less globose to reniform, and dehiscent distally along the margins; xylem strand slender or large, endarch or exarch; tracheids scalariform when well defined.

A single vascular plant, *Cooksonia,* is known from the close of Silurian time, about 405 million years ago. Fossils have been found in Wales, and others from various parts of the world may be of the same genus. *Cooksonia* was in the Class Rhyniopsida, from which nearly all the living and most of the Carboniferous vascular plants were derived. Soon afterward (geologically), at the beginning of Devonian time, another group of genera appeared, the best-known being *Zosterophyllum,* initiating the line of development leading to the club mosses. It was a member of the class Zosterophyllopsida, also in the Rhyniophyta. Despite divergence, there were so many characters in common that an underlying remote relationship is indicated, even though each class represents the beginning of a major line of evolution. Although a common ancestor is to be sought in Silurian rocks, none is known, and solid facts upon which to base a classification are few.

Class RHYNIOPSIDA

Stems naked or with papillae or prickles, branching dichotomously or the main axis branching dichotomously and spirally and the lateral axes di- or trichotomously; fertile branches either merely some of the terminal ones on the forking axis and with narrowly ovoid terminal sporangia or much branched and bearing a mass of terminal, paired, erect or drooping, ellipsoid (or narrowly so) sporangia, the sporangia dehiscent usually lengthwise; xylem strand slender or stout, centrarch, the tracheids scalariform.

Order RHYNIALES; Family RHYNIACEAE

The genera discussed below are those clearly of the Rhyniaceae. Others probably could be included, but those proposed for the family are based on incomplete specimens or are of debatable relationship.

Cooksonia. This is the simplest and oldest known vascular plant, as indicated above. It was only a few inches high, and there were

* A difficulty exists in choice of a name for the Division, other authors having considered the highest level of classification of the Devonian vascular plants as subdivision—under an inclusive Division Tracheophyta for all the vascular plants. The logical choice would be Psilophyta, but this is the name for a division of living plants, based nomenclaturally on *Psilotum,* modified into the normal ending for a division. The fossil and the living groups have been classified together, but this procedure is not accepted. Furthermore, the origin and classification of the Psilotaceae are the center of a controversy (pp. 679, 686). For the Devonian plants, Rhyniophyta is chosen to represent *Rhynia, Psilophyton, Zosterophyllum,* and the relatives of each. It is based on the Class Rhyniopsida, Order Rhyniales, Family Rhyniaceae, and Genus *Rhynia,* long the best-known plant in the Division, and, despite recent studies, as well understood as *Psilophyton.*

neither leaves nor roots, but only a slender forking axis, or stem. Some of the aërial branches were sterile (nonreproductive) and probably photosynthetic; others ended in more or less globelike or slightly egg-shaped sporangia, or spore cases, still containing the typical vascular plant spores. Little more is known about the plant, but presumably the details of structure were similar to those of the related genus *Rhynia,* of about 375 million years ago during the early part of Devonian time. *Rhynia,* along with two other genera of or related to the Rhyniophyta, *Horneophyton* and *Asteroxylon,* was preserved in a silicified form in the Rhynie Chert of Scotland. All these genera are well known from clearly preserved fossils in which the microscopic cellular structure can be studied in the patterns of cell walls in the rocks.

Rhynia. The sporophyte consisted of little besides a slender forked axis a few inches long. This was not differentiated into stem and root. The erect aërial portion of the axis was without leaves, and it branched dichotomously. The internal tissues of the entire axis were arranged essentially as in the roots of modern plants, for the xylem and phloem formed a solid cylinder (protostele) with the phloem on the outside. This protostele was of a very primitive type, with the first-maturing xylem at the center (technically, with the protoxylem centrarch). The sporangia were relatively large, thick-walled, terminal structures (i.e., each small branch of the axis ended in a sporangium) with many spores. The sporangia opened irregularly or by a simple split and there was no annulus (p. 659) or other surface structure. Figs. 21-2, 3, 6.

From all indications *Rhynia* was primitive, and it is possible to follow step by step through a series, first of fossils and then of living plants, from *Rhynia* to any of the major groups of more complicated vascular plants, except the line of development culminating in the club mosses. Because the ancestors of all groups disappeared more than 300 million years ago, the record is imperfect, and there are many gaps between groups of organisms, both living and fossil. However, during the

A list of the taxa of Silurian and early to middle Devonian. Names in parentheses are among the principal differing ones used by Banks. The system employed by Bierhorst (*Morphology of Vascular Plants.* 1971. Macmillan Co., New York) is similar but slightly different in levels of rank. The classification presented here is much like that of Kasper, Andrews, and Forbes (American Journal of Botany 61:357. 1974).

Division	RHYNIOPHYTA		
(*Subdivisions*)	(Rhyniophytina)	(Trimerophytina)	(Zosterophyllophytina)
Classes	RHYNIOPSIDA		ZOSTEROPHYLLOPSIDA
Orders	RHYNIALES	PSILOPHYTALES (Trimerophytales)	ZOSTEROPHYLLALES
Families	RHYNIACEAE	PSILOPHYTACEAE	ZOSTEROPHYLLACEAE
Some genera	*Cooksonia* *Rhynia* Horneophyton ?	*Psilophyton* *Trimerophyton* *Pertica*	*Zosterophyllum* *Sawdonia* *Gosslingia ?* *Crenaticaulis ?*

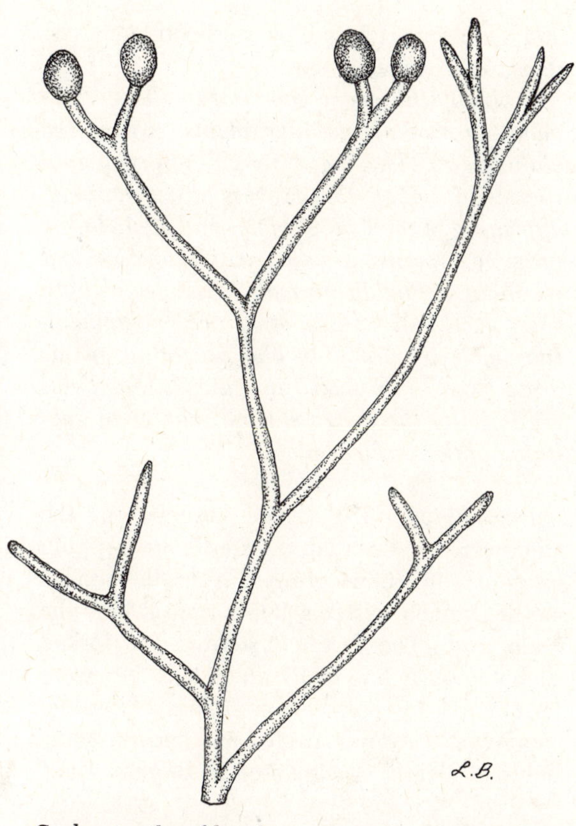

Cooksonia, the oldest known vascular plant; end of Silurian, about 405 million years ago. The stem was a forking, cylindroidal axis a few inches long; there were no leaves or roots; the gametophytes are not known. The terminal branchlets were either only photosynthetic (vegetative or "sterile") or terminated by sporangia (reproductive or "fertile").

last twenty years a number of significant organisms have come to light, and knowledge of others has increased. Thus, some major gaps have been filled or partly so, and plausible outlines of interrelationships of the major groups of vascular plants are emerging. The ferns, gymnosperms (seed ferns, cone-bearing trees, cycads), and flowering plants appear to have been derived from the Psilophytales (Trimerophytyales), an offshoot of the Rhyniales that appeared about 390 million years ago, early in Devonian time.

Horneophyton. Horneophyton, first known as *Hornea,* an invalid name, occurred with *Rhynia* in the Rhynie Chert, and it has been supposed to be a close relative of *Rhynia.* However, doubt has been cast by the thorough study of Eggert (American Journal of Botany 61: 405–413. 1974), who obtained a piece of the chert at Rhynie and sectioned it from various angles. The plant differs from *Rhynia* (1) in having a columella continuing (or at least as a structure beyond) the stele and forming a rod of tissue running up the central part of the sporangium, (2) in formation of elongate sporangia running through one or two forks continuous with the stem and thus having 2 or 4 lobes, (3) in having several swollen, tuberous underground structures giving rise to the stems and many rhizoids resembling roothairs, and (4) in having in the cells presumed to be phloem well-developed thickenings in the angles of their walls (as do collenchyma cells of flowering plants). Thus, inclusion of *Horneophyton* in the Rhyniales seems doubtful.

As indicated by Eggert, the columella of the sporangium of *Horneophyton* is slightly reminiscent of the columella of *Sphagnum,* or peat moss, and it is less so of the columella of a true moss. If there is a similarity to one of the Bryophyta (liverworts, hornworts, and mosses), it is to the hornworts, or Anthocerotae, rather than to the Musci (mosses and peat mosses). In *Anthoceros* almost the whole stem is an elongate sporangium with a continuous columella. However, the sporangium is unbranched, it splits open from apex to almost the base, where a meristem produces spores and elaters (sterile, moisture-sensitive cells that push the spores loose). If there is any connection between the hornworts and *Horneophyton,* certainly one was not derived from the other, though there may have been a remote common ancestary. No clear connection between the vascular plants and the Bryophytes is indicated by this meager evidence, but the possibility should not be overlooked. On the basis of similarities there may be a distant common ancestry of the Bryophyta and the divisions of the vascular plants, but the origin of the vascular plants is not to be sought in the bryophytes.

Order PSILOPHYTALES; Family PSILOPHYTACEAE *

(Alternate: Class Trimerophytopsida; Order Trimeophytales; Family Psilophytaceae)

This order differs from the Rhyniales as follows: (1) main axis sometimes branched spirally, as well as dichotomously; (2) smaller branches forked trichotomously, as well as dichotomously; (3) fertile branches much forked and terminating in a mass of numerous sporangia, these paired, more or less ellipsoid; and (4) xylem strand larger than in *Rhynia*. The fertile branches tend to be much rebranched and to be overtopped by sterile branch systems forming a main axis.

Probably this order gave rise to all the later major groups of vascular plants, except the club mosses. This was through the progymnosperms (p. 642). The process of overtopping, appearing in the Psilophytales and again in the progymnosperms, together with planation and webbing, probably formed the leaves of both the Cycad and Conifer lines of development. Inasmuch as probably the flowering plants arose from the Cycad Line, the Psilophytales are basic to the entire current scheme of evolution of the vascular plants.

Psilophyton. The fertile branches of the stem were smaller than the sterile overtopping ones or sometimes groups of fertile and of sterile branch systems alternating along the main axis. The fertile systems were forked dichotomously 6 to many times, and they were terminated by pairs of drooping, ellipsoidal sporangia. Several species are known, some being as clearly documented as *Rhynia.*

Trimerophyton. The genus is known only from the fertile branches, which were arranged spirally as lateral branches in three vertical series. Further branching was dichotomous and terminated by paired, erect sporangia.

Pertica. The genus bore lateral branches in four vertical series. The sterile or fertile lateral branches produced compact masses of sporangia. The sporangia were erect in dense clusters of pairs.

Class ZOSTEROPHYLLOPSIDA

Stems bare to scaly or prickly; branching dichotomous to unequal and with the larger segments forming a main axis (pseudo-monopodial); sporangia lateral, stalked to nearly sessile, more or less globose to reinform, dehiscent distally along the margins; xylem strand large, exarch so far as known, the tracheids scalariform.

---

* Until the status of *Psilophyton* was clarified by Banks and Hueber (Taxon 16: 81–85. 1967) by removal of discordant elements and designation of a lectotype and until the species were clarified in subsequent papers by Banks, Leclerq, and Hueber (Paleontographica Am. 8: 73–127. 1975 and several other papers), the genus was a shaky basis for the names of higher taxa. However, the excellent work of Banks, Andrews, and their associates has made *Psilophyton* one of the best known and defined genera of Devonian fossil plants.

Inasmuch as the name of the Family Psilophytaceae is of long standing and based securely on *Psilophyton,* it cannot be replaced by the recent name Trimerophytaceae, based on *Trimerophyton,* a name published by Hopping in 1956. However, above the rank of family the rule of priority does not apply, and the designation of a name for the division, class, or order is subject to the choice of the author. An ordinal name based on *Psilophyton* seems preferable to substitution on one based on *Trimerophyton* in the same family, because it is in harmony with the mandatory use of Psilophytaceae for the family and because (1) the names are long established, and they are known to nearly all botanists and a host of botany students, and (2) *Trimerophyton* is known to few, and it is based upon incomplete material.

The assemblage of plants appearing in the books and papers of many years under an inclusive order Psilophytales or class Psilopsida, of course, could not stand in the light of recent discoveries, because too many plants of no close relationships were included in the order. In taxonomy, the standard method of dealing with such cases is restriction of the taxon to the natural unit including the nomenclatural type, which then continues to bear the name. Families and lower taxa *must* be named according to the rule of priority. Thus, the family including *Psilophyton* must be known by the oldest valid name, Psilophytaceae. Properly restricted, the higher taxa based on *Psilophyton* are still valid, regardless of the diversity of the plants covered by the order as interpreted by past authors or the literary confusion concerning their nature and relationships. Once a classification system has been devised, nomenclature follows automatically up to the family level. Above that level there is latitude, but the best practice is to follow the same rules if there is no obstacle to employing them.

The nature of *Trimerophyton,* selected as a substitute for *Psilophyton* as the type of genus before the status of the latter was clarified [Banks, in E. Drake (editor), *Evolution and Environment* . . . , 1968. Yale University Press. 73–107] is not clear. According to Banks (Taxon 24: 410. 1975), "*Trimerophyton* is known only from the fertile portion of the plant . . . ," and it seems now to be an inadequately known taxon. It is not a good substitute for *Psilophyton,* in view of the thorough recent work of Banks and his associates on the nomenclatural status and the structure of *Psilophyton.*

Primitive vascular plants along the shores of a Devonian sea; *left and right, Rhynia; left-center, Baragwanathia; center,* (shorter) *Asteroxylon;* (taller) *Psilophyton.* Cf. also **Figure 13-9.** Since this illustration was prepared in 1945, much new information has become available. The *Psilophyton* in the figure was based on incomplete material, and the common overtopping of one axis by another occurring in at least the more advanced species of *Psilophyton* was unknown. In the main a vegetative branch assumed dominance, and the sporangium-bearing branches were clustered and seemingly lateral. In *Psilophyton dapsile* the axes were dichotomous, but the fertile branches were in clusters. Unknown primitive species may have had the form of branching shown here.

The Zosterophyllopsida lived from about 395 to 355 million years ago, from the beginning to nearly the end of the Devonian Period. *Strong evidence indicates them to be the precursors of the line of* development leading to the Carboniferous Lepidodendrales and the *present club mosses* (pp. 698–709).

The enations along the stem included prickles ranging from one-celled to many-celled or there were multicellular projections or scales. According to the evidence from series of later plants, these were the forerunners of the small simple leaves of the Club Moss Line of development. The sporangia, borne on short lateral branches without vascular tissues, probably were the forerunners of those aggregated into strobili (cones) of sporangia in the club mosses. The dehiscence of the sporangium along a distal line following the rim is in contrast to the indehiscence of the sporangia of the Rhyniopsida. The exarch, rather than centrarch, protostele of the stem is a feature of the line of development.

Order ZOSTEROPHYLLALES; Family ZOSTEROPHYLLACEAE

Zosterophyllum. Usually the plants have been drawn as though they lived in water and the upper branches floated on the surface. These have been considered to have been flattened, though this may have been due to compression of the stems as they were preserved.

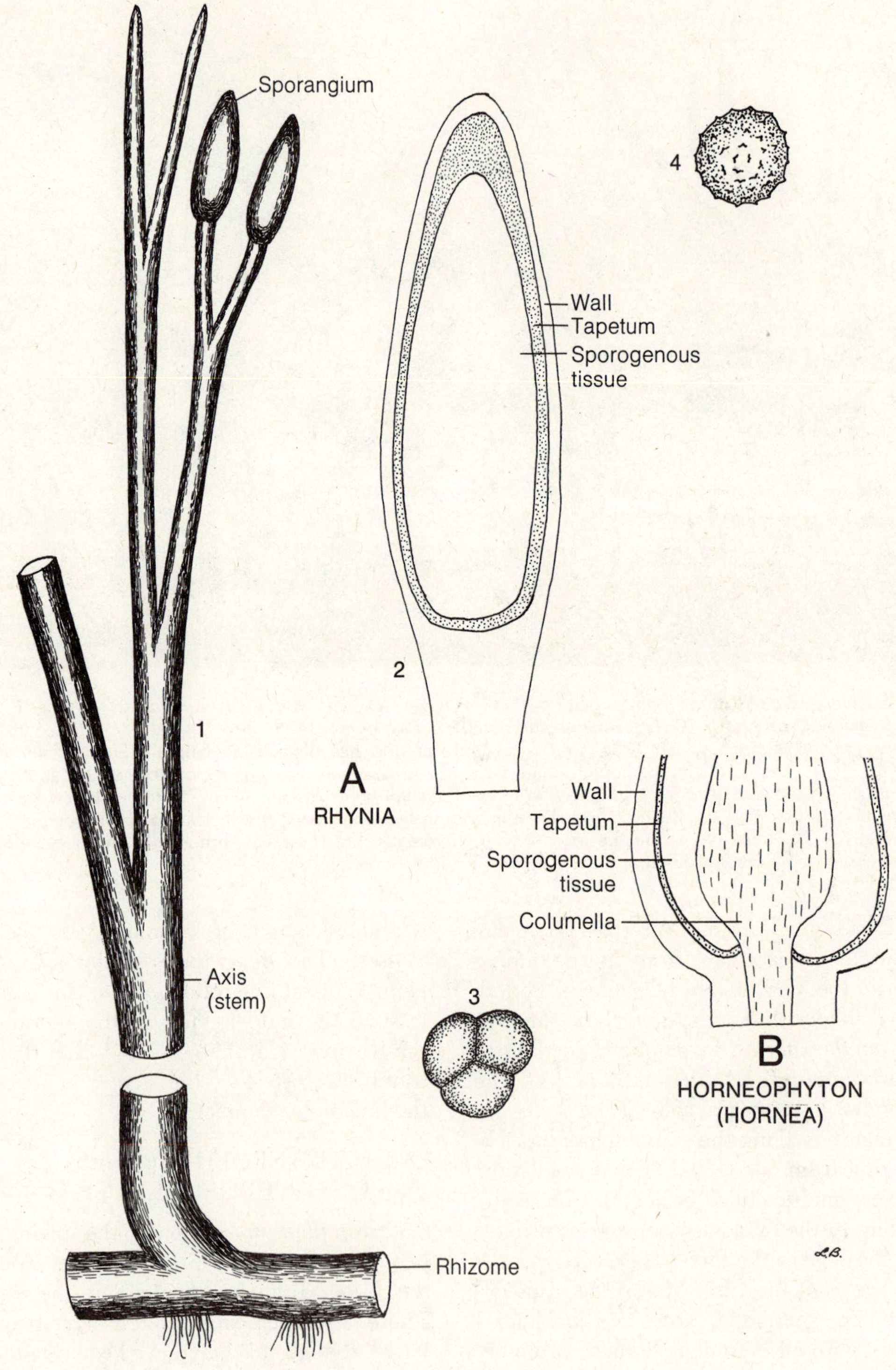
Sporangium
Axis
(stem)
Rhizome
1
2
Wall
Tapetum
Sporogenous
tissue
3
4
A
RHYNIA
Wall
Tapetum
Sporogenous
tissue
Columella
B
HORNEOPHYTON
(HORNEA)
L.B.

The rhizome gave rise to ascending branches, and the branching of both probably was dichotomous below as well as above ground. Scars suggest possibly the presence of rhizophores similar to those of the modern little club mosses (*Selaginella*), that is, branches or root trusses bearing the roots, which are adventitious rather than continuing the axis of the stem. The sporangia were essentially of the form of the club mosses, flattened a little with the faces parallel to the axis or stem and to the apical line of opening.

Horneophyton Lignieri: 1 Reconstruction of the entire plant, about ½ natural size, showing the dichotomous branching of the stem, 2 enlargement and a sectional view of a once-dichotomous sporangium, the stem bases bulbous. From DONALD A. EGGERT, American Journal of Botany 61: 411. 1974. Used by permission.

Evolution of other Vascular Plants

So far as can be determined from the meager fossil record known now, the more recent vascular plants were derived from plants similar to those described in this chapter. The outlines of relationships and the indications of evolution now available are summarized in the chart on p. 596. As each major taxon is studied, its more primitive members should be checked for similarities to the Rhyniophyta or to the Aneurophytopsida in the chart. Those appearing nearest are the probably related taxa.

Lines of Descent: Radiation

Historical Note. During the Nineteenth Century, while systems of angiosperm classification were being devised, information was accumulating not only from studies of the characters of species, genera, and families of flowering plants, but also from brilliant research on the anatomy and the life histories of other plant groups. These divisions, which had received little attention in earlier classifications, were raised toward their proper proportions. In 1843 Brongniart proposed two now classical major divisions for the Plant Kingdom

Psilophytales; *Rhynia* and *Horneophyton* (*Hornea*): **A** *Rhynia,* 1 entire plant, illustrating the parts and the dichotomous branching of the axis, 2 sporangium in longitudinal section, 3 spores in a group of four (three showing) just after formation by the inferred reduction divisions, 4 single spore; **B** *Horneophyton,* lower part of the sporangium in longitudinal section, showing the columella of phloem continuous with the stele of the axis. Based upon R. KIDSTON and W. H. LANG, Royal Society of Edinburgh. Transactions 51: 761–784. plate 9, fig. 62, 1917; 52: 603–627. plate 10, fig. 69, 1920; 52: 831–854. plate 1–2. 1921. Redrawn by permission.

—Cryptogamae (without flowers) and Phanerogamae (with flowers). The Phanerogamae were divided into monocotyledons and dicotyledons, but the gymnosperms were included in the dicotyledons. The terms proposed by Brongniart are still in informal use, as cryptogams and phanerogams (the seed plants). In 1883 Eichler presented an arrangement of the Plant Kingdom dominant in textbooks until recently —Thallophyta, Bryophyta, Pteridophyta, and Spermatophyta. The first three were grouped under Cryptogamae and the fourth constituted the Phanerogamae.

In the present systems of classification the Thallophyta are divided into major taxa, including several divisions, and the bacteria and the bluegreen algae are excluded from the Plant Kingdom. The Bryophyta and the Spermatophyta remain as divisions. The Pteridophyta of Eichler included not only the restricted Pteridophyta (embracing only the ferns) but also the Psilophyta, Sphenophyta, and Lycophyta. Under the broad usage, the pteridophytes, or Pteridophyta, included all vascular plants not reproducing by seeds. The best-known feature in common of their repro-

Psilophyton, showing gradual overtopping (exceeding in length) of the sporangium-bearing branchlet systems by the larger and more elongate sterile (vegetative) branches, probably following development from a more primitive ancestor of the hypothetical type in Figure 27-4. 2 *Psilophyton dapsile* (slightly less than natural size), the sporangia clustered but on branchlet systems nearly as long as the vegetative photosynthetic systems. 1 *Psilophyton microspinosum*, with less differentiation of the branch systems, and 3 *Psilophyton Dawsonii* with more differentiation and with formation of thicker vegetative stems. 1, 2 ANDREW E. KASPER, HENRY N. ANDREWS, and WILLIAM H. FORBES, American Journal of Botany 61: 343, 348. 1974; 3 HARLAN P. BANKS, Bioscience 25: 735. 1975. Used by permission.

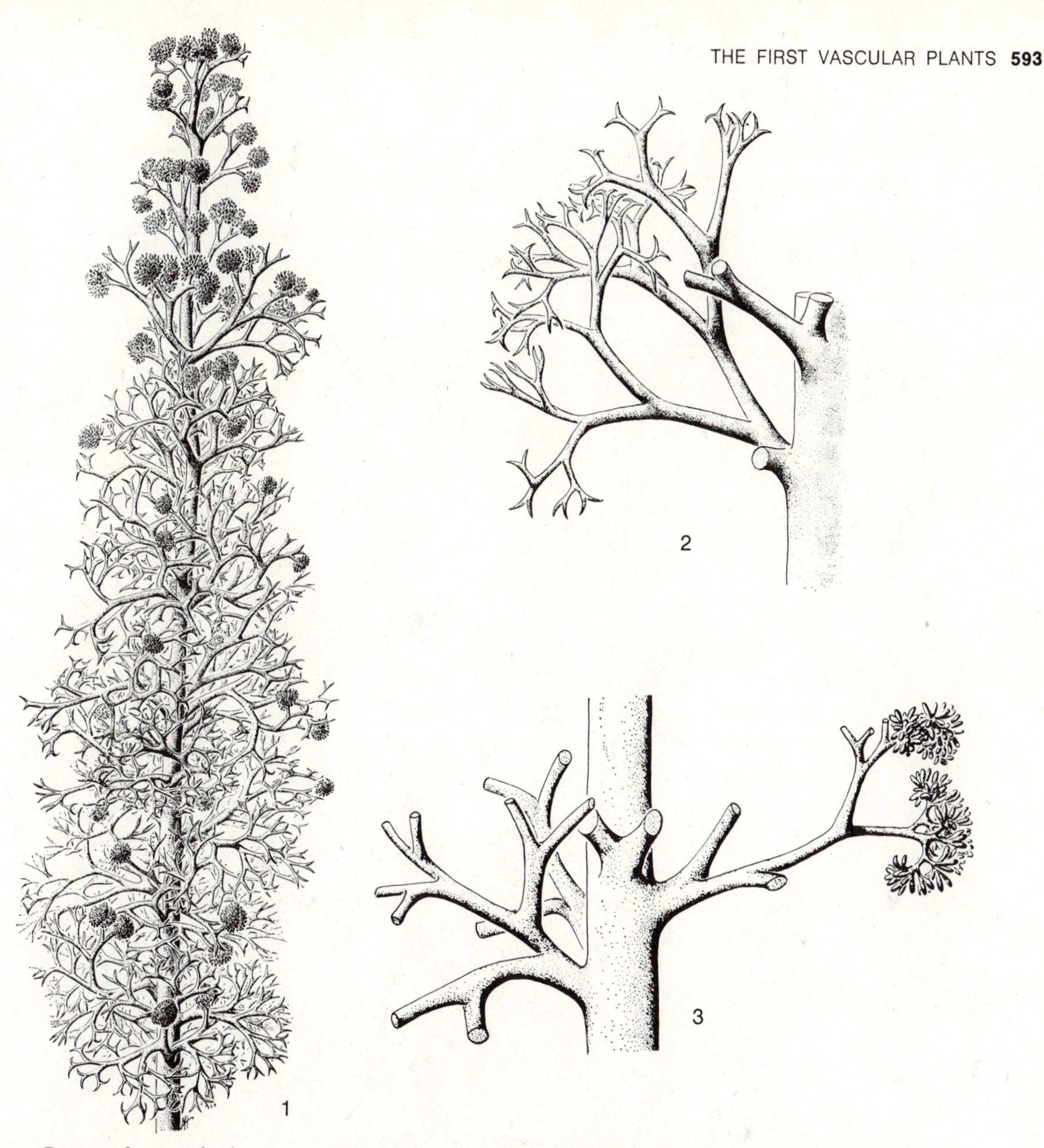

Pertica, showing further overtopping of the sporangium bearing parts of the branch system by the vegetative parts and the formation of both dense clusters of sporangia and an elongate main axis of the stem. 1 reconstruction of *Pertica quadrifolia,* showing the elongate axis perhaps up to 1 m. long and the more open masses of small vegetative branches and compact and much smaller masses of branches bearing sporangia, 2 vegetative branches, 3 small branches terminated by sporangia, these in a relatively small, tight cluster. ANDREW E. KASPER and HENRY N. ANDREWS, American Journal of Botany 59: 905, 908, 909. 1972. Used by permission.

ductive cycles is spore formation.

The most widely known pteridophytes (in this broad sense) are the ferns, but three other groups are differentiated sharply in the living flora. These are the tropical and subtropical genera *Psilotum* and *Tmesipteris* (of debated relationship and degree of evolutionary development) and the widely distributed horsetails (*Equisetum*) and club mosses (primarily *Lycopodium* and *Selaginella*). These plants

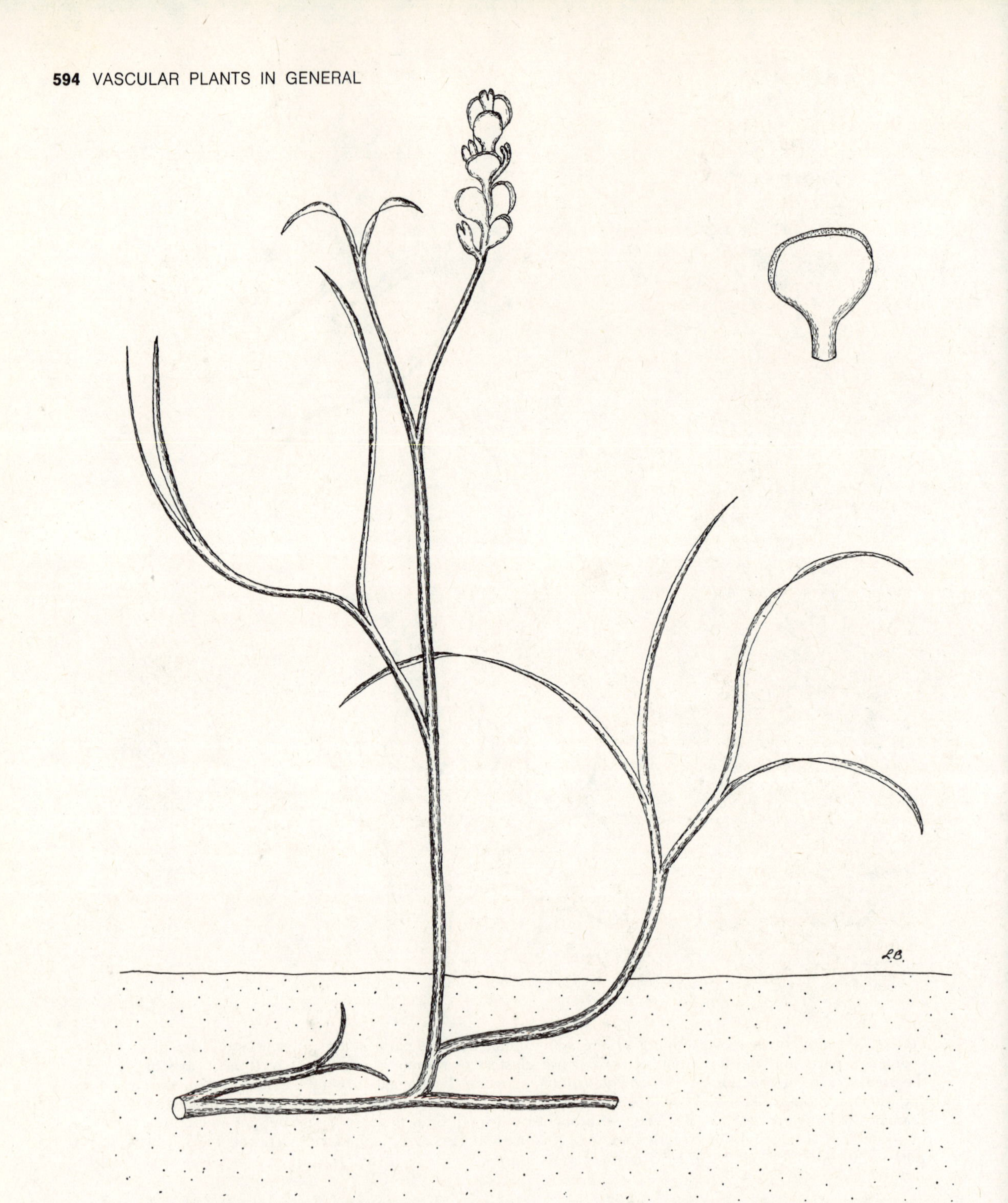

Zosterophyllum: 1 reconstruction of a part of a plant with the axis composed of cylindroid rhizomes, erect branches, flattened photosynthetic branches (reminiscent of the leaves of *Zostera,* the eel-grass of brackish water along coasts), and an axis bearing sporangia on short lateral branches forming a racemelike cluster, 2 single sporangium, showing its dehiscence along a terminal slit.

are interpreted as representing ancient strongly differentiated lines of evolution, separate since Devonian and Carboniferous. The formal divisional name Pteridophyta is retained for the ferns alone, and the common name pteridophytes is restricted to them. The other three groups constitute the divisions Psilophyta, Sphenophyta, and Lycophyta, but divisional status is assigned to the Psilophyta with reservations because their nature and status are a center of controversy.

As plants invaded the land, they became diverse, adapting in various ways to the existing habitat niches and creating new ones by their presence. Bleak landscapes were transformed by the cover of plants, and in time forests, grasslands, and the other ecosystems of the present evolved, but during Devonian probably only simple systems were possible because the plants were small and perhaps not diverse. The few known fossils do not provide adequate evidence for evaluation of the ecosystems, though doubtless they were more complex than the few remnants indicate. By late Devonian and Carboniferous, forests had developed, and there must have been many new habitats not only in the swampy areas where records remain, particularly in coal, but probably in upland or drier areas as well.

As habitats became more diverse, the plants adapted to them, radiating into the niches available and ultimately occupying all of them. The combination of mutation and selection produced many kinds of plants, but the living ones and the known fossil ones can be organized into relatively few major groups, as shown in the figure on p. 596. Adaptive radiation must have occurred on a grand scale, but, once certain groups of plants had arisen, it was quicker and more effective for new plants to arise from the existing groups than for evolution to start all over again, repeating the course of development from the algae. This has tended to keep development within certain major channels that appeared during Devonian. These "lines" or trends of development maintained a head start of millions and ultimately four-tenths of a billion years. Great innovations, like the beginning of flowering plants during Cretaceous, were based upon modification of existing organisms, not upon new groups arising directly from the ancestral one-celled plants.

The slowness of recognition of some major groups in classification was due to difficulty of study with need for microscopes and for equipment and methods for preparation of materials for microscopic study. For this reason the one- or few-celled organisms were late in being studied and the life histories of the "pteridophytes" in the broad sense were slow in being understood. Even the fertilization of the pteridophytes was shown by a professor (Douglas Houghton Campbell) who retired the spring before I was a freshman, fertilization in angiosperms having been demonstrated a little earlier by Strasburger. If even the fundamentals of life history were unknown, the classification of major groups could not be far along. Recent changes in classification reflect additions to and corrections of basic knowledge. In the light of the information accumulated recently, there are three or four major lines of development among the vascular plants, but one overshadows the others. Nearly all plants of today are in the Fern Line; only a handful of genera of living plants represent the *Psilotum* line (which is of debated status), the Horsetail Line, and the Club Moss Line together. It is no wonder that the significance of these three lines has been overlooked. They are remnants of developmental lines that once were of great importance in the floras. If the Psilotaceae are primitive, they are only a step from the Rhyniophyta of late Silurian and Devonian time; otherwise they are holdovers from the late evolution of the living pteridophytes. The Horsetail Line was a major element in formation of the Carboniferous coal deposits from the lowland forests of the time. So also were the giant forerunners of the present club mosses. However, these lines of development

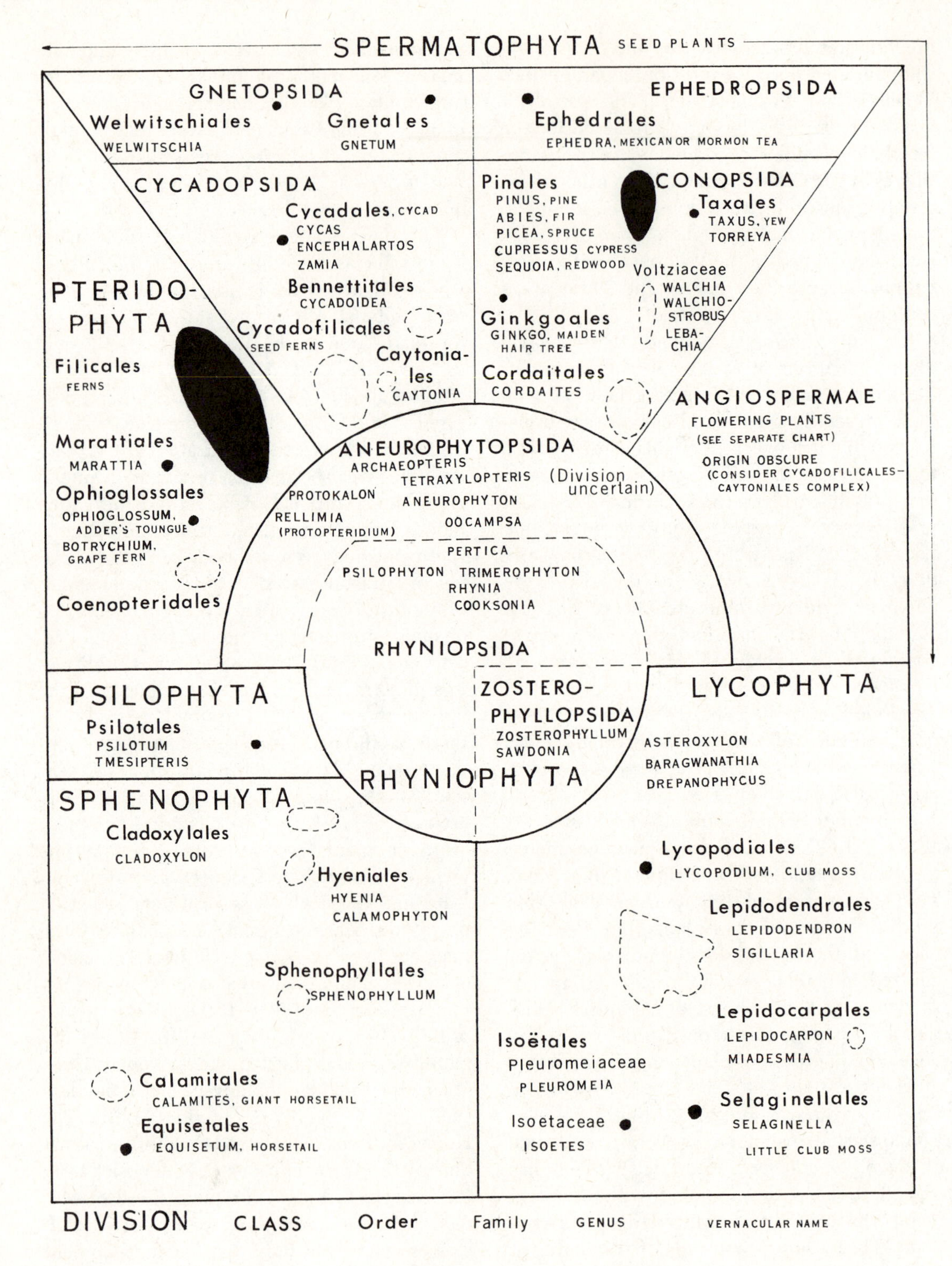
SPERMATOPHYTA SEED PLANTS
GNETOPSIDA
Welwitschiales
WELWITSCHIA
Gnetales
GNETUM
EPHEDROPSIDA
Ephedrales
EPHEDRA, MEXICAN OR MORMON TEA
CYCADOPSIDA
Cycadales, CYCAD
CYCAS
ENCEPHALARTOS
ZAMIA
Bennettitales
CYCADOIDEA
Cycadofilicales
SEED FERNS
Caytoniales
CAYTONIA
Pinales
PINUS, PINE
ABIES, FIR
PICEA, SPRUCE
CUPRESSUS CYPRESS
SEQUOIA, REDWOOD
CONOPSIDA
Taxales
TAXUS, YEW
TORREYA
Voltziaceae
WALCHIA
WALCHIOSTROBUS
LEBACHIA
Ginkgoales
GINKGO, MAIDEN HAIR TREE
Cordaitales
CORDAITES
PTERIDOPHYTA
Filicales
FERNS
Marattiales
MARATTIA
Ophioglossales
OPHIOGLOSSUM, ADDER'S TOUNGUE
BOTRYCHIUM, GRAPE FERN
Coenopteridales
ANEUROPHYTOPSIDA
ARCHAEOPTERIS
TETRAXYLOPTERIS
ANEUROPHYTON
OOCAMPSA
PROTOKALON
RELLIMIA
(PROTOPTERIDIUM)
(Division uncertain)
ANGIOSPERMAE
FLOWERING PLANTS
(SEE SEPARATE CHART)
ORIGIN OBSCURE
(CONSIDER CYCADOFILICALES–CAYTONIALES COMPLEX)
PERTICA
PSILOPHYTON TRIMEROPHYTON
RHYNIA
COOKSONIA
RHYNIOPSIDA
PSILOPHYTA
Psilotales
PSILOTUM
TMESIPTERIS
ZOSTEROPHYLLOPSIDA
ZOSTEROPHYLLUM
SAWDONIA
RHYNIOPHYTA
LYCOPHYTA
ASTEROXYLON
BARAGWANATHIA
DREPANOPHYCUS
SPHENOPHYTA
Cladoxylales
CLADOXYLON
Hyeniales
HYENIA
CALAMOPHYTON
Sphenophyllales
SPHENOPHYLLUM
Calamitales
CALAMITES, GIANT HORSETAIL
Equisetales
EQUISETUM, HORSETAIL
Lycopodiales
LYCOPODIUM, CLUB MOSS
Lepidodendrales
LEPIDODENDRON
SIGILLARIA
Lepidocarpales
LEPIDOCARPON
MIADESMIA
Isoëtales
Pleuromeiaceae
PLEUROMEIA
Isoetaceae
ISOETES
Selaginellales
SELAGINELLA
LITTLE CLUB MOSS
DIVISION CLASS Order Family GENUS VERNACULAR NAME

are minor today, and the flowering plants have all but replaced even the ferns and the various gymnosperms. The species of ferns and gymnosperms fill only a few pages of any flora or manual; the flowering plants cover hundreds or even thousands of pages.

THE MAJOR LINES OF DEVELOPMENT OF VASCULAR PLANTS

1. *The Fern Line.* Divisions **Pteridophyta** and **Spermatophyta.** Beginning with *Rellimia* (*Protopteridium*), a relative of the Rhyniophyta and culminating in the ferns and the seed plants.

The Fern Line is characterized by large complex leaves each derived presumably from a major flattened branch system, and sporangia are produced on the backs or the margins of the leaves. A branch or a bud develops characteristically in each leaf axil in many members of this line of development.

The Fern Line includes not only the ferns but also two groups which have diverged sufficiently to be considered as a separate Division, the Spermatophyta, including the angiosperms and gymnosperms. The direct origin of the gymnosperms, or at least of some of them, from ancestors also near the beginning of the Fern Line is fairly clear. Origin of the angiosperms is controversial, and their relationship is obscured by lack of fossil connecting links. The most likely ancestry is in the Cycadofilicales (seed ferns) or the related Caytoniales, extinct gymnosperms.

2. *The Psilotum Line* (if it is interpreted as primitive; see Chapter 22). Division **Psilophyta.** A relict group of plants with the fundamental organization of the Rhyniales, or a family of the living order Filicales according to the opposed interpretation.

There are no roots or leaves, unless the scales on the stem are interpreted as leaves; the sporangia are fundamentally terminal, and they are thick-walled.

3. *The Horsetail Line.* Division **Sphenophyta.** Beginning with *Psilophyton* (Rhyniophyta) and continuing through the Hyeniales to the giant horsetails (*Calamites*) of Carboniferous and the living horsetails (*Equisetum*).

The leaves were derived from minor systems of flattened branches (in contrast to those from major branch systems in the Fern Line); they are whorled at the nodes of strongly jointed stems; the sporangia are in fundamentally drooping pairs (but in the living horsetails appearing otherwise because of union of the sporophylls into peltate structures).

4. *The Club Moss Line.* Division *Lycophyta.* Beginning with *Asteroxylon,* near the Rhyniophyta, and culminating in the Carbon-

---

Chart of relationships of the vascular plants, the living major taxa represented by black areas, the fossil taxa by areas enclosed by dotted lines. The less specialized and presumably more primitive taxa are at the center of the chart, the more specialized taxa toward the periphery. The known taxa are not necessarily presumed to have been derived directly from the known related taxa; in most instances both arose from a common ancestor most likely undiscovered. Because fossils are relatively few, the record is meager, but the main outlines of evolution of the vascular plants are emerging. This chart should be consulted as each group of vascular plants is studied, and each group should be correlated with the Rhyniophyta at the center of the chart. Thus, each group of vascular plants may be correlated with the Silurian and Devonian simple primitive types discussed in this chapter.

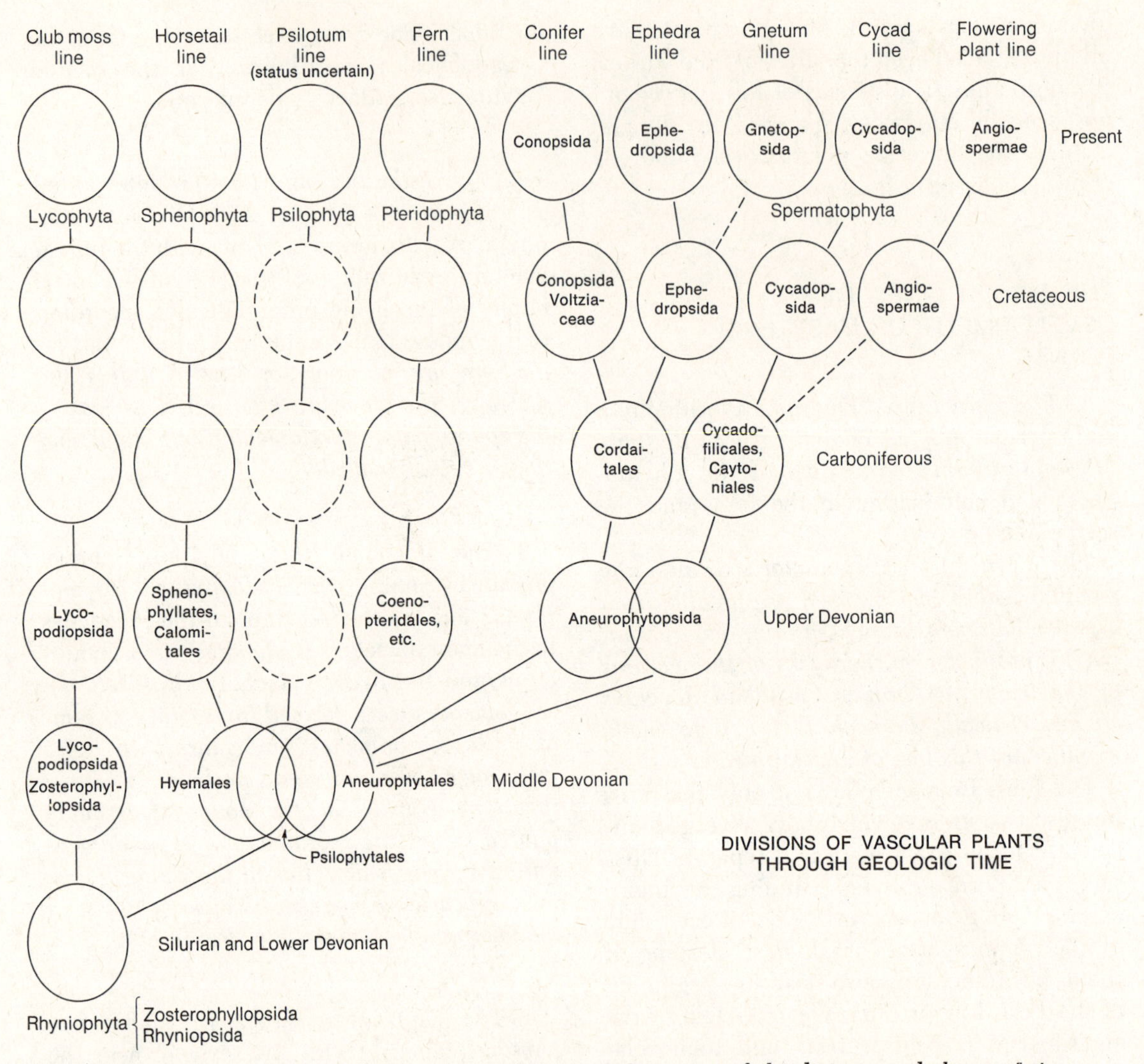

Diagram illustrating the gradual divergence through geologic time of the divisions and classes of the vascular plants. (The nature and status of the Psilophyta are in dispute.)

iferous Lepidodendrales and the living club mosses (*Lycopodium*), little club mosses (*Selaginella*), and quillworts (Isoëtes).

In contrast to development of leaves from flattened branch systems, the leaves are flattened outgrowths of the superficial layers of the stem (enations); there is only one leaf vascular bundle, the midrib; the sporangia are lateral to the stem, in the living plants with one on the ventral side of the base of each sporophyll, but in the extinct primitive plants not associated with the leaves.

Gymnosperms

Division *SPERMATOPHYTA*

Classes *CONOPSIDA, EPHEDROPSIDA, GNETOPSIDA,* and *CYCADOPSIDA*

As pointed out in Chapter 1, the gymnosperms, like the angiosperms, are vascular plants characterized by seeds (Spermatophyta), but they differ from the angiosperms in the following obvious and many microscopic characters:

1. Absence of flowers, the pollen and seeds being produced usually in cones.
2. Absence of a closed pistil, the seeds being naked on a specialized leaf or a scale of the cone.

Other more technical distinguishing characters are seen only on prepared slides by means of a compound microscope.

In North America north of Mexico, there are three major groups of gymnosperms, the **conifers** (pines, firs, Douglas fir, spruce, hemlock, cypress, juniper, bald cypress, redwood), **Ephedra** (Mexican tea or Mormon tea), and the **cycads** (represented in Florida by *Zamia*).

18

The Vocabulary Describing Gymnosperm Characteristics

Conifers

Through most of the Northern Hemisphere pines are the best-known gymnosperms, and the genus *Pinus* is described below to illustrate the group. Since in *Pinus* the arrangement of leaves in fascicles on the branchlets and the often minute bracts of the cone are unusual, *Abies,* including the true firs of northern and mountainous regions, and *Pseudotsuga,* including the so-called Douglas fir of the western United States, will be used occasionally as supplementary examples.

THE PLANT BODY

Nearly all conifers are trees, the few exceptions being shrubs. The roots and stems are similar externally to those of dicotyledons, although on the average there is a greater tendency toward a straight tall trunk. Increase in girth is by a cambium as in the dicotyledons. The leaves differ more markedly from those of angiosperms. Usually they persist for more than one season and consequently the plant is evergreen, but this is true also of many dicotyledons. In many species the leaves are needlelike, but in others (e.g., cypresses and junipers) they are scalelike and small and in some conifers they are lanceolate and fairly large (e.g., monkey-puzzle tree and *Podocarpus*). In a related group they are broader than long (maidenhair tree).

A pine tree has four kinds of leaves, as follows: **cotyledons** (**primary** or **seed leaves**), **juvenile leaves, scale leaves,** and **foliage leaves.**

The last three types are secondary leaves; only the last two occur normally on the adult plant.

Cotyledons. The first leaves of any spermatophyte are those formed in the embryo while it is enclosed in the seed. These are the cotyledons. The pine has usually from six to ten, but in the pine family the number may run as high as seventeen. Other gymnosperms have only two. When the seed germinates, the cotyledons become green; and they form a leaf whorl on the stem of the seedling.

Juvenile Leaves. As the stem elongates above the cotyledons it gives rise first to spirally arranged solitary needlelike juvenile leaves unlike those of the adult plant.

Scale Leaves. In the adult pine, small flat scale leaves occur singly at the nodes of the branchlets. They are brown or tan, short, and more or less triangular.

Foliage Leaves. In the mature pine the needlelike foliage leaves are developed in compact clusters at the ends of very short branchlets in the axils of the scale leaves. Each cluster includes a definite small number of leaves characteristic of the species—usually two, three, or five, but rarely one or four. The sheath of membranous scales surrounding the base of the young cluster makes the peculiar arrangement recognizable even when there is only one leaf in a sheath.

The arrangement of foliage leaves in most conifers is simpler than in the pines. The true firs and the Douglas fir have solitary needles arranged spirally on the branchlets. They are without basal sheaths and are not subtended by scale leaves.

THE CONES

The pollen- or seed-bearing structures of the pine compose **strobili** (singular, **strobilus**) or **cones.** A single tree produces two kinds. Small cones, in the average species perhaps half an inch long, are pollen-producing (**polliniferous**); larger ones, ranging from an inch or two to about eighteen inches long according to the species, are seed-bearing (**ovuliferous**). Thus the pine is **monoecious,** that is, with both kinds of strobili on the same individual.

The polliniferous cones appear in the spring in the northern and mountainous parts of North America, but in the California lowlands and other areas with warm winters trees in cultivation may reach the height of pollen development in January.

The pollen-bearing cone is a simple structure. It consists of an **axis** (branchlet) bearing spirally arranged **microsporophylls** corresponding to the stamens of flowering plants. A microsporophyll has no blade, and it is made up of a short flattened petiole continuing into a midrib with two pollen sacs lying parallel and dorsal to it. Each pollen sac is a microsporangium, that is, literally a sac or a **sporangium** producing small spores.* The microsporangia on the back of the sporophyll correspond to sporangia on the backs of the fern leaves (p. 659) or to the anther sacs or pollen chambers of flowering plants (cf. p. 52).

* For many years the specialized leaves have been called *microsporophylls,* their sporangia *microsporangia,* and their spores *microspores.* Recently Thomson has shown the microspores and megaspores to be actually the same size. However, the classical terminology is retained, and the prefixes *micro-* and *mega-* are used as though the spores were of two sizes. In any event they produce gametophytes (small sexual plants) of different sizes.

The prefix *micro-* (small) may be opposed by either *mega-* (great) or *macro-* (long in extent or duration). In biology *macro-* may be used in the sense of some particular parts being unusually large, and it is used sometimes in the place of *mega-*.

Figure 18-1. Leaves of conifers: **A** pine (*Pinus canariensis*), abnormal branch from near an injury to the trunk, showing (drooping) two fascicles of needlelike foliage leaves, the brownish scale leaves on the stem (the fascicles being axillary to these), and a short branch (above) with juvenile needlelike leaves, these arranged spirally along the branch instead of in fascicles; **B** spruce (*Picea*) with needlelike leaves arranged spirally on the branch; **C** deodar (*Cedrus Deodara*) with the needlelike leaves restricted (except

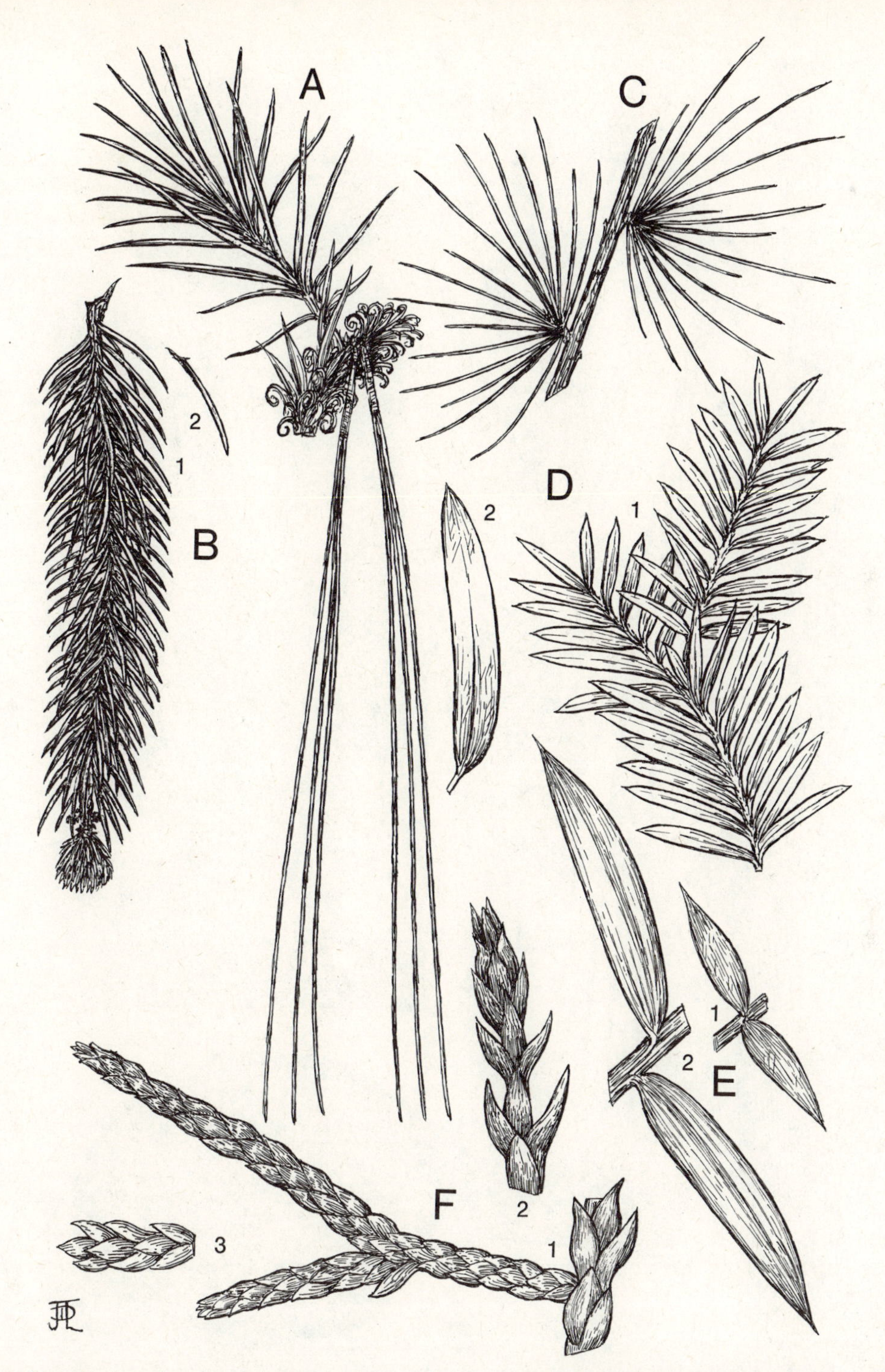

on young branches) to dwarf spur branches; **D** redwood (*Sequoia sempervirens*) with flattened leaves arranged spirally along the branch, but the blades deflected into two rows or ranks; **E** Bidwill "pine" of Australia (*Araucaria Bidwillii*), leaves (arranged as in the redwood [**D**]); **F** Italian cypress (*Cupressus sempervirens*), scalelike green leaves, mostly appressed against the branchlets. Under each letter, **1** branchlet and leaves, **2** and **3** detail.

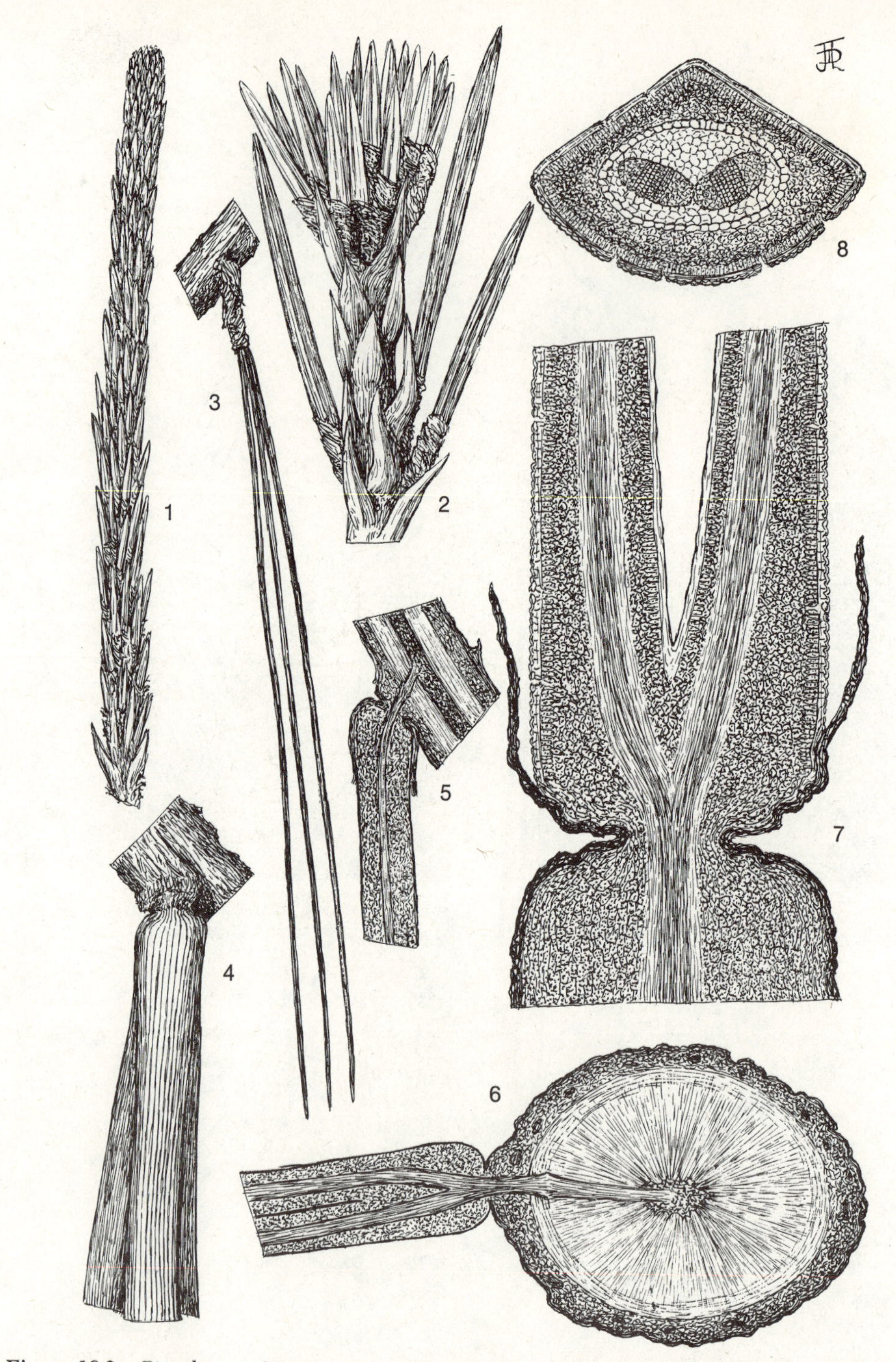

Figure 18-2. Pine leaves; Monterey pine (*Pinus radiata*): **1** rapidly growing shoot with foliage leaves emerging, **2** fascicles of young foliage leaves, each subtended by a scale leaf, **3** fascicle of three adult foliage leaves, **4** base of a fascicle emerging from the stem (the sheath removed), **5–7** longitudinal sections of the base of the fascicle and of the vascular connection from the dwarf branch of the fascicle to the stele of the twig, **8** cross section of a leaf, in this species with two vascular strands.

Figure 18-3. Polliniferous cones of a pine; Monterey pine (*Pinus radiata*): **1–3** polliniferous cones in **1** old cones with the shoot developing foliage stems beyond them), **4–5** longitudinal sections showing the microsporophylls, **6–8** views of a microsporophyll, **9** cross section of a microsporophyll showing the two pollen chambers, **10** pollen grain in the two-celled stage, the upper structures being air bladders ("wings") in the wall of the pollen grain.

The microspores of the gymnosperms are retained for a time in the microsporangium, where each one develops by internal cell division into a pollen grain.

The ovuliferous (later *seed-bearing*) *cones* of the pine appear at the same time as the polliniferous ones, but they are so inconspicuous that ordinarily they are overlooked during the earlier stages when they are small. A year or more after pollination the cones of conifers reach maturity and a much larger size.

The cone is made up of an axis and spirally arranged appendages, but the appendages are of two kinds—**bracts** and **ovuliferous scales** (usually shortened to **scales** or **cone scales**).

A bract is a specialized leaf subtending a reproductive structure. It is not seen readily in the mature cones of many pine species, because often it does not increase in size as the cone scale enlarges after the time of fertilization. The bract is visible at maturity in the true firs and is conspicuous in the Douglas fir.

The ovuliferous scale is a complex structure derived from the bud (embryonic branch) in the axil of the bract (specialized leaf). Occurrence of a bud in a leaf axil corresponds with the universal relative arrangement of leaves and branches among the seed plants, but the method of development of the cone scale from the structures produced by the axillary bud has been the subject of much controversy.

On the upper side of each scale there are two ovules. Each ovule is a megasporangium (**nucellus**) enclosed in an extra structure, the **integument.** Inside the ovule a single functional megaspore is developed, and it gives rise to a special tissue composed of hundreds of cells. About two or three cells, each part of a special structure, are eggs. These are near the **micropyle,** a small opening in the integument of the ovule.

POLLINATION AND FERTILIZATION

Pine pollen is so abundant that it may blanket an entire region, and almost surely some will be carried by the wind to the micropyle. Having landed in a favorable situation, the pollen grain grows into an elongated pollen tube, which digests its way through the intervening megasporangium (nucellus) tissue to an egg. Two male gametes are carried toward the megagametophyte by growth of the pollen tube; one unites with the egg, effecting fertilization.

DEVELOPMENT OF THE EMBRYO

After fertilization the resulting zygote goes through an elaborate developmental procedure resulting in formation of four embryos, only one of which ordinarily survives and develops to maturity. Ultimately, besides the cotyledons, the embryo consists of an axis differentiated into a **hypocotyl** and an **epicotyl,** the hypocotyl being the portion below the cotyledons and the epicotyl that above. The extremity of the hypocotyl forms a radicle or root tip; the epicotyl is a small mass of embryonic tissue which develops into the stem and its attached juvenile and later types of leaves.

THE SEED

The ovule and its enclosing integument become a seed, which consists of the following: (1) the **seed coat,** a hard outer covering formed from the integument; (2) the megasporangium or nucellus, persisting as a thin, papery inner layer; (3) a food storage reservoir formed from the large tissue (megagametophyte) which produced the eggs; and (4) the embryo.

Ephedra

The genus *Ephedra,* which includes the Mexican tea and Mormon tea of Southwestern Deserts, is composed of bushy, diffusely branched desert shrubs usually one to four feet high. The leaves are opposite or in whorls of three. They are small and scalelike and without chlo-

Figure 18-4. Ovuliferous cones of a pine; Monterey pine (*Pinus radiata*): **1** young cones (after pollination), **2–3** longitudinal section, **3** showing an ovule (above a scale), and a bract, **4** mature cone, external view, **5** longitudinal section, **6** dissected cone showing the two wingéd seeds on each ovuliferous scale, **7–9** views of the scale with seeds on one side and the bract on the other, the scale now much larger than the bract.

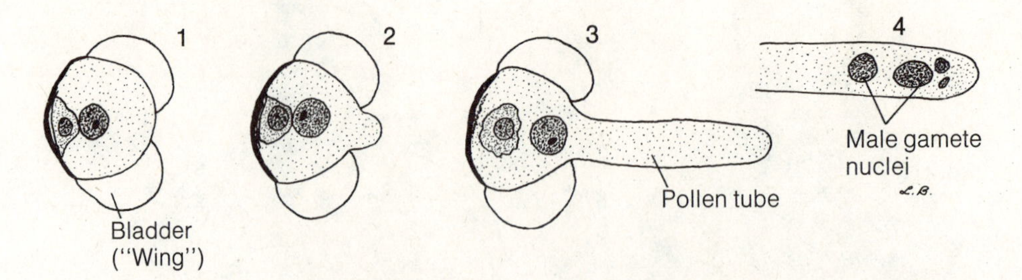

Figure 18-5. Development of the pollen grain of a pine: **1** pollen grain consisting of a tube cell (nucleus large), an enclosed generative cell, and remains of two prothallial cells (this being the microgametophyte or pollen grain as it is carried by wind), **2–3** growth of the pollen tube, **4** tip of the pollen tube before fertilization, the male gametes having been formed from the generative cell.

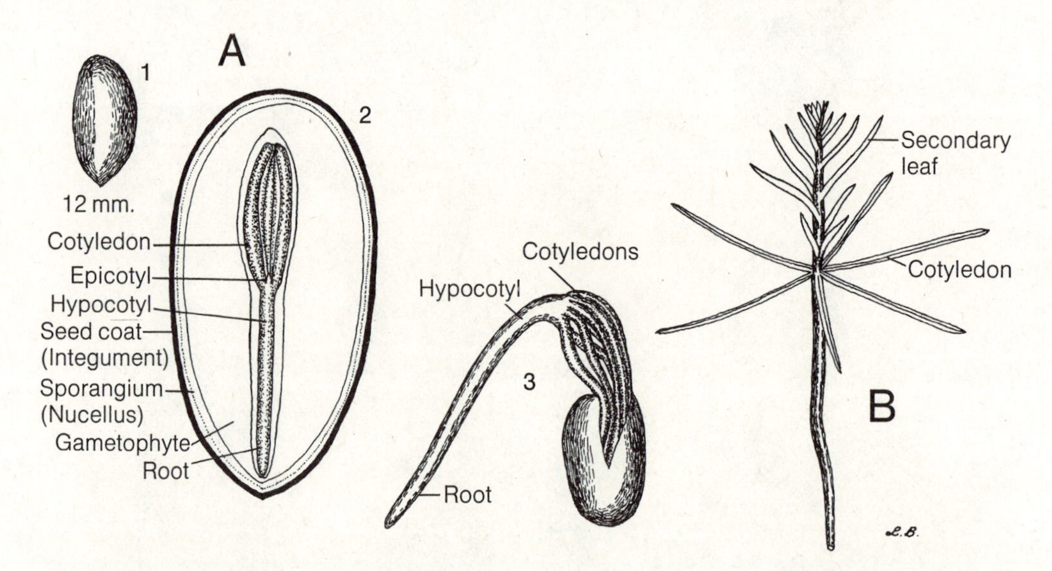

Figure 18-6. Seed and seedling; conifers: **A** seed of the one-leaf pinyon (*Pinus cembroides* var. *monophylla*), **1** seed, **2** longitudinal section (enlarged), **3** embryo emerging from the seed coat; **B** seedling of red fir (*Abies magnifica*).

rophyll, and photosynthesis is restricted to the green twigs.

The plants are dioecious. The polliniferous cone consists of pairs of opposite scales subtending stalklike microsporophylls, each with five or six sporangia. The ovules are solitary or in pairs, and they are subtended by pairs of scales. The ovule has an elongated tube around the micropyle. This appears like a style, and the ovule may be mistaken for a pistil.

Cycads

The cycads are represented by *Cycas revoluta,* the sago "palm," or by *Zamia,* a genus native in Florida.

Most cycads appear to be dwarf or small palm trees, the usually unbranched trunk having a dense crown of spirally arranged leaves at the top. However some cycads (e.g., *Zamia*) have no elongated trunk, but only a tuberous structure. The leaves are pinnate in all the genera except *Bowenia,* in which they are bipinnate. They are large and somewhat like fern leaves, but usually thicker and more leathery.

The strobili are produced at the apex of the plant. So far as can be established, the plants are strictly dioecious, that is, having microsporophylls on one plant and ovules on an-

Figure 18-7. *Ephedra californica* growing on a small sand dune near Palm Springs, California. As the plant grows upward and outward the sand dune does, too. This is in the Colorado Desert. The leaves are small, whorled, and scalelike; nearly all the photosynthesis occurs in the slender stems.

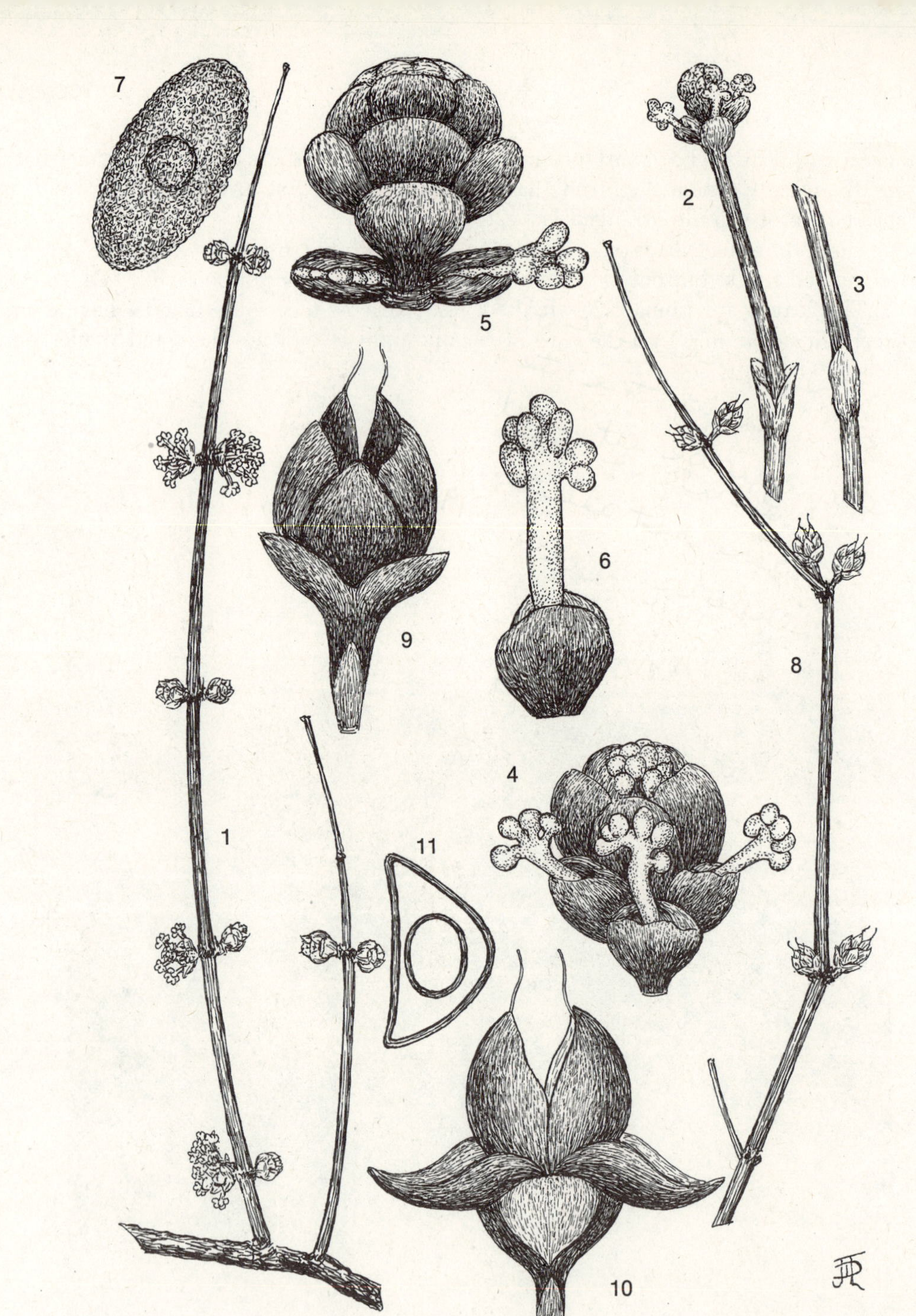

Figure 18-8. *Ephedra nevadensis:* **1** branch with polliniferous cones, **2** enlargement, showing a polliniferous cone and the paired leaves (scalelike and brown), **3** another view of the pair of leaves, **4** polliniferous cone of paired bracts and microsporophylls, each microsporophyll with several microsporangia (pollen sacs), **5–6** other views of bracts and microsporophylls, **7** pollen grain, **8** branch with ovuliferous cones, each with (in *Ephedra nevadensis* var. *nevadensis*) two ovules (only one in var. *aspera* and in various other species), **9–10** views of the ovulate cone, showing the bracts and the ovules (**9**) and seeds (**10**), each with an elongated tube around the micropyle, this appearing superficially like the style of a flowering plant pistil, **11** diagrammatic cross section of the seed (seed coat and outline of the embryo).

Figure 18-9. A cycad, *Cycas revoluta,* in cultivation: *upper left,* crown of an ovulate plant, with the megasporophylls forming a dome at the stem apex, these surrounded by vegetative leaves; *lower left,* mature plant about 15–20 feet tall, whorls of megasporophylls bearing old seeds between the whorls of leaves; *right,* polliniferous cone surrounded by leaves.

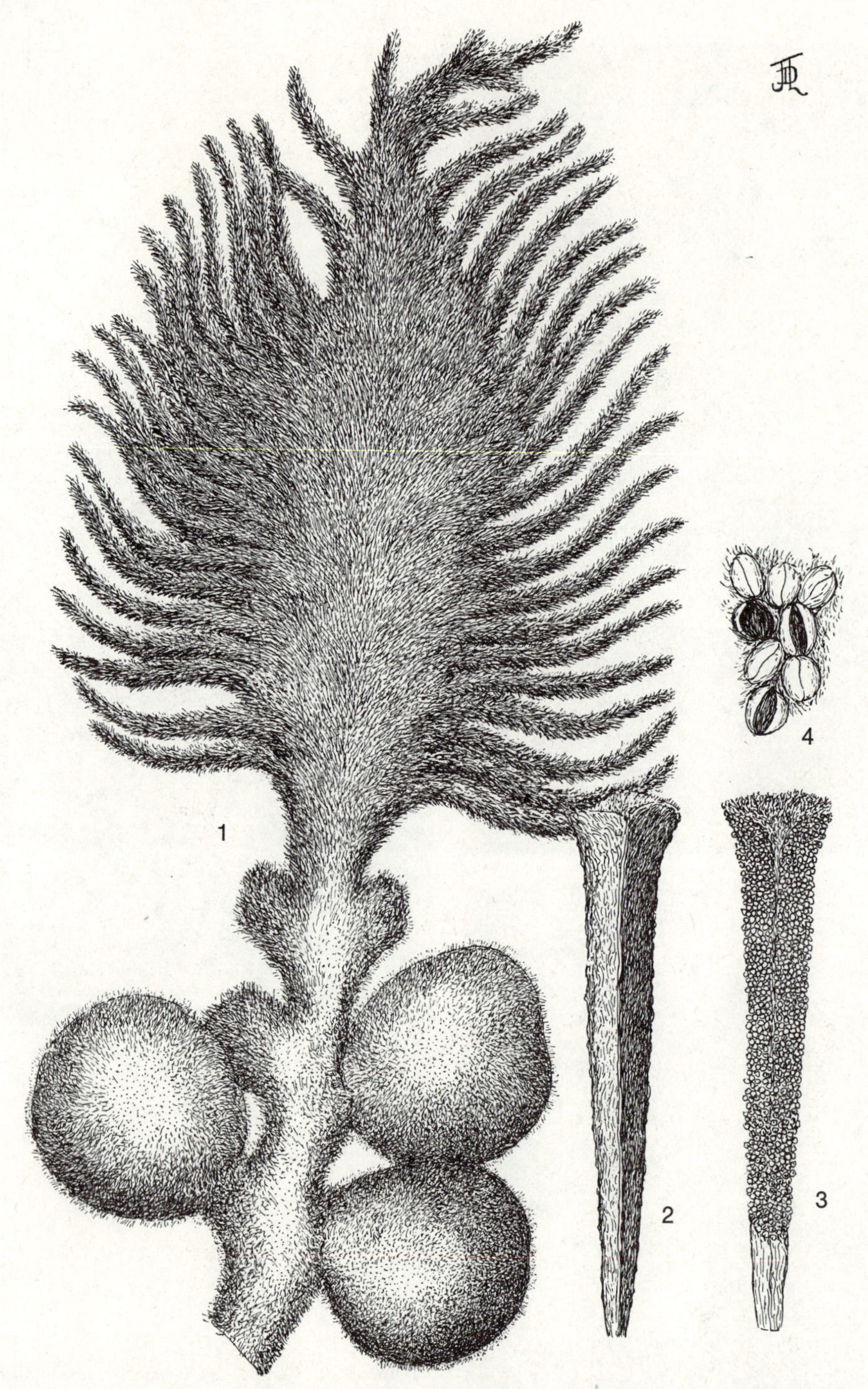

Figure 18-10. Sporophylls of *Cycas revoluta:* **1** megasporophyll with three developed seeds and three sterile ovules and with nonfunctional divisions above, these not green and covered densely by hairs, **2–3** ventral and dorsal views of a microsporophyll from a polliniferous cone, the dorsal surface bearing numerous microsporangia, **4** closed and open microsporangia.

other. The polliniferous cone of *Cycas revoluta* is very large, resembling superficially the ovuliferous cone of a sugar pine. Each microsporophyll has many microsporangia on its surface. In the more primitive cycads, such as *Cycas revoluta,* the megasporophylls are of the same outline as vegetative leaves, but they are relatively small, yellowish, and covered with dense hair. Several large seeds are developed along each margin. Among the more specialized species of *Cycas* there is a gradual reduction of the size and complexity of the megasporophyll. In the species at the opposite extreme in development from *Cycas revoluta,* the sporophyll has only a toothed margin and two ovules instead of as many as eight. In the other genera the sporophylls are hard and woody and are arranged in a compact ovuliferous cone. Each produces only two ovules as does the ovuliferous scale of the pine.

19

The Process of Identification

Manual of the Major Taxa of Living Gymnosperms

TABLE OF THE MAJOR TAXA

Class	*Order*	*Family*	*Examples*
1. Conopsida	1. Ginkgoales	1. Ginkgoaceae	Maidenhair tree
	2. Pinales	*1. Pinaceae	Pine, fir
		*2. Taxodiaceae	Bald cypress, redwood
		*3. Cupressaceae	Cypress, juniper
		4. Araucariaceae	Monkey-puzzle, Kauri
		5. Podocarpaceae	*Podocarpus*
		6. Cephalotaxaceae	Plum-yew
	3. Taxales	*1. Taxaceae	Yew, *Torreya*
2. Ephedropsida	1. Ephedrales	*1. Ephedraceae	Mormon or Mexican tea
3. Gnetopsida	1. Gnetales	1. Gnetaceae	*Gnetum*
	2. Welwitschiales	1. Welwitschiaceae	*Welwitschia*
4. Cycadopsida	1. Cycadales	*1. Cycadaceae	Cycads, including *Zamia*

* As with the Angiospermae, families occurring without the intentional aid of man in North America north of Mexico are marked in the table and in the text by the asterisk. However, all families are included in the keys.

NOTE: The classes of the Gymnosperms are keyed out on page 7 in Chapter 1.

Class 1. **CONOPSIDA.** CONIFER CLASS

Shrubs or usually tall trees; trunk commonly long and straight and much longer and larger than the numerous branches, but the plants sometimes diffusely branched; stem with a very small central pith, much wood (increased by annual cylinders formed by the cambium) and little pith or cortical tissue (external to the wood); bark present, relatively thick but scaling away on the outer surface. Leaves relatively small, simple, usually linear, needlelike, sometimes broader, then lanceolate or broadly fan-shaped. Plants usually monoecious. Polliniferous cones small (usually 2–3 centimeters or less in length); each microsporophyll usually with 1–several pollen sacs. Seed coat of usually 2 layers (3 in the Ginkgoales), usually without vascular bundles; embryo with a resting stage in the seed, germinating after an inactive period, this sometimes of several years' duration.

Key to the Orders

1. Leaves 2–3 or more cm. in length, reniform to fan-shaped, usually broader than long, usually with one apical notch, annually deciduous; ovules not in cones, in pairs, terminating long stalks arranged spirally on short spur branches; trees dioecious, much-branched, resembling dicotyledons.

   Order 1. **Ginkgoales,** below.

1. Leaves elongated and flat or needlelike or small and scalelike and only 1 to a few mm. long, entire, evergreen (or usually so); ovules in cones (or rarely solitary and not in cones, then sessile); trees or sometimes diffusely branched shrubs or small trees, *not* closely resembling the dicotyledons.
   2. Seed *not* enclosed or partly enclosed in a fleshy or leathery aril; seeds usually (always in the species occurring in North America north of Mexico) enclosed in cones (these sometimes berrylike and with only 1–3 seeds, the seeds then being much smaller than the cone as a whole, which is less than 1 cm. in diameter).

      Order 2. **Pinales,** below.

   2. Seed enclosed or partly enclosed in a fleshy aril, this either brilliantly colored (red) at maturity or green with purplish streaks or blotches (and then 2–4 cm. long).

      Order 3. **Taxales,** page 627.

Order 1. ***GINKGOALES.***

MAIDENHAIR TREE ORDER

Diffusely branched trees resembling dicotyledons; wood containing mucilage. Leaves mostly broader than long, reniform to fan-shaped, usually with a single apical incision, annually deciduous after turning yellow in autumn. Plants dioecious. Microsporophyll resembling an epaulet, that is, umbrellalike, with the 2 pollen sacs pendent from the apical enlarged structure. Ovules produced in pairs, terminating long stalks arranged spirally on stubby spur branches; seed coat 3-layered, with 2 or rarely 3 vascular bundles in the inner layer.

Family 1. **Ginkgoaceae.**

MAIDENHAIR TREE FAMILY

Permian Period to the present. The order is known from extinct genera, including *Baiera,* and a living genus, *Ginkgo,*[*] which existed also in Mesozoic time, the single living species, *Ginkgo biloba,* having been preserved through cultivation in the Orient. Occasionally travelers report it as native in western China, but specimens are lacking. The species is cultivated commonly in the United States. It is a beautiful street and park tree, beloved for its habit of turning yellow in the fall and the sudden loss of all its leaves. The ovulate trees are not so beloved for the rank butyric acid odor of the seeds when they become rancid.

Order 2. ***PINALES.*** CONIFER ORDER

Mostly straight-trunked tall trees with whorls of much smaller lateral branches, but sometimes diffusely branched and varying from small trees to prostrate shrubs, not closely resembling dicotyledons, with resin (pitch) in the wood and leaves. Leaves needlelike and elongate or scalelike and only 1–few mm. long, entire, evergreen (i.e., deciduous after a year or more) in all but a few genera (e.g., *Larix,* the larches or tamaracks). Plants usually monoecious, rarely dioecious (e.g., in some species of *Juniperus*); sporophylls or ovuliferous scales usually in cones. Microsporophylls

---

[*] This name and those of the higher taxa based upon it may have to be altered in spelling, substituting *y* for the second g.

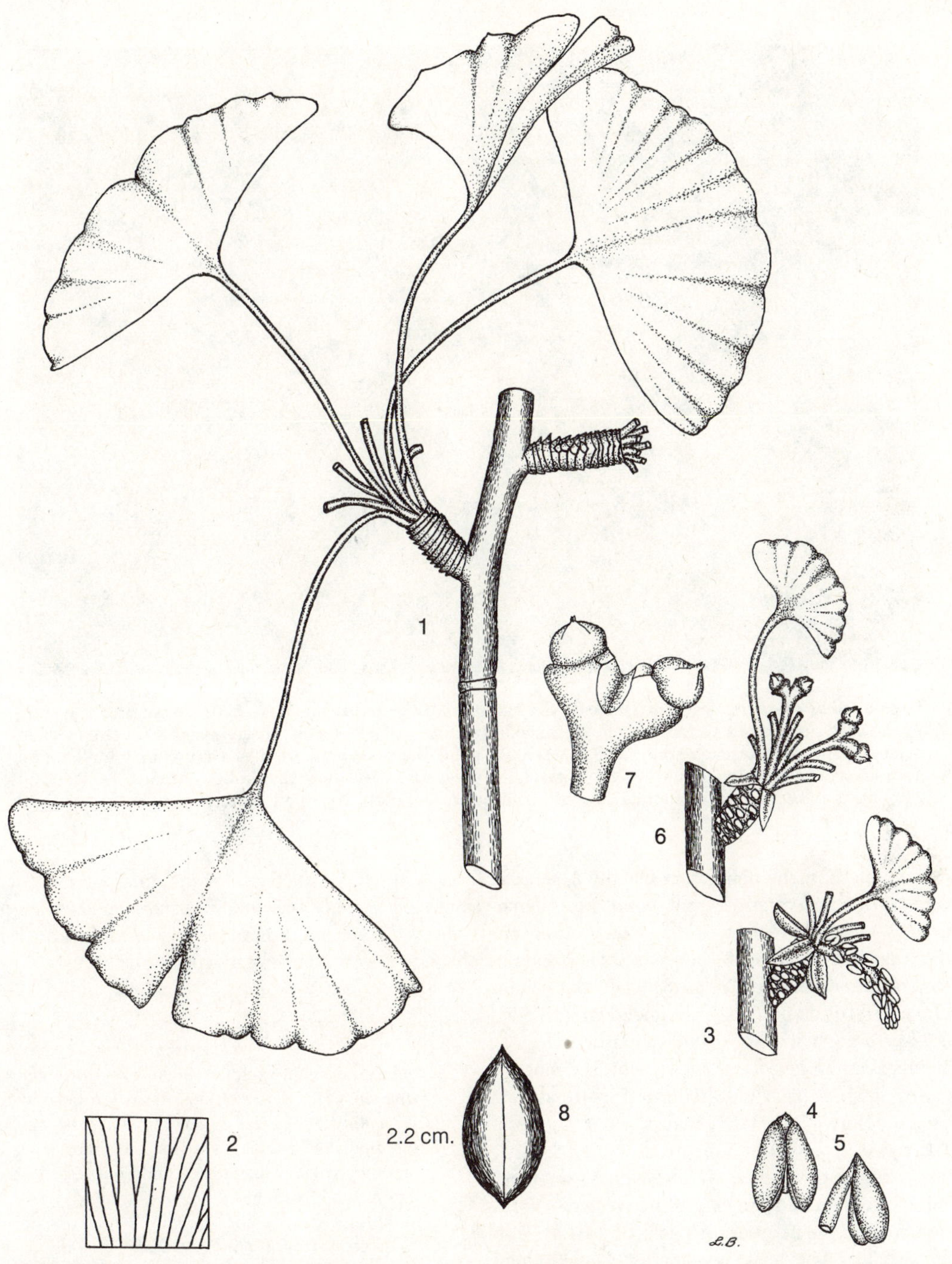

Figure 19-1. Ginkgoales; Ginkgoaceae; *Ginkgo biloba,* the maidenhair tree, a "living fossil," known only from cultivation, originally in the Orient: **1** branch bearing the characteristic spur branches which produce a new tuft of leaves each year, **2** the dichotomously branched veins, **3** polliniferous strobilus, **4–5** views of a microsporophyll showing the two pollen sacs, **6** the stalked pairs of ovules, **7** pair of ovules enlarged, **8** seed after decay of the outer fleshy layer of the seed coat.

Figure 19-2. Ginkgoales; Ginkgoaceae; *Ginkgo biloba,* the maidenhair tree: twigs bearing leaves in apical tufts, new ones forming each year; seeds the size of large olives, these naked and solitary or in a pair at the apex of each stalk. Seeds were not available for preparation of **Figure 19-1** for Edition 1, because the female trees are planted rarely. The seeds, after they have fallen, become rancid, and they have a strong, and unpopular, butyric acid odor, like that of spoiled butter.

rarely epauletlike, the pollen sacs usually 2–several. Ovules 1–many (often 2) per cone scale (ovuliferous scale), the scale subtended by a bract, but often fused with it and the two structures not distinguishable; in the *Podocarpaceae* the ovules solitary, but fundamentally in reduced cones. Seed coat 2-layered, without vascular bundles.

Pennsylvanian Period to the present, the modern families having been differentiated gradually in Mesozoic time. The tall Pinales occur throughout the world, and they are the dominant lumber trees of temperate regions. The richest development of genera and species is in western North America and the temperate parts of eastern and central Asia. The areas of poorest development are the Tropics and South Africa. The group is common in eastern North America, but it has relatively few representatives in northern Europe. Some of the most extensive forests, as, for example, those in Scandinavia, include only a few species. Only two medium-sized species, a spruce and a pine, are common in Scandinavia, a region once covered largely by coniferous forests.

Key to the Families

1. Seeds borne in cones, attached to the cone scales (the cone itself sometimes fleshy or berrylike and with only 1 to 3 seeds, and in that case the plant distinguished by leaves which are minute, scalelike or somewhat elongated, and appressed tightly against the branchlets).
   2. Ovules (and seeds) usually 2 or more per cone scale, rarely 1 on some scales or only 1–3 per entire (berrylike) cone, the seed *not* embedded in the scale; leaves either linear, needlelike, or scalelike.
      3. Leaves and cone parts one at each node, that is, arranged alternately or in spirals (in one

genus the adult foliage leaves in fascicles of usually 2, 3, or 5 or rarely 1 or 4 and the fascicle enclosed at the base by a membranous sheath); seed-bearing cones with woody scales, the scales numerous.

4. Leaves *either* (1) jointed (later disjointing) at the bases and after several years individually deciduous *or* (2) deciduous in sheathed clusters of a constant small number (usually 2, 3, or 5 or rarely 1 or 4); bud scales present, membranous; pollen sacs 2 to each microsporophyll; seeds 2 to each cone scale.

Family 1. **Pinaceae** (pine family).

4. Leaves not jointed at the bases, not deciduous either individually or in sheathed clusters but persistent on elongated branchlets which are finally deciduous; bud scales none; pollen sacs several to each microsporophyll; seeds several to each cone scale.

Family 2. **Taxodiaceae** (bald cypress family).

3. Leaves and cone scales more than one at a node, arranged in cycles of two or three; bud scales none; seed-bearing cone scales hard and woody or sometimes papery, pulpy, or juicy, the seed-bearing cones compact and the scales not more than 14.

Family 3. **Cupressaceae** (cypress family).

2. Ovule (and seed) 1 per cone scale, the seed often embedded in the scale, the scale fused with the (larger) bract; leaves usually lanceolate but sometimes narrow and then thicker than broad.

Family 4. **Araucariaceae** (monkey-puzzle tree family).

1. Seeds not enclosed in a cone (except sometimes in the extra-territorial family Podocarpaceae), solitary, with either scalelike subtending bracts *or* a conspicuous aril of fleshy tissue, this often a colored cup covering much or nearly all of the seed; leaves elongate, not scalelike.

2'. Pollen sacs 2 per microsporophyll; seed solitary.

Family 5. **Podocarpaceae** (*Podocarpus* family).

2'. Pollen sacs 3–7 per microsporophyll; ovules (and often seeds) in pairs.

Family 6. **Cephalotaxaceae** (plum-yew family).

*Family 1. **Pinaceae.** PINE FAMILY

(Synonyms: *Abietaceae* and *Coniferae*)

Large or sometimes rather small trees; main trunk usually elongate, commonly unbranched. Buds enclosed by special scalelike leaves (bud scales); leaves slender, needlelike, arranged spirally, deciduous after a few years, falling singly *or* in *Pinus* (pine) in clusters of a definite small number, usually 2, 3, or 5, or rarely 1 or 4, and the cluster enclosed basally by a membranous sheath. Microsporophyll with two embedded

Figure 19-3. Pinales; Pinaceae; *Pinus ponderosa*, western yellow pine, with a trunk diameter of about five feet. The broad scales of the bark are characteristic. This is the most widely distributed pine in North America, occurring in various forms and varieties from British Columbia to Baja California and western Texas. The open, parklike stands of large trees are characteristic. Photograph from Hat Creek, north of Mt. Lassen, Cascade Mountains, California.

Figure 19-4. Fir cones; Pinales; Pinaceae; *Abies magnifica,* red fir. Each cone of *Abies* and of *Cedrus* is erect on the branch and it falls a scale at a time, leaving the axis on the tree. Photographs from Mammoth Lakes, Sierra Nevada, California.

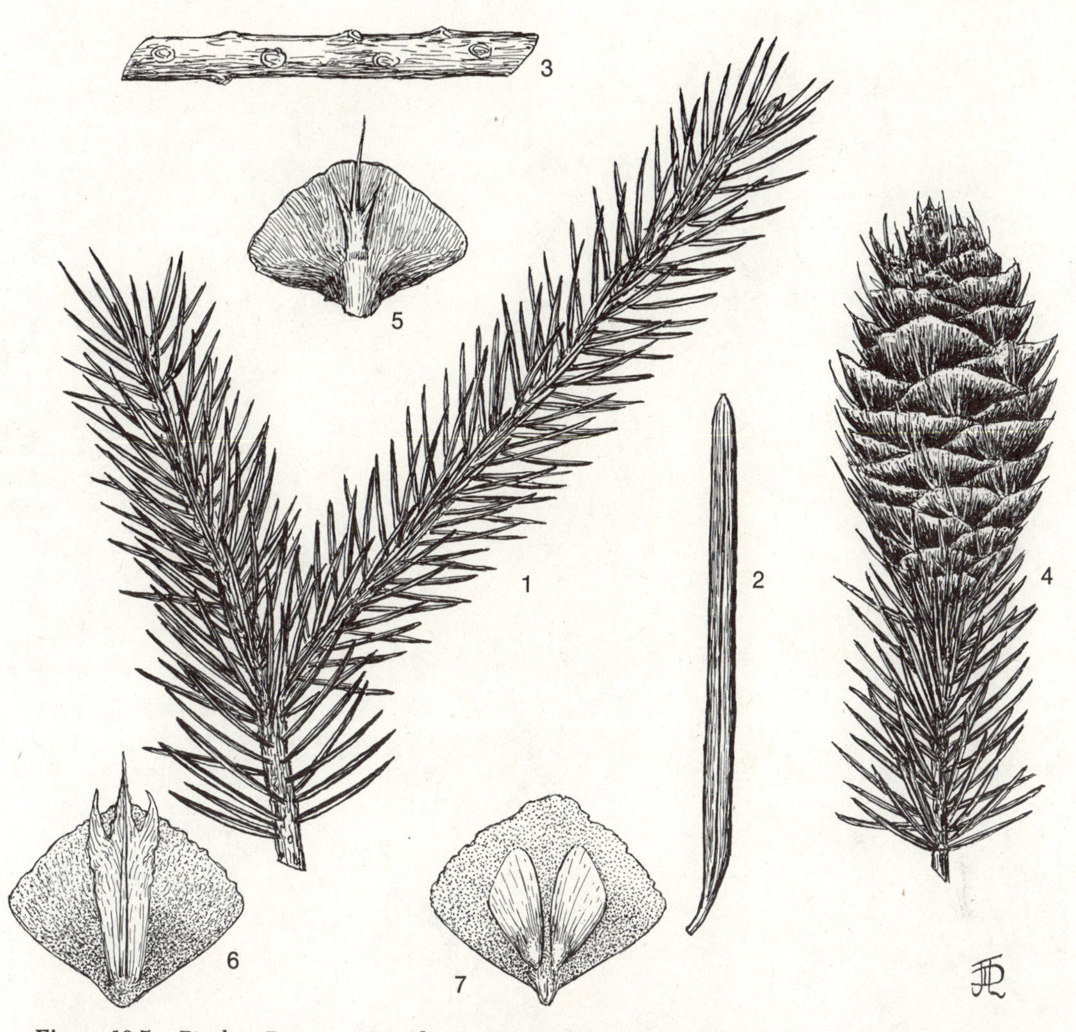

Figure 19-5. Pinales; Pinaceae; *Pseudotsuga Menziesii,* the Douglas-fir: **1** leafy branch, **2** leaf, **3** twig bearing leaf scars, **4** twig with a terminal cone, the 3-pronged bracts exserted beyond the scales, **5–6** dorsal views of cone scales and the subtending bracts, **7** ventral view of a cone scale and the two wingéd seeds.

sporangia. Ovuliferous scale axillary to a bract, developed from an axillary bud; ovules two on the upper surface of each scale.

Lower Cretaceous Period to the present. The following are among the best-known living genera of this Northern Hemisphere family.

Pinus (pine), including about 80 to 90 species, is one of the principal lumber-producing genera. Some species, for example, the western yellow pine of North America (*Pinus ponderosa*) and the sugar pine of California (*Pinus Lambertiana*), become as much as 200 feet tall and 8 or rarely up to 13 feet in trunk diameter.

Abies (fir), including about 40 species. *Abies magnifica,* the red fir of California, may be 200 feet high and with an 8-foot trunk diameter. The common northeastern fir is *Abies balsamea,* the balsam fir.

Picea (spruce), including also 40 species. Both spruces and firs predominate in the forests of the far north and of the high mountains near timber line.

1
2
3
4
5
6
7
8
9
10
11
12
13
14
15

Pseudotsuga, including about seven species, the best known being the Douglas fir of western North America, one of the most important lumber trees of the world and the mainstay of lumbering in the Pacific Northwest. Some trunks are 200 feet high and 10 feet or more in diameter. The wood, sold as Douglas fir or as "Oregon pine" or "O.P." is straight-grained, strong, durable, and exceedingly tough.

Tsuga (hemlock) including about 14 species and *Larix* (larch or tamarack) including about 10. Both are represented by two species in the Pacific Northwest and one in the northeast.

*Family 2. **Taxodiaceae.**

BALD CYPRESS FAMILY

(Sometimes included in *Pinaceae*)

Large or often enormous trees with massive trunks. Buds not scaly, the embryonic leaves exposed; leaves narrow or scalelike, but not markedly needlelike, arranged spirally, not individually deciduous but persistent on elongated later deciduous branchlets.* Microsporangia several per microsporophyll. Ovules several to many per ovuliferous cone scale, arranged on all sides of its basal portion; cone scale not evidently subtended by a bract, probably fused with it; ovules 2–several per cone scale.

Jurassic Period to the present. Three living genera, *Taxodium, Sequoia*, and *Sequoiadendron*, occur in the United States.

Taxodium, includes the bald cypress (*Taxodium distichum*) characteristic of the river swamps of the Southern States, and the pond cypress (*Taxodium ascendens*). The bald cypress is noteworthy for the irregular trunk base swellings known as cypress "knees." The Big Tree of Tule in Mexico (*Taxodium mucronatum*) has a trunk diameter of over 50 feet, but hardly longer than its diameter. According to Boone Hallberg, it is formed by fusion of enlarged crown sprouts or of several trees or of an ancient stump.

Sequoia consists of a single species, the coast redwood (*Sequoia sempervirens*) restricted to the ocean fog belt from the extreme southwestern corner of Oregon to about 100 miles south of San Francisco, California. These trees attain heights up to 360 feet, and they are the tallest in the world. Maximum trunk diameter is about 20 feet.

Sequoiadendron includes a single species the Big Tree (*Sequoiadendron giganteum**) of the California Sierra Nevada. It is restricted largely to groves occurring on ridges with good cold air drainage during the winter, but at the southern end of its range in and near Sequoia National Park it forms an extensive and continuous forest. Although the Big Tree is not quite as tall as the coast redwood, it has a more massive trunk. Just above the irregular swellings at ground level the stem measures up to 33 feet in diameter, that is, about the length of an average classroom. The trees are of great age, and many of the living ones were a thousand to eighteen hundred years old at

* This character may be detected by examination of the debris beneath the tree.

* The Big Tree, long known as *Sequoia gigantea*, has been segregated recently as *Sequoiadendron giganteum*. The evidence upon which this generic segregation rests is strong, because the two redwoods differ in many characteristics, and they are completely isolated at least geographically. A fortunate by-product of segregation (but not an argument in favor of it) is the probable solution to a difficult nomenclature problem. Although the Big Tree is one of the best known plants in the world, the name combination *Sequoia gigantea*, by which it was well known for almost a century, is not valid in its application to the Big Tree because an exactly duplicating though invalid name combination was used earlier for the coast redwood, which is known by the still earlier name *Sequoia sempervirens*. (Cf. J. T. Buchholz, *The Generic Segregation of the Sequoias*. American Journal of Botany 26: 535–538. 1939.) *Americus*, an earlier generic name, has been taken up by some authors. However, its effective publication is dubious (cf. Taxon 4: 40–42. 1955).

Figure 19-6. Pinales; Taxodiaceae; *Sequoia sempervirens*, the coast redwood of California and southwestern Oregon: **1** branchlets with spreading, spirally arranged leaves, the tips deflected into two ranks, the leaves at the apices of these branchlets scalelike and resembling those of the big tree, *Sequoiadendron giganteum*, the twig ending in a young ovuliferous cone, **2** twig with polliniferous cones, **3** polliniferous cone, **4** polliniferous cone dissected, revealing the microsporangia and the axis, **5–6** views of the microsporophyll, showing the pollen sacs, **7** young ovuliferous cone, **8** longitudinal section, showing the megasporophylls and ovules, **9–10** views of the megasporophyll, **11** mature seed-bearing cone, **12** cone scales spreading apart and revealing the seeds, **13** cone opened after the seeds are shed, **14–15** views of the seed. Cf. **Figures 25-7** redwood, **25-22** bald cypress, **25-43** (above) pond cypress.

Figure 19-7. Pinales; Taxodiaceae; *Sequoiadendron giganteum,* the big tree. The tree at the right is about twenty-five feet in diameter above the basal swellings of the trunk; the trees in the background are large forest trees five to ten feet in diameter. Wawona Grove, Yosemite National Park, California.

Figure 19-8. Pinales; Taxodiaceae; *Sequoiadendron giganteum,* giant sequoia, big tree, Sierra redwood; in winter; Wawona Grove, Yosemite National Park. The end wall of the cabin at the right measured 20 feet across, exclusive of the eaves.

the beginning of the Christian era. The ages of the largest trees have not been determined accurately. An 11-foot section of a log of *Sequoiadendron* sent to the Riksmuseet in Stockholm in 1952 is 2,800 years old. The age of a 33-foot tree must be far greater, but there is no accurate boring equipment for trees of such size. Pines known to be 4,600 years old may or may not be older. According to Muller, oaks growing in Texas sand dunes may have grown partly beneath the sand since the close of Pleistocene 10,000 years ago.

The bald cypress loses its leafy branchlets every year, the leaves turning yellow or slightly reddish in the autumn. The branchlets of both the redwoods fall away (without coloring) after more than one season rather than annually, so the trees remain evergreen.

Recently another genus, *Metasequoia,* known first from fossils, has been discovered in the interior of China (eastern Szechuan and western Hupeh). In mid-Tertiary time this plant and the coast redwood occurred through much of the Northern Hemisphere. Now one is making a last stand in China, the other in California.

*Family 3. **Cupressaceae.**

CYPRESS FAMILY

(Sometimes included in *Pinaceae.* Segregate: *Juniperaceae*)

Small trees to prostrate shrubs, diffusely branching, the main trunk usually short. Buds covered by only the vegetative leaves, without specialized bud scales; leaves scalelike, usually triangular or sometimes elongated and narrow but not clearly needlelike, arranged in cycles of 2 or 3, not individually deciduous, some of the branchlets falling away each season and carrying the attached leaves with them. Microsporophylls somewhat peltate, each with 2–6 pollen sacs. Ovules usually several or many per cone scale, but in some genera only 1 or 2, if numerous, arranged on all sides of the base of the cone scale; cone scale not evidently subtended by a bract (probably fused with it).

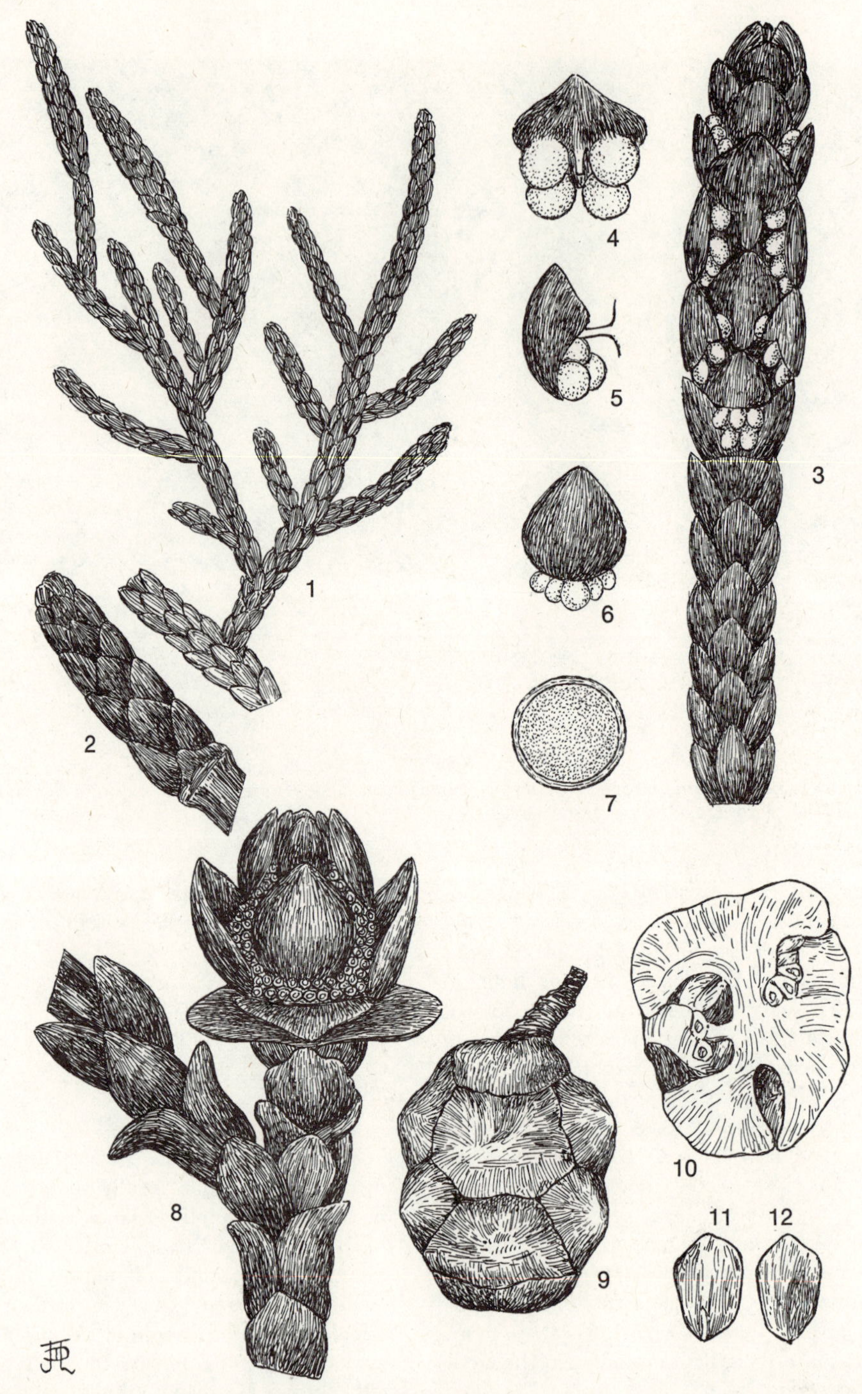

Figure 19-9. Pinales; Cupressaceae; *Cupressus sempervirens*, Italian cypress: **1** twigs with opposite appressed scalelike green leaves, **2** enlargement, **3** polliniferous cone with some of the microsporangia (pollen sacs) exposed, **4–6** views of a microsporophyll, **7** pollen grain, **8** ovuliferous cone at the time of pollination, the numerous ovules exposed, **9** mature cone, **10** longitudinal section showing the numerous seeds around the ovuliferous scales, **11–12** views of seed.

Upper Cretaceous Period to the present. The cypress family is represented in North America by a number of living genera, the most common being *Juniperus* (juniper). Some junipers are trees, some large shrubs, and others small prostrate shrubs. Like a number of other North American plants, some species are known as "cedars." Several taxa of *Cupressus* (cypress) occur in California, two in Arizona, and others in Mexico. The largest trees in the family and the only North American species which make good lumber are the incense cedar (*Libocedrus decurrens**) of California and southwestern Oregon, the giant cedar (*Thuja plicata*) of the Pacific Northwest and northwestern California, and the Lawson cypress (*Chamaecyparis Lawsoniana*) of south coastal Oregon and northwestern California. All these three trees have excellent wood with a fine straight grain. Some junipers are used for fence posts.

Family 4. **Araucariaceae.**

MONKEY-PUZZLE TREE FAMILY

Large, sometimes gigantic, trees. Buds naked; leaves, except in a few species, not needlelike but broad and lanceolate, arranged spirally on the branchlets, not deciduous individually, the branchlets deciduous and carrying the attached leaves with them. Microsporangia numerous on each microsporophyll. Ovule 1 per ovuliferous scale; seed often embedded in the tissues of the scale fused with the larger bract.

Triassic Period to the present. This is the dominant family of living conifers in the Southern Hemisphere. Many of its members are known erroneously as "pines." The Kauri (*Agathis*) of New Zealand is one of the largest species of the family, and often it may be 160 feet tall and with a trunk 6 or 7 or even 12 feet in diameter. Several species of *Araucaria* are common in cultivation on the American Pacific Coast, in the Southern States (especially Florida), and in other areas where the winters are not too cold. The most striking cultivated species is the monkey-puzzle tree, *Araucaria imbricata*, of the southern Andes. This tree is noteworthy for its whorls of branches each densely covered with rather broad lanceolate leaves. Its appearance is somewhat like that of *Lepidodendron*, the scale tree, a spore plant of the Carboniferous Period.

Family 5. **Podocarpaceae.**

PODOCARPUS FAMILY

Small to large trees or rarely shrubs. Leaves alternate or opposite, needlelike to lanceolate, oblanceolate, linear, or narrowly oblong. Plants either monoecious or dioecious. Microsporophylls with 2 pollen chambers. Ovules solitary (in reduced cones) or when in strobili 1 per ovuliferous cone scale.

Jurassic Period to the present. The family is restricted largely to the Southern Hemisphere and the Orient. A few species of *Podocarpus* are cultivated in the southern portions of the United States.

The family may be related to the Taxaceae, but, if so, the relationship is remote. According to recent data the lines of development have been long separated.

Family 6. **Cephalotaxaceae.**

PLUM-YEW FAMILY

Shrubs or trees; branching opposite. Leaves arranged spirally, linear, small. Plants usually dioecious, sometimes monocious. Microsporophyll with 3–8 pollen sacs. Ovules in pairs, not in cones, with short stalks, large, drupelike at maturity, about 2.5 cm. long and resembling plums.

The family includes two genera restricted to eastern Asia. Some species of *Cephalotaxus* are cultivated in the United States.

Order 3. ***TAXALES.*** YEW ORDER

Shrubs or small trees. Leaves linear, arranged spirally. Microsporophyll resembling an epaulet, with 3 to 8 pollen sacs. Ovules not in cones, the ovule or seed solitary, *either* (in *Taxus*, yew) surrounded and largely enclosed by a cuplike structure (aril) which turns red when the seed is mature *or* (in *Torreya*, "nutmeg") completely enclosed by a green aril and subtended by scalelike leaves; seed coat with a vascular bundle in the outer layer.

---

* According to Li, *Libocedrus* is restricted to the Southern Hemisphere, and the North American plant, if so, should be *Calocedrus* (*Heyderia*).

Figure 19-10. Pinales; Araucariaceae; *Araucaria Bidwillii,* Bidwill "pine," bunya-bunya; the sharp-pointed lanceolate leaves actually spiral along the twigs but displaced so that they appear to be in two lateral ranks; cone 10 or 12 inches long, single scale enclosing the solitary seed and the seed dissected out. Native of Australia.

*Family 1. **Taxaceae.** YEW ORDER

Jurassic Period to the present. Two of the three living genera, *Taxus* and *Torreya*, are native in North America; the third, *Austrotaxus*, is restricted to New Caledonia. One species of *Taxus* (yew) is native on the Pacific Coast, another in the Middle West and East. One species of *Torreya* is a narrow endemic in northern California, as is another in northern Florida. Three species are native in the Orient, and evidently *Torreya* is made up of survivors of a once widely distributed genus. The Californian species is noteworthy for its exceedingly sharp leaf tips.

The wood of the yew family is noted for its resiliency, and in the Middle Ages it was a favorite for making bows. Its springiness is due to extra spiral thickenings in the walls of the xylem cells.

Figure 19-11. Pinales; Araucariaceae; *Agathis robusta,* Queensland Kauri, Dammar "pine"; a gigantic tree as much as 160 feet tall and with a trunk up to 12 feet in diameter; twigs with lanceolate leaves; a cone about 8 inches long. Native of Queensland, northeastern Australia. Another of the "pines" of the Southern Hemisphere, where there are no real pines (*Pinus*).

Class 2. **EPHEDROPSIDA.** EPHEDRA CLASS

Scraggly, diffusely branched shrubs usually 1–4 feet high; stems green and photosynthetic. Resin canals none. Leaves in cycles of two or three at each node, without chlorophyll, reduced to dry, brown or tan scales. Plants dioecious. Polliniferous strobili consisting of opposite scales subtending stalklike microsporophylls each with about five or six sporangia. Ovules solitary or in pairs, subtended by several pairs of scales, the cone being reduced, the micropyle within an extended tube.

Order 1. ***EPHEDRALES.*** EPHEDRA ORDER

*Family 1. **Ephedraceae.** EPHEDRA FAMILY

The single relict genus *Ephedra* is represented by a number of species occurring in arid portions of both the Eastern and the Western hemispheres. Several are native in (especially the deserts of) the southwestern United States, where they are known as Mexican or Mormon tea.

Class 3. **GNETOPSIDA.** GNETUM CLASS

Shrubs, woody vines, or thick, fleshy, low herblike plants. Resin canals none. Leaves opposite, either broad (as in dicotyledons) or up to 2 m. long and ribbonlike. Plants dioecious or rarely monoecious. Polliniferous cones more or less conelike or elongate and interrupted. Ovules and seeds at several levels in an ovulate strobilus; seed with two integuments, the micropyle within an extended tube resembling the style of a flowering plant (as also in *Ephedra*) but not homologous with it.

Figure 19-12. Taxales; Taxaceae; *Taxus baccata*, English yew: **1** leafy twig with polliniferous cones, **2** enlargement, **3–4** polliniferous cone, the bracts removed in **4**, **5** twig with an ovule enclosed in the red aril, **6–7** enlargements, **8** aril opened, exposing the seed, **9** seed, **10** longitudinal section, exposing the embryo.

Key to the Orders

1. Leaves in numerous pairs, ovate to oblong, the veins netted as in the dicotyledons; tropical or subtropical shrubs or woody vines; microsporophylls in each unit of the strobilus *not* surrounding a sterile ovule; strobilus elongate, with somewhat separated whorls of microsporophylls or of ovules (these sometimes with sterile microsporophylls below them).

   Order 1. **Gnetales,** below.

1. Leaves in a single pair, linear and elongate (up to 2 m. long); desert (South West African) bulbous plants; microsporophylls in each unit of the strobilus surrounding a sterile ovule; strobilus relatively short, the bracts in overlapping pairs, the groups of microsporophylls or of ovules not conspicuously separated.

   Order 2. **Welwitschiales,** below.

Order 1. ***GNETALES.*** GNETUM ORDER

Shrubs or woody vines resembling dicotyledonous flowering plants. Leaves simple, in opposite pairs at the nodes, green, ovate to oblong, the veins netted. Plants usually dioecious, but sometimes with microsporophylls and ovules in the axils of the same bract. Polliniferous strobilus long and slender, interrupted. Ovuliferous strobilus with 5 or 6 somewhat separated whorls of ovules.

Family 1. **Gnetaceae.** GNETUM FAMILY

The single relict genus *Gnetum* occurs in tropical regions of both the Eastern and the Western hemispheres. About a dozen species are native in the West Indies and northern South America.

Order 2. ***WELWITSCHIALES.*** WELWITSCHIA ORDER

Stem tuberous, resembling an inverted cone or a turnip or somewhat headlike, rarely protruding more than 1–1.5 dm. above ground level. Leaves 2 in an opposite pair, ribbonlike or straplike, persisting throughout the life of the plant (for several centuries), eventually frayed into narrow ribbons by the action of the wind whipping them across the desert soil, each with six leathery parallel veins. Plants dioecious. Microsporophylls in each unit of the strobilus surrounding a sterile ovule (arranged like stamens around a pistil), enclosed in 2 pairs of bracts and a single larger bract. Fertile ovule not associated with microsporophylls, similarly enclosed in bracts within the ovulate strobilus.

Family 1. **Welwitschiaceae.** WELWITSCHIA FAMILY

The family consists of the single relict genus *Welwitschia,* which occurs only in the area from Latitude 15° S. to the Tropic of Capricorn near Walvis Bay on the west coast of South West Africa. The plant grows under extreme desert conditions in a region where rainfall is irregular and sometimes lacking for four or five years. This is one of the most weird plants ever discovered. There is only one species, known correctly as *Welwitschia mirabilis* (cf. L. Benson, Cactus and Succulent Journal 42: 195–200. 1970).

Class 4. **CYCADOPSIDA.** CYCAD CLASS

Small or dwarf trees or bulbous-stemmed large herbs; trunks (when usually present) ordinarily simple, sometimes slightly branched; stem with a thick central pith and a thick cortex and with little wood. Leaves large, pinnate or in (*Bowenia*) bipinnate. Polliniferous cones relatively very large; each microsporophyll with numerous pollen sacs. Plants dioecious. Seeds either (1) several and marginal on leaflike megasporophylls not aggregated in cones *or* (2) reduced to a single pair marginal on each hard woody scalelike megasporophyll of a compact large cone. Seed coat with 3 layers, the outer and inner each with a vascular bundle; embryo with no resting stage in the seed, germinating at once (perhaps a function of the tropical character of these plants).

Order 1. ***CYCADALES.*** CYCAD ORDER

* Family 1. **Cycadaceae.** CYCAD FAMILY

Mesozoic Era to the present. There is some evidence that cycads occurred as early as Carboniferous time or the Permian Period. The nine living genera are distributed as follows: *Cycas* in the Eastern Hemisphere; *Stangeria* in Africa; *Bowenia* in Australia; *Macrozamia* in Australia; *Encephal-*

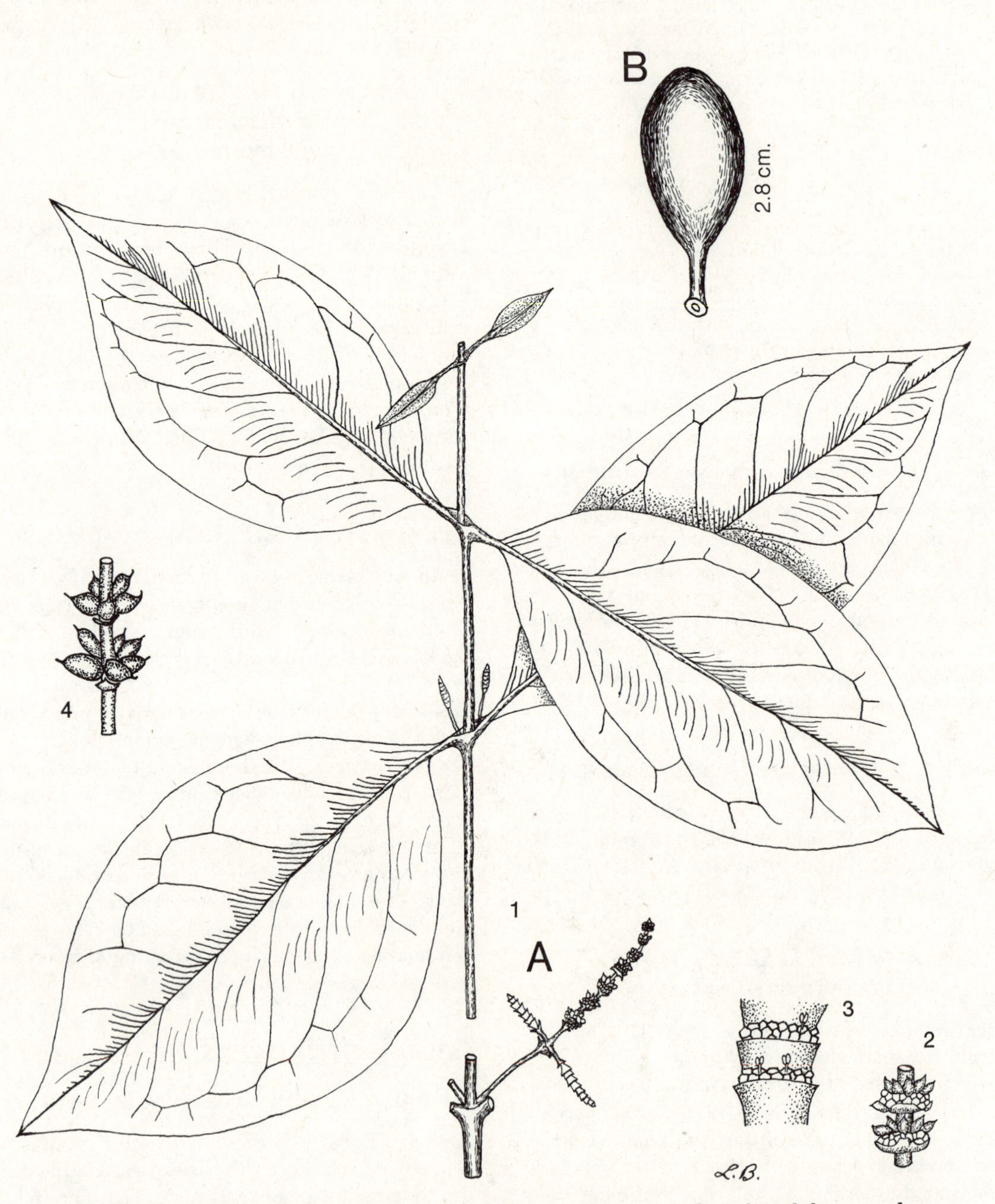

Figure 19-13. Gnetales; Gnetaceae; *Gnetum:* **A** *Gnetum Gnemon,* **1** branch with leaves and strobili, **2** ovulate strobilus with (as sometimes) at least sterile microsporophylls below the ovules, **3** polliniferous strobilus, with a few microsporophylls (stamenlike) projecting, **4** ovuliferous strobilus with whorls of ovules; **B** *Gnetum latifolium,* mature seed.

Figure 19-14. Wilwitschiales; Welwitschiaceae; *Welwitschia mirabilis,* the only species and a good candidate for most weird plant in the world. The stem is up to four feet in diameter and only to 8 or 10 inches long; it produces only the cotyledons and then two leaves, which must last for hundreds of years, growing continually from the bases and wearing away at the apices by being swept by the wind against the sand about 4 feet out from the trunk of an adult plant. *Left,* plants growing in the Namib Desert in South West Africa (Namibia), occupying an area here about 2 or 3 miles long and one-quarter mile wide, the plants scattered and few. The sand and gravel of the soil surface are underlain by caliche (as we call it in the deserts of the United States), a calcareous hardpan like soft concrete. The root, 3 or 4 inches in diameter, grows downward, and, because of the difficulty of digging for removal, by permit, of one plant for scientific purposes in southern California, there was some speculation as to whether the root might reach there already. *Below,* "young" plant perhaps 50 years old with leaves about 2 feet long, these frayed and dead apically but almost intact otherwise. Age is difficult to judge, because growth is slow and intermittent. Rainfall is less than ½ inch per year, and many years there is no rain; growth occurs mostly during the few years when rainfall is significant. As the plants become older, the leaves split lengthwise. This occurs readily, but the leaves are very strong crosswise and reminiscent of sole leather. Junction of the Kahn and Swakop rivers, great omurambas or dry washes most of the time, "back of beyond." LYMAN BENSON, Cactus and Succulent Journal 42: 195–200. 1970. Used by permission.

Figure 19-15. Wilwitschiales; Welwitschiaceae; *Welwitschia mirabilis.* Nearer view of a plant like the one in **Figure 19-14:** *above,* the leaves divided lengthwise by the sweeping of the wind but not divided into ribbons as with some plants. Each leaf has been split into 4 or more segments, and the spreading of all the segments gives the appearance referred to as like an *Octopus.* LYMAN BENSON, Cactus and Succulent Journal 42: 195–290. 1970. Used by permission.

Figure 19-16. Welwitschiales; Welwitschiaceae; *Welwitschia mirabilis* (*mirabilis* = wonderful), plants in the desert at Brandberg, South West Africa: *above,* mature ovuliferous plant with young strobili and with fallen mature ones; *below,* large polliniferous plant with numerous polliniferous strobili and with the dead peduncles of old strobili (fallen into the center). Photographs by ROBERT J. RODIN, American Journal of Botany 40: 282–3, figs. 2, 4. 1953. Used by permission.

Figure 19-17. Welwitschiales; Welwitschiaceae; *Welwitschis mirabilis:* above, plants with disintegrated seed cones, the wingéd seeds carried by the common high winds and distributed widely, but few having the benefit of the right weather for germination and growth of the seedling; many remaining within the windbreak formed by the female parent plant, as here; *below,* close-up view of parts of a leaf, the axes of cones after shedding of seeds, and the wingéd seeds. The bug (*upper right*) occurs exclusively on *Welwitschia.* It lives on the liquid materials of the plant, gotten with the typical sucking mouthparts of the Hemiptera, and it carries the pollen. *Welwitschia* probably has lived in isolation in the colonies occurring from Angola to the Swakop River in South West Africa for millions of years, and likely the relationship with the bug is very old. LYMAN BENSON, Cactus and Succulent Journal 42: 195–200. 1970. Used by permission.

Figure 19-18. Cycadales; Cycadaceae; *Zamia;* leaves and cones from subterranean tubers: *above,* plant with a polliniferous cone; *below,* plant with an ovuliferous cone. Ocala National Forest, Florida.

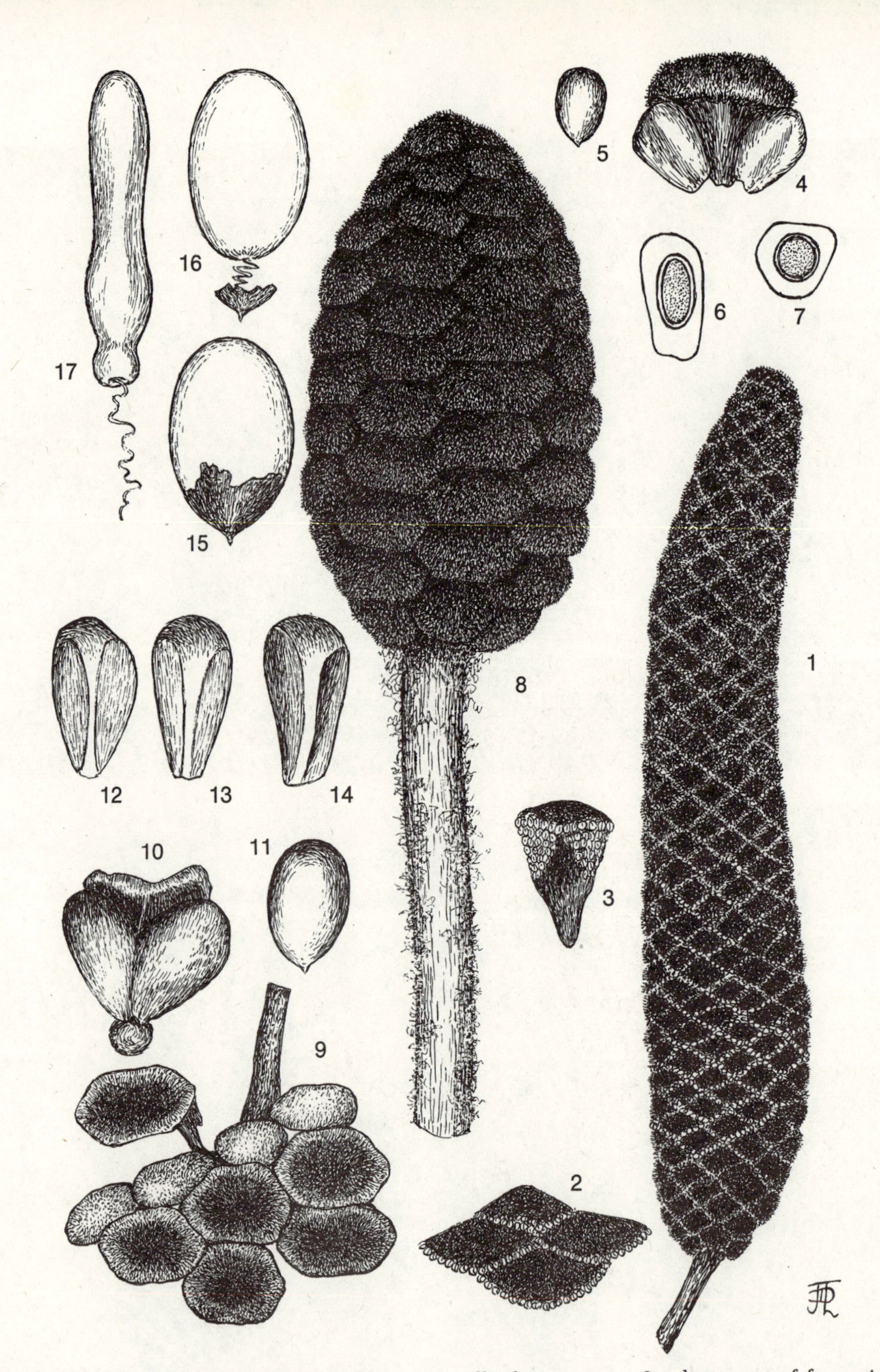

Figure 19-19. Cycadales; Cycadaceae; *Zamia:* **1** polliniferous cone, **2** enlargement of four microsporophylls, **3** microsporophyll with microsporangia, **4** megasporophyll and two ovules from the ovuliferous cone, **5** ovule, **6–7** sections of the ovule, **8** seed-bearing (older) cone, **9** scales and ovules from the seed-bearing cone, **10** megasporophyll and two seeds, **11** seed, **12–14** views of a seed angled by compression in the cone, **15** seed with the integument removed except at the micropylar end, **16** remnant of the integument removed revealing the coiled suspensor of the embryo, **17** embryo.

artos in Africa; *Dioön* in Mexico and Honduras; *Microcycas* in Cuba; *Zamia* in Cuba and Florida and from Mexico to Bolivia; *Ceratozamia* in Mexico.

The single genus (*Zamia*) native in North America north of Mexico has a bulbous subterranean stem base; and the leaves, which stand up two or three feet high, resemble those of a palm tree. Most cycads have trunks, and the plants resemble dwarf palms. The commonest species in cultivation is *Cycas revoluta,* known as sago "palm."

The structure of the stele in the stem is unlike that of any other group of plants; there is no elongated stele but only a horizontal square plate of xylem and phloem from which vascular bundles extend upward through pithy tissue to the leaves.

20

The Basis for Classification of Gymnosperms

Derivation and Starting Points

The genera of the gymnosperms are few, and, clearly now, they are relics of a much larger group of dominant plants of long ago. Only the conifers are of marked importance in the modern vegetation, and the relatively few genera of even this group belong to a small number of well-differentiated families. Their position of ecological dominance in many temperate forests is due to the great height of the individuals rather than to number of plants or of species. They are far outnumbered by the associated angiosperms.

The living gymnosperms belong to four clearly distinct groups. The major problem in classifying them is not their disentanglement, as with the orders of angiosperms, but an evaluation of the degree of remoteness of their relationships. Different authors have considered the chief taxa to be of all ranks from family to division. The Pinales (Coniferales) have appeared in books and papers as a single family and as eight families in four orders of two classes.

The living and extinct developmental lines* of the gymnosperms are as follows:

1. *The Conifer Line,* beginning with the Cordaitales and culminating in the modern Ginkgoales, Pinales, and Taxales.
2. *The Ephedra Line,* beginning with plants which have left a long though meager fossil record and surviving in *Ephedra*.
3. *The Gnetum Line,* beginning with plants which have left no fossil record and surviving in *Gnetum* and *Welwitschia*.
4. *The Cycad Line,* beginning with the Cycadofilicales and culminating in the extinct Bennettitales and the living Cycadales.

Little is known of the origin of the second and third lines listed above, but there is new information concerning the origin and relationships of the first and fourth. This is based on the recent research of Charles B. Beck,

* The word "line" is not wholly satisfactory because it connotes a simple linear sequence. "Trend" denotes a particular tendency; however, tendencies change at any time. A more satisfactory word has been sought, but so far one has not come to mind. Cf. p. 597.

Chester A. Arnold, H. N. Andrews, Harlan P. Banks, and their associates and students. The apparent great disparity of the Conifer Line and the Cycad Line has been the basis of many theories, including the contention that the two lines lack close relationship. This question is resolved by studies of the progymnosperms of middle Devonian to Carboniferous, forming a link between the Psilophytales, or Trimerophytales (p. 588), and the gymnosperms of both major groups. Among the key organisms are *Rellimia*, or *Protopteridium* (to be discussed with the pteridophytes in Chapter 21), *Aneurophyton*, *Tetraxylopteris*, and *Archaeopteris*.

Aneurophyton was a 40-foot tree, as were the members of the group in general, except *Tetraxylopteris*, a shrub. The branch systems from the trunk were up to a yard long, and the major branching was spiral and hence in three dimensions. The ultimate branching was dichotomous forking, as in the Rhyniales and Psilophytales. There were no clearly differentiated leaves; there were few leaves or many, according to interpretation. Either the whole system of small branches was one leaf, following the pattern of probable origin of fern leaves (Chapter 21), or the last small divisions were leaves. Neither gave the impression of leaves. The fusiform sporangia were terminal on clusters of recurved branchlets. Thus the plant body was essentially like a giant plant of the Psilophytales (p. 588), but the woody stem bears a striking resemblance to those of the gymnosperms, especially in that there was a cambium that produced secondary wood. A genus *Oocampsa* (see Andrews, Gensel, and Kasper, Canadian Journal of Botany 53: 1719–1728. 1975) tends to bridge the gap between *Aneurophyton* and the Psilophytales. This plant occurred during early Devonian. In it there was "overtopping" of one member of a dichotomy by the other, the former bearing sporangia and the latter vegetative photosynthetic branches. This tendency was the beginning of the development of special reproductive branches in the progymnosperms and ultimately the cones of the conifers.

Tetraxylopteris was similar to *Aneuropteris*, but the branching was opposite rather than alternate (spiral), and there was in part the same problem as to what is a leaf. The ultimate divisions of the branching system were internally of less complex structure than the larger branchlets, and thus there was a somewhat greater differentiation from the branchlets that bore them. The fertile (sporangium-bearing) branchlets were more leaflike than the vegetative ones, being branched dichotomously and further subdivided, as well as being flattened and appearing somewhat like small fern leaves. Thus, perhaps the ultimate branchlets may have had the potentiality for developing into complexly indented leaves like those of ferns and seed ferns. *Protokalon* had a more highly developed branching system, possibly evolving in the direction of having leaves of the fern type. Derivation of both simple leaves like those of conifers and more complex leaves like those of the seed ferns may have been possible starting with the plants of this group.

Archaeopteris was a forest tree with branch systems resembling the leaves of ferns and seed ferns. The branch system was flattened in one plane, and the divisions were joined together by soft tissue, forming fronds like the leaves of the ferns and fernlike plants occurring later on during the Carboniferous Period. The stem had pith encircled by the cylinder of "vascular bundles," and the cylinder gave rise to branch traces. In at least some species there was differentiation of micro- and megaspores, as in the gymnosperms of all groups but rarely the ferns.

On the basis of the interpretation of the gymnosperms given here, they form a large natural group with four small divergent living classes, the connecting links have disappeared long ago from the living flora. However, the old class name Gymnospermae, applied under Spermatophyta as a division,

Figure 20-1. Progymnosperms: **1** *Tetraxylopteris,* reconstruction of the stem and part of the sterile branch systems, or fronds, with only the basal parts shown, the branching dichotomous, **2** fertile branch with numerous sporangia, **3** enlargement of a small portion of **2,** **4** *Triloboxylon,* reconstruction of a part of the plant, the large branch bearing systems of vegetative flattened branchlets above and below and systems with sporangia along the major branch between them, the sporangia removed from two branches (indicated by arrows) to show the basic dichotomous branching. **1** Charles B. Beck, American Journal of Botany 44: 362. 1957; **2, 3,** P. M. Bonamo, and H. P. Banks, American Journal of Botany 54: 755–768. 1967; **4** Stephen E. Scheckler, American Journal of Botany 62: 924. 1975. Used by permission.

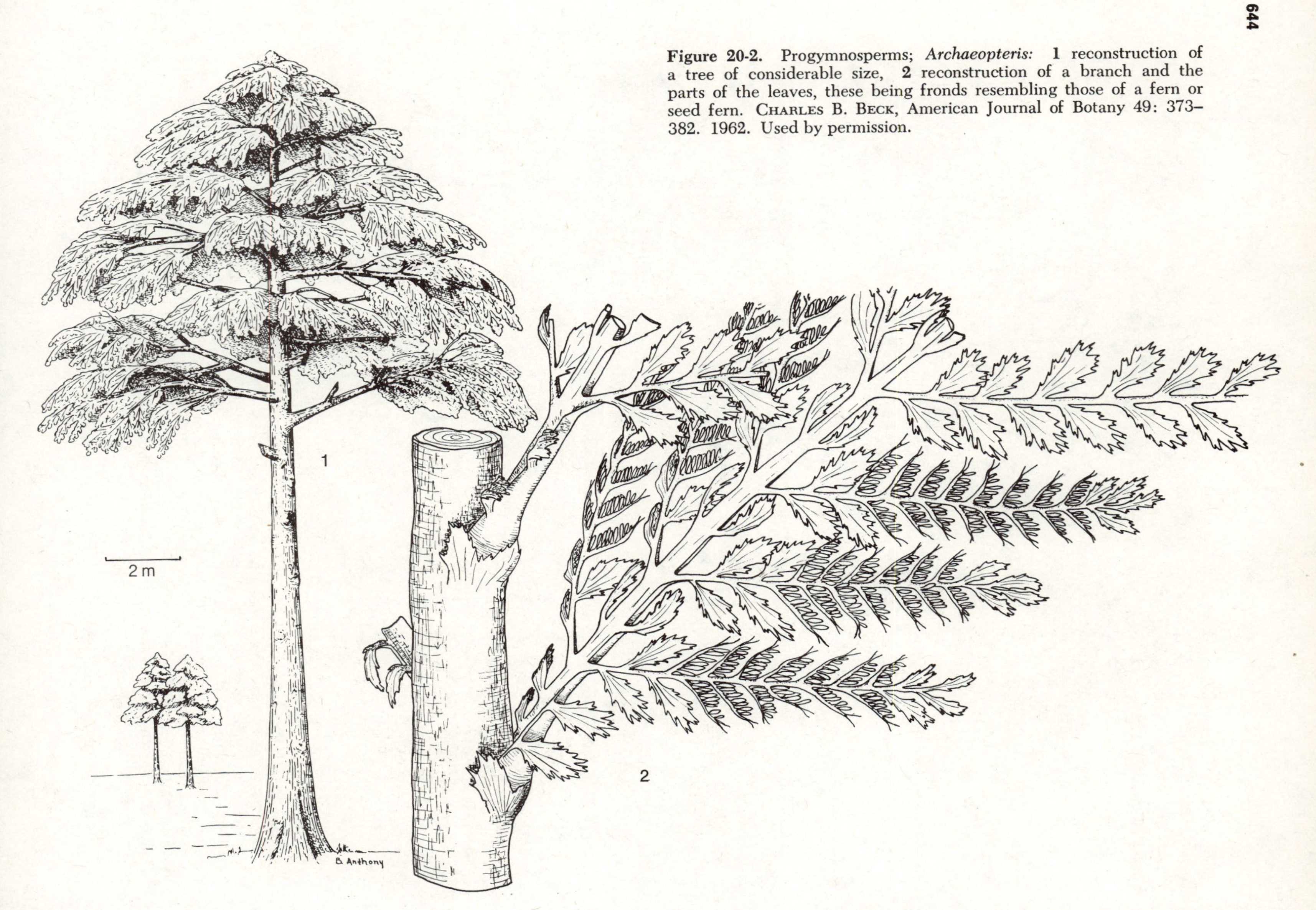

Figure 20-2. Progymnosperms; *Archaeopteris:* **1** reconstruction of a tree of considerable size, **2** reconstruction of a branch and the parts of the leaves, these being fronds resembling those of a fern or seed fern. CHARLES B. BECK, American Journal of Botany 49: 373–382. 1962. Used by permission.

should not be revived. The rank is higher than class, and the taxa of today do not include the plants of other units of different stages in geologic time (p. 551). The taxa of the present day include only the plants now living, and the plants of the past do not affect the status in classification of the Conopsida or Cycadopsida or of the Ephedropsida and Gnetosida. The gap between the taxa, which is one basis for recognizing units of classification, is expected to be filled by the organisms of the present from which they arose or to which they were related, but the extinct taxa belonged to another age.

The Conifer Line

Probably the first great forests of the world were dominated by the members of the order Cordaitales. Cordaitales were a major element in the swamp forest remains which compose the coal fields formed in several parts of the world in Carboniferous time, and they persisted at least until the termination of the Permian Period. Probably the Cordaitales bore the same relationship to other trees in the Paleozoic forests as do the tall conifers to those in the temperate forests of today.

The stems were usually straight, tall, and slender, with branches only near the tops. The leaves were in dense clusters on the upper branches, and probably the lower leaves and branches of the tree were shaded out gradually as they are in modern coniferous forests. However, the stem of *Cordaites* was less woody than the trunks of modern conifers, and at least some species would have been poor lumber trees.

The leaves of the Cordaitales were larger than those of modern conifers but much smaller than those of the Cycad Line. Although leaves of a few species were a yard long and about eight inches wide, most were much smaller. Some were even needlelike, less than an inch long, leathery, and apparently adapted to dry climates. Most leaves resembled those of monocotyledons in that the veins did not fork much except at the bases and that the principal ones were parallel.

Most Cordaitales were monoecious like the pine. The polliniferous strobilus consisted of an axis with numerous bracts and fewer sporophylls, each with one, two, or four pollen sacs (microsporangia). The ovuliferous strobilus had a relatively stout axis with many bracts and with a few ovules among them. The ovule was produced on a usually short structure developed from a bud axillary to a bract, fundamentally as it is in the pine. In some genera the sporophyll bearing the ovule was elongated and drooping, protruding beyond the strobilus; in others it was reduced.

One probable derivative and fairly close relative of the Cordaitales is the order Ginkgoales, represented in the living flora by the single species, *Ginkgo biloba*, the maiden-hair tree (pp. 616 and 617). The time of origin of the Ginkgoales is not certain, but they were present during the earliest part of the Permian Period. Although many of the fossil leaves discovered so far could be referred to the several forms occurring on *Ginkgo biloba*, they may have belonged to other species of *Ginkgo*, and there were related genera (e.g., *Baiera*). Since Jurassic time, only the one species has survived; and recent extinction of this one may have been prevented for many centuries by cultivation about Oriental temples.

The evidence available is taken to indicate origin of the Ginkgoales from the Cordaitales, or the origin of both groups from a common ancestor.

The characteristic group of living gymnosperms is the widespread order Pinales. There are several lines of development within the order, as indicated in the families described in Chapter 19. These were differentiated gradually during Mesozoic time.

Conifers transitional between the Cordaitales and the modern Pinales occurred during

Figure 20-3. A forest of members of the order Cordaitales in Carboniferous time; France. Reconstruction by M. S. CYRILLE GRAND'EURY, Academie des Sciences. Memoires 24: 1–624. tab. veg. D. 1877. Used by permission.

Figure 20-4. Twigs of members of the Cordaitales with strobili, e.g. the figure at the *upper left* with ovuliferous strobili, the one in the *lower center* with a polliniferous strobilus on the *left* and an ovuliferous one on the *right*. From M. S. CYRILLE GRAND'EURY, Academie des Sciences. Memoires 24: 1–624. Pl. 25. 1877. Used by permission.

the long course of evolution from Carboniferous (Paleozoic) time through Triassic (Mesozoic) from 300 million years ago to 200 million. The modifications of the characters of the pollen cones, microsporangiosphores or pollen cone scales, pollen grains, wood, and leaves were not great, and the transitions present no major problems. The long unanswered, and consequently much theorized and hotly disputed, question was the origin, nature, and composition of the Pinales cone scale (Figure 18-4). The bract subtending the scale has been conceded to be just that—a specialized leaf subtending a reproductive part, the cone scale and its attached ovules. The scale has been interpreted as (1) a leaf bearing ovules, (2) a branchlet bearing ovules, or (3) a shoot with specialized fused leaves forming the body of the scale. Research by Florin on the transitional fossil plants (placed, at least for convenience, in the single family Voltziaceae) has provided a satisfactory solution to the problem. It confirms the third alternative.*

In *Lebachia* (late Carboniferous, about 300 million years ago) both pollen and seed cones were borne on the tips of the twigs of the same tree. In the compact seed cones, short "fertile," or ovule-bearing, shoots were developed in the axils of major bracts with bifid tips. These shoots were very short, and the axis bore scalelike smaller bracts and terminated in a solitary stalked ovule. They were similar to the short fertile shoots of the strobili of *Cordaianthus* (Cordaitales), which had 1–3 ovules on the short shoot. Other genera developed later had more elaborate development of the scale. *Walchiostrobus* (or perhaps *Ernestiodendron*) of Permian had 3–7 ovules on stalks at the end of the shoot, and these were inverted as in modern conifers. There were at least 20 scales of varying size on the fan-shaped shoot. In *Walchia* (or perhaps *Ernestiodendron*) there were 3–7 stalked ovules but only a few scales below them.

In *Pseudovoltzia,* upper Permian, the shoot and its appendages approached the character of those of the living Pinales more closely, because the shoot (now rudimentary) was subtended by an entire bract and it bore 5 scales toward the apex and 3 inverted ovules. In *Glyptolepis,* upper Permian and Triassic, there were only 2 ovules and a fan-shaped structure formed from 5 fused scale leaves. If the scales were fused into one, the cone scale of *Pinus* would be essentially duplicated. It consists of a single scale body bearing 2 inverted ovules toward the base of the upper (ventral) side, and this is subtended by the major bract. In *Pinus* usually the bract is small and simple, but in other genera such as *Abies* it is larger and with a terminal point, and in *Pseudotsuga* there are 3 terminal prongs, the bract protruding from the cone.

The Ephedra Line and the Gnetum Line

The origin and historical development of these groups are obscure. Presence of the peculiar pollen indicates occurrence as far back as during Triassic or Permian for *Ephedra,* but there is little fossil evidence available for Gnetum or *Welwitschia.* Evidently all are survivors from old groups once much more common and variable. As shown in Chapter 19, these lines of development are known from only three peculiar living genera, which combine some features of the gymnosperms and others paralleling but not homologous with those of the angiosperms. There is no evidence of their close affinity to any other major groups of gymnosperms, though *Ephedra* resembles somewhat the Cordaitales and *Gnetum* and *Welwitschia* the Bennettitales (Cycad Line).

* Florin, R. 1951. *Evolution in Cordaitales and Conifers.* Acti Horti Bergiani 15: 255–388; 1955. *The Systematics of Gymnosperms* in *A Century of Progress in the Natural Sciences* 1853–1953.

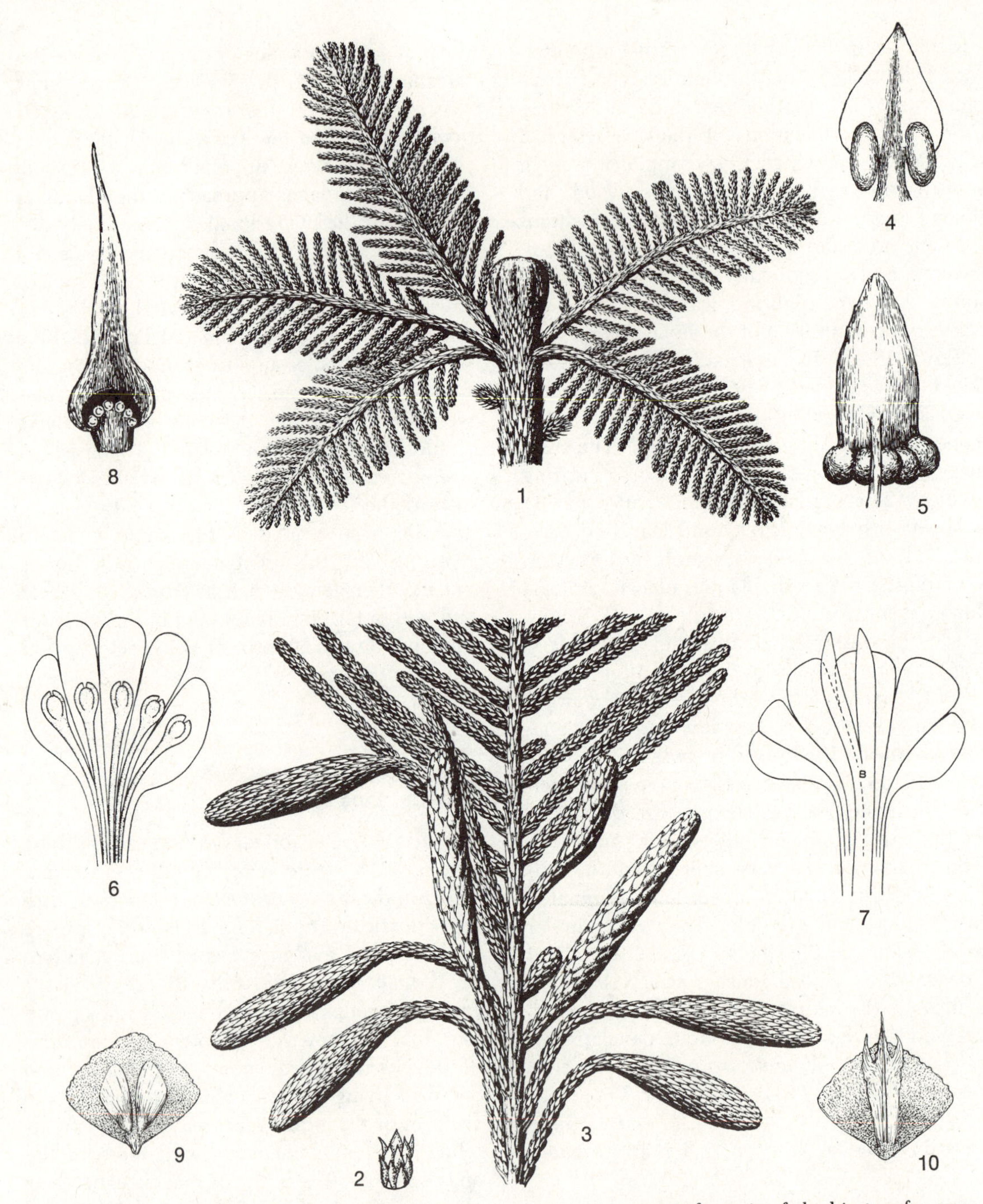

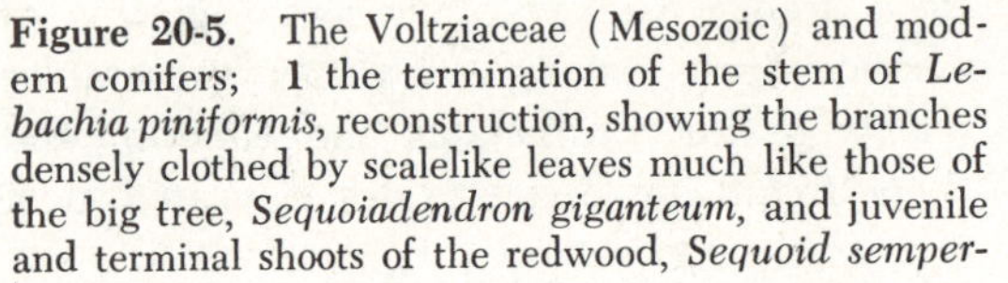

Figure 20-5. The Voltziaceae (Mesozoic) and modern conifers; **1** the termination of the stem of *Lebachia piniformis,* reconstruction, showing the branches densely clothed by scalelike leaves much like those of the big tree, *Sequoiadendron giganteum,* and juvenile and terminal shoots of the redwood, *Sequoid sempervirens,* **2** segment of a twig of the big tree for comparison, **3** branches of *Lebachia* with spreading to pendent pollen cones (below) and (just above) erect seed cones, both cone types terminating short branchlets, **4** *Lebachia hypnoides,* microsporophyll with two microsporangia, **5** *Sequoia sempervirens,* microsporo-

Segregation of these classes must have occurred far back in geological history. Both *Gnetum* and *Welwitschia* superficially approach the angiosperms in some respects, but they lack many flowering plant characters, including an ovary enclosing the ovules, and they have a different sequence of development of thickening of the walls of the xylem cells (cf. Chapter 17). The Ephedra Line may have originated from ancestors near the Cordaitales, the Gnetum Line from relatives of the Bennettitales.

The Cycad Line

The Cycadofilicales and related orders were at least mostly woody plants composing one of the most important species groups forming the coal deposits of Carboniferous time.

The leaves of seed ferns resembled those of ferns so closely that for many years the Carboniferous was thought to have been an age of ferns. Ultimately seeds were found attached to the leaf margins or to branchlets of these plants, and they were reclassified as gymnosperms related to but more primitive than the cycads. Details of structure show relationship to forerunners of the ferns. The Cordaitales also had features in common with the apparent ancestors of the ferns, but the relationship was less obvious because the leaves were not fernlike but somewhat similar to those of monocotyledons. The Ephedropsida and Gnetopsida do not resemble ferns, but their life history includes features in common with the other gymnosperms.

Both microsporangia and ovules were produced in various positions—in some genera on the backs or margins of leaves and in others on short specialized branches. The integument of the ovule often was completely free from the megasporangium (nucellus) instead of adnate with it as in the flowering plants or the more advanced gymnosperms. Andrews (Science 142: 925–931. 1963) illustrates a sequence of steps through which the single integument, or seed coat, may have developed through stages similar to those of known Cycadofilicales. This sequence proceeds from surrounding of the megasporangium with separate integument lobes through formation of a cup from the lower parts of the lobes and on to ultimate complete coalescence around a micropylar apical opening. Among the seed ferns the apex of the nucellus, or sporangium, was modified in various ways making it more effective as a pollen-receiving organ. However, ultimately in the gymnosperms this was superseded by secretion of a pollen drop of a secreted sticky solution, which is far more effective. According to Rothwell (Science 198: 1251–1252. 1977), probably some of the Cycadofilicales developed pollen drops. An immature ovule of *Callospermarion* (or a relative) from middle Pennsylvanian has a body of noncellular material in the right place for a pollen drop, and several probable pollen grains or spores are either embedded in the material or adhering to it.

The seed ferns had the following and other characters in common with the fern line of development: (1) similar leaf forms, (2) uncoiling of the leaf blade as it develops, and (3) leaf gaps (gaps in the xylem and phloem of the stem above the points of departure of the vascular traces leading to the leaves).

---

phyll with several microsporangia, **6** *Voltziopsis africana,* ovuliferous scale with 5 erect ventral ovules and with the upper parts of the scale separate, **7** dorsal view of the scale and the subtending bract (**B**), **8** *Sequoia sempervirens,* ovuliferous scale with several erect ventral ovules, **9** Douglas fir, *Pseudotsuga Menziesii,* ovuliferous scale, ventral side, with 2 ovules, **10** dorsal side, showing the 3-lobed bract. **1, 3, 4** (redrawn), **6, 7** from Rudolf Florin, Palaeontographica 85B (6): 365–664. 1944. Used by permission; **2, 4** Lyman Benson; **5, 8–10** J. D. Laudermilk.

Figure 20-6. Seeds of Carboniferous gymnosperms (Cycadofilicales or seed ferns, with the appearance of ferns) varying in degree of coalescence of the lobes of the integument and in the tendency toward forming a micropyle: 1 *Genomosperma Kidstonii,* the seed about 1.5 cm. long, 2 *Genomosperma patens,* about 8 mm., 3 *Salpinostoma dasu,* 5 cm., 4 Physostoma elegans, 6 mm., 5 *Eurystoma angulare,* 8 mm., 6 *Stamnostoma huttonense,* 3.7 mm. The integument investing the sporangium became adnate with it, forming the ovule. HENRY N. ANDREWS, Science 142: 927. 1963. Copyright 1963 by the American Association for the Advancement of Science. Used by permission.

The Cycad Line (present in middle Devonian time) continued to develop after the close of the Paleozoic Era. It reached its peak in Mesozoic time, and from the standpoint of plants the Mesozoic is known as the "Age of Cycads," although from the standpoint of animals it was the "Age of Reptiles," especially dinosaurs.

The predominant order of the Mesozoic was the Bennettitales* (Cycadeoidales). Some Bennettitales were dwarf trees reaching a height of about six feet, and they had columnar unbranched trunks, each terminated by a crown of spirally arranged leaves. Other species were branched profusely and perhaps were shrubs. Still others developed gigantic basal tubers protruding somewhat above ground level. The leaves were pinnate, and in most characters they resembled those of ferns. In some species they were as much as nine or ten feet long, but a foot or two was the common length.

The strobili were the characteristic feature of the order. They were produced on the upper portions of the plant, and sometimes they occurred by the hundreds in the axils of the branches. These strobili are of particular interest because each one bore both microsporangia and ovules. The base of the strobilus was made up of many spirally arranged scalelike bracts (leaves). Above them were spirally arranged bipinnate microsporophylls with leaflets bearing no blades but only numerous microsporangia developed as **synangia** (coalescent sporangia). The apex of the strobilus was a compact structure bearing stalklike megasporophylls each terminated by a single ovule and probably representing no more than a petiole and midrib. Apparently the ovules borne on the sides of the megasporophyll in ancestral types disappeared in the course of evolution of the Bennettitales, and only the terminal ovule remained. Among the mega-

* The principal genus is *Cycadeoidea* (1827), this name having priority over *Bennettites* (1867). The familiar name Bennettitaceae and the ordinal name Bennettitales have priority over Cycadeoidaceae and Cycadeoidales, however.

Figure 20-7. Restoration of the seed fern or pteridosperm *Medullosa* (Cycadofilicales); upper Carboniferous (Pennsylvanian) time; Iowa. Plate supplied through the courtesy of HENRY N. ANDREWS, JR., *Botanical Gazette* 110: 15. fig. 1. 1948. Used by permission of the University of Chicago Press.

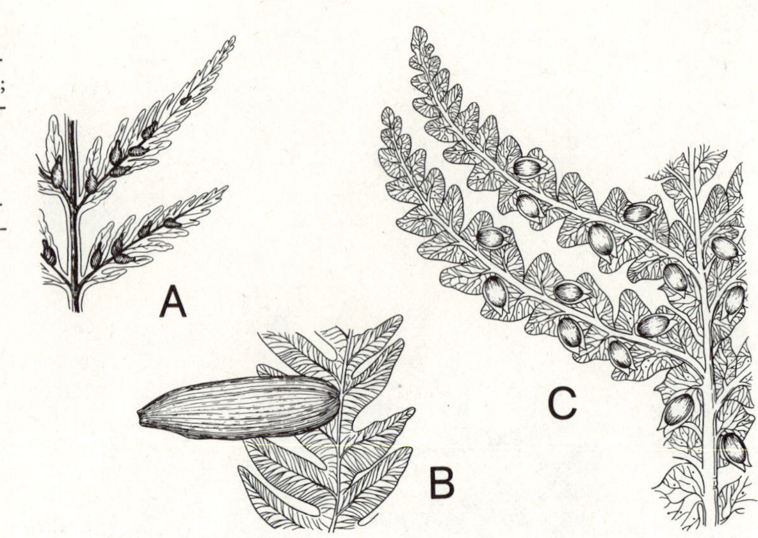

Figure 20-8. Seed ferns or pteridosperms with leaves bearing seeds; order Cycadofilicales; Permian time; China: **A** *Sphenopteris*, **B** *Alethopteris*, **C** *Emplectopteris*. Drawings furnished through the courtesy of HENRY N. ANDREWS, JR., Botanical Gazette 110: 26. fig. 16–18. 1948. Used by permission of the University of Chicago Press.

Figure 20-9. Restoration of the bisexual sporangium of *Cycadeoidea ingens*, order Bennettitales. The conelike central portion of the strobilus bore ovules and sterile scales (cf. **Figure 20-10**); the surrounding microsporophylls were bipinnate; the outer (hairy) leaves were sterile. From G. R. WIELAND, *American Fossil Cycads*. Carnegie Institution of Washington Publication No. 34. 1: 106. fig. 54. 1906. Used by permission.

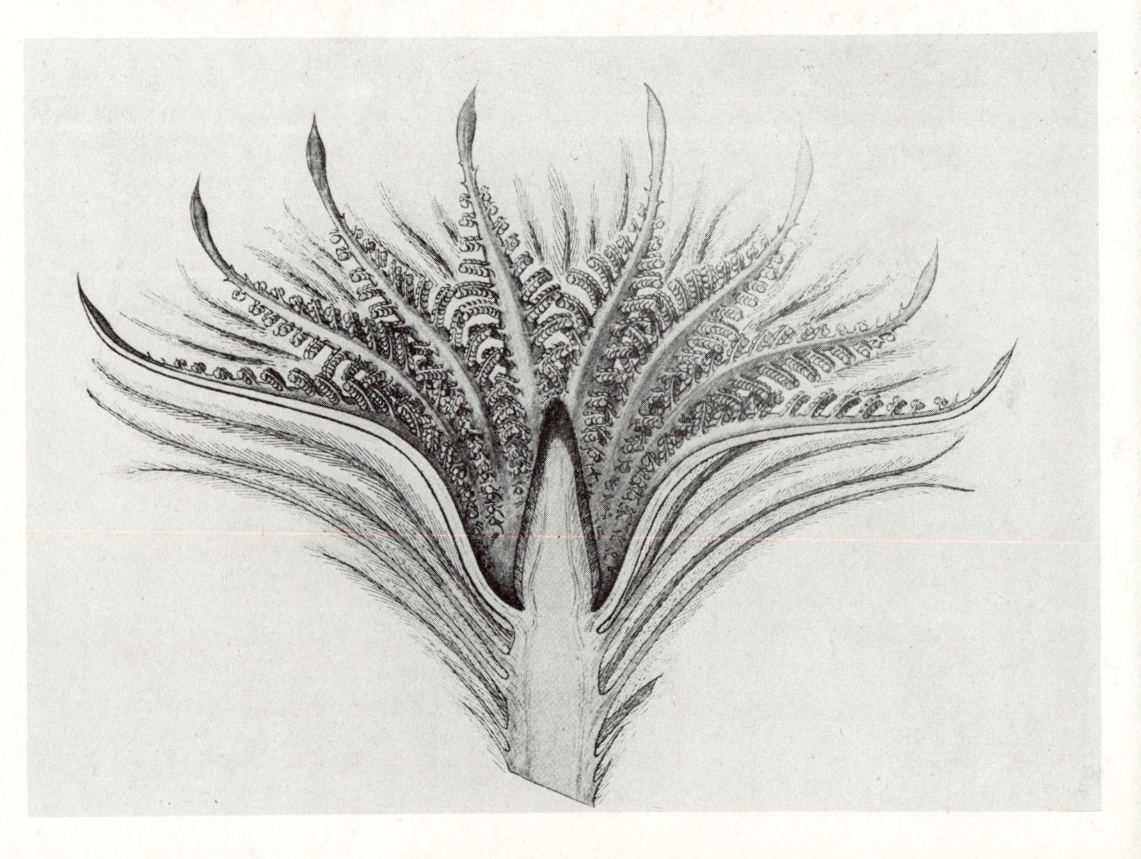

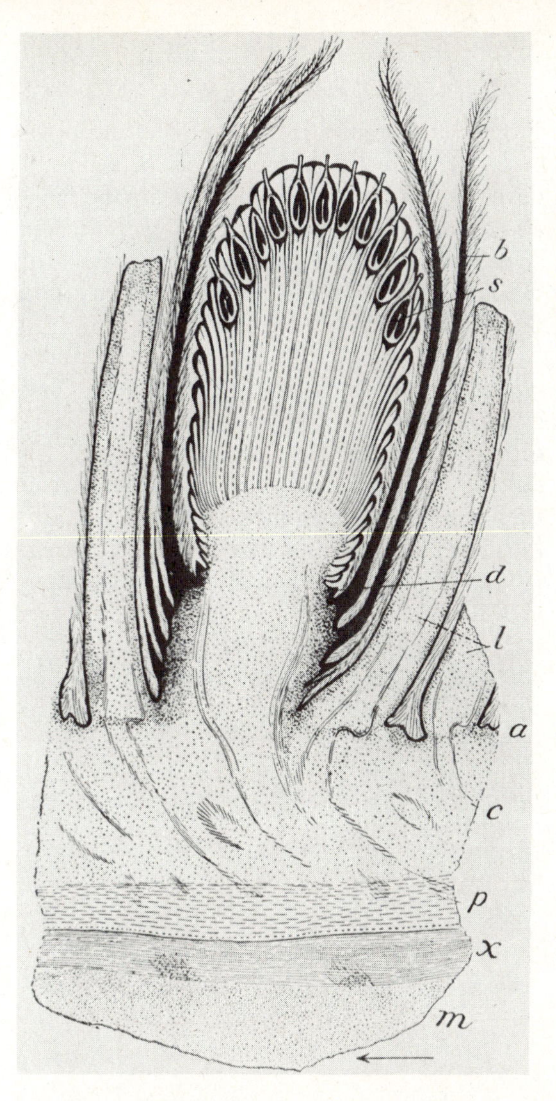

Figure 20-10. Restoration of an ovulate strobilus of *Cycadeoidea,* s seed, borne at the summit of a long stalklike megasporophyll (the megasporophylls intermingled with sterile sporophylls), b bract (hairy), d insertion (attachment) of a disc of tissue, l old leaf bases, a insertion of surface covering of the stem, c cortex (outer soft tissue) of the stem, p phloem, x xylem, m pithy inner tissue of stem. From G. R. Wieland, *American Fossil Cycads.* Carnegie Institution of Washington Publication No. 34. 1: 110. fig. 56. 1906. Used by permission.

Figure 20-11. Characteristic forms of the megasporophylls of the living cycads (Cycadaceae), showing the trend from several lateral pairs of ovules to one lateral pair (in the Bennettitales the trend having been from several lateral pairs to a single terminal ovule): **A** *Cycas revoluta,* **B** *Cycas circinalis,* **C** *Cycas Normanbyana,* **D** *Dioön edule,* **E** *Macrozamia Fraseri,* **F** *Zamia integrifolia,* **G** *Ceratozamia mexicana.* From G. R. Wieland (from Engler and Prantl, **A** after Sachs, **C** after F. V. Müller, **E** after Miquel, **F** after Richard), *American Fossil Cycads.* Carnegie Institution of Washington Publication No. 34. 1: 217. fig. 126. 126. Used by permission.

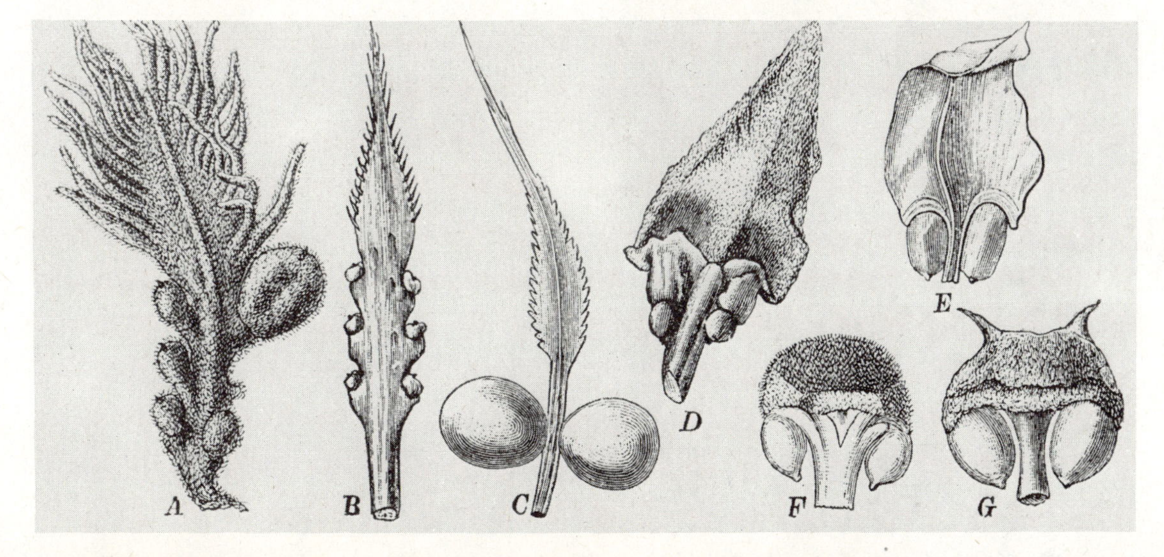

sporophylls were sterile structures (either sterile megasporophylls or bracts) of slightly greater length. The expanded tips of these more or less covered the tops of the ovules, and it was necessary for the pollen to sift between them before reaching the micropyles.

The spirally arranged sterile leaves, microsporophylls, and megasporophylls of the Bennettitales occurred in the same order on the axis of the strobilus as do perianth parts (sepals and petals), stamens, and pistils in the flowering plants. This sequence within the strobilus of the Bennettitales suggested to Arber and Parkin and to Bessey the flower structure of the magnolia family, and they postulated the origin of the flowering plants from the Bennettitales or their relatives during Mesozoic time (cf. p. 537).

In Mesozoic time the Cycadales (cycads) were abundant and widely distributed, but today they are restricted to limited areas in the Tropics and adjacent warm regions and obviously they are the last survivors of a line of development almost crowded out by the flowering plants. The only survivors of the whole Cycad Line are the nine genera (p. 631) of the family Cycadaceae, which now constitute the entire order Cycadales.

The differences between the Bennettitales and the Cycadales (described above) and especially those of the megasporophyll, indicate two parallel lines of development. Although probably each continued the Cycad Line from a common ancestor in the Cycadofilicales, neither was derived from the other. It is to be noted particularly that in the Bennettitales reduction of the megasporophylls resulted in retention of a single terminal ovule, whereas in the Cycadales reduction of the sporophyll resulted in retention ultimately of a single pair of lateral ovules.

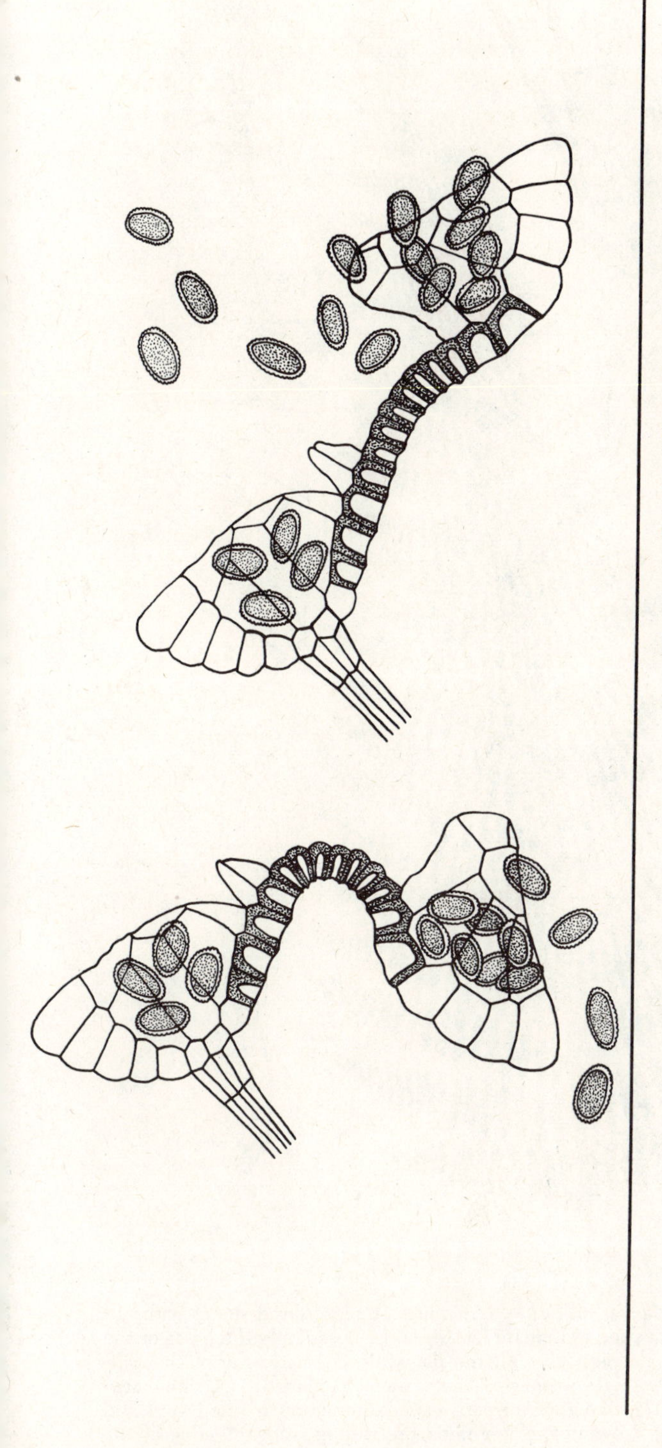

Ferns

(or pteridophytes, interpreted strictly, as the ferns)

Division *PTERIDOPHYTA*

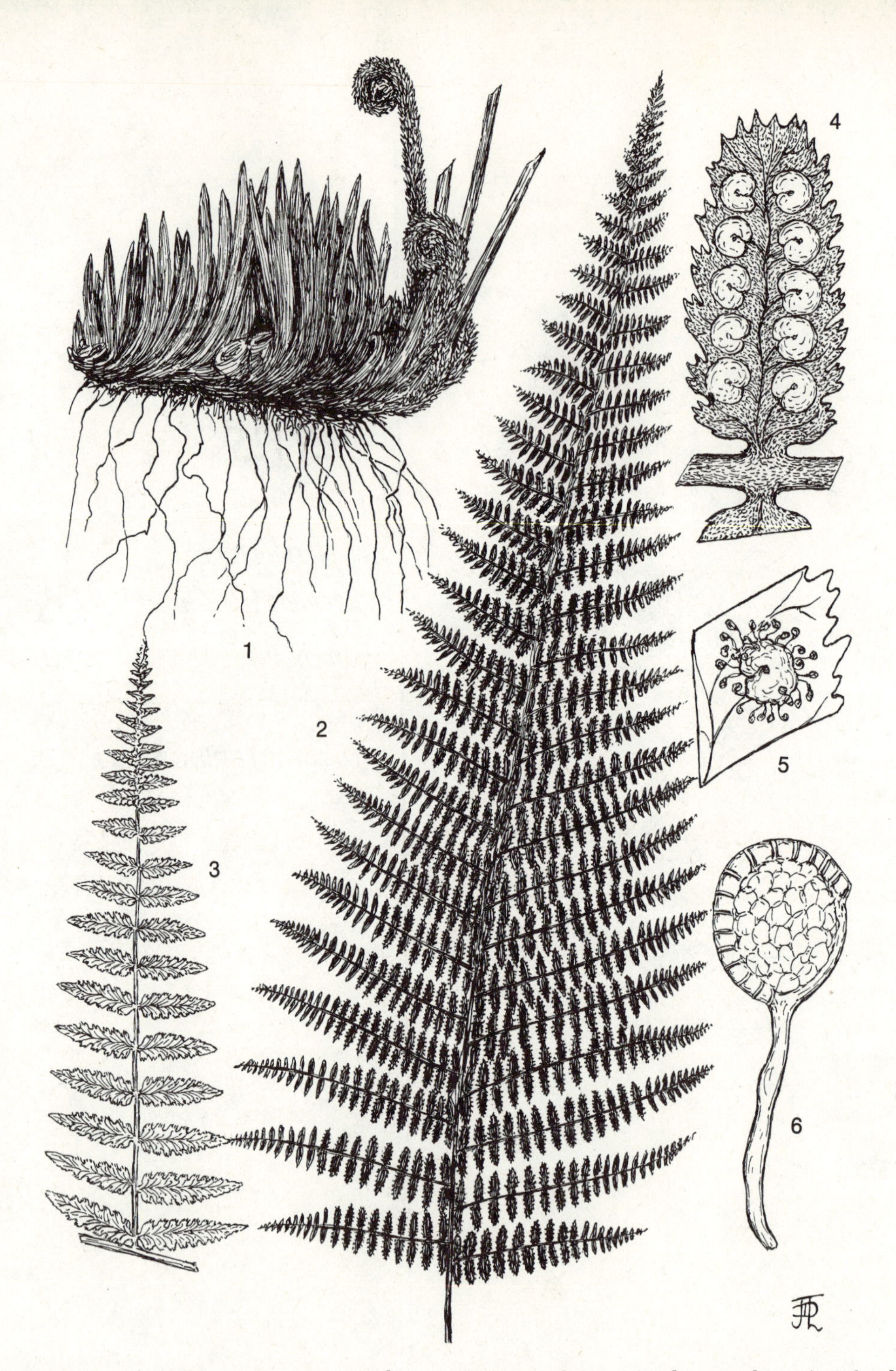

Figure 21-1. A wood fern, *Dryopteris rigida* var. *arguta:* **1** rhizome (underground stem) with adventitious roots and (above) old leaf bases, young circinate leaves, and leaves (only the bases of the petioles or stipes shown), the young leaves and petioles with the flat scale characteristic of the Filicales, **2** bipinnate leaf blade (ventral), **3** single primary leaflet (pinna) enlarged, **4** single secondary leaflet (pinnule) enlarged, showing the two rows of sori, each terminating a small vein and covered by an indusium, **5** sorus enlarged, showing the sporangia protruding from the edge of the indusium, **6** single sporangium (detail shown in **Figure 21-2**).

21

The Ferns

THE VOCABULARY DESCRIBING FERN CHARACTERISTICS

Alternation of Generations

The characteristic feature in the life histories of all the advanced members of the plant kingdom is an alternation of unlike generations. In the seed plants only one generation is evident because the other is minute and not free-living as a macroscopic independent plant. In the ferns and their relatives both generations are visible, one, the **sporophyte** or *spore-producing* plant, being the large conspicuous type, the other, the **gametophyte** or *gamete-producing* plant, being often only about one-quarter inch or less in length. The larger plant produces spores, which, germinating on moist soil, grow into gametophytes, which produce gametes. After union of gametes, the fertilized egg grows into a sporophyte, thus completing the cycle. Each cell of the sporophyte contains 2n chromosomes (the diploid number), each cell of the gametophyte n (the haploid number).

The Life History of a Fern

A fern with a relatively common leaf form is *Dryopteris* or wood fern. Since the wood ferns are representative of most of the features of common ferns, one of them is used as an example of the group. *Dryopteris* has appeared in some books as *Aspidium.*

THE SPOROPHYTE

Stem. The stems of most ferns are rhizomes. They may be either straight and unbranched or crooked and slightly to intricately branched. One end of the stem as a whole or the end of each of the branches is a growing point giving rise every favorable season to one or more leaves. The other end remains static or gradually dies and decays away. If death of the stem proceeds past a point of branching, the plant becomes separated into two individuals. Similar "reproduction by death" is not an uncommon feature in the plant kingdom. At various points along the stem adventitious roots are produced; there is no primary root or taproot in the adult fern. Often the stem is covered by flat brown scales which are appendages of its surface.

Leaves. The young leaves of ferns uncoil gradually, and this uncoiling is carried out by each segment as well. The arrangement of the young leaves (or leaf) within the bud is known as **vernation,** and a coiled arrangement is **circinate.**

Mature leaves (**fronds**) of ferns are divided into two principal parts: the large expanded portion above ground is the blade; the supporting stalk attached to the growing end of the stem is the **petiole** or **stipe.** Frond and stipe are special terms applied to fern leaves since long before their nature was understood.

Leaf types have been discussed under the flowering plants (Chapter 3). The amount of division or indentation of the leaf is used as a key character in each group of ferns, and this subject should be studied before attempting classification.

Sporangia. The sporophyte plants of the common ferns such as *Dryopteris* reproduce by spores developed in spore cases, or **sporangia,** usually aggregated into clusters forming conspicuous brown dots on the backs of

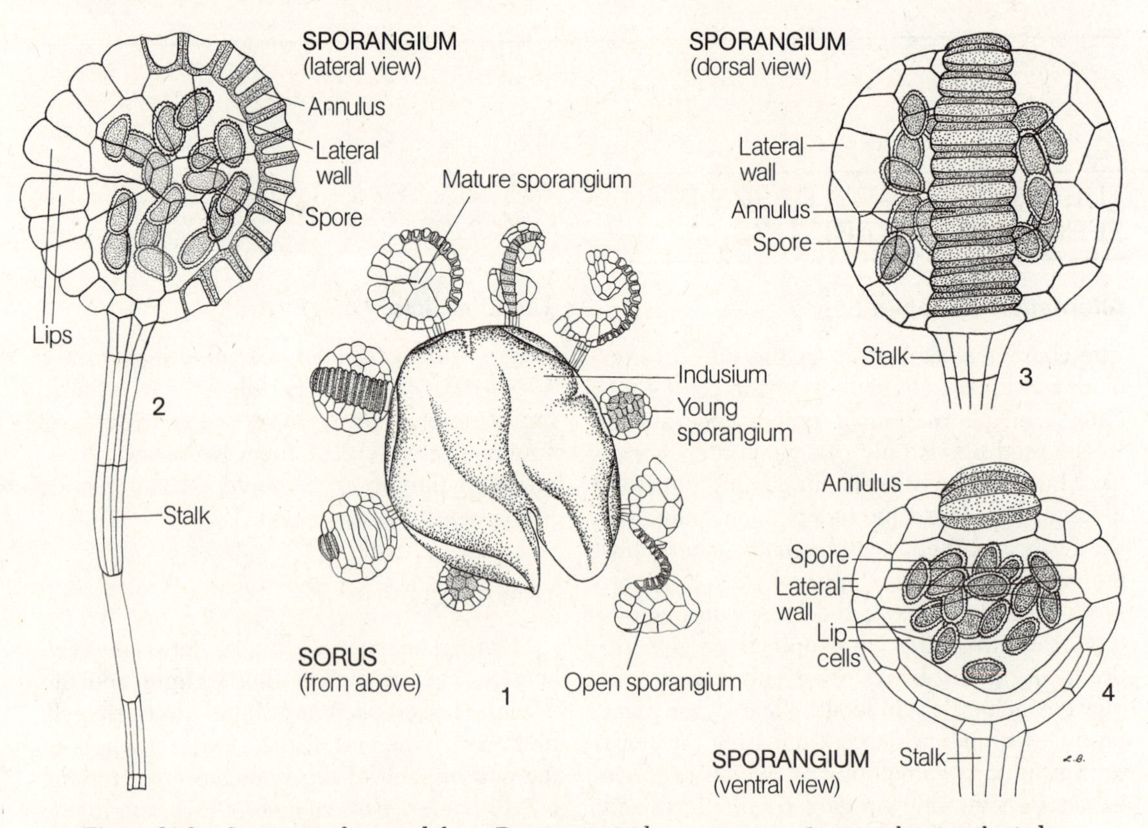

Figure 21-2. Sporangia of a wood fern, *Dryopteris rigida* var. *arguta:* **1** sorus, showing the indusium (notched on one side in this genus) and sporangia of various ages (irregularity within the sorus in the time of maturing being a characteristic of the Polypodiaceae), **2** sporangium just beginning to open (side view), **3** dorsal view, **4** ventral view.

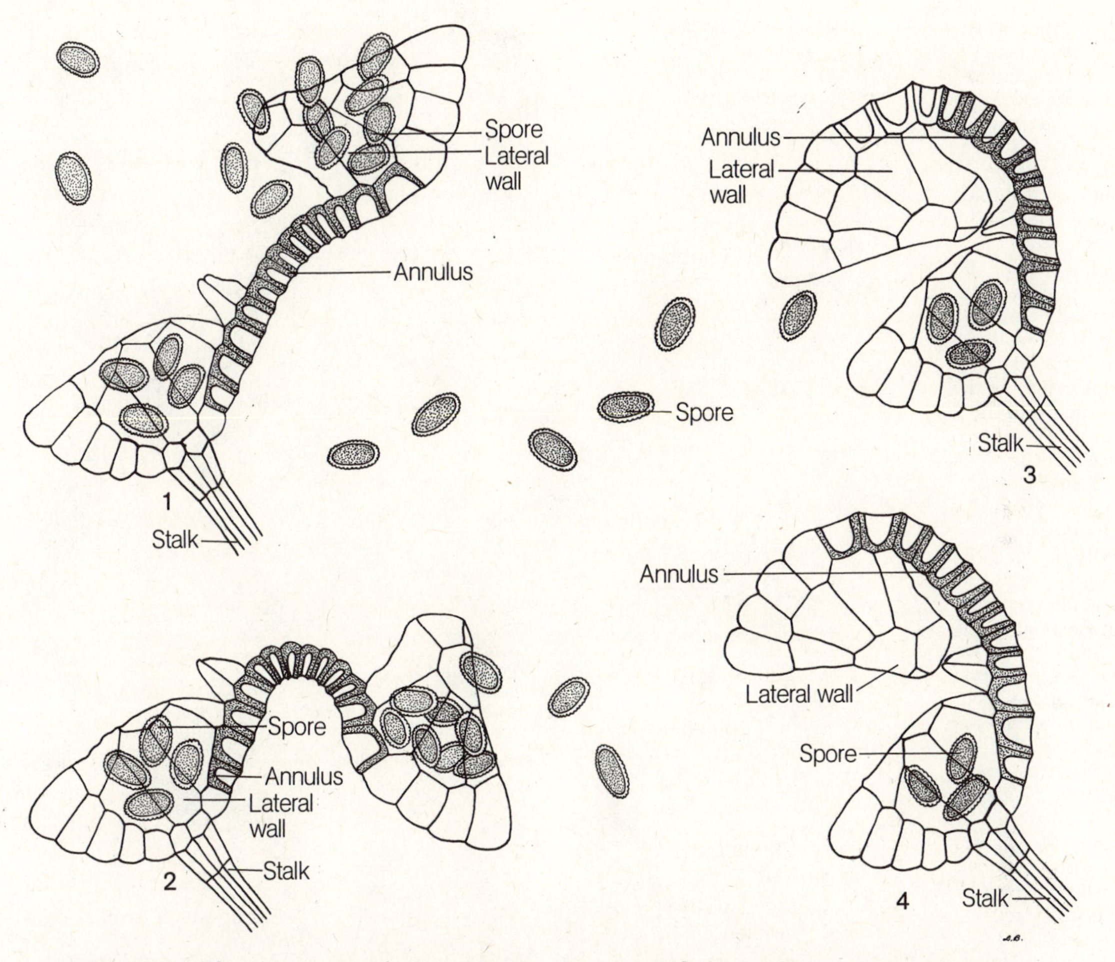

Figure 21-3. The sporangium of **Figure 21-2,** as it opens and throws the spores into the air: **1** annulus straightened out, **2** annulus recurved, as it is for only an instant, **3** annulus snapped back almost to the original form, the motion having thrown out most of the spores, **4** annulus assuming an intermediate position which is maintained until it becomes moist once more. The opening of the sporangium by the annulus is due to drying; moistening followed by drying initiates repetition of the movements in opening the capsule (spore sac) and throwing the spores into the air.

the leaves. A cluster of sporangia is a **sorus,** (plural, **sori**). In *Dryopteris* and the majority of other common ferns the young sorus is covered by a scale called an **indusium.** At maturity usually the sorus is covered only in part.

Each sporangium is a spore-containing ball usually on a stalk. The rupturing of the ball and scattering of the spores is accomplished as described below. A single row of specialized cells forms a crest (the **annulus**) beginning at the top of the stalk of the sporangium and arching over the ball and sometimes part way down the other side (then being connected to the stalk like the top of a question mark). The inner and radial * walls of the annulus cells are thickened and hygroscopic (sensitive to moisture). When these cells reach a particular stage of dryness, the annulus reverses its direction of curvature, rupturing the side of the

* Any wall with its axis on a radius of the spheroid formed by the upper portion of the sporangium.

Figure 21-4. Life history of a fern, the inconspicuous stages, i.e. the gametophyte and the embryo: **1** spore, **2** spore germinating, **3** sporeling, **4** mature gametophyte generation plant (prothallium), **5** archegonium (female sex organ) enlarged (longitudinal section), **6** antheridium (male sex organ) enlarged (longitudinal section), **7** male gamete (antherozoid) enlarged, this being the cell which swims into the archegonium and unites with the egg (female gamete), **8** four-celled stage of the embryo (young sporophyte), formed by division of the fertilized egg (zygote), each of these four primary cells giving rise to an organ by further divisions, i.e. to the primary root, primary leaf, foot (an organ which absorbs food for a time from the gametophyte which bore the archegonium), and stem (which in time gives rise to the adult plant or sporophyte by becoming the rhizome), **9** old embryo with a well-developed primary root and primary leaf, **10** young sporophyte plant with secondary leaves developing from the still very short stem, the primary leaf having disappeared. See figure, p. 580.

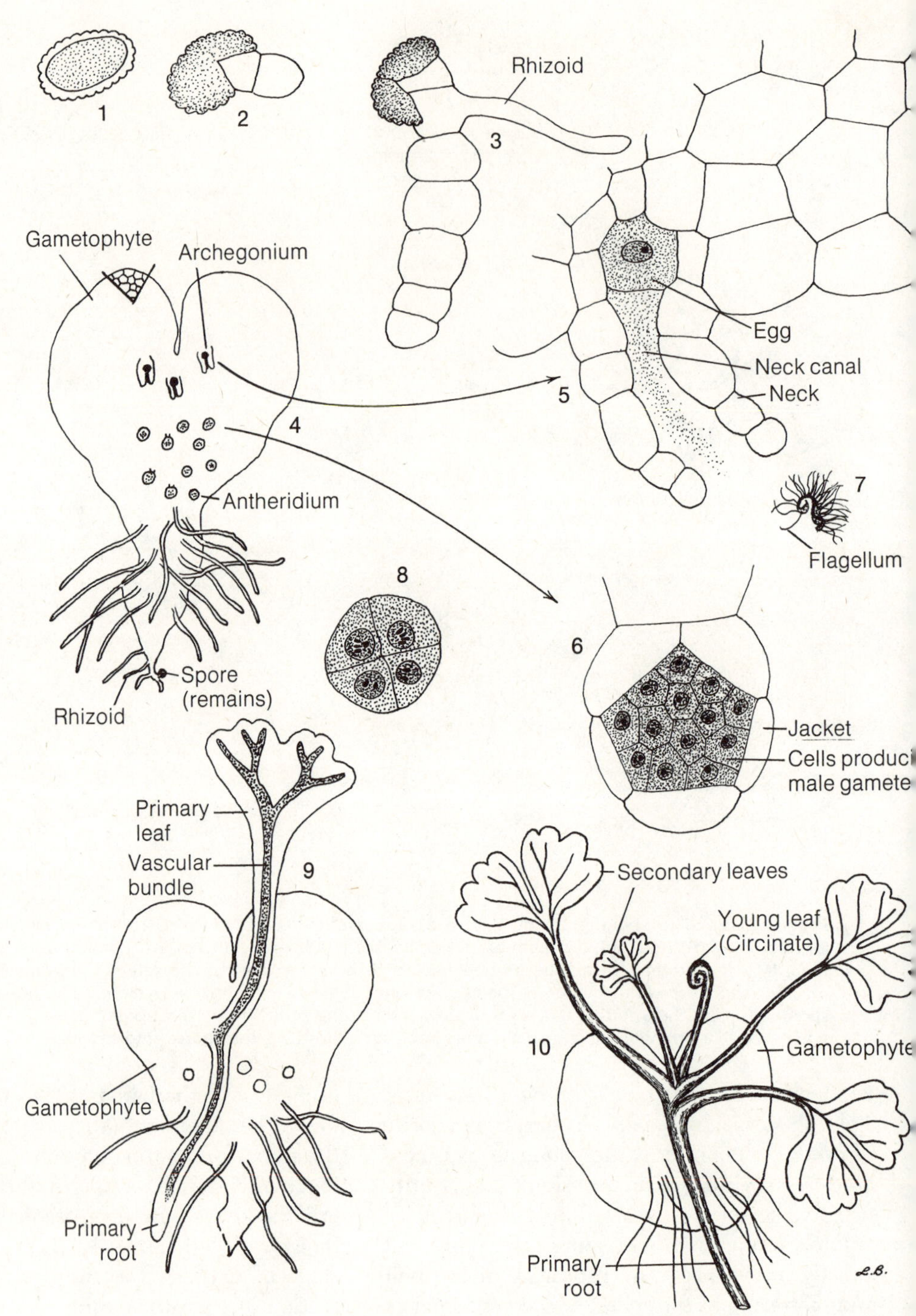

ball-like part of the sporangium and throwing some of the spores out into the air. Then it snaps back nearly to a closed position throwing more spores into the air. Drying of the cells of the annulus occurs after the spores are mature and when the moisture content of the air is low or the wind is blowing, and these atmospheric situations are favorable also for dispersal of the spores. The cells of the annulus are hygroscopic even after the spores have been shed, and when a moist fern leaf is placed on the stage of a dissecting microscope or under a hand lens, the movements of the annuli may be observed as drying occurs.

THE GAMETOPHYTE

As stated above, the sporophyte generation is characterized by its ability to produce spores. Equally important is presence of the diploid (2n) or unreduced number of chromosomes in each cell. Since the spores are formed by reduction divisions (cf. p. 581) and they contain only n chromosomes, they are the first cells of the gametophyte generation.

In a suitable moist, shady situation the spore may germinate and form a heart-shaped structure, the gametophyte or **prothallium,** which consists of a flat, green, cellular sheet three or four or sometimes more cells thick and usually an eighth of an inch or more in diameter. The sex organs are borne on the lower side of the prothallium. Fertilization is followed by development of an embryo or young sporophyte plant consisting at first of a primary root, a primary leaf, the rudiment of the new stem, and the "foot," an organ which for a time absorbs food from the gametophyte.

THE PROCESS OF IDENTIFICATION

Manual of the Major Taxa of Living Pteridophyta

(North America north of Mexico)

TABLE OF THE MAJOR TAXA

Division	*Class*	*Order*	*Family*	*Examples*
Pteridophyta	1. Pteropsida (Appearance of a feather, the form of a fern leaf)	1. Filicales	*1. Schizaeaceae	Curly grass, climbing fern
			*2. Gleicheniaceae	*Gleichenia*
			*3. Cyatheaceae	Tree fern
			*4. Osmundaceae	Royal fern
			*5. Hymenophyllaceae	Filmy fern
			*6. Polypodiaceae	Common ferns
			*7. Salviniaceae	*Salvinia, Azolla*
			*8. Marsileaceae	*Marsilea, Pilularia*
		2. Marattiales	1. Marattiaceae	*Marattia*
		3. Ophioglossales	*1. Ophioglossaceae	Adder's-tongue, grape fern

NOTE: The divisions of the Plant Kingdom are segregated in a key on page 6 in Chapter 1.

PRELIMINARY LIST FOR ELIMINATION OF EXCEPTIONAL CHARACTER COMBINATIONS

Before using the keys in this chapter, it is necessary to eliminate plants with the key characteristics obscure or often not available and plants with characters not obviously similar to the ones marking their relatives. **The plant must correspond with every character in the combination.**

1. Plants *either* (1) aquatic, floating, and with entire or 2-lobed small leaves *or* (2) resembling a "four-leaf" clover or young grass; sporangia completely enclosed in sporocarps (special sacs). Families *Marsileaceae* and *Salviniaceae*.

   Class 1. ***PTEROPSIDA,*** below.

2. Stem minute, producing a cluster of elongated grasslike leaves; leaves, or some of them, with large sporangia embedded in their bases. Family *Isoëtaceae.*

   Class 3. ***LYCOPSIDA*** **(LYCOPHYTA)**, Page 701.

Division **PTERIDOPHYTA.** Ferns

The Division Pteridophyta includes only ferns and not (as traditionally) the other divisions. The English name "pteridophytes" is used in a restricted sense to mean only the ferns or in a broad sense to include all vascular plants with spores and without seeds (not used in this book).

Class 1. **PTEROPSIDA.** FERN CLASS

Stem usually subterranean but sometimes growing above ground, not jointed, the division into nodes and internodes internal. Leaves large, usually deeply parted or divided or compound, formed (historically) by coalescence of the flattened branchlets in a major branch system; (leaf gaps present in the stele of the stem). Sporangia borne usually in clusters (sori), these on the back (dorsal side) of the leaf, sometimes near or on the margin.

Key to the Orders

1. Sporangia enclosed in sacs (sporocarps); plants *either* (1) floating and with small single or 2-lobed leaves *or* (2) resembling a "four-leaf" clover or young grass. Families *Marsileaceae* and *Salviniaceae.*

   Order 1. **Filicales,** below.

1. Sporangia not enclosed in sacs.
   2. Annulus present, but varying in form (a vertical crest, an oblique or horizontal band, or a shield-shaped lateral patch of cells); stems and leaves with simple or complex hairs or many-celled scales or both; wall of the sporangium thin, composed of a single layer of cells; stipules none.

      Order 1. **Filicales,** below.

   2. Annulus none (except in two extraterritorial genera occurring in the region of the western Pacific Ocean); stems and leaves either with no hairs or scales or with only short simple (unbranched) hairs; walls of the sporangia thick, formed from several layers of cells; stipules present.
      3. Sporangia borne on the backs of the ordinary vegetative portions of the green leaves, those of each sorus usually coalescent into a synangium; stipules thick and not sheathing.

         Order 2. **Marattiales,** page 673.

      3. Sporangia borne on specialized leaflets or on the terminal portion of the leaf, separate; stipules thin and sheathing.

         Order 3. **Ophioglossales,** page 673.

Order 1. ***FILICALES.*** FERN ORDER

Stems and leaves firm, containing much internal strengthening tissue, with simple or complex hairs or scales or both. Young leaves circinate; stipules none. Sporangia borne on the backs of the leaves, usually on ordinary vegetative leaves but sometimes on specialized leaves or specialized segments of leaves, the walls 1 cell thick; annulus present but variable in type, not necessarily vertical, sometimes oblique or horizontal or consisting of only a shield-shaped lateral patch of cells. Gametophyte usually flattened, rarely a branching filament of cells (*Schizaea*), usually ceasing growth after attaining a particular size.

The order Filicales is **leptosporangiate,** and, although this character cannot be used in the keys, it is a primary distinguishing feature of the group. Leptosporangiate refers to the origin of the indi-

Figure 21-5. Sori of various fern families: **A** Marattiaceae, (Marattiales, the others being Filicales), *Marattia,* **1** portion of a leaf, **2** leaflet with synangia formed by coalescence of several sporangia, **3–5** views of a synangium; **B** Osmundaceae, *Osmunda regalis* var. *spectabilis,* royal fern, **1–3** photosynthetic and sporangium-bearing leaflets from the same large, complex leaf, **3** being enlarged, **4** sorus, enlarged, each sporangium opening along a vertical slit, the indusium broad and lateral, not a crest; **C** Cyatheaceae, *Alsophila,* a tree fern, the sori as in the Polypodiaceae, but each sporangium with the crest running on the diagonal instead of vertically over the top of the sporangium; **D** Schizaeaceae, *Lygodium palmatum,* the climbing fern, **1** photosynthetic and sporangium-bearing leaflets from the same elongated climbing leaf blade, **2–3** sporangia in rows along the midrib, the annulus a horizontal ring around the apex of the sporangium; **E** Hymenophyllaceae, *Trichomanes,* a filmy fern, the sorus enclosed by a basal cup, and the sporangia developed on the base of an elongated axis a sterile portion of which protrudes beyond the cup.

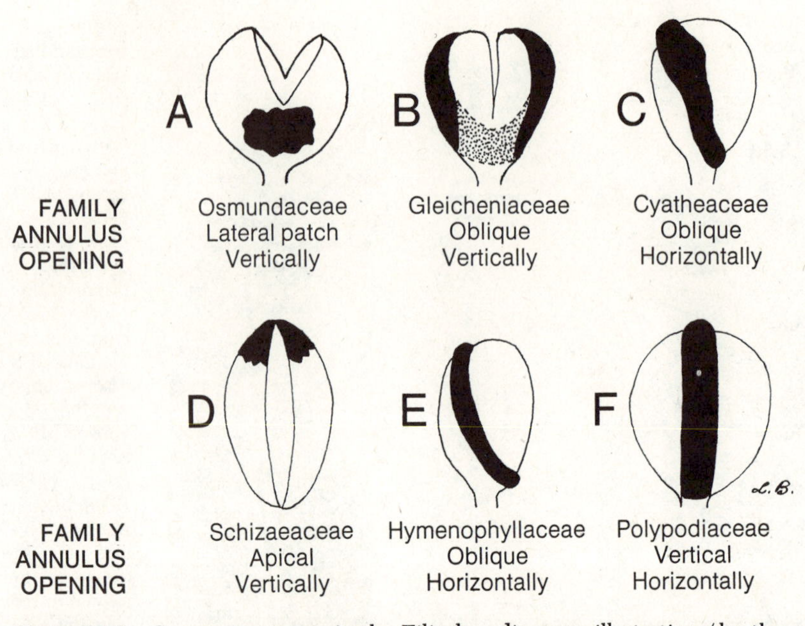

Figure 21-6. Sporangium types in the Filicales: diagrams illustrating (by the dark areas) the positions occupied by the thick-walled cells of the annulus of the sporangium. Stippling in **B** indicates the portion of annulus on the other side of the sporangium.

vidual sporangium, which may be traced back to a single cell which gave rise to all the others forming the sporangium. The Ophioglossales and Marattiales are **eusporangiate,** that is, the origin of the sporangium cannot be traced back to a single cell but only to a group of cells. In technical work frequently the Filicales are referred to as the leptosporangiate ferns and the other two orders as the eusporangiate ferns. The family Osmundaceae in the Filicales is an exception, for the stalk of the sporangium is developed from cells other than the single one which forms the body.

Key to the Families

(North America north of Mexico)

1. Sporangia not enclosed in baglike or "seedlike" sporocarps.

(NOTE: If no sporangia are present, the characters described in leads 2', under lower lead 1, should be checked.)

2. Annulus not a vertical crest on the sporangium, either oblique, horizontal, apical, or lateral; sporangia either all maturing at once or those within a sorus maturing in regular gradation from one area to another.

3. Sporangium *not* borne on a special axis, the sorus *not* enclosed basally in a cup or a two-lipped indusium; sporangium opening vertically, except in the Cyatheaceae (tree ferns); annulus oblique or lateral.

4. Sporangium solitary, sessile or stalked; annulus a horizontal ring around the extreme apex of the sporangium; sporangium-producing and sterile leaves or segments of the leaves markedly differentiated in most but not all species.

Family 1. **Schizaeaceae** (climbing fern family).

4. Sporangia clustered in sori, the sori with or usually without indusia; annulus not a horizontal ring of cells around the extreme apex of the sporangium.

5. Annulus an obliquely horizontal ring; sporangium with only a very short stalk

(subsessile); sporangia on the backs of ordinary vegetative leaves.

6. Sporangia splitting lengthwise, sessile; annulus oblique-horizontal; sorus with the sporangia all maturing at once; plants not trees.

Family 2. **Gleicheniaceae** (*Gleichenia* family).

6. Sporangia opening horizontally, stalked; annulus oblique-vertical; sorus with the central (upper) sporangia maturing first; plants usually trees.

Family 3. **Cyatheaceae** (tree fern family).

5. Annulus a shieldlike plate on one side of the sporangium or sometimes a broad horizontal band forming an arc part way around the sporangium; sporangium stalked; the North American genus with specialized nongreen sporangium-producing leaves or portions of leaves.

Family 4. **Osmundaceae** (royal fern family).

3. Sporangia borne along a stalk which forms an elongated axis of the sorus, the sorus surrounded by a cup or by a two-lipped indusium; sporangia maturing successively, those at the apex of the axis developing earlier and those toward the base last; sporangia splitting open horizontally; annulus oblique.

Family 5. **Hymenophyllaceae** (filmy fern family).

2. Annulus a vertical crest on the sporangium (not completely encircling the sporangium, the cells in the gap thin-walled and permitting dehiscence), the sporangium opening horizontally; sporangia within a sorus maturing at different times, mature and immature ones mixed irregularly. (**See note under the treatment of the family and Figure 21-2-1.**)

Family 6. **Polypodiaceae** (fern family).

1. Sporangia enclosed in sporocarps (hard or leathery sacs appearing more or less like seeds), upon germination the enclosed sporangia pushed out by the swelling of enclosed gelatinous tissues; plants aquatic or of muddy shores; sporangia of 2 kinds, some producing microspores, other megaspores.*

2'. Rhizomes floating; leaves small or 1–1.5 cm. long, simple, entire or 2-lobed, appearing to be produced in two rows along the stem; sporangia in soft sporocarps on the lobes of submersed leaves.

Family 7. **Salviniaceae** (water fern family).

2'. Rhizomes growing in mud on wet shores or beneath shallow water, bearing roots; leaves resembling those of either a "four-leaf" clover or young grass; sporangia in hard sporocarps produced at or beneath ground level.

Family 8. **Marsileaceae** (*Marsilea* family).

*Family 1. Schizaeaceae.

CLIMBING FERN FAMILY

Leaves or their leaflets usually of two types, one producing sporangia, the other not, the two markedly differentiated from each other (in *Lygodium* the elongated leaf resembling a climbing vine and the palmately cleft primary leaflets appearing to be individual leaves, the upper leaflets smaller and bearing sporangia). Sporangium solitary, stalked; annulus a horizontal ring around the apex of the sporangium.

Jurassic Period (or possibly earlier) to the present. Of the several living genera, only *Schizaea, Lygodium,* and *Anemia* are native in the forests of eastern North America. *Anemia* is restricted to peninsular Florida and central and southwestern Texas.

*Family 2. Gleicheniaceae.

GLEICHENIA FAMILY

Leaves not differentiated into unlike sterile and sporangium-bearing portions. Sporangia in sori, those of each sorus maturing simultaneously, the sorus with no indusium; sporangium subsessile, annulus obliquely horizontal, the sporangium wall opening vertically.

* The prefix *micro-* (small) may be opposed by either *mega-* (great) or *macro-* (long in extent or duration). In biology *macro-* is used in the sense of some particular parts being unusually large, and it is used sometimes in the place of *mega-*.

Cretaceous Period to the present (but the maximum development in Cretaceous time). The genera include *Gleichenia,* common and widely distributed in the Tropics, and *Hicriopteris.* A single colony of *Gleichenia flexuosa* grew for some years on Mons Louis Island, near Mobile, Alabama, but it was eliminated by the intervention of man. This species was one of a group known sometimes as *Dicranopteris.*

The leaves of many species have a peculiar habit of continuous growth and branching which gives the impression of a dichotomously forked stem.

*Family 3. **Cyatheaceae.**

TREE FERN FAMILY

Plant usually treelike and the stem growing above ground (in some plants, however, the stem rhizomatous and underground). Leaves not differentiated into sterile and sporangium-bearing forms. Sporangia in sori, those of the inner (upper) part of the sorus maturing first; indusium present or absent, a cup-shaped or a lateral flap; sori dorsal on the veins; sporangium stalked; annulus oblique, the sporangium opening horizontally.

Jurassic Period to the present. The family is largely tropical and subtropical. It is not represented clearly in North America north of Mexico.

One species of *Dennstaedtia,* included in this family by some authors, is common in eastern North America and at one time another species grew either as a native or as an introduced plant near Lake Okeechobee, Florida. *Dennstaedtia globulifera* occurs in limestone sinks and in caves north of Comstock, Texas.

*Family 4. **Osmundaceae.**

ROYAL FERN FAMILY

Some leaves or segments of them nongreen and specialized in production of sporangia, others green and not producing sporangia. Sporangia in sori, those of each sorus maturing simultaneously, the sorus with no indusium; sporangium stalked; annulus at the side of the sporangium, either a shield-shaped group of cells or a broad horizontal band partly encircling the sporangium.

Mesozoic time to the present. There are three living genera, *Osmunda, Todea,* and *Leptopteris.* The cosmopolitan genus *Osmunda* is represented in eastern North America by three species; *Todea* occurs in Africa, Australia, and New Zealand; *Leptopteris* is restricted to areas about the southwestern Pacific Ocean.

In *Osmunda cinnamomea* (cinnamon fern) the photosynthetic and sporangium-bearing leaves are different; in *Osmunda Claytoniana* (interrupted fern) the middle primary leaflets of some of the leaves are "fertile"; in *Osmunda regalis* var. *spectabilis* (royal fern) the terminal leaflets of only certain leaves are fertile.

*Family 5. **Hymenophyllaceae.**

FILMY FERN FAMILY

Leaves one or only a few cells thick, more or less translucent, not differentiated into unlike sterile and sporangium-bearing forms. Sporangia produced on a stalk surrounded at least basally by a cup or a two-lipped indusium, those at the apex of the stalk maturing first, those successively lower on the stalk gradually maturing later; annulus oblique, the sporangium opening horizontally.

Upper Carboniferous (Pennsylvanian) Period to the present. The two genera, *Trichomanes* and *Hymenophyllum,* are common in warm regions. Six species of *Trichomanes* occur in eastern North America, mostly in the Southern States and especially in southern Florida. One occurs as far northward as Ohio, Illinois, and West Virginia; another has been recorded from Illinois. One species of *Hymenophyllum* occurs in South Carolina. *Sphenomeris* (subfamily Lindsaeaoideae) ocurs in southern Florida.

*Family 6. **Polypodiaceae.** FERN FAMILY

First, see important note on p. 667.

Vegetative structure variable. Sporangia usually in sori or concentrated along the leaf margin, sometimes distributed irregularly or following the veins; sorus composed of sporangia in various stages of development, these distributed irregularly; sporangium stalked; annulus vertical on the sporangium, arching almost completely over it, the sporangium opening horizontally; spores not exceeding 64, usually fewer (in the preceding families usually much more numerous).

Tertiary Period to the present; possibly represented in Jurassic time. This is the common fern family. There are about 170 genera and between 7,000 and 8,000 species, including approximately 36 genera and over 200 species in Canada and the United States.

Ferns are thought of as being abundant only

IMPORTANT NOTE: Recently some fern specialists have favored division of the Polypodiaceae into a number of families. Probably these are natural small units, but family status is debatable in at least most cases. The question is whether the units represent needed rearrangement or merely a shift to liberal policy or in some instances one and in some the other. The policy of the author is conservative, and tentatively the segregations, or all but a few of them, are rejected, though this does not represent a hard and fast rejection. It is hoped that the problems involved can be studied more deeply. For a general discussion of the problems of conservative or liberal policy, the following references are suggested: (1) Lyman Benson, *Plant Taxonomy, Methods and Principles*, 1962, Ronald Press Co., New York, 1978, reprinted, John Wiley & Sons, New York; (2) Lyman Benson, *The Cacti of the United States and Canada,* Stanford University Press, in press.

The recent segregations of families from the Polypodiaceae appears in some books, though most manuals and floras will not include them. In the book to be used for determination of the genus and species, the content of the keys and family descriptions should be checked for inclusion of such families as Pteridaceae, Grammitaceae, Adiantaceae, Dryopteridaceae, Aspidiaceae, Blechnaceae, Aspleniaceae, Vittariaceae, Davalliaceae, Parkeriaceae, and Lomariopsidaceae. If these are found, they represent division of the Polypodiaceae as interpreted here. The key to the families of the Filiacales (sometimes as Polypodiales) in the local or regional manual or flora should be used to find the segregated family.

An astute student of the ferns, Warren H. Wagner, Jr., has presented a middle-of-the-road interpretation of the fern families, because the arguments over classification are concerned mostly with the rank to be assigned each group of genera rather than with the outlines of their classification. The same group may appear as merely genera within the Polypodiaceae, as a subfamily, or as a family. Wagner recognizes three families, each composed of subfamilies. The following relatively simple key will segregate them:

1. Sorus without an indusium, naked on the back or margin of the leaf or in a groove or covered by the rolled-back leaf margin.
   2. Sorus without an indusium, round.

      Family **Polypodiaceae.**

   2. Sorus in a line along a vein on the back of the leaf or at the tip of the vein at the margin of the lower side, the shape variable.

      Family **Adiantaceae.**

1. Sorus on the back of the leaf, covered by an indusium, this oblong to round or linear, reniform, or peltate, but sometimes vestigial.

   Family **Aspleniaceae.**

The proposed families include the following subfamilies:

Family Polypodiaceae
  Subfamilies Polypodioideae, Grammitoideae
Family Adiantaceae
  Subfamilies Adiantoideae, Ceratopteridoideae (Parkeriaceae)
Family Asplenioideae (the largest by far)
  Subfamilies Dryopteridoideae, Blechnoideae, Asplenioideae

A brief readable summary of this system is in the 1974 edition of the Encyclopaedia Brittanica, under "Fern," Macropaedia volume 7.

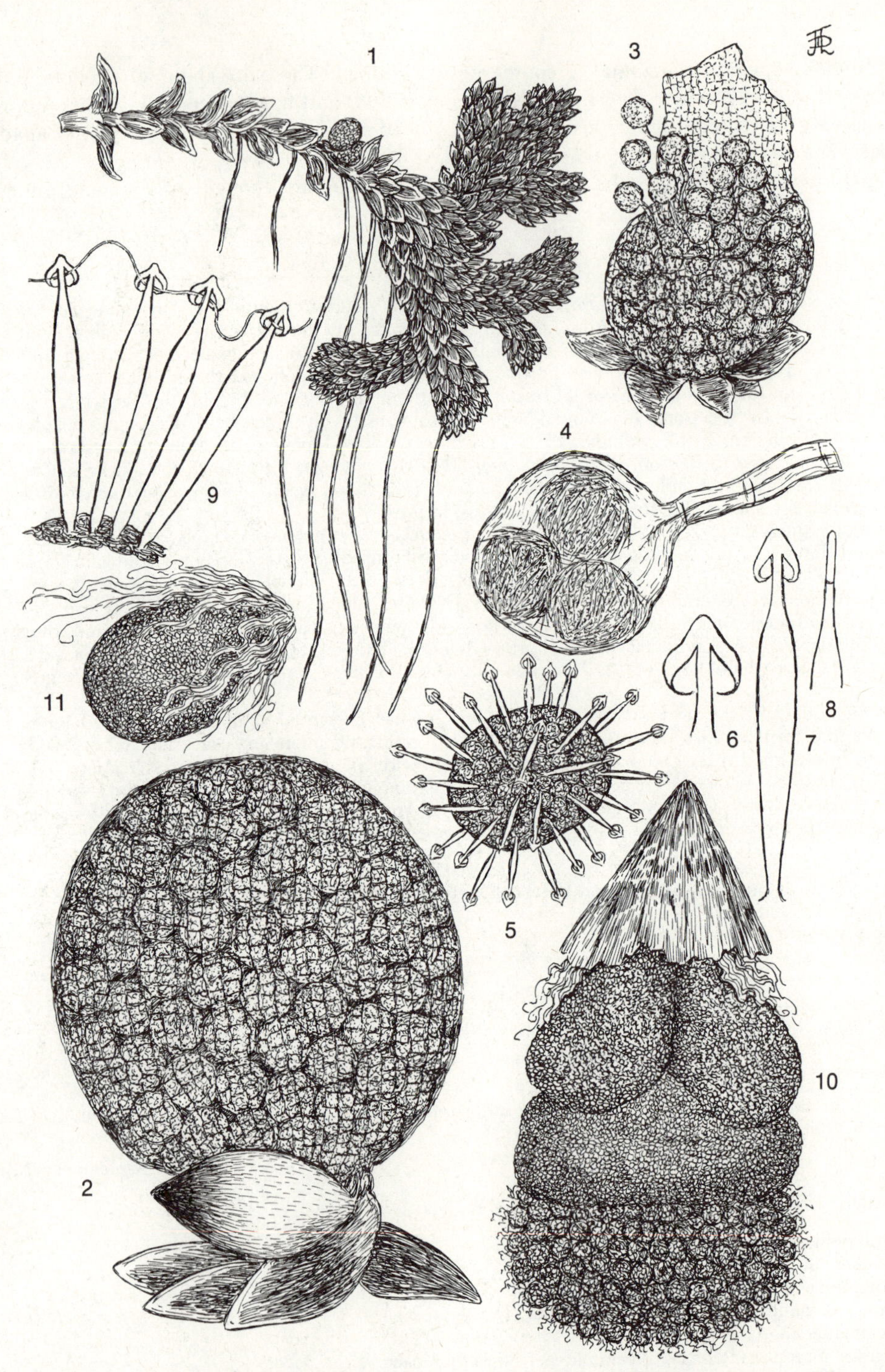

Figure 21-7. Filicales; Salviniaceae; *Azolla filiculoides*, a water fern: **1** plant showing the stem, leaves, roots, and a sporocarp, **2** pair of sporocarps enlarged, the large upper globular one a microsporocarp, the lower ellipsoidal one (shaped like an American football) a megasporocarp, the lower structures leaves, **3** microsporocarp containing microsporangia, **4** microsporangium containing massulae, complex

in moist woods, but this is not true because there are numerous species in even the deserts of Arizona. In the moist subtropical forests of Florida, where they occur with subtropical members of other families of the Filicales, the *Polypodiaceae* are the predominant fern family.

In older books, this family is named *Filices.* Perhaps it needs division and this has been done by some authors, notably Copeland, but this is a matter requiring much further study.

*Family 7. **Salviniaceae.**

WATER FERN FAMILY

Roots present but simple in *Azolla,* absent in *Salvinia;* rhizomes floating on the surface of the water or the plants sometimes stranded in mud along the shore. Leaves seemingly in two rows along the stem and its branches, small, simple, entire (*Salvinia*) or 2-lobed (*Azolla*). Sporangia enclosed in relatively soft sporocarps on the lobes of submersed leaves, the microsporangia and megasporangia in different sporocarps.

Miocene Epoch to the present. The family is distributed widely especially in warm regions. Both genera, *Salvinia* and *Azolla,* occur in North America. One species of *Salvinia* grows in ponds and slow-moving streams in peninsular Florida. Another was collected once in Missouri, but it may have been an escape from cultivation. There are two native species of *Azolla,* one of the Pacific Coast and the other occurring from Alaska to Ohio, Arizona, North Carolina and Florida.

Azolla resembles aquatic mosses, and often it is overlooked by collectors of vascular plants.

Sometimes the family is separated as an order Salviniales composed of two families, Salviniaceae and Azollaceae.

At some times of the year, especially in the fall, the plants turn red. If the leaves are crushed on a slide and mounted in water under a compound microscope, filaments of the blue-green alga *Anabaena azollae* may be seen. This alga is beneficial to its associate, *Azolla,* because it fixes free nitrogen and makes it available to the larger plant which gives *Anabaena* shelter (according to unconfirmed reports).

*Family 8. **Marsileaceae.**

MARSILEA FAMILY

Plant resembling either a "four-leaf" clover or young grass, the leaves either with four leaflets at the summit of an elongated petiole or consisting of only the petiole. Sporangia enclosed in a hard sporocarp formed from a specialized leaf, of two kinds, some producing microspores, others megaspores, attached to a gelatinous ringlike structure which swells when water enters the sporangium, the swelling of this gelatinous layer pushing the sporangia from the sporocarp.

The family is not known definitely from fossils. The three genera are distributed widely. In North America north of Mexico there are five species of *Marsilea* and one of *Pilularia.* Both genera grow near or in water, usually inhabiting the muddy shores of lakes and streams or the margins of ponds or vernal pools. *Marsilea* may be aquatic, and the petioles may rise to the surface from a depth of a foot or two.

Sometimes the family is considered as composing a separate order, Marsiliales.

---

sporangium, showing the following structures: (*above*) an apical cap of megasporangium wall tissue, next two of the four structures homologous with massulae of the microsporangium but not containing spores, next structures developed from the contents of the microsporangium including the spores, **5** liberated massula with external structures, the glochidia, **6–8** views of a glochidium, showing the two hooklike apical structures, **9** glochidia in which the hooks have snapped onto a thread from the megaspore or its associated structures, **10** megaspore after opening of the megaspore, a columella of material from the megaspore, (*below*) the megaspore, its wall roughened and covered with long hairs, **11** (reduced scale) the megaspore and associated structures free in the water. The hairs associated with the megaspore are caught by the glochidia of the massulae from the microsporangium, and this insures development of micro- and megagametophytes side-by-side, favoring ultimate fertilization of the eggs. This occurs at the bottom of the pool, lake, or stream.

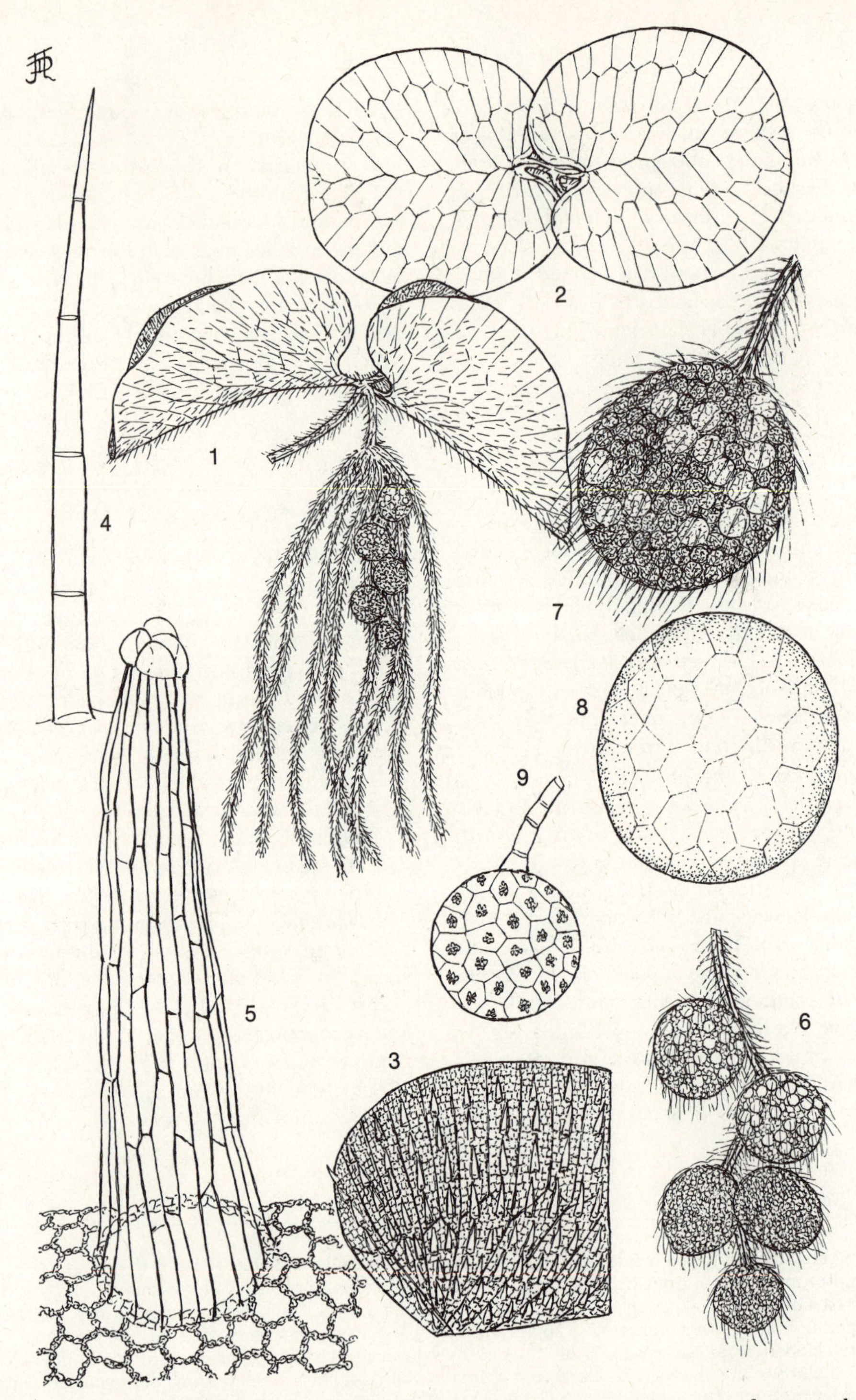

Figure 21-8. Filicales; Salviniaceae; *Salvinia natans,* a floating fern: **1** stem and a typical whorl of three leaves, two simple and floating, one dissected and submersed, this bearing sporocarps, **2** the floating leaves, **3** hairy lower surface of a submersed leaf, **4, 5** hairs of a floating leaf, **6** sporocarps (those with micro- and megasporangia together and similar in appearance), the first one or two formed in each cluster of four to twenty being megasporocarps, **7,** single sporocarp, **8,** megasporangium from external view, **9,** microsporangium.

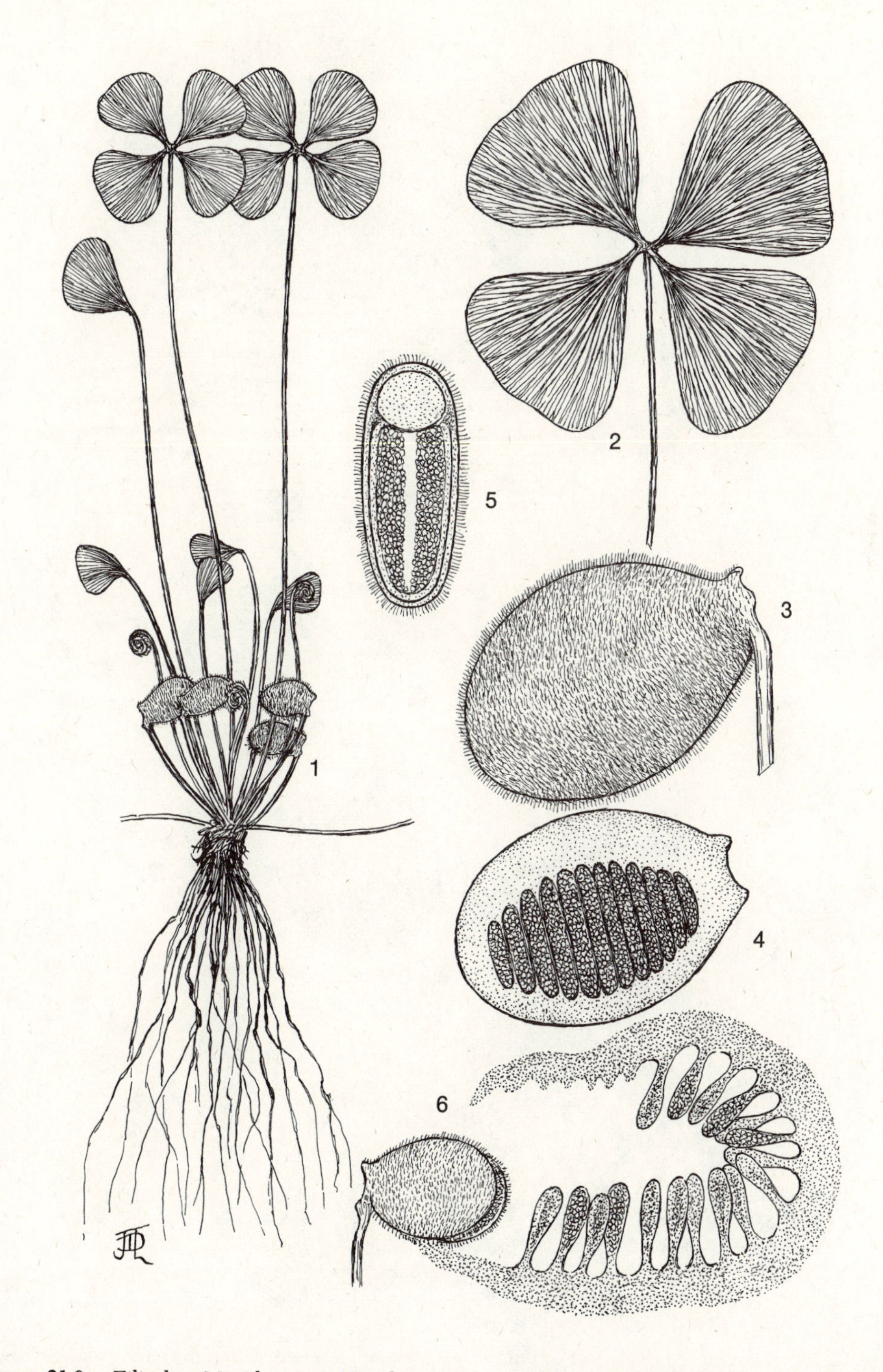

Figure 21-9. Filicales; Marsileaceae; *Marsilea vestita:* **1** plant with a slender rhizome, this bearing a cluster of leaves, sporocarps, and roots, **2** leaf blade divided into four leaflets, **3** sporocarp, **4** sporangia within the sporocarp, **5** the same in cross section, **6** germinating sporocarp with the swollen gelatinous structure carrying the sporangia from the sporocarp.

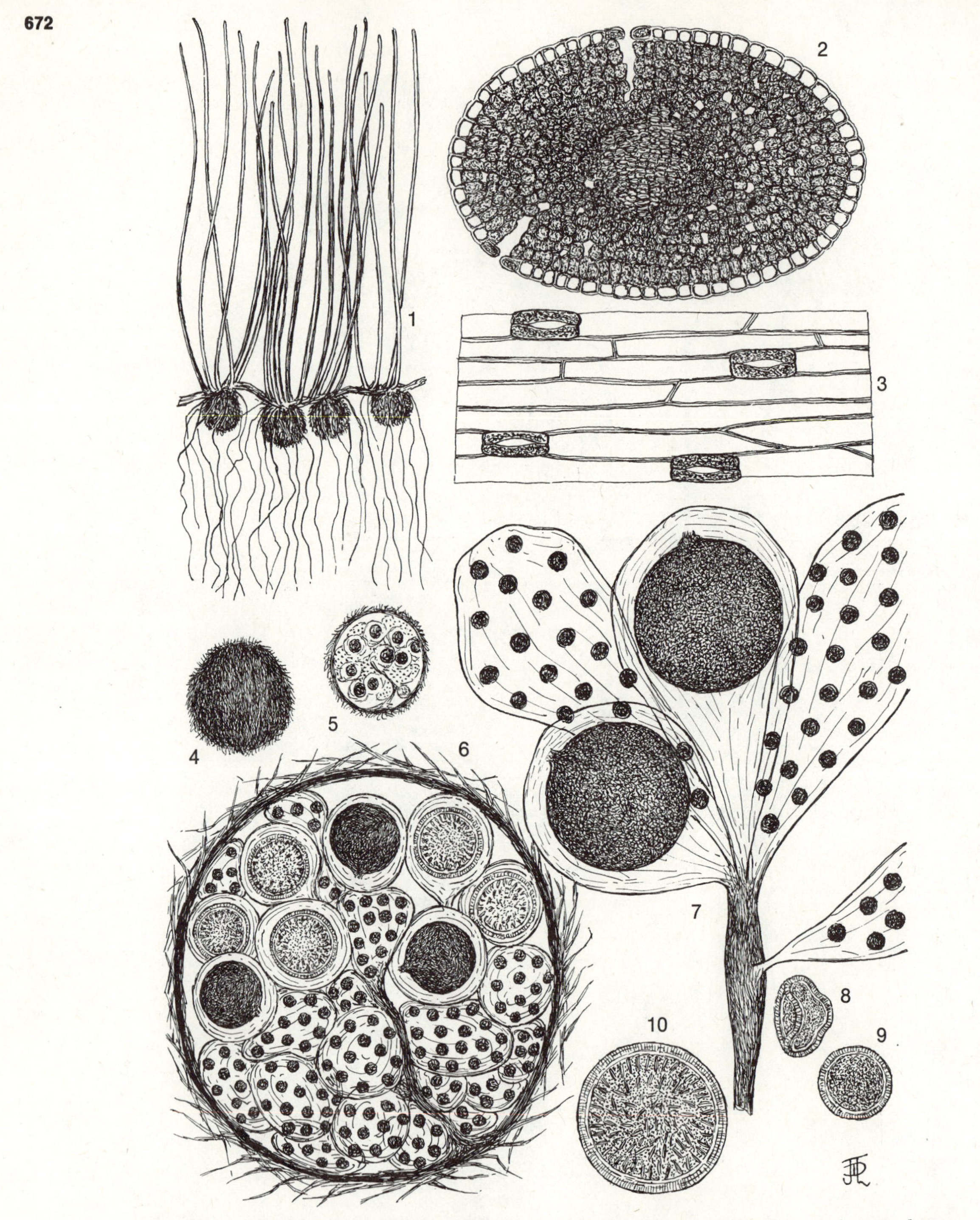

Figure 21-10. Filicales; Marsileaceae; *Pilularia americana:* **1** plant with a rhizome bearing leaves (composed of only the petioles), roots, and sporocarps, **2** cross section of the leaf, **3** stomata (pores) in the leaf epidermis, **4** sporocarp, **5** internal view, **6** this enlarged, showing the micro- and megasporangia borne on a stalk, **7** stalk removed, showing the attached micro- and megasporangia, **8–9** microspores, **10** megaspore.

Order 2. ***MARATTIALES.***
MARATTIA ORDER

Stems and leaves with simple hairs, with more strengthening tissue than in the Ophioglossales but less than in the Filicales. Young leaves circinate; stipules thick and not sheathing. Sporangia borne dorsally on the ordinary foliage leaves, not at the tips of vascular bundles, those of each sorus coalescent into a synangium, the walls thick, each sporangium with numerous spores.

Family 1. **Marattiaceae.**
MARATTIA FAMILY

Most members of this family occur in Taiwan (Formosa), the southern Peoples Republic of China, and thence across the islands of the southwest Pacific to Australia. Two genera, *Marattia* and *Danaea,* occur in the Tropics of the Western Hemisphere. (Cf. Figure 21-6A.)

Order 3. ***OPHIOGLOSSALES.***
ADDER'S-TONGUE ORDER

Hairs none or short and simple; scales none; stems and leaves soft, containing little strengthening tissue. Young leaves erect, not circinate in *Ophioglossum* but circinate in *Botrychium;* stipules thin and sheathing. Sporangia not produced on the backs of the vegetative segments of the leaf but on a segregated terminal portion apart from the green section of the blade; sporangium with no annulus, fundamentally terminal, formed at the tip of a vascular bundle, the wall thick. Gametophyte cylindroidal and subterranean, colorless and dependent upon fungi for absorbing food from the surrounding leaf mold, producing sex organs over its whole surface (resembling the gametophyte of *Psilotum,* cf. p. 687). See Figures 21-11 to 21-13, pp. 674, 675, and 676.

*Family 1. **Ophioglossaceae.**
ADDER'S-TONGUE FAMILY

The family is composed of three genera, two of which are world-wide. Those represented in North America north of Mexico are *Ophioglossum* (adder's-tongue) and *Botrychium* (moonwort, or grape fern). Several species of each occur in this region.

THE BASIS FOR CLASSIFICATION OF FERNS

The Fern Lines

The ferns, constituting the Division Pteridophyta, are related, more broadly, to the Spermatophyta including the gymnosperms and flowering plants. As nearly as can be discerned from the fragmentary fossil record, the entire great assemblage arose in one way or another from the Devonian Rhyniales, through the Psilophytales (see p. 641). However, the origins of the individual divisions and classes must have been complex, and only a little of the various lines of development can be made out. The few fossils shed some light on the subject, but knowledge is irregular, and there are many blank spots in the record. Recently, enough information has been acquired to point the way toward understanding some of the main outlines of evolution, but the origin of the ferns and of the flowering plants is vague. That of seed ferns and cycads and of the coniferous trees is becoming clarified (p. 642). See Figure 21-14, p. 677.

The pteridophytes that might be called incipient ferns appearing to have been near the lines of development of the more primitive living ferns are discussed below. Some trends of their evolution and, doubtless to a greater extent, that of their extinct Paleozoic relatives are indicated but not proved.

Rellimia (*Protopteridum*). The axis of this genus was similar to that of *Psilophyton,* but there were two types of branch systems: one sterile, the other fertile. The sterile branch

Figure 21-11. Ophioglossales; Ophioglossaceae; *Botrychium virginianum,* a grape-fern: **1** leaf with photosynthetic and spore-bearing segments, **2–5** enlargements of segments, in **3–5** showing the sporangia, **6–8** views of spores. Compare with Rellimia (*Protopteridium*), **Figure 23-3 A.**

Figure 21-12. Ophioglossales; Ophioglossaceae; ***Botrychium virginianum,*** a grape fern, showing the circinate vernation or gradual uncoiling of the young leaf as in other ferns, but not in *Ophioglossum,* the genus reported for the vernation of the family.

system was composed of many small branchlets all flattened in the same plane, the whole resembling a fern leaf more than once pinnate. In early development these structures were circinate as are the young leaves of ferns, and the smallest branchlets bore terminal drooping pairs of sporangia. The fertile branches were cylindroidal like those of *Rhynia.* It is possible that *Rellimia* or one of its relatives, was a forerunner of the ferns and probable that the fern leaf originated from a major flattened branch system. This theory of origin of the leaves in the Fern Lines was proposed by Jeffrey in 1907.

Coenopteridales. From late Devonian time until the Permian Period and particularly in the early part of the Carboniferous Period, plants of this extinct order developed flattened branch systems like those of *Rellimia.* However, in some species they were more elaborate, branching in two or more planes. Often the young branch systems were circinate. There was no definite demarcation of them from the cylindroidal stems and branches, and the internal structure of the stem was of a primitive type as in *Rhynia.* The sporangia were similar also to those of the primitive Psilophytales, p. 588, but they were in sori like

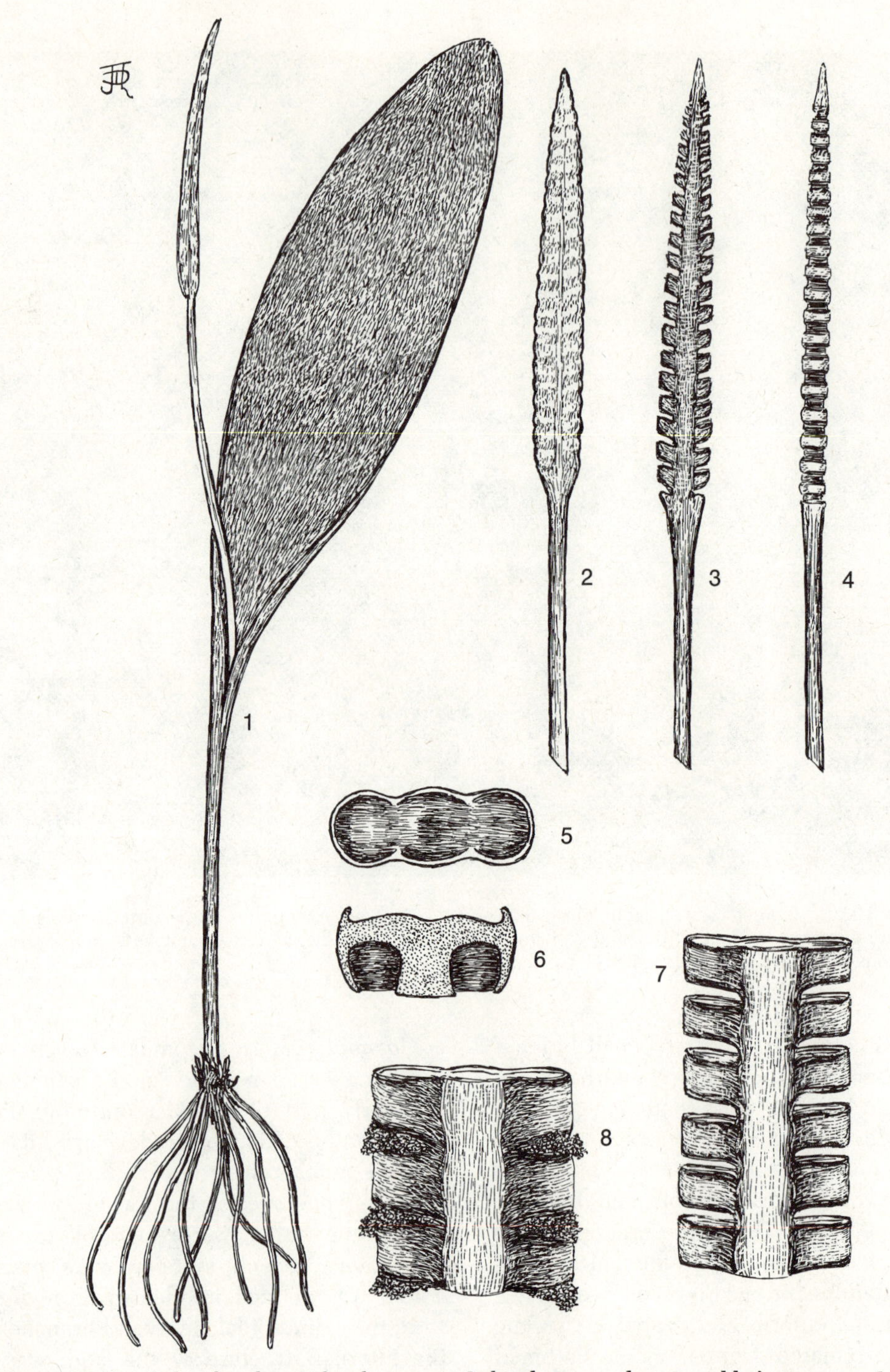

Figure 21-13. Ophioglossales; Ophioglossaceae; *Ophioglossum vulgatum,* adder's tongue: **1** plant, showing the photosynthetic and sporangium-bearing portions of the leaf, **2–8** views of the sporangia (forming a synangium), the spores being released in **8. 7** is the synangium after release of the spores.

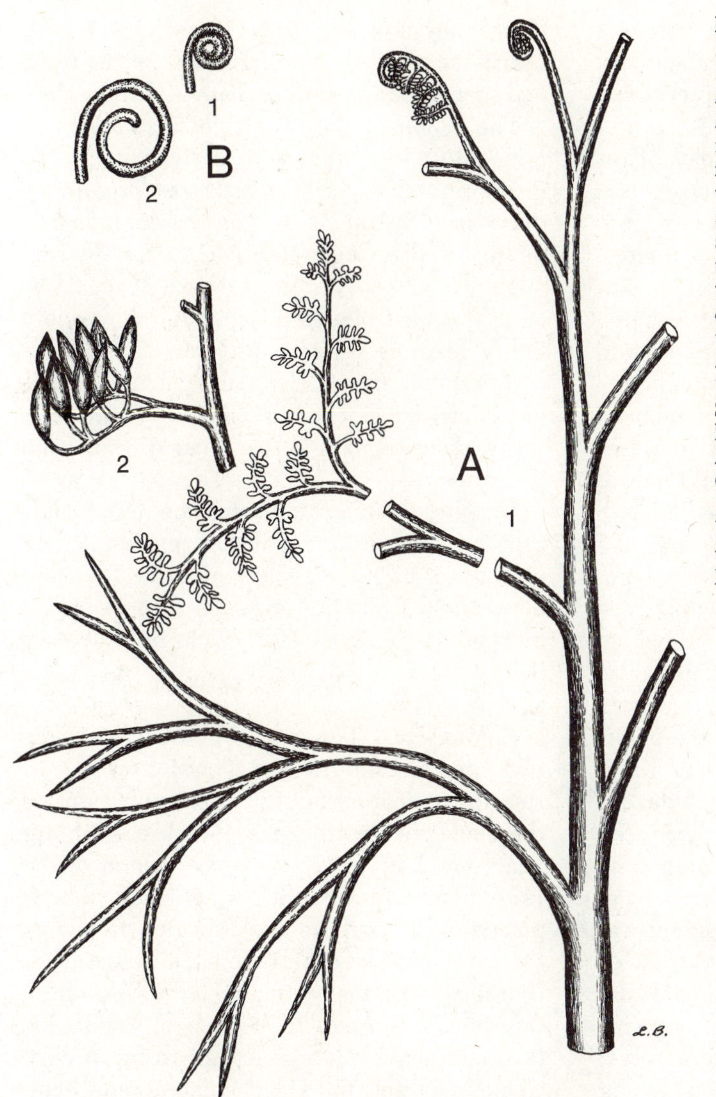

Figure 21-14. The Fern Line (primitive Devonian members) (Psilophytales; *Rellimia*, long known by another name, *Protopteridium*, and *Psilophyton*): **A** *Rellimia*, **1** dichotomous stem with major portions of the branch system flattened, and similar to a fern leaf, the vernation circinate as in the ferns, **2** sporangium-bearing portion of the branch system; **B** *Psilophyton*, circinate vernation of the branches of the axis. Circinate vernation occurred also in *Asteroxylon elberfeldense*. *Protopteridium* based upon R. KRÄUSEL and H. WEYLAND, *Palaeontographica* 78, Abt. B: 1–47. *Textabbildungen* 3–4, 8. 1933. E. Schweizerbart'sche Verlagsbuchhandlung, Stuttgart. Used by permission. *Psilophyton*, after J. W. DAWSON, Geological Society of London. Quarterly Journal 15: 477–488. fig. 1c, 1d. 1859. Used by permission. Compare *Rellimia* with the living genus *Botrychium*, **Figures 21-11** and **21-12.**

those of some of the ferns. Sometimes the sporangia of a single sorus were coalescent, forming a synangium as in some groups of living ferns (e.g., cf. Marattiaceae, p. 673). Possibly the Coenopteridales formed a connecting link between *Rellimia* and the modern ferns, but probably they were not in the direct line of ancestry for most ferns. More likely they constituted a group that has left no descendants since it became extinct in the middle of Carboniferous time, when it formed part of the great Paleozoic complex including ferns, seed ferns, Cordaitales, and other extinct groups. These were its advanced relatives but probably not its descendants. The one possible group of living descendants is the

Opioglossales, which certainly are relatives though likely derived from a common ancestry rather than directly from the Coenopteridales.

Ophioglossales. Among present-day plants the nearest approach to *Rellimia* is *Botrychium* (p. 673, Figure 21-1). *Botrychium* has circinate leaves, and the other external characters are similar. The leaf has a sterile portion similar to the flattened branch system of *Rellimia* and a similar dichotomously branched section ending in terminal thick-walled sporangia.

The sporangium has no annulus, and it is similar to that of *Rhynia* and *Psilophyton.* The development of more specialized types of sporangia is gradual through the modern families of the Filicales beginning with the primitive ferns and culminating in the Polypodiaceae. (Cf. Figure 21-6). The original type of sporangium in the Fern Lines of development may have been that of the Psilophytales (p. 588). Perhaps this extinct family of lower Devonian may have given rise to all the plants in this group of Divisions of the Plant Kingdom. The sporangia of the Psilophytaceae were terminal on the branchlets of a reduced lateral system, with a simple thick wall and no annulus, and dehiscent by a lengthwise slit as in nearly all Rhyniales. The sporangium was elongate.

The gametophytes of the Ophioglossales and of some other primitive ferns (some species of *Schizaea* and the Stromatopteridaceae) are subterranean, simple, and rodlike or similar and branched, and with antheridia and archegonia arising over the surface at almost any point, as also in the Psilophyta and *Lycopodium.* This type of gametophyte could be the primitive one in the vascular plants. However, transitions to green surface-living gametophytes occur in the Schizaeaceae, some species having either subterranean or aërial haploid generations, according to the water content of the soil, going underground on drier substrates. In *Lycopodium* (Club Moss Line) the gametophytes of the temperate species are radishlike and subterranean, those of the tropical species on the surface, flattened and green.

The origin of the Ophioglossales is uncertain. Most likely they are derivatives of the Protopteridales (order based on *Protopteridium* i.e., *Rellimia*), having arisen independently of the Coenopteridales. The internal anatomy, as well as the external form, indicates this possible relationship and derivation. Other features suggest origin from the Coenopteridales, but derivation of both Ophioglossales and Coenopteridales from the Protopteridales is more likely than of one from the other.

Despite their relationship, the Ophioglossales probably were not of the same immediate origin as the Marattiales and the Filicales. Separation from an original common stock must have occurred far back in Paleozoic time.

Psilotaceae. This family, consisting of only two genera, *Psilotum* and *Tmesipteris,* occurring in tropical and adjacent warm regions, is the center of controversy, and discussion appears in Chapter 22 after introduction of the Psilophyta. Recently the Psilophyta have been placed as a family of the Filicales, following the very extensive and detailed research of Bierhorst, who has postulated a transition from this family through the Stromatopteridaceae (a single genus and species occurring in New Caledonia) and the Gleicheniaceae and Schizaeaceae to the other families of ferns. The basis for this has been debated by various botanists, including Bierhorst, Gensel, Kaplan, and Wagner, and an excellent summary of both sides of the argument has been published in their four symposium papers (Brittonia 29: 1–68. 1977.) The problem is summarized after the treatment of the Psilophyta (pp. 686–688).

Insofar as the taxonomic arrangement of the plants in this book is concerned, the principle involved is that discussed elsewhere, as for the flowering plants (p. 551). A book on the living flora is devoted to classification of the plants on earth today. Those of the past, if enough is known about them, may indicate the presence of a common origin and therefore the relationship of living groups. However, even if the complete fossil record of the living plants were known, it would not determine the limits of taxa. The plants of the present are classified into groups as they are, not according to what they or their ancestors were. Long ago, a popular radio character put it this way, "It ain't what you was has been; it's what you now am is!"

The research on the morphology of all the groups involved in the controversy has been primarily by Bierhorst and, with differing results, by Siegert. It raises two questions of importance in taxonomy: (1) Are the Psilotaceae related more than superficially or very distantly to the Filicales? (2) Are several developmental and structural features occurring in the Psilotaceae homologous with those in the more primitive Filicales and are they transitional between those of the ferns and the Psilotaceae and those of some extinct group probably derived directly or indirectly from the Psilophytales (p. 588)? The answer to the first question depends upon that to the second. If a degree of relationship between the Psilophytaceae and the ferns can be worked out, the family may constitute (1) a class, Psilopsida, and an order, Psilotales, within the division Pteridophyta; (2) an order, Psilotales, within the class Pteropsida; or (3) only a family, Psilotaceae, within the order Filicales, as proposed by Bierhorst. If no particularly close relationship can be determined, the Psilophyta will remain as a division of the Plant Kingdom, as that kingdom exists today and as it is interpreted here and as it has been interpreted during much of the Twentieth Century. However, there can be no commitment to one interpretation or another until further research clarifies the problem.

The Fern Lines are characterized by large complex leaves thought to be (historically) formed from a major flattened branch system, and sporangia are produced on the backs or margins of the leaves. In the Spermatophyta a branch or bud develops characteristically in each leaf axil, but this is not usual in the ferns.

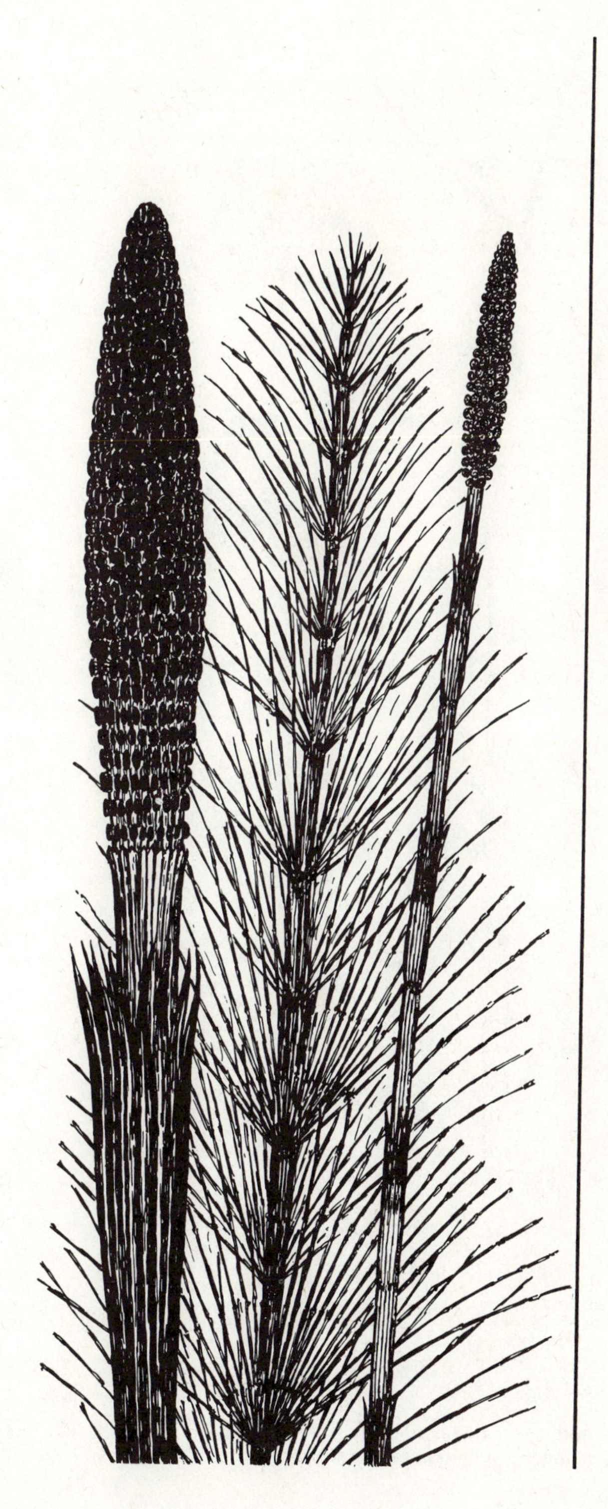

Psilophytes

Division *PSILOPHYTA*

and Horsetails

Division *SPHENOPHYTA*

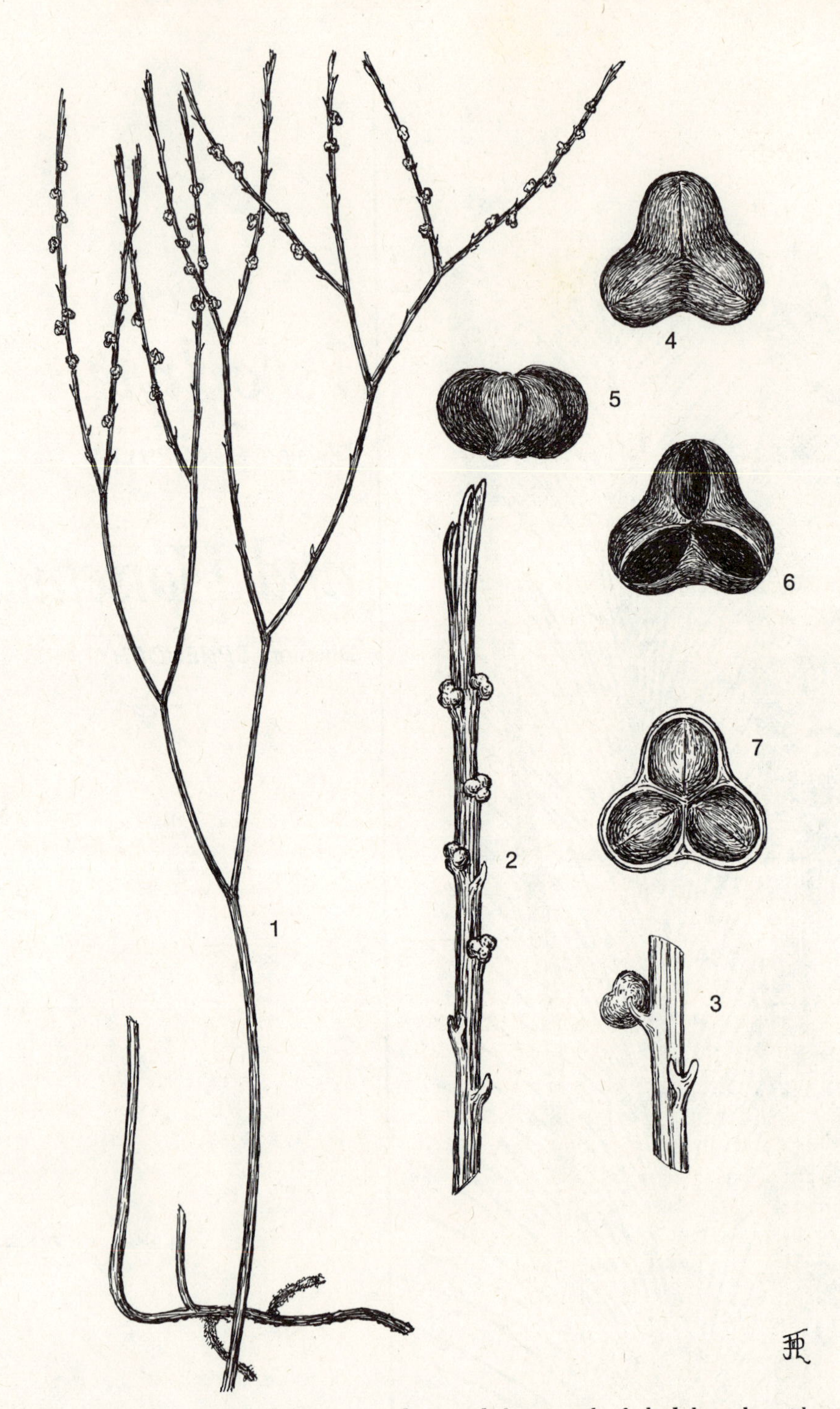

Figure 22-1. *Psilotum nudum:* **1** rhizome and erect dichotomously forked branch, with scalelike flattened branchlets, **2–3** enlarged branches showing synangia (each composed of 3 fused sporangia), **4–7** views of synangia, in **6** opened.

22

Psilophytes (Psilophyta) and Horsetails (Sphenophyta)

Psilophytes

THE VOCABULARY DESCRIBING PSILOPHYTE CHARACTERISTICS

Psilotum is a tropical and subtropical genus occurring in the United States only in southernmost Arizona and from Texas to South Carolina, and Florida but cultivated occasionally in greenhouses farther northward; abundant in Hawaii.

THE SPOROPHYTE

(According to the prevailing interpretation of most of the Twentieth Century, but see pp. 679 and 686–688.)

Stem. The axis is not differentiated into a root and stem. Its repeated forking is **dichotomous,** i.e., into two equal branches, and the plant has a bushy appearance despite the fact that commonly it is less than a foot high. There are no leaves and only scalelike structures. The related genus *Tmesipteris,* occurring in the lands in the region of the southwestern Pacific Ocean, has leaflike flattened structures, homologous with the scales of *Psilotum.* These have been interpreted as enations (p. 589 and Chapter 23), small leaves, small leaflets, and branches.

Sporangia. In *Psilotum* the sporangia are divided internally into three chambers (or probably coalescent into threes forming **synangia**). They are in the axils of scales. In *Tmesipteris* the sporangia are two-chambered and borne on the leaflike branches. The sporangia of both genera are relatively thick-walled (4–6 cells thick). Dehiscence is by two or three radiating slits, one in the apex of each chamber. There is no annulus.

THE GAMETOPHYTE

The gametophytes are rodlike but not necessarily straight. They grow buried in humus in the soil or on tree trunks, and often they are mistaken for fragments of the rhizome of the sporophyte. The entire surface may bear sex organs and there is continuous indeterminate growth from an apical cell with three cutting faces.

Figure 22-2. Psilotales; Psilotaceae; *Psilotum nudum*, plant from Hawaii (tropical rainforest), showing the dichotomously forked stems (or fronds?) with only scalelike, very small appendages interpreted either as leaves or as leaflets and with apparently lateral but fundamentally terminal clusters of 3 sporangia coalescent into synangia.

THE PROCESS OF IDENTIFICATION

Manual of the Major Taxa of Living Psilopsida

(See discussion of status and relationships, pp. 686–688.)

NOTE: The divisions of the Plant Kingdom are segregated in a key on page 6 in Chapter 1.

Division **PSILOPHYTA.** Psilotum

Class 1. **PSILOPSIDA.** PSILOTUM CLASS

Roots none; branching of the stem (axis) dichotomous. True leaves with veins none (*Psilotum* with scales, *Tmesipteris* with flattened structures resembling leaves). Sporangia fundamentally terminal on reduced branchlets but apparently lateral on the stem or flattened branchlets, each supplied with a vascular bundle, with 2 or 3 internal chambers, large, thick-walled, with many spores, opening irregularly or splitting simply. Gametophyte rodlike, growing continuously at one end, producing sex organs over its entire surface.

Order 1. ***PSILOTALES.*** PSILOTUM ORDER

*Family 1. **Psilotaceae.**
PSILOTUM FAMILY

Known only from living plants. The family occurs in the Tropics and adjacent warm regions. A single species, *Psilotum nudum* grows in Santa Cruz County, southern Arizona, and in forests from eastern Texas to southern South Carolina and Florida and Hawaii. The other genus, *Tmesipteris*, is confined to the southwestern area of the Pacific Ocean.

The plants grow on soil or on tree trunks, especially those of tree ferns in the Hawaiian forests. Both the rhizomes of the sporophytes and the similar gametophytes are submerged in the leaf-mold of the forest.

Psilotum nudum is the most common species of two to a number, *Psilotum complanatum* of Hawaii being the other one of certain status. As whisk fern, it is cultivated frequently in greenhouses as a curiosity and particularly because of its apparent similarity to the Devonian genus *Rhynia* (p.586), to which it at least seems to be related. The presence of larger and more leaflike appendages makes *Tmesipteris* seem more like a fern than does *Psilotum*.

Figure 22-3. Psilotales; Psilotaceae; *Psilotum complanatum*, an Hawaiian species of tropical rainforests; vegetative branches (or frond segments) broader and thinner than those of *Psilotum nudum*, i.e., flattened, with scalelike appendages interpreted either as leaves or leaflets.

THE BASIS FOR CLASSIFICATION OF PSILOPHYTES

The Psilotum Line

The sporophyte plant body of *Psilotum*, if interpreted as a living relative of *Rhynia*, is a forked axis not differentiated into roots and stems and bearing no leaves. The sporangia are large and thick-walled. They are coalescent in 3's forming synangia apparently borne on the sides of the branches but thought technically terminal on very short branchlets. This is indicated by the forking of the vascular traces. Each sporangium produces numerous spores, and at maturity it splits irregularly or simply along three lines.

Although *Psilotum* is almost identical with *Rhynia* in form, it has scattered small scales along the axis, and both the sporangia and the axis anatomy have special features not to be found in the early Devonian Rhyniales, for which it seems to be the reincarnation. Especially in Japan, numerous races of *Psilotum nudum* have been long in cultivation. Rouffa (American Fern Journal 61: 75–86. 1971) has called to attention and analyzed a form having no scalelike or other stem appendages and terminal (usually 3 but sometimes more) sporangia (in synangia). The significance is difficult to determine. The form of the plant may be held over from Paleozoic, but it may be due to recent mutation.

The related genus *Tmesipteris* is similar in most respects, but with structures interpreted by some as flattened branches and by others as leaves and with 2-chambered sporangia.

According to this tentative interpretation, the Psilotum Line retains the fundamental organization of the Psilophytales. There are no leaves, though there may be scalelike or leaflike flattened structures. The sporangia are fundamentally terminal, and they are thick-walled.

An Alternate Interpretation—The Psilotaceae and Origin of the Ferns

The riddles of the nature, origin, and relationships of the Psilotaceae and of the origin of the ferns have been given a proposed revolutionary solution through the extensive and meticulous work of David W. Bierhorst. The genera *Psilotum* and *Tmesipteris* are indicated to be near relatives of the more primitive ferns, such as *Stromatopteris* (one species in New Caledonia) and members of the Schizaeaceae and Gleicheniaceae. At the center of this interpretation is a new theory of the aërial body of the members of the Psilotaceae as a single incompletely differentiated leaf formed from a branch system not clearly separable from or an appendage of the rhizome. This interpretation is given to the leaves of such ferns as *Stromatopteris*, in particular, as well. The other features of the Psilotaceae are considered to link them also with the ferns. The following characters are included:

The Stem and Leaf. One to six stem apices occur near ground level, terminating forks of the dichotomous underground stem, but those within the green region do not develop. One apex divides, and one half of it develops the frond. This procedure occurs also in various ferns, including those thought primitive (Stromatopteridaceae, *Gleichenia*, Hymanophyllaceae), as well as even *Pteridium*. Thus, the aërial part of either *Psilotum* or *Tmesipteris* is interpreted as a leaf of the pteridophyte type but in transition from a branch system, and this interpretation is given also to the leaves of the primitive ferns. The scales on the frond of *Psilotum* and the broader appendages of *Tmesipteris* are interpreted as leaflets, as are the branches of the system. Structures in *Schizaea dichotoma* and *Gleichenia* are considered homologous. The implication is not the origin of the Psilotaceae from the Psilophytales or some other extinct group and of the ferns from the Psilotaceae but instead the origin of the Psilotaceae from some group of primitive ferns within the complex of the Filicales and their extinct near relatives.

Anatomy of the Stems and Frond Bases. The smaller underground axes of the stems of both the Psilotaceae and Stromatopteridaceae have simple protosteles without protoxylem and no distinction of formation of early or late metaxylem (that maturing after stem formation). The smallest branches have no formation of a stele. The larger axes have no protoxylem but formation of considerable early and late metaxylem. The base of the frond, where it joins the stem, has a somewhat more complex protostele, but higher up it changes organization to fundamentally that of a petiole of the Filicales. The cortex cells are impregnated with dark substances (polyphenolic compounds), as they are also only in the Filicales.

The Root. Lack of roots is a major feature of the Psilotaceae. Young plants of *Stromatopteris*, up until there are several leaves and many rhizome branches, have no roots. Later the older plants develop a few roots; these are mostly unbranched and glabrous. There are only stubby vestiges of root hairs. The roots are apparently vestigial, and this is considered further evidence of reduction of the structures of the Psilotaceae from those of more elaborate primitive fern ancestors. There is no special cell or group of cells in the embryo giving rise to a root, and this may be primitive, but in *Stromatopteris* and *Actinostachys* (if segregated from *Schizaea*) there are roots, even though there is no distinction of a root in the embryo.

The Fertile Pinnae, Sorus, and Sporangia. The axis of the sorus is interpreted as bearing two scales of the fertile pinna below and the two or three (rarely four or more) sporangia above (these forming a synangium through coalescence). Rarely the axis continues to grow and develops more fertile pinnae and sporangia above. The part of the fertile pinna below the two scales also may become more elaborate. The sporangium has a small group of apical cells suggestive of the annulus of some primitive ferns, especially *Osmunda* and *Mohria.* The sori of the Psilotaceae are likened to simpler versions of those of *Gleichenia,* in which 1–12 sporangia are arranged on a mound or axis. The wall of the sporangium, 4–6 cells deep in the Psilotaceae, is matched by almost identical walls found recently in *Stromatopteris, Schizaea* (*Actinostachys*), *Gleichenia,* and *Osmunda.* In all these plants, the inner wall layers break down in part. The "massive" sporangia of the Psilotaceae are of polyploids (most are high polyploids), and those of the diploids are of about the same size as the sporangia of ferns.

Spores. The spores of the Psilotaceae are similar to those of the Filicales.

Gametophytes. The gametophytes of the Psilotaceae are identical with those of *Stromatopteris* and *Schizaea* (*Actinostachys*), as well as of *Botrychium.* The gametophyte grows underground, becomes rodlike, grows indeterminately from an apical cell with three cutting faces, and bears antheridia and archegonia at any point on its surface. The rhizoids of all these gametophytes are septate (with internal cross walls), as in no other vascular plants (those of *Botrychium* not homologous). The Psilotaceae, Stromatopteridaceae, and *Schizaea* (*Actinostachys*) have (1) antheridia of the Filicales type; (2) archegonia with thickened walls; and (3) cutinized outer cells of the plant body, these restricting absorption from the soil to the hyphae (filaments of cells) of symbiotic fungi.

Embryo. The embryo in each of the four genera just mentioned is composed of only a foot and a stem apex.

Thus, according to Bierhorst's interpretation, placing the Psilotaceae in the Filicales is an obvious consequence of investigation. His work is to be taken with the utmost seriousness. He has made an enormous contribution to study of details of obscure organisms others have passed by or not seen. The summary above is necessarily brief, but it is based on much of a book (Bierhorst, 1971) and a brief summary (Bierhorst, 1977), as well as other technical papers. The research is a model of thoroughness, and the interpretations are made cautiously.

In view of the paragraph above, the question would seem to be settled, but there is sharp disagreement on interpretation of the structures, and unfortunately there is no fossil evidence concerning the problem. An account of the case for and against Bierhorst's interpretation of the Psilotales is published in a symposium (Brittonia 29: 1–68. 1977). Papers by David Bierhorst, Patricia G. Gensel, Donald R. Kaplan, and Warren H. Wagner present both sides of the case.

Kaplan rejects the interpretation of the fernlike leaf of *Stromatopteris* as a continuous development from the rhizome and considers the leaf to be an appendage of the stem. Basing much of his argument on the investigation of Siegert, he considers the forked axes of *Psilotum* and *Tmesipteris* to be aërial stems and the scalelike appendages of *Psilotum* and the more leaflike ones of *Tmesipteris* to be leaves rather than leaflets. If these interpretations are correct, of course, Bierhorst's principal point is negated and some lesser points fall as corollaries. Kaplan calls for further research directed to solving the problem.

Wagner considers the proposal of Bierhorst from a taxonomic point of view, and he points out the following opposed interpretations:

The Fertile Pinnae, Sorus, and Sporangia. The nature of the structures bearing the sporangia is considered undetermined, except that the sporangial groups in *Psilotum* and *Tmesipteris* are borne on forked appendages, and these are designated by an uncommittal term, sporangiophores, rather than sporophylls (specialized spore-bearing leaves). The sporangium position of *Stromatopteris* is considered exactly as in other ferns. The two or three sporangia of the Psilotaceae are coalescent into synangia; those of *Stromatopteris* are separate and up to about 25 in a cluster, or sorus, and with stalks as in other Filicales. Presence of an annulus in the Psilotaceae is rejected, and presence of a distinct "bow" of annulus tissue in *Stromatopteris* is pointed out. The latter genus also has paraphyses (sterile structures) associated with the sporangia, as do many Filicales, while the Psilotaceae are completely glabrous.

Spores. The spores of the Filicales and other ferns are so variable that probably similarity of those of the Psilotales to some of them is of no great significance.

Gametophytes. The significance of aërial green and subterranean nongreen gametophytes is questioned. The occurrence of both types among the various families of the Filicales is pointed out, and the presence of three types in *Schizaea* and four in *Lycopodium* is called to attention.

Trichomes (hairs). These are surface appendages (of especially the rhizomes) almost universal in the ferns; the Psilotaceae are wholly glabrous.

The little known about the chemical characters of the Psilotaceae and the Filicales does not support the inclusion of the family in that order. Caldicott, Simoneit, and Eglinton (Phytochemistry 14: 2223–2228. 1975) found trihydroxy hexadeconols otherwise unknown in the cutin of the stems of the Psilotaceae; the Pteridophyta and Spermatophyta have dihydroxyhexadecanoic acids instead. Cooper-Driver (Science 198: 1260–1262. 1978) found only biflavonyls (amentoflavone) in the Psilotaceae and flavonols (kaempferol and quercetin) and Proanthocyanidins (procyanin and prodelphinidin) in the Filicales of the primitive group including Stromatopteridaceae, Gleicheniaceae, and Schizaeaceae. Wallace and Markham (American Journal of Botany 65: 965–969. 1978) found no flavonoids in common between the primitive ferns and the Psilotaceae. In view of the ability of major divisions of the Plant Kingdom to produce only certain flavonoids, this may be considered strong though by no means infallible evidence of the separateness of the Psilotaceae and the Filicales.

Obviously, there is a problem. Too little is known, and what is known may be interpreted differently by various authors. Consequently, the nature, organization, structures, origin, and evolution of the Psilotaceae must be subjected to further investigation and discussion before a solution to the questions about them will emerge. Their classification is still undetermined (pp. 679–680), and the effect upon classification, as well as ideas about origin of the ferns, must be held in abeyance. For this reason, in Chapter 21 the family Psilotaceae is not included in the Filicales, and here it is retained as forming a division of the Plant Kingdom, Psilophyta. However, this does not reflect a conclusion; it merely indicates caution concerning changing a well-established classification until the supporting data are known to be conclusive.

If the proposals of Bierhorst are accepted, the description of the family should appear as follows, as a substitute for the one on p. 684.

*Family 1. **Psilotaceae.**
PSILOTUM FAMILY

Root none. Stem underground, forking dichotomously like that of the Rhyniales, continuous with the whiskbroomlike or somewhat leaflike aërial structure, which forms fundamentally a large leaf, though this does not necessarily appear like one and it is not clearly appendicular to the stem (the leaf being transitional between a system of branches and the complex leaf of a fern formed from a flattened branch system), the leaf bearing scalelike leaflets in *Psilotum* and flattened corresponding leaflets in *Tmesipteris*. Sporangia fundamentally terminal (each sometimes supplied with a vascular bundle), though appearing lateral, on the upper axis above 2 scales on the fertile leaflet, forming synangia, coalescent in 3's (*Psilotum*) or 2's (*Tmesipteris*), usually large, with thick walls 4–6 cell layers deep, with many spores, each opening by a longitudinal slit from the center of the synangium. Gametophyte rodlike, growing continuously from an apical cell with 3 cutting faces, not determinate as are the gametophytes of other ferns, producing antheridia and archegonia over the entire surface, often confused with the stems of sporophytes growing with it.

SELECTED REFERENCES

BANKS, HARLAN P. *Evolution and Plants of the Past.* 1964, 1970. Wadsworth Publishing Co., Belmont, California.

BIERHORST, DAVID W. *Morphology of Vascular Plants.* 1971. MacMillan Co., New York.

———. *The Systematic Position of Psilotum and Tmesipteris.* Brittonia 29: 3–13. 1977.

GENSEL, PATRICIA G. *Morphologic and Taxonomic Relationships of the Psilotaceae Relative to Evolutionary Lines in Early Land Vascular Plants.* Brittonia 29: 14–29. 1977.

KAPLAN, DONALD R. *Morphological Status of the Shoot Systems of Psilotaceae.* Brittonia 29: 30–53. 1977.

WAGNER, HERBERT WARREN, JR. *Systematic Implications of the Psilotaceae.* Brittonia 29: 34–63. 1977.

Horsetails

THE VOCABULARY DESCRIBING HORSETAIL CHARACTERISTICS

The only living genus is *Equisetum,* known as horsetail or scouring-rush. The stems and branches of some plants are bushy and with a slight resemblance to a horse's tail. Because the surface includes cells with glasslike deposits of silicon dioxide (silica), the plants were used once for scouring pots and pans.

THE SPOROPHYTE

The widely occurring species *Equisetum arvense* and the larger similar Pacific Coast species *Equisetum telmateia* are characteristic of the genus.

Stems and Leaves. The stems are partly rhizomatous, and the aërial portions may be up to about one foot (3 dm.) in height. In this species they are of two kinds, those appearing early in the spring being yellowish and each terminating in a conelike strobilus, the later ones being green and with whorls of branches but no sporangia. Both types are conspicuously "jointed," that is, with obvious nodes and internodes. Each node bears a whorl of scalelike nongreen leaves, which are joined into a sheath enclosing a small portion of the stem. On the green stems a whorl of branches is produced just below the leaves at each upper node.

Sporangia. The strobili appear like small cones of gymnosperms. Each is composed of an axis bearing closely packed peltate sporophylls, which are hexagonal from the external view of the strobilus. Several sporangia droop toward the axis of the strobilus from the mar-

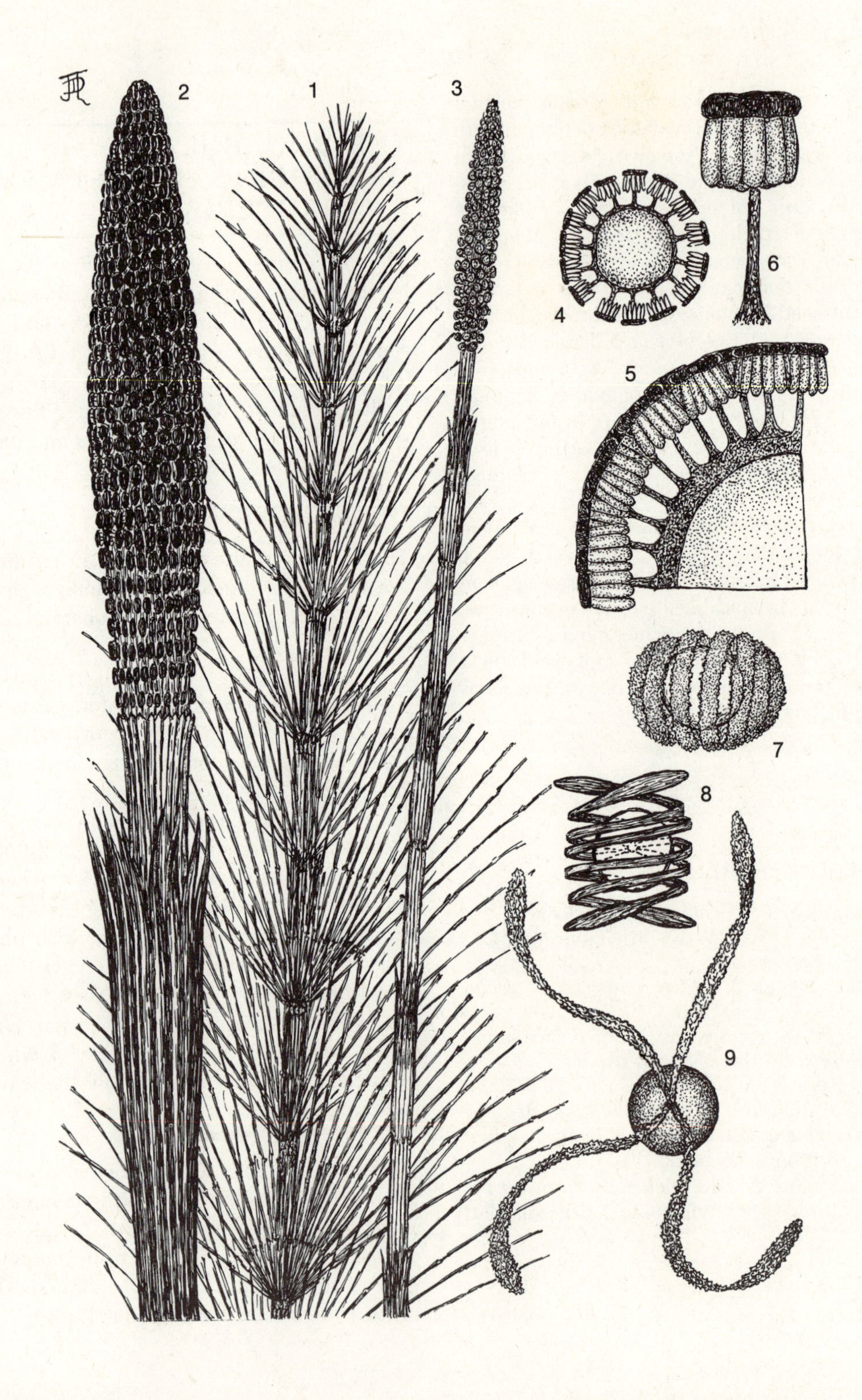
2
1
3
4
5
6
7
8
9

gin of the sporophyll. Each sporangium opens by a longitudinal slit, releasing the spores. There is no annulus.

THE GAMETOPHYTE

The spore has four attached threadlike "elaters," which coil and straighten rapidly as the moisture content of the air fluctuates. Their action in moving the spores may be observed under a dissecting microscope as the humidity is changed by human breath. The movements aid in releasing the spores from the sporangium. The mature gametophyte resembles that of a fern, but it is lobed irregularly.

THE PROCESS OF IDENTIFICATION

Manual of the Major Taxa of Living Horsetails

NOTE: The divisions of the Plant Kingdom are segregated in a key on page 6 in Chapter 1.

Division **SPHENOPHYTA.** Horsetails

Class 1. **SPHENOPSIDA.**
HORSETAIL CLASS

Roots present, adventitious, growing from the stem and the subterranean branches; stems with conspicuous nodes and internodes; branches in whorls at the nodes. Leaves in whorls above the branches at the nodes, numerous in each whorl; (leaf gaps none in the stele of the stem). Sporangia borne in a strobilus, the sporophylls peltate, each with several sporangia drooping inward from the margin; each spore with four attached "elaters." Gametophyte a flat or irregularly lobed thallus.

Order 1. ***EQUISETALES.***
HORSETAIL ORDER

*Family 1. **Equisetaceae.**
HORSETAIL FAMILY

Upper Carboniferous (Pennsylvanian) Period to the present. The single genus, *Equisetum,* occurs in all continents but Australia. About fifteen species are native in North America north of Mexico.

THE BASIS FOR CLASSIFICATION OF THE HORSETAILS

The Horsetail Line

The Horsetail Line of development may be traced from *Rhynia* through *Psilophyton,* an early Devonian genus of the Psilophytales, as follows.

Psilophyton. Except in the characters of its sporangia and the greater length of the sterile branches, the genus resembled *Rhynia.* At the last fork of the axis the two slender branchlets turned downward, and the two sporangia, each terminating one of the branchlets, were therefore in a drooping position. Similar

Figure 22-4. Horsetail, *Equisetum telmateia* (Pacific Coast): **1** sterile branch with a whorl of branchlets at each node (just below the whorl of scalelike nongreen leaves), **2** fertile branch terminated by a strobilus, an enlarged sheath formed from a whorl of many coalescent leaves shown below, **3** the fertile branch of another species, showing the sheath of leaves at each node, **4** cross section through a strobilus, showing the peltate sporophylls, **5** enlargement, **6** single sporangiophore with the drooping sporangia, **7–9** spores with the attached "elaters" which straighten and coil with changes in moisture content of the surrounding air, as under human breath. The movements are sudden and violent, and the squirming of the elaters throws the spores into the air, where they may be picked up by a breeze.

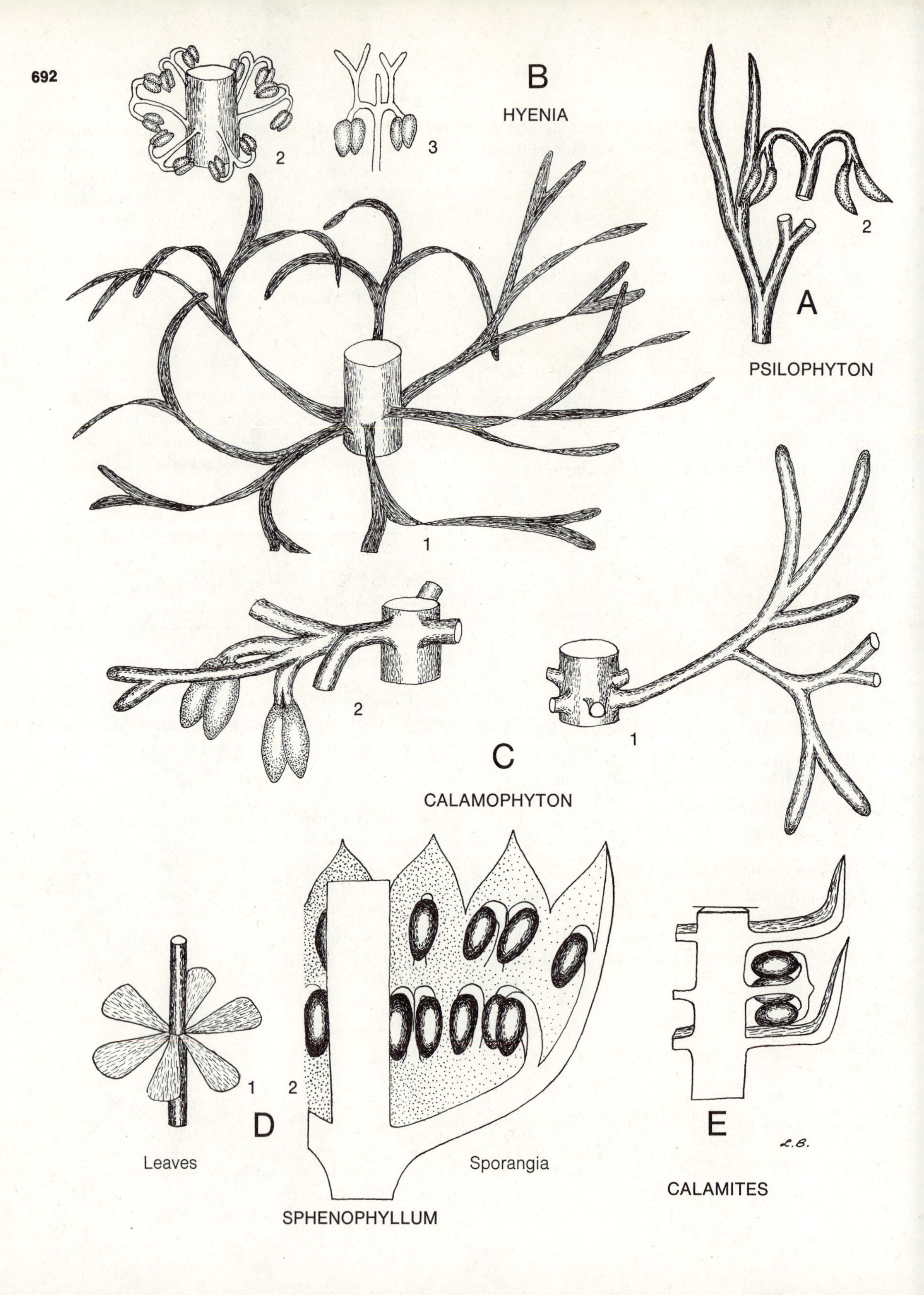
B
HYENIA
2
3
A
2
PSILOPHYTON
1
2
C
1
CALAMOPHYTON
1
2
D
Leaves
Sporangia
SPHENOPHYLLUM
E
CALAMITES
L.B.

drooping sporangia usually in pairs or fundamentally so are characteristic of the Horsetail Line. They occurred in the following two genera of Sphenopsida composing the Devonian order Hyeniales, probably derived from a *Psilophyton*-like ancestor.

Hyenia. In these small plants the appendages of the stem were arranged in imprecise whorls of 6 foreshadowing a characteristic of all members of the Horsetail Line. They were flattened, and they may be interpreted as leaves or not. The main stem was unbranched or it forked only once, and many stems arose from a rhizome. The stem was not jointed. The sporangiophores, borne on separate branches from those bearing the sterile appendages, were in 3 pairs forming an irregular whorl of 6.

Calamophyton. This genus of the Hyeniales resembled *Hyenia*, but its stem has been interpreted as jointed, a characteristic feature of the more advanced members of the Horsetail Line. However, Leclerq and Andrews have pointed out the irregular occurrence of the joints and suggested that they may be cracks formed in the specimens during fossilization. The appendages, or leaves, were not flattened as in *Hyenia*.

In the Hyeniales often many branches bearing pairs of drooping sporangia were aggregated into a sort of loose strobilus with 6 branches in a whorl or these united in pairs. The sporangiophores (sporangium-bearing branches or leaves) were forked once, and each fork bore 3 branches, two terminating in pairs of elongated sporangia resembling those of *Psilophyton* (i.e., Psilophytaceae). These were terminal and curved backward as in *Psilophyton*. The middle branch of the sporangiophore was vegetative and weakly branching.

Development of taxa probably from the Hyeniales in late Devonian, Carboniferous, Permian, and Triassic time is illustrated by the following plant groups.

Sphenophyllum. This genus was composed of slender herbs characterized by whorls of well-developed but variable leaves at each node of the stem. Probably the plants resembled the present genus *Galium*. The leaf appears to have originated by fusion of the small flattened branchlets of a minor branch system. The leaves varied from deeply lobed or divided to apically toothed or entire, but in all

Figure 22-5. The horsetail (*Equisetum*) ancestry. **A** *Psilophyton* (early Devonian), **1** the vegetative axis, forking dichotomously, **2** a reproductive branchlet terminated by a pair of drooping sporangia; **B** *Hyenia* (middle Devonian), **1** with flattened minor branchlet systems attached to the main axis, these doubtless photosynthetic, occurring in 6's at each level (incipient node), **2** a reproductive branch with each of 6 branchlets bearing two pairs of drooping sporangia, **3** development of small vegetative branches beyond those bearing sporangia, this present and variable or absent; **C** *Calamophyton* (commonly figured with strongly differentiated nodes and internodes, but this perhaps due to cracking during fossilization), **1** with terete minor branchlet systems attached to the main axis, these presumably photosynthetic, **2** three coalescent pairs of sporangium-bearing branchlets at a node on a fertile stem, the vegetative axis continued beyond the branchlets bearing sporangia; **D** *Sphenophyllum* (Carboniferous), an abundant genus in the coal fields, **1** whorled leaves at each node, **2** the leaves of the strobilus like the branchlets of *Calamophyton* (bearing sporangia) in **C 2** but fused into a cup and bearing alternately paired and solitary sporangia; **E** *Calamites*, giant horsetail (Carboniferous), portion of the strobilus with the bracts not fused and with pairs of sporangia fused into sets of four. **A** based upon J. W. Dawson, *The Geological History of Plants* 64. fig. 19. 1896. Appleton-Century-Crofts, Inc.; **B, C** upon R. Kräusel and H. Weyland, *Senckenbergische Naturforschende Gesellschaft.* Abhandlungen Band 4, Heft 2: 115–155. plate 16–17. 1926, with ideas from recent literature; **D 2** upon Max Hirmer, *Handbuch der Paläobotanik* 358–359. fig. 418–21. 1927, Verlag von R. Oldenbourg, München; **E** *ibid.* 402. fig. 483. Used by permission.

there was a clearly dichotomously forked system of veins. The sporangia were produced at the tips of leaf veins corresponding to the vascular bundles of the component flattened branchlets. In some species the sporophylls were arranged in an elongated, slender strobilus; in others they were scattered on the stem among the sterile leaves. A wide variety of strobilus types is known. Each sporophyll had two lobes; in some cases both were fertile, but often the upper was fertile and the lower sterile. On the fertile lobe the sporangia were borne either singly or in pairs on more or less umbrella-shaped tips, where they were bent backward toward the axis of the strobilus rather than downward as in *Psilophyton*. Because of branching of the veins of the consolidated branchlets forming the fertile lobe, a single sporophyll had altogether as many as three to six or even more sporangia. Thus, *Sphenophyllum* carried on the arrangement of drooping sporangia found in *Psilophyton* and *Hyenia*, but the details were more elaborate. Forms of the same pattern characterize the rest of the line of development.

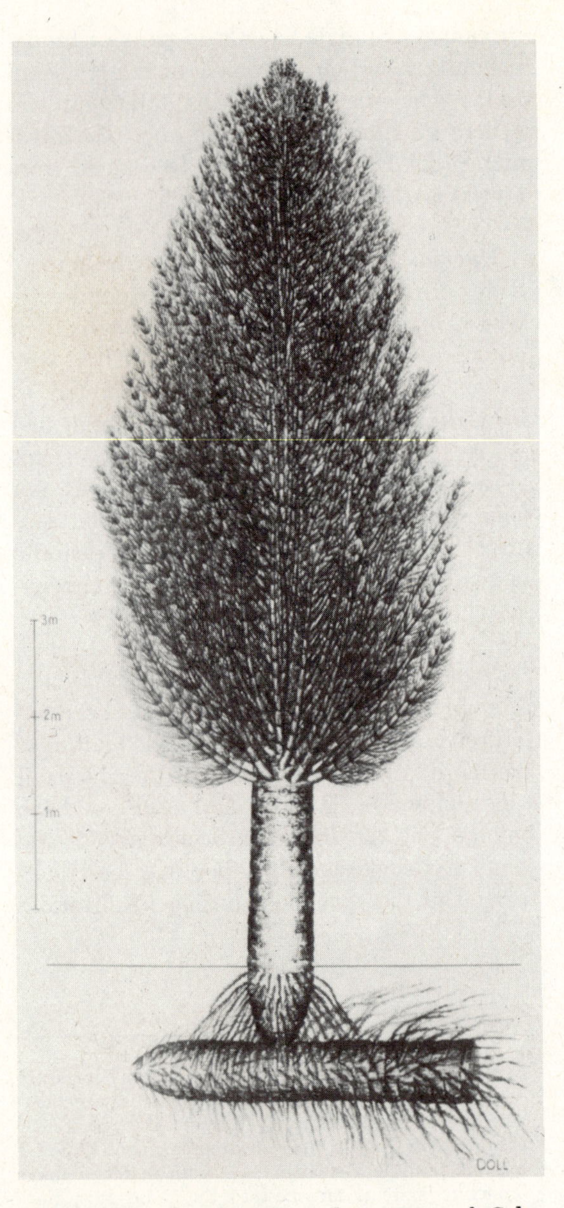

Figure 22-6. Reconstruction of a species of *Calamites,* giant horsetail. From MAX HIRMER, *Handbuch der Paläobotanik* 435. fig. 527. 1927. Verlag von R. Oldenbourg, München. Used by permission.

Calamites and its Relatives. The extinct family Calamitaceae probably included a number of genera, but most of the fossils have been referred to as "*Calamites*" or giant horsetail. In many respects these trees (and probably some herbs) resembled the present-day horsetails in vegetative structure. The stems were markedly jointed and longitudinally ridged and grooved. Whorls of leaves and branches grew at each node. There were several types of strobili, but all types included whorls of more or less peltate sporophylls with sporangia "drooping" inward from their margins toward the axis of the strobilus. Fundamentally, probably these sporangia were paired, but with the pairing obscured by fusion of the branchlets of the flattened branch system forming the sporophyll.

The Horsetail Line is represented today by a single common and widely distributed genus, as follows.

Equisetum. The horsetails carry out the fundamental features of the fossil line of development described above, for the stems are jointed, the leaves and branches are in whorls at the nodes, and the sporangia droop from the edge of a peltate sporophyll as in the giant horsetails, being fundamentally in pendent pairs.

In contrast to development of leaves in the Fern Line, the leaves of the Horsetail Line originated from a minor system of flattened branches, and those of the living horsetails are small, scalelike, and not green. The stems are jointed, the leaves whorled. The sporangia of the living plants are on peltate sporophylls in compact strobili, being fundamentally in drooping pairs but not appearing so.

Figure 22-7. The Horsetail Line; *below, Sphenophyllum,* piece of coal with many branches and whorls of leaves similar to those in **Figure 22-4 D 1**; *above, Calamites,* surface ridges and nodes of the stem in coal. Photograph by Frank B. Salisbury.

Club Mosses

Division *LYCOPHYTA*

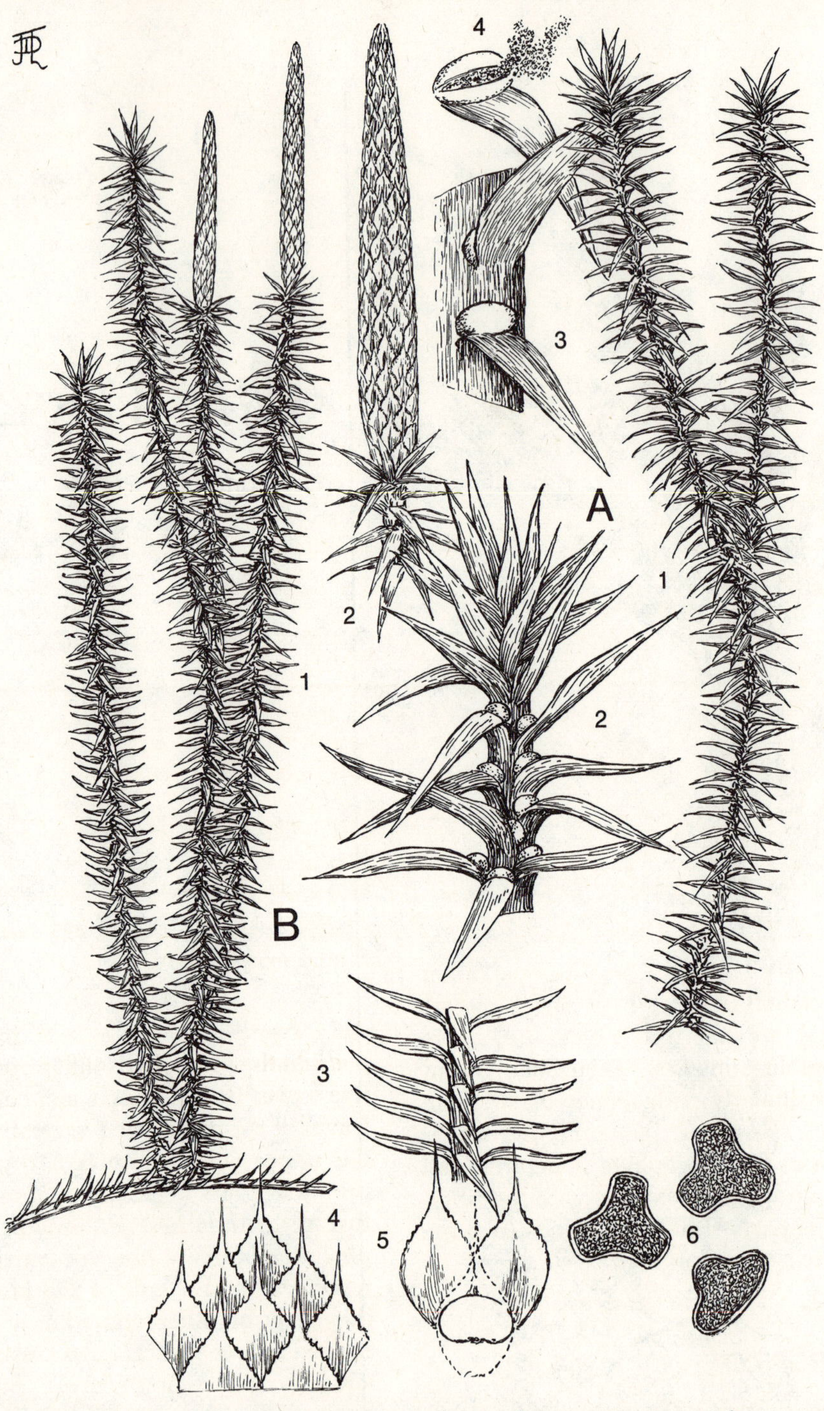

Figure 23-1. Club moss, *Lycopodium:* **A** *Lycopodium Selago* var. *patens,* **1** bearing sporangia in the upper leaf axils, **2** enlargement, showing the sporangia, **3** greater enlargement, **4** sporangium opening and releasing the spores; **B** *Lycopodium annotinum,* **1** rhizome or stolon and erect branches bearing terminal strobili, **2** strobilus enlarged, **3** leaves, **4** sporophylls, **5** sporophylls and a sporangium, the sporophyll subtending the sporangium removed but represented by dotted lines, **6** spores. Note similarity of sporangium in **4** to that of *Zosterophyllum*, p. 594.

23

The Club Mosses

THE VOCABULARY DESCRIBING CLUB MOSS CHARACTERISTICS

The club mosses (*Lycopodium*) and the rather distantly related little club or spike mosses (*Selaginella*) are not true mosses but spore plants resembling mosses. Their small stems are clothed similarly with numerous scalelike green leaves. However, the stems and leaves of club mosses are sporophyte structures, and they are in no way homologous with the similar-appearing parts of mosses, which are of the gametophyte generation.

THE SPOROPHYTE

Stems and Leaves. The stems branch considerably, and they are a few inches or sometimes several feet long and usually an eighth of an inch or less in diameter. They are not jointed, and the leaves are borne in dense spirals. The roots are adventitious and on the rhizomes or sometimes along the stems. In *Selaginella* they terminate threadlike roots, or special branches, known as rhizophores (their fundamental nature debated).

Sporangia. In one species group of *Lycopodium* the sporophylls differ from the vegetative leaves in size, shape, and color, the lower leaves of the plant being vegetative and green, the upper ones (sporophylls) yellow, spore-producing, and aggregated into strobili at the tips of the branches. In *Selaginella* the sporophylls are usually like vegetative leaves, but restricted to the ends of the branches where they may be differentiated only by being arranged in four vertical rows (instead of apparently irregular) and more tightly appressed against the stem. In this genus the sporangia are of two kinds, one producing large spores (megaspores), these often being yellow, and the other small spores (microspores), these often being orange.

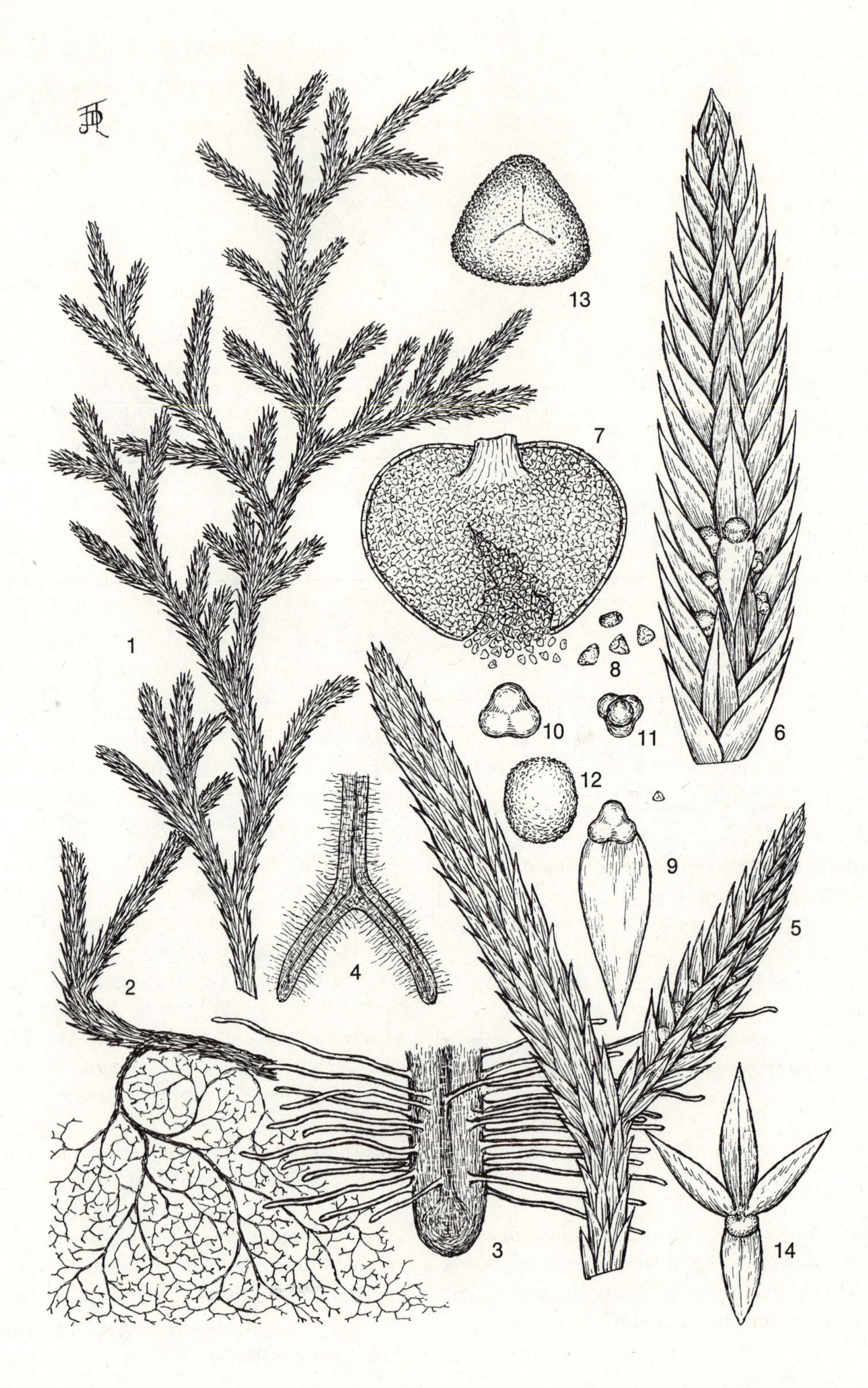

1
2
3
4
5
6
7
8
9
10
11
12
13
14

THE GAMETOPHYTES

The gametophytes of the species of *Locopodium* occurring in northern and temperate regions resemble turnip roots, and they are subterranean. Those of most tropical species are flattened thalli somewhat like the prothallia of the ferns, but they are larger and they continue to grow indefinitely instead of reaching a certain size and stopping. The gametophytes of *Selaginella* are formed within the walls of the spores, and they are of microscopic size. They are of two greatly differing sizes. The microspores produce internal microgametophytes each consisting of a single prothallial cell, representing the thallus, and a solitary antheridium. The megapore produces an internal, but slightly exposed apically, larger megagametophyte consisting of an enormous basal many-nucleate cell containing reserve food, a cushion of cells partly exposed at the apex, and archegonia in the cushion. Rhizoids also are developed.

THE PROCESS OF IDENTIFICATION

Manual of the Major Taxa of Living Club Mosses

NOTE: The divisions of the Plant Kingdom are segregated in a key on page 6 in Chapter 1.

Division **LYCOPHYTA.** Club Mosses

Class 1. **LYCOPSIDA.** CLUB MOSS CLASS

Roots adventitious on the stem or main branches or on special branches from them; stems dichotomous; (xylem and phloem of the stem almost always in a solid cylinder, i.e., protostele, but this sometimes divided into segments, the stem with no leaf gaps). Leaves simple, scalelike, clothing the stem densely, each with only a midrib. Sporangia in the leaf axils, attached either to the stem or to the base of the upper side of the leaf; sporophylls often differing more or less from vegetative leaves and then arranged in strobili. Gametophytes of various types.

Key to the Orders

1. Stems elongated and much branched; leaves numerous, scalelike, closely clothing the branches and overlapping each other like shingles, often not more than three or four times as long as broad; sporophylls often in terminal strobili; sporangia produced in the axils of the leaves, not embedded in the leaves, not divided internally by crosswalls.
   2. Spores all of one kind, minute, numerous in the sporangia.

      Order 1. **Lycopodiales,** below.

   2. Spores of two sizes, some sporangia producing numerous minute spores, others only four large spores, the spores (and their sporangia) often differentiated by color, as, for example, microsporangia and microspores orange or red and the megasporangia and megaspores yellow or white.

      Order 2. **Selaginellales,** below.

Figure 23-2. Selaginellales; Selaginellaceae; *Selaginella Bigelovii:* **1** aërial branch clothed densely with green scalelike leaves, **2** lower and underground portion of the branch with a rhizomorph (rootlike branch bearing roots), **3** tip of a rhizomorph enlarged, **4** dichotomous branching of the root, **5** sterile branchlet (*left*) and fertile branchlet or strobilus (*right*), the former with (usually spreading) leaves not arranged precisely in four vertical rows, the latter with (usually appressed) leaves (sporophylls) arranged precisely in four vertical rows, each leaf with an axillary sporangium, in this species the (mostly) upper leaves with microsporangia (orange) and the (mostly) lower with megasporangia (yellow), **6** strobilus showing sporangia, the megasporangia indented around the four spores (the bulges of three visible), **7** microsporangium opened, releasing the microspores, **8** microspores enlarged, **9** megasporophyll, showing the megasporangium, **10–11** megaspore tetrads in two views, **12** megaspore, **13** megaspore in another view, showing the triradiate crest (Chapter 17, p. 566), **14** microsporophyll, showing the microsporangium.

1. Stem minute, short, completely enclosed in a rosette of leaves, its presence not obvious to the naked eye; leaves greatly elongated, many times as long as broad, usually at least 2–3 cm. long; sporophylls aggregated about the short stem; sporangia each embedded in the broad base of a leaf, divided horizontally into compartments; spores of different sporangia of two markedly differing sizes.

Order 3. **Isoëtales**, page 703.

Order 1. *LYCOPODIALES.*

CLUB MOSS ORDER

Roots adventitious, produced directly from the stem and branches; stem elongated and usually much branched. Leaf small, often not more than three or four times as long as broad, with *no* scale (ligule) on the upper side near the base. Sporangia in the axils of the leaves, each sporangium usually attached to the base of the leaf, not divided internally, the spores of only one kind, small. Gametophyte either a turnip-shaped more or less subterranean structure of a flattened thallus growing on the surface.

*Family 1. **Lycopodiaceae.**

CLUB MOSS FAMILY

Upper Carboniferous (Pennsylvanian) Period to the present. The family is composed of one large world-wide genus and another of a single species restricted to Australia and New Zealand. About twenty species of *Lycopodium* (club moss) occur largely in the northern and eastern portions of North America.

Commonly plants of *Lycopodium* are only a few inches high, but in some species the stems may be several feet long.

In one subgenus of *Lycopodium* the sporophylls are identical with vegetative leaves; in the other the sporophylls differ from the vegetative leaves in size, shape, and color. The lower leaves of the plant are vegetative and green; the upper ones are yellow, spore-producing, and aggregated into strobili at the tips of the branches.

The gametophytes of the species occurring in northern and temperate regions resemble a turnip root, and they are subterranean. Those of most of the tropical species are flattened thalli somewhat like the gametophytes (prothallia) of the ferns, but they are larger and continue to grow indefinitely instead of stopping at a certain size.

Order 2. *SELAGINELLALES.*

SELAGINELLA ORDER

Roots usually adventitious, at the ends of special branches called **rhizomorphs;** stems much branched. Leaves small, clothing the stem densely, overlapping each other like shingles, usually not more than 2 to 4 times as long as broad, each with a membranous scale, the ligule, on the upper side near the base. Sporangium attached to the stem above the sporophyll, not divided internally by partitions; spores of two kinds, some sporangia containing many microspores; others with only four large spores (megaspores), these usually of a different color, for example, the microspores orange or red and the megaspores yellow or white. Gametophyte produced within or almost within the wall of the spore, living at least largely on food stored in the spore, the gametophytes differentiated into two types, the microspores producing microgametophytes with male sex organs, the megaspores megagametophytes with female sex organs.

*Family 1. **Selaginellaceae.**

SELAGINELLA FAMILY

Carboniferous Period to the present. The family is composed of a single world-wide genus, *Selaginella,* with 30 or 35 species in North America north of Mexico. Most of these species occur on the Pacific Coast and across the southern portions of the United States but a few are of northern distribution.

The adult sporophyte of *Selaginella* resembles that of *Lycopodium,* but the genera differ radically in life history from the time of spore production on through the gametophyte generation and development of the embryo. Despite great differences in life history, there are evidences of relationship.

Plants of *Selaginella* are usually only a few inches high, but some tropical and other species growing on trees may be more elongate, some vinelike types being as much as sixty feet long.

The species growing in the Southwestern Deserts and the drier lowlands of California are in varying degree "resurrection plants." In the dry season the leaves are appressed tightly against the stem, the branches tend to curl together (strongly in some species but little in others), and all parts turn brown. If the plant is immersed in water, it may open out and turn green within an hour or two.

Order 3. *ISOËTALES.* QUILLWORT ORDER

Roots from a basal rhizomorph, this stem base usually 4-lobed and with roots growing from each lobe; stem minute, not visible without dissection or sectioning, enclosed or largely so in the tuft of leaves. Leaves elongated, many times as long as broad, usually at least 2 or 3 cm. long. Sporangium large, fundamentally on the upper side of the leaf base but sunken within its tissue, divided internally by horizontal partitions (septa); spores of two kinds, microspores and megaspores—produced in microsporangia and megasporangia. Gametophytes small, of two types, microgametophytes and megagametophytes, each developed wholly or almost wholly within the wall of the spore and dependent upon food stored in it.

*Family 1. **Isoëtaceae.** QUILLWORT FAMILY

The family is distributed widely in temperate regions. The only genus, *Isoëtes,* is represented in North America by about 20 species occurring mostly in the northern part of the Continent. Some of these grow in lakes, and they may be submerged to depths equal to several or many times their height; others grow in very wet soil along the shores. Some Californian species occur in temporary vernal pools.

THE BASIS FOR CLASSIFICATION OF CLUB MOSSES

The Club Moss Line

This line of development began with early Devonian the Zosterophyllales (p. 589) and reached its greatest development in Carboniferous time. There are three common modern genera, *Lycopodium, Selaginella,* and *Isoëtes,* and a rare genus, *Phylloglossum,* which includes a single species occurring in Australia and New Zealand.

The early Devonian (and Silurian?) beginning of this series is found in the following three genera related to *Zosterophyllum* and included also in the extinct order Asteroxylales of the Lycopsida.

Asteroxylon. *Asteroxylon* resembled *Rhynia,* but the aërial axis was clothed with small scalelike structures resembling leaves. These were arranged irregularly or possibly spirally on the stem, and they overlapped as do shingles on a roof. These "leaves" have been interpreted as **enations,** that is, outgrowths of the superficial (cortical) tissues of the stem. In one of the known species the leaves clothed the aërial portions of plant axis to the tips of the branches. The stalked sporangia were borne laterally in random positions on the stem or at least not clearly above leaves. A vascular trace from the central stele ran to the base of each leaf but did not enter it. The vascular system of the underground portions of the axis was a simple protostele like that of *Rhynia.* The sporangia were borne along the side of the stem, terminating short lateral branches and not being associated with the leaves (enations). They were dehiscent by simple lengthwise splitting.

Baragwanathia. *Baragwanathia* was similar to *Asteroxylon.* The leaves were considerably larger, being as much as 4 cm. (over 1½ inches) long, and they were arranged in spirals on the stem. Each one had a single vascular bundle forming a midrib. In this genus the sporangia were nearly sessile and **lateral,** that is, on the side of the axis. Their position bore some relationship to the leaves, but the correlation was not definite. The sporangia, as well as the leaves, were in spiral series. In some cases each sporangium seemed to be in a leaf axil, but in others apparently each sporangium was attached just below a leaf.

Drepanophycus. This genus resembled *Baragwanathia.* Some species differed in the

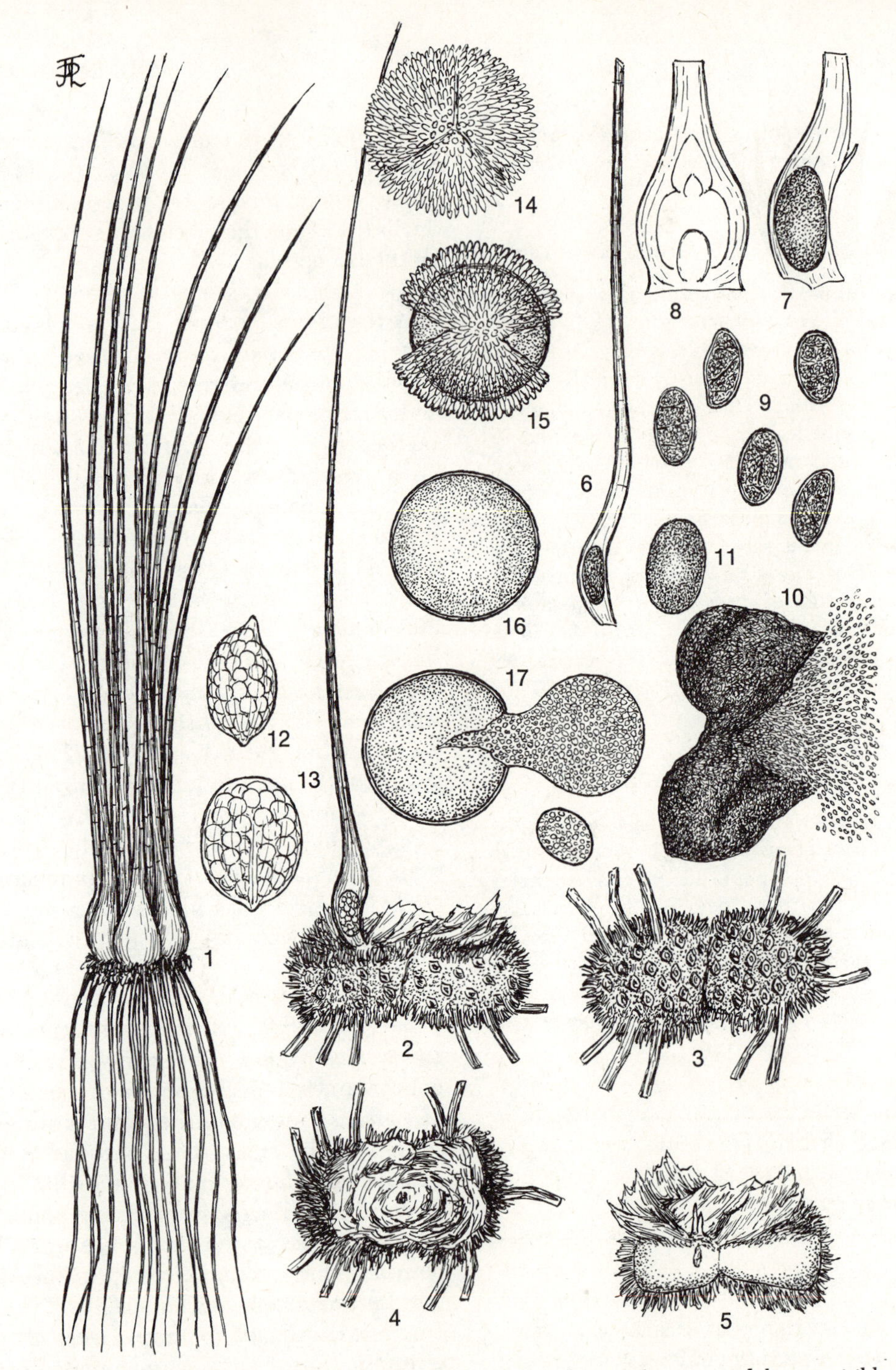

Figure 23-3. Isoëtales; Isoëtaceae; *Isoëtes,* quillwort: **1** plant with roots and leaves visible, **2–4** stem with most of the attached structures removed, side, botom, and top views, **2** with the leaf in longitudinal section, showing the megasporangium in the base, **5** stem with all the attached structures removed, **6** leaf in longitudinal section, showing the embedded basal microsporangium, **7** enlargement, **8** ventral view, **9** view showing the internal division, **10** release of microspores, **11** microspore, **12–13** views of a megasporangium, **14** megaspore, enlarged, **15** part of the wall removed, **16** wall removed, **17** contents being squeezed out.

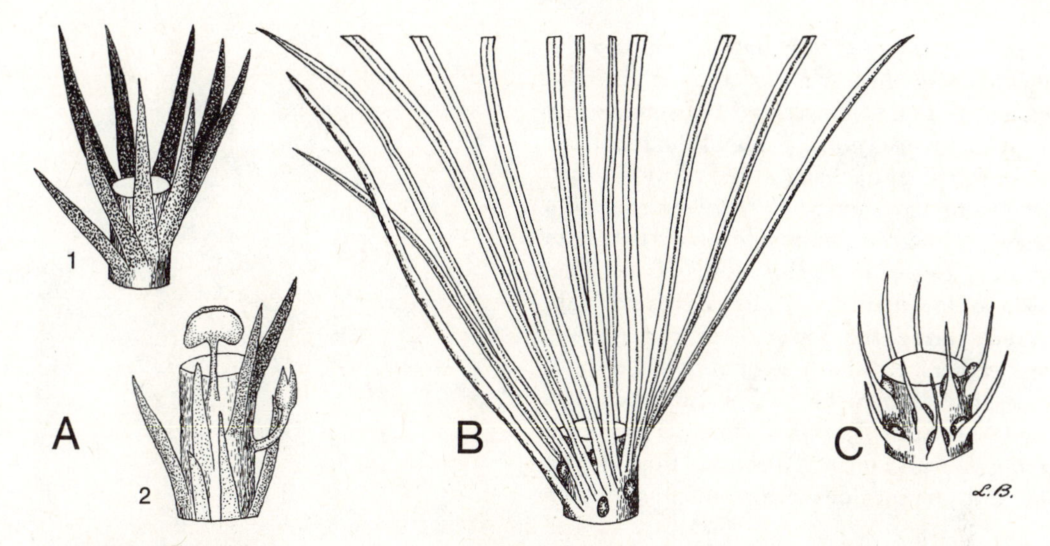

Figure 23-4. The Club Moss Line (early Devonian, see *Zosterophyllum,* pp. 589–594): **A** *Asteroxylon,* **1** sterile branches with leaflike outgrowths of the outer tissues (enations), these with no veins or connections to the vascular system of the axis, or stem, **2** short lateral branches terminated each by a reniform sporangium dehiscent along an apical marginal line, the fertile branchlets not correlated with the positions of adjacent enations (or leaves); **B** *Baragwanathia,* the leaves long and each with one vein from the vascular system of the axis; the sporangia lateral on the stem among the leaves but of uncertain relationship to them; **C** *Drepanophycus,* the leaves (enations) of this species spinelike, in others up to 2 cm. long (stems then up to 4 cm. in diameter) the sporangia in their axils. **A 1** based on R. KIDSTON and W. H. LANG, Royal Society of Edinburgh. Transactions 52: 831–854. pl. 2. 1921; **2** based partly on HARLAN P. BANKS. *Evolution and Plants of the Past* 69. fig. 4–6 (middle). 1968; **B** upon W. H. LANG and ISABEL COOKSON, Royal Society of London. Philosophical Transactions. Series B, 224: 421–449, plate 29. fig. 1, 6–7. 1935; **C** upon R. KRÄUSEL and H. WEYLAND, Palaeontographica 78, Abt. B: 1–47. Textabbildung 19–33, plate 6, fig. 7–16, 7, fig. 1. 1933. Used by permission.

form of the stem appendages which were not leaflike but spinelike. The leaves probably corresponded to the leaves of *Baragwanathia,* for each had a vascular bundle. The sporangia were borne in the axils of the spines.

The three genera, *Asteroxylon, Baragwanathia,* and *Drepanophycus,* apparently formed a connecting link between *Zosterophyllum* and the Lycopsida, for they had features also characteristic of the Lycopsida as follows: (1) all three genera had appendages resembling the leaves of the club mosses and similarly arranged irregularly or spirally along the aërial portions of the axis; (2) the sporangia were lateral; (3) in *Drepanophycus* the sporangia were not only lateral but clearly axillary, that is, in the angles above the stem appendages.

The living plants most closely resembling the three extinct genera discussed above are the club mosses and particularly certain species, such as *Lycopodium Selago* and *Lycopodium lucidulum.* In these small plants most of the leaves resemble those of *Baragwanathia,* but those of the first-formed portions of the stem lack a vascular bundle as did those of *Asteroxylon.* The sporangia are produced on many of the leaf bases just above the axils of the leaves. The only clearly significant differences from the Zosterophyllales are as follows:

(1) presence of roots (though in this line the divergence of stem and root appears to be ill defined); (2) development of the sporangium from the upper (ventral) side of the leaf base, the stalk being attached to the leaf rather than directly to the stem; (3) development of a somewhat more complex internal structure in the stem, especially in the details of the stele.

These forerunners of the Lycopsida appeared during the lower Devonian Period, even earlier than the time of *Rhynia*. Several lines of development within the class had arisen by late Devonian time. The major groups occurring in Paleozoic time represented divergent lines of evolution as follows.

Protolepidodendron. This early and middle Devonian plant resembled both the genera above and the club mosses. Its leaves were deciduous, and attachment scars were left on the stem.

Lepidodendrales. Scale Tree Order. In the great Coal Age, or Carboniferous time, the Lycophyta reached their greatest development. The principal order was the Lepidodendrales, which included two large genera of trees which were dominant features of at least the swamp landscape, where the plants were preserved as fossils. These trees, *Lepidodendron* and *Sigillaria*, are known to the geologist as scale trees because of their scalelike leaves resembling but very much larger than those of the club mosses. The leaves were deciduous, and each one left a characteristic attachment scar when it fell from the trunk or a branch. The sporophylls were arranged in strobili. There were microspores and megaspores, and the microscopic phases of life history were similar to those of *Selaginella*. The order lived from the upper Devonian Period until the Triassic Period.

Lepidodendron differed from *Sigillaria* as follows: (1) leaf scars arranged in obvious spirals (rather than along vertical ridges of the

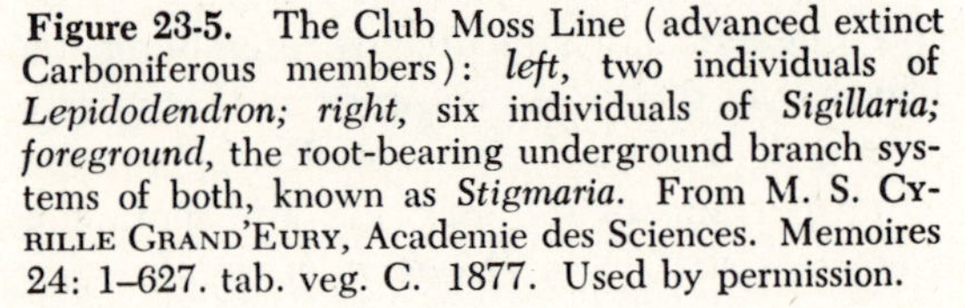

Figure 23-5. The Club Moss Line (advanced extinct Carboniferous members): *left,* two individuals of *Lepidodendron; right,* six individuals of *Sigillaria; foreground,* the root-bearing underground branch systems of both, known as *Stigmaria*. From M. S. CYRILLE GRAND'EURY, Academie des Sciences. Memoires 24: 1–627. tab. veg. C. 1877. Used by permission.

Figure 23-6. The Club Moss Line (advanced extinct Carboniferous members): coal showing the patterns of leaf-attachment scars on the trunks: *above, Lepidodendron,* two fossils showing the spiral arrangement of leaf-scars (not in vertical rows) (the fossil at the right inverted); *below, Sigillaria,* showing the spirally arranged leaf scars also in vertical ranks with grooves between the ranks.

stem), vertically elongated; (2) leaves relatively short; (3) stems much branched above; (4) strobili terminal on the leafy branches instead of on special short branches from the trunk below the upper leafy branches.

Lepidodendron and *Sigillaria* cannot be distinguished by the roots or the rhizomorph (four-lobed rootbearing structure) at the base of the stem, and fossil rhizomorphs and roots of both genera are called "*Stigmaria.*" Often these fossils are found upright in sandstone or shale just below deposits of coal containing horizontal trunks of *Lepidodendron* and *Sigillaria.* Evidently the tree trunks fell into the swamp or moist ground in which they grew, leaving the rhizomorphs and roots in the silt

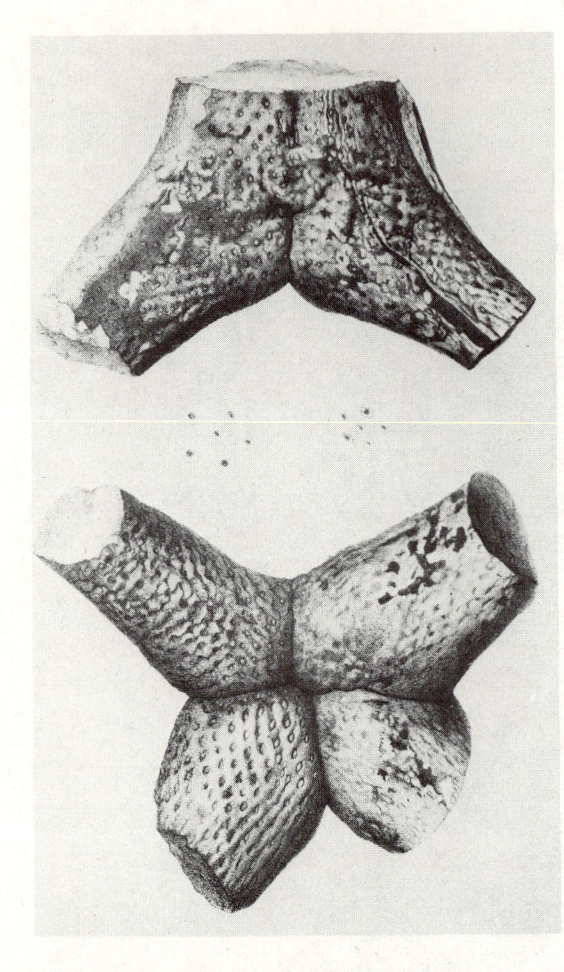

Figure 23-7. The rhizomorph (group of four root-bearing branches) of *Lepidodendron, Sigillaria,* and their relatives, known as *Stigmaria;* Carboniferous time. The scars represent the attachment points of the small roots. From W. C. WILLIAMSON, Palaeontological Society of London. Monographs 1886: i–iv, 1–62. plate 3. 1887. Used by permission.

beneath. In time, the vegetation in the upper layer became coal and the silt at the lower level was cemented into sandstone or shale.

Another genus, *Bothrodendron,* occurred in the latter part of the Devonian Period and about the first half of Carboniferous time (Mississippian Period), and it is the earliest known member of the order.

Figure 23-8. *Pleuromeia,* a fossil relative of *Isoëtes,* the plant about four feet high. The rhizomorph at the base of the stem is the near equivalent of the whole stem of *Isoëtes* (**Figure 23-7**), wherein the leaf-bearing part is very short. Triassic time. From MAX HIRMER, Palaeontographica 78, Abt. B: 47–56. Textabbildung 5. 1933. E. Schweizerbart'sche Verlagsbuchhandlung, Stuttgart. Used by permission.

Lepidocarpales. Lepidocarpon Order. The order included two genera, a tree (*Lepidocarpon*) resembling the *Lepidodendrales* and an herb (*Miadesmia*) resembling *Selaginella.* The plants were similar to those of the *Lepidodendrales.* They are of interest because the order parallels the seed plants in retention within the original megasporangium of a single functional megaspore, the megagametophyte formed from it, and finally the early stages of the embryo. The megasporangium was enclosed permanently within a "seed coat" formed by the entire megasporophyll rather than by an integument or special covering developed from a small part of the sporophyll as in the seed plants. Access of the "pollen" (microgametophyte) to the megasporangium and to the enclosed megagametophyte was through a slit or hole forming a passage into the megasporophyll.

The production of seedlike structures by both the Lepidocarpales and the Spermatophyta is a case of parallel evolutionary development, and the Lycopsida are not possible ancestors of the seed plants because the "seeds" are not homologous (of corresponding origin) with those of Spermatophyta.

Pleuromeiaceae. Pleuromeia Family. This family was a Triassic member of the order Isoëtales related also to the Carboniferous Lepidodendrales. *Pleuromeia* resembled a yucca; it had an erect trunk as much as two meters high arising from a massive basal rhizophore. The elongated leaves were deciduous. The terminal strobilus bore sporophylls unlike the vegetative leaves; these bore dorsal sporangia.

In contrast to development of leaves in the preceding two lines, the leaves of the Club Moss Line originated as outgrowths (enations) from the superficial tissues of the stem. They have but one vascular bundle, the midrib. The sporangia are solitary, one on the ventral side of the base of each sporophyll.

Association of Species in Natural Floras

24

Classification of Natural Floras

Early Systems of Vegetation Classification

Early attempts to classify floral and faunal regions were based wholly upon elaboration of the climatic belts long recognized in geography, that is, the Arctic, Temperate, and Tropical zones. This resulted in systems of zoning or dividing these regions, and particularly the Arctic and Temperate ones, in accordance with temperature.

Temperature played the chief part in formation of several early classifications of vegetation, as for example that of De Candolle. The principal system of zoning North America according to temperature was published (in 1898) by C. Hart Merriam, Chief of the United States Biological Survey. This system has had wide use in systematic botany, especially in the western United States, and it has been employed extensively by zoölogists, particularly western ornithologists and mammalogists. Although the Merriam system of life zones is still in use for certain purposes, for many years most ecologists* have accorded it no more than historical significance. It has the element of an artificial system of classification, for it represents an attempt to arrange groups of plants and animals on the basis of predetermined grounds, that is, variation in a single factor of the environment—temperature. The northernmost zones proposed by Merriam stand scrutiny better than do the more southerly ones of the series. Probably this is because in the far north temperature is the predominant factor. An analysis of the Austral (southern) Zones is far from reassuring. For example, matching up in one "zone" the flora of the Southwestern Desert Region with that of the Gulf Coast because they have about the same average temperature is like harnessing a mule and a jack rabbit into a team because they have long ears. The results are not satisfactory.

* Ecology is the study of the relationship of the plant or animal to its environment (e.g., physical factors, soil, and living organisms).

There have been other attempts to classify vegetation upon the basis of a single factor. Even though (in 1909) Warming developed the chief foundations of modern ecology, he erred as did Merriam in assigning responsibility for the presence of plant communities to a single factor, though he chose water supply rather than temperature. In the days when there were many followers of both the temperature and water schools of thought, one camp was known as the "hot dogs," the other as the "water dogs."

A. F. W. Schimper, author of a thoroughly interesting and readable large volume translated as *Plant Geography upon a Physiological Basis* (1903), must be given credit for a much deeper insight into the complexity of the problem. Schimper realized that the difficulties in analysis of plant distribution are due to complications resulting from interaction of many factors, no one of which is predominant in all situations. These factors include temperature (not just average temperature, but temperature during growing and dormant seasons, in its relationship to seed germination, in its cycles with regard to day and night, in its relationship to enzymatic action, etc.), radiant energy, water (in respect to its availability to the plant in various seasons and stages in life history), evaporation and atmospheric conditions, length of day, soil types, interrelationship with other organisms, and the history of past barriers to or facilities aiding migration.

Raunkiaer has added a significant vegetation classification system based upon life forms of plants. Primary emphasis is upon the location and protection of the organs (buds and shoot apices) enabling the plant to survive through such unfavorable seasons as cold winters or dry, hot summers. These may be seeds (as in annuals), rhizomes, or stems of varying heights above or below ground level. Raunkiaer classified plants according to the position of the buds and shoot apices (right column) as follows:

1. *Phanerophytes*	Projecting into the air
2. *Chamaephytes*	Near ground level
3. *Hemicryptophytes*	On the soil surface
4. *Cryptophytes*	Below ground level
5. *Therophytes* (*Annuals*)	The plant growing only in favorable seasons; living over as a seed or a spore

The form of the woody or other perennial plant varies in accordance with the position of the buds and shoot apices, i.e., it may be tall and erect or ascending, low and bushy, carpeting the ground, or tightly matted against the soil. Vegetation was classified according to the percentage of plants in each of the five categories and their subdivisions, this bearing a relationship to the type and severity of the climate during the unfavorable season.

In order to formulate a workable classification of types of vegetation and of the floristic elements or groups of species composing them, the principles of *plant equilibrium and succession, kinds of succession, climax,* and *world climatic belts* must be introduced.

Plant Equilibrium and Succession

No association of plants and animals is static; changes are going on perpetually. In a particular lawn in southern California the dynamic equilibrium of the components shifts from season to season and from time to time during each growing period. Three dominant plants —Bermuda grass, a small sorrel, and *Dichon-*

NOTE: The following are earlier and contemporary publications by the author developing this subject: *The Relationship of Ranunculus to the North American Floras,* American Journal of Botany 29: 491–500. 1942; *A Treatise on the North American Ranunculi,* American Midland Naturalist 40: 1–261. 1948; *Plant Classification,* ed. 1, 1957; *Plant Taxonomy, Methods and Principles,* Ronald Press Co., New York, 1962; *The Cacti of Arizona,* ed. 3, University of Arizona Press, 1969; *The Native Cacti of California,* Stanford University Press, 1969; *The Trees and Shrubs of the Southwestern Deserts* (with Robert A. Darrow), ed. 3, University of Arizona Press, in press; *The Cacti of the United States and Canada,* Stanford University Press, in press (with considerable development of this subject); *Evolution of the North American Floras* (in preparation).

dra (a highly desirable lawn plant of the morning-glory family)—are in mortal combat, and a number of other species is involved as well. At first *Dichondra* was planted in only one corner, but, because it produces many seeds and scatters them effectively, it has spread through the entire lawn and has displaced a large part of the Bermuda grass. At times conditions change slightly, and Bermuda grass wins out. In the shifting conditions of one or two summers, the sorrel has defeated the *Dichondra* as a result of growing a little higher and shading it. The same kind of competition prevails in natural vegetation. First one set of species, then another gains the advantage in any locality, and the result is yearly or seasonal shifts in the prevalence of particular herbaceous species. Also, in both the lawn and natural vegetation areas the balance among various elements shifts from place to place so that vegetation varies greatly according to locality or the exceedingly localized conditions in one spot or the next. In forests and in associations of perennial and shrubby species the struggle is no less real although perhaps not so apparent. Fluctuations are slower, because the dominant plants are longer-lived.

The complex interrelationships underlying the dynamic equilibrium of a plant and animal community are destroyed when the area is denuded, as by fire. The year following a very hot forest fire the ground may be completely bare at the beginning of the growing season—not only has the permanent vegetation been removed but also the accumulation of leaves and leaf mold has been burned away. Soil temperature may change rapidly in either direction, for the sun shines directly upon the surface, and at night heat radiates directly into the atmosphere. Certain salts have been released by burning of the vegetation, and there is at least a slight change in the pH (alkalinity or acidity) of the soil. For lack of humus, the soil does not absorb and hold water as readily as it did before the fire. Every organism which lived there is affected—many of the animals, such as the earthworms, are dead, others have gone to better environments. Many cannot return before the restoration of shade and leaf mold. Some animals cannot live in the area until their food plants are replaced and have become large enough to bear seeds or other edible parts; other animals with different food and shelter requirements replace them.

Heavy rains or wind may carry away much of the top soil, allowing gully erosion to start, and, if the vegetation does not return quickly, deep ditchlike ravines may be formed.

Fires do not kill all the seeds in the ground. Even an exceedingly hot fire may not accomplish this because heated air moves upward. The introduced weedy plants which come up the first year crowd out the seedlings of many native species, but they provide a certain amount of humus and they reduce erosion; consequently, the character of the soil begins to change. Presence of the weeds changes the environment gradually so that there is a succession of weeds and later of native plants, and eventually the native trees become established as seedlings. As these woody plants grow they provide shade and alter the environment further, favoring return of their small associates. The eventual restoration of the original ecological balance is complicated, and it requires many years and many shifts from one set of plants to another. However, if there is no further disturbance, after many years the original vegetation becomes predominant, and ultimately a stable association is developed. The term **succession** is applied to the sequence of vegetation changes as one type gradually replaces another until finally a **climax** is reached.

Kinds of Succession

When a bare area, such as a newly formed island or a locality exposed by the melting of a great ice sheet, becomes available for plant growth, **primary succession** is instituted. The first few **pioneer** plants to establish themselves

Figure 24-1. Plant succession following a forest fire on the Grimes Creek Drainage, midway between Grimes Creek and the South Fork of Payette River, Idaho. The underbrush and litter were burned completely by a hot fire in 1931, and most of the trees were killed (*above*). In 1950 (*below*) the forest had been replaced by brush, a stage in the long succession leading ultimately back to forest. Photographs, courtesy of the United States Forest Service.

on such an area may be mosses, algae, lichens, or other groups with special features, such as ability to retain water, enabling them to survive in such environments. After the pioneers have become established the way is open for others because of the meager humus and soil provided by the "first settlers" and the dust which accumulates among them.

A **secondary succession** follows interruption of the normal or primary succession or equilibrium by factors such as fire, clearing for cultivation, lumbering operations, or violent wind storms. The kinds of plants capable of starting a secondary succession depends both upon the degree to which the area becomes denuded and upon the effect which the disaster has had upon individual plant species in its path. For example, native species may have special features enabling them to withstand fire. Many shrubs characteristic of the chaparral (brush) in California have large tuberous subterranean swellings from which sprouts shoot up rapidly, restoring the original woody vegetation. The final stages of a primary or a secondary succession are the same, that is, they represent a gradual approach to the climax type for the particular region.

In Canada and the United States today few places are not in secondary succession, for the coming of white men to North America wrought profound disturbance in the natural vegetation through cultivation, lumbering, fire, and overgrazing. Even in the remote portions of the deserts and mountains fire or overgrazing has disturbed the climax vegetation and instituted various secondary successions.

Figure 24-2. An early stage in plant succession: *Phlox Hoodii* growing in the cracks of north-facing bare sandstone above Miles City, Montana. Photograph, courtesy of the United States Forest Service.

Climax

In any given region the ultimate trend of succession is toward a type of vegetation adapted to withstanding even the extremes of local climate (dry and wet years, warm and cold winters, etc.); consequently, the climax vegetation is relatively stable. As individuals within this community mature and die, they are replaced by their own offspring. Thus the character of the community tends to remain unchanged, unless there is a disturbance or a gradual trend of change of the regional climate. Within any of the major vegetation regions of North America there is local variation within a climax. However, these local variations as a whole constitute a set of associations characteristic of the region and different from the corresponding set for other regions.

World Climatic Belts

The well-known belts of temperature and climate are the primary factor in world plant distribution. In North America north of Mexico these belts or zones are as follows:

1. *The Arctic Zone and the Alpine Areas*—beyond timber line in the far north or above it in the western mountains.
2. *The Subarctic Zone*—coniferous forests south of timber line across central Alaska and the northern part of Canada; barely in the northeastern United States.

Figure 24-3. Succession in a forest in the Sierra Nevada at Mammoth Lakes, California. The older trees in the background are lodgepole pine (*Pinus contorta* var. *latifolia*), probably replacing the climax forest of red fir (*Abies magnifica*) and other species as the result of a fire many years earlier. The trees coming up under the temporarily more or less stabilized lodgepole pine forest are exclusively red fir (Christmas tree type). The dense stand of young lodgepole pines on the disturbed soil of the road embankment in the left foreground includes no red firs. Ultimately, barring fire or other disturbances, the fir trees will overtop and crowd out the lodgepole pines in the background. Cf. **Figure 25-10.** Photograph, 1954.

3. *The Temperate Zone*—including (a) the coniferous forests of the West; (b) the deciduous forests of the Middle West, East, and South; (c) the grasslands of the Great Plains and near the Mississippi River system; and (d) the dry woodlands, brushlands, Sagebrush Desert and similar deserts, and some grasslands of the West.

4. *The Horse Latitudes* or *Desert Zone*—the creosote bush deserts ranging from Palm Springs, California, to Texas west of the Pecos River, and northern Mexico.

5. *The Torrid Zone*—The tropical forests and swamps at the southernmost tip of Texas on the Gulf of Mexico and in southern Florida.

Each climatic belt encircles the Earth at its own latitude, and each is determined by relationship to various factors but primarily to the global or planetary winds and calms.

The Horse Latitudes were dreaded by the ancient mariners, because there a ship might be becalmed for many days "...'til all the boards did shrink." In this belt of calms there is no prevailing wind, only a great sinking air mass from the upper atmosphere, and it drifts away only slowly.

The descending air determines the climate of the region, because as it comes downward it is compressed and it becomes warmer. This increases its moisture-holding capacity, and it tends to take up water and rarely to deposit any. Consequently, great land areas of the Horse Latitudes are primarily deserts. However, the climate of uplands and mountains in the desert regions is cooler and moister, and, if the air descends over water, moisture is taken up from the warm seas and deposited on the adjacent land, as in Florida and along the Gulf Coast.

The air mass descending upon the Horse Latitudes spreads out in all directions, but, because the belt encircles the Earth between 23° and 34° latitude, N. or S., for practical purposes this means to the north and south. Thus, at the Horse Latitudes in both hemispheres the air mass divides near ground level and begins to drift toward the equator and the poles. The planet rotates from west to east, and the air is carried along at about an equal speed in the same direction because of friction with the ground beneath. The rotation speed of the Earth at the equator is calculated easily. The circumference of the planet there is roughly 24,000 miles, and, since each revolution requires 24 hours, the ground speed is approximately 1,000 miles per hour. However, because the circumference of the world is less in the Horse Latitudes, the rotation speed there averages only about 900 miles per hour. The air moving from the Horse Latitudes into the Tropics tends to lag to the westward behind the faster and faster moving surface, and in the Northern Hemisphere the actual air stream is deflected so that it does not flow from north to south but from northeast to southwest. This mass of moving air is called the trade winds, and the sailing ships from Europe were carried by them from the northwestern edge of Africa to the New World Tropics.

The trade winds north of the equator blow from the northeast, but those south of the equator from the southeast, and the two meet on the diagonal at the heat equator. There the air can go in only one direction—up, resulting in the belt of tropical calms or doldrums. The atmosphere, saturated with water it absorbed in passing over warm seas, expands, cools, and loses moisture-holding capacity as it rises. Consequently, there are heavy rains in the region of the heat equator. As the heat equator moves north and south with the seasonal tilting of the axis of the Earth, the belt of

Figure 24-4. Disturbance of natural vegetation by overgrazing; California Grassland. The area between the fences was grown high with grass, the ground on both sides bare and beaten to dust. San Emigdio Canyon, California, 1949. Photograph by Avery H. Gallup.

Figure 24-5. Succession following overgrazing: *above,* Sagebrush Desert near Church Buttes, Uinta County, Wyoming, August 28, 1937, the dominant plants sagebrush (*Artemisia tridentata*) and rabbit-brush (*Chrysothamnus nauseosus*), with a few individuals of the species characteristic of the Palouse Prairie scattered among them, the stream not clear and at a low level; *below,* the camp of the Hayden Expedition at the same spot, September 10 and 11, 1870, the vegetation type, Palouse Prairie, the stream clear and eroding its banks more slowly. Photographs, courtesy of the Soil Conservation Service, United States Department of Agriculture.

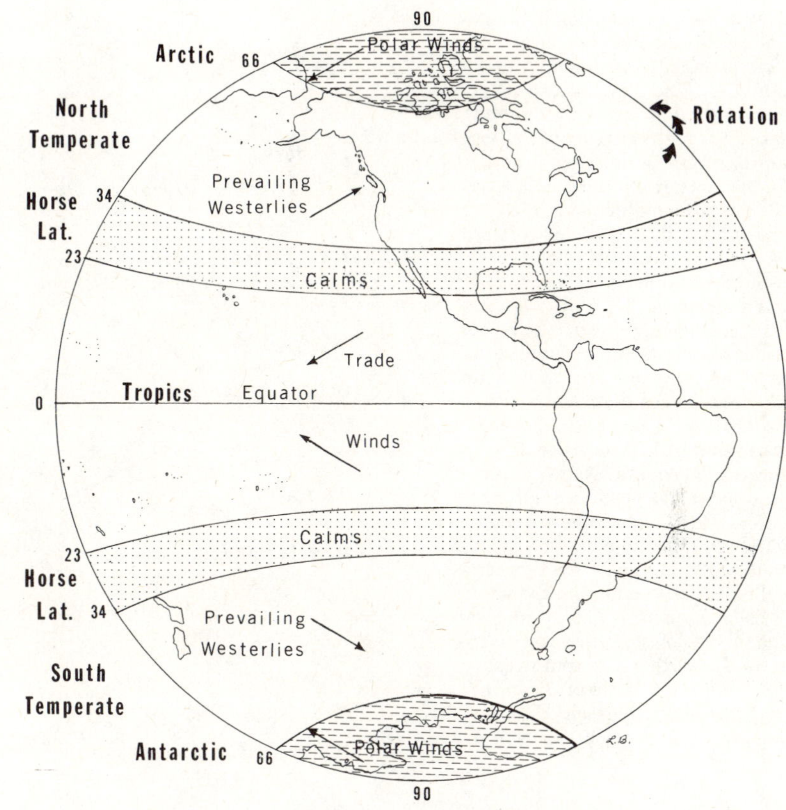

Figure 24-6. The planetary or global winds for the Western Hemisphere, these determining the climatic zones. A great air mass descends over the Horse Latitudes, and some then moves north and some south in either hemisphere, creating the trade winds blowing toward the equator and the prevailing westerlies blowing toward but not to the poles. The Horse Latitudes, where the winds start, are belts of calms. In the cold arctic zones the dense air (contracted at low temperatures) tends to expand, forming the polar winds blowing in the only possible direction—toward the equator. Each global wind or calm is associated with a cold, temperate, or hot major climatic region. (From LYMAN BENSON, *The Cacti of the United States and Canada.* Stanford University Press, in press.)

heavy rain moves also across the Tropics to the northward in the summer of the Northern Hemisphere and southward in that of the Southern. The tropical winter of each hemisphere is the dry season. Because the heat equator crosses the equator twice each year, there are two rainy seasons there.

Meanwhile, the air moving northward from the Horse Latitudes of the Northern Hemisphere enters the North Temperate Zone. However, there the ground is moving more slowly than the 900 miles per hour picked up by the air mass while it was in the Horse Latitudes. At the 40th parallel (Salt Lake City to Philadelphia) the ground speed is roughly 800 miles per hour; at the 48th (Seattle to north of the city of Quebec) about 700. Consequently, the atmosphere tends to forge ahead of the ground, and the air movement takes a trend to the east as well as north. These winds come from the west (actually southwest), and therefore they are called the prevailing westerlies. Because the diameter of the Earth decreases rapidly toward the poles, the intensity of the prevailing westerlies increases as they move northward. Consequently, especially in the Southern Hemisphere, where there is little land in the way of the winds, the Temperate Zone at latitudes between 40° and 50° was known to sailors as the "roaring forties."

Sailing ships returning to Europe from the Americas picked up the prevailing westerlies to carry them across the Atlantic Ocean from near Cape Hatteras in North Carolina. To sail across the Horse Latitudes from the tropical Caribbean Sea, they had used local breezes,

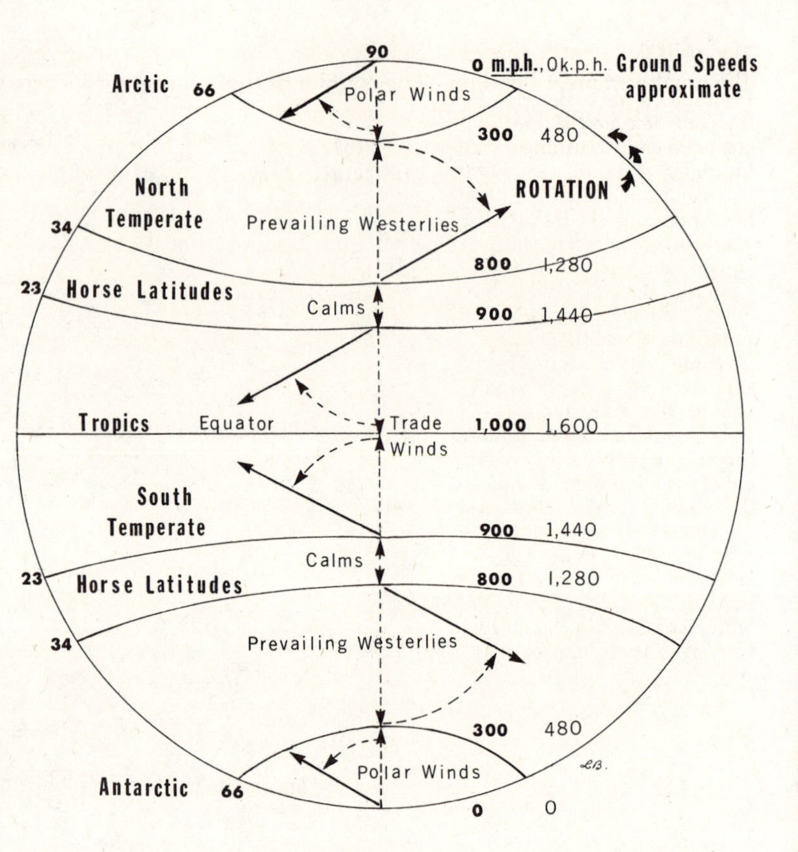

Figure 24-7. Deflection of the global winds according to relative rotation speeds of the Earth at different latitudes. The two columns at the right indicate roughly the ground speeds or rotation velocities at various latitudes, in miles per hour and kilometers per hour. As air moves toward the equator the ground speed beneath it becomes greater; as it moves toward the poles the ground speed becomes less because the diameter of the Earth becomes less. Air moving toward the equator moves more slowly than the surface and therefore westward; winds blowing toward the poles move faster than the surface and therefore eastward. Thus any wind toward the equator is toward the west as well as north or south (according to the hemisphere), and a wind toward a pole is also deflected to the east. A wind from the east is called easterly; one from the west is called westerly. This is true even though it blows on the diagonal, as from the northeast or southwest. (From Lyman Benson, *The Cacti of the United States and Canada.* Stanford University Press, in press.)

created by the temperature differences of sea and land, to carry them past the pirates lurking along the channel east of Florida. This was the only route crossing the calms near land. To have tried to cross the Horse Latitudes in the open ocean would have provided only an alternate type of death—from being becalmed. Against pirates there was a fighting chance.

Storms form over the Temperate Zone, but they may be understood better in the light of the polar winds, as well as the prevailing westerlies.

In the Arctic (and Antarctic Zone) the air mass is cold and contracted, and it tends to expand. In view of the gas laws, this seems strange, but Boyle's and Charles' laws commonly are thought of in relation to closed systems. The air above the Earth is an open system wherein, if it is cold, the mass already is contracted, and, if it is warm, the mass already is expanded. In the Arctic the cold atmosphere is dense, but that adjacent to and surrounding it is a little warmer and less dense. The cold, dense, contarcted air tends to expand into the area of the less dense air around it and then into the still less dense air beyond. This movement is the beginning of the cold polar winds, blowing southward in the Northern Hemisphere. However, the polar winds move from an area where the circumference of the Earth is very small, approaching zero at the pole, to one where it is greater and where consequently the ground speed of rotation is greater. Therefore, the polar winds lag behind the surface, and they are deflected to the westward, becoming northeasterly winds (i.e., from the northeast).

Along the northern margin of the Temperate Zone the warmer prevailing westerlies from the southwest and the cold polar winds from the northeast meet head on. The conflict of the two directly opposed wind systems produces an area of turbulence, where the lighter air of the westerlies comes to lie above the colder polar air. This is a belt of storms because of the initial chilling and the rise and further chilling of the warmer and relatively moist westerlies and of the general turbulence. The area of meeting, called the *polar front,* is a spawning area for swirling low- and high-pressure areas which move from west to east, as well as for the jet stream generated above. The low-pressure areas, or cyclones (not tornados), become storm centers, because, as air rushes into them from all directions, it rises, expands, cools, and loses moisture-holding capacity. On the other hand, where there are high-pressure areas, the air flows outward in all directions along the surface, and there is

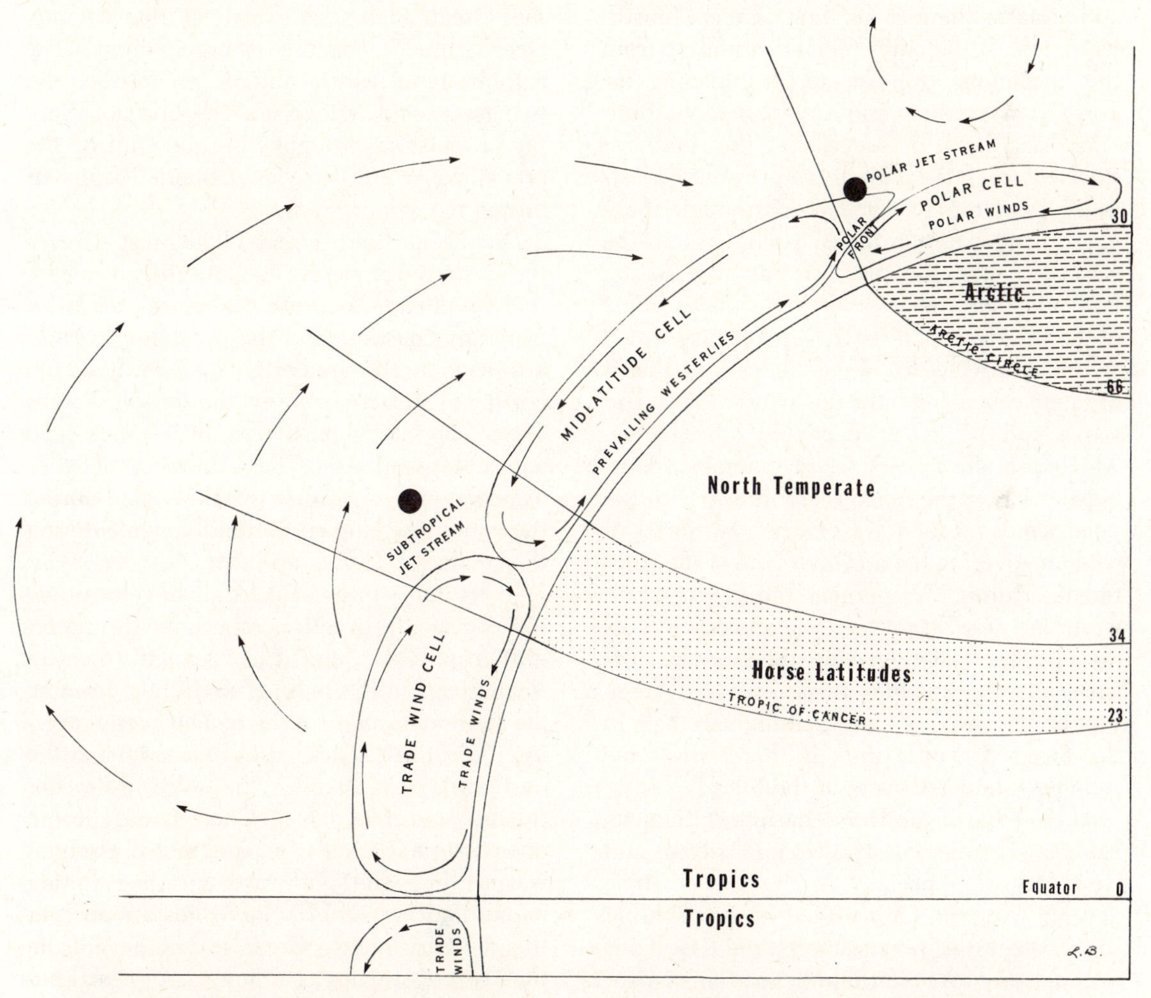

Figure 24-8. Cells of circulation of the Earth's atmosphere. The periphery of each major climatic zone and the circulation of air in a "cell" above it over a little more than one-fourth the surface of the world. Only the movements from the north or south are shown, these being modified actually by deflection of each wind system to the east or west. (LYMAN BENSON, *The Cacti of the United States and Canada.* Stanford University Press, in press.)

fair weather. The polar front meanders or undulates across North America, changing position irregularly between north and south from day to day, and the cyclonic storms emanating from it move along from west to east.

In most of the West, the prevailing westerlies originate over the Pacific Ocean. On the Pacific Slope, at times when the land is colder than the ocean, moisture is deposited on the coastal lowlands. Thus, in the three Pacific Coast States winter rain is common, but in the summer, the land is warmer than the ocean, and usually there is no rain. Except northward, almost the only summer rainfall is from the uncommon tropical storms traveling far north. The northern and central Rocky Mountains, and to a lesser extent the southern Rockies, are in the path of the prevailing westerlies from the Pacific Ocean, though these have lost moisture to the coastal areas and the Cascade-Sierra Nevada Axis, as well as to, here and there, minor mountain ranges. East of the Rockies a rain shadow (p. 734) is formed on the relatively dry Great Plains, but this is of significance for only the winter rains and snows, and the area is affected by other factors which also affect the Rocky Mountains. During the winter the Rockies are in line with the polar winds emanating from the great mass of cold air covering the northern part of the Continent. During the summer the great humid warm air mass lying over the eastern half of the Continent may envelop the Rockies, too. Consequently, summer rains, especially convection showers, are more abundant than in the Sierra Nevada, though there are some summer thundershowers in the Sierra.

At the edge of the Horse Latitudes, the prevailing westerlies most affecting Arizona and New Mexico, especially in the summertime, develop over the Gulf of California. During the summer the very warm, humid Gulf air mass yields convection showers as the air rises over the hot ground in the deserts or over the mountains. As the temperate region shifts southward, attenuated Pacific winter storms are brought from that area by the prevailing westerlies. However, the winter storms are lighter than those of July and August, which is the principal rainy season.

The segment of the prevailing westerlies most affecting eastern North America is that originating in the Horse Latitudes over the Gulf of Mexico. From there most of the eastern region lies northward. The air descending on the Gulf becomes both warm and moist, and, where ultimately it encounters the polar winds along an irregular, fluctuating line from the Great Plains eastward to the Atlantic Ocean, much moisture is precipitated. The extreme temperature differences between the two great wind systems and the enormous supply of moisture brought from the Gulf by the prevailing westerlies yield abundant rain or, during the winter, snow.

On the northern Great Plains, just west of the Eastern Temperate Region, there is a *continental climate,* because the entire area is far from any coastline and the prevailing westerlies reaching this area originate over deserts in northern Mexico or over the far-off Pacific Ocean, and they must pass over much land and high mountains. The stabilizing influence of water on temperature is lacking, and during the winter the land cools rapidly, while during the summer it builds up heat.

In *winter* a mass of cold air develops over the continental interior, especially the northern Great Plains, and it merges with the seasonally expanding mass of arctic air, forming the enormous *polar continental air mass,* covering the entire Arctic, most of Canada, and the northern Great Plains. The cold air is contracted, and it is dense. Consequently, in an open system it tends to expand into the lighter, warmer air to the southward, and the resulting strong and very cold polar winds sweep over the southern areas. Great storms develop in the low pressure areas around the zone of contact with the warmer and moister prevailing westerlies.

In *summer* the warm, moist air mass of the

prevailing westerlies moves northward from the Gulf of Mexico over the eastern half of the Continent under the pressure resulting from descent of more air in the Horse Latitudes. As the polar continental air mass weakens and shrinks back toward the Arctic, it is pushed northward and replaced by the warm, humid Gulf air. Some of the summer storms of the eastern part of the Continent are cyclonic, and others are due to local convection over the heated ground. Most of the convection showers are accompanied by lightning and thunder.

The arctic and subarctic climates are controlled by the cold air mass of the north, which is not only at low temperature but also dry, partly because cold water evaporates slowly. The region is much less exposed to the warmth of the sun than places farther southward, because the sun is always at a low angle and it disappears, or nearly so, through the middle of the far northern winter. The descending air in the north does not release much moisture, and precipitation is as low as in the deserts farther south, perhaps 4 to 6 inches per year. The water is frozen during all but a few weeks to two or three months, depending on latitude, but during the brief growing season it is released from the melting snow and ice. Because the relatively warm season is short, there is enough water to keep the soil moist during the critical season for plants.

Vegetation—Ecological Formations and Associations

The vegetation of an area is composed of the plants that grow together. Commonly they compose a climax characterized by the form of the conspicuous ecologically dominant plants, and this is called an ecological **formation.** For example, if the dominant plants are cone-bearing trees, the vegetation is said to be a coniferous forest formation. Commonly the formation includes regional or local manifestations of the climax, and these are ecological **associations.** Examples might be yellow-pine forest, redwood forest, or Douglas fir forest, designating climax communities of lesser rank. The ecological formations occurring in North America include tundra (in the frozen north and on high mountains), coniferous forests (across the far north but south of timberline and in the West), deciduous forests (in the eastern third of the Continent), grasslands (in the mid-Continent), relatively dry woodlands (in the West), dry brushlands (including the chaparral of the Southwest), deserts (largely in the Southwest), and tropical forests (southernmost Texas and southern Florida).

Each formation is the dominant one at the lower altitudes in one of the major climatic belts (or a northern or southern segment of it) around the Earth. It tends to occur nearer the equator in the adjacent climatic zone but at higher elevations. Each formation is dictated by the climate, and it is correlated with the North American climatic belts as follows:

Arctic Zone and Alpine Areas—tundra

Subarctic Zone—coniferous forest

Temperate Zone—forest, woodland, brushland, or grassland, as modified by other factors than temperature

Horse Latitudes or Desert Zone—deserts

Tropics—tropical rain-forest (or toward the edges savannah [grassland] or woodland savannah)

Floras and Floristic Associations

The ecological units of vegetation occur around the world without reference to the species composing them in one area or another or to their geological history or relationships. However, the plants composing the formations of vegetation on different land masses at the same latitude and in a similar climate are not the same. The continents within each climatic belt have been separated from each other through continental drift for periods ranging up to 130 million years, and, since separation, on each land mass evolution of plants and animals has taken an independent course. Thus the **floras** of the continents and other large land masses are largely different, and the species and higher taxa composing each one tend to be endemic, that is, to occur nowhere else in the world. Each flora is composed of other species, genera, and families than those occurring in similar vegetation formations at about the same latitude on other land masses, and essentially they are of different origins. However, a few plants may share a remote common ancestry, and of course all major groups were derived from exceedingly remote common ancestors of Cretaceous or early Cenozoic time.

An example of the similar appearance of the little- or unrelated vegetation occurring in the same type of climatic belt is drawn from the Mediterranean climates of the world. These occur at the edge of each Temperate Zone next to the Horse Latitudes in both hemispheres—in the Northern Hemisphere, on the hills about the Mediterranean Sea and west of the Cascade-Sierra Nevada Axis in California; in the Southern Hemisphere, on the hills of central Chile, those of western Cape Province in South Africa, and those of parts of western Australia. Brushland, known in California as chaparral, is abundant in all five areas, but (except in rare instances in the Mediterranean region and California or in California and Chile) no species are the same. In the brushlands of the two hemispheres not only the species but also the genera are largely different, as are many of the families.

Similarly, in the Namib Desert of South West Africa, or Namibia, the canyon of the Kahn River has the appearance of the more extreme Southwestern Deserts of the United States, as in Death Valley, but the only species in common is the tree tobacco, *Nicotiana glauca,* introduced into both deserts from South America. Thus, although a particular vegetation type, or formation, is predominant in each climatic belt around the world, on each continent or large island it is developed from different species. Consequently, classification by ecological formations and associations alone is inadequate.

Natural floras are limited to the regions or areas they have been able to occupy. For the most part, each is restricted to a single continent by some barrier. Thus, the Temperate Regions, Horse Latitudes, and Tropics of the Old and New Worlds have been separated for many millions of years, and there has been little interchange of species or other taxa. The only exchange having been possible was through mechanisms of long-distance dispersal, such as carrying by birds, floating in the ocean, rafting there on debris from floods or storms, or (for light seeds and spores) air currents. The North American floras discussed here include five limited to the Continent, one (tropical) shared with South America, and two occurring across the northern part of North America and Eurasia where separation has been through a relatively shorter period (cf. pp. 741–742 and 746 to 747).

Thus, each flora is characterized by some endemic races, varieties, species, genera, and even families; each embraces a characteristic species composition unapproached in other floras; and each includes a considerable group of species long associated during geological history.

Although natural floras are more or less stable units of plants that have remained together through one, two, or more epochs of geologic time, over long periods some of their elements have become adapted to different regional environments, or even to different but adjacent major climatic zones. These elements have become distinguishable units of the flora, that is, **floristic associations.** For example, the Western Forest Flora of North America is composed of six floristic associations—the Pacific Lowland Forest, the Pacific Montane Forest, the Pacific Subalpine Forest, the Palouse Prairie, the Rocky Mountain Montane Forest, and the Rocky Mountain Subalpine

Forest. All but one of these coincide with the prevailing ecological formation, coniferous forest; the other is a grassland (see below). The floristic associations are correlated with the complex and mountainous geographical patterns, the resulting climatic complexity, and the diverse geological history of the two main mountain axes of the West and the areas between and near them. As at a higher level with floras, each of them is marked by many characteristic native taxa, including varieties, species, and even genera occurring nowhere else. Most of the forebears of these taxa have been associated for a very long time, as since Miocene.

Although commonly each flora is expressed in a characteristic ecological formation (e.g., forest or grassland), every subdivision being manifested in the same formation, as indicated above, this is not necessarily true. For example, local selection of species included in a forest may yield small patches dominated by grass, small trees, or brush. Extensive development of one of these may produce a formation different from the original forest, e.g., a grassland, but, from the standpoint of species competition and historical development, the relationships of this new floristic association are to the forest and not to similar formations derived from different sources—it is part of the forest flora.

Each floristic association, then, ordinarily coincides with an ecological formation, because it develops according to the combination of factors dictated by the climate in which it lives. However, it may have no floristic (i.e., genetic) relationships to the plants of the same ecological formation as it occurs in other areas beyond a barrier to migration, such as an ocean or an environmental barrier. For plants of the Arctic, Subarctic, or Temperate Zones or the Horse Latitudes the climate of the Tropics is a barrier or deterrent to passage across the equator. Similarly, plants of tropical rainforests are limited by ecological barriers, such as drought or frost.

Evolution of the Ecosystem

Floras and the associated faunas constitute superecosystems composed of living organisms together with the environment to which they are adapted. These great ecosystems or their subdivisions and their included floral elements may be more or less momentary associations derived from diverse species able to live together in the time and place. These elements may dissociate if the climate or the physiography changes, and their surviving members may take up new alliances. However, ordinarily only a small percentage of the members of a major ecosystem, like a flora, such as those composing a floristic association which is part of it, is likely to be lost, and only a few new members are likely to be incorporated unless the environmental change is drastic.

The changes of the environment during geologic time may be gradual, but they may be rapid, as during the glacial periods of Pleistocene—nearly all of the last one to three million years, according to different estimates. Four times ice sheets up to two miles thick covered southern Alaska, Canada, and the northern and especially the northeastern United States. At the same time the climates occurring now farther south in the Temperate Zone were shifted still farther southward by the lower prevailing temperatures and the greater precipitation. This affected and modified all the ecosystems.

When conditions are stable, species, varieties, and other taxa are relatively stable, too. The species of one ecological niche are replaced by their own offspring, as are those of the surrounding niches. Hybrids between species rarely survive, except where there is local disturbance or an intermediate habitat, because the members of their own age group resembling the female parent that produced the seeds have a strong advantage in the environment in which the female parent grows. This is where the seeds are most likely to sprout. Nonhybrid seedlings have inherited gene combinations that have withstood the prevailing ecological conditions for countless generations. The hybrid has an untried combination, and so long as the environment does not change, its trial nearly always results in failure, unless somehow it grows in a different niche from that of either parent. However, when the environment changes, often a few of the hybrids turn out to be better adapted to the new conditions than their parents. Thus, after natural selection, the progeny of a cross may yield a few plants that give rise to a race or a new variety or species. These are adapted to the new or "hybridized" habitat. Survival of such taxa resulting from hybridization alters the minor ecosystem, and it, as well as its included species, evolves. Alteration of many minor ecosystems affects the major ones.

As the environment changes, the plants change. Ordinarily the differences accumulate slowly over millions of years, and the trends of change may be gradual even in terms of geologic time. Thus, there has been a slow but irregular shift in climate and the character of the flora in western North America from Eocene to the present. Prevailingly, this has been in the direction of a cooler and drier climate, either because the oceans have become cooler or perhaps because of some drift of the continent to the northward. Whenever the climate changes, the boundaries between floristic associations shift, and if the trend of change is great, some associations may move into wholly new territory to the north or south and up and down the mountains. When shifts occur there are unstable habitats not only on the border zones but also throughout the other areas. As the habitats become unstable, hybrids and mutants tend to survive, forming hybrid swarms from which some individuals better adapted to the changed conditions tend to persist. Often some of them replace the parental types of plants.

The pace of evolution of the habitat is not necessarily slow; it may be rapid, as it was during Pleistocene with the changes back and forth between cold, wet glacial and warmer, drier interglacial times. Then climate changed rapidly as did the habitats. During much of the time nothing was stable, and the gene pools of all the species were upset, favoring hybridization, survival of new genetic combinations, and rapid evolution of not only taxa but also of both the lesser ecosystems and the floristic associations and floras.

Classification According to Natural Floras

Natural historic floras and floristic associations are an important basis for classification of the major systems of plant communities encountered in the field, because some species *do* remain together through long periods of geologic time. If the ecological factors change, the varieties, species, and even higher taxa must change, also, but those living together in the same environment encounter the same trend of changes. If they are to survive, all or some must adapt. Some of the adaptations are parallel and the rest, although maybe different, produce the same results. In any event, some or many of the species are able to persist in the new environment, though commonly with evolutionary changes. This factor tends to keep the association together, though usually it is modified.

The Choice of a Classification System

In view of the foregoing discussion, the validity of historic floras as part of the basis for a method of classification of plant associations is dependent upon the following consideration. Climate in an *isolated* part of a floristic region may remain without major change for even one or more geological epochs; and consequently, the plants may persist with relatively minor modification. In the meantime, because of such factors as mountain formation, the climate of another isolated part may undergo change, and its plants may be modified into a different association. However, the second floristic association is related to the first, and some of its species having broad ecological tolerance are likely to be identical. Likewise, differing modifications of climate within two or more geographical portions of a floral unit may produce eventually a like number of associations resembling but also differing from the original one.

The interest of the systematic botanist may center more strongly in the selection of species that composes a floristic unit and in the origin and relationships of these species than in the formation (or other ecological considerations), although of course he is concerned with both. The ecologist may have any of several primary objectives in his study of natural vegetation, and it does not necessarily center in the formations as with many recent American students of the subject. His primary interest may be in such subjects as the relationships of plants to the physical and chemical factors affecting them, the life forms as described by Raunkiaer, or the interrelationships of all organisms (both plants and animals) constituting the community of organisms. Consequently, his classification system for living vegetation may be altered in one direction or another, and his interpretation is not necessarily the one of primary interest to the taxonomist. Until all objectives can be met by one system, more than one is needed.

Other Major Factors Affecting Climate

Variation in climate and consequently in vegetation is the most striking feature of North America. For example, the desert landscape of Southern Arizona resembles that of the East and Middle West little more than it does the moon. Primarily this is due to occurrence of the major zones of climate, but there are important modifying factors within each zone. The chief ones are adjacent *ocean currents, elevation* above sea level, and *rain shadows.*

Ocean Currents. As indicated above, the ecological formation occurring at the lower levels of altitude in a given climatic belt is dictated by the environment produced by the climate of the latitude. This may be changed by various factors, such as the ocean currents upwind from the region or adjacent to it.

For example, the coast of western Europe has a much milder climate than that occurring at the same latitudes in eastern North America, because the Gulf Stream, coming northeastward from the Gulf of Mexico, warms it. Consequently, it is inhabited up to North Cape, Norway, the northernmost point in Europe. Even at North Cape the Tundra is thin, because during the summer the temperatures are too high for it to build up to great thickness. The dead and undecayed or only partly decayed bodies of plants and animals do not accumulate in continuous deep freeze as they do at the same latitude at Point Barrow, Alaska. Among the islands and fjords of the northern Norwegian Coast there is much irregular variation between arctic and subarctic local climate, as shown by the occurrence of local patches of tundra and of the Eurasian Northern Forest distributed here and there according to the eddies of the Gulf Stream into some of the complex ocean inlets more than others. Tundra appears at the Arctic Circle, but a little farther north it is displaced by forest and each formation reappears a number of times along the fjords at different latitudes and altitudes. In Alaska forest of a similar type appears only to just above the Arctic Circle, south of Nome and in the Yukon Valley, at least 4 degrees of latitude (about 267 miles) farther south than in Norway, where it occurs up to about 70° N.

Similarly, along the northern coast of South West Africa (or Namibia) the ocean current along the shore is very cold, and there is much fog, but rain is rare in the coastal region, which is the Namib Desert (where *Welwitschia* grows; pp. 631–636). There the prevailing winds are the trade winds coming from the southeast across the Indian Ocean east of Africa. Since across Africa they have come a long distance over land, they bring little moisture by the time they have reached the western coast along the Atlantic Ocean. There is not much evaporation from the cold water of the

Figure 24-9. Succession in a lake in the glaciated area of northern Minnesota. Along the shore and outward from the island, shore plants and farther out aquatic plants are filling in the shallow portions of the lake. The land surface is carried out gradually into the areas once occupied by deeper and deeper water, as the remains of dead plants, silt deposited from the water, and dust from the air fill in the shallow areas. In time the lake will become a bog and later the bog will become land occupied by forest. Note the young trees at the edge of the advancing forest. Photograph courtesy of the United States Forest Service.

Atlantic, and the little moisture coming on shore is that condensed from fog. The fog is the chief source of water for plants growing in the coastal region, except during the rare downpours of desert convection storms. In South West Africa the deserts, characteristic of the Horse Latitudes, are developed near the coast northward well into the Tropics.

Elevation. Uplands and mountains are colder and moister than the adjacent areas at lower elevations. The temperature is about 4° or 5°F lower for each 1,000 feet of altitude, and in the climate of California, for example, annual precipitation increases about 5 inches for each similar change in elevation. For this reason, in the Northern Hemisphere small or large "islands" of floristic associations normally nearer the pole occur farther south. These are surrounded at lower altitudes by the ecological formation characteristic of the climatic zone appearing near sea level.

Extreme differences may be expressed in zonation of floristic associations and ecological formations on high mountains. One of the most extreme ranges of altitude in the United States is in Arizona. Along the Colorado River on the western boundary of the State the elevation above sea level is only about 200 feet; at the summit of San Francisco Peaks just north of Flagstaff it is 12,611 feet. The Colorado Desert* with its characteristic flora is adjacent to the Colorado River and at altitudes up to approximately 1,500 feet; the Arizona Desert begins at that level and continues to approximately 2,500 or 3,000 feet. These are

* The Colorado Desert occurs at lower altitudes along the Colorado River in California and Arizona and the Arizona Desert at slightly higher levels in central and southern Arizona. The Colorado Desert is named for the Colorado River.

subdivisions of the Mexican Desert Flora. The Sierra Madrean Flora occurs from 2,500 feet to nearly 6,000 feet, and it is replaced at the next higher level by the yellow-pine forest or Rocky Mountain Montane Forest of the Western Forest Flora. Spruce-fir forest (Rocky Mountain Subalpine Forest) of the same region replaces the yellow pines on the slopes of San Francisco Peaks at about 8,500 feet. Timberline is at about 11,000 feet, and the plants on the highest peaks form an alpine version of the tundra of the Arctic Zone and the Alpine Areas. (Cf. Figure 24-12.)

The number of floristic associations in the mountain ranges to the northward is reduced according to latitude. Those represented between the Colorado River and the summit of San Francisco Peaks drop out in the order in

Figure 24-10. Tundra thickness, exposed by rock removal. North Cape, Norway, a little more than 71° N. latitude, about the same latitude as Point Barrow, Alaska, each place being at the northern tip of a continent. At North Cape the tundra is only a few inches thick and not frozen during the summer, because of the warmth of eddies of the Gulf Stream. At Point Barrow (**Figures 25-1 and 25-2**) the depth of the tundra has been determined only by drilling oil wells; it may run to hundreds or even thousands of feet. There the summer temperatures are only a little above freezing, and the great bulk of dead plant and animal material goes immediately into "deep freeze" for an indefinite period. Only the top 20 inches thaws for 3 to 6 weeks in July and August.

Figure 24-11. The Norwegian fjords, deep glacial valleys formed during Pleistocene when about 350 feet of water was withdrawn from the ocean and when great glaciers moved down the stream channels to the sea. When the ice sheets melted, the ocean rose into the valleys cut by the glaciers. *Above,* the Black Glacier with tundra on the north-facing slope at the left and with forest on the south-facing slope at the right; *below,* wall of a fjord above the Arctic Circle with forest on the lower part of the south-facing slope and tundra on top, where it is colder. Although at the Arctic Circle tundra came down to the edge of the water, here the climate is tempered by a warm eddy from the Gulf Stream.

Figure 24-12. The effect of altitude upon vegetation; three levels of altitude in Arizona. *Above,* San Francisco Peaks from the vicinity of Flagstaff, the West American Alpine Tundra at the summit (12,611 feet) being replaced on the upper half of the mountain by Rocky Mountain Subalpine Forest and at the base (as at 7,000 feet where the picture was taken) by Rocky Mountain Montane Forest (dominated by western yellow pines, *Pinus ponderosa*). *Middle,* Southwestern Oak Woodland and (in other instances) Chaparral (brushland), prevalent at middle altitudes (5,500 to 6,500 feet). *Below,* Arizona Desert, occurring at mostly 1,500 to 3,000 feet. Photographs, courtesy of the United States Forest Service.

which they were listed, that is, from the lowest to the highest. On the south side of the Brooks Range north of Fairbanks, Alaska, only two floristic associations remain. Plants characteristic of the Northern Forest Flora (Subarctic Region) occur at lower levels and those of the Arctic Zone (Tundra Flora) at higher levels. North of the Brooks Range, the Tundra Flora of the Arctic Zone is the only flora. It occurs at sea level along the coast of the Arctic Ocean.

Zonation resembling but less striking than that throughout the mountains of the West occurs in the East, as well. The strictly deciduous forests cover the valleys, and the Laurentian Forest, which bears some affinity to the Northern Forest Flora, occupies the higher parts of the Appalachian System. In New England the Deciduous Forests of the Eastern Forest Flora are present on the lowlands, the Northern Forest Flora occurs on the higher northernmost mountains, and the Tundra Flora of the Arctic Zone occupies small "islands" at the summits of the highest of the White Mountains in New Hampshire (particularly Mt. Washington).

Rain Shadows. When a mountain range lies across the path of a global wind, such as the prevailing westerlies, it forces the air mass upward as it crosses. If the mountain system or range is long, like the Cascade-Sierra Nevada Axis on the Pacific Coast of North America, the climate of a large area in the lee of the mountains may be altered profoundly. The State of Washington is considered to be the land of the tall timber of temperate rainforests, and this is correct for the area west of the Cascade Mountains, where in some places the annual rainfall may be as much as about 8 to 13 feet (100 to 160 inches). However, at the same altitude only 60 or 70 airline miles away, east of the Cascade Mountains at Wenatchee or Ellensburg, it is only about 8 or 10 inches, and the country is Sagebrush Desert. The air mass from the Pacific Ocean must rise over the mountain axis, and in doing so it expands, cools, and loses moisture as rain or snow falling on the mountains. As the air comes down on the other side, it becomes compressed, warmer, and with the ability to hold more water. Consequently, the climate on the leeward side of the range is much drier, and the area is said to be in a rain shadow. The obvious part of this rain shadow extends for about 200 miles northward into British Columbia and southward through Oregon and California to northern Baja California—about 1,800 additional miles. The effect of the rain shadow extends far eastward across the Columbia basin, the Great Basin, and the lower Colorado Basin to the Rocky Mountain System. In the rain shadow there are various deserts, grasslands, and (at slightly higher levels) dry woodlands and brushlands.

Similar rain shadows develop in many other places, as on the dry, nearly desert sides of the islands of the tropical Hawaiian Chain. On the lee (southwestern) side of each mountainous island rainfall is very low, but on the trade wind (northeastern) side of the same small island there is tropical rain forest supported by 200 to 700 inches of rainfall.

In western South America the deserts formed normally in only the Horse Latitudes are extended to the north and south along the lee of the Andes. North of the Horse Latitudes, they follow along the western (Pacific) coast toward the equator, because the trade winds, coming from the southeast, must pass over the high Andes. (There is a cold current along the coast, as well.) South of the Horse Latitudes, deserts occur along the eastern flank of the Andes, these being there in the path of the prevailing westerlies of the Southern Hemisphere, which blow across the Andes from the northwest.

The North Temperate Zone: North American Subdivisions

The physiographic complexity of North America affects the climate of the Temperate Zone more than that of any other climatic zone on the Continent. In each other climatic zone

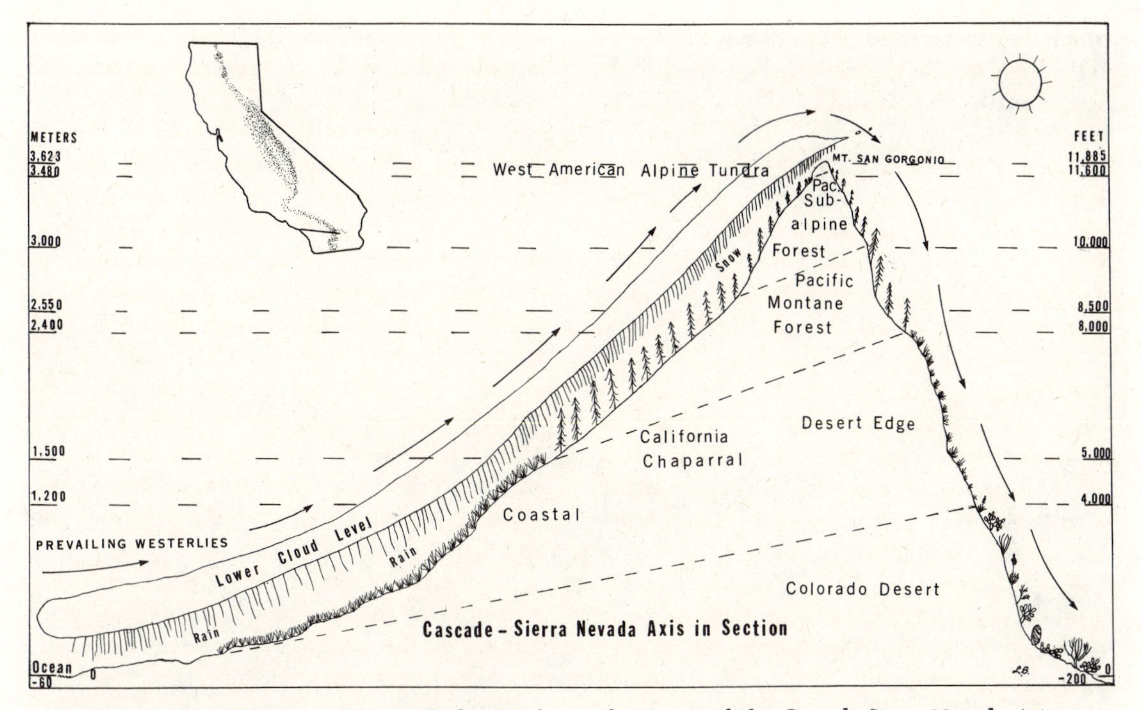

Figure 24-13. Diagram of a rain shadow in the southern part of the Cascade-Sierra Nevada Axis, in southern California, running from the Pacific Ocean to the Salton Sea Basin. The section is not quite straight, in order to include all features in a single section, and the almost flat areas west of the mountains are shortened in order to save space. During the cool season of winter, the prevailing westerlies from over the ocean deposit moisture on the lowlands, then cooler than the ocean, and, as they rise in going over the mountains, they deposit more moisture on the western sides and the summit, the air mass rising, cooling, and losing water-holding capacity. During the descent on the east, the air is warmed, and moisture-holding capacity is increased. No rain falls and often some is taken up in the usual course of events. Thus, deserts occur east of the mountains, not only in southern California but also along most of the entire 3,000-mile length of the mountain axis. (From LYMAN BENSON, *The Cacti of the United States and Canada.* Stanford University Press, in press.)

there is only one flora, but in the Temperate Zone each of four regions is occupied by a distinctive flora. The principal climatic subdivisions are the Western, Eastern, Midcontinental, and Southwestern temperate regions, and the floras are the Western Forest, Eastern Forest, Prairie, and Sierra Madrean floras. The occurrence of these is correlated with regional topography and climate.

One of the two great mountain systems of the West is the Cascade-Sierra Nevada Axis, slightly higher than the Rocky Mountains. The Axis runs from central Alaska to the tip of Baja California, as an essentially continuous great stone wall more than 3,000 miles long. This stands crosswise to the prevailing westerlies coming in from the Pacific Ocean. The lowlands along the coast and the western sides of the mountains are wet during the winter, and, except during the brief rainy seasons, the leeward side to the east is very dry, more so southward. The Rocky Mountain System farther inland to the east is a discontinuous group of north-south mountain axes also lying across the prevailing westerlies but not affecting climate so strongly. Between the two great mountain systems there is a series of inland basins and upland plains. The Columbia Basin is in southern British Columbia, eastern Washington, northeastern Oregon, northern Idaho, and

Figure 24-14. Rain shadow in Hawaii. The southeastern, dry side of western Maui, an area resembling the deserts of the southwestern United States on the mainland. Directly opposite, on the part of western Maui exposed to the trade winds from the northeast, the rainfall is about 400 inches per year, here, across the mountains, perhaps 15. Clouds, as usual, are on the eastern and northern sides of the mountain mass and on its summit, which is only about 4,000 feet high. (From LYMAN BENSON, *The Cacti of the United States and Canada.* Stanford University Press, in press.)

a part of western Montana. The Columbia Plateau occupies the lower elevations of the Snake River drainage in southeastern Oregon, southern Idaho, and the northern edges of Nevada and Utah. For botanical purposes it is a part of the Great Basin just south of it. The Great Basin runs from the east side of the Sierra Nevada in California to the Rocky Mountain System in western Utah and to the Colorado Plateau in the southern part of the state and the adjacent parts of Colorado, Arizona, and New Mexico. The Colorado Basin is the low-lying watershed of the part of the Colorado River below the Grand Canyon in southeastern California, southern Nevada, southwesternmost Utah, and the lower altitudes of western, central, and southern Arizona and adjacent Baja California and Sonora in Mexico.

Since late Miocene the Cascade-Sierra Nevada Axis has risen out of the Miocene plains and hills that covered the entire West, and the second generation of the Rocky Mountain System has been elevated from the peneplane to which the first generation had been worn down. This mountain building has produced the great diversity of habitats in the West, resulting in the combination of radically differing elevations and both wide-spreading and local rain shadows. The climate has favored a forest flora in the moist northern lowlands and, in modified forms, in the mountains. From this forest flora the present six floristic associations have arisen. The areas with less rainfall, generally southward and at lower elevations, but almost throughout the West, favor the derivatives of the dryland Madrotertiary Geoflora, an oak woodland, which, according to Axelrod, during late Tertiary migrated northward and northwestward from the present site of the Sierra Madre Occidentál, the great western mountain axis of Mexico. The derivatives constitute the Sierra Madrean Flora, composed in the United States of eight floristic associations and several important groupings of lesser rank. For the most part, these occupy California west of the Cascade-Sierra Nevada Axis, the lowest elevations of the Columbia Basin, the Columbia

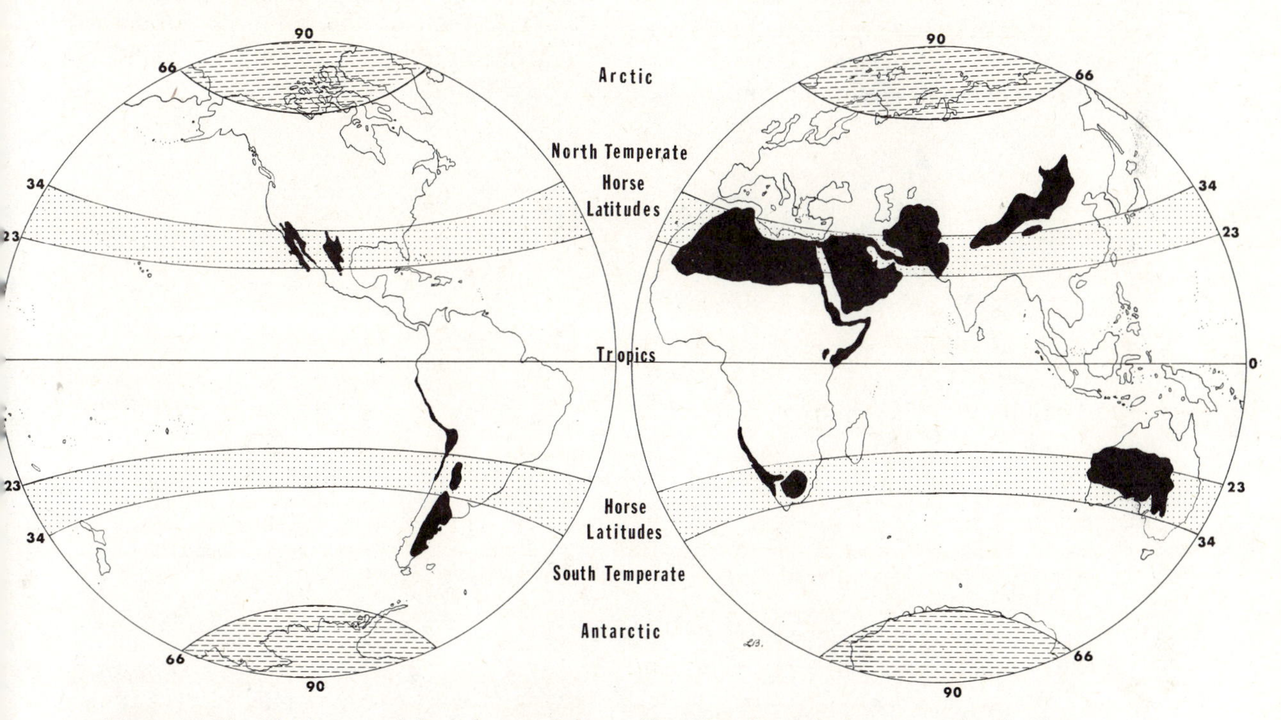

Figure 24-15. The larger and drier deserts of the world. Although mostly these are within the Horse Latitudes, other factors affect climate and the deserts may extend to the north or south. Some deserts, for example, those near the high Andes of South America, continue to the north or south for considerable distances. The deserts there extend northward into the tropics west of the Andes, where the trade winds blow from the southeast and the Pacific side of the range is to the leeward. Farther south the desert extends into the southern temperate region but on the Atlantic or eastern side of the mountains in the lee of the prevailing westerlies from the northwest. The northern extension of the deserts along the Pacific Coast of Chile is favored also by a very cold offshore current over which there is little evaporation and which is almost always colder than the adjacent land. In eastern Asia the Gobi Desert extends northward to 50°, the latitude of Vancouver, British Columbia, because it receives no moisture-laden winds. Those to the west must pass over the largest land mass in the world, those from the south over the highest. The north winds originate over the Arctic ice pack from which there is little evaporation. The east is downwind. Small patches of desert occur downwind on high islands in the pathway of prevailing winds, low islands and headlands where the wind is not uplifted, or on "desert" islands lying in cold ocean currents, e.g., Galapagos. (From LYMAN BENSON, *The Cacti of the United States and Canada.* Stanford University Press, in press.)

Plateau, the Great Basin, the Colorado Plateau, and the lower elevations of the southern area of the Rocky Mountain System, as well as uplands and hills in northeastern Sonora and northern Chihuahua, Mexico.

Because the topography of Eastern North America is relatively simple, the climatic variation and the regions of vegetation and floras are less complex than in the West. The great mountain axis of the region, the Appalachian System, was both long and gigantic during the Paleozoic time of great uplift—Carboniferous (Pennsylvanian) and Permian, and the ranges were high during Mesozoic. Through Cenozoic they have been lower, and the uplifts have been relatively minor. The ranges lie mostly parallel or more or less so to the direction of the prevailing westerlies, and consequently rain shadows are minor. There are great and irregular gradients of available moisture, increasing from west to east, and of temperature, rising from north to south. Over the great areas of plains and low valleys, the climate is determined chiefly by interaction of these two varying factors, and plant responses and the compositions of floras follow accordingly. Two floras occur—in the drier mid-Continental region the Prairie Flora and in the moister eastern one-third of the Continent the Eastern Forest Flora.

25

The Floras of North America

THE NORTH AMERICAN FLORAS AND THEIR SUBDIVISIONS

NOTE: Cf. Appendix: Favorable Seasons for Plant Collection in Various Parts of North America, pages 851–856.

The Tundra Flora

(The Arctic Zone and the Alpine Areas)

The Boreal Flora has had a nomadic existence. In the Pleistocene Epoch enlarging ice sheets pushed it southward four times. Each time the ice melted many plants returned to the north, but some were left stranded on the peaks of the high Western mountain chains. Being isolated, some evolved into new regional or local races, varieties, or species. Other species occurring in or near the mountains already were adapted or they became so to the conditions under which the Tundra Flora lives, and some of these alpine species migrated northward in the wake of the ice, too. In North America the Boreal Flora has come to have two major subdivisions—one beyond timber line in the Arctic, the other above timber line in the mountains of the West. Typically the plants grow either among loose rock fragments or on sand or in tundra, that is, masses of plant remains which decay only slowly at the prevailing low temperatures. The tundra community is the stable one, that is, the climax.

1. THE ARCTIC TUNDRA

Occurrence. Circumpolar. In North America from the Alaskan coast of the Bering Sea to the mountains of northern Alaska, the islands of the Arctic Ocean, and the far northern plains and mountains from the Yukon eastward to Labrador and southward to James Bay; Greenland; mountains of the Gaspé Peninsula, Quebec; White Mountains, New Hampshire.

Characteristic Species. Many herbaceous plants are circumpolar and endemic to the Arctic; some are endemic to individual regions. There are no large woody species, and dwarf willows forming low mats in the tundra are the chief small ones. Many species are circumpolar, and it is to be noted that withdrawal of ocean water into the Pleistocene ice sheets

Figure 25-1. Arctic Tundra: plains and lake near the Arctic Ocean south of Point Barrow, Alaska, the tundra turning brown at the end of the flowering season, August 19, 1950.

probably lowered the northern Bering Sea sufficiently to make a broad temporary arctic land connection between Asia and northern Alaska, which, because of low precipitation, had no ice sheets.

Discussion. Probably the Arctic species of plants were derived from those of adjacent regions with some members able to adapt to the severe climate of the far north. If they are interpreted conservatively, the few genera and species in the Arctic do not include many endemics, but there are numerous endemic varieties and races of species. Even the genera with the most species in the Arctic, such as *Carex, Saxifraga,* and *Ranunculus,* have far more in the regions to the southward. According to Savile, "The Arctic flora is a depauperate miscellany from various regions." Derivation of the Subarctic and Temperate floras from the very small Arctic flora is unlikely, and probably the reverse is true.

The Tundra Flora is of relatively recent origin, as indicated by presence of endemism mainly in the lesser taxa. It developed either after the oceans cooled or after the continents drifted far enough north, whichever also caused the tropical forests to disappear from the present Temperate Regions at the close of Oligocene.

Pleistocene was not necessarily devastating to the Arctic flora. Whatever type of tundra flora may have been present in the north before Pleistocene was not eradicated by glaciation, because some areas of the far north were not glaciated for lack of sufficient precipitation to form ice sheets. Thus, there was also a refugium for Arctic species in the north, as many species moved southward in front of the ice.

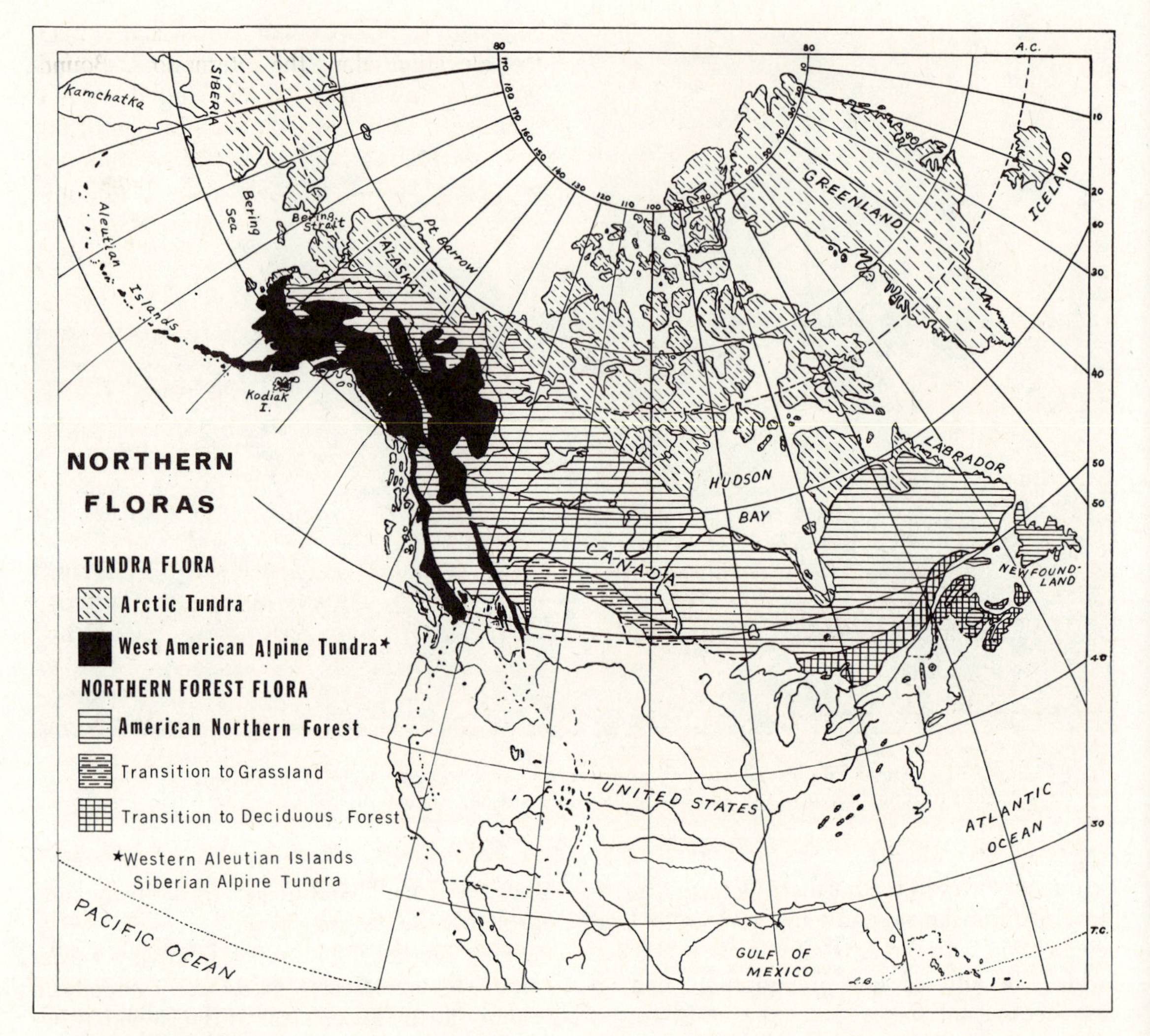

Distribution Map 1

During Pleistocene the oceans were lowered by about 350 feet by withdrawal of water into the ice sheets. The shallow North Bering Sea was a plain across which Arctic species might migrate between continents. Thus, the Arctic Regions of the New and Old Worlds have been separated for only about 10,000 years. The postulated basaltic land mass, Scandia, in the North Atlantic may have been an earlier avenue between continents.

North of timberline or above it in the mountains the winter winds are a major ecological factor. They dry the plants at a time when the soil water is frozen and unavailable, and they bombard the exposed stems and leaves with sharp ice particles, which tear them to pieces. A plant rising more than a few inches above the ground has no chance of survival. The plants are perennials, and the underground parts live over while those above the surface of the tundra die. Underground or with the protection of snow the surviving portion of each plant may live through the winter, carrying on only basal metabolism.

Figure 25-2. Arctic Tundra: lake region south of Point Barrow, Alaska; area of hummocks formed by ice-heaving in the winter, soil exposed in the cracks and supporting species characteristic of disturbed soil in the Arctic (but no weeds, none being able to withstand the climate); the pool supporting various species growing within one degree of the freezing point, *Ranunculus Gmelinii* having flowered and fruited under water; the area underlain at about 15–20 inches by permafrost, August 19, 1950.

2. THE WEST AMERICAN ALPINE TUNDRA

Occurrence. Above timber line in the mountains of western North America from the Aleutian Islands and western and central Alaska to Alberta and southward to California, Arizona, and New Mexico. Whereas in the Aleutian Islands the theoretical timber line is below sea level, at Kodiak, Alaska, it is at 2,000 feet elevation; along the International Boundary between Canada and the United States it is between 5,000 feet (Washington) and 7,000 feet (Montana), and in southern California and New Mexico it is between 11,000 and 12,000 feet, according to local conditions.

Characteristic Species. These are herbaceous, except for dwarf woody plants (particularly specialized willows). Some boreal species occur without obvious modification, but in many groups the species are related to the boreal ones but different from them.

Discussion. Although the severe conditions (extreme cold in winter and prevailing strong winds) at high altitudes in the western mountains resemble those in the Arctic, they do not duplicate them. In the growing season the mountains have alternation of day and night; the Arctic has continuous daylight for several weeks. At high altitudes in the southern mountains the daily range of temperature is extreme because the thin atmospheric blanket allows the sun's rays to penetrate readily in the daytime but permits rapid radiation of heat at night. Furthermore, the sun is more nearly overhead instead of at a low angle to the surface of the Earth, and the quality of the light is different because less ultraviolet light is filtered out. Even the continuous daylight in the Arctic may not provide much warmth; consequently, the tundra there is underlain by permafrost, that is, permanently frozen soil or tundra. At Point Barrow, the extreme northern tip of Alaska, permafrost is only fifteen or twenty inches below the surface of the tundra, and even in summer the temperature hovers near the freezing point, occasionally on "hot" days climbing into the forties (Fahrenheit) and rarely into the sixties. At high altitudes the evaporation rate is high, and both plants and the soil dry quickly. If they are to survive, the plants not occurring in meadows must

Figure 25-3. West American Alpine Tundra: rivulets at 2,000 feet elevation in the mountains of Kodiak Island, Alaska, the perennial species (left foreground) growing rapidly in the short season just after melting of the snow, July 13, 1952.

have structural and physiological features similar to those of desert species. Such extreme adaptations are unnecessary for arctic plants, and the factors discussed above have contributed to divergent evolution of alpine vs. arctic species.

Relative difference in behavior of arctic and alpine races is illustrated by the experimental studies of Clausen, Keck, and Hiesey. An arctic race of yarrow, *Achillea,* was secured at Abisko, Swedish Lapland, about 2° north of the Arctic Circle, where it grew under low arctic conditions. Divisions of individuals were planted at the experimental stations of the Carnegie Institution of Washington at various altitudes ranging from near sea level at Stanford University to timber line at about 10,000 feet elevation in the Sierra Nevada. At timber line, the Californian alpine forms developed vigorously and flowered abundantly during the short summer growing season, but the plants from Lapland remained low and did not flower, barely surviving as slowly spreading mats of vegetative rosettes of leaves and caudices. In four years only two individuals out of twenty showed any tendency to produce flowering stems, and these were still immature and inconspicuous at the end of the growing season. Divisions of the same plants flowered freely in April at Stanford University at a time when in Lapland they would have been covered with snow. Individuals grown under controlled conditions in a specially constructed greenhouse were found to remain vegetative at low temperatures. However, with increased warmth, and more so with addition of artificial light during the night, they flowered. At Abisko, the snow is gone about May 10, and in June and July there is continuous daylight. With the sun present continuously, the weather is relatively warm, and the yarrows flower and fruit. They are genetically adapted to survival in a region where there is no hurry concerning development.

3. THE SIBERIAN ALPINE TUNDRA

Occurrence. Northern Asia; mountains of the Aleutian Islands mostly from Adak westward, there being on Adak a minority of alpine species from the Alaskan mainland. East and west of Adak the species balance shifts in opposite directions.

Discussion. The weather of the Aleutian Islands is far different from that of the Arctic. There is much rain and fog throughout the year, but there is not the extreme cold of the north. It is always cold, but commonly not far below or above freezing, the temperature

Figure 25-4. West American Alpine Tundra in the Rocky Mountains in Colorado: *above,* Conejos River region; *below,* near Jones Pass, timber line showing below. Photographs, courtesy of the United States Forest Service.

being controlled by the surrounding ocean. The frozen plant and animal bodies composing the tundra do not build up to great thickness, because freezing weather is not the rule. The tundra is about 9 inches to 1 foot thick.

The Northern Forest Flora

(Subarctic Zone)

During Pleistocene plants of the Arctic managed to survive in refugia north of the ice sheets, where there was too little precipitation for formation of a continuous ice cover, and the brunt of the glaciations fell upon the present Subarctic, areas now covered by the Northern Forest Flora. There the ice became up to two miles thick, and its great weight caused it to spread to the edges of the oceans, where great icebergs broke off, and southward on the land. All was scoured before the moving ice mass, and, as finally it melted, some areas were left as bare rock, others as sand and gravel, none as good soil.

While the ice was melting, great streams and lakes were formed, and there were devastating floods. One of the largest occurred when the ice jam that had maintained Lake Missoula, covering a great area in western Montana and northern Idaho, suddenly gave way, and the soil, wind-blown over vast areas of lava in eastern Washington, was washed away leaving many square miles of "scablands."

Only about 10,000 years ago, a short time geologically speaking, the climate warmed, and the glaciations were over. If the age of the Earth (about 4.5 billion years) is represented by the distance of a race (sprint) (100 yards), the 10,000 years since the close of the last glacial period is equal to one-half the thickness of the tape each runner hopes to break. Thus, geologically, the time since the ice sheets is short.

The devastated region has been occupied through an "Oklahoma land rush" of plants to invade the newly available territory in the north. At the same time there has been movement of all other floristic associations northward and/or up the mountains.

In the newly reconstituted American Northern Forest, modified from the type pushed southward during glacial times, there is only a subclimax composed of the plants able both to migrate rapidly and to withstand the northern climate. If the present climate persists, eventually many of the current species will be replaced by slower-moving ones better adapted to the ultimate climax for the Subarctic Zone. To a considerable extent the species now in the north and those moving there are the ones that occupied the area during the interglacial times and before Pleistocene during the warm Pliocene Epoch, but there has been much evolutionary change and the gain and loss of taxa.

The Northern Forest Flora spans the Atlantic Ocean in the Subarctic Zone just south of timber line, but its floristic associations in the Eastern and Western Hemispheres are different. Parallel but different changes occurred in northern Eurasia, and the Eurasian Northern Forest, or Taiga, resembles the American. However, the species composition is largely different, even though, expectedly, the formation is similar. The ecologically dominant trees, such as spruces, firs, and northern types of pines, are largely of the same genera but different species. On the average, differentiation is at only about the level of species, because continential drift separated the northern areas of Eurasia and North America later than those farther south. Until about Pliocene, in the Subarctic considerable plant migration was possible across volcanic islands similar to Iceland and even, as mentioned above, probably a postulated larger igneous land mass, Scandia, occurring for a time. These resulted from outpourings of molten rock, much of it along the Midatlantic Ridge, formed along a fissure in the crust of the Earth extending from the Arctic to the southern end of the ocean.

Floristic differences on the great land masses of the Northern Hemisphere are in rough proportion to the passage of time since the last effective pathways of migration of plants disappeared. This becomes gradually greater southward. In the Arctic the circumference of the Earth is so small that there never has been separation of the land masses by great distances, and nearly the same Tundra Flora spans the Old and New Worlds. In the Subarctic, the separation of floras is mostly into different but related species, varieties, and races, but farther south separation is into higher and higher taxa. In the Temperate Zone independent evolution has produced mostly different species and many different genera, and in the Horse Latitudes and the Tropics the separated portions of the climatic belts are occupied mostly by different genera or even families. (See pp. 477–478.)

The species that reoccupied the Subarctic are necessarily aggressive types—rapid in spreading from one place to another and vigorous in competition. Because such aggressive species do not remain static in a single locality, the forests of the glaciated areas include few local endemics.

1. THE AMERICAN NORTHERN FOREST

Occurrence. From the Matanuska and Yukon valleys of Alaska eastward to southern Labrador, parts of Nova Scotia, the Great Lakes, and northern Maine; southward in modified composition on the highest peaks and ridges of the northern edge of the eastern United States in New York and northern New England.

Elements of the American Northern Forest (or spruce-fir forest) occur in the subalpine forests of both the Rocky Mountain System and the Cascade–Sierra Nevada Axis on the Pacific Coast. Especially in the Rockies they are represented significantly near timber line.

CHARACTERISTIC WOODY* SPECIES

Gymnosperms

- *Abies balsamea,* balsam fir
- *Picea glauca,* white spruce
- *Picea mariana,* black spruce
- *Larix laricina,* black larch or tamarack
- *Pinus Banksiana,* Jack pine (in the Eastern Forest Flora)
- *Thuja occidentalis,* white cedar

The Western Forest Flora

(Temperate Zone: Western Temperate Region)

Until Pliocene there was no great differentiation of the surface of the West into mountain axes, and during Miocene it was a region of low hills and rolling plains dotted by relatively short and disconnected mountain ranges. Because both the physiography and the climatic divisions of the West were less complex than now, so were the forests covering the region. They occupied roughly the area north of San Francisco, Salt Lake City, and Colorado Springs, though the boundaries followed the existing topography and they were irregular.

During Miocene a major system of versions of the temperate forests of the Northern Hemisphere, dominated by both conifers such as the coast redwood, *Sequoia sempervirens,* and deciduous trees, occurred on the great plain of the West. As with modern forests, there must have been great variability between large or small regions or from one continent to another. However, during Miocene doubtless there must have been less differentiation between Eurasian and North American temperate forests than now, because there had been less time for evolutionary divergence after separation of the land masses.

In some form, the forerunners of the elements composing the Miocene forest flora had been on both continental land masses before

* Woody plants are chosen for lists such as this because they are available at all seasons. Herbs are equally significant, except that presence of woody plants may affect the conditions under which the herbs grow.

Figure 25-5. American Northern Forest: *above,* swamp forest of white cedar, *Thuja occidentalis,* northern Michigan; *below,* spruce forest at Horseshoe Lake, Mt. McKinley National Park, Alaska. Photographs, above, courtesy of the United States Forest Service; below, courtesy of the United States National Park Service.

their separation by continental drift during late Mesozoic and early Cenozoic. However, the species may not have been associated in extensive temperate forests until Miocene, because through Eocene and Oligocene tropical forests dominated the now temperate lowland areas very far north, as along the Alaskan coast. The origin of the widespread Miocene temperate forests is not clear. It may have been through extension and modification of existing localized upland temperate forests of the preceding epochs or through bringing together assemblages of plants from other sources but adaptable to temperate forest conditions, or both. Probably the forests developed from a complex combination of sources.

The similarities of the Miocene temperate forests of Eurasia and North America are due to the relatively late separation by continental drift of both the north Pacific and the north Atlantic areas. In the western part of North America, land separation is by only the narrow Bering Strait, and the northern Bering Sea is so shallow that it was a connecting plain for migration of Arctic plants between continents during the Pleistocene glaciations, when the withdrawal of water into the ice sheets lowered the oceans by 300 feet. According to Wolfe, during Miocene there was an apparently continuous temperate forest extending from northern China and Japan to Alaska and the Pacific Northwest. In the North Atlantic, until Pliocene there were also migration routes for plants along volcanic island stepping stones and perhaps temporary volcanic land masses. During the later period of the connection, these were restricted partly or altogether to the plants of the Arctic and Subarctic zones, but earlier they may have had a more temperate climate.

At present, mostly the forests of the West occupy three major climatic and physiographic areas: (1) the lowlands west of the Cascade–Sierra Nevada Axis; (2) the Cascade–Sierra Nevada Axis; and (3) the Rocky Mountain System. The mountains composing the Cascade–Sierra Nevada Axis began to form during late Miocene, but their chief development occurred during Pliocene and Pleistocene. The Cascades are a volcanic chain running from the southern edge of British Columbia to Mt. Lassen in northern California. The Sierra Nevada is a gigantic tilted fault block continuing southward through the rest of northern California. A series of short granitic ranges continues the axis through southern California and on to the tip of Baja California. The forest floristic associations comprising the Western Forest Flora are developed in the principal climatic regions of the forest, one in the lowlands of the Pacific Northwest, and at two levels in each of the great mountain systems.

The Miocene forest complex of the Pacific Northwest included three major descendent elements. According to Chaney, Condit, and Axelrod, these are as follows:

The West American Element. Near equivalents of modern western species of fir (*Abies*), maple (*Acer*), alder (*Alnus*), serviceberry (*Amelanchier*), birch (*Betula*), chinquapin (*Castanopsis*), dogwood (*Cornus*), ash (*Fraxinus*), salal (*Gaultheria*), incense cedar (*Libocedrus,* perhaps to be segregated as *Calocedrus*), tan oak (*Lithocarpus*), barberry (*Berberis*), *Rhododendron,* rose (*Rosa*), willow (*Salix*), redwood (*Sequoia*), mountain-ash (*Sorbus*), giant cedar (*Thuja*), hemlock (*Tsuga*), and huckleberry (*Vaccinium*).

The East American Element. Trees of genera no longer occurring in the West, including hickories (*Carya*), hornbeam (*Carpinus*), chestnut (*Castanea*), beech (*Fagus*), sweet gum (*Liquidambar*), tupelo gum (*Nyssa*), bald cypress (*Taxodium*), and elm (*Ulmus*).

The East Asian Element. Trees of genera no longer occurring in North America, including tree-of-Heaven (*Ailanthus*), *Cercidiphyllum,* maidenhair tree (*Ginkgo*), *Keteleeria, Metasequoia* (a recently discovered relative of the redwoods), and Chinese wingnuts (*Pterocarya*). These occur today only in eastern Asia.

These plants lived in a climate of ample year-round rainfall and moderate temperature. Gradual elimination of summer rainfall during the Pliocene Epoch left during the warm, arid climate of middle Pliocene time only the West American Element, which consists of species capable of survival without summer rainfall. Some species requiring summer showers persisted in California until middle Pliocene or as last survivors until late Pliocene.

In view of their long and complicated geological history, the present subdivisions of the surviving West American Element forming part of the Western Forest Flora compose a fairly homogeneous major unit, but each subdivision has been intruded by wide-ranging species and by members of adjacent floras. The individual divisions are "frayed around the edges" from contact with their neighbors. During late Cenozoic time, each subdivision gained and lost some members, and some species were replaced wholly or partly by unusual forms selected from among their previous components. Association of species has varied somewhat from time to time because no two had exactly the same tolerance for the complicated ecological conditions under which they lived. Gradual uplift of the Cascade–Sierra Nevada Axis from Washington to California beginning in late Miocene and early Pliocene gave rise to mountain areas with new conditions. However, differentiation of the residual West American Element into coastal and mountain forests was slow and not far along even in upper Pliocene. More rapid elevation occurred in Pleistocene.

The surviving redwood forest of the northwest coast of California is modified considerably from the redwood forest which occupied the same and a much larger area in Miocene time, but its modification is relatively less than for other elements of the current Western Forest Flora. It approaches the Pliocene archetype of the present flora except for disappearance of species now restricted to the mountain chains. Many plants occurring in the Pacific Lowland Forest may be remnants of the prototypes of wide-ranging species complexes, and the subalpine and montane species and varieties may be more greatly modified descendants of populations once occurring on the forested Miocene plains, valleys, and hills. For example, each of four variable species of *Ranunculus* is represented by a single variety in the somewhat disjunct surviving relics of the Miocene forest. A variety of each species (*R. occidentalis, R. uncinatus, R. orthorhynchus,* and *R. alismaefolius*) is replaced by other varieties or closely related species in the Cascade–Sierra Nevada Axis, the mountains within and about the Columbia Basin and the western and northern Great Basin, and the Rocky Mountains.

With the emergence of the Cascade–Sierra Nevada Axis, the climate of most of the Columbia Basin and of the entire area east of the mountains changed profoundly. Rainfall and winter temperature were reduced, and summer temperature was increased. Consequently, the forest disappeared, being replaced by more **xerophytic** (drought resistant) floras. This was due to factors affecting the climate of the entire Pacific Coast, as described in the next paragraphs.

In winter the land often is cooler than the sea, and the Prevailing Westerly Winds from the Pacific Ocean deposit moisture along the coastal lowlands. Consequently, the Pacific Coast is an area of winter rainfall. As the air mass moves inland across the mountains, more rainfall is precipitated along the Cascade–Sierra Nevada Axis, and the dry air descending along the eastern side of the range creates a rain shadow. Opposite Puget Sound, there is wet Pacific Lowland Forest; on the east side of the Cascades there is dry Sagebrush Desert. Similarly, at Claremont on the western side of the main southern California mountain axis the average rainfall is about eighteen inches, but only forty airline miles away on the east slope of the mountains the average is two or three inches. The first area is California Chaparral, a dense brushland of the Sierra Madrean Flora; the second is one of the driest parts of the Southwestern Deserts.

The woodland gilias provide an example of evolution of both mountain and desert species from a prototype which occurred formerly in the coastal Pacific Lowland Forest of California. Grant reports that the hypothetical original species of the group has disappeared, but it is approached by living races of two species—one linking it with derivatives in the high mountains, the other with descendants in the desert region. Some races of *Gilia leptalea* of the California North Coast

Ranges (north of San Francisco Bay) are similar to the hypothetical prototype, while others farther inland tend toward *Gilia capillaris* of the Sierra Nevada. Some populations of *Gilia splendens* of the South Coast Ranges have character combinations approaching the original type, while others farther inland pass into a species characteristic of drier habitats to the southeastward. This species, *Gilia caruifolia,* forms a connecting link to *Gilia australis,* an ephemeral annual of dry areas in the Mojave Desert region as far east as the Old Dad Mountains. The races of both *Gilia leptalea* and *Gilia splendens* more or less similar to the prototype of the species group have large flowers, which permit pollination by various insects. On the other hand *Gilia capillaris* and *Gilia australis* at the other ends of the mountain and desert series have small flowers which are self-pollinated. The open pollination system of the first pair of species permits crossing of various races and formation of local population complexes with new combinations of characters and the ultimate development of new species. The closed pollination system of the second pair restricts recombination of characters and development of new races or species. Consequently, evolution could have proceeded more easily from *leptalea* to *capillaris* or from *splendens* to *australis* than in the reverse direction. Uplift of the Sierra Nevada and the continuing ranges to the southward created new montane and subalpine as well as desert habitats to which the new derivative species are adapted.

On the Pacific Slope in summer the land is warmer than the ocean, and in the south summer rain is rare. The period of summer drought decreases gradually toward the northward, and about Puget Sound there may be occasional rain during the summer. In coastal Alaska and British Columbia the amount of summer rain is considerable. In Alaska the spring growing season is long delayed by cold, and it merges into the brief summer.

The original forests of the great Miocene plain covering the West included both coniferous and deciduous dominant trees, but after Miocene reduction of total precipitation and disappearance of summer rainfall in the coastal region and much of the interior changed the ecological formation to coniferous forest. The broad-leaved trees, or dicotyledons, require water during the summer. Today in Western North America there is no area with enough summer rainfall to support a cover of deciduous forest, and the more drought-resistant conifers are dominant. The relatively small number of deciduous trees and shrubs includes few species but those restricted to places with extra moisture—low land near streams, lakes, and boggy meadows or spots where there is moisture below the surface. However, deciduous trees do occur in the coastal fog belt and in the areas of greatest summer precipitation in temperate rain forests like that west of the Olympic Mountains in Washington.

The area of greatest abundance of species requiring extra water is in the Siskiyou Mountains near the coast in the adjoining corners of Oregon and California. This area combines the rainfall occurring on cool land along the ocean with that derived from uplift of the air mass as it passes over the coastal mountains, and during the summer there are coastal fogs as well. Consequently species from the Miocene forest have persisted and some deciduous as well as coniferous plants have been retained through two epochs of geologic time while they died out elsewhere.

The plants of the forests in the Rocky Mountain System are related not only to those in the mountains farther west, but also, though less so, to those in the American Northern Forest and to a still lesser extent in the Tundra Flora. Evidence of a relationship to the floras of boreal regions is provided, for example, by the genus *Ranunculus.* The species occurring in the Arctic Tundra, the West American Alpine Tundra, and the Rocky Mountain Forests are primarily in a single section, *Epirotes,* which is absent or represented by only a few aberrant species in the rest of North America.

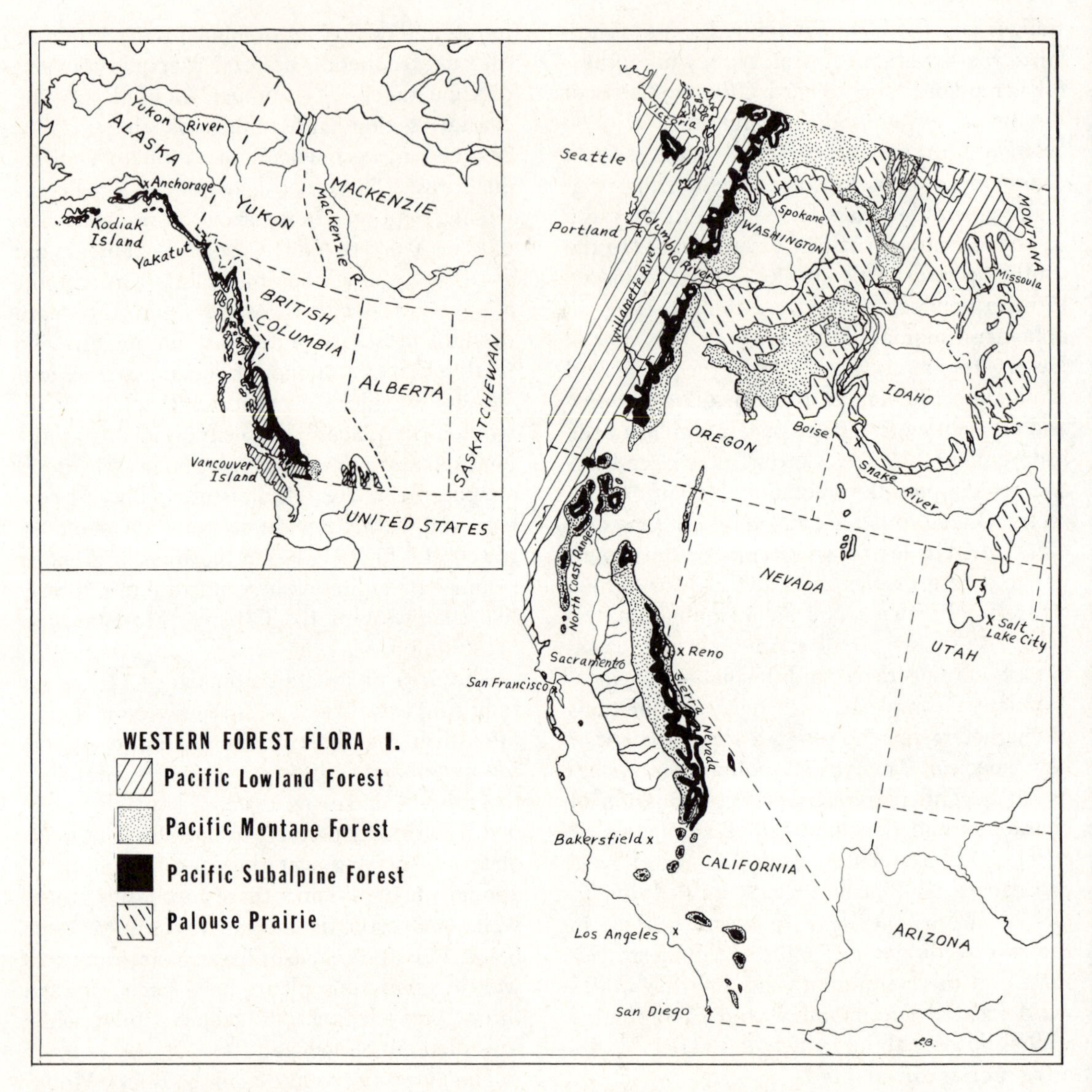

Distribution Map 2

Epirotes includes nine of the eleven species and four of the five endemics occurring in the Arctic Tundra, twelve of the thirteen species and varieties and all ten endemics in the West American Alpine Tundra, and twelve of the seventeen species and varieties and nine of the twelve endemics in the Rocky Mountain Forests. In keeping with having been denuded by ice and reinvaded by relatively few aggressive species, the American Northern Forest includes no characteristic terrestrial *Ranunculi* but only wide-ranging aquatic and marsh species. However, evidence from other plant groups indicates some degree of relationship of the plants of the forests of the Rocky Mountains and especially of their subalpine phase to the Northern Forest Flora.

According to species composition, the western mountain forest types correlate more clearly with the principal north and south

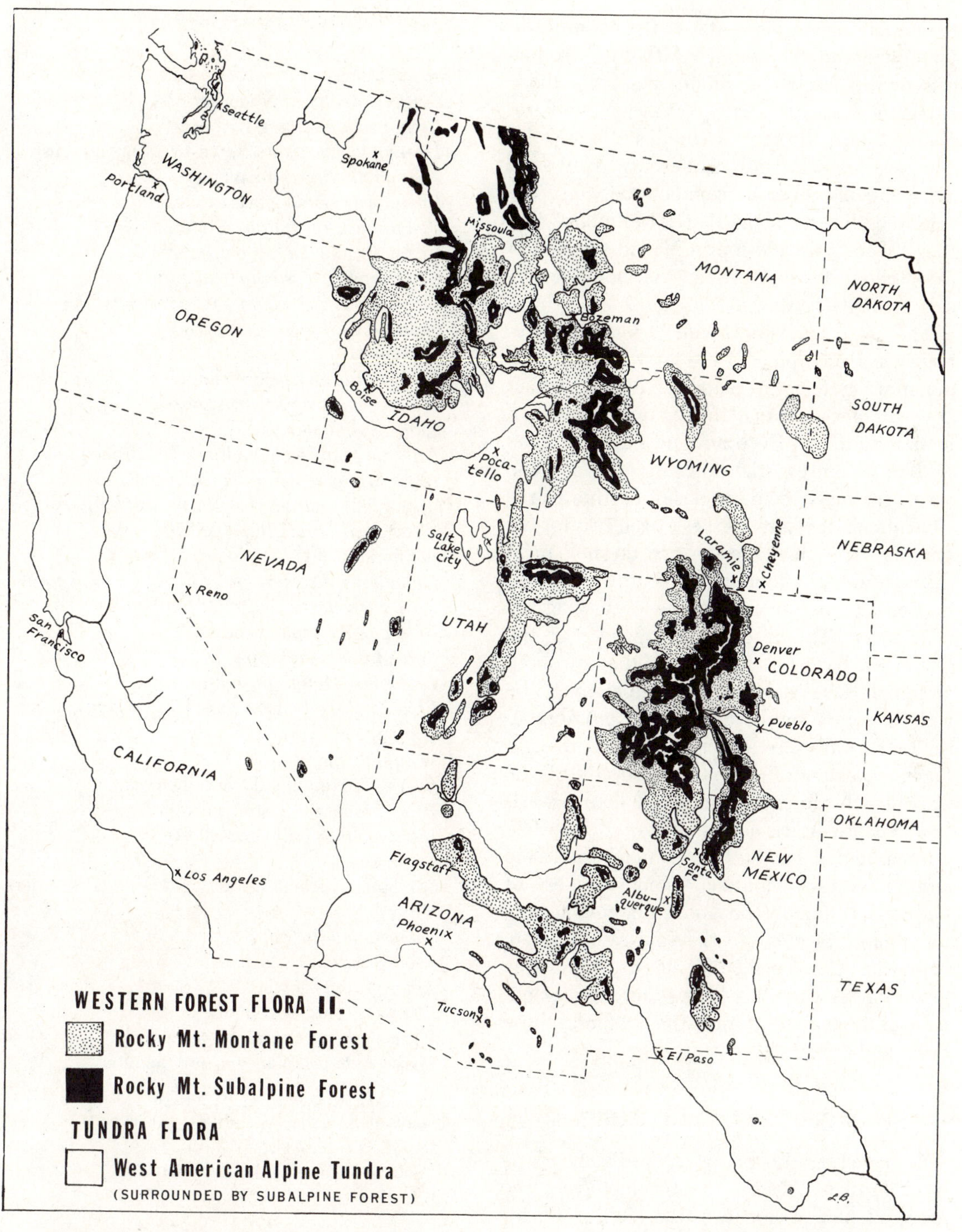

Distribution Map 3

mountain axes than with belts of altitude (montane and subalpine). Although the forests of the Rocky Mountains resemble those along the Cascade–Sierra Nevada Axis of the Pacific Coast, the species composition and the climates are markedly different. However, there are elements in common between the subalpine forests of the Rocky Mountains and those of the Cascade–Sierra Nevada Axis, and also, though fewer, between the montane forests of the two systems.

The present possible route of migration between the subalpine forests of the two great mountain systems presents less difficulty than any still perhaps available between the two montane forests. Even now passage from one subalpine forest to the other may be possible across the moist, cold regions of northern British Columbia, where species of the two forests mingle. The montane forests do not meet through any continuous easy pathway of migration, at least for the species occupying the drier sites. This is because the floor of the Columbia Basin is occupied by the still more xeric Palouse Prairie and because many of the valleys across southern British Columbia to the northward are covered by the moister Pacific Lowland Forest. Since uplift of the Cascades this forest has persisted along a west-to-east storm track across eastern British Columbia, and it has outliers in northern Idaho and in western Montana. The members of both pairs of montane and subalpine forests differ most in areas to the southward where migration has been difficult to impossible much longer and where the differences of climate of the mountains near the coast and those far inland are greater.

1. THE PACIFIC LOWLAND FOREST

Occurrence. Lowlands of the Pacific Slope from Kodiak, Alaska, to and including the redwood belt of northwestern California; inland north of the Cascade Mountains across British Columbia to northwestern Montana and southward to parts of northern Idaho.

CHARACTERISTIC WOODY SPECIES

Thalamiflorae
- *Berberis nervosa*, Oregon-grape
- *Berberis Aquifolium*, barberry
- *Euonymus occidentalis*, western burning-bush
- *Acer circinatum*, vine maple
- *Rhamnus Purshiana*, cascara sagrada
- *Ceanothus thyrsiflorus*, mountain-lilac
- *Ceanothus incanus*, mountain-lilac
- *Castanopsis chrysophylla*, chinquapin
- *Lithocarpus densiflora* var. *densiflora*, tan oak
- *Quercus Garryana*, Garry oak

Corolliflorae
- *Rhododendron californicum*
- *Menziesia ferruginea*, rusty-leaf
- *Gautheria Shallon*, salal
- *Arbutus Menziesii*, madrone (madroño)
- *Arctostaphylos columbiana*, manzanita
- *Vaccinium ovatum*, California huckleberry
- *Vaccinium parvifolium*, red bilberry

Calyciflorae
- *Holodiscus discolor*, cream-bush (typical form)
- *Rosa nutkana* (at least var. *muriculata*)
- *Rosa gymnocarpa*, wood rose
- *Pyrus fusca*, crab-apple
- *Crataegus Douglasii*, western black haw
- *Philadelphus Lewisii* var. *Gordonianus*, mock-orange
- *Whipplea modesta*
- *Ribes sanguineum*, flowering currant
- *Ribes bracteosum*, stink currant
- *Ribes Lobbii*, Lobb gooseberry

Ovariflorae
- *Sambucus racemosa* var. *callicarpa*, red elderberry
- *Viburnum ellipticum*

Gymnosperms
- *Abies grandis*, lowland fir
- *Pinus contorta*, beach pine
- *Picea sitchensis*, tideland spruce
- *Tsuga heterophylla*, western hemlock

Pseudotsuga Menziesii var. *Menziesii*, Douglas fir
Sequoia sempervirens, coast redwood
Thuja plicata, giant cedar
Chamaecyparis Lawsoniana, Lawson cypress

Discussion. The chief elements of the Pacific Lowland Forest occur as relics in valleys of eastern British Columbia, northeastern Washington, northern Idaho, and northwestern Montana, forming a fringe north and northeast of the more xerophytic floras of the bulk of the Columbia River Basin. This forest west of the Continental Divide in Glacier National Park is in striking contrast to the Rocky Mountain Forests on the east. The coastal character of the vegetation is evident particularly about Avalanche Creek above Lake McDonald, where similarity to the California redwood belt and particularly to the Washington and Oregon coastal forest is outstanding. This is even more striking at Linden Mill on the western edge of the Swan River Valley.

In other areas in Montana plants characteristic of the Pacific Lowland Forest and of the Rocky Mountain Montane Forest are mixed in various proportions. For example, on the east shore of Flathead Lake, a plurality of Pacific Lowland Forest plants is mingled with an almost equally strong representation of Rocky Mountain Montane Forest species, and there are smaller numbers of plants characteristic of the American Northern Forest and of the Rocky Mountain Subalpine Forest. The area was glaciated in the Pleistocene Epoch, and the Pacific Lowland Forest must have migrated northward and southward during and between glaciations, thus becoming mixed with other elements.

The principal storm tracks from the Pacific Ocean to the Rocky Mountains are located near and chiefly just north of the boundary between the United States and Canada. Probably these are responsible for long maintenance of a climate suitable for the Pacific Lowland Forest, requiring more moisture than the Rocky Mountain Forests. Currently, the Pacific Lowland Forest is continuous, except for one or two gaps, in the lower valleys across British Columbia from the coast to the vicinity of the Selkirk Mountains.

In the Pacific Northwest the Lowland Forest includes an Alaskan phase, a coastal phase, a local grassland phase, and Columbia Basin phases with local segregation of dominant species; in California it includes a redwood phase, a Douglas fir phase, a tan-oak-madrone phase, a soft chaparral phase, and a local grassland phase.

2. THE PACIFIC MONTANE FOREST

Occurrence. Eastern slopes of the Cascade Mountains and adjacent ranges from the vicinity of Kamloops, British Columbia, to Klamath Lake, Oregon; higher levels in eastern Washington and the mountains of eastern Oregon; mountains of California at about 2,000 to 5,000 feet elevation northward and 5,000 to 7,000 or 8,000 feet southward; mountain ranges of western Nevada; high mountains of northern Baja California.

CHARACTERISTIC WOODY SPECIES

Thalamiflorae
- *Staphylea Bolanderi,* bladdernut
- *Ceanothus prostratus,* mountain-lilac

Corolliflorae
- *Penstemon Jaffrayanus*

Calyciflorae
- *Chamaebatia foliolosa,* mountain-misery
- *Ribes nevadense,* Sierra currant
- *Ribes Roezlii,* spiny gooseberry

Amentiferae
- *Quercus Kelloggii,* California black oak
- *Quercus Garryana* var. *Breweri,* Brewer oak
- *Quercus Garryana* var. *semota,* Kaweah oak (at low altitudes)

Gymnosperms
- *Pinus Lambertiana,* sugar pine
- *Pinus ponderosa,* Western yellow pine
- *Abies concolor,* white fir (Californian form)
- *Sequoiadendron giganteum,* big tree, Sierra redwood (restricted to ridges on the west slope of the Sierra Nevada)
- *Libocedrus decurrens,* incense cedar (perhaps properly classified as *Calocedrus* [synonym, *Heyderia*] *decurrens*)

The following woody species or varieties are characteristic of both the Pacific Lowland and the Pacific Montane forests and are restricted to them.

Figure 25-6. Pacific Lowland Forest: *above,* road to Paradise Valley, Mt. Rainier National Park, Washington; virgin stand of timber; white cedar, *Thuja plicata;* Douglas fir, *Pseudotsuga Menziesii;* western hemlock, *Tsuga heterophylla;* and lowland fir, *Abies grandis;* about 1,000 feet elevation; *below,* forest near sea level, near Tacoma, Washington, in 1926, primarily of gigantic Douglas firs, *Pseudotsuga Menziesii,* but with a number of other trees in association. The devil's club, *Oplopanax horrida,* is the large-leaved plant in the left foreground. (*Below,* by Homer L. Shantz, courtesy of the Herbarium of the University of Arizona and the Curator, Charles T. Mason, Jr.)

Figure 25-7. Pacific Lowland Forest: redwood forest; coast redwood, *Sequoia sempervirens;* Pepperwood Flat, Avenue of the Giants, Humboldt County, California. The ground cover includes swordfern, *Polystichum munitum,* and *Oxalis oregana.* The large tree is at least fifteen feet in diameter. Photograph by Moulin, for the Save-the-Redwoods League, used through the courtesy of Newton B. Drury, Chief, Division of Beaches and Parks, State of California and the Moulin Studio.

Figure 25-8. Pacific Montane Forest: relatively young big trees, *Sequoiadendron giganteum;* sugar pine, *Pinus Lambertiana;* white fir, *Abies concolor* (California form); Ten Mile Grove, Sequoia National Forest, Sierra Nevada, California. Cf. also **Figures 19-3, 7,** and **8.** Photograph, courtesy of the United States Forest Service.

Thalamiflorae
- *Acer macrophyllum,* big-leaf maple (also following streams to drier areas at much lower altitudes)

Corolliflorae
- *Rhododendron occidentale,* western azalea

Calyciflorae
- *Osmaronia cerasiformis,* osoberry
- *Cornus Nutallii,* western dogwood
- *Corylus cornuta* var. *californica,* California hazelnut

Gymnosperms
- *Taxus brevifolia,* western yew

3. THE PACIFIC SUBALPINE FOREST

Occurrence. Coastal southeastern Alaska under 2,000 feet; characteristic of higher elevations in the Cascade Mountains and the Sierra Nevada, but represented in the high neighboring ranges immediately to the west and the east. In Washington at 3,000 to 5,500 feet elevation, in California at 7,000 to 9,000 or 11,000 feet.

CHARACTERISTIC WOODY SPECIES

Thalamiflorae
- *Ceanothus cordulatus,* mountain-lilac (lapping into the Sierran Montane Forest)
- *Castanopsis sempervirens,* Sierra chinquapin
- *Quercus Sadleriana,* Sadler oak (Siskiyou Mountains)
- *Quercus vaccinifolia,* huckleberry oak

Corolliflorae
- *Arctostaphylos nevadensis,* mat manzanita
- *Penstemon Newberryi* (occurring also at somewhat lower altitudes)

Calyciflorae
- *Holodiscus discolor* var. *dumosus*
- *Sorbus* (or *Pyrus*) *sitchensis* var. *californica,* mountain-ash
- *Ribes leptanthum* var. *lasianthum*

Gymnosperms
- *Tsuga Mertensiana,* mountain hemlock
- *Pinus monticola,* western white pine
- *Pinus albicaulis,* white bark pine (at timber line)
- *Pinus Balfouriana,* foxtail pine
- *Pinus ponderosa* var. *Jeffreyi,* Jeffrey pine
- *Pinus contorta* var. *latifolia,* lodgepole *pine*
- *Picea Breweriana,* weeping spruce (Siskiyou Mountains)
- *Albies magnifica,* red fir
- *Abies nobilis*
- *Juniperus occidentalis* var. *australis,* mountain juniper
- *Chamaecyparis nootkatensis,* Alaska cedar

Discussion. The mountain forests are divisible according to any of several criteria, and the designations Pacific Subalpine Forest and Pacific Montane Forest used here may be considered arbitrary. The forests of the extreme upper levels are far different from those of lower elevations, and the chief difficulty in separation is the extensive zone of intergradation at middle altitudes and the complex composition of the forest as a whole. In California the subalpine forest occurs at mostly 7,000 to 9,000 feet elevation. It is characterized at lower levels by red fir, Jeffrey pine, and western white pine and at higher levels by white bark pine and mountain hemlock. Probably at various times plants from the American Northern Forest have crept in and survived, with or without modification, in the higher mountains especially near timber line. Others have been derived with modification from types existing in moist situations at lower altitudes before uplift of the mountain axis. Some species now characteristic of the Rocky Mountains have persisted along the Cascade–Sierra Nevada Axis since Pliocene at higher altitudes where there are significant summer thunder showers.

At relatively higher elevations than for the same species in the Rocky Mountains there are large areas occupied by lodgepole pine and often a characteristic selection of associated species. Classification of this vegetation is difficult, because, as mentioned on p. 759, in the Rocky Mountain region lodgepole pine forest often develops after fire destroys the climax vegetation. This is true also in the Sierra Nevada, but lodgepole pines are also a part of each climax forest.

Figure 25-9. Pacific Subalpine Forest; Mt. Adams, Cascade Mountains, Washington, from Council Bluff with Council Lake in the foreground. Photograph, courtesy of the United States Forest Service.

Figure 25-10. Pacific Subalpine Forest: *above,* forest near timber line at Picture Lake and Mt. Shuksan, Mt. Baker Region, Washington; *below,* typical red fir (*Abies magnifica*) forest in the Sierra Nevada between Park Ridge and Bacon Meadow, Sequoia National Forest, California. Cf. also **Figure 19-4.** Photographs, courtesy of the United States Forest Service.

4. THE PALOUSE PRAIRIE

Occurrence. Centering in the Columbia River Basin. Areas below the forests and above the Sagebrush Desert at about 2,000 to 3,000 feet elevation in eastern Washington, northern Idaho, northwestern Montana, and northeastern Oregon and at 4,500 to 5,500 feet in central Idaho and mid-northern Utah; western edge of Wyoming.

CHARACTERISTIC SPECIES

The best-known ones commonly considered as indicators are the following grasses.

Monocotyledoneae

Festuca idahoensis var. *idahoensis;* Idaho fescue or blue bunch-grass

Agropyron spicatum, wheat grass or bunch grass

Discussion. Essentially, the Columbia Basin is a now dry spot in the former plains area once occupied by the Miocene forests. It became dry as the Cascade Mountains arose beginning in Miocene but chiefly during Pliocene and Pleistocene, creating a rain shadow to the eastward. During the early periods of mountain building only drier and drier forest could have originated in the Columbia Basin, and during the glacial periods there must have been forests on the areas free from ice. Development of extensive grassland probably did not occur until after Pleistocene, then being stimulated by fires set by the Indians as well as by the drying climate.

The Palouse Prairie is composed of a hodgepodge of species able to adapt to the conditions developed during the last 10,000 years, but probably most were derived with modification from species in the surrounding temperate forests. The grassland includes species identical with or related to those in the adjacent Pacific Montane and Pacific Lowland forests, the Rocky Mountain Montane Forest, and the Sagebrush Desert. Adapted or adaptable species must have migrated to the Columbia Basin from all directions, and, though some are very wide-ranging, most occur in the neighboring floristic associations listed above. Even the two dominants are of wide geographical range and not confined to the Palouse Prairie. *Festuca idahoensis* var. *idahoensis* occurs from British Columbia to Alberta and southward to the Sierra Nevada of California and to Montana and Colorado. *Agropyron spicatum* occurs from Alaska to California, Alberta, the Dakotas, and New Mexico. Consequently, these plants are dominants from only an ecological point of view, and they are insignificant from the floristic view.

Thus, the Palouse Prairie probably has arisen only since Pleistocene and then partly in response to fire, which has disappeared as a major factor. The included species nearly all occur elsewhere, as well, and this does not indicate an established and highly evolved floristic association but, rather, an incipient one.

Unfortunately, the Palouse Prairie is nearly extinct, and there are only a few good patches of it left, mostly in southeastern Washington and adjacent northern Idaho and in northwestern Montana. The fine windblown soil overlying the great lava flows of the Columbia Basin is excellent wheat land, and most of the Prairie did not last long after coming of the white man.

5. THE ROCKY MOUNTAIN MONTANE FOREST

Occurrence. The lower half of the forest belt: mountains from north-central Idaho and northwestern Montana (3,000 to 5,000 feet or lower or higher, depending upon locality) to Arizona and New Mexico (6,000 to 8,500 feet) and the Black Hills.

CHARACTERISTIC TREES

Thalamiflorae

Quercus Gambelii, Gambel oak

Calyciflorae

Robinia neo-mexicana, New Mexico locust

Figure 25-11. Palouse Prairie. The most prominent grasses are *Festuca idahoensis* and *Agropyron spicatum;* 5 miles west of Colton, Washington. (By ROBERT L. BENSON, from LYMAN BENSON, *The Cacti of the United States and Canada.* Stanford University Press, in press.)

Gymnosperms

Pseudotsuga Menziesii var. *glauca,* Rocky Mountain Douglas fir
Abies concolor, white fir
Pinus flexilis var. *reflexa,* Mexican white pine
Pinus ponderosa var. *ponderosa,* western yellow pine (Rocky Mountain form)
Pinus ponderosa var. *scopulorum,* Rocky Mountain yellow pine
Pinus ponderosa var. *arizonica,* Arizona yellow pine
Pinus ponderosa var. *Mayriana,* Apache pine
Juniperus scopulorum, Rocky Mountain juniper or red cedar

Discussion. The relationships of this forest are diverse. The affinities of some species to those in the American Northern Forest are evident, but most connections are with the montane forest of the Cascade–Sierra Nevada Axis, the Prairie Flora, and the Eastern Forest Flora. As indicated by *Ranunculus,* there is a clear similarity to the species in the pine forests of the higher Sierra Madre Occidental of northwestern Mexico, where also the section *Epirotes* is dominant. Although the Mexican species are related to those in the United States, they are different from them and endemic.

Throughout the Rocky Mountain system extensive forests are not climax; they develop in the wake of forest fires. These are dominated by lodgepole pines (*Pinus contorta* var. *latifolia*) or quaking aspens (*Populus tremuloides*). Consequently, the normal climax vegetation is obscure, and distinction of local subalpine and montane forests may be difficult or impossible.

6. THE ROCKY MOUNTAIN SUBALPINE FOREST

Occurrence. The upper half of the forest belt: mountains from eastern British Columbia (3,500 to 6,000 feet elevation) to Arizona and New Mexico (9,000 to 11,000 or 12,000 feet).

Figure 25-12. Rocky Mountain Montane Forest (cf. also **Figure 24-12,** *upper,* p. 733): stand of western yellow pine, *Pinus ponderosa,* in the Kaibab Forest, Arizona; other coniferous species in the canyon in the background. The associates of the yellow pine here are almost wholly different from those in the forests of the Pacific States. Photograph, courtesy of the United States Forest Service.

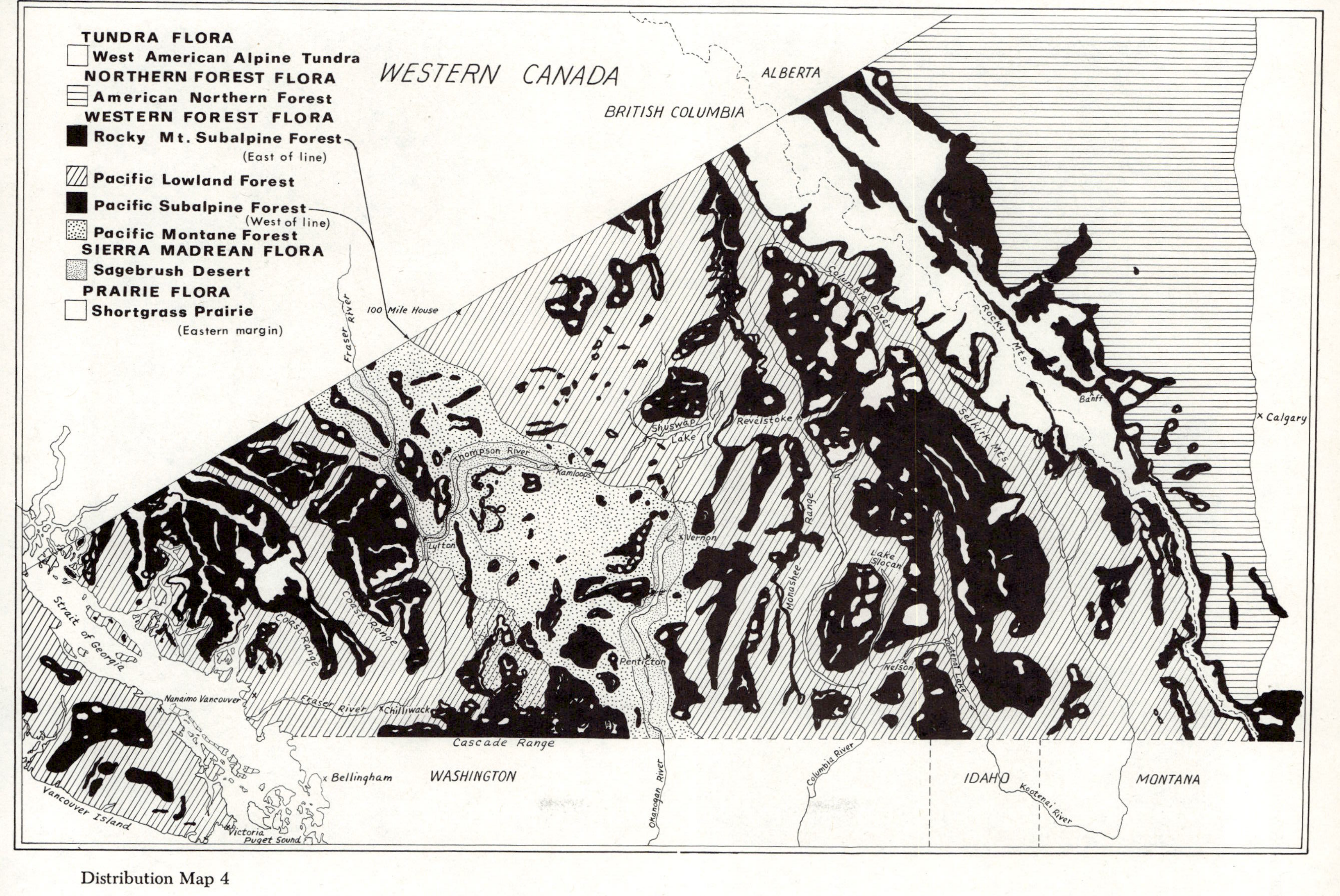

Distribution Map 4

Figure 25-13. Timber line in the Rocky Mountains: *above,* wind-trimmed timber on the road to Mt. Evans, Colorado, branches growing only away from the wind, the terminal buds of those on the windward side dried rapidly by the prevailing high winds in combination with the high evaporation rate at upper altitudes; *below,* wind-trimmed timber at timber line on Cove Mountain, Utah, the last advance of tree branches at low levels where they are somewhat sheltered from the winds coming over the mountaintop, the rocky soil above with many alpine plant species growing in the crevices. Photographs, courtesy of the United States Forest Service.

CHARACTERISTIC TREES

Thalamiflorae

Salix Dodgeana, willow
Salix Farriae
Salix Tweedyi
Salix monticola (also in the American Northern Forest)
Salix Wolfei vars. *Wolfei* and *idahoense*
Salix phylicifolia var. *planifolia*

Corolliflorae

Cassiope Mertensiana var. *gracilis*

Calyciflorae

Currant, *Ribes mogollonicum*

Gymnosperms

Picea Engelmannii, Engelmann spruce
Picea pungens, Colorado blue spruce (also in the upper part of the montane forest)
Abies lasiocarpa, alpine fir
Abies lasiocarpa var. *arizonica,* corkbark fir
Pinus flexilis, limber pine (also in montane forests)
Pinus aristata, bristle-cone pine

Discussion. The characteristic species are allied more closely to those of the Boreal-Alpine and Northern Forest regions than to the floras of other areas in North America. Many of the species occurring at timber line in both the Rocky Mountain Subalpine Forest and the high mountains of the Pacific Coast are wide-ranging in the far north.

The Eastern Forest Flora

(Temperate Zone: Eastern Temperate Region)

The (primarily) Deciduous Forests composing the Eastern Forest Flora occupy one of the most extensive floristic regions in North America. In the United States they cover most of the Middle West, East, and South, about two-fifths of the contiguous 48 states. In Canada they occur along the St. Lawrence River watershed.

Throughout eastern North America there is an abundance of precipitation the year around, but the "four seasons" of the region are markedly different. Winter is dominated by the cold northern air and summer by the warm, humid southern air. Not infrequently the average temperature difference between these two seasons is 50° to 60° F, and the changes of seasons are sudden. The rapid plant growth and flowering during the transitional season of spring produces a period far different from winter, and the quick appearance of the transitional season of autumn is marked by the beautiful coloring and the fall of the foliage of the deciduous trees. The fall coloring is matched only in the temperate interiors of eastern Europe, eastern Asia, and a small mountainous area in eastern Mexico.

If they are to survive the cold winter of the East, Middle West, and South, the broad-leaved dicotyledonous trees must eliminate the evaporative surface of their foliage. During such a winter, at least at times much of the ground water is frozen and unavailable, and, if the plant were to lose moisture to the atmosphere as it does during the rest of the year, it would dry out and die. Thus, the broad-leaved deciduous trees must lose their leaves, while the more drought-resistant conifers may retain most of theirs.

In western Europe, fall coloring does not match that in eastern North America, because the warm Gulf Stream is up wind and the chill of winter is much less. The loss of leaves is less urgent and less spectacular. European trees transplanted to eastern North America retain their inbred slower reactions to the coming of fall.

In addition to their water relations, broad-leaved dicotyledonous trees, if they retained their leaves during the eastern winter snows, would accumulate too great a weight of snow or ice. This would break off most of the branches and even split the trunks. Conifers have small leaves, and those of very cold regions around the world are characteristically with short, flexible branches. These bend downward and shed the snow, then spring upward. Because of their short branches, the conifers, especially those of northern and sub-

Figure 25-14. Rocky Mountain Subalpine Forest: *above,* about mountain meadows near timber line, Sacramento Creek near Mt. Sherman, Colorado; *below,* Engelmann Spruce (*Picea Engelmannii*) and alpine fir (*Abies lasiocarpa*) forest, Wellington Creek, Kaniksu National Forest, Idaho. The narrowly conical form of the trees is characteristic of subalpine forests around the world. Photographs, courtesy of the United States Forest Service.

Figure 25-15. Dense stand of subclimax lodgepole pine developed after a fire in the Lewis and Clark National Forest, Montana (cf. succession in **Figure 24-3**). This type of forest is abundant in the areas occupied normally by Rocky Mountain Subalpine or Montane forests. Similar areas of lodgepole pines occur in the Pacific Montane and Subalpine forests. Photograph, courtesy of the United States Forest Service.

alpine forests, have a narrowly conical form, far different from the outlines of the deciduous trees of eastern North America.

During Paleozoic time the Appalachian chain was very long and high, but since Mesozoic the geological history of the eastern part of the Continent has been unspectacular, except during the glacial periods. The topography has been low, except for the moderately high ridges of the Appalachians.

Despite the low relief, there have been subtle trends of climate, beginning with the tropical environment of the eastern lowlands during Mesozoic and on into Eocene and Oligocene. During Miocene versions of the temperate forests appeared in extensive lowland areas in the East, as well as the West, then reduced to a peneplane, and the forests of both regions featured a combination of conifers and deciduous trees as dominants. In the East the temperate forests must have had forerunners in the Appalachian System during the first part of Tertiary time, and later the Miocene forests must have included both temperate elements from the mountains and modified tropical taxa from the lowlands. During Pleistocene the temperate forests must have moved north and south and up and down the mountainsides as the climate changed from cold to mild and the reverse during and between glacial times.

During and since Miocene the isolation of the temperate part of eastern North America has been effective—(1) on the *east* and *south* essentially complete isolation because of the Atlantic Ocean and the Gulf of Mexico; (2) on the *west* and *southwest* intermittently complete isolation, as now while migration across the Great Plains and Texas is blocked by dry areas, though these disappeared during the glacial periods; (3) on the *north* essentially complete isolation, except for the mostly temporary introduction of subarctic species during

Figure 25-16. Grove of subclimax quaking aspen, *Populus tremuloides,* near Panguitch Lake, Utah. Groves like this are characteristic of large moist areas particularly in the Rocky Mountain Forests. Engelmann spruce (right foreground) and other conifers ultimately come up among the aspens and take over the area. The coniferous forest is the climax. Similar areas of quaking aspens occur in the Pacific forests. Photograph, courtesy of the United States Forest Service.

the glaciations, followed by retention of some of them in favorable spots. Thus, at times, especially during the moist periods of Pleistocene, there has been invasion by new species or species groups, but in general during the latter part of Cenozoic time, even during Pleistocene, the temperate flora of eastern North America has developed chiefly from the plants present there in the Miocene forests.

The relationship of the temperate forests of North America and Eurasia is clear but not close. According to A. Löve (Taxon 16: 327. 1976) about 2,000 taxa in many families occurring in the temperate forests of North America and Eurasia now exist in "... all stages of differentiation, from that of genera and sections through species, subspecies, and varieties down to populations which still seem to remain practically identical in at least two of these vastly separate[d] areas." This indicates a floristic relationship of the temperate forests of the Northern Hemisphere occurring in the New and Old Worlds. Probably it reflects the existence of their forerunners in favorable areas on both land masses during the first part of the Tertiary Period when migration still was possible. During early Cenozoic the lowlands were tropical, and the temperate forests of North America and Eurasia were contracted in area, though they must have included complex series of forests on the mountains of the two continental masses. Gradually these forests became differentiated further with the passage of time and with complete separation geographically by further continental drift.

Since Miocene, during the long period of isolation of the plants occurring in temperate eastern North America, a remarkably rich forest flora has developed, and it includes an enormous number of endemic species. This reflects the antiquity of the flora and the adaptations of many plants to strong competition with others in a climate favorable to any of many temperate species. There have been adaptations to small differences in local habitat niches in each of which a different competitor may have a slight advantage over other taxa, as similarly in a tropical rainforest where there is a more or less universally favorable physical and chemical environment.

There are several floristic associations in the Eastern Forest Flora, but this flora covers an immense region with relatively minor relief and with no sharp boundaries of climatic areas. The only significant mountain system trends from southwest to northeast, parallel to the direction of both the prevailing westerlies and the polar winds. Thus, there are only localized uplifts of air over mountains and consequently only minor rain shadows beyond. From place to place the climate varies within relatively narrow limits, and the transitions between environments occur only gradually over long distances. This is reflected in the flora; there are distinctive units insofar as extreme development is concerned, but intergradations are mostly gradual to imperceptible. Consequently, classification into floristic associations is difficult.

THE DECIDUOUS FORESTS AS A GROUP

Occurrence. Lowlands of southeastern Canada chiefly near the St. Lawrence River and the Great Lakes; throughout the lowlands and the mountains of the Middle Western, Eastern, and Southern states east of the area of common occurrence of the Prairie.

Discussion. From the standpoint of ecology, the interpretation of deciduous forests employed here is loose. Strictly, a temperate forest formation has either coniferous or deciduous (dicotyledonous) dominants or the dominants are "mixed"—some coniferous and some deciduous. Ecologically this is important, because the dominants affect the lower stories of forest vegetation.

From a floristic point of view the importance of the ecological dominants is much less. The species in the forest are associated with each other not only with respect to forest ecology but also according to their origin and relationships, connecting them with a particular flora. For floristic study the whole combination of taxa, not just the dominants, is of primary concern. Even a "coniferous" forest is composed overwhelmingly of dicotyledons, and it includes many more species of even monocotyledons than of conifers. Even though they may be large and ecologically dominant, the present-day conifers are of only a few species. The flowering plants are legion, and regardless of the formation, floristically the gymnosperms are a drop in the bucket.

Some forests of eastern North America, especially the Cumberland Forest and the Laurentian Forest, include both coniferous and dicotyledonous dominants, a condition carried over from the Miocene forests. However, here the entire group of forests is referred to as the Deciduous Forests, because all the forests are interrelated so closely that no clear separation is possible. The deciduous trees are the characteristic ones for the entire area, and they form the forest canopy. The conifers, among them at some points, are not an ecologically dominant group as they are in the West, where deciduous trees have been eliminated because of a radically different climate developed since Miocene. In the West after Miocene there was much alteration of the species composition of the forests. Elimination of the broad-leaved

Figure 25-17. Deciduous Forest: *above,* in winter, Pisgah National Forest, North Carolina, the trees mostly sugar maple (*Acer saccharum*); *below,* in summer, near Holly Bluff, Mississippi, Delta hardwoods, the large trees, red gum (*Liquidambar styraciflua*). Photographs, courtesy of the United States Forest Service.

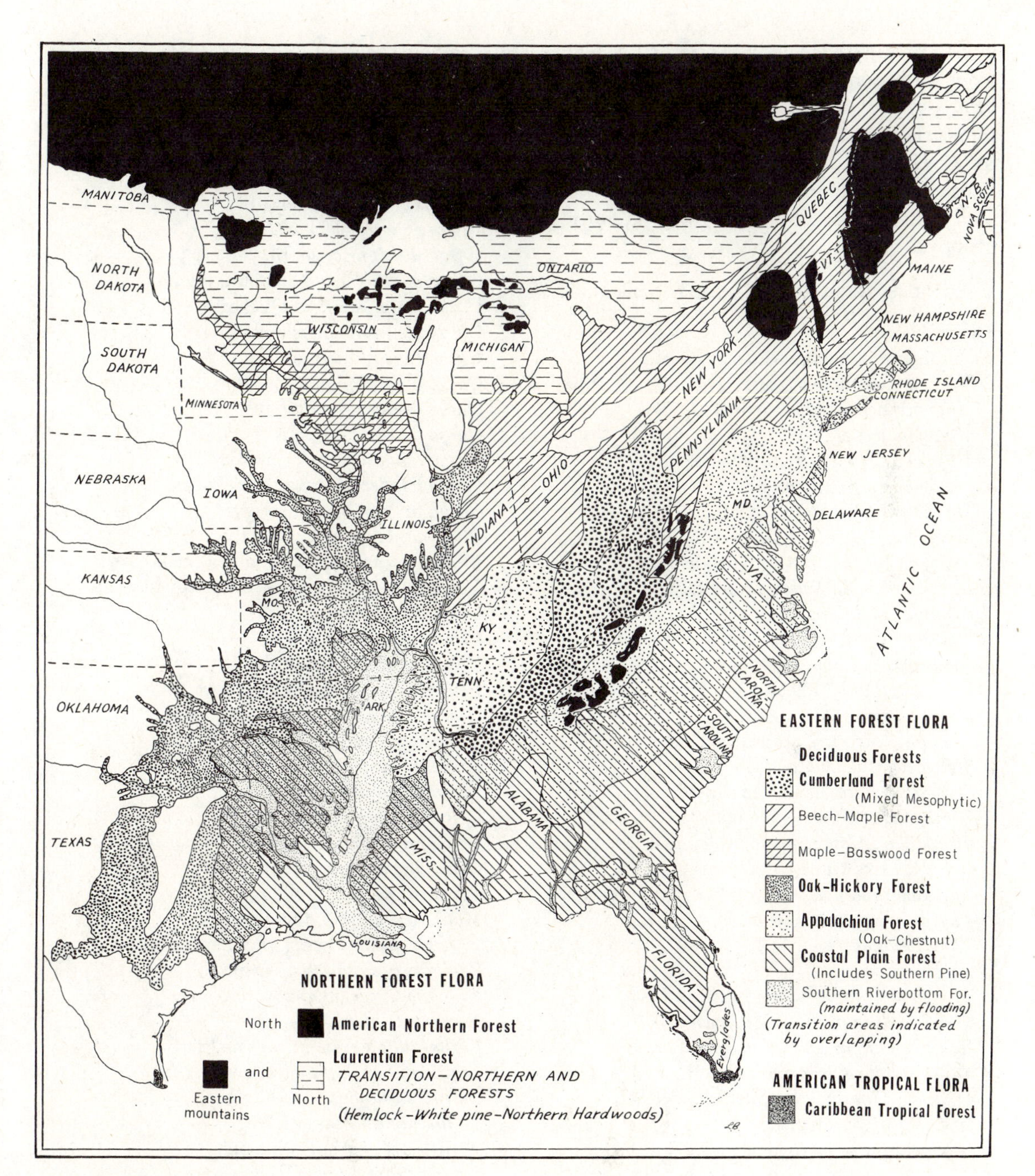

Distribution Map 5

trees forming the canopy and their replacement with conical conifers that in the drier forests do not form a canopy has had a profound effect on the plants growing beneath the trees. However, it is to be noted that in the moister coniferous forests, like the Pacific Lowland Forest, the coniferous trees do form a canopy.

SOME CHARACTERISTIC TREES AND LARGE SHRUBS

Thalamiflorae

Schisandra coccinea, wild sarsaparilla
Magnolia virginiana, laurel magnolia
Magnolia acuminata, cucumber-tree
Magnolia macrophylla
Magnolia tripetala, umbrella-tree
Magnolia grandiflora, southern magnolia
Magnolia pyramidata
Magnolia Fraseri, ear-leaved umbrella-tree
Calycanthus fertilis, Carolina allspice
Liriodendron Tulipifera, tulip-tree
Asimina triloba, pawpaw
Asimina parviflora, dwarf pawpaw
Sassafras albidum, sassafras
Persea Borbonia, red bay
Litsea aestivalis, pond-spice
Lindera Benzoin, spicebush
Stewartia Malachodendron, silky camellia
Stewartia ovata, mountain camellia
Gordonia lasianthus, loblolly bay
Franklinia Alatamaha, lost camelia (long extinct, except in horticulture)
Tilia heterophylla, basswood
Tilia americana, basswood
Tilia floridana
Tilia neglecta
Tilia heterophylla, white basswood
Tilia caroliniana, Carolina basswood
Xanthoxylum americanum, prickly-ash
Xanthoxylum Clava-Herculis, Hercules'-club
Cliftonia monophylla, buckwheat tree
Ilex, holly (many species)
Nemopanthus mucronata, catberry
Euonymus atropurpureus, burning-bush
Euonymus americanus, strawberry-bush
Staphylea trifolia, bladdernut
Aesculus octandra, buckeye
Aesculus (several other species)
Cotinus obovatus, American smoke tree
Cyrilla racemiflora
Acer saccharum, sugar maple
Acer rubrum, red maple
Acer nigrum, black maple
Acer barbatum, southern sugar maple
Acer (several other species)
Juglans cinerea, butternut
Juglans nigra, black walnut
Carya illinoensis, pecan
Carya cordiformis, butternut, pignut
Carya ovata, shagbark hickory
Carya laciniosa, big shellbark
Carya tomentosa, mockernut
Carya glabra, pignut
Carya ovalis, sweet pignut
Carya myristicaeformis
Carya aquatica, water hickory
Carya leiodermis, swamp hickory
Carya texana, black hickory
Rhamnus lanceolata, buckthorn
Rhamnus caroliniana, Carolina buckthorn
Ceanothus americanus, New Jersey tea
Ceanothus ovatus
Ulmus americana, elm
Ulmus rubra, slippery elm
Ulmus alata, winged elm
Ulmus Thomasii, cork elm
Ulmus serotina, September elm
Ulmus crassifolia, sugar elm
Planera aquatica, water elm
Celtis occidentalis, hackberry
Celtis tenuifolia, hackberry
Celtis laevigata, hackberry
Morus rubra, red mulberry
Maclura pomifera, Osage-orange
Prunus, plum, cherry (numerous species)
Platanus occidentalis, sycamore
Hamamelis virginiana, witch-hazel
Hamamelis vernalis
Fothergillia Gardenii, witch-alder
Liquidambar Styraciflua, sweet gum
Ostrya virginiana, hop hornbeam
Betula nigra, black birch
Alnus serrulata, smooth alder
Carpinus caroliniana, hornbeam
Fagus grandifolia, beech
Castanea dentata, chestnut
Castanea alnifolia, downy chinquapin
Quercus alba, white oak
Quercus rubra, red oak

Quercus prinoides, chinquapin oak
Quercus imbricaria, shingle oak
Quercus coccinea, scarlet oak
Quercus Phellos, willow oak
Quercus Prinus, chestnut oak
Quercus velutina, black oak
Quercus stellata, post oak
Quercus macrocarpa, bur oak
Quercus lyrata, overcup oak
Quercus Muehlenbergii, chinquapin oak
Quercus Margaretta, sand post oak
Quercus Drummondii
Quercus sinuata
Quercus virginiana, eastern live oak
Quercus hemisphaerica, laurel oak
Quercus laurifolia, swamp laurel oak
Quercus nigra, water oak
Quercus Shumardii, Shumard red oak
Quercus falcata, southern red oak
Quercus marilandica, blackjack oak
Quercus cinerea, bluejack oak

Corolliflorae
Bumelia lyciodes, southern buckthorn
Bumelia Smallii
Halesia diptera, snowdrop tree
Diospyros virginiana, persimmon
Symplocos tinctoria, horse-sugar
Halesia carolina, silverbell-tree
Styrax grandifolia
Styrax americana
Clethra alnifolia, white alder
Clethra acuminata
Rhododendron, rhododendron and azalea (numerous species)
Kalmia latifolia, mountain laurel
Pieris floribunda, fetterbush
Oxydendrum arboreum, titi
Vaccinium arboreum, sparkleberry
Vaccinium, blueberry (numerous species)
Fraxinus americana, white ash
Fraxinus tomentosa, red ash
Fraxinus pensylvanica, red ash
Fraxinus caroliniana, Carolina ash
Fraxinus quadrangulata, blue ash
Forestiera acuminata, swamp privet
Forestiera ligustrina
Chionanthus virginiana, old man's beard
Osmanthus americana, wild olive
Callicarpa americana, French mulberry
Catalpa speciosa, catawba tree

Calyciflorae
Gymnocladus dioica, Kentucky coffee-tree
Gleditschia triacanthos, honey-locust
Gleditschia aquatica, water-locust
Cercis canadensis, redbud
Robinia Pseudo-Acacia, black-locust
Physocarpus opulifolius, ninebark
Pyrus angustifolia, crab-apple
Pyrus coronaria, crab-apple
Pyrus ioensis, crab-apple
Pyrus arbutifolia, floribunda, and *melanocarpa,* chokeberry
Pyrus (or *Sorbus*) *americana* and *decora,* mountain-ash
Amelanchier, serviceberry (numerous species)
Crataegus, hawthorn (numerous species)
Philadelphius inodorus, hirsutus, and *pubescens,* mock-orange
Hyrangea arborescens, hydrangea
Itea virginiana, Virginia willow
Ribes, currant, gooseberry (numerous species)
Cornus florida, flowering dogwood
Cornus, dogwood (several species)
Nyssa sylvatica, black gum
Nyssa aquatica, Tupelo
Aralia spinosa, Angelica tree, Hercules' club

Ovariflorae
Viburnum (numerous species)
Sambucus canadensis, elderberry

As the list indicates, the flora is by far the richest in deciduous woody species. It has also a wealth of herbaceous species far exceeding any other North American flora.

Discussion. The forests are rich in species at their center in the geologically old Cumberland Plateau and the southern Allegheny Mountains. In every direction the central forest gives way gradually to a special selection of its component species. To the northward in the areas denuded by the great Pleistocene ice sheets beech and sugar maple become more and more important in the association. Sugar maple and basswood dominate a large (driftless) area untouched by the ice. In the vicinity of the Great Lakes and at higher levels in the Appalachian System where the forest

Figure 25-18. Deciduous Forests: stand of virgin hardwoods, Spring Mill State Park, Indiana, the large trees mostly white oak (*Quercus alba*). Photograph, courtesy of the United States Forest Service.

has both dicotyledons and conifers as dominants, as is true in the Cumberland Forest (see below). On the drier margins to the west and south the deciduous forest becomes an association dominated primarily by oaks and hickory. Until disturbance by man the uplands east and southeast of the Cumberland-Allegheny area were dominated by forests featuring oaks and chestnuts, but in recent years blight has almost eliminated the chestnut. Another forest dominates the Atlantic and Gulf coastal plains.

According to Braun, the Deciduous Forests include the subdivisions mentioned above, and the following discussion is adapted partly from her work. As indicated in the general discussion of the Eastern Forest Flora, because the area is without sharp climatic divisions and is of low topography, segregation of the intergrading phases is difficult and inconsistent. However, in its extreme form each forest is distinctive, and it is characterized by a wealth of woody and herbaceous endemic species.

Although the following tentative floristic associations represent only the beginning of classification of the Eastern Forest Flora, they do provide a starting point. Ecological classifications abound, but they represent different approaches and emphases on different problems than those of primary concern here. Construction of a better floristic classification for the region is a project for many years of research.

1. THE CUMBERLAND FOREST

Occurrence. Southeastern Ohio and western Pennsylvania to West Virginia, the western tip of Maryland, eastern Kentucky, eastern middle Tennessee, and northern Alabama; best developed on the Cumberland and Allegheny plateaus and on the mountains of the same names. The Cumberland Forest is almost absent from the ridges and valleys of the eastern part of the Appalachian System.

SOME CHARACTERISTIC WOODY SPECIES

Thalamiflorae

Liriodendron Tulipifera, tulip-tree
Tilia neglecta, basswood
Tilia heterophylla, white basswood
Tilia floridana, basswood
Acer saccharum vars. *saccharum, nigrum,* and *Rugelii,* sugar maples, some extending beyond this forest
Aesculus octandra, sweet buckeye
Fagus grandifolia, at lower elevations, the white beech and, at higher elevations, the red beech of Camp (p. 780).
Castanea dentata, chestnut
Quercus alba, white oak
Quercus rubra, red oak

Gymnosperms

Tsuga canadensis, hemlock. (The species is widespread in occurrence, but it has been a dominant member of this floristic association and its predecessor probably since early Tertiary.)

Discussion. According to Braun, the Cumberland Forest, or Mixed Mesophytic Forest, is the richest deciduous forest occurring in eastern North America; it is composed of more species than any other. It is the heart of the Eastern Forest Flora and the closest living approach to the widespread Mixed Tertiary Forest, members of which persist mostly in the Cumberland-Allegheny region. The floristic association has had greater stability than its neighbors, many of the taxa having been together since Miocene or before. Temperate habitats probably have existed at least in the mountains of the region during most or all of Cenozoic time and perhaps longer. This forest was not in the path of the Pleistocene ice, and therefore it was not subject to the catastrophic change that affected the other eastern forests.

As indicated in the discussion of the Deciduous Forests as a group, the Cumberland Forest is related closely to all the other floristic associations composing the Eastern Forest Flora, and it intergrades with all of them. In the other floristic associations there are surviving pockets of Cumberland Forest or modifications of it. Within the Cumberland Forest there are numerous subclimaxes resembling the other Deciduous Forests, and succession in these areas goes ultimately to Cumberland Forest as a climax. Outward in every direction the component species of the central Cumberland Forest thin out in accordance with local emphasis on various environmental factors and with the recent events of geological history.

The interior low plateau area from the Cumberland Forest westward to the bluffs of the Mississippi River commonly is distinguished as a western phase of this forest. It is a transitional forest including a mosaic of minor locally segregated floristic types tending to resemble the Oak-Hickory Forest, as well as the Cumberland.

2. THE LAURENTIAN FOREST

Occurrence. Primarily in the mountains and on the direct watershed of the St. Lawrence River, from which the name is derived, and including the northern part of the Great Lakes region; Canada from southeastern Manitoba (in "islands") to Nova Scotia and southward to central and northeastern Minnesota, northern Wisconsin, and northern Michigan; in scattering highland areas along the Appalachian chain at higher elevations, ranging up to the summits of the southern mountains.

CHARACTERISTIC WOODY SPECIES

Thalamiflorae

Tilia americana, northern basswood
Acer spicatum, mountain maple
Acer pensylvanicum, striped maple
Acer saccharum var. *saccharum,* common sugar maple
Betula lutea, yellow birch
Fagus grandifolia, beech (in the local beech-maple communities, the red beech, or, near the American Northern Forest and along the higher parts of the Appalachian System, the gray beech of Camp [see p. 780]).

Figure 25-19. The Laurentian forest of the Great Lakes, St. Lawrence River region, and higher Appalachians: *above,* white pine (*Pinus Strobus*) and hemlock (either *Tsuga canadensis* or *Tsuga caroliniana*), Laurels Recreation Area, Cherokee National Forest, Tennessee; *below,* Laurentian forest in the Great Lakes region, white pine (*Pinus Strobus*) and red pine (*Pinus resinosa*), understory of young deciduous trees which may replace many of the conifers, Norway Beach, Chippewa National Forest, Minnesota. Photographs, courtesy of the United States Forest Service.

Corolliflorae

Fraxinus americana, white ash

Gymnosperms

Thuja occidentalis, white cedar or arbor vitae (most commonly of the American Northern Forest)

Abies balsamea, balsam fir (most commonly of the American Northern Forest)

Tsuga canadensis, hemlock

Tsuga caroliniana, Carolina hemlock (southern Appalachians)

Larix laricina, larch or tamarack (also in the American Northern Forest)

Pinus Banksiana, Jack pine

Pinus resinosa, red pine

Pinus pungens, table mountain pine

Pinus Strobus, white pine

Picea mariana, black spruce (relict in bogs; mostly in the American Northern Forest)

Picea rubens, red spruce

Picea glauca, white spruce (mostly of the American Northern Forest)

Discussion. In ecology this forest is known by the elongate name, Hemlock-White Pine-Northern Hardwoods Association, based on dominance of some coniferous and some dicotyledonous trees. Except in the Appalachian System, it occurs on glaciated areas of the north, from whence the first claimant after the glaciations, the American Northern Forest, has moved still farther northward. It intergrades with the American Northern Forest, and with the strictly deciduous forests to the south or at lower altitudes in the mountains.

The northern lowland areas in which the Laurentian Forest occurs have a milder climate than those just northward now occupied by the American Northern Forest. The territory was released from the ice sooner than that farther north, and during repopulation and occupation first by the Northern Forest and then by the Laurentian there has been time to build up a deeper, better forest soil. Even though it is not yet climax, the Laurentian Forest is nearer climax than the Northern Forest, and it includes more species. Along the southern edge, the present Laurentian Forest is in competition with advancing species from the south still moving northward into the glaciated areas. During Pleistocene the forest was invaded by some northern species, and some of these linger within it, especially in such local habitats as those about cold bogs. In the middle and southern Appalachian System this forest was not destroyed or pushed out by glaciation, but it descended to lower altitudes. Probably since Pleistocene some readjustments similar to those in the glaciated north have been necessary.

THE MAPLE-BASSWOOD FOREST

Occurrence. Central and southeastern Minnesota, southern Wisconsin, northeastern Iowa, and northwestern Illinois; on the patchwork of "driftless areas," unglaciated during Pleistocene; also on the adjacent glaciated areas into which some of the species have migrated since Pleistocene.

CHARACTERISTIC WOODY SPECIES

Thalamiflorae

Tilia americana, basswood

Acer saccharum, sugar maple

Quercus rubra, red oak

Quercus rubra var. *borealis,* northern red oak

Discussion. Probably the Maple-Basswood Forest is a depleted remnant of the more widely distributed forerunner of the Cumberland Forest of preglacial times. During Pleistocene it persisted in nonglaciated "islands" near the southern limit of the last glaciation. In some ways it is similar to the Laurentian Forest but the beech and some other species are missing. The islands of Maple-Basswood Forest surviving the glaciers have served as centers from which the forest has spread to surrounding areas, including recently parts of the Prairie formerly sustained through annual burning by the Indians but replaced by forest when the fires ceased. The forest is a group of subclimaxes, and they do not constitute a floristic association.

THE BEECH-MAPLE FOREST

Occurrence. Across the Middle Western States from the southern Great Lakes to Indiana and Ohio and barely into southern Ontario. North of the southern limits of the last (Wisconsin) glaciation; the area of occurrence ice free for only about 10,000 years.

CHARACTERISTIC WOODY SPECIES

Thalamiflorae

Acer saccharum, sugar maple
Fagus grandifolia, beech

Discussion. After disappearance of the Pleistocene glaciers, the northward movement of the Deciduous Forests was rapid. Beech-Maple Forest developed on the glacial sand and gravel in the areas bared first by the melting of the ice sheets, but this was as they were deserted by the northward-moving American Northern Forest, which occupied them first.

The Beech-Maple Forest consists of subclimax stages similar to some in disturbed areas within the Cumberland Forest just to the south, and it must be regarded as a temporary association of species. There are only two important dominants, the beech and the common sugar maple, both occurring abundantly in various floristic associations, rather than being endemic or nearly so to this forest.

The beech, *Fagus grandifolia,* has been taken to be a major indicator of ecological formations, but the species as a whole does not really indicate floristic associations. Camp (personal discussions and in Braun) detected presence of three strains, or perhaps varieties, of the species in eastern North America, and these tend to grow as follows:

Gray beech. In more severe climates; lower levels from the Great Lakes to Nova Scotia and at higher elevations southward along the Appalachian System to eastern Tennessee and North Carolina. Primarily in the Laurentian Forest.

Red beech. In medium climates; Appalachian System at middle altitudes and intergrading with the white beech in the Beech-Maple Forest.

White beech. In general, where the climate is milder; low altitudes west of the Appalachians, on the Coastal Plain, and in the Piedmont region.

Thus, the floristic significance of the beech is not at the level of the species but at an infraspecific level, and the term "beech forest" means little. In the Beech-Maple Forest, the red and white beeches often occur together, and they intergrade.

The Beech-Maple Forest cannot be considered a floristic association; it is a subclimax appearing in the glaciated area being repopulated primarily from the Cumberland Forest just to the south. It is an impoverished forest with little floristic character of its own.

3. THE OAK-HICKORY FOREST

Occurrence. Unglaciated areas: centering in the Ozark region of Missouri, Arkansas, and eastern Oklahoma, but occurring northward to the eastern parts of southwestern Michigan and eastward to the Piedmont of the Appalachian System and thence northeastward from Alabama to New Jersey.

CHARACTERISTIC WOODY SPECIES

Thalamiflorae

Quercus stellata, post oak
Quercus lyrata, overcup oak
Quercus macrocarpa, mossy-cup oak
Quercus Shumardii, Shumard red oak
Quercus ellipsoidalis, Jack oak
Quercus marilandica, blackjack or Jack oak
Quercus nigra, water oak
Quercus alba, white oak, also occurring in other forests
Carya illinoensis, pecan
Carya aquatica, water hickory or bitter pecan
Carya cordiformis, pignut
Carya ovata, shagbark hickory
Carya texana, black hickory

Discussion. Most of the Oak-Hickory Forest is a response to the drier climate intermediate in position and character between that of the Great Plains in the middle of the Continent and the moister areas of the eastern one-third. It occurs also in the drier climate of the faint rain shadow of the Piedmont region on the eastern side of the Appalachian System. Inasmuch as there is no sharp boundary between the climatic areas, there is none between the Oak-Hickory Forest and other deciduous forests. Intergradation is the rule, and islands of one forest occur within the area of the other. Within the Cumberland Forest, some subclimaxes duplicate the Oak-Hickory Forest.

The trend from the Great Plains eastward is toward moister climate. To the eastward the dry glassland gives way gradually to drier, then moister, forest. On the plains the grasses are dominant, but they are accompanied by wavy bands of lush forest along the watercourses. A little farther east the areas between streams have patches of post oak, *Quercus stellata,* the first outpost of the Oak-Hickory Forest, and beyond these the forest fills in most of the areas between streams with a relatively dry version of the Eastern Deciduous Forests. This transition occurs from the Dakotas, where the forest forms a thin marginal band, to Texas. Although, where there is no disturbance, the trees may form a canopy, much of the forest is dry and open.

In the Piedmont region, the following woody plants are added to the list of those characteristic of most of the Oak-Hickory Forest, its features to some extent being altered:

Corolliflorae
- *Oxydendrum arboreum,* sourwood
- *Liquidambar styraciflua,* liquidambar or sweet gum
- *Fagus grandifolia,* beech (the white beech of Camp, p. 780)
- *Carya myristiciformis*

Gymnosperms
- *Pinus Taeda,* loblolly pine
- *Pinus australis,* longleaf pine
- *Pinus virginiana,* scrub pine
- *Pinus echinata,* eastern yellow pine

4. THE APPALACHIAN FOREST

Occurrence. Lowlands from southeastern New York to southern New England, New Jersey, and westcentral Maryland; in the Appalachian System southward to southeastern Tennessee, northeastern Georgia, and northwesternmost South Carolina. Mostly on the ridges of the valley-and-ridge system of the Appalachian ranges, where the lower valleys between the many parallel ridges support a version of the Oak-Hickory Forest, in which the white oak, *Quercus alba,* is prominent.

CHARACTERISTIC WOODY SPECIES

Thalamiflorae
- *Liriodendron Tulipifera,* tulip-tree
- *Lindera Benzoin,* spicebush
- *Stewartia ovata,* mountain camellia
- *Ilex montana,* mountain holly
- *Parthenocissus quinquefolia,* Virginia creeper (a wide-ranging species)
- *Hamamelis virginiana,* witch-hazel
- *Castanea dentata,* chestnut
- *Quercus alba,* white oak
- *Quercus Prinus,* chestnut oak
- *Quercus rubra,* red oak
- *Quercus coccinea,* scarlet oak
- *Quercus velutina,* black oak

Calyciflorae
- *Cornus alternifolia,* pagoda dogwood

Discussion. This forest, known in ecology as Oak-Chestnut Forest, occurs where summer drought is more frequent than in most of the eastern part of the Continent. Its character has changed greatly during the Twentieth Century, because essentially the chestnut blight has eliminated the chestnut. This tree was one of the primary parts of the forest canopy, and its disappearance affected the lower stories of vegetation. The entire region of the forest is in various stages of succession leading ultimately to a new and undetermined climax.

The equilibrium has been changed, and it cannot be restored. The remaining hulks of dead chestnuts, now few, are decaying away, and other trees are replacing this beautiful tree of former times.

5. THE COASTAL PLAIN FOREST

Occurrence. Adjacent parts of easternmost Texas and western Louisiana and east of the Mississippi River from the eastern edge of Louisiana and southern Mississippi along the Gulf and Atlantic coastal plains to New Jersey and central Florida.

CHARACTERISTIC WOODY SPECIES
OF DRIER INLAND AREAS

Thalamiflorae
- *Liquidambar styraciflua,* liquidambar or sweet gum
- *Quercus laevis,* turkey oak
- *Quercus alba,* white oak
- *Quercus cinerea,* bluejack oak
- *Quercus marilandica,* blackjack oak
- *Quercus laurifolia*
- *Quercus falcata*
- *Fagus grandifolia,* beech (the white beech of Camp)
- *Carya glabra,* pignut
- *Carya tomentosa,* mockernut

Calyciflorae
- *Nyssa sylvatica,* black gum

Gymnosperms
- *Pinus australis,* longleaf pine
- *Pinus Taeda,* loblolly pine
- *Pinus echinata,* eastern yellow pine
- *Pinus virginiana,* scrub pine (more wide-ranging)

Discussion. The dominant trees of the inland parts of the Coastal Plain Forest are mostly oaks, including the wide-ranging white oak, *Quercus alba;* the blackjack oak, *Quercus marilandica;* the turkey oak, *Quercus laevis;* the bluejack oak, *Quercus cinerea; Quercus laurifolia;* and *Quercus falcata.* The ecological dominants also include the sweet gum, *Liquidambar styraciflua;* several hickories, *Carya glabra* and *Carya tomentosa* being most characteristic; the beech, *Fagus grandifolia* (the white beech of Camp); and the black gum, *Nyssa sylvatica.* The areas near the coast have more moisture and deeper soil, and they support a wider variety of trees, shrubs, and other plants, especially in Florida. Typical trees include, in addition to some of those mentioned above, the eastern live oak, *Quercus virginiana;* the water oak, *Quercus nigra;* and the southern magnolia, *Magnolia grandiflora.* Festoons of Spanish moss, *Tillandsia* (Bromeliaceae) occur on various trees.

The climax Coastal Plain Forest is dominated by deciduous trees, especially oaks, but pines are common in the floristic association. Unfortunately, there has been disturbance, particularly by fire and flooding, and much of the area of occurrence is occupied by subclimaxes.

The drier inland parts of the forest are dominated by oaks or oaks and hickories, but after disturbance, at first the pines dominate. Eventually the young oaks rise above the pines and shade them out. The inland areas are dry, and the soil is poor, as indicated by the pitiful cotton fields growing where the ground has been cleared, but the areas along the coast have more water and richer soil.

ADDITIONAL CHARACTERISTIC TREES
OF MOISTER COASTAL AREAS

Thalamiflorae
- *Magnolia grandiflora,* southern magnolia
- *Ulmus,* elms
- *Celtis,* hackberries
- *Quercus nigra,* water oak
- *Quercus virginiana,* eastern live oak

Toward the coast the additional trees listed above enter the floristic association. The coastal areas are both richer in species and more lush, the trees often being festooned with Spanish moss, in greater quantity.

Figure 25-20. Deciduous Forests; Appalachian or Oak-Chestnut Forest: Chestnuts on Cold Mountain near Natural Bridge, Virginia, in 1925. Photograph, courtesy of the United States Forest Service. A blight brought from the Old World has all but eliminated the chestnut.

THE SOUTHERN PINE FOREST

Subclimax pine forest is abundant from Louisiana and adjacent Texas to the Carolinas and central Florida and in lesser areas northward along the Atlantic Coast to Long Island. Various species of pine, including pitch pines in the New Jersey pine barrens and longleaf and loblolly pines in the Southern States, are subclimax dominants of the forest, but there is evidence that they would be replaced by deciduous trees such as oak and hickory if the area were protected from fire. Thus, the "piney woods" of the South are ephemeral.

Some characteristic species include the following.

Gymnosperms

Pinus australis, longleaf pine
Pinus Taeda, loblolly pine
Pinus echinata, shortleaf pine
Pinus virginiana, scrub pine
Pinus serotina, pond pine
Pinus palustris, longleaf pine

THE SOUTHERN SWAMP FOREST

In the numerous marsh areas in the Southern States the outstanding tree is the swamp cypress or bald cypress (*Taxodium distichum*) of river swamps. Other characteristic species are associated with it, e.g., the cotton gum, *Nyssa aquatica; Nyssa sylvatica* var. *biflora;* the pond cypress, *Taxodium ascendens,* of

Figure 25-21. Special forest types, Coastal Plain Forest area—I; Southern Pine Forest: *above,* virgin stand of longleaf pine (*Pinus australis*); Crosby Lumber Company property, Mississippi; *below,* young longleaf pine forest with a nearly pure stand of a grass, slender bluestem (*Andropogon tener*), which burns almost every year. The Southern Pine Forest is not a climax type, and it is maintained by fire. Probably the area would be occupied by Deciduous Forest if fire were not a factor. Photographs, courtesy of the United States Forest Service.

Figure 25-22. Special forest types, Coastal Plain Forest area—II; Southern Riverbottom Forest: *above,* cypress swamp with bald cypress (*Taxodium distichum*), showing the swollen bases of the trees, the water completely covered with water-hyacinth (*Eichornia crassipes*); near Round Lake, St. Martin Parish, Louisiana; *below,* forest along a creek, the water level low, revealing the cypress knees, through which air reaches the roots when the water level is higher; Wadboo Creek, Francis Marion National Forest, South Carolina. The Southern Riverbottom Forest is dependent upon persistence of large flooded areas. Photographs, courtesy of the United States Forest Service.

small ponds or swamps; and the pond pine, *Pinus serotina.* This apparently distinctive floristic type is maintained as a result of flooding, a local condition perhaps likely to disappear in the course of time.

The Prairie Flora

(Temperate Zone: Midcontinental Temperate Region)

East of the Rocky Mountain System from Alberta to New Mexico and Texas; westward into the high plains of northern and eastern Arizona; eastward to parts of Wisconsin, Ohio, and northernmost Tennessee; occurring in patches along the Gulf Coast and the Atlantic Coast, appearing there as far north as Montauk Point, Long Island.

The original nature of grasslands is difficult to ascertain, because this vegetation formation lends itself readily to *plowing* and immediate destruction of the native flora and fauna or to *grazing* and in most areas overgrazing and ultimate destruction or great alteration of the normal ecosystem. In agricultural areas floristic study must be based on a few small remnants on steep or otherwise poor land for crops, in old cemeteries, or along road or railroad rights of way. Rarely a plot of ground has been preserved by the owners, by a research or teaching institution, or by a government agency. In grazing country preservation is better, but areas free from overgrazing are difficult to find and to evaluate for degrees of modification from the original flora.

The present grasslands have been formed since Pleistocene, but some prairies may have been present before or between the Pleistocene pluvial periods. Evidence is lacking, but, ever since Miocene, if a local climate dictated such a formation, some kind of at least limited grassland may have developed, possibly with the aid of fires set by lightning. During glacial times, much, if not all, of the Great Plains region was forested, but probably some of the forest was open and grassy. The grasslands, if they existed, must have been forced southward and probably contracted and altered and perhaps eliminated.

According to Wells, the presence of prairie and the absence of either forest or of a mosaic of forest, woodland, and grassland on the Great Plains is due to the presence of a relatively dry climate, borderline for forest, and to the raging Indian-set prairie fires. All the grasslands of the United States were maintained in that formation by the annual fires the Indians started to drive out or to roast game on the hoof or to improve hunting in future years by killing the seedlings of trees and shrubs that otherwise would turn the grassland into forest or woodland. Large-scale burning by the Indians in Texas was reported by Cabeza de Vaca in 1528 and observed on the northern Great Plains by Lewis and Clark in 1804 to 1806. Ruhe reports organic carbon in the fossil remnants of grasslands of 1,000 years ago. Since the coming of the white man, fire has been reduced or eliminated, and much grassland has been replaced as woody plants have taken over. Thus, probably the Prairie Flora represents a system of subclimaxes.

According to Wells, even during late Pleistocene, forests in borderline dry habitats on the plains must have been killed by the great fires set by the Indians. As the areas of a drying climate became barely able to support the forests earlier developed within them, they became more vulnerable to fire. Thus, the recent origin of the Prairie Flora on sites burned over frequently is plausible, particularly so in view of the present persistence of patches of dryland forest in the prairie region downwind from bodies of water or on rocky slopes about the bluffs along rivers, or wherever else there is a natural barrier to fire.

Doubtless species from other floras have invaded the grasslands since whenever the prairies came into being. The Great Plains have no physiographic or other protection, and the only limitation to entry of plants is that they must be able to survive the climate and

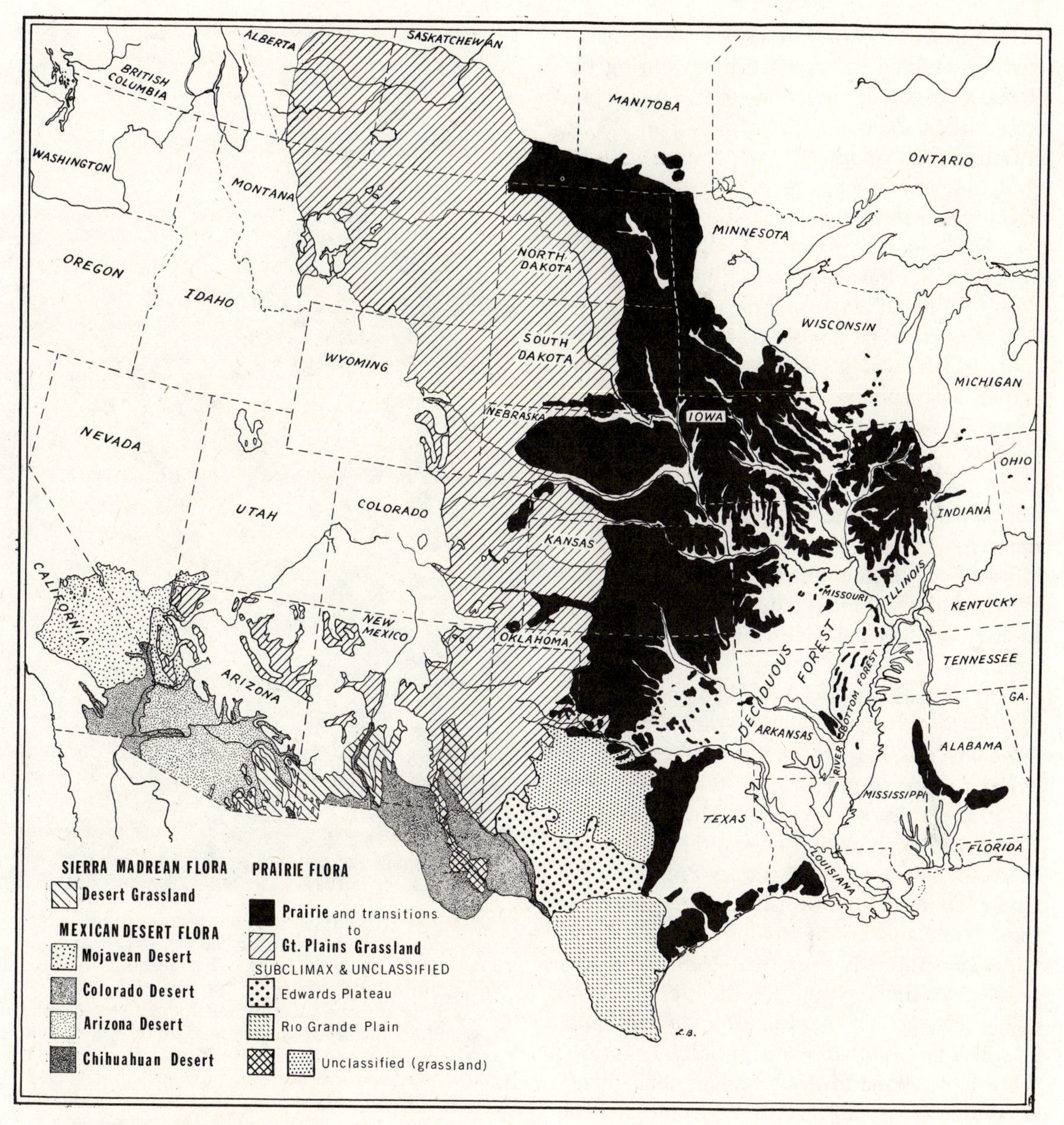

Distribution Map 6

to compete with other species so adapted. Thus, migration may occur at any time, and reverse movement is restricted by only the necessity for the ability to cope with other climates and competitors. In the prairies this is reflected not only in the paucity of endemic taxa of both plants and animals but also in the occurrence of all of even the dominant species of the grasslands in one to several other floristic associations in other floras. The only marker of the Prairie Flora is the unique local combination of taxa and the absence of trees.

The present Prairie Flora shows evidence of hasty assemblage or reconstitution during the 10,000 years since the ice melted. The species composition indicates this rather than a long period of development in company with gradual climatic change or a great time of stability during which characteristic endemic taxa might have evolved and become adapted to various habitat niches in a shielded region. Thus, the assemblage of prairie species must be regarded as composing only an incipient flora, which, had it been given enough time before disturbance, might have developed more of its own species and other taxa.

Although the Prairie Flora has few features of its own, it does not fit clearly as a subdivision of another flora. It shows affinity to the Eastern Forest Flora, the Western Forest Flora (Rocky Mountain Montane Forest), the Sierra Madrean Flora, and the Mexican Desert Flora, but it is not clearly a part of any of them.

The floristic associations composing the Prairie Flora are difficult to define, because there are no sharp or natural internal boundaries of topography or climate. The true or tall grass Prairie occurs in the Middle West in an area of several times the precipitation in that occupied by the Shortgrass Prairie on the Great Plains. The transition area between these grasslands is up to hundreds of miles wide, and there is a gradual trend from tall grasses to short ones. Division of such an assemblage of local grasslands is difficult, and, as between montane and subalpine forests, there is no place to draw a sharp line of demarcation. An alternative is to divide the grasslands into three or more divisions, designating the intermediate areas as floristic associations. However, little is to be gained by this, and confusion is promoted. If only two divisions are recognized, even though the boundary between them is vague, each has a relatively sharp line of demarcation with the forest on one side. However, if a third division is made in the middle, it has no clear boundary anywhere, and it is much more difficult to differentiate. Ecologists have attempted to define a formation of "midgrasses" between the tall and short grasses. This is useful for their purposes and for range management, but the degree of floristic significance has not been worked out.

Figure 25-23. Outpost of the western yellow pine, *Pinus ponderosa,* on the Great Plains in Nebraska, surrounded by Shortgrass Prairie. The seedlings of trees and shrubs are vulnerable to fire, including those set during thousands of years by the Indians to keep the grassland dominant for grazing land and to cook game animals as they ran before the fire. Trees, like the yellow pine, and shrubs have survived only in places protected from intense fire, like those shown here growing on rock outcrops where the grass is sparse. South of Chadron.

1. THE PRAIRIE

(The true or tall grass prairie)

Occurrence. Small remnants in patches through the Middle West from the eastern edges of the Dakotas to Wisconsin, parts of Ohio, northcentral Texas, and the northern edge of Tennessee; in modified form here and there on the Gulf and Atlantic Coastal plains.

CHARACTERISTIC SPECIES

Monocotyledoneae (numerous grasses, including the following ecological dominants, all occurring also far beyond the Prairies)

Andropogon Gerardii, big bluestem (up to 10 feet tall) (Saskatchewan to Quebec, northern Mexico, and Florida)
Andropogon scoparius, little bluestem (Missouri to Massachusetts, Texas, and Florida)
Stipa spartea, porcupine grass (British Columbia to Ontario, New Mexico, and Indiana)
Sorghastrum nutans, Indian grass or wood grass (Manitoba to Quebec, Florida, and Mexico)

Discussion. The (tallgrass) Prairie once covered a large portion of the Middle West west of the Deciduous Forests and patches eastward to Ohio and northern Tennessee and

Figure 25-24. Tallgrass (or true) Prairie near Lake Minnetonka, Minnesota, the area being invaded by young oak trees, the seedlings of which are no longer killed by Indian-set fires.

the coastal plains. However, most of it was plowed up because this was easy, and the soil it occupied was rich, well-watered, and perfect for cornfields. The Prairie was not dry, like the Shortgrass Prairie to the westward, where snow and rainfall are much less. The grass on the Prairie was unbelievably high, up to 10 feet, and farmers rushed to the area. Since the cessation of set fires, most of the remaining Prairie has reverted to forest. Now the remnants are few. An outstanding example of the tall grasses is adjacent to Brownfield Woods of the University of Illinois at Urbana.

2. THE SHORTGRASS PRAIRIE

Occurrence. Typically on the plains east of the Rocky Mountain System; 2,000 to 5,000 feet. Shading off imperceptibly and irregularly into the Prairie occurring in the moist Mississippi Valley; Alberta to Manitoba and the Dakotas and southward to New Mexico and Texas; westward to the higher plains of northern and eastern Arizona.

CHARACTERISTIC SPECIES

A large number of grasses and other herbs, nearly all occurring also in other floristic associations; the woody plants restricted to bottomlands or spreading out onto the plains if they are not checked by fire. The plains yucca, *Yucca glauca*, forms protruding mounds of leaves among the grasses, and in season its elongate panicles of flowers or fruits rise high above the leaves.

Discussion. The Great Plains have both spring and summer rainfall, but the total is not great. During the rainy season, there are miles of green grass and herbs, but during the rest of the year the grass is dry and gray or brown. Flowering of various species is during the spring or summer or both, according to genetic makeup.

The Great Plains were the hunting grounds of a number of tribes of Indians, and once the rolling country was covered by great herds of buffalo, or bison. However, the hills and plains have been plowed up for dry farming, i.e., growing of wheat or other cereals without irrigation. Other farming was started, especially to the eastward, where there were more moisture and better soil. However, vast areas have remained as range land for livestock, especially cattle and sheep. On the ranges the poorer land of the hills and bluffs has had less of the white man's devastation than has the Prairie in the Middle West, and much larger, though modified, remnants of the original vegetation are available. The native plants have not been replaced to as great an extent by weeds from other continents.

The native plants of the Shortgrass Prairie are drawn from every direction, but the strongest relationships are with those of the Deciduous Forests, the Rocky Mountain Montane Forest, some elements of the Sierra Madrean Flora, and the Mexican Desert Flora. The dry habitat is suitable for some plants from each of these sources.

THE EDWARDS PLATEAU

Occurrence. The central Plateau of Texas, including the hills along its escarpment.

Discussion. This is an extension of the Shortgrass Prairie, but since cessation of Indian fires it has been invaded by trees and shrubs, and it appears as woodland and chaparral. The trees include shinnery oak, *Quercus Havardii;* eastern live oak, *Quercus virginiana;* junipers (known as "cedars"), *Juniperus Pinchottii* and *Ashei;* and honey mesquites, *Prosopis juliflora* var. *glandulosa.* These species have become dominants.

THE RIO GRANDE PLAIN

Occurrence. Southern Texas from Val Verde and Webb counties to the Gulf of Mexico near Corpus Christi and the south to the Rio Grande; thence southward into adjacent Mexico; low elevations, mostly under 700 feet; within the Horse Latitudes.

Figure 25-25. Shortgrass Prairie: *above,* on rolling hills near the Bearlodge Mountains, northeastern Wyoming; *below,* plains covered primarily with blue grama (*Bouteloua gracilis,* grass tribe Chlorideae). Photographs, courtesy of (*above*) United States Forest Service and (*below*) United States Soil Conservation Service. Department of Agriculture.

Discussion. Originally this was a grassland or perhaps with scattered trees. The area has been overgrazed for more than two centuries, and set fires have been reduced or essentially eliminated. Because of these changes, woody plants have invaded the grassland or have become more prominent in it. Common dominant trees include most commonly the honey mesquite, *Prosopis juliflora* var. *glandulosa*, and in other places the post oak, *Quercus stellata*, and the eastern live oak, *Quercus virginiana*. The Rio Grande Plain in Texas and adjacent Mexico includes 14 taxa of endemic cacti, an indication that it may be a floristic association in its own right. Study is needed.

The Sierra Madrean Flora

(Temperate Zone: Southwestern Temperate Region)

Occurrence. At mostly low or moderate elevations on the Pacific Slope in southwestern Oregon and in all of lowland western California in the rain shadow of the Cascade–Sierra Nevada Axis from Thompson River, British Columbia, to the Inland Basins between the great western mountain systems (Columbia, Great, and Colorado [River] basins) and the adjacent Columbia Plateau and to the uplands of the Colorado Plateau and the entire southern part of the Rocky Mountain System; northwestern Baja California, eastern Sonora, and northern Chihuahua, Mexico; more common and better developed southward than northward.

The complexity of the western geography and climate is reflected in the intricate pattern of ecological and floristic associations. The Southwestern Temperate Region, supporting the Sierra Madrean Flora, is intermediate in position and climate between the following adjacent regions and floras:

The Western Forest Region—Western Forest Flora
The Horse Latitudes or Desert Zone—Mexican Desert Flora

The former occurs either northward or at higher elevations, the latter southward or at lower elevations.

The present distributional pattern of these three regions and floras is reminiscent of a jigsaw puzzle. The Mexican Desert Flora occurs only near the Mexican Boundary from California to western Texas, but the Sierra Madrean and the Western Forest floras, following the complex outlines of mountains and valleys, occur in an intricate, mosaiclike geographical pattern in western Canada, in all eleven contiguous Western States, and in Texas west of the Pecos River. Various floristic associations of both floras are in contact along each zone of meeting, and there are complex intergradations where species of the two floras having overlapping tolerances to the environment occur together. Furthermore, in such places related species hybridize, and there may be intergrading of those of one flora with those of the other.

Today the Southwestern Temperate Region is dominated by either open, parklike woodlands with well-spaced, small, rounded trees or by xerophytic brushlands. However, there are grasslands as well—the California Grassland of the Great Valley (Sacramento-San Joaquin) of California and the Desert Grassland ranging from central Arizona to western Texas.

Despite its present diversity, the flora was derived largely from a Mexican oak woodland prototype, the Madrotertiary Geoflora (Axelrod). This woodland migrated northward and westward during Oligocene from the region now occupied by the Sierra Madre Occidentál, the great mountain range of western Mexico. During early and middle Miocene it dominated the southern half of the plain then covering the West. It occurred in much of California, the Great Basin, and the areas about the present site of the southern Rocky Mountain System, from about the latitude of San Francisco, Salt Lake City, and Colorado Springs southward into northern Mexico.

Distribution Map 7

Since Miocene, the two great mountain axes have arisen and created cooler, moister places at higher altitudes and rain shadows downwind beyond, and simultaneously rainfall and temperature have become lower throughout the West. Also, summer rainfall has disappeared almost completely from the Pacific Slope (except its northernmost part) and from the Columbia Basin, the Columbia Plateau, and the Great Basin. It persists in the Rocky Mountain Region, including the areas at moderate elevations around its southern end in Arizona and New Mexico and the western tip of Texas. Consequently, from place to place there is great variation in amount of rainfall and in its proportion at different seasons, the northwestern and western regions having little but winter rain and the southern and southeastern having relatively light precipitation during both winter and summer.

Since in Arizona, New Mexico, and northwestern Mexico there has been less climatic change than in other areas, the Southwestern Oak Woodland has been modified relatively less than other Sierra Madrean floristic associations, and it is the nearest modern approach to the archetype of the group. In Pliocene time, as summer rain ceased in California, independent development produced a vegetation almost wholly distinct in species from the descendants of the same original type in Arizona, New Mexico, and Mexico. Only the following woody species occur unmodified in both the Southwestern Oak Woodland and Chaparral and the California Oak Woodland or the California Chaparral.

Thalamiflorae
- *Rhus ovata,* sugarbush
- *Ceanothus Greggii*

Corolliflorae
- *Arctostaphylos pungens,* manzanita

Calyciflorae
- *Cercis occidentalis,* western redbud
- *Ribes quercetorum,* oak gooseberry

Ovariflorae
- *Lonicera interrupta,* chaparral honeysuckle

With the slow uplift of the Cascades and the Sierra Nevada in late Miocene and early Pliocene and the more rapid rise in Pleistocene and the resulting development of a dry interior climate to the eastward, characteristic present-day California Oak Woodland species disappeared from the Great Basin, and the Sagebrush Desert and the Great Basin Juniper-Pinyon Woodland developed. With the more recent elevation of the Coast Ranges, grassland of a dry land type developed in the Great Valley of California.

During Pleistocene the complexes of species and varieties of the Sierra Madrean Flora were connected across the Colorado Basin, where now at lower elevations there are great gaps filled by the Mexican Desert Flora. Wells and Berger found 17 ancient middens of woodrats, or packrats, protected in small caves, crevices, and recesses in rocks and cliffs in the Mojavean Desert. By carbon dating, the ages of the nests were determined to range from 7,400 to 19,500 years. Inasmuch as woodrats bring in plants and other objects from usually less than 100 feet from the nest, the remains represent the vegetation of the immediate surroundings at the time when the midden covering the burrow was made or during even thousands of years of its occupation. During Pleistocene the floristic associations occurred about 1,000 to 2,000 feet lower than now, for the woodrat middens show the flora of areas now Mojavean Desert to have been Great Basin Juniper-Pinyon Woodland during the later stages of the last (Wisconsin) glaciation, when the weather was cooler and moister much farther south. The Sierra Madrean Flora was continuous across the low altitudes of the upper part of the Colorado Basin in southeastern California, southern Nevada, the southwestern corner of Utah, and northwestern Arizona where now there is Mojavean Desert.

Since Pleistocene drying and desert formation have separated the Sierra Madrean Flora into eastern, northern, and western segments in Arizona, Nevada, and California. The floristic associations on the opposite sides of the Colorado River and north of the bend of the river no longer are connected, and the present isolated local taxa occur in place like the fingers of a glove but without the hand. The pinyons or piñons, i.e., nut pines, *Pinus cembroides* vars. *cembroides, edulis, monophylla,* and *Parryana,* occurring on the margins of the Colorado Basin, are an example of closely related taxa isolated in this way. Although some authors have considered these taxa to be species, chiefly because they grow in different places and because each has a different prevailing number of needles per bundle, field studies by Pomona College students* have shown that the combination of characters is unstable. It fluctuates from individual to individual plant and population to population and even on the same individual, as might be expected in the recently separated geographical extremes of a complex system of intergrading populations. Another example is the varieties of the Arizona cypresses, *Cupressus arizonica,* which rim the Colorado Basin. Vars. *arizonica* and *glabra* occur in Arizona at different points on the eastern rim of the Colorado Basin, var. *nevadensis* in California just south of the southern tip of the Sierra Nevada, var. *Stephensonii* in California on Cuyamaca Peak in San Diego County, and var. *montana* in Baja California. Each of these has been considered, also, to be a separate species. This pattern of distribution is paralleled in other genera, some of the woody ones appearing in the list on pp. 796 and 797.

1. THE SOUTHWESTERN OAK WOODLAND AND CHAPARRAL

Occurrence. Southern edge of Utah in Zion Canyon (poorly developed, but some characteristic taxa present); Arizona below the Mogollon Rim and southeastern Arizona; southwestern New Mexico; Transpecos Texas, northwestern Mexico; 4,000 or mostly 5,000 to 6,500 feet.

* In 1973–1974, by Carol Larkin, Kathy Martinez, Abigail Miller, Carol Nimick, and Greg Wright.

CHARACTERISTIC WOODY PLANTS

Thalamiflorae

Rhamnus crocea var. *ilicifolia,* redberry (shrub)
Ceanothus Greggii, mountain lilac (shrub)
Quercus diversicolor, netleaf oak (also in the Rocky Mountain Montane Forest)
Quercus oblongifolia, Mexican blue oak
Quercus Emoryi, Emory oak, bellota
Quercus hypoleucoides, white-leaf oak
Quercus arizonica, Arizona oak
Quercus turbinella, desert scrub oak (shrub or small tree)

Corolliflorae

Arbutus arizonica, Arizona madrone
Arctostaphylos pungens, manzanita (low shrub)
Arctostaphylos Pringlei var. *Pringlei,* manzanita (tall shrub)

Gymnosperms

Pinus leiophylla var. *chihuahua,* Chihuahua pine
Pinus cembroides var. *cembroides,* pinyon or piñon
Cupressus arizonica vars. *arizonica* and *glabra,* Arizona cypresses
Juniperus Deppeana, alligator bark juniper
Juniperus monosperma, juniper (not common)

Discussion. The climate is dominated by summer rain, which occurs during July and August, mostly as torrential convection storms. During the hot weather growth is rapid, and the country becomes lush and green. Species of perennial grasses and numerous other herbs form the groundcover, and they stand knee-deep beneath and between the trees. Usually the winter rain and snow are minor, and the weather tends to remain cold for so long that often spring growth and flowering are of little consequence. The water has evaporated too soon, and often by April or May the soil has become unbelievably dry. On the other hand, when there is more than usual winter rain, the spring growth may be extensive, and the flowering season good.

Woody species composed of varieties occurring in both the Southwestern Oak Woodland and Chaparral and/or the Southwestern Juniper-Pinyon Woodland and in the California Oak Woodland and/or the California Chaparral.

SOUTHWESTERN OAK WOODLAND AND CHAPARRAL and/or SOUTHWESTERN JUNIPER-PINYON WOODLAND	CALIFORNIA OAK WOODLAND and/or CALIFORNIA CHAPARRAL
Thalamiflorae	
Fremontodendron californicum, Fremontia	
Var. *californicum*	Vars. *californica, napensis,* and (not, so far, recombined as varieties) *F. obispoensis* and *F. crassifolia*
Rhamnus californica, coffee berry	
Var. *ursina*	Vars. *viridula, crassifolia, tomentella,* and *occidentalis*
Rhamnus crocea, redberry	
Var. *ilicifolia*	Vars. *crocea* and *ilicifolia*
Ceanothus integerrimus, mountain-lilac or deer brush	
A single form	Numerous intergrading forms, these named but probably not really distinguishable as varieties
Ceanothus Greggii, mountain-lilac	
Var. *Greggii*	Vars. *vestitus* and *perplexans*
Corolliflorae	
Arctostaphylos Pringlei, manzanita	
Var. *Pringlei*	Var. *drupacea*
Calyciflorae	
Cercocarpus montanus, Mountain mahogany	
Vars. *montanus, paucidentatus,* and *glaber*	Vars. *glaber* (as *C. betuloides*) and *minutifolius*
Amorpha californica, mock locust	
Var. *californica*	Vars. *californica* and *napensis*
Quercus turbinella, desert scrub oak	
Vars. *turbinella* and *ajoensis*	Var. *californica*
Quercus chrysolepis, canyon live oak	
Var. *Palmeri*	Vars. *chrysolepis* and *Palmeri*
Gymnosperms	
Pinus cembroides, pinyon or piñon	
Vars. *cembroides* and *edulis*	Vars. *monophylla* and *Parryana*
Cupressus arizonica, arizona cypresses	
Vars. *arizonica* and *glabra*	Vars. *nevadensis, Stephensonii,* and *montana*

Pairs or groups of closely related, or vicarious, woody species, occurring one or more in the Southwestern Oak Woodland and Chaparral and/or the Southwestern Juniper-Pinyon Woodland and also one or more in the California Oak Woodland and/or the California Chaparral.

SOUTHWESTERN OAK WOODLAND AND CHAPARRAL and/or JUNIPER-PINYON WOODLAND	CALIFORNIA OAK WOODLAND and/or CALIFORNIA CHAPARRAL
Thalamiflorae	
Berberis haematocarpa, red barberry	*Berberis Nevinii*, Nevin barberry
Vitis arizonica, Arizona grape	*Vitis californica*, California grape; *Vitis Girdiana*, desert grape (also along the western edges of the deserts)
Alnus oblongifolia, alder	*Alnus rhombifolia*, white alder
Quercus oblongifolia, Mexican blue oak	*Quercus Douglasii*, white or blue oak; *Quercus Engelmannii*, Engelmann oak

There are two distinctive phases of the floristic association, and these compose different ecological formations, though they are of incomplete floristic distinction. Under ecologically somewhat different conditions these phases arise through emphasis of some species and suppression or extinction of others. They are not differentiated by characteristic endemic species, and apparently their segregation is transitory, each formation arising here or there according to local environmental factors and being composed of species occurring also in the general floristic association.

THE SOUTHWESTERN OAK WOODLAND

Occurrence. Middle altitudes of hills below the Mogollon Rim in central Arizona and common in southeastern Arizona; southwestern corner of New Mexico; Sonora and Chihuahua, Mexico; 4,000 or 5,000 to 6,500 feet.

Discussion. The oak woodland is partly open and parklike, but in places it tends to be more dense, and it merges into the locally occurring Southwestern Chaparral. This woodland is the nearest approach in the living flora to the archetype Madrotertiary Geoflora, which, under various climatic modifications, gave rise to all the present floristic associations of the Sierra Madrean Flora. The oak woodland is rich in species.

THE SOUTHWESTERN CHAPARRAL

Occurrence. Mostly dispersed into islands in the Southwestern Oak Woodland, but also sometimes well developed in other areas, especially northwestward nearer California, as on Hualpai Mountain in northwestern Arizona and along the Mogollon Rim southwest of Prescott.

Discussion. The shrubs have the appearance of the California Chaparral, which is much more extensive and better differentiated from the adjacent California Oak Woodland, being characterized by a host of endemic species. In the Southwestern Chaparral, the char-

Figure 25-26. *Above,* Southwestern Oak Woodland and Desert Grassland: Southwestern Oak Woodland on hills just above Desert Grassland (foreground); southeastern Arizona; *below,* Southwestern Oak Woodland near the summit of Kitt Peak, Baboquivari (Quinlan portion) Mountains, Pima County, Ariz.: bellota or Emory oak, *Quercus Emoryi,* Mexican blue oak, *Quercus oblongifolia,* etc. Photographs, courtesy of the United States Forest Service.

Figure 25-27. Chaparral or brushland area in the Southwestern Oak Woodland and Chaparral (for the associated oak woodland cf. **Figure 24-1,** *middle,* p. 716; and **Figure 25-17**), west of Globe, Arizona.

acteristic species are the five shrubs appearing in the list for the floristic association and occurring also in the adjacent oak woodland. Three of these (*Rhamnus crocea* var. *ilicifolia, Ceanothus Greggii,* and *Arctostaphylos pungens*) are prominent in the chaparral of southern California, as well. *Arctostaphylos Pringlei* var. *Pringlei* is replaced there by var. *drupacea, Quercus turbinella* var. *turbinella* by var. *californica.* The wide-ranging *Quercus turbinella* var. *turbinella* also occurs disjunctly in California in the New York Mountains (surrounded by the Mojavean Desert), in southern Nevada, and in Baja California. In these locations it must be a relic. Thus, probably the chaparral east of the Colorado Basin, disjunct since Pleistocene, represents a less differentiated phase of the chaparral derived, like the Californian type, from the changing oak woodland, but in a situation where the climate has been modified relatively less. Perhaps this incompletely differentiated chaparral represents only the persistence of a less rich tongue of this floristic type cut off at the end of Pleistocene and surviving in favorable sites along the eastern rim of the Colorado Basin. If so, both chaparrals are parts of a modern general type of chaparral developed since Pleistocene and preceded by a Pliocene version, which was more extensive.

2. THE SOUTHWESTERN JUNIPER-PINYON WOODLAND

Occurrence. Southern edge of Utah, where there is intergradation with the Great Basin Juniper-Pinyon Woodland, to the foothills of the Rocky Mountain System from southern Colorado and northern and eastern Arizona, and to New Mexico and Transpecos Texas; adjacent Mexico; 5,000 to 7,000 feet.

CHARACTERISTIC TREES (MARKERS)

Gymnosperms

Pinus cembroides var. *edulis*, pinyon or piñon
Juniperus Deppeana, alligator bark juniper
Juniperus monosperma, juniper

Discussion. This coniferous woodland has much in common with the Southwestern Oak Woodland, but it occurs exclusively through a great region and there is considerable endemism of herbaceous species. In central and southeastern Arizona, as west of Camp Verde and around the base of Mt. Graham (Swift Trail), there are intermediate woodlands in which both oaks and conifers are prominent.

The Great Basin and Southwestern Juniper-Pinyon woodlands are far different floristically, and they differ greatly in relationship to climatic and ecological factors. Because the coniferous woodland formation is the same, the woodlands have not been differentiated by ecologists. However, even the dominant taxa, the junipers and pinyons, which govern the appearance of the ecological formation, are different. In the Great Basin Juniper-Pinyon Woodland they are *Pinus cembroides* var. *monophylla*, pinyon, and *Juniperus californica* var. *osteosperma*, Utah juniper, neither of which occurs in the Southwestern Juniper-Pinyon Woodland. The plants, especially the herbaceous ones, are adapted to far different climates, and most varieties, species, and genera are different. The Southwestern Juniper-Pinyon Woodland occurs in a region of summer rain and of relatively light winter precipitation; the Great Basin Juniper-Pinyon Woodland occurs in a region with only winter rain and snow. This has a profound effect on the presence of herbaceous species. In the Great Basin Juniper-Pinyon Woodland they must grow, flower, and fruit on the basis of the water held over from winter until the weather is warm enough for development. In the meantime much has evaporated. Farther south where there is summer rain concentrated in two months, July and August, the water may be used immediately while the weather is warm. The vastly different environmental conditions favor an almost wholly different group of species, especially the herbaceous ones.

3. THE DESERT GRASSLAND

Occurrence. This dry type of prairie is developed best in southeastern Arizona, southwestern New Mexico, westcentral and western Texas (in clear form west of the Pecos River), and northern and northcentral Mexico. In Arizona and New Mexico the Desert Grassland is chiefly at about 3,500 or 4,000 to 5,000 feet elevation, occurring just above the Southwestern Deserts and below the Southwestern Oak Woodland and Chaparral. Cf. Figure 25-26, *above*.

Discussion. The growing season is in the summer (July and August), and, as with the adjacent floristic associations, the spring period is very dry by the time the weather becomes warm enough for plant growth. The total precipitation is about 12 to 18 inches, mostly during July and August, when the water is usable immediately. The hills and plains become lush during August and early September. Sometimes during wet winters enough rain may fall to last over into the spring and to produce a good growing and flowering season then.

The relationships of the Desert Grassland have been assumed to be to the other grass-

Figure 25-28. Desert Grassland with many perennial grasses and annual herbs. Two young oaks have reached significant size, and a few others are coming up. Mesquites are growing along the small wash. The most conspicuous grasses are various grama grasses (blue, *Bouteloua gracilis;* side-oats, *Bouteloua curtipendula;* hairy, *Bouteloua hirsuta,* and sprucetop, *Bouteloua chondrosioides*), wolf-tail (*Lycurus phleoides*), and curly mesquite (*Hilaria Belangieri*), all important range grasses. (Photograph, courtesy of the United States Forest Service.)

lands to the east and north. However, despite some ecological similarity, the plants, including grasses, show more affinity to those of the Southwestern Oak Woodland. Further floristic analysis is needed.

In Texas the Desert Grassland intergrades with the Shortgrass Prairie, as may be expected of two dry grasslands having no sharp physiographic or climatic boundary between them. Probably during Pleistocene, some elements now forming the Shortgrass Prairie were pushed southward because of the change of climate. Later, as they returned northward after the pluvial period, taxa of the Desert Grassland and the Southwestern Oak Woodland must have gone with them. These plants occur in the Shortgrass Prairie today.

Since the advent of overgrazing and the cessation of Indian-set fires, both the character of the Desert Grassland and its extent have changed. Woody species from the Southwestern Oak Woodland, on the hills just above the Desert Grassland, have penetrated to lower levels as their seedlings were no longer burned and as some areas became overgrazed. For the same reasons species of trees, shrubs, and herbs have moved up from the Arizona and Chihuahuan deserts into the grassland. In addition, the velvet mesquite, *Prosopis juliflora* var. *velutina,* and sometimes the western honey mesquite, var. *Torreyana,* have been carried into the grasslands from the neighboring bottomlands and swales by cattle, which eat the pods. Some seeds are not digested, and they are distributed on the plains and hills with moisture-holding manure containing fertilizer. This has extended the range of mesquites to vast dry areas in which seedlings could not become established before the coming of cattle. Because of encroachment of woody plants from above and below, the Desert Grassland has become 1,000 to 2,000 feet narrower in altitude than it was a century ago. Except on plains and low, rolling hills spread out at the proper elevation, it forms a thin and irregular zone.

4. THE CALIFORNIA OAK WOODLAND

Occurrence. Represented by a few species in the Rogue River Valley, southwestern Oregon; *the predominant vegetation in the lowlands west of the Sierra Nevada in Northern California;* restricted to "islands" in the chaparral of Southern California; faintly represented by a few species in northwestern Baja California. Elevation: Coast Ranges, 200–2,000 feet; Sierra Nevada foothills, northward 200–2,000 feet, southward 1,600–4,500 feet.

Most of the numerous characteristic species are herbs, these being associated with relatively few species of trees or shrubs.

CHARACTERISTIC WOODY SPECIES

Thalamiflorae
- *Aesculus californica,* California buckeye (primarily in oak woodland, but frequently in chaparral)
- *Quercus Engelmannii,* Engelmann oak
- *Quercus lobata,* valley oak
- *Quercus Douglasii,* blue oak
- *Juglans californica* var. *californica,* southern California black walnut
- *Juglans californica* var. *Hindsii,* northern California black walnut

Calyciflorae
- *Ribes quercetorum,* oak gooseberry

The following woody species or varieties occurring in California are characteristic of both the California Oak Woodland and the California Chaparral, and they are restricted to these floristic associations.

Thalamiflorae
- *Quercus Wislizenii,* interior live oak
- *Quercus agrifolia,* coast live oak

Corolliflorae
- *Arctostaphylos manzanita* var. *manzanita,* manzanita
- *Arctostaphylos manzanita* var. *elegans,* Lake County manzanita
- *Arctostaphylos viscida,* manzanita

Calyciflorae
- *Cercis occidentalis,* western redbud
- *Prunus ilicifolia,* holly-leaved cherry
- *Photinia arbutifolia,* Christmasberry

Gymnosperms
- *Pinus Sabiniana,* Digger pine

Discussion. Although the California Oak Woodland resembles the Southwestern type in growth form of the dominant plants, the species composition is almost completely different. This is to be expected in view of the markedly different climate. The California Oak Woodland supports a rich growth of annual grasses and of endemic annual and perennial herbs during the moist lush winter and spring growing season. All herbaceous plants except those near permanent water dry up late in May and remain brown until October or November when the winter rains begin. Obviously this favors a different set of plants from those in Arizona and New Mexico where summer rains predominate. There the emphasis is on perennial grasses and other perennial herbs.

Figure 25-29. California Oak Woodland, lower foothill phase dominated by the blue oak, *Quercus Douglasii;* inner North Coast Ranges west of Williams, Colusa County, California.

The California Oak Woodland is the most characteristic feature of northern California, but it is uncommon in southern California. It includes a number of somewhat differentiated subordinate types. This is to be expected in view of the complicated topography even west of the principal mountain axis.

5. THE CALIFORNIA CHAPARRAL

Occurrence. In "islands" at 100 to 3,000 or sometimes 4,000 feet elevation in northern California from Trinity and Shasta counties southward in the Coast Ranges and in the foothills of the Sierra Nevada; *the prevailing vegetation of southern California below 6,000 feet elevation;* northwestern Baja California.

Numerous shrubs and a few trees are endemic in the California Chaparral, and some occur also in the California Oak Woodland, as listed above. Although there are many characteristic herbaceous species, these are not as abundant as in the California Oak Woodland.

CHARACTERISTIC WOODY SPECIES

Thalamiflorae

Berberis dictyota, barberry
Berberis Nevinii, Nevin barberry
Dendromecon rigida, bush poppy
Rhus laurina, laurel sumac
Rhus integrifolia, lemonade-berry
Rhamnus crocea vars. *crocea* and *ilicifolia*, redberries
Ceanothus foliosus, mountain-lilac

Figure 25-30. California Chaparral forming a dense covering of the hills with the appearance of a shag rug. The conspicuous plants are leather oak, *Quercus durata,* and scrub oak, *Quercus dumosa,* here forming most of the cover; chamise brush, *Adenostoma fasciculatum,* lower right corner; and knobcone pine, *Pinus attenuata,* which grows in the chaparral, the cones tightly sealed by pitch for many years until there is a brush fire, then opening and releasing the seeds, which will germinate only after heating. The adult pines are killed by the fire, but seedlings grow rapidly. Hopland Grade, Lake County, California, between Lakeport and Hopland. Robert L. Benson.

Ceanothus spinosus, red-heart
Ceanothus divaricatus, mountain-lilac
Ceanothus tomentosus, mountain-lilac
Ceanothus oliganthus, mountain-lilac
Ceanothus verrucosus, mountain-lilac
Ceanothus megacarpus, mountain-lilac
Ceanothus crassifolius
Ceanothus cuneatus, buck brush
Ceanothus Jepsonii, musk brush (serpentine)
Eriogonum fasciculatum (especially var. *foliolosum*), wild buckwheat (coastal sagebrush stage)
Opuntia prolifera, coastal cholla
Opuntia littoralis vars. *littoralis, Vaseyi,* and *austrocalifornica,* prickly pears
Opuntia oricola, coastal prickly pear
Cereus Emoryi
Quercus dumosa, scrub oak
Quercus durata, leather oak (serpentine)
Garrya Veatchii, silktassel-bush
Garrya Fremontii, silktassel-bush
Garrya flavescens, silktassel-bush

Corolliflorae

Styrax officinalis var. *californica,* California styrax
Arctostaphylos glauca, great-berried manzanita

Arctostaphylos Stanfordiana, Stanford manzanita (serpentine phase)
Arctostaphylos canescens, manzanita
Arctostaphylos glandulosa, Eastwood manzanita
Arctostaphylos tomentosa
Fraxinus dipetala, foothill ash
Turricula Parryi
Eriodictyon trichocalyx, southern yerba santa (coastal sagebrush stage)
Eriodictyon californicum, yerba santa
Eriodictyon crassifolium, white yerba santa
Salvia mellifera, black sage
Salvia apiana, bee sage (coastal sagebrush stage)
Salvia leucophylla, purple sage (coastal sagebrush stage)
Penstemon cordiofolius
Penstemon corymbosus
Penstemon antirrhinoides
Penstemon Lemmonii
Penstemon spectabilis (coastal sagebrush stage)
Penstemon heterophyllus
Mimulus (*Diplacus*) *Clevelandii,* bush monkey-flower
Mimulus (*Diplacus*) *longiflorus,* bush monkey-flower
Mimulus (*Diplacus*) *puniceus,* red bush monkey-flower

Calyciflorae
Pickeringia montana, chaparral pea
Cercocarpus montanus vars. *glaber* and *minutiflorus,* mountain-mahogany
Adenostoma fasciculatum, chamise-brush
Adenostoma sparsifolium, red-shank
Ribes malvaceum vars. *malvaceum, viridifolium,* and *indecorum,* flowering currants

Ovariflorae
Lonicera subspicata, moronel
Lonicera interrupta, chaparral honeysuckle
Haplopappus pinifolius (coastal sagebrush stage)
Haplopappus ericoides, mock-heather
Haplopappus Parishii
Artemisia californica, coastal sagebrush (coastal sagebrush stage)
Senecio Douglasii, bush groundsel (coastal sagebrush stage)
Lepidospartum squamatum, scale broom (coastal sagebrush stage)

Monocotyledoneae
Yucca Whipplei var. *Whipplei,* yucca (partly coastal sagebrush stage)

Gymnosperms
Pinus Sabiniana, Digger pine (in oak woodland, as well)
Pinus attenuata, knob-cone pine
Cupressus Sargentii, Sargent cypress (serpentine phase)
Cupressus Macnabina, McNab cypress (serpentine phase)

Discussion. The oak woodland and the chaparral occur one above the other or sometimes at the same elevation, the oak woodland often growing on the north slopes and the more drought resistant chaparral on the south. In the dry, mild climate of middle Pliocene time, the chaparral attained its maximum development, and many species now restricted to southern California occurred farther north. In northern California today the chaparral is a discontinuous relict phase occurring primarily as "islands" surrounded by oak woodland. In coastal southern California (south of the Tehachapi Mountains) this relationship is reversed, and chaparral is the continuous phase while the oak woodland occurs only in relict patches surrounded by brush. These islands of oak woodland occur near the so-called "Ridge Route," which is the principal highway from Bakersfield southward to Los Angeles, in the valleys south of the San Gabriel Mountains from Pasadena eastward to San Bernardino County, and southward to interior San Diego County.

A successional stage of the chaparral featuring sages and sagebrush (a sagebrush far different from that in the Sagebrush Desert) is common especially in southern California on well-drained, usually sandy soils. Some authors have given the coastal sagebrush formation an ecological status equal to that of the California Chaparral, because physical appearance and association of species differ from those of the other phases of the chaparral. However, the coastal sagebrush occurs only on disturbed soils, and commonly it is only transitory. Another important transitory phase is a chamise-brush stage in succession following the coastal sagebrush stage.

In coastal northern California a special phase of the chaparral, with a considerable number of endemic to nearly endemic species or varieties or races, occurs on soil from serpentine rocks. This soil is sterile for most plants. The taxa and races adapted to serpentine are able to survive there for lack of competition, but many of them are less vigorous elsewhere than their competitors and therefore unable to survive away from serpentine.

6. THE CALIFORNIA GRASSLAND

Occurrence. California; almost restricted to the Great Valley (Sacramento and San Joaquin valleys); originally in patches among oak woodland and some chaparral in the Sacramento Valley, but the prevailing vegetation of the San Joaquin Valley. At lower latitudes the grassland rises to higher and higher elevations, and on the north side of the Tehachapi Mountains it reaches about 1,600 feet. A well-developed disjunct area of Pacific Grassland occurs, also, in dry areas at about 3,000 to 4,000 feet just southeast of the San Joaquin Valley in and near Antelope Valley, where it merges into the Mojavean Desert.

This vegetation is remarkable for its lack of woody plants, except in the washes and alkali flats and the few sand dunes (recently destroyed), and annuals are abundant and dominant. The few woody species include the following endemic ones.

CHARACTERISTIC WOODY SPECIES

Thalamiflorae

Atriplex spinifera, spiny saltbush

Opuntia basilaris var. *Treleasii,* prickly pear

Ovariflorae

Eastwoodia elegans

Discussion. The area is one of low rainfall (5–10 inches), relatively mild winters, and hot, dry summers.

Throughout the California Grassland changes have been extensive because of overgrazing, cultivation, and introduction of numerous weeds, particularly from the Mediterranean region. This, and the inherent complexity of the original flora, make settling the question of origin difficult.

Before the coming of European man the California Grassland may have been bunch-grass prairie, and traces of this formation occur in eastern San Luis Obispo County and in the lower (northern) San Joaquin Valley. Currently it includes, in addition to many annual grasses and other plants introduced from the Mediterranean Region, partly a selection of annual grasses and annual herbs from the adjacent California Oak Woodland; partly a selection of Sagebrush Desert plants (having migrated through the gap of Pit River) and Mojavean Desert plants (having migrated through Walker, Tehachapi, and Tejon passes); and partly a small but distinctive group of *endemic herbaceous species,* now restricted by agriculture largely to canyons in the foothills. Probably these endemic herbs are the best key to the nature of the original flora.

The presence of the endemics is remarkable in view of the probable origin of the California Grassland since Pleistocene, that is, during the last 10,000 years. Most of the present native grassland species are present also in the California Oak Woodland or the California Chaparral surrounding the grassland, and this is in harmony with the near certainty that the Great Valley was occupied by these two Sierra Madrean Flora floristic associations during Pleistocene. Since Pleistocene the California Grassland, like the other North American prairies, has been maintained in part through the dryness of the climate and in part by fires set by the Indians. This combination has not produced many endemic taxa in the other grasslands. However, the California Grassland emphasizes annual species, mostly dicotyledons. Originally there may have been more bunch grasses than now, especially in the northern parts with greater precipitation, but

Figure 25-31. California Grassland: *above,* in the springtime, this particular area dominated by California poppies, *Eschscholtzia californica* var. *crocea;* Antelope Valley between Lancaster and Gorman; *below,* in early fall, the hills dry and brown, the shrubs in the foreground *Haplopappus venetus;* San Luis Obispo County, between Soda Lake and McKittrick, Kern County. *Above,* Burton Frasher, Sr.; courtesy Burton Frasher, Jr., and the Pomona Public Library.

the surviving native plants, including the endemic taxa, are almost exclusively annuals of characteristically Californian species groups. Probably the unusual percentage of endemic taxa in the California Grassland is due to the rapid rate of evolution possible among annuals, which produce enormous numbers of individuals, the vast majority of which cannot live to maturity and none of which will live until the next year. These plants undergo rapid natural selection each year according to a rigorous climate. Thus, production of endemic taxa takes much less time than in other floristic associations, including the other grasslands, all of which are dominated by long-lived perennial grasses and dicotyledons.

The great flower fields of the San Joaquin Valley have produced a display equal to any in the world, including that in Namaqualand in South Africa, and they are far more colorful on good years than those in the deserts or elsewhere in North America. Years of good growth and flowering are rare—recently 1932, 1935, 1937, 1941, 1952, 1958, 1973, and 1978.* They follow relatively wet winters and springs during which the rains have been frequent, though not necessarily heavy. In the 1930's the entire upper end of the San Joaquin Valley, south of Bakersfield, was a sea of many vivid colors during late March and the first week of April. Then it dried during a few days, and summer came. (See Appendix, p. 854.) In 1973 because of cultivation of the valley floor, the largest flower fields were restricted to a horseshoe on the great alluvial fans, mostly on the Tejon Ranch. This vividly colored area was about 25 miles long and 2 to 3 miles wide.

According to Marvin Dodge, California Department of Forestry, early Spanish accounts indicate former presence of more grassland and forest in southern California before disturbance by grazing and agriculture. During the 200 years of disturbance since founding of the Mission San Diego in 1769, the chaparral, once mostly in more localized niches, has spread over most of the lower areas. Grassland remnants are preserved in the area set aside for the Camp Pendleton Marine Base. The relationships of this nearly extinct grassland are difficult to determine. Whether it bore much floristic relationship to the California Grassland is undetermined.

7. THE CLOSED-CONE PINE FOREST

Occurrence. Along the Californian coast in patches and strips from Trinidad, Humboldt County, to northern Santa Barbara County and parts of the northern group of southern Californian islands; mostly from near sea level to 200 feet.

CHARACTERISTIC WOODY SPECIES

Corolliflorae

Rhododendron californicum, California rhododendron
Arctostaphylos Hookeri, Monterey manzanita
Arctostaphylos Nummularia, Fort Bragg manzanita
Arctostaphylos pumila, dune manzanita

Gymnosperms

Pinus muricata, Bishop pine
Pinus radiata, Monterey pine
Pinus contorta var. *Bolanderi,* pygmy beach pine
Cupressus Goveniana, Gowen cypress
Cupressus Goveniana var. *pygmaea,* pygmy cypress
Cupressus macrocarpa, Monterey cypress

Discussion. The Closed-Cone Pine Forest is not composed of forest trees, and it may be looked upon as a dense coniferous woodland dominated by trees with affinities to the Sierra Madrean Flora. The characteristic plants are species that occurred farther inland during the warm epoch of Pliocene and which have been

* In the winter of 1977–1978 an enormous sandstorm covered many square miles on the Tejon Ranch and adjacent farms with varying depths of sand, and the area included the best flower fields. Hillsides were sandblasted. Succession will require time.

Figure 25-32. Closed-Cone Pine Forest on rapidly eroding land above an ocean inlet, where the forest forms a subclimax maintained only because the disturbed ground is not suitable for redwood forest (Pacific Lowland Forest); mouth of Russian Gulch, south of Fort Bragg and just north of Mendocino, Mendocino County, California. The pines are mostly Bishop pine, *Pinus muricata,* with some beach pines, *Pinus contorta,* and with some lowland fir, *Abies grandis.*

able to survive in a maritime climate where there are no killing frosts, as likewise with those of the Insular Association on the islands of southern California. The plants of the Closed-Cone Pine Forest are adapted to a moister, cooler, but essentially frostless climate farther north, those of the Insular Association to a drier, warmer, similarly frost-free environment farther south.

The forms of trees are affected by local environment. The pines and cypresses of the north coast may be spaced rather widely, forming woodlands, though in many places they are dense and they compose forests of low stature. Commonly in dense young forests many small trees and the lower branches of the rest are killed by shading. This affects the form of the tree, yielding a bare lower trunk and a conical crown. Wider spacing of trees favors growth of the lower branches spreading at right angles to the trunk, and this produces ovoid or globose trees. On ocean bluffs and headlands the prevailing wind produces weird tree shapes through wind trimming, that is, the exposed buds up wind are dried and killed, while those down wind are sheltered and the branches on that side become long. One-sided trees are common along the Pacific Coast, because the prevailing westerlies come from the open ocean.

In general the climate is that of the adjacent redwood forest phase of the Pacific Lowland Forest. There prevailing climatic features are a small, nearly constant temperature range and much rain. The ocean temperature varies only a little, and commonly that of the adjacent land is about 50° to 60° F, winter and summer. Rainfall is heavy, about 60 to 100 inches annually. Fog is common, and the plants absorb water from it. The fog condenses on the trees, which may absorb water directly, and the soil beneath them is moistened even in summer by the drip from the narrow, sharply pointed coniferous leaves.

On the other hand, the Closed-Cone Pine Forest lives under environmental factors to which the adjacent redwood forest is not subjected. The wind velocity is high, because the prevailing westerlies have had no obstruction for thousands of miles. The plants in the low elevations on the immediate coast are dried by the wind and the salt spray whipped by it from the waves deposited on both the vegetation and the soil beneath. Some of the land on which the coastal plants grow in undercut by waves, and this creates a zone in which plants able to adapt to unstable soil may be free from competition. The plants of the Closed-Cone Pine Forest have been selected for this ability, and probably this accounts for the survival of this floristic association in an area where otherwise it is shaded out by the taller trees of the Pacific Lowland Forest.

The Closed-Cone Pine Forest in most of its areas of occurrence in Mendocino County is a subclimax in succession, leading, slightly back from the coast, to redwood forest. This is apparent wherever man has cleared the forest and later abandoned use of the land. In some places there are open fields dotted by young Bishop pines; in others stands of maturing pines and some of their usual associates, including various shrubs and especially rhododendrons; in still others mature or maturing Closed-Cone Pine Forest with Bishop pines 30 feet tall and with about two-foot trunks. At the inner edge of this forest the redwoods and Douglas firs soon overtop the closed-cone pines, and these and their associates are doomed.

An amazing pygmy forest occurs on old sea beaches in Mendocino County, these now elevated far above the ocean. The plants grow on extremely acid, nearly sterile soil formed from the beach sand. They are underlain by hardpan, resulting in an excess of water just under the surface nearly the whole year. The soil, itself, is deficient in calcium, potassium, phosphorus, and magnesium. Despite the heavy rainfall, this combination of local environmental conditions can support only dwarf, specialized trees. These may be mature or even old, and they may bear the cones of many seasons when they are only a few feet high. They are amazingly adapted and ecologically specialized small races or varieties of a species the other variety of which occurs commonly in the rest of the Closed-Cone Pine Forest. They have been named *Pinus contorta* var. *Bolanderi*, pygmy beach pine, and *Cupressus Goveniana* var. *pygmaea*, pygmy cypress.

8. THE INSULAR ASSOCIATION

Occurrence. Limited to two groups of islands off the coast of southern California—northern group: San Miguel, Santa Rosa, Santa Cruz, and Anacapa—southern group: San Nicolas, Santa Catalina, Santa Barbara, and San Clemente.

CHARACTERISTIC WOODY SPECIES

Thalamiflorae

- *Dendromecon Harfordii*, island bush poppy
- *Lavatera assurgentiflora*, tree-mallow
- *Ceanothus arboreus*, mountain-lilac
- *Ceanothus insularis*, island mountain-lilac
- *Quercus MacDonaldii* (status doubtful), McDonald oak

Corolliflorae

- *Arctostaphylos insularis*, island manzanita
- *Arctostaphylos (Xylococcus) bicolor*, manzanita
- *Mimulus (Diplacus) parviflorus*, bush monkey flower

Calyciflorae

- *Crossosoma californicum*, island crossosoma
- *Lyonothamnus floribundus* vars. *floribundus* and *asplenifolius*, Catalina ironwoods
- *Ribes viburnifolium*, island currant
- *Prunus ilicifolia* var. *occidentalis*, Catalina cherry
- *Cercocarpus montanus* vars. *Traskiae* and *alnifolius*, mountain mahoganies

Ovariflorae

- *Haplopapus canus*

Gymnosperms

- *Pinus Torreyana*, Torrey pine (also on the mainland at Del Mar, San Diego County)

Discussion. The plants listed above are a remarkable group of island endemics. The species are well distinguished, and some are without very close relatives. *Lyonothamnus*, for example, is a distinctive monotypic genus, the single species consisting of two island varieties that are cultivated widely. The island species of *Crossosoma* and *Dendromecon* are considerably different from the one other species in each genus, in the first case a desert shrub and in the second a Californian mountain chaparral shrub. In addition to the woody endemics, there are many herbaceous ones. A few of the essentially endemic insular species occur also at one or two points on mainland coastal bluffs, as with the Torrey pine, *Pinus Torreyana*.

Along with the endemic species, which form a considerable part of the flora of the islands, there are others common in the California Chaparral or less commonly the California Oak Woodland on the mainland, and often those predominate. Thus the relationships of the Insular Association to the Sierra Madrean Flora are clear, as they are also from noting the genera and the species groups to which the endemics belong.

According to Axelrod, the insular flora, like the Closed-Cone Pine Forest farther north, is highlighted by endemics that have persisted there since late Pliocene. Severe cooling of the climate during Pleistocene restricted the plants to the islands, because the insular climate is free from killing frosts. Severe frosts occur sometimes on the mainland, and the island endemics cultivated at the Rancho Santa Ana Botanic Garden in Claremont show frost sensitivity. During the warm Pliocene Epoch at least some of the island endemics occurred much farther inland, even in the area now occupied by the Southwestern Deserts.

In the past the northern group of islands may have been connected to the mainland in the Santa Barbara region, and the southern group may have been joined to the land in San Diego County. At any rate, chaparral and oak woodland species occurring on the adjacent mainland tend to be also on the nearer islands, and this would indicate that in the past there was a period of easier migration in each of the two areas.

9. THE GREAT BASIN WOODLAND AND DESERT

Occurrence. Vicinity of Thompson River, British Columbia, to the lower altitudes of the Columbia Basin and to the floor, hills, and lower mountains of the Great Basin; 3,000 to 6,000 or 8,000 feet, depending on the latitude and the height of the mountains creating the rain shadow.

Discussion. On the valley floors of the Great Basin and adjacent areas, Sagebrush Desert predominates, covering many square miles. The Great Basin Juniper-Pinyon Woodland is on the hills above the Sagebrush Desert, because it is adapted to the cold and additional precipitation of higher altitudes and to the rocky soils of the hills. The appearance of these two phases is more sharply differentiated than the floristic composition of the plant associations, and in its lower parts the woodland appears to be Sagebrush Desert with the addition of scattered small trees. The Navajoan Desert shows close relationship to the other two phases, but it differs considerably and it is marked by endemic species.

THE GREAT BASIN JUNIPER-PINYON WOODLAND

Occurrence. The dominant minor floristic association of the numerous discontinuous small mountain ranges of the Great Basin region from eastern Oregon and southern Idaho southward through the eastern edge of northern California, Nevada, Utah, and western Colorado; mostly from 3,000 to 4,000 feet in the north and from 5,000 to 7,000 feet in the south.

CHARACTERISTIC WOODY SPECIES

Ovariflorae

Chrysothamnus nauseosus (several of the numerous varieties), rabbit-brush

Figure 25-33. Great Basin Juniper–Pinyon Woodland. An open, parklike woodland typical of the Great Basin but in this area with some plants more characteristic of the Southwestern Juniper–Pinyon Woodland. This is a transitional area, with some summer rain, on the border zone between the ranges of the two woodlands. About 4,000 feet in Zion Canyon, Zion National Park, Utah. The trees are one-leaf pinyon, *Pinus cembroides* var. *monophylla*, and Utah juniper, *Juniperus californica* var. *osteosperma*. Some of the pines may have two needles in a bundle, instead of one, and this is a character of var. *edulis*, more common from here eastward and southward.

Figure 25-34. Sagebrush Desert and Juniper-Pinyon Woodland: *above,* sagebrush, *Artemisia tridentata,* in the Sagebrush Desert near Thompson River west of Kamloops, British Columbia, about 200 miles north of the International Boundary (two species of cacti occurring with the sagebrush at even this latitude); *below,* Juniper-Pinyon Woodland on the mountains north of St. George, Utah, dominated by Utah juniper, *Juniperus californica* var. *osteosperma,* and the one-leaf pinyon, *Pinus cembroides* var. *monophylla;* foreground, transition to Sagebrush Desert, young junipers invading the sagebrush.

Gymnosperms

Pinus cembroides Zucc. var. *monophylla*, one-leaf pinyon
Juniperus californica var. *osteosperma*, Utah juniper

Discussion. The small number of woody plants differentiating the Great Basin Juniper-Pinyon Woodland from the Sagebrush Desert points out the need for a thorough reëxamination of the total floristic composition of both floristic associations. The woodland may not be as well segregated from the desert as the presence of the large species (junipers and pinyons) seems to indicate.

The relationship of the two juniper-pinyon woodlands is discussed under the Southwestern Juniper-Pinyon Woodland, above.

THE SAGEBRUSH DESERT

Occurrence. From the vicinity of Kamloops, British Columbia, southward just east of the Cascade Mountains to low canyons and coulees in the Columbia River Basin in eastern Washington, northern Idaho, and northeastern Oregon; the Columbia Basin and throughout the lower areas of the Columbia Plateau and the Great Basin (southeastern Oregon; eastern California, Nevada, southern Idaho, and Utah); high plateau country in northwestern Arizona; and plains of southwestern Wyoming. Represented east of the Continental Divide by some elements occurring on the Great Plains as far north and east as Saskatchewan and along the western edges of the Dakotas. Probably these have migrated through the gap in the Rocky Mountains in southern Wyoming since disturbance of the vegetation by grazing of cattle and sheep.

CHARACTERISTIC WOODY SPECIES

Thalamiflorae

Atriplex Nuttallii, salt sage
Opuntia fragilis, little prickly pear
Pediocactus Simpsonii

Corolliflorae

Salvia Dorrii var. *carnosa*, purple sage

Calyciflorae

Purshia tridentata, antelope-brush
Prunus Andersonii, desert peach

Ovariflorae

Tetradymia axillaris var. *axillaris*, spiny horse-brush
Artemisia tridentata, sagebrush
Artemisia arbuscula, black sage
To some extent the following species:
Artemisia trifida
Artemisia cana
Artemisia nova

The following woody species are characteristic of the Sagebrush Desert, but they occur also in the Mojavean Desert.

Thalamiflorae

Grayia spinosa, hop-sage
Atriplex confertifolia, sheep-fat, spiny saltbush
Atriplex lentiformis var. *Torreyi*, water sage
Atriplex canescens, shad-scale

Calyciflorae

Dalea Fremontii, indigo-bush
Dalea polyadenia

Ovariflorae

Tetradymia glabrata, horse-brush
Tetradymia axillaris var. *longispina*, Mojave horse-brush
Artemisia spinescens, bud-sage

Discussion. This desert is the characteristic feature of the Great Basin Region. It occurs typically at elevations just below the Juniper-Pinyon woodland, often covering the desert floor while the woodland covers the low desert mountain ranges and the foothills of the higher ones.

Association of the Sagebrush Desert primarily with the Sierra Madrean Flora has been questioned because some of the dominant species were derived from the Western Forest Flora. However, the Juniper-Pinyon Woodland shows affinity to the Sierra Madrean Flora, and it and the Sagebrush Desert probably are difficulty separable. Some species of both are of northern origin, as might be expected.

The Sagebrush Desert includes two phases. Since the typical sagebrush phase has a higher water requirement, it is restricted to better and usually deeper soils. The shad-scale phase, characterized by spiny saltbush and bud-sage, is developed on alkaline or dry, rocky soil.

THE NAVAJOAN DESERT

Occurrence. Colorado Plateau in southern Utah, southwesternmost Colorado, northern Arizona, and northwestern New Mexico; about 4,000 to 5,000 feet.

CHARACTERISTIC SPECIES

Four of the eight species of *Sclerocactus* and five of the seven species and their varieties of *Pediocactus* are restricted to the Colorado Plateau and the Navajoan Desert. According to Welsh 182 endemic plant taxa occur below 6,500 feet elevation on the Colorado Plateau. Many of these are restricted to the Navajoan Desert. A thorough study is needed to determine the real floristic status of the desert.

Discussion. The Navajoan Desert occupies the heart of one of the most beautiful areas on Earth, the redrock country of the four corners region of Utah, Colorado, Arizona, and New Mexico, mostly on the Navajo Indian reservation. This is an area of pastel or sometimes vivid colors and lighting effects, mixed with grays, browns, and greens. The colorful cliffs, canyons, and plains stretch on and on, and visibility in the clear, blue desert air is expanded to 50 or 100 miles. There are many National Parks and National Monuments, but an untold number of other spots would be in that category if they occurred anywhere else. Here, each one is lost among its many equals. Hour by hour the color changes of a single day are amazingly beautiful.

Hot summers and cold winters are the rule, and the rainfall is light at both times. The air is dry, and evaporation is rapid; thus 100° F, about the maximum, is more comfortable than 80° in the eastern half of the Continent. Spring flowering occurs in April and May, following the light winter snows and spring rains. Summer flowering of a different set of species follows the rains of July and August, and, at the high temperatures, any water available then is used to advantage for plant growth.

The Navajoan Desert includes elements appearing also in both the Great Basin and Southwestern Juniper-Pinyon woodlands.

The Mexican Desert Flora

(Horse Latitudes or Desert Zone)

NOTE: See map 6 page 787.

Southwestern California to southern Nevada, southwesternmost Utah, southern Arizona, southern New Mexico, and Texas west of the Pecos River; Mexico as far south as Sinaloa and San Luis Potosí; disjunct in Querétaro.

The Horse Latitudes occur between latitudes 23° and 34° N. or S. but above 30° deserts may not develop unless there are additional factors causing dryness. For example, even though deserts develop in the Horse Latitudes of the West, they do not in those of the Southern States, because at that longitude the air mass descends over the Gulf of Mexico, and there it becomes both warm and moist (p. 724). The 34th parallel runs about through Los Angeles in coastal southern California, but the coastal strip is not desert but largely California Chaparral. At the same latitude east of the Cascade-Sierra Nevada Axis there are deserts, where the rainfall is only about one-third to one-eighth that at the same altitude a few miles away on the west side of the mountains. The intervening mountains range up to 11,845 feet, and the rain shadow to the east is remarkably effective in southeastern California, southern Nevada, southwestern Utah, and much of Arizona. The Colorado Desert lies at a low altitude, mostly below 1,500 feet, in the lee of the highest mountains of southern California, and it is the driest of the Southwestern Deserts, mostly with only 2 to 5 inches of annual rainfall. The Mojavean Desert is higher, commonly 2,000 to 3,000 or 5,000 feet, and the

Figure 25-35. Navajoan Desert with Zuñi Mountain in the background; western edge of New Mexico. The mountain is the ancestral home of the Zuñi Indians, who live now along the river at the base of the mountains, the roving Apaches and Navajos having been pacified and the stronghold above being unnecessary. The fantastic and beautiful coloring of the rocks in the Navajoan Desert is unrivaled in the United States and Canada, at least.

weather is cooler and the rainfall greater, about 8 to 12 inches.

According to Axelrod, occurrence of deserts around the world in the Horse Latitudes is a new feature of the Earth. About 20% of the present land surface is desert, but this was not always so. After the beginning of flowering plants sometime in Mesozoic, there were no large lowland drought areas until the latter half of Eocene, and true deserts appeared only recently, since Pleistocene.

Evolution in extreme climates is rapid because of the rigid natural selection, and in small areas it proceeds quickly because the balance of prevailing genetic types in a small population system can be changed in a relatively short time. According to Axelrod, during much of Cenozoic time plants adapted to relatively dry conditions developed in local drought areas. These included rain shadow pockets on mountains, or nearby in the downwind valleys, and on domes of granite or gneiss, the meager soil on outcrops of these hard crystalline basement rocks drying rapidly, as on the granite domes of the southern Appalachians or Sugarloaf at Rio de Janeiro. Ultimately the plants of these dry localities and niches spread into the deserts as they were formed, perfecting their evolutionary adaptations as the climate became drier. The incipient desert taxa included individuals capable of further selection in the same direction, in this case toward becoming xerophytes, or drought-tolerant plants.

The deserts as we know them developed after Pleistocene as rainfall subsided and temperatures rose. The deserts of the United States were populated by migration of some dryland species northward and movements of others from rain shadow pockets nearby. Doubtless in both cases, the plants completed their adaptation as the deserts developed. The uplift of the Cascade-Sierra Nevada Axis at the western edges of the deserts during Pliocene and Pleistocene and afterward added impetus to desert formation, and the combination of desert species, assembled so rapidly, is remarkable in its adaptations to a new desert climate.

Today deserts occur on the large and relatively low land masses in the Horse Latitudes, as in the Sahara; on the leeward sides of mountains on both the mainland and islands; on low islands, even like Curaçao in the Tropics, where the prevailing winds are not lifted and they pass over without losing much water; on low headlands of islands where the wind similarly is not yet lifted; in coastal areas or on tropical desert islands where the surrounding water is cold, and it yields little moisture to the atmosphere; and in interior continental areas cut off from any source of water, as with the Gobi Desert. The Gobi Desert receives the prevailing westerlies only after they have passed over the greatest land mass in the world, the polar winds coming from the frozen north, and winds from the south after they have crossed the Himalaya range, the highest in the world. In the fourth direction, the Pacific Ocean is downwind and far across land.

Desert soils are of many types, and some less known from other areas include alluvial fans, desert pavement, sand dunes, and playas.

Figure 25-36. A desert island, Curaçao in the Dutch West Indies near the northern edge of South America. Despite the presence of the trade winds from warm seas, there is little rain because the island is so low that the air mass is not lifted and it passes over without dropping moisture. The city, Willemstad, with a population of 130,000, is dependent wholly upon conversion of sea water to fresh. The cactus is *Melocactus*, the species of which are in a confused state because of much splitting off of minor forms as presumed species. The apical turret of each stem is flower-bearing, as well as woolly.

An *alluvial fan* or bajada forms in the mouth of each desert canyon, and often the fans of several canyons merge into a single slope, often many miles across. The rocks, gravel, and sand washed from the canyons by the infrequent torrential rains form the fans, which vary irregularly in soil consistency, though in general the larger rocks and fragments are deposited nearer the mountain and the finer ones and gravel and sand are carried toward the edge of the fan. The alluvial fans are excellent places for desert plants of many species, and they are populated heavily.

The floor of the desert often is covered with *desert pavement*, a mosaic of rock fragments covering fine, flourlike soil and preventing its further erosion. The wind and rain have carried away the upper fine soil, leaving the rocks it contained as a thin pavement, which protects the rest of the soil.

Sand dunes, occurring extensively in the Sahara and appearing in legends and stories, are thought to be characteristic of deserts, but they are uncommon, and they cover only a tiny fraction of the surfaces of most desert areas. The most spectacular in the Southwestern Deserts of the United States are in California west of Yuma, Arizona, and they have been the sets for "Foreign Legion" moving pictures. Extensive areas in the southern Mojave Desert in California have blowing sand, which buries hills or mountains.

A *playa* is the floor of an undrained desert basin. Because of hardpan formation, often it forms a dry lake, level as a floor and centrally without vegetation. There is shallow water only briefly after heavy storms, and its loss is wholly by evaporation. Usually the playa becomes strongly alkaline, as does the soil in the low areas around it, and these soils support a distinctive vegetation, emphasizing various Chenopodiaceae, like the salt-bushes, adapted to high concentrations of salts.

Desert plants are adapted in many ways to conservation of water. Cacti have succulent stems with much peripheral and central water storage tissue and a heavy covering epidermis nearly impervious to water. The plants blot up any moisture available through shallow, widespreading root systems, adapted to securing water from even light rains. Cacti and other succulent plants of various families close their stomata (pores) during the daytime but open them at night when water loss is low. Carbon dioxide is absorbed then and combined with organic acids until the next day. Photosynthesis starts at daylight, and the carbon dioxide is released for use during the day. Thus, there is no necessity for having the stomata open in the daytime when water loss would be rapid. Some desert plants like the mesquites have deep roots as well as those nearer the surface, and this enables them to endure long dry periods. Most desert bushes, shrubs, and trees reduce transpiration by losing their leaves during the dry season, but they rapidly develop new ones after the next rains. During drought, perennial herbs die down to or near ground level, and the tops are regrown during the next moist season. Annual herbs die after flowering, but the seeds live in dormancy until rain and the proper temperature come again. In areas like Tucson, Arizona, having both winter and summer rain, the annuals growing during the cool spring and those during the hot summer are of different species, being physiologically adapted to different conditions. "Resurrection plants," like desert ferns and *Selaginella*, curl up and turn brown or tan during the dry season, but when water comes, naturally or artificially, they resume their normal shape and turn green within a fraction of an hour to a few hours.

The Horse Latitudes of North America are chiefly in northern Mexico, and the deserts in the United States are primarily those along the fringes of the Desert Zone, where addition of rain shadows is necessary to produce deserts. The Mexican Desert Flora centers south of the border, but it is represented well in the United States, roughly from Palm Springs in southeastern California to the Pecos River in western Texas. The deserts of the United States are in an area of very low rainfall, mild win-

Figure 25-37. Specialized desert habitats: *above,* sand dunes, including those forming about mesquite trees (*Prosopis juliflora* var. *Torreyana*); near Stovepipe Wells, Death Valley, California (sand dunes being occasional in occurrence in the deserts but covering only a minute percentage of desert areas); *below,* an alkali flat at the edge of a playa (dry lake), with a specialized flora including the iodine bush (*Allenrolfea occidentalis*), black greasewood (*Sarcobatus vermiculatus*), alkali blight (*Suaeda Moquinii* and *Suaeda suffrutescens*), and in less alkaline soil various species of saltbush, *Atriplex;* near Bad Water, Death Valley, California.

ters, and hot summers. In the western portion rain is confined ordinarily to winter; in the eastern portion it falls in both winter and summer, but summer rain is more important.

The entire Mexican Desert flora is marked by the creosote-bush, *Larrea tridentata.* Various other plants each occur through a part of the Mexican Desert Flora, but no other species occurs through even a high percentage of the desert region of North America. Consequently, the desert is defined, for convenience, in terms of the creosote-bush, which has a broad ecological tolerance for desert environments but which either cannot endure other dryland conditions or cannot compete with other plants better adapted to them.

Nearly all the species and a great many of the genera occurring in the Southwestern Deserts or creosote-bush deserts of the United States are different from those of the Sagebrush Desert, and the flora is in sharp contrast to that of any other floristic association occurring in the United States. The affinities of most of the species are with Mexican plant groups.

There are four desert floristic associations in the United States: the Mojavean, Colorado, Arizona, and Chihuahuan deserts. Commonly the Colorado and Arizona deserts are grouped under the Sonoran Desert or Deserts in Baja California and Sonora, Mexico, just to the southward. However, a tabulation of trees and shrubs made during preparation of the third edition of *The Trees and Shrubs of Southwestern Deserts* (University of Arizona Press, in press) indicates the floristic interrelationships of the four deserts to be about equal.

1. THE MOJAVEAN DESERT

Occurrence. The northern part of southeastern California; southern Nevada; extreme southwestern Utah; northwestern Arizona and southward to a line just south of the Bill Williams River; 2,000 to 3,500 or 5,000 feet.

The term Mojavean Desert describes the floristic association, which is not quite coextensive in California with the Mojave (or Mohave) Desert (a geographical term). The Mojavean Desert area is penetrated at lower levels of altitude by the exceedingly dry Colorado Desert, as, for example, along the lower Mojave River and in low-lying country just southwest of it.

CHARACTERISTIC WOODY SPECIES

Thalamiflorae

Atriplex Parryi, Parry salt-bush
Mortonia scabrella var. *utahensis*
Canotia Holacantha, crucifixion-thorn
Opuntia acanthocarpa var. *coloradensis,* staghorn cholla
Opuntia erinacea
Echinocereus triglochidiatus var. *mojavensis,* Mojave hedgehog cactus
Echinocactus polycephalus, a small barrel cactus
Echinocactus polyancistrus, fishhook barrel cactus
Echinocactus Johnsonii
Mammillaria vivipara var. *deserti,* desert-pincushion

Corolliflorae

Mendora spinescens
Lycium pallidum var. *oligospermum*
Lycium Parishii
Salvia funerea, Death Valley sage
Salvia mohavensis, Mojave sage
Penstemon albomarginatus

Calyciflorae

Cassia armata, spiny senna (chiefly of this desert)
Dalea arborescens
Dalea Fremontii vars. *pubescens, Saundersii,* and *minutifolia,* indigo-bushes

Ovariflorae

Amphiachyris Fremontii
Acamptopappus Shockleyi
Haplopappus linearifolius var. *interior*
Chrysothamnus teretifolius
Lepidospartum latisquamum
Tetradymia stenolepis, horse-brush
Viguiera reticulata
Ambrosia eriocentra, woolly bur sage

Figure 25-38. The creosote-bush, *Larrea tridentata,* the species most characteristic of the Mexican Desert Flora and the one outlining it almost exactly. Widely spaced individuals, as shown here, are characteristic. (Frank P. McWhorter.)

Monocotyledons
Yucca brevifolia, Joshua-tree
Gymnosperms
Ephedra viridis, green ephedra
Ephedra californica var. *funerea,* Death Valley ephedra

Discussion. According to Axelrod the southern elements of the flora of the Mojavean Desert had not mingled with those of the Sagebrush Desert of the Great Basin before middle Pliocene. Invasion of elements from the Great Basin occurred probably during late Pliocene and Pleistocene, when the winter climate became cold and the Mojave region was elevated and when mountains arose rapidly along its western border.

At the close of Pleistocene plants must have migrated from local rain shadow pockets and from the drier areas of the Horse Latitudes into the Mojavean Desert region, as well as into the present Colorado Desert. Even during Miocene the short mountain ranges had local rain shadows in their lees. These did not form deserts, but they were dry.

The rain falls in the wintertime, and summer rains occur only when a tropical storm goes farther north than usual. In moist years, during April and sometimes the first week of May there are numerous species of annuals and of flowering shrubs. However, in most years a visitor may drive clear across the Mojavean Desert when it is in full bloom without knowing it is the flowering season, because in most areas the annual plants of dry years tend to occur singly under bushes rather than in dense masses. However, in some places, especially during moist years, there are colorful fields of flowers.

Figure 25-39. Mojavean Desert: *above,* the upper part of the desert at the level of Joshua trees, *Yucca brevifolia,* Clark Mountain, Mojave Desert, California; *below,* the floor of the lower desert with bur sage, *Ambrosia dumosa,* predominating but with some creosote-bush, *Larrea tridentata;* near the Valley of Fire, north of Lake Mead, Nevada.

In both the Mojavean and the Colorado deserts, a combination of late summer tropical storms may bring about seed germination of species occurring mostly farther east in Arizona, where there is annual summer rain. In southeastern California this occurs only once in five to twenty years, and the younger woody plants occur in age classes according to the years of the more recent abundant summer rains.

The higher levels of the Mojavean Desert, above about 3,000 feet, are characterized by the fantastic Joshua Tree, *Yucca brevifolia* (Figure 25-39). This species serves as a marker of the outlines of the Mojavean Desert, and it does not occur much beyond the floristic association.

2. THE COLORADO DESERT

Occurrence. Low-lying areas on the drainage of the lower Colorado River; California in the Salton Sea Basin and near the Colorado River at −200 to 1,500 feet, but in the lee of the high and steep San Jacinto Mountains up to 4,200 feet; extreme southern tip of Nevada; Arizona along the Colorado and Gila Rivers up to about 1,500 feet; northeastern Baja California and northwesternmost Sonora, Mexico.

CHARACTERISTIC WOODY SPECIES

Thalamiflorae
- *Ayenia compacta*
- *Tetracoccus Hallii* (chiefly in this desert)
- *Castela* (*Holacantha*) *Emoryi,* desert crucifixion-thorn
- *Condalia Parryi*
- *Opuntia ramosissima,* pencil cholla
- *Ferocactus acanthodes,* barrel cactus (primarily in this desert)
- *Mammillaria tetrancistra,* fishhook cactus (chiefly in this desert)

Corolliflorae
- *Asclepias subulata,* desert milkweed
- *Asclepias albicans,* desert milkweed
- *Lycium Andersonii* var. *deserticola*
- *Lycium brevipes*
- *Salvia Greatae,* sage
- *Salvia eremostachya,* sage
- *Justicia californica*

Calyciflorae
- *Hoffmanseggia microphylla*
- *Olneya Tesota,* desert ironwood
- *Dalea spinosa,* smoke tree
- *Dalea Schottii*
- *Dalea Emoryi*
- *Chrossosoma Bigelovii* var. *glaucum*
- *Prunus Fremontii,* desert apricot

Ovariflorae
- *Brickellia frutescens*
- *Gutierrezia californica* var. *bracteata,* match-weed
- *Haplopappus propinquus*
- *Ambrosia ilicifolia,* bur sage

Monocotyledons
- *Washingtonia filifera,* California fan palm
- *Agave desertii,* desert century-plant

CHARACTERISTIC WOODY SPECIES OF BOTH THE MOJAVEAN AND THE COLORADO DESERTS

Thalamiflorae
- *Atriplex hymenelytra,* desert holly
- *Thamnosma montana*
- *Opuntia basilaris,* beavertail cactus

Calyciflorae
- *Prunus fasciculata,* desert almond (chiefly Mojavean)

Ovariflorae
- *Hofmeisteria pluriseta*
- *Brickellia desertorum*
- *Haplopappus Cooperi*
- *Chrysothamnus paniculatus*

Discussion. In the Colorado Desert the winter rains (October to March) are spaced widely, irregular in distribution, and light, their total being only two to five inches. Nonetheless, they support many species of vernal annuals. This spring vegetation is fairly homogeneous in the Colorado Desert, and many of the same species occur also in the Arizona Desert as a part of the spring flora there, where there are both winter and summer rain. However, the Arizona Desert species dependent on summer rain are absent

Figure 25-40. Colorado Desert: *above,* Palm Canyon south of Palm Springs, California, with many desert shrubs on the hillsides and with palms, California fan-palm, *Washingtonia filifera,* along the creek coming down from the higher mountains near by; *below,* desert hills in Mason Valley near the Anza-Borrego State Park in San Diego County, California; barrel cactus, *Ferocactus acanthodes;* chollas, *Opuntia acanthocarpa* var. *Ganderi* (center); desert century plant, *Agave desertii;* a young ocotillo, *Fouquieria splendens* (just left of center).

from California or they are rare and dependent for seed germination on the unusual combination of several summer tropical storms. Occasionally, after unusually heavy winter rains, the Colorado Desert or parts of it may become a flower garden in the spring, that is, during February and March or sometimes the first week of April. Summer comes in April. Temperatures of 110°F to 115°F are common during the summer, and 120° is unusual. Ordinarily the humidity is low, commonly very low.

Much of the Colorado Desert occupies the site of a prehistoric northward extension of the Gulf of California later covered for a long period by a freshwater lake only 40 feet above sea level. This left a beach line still obvious for many miles along the hills west of the Salton Sea. Certain characteristic species of the seacoast of lower California and southern California persist along this beach line; one of these is the shrub *Euphorbia misera.* The California fan palm (*Washingtonia filifera*) is a relict endemic species occurring in the canyons above the beach line, perhaps having persisted there since a time like late Pleistocene when precipitation was more plentiful and distribution more nearly continuous.

3. THE ARIZONA DESERT

Occurrence. Just south of the Bill Williams River and of the Mogollon Rim in western, central, and southern Arizona; near the Gulf of California in Sonora, Mexico.

CHARACTERISTIC WOODY SPECIES

Thalamiflorae

- *Berberis Harrisoniana,* Kofa Mountain barberry
- *Gossypium Thurberi,* desert cotton
- *Abutilon Pringlei*
- *Acalypha Pringlei*
- *Jatropha cardiophylla*
- *Opuntia arbuscula,* pencil cholla
- *Opuntia Kleiniae* var. *tetracantha*
- *Opuntia acanthocarpa* vars. *major* and *Thornberi,* cane chollas
- *Opuntia versicolor,* staghorn cholla
- *Opuntia fulgida* vars. *fulgida* and *mamillata,* jumping chollas
- *Opuntia Stanlyi,* devil cholla
- *Opuntia Stanlyi* var. *Kunzei,* Kunze cholla
- *Opuntia Stanlyi* var. *Peeblesianus*
- *Opuntia phaeacantha* var. *flavispina,* prickly pear
- *Cereus Thurberi,* organ-pipe cactus
- *Cereus Schottii,* senita
- *Cereus giganteus,* saguaro, giant cactus
- *Echinocereus fasciculatus,* vars. *fasciculatus* and *Boyce-Thompsonii,* hedgehog cacti
- *Echinocereus Engelmannii* vars. *acicularis* and *Nicholii,* hedgehog cacti
- *Ferocactus acanthodes* var. *Eastwoodiae,* Eastwood barrel cactus
- *Ferocactus Covillei,* Coville barrel cactus
- *Echinocactus horizonthalonius* var. *Nicholii*
- *Neolloydia erectocentrus* var. *acunensis*
- *Mammillaria microcarpa,* fishhook cactus
- *Mammillaria Thornberi,* fishhook cactus

Corolliflorae

- *Haplophyton Crooksii,* Arizona cockroach plant
- *Forestiera Shrevei,* desert olive

Calyciflorae

- *Cercidium microphyllum,* foothill palo verde
- *Sophora arizonica,* Arizona sophora
- *Sophora formosa,* Gila sophora
- *Coursetia glandulosa*
- *Crossosoma Bigelovii* var. *Bigelovii*
- *Cowania subintegra,* Burro Creek cliffrose

Ovariflorae

- *Brickellia Coulteri*
- *Haplopappus tenuisectus,* burro-weed
- *Baccharis sarothroides,* desert-broom
- *Ambrosia ambrosioides,* canyon ragweed
- *Ambrosia deltoidea,* bur sage
- *Ambrosia cordifolia*

Monocotyledoneae

- *Yucca baccata* var. *brevifolia,* Thornber yucca

CHARACTERISTIC WOODY SPECIES OF BOTH THE COLORADO DESERT AND THE ARIZONA DESERT

Thalamiflorae

- *Simmondsia chinensis,* jojoba, goat nut
- *Colubrina texensis* var. *californica*
- *Condalia globosa,* bitter condalia
- *Condalia Warnockii* var. *Kearneyana*
- *Ziziphus Parryi,* California lote bush
- *Opuntia Bigelovii,* Teddy-bear cactus

Figure 25-41. The fantastic Arizona Desert (cf. also **Figure 24-1,** *lower,* p. 716). *Opposite, upper photo,* grove of saguaros (giant cacti, *Cereus giganteus*), the associated plants including burro-weed (*Haplopappus tenuisectus*), left foreground, *Condalia Warnockii* var. *Kearneyana,* right foreground, a cholla (*Opuntia versicolor,* the cactus at the extreme left), desert hackberry (*Celtis Tala* var. *pallida*) beneath the two young saguaros at the left, foothill palo verde (*Cercidium microphyllum*) at the left of the large group of saguaros, and velvet mesquite (*Prosopis juliflora* var. *velutina*) at the extreme right; Saguaro National Monument, east of Tucson, Arizona; *opposite, lower photo,* desert near Florence, Arizona, the smaller cacti being the jumping cholla (*Opuntia fulgida*) and a prickly pear (*Opuntia phaeacantha*); *above,* desert hillside in the Organ Pipe Cactus National Monument, western Pima County, Arizona; saguaros, *Cereus giganteus,* the older ones branching above, the younger still unbranched; organ pipe cactus, *Cereus Thurberi,* with branches from the base; a barrel cactus, *Ferocactus Covillei,* two large plants growing in the center of the major slope; prickly pears, *Opuntia phaeaacantha* var. *flavispina,* in the foreground; foothill palos verdes, *Cercidium microphyllum,* the large shrubs. Photographs, courtesy of the United States National Park Service.

Corolliflorae

Fouquieria splendens, ocotillo (also in the Chihuahuan Desert)

Menodora scabra var. *glabrescens* (chiefly in these deserts)

Lycium Cooperi

Lycium macrodon

Lycium Fremontii (chiefly in these deserts)

Hyptis Emoryi, desert lavender

Calyciflorae

Acacia Greggii, cat-claw

Calliandra eriophylla, fairy-duster (chiefly Arizona Desert)

Cercidium floridum, blue palo verde

Ovariflorae

Trixis californica

Peucephyllum Schottii

Hymenoclea Salsola var. *pentalepis*, burro-bush

Discussion. The Arizona Desert is the most picturesque floristic association in North America. Commonly the visitor from a well-watered country is shocked at first by the dryness of the landscape of any desert and the absence of green grass, blue water, and white snow, representing the colors he has been taught to consider beautiful in nature. Slowly he becomes enthralled with the many other colors that overshadow the stock three, and he sees beauty not to be found in his previous world.

The dominant feature of the landscape is the gigantic saguaro, standing like a candelabrum on the desert and towering above all other plants—up to 50 feet high. Along with the saguaro is a host of other cacti and trees and shrubs like no others in the world.

In the Arizona desert there are two rainy seasons. The average winter rainfall at Tucson is about 4 or 5 inches and the summer 6 to 7 inches. The woody and large succulent plants are present the year around, and most of them bloom in May, but the herbs have two flowering seasons—March and April and late July to early September. This reflects the two rainy seasons, the winter rains occurring from October to April (or sometimes only part of this period) and the summer rains in July and August. One set of herbs and some other plants depend primarily on the winter rains; another set of herbs is dependent primarily on the summer rains. In one group the seeds germinate only when it is cool and moist, in the other when it is warm and moist. Most trees, shrubs, and cacti are dependent on summer rain for seed germination. Thus, to a considerable extent, one flora is superimposed upon the other, and two crops of native plants grow on the same soil. The summer rain, coming during the hot season, is particularly effective, and it is concentrated into only two months. Thus, during the summer growth both can and must be rapid.

4. THE CHIHUAHUAN DESERT

Occurrence. Arizona (represented by a few species in the southeastern corner of the State); southern New Mexico, as along the Rio Grande, from the Tularosa Valley southward and southeastward, and in the lower portion of the Pecos River Valley; Texas west of the Pecos River and southward along the Rio Grande; Mexico as far south as San Luis Potosí and disjunct in Querétaro.

CHARACTERISTIC WOODY SPECIES IN THE UNITED STATES

Thalamiflorae

Berberis trifoliolata, red barberry

Opuntia Schottii var. *Grahamii*, cholla

Opuntia arenaria

Opuntia rufida, blind prickly pear

Opuntia violacea vars. *macrocentra* and *Castetteri*, prickly pears

Opuntia strigil var. *strigil*

Cereus Greggii var. *Greggii*, desert night-blooming cereus

Echinocereus triglochidiatus var. *paucispinus*, red-flowered hedgehog cactus

Echinocereus enneacanthus vars. *stramineus* and *dubius*, hedgehog cacti

Echinocereus pectinatus vars. *Wenigeri* and *minor*

Echinocereus Reichenbachii var. *chisosensis*

Figure 25-42. Chihuahuan Desert: *above,* hillside stand of a century plant, *Agave Lechuguilla,* cat-claw, *Acacia Greggii,* and other large shrubs in the wash (draw); Big Bend National Park, Texas, the Chisos Mountains in the background; *below,* the creosote-bush, *Larrea tridentata,* the characteristic species of the Mexican Desert Flora in a typical almost pure stand but with a few lechuguillas and western honey mesquites, *Prosopis juliflora* var. *Torreyana;* soil formation, *desert pavement,* a common mosaic of surface pebbles and rock fragments underlain by fine dust, the rock having been left behind by wind and water erosion of several feet of soil until a continuous surface rock layer was formed and the rapid erosion was stopped; Big Bend National Park north of the Chisos Mountains.

Lophophora Williamsii, peyote (also Rio Grande Plain)
Echinocereus horizonthalonius var. *horizonthalonius*
Epithelantha Bokei
Neolloydia conoidea
Neolloydia Warnockii
Neolloydia mariposensis
Ancistrocactus uncinatus var. *Wrightii*
Coryphantha macromeris var. *macromeris*
Coryphantha Scheeri vars. *Scheeri* and *uncinatus*
Coryphantha ramillosa
Coryphantha Sneedii vars. *Sneedii* and *Leei*
Coryphantha dasyacantha var. *varicolor*
Ariocarpus fissuratus
Mammillaria Pottsii

Corolliflorae
Bumelia lanuginosa var. *rigida,* chittamwood
Chilopsis linearis var. *linearis,* desert-willow
Chilopsis linearis var. *glutinosa,* sticky desert-willow
Tecoma stans var. *angustatum,* yellow trumpet-flower
Anisacanthus Thurberi, chuparosa

Calyciflorae
Acacia constricta var. *vernicosa*
Cassia Wislizenii, shrubby senna
Koeberlinia spinosa, desert crucifixion-thorn

Ovariflorae
Flourensia cernua, tar-brush or black brush
Parthenium incanum, mariola

Monocotyledons
Dasylirion leiophyllum, sotol
Agave Lechuguilla, lechuguilla

There are many other characteristic and endemic species in Transpecos Texas and in Mexico, where the Chihuahuan Desert extends over most of the great Mexican Plateau. These species have not been evaluated sufficiently to be certain which are really characteristic of this desert and which occur in other floristic associations as well.

Discussion. The Chihuahuan Desert is at the core of the Mexican Desert Flora, and it is the most extensive of the North American deserts. The part of this desert occurring in the United States is only a northern fringe. The center of development north of the border is in the Big Bend of the Rio Grande in western Texas. The Big Bend and El Paso County are rich in species characteristic of the Chihuahuan Desert, and they afford the best example of this type of desert in the United States.

The climate differs from that of the Arizona Desert in the colder winters and the even greater emphasis on summer rain. The winter is not continuously cold, but it is in the pathway of "northers," the polar winds from the western part of the polar continental air mass, and at times these cause severe chilling.

The Chihuahuan Desert seems drier than those of the Colorado Basin, and commonly, at least in the United States, the creosote bushes are smaller and with a tendency to be yellow green.

Cacti are abundant, and the species are as numerous as in Arizona, but mostly the plants are smaller and less conspicuous from a distance, though a few species are of up to about 4 feet tall. Thus, in the United States cacti are abundant in but not the primary feature of the Chihuahuan Desert, as they are of the Arizona Desert. Nevertheless, they are an important part of the landscape, but more through appearing on a considerable amount of desert floor and through their variety than through towering above the other desert plants or being more conspicuous than they. Farther north, in Mexico, cacti, as well as yuccas and century plants (*Agave*: mescal or maguey), become more prominent.

The American Tropical Flora

(Torrid Zone)

The Tropics span 46° of the 180° of latitude—in the region of the greatest circumference of the Earth. They lie between the Tropic of Capricorn and the Tropic of Cancer, each about 23° of latitude from the equator. They include the two belts of trade winds and the belt of tropical calms or doldrums, where the winds meet at the heat equator. There the air rises over the heat equator, resulting in the torrential rains and the seasons described in broad terms earlier (p. 721). The rainy belt of the heat equator moves north and south each year as the axis of the Earth shifts. Consequently there is heavy summer rain in both the northern and southern parts of the Tropics, and twice a year the equatorial region is drenched as the heat equator crosses, once going north and again going south. During the winter of either hemisphere, when the heat equator is in the other hemisphere, the dry season is pronounced—the trees become leafless or with dry leaves, the water holes dry up, mosquitoes and malaria disappear, and the large animals stay in the vicinities of waterholes.

The Tropics are prevailingly warm and moist, and there are no frosts. Consequently most of the plants on the Earth can grow there, provided their physiological cycle does not require a cold period, they can withstand an abundance of moisture, and they can compete effectively with many other species meeting about the same requirements. Since severity of climate does not eliminate species, many tend to survive, and the competition is intense. Survival against competition requires plants to be adapted to special habitat niches in which they have a slight advantage. Usually there is no great mass of members of the same species; one or a few individuals may occur here or there, wherever the species has a slightly better adaptation than the others. Innumerable other species are similar in their adaptations to specific niches, and the flora is exceedingly rich in taxa.

During the first part of Mesozoic (Triassic and Jurassic) the single great supercontinent, Pangaea, began its separation into the present continents, but they were not far apart. During Cretaceous separation became more and more significant. When flowering plants appeared during middle Mesozoic, there were only tropical lowlands and uplands. According to Axelrod, angiosperms originated in the warm temperate uplands and then invaded the lowlands, where they replaced the forests of gymnosperms, pteridophytes of all the groups, and byrophytes.

In the meager extension of this flora into the southeastern United States, subdivisions are difficult to classify.

1. THE CARIBBEAN TROPICAL FOREST

Occurrence. This floristic association is represented in the United States chiefly along the shores and in the "hammocks" and Everglades of southern Florida and to a slight extent in swampy regions along the Gulf of Mexico, particularly near Brownsville, Texas.

CHARACTERISTIC WOODY SPECIES

Thalamiflorae

Annona glabra, custard-apple or pawpaw
Annona squamosa, sugar-apple
Ficus aurea, strangler fig, golden fig
Ficus citrifolia, wild banyan
Trema floridana, Florida nettle tree
Trema Lamarckiana, West Indian nettle tree
Opuntia triacantha, prickly pear
Opuntia cubensis, prickly pear
Opuntia stricta vars. *stricta* and *Dillenii,* prickly pears
Rhipsalis baccifera, pencil cactus
Cereus Robinii vars. *Robinii* and *Deeringii*
Cereus eriophorus var. *fragans*
Cereus gracilis vars. *Simpsonii* and *arboriginum*

Corolliflorae

Ardisia escalonioides, marlberry
Jacquinia keyensis, Joe-wood
Myrsine guayenesis, myrsine
Chrysophyllum olivaeforme, satinleaf
Mastichodendron foetidissimum, mastic
Dipholis salicifolia, bustic, cassada
Bumelia celastrina var. *angustifolia,* ant-wood
Mankilhara bahamensis, wild sapodilla
Avicennia germinans, black mangrove

Figure 25-43. The Caribbean Tropical Forest: *above,* palm forest southeast of Fort Myers, Florida, this type of forest occurring only in a narrow zone somewhat south of the middle of the state; *below,* second growth subtropical forest on Key Largo, Florida Keys. The forest comes back rapidly after clearing, this area having been a pineapple plantation.

Calyciflorae
Lysiloma bahamensis, wild tamarind
Pithecellobium Unguis-Cacti, cat-claw
Pithecellobium keyense
Caesalpinia pauciflora
Caesalpinia crista, gray nicker
Caesalpinia Bonduc, yellow nicker
Conocarpus erecta, buttonwood
Bucida Buceras, black olive
Laguncularia racemosa, white mangrove
Myrtus verrucosa, stopper
Myrtus bahamensis, stopper
Eugenia myrtoides, Spanish stopper
Eugenia axillaris, white stopper
Eugenia rhombea, red stopper
Eugenia confusa, ironwood
Calyptranthes pallens, spicewood
Calyptranthes Zuzygium, myrtle-of-the-river
Rhizophora Mangle, mangrove
Ovariflorae
Exostema caribaeum, princewood
Casasia clusiifolia, seven-year apple
Psychotria undata, wild coffee
Psychotria ligustricifolia, Bahaman wild coffee
Psychotria Sulzneri
Monocotyledons
Pseudophoenix Sargentii, cherry cabbage palm
Roystonea elata, Florida royal palm
Thrinax parviflora, Florida thatch palm
Thrinax microcarpa, brittle thatch palm
Coccothrinax argentea, silver palm
Acoelorraphe Wrightii, Everglades saw-cabbage palm
Gymnosperms
Pinus caribaea
Zamia floridana, cycad

Discussion. The tropical areas around the Gulf of Mexico and the Atlantic Ocean, including the tip of Florida and the Keys, have heavy rains during the summer and dry weather during the winter. Thus Florida has wet summers and dry winters, while California has the reverse.

Florida arose from under the ocean during Pleistocene, and it is only slightly above the sea. In the tropical Everglades the land and sea are at so nearly the same level that only a few inches to a yard may make the difference between saw-grass (*Mariscus*) swamp (everglades), land, or hammocks. A hammock is an area slightly higher than the swamp, often submerged or nearly so during the summer and emerging from the water during the winter dry season. It is a patch of forest in the swamp, and the forest includes not only tropical trees and shrubs but also a variety of epiphytes, such as ferns, orchids, and bromeliads, living on the tree trunks and branches but not parasitic. The tropical flora as a whole is rich in epiphytes, including ferns and orchids, and various genera and species are not to be found elsewhere in the United States.

The Caribbean Tropical Forest occurs in various forms on the islands and shores of the Caribbean Sea. The migration of the largely insular plants to southern Florida has occurred over water during the short period of existence of Florida. According to Richard A. Howard, the representation of the tropical flora in Florida is weak, and special emphasis is on strand plants, ferns, epiphytes, and weeds, the species carried most readily across water from the Antilles.

The Hawaiian Flora

On the Hawaiian Chain, the islands extending about 2,000 miles from Hawaii to Midway; sea level to 13,784 feet; tropical at the lower levels and temperate at the higher altitudes.

Almost all the native plants, about 2,000 species, are endemic to the islands, the few not endemic being mostly strand plants appearing on many tropical shores. This is the most highly endemic flora considered in this book, and it has been described as the most strongly endemic in the world.

The Hawaiian Flora is not one of the North American Floras, and it is discussed here only because Hawaii is one of the United States and because many people from the other states visit there. For the most part, the flora is of Southeastern Asian insular origin, but it cannot be assigned to a single source. The islands

Figure 25-44. Caribbean Tropical Forest: the Everglades, Florida: *above,* in the foreground the typical surface of the Everglades, featuring saw-grass, a sedge (*Mariscus jamaicensis*), in the background a cypress hammock, the principal tree being the pond cypress (*Taxodium ascendens*); *below,* epiphytes, including bromeliads and orchids, growing on the pond cypresses of a hammock. Everglades National Park. (Cf. also the photographs of mangroves, which occur in shallow water along or near the Gulf of Mexico and the Atlantic Ocean, **Figure 10-87.**) Photographs, courtesy of the United States National Park Service.

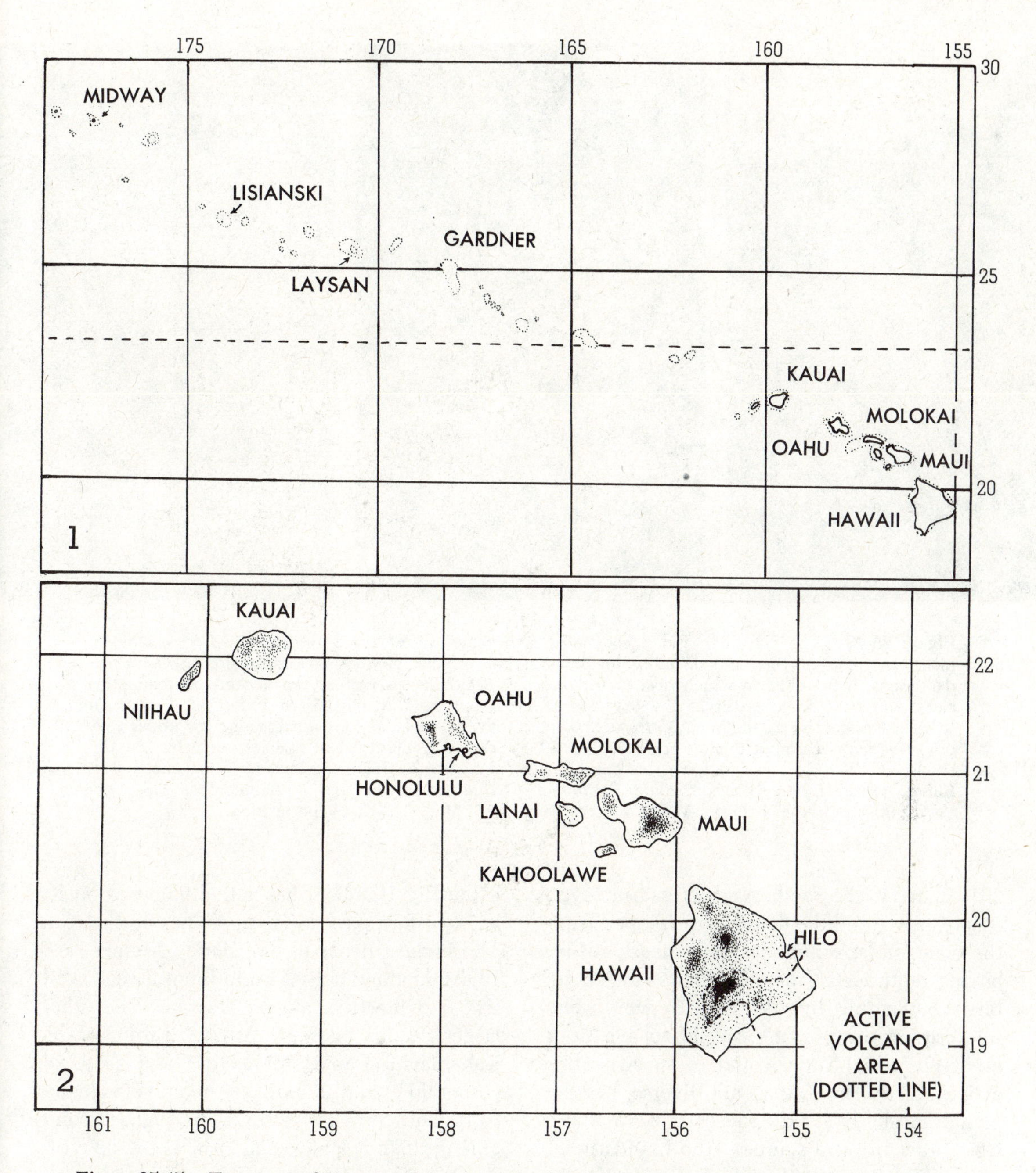

Figure 25-45. Two maps of Hawaii: Above the entire Hawaiian Chain of islands, reaching from Midway on the northwest to the "Big Island" of Hawaii on the southeast, the islands having been formed over an opening in the crust of the earth under the southeastern part of the Big Island where there is current volcanic activity, as there has been for 5 million years. The newly formed islands have been carried slowly westward and northward by seafloor spreading, eroding as they went, those west of the well-known cluster of larger islands being eroded to base level insofar as lava is concerned, but covered with calcareous material from corals and other marine organisms. LYMAN BENSON, *Plant Taxonomy, Methods and Principles.* © Ronald Press Co., New York, 1962; reprinted by John Wiley & Sons, New York, 1978. Used by permission.

Figure 25-46. The Hawaiian Flora, ecological niches and forest types in Hawaii: *Left* ecological niches in steep canyons and on cliffs, on Nuuanu Pali on the windward or rainy side of Oahu; indentations in the cliffs in which taxa may develop in isolation, particularly in pockets on mountainsides running from the wet to the dry side of an island; *middle* Ohia-Fern Forest on Hawaii in Kilauea National Park, the ohia, *Metrosideros* spp., with its trunks and crowns forming the upper canopy, tree ferns forming the lower (at about 10 to 15 feet); many plant species occurring on the trunks of the tree ferns and on the ground; *right* Koa Forest, featuring *Acacia Koa;* near Kokee, Kauai. (From LYMAN BENSON, *Plant Taxonomy, Methods and Principles.* © Ronald Press Co., New York, 1962; reprinted by John Wiley & Sons, New York, 1978. Used by permission.

are isolated in the Pacific Ocean, far from even other islands, and the flora has developed from the plants able to migrate there, nearly all by having reproductive or other parts carried by birds, by floating, by rafting, or by wind. The sole criteria for entering the flora have been ability to reach Hawaii and to survive after arrival. Thus, the sources are diverse, but the plants most likely to succeed were those from other tropical sources—the islands nearest to Hawaii and with similar climates.

The Hawaiian Chain of islands is composed of the well-known volcanic islands on the southeast and also of many little-known atolls and other small islands stretching westward to Midway. The southeasternmost island, Hawaii ("the Big Island"), has active volcanos on its southeastern part and extinct ones on the rest. The highest peaks in the islands, Mauna Loa, 13,680 feet and active, and Mauna Kea, 13,784 feet and inactive, are on Hawaii. The other larger islands (going northwestward), Maui, Kahoolawe, Lanai, Molokai, Oahu (site of Honolulu), and Kauai, are inactive volcanically, and they are lower. The highest point among them is Haleakala, 10,029 feet, on eastern Maui near Hawaii. The volcanic islands have originated one at a time over an opening in the crust of the Earth, where lava erupts. They have moved northwestward through seafloor spreading due to wedging the crust apart by lava arising through a fissure. Each of the islands has been worn down gradually by rain, those west of Kauai having been worn to or near sea level and, except rarely, having the entire surface covered with coral and coral sand lying over the remaining lava now lower than the level of the ocean.

Although the islands are small, Oahu being only 40 miles long and even Hawaii only 90, the differences in altitude, the effect of the trade winds, the steep topography, and the two far different major types of soils produce a great variety of habitats, providing many possible niches to be occupied by invading plants. Especially in the beginning, about 5 million years ago, any new habitat was available for the taking. The first plants reaching the Hawaiian Chain had no competition, and, if they could adapt to even one physical and chemical habitat, they survived. Those that were successful proceeded into evolutionary adaptive radiation; that is, they or their modified descendants occupied all the niches to which plants with any gene combinations they could produce might adapt. As time went on and other migrants arrived and occupied niches, competition developed. When all the habitats were taken, competition became more intense, but much less so than elsewhere in the tropics.

The habitats are produced by the factors mentioned above. The *differences of altitude* provide both tropical and temperate niches. The *trade winds* of the Northern Hemisphere, coming from the northeast, are elevated in passing over the main islands, and abundant rain falls and produces warm rainforests on the windward (northeastern) sides, but the rain shadows result in deserts on the leeward (southeastern) sides. The *steep topography* with great cliffs and gashlike canyons on the mountains provides a series of isolated habitats ranging from wet to dry around the sides of each peak or high island, and different species and varieties of the same genus occur in each of these climatic pockets. The contrasting soils —volcanic and coral-limestone—provide radically different habitat series, as well. Thus, the Hawaiian Flora is rich in endemic species of only 86 families, and they have been derived from probably only about 300 chance natural invasions of the islands.

The coming of European man changed the island flora. Today on the lowlands of Oahu and most low parts of the other islands there are almost no native plants. Nearly all have been introduced from the lowlands of other parts of the Tropics, where they have developed under intense competition. When these plants reached the homeland of the less competitive native Hawaiian flora, the area was being disturbed by man, and the invaders soon crowded out the natives. Consequently, the surviving native plants are largely those growing on the high hills and the mountains, though there are some relatively little disturbed lower areas here and there, even near the city of Hilo on Hawaii.

Further floristic study in Hawaii is needed. However, the following appear to be worthy of consideration as possible floristic associations. Obviously each intergrades with the others.

THE OHIA-FERN FOREST

Occurrence. Windward sides of all the islands, but best developed on Hawaii and Maui; mostly replaced by invading species on the Oahu lowlands and much of lower parts of the other islands.

Discussion. In the rainforest tree ferns form a canopy 10 to 20 feet above the ground. The Ohia (*Metrosideros* spp., Myrtaceae) forms an upper canopy, the trunks of the trees appearing like pillars supporting the roof of a large building.

THE KOA FOREST

Occurrence. Middle elevations in the mountains of all the larger islands.

Discussion. The characteristic tree, the koa, *Acacia Koa,* is considered to be a marker of the forest. However the makeup of the floristic association as a whole is in need of further study.

THE LEEWARD ASSOCIATION

Occurrence. Dry areas on the leeward sides of all the islands.

Discussion. The areas at low elevations in the rain shadows are very dry, and some are strongly reminiscent of the Southwestern Deserts of the continental United States.

THE SUMMIT ASSOCIATION

Occurrence. Near the summits of the higher volcanic peaks, which tend to come to sharp points; eastern Maui and Hawaii.

Discussion. The summits of the higher mountain peaks are so narrow that they are said to "part the clouds." The trade winds are elevated in passing over the broader bases of the mountains, but they divide and go around the summits, leaving them dry. Thus, there is a special floristic association at the higher levels. In at least some places, as on Haleakala, it forms a chaparral appearing like the other brushlands or chaparrals of the world but with no species and few genera in common with them.

THE STRAND ASSOCIATION

Occurrence. To some extent on all the islands, but chiefly on the small islands west of the main group of Hawaiian islands.

Discussion. The small islands are composed largely or wholly of coral rocks, sands, and soils. Often there is limestone, cemented together from the residue of corals and other marine organisms. The small islands rise only a little above sea level, according to uplift of the rocks or to changing levels of the ocean over geologic time. The few species are nearly all migrants characteristic of many shores, but the association varies from place to place according to local ecology in the South Pacific.

Specialized Floral Elements

THE COASTAL ELEMENTS

THE STRAND VEGETATION

This flora occurs just back of the sandy beaches along the ocean. It is only in part a natural grouping because it varies from coast to coast, but certain species are found on many widely separated shores.

THE BLUFF VEGETATION

This flora is of doubtful designation, for the sea bluffs of North America include many diverse elements. The thread of relationship among them may be only a minor one.

THE SALTMARSH VEGETATION

This flora varies from coast to coast, and on the southern portion of the West Coast there is a close resemblance to the vegetation of desert alkali sinks or playas. As a vegetation element it is much more homogeneous than either the strand or the bluff vegetation.

The following is quoted from W. S. Cooper (University of California Press, 1936):

> A province such as the one here treated [strand and dune], linear in form, frequently interrupted, unstable in the extreme, cannot be expected to be a center of evolution, to give rise to a distinctive flora, to become a focus for centrifugal migration [migration outward from a center]. Instead, the population of such a province is made up in part of species differentiated from neighboring inland stock in response to special conditions of the shore, and in part of species that have migrated mainly along the shore, from other maritime provinces where they have differentiated from inland stock. (Quoted by permission.)

Figure 25-47. The Strand Vegetation on the Atlantic coast of northern Florida: *above,* palmetto (*Serenoa repens,* glaucous form), in a sand dune, and sea-oats; *below,* beach with grasses and sedges at the edges of the dunes and with railroad vine (*Ipomaea Pes-Caprae*) growing out toward the shore.

THE AQUATIC, PALUSTRINE, AND RIPARIAN ELEMENTS

THE AQUATIC VEGETATION

Distribution of water plants usually is not controlled in the same ways as occurrence of the plants growing in adjacent terrestrial habitats. Many aquatic species are common through large areas, appearing in lakes, ponds, or streams throughout a continent, a hemisphere, temperate regions, the Tropics, or most of the world. Yet some are restricted to small areas or to occurrence in a few special kinds of lakes many miles apart. Only a few are restricted to lakes or streams within the area of a particular land flora or one of its subdivisions, as with *Ranunculus aquatilis* var. *hispidulus,* which is limited almost completely to the Pacific Lowland Forest. Some, like *Ranunculus aquatilis* var. *Harrisii,* are strictly within a particular drainage area, in this instance the eastern side of the southern half of the Cascade Mountains and the rivers which cut through them from the Great Basin to the coast of southwestern Oregon and adjacent California.

Aquatic plants are affected by various physical and chemical factors, such as the following.

Temperature of the Water at Various Seasons. In the Arctic, several aquatic species flourish in water frozen solid all winter and barely above freezing for a few weeks in summer. Other species may be restricted to the Tropics or to various latitudes by inability to withstand winter cold (usually accomplished in other plants by becoming dormant). Certain blue-green algae grow only in the superheated water of hot springs, such as those at Yellowstone National Park.

Light. Color or murkiness of the water may determine which species can survive. Depth may be correlated with light as well as, of course, the length of the stem.

Physical Properties of the Bottom. Mud or sand bottoms favor different species; gravel bottoms support only a few aquatics, such as species of *Isoëtes* growing in mountain lakes.

The pH and the Saltiness of the Water. Large numbers of aquatics are restricted to acid waters or to limestone areas where the water is basic or to alkaline or brackish waters of deserts or of ocean inlets. Not only the pH of the water, but the kinds and quantities of salts may be important in governing the distribution of aquatic plants. A few are restricted to the ocean. Eel grass, *Zostera marina,* occurs in ocean inlets where the salt water may be somewhat diluted. In Florida three other monocotyledons, *Cymodocea manatorum* (manatee-grass), *Halodule Wrightii,* and *Thalassia testudinum* (turtle-grass) occur in coastal creeks or bays or on submarine sandy bottoms. The leaves are washed ashore especially after storms. On the Pacific Coast two monocotyledonous species, the surf-grasses, *Phyllospadix Torreya* and *Phyllospadix Scouleri,* occur only attached to rocks of the ocean near and below low tide level.

Motion of the Water. Some species such as the riverweeds, are restricted to running streams where they remain attached to sticks or stones. Others grow only in quiet waters.

Organic Matter. Some species must live in the purest water of mountain lakes; others tolerate great quantities of decaying organic matter or of floating or swimming algae.

Other factors determining the occurrence of species are their aggressiveness in competition with others, their means of seed or vegetative fragment dispersal, and the herbivorous animals living in the water or visiting it.

Some dicotyledonous families with aquatic species include the Ranunculaceae, Nymphaeaceae, Ceratophyllaceae, Callitrichaceae, Podostemaceae, Lentibulariaceae, Onagraceae, Hydrocaryaceae, Haloragidaceae, Hippuridaceae, Umbelliferae, and Compositae. Some monocotyledonous families are the Lemnaceae, Typhaceae, Sparaganiaceae, Zosteraceae, Naia-

Figure 25-48. Riparian Vegetation: cottonwoods, willows, and other streambank plants along the North Platte River in Wyoming. The hills are in the Shortgrass Prairie, but it has been disturbed and replaced by brush. Photograph, courtesy of the United States Forest Service.

daceae, Juncaginaceae, Alismaceae, Butomaceae, Araceae, and Pontederiaceae.

THE PALUSTRINE VEGETATION

Aquatic vegetation shades off into marsh vegetation, and marsh plants are not always to be distinguished from those of the shore. Marsh or palustrine species occur in many genera.

Dicotyledonous genera with palustrine species occur in the Lythraceae and the families having aquatic species (cf. above).

Some of the more common monocotyledonous genera are *Scirpus* (e.g., the tules and bulrushes), *Carex* (sedge), *Cyperus* (galingale), *Eleocharis* (spike rush), *Eriophorum* (cotton grass), *Juncus* (rush), *Typha* (cattail), *Sparganium* (bur-reed), *Alisma* (waterplantain), *Echinodorus* (burhead), *Sagittaria* (arrowhead), some grass genera, and various genera of the families listed above as including aquatic species.

THE RIPARIAN VEGETATION

The vegetation occurring on the banks of streams, ponds, and lakes is limited partly by some of the factors affecting the plants in the water, as for example the pH or alkalinity of the water about their roots. On the other hand, the factors affecting the strictly terrestrial plants of the region have more influence upon riparian than upon aquatic or palustrine plants, and the shore and bank species must be considered intermediate or transitional in their requirements. Certain genera are characteristic of riparian habitats, and some are restricted to them. Some of the more common woody genera are *Salix* (willow), *Populus* (cottonwood, poplar, aspen), *Betula* (birch), *Alnus* (alder), *Fraxinus* (ash), and *Cephalanthus* (button-willow). Most of these are plants long classified in the Amentiferae.

In all areas the riparian vegetation is distinctive, though its species composition varies from place to place. In the mountains the stream bank species follow the water and the cool air down the canyons far below their associates at higher levels, accentuating the topography. In the deserts and on the Great Plains riparian species form a thread of green along each stream descending from the mountains or along each dry wash which carries only flood water. On even forested plains, the vegetation adjacent to streams and in wet bottomlands is in contrast to that on higher ground.

SELECTED REFERENCES

The materials in this chapter are derived from field studies of the regions of natural vegetation and floras, from research on special plant groups, and from the botanical literature. The following references are suggested for background reading or for the original investigations they represent. (Chapters 24 and 25.)

ADAMS, CHARLES C. 1902. *Postglacial Origin and Migration of the Life of the Northeastern United States.* Journal of Geography 1: 300–310, 352–357.

ADAMS, JOHN. 1946. *The Flora of Canada.* Canada Year Book, 1938, as revised in 1945. Edmond Cloutier, King's Printer, Ottawa.

AXELROD, DANIEL I. 1940. *Late Tertiary Floras of the Great Basin and Border Areas.* Bulletin of the Torrey Botanical Club 67: 477–488.

———. 1948. *Climate and Evolution in Western North America During Middle Pliocene Time.* Evolution 2: 127–144.

———. 1950a. *Classification of the Madro-Tertiary Flora.* Carnegie Institution of Washington Publication 590: 1–22.

———. 1950b. *Evolution of Desert Vegetation in Western North America.* Carnegie Institution of Washington Publication 590: 215–306.

———. 1952. *Variables Affecting the Probabilities of Dispersal in Geologic Time.* Bulletin of the American Museum of Natural History 99: 177–188.

———. 1956. *Mio-Pliocene Floras from West-Central Nevada.* University of California Publications in Geological Sciences 33: 1–322.

———. 1957. *Late Tertiary Floras and the Sierra Nevadan Uplift.* Bulletin of the Geological Society of America 68: 19–45.

———. 1958. *Evolution of the Madro-Tertiary Geoflora.* Botanical Review 24: 433–509.

———. 1959. *Late Cenozoic Evolution of the Sierran Bigtree Forest.* Evolution 13: 9–23.

———. 1966a. *A Method for Determining the Altitudes of Tertiary Floras.* The Paleobotanist 14: 144–171.

———. 1966b. *Origin of Deciduous and Evergreen Habits in Temperate Forests.* Evolution 20: 1–15.

———. 1966c. *Potassium-Argon Ages of Some Western Tertiary Floras.* American Journal of Science 264: 497–506.

———. 1967a. *The Evolution of the Californian Closed-Cone Pine Forest.* Proceedings of the Symposium on the Biology of the California Islands. Santa Barbara Botanic Garden: 93–149.

———. 1967b. *Geological History of the Californian Insular Flora.* Proceedings of the Symposium on the Biology of the California Islands. Santa Barbara Botanic Gardens: 267–315.

———. 1967c. *Drought, Diastrophism, and Quantum Evolution.* Evolution 21: 201–209.

———. 1967d. *Quarternary Extinctions of Large Mammals.* University of California Publications in Geological Sciences 74: 1–42.

———. 1968. *Tertiary Floras and Topographic History of the Snake River Basin, Idaho.* Geological Society of America Bulletin 79: 713–734.

———. 1972a. *Edaphic Aridity as a Factor in Angiosperm Evolution.* American Naturalist 106: 311– 320.

———. 1972b. *Ocean-Floor Spreading in Relation to Ecosystematic Problems.* University of Arkansas Museum Occasional Paper No. 4: 15–76.

———. 1973. *History of the Mediterranean Ecosystem in California.* In F. Di Castri and H. A. Mooney (editors), *Ecological Studies, Analysis and Synthesis.* Springer-Verlag, Berlin–Heidelberg–New York. 225–277.

———. 1974. *Revolutions in the Plant World.* Geophytology, Paleobotanical Society, Lucknow, India.

———. 1975a. *Evolution and Biogeography of Madrean-Tethyan Sclerophyll Vegetation.* Annals of the Missouri Botanical Garden 62: 280–334.

———. 1975b. *Plate Tectonics and Problems of Angiosperm History.* Mémoires du Muséum d'Historie Naturelle. N. S. Série A, Zoologie 88: 72–86.

———. 1976. *History of the Coniferous Forests. California and Nevada.* University of California Publications in Botany 70: 1–62.

———, and Harry P. Bailey. 1969. *Paleotemperature Analysis of Tertiary Floras.* Palaeogeography, Palaeoclimatology, and Palaeoecology 6: 163–195.

——— and ———. 1976. *Tertiary Vegetation, Climate, and Altitude of the Rio Grande Depression, New Mexico–Colorado.* Paleobiology 2: 235–254.

Bailey, Harry P. 1964. *Toward a Unified Concept of the Temperate Climate.* Geographical Review 54: 516–545.

———. 1966. *The Climate of Southern California.* University of California Press, Berkeley and Los Angeles.

Baldwin, Henry I. 1977. *The Induced Timberline of Mt. Monadnock, New Hampshire.* Bulletin of the Torrey Botanical Club 104: 324–333.

Barbour, Michael G., and Jack Major. 1977. *Terrestrial Vegetation of California.* Wiley-Interscience, Somerset, New Jersey.

Beaty, Chester B. 1978. *The Causes of Glaciation.* American Scientist 66: 452–459.

Benson, Lyman. 1942. *The Relationship of Ranunculus to the North American Floras.* American Journal of Botany 29: 491–500.

———. 1948. *A Treatise on the North American Ranunculi.* American Midland Naturalist 40: 1–261. Supplement 52: 328–369. 1954.

———. 1953. *Relationships of the Ranunculi of the Continental Divide and of the Pacific and Eastern Forests of North America.* Proceedings of the Seventh International Botanical Congress, Stockholm, 1950: 862–863.

———. 1955. *The Ranunculi of the Alaskan Arctic Coastal Plain and the Brooks Range.* American Midland Naturalist 53: 242–255.

———. 1957. *Plant Classification.* D. C. Heath, Lexington, Massachusetts. (Especially 565–647.)

———. 1962. *Plant Taxonomy, Methods and Principles.* Ronald Press Company, New York. 1978, reprinted by John Wiley & Sons, New York. (Especially Chapter 5.)

———. 1969. *The Native Cacti of California.* Stanford University Press, Stanford, California.

———. In press. *The Cacti of the United States and Canada.* Stanford University Press, Stanford, California.

———, and Robert A. Darrow. 1945, 1954. Ed. 1, *A Manual of Southwestern Desert Trees and Shrubs.* University of Arizona Biological

Science Bulletin 5 (6): 1–411; Ed. 2, 1954. *The Trees and Shrubs of the Southwestern Deserts.* University of New Mexico Press, Albuquerque. Ed. 3 in press.

BILLINGS, W. D. 1945. *The Plant Associations of the Carson Desert Region, Western Nevada.* Butler University Studies. Botany 7: 89–123.

BIRKS, H. J. B., and R. G. WEST. 1974. *Quaternary Plant Ecology.* Halstead (Wiley), New York.

BOCHER, TYGE W. 1950. *Distributions of Plants in the Circumpolar Area in Relation to Ecological and Historical Factors.* Journal of Ecology 39: 376–395.

BRAUN, E. LUCY. 1938. *Deciduous Forest Climaxes.* Ecology 19: 515–522.

———. 1955. *The Phytogeography of the Unglaciated Eastern United States and Its Interpretation.* Botanical Review 21: 297–375.

———. 1956. *The Development of Association and Climax Concepts: Their Use in Interpretation of the Deciduous Forest.* American Journal of Botany 43: 906–911.

———. 1967. *Deciduous Forests of Eastern North America.* Hafner Publishing Company, New York.

BRAY, J. R. 1977. *Pleistocene Volcanism and Glacial Initiation.* Science 197: 251–254.

BRITTON, MAX E. (editor). 1973. *Alaskan Arctic Tundra* (symposium). Arctic Institute of North America, Washington, D.C.

BROOKS, C. E. P. 1949. *Climate Through the Ages.* McGraw-Hill Book Company, New York. Ed. 2. Reprinted, Dover Publications, New York. 1970.

BROWN, R. W. 1962. *The Paleocene Flora of the Rocky Mountains and Great Plains.* United States Geological Survey Professional Papers 375.

BRYAN, E. H., JR. 1954. *The Hawaiian Chain.* Bishop Museum, Honolulu.

BRYSON, R. A. 1974. *A Perspective on Climate Change.* Science 184: 753–760.

CAIN, STANLEY A. 1943. *The Tertiary Character of the Cove Hardwood Forests of the Great Smoky Mountains National Park.* Bulletin of the Torrey Botanical Club 70: 213–235.

———. 1944. *Foundations of Plant Geography.* Harper & Brothers, New York.

CAMP, W. H. 1947. *Distribution Patterns in Modern Plants and the Problems of Ancient Dispersals.* Ecological Monographs 17: 123–126, 159–183.

CANDOLLE, ALPHONSE DE. 1855. *Geographie Botanique Raisonee.* Paris.

CHABOT, BRIAN F., and W. DWIGHT BILLINGS. 1972. *Origins and Ecology of the Sierran Alpine Flora and Vegetation.* Ecological Monographs 42: 163–199.

CHANEY, RALPH W. 1925. *A Comparative Study of the Bridge Creek Flora and the Modern Redwood Forest.* Carnegie Institution of Washington Publication 349: 1–22.

———. 1936. *The Succession and Distribution of Cenozoic Floras Around the North Pacific Basin.* Essays in Geobotany in Honor of W. A. Setchell, edited by T. H. GOODSPEED: 55–85.

———. 1938. *Paleoecological Interpretation of Cenozoic Plants in Western North America.* Botanical Review 4: 371–396.

———. 1940. *Teritary Floras and Continental History.* Bulletin of the Geological Society of America 51: 469–488.

———. 1947. *Tertiary Centers and Migration Routes.* Ecological Monographs 17: 139–148.

———. 1948. *The Bearing of Living Metasequoia on Problems of Tertiary Paleobotany.* Proceedings of the National Academy of Sciences 34: 503–515.

———. 1951. *A Revision of Fossil Sequoia and Taxodium in Western North America Based upon the Recent Discovery of Metasequoia.* Transactions of the American Philosophical Society 40: 171–263.

———. 1952. *Conifer Dominants in the Middle Tertiary of the John Day Basin, Oregon.* Paleobotanist 1: 105–113.

———, and DANIEL I. AXELROD. 1959. *Miocene Floras of the Columbia Plateau.* Carnegie Institution of Washington Publications 617: Part II, 1–134, by Ralph W. Chaney; Part II, Systematic Considerations, 135–237, by Ralph W. Chaney and Daniel I. Axelrod.

———, CARLTON CONDIT, and DANIEL I. AXELROD. 1944. *Pliocene Floras of California and Oregon.* Carnegie Institution of Washington Publication 533.

CLARK, THOMAS H., and COLIN W. STEARN. 1960. *The Geological Evolution of North America.* Ronald Press Company, New York.

CLAUSEN, JENS, DAVID D. KECK, and WILLIAM M. HIESEY. 1948. *Experimental Studies on the Nature of Species III. Environmental Responses of Climatic Races of Achillea.* Carnegie Institution of Washington Publication 581: i–iii, 1–129.

CLEMENTS, FREDERICK E. 1920. *Plant Indicators.* Carnegie Institution of Washington Publication 290: 1–388.

———, E. V. MARTIN, and F. L. LONG. 1950. *Adaptation and Origin in the Plant World; the Role of Environment in Evolution.* Chronica Botanica, Waltham, Massachusetts. Review by GERALD OWNBEY. Ecology 33: 431–433. 1952.

CLIMAP PROJECT MEMBERS. 1976. *The Surface of the Ice-Age Earth.* Science 191: 1131–1137.

CONSTANCE, LINCOLN, L. R. HECKARD, KENTON L. CHAMBERS, ROBERT ORNDUFF, and PETER H. RAVEN. 1963. *Amphitropical Relationships in the Herbaceous Flora of the Pacific Coast of North and South America.* Quarterly Review of Biology 38: 109–177.

COOPER, W. S. 1922. *The Broad-Leaf Sclerophyll Vegetation of California.* Carnegie Institution of Washington Publication 319: 1–124.

———. 1936. *The Strand and Dune Flora of the Pacific Coast of North America.* In *Essays in Geobotany,* edited by T. H. GOODSPEED. 141–187. University of California Press, Berkeley.

DALRYMPLE, BRENT G., ELI A. SILVER, and EVERETT D. JACKSON. 1973. *Origin of the Hawaiian Islands.* American Scientist 61: 294–308.

DAUBENMIRE, REXFORD F. 1940. *Plant Succession Due to Overgrazing in Agropyron Bunchgrass Prairie in Southeastern Washington.* Ecology 21: 55–64.

———. 1943. *Vegetational Zonation in the Rocky Mountains.* Botanical Review 9: 325–394.

———. 1954. *Alpine Timberlines in the Americas and Their Interpretation.* Butler University Botanical Studies 11: 119–136.

———. 1969. *Ecologic Plant Geography of the Pacific Northwest.* Madroño 20: 111–128.

———. 1978. *Plant Geography, with Special Reference to North America.* Academic Press, New York.

DICE, LEE R. 1943. *The Biotic Provinces of North America.* University of Michigan Press, Ann Arbor.

DORF, ERLING. 1955. *Plants and the Geologic Time Scale.* Geological Society of America Special Papers 62: 575–592.

———. 1959. *Climatic Changes of the Past and Present.* Contributions from the Museum of Paleontology, University of Michigan 13: 181–210.

———. 1960a. *Tertiary Fossil Forests of Yellowstone National Park, Wyoming.* Billings Geological Society, Eleventh Annual Field Conference: 253–260.

———. 1960b. *Paleobotany.* McGraw-Hill Encyclopedia of Science and Technology: 499–506.

———. 1969. *Paleobotanical Evidence of Mesozoic and Cenozoic Climatic Changes.* Proceedings of the North American Paleontological Convention, September 1969. Part D: 323–346.

DORT, WAKEFIELD, JR., and J. KNOX JONES (editors). 1968. *Pleistocene and Recent Environments of the Central Great Plains.* University of Kansas Department of Geology Special Publication No. 3. University of Kansas Press, Lawrence.

EMILIANI, CESARE. 1972. *Quaternary Paleotemperatures and the Duration of the High-Temperature Intervals.* Science 178: 398–401.

EVERDE, J. F., and G. T. JAMES. 1964. *Potassium-Argon Dates and the Tertiary Floras of North America.* American Journal of Science 262: 945–974.

FERNALD, M. L. 1911. *A Botanical Expedition to Newfoundland and Southern Labrador.* Rhodora 13: 135–162.

———. 1925. *Persistence of Plants in Unglaciated Areas of Boreal America.* Memoirs of the American Academy of Arts and Sciences 15: 239–342.

———. 1929. *Some Relationships of the Floras of the Northern Hemisphere.* Proceedings of the International Congress of Plant Sciences at Ithaca, New York 2: 1487–1507.

———. 1931. *Specific Segregations and Identities in Some Floras of Eastern North America and the Old World.* Rhodora 33: 25–63.

FISHER, R. V. 1964. *Resurrected Oligocene Hills, Eastern Oregon.* American Journal of Science 262: 713–725.

FLINT, RICHARD FOSTER. 1977. Glacial and Quaternary Geology. John Wiley & Sons, New York.

FOSBERG, F. RAYMOND. 1948. *Derivation of the Flora of the Hawaiian Islands.* In E. C. ZIMMERMAN (editor). *Insects of Hawaii,* Vol. 1. University of Hawaii Press, Honolulu.

———. 1961. *A Classification of Vegetation for General Purposes.* Tropical Ecology 2: 1–28.

———. 1962. *Qualitative Description of the Coral Atoll Ecosystem. Proceedings of the Ninth Pacific Science Congress, 1957.* 4: 61–167.

———. 1966. *Restoration of Lost and Degraded Habitats.* In F. Fraser Darling and J. P. Milton. *Future Environments of North America.* Natural History Press, Garden City, New York.

———. 1967. *Opening Remarks: Island Ecosystem Symposium.* Micronesia 3: 3–4.

———. 1974. *Phytogeography of Atolls and Other Coral Islands.* Proceedings of the Second International Coral Reef Symposium I. Great Barrier Reef Committee, Brisbane, Australia.

———. 1976. *Geography, Ecology, and Biogeography.* Annals of the Association of American Geographers 66: 117–128.

Geological Survey. 1973. *The Channelled Scablands of Eastern Washington, The Geologic Story of the Spokane Flood.* U.S. Department of the Interior, Superintendent of Documents, U.S. Government Printing Office, Washington, D.C.

Gleason, Henry Allen. 1912. *An Isolated Prairie Grove and Its Phytogeographical Significance.* Botanical Gazette 53: 38–49.

———. 1913. *Relation of Forest Distribution and Prairie Fires in the Middle West.* Torreya 13: 173–181.

———. 1917. *The Structure and Development of the Plant Association.* Bulletin of the Torrey Botanical Club 44: 463–481.

———. 1923. *The Vegetational History of the Middle West.* Association of American Geographers, Annals. 12: 39–85.

———. 1926. *The Individualistic Concept of the Plant Association.* Bulletin of the Torrey Botanical Club 53: 7–26.

———, and Arthur Cronquist. 1964. *The Natural Geography of Plants.* Columbia University Press, New York.

Good, R. D'O. 1964. *The Geography of Flowering Plants.* John Wiley & Sons, New York.

Gould, Frank W. 1962. *Texas Plants—A Checklist and Ecological Summary.* The Agricultural and Mechanical College of Texas; Texas Agricultural Experiment Station, College Station.

Grant, Verne, and Alva Grant. 1954. *Generic and Taxonomic Studies in Gilia VII. The Woodland Gilias.* El Aliso 3: 59–91.

Griggs, Robert F. 1934. *The Edge of the Forest in Alaska and the Reason for Its Position.* Ecology 15: 80–95.

Grinnell, Joseph. 1935. *A Revised Life-Zone Map of California.* University of California Publications in Zoology 40: 327–330.

Grootes, P. M. 1978. *Carbon-14 Time Scale Extended: Comparison of Chronologies.* Science 200: 11–15.

Gulick, John Thomas. 1905. *Evolution, Racial and Habitudinal.* Carnegie Institution of Washington, Washington, D.C.

Halliday, W. E. D. 1937. *A Forest Classification for Canada.* Forest Research Division Bulletin 89: 1–50. Department of Resources and Development, Ottawa.

———, and A. W. A. Brown. 1943. *The Distribution of Some Important Forest Trees in Canada.* Ecology 24: 353–373.

Hamilton, W. 1968. *Cenozoic Climatic Change and Its Cause.* Meteorological Monographs 8: 128–133.

Hammond, Allen L. 1971a. *Plate Tectonics: The Geophysics of the Earth's Surface.* Science 173: 40–41.

———. 1971b. *Plate Tectonics (II): Mountain Building and Continental Geology.* Science 173: 133–134.

———. 1976a. *Paleoceanography: Sea Floor Clues to Earlier Environments.* Science 191: 168–170, 208.

———. 1976b. *Paleoclimate; Ice Age Was Cool and Dry.* Science 191: 455.

Harrison, A. T., E. Small, and H. A. Mooney. 1971. *Drought Relationships, and Distribution of Two Mediterranean-Climate California Plant Communities.* Ecology 52: 869–875.

Hays, J. D., J. Imbrie, and N. J. Shackleton. 1976. *Variations in the Earth's Orbit: Pacemaker of the Ice Ages.* Science 194: 1121–1132.

Hernandez-X, E., E. H. Crum, W. B. Fox, and A. J. Sharp. 1951. *A Unique Vegetational Area in Tamaulipas.* Bulletin of the Torrey Botanical Club 78: 458–463.

Hooker, J. D. 1862. *Outlines of Distribution of Arctic Plants.* Transactions of the Linnaean Society 23: 251–348.

Hopkins, D. M. 1959. *Cenozoic History of the Bering Land Bridge.* Science 129: 1519–1528.

——— (editor). 1967. *The Bering Land Bridge.* Stanford University Press, Stanford, California.

Hultén, Eric. 1937. *Outline of the History of the Arctic and Boreal Biota During the Quaternary Period.* Bokförlags aktiebolaget Thule, Stockholm.

———. 1958. *The Amphi-Atlantic Plants.* Kungl. Vetenskapsakademiens Handlingar Fjärde Serien. Band 7, Nr. 1. Stockholm.

———. 1963. *Phytogeographical Connections of the North Atlantic.* In A. Löve and D. Löve (eds.). *North Atlantic Biota and Their History*. Pergamon Press, Oxford.

HUMPHREY, R. R. April 1958. *The Desert Grassland: A History of Vegetational Change and an Analysis of Causes*. Botanical Review 24: 193–252. Republished by the Agricultural Experiment Station, University of Arizona, Tucson, December 1958.

———. 1963. *Arizona Natural Vegetation*. Arizona Agricultural Experiment Station Bulletin A-45 (map).

ILTIS, HUGH H. 1969. *A Requiem for the Prairie*. Prairie Naturalist 1: 51–57.

———. 1973. *Long-Distance Dispersal (LDD) within the Arcto-Tertiary Geoflora: Eastern North America as a Floristic Oceanic Archipelago*. Abstract of a paper presented at the First International Congress of Systematic and Evolutionary Biology, Boulder, Colorado.

JACKSON, JAMES P. 1973. *Visions of Prairie Parks*. National Parks and Conservation Magazine 47: 23–26.

JOHNSTON, IVAN M. 1940. *The Floristic Significance of Shrubs Common to the North and South American Deserts*. Journal of the Arnold Arboretum 21: 356–363.

KERR, RICHARD A. 1978. *Climate Control: How Large a Role for Orbital Variations?* Science 201: 144–146.

KING, JAMES E., and THOMAS R. VAN DEVENDER. 1977. *Pollen Analysis of Fossil Packrat Middens from the Sonoran Desert*. Quaternary Research 8: 191–204.

KUCHLER, A. W. 1949. *Natural Vegetation of the World*. Map in Goode's School Atlas. Rand-McNally, Chicago.

———. 1964. *Potential Natural Vegetation of the Coterminus United States*. American Geographical Society, New York.

———. 1967. *Vegetation Mapping*. Ronald Press Comany, New York.

KUKLA, G. J., and R. K. MATTHEWS. 1972. *When Will the Present Interglacial End?* Science 178: 190–191.

LANNER, RONALD M., and THOMAS R. VAN DEVENDER. 1974. *Morphology of Pinyon Pine Needles from Fossil Packrat Middens in Arizona*. Forest Science 20: 207–211.

LARSEN, J. A. 1930. *Forest Types of the Northern Rocky Mountains and Their Climatic Controls*. Ecology 11: 631–672.

LEOPOLD, E. B., and H. D. MACGINITIE. 1972. *Development and Affinities of Tertiary Floras in the Rocky Mountains*. In A. GRAHAM (editor). *Floristics and Paleofloristics of Asia and Eastern North America*. 1–278. Elsevier Publishing Company, Amsterdam.

LEOPOLD, LUNA B. 1951. *Vegetation of Southwestern Watersheds*, in the Nineteenth Century. Geographical Review 41: 295–316.

LI, HUI-LIN. 1952. *Floristic Relationships Between Eastern Asia and Eastern North America*. Transactions of the American Philosophical Society, II. 42: 371–429.

LIVINGSTON, B. E., and FORREST SHREVE. 1921. *Climatic Areas of the United States as Related to Plant Growth*. Carnegie Institution of Washington Publication 284: 1–590.

LONG, ROBERT W. 1974a. *The Vegetation of Southern Florida*. Florida Scientist 37: 33–45.

———. 1974b. *Origin of the Vascular Flora of Southern Florida*. In P. J. GLEASON (editor). *Environments of South Florida*. Memoirs of the Miami Geological Society.

———, and OLGA LAKELA. 1971. *Geology of Southern Florida*. In *A Flora of Tropical Florida*. 11–26. University of Miami Press, Coral Gables.

LÖVE, ÅSKELL. 1959. *Origin of the Arctic Flora*. McGill University Museums Publication 1: 82–95.

———, and DORIS LÖVE (editors). 1963. *North Atlantic Biota and their History*. Oxford University Press, Oxford.

——— and ———. 1967a. *Continental Drift and the Origin of the Arctic-Alpine Flora*. Revue Roumaine de Biologie Série Bot. 12: 163–169.

——— and ———. 1967b. *The Origin of the North Atlantic Flora*. Department of Biology and Institute for Arctic and Alpine Research, University of Colorado, Boulder. Aquilo, Ser. Botanica 6: 52–66.

LÖVE, DORIS. 1962. *Plants and Pleistocene—Problems of Pleistocene and Arctic*. McGill University Museums Publication 2: 17–39.

———. 1970. *Subarctic and Subalpine: Where and What?* Arctic and Alpine Research 2: 63–73.

MACBRIDE, J. FRANCIS. 1950. *Natural Landscapes of the United States*. Chicago Museum of Natural History, Popular Series, Botany No. 27.

MACDOUGAL, D. T. 1913. *North American Deserts*. Geographical Journal 39: 105–120.

MACGINITIE, H. D. 1933. *Redwoods and Frost*. Science 78: 190.

———. 1953. *Fossil Plants from the Florissant Beds, Colorado*. Carnegie Institution of Washington Publication 599.

McKenna, M. C. 1972. *Possible Biological Consequences of Plate Tectonics.* Bioscience 22: 519–525.

Map of the Dominion of Canada Indicating Vegetation and Forest Cover. National Development Bureau, Department of the Interior, Dominion of Canada, Ottawa.

Mason, Herbert L. 1934. *The Pleistocene Flora of the Tomales Formation.* Carnegie Institution of Washington Publication 415: 83–180.

———. 1947. *Evolution of Certain Floristic Associations in Western North America.* Ecological Monographs 17: 201–210.

Mayer-Oakes, W. J. (editor). 1967. *Life, Land, and Water. Proceedings of the 1966 Conference on Environmental Studies of the Glacial Lake Agassiz Region.* Occasional Papers, Department of Anthropology, University of Manitoba, No. 1. University of Manitoba Press.

Melville, R. 1966. *Continental Drift, Mesozoic Continents, and the Migrations of the Angiosperms.* Nature 211: 116–120.

Merriam, C. Hart. 1890. *Results of a Biological Survey of the San Francisco Mountain Region and Desert of the Little Colorado in Arizona.* United States Division of Ornithology and Mammalogy. North American Fauna 3: 1–34.

———. 1893. *The Geographical Distribution of Life in North America, with Special Reference to Mammalia.* Smithsonian Institution Annual Report to July 1891: 365–415.

———. 1894. *The Geographical Distribution of Animals and Plants in North America.* United States Department of Agriculture Yearbook 18: 203–214.

———. 1898. *Life Zones and Crop Zones in the United States.* United States Division of the Biological Survey Bulletin 10: 1–79.

Miranda, Faustino, and A. J. Sharp. 1950. *Characteristics of the Vegetation in Certain Temperate Regions of Eastern Mexico.* Ecology 31: 313–333.

Mooney, H. A., J. Ehleringer, and J. A. Berry. 1976. *High Photosynthetic Capacity of a Winter Annual in Death Valley.* Science 194: 322–324.

Nichol, A. A. 1931. *The Natural Vegetation of Arizona.* University of Arizona College of Agriculture. Agricultural Experiment Station Technical Bulletin 68: 181–222.

Oosting, H. J. 1948, 1956. *The Study of Plant Communities.* W. H. Freeman, San Francisco.

Parish, S. B. 1930. *Vegetation of the Mojave and Colorado Deserts of Southern California.* Ecology 11: 481–499.

Piper, Charles V. 1906. *Flora of the State of Washington.* Contributions from the United States National Herbarium 11.

Polunin, Nicholas. 1951. *The Real Arctic: Suggestions for Its Delimitation, Subdivision, and Characterization.* Journal of Ecology 39: 308–315.

Porsild, A. Erling. 1958. *Geographical Distribution of Some Elements in the Flora of Canada.* Geographical Bulletin 11: 57–77.

Quarterman, Elsie, and Catherine Keever. 1962. *Southern Mixed Hardwood Forest: Climax in the Southeastern Coastal Plain: U.S.A.* Ecological Monographs 32: 167–185.

Raunklaer, C. 1934. *The Life Forms of Plants and Statistical Plant Geography; Being the Collected Papers of C. Raunkiaer.* Clarendon Press, Oxford.

Raup, Hugh M. 1941. *Botanical Problems in Boreal America.* Botanical Review 7: 147–248.

Raven, Peter. H., and Daniel I. Axelrod. 1974. *Angiosperm Biogeography and Past Continental Movements.* Annals of the Missouri Botanical Garden 61: 539–673.

——— and ———. 1978. *Origin and Relationships of the California Flora.* University of California Publications in Botany: 72.

Reichle, David E. 1970. *Analysis of Temperate Flora Ecosystems.* Springer-Verlag, New York.

Richards, P. W. 1952. *The Tropical Rain Forest.* Cambridge University Press. Review by A. G. Tansley. Journal of Ecology 41: 398–399. 1952.

Richmond, Gerald M., et al. 1965. *The Cordilleran Ice Sheet of Northern Rocky Mountains and Related Quaternary History of the Columbia Plateau.* In H. E. Wright and David G. Frey (editors). *The Quaternary of the United States.* Princeton University Press, Princeton, New Jersey.

Rowe, J. S. 1959. *Forest Regions of Canada.* Canada, Department of Northern Affairs and National Resources, Forestry Branch. Bulletin 123. Minister of Northern Affairs and National Resources, Ottawa.

St. John, Harold. 1946. *Endemism in the Hawaiian Species of Gunnera (Haloragidaceae). Hawaiian Plant Studies, 11.* Proceedings of the California Academy of Sciences IV 25: 377–420.

Saville, D. B. O. 1972. *Arctic Adaptations of Plants.* Research Branch, Canada Department of Agriculture, Monograph No. 6.

Schimper, A. F. W. (translated by W. R. Fisher).

1903. *Plant Geography upon a Physiological Basis,* 1–839. Oxford University Press, Oxford. Reprinted 1960 by Lubrecht & Cramer, Monticello, New York.

SCHUCHERT, CHARLES, and CARL O. DUNBAR. 1941. *Textbook of Geology. Part II. Historical Geology.* Ed. 4. John Wiley & Sons, New York.

SCOGGAN, H. J. 1950. *The Flora of Bic and the Gaspé Peninsula,* Quebec. National Museum of Canada, Ottawa.

SELLERS, E. D. 1965. *Physical Climatology.* University of Chicago Press, Chicago.

SENN, HAROLD A. 1951. *A Bibliography of Canadian Plant Geography. IX.* 1941–45. Department of Agriculture, Ottawa, Publication 863.

SHANTZ, HOMER L., and RAPHAEL ZON. 1924. *Natural Vegetation.* Atlas of American Agriculture. United States Department of Agriculture.

SHARP, AARON J. 1946a. *A Preliminary Report on Some Phytogeographical Studies in Mexico and Guatemala.* (Abstract.) American Journal of Botany 33: 844.

———. 1946b. *Some Fungi Common on the Highlands of Mexico and Guatemala and Eastern United States.* (Abstract.) American Journal of Botany 33: 844.

———. 1950. *The Relation of the Eocene Wilcox Flora to Some Modern Floras.* Evolution 5: 1–5.

———. 1952. *Notes on the Flora of Mexico: World Distribution of the Woody Dicotyledonous Families and the Origin of the Modern Vegetation.* Journal of Ecology 41: 374–380.

———. 1969. *Studies of Vegetation in Two Areas of Mexico* (Review). Ecology 50 (5): late summer, 1969.

———. 1970. *Different Ideas Concerning the Migration of Vascular Plants and Bryophytes from the Old World to the New.* ASB Bulletin 17: 63.

———. 1971. *Epilogue.* In PERRY C. HOLT (editor). *The Distributional History of the Biota of the Southern Appalachians.* Virginia Polytechnic Institute and State University, Blacksburg. Research Division Monograph 2.

SHELFORD, VICTOR E. 1963. *The Ecology of North America.* University of Illinois Press, Urbana.

SHREVE, FORREST. 1917a. *The Establishment of Desert Perennials.* Journal of Ecology 5: 210–216.

———. 1917b. *A Map of the Vegetation of the United States.* Geographical Review 3: 119–125.

———. 1917c. *The Physical Control of Vegetation in Rain-Forest and Desert Mountains.* Plant World 20: 135–141.

———. 1922. *Conditions Indirectly Affecting Vertical [Plant] Distribution on Desert Mountains.* Ecology 3: 269–274.

———. 1925. *Ecological Aspects of the Deserts of California.* Ecology 6: 93–103.

———. 1941. *Forest Climatology.* Plant World 18: 150–151.

———. 1942. *The Desert Vegetation of North America.* Botanical Review 8: 195–246.

———. 1944. *Rainfall in Northern Mexico.* Ecology 31: 368–372.

———, and IRA WIGGINS. 1951, 1964. *Vegetation and Flora of the Sonoran Desert,* Volume I. *Vegetation of the Sonoran Desert* by Forrest Shreve. Carnegie Institution of Washington Publication 591. Also in the complete book, Stanford University Press, Stanford, California. 1964.

SMILEY, TERAH L. 1958. *Climate and Man in the Southwest.* Program in Geochronology Contribution No. 6. University of Arizona, Tucson.

SPURR, STEPHEN H. 1952. *Origin of the Concept of Forest Succession.* Ecology 33: 426–427.

STEBBINS, G. LEDYARD, JR. 1941. *Additional Evidence for a Holarctic Dispersal of Flowering Plants in the Mesozoic Era.* Proceedings of the Sixth Pacific Science Congress: 649–660.

———, and JACK MAJOR. 1965. *Endemism and Speciation in the California Flora.* Ecological Monographs 35: 1–35.

STODDART, L. A. 1941. *The Palouse Grassland Association in Northern Utah.* Ecology 22: 158–163.

STONE, KIRK H. 1948. *Aërial Photographic Interpretation of Natural Vegetation in the Anchorage Area.* Geographical Review 38: 465–474.

SUTTON, A., and M. SUTTON. 1966. *The Life of the Desert.* McGraw-Hill, New York. 1–232.

TAYLOR, R. L., and R. A. LUDWIG. 1966. *The Evolution of Canada's Flora.* University of Toronto Press.

THOMPSON, PETER, and HENRY P. SCHWARTZ. 1947. *Continental Pleistocene Climatic Variations from Speleothem* [cavern] *Age and Isotopic Data.* Science 184: 893–894.

TISDALE, E. W. 1947. *The Grasslands of Southern Interior British Columbia.* Ecology 28: 346–382.

TROUGHTON, JOHN H., P. V. WELLS, and H. A. MOONEY. 1974. *Photosynthetic Mechanisms* *and Paleoecology from Carbon Isotope Ratios in Ancient Specimens of C_4 and CAM Plants.* Science 185: 610–612.

Van Steenis, C. G. G. J. 1962. *The Land-Bridge Theory in Botany.* Blumea 11: 235–372.

Weaver, J. E. 1917. *A Study of the Vegetation of Southeastern Washington and Adjacent Idaho.* University of Nebraska Studies 17: 1–133.

———, and F. W. Albertson. 1956. *Grasslands of the Great Plains.* Johnson Publishing Company, Lincoln, Nebraska.

———, and F. E. Clements. 1938. *Plant Ecology.* Ed. 2.

Weber, William A. 1965. *Plant Geography in the Southern Rocky Mountains.* In H. E. Wright, Jr., and David G. Frey (editors). *The Quaternary in the United States.* Princeton University Press, Princeton, New Jersey.

Wells, B. W. 1928. *Plant Communities of North Carolina and Their Successional Relations.* Ecology 9: 230–242.

Wells, Philip V. 1966. *Late Pleistocene Vegetation and Degree of Pluvial Climatic Change in the Chihuahuan Desert.* Science 153: 970–975.

———. 1970a. *Postglacial Vegetational History of the Great Plains.* Science 167: 1574–1582.

———. 1970b. *Vegetational History of the Great Plains: A Post-Glacial Record of Coniferous Woodland in Southeastern Wyoming.* In *Pleistocene and Recent Environments of the Central Great Plains.* Department of Geology, University of Kansas, Special Publication 3. University of Kansas Press, Lawrence. 185–202.

———. 1976. *Postglacial Origin of the Chihuahuan Desert Less than 11,500 Years Ago.* In Roland H. Wauer and D. H. Riskind (editors). Transactions of a Symposium on the Biological Resources of the Chihuahuan Desert, U.S. and Mexico. National Park Service, Washington, D.C.

———, and Rainer Berger. 1967. *Late Pleistocene History of Coniferous Woodland in the Mohave Desert.* Science 155: 1640–1647.

———, and Clive D. Jorgenson. 1964. *Pleistocene Wood Rat Middens and Climatic Change in the Mohave Desert: A Record of Juniper Woodlands.* Science 13: 1171–1174.

Welsh, Stanley L. 1976. *Problems in Plant Endemism on the Colorado Plateau. Intermountain Biogeography, a Symposium.* Brigham Young University, the Intermountain Forest and Range Experiment Station of the U.S. Forest Service, the Botanical Society of America, and the Biological Science Section of the Pacific Section of the A.A.A.S. Missoula, Montana, June 1976. (To be published.)

Whitford, H. N., and Roland D. Craig. 1917. *A Report on the Forests of British Columbia* (accompanied by map, Climatic Forest Types of British Columbia). Commission on Conservation, Canada. Ottawa.

Wilson, J. Tuzo (editor). 1972. *Continents Adrift.* [Fifteen articles on continental drift and the history of the Earth, published in the Scientific American from 1952 to 1972.] W. H. Freeman and Company, San Francisco.

Wolfe, Jack A. 1966. *Tertiary Plants from the Cook Inlet Region, Alaska.* United States Geological Survey Professional Papers 398–B.

———. 1969. *Neogene Floristic and Vegetational History of the Pacific Northwest.* Madroño 20: 83–110.

———, and D. M. Hopkins. 1967. *Climatic Changes Recorded by Tertiary Land Floras in Northwestern North America.* In K. Hatai, *Tertiary Correlation and Climatic Changes in the Pacific.* Sendai, Japan.

———, ———, and E. B. Leopold. 1967. *Neogene and Early Quaternary Vegetation of Northeastern Eurasia.* In D. M. Hopkins (Editor). *The Bering Land Bridge,* Stanford University Press, Stanford.

Wood, Carroll E., Jr. 1971. *Some Floristic Relationships between the Southern Appalachians and Western North America.* Research Division Monograph 2. Virginia Polytechnic Institute and State University, Blacksburg.

Wright, H. E., Jr. 1971. *Late Quaternary Vegetational History of North America.* In K. K. Turekian (editor). *The Late Cenozoic Ice Ages.* 425–464. Yale University Press, New Haven, Connecticut.

———, and D. G. Frey (editors). 1965. *The Quaternary of the United States.* Princeton University Press, Princeton, New Jersey.

Wynne-Edwards, V. C. 1937. *Isolated Arctic-Alpine Floras in Eastern North America: A Discussion of Their Glacial and Recent History.* Transactions of the Royal Society of Canada II (5) 31: 1–26.

———. 1939. *Some Factors in the Isolation of Rare Alpine Plants.* Transactions of the Royal Society of Canada III (5) 33: 35–42.

Appendix

Favorable Seasons for Plant Collection in Various Parts of North America

It is natural to assume that the cycle of seasons everywhere is much the same as at home. This is not true, and nearly every plant collector who has strayed far in search of plants has been disappointed upon reaching the goal of a long anticipated collecting trip only to find that it was the wrong season of the year. This appendix is written with the hope of reducing such disappointments.

If a long trip is contemplated, it is well to secure from agricultural county agents, the Weather Bureau, Chambers of Commerce, the National Park Service, the Forest Service, or other sources of data on the particular season because unusual weather is not restricted to the environment in which this book is written. Many seasons vary one way or another from normal timing or toward the local extremes of wetness or dryness, heat or cold. No two years are exactly alike anywhere, and variation within broad limits is to be expected. In some areas, especially in dry or cold regions, fluctuations are great. Unusual years do not represent a change in climate, as the "old timers" of ten years' standing proclaim, but simply another phase of a variable local pattern of seasons.

Since existence of the major floras of the Continent is correlated with climate, the most feasible arrangement for a list of collecting seasons is in accordance with floras and floristic associations.*

The Tundra Flora

This region is characterized by a short flowering season and by much variation from year to year. For example, at timber line on Mt. Rainer, Washington, in the summer of 1929 plant collecting was at its peak on July 22. The meadows about timber line were in full bloom, and there was a colorful display of wild flowers everywhere. On August 10, 1933, the same area was still covered by three to four feet of snow. In general, in the regions above and beyond timberline the flowering season reaches its height in the latter part of July, although the earliest flowers may come out before the end of June. Seasonally or locally the peak of the season may range into the first or second week of August, but, most frequently, after the first week of August most of the

* In general the data presented here are drawn from the field experience of the author, and he takes responsibility for any errors. Gratitude for data on special regions is expressed to Dr. Arthur Cronquist, Dr. Robert A. Darrow, Dr. George J. Goodman, Dr. Gerald Ownbey, and Dr. A. Erling Porsild and to the late Mr. J. P. Anderson, Mr. V. L. Cory, Dr. Marion Ownbey, Dr. Walter S. Phillips, and Dr. Erdman West.

NOTES ON PROGRESS OF SEASON IN THE ARCTIC AND IN CANADA
by A. E. Porsild, National Museum of Canada

	First flowers	*Height of Season*	*Killing frost*
Western Arctic	June 10–20	July 15–25	Aug. 20–25
Eastern Arctic	June 24–July 15	July 25–Aug. 10	Sept. 15
West Greenland 60°–72° N. Maritime	June 1–15	July 25–Aug. 10	Sept. 15–25
Interior Coastland	End of May–June 5	July 15–25	Sept. 1
Central Yukon, elev. 2,500′–4,000′	May 5–10	June 25–July 15	Aug. 25–Sept. 1
Upper Mackenzie Basin	May 15	End of June to early July	Middle Sept.
Lower Mackenzie Basin	End of May	July 10–20	Aug. 25–Sept. 10
Banff-Jasper Parks in Can. Rocky Mts., Alta., 2,500′–6,000′	May 1–June 25	June 10–July 10	Aug. 20–Sept. 1

plants are in fruit, and only occasional individuals are found still with flowers. In some seasons there is no good flowering period.

The Northern Forest Flora

The flowering season is longer than in the Boreal-Alpine region, and it may begin earlier. The earliest plants may flower in May or early June, and the peak of the flowering season is reached in June or July. If there is summer rain, some plants may continue to bloom into August, and special groups, such as composites, flower in the fall.

The Western Forest Flora

The flowering seasons vary according to the subdivisions, as follows.

The Pacific Lowland Forest. At Seattle, Washington, the earliest flowers appear late in March before development of the leaves on the deciduous trees, but the height of spring is in May. Since in some seasons there are occasional summer rains, certain plants are likely to be collectible through June and July and sometimes into August. Most species available in July and August are in fruit. About the same sequence applies to the extension of this floristic association eastward to northwestern Montana. Flowering in south coastal Alaska is nearly one month later than at Seattle; on the other hand, flowering in the redwood belt of northwestern California begins somewhat earlier, and the peak is reached in late April or in May. Usually good collecting of flowering or fruiting specimens is available through May, June, and early July. By August the woods are dry.

The Pacific Montane Forest. Flowering is maximum in June, although early plants flower through May, and in some areas a great many plants flower during the last two weeks of that month. In July the mountains dry out rapidly, and in August the soil is powdery because of the high evaporation rate in the mountains.

The Pacific Subalpine Forest. In the higher forests of the Cascades and the Sierra Nevada the peak of the flowering season is the last of July or the first of August, shifting a little earlier in some seasons. The earliest flowers appear at the end of June, and drying occurs abruptly in August.

The Palouse Prairie. Flowering is most abundant in May and the first part of June, but early flowering plants are present in April. During June the prairie dries rapidly, and, except in swales, there are relatively few flowering species during July and August. In mild winters a few early species, such as *Ranunculus glaberrimus,* may produce a few flowers all winter.

The Rocky Mountain Montane and Subalpine Forests. In the northern and central mountains the montane forests reach their best flowering in early June, though in some areas the earlier plants are available in May. The Subalpine Forest flowers mostly in July.

In the southern Rocky Mountains, especially in Arizona and New Mexico, there are two flowering seasons: a spring season late in May and June, and a late summer season in August and early September following the summer rains. In the montane forest the ground may become unbelievably dry by the time in May or early June when the weather is warm enough for plant growth, and the spring season is minor; in the subalpine forest there is a good spring season in June. During the summer season the growth of plants and the abundance of species in the southern Rocky Mountain Forests is amazing, as it is also in the Southwestern Oak Woodland and Chaparral and the Desert Grassland at lower elevations.

The Eastern Forest Flora

The arrival of spring depends upon latitude. At Minneapolis, Minnesota, early species appear about April 15, and the best flowering is about May 15. At the latitude of St. Louis, Missouri, some plants are in bloom in the latter part of March, and spring flowering is best developed in late April or early May. Spring comes relatively early in the Southern States, and plants are abundant on the coasts in March and in the Piedmont region of the Atlantic Coast in April. In eastern Oklahoma the first plants appear in February or March and maximum spring flowering is in April and May.

About one month after maximum flowering the spring flora is replaced almost completely by summer plants, and these give way from time to time to a series of replacements until the first killing frost in October or November. In some areas the season may end in early September.

Near the Gulf Coast, as for example in northern Florida, the early plants bloom in January, flowering is significant in February, and there is a good display of flowers in March and April unless the year is dry. Another season of conspicuous flowering with emphasis upon Compositae during late August, September, and October. Some native plants are in flower in every month of the year.

The Prairie Flora

The Prairie. The true Prairie region has about the same flowering seasons as in the portion of the Eastern Forest Flora at the same latitude.

The Shortgrass Prairie. The Great Plains are primarily an area of both spring and late summer rain, and flowering occurs at both seasons. To the northward spring is at its best in May and to the southward (Texas) in April; the late summer season is mostly in September.

The Sierra Madrean Flora

The flowering season varies according to the subdivisions of the region, as follows.

The Southwestern Oak Woodland and Chaparral. In this area the spring flowering season is relatively minor. The meager winter precipitation evaporates rapidly with the ceasing of rainfall and snowfall in the spring, and few plants mature and flower before the soil becomes too dry for their growth. Nonetheless, some species do flower, and the region is not to be overlooked for collecting plants in April. During unusually rainy springs there are growth and flowering and green hills. The summer rains may begin late in June and may lap over into early September, but ordinarily they coincide closely with the months of July and August. This is the lush time of year and a very rich one for plant collecting, particularly in August after the plants started by the first rains have matured. General collecting may remain good through September, and there is usually an abundance of grasses and of some other plants still remaining in collectible condition, despite the fact that drying is rapid with the usual abrupt cessation of summer rain early in the month. Some grasses persist in October.

The Desert Grassland. In Arizona, New Mexico, and West Texas (west of the Pecos River) flowering occurs in late March and in April and again in the late summer or early fall. The winters are milder than in the Southwestern Oak Woodland and Chaparral (at the next higher level of elevation) and more moist than in the Southwestern Deserts (just below). Since warm weather comes earlier than in the adjacent oak woodland and at a shorter interval after the end of the winter rains, there is a modest spring flowering season, but the summer season produces the principal growth. The summer rains are heavier and more consistent in occurrence than in the desert at Tucson, and there is a good flowering season every year.

The California Oak Woodland and the California Chaparral. In northern California, where the oak woodland is predominant, the rains begin usually in October, but in southern California they may begin in November or even January. The seeds of herbaceous plants are sprouted by the first rains, and at the same time perennials dormant over the dry summer are stimulated to activity. Consequently, the landscape turns more or less green and remains that way through the winter, except at the upper levels toward the northern extremity of the oak woodland where development may be relatively slower and a truly green blanket does not appear on the hills until toward spring. Spring comes early. Some of the earlier shrubs, such as the manzanitas (*Arctostaphylos*), currants, gooseberries, and silk-tassel-bushes (*Garrya*), begin flowering in December. A few other early plants appear in January, and many are available in February. March and April are the height of the flowering season. In middle or late May the fields begin to dry up and, except at higher levels, the season for herbs is over in June, although many shrubs bloom then. However, in late May or June herbaceous plants are still to be found in the more moist locations, and there is a wealth of species at the edges of the drying vernal pools and along the drying sand and gravel bars of creeks which run through the winter but become dry washes in the summer. Only a few specialized plants are in flower in July and August and through the fall months, but there are some shrubs and other perennials in flower at every season of the year. Nonetheless, a first visit to the California lowlands in summer or early fall (i.e., from late May to October) is likely to be disappointing because this is the dormant season for plants and their remains are dry and brown at just the time when the East is green.

The Pacific Grassland. The most magnificent flower displays in North America possibly equalled only by some in Texas have occurred in the Great Valley of California and particularly at its southern end near Bakersfield, where the Pacific Grassland is developed best. A few years ago, before a considerable area of land was plowed up for planting of wheat, potatoes, and cotton, it was possible to see a continuous stretch of five to ten miles of brilliantly colored wild flowers packed as densely as though they were in a garden. Much of this was ended just before and during the Second World War, and in even the rainy spring of 1952 only one area of half a section matched the display of the once enormous fields of wild flowers. Nevertheless, on rainy springs the adjacent hillsides offer both beautiful floral displays on a smaller scale and excellent specimen collecting. In 1973, following 14 consecutive poor years, a gigantic horseshoe of color, curving for 25 miles and 3 miles wide, appeared at the head (south end) of the San Joaquin Valley, mostly on the Tejon Ranch. The average rainfall nearby at Bakersfield is five and one-half inches, and there are more poor years with two or three inches of rainfall or with only a few big storms than good ones with seven to ten inches coming at regular intervals through the winter and especially the spring. The flowering season in the San Joaquin Valley and particularly the southern end is early. It begins in February, reaches its peak the latter part of March, and ends usually by the first of April, although it may continue through the first two weeks of April if the rains last longer than usual. Ordinarily summer has come by the beginning of April and the fields are brown from then until the first fall rains in October or November. Approximately the same schedule occurs in the Sacramento Valley, but there is a little more rainfall and the season is slightly later.

The Juniper-Pinyon Woodland and the Sagebrush Desert. In general, flowering begins in April, and the peak month for this vast region centering in the Great Basin is May. Only the late-flowering shrubs are available for collecting through the summer and fall.

The Mexican Desert Flora

In the United States the creosote-bush deserts are developed best in southern and central Arizona, where there are two rainy seasons each year. At Tucson, which is typical, the average rainfall is about eleven inches. Ordinarily four or five inches occur as winter rain, falling from October through April, but some years the amount may be considerably more or less, ranging up to ten inches or down to two. Since these rains are spaced over a long period, they are effective only on wet years when they bring forth an abundance of wild flowers, making the desert colorful, though not nearly as colorful as the southern San Joaquin Valley in its good years. Few plants are to be found along the main railways and highways which, for the sake of speed and ease of travel, cross the flattest, least interesting parts of the desert; but they are abundant near secondary roads in the small rocky desert mountain ranges and in the foothills of the higher ranges. The height of spring flowering of desert annuals is reached toward the end of March. The annuals are followed by trees, shrubs, and cacti, which flower in April and May. The height of cactus blooming is likely to be in the month of May, although many species flower earlier, some only in June, and others in the summer. For example, the night-blooming cereus flowers largely on a single night, usually in June. The desert is hot and dry in May and June, but, with the coming of summer rain at the end of June or usually early in July, it blooms again. Some of the trees and shrubs bloom twice a year, though most profusely in the spring. Most of the summer herbaceous plants are different from those of the spring. In particular, there are numerous perennial grasses. The peak of the summer flowering is in August, and drying in early or middle September is abrupt. North and west of Tucson the proportion of winter to summer rain increases, but southward and eastward the trend is reversed, the summer rain being predominant and the winter rain more or less negligible, coming as only the tail end of the California winter storms.

The Mojavean Desert. Usually (except on very wet years) maximum flowering is in April; after the first week of May only the higher hills are in bloom. Ordinarily the California deserts do not have a fall flowering season because there are rains in the late summer or early fall only once in ten or twenty years.

The Colorado Desert. In the low-lying desert of southeastern California and the western edge of Arizona the height of the flowering season is in March, and ordinarily the plants are dry in April. A few early species may appear during February, but this time is unsatisfactory for collecting specimens because few fruits are available. Ordinarily, there is no summer flowering season, although once in several years unusual fall rains may promote flowering of some plants near the mountains.

The Arizona Desert. This area has the best-developed desert flora in the United States. It has been discussed above under the Mexican Desert Flora as a whole.

The Chihuahuan Desert. There is a spring flowering season in April, but the important season is late summer, particularly September.

The Caribbean Tropical Forest

In southern Florida the flowering periods are determined by rainfall. Since winter is dry and summer wet, the best growing season is summer (the opposite of California where it is winter). Flowering begins in May or June according to the dryness of the season, and the peak is in the early summer. In the jungle-like hammocks of Southern Florida and the Keys some relatively inconspicuous epiphytes and small herbs flower the year around. From Lake Okeechobee southward certain plants may be collected at any season.

The Hawaiian Flora

The Hawaiian Islands do not have marked wet and dry or warm and cool seasons. The average temperature is about 75°F, and in Honolulu the record high is 88° and the low 57°. The humidity is high, but the trade winds, when they are blowing, as they nearly always are, make the weather pleasant. Many plants are in bloom and collectible at any time of the year. Because the humidity is high, plant presses must have heat and air movement for drying specimens.

Aquatic Plants of All Regions

In almost every region aquatic plants or those occurring in wet situations flower later than do the terrestrial species. They continue to flower through a considerable part of the summer or far into the fall.

NEW NAME COMBINATIONS

The following name combinations are validated here so that they may be used in the text:

Quercus turbinella Greene var. **californica** (Tucker) L. Benson, comb. nov. *Quercus turbinella* subsp. *californica* Tucker, Madroño 11: 240. 1952.

Quercus turbinella Greene var. **ajoensis** (C. H. Muller) L. Benson, comb. nov. *Quercus ajoensis* C. H. Muller, Madroño 12: 140. *f. 1, above.* 1954. *Quercus turbinella* Greene subsp. *ajoensis* Felger & Lowe, Jour. Ariz. Acad. Sci. 6: 83. 1970.

Juniperus californica Carr. var. **osteosperma** (Torrey) L. Benson, **comb. nov.** *Juniperus tetragona* Schlecht. var. *osteosperma* Torrey, Pac. R. R. Rept. 4: 141. 1857.

These combinations are discussed fully in *The Trees and Shrubs of the Southwestern Deserts,* ed. 3., in press, University of Arizona Press, which is likely to appear later than this book.

Glossary

Some Common Technical Terms

(Largely of Taxonomic Terms as Applied to Vascular Plants)

Abaxial. The side of an organ turned away from the axis, for example, the underside of a leaf; the dorsal side.

Aberrant. Different from the normal.

Abnormal. Different from the usual or prevailing form or structure.

Abortion. Imperfect development or failure of development.

Abortive. Imperfectly developed or undeveloped.

Abruptly pinnate. Pinnate, but without a terminal leaflet. See *even pinnate.*

Acaulescent. Apparently without a stem, that is, with the main stem underground and only basal leaves and slender, leafless flowering stems appearing above ground level.

Accessory. Additional beyond the normal; extra.

Accrescent. Increasing in size after flowering; enlarging with age.

Accumbent. Applied to cotyledons with their edges turned toward the main axis of the embryo. See *incumbent.*

Aceriform. Similar to the leaf of a maple.

Acerose. Needlelike.

Achene. A dry, one-seeded fruit with a firm close-fitting wall which does not open by any regular dehiscence.

Achene beak. The persistent and hardened style of an *achene.*

Achlamydeous. Having neither a calyx nor a corolla.

Acicular. Needle-shaped and very slender.

Acorn. The leathery fruit of an oak, containing a single large seed and enclosed basally in a cup formed from bracts.

Actinomorphic. *Radially symmetrical,* that is, capable of division by straight boundaries into three or more similar sections.

Active bud. A growing bud.

Aculeate. Covered with prickles.

Acuminate. With a long, tapering point set off rather abruptly from the main body (e.g. of a leaf).

Acute. With a pointed end forming an acute angle, that is, less than a right angle.

Adaptive radiation. The filling of all the environmental niches available to a taxon with either existing or newly evolving lesser taxa. This occurs rapidly when a new area becomes available, as by uplift of new islands.

Adaxial. On the side toward the axis, as for example, the upper side of a leaf; the ventral side.

Adherent. Grown fast to a structure of a different kind, as for example, a stamen grown fast to a petal.

Adnate. Adjective. See *adnation.*

Adnation. Fusion with or attachment to another structure from the beginning of development.

Adventitious. A structure occurring in an unusual position as, for example, an adventitious root formed from a stem or a leaf.

Adventitious bud. A bud produced in an unusual or unexpected place, as for example, near a point of injury of a stem or on a leaf or a root.

Adventive. Said of an introduced plant beginning to spread into a new locality or region.

Aërial root. Roots exposed to the air, as those of tropical plants growing on trees.

Aestival. Of the summer.

Aestivation. Arrangement of the young flower parts in the bud.

Agglomerate. Crowded into a dense cluster but not joined.

Aggregate. See *agglomerate.*

Alate. With a wing.

Albumen. A deposit of reserve food material accompanying the embryo. Such reserves often are in the endosperm.

Alga. A seaweed or one of the essentially microscopic pond scums. Plural, **algae.**

Alliaceous. Having the odor or taste of garlic or onions.

Alpine. Above timberline in the mountains.

Alternate. With a single structure of each kind occurring at each level of the axis, e.g., stem. These structures (e.g., leaves and branches) appearing to alternate on opposite sides of the axis, but actually in a spiral arrangement.

Alveolate. With angular depressions forming a pattern like a honeycomb.

Alveolus. A depression within an *alveolate* pattern.

Ament. A soft, usually scaly spike of small apetalous, unisexual flowers, this usually falling as a single unit. See *catkin.*

Amentiferous. With the flowers in *aments,* i.e., catkins.

Amorphous. With no definite form.

Amphitropous. Describing an ovule bent back along and adnate to the funiculus, but with the micropyle not bent all the way back to the funiculus.

Amplexicaul. Clasping the stem, that is, the base nearly surrounding it.

Ampliate. Enlarged.

Anastomosing. Joining each other and forming a network, the term being applied to veins, as the veins of a leaf.

Anatropous. Descriptive of an ovule in which the body is bent backward along the funiculus and adnate with it. The micropyle is against the funiculus.

Ancipital. Two-edged.

Androecium. The male reproductive organs; in the flowering plants the stamens—a collective term for all those within a flower.

Androgynous. A term applied to an inflorescence which includes both staminate and pistillate flowers, the staminate ones being apical, the pistillate basal.

Androphore. A stalk or other supporting structure under stamens raised above their normal position in the flower.

Anemophilous. With the pollen carried by wind.

Annual. A plant completing its life cycle in a year or less.

Annular. Ringlike.

Annular thickening. Thickening of the wall of a xylem cell laid down in the form of a ring.

Annulus. A ring; in the ferns a crestlike or other special structure on the sporangium.

Anther. The portion of the *stamen* which produces pollen.

Anther tube. A tube formed by coalescent anthers, as in the sunflower family.

Antheridium. A structure producing antherozoids, that is, male gametes or sex cells.

Antherozoid. A male *gamete* (sex) cell of a plant.

Anthesis. The time when the flower expands and opens or the process of expansion and opening.

Anthocyanin. Any of a common class of pigments having colors ranging from lavender to purple. These pigments are affected by the acidity or alkalinity of the cell sap, and they change color in approximately the same way as litmus paper, tending toward red in an acid medium and blue in a basic (alkaline) medium. These pigments are water-soluble and not associated with plastids.

Anthoxanthin. Flavonoid pigment related to the anthocyanins, but yellow to reddish orange or orange.

Anticyclone. An area of high pressure; thus air does not move upward, as with a cyclone, but outward in all directions; since no air is elevated, there is no rain but fair weather.

Antipodal cells. Three (or after development more) cells at the end of the flowering plant megagametophyte opposite the egg.

Antrorse. Directed upward or forward.

Apetalous. Without petals.

Aphyllous. Without leaves.

Apical. Of the apex.

Apical bud. See *terminal bud.*

Apical placenta. A placenta at the distal (apical) end of the ovary.

Apiculate. Terminated by an abrupt, short, flexible point.

Apiculation. An abrupt, short, flexible point.

Apocarpous. With the carpels separate.

Apogamous. Developed without the joining of gametes (sex cells), that is, asexually.

Apophysis. A swelling on the surface of an organ as, for example, the somewhat swollen exposed portion of the cone scale of a gymnosperm.

Appendiculate. With a basal appendage.

Appressed. Lying tightly against another (usually larger) organ.

Apterous. Without wings.

Aquatic. Growing in the water.

Arachnoid. With slender, tangled hairs resembling the threads of a spiderweb.

Arboreous. Of treelike form, that is, with a main woody trunk.

Arborescent. Of large size and more or less treelike but without the clear distinction of a single trunk.

Archegonium. A flask-shaped organ containing an egg, that is, a female gamete or sex cell.

Arctic. The area beyond timber line at high latitudes.

Arcuate. Forming a moderate curve or arc.

Areolate. Divided into small, marked-off spaces. See *reticulate.*

Areole. Diminutive of area; a small, clearly marked space. The term is used most frequently in reference to the small, special spine-bearing areas on the stem of a cactus.

Aril. A large appendage of the funiculus at the hilum (attachment area) of a seed. It tends to envelope the seed.

Arista. A stiff bristle.

Aristate. With a stiff bristle.

Articulate. With conspicuous segments or joints.

Artificial system (of classification). A system made up according to predetermined supposedly "important" characters.

Ascending. Arising at an oblique angle (or on a curve).

Asperous. Rough.

Association (ecological). A local or regional manifestation of a climax formation, such as a forest, marked by one or more dominant species, e.g., Douglas fir or redwood forest.

Assurgent. See *ascending.*

Attenuate. With a long, tapering point, this usually set off rather abruptly from the main body of the object (for example, a leaf blade).

Auricle. An appendage shaped like the lobe of a human ear (that is, like the type with a rounded lobe).

Auriculate. Having an *auricle.*

Austral. Southern.

Awl-shaped. With a narrow flattened body tapering very gradually upward into a point. See *subulate.*

Awn. A long stout or stiff bristle.

Awned. With an *awn.*

Axil. The adaxial angle between two organs, particularly the angle between the upper side of a leaf and the stem.

Axile. On the axis, as for example, said of placentae at or near the center of an ovary.

Axillary. In the *axil.*

Axillary bud. A bud in a leaf *axil.*

Axillary flower. A flower in the *axil* of a leaf or a bract.

Baccate. Berrylike, that is, fleshy or pulpy.

Banner. The upper and usually largest petal in the *papilionaceous* corolla of a plant of the pea family (of the common type having markedly bilaterally symmetrical flowers). Known also as a standard or a vexillum.

Barbate. Bearded. (Diminutives, *barbellate, barbellulate.*)

Barbed. With a rigid barb like the barb of a fish hook.

Barbulate. With a beard of fine hairs.

Bars of Sanio. Special thickenings on the walls of some xylem cells.

Basal leaves. Leaves at the base of an herbaceous plant, arising from several nodes separated by exceedingly short internodes occurring at about ground level.

Basal placenta. A placenta at the basal end of the ovary.

Basifixed. Attached at the base.

Beaked. Ending in a firm elongated slender structure.

Bearded. With long or stiff hairs.

Berry. A fleshy or pulpy fruit with more than one seed and formed from either a superior or an inferior ovary. The seeds are embedded in pulpy tissue.

Betacyanin. A pigment occurring in the Cactales and some Caryophyllales; not related to but taking the place of anthocyanins in other flowering plants and paralleling them in color (toward red in an acid medium and toward blue in a basic medium); water-soluble and not associated with plastids; the molecule includes nitrogen, but anthocyanins do not.

Betalain. A betacyanin or a betaxanthin.

Betaxanthin. Pigment similar to a betacyanin but yellow to reddish orange or orange.

Bidentate. With two teeth.

Biennial. Completing the life cycle in two years. Biennial plants usually produce only basal leaves above ground the first year and both basal leaves and flowering stems the second.

Bifid. Forked, that is, ending in two parts.

Bilabiate. Two-lipped. A bilabiate corolla has petals in two sets, commonly with two in the upper and three in the lower.

Bilateral. Two-sided.

Bilaterally symmetrical. Capable of division into only two similar sections, which are mirror images of each other.

Bilocular. With two cavities, as an ovary with two seed chambers.

Bipinnate. Pinnate and the primary leaflets again pinnate.

Bipinnatifid. Pinnatifid with the primary divisions again pinnatifid.

Bisexual. With both sexes represented in the same individual or organ, e.g., with stamens and pistils in the same flower.

Blade. The broad, usually flat, part of a leaf.

Bloom. A usually waxy, whitish or bluish powder covering the surface of a leaf, stem, fruit, or other organ.

Blossom end. The end of an inferior ovary and the surrounding floral cup which supports the flower parts or at fruiting time the calyx and remains of other flower parts.

Boreal. Northern.

Boss. A protrusion, as for example, a protuberance near the center of the *apophysis* of a scale of a gymnosperm cone (or of a shield).

Bract. A leaf subtending a reproductive structure, such as a *flower* or a cluster of flowers or an *ovuliferous scale.* Usually the leaf is specialized and at least somewhat dissimilar to the foliage leaves.

Bracteate. With *bracts.*

Bracteolate. With bractlets.

Bracteole. Diminutive of *bract,* a mere scale.

Bractlet. A minor *bract.*

Branchlet. A small branch, that is, one of the smallest degree.

Branch primordium. The hump of tissue which develops into a young branch.

Bristle. A stiff hair.

Bryophyta. The mosses, liverworts, and *Anthocerotae.*

Bud. A growing structure at the tip of a stem or branch, with the enclosing scale leaves or immature leaves; a young flower which has not yet opened. There are *vegetative buds* and *flower buds,* as described in the definition above.

Bulb. An underground bud covered by fleshy scales, the coating formed from the bases of leaves.

Bulbiferous. With *bulbs.*

Bulbil. A small *bulb.* The word is applied to bulblike structures produced on the stems, usually in the axils of leaves, or sometimes (as in some onions and some century plants) in the places where flowers ordinarily occur.

Bulblet. A small underground bulb.

Bulbous. Bulblike.

Bullate. Blistered or puckered.

Caducous. Falling very early. In the poppy family, the caducous sepals fall away when the flower opens.

Caespitose. Growing in tufts or mats. (Also spelled *cespitose.*)

Calcarate. With a spur.

Callosity. A thickened hard structure.

Callous. With the texture of a *callus.*

Callus. A hard or tough swollen area.

Calyculate. With small bracts in a series below the flower, these resembling a calyx.

Calyptra. A cap, cover, or lid.

Calyx. A cup; the sepals of a flower; the outermost series of flower parts. Each sepal has the same number of vascular traces as a leaf of the species. If there is doubt, it may be distinguished by this character. Cf. *petal.*

Calyx tube. A tube formed from the lower portions of the sepals. The term has been used loosely for (a) a floral cup or floral tube regardless of its origin, (b) a floral cup or tube formed by coalescence and adnation of the bases of the sepals, petals, and stamens, (c) a perianth tube of the type in the monocotyledons formed by edge to edge adnation of the adjacent sepals and petals together with the bases of the stamens, and (d) the portion of a floral cup or tube above the area of adnation to an inferior ovary.

Cambium. A layer of dividing cells which add during each growing season a layer of woody material (largely xylem) on their inner side toward the center of the stem or root and a layer of bark (phloem and associated tissues) on the outer side. Known also as a **vascular cambium.**

Campanulate. Bell-shaped, that is, the shape of an inverted church bell, i.e., rounded at the attachment and with a broad flaring rim.

Campylotropous. Descriptive of an ovule which curves in such a way that the micropyle (opening) is near the funiculus or stalk, but the side of the ovule is not adnate to the funiculus.

Canaliculate. With lengthwise furrows or channels.

Cancellate. Appearing to be a lattice-work.

Canescent. Grayish-white or hoary, densely covered with white or gray fine hairs, these usually short.

Capillary. Like an elongated delicate hair or thread.

Capitate. In a dense cluster or head.

Capitulate. Diminutive of *capitate.* Actually this word should be used instead of capitate because the reference is to a *capitulum.*

Capitulum. A small head.

Capsule. A dry, many-seeded fruit made up of more than one carpel and splitting open (dehiscent) lengthwise at maturity.

Carina. A keel, literally like the keel of a boat. Sometimes said of a structure which simply is folded and creased and without a projecting keel on the crease.

Carinate. With a keel.

Carpel. A specialized leaf which forms either all or part of a pistil.

Cartilaginous. With the texture of cartilage, that is, tough and firm but somewhat flexible.

Caruncle. An appendage at the attachment point (*hilum*) of a seed.

Carunculate. With a *caruncle.*

Caryopsis. The fruit of a grass, that is, a one-seeded, indehiscent fruit with the pericarp adnate with the seed.

Castaneous. Chestnut-colored.

Catkin. A soft, usually scaly spike or raceme of small apetalous unisexual flowers, the inflorescence usually falling as a single unit. See *ament.*

Caudate. With a taillike structure.

Caudex. A largely underground stem base which persists from year to year and each season produces leaves and flowering stems of short duration.

Caulescent. Having a well-developed stem above ground as opposed to having only what appears to be the stalk of the individual flower or cluster of flowers.

Cauline. Of or on the stem.

Cell. The structural and functional unit of living organisms. This unit is surrounded by a membrane which encloses at least the living substance (nucleus and cytoplasm) as well as frequently other structures.

Cell (atmospheric). A unit of the atmosphere or especially of its circulation.

Centrifugal. Developing first at the center and then gradually toward the outside.

Centripetal. Developing first at the outside and then gradually toward the center.

Cernuous. Nodding.

Cespitose. See *caespitose.*

Chaff. Dry, membranous scales or bracts, that is, similar to the chaff which comes from a thrashing machine.

Chaffy. Resembling chaff.

Chamber. A room. Applied to the cavities (locules or "cells") of an anther or an ovary.

Channelled. With lengthwise grooves, the grooves usually deep.

Chartaceous. Like writing paper.

Chlorophyll. The green coloring of most plants. A substance which aids in making the energy of light available to *photosynthesis* (manufacture of a simple carbohydrate, formed [in effect] from carbon dioxide and water).

Chloroplast. Literally a green body, actually a solid body with chlorophyll in its outer part.

Choripetalous. With the petals separate (distinct) or at least with some of them separate from the others.

Chromosome. One of two to several units in the nucleus of a cell; composed of a gigantic DNA molecule with two strands (spiral) of deoxyribose and a phosphate group alternating in the sequence and with paired bases joining the spiral elements into a ladder; the pattern of the units in the chromosome constitutes the genetic code (see genetic code and DNA).

Cilia. Hairs along the margin of a structure, placed like the eyelashes on a human eyelid.

Ciliate. With *cilia* along the margin.

Ciliolate. Diminutive of *ciliate.*

Cinereous. The color of ashes.

Circinate. Coiled at the tip, as the young coiled leaf of a fern, which uncoils gradually as it develops.

Circumscissile. Opening by a horizontal circular line, the top coming off like a lid.

Class. The next taxon below division; a group of related orders or sometimes a single order. Usual ending: -opsida.

Clavate. Gradually enlarged upward after the manner of a baseball bat or the traditional giant's club, that is, either tapering gradually upward or with an enlarged knob at the summit.

Clavellate. Diminutive of *clavate.*

Claw. The narrow stalk at the base of a petal, this resembling the petiole of a leaf.

Cleft. Indented about half way or a little more than half way to the base or to the midrib.

Cleistogamous. Self-fertilized in the bud stage, or at least without opening.

Climatic zone. A belt around the earth having similar climate, as in the arctic, subarctic, temperate, horse latitude, or tropical zone. The vegetation is similar around the world in such a belt, but the flora is not.

Climax (ecological). The end stage in succession; a type of vegetation and flora adapted to withstanding the extremes of local climate and therefore relatively stable until there is a major change of climate. Mature individuals tend to be replaced by their own offspring.

Climbing. Supported by clinging. See *scandent*.

Coalescence. The union of similar parts, as for example, the petals of a flower.

Coalescent. Adjective. See *coalescence*.

Cochleate. Spiral, like a turban snail.

Codon. A group (triplet) made up of any 3 of guanine, cytosine, adenine, and/or uracil with or without repetition of any of these. This triplet, carried in DNA, then RNA, calls for a particular amino acid to enter at a particular point in the linear series making up an enzyme (protein).

Coherent. Grown fast together (similar structures joined to each other).

Colonial. In colonies. For the angiosperms, used primarily in reference to plants occurring in clumps connected by rhizomes.

Column. A slender aggregation of coalescent stamen filaments, as in some members of the mallow family; the adnate style and filaments or filament (*gynandrium*) of the orchid family.

Coma. A tuft of hairs on the end of a seed.

Commissure. The surface along which two or more *locules* are joined to each other.

Comose. With a *coma*.

Companion cell. A component of the phloem; small complete cells between the sieve tube cells.

Complete flower. A flower with all the four usual series, that is, sepals, petals, stamens, and pistils.

Compound. Composed of two or more similar elements. For example, a panicle is a compound of spikes, racemes, or corymbs, that is, composed of two or more of any one of these elements; a **compound leaf** is composed of two or more leaflets; a **compound pistil** is composed of two or more coalescent carpels.

Compressed. Flattened, particularly from side to side (i.e., **laterally compressed**). See *obcompressed*.

Conduplicate. Folded together lengthwise.

Cone. A reproductive structure composed of an axis (branch) bearing sporophylls or other seed- or pollen-bearing structures. See *strobilus*.

Cone scale. See *ovuliferous scale*.

Conglomerate. Densely aggregated; a cluster or a heap.

Coniferous. Cone-bearing.

Conjugate. Lying together in pairs.

Connate. Joined from the beginning of development.

Connate-perfoliate. Both connate and perfoliate; e.g., two opposite sessile leaves together encircling the stem, growing together, and appearing as one.

Connective. The middle part of an anther connecting the two pollen sacs or two pairs of pollen sacs.

Connivent. Standing together. Stamens with their tips ending against each other are connivent.

Continental climate. The climate of the interior of a temperate continent; for lack of contact with the ocean, the temperature runs high in summer and low in winter; for lack of a water source, the climate tends to be dry.

Contorted. Twisted out of the usual form. Petals may be contorted in the bud.

Convolute. Rolled up lengthwise. Petals may be convolute in the bud.

Cordate. Of a conventional heart-shape, the length greater than the width, the petiole attached in the basal *sinus*, the apex acute; applied also to the basal indentation of a leaf or other structure.

Coriaceous. Leathery.

Corm. A bulblike structure formed by enlargement of the stem base. It is sometimes coated with one or more membranous layers.

Corneous. Horny.

Corniculate. With a small horn or with small horns.

Corolla. The petals of a flower, that is, the inner series (or several series) of the perianth. Each petal has usually a single vascular trace as does a stamen.

Corolla division. See *corolla lobe*. A division is longer than a lobe or a part.

Corolla lobe. See *corolla teeth*. A lobe is longer than a tooth. Lobe may be applied in a broad sense to cover lobes, parts, or divisions.

Corolla part. See *corolla lobe*. A part is longer than a lobe.

Corolla teeth. The separate tips of the individual petals of a sympetalous corolla.

Corolla tube. The hollow cylinder formed by coalescence of petals.

Corona. A small crown, the term being applied to special appendages of the corolla. These sometimes form a tube, e.g., daffodil.

Coroniform. Shaped like a crown.

Corrugated. With many small folds or wrinkles.

Cortex. The layers of living cells outside the central stele (which includes the conducting tissues) but inside the outermost single layer of cells (the epidermis) of a stem or a root.

Cortical. Of the *cortex*.

Corymb. A flat-topped cluster of flowers; fundamentally like a raceme but with the pedicels of the lower flowers longer and the pedicels of the upper flowers gradually shorter. As a result of this arrangement, at a middle stage of development there may be fruits on the outside, buds in the center, and flowers in between. (The opposite of the arrangement in a *cyme*.)

Corymbiform. In the form of a *corymb*.

Corymbose. Arranged in corymbs.

Costa. A rib or a prominent nerve.

Costate. Ribbed lengthwise.

Cotyledon. One of the first leaves developed in the embryo in the seed at the joining point of the hypocotyl and the epicotyl. These leaves are present in the seed and they may or may not enlarge and become green when the seed germinates. Often food materials are stored in them. In the flowering plants there are one or two or very rarely more. Among the gymnosperms usually there are two, but in the pine family there may be as many as seventeen.

Creeping. The stem growing along the ground and producing adventitious roots.

Crenate. With rounded teeth projecting at right angles to the edge of the leaf.

Crenulate. Diminutive of *crenate.*

Crested. With a crest, that is, a projection at a prominent position such as the apex or the midrib of an organ.

Crisped. Ruffled, that is, with the same kind of winding as in *sinuate,* but in the vertical plane instead of the horizontal.

Cristate. Crested.

Cristulate. The diminutive of *cristate.*

Cross section. A slice cut across an object.

Crown. See *corona.*

Cruciate, cruciform. Crosslike.

Crustaceous. Hard and brittle in texture.

Cryptogams. An old term applied to the thallophytes, bryophytes, and the pteridophytes, that is, the plants which regularly reproduce without formation of seeds.

Cucullate. Hood-shaped.

Culm. The hollow stem of a grass, the term being applied sometimes to sedges also.

Cuneate, cuneiform. Wedge-shaped; essentially a narrow isosceles triangle with the distal corners (those away from the petioles) rounded off. The petiole of a cuneate leaf is attached at the sharp angle.

Cupule. A little cup, the term being applied particularly to the cup of an acorn (which is an involucre formed from coalescent bractlets).

Cuspidate. With a sharp, firm point at the tip. As applied to leaves, the term indicates a point of firmer texture than the rest of the blade.

Cuticle. The waxy, more or less waterproof coating secreted by the cells of the epidermis (outer layer) of a leaf, stem, or flower part.

Cyathiform. Cup-shaped.

Cyathium. A cuplike involucre enclosing flowers. In the genus *Euphorbia,* the cyathium includes one pistillate and a number of staminate flowers each of a single stamen. A joint in the "filament" is the receptacle terminating a slender pedicel and bearing the true filament. There are glands on the edges of the cyathium, and frequently each of these has a petal-like appendage. Consequently the cyathium and the flowers appear to be a single flower. Some authors use cyathium to include the contents as well as the cup.

Cycle. A circle. Leaves or flower parts arranged in a single series at a single node are in a cycle.

Cyclic. Arranged in a *cycle.*

Cyclone. An area of low pressure; the air tends to rush in, rise, expand, cool, and drop moisture; hence this is storm area. Not to be confused with tornado.

Cylindroidal. In approximately the shape of a cylinder.

Cyme. A broad, more or less flat-topped cluster of flowers. If the branches of the inflorescence are alternate, the form is like that of a *corymb,* the lower pedicels being elongated and the upper gradually shorter. However, since the terminal flower blooms first, at a middle stage of development the fruits are in the center; the buds are on the outside; and flowers are between. In plants with opposite branching in the inflorescence, a flower terminates the major branch. The two buds developing from opposite sides of the nearest node next produce flowers, as then do two buds below each of these, the process being repeated a variable number of times. In both types of cymes increase in length of the stem is stopped by conversion of the terminal bud into a flower. The length of the inflorescence is predetermined or determinate, and in a cyme with alternate branching, the number of flowers is predetermined when the first flower blooms. In a raceme, spike, or corymb, the increase in the growth of the stem is indeterminate because the apical bud continues to increase indefinitely the length of the stem (axis of the flower cluster).

Cymose. With *cymes* or with cymelike clusters of flowers.

Cymule. Diminutive of *cyme;* sometimes applied to a portion of a cyme.

Cypsela. An achene which is attached to the enclosing floral cup or tube, that is, an achene formed from an inferior ovary. E.g., dandelion fruit.

Cytology. The study of the cell. Often particular emphasis is placed upon the chromosomes.

Cytoplasm. The living substance outside the nucleus and inside the enclosing membrane.

Deciduous. Falling off at the end of each growing season.

Declinate. Bending or curving downward or forward.

Declined. Bending over in one direction.

Decompound. More than once compound (or divided).

Decumbent. Reclining except at the apex.

Decurrent. Leaf bases which continue along the stem as wings or lines.

Decussate. In opposite pairs (as opposite leaves) with the alternate pairs projecting at right angles to each other.

Deflexed. Abruptly bent or turned downward.

Dehisce. To split open along definite lines.

Dehiscence. The process of splitting open at maturity.

Dehiscent. Splitting open along definite lines.

Deltoid. Of the shape of the Greek letter delta, that is, an equilateral triangle, the attachment being in the middle of one side.

Dentate. With angular teeth projecting at right angles to the edge of the structure (e.g., a leaf).

Denticulate. Diminutive of *dentate.*

Depauperate. Stunted, that is, very small. The term is applied particularly if the plant is undeveloped as compared with others of the species, but also if it is a representative of a small species.

Depressed. Flattened from above as if pushed downward.

Determinate. With a definite predetermined number of structures (e.g., flowers). See *cyme.*

Diadelphous. In two brotherhoods, the term being applied to stamens coalescent in two sets. The common usage is for members of the pea family in which nine of the ten stamens are coalescent and the other stands alone.

Diandrous. With two stamens.

Dichasium. A cyme with two axes running in opposite directions, that is, the type of cyme formed in plants with opposite branching in the inflorescence. See *cyme.*

Dichlamydeous. With both calyx and corolla.

Dichotomous. Forking, with two usually equal branches at each point of forking.

Dicotyledonous. With two *cotyledons.*

Didymous. Twinlike, that is, occurring in pairs.

Didynamous. In two pairs, the pairs not being of the same length. The term is applied to stamens.

Differentiation. Accumulation of differing characters in different populations or taxa.

Diffuse. Spreading widely and diffusely in all directions.

Digitate. Resembling the fingers of a human hand, that is, with several similar structures arising at a common point. See *palmate.*

Digynous. With two pistils.

Dimerous. With two members; the flower parts in twos.

Dimidiate. With the appearance of being only half a structure.

Dimorphous. With two forms.

Dioecious. The flowers unisexual and the staminate on one individual and the pistillate another. The term is used also for gymnosperms with pollen and seed cones on different individuals.

Diploid. With 2n chromosomes per cell.

Diplostemonous. The stamens in two series, those of the outer series alternating with the petals.

Dipterous. With two wings.

Disc, disk. In the sunflower family or *Compositae,* the central portion of the compound receptacle bearing the *disc flowers.*

Disc flower. In the sunflower family or Compositae, one of the flowers with tubular corollas.

Discate. In the form of disc, i.e., circular and flattened.

Disciform. Circular and flattened like a disc.

Discoid. Resembling a disc. In the sunflower family or *Compositae* a **discoid head** is one having only *disc flowers* and no *ray flowers.*

Discrete. Separate.

Dissected. Divided into narrow segments.

Dissepiment. A partition dividing an ovary or a fruit into chambers.

Distichous. In two vertical series or ranks.

Distinct. Separate.

Diurnal. Occurring in the daytime. The term is applied to the flowers which open in daylight as opposed to the nocturnal flowers which open at night.

Divaricate. Spreading widely, that is, *divergent.*

Divergent. Spreading away from each other.

Divided. Indented essentially to the base or midrib.

Division. The highest rank of taxon in the plant kingdom; a group of related classes or sometimes a single class. Usual ending: -phyta.

Division. A segment of a structure, such as a leaf.

DNA or **deoxyribonucleic acid.** The genetic material determining indirectly the inherited characters of living organisms; composed of 2 helical elements of deoxyribose (a sugar) and of a phosphate group (these alternating), with paired bases joining the two spiral elements into a ladder.

Dorsal. On the outer surface of an organ, that is, the side away from the axis (e.g., the back of a leaf); the *abaxial* surface of a structure such as a petal or a leaf. See *ventral.*

Dorsiventral. A structure having a clear differentiation of a back and front or upper and lower side.

Dorsoventral. (A zoölogical term sometimes applied in botany.) The distance or measurement from the back of a structure to the front, that is, from dorsal to ventral, as opposed to lateral (side to side) distance.

Double samara. A *samara* with two locules and two wings.

Downy. Finely and softly *pubescent.*

Drainage area. The watershed or area into which the excess precipitation of a region drains.

Drupaceous. Drupelike.

Drupe. A fruit with a fleshy exocarp and a hard, stony endocarp about each seed. The classical example is one of the pitted fruits, that is, a plum, cherry, apricot, peach, or almond. The term is used also for other fruits which are fleshy or pulpy on the outside and with one or more stones inside.

Drupelet. Diminutive of *drupe.* The cluster of fruits of a blackberry or a raspberry is composed of drupelets which seem to form a single fruit.

Echinate. Covered with prickles.

Ecology. The study of the interrelationships of organisms with their physical and chemical environment and with each other.

Ecosystem. The complex group of living organisms occurring together and interdependent, together with the physical and chemical factors affecting them.

Egg. A female *gamete* (sex) cell of a plant.

Ellipsoid, Ellipsoidal. Elliptical in outline with a three-dimensional body. See *elliptic.*

Elliptic, elliptical. In the form of an ellipse, that is, about one and one half times as long as broad, widest at the middle, and rounded at both ends. A two-dimensional figure.

Emarginate. With a shallow broad notch at the apex.

Embryo. The new plant enclosed in the seed. Its formation follows union of gametes (sex cells). It consists of an axis and the attached cotyledons and young secondary leaves. The embryos of pteridophytes (and of nonvascular plants) are not enclosed in seeds, and they have no resting stage.

Emersed. Above water.

Enation. An outgrowth from the superficial (surface) tissues of the stem.

Endemic. Restricted in occurrence to a particular geographical area.

Endocarp. The inner portion of the *pericarp.* See *exocarp* and *mesocarp.*

Endosperm. A cell layer occurring

in at least the immature seeds of flowering plants. If the endosperm persists until the seed is mature, usually it becomes a food-storage area. As a result of fusion of two nuclei of the original megagametophyte and later addition of a third male nucleus, the nucleus of the original cell of the endosperm and also the nucleus of every later-formed cell includes 3n chromosomes.

Ensiform. Swordlike; in the form of a sword.

Entire. Without division or toothing of any kind.

Entomophilous. Insect-pollinated.

Enzyme. A substance promoting a chemical reaction but not entering into it.

Epeiric seas. Seas formed in depressions of the mainland (of a continent).

Ephemeral. Lasting for a brief period, as for example, one day.

Epicotyl. The portion of the embryo of a seed plant just above the cotyledon(s); the young stem.

Epigynous. The sepals, petals, and stamens apparently upon the ovary, but actually growing from the edge of the floral cup (or tube), which is adnate with the ovary. An **epigynous flower** is one in which the floral cup (or tube) is adnate with the ovary.

Epigynous disc. A disc within the floral cup and on top of an *inferior ovary.* See *hypogynous disc* and *perigynous disc.*

Epiphyte. A plant growing upon another plant but not parasitic upon it.

Equal. See special use in the latter part of Chapter 9.

Equator. An imaginary line encircling the earth midway between the poles.

Equitant. With the leaves folded around a stem after the manner of the legs of a rider around a horse, or sometimes with the leaves folded around each other in rows, as, for example, in *Iris.*

Erect. Standing upright.

Erose. With the margin appearing to have been gnawed.

Etiolated. White through failure to develop chlorophyll. Etiolated stems (e.g., those developed in darkness or weak light) are elongated and spindly as well.

Even pinnate. Without a terminal leaflet. See *odd pinnate.*

Evolution. The process by which new living organisms originate or have originated through modification of existing or extinct organisms.

Excentric. Off-center.

Excurrent. Projecting, as, for example, a leaf base which projects beyond the margin of the blade.

Exfoliating. Separating into thin layers.

Exocarp. The outer layer of the *pericarp.* See *mesocarp* and *endocarp.*

Explanate. Flattened and spread out into divergent structures.

Exserted. Projecting beyond the usual containing structure, as, for example, stamens projecting beyond the rim of a sympetalous corolla.

Extraterritorial. Occurring beyond the specific geographical range of some features of this book, i.e., beyond North America north of Mexico.

Extrorse. Facing outward.

Falcate. In the shape of a scythe, that is, curving, flat, and tapering gradually to a point.

False partition. An ovary partition not formed by infolding of the edges of the carpel, growing from the middle of the placenta on one side to the middle of the placenta on the opposite side or formed by ingrowth of placentae.

Family. A group of related genera or single genus; a unit in classification. Usual ending: -aceae, but there are several exceptions.

Farinose. Mealy, that is composed of mealy granules (e.g., the *endosperm* of a seed) or with granules on the surface (e.g., a leaf or stem).

Fasciate. Literally, in a bundle or bundled together, the term being applied, also, as **fasciated,** to branches remaining parallel and grown abnormally together.

Fascicle. A bundle or cluster.

Fascicled. In bundles or clusters.

Fascicled root system. Clustered *storage roots.*

Fasciculate. In bundles or clusters.

Fastigiate. Erect and close together.

Faveolate or **favose.** Resembling a honey-comb.

Female gamete. An egg; a female gamete cell. See *gamete.*

Fenestrate. Perforated.

Ferruginous. Rust-colored.

Fertile. Productive. The term is used to mean capable of producing fruit or spores. Sometimes a staminate flower is referred to as *infertile* or *sterile* because it produces no seeds.

Fibrillose. With fine fibers.

Fibrous root system. A root system with several major roots about equal and arising from approximately the same point.

Filament. A thread; the stalk of a *stamen.*

Filamentous. Composed of threads or like a thread.

Filiform. Threadlike, that is, long, slender, and cylindroidal.

Fimbriate. Fringed, that is, resembling the fringe on the sleeve of an early American buckskin shirt.

Fimbrillate. The diminutive of *fimbriate.*

Fimbriolate. With a very fine fringe.

Fistulose. Hollow and cylindroidal.

Flabellate or **flabelliform.** Fan-shaped.

Flaccid. Weak.

Flagelliform. Whiplike or lashlike.

Flagellum. A whiplike thread of cytoplasm propelling a vegetative or reproductive cell through water. In the vascular plants flagella occur only in the male gamete cells.

Fleshy fruit. A fruit with soft, juicy tissues.

Flexuous. Curved in first one direction, then the opposite.

Floccose. With tufts of woolly hair.

Flocculent. Diminutive of *floccose.*

Flora (natural). A group of taxa (or their descendants), such as genera, species, and varieties, that have tended to remain together through one or more epochs of geologic time, having similar (though not identical) tolerances to environmental conditions. A flora is characterized by its taxonomic composition, i.e., the species or other taxa of which it is made up. See vegetation, formation, and floristic association.

Floral bud. See *flower bud.*

Floral cup. A cup bearing on its rim the sepals, petals, and stamens; originating as (a) a *hypanthium,* (b) a *"calyx tube"* formed by coalescence and adnation of the bases of the sepals, petals, and stamens, or (c) a *perianth tube* of the type in the monocotyledons, the sepals and adjacent petals being adnate edge to edge.

Floral envelope. See *perianth.*

Floral tube. An elongated, slender *floral cup.*

Floret. A small flower; the flower of a grass and the two immediately enclosing bracts, that is, the *lemma* and *palea.*

Floricane. A canelike stem producing flowers. See *primocane.*

Floriferous. Bearing flowers.

Floristic. Pertaining to the taxa composing a flora.

Floristic Association. A segment or subunit of a natural flora; usually corresponding also to an ecological formation or to a local or regional part of it.

Flower. A complex *strobilus* formed at the end of a branch, including the receptacle (*thalamus* or *torus*) and bearing *sepals, stamens, petals,* and *pistils* or some of these.

Flower bud. A bud which will develop into a flower, actually the young flower.

Foliaceous. Leaflike.

Foliage bud. A bud containing new young leaves.

Foliage leaf. The green leaf of the mature plant.

Foliar. Relating to a leaf.

-foliate. -leaved, an example being *3-foliate,* or *trifoliate,* that is, with 3 leaves or 3-leaved.

-foliolate. With leaflets, an example being 3-foliolate or trifoliolate, that is, with 3 leaflets.

Foliose. Bearing many leaves; leafy.

Follicle. A dry fruit formed from a single carpel, containing more than one seed, and splitting open along the suture. The term is applied also sometimes to similar fruits splitting only along the midrib.

Follicular. Of the nature of a *follicle* or similar to a follicle.

Forked. Dividing into branches which are nearly equal.

Formation (ecological). A climax vegetation characterized by the form of the dominant plants, e.g., a coniferous forest, a deciduous forest, a woodland, a brushland, a grassland, or a tundra.

Foveolate. Pitted.

Free. Separate from other organs. Botanical equivalent, *distinct.* Cf. special use, latter part of Chapter 9.

Frond. The leaf blade of a fern.

Frondose. Leafy, that is, with frondlike leaves.

Fruit. A matured ovary with its enclosed seeds and sometimes with attached external structures (as with an inferior ovary, the floral cup or tube).

Frutescent. Shrubby or becoming so at length.

Fruticose. Distinctly woody, living over from year to year, and attaining considerable size, with several main stems instead of a single trunk. The plant a *shrub.*

Fugacious. Falling away early.

Fulvous. Tawny.

Fungus. A plant without chlorophyll and with no vascular tissue or roots, stems, or leaves. Examples include mushrooms and bread mold. Plural, **fungi.**

Funiculus or **Funicle.** The stalk of an ovule or a seed.

Funnelform. In the shape of a funnel.

Furcate. Forked.

Fuscous. Grayish-brown.

Fusiform. In the shape of a spindle, that is, widest at the middle and tapering gradually to each pointed end, the body being circular in cross section.

Galea. A hood formed from a portion of the perianth derived, for example, from the two upper petals coalescent indistinguishably into one (in some members of the mint and snapdragon families) or from the upper sepal (in the monkshood).

Galeate. With a *galea.*

Gamete. A sex cell, i.e., one which unites with a cell of the opposite sex, resulting in the first cell of a new individual (the fertilized egg).

Gamopetalous. With all the petals coalescent. See *sympetalous.*

Gametophyte. The gamete-producing generation, each cell with n chromosomes.

Geminate. Equal and arranged in pairs.

Gene. A segment of the molecule of DNA carrying the pattern of inheritance of a particular character; the gene indirectly but precisely controls the synthesis of an enzyme (protein) and thus the development of a structural or physiological character of the organism.

Generative cell. A cell in the microgametophyte (or pollen grain) which may give rise directly or indirectly to male gamete cells (antherozoids) or nuclei.

Genetic code. The arrangement of codons (see codon) along the gigantic molecule of DNA in a chromosome; this specifies formation of particular amino acids at definite locations in the enzymes (proteins) formed in the cell. The pattern is carried to these sites by RNA. The pattern of heredity occurs in every cell of the organism, and it is transmitted to later generations.

Genetics. The study of heredity.

Geniculate. Bent like the human knee.

Genus. A group of related species or sometimes a single species. Plural, **genera;** a unit of classification. Usually a Latin or latinized noun.

Gibbous. Swollen or distended on one side.

Glabrate, glabrescent. At first hairy but later becoming *glabrous.*

Glabrous. Not hairy.

Gladiate. Swordlike.

Gland. A secreting organ. Usually glands are recognized by their secretion which accumulates into droplets or lumps. Often plant glands are on the tips of hairs.

Glandular. Bearing glands, small cellular organs secreting oils, tars, resins, etc. Often only the secretion is visible.

Glaucescent. More or less *glaucous.*

Glaucous. Covered with a white or bluish powder or *bloom,* this often composed of finely divided particles of wax. Example: the bloom on a plum.

Globose, globular. Spheroidal.

Glochid. A sharp hair or bristle tipped with a barb.

Glochidiate. Barbed at the tip.

Glomerate. In compact clusters.

Glomerulate. Diminutive of *glomerate.*

Glomerule. A single *cyme* of sessile flowers forming a compact cluster. It may be distinguished from a *head* by blooming of the central flower first.

Glumaceous. Resembling a *glume* of a grass spikelet.

Glume. One of the two chafflike bractlets at the base of a grass spikelet. The glumes do not enclose flowers. The term has been used for other bractlets, as those of the sedges.

Glutinous. Covered with sticky material.

Granulose. Covered by minute grains of hardened material.

Gregarious. In large colonies.

Gymnospermous. With the seeds naked, that is, not enclosed in an ovary.

Gynandrium. A structure formed by adnation of stamen(s) and pistil.

Gynandrous. With the stamens adhering with the pistils.

Gynecandrous. With staminate and pistillate flowers in the same inflorescence, the pistillate above and the staminate below.

Gynobase. An enlarged or elongated portion of the receptacle bearing the pistil.

Gynoecium. The pistil of a flower or its group of pistils.

Gynophore. A special stalk under a pistil.

Habit. The general appearance of a plant.

Habitat. The type of locality or the set of ecological conditions under which the plant grows.

Hair. A slender cellular projection. See *trichome.*

Halberd-shaped. See *hastate.*

Halophyte. A plant growing in salty soil as, for example, an alkali flat or a salt-water marsh.

Hamate. With a hook at the tip.

Haploid. With n chromosomes per cell.

Hastate. More or less *sagittate* (arrow-head shaped) but with divergent basal lobes.

Head. A cluster of sessile or essentially sessile flowers or fruits at the apex of a *peduncle.* Essentially a spike with a very short axis. The marginal flowers bloom first, the central last. See *glomerule.*

Heat equator. Area parallel to the equator but moving northward or southward as the area affected most by the sun moves north and south seasonally.

Helicoid. Spiral, like the shell of a turban snail.

Helicoid cyme. A coiled inflorescence. The main stem terminates in a flower and a single bud just below grows out as a stem but terminates also in a flower, the process being repeated many times and always in the same direction.

Herb. A non-woody plant, at least one which is not woody above ground level.

Herbaceous. Not woody.

Herbarium. A collection of pressed plant specimens.

Hermaphrodite. Bisexual, that is, with stamens and pistils in the same flower. See *perfect.*

Heterocarpous. Producing more than one kind of fruit.

Heterogamous. With more than one kind of flower.

Heterogeneous. Of various kinds, that is, not all individuals the same.

Heterozygous. With unlike genes in each pair under consideration, e.g., one gene of the pair promoting presence of red pigment in the petals, the other blue.

Hilum. The point of attachment of the *funiculus* (stalk) to the seed; the scar where the funiculus was attached (after the seed has separated from its stalk).

Hip (of a rose). A floral cup which usually becomes enlarged and fleshy at fruiting time. The true fruits are the achenes inside.

Hippocrepiform. Horseshoe-shaped.

Hirsute. With fairly coarse more or less stiff hairs.

Hirsutulous. Slightly hirsute.

Hirtellous. Minutely hirsute.

Hispid. With rigid or stiff bristles or bristly hairs.

Hispidulous. Diminutive of *hispid.*

Hoary. Covered with short, dense, grayish-white hairs, the surface of the stem, leaf, or other structure therefore appearing white.

Homogamous. With only one kind of flower.

Homogeneous. With all members similar.

Homozygous. With like genes in each of the pairs under consideration, e.g., with both genes promoting presence of red pigment in the petals, as opposed to blue.

Hood. A hoodlike structure, often formed from a petal or a sepal or more than one of either.

Humifuse. Spreading on the ground.

Hyaline. Thin and membranous, being transparent or translucent.

Hybrid. Produced by dissimilar parents.

Hydrophyte. An aquatic plant. See *mesophyte* and *xerophyte.*

Hygroscopic. Changing form with changes of moisture content.

Hypanthium. A floral cup or tube developed by extra growth of the margin of the receptacle. An alternative adopted by some authors is use of hypanthium with the meaning of floral cup as used in this book.

Hypocotyl. The portion of the axis of an embryo of a seed plant just below the cotyledon(s).

Hypogaeous. Growing underground.

Hypogynous. Below the ovary, that is, referring to flower parts which come directly from the receptacle and not from a floral cup or tube. A *hypogynous flower* is one which does not have a floral cup or tube. See special use, latter part of Chapter 9.

Hypogynous disc. A fleshy cushion of tissue growing from the receptacle below the ovary, often the stamens and sometimes the petals being attached to it. Flowers having a hypogynous disc are *not* interpreted here as perigynous but rather as hypogynous. In some works they are considered to be perigynous, that is, to be with only the stamens and sometimes the petals perigynous.

Imbricate. Overlapping like the shingles on a roof.

Impari-pinnate. See *odd pinnate.*

Implexed. Tangled; interlacing.
Implicate. Woven in.
Inactive bud. A bud dormant either before the beginning of the growing season or through a longer period.
Incanous. Covered with white (or grayish) hairs.
Incised. Cut deeply and sharply into narrow, angular divisions.
Included. Not protruding beyond the normal surrounding structures.
Incomplete flower. A flower lacking one or more of the four usual series of structures, that is, sepals, petals, stamens, or pistils.
Incrassate. Thickened.
Incumbent. Applied to a pair of cotyledons which lie with the back of one against the axis of the embryo. See *accumbent.*
Indehiscent. Not opening by splitting along regular lines or not opening at all.
Indeterminate. Of indefinite growth, i.e., the size or number of organs to be produced not predetermined. See *determinate* and *cyme.*
Indigenous. Native in a particular region.
Indument. A covering of hairs.
Induplicate. With the edges folded inward.
Indurate. Hardened.
Indusium. The membranous covering of a sorus (sporangium cluster) of a pteridophyte.
Inferior. Below. An *inferior ovary* is distinguished by being adnate with the floral cup or tube. Therefore it appears to be embedded in the pedicel below the other flower parts.
Inferior floral cup (or **tube**). The portion of a floral cup or tube adnate with the wall of an inferior ovary.
Inflated. Appearing as if inflated with air.
Inflexed. Turned or bent inward, incurved.
Inflorescence. The flowering area or segment of a plant.
Innate. Attached at the apex of the supporting structure.
Inserted. Attached upon another structure.
Integument. The outer coating of an ovule, a structure grown up from the sporophyll and surrounding the sporangium (nucellus). The integument becomes the seed coat. In gymnosperms there is only 1 integument, in the angiosperms 1 or 2.
Internode. The portion of the stem between two *nodes.*
Introduced. Brought in from another area.
Introrse. Turned inward, that is, toward the axis.
Involucel. The *involucre* of a secondary *umbel* (of a compound umbel).
Involucellate. With an *involucel.*
Involucral. Pertaining to an *involucre.*
Involucrate. With an *involucre.*
Involucre. A series of bracts surrounding a flower cluster or sometimes a single flower.
Involute. Rolled inward.
Irregular. Used in botany to indicate a *bilaterally symmetrical* structure, as, for example, a bilaterally symmetrical flower.
Isolation. Keeping of plants or their taxa apart—as a result of hereditary, geographic, or ecological factors.
Isomerous. With the members of the various series of flower parts of equal number.

Juvenile leaves. Secondary leaves appearing at the beginning of growth of the young plant and differing somewhat from the later-produced secondary leaves.

Keel. A ridge along the outside of a fold, like the keel of a boat. The term is applied to the two coalescent lower petals of a *papilionaceous corolla* of the pea family.
Key. An outline prepared for use in identifying plants (or animals, etc.) by a process of elimination.

Labiate. With lips, that is, two opposed structures as, for example, the upper two petals as opposed to the lower three petals in the mint or the snapdragon family.
Lacerate. Cut irregularly as if the structure had been slashed.
Laciniate. Slashed, that is, cut into narrow, pointed segments.
Lacunose. Perforated or with depressions.
Lacustrine. Of lakes.
Lamella. A small thin sheet or plate.
Lamellate. Made up of small thin plates.
Lamina. A sheet or plate; the flat, expanded portion of a structure such as a leaf or a petal.
Lanate. With long woolly hair.
Lanceolate. Lance-shaped; four to six times as long as broad, broadest toward the basal (attachment) end, sharply angled at both ends and especially the apical end, the sides being curved at least along the broad part.
Lanuginous. Covered with hair appearing like cotton or wool.
Latent bud. One remaining dormant for several years.
Lateral. At the side or sides, i.e., along the margins.
Lateral bud. One in a leaf axil, i.e., on the side of the stem.
Latex. Milky juice.
Leaf base. The usually expanded portion of the leaf attached to the stem.
Leaflet. A leaflike segment of a compound leaf.
Leaf primordium. The hump of tissue which develops into a young leaf.
Legume. A dry, several-seeded fruit formed from a single carpel and dehiscent on both margins (i.e., the suture and the midrib). This fruit occurs only in the pea family. The members of the family are referred to as *leguminous plants.*
Leguminous. Of the pea family; referring to a leguminous plant.
Lemma. The lower member of the pair of bracts surrounding a grass flower, this bract enclosing not only the flower but the other bract (palea).
Lenticular. Lens-shaped, that is, biconvex.
Lepidote. Covered with dandrufflike scales or hairs.
Liana. A woody twining vine.
Ligneous. Woody.
Ligulate. With a ligule, that is, a strap-shaped body.
Ligule. A flat, straplike body; the flattened corolla of a *ray flower* (of the sunflower family); a flat membranous scale produced at the point of joining of the sheath and the blade of a grass leaf.
Liguliform. Strap-shaped.
Limb. The expanded and spread-

ing parts of a sympetalous corolla or the broad portion of a petal; a branch.

Line. One-twelfth of an inch.

Linear. Long and narrow, the sides being parallel and the length at least eight times the width, tending to be an elongated rectangle.

Lineolate. Marked with fine lines.

Lingulate. Tonguelike.

Lip. See the explanation under *labiate.*

Litoral or **littoral.** Occurring on the sea shore.

Liverwort. A green nonvascular land (or fresh-water) plant usually with a flat *thallus* (sometimes appearing leafy).

Livid. Pale lead-colored.

Lobe. A short segment of a leaf or other organ.

Lobed, lobate. In the broad sense, from moderately to very deeply indented toward the base or midrib; in the restricted sense, indented significantly but less than half way to the base or midrib. The broad sense includes *lobed* (restricted sense), *cleft, parted,* and *divided.*

-locular. Divided into locules. A *3-locular* or *trilocular* structure has 3 locules as, for example, a 3-chambered ovary.

Locule. The cavity of an anther or an ovule, that is, a pollen chamber or a seed chamber.

Loculicidal. Dehiscent along the midrib of a carpel of an ovary containing more than one carpel and chamber, i.e., along the middle of the back of each chamber.

Lodicule. One of the two small scales at the base of a grass flower, these apparently corresponding to either two sepals, two petals, or a sepal and a petal.

Loment. A legume divided by constriction into a linear series of segments, each containing one seed.

Longitudinal. Lengthwise.

Longitudinal section. A slice cut lengthwise.

Lunate. Half-moon-shaped or crescent-shaped.

Lurid. Dingy.

Lyrate. Pinnatifid, with the terminal lobe rounded and much larger than the others.

Maculate. Blotched or mottled.

Male gamete. An *antherozoid* or a *male gamete cell* or *nucleus.* See *gamete.*

Male nucleus, male gamete nucleus. The male gamete of a flowering plant, that is, a single nucleus with the associated cytoplasm but with no cell wall.

Malpighiaceous. An adjective used to describe a hair with two branches and almost no stalk, these appearing to be a single straight hair with the attachment point at the middle.

Marcescent. Persisting on the plant after withering.

Marginate. With a distinct margin.

Maritime. Of the sea.

Median. Middle.

Megagametophyte. A female *gametophyte;* the usually larger gametophyte developed from a *megaspore,* as opposed to the smaller from a microspore.

Megasporangium. A *sporangium* producing *megaspores.*

Megaspore. The larger type of spore developed by a plant with spores of two sizes or kinds.

Megasporophyll. A sporophyll bearing one or more *megasporangia.*

Meiosis. The reduction divisions which result in production of four cells from a single one, the number of chromosomes per cell being reduced from 2n to n.

Membranaceous, membranous. Thin, soft, pliable, and translucent.

Mericarp. A portion of a fruit which appears to be a whole fruit.

Meristem. A tissue composed of cells which retain the power of division.

Meristematic. Of the nature of a *meristem.*

-merous. Composed of parts, for example, *3-merous* or *trimerous,* meaning with 3 parts in each series, e.g., 3 sepals, 3 petals, etc.

Mesocarp. The middle layer of a *pericarp.* See *endocarp* and *exocarp.*

Mesophyte. A plant occurring under medium conditions, that is, average moisture, light, temperature, etc. See *hydrophyte* and *xerophyte.*

Metric system (of measurement). See explanation at the end of Chapter 1.

Microgametophyte. A male *gametophyte;* the smaller gametophyte developing from a *microspore.* See *megagametophyte.*

Micropyle. A small opening in the *integument* which forms the outer layer of an *ovule* or a *seed.* The *pollen tube* enters the ovule through this opening.

Microsporangium. A *sporangium* producing *microspores.*

Microspore. The smaller type of spore in a plant producing spores of two sizes.

Microsporophyll. A *sporophyll* bearing one or more *microsporangia.*

Midrib. The middle vein of a leaf or other structure.

Mixed bud. A *bud* containing both leaf primordia and rudimentary flowers.

Monadelphous. In one brotherhood; referring to stamens with their filaments coalescent into a single tube.

Moniliform. Like a string of beads.

Monochlamydeous. With a single series in the *perianth,* that is, only the sepals.

Monocolpate. With a single furrow, e.g., on a *pollen grain.*

Monocotyledonous. With only one *cotyledon.*

Monoecious. With unisexual flowers, the staminate and the pistillate occurring on the same individual.

Montane. Of the mountains, the term referring to areas of moderate elevation in the mountains.

Moss. A non-vascular green plant of the division *Bryophyta* (Class *Musci*). The plant has leaf- and stemlike structures belonging to the gametophyte generation. The antheridia and archegonia are on the leafy branches. Ultimately a sporophyte is produced still attached to the archegonium.

Mucilaginous. Slimy, as if covered with mucilage.

Mucro. A short pointed structure terminal upon the organ (e.g., leaf) which bears it and of about the same texture as the supporting organ (e.g., leaf blade).

Mucronate. With a *mucro.*

Multifid. Divided into many lobes or segments.
Muricate. Roughened by the presence of short, hard points.
Muriculate. The diminutive of *muricate.*
Muticous. Without a point; blunt.

Naked bud. A bud not covered by special scales but only by the outer leaves.
Napiform. Turnip-shaped.
Natural system (of classification). A system made up on the basis of relationships, or, practically, as indicated by many characters in common as opposed to few differences.
Naturalized. Established thoroughly after introduction from another region.
Navicular. Boat-shaped.
Nectar. A sugar solution attracting insects.
Nectariferous. Producing *nectar.*
Nectary. A gland producing *nectar.*
Nectary scale. A flap covering or partly surrounding a nectar gland.
Nerve. A simple vein or rib.
Netted-veined. With the veins forming a network.
Nigrescent. More or less black.
Niveus. White; like snow.
Nocturnal. Occurring at night. The term is applied to flowers which open at night.
Node. A "joint" of the stem, that is, the area which bears a leaf or a pair of or several leaves, usually each with a branch or a bud in its axil.
Nodose. Knobby.
Nodule. Diminutive of *node.* The term is applied to swellings produced on the roots of leguminous plants (members of the pea family) by partly parasitic or symbiotic bacteria which combine nitrogen from the air into chemical compounds.
Nucellus. The sporangium enclosed in the ovule of a seed plant.
Numerous. See special use, latter part of Chapter 9.
Nut. A hard, relatively large, indehiscent, 1-seeded fruit.
Nutlet. Diminutive of *nut.*

Ob- A Latin prefix meaning inverted; reversed from the usual arrangement.
Obcompressed. Compressed *dorsoventrally.*
Obconical. A cone attached at the pointed end.
Obcordate. A conventional heart-shape but with the attachment at the point instead of at the indentation.
Obdeltoid. *Deltoid* but with the attachment of the petiole at one of the angles instead of on a side of the triangle.
Obdiplostemonous. The stamens in two series, those of the outer series opposite the petals.
Oblanceolate. *Lanceolate* but with the widest part toward the attachment.
Oblate. Nearly spherical but compressed at the poles.
Oblique. With the sides unequal or slanting.
Oblong. Rectangular and longer than broad, the length roughly two to three times the breadth.
Obovate. *Ovate* but with the narrow part toward the attachment.
Obovoid. *Ovoid,* but with the attachment at the small end.
Obsolescent. Rudimentary or having nearly disappeared; becoming extinct.
Obsolete. Rudimentary or having practically or wholly disappeared.
Obtuse. Blunt, that is, forming an obtuse angle, or somewhat rounded instead of strictly angular.
Ochroleucous. Yellowish-white.
Ocrea. A tube formed by coalescence of a pair of stipules around the stem.
Ocreate. Having a tubular pair of coalescent stipules.
Odd pinnate. With a terminal leaflet.
Offset. A young plant produced at the end of a *stolon* with only one *internode.*
Olivaceous. Olive-green.
Opaque. Impervious to light; dull.
Operculate. With a lid or cap.
Operculum. A lid or cap.
Opposite. Two organs (e.g., leaves) occurring at the same level and on the opposite sides of the supporting structure (e.g., stem).

Orbiculate. Circular.
Order. A taxon composed of related families or sometimes of a single family. Ending: -ales.
Organ (of a plant). A *vegetative organ* that is, a root, stem, or leaf, or a *reproductive organ,* that is, a flower, fruit, or cone, or the sporangia borne on leaves.
Orifice. Opening.
Orthotropous. Descriptive of an ovule which continues the direction of the funiculus, that is, one which is not bent or curved.
Oval. Broadly elliptic.
Ovary. The lower part of the pistil, which contains the ovules or later the seeds.
Ovate. Egg-shaped; a two-dimensional object about one and one-half times as long as broad, with rounded ends, and widest toward the base or attachment point.
Ovoid. *Ovate* but a three-dimensional figure.
Ovulate. Producing ovules.
Ovule. The structure which (after fertilization) develops into a seed. Essentially it is at first a *megasporangium* enveloped by a special structure, the *integument.* The (ordinarily) solitary *megaspore* develops into a *megagametophyte,* which produces an *egg.*
Ovuliferous. Bearing ovules.
Ovuliferous scale. A complex structure formed from a leaf or possibly partly from a leaf or leaves and partly (at least basally) from a branch. The entire structure, whatever its origin, bearing naked ovules.

Paedomorphosis. Arresting of development in a juvenile stage and occurrence of divergent evolution from there.
Palate. A projection near the lower lip of a more or less tubular corolla, the projection closing or nearly closing the opening.
Palea. The upper bract of the pair which encloses the grass flower, this bract being enclosed by the larger lower bract or *lemma.*
Paleaceous. *Chaffy.*
Palmate. Descriptive of compound leaves in which leaflets arise at the same point, that is, at the apex of the petiole.

Palmately. In a *palmate* manner. A leaf may be *palmately lobed, cleft, parted,* or *divided.*

Palmatifid. Palmately divided and almost but not quite palmate.

Paludal, paludose, palustrine. Growing in marshes.

Panicle. A cluster of associated spikes, racemes, or corymbs.

Paniculate. In a *panicle;* similar to a panicle.

Pannose. Feltlike or like closely woven woolen cloth.

Papilionaceous. Butterflylike; descriptive of the corolla found only in one (the largest) subfamily of the pea family, this composed of a banner, two wings, and a keel, the latter formed by the coalescence of the two petals which enclose the stamens and the pistil.

Papillose. Bearing minute rounded projections.

Pappus. The specialized calyx of members of the sunflower family; composed of bristles or scales.

Parallel evolution. Duplication of some features in unrelated or only distantly related organisms through independent evolution. The features are usually only analogous, not homologous. They resemble each other in form and/or function but not in basic development or even structure.

Parallel veined. With the principal veins parallel and usually close together.

Parasitic. Living upon food or water derived from another individual, ordinarily of another species. Noun, **parasite.**

Parenchyma. Thin-walled cells forming a soft tissue which remains alive.

Parietal. On the outer wall.

Parted. Indented more than half way or nearly all the way to the base or midrib.

Parthenogenetic. Developing without fertilization.

Partly inferior (ovary). With only the basal portion of the ovary adnate with the floral cup.

Patelliform. Disc-shaped.

Pectinate. Resembling a comb.

Pedate. Palmately divided or parted with the lateral segments forked.

Pedicel. The internode below a flower.

Pedicellate. With a *pedicel.*

Peduncle. The stalk of a cluster of flowers or the next to the last internode below a single flower.

Pedunculate. With a *peduncle.*

Pellucid. Clear or transparent.

Peltate. Shieldlike; supported by a stalk attached near the center of the lower surface (thus resembling a coin balanced on the end of a pencil).

Pendulous. Hanging downward.

Penicillate. Brushlike.

Pentagynous. With five pistils.

Pentamerous. With five members; the flower parts in fives.

Pentandrous. With five stamens.

Pepo. A fruit of the type in the gourd family, that is, with a hard or leathery rind on the outside and fleshy placental tissue inside. The seeds are numerous and there is only one chamber (except sometimes by intrusion of placentae).

Perennial. Continuing to grow year after year.

Perfect. Description of a flower having both functional pistils and functional stamens.

Perfoliate. Descriptive of a sessile leaf encircling the stem.

Perianth. A collective term for the *calyx* and the *corolla.*

Perianth tube. A tube formed by coalescence and adnation of the lower portions of the sepals and petals. In the monocotyledons, the sepals and petals are adnate edge to edge, and the bases of the stamens are also adnate.

Pericarp. The wall of a matured ovary, that is, the wall of the fruit or the inner wall if the ovary is inferior.

Perigynium. A special sac which encloses the ovary of *Carex* (sedge family).

Perigynous. Descriptive of the parts of a flower (sepals, petals and stamens) attached to a floral cup or tube which is not adnate with the ovary. A *perigynous flower* has a floral cup but the ovary is not adnate with it.

Perigynous disc. A disc adnate to the floral cup of a *perigynous flower.* See *hypogynous disc* and *epigynous disc.*

Peripheral. On the margin.

Persistent. Remaining attached longer than might be expected, for example, a calyx which remains on the receptacle or floral cup until fruiting time.

Petal. One member of the series of flower parts forming the *corolla.* See *corolla.*

Petaloid (adj.). Resembling a *petal* in color and texture.

Petaloid (adj. used as a noun). One of the petal-like structures of a cactus flower or any other not formed in the usual way from a stamen.

Petiolate. Having a petiole.

Petiole. The stalk of a leaf supporting the expanded portion or blade.

Petioled. With a *petiole.*

Petiolulate. With a *petiolule.*

Petiolule. The stalk of a leaflet.

Phanerogam. A member of the *Spermatophyta* or seed plants.

Phloem. A conducting tissue which usually carries manufactured food downward to places of use or storage. It includes two elements: *sieve tubes,* without nuclei and forming elongate conducting elements composed of many cells with connecting perforated sieve plates and *companion cells* with nuclei apparently supplying the functions of nuclei to the portions of the sieve tubes.

Photosynthesis. The process of combining in the presence of chlorophyll and light of (in net effect) carbon dioxide and water yielding a carbohydrate and oxygen.

Photosynthetic. Descriptive of a green portion of a plant, capable of carrying on *photosynthesis.*

Phyllary. An *involucral bract,* the term being used for the *Compositae* or sunflower family.

Phyllodium, Phyllode. A broadened petiole which takes the place of a leaf blade.

Phylum. In zoölogy the approximate equivalent of division; not applicable to the Plant Kingdom.

Pilose. Hairy, the hairs being elongated, slender, and soft.

Pinna. A primary leaflet of a pinnately compound leaf.

Pinnate (leaf). Compound, with the leaflets arising along an axis. (Axis fundamentally the midrib.)

Pinnately. In a *pinnate* manner. A leaf may be *pinnately lobed, parted, cleft,* or *divided.*

Pinnatifid. Deeply pinnately divided, the segments being not quite separate from each other.

Pinnule. A secondary, tertiary, or quaternary leaflet, whichever is the smallest degree in the compound leaf.

Pisiform. Pea-shaped.

Pistil. The organ of a flower which bears ovules and later seeds. It is composed of at least one ovary, stigma, and style and formed from one or more *carpels.*

Pistillate. Having *pistils,* that is, a flower which has pistils but no stamens.

Pit. A depression; a gap in the extra thickening formed on the wall of a xylem cell. Only the middle layer (lamella) between two adjacent cells serves as a wall at a point opposite a pit. This type of xylem thickening is called **pitted thickening.**

Pith. The central soft tissue of a stem.

Pitted. With small depressions.

Placenta. The tissue of an ovary which bears ovules or seeds.

Plane. With the surface flat.

Plastid. A solid body in the cytoplasm of a cell, often bearing a pigment.

Plastid pigment. Coloring matter deposited in a *plastid,* the pigment being insoluble in water, therefore remaining in or on the plastid and not being dissolved in the cell sap.

Plicate. Folded (usually lengthwise) into plaits.

Plumose. Like a feather, the term being applied to hairs which have finer hairs attached along each side.

Plumule. A tuft of embryonic secondary leaves surrounding the epicotyl of an embryo.

Pollen, pollen grains. The spheroidal structures produced in an anther of a flower or in the microsporophyll of a gymnosperm. The pollen grains are *microgametophytes* developed from *microspores.*

Pollen chamber. A pollen-bearing cavity or *locule.*

Pollen sac. See *pollen chamber.*

Pollen tube. A tube developed by a *pollen grain* and growing through the micropyle of an *ovule* where its contained cells or nuclei effect fertilization of the egg. The pollen tube is developed on the *stigma* of an angiosperm or in the *micropyle* of a gymnosperm.

Polliniferous. Bearing pollen.

Pollinium. The waxy mass of pollen produced in the milkweed family or the orchid family.

Polyadelphous. In several brotherhoods; with several groups of coalescent stamens.

Polyandrous. With an indefinite large number of stamens.

Polyembryony. Production of more than one embryo in a single ovule.

Polygamo-dioecious, polygamodioecious. *Polygamous,* but in the main *dioecious.*

Polygamo-monoecious, polygamomonoecious. *Polygamous,* but in the main *monoecious.*

Polygamous. With bisexual and unisexual flowers on the same or different individuals of the species.

Polygynous. With a large indefinite number of pistils.

Polymorphous. With several or many forms.

Polypetalous. See *choripetalous.*

Polyploid. With several to many times n chromosomes per cell.

Pome. A fleshy fruit with several seed chambers, this formed from an inferior ovary, the fleshy tissue being largely the floral cup, the seeds not embedded in pulp. Apples and pears are classical examples.

Porrect. Directed outward and forward.

Praemorse. Appearing to have been bitten off.

Precocious. Appearing early in the season, as for example, flowers occurring before the leaves.

Prickle. A sharp, pointed outgrowth from the superficial tissues of the stem, that is, from the epidermis or the cortex, as in a rose prickle. The structure is not associated with the conducting tissues at the center of the stem.

Primary endosperm cell. The original cell of the endosperm of a flowering plant seed. This has at first two nuclei which fuse and later join with a male gamete nucleus, thus giving rise to a tissue of 3n chromosomes. See *endosperm.*

Primary leaf. A leaf present in the embryo, that is, in seed plants, a *cotyledon.*

Primary leaflet. A leaflet of the first degree in a *bipinnate* or more complex compound leaf.

Primary root. The root which continues the main axis developed in a seedling.

Primocane. A cane in its first season of growth, as in a bramble or blackberry.

Prismatic. The shape of a prism, that is, angular but with flat sides.

Procumbent. Lying on the ground, but not rooting at the nodes.

Proliferating, proliferous. Producing buds or small plants from the leaves or as offshoots from the stem.

Prostrate. Flat upon the ground.

Protected bud. A bud with a protective layer of scales and resin.

Proterandrous. With the anthers of the flowers producing pollen before the pistils are receptive to it.

Proterogynous. With the stigma receptive to pollen before the pollen of the same flower is released.

Prothallium. The gametophyte of a fern or other pteridophyte. A prothallium is determinate (it grows to a certain size and degree of complexity and stops); a thallus is indeterminate (it keeps on growing indefinitely).

Protoplasm. The living substance of a cell.

Pruinose. Covered with a bloom, that is, finely divided particles of a waxy powder. See *glaucous.*

Pteridophyte. A fern or fern relative. Any member of the divisions *Pteridophyta, Sphenophyta, Lycophyta,* or *Psilophyta.*

Puberulent. Finely and minutely pubescent, the hairs being very short.

Pubescent. Hairy or downy, usually with fine soft hairs. Commonly the term is used to indicate hairiness of a generalized instead of a specialized type, and it is used loosely to cover any kind of hair. The noun is *pubescence.*

Pulvinus. An enlargement at the leaf base where it is attached to the stem.

Punctate. Dotted with depressed glands or colored spots.
Puncticulate. Diminutive of *punctate.*
Pungent. Sharp-pointed.
Pustular, pustulate. Blistered.
Pyramidal. Pyramid-shaped.
Pyriform. Pear-shaped.
Pyxidate. With a lid.
Pyxis. A *circumscissile* capsule.

Quadrate. Nearly square.
Quadripinnate. Four times pinnate, that is, with primary, secondary, tertiary, and quarternary leaflets.
Quarternary leaflet. A leaflet of the fourth degree; a leaflet of a *tertiary leaflet.*

Raceme. An inflorescence composed of pedicelled flowers arranged along an axis which elongates for an indefinite period. The lower flower blooms first and eventually the terminal bud forms the last flower.
Racemiform. In the form of a *raceme.*
Racemose. In *racemes;* similar to racemes.
Rachilla. A secondary axis. See *rachis.*
Rachis (or **Rhachis**). The axis of a pinnate leaf or of an inflorescence.
Radiate. Spreading from a common center; descriptive of a head which includes *ray flowers* (sunflower family).
Radially symmetrical. Capable of division into three or more similar sections.
Radical leaves. Literally root leaves, actually *basal leaves.*
Radicant. Rooting; descriptive of a stem which produces adventitious roots.
Radicle. The portion of an embryo which grows into a root.
Rain shadow. After the air mass of a moist prevailing wind is lifted in going over a mountain range, its moisture has been much reduced; as it descends on the leeward side of the mountains, there is increased moisture holding capacity, and little rain falls. The dry area in the lee of the mountains is a rain shadow.
Ramification. Branching.
Ramose. Branching, that is, with many branches.
Ramulose. With many branchlets.
Raphe. The portion of the funiculus grown fast to the edge of an *anatropous ovule.*
Ray. A pedicel within an *umbel;* a *ray flower* or its corolla.
Ray flower. In the sunflower family or *Compositae,* one of the flowers with a ligulate (strap-shaped or flattened) corolla. The "flattening" is due to failure of complete coalescence between two petals of the sympetalous corolla.
Receptacle. The apical area beyond a pedicel, that is, the portion which bears flower parts. The receptacle consists of several or many nodes and short internodes.
Reclinate. Turned or bent downward.
Reclining. Sprawling or lying down.
Recurved. Curved downward or backward.
Reduced. Small but probably derived from larger forerunners.
Reflexed. Bent or turned abruptly downward.
Regular. Uniform in shape or structure; *radially symmetrical* (especially as applied to a corolla).
Relic. A plant persisting in relatively small portion or portions of its previous range.
Relict. Descriptive of a relic, that is, the adjective form.
Reniform. Kidney-shaped or bean-shaped, with the attachment in the indentation, the width greater than the length, the apex rounded.
Repand. With a somewhat uneven *sinuate* margin. See *undulate.*
Repent. Lying flat on the ground and rooting adventitiously.
Replicate. Folded backward.
Reproductive bud. A *bud* developing into a flower, cone, or other pollen-, seed-, or spore-bearing structure.
Reproductive organ. See under *organ.*
Resiniferous. Bearing resin.
Resting bud. An inactive *bud.*
Resupinate. Turned upside down.
Reticulate. In a network or a pattern which appears like a network.
Reticulum. A network.
Retrorse. Turned back or downward.
Retrorsely toothed. Similar to *serrate* but with the teeth projecting backward toward the base of the leaf.
Retuse. With a shallow and rather narrow notch in a broad apex.
Revolute. With the margins or the apex rolled backward.
Rhachilla. See *rachilla.*
Rhachis. See *rachis.*
Rhizomatose, rhizomatous. With a *rhizome;* rhizomelike.
Rhizome. A horizontal, underground stem.
Rhombic. More or less diamond-shaped; an equilateral parallelogram with oblique angles. The petiole is attached at one of the sharper angles.
Rhomboid, rhomboidal. A three-dimensional *rhombic* figure.
Rib. A prominent raised nerve or vein.
Ringent. Gaping open.
Riparian. Of rivers or streams.
RNA, or **ribose nucleic acid.** A substance allied to DNA (see DNA) to which the genetic code (see genetic code) is transferred and by which it is carried to the sites of enzyme (protein) synthesis. There are three types of RNA—messenger, transfer, and ribosomal.
Root. The underground portion of the main axis of a plant or the branches of the axis. There are no nodes and internodes or leaves and usually there is a solid core of xylem in the center (i.e., no pith).
Root-hairs. Projections of the epidermal cells of a root, these absorbing water and salts from the soil.
Rootstock. An old, inaccurate term for *rhizome.*
Root tip. The apex of a young root; the absorbing area (a centimeter or a few centimeters long).
Rosette. A circular cluster.
Rostellate. The diminutive of *rostrate.*
Rostellum. A small beak.
Rostrate. With a beak.
Rostrum. A beak.
Rotate. Spreading; wheel-shaped or saucer-shaped.

Rotund. Rounded.
Rounded. Gently curved.
Ruderal. Growing in waste places or in rubbish heaps.
Rudiment. A vestigial organ.
Rufous, rufescent. Reddish-brown.
Rugose. Wrinkled.
Ruminate. Appearing to have been chewed.
Runcinate. With the sharply cut divisions directed toward the base, that is, backward.
Runner. A slender *stolon.*
Rupturing. Bursting open along irregular lines.

Sac, sack. A pocketlike or baglike indentation.
Saccate. Forming a sac.
Sagittate. Arrowhead shaped, that is, an isosceles triangle in outline, but with an angular indentation on the short side and the attachment in the indentation.
Salverform. Descriptive of a *sympetalous* corolla with the slender basal tube abruptly expanded into a flat or saucer-shaped upper portion. (Strictly with a constriction at the upper end of the tube.)
Samara. A dry, indehiscent fruit with a wing.
Saprophyte. A plant which lives upon dead organic matter such as the leaf mold on forest floors. Adjective, *saprophytic.*
Sarmentose. With long slender *stolons.*
Savannah. Tropical grassland occurring in areas where there are marked wet and dry seasons, primarily along the northern and southern margins of the trade wind belt; woodland savannah includes some trees.
Scabrous. Rough to the touch, with minute rough projections.
Scalariform. Ladderlike.
Scalariform thickening. Thickening of the wall of the xylem cell laid down over the inner surface of the wall except in large windowlike openings which are arranged in such a way that the side of the cell looks somewhat like a ladder.
Scale. A thin, membranous structure; a small more or less triangular leaf; a chafflike bract; a flattened hair.
Scale leaves. Small leaves resembling *scales.*
Scandent. See *climbing.*
Scape. A flowering stem which bears no leaves, or only a small bract or a pair or whorl of bracts.
Scapiform. In the form of a *scape.*
Scapose. With a *scape* or in the form of a scape.
Scarious. Dry, thin, membranous, non-green, and translucent.
Schizocarp. A fruit which splits into one-seeded sections (mericarps).
Sclerenchyma. A tissue composed of cells with thick hard walls (a strengthening tissue).
Scorpioid. Often used as descriptive of an inflorescence which is coiled in the bud stage as in the forget-me-not or fiddle-neck. This is a *helicoid cyme* although it appears to be a spike or a raceme.
Scurfy. With scalelike particles on the surface (the scales resembling human dandruff).
Secondary leaf. One produced above the cotyledon(s) or primary leaf or leaves.
Secondary leaflet. A leaflet of the *primary leaflet.*
Secondary root. A branch root arising directly from the primary root. There are also tertiary roots, quaternary roots, etc.
Secund. Turned to one side usually by twisting.
Seed. A matured ovule consisting of an integument (seed coat), an enclosed nucellus (sporangium), the remains of the megagametophyte, the endosperm (in flowering plants), and the embryo.
Seed coat. The outer coating of a seed developed from the integument of the ovule.
Sepal. One of the flower parts of the outer series, the sepals forming a *calyx.* See *calyx.*
Sepaloid (adj.). Similar in appearance and/or position to a sepal; sepallike.
Sepaloid (adj. used as a noun). One of the sepallike structures of a cactus flower or of other flowers in which they are formed in the same manner.
Septate. Divided by partitions.
Septicidal. Descriptive of a fruit dehiscent through the partitions between the seed chambers.
Septum. A partition.
Sericeous. Silky.
Serotinous. Developing late in the summer; in willows the catkins developing later than the leaves (as in certain species).
Serrate. With teeth like sawteeth, that is, angular and directed forward.
Serrulate. Diminutive of *serrate.*
Sessile. Without a stalk, that is, "sitting."
Seta. A bristle.
Setaceous, setiform. Bristle-shaped.
Setiferous. With bristles.
Setose. Covered with bristles.
Setulose. With minute bristles.
Sex cell. See *gamete.*
Sheath. A tubular cover. An example is the basal portion of a grass leaf, which surrounds the stem.
Sheathing. Covering or enclosing.
Shrub. A woody plant not having a main trunk but several main branches. In general, shrubs are smaller than trees.
Sieve tube. An element of the phloem composed of the walls and cytoplasm but not the nuclei of specialized cells; the original cells connected by perforated sieve plates.
Sigmoid. Resembling the letter S or the Greek letter sigma; with a double curve.
Siliceous, silicious. Composed of or covered with silica (silicon dioxide), the principal ingredient of sand or glass.
Silicle. A short *silique,* that is, one usually not more than two or three times as long as broad.
Silique. The elongate capsular fruit of the mustard family, which has two seed chambers separated by a false partition from the middle of one placenta to the middle of the other. The "false partition" in this case results from presence of some fertile and some sterile carpels in the pistil.
Silky. Covered densely with apressed, soft, straight hairs.
Sinistrorse. Turned to the left.
Sinuate, sinuous. With a wavy margin, the margin winding strongly inward and outward.
Sinus. A cleft, recess, or embayment.
Smooth. Not rough. Some authors have employed this term inac-

curately to mean glabrous.

Solitary. Alone.

Sordid. Appearing dirty as opposed to white.

Sorus. A cluster of sporangia. Plural *sori.*

Spadix. A spike with a fleshy or succulent axis, the flowers often partly embedded in the axis.

Spathe. A large bract enclosing an inflorescence (often a spadix) at least when it is young. Spathes are either white or highly colored but usually not green.

Spathulate, spatulate. An oblong or somewhat rounded structure with the basal end long and tapered; the shape of a spatula.

Species. A group of related varieties or often a single unit. Plural, *species.* See *taxon.* The definition of taxon applies to categories of all ranks. Cf. discussion in Chapter 14.

Sperm. The male gamete of an animal (not applicable to plant gametes); seed.

Spermatophyte. A plant which produces seeds. A member of the Division Spermatophyta.

Spheroidal. In approximately the shape of a sphere.

Spicate. In the form of a *spike.*

Spiciform. Spikelike.

Spiculose. With a surface covered with fine spikelike points.

Spike. An inflorescence in which the sessile flowers are arranged along an axis. The basal flower blooms first; the last one formed is at the apex.

Spikelet. Diminutive of spike; the small densely bracteate spike of a grass or a sedge.

Spindle-shaped. A three-dimensional object, circular in any cross-section and broadest at the middle and tapering to both pointed ends. See *fusiform.*

Spine. A sharp more or less woody or horny outgrowth from a leaf or a part of a leaf, sometimes representing the entire leaf. The term has been applied sometimes to similar outgrowths from the stem. As interpreted here these are *thorns.*

Spinescent. With spines; spiny.

Spinose. With spines; spinelike. The word is used in compounds, as spinose-dentate.

Spinous. See *spinose.*

Spinule. The diminutive of *spine.*

Spiral. Arranged in a spiral. See special use, latter part of Chapter 9.

Spiral thickening. Thickening of the walls of a xylem cell laid down in the form of a spiral.

Sporadic. Of irregular occurrence here and there; not forming a continuous population.

Sporangium. A spore-case.

Spore. A simple, one-celled reproductive structure, in the higher plants containing n chromosomes and formed directly as the result of the reduction divisions (meiosis).

Sporocarp. A special hard or leathery structure containing sporangia or a sporangium.

Sporophyll. A *sporangium-bearing* leaf.

Sporophyte. The spore-producing (also often seed-producing) generation, each cell with 2n chromosomes.

Spur. An elongated sac produced from a part of the flower, as in the larkspurs from a sepal or in the columbines from a petal.

Squamellate. With small *scales.*

Squamulose. With *scales.*

Squarrose. The parts with widely spreading or recurved tips.

Stalk cell. The cell supporting a *generative cell* of a gymnosperm.

Stamen. The pollen-producing structure of a flowering plant, consisting of an *anther* (which includes pollen sacs) and of a *filament* (stalk).

Staminate. With *stamens* but not *pistils.*

Staminodium. A sterile stamen, that is, without an anther or at least not producing pollen.

Standard. See *banner.*

Stellate. Star-shaped; descriptive of hairs branched so that the hair appears like a star. However, often the term is applied to hairs with only two or three branches.

Stem end. The basal portion of an inferior ovary at fruiting time; the point of attachment of the *pedicel.*

Sterile. Producing nothing, e.g., not producing pollen or seeds. However, cf. under *fertile.*

Stigma. The apical portion of a pistil, i.e., the portion receptive to pollen. It is covered commonly with minute papillae, and it is sticky through production of a sugar solution in which the pollen grains germinate.

Stigmatic crest. Pollen-receptive surface on a portion of the margin of the apical part of the carpel. The forerunner of a stigma.

Stigmatic line. A line along which the pollen-receptive surface occurs, as on the stigma of a member of the sunflower family.

Stipe. A stalk; the petiole of a fern leaf. The term is applied to a special stalk under the pistil of a flower, a stalk developed from either ovary tissue or the receptacle. A stipe is not to be confused with the pedicel of the flower.

Stipitate. With a *stipe.* Stipitate *glands* are glands with basal stalks. Usually they terminate hairs.

Stipular. In reference to *stipules.*

Stipulate. With stipules.

Stipule. One of a pair of appendages at the base of the petiole or the leaf base at the point of attachment to the stem. These structures may be either thin and scalelike, thickened and hard, green and leaflike, or reduced to mere glands. Occasionally they are specialized as spines.

Stolon. A runner, that is, a branch which grows along the ground and produces adventitious roots.

Stoloniferous. With *stolons.*

Stóma. A minute opening in the surface of a leaf or a stem, the opening surrounded by a pair of guard cells, that is, specialized cells of the epidermis. The gas exchange between the cells of the leaf and the surrounding atmosphere occurs through the stóma. Plural, *stómata.*

Storage root. A thickened root storing reserve food.

Stramineous. Straw-colored.

Striate. With fine longitudinal lines or streaks.

Strict. Standing upright; straight.

Strigose. Covered with depressed, sharp, thin, straight hairs.

Strobilus. A conelike reproductive structure composed of a central axis or branch bearing sporophylls.

Strophiole. A special appendage at the hilum in certain kinds of seeds.

Style. The tubular upper or middle part of a pistil connecting the stigma and the ovary.
Stylopodium. A special swelling on the base of the style as in the members of the parsley family.
Sub-. Almost or nearly.
Subclass. A taxon of a rank between class and order (not necessarily used).
Subclimax. A temporary stage in succession, though often a striking one; representing only one of the transitory phases appearing before climax is attained.
Subfamily. A group of closely related genera or sometimes a single one forming a unit within a family.
Suborder. A group of closely related families or sometimes a single one within an order.
Suberose. Corky.
Subglabrous. Nearly glabrous, i.e., with few hairs.
Submersed. Submerged.
Subspecies. A taxon of a rank between species and variety (used by some authors); a group of related varieties. Sometimes substituted for variety when only one infra-specific category is employed.
Subulate. Awl-shaped, that is, more or less flat, very narrow, tapering gradually from the base to the sharp apex.
Succession. Progression of vegetation toward a climax, or state in which the species are able to withstand the extremes of the climate and therefore relatively stable in their association and in which the mature members of the community tend to be replaced by their own offspring. **Primary succession** begins from no vegetation or soil. **Secondary succession** begins after disturbance, as by fire or flood, there being some soil and some reproductive parts of plants in it.
Succulence. Fleshiness or juiciness; see succulent.
Succulent. Fleshy and juicy like a branch of a cactus or the leaf of the garden hen-and-chickens, the structure (leaf or stem) much thicker than in most plants; applied as an adjective to fruits; applied as a noun to a plant with succulent parts, especially stems or leaves.
Sucker. A rhizome or a branch of a rhizome which comes to the surface and grows into a leafy shoot.
Suffrutescent. With the lower part of the stem just above ground level somewhat woody and living over from year to year.
Suffruticose. The next degree above *suffrutescent,* that is, with woody permanent stems extending a few inches above ground level, these living over from year to year.
Sulcate. Grooved or furrowed.
Super-, supra-. Above.
Superior. Above. A *superior ovary* is not adnate with the *floral cup.*
Superior floral cup (or **tube**). The portion of a floral cup or tube above and free from an inferior ovary.
Surculose. Producing suckers.
Suspended. Hanging downward. A *suspended ovule* hangs downward from the very summit of the seed chamber of the ovary.
Suture. A line of joining; a seam. A *carpel* shows usually both a *midrib* and a *suture* (line of joining of the margins).
Syconium. An enlarged hollow receptacle bearing flowers and ultimately fruits inside, e.g., the fruitlike structure of a fig.
Symbiosis. Dissimilar organisms living together with mutual benefit.
Symbiotic. Referring to *symbiosis.*
Symmetrical. Balanced, the parts similar to each other. The flower, for example, may be *radially symmetrical,* that is, capable of division into three or more similar parts, or *bilaterally symmetrical,* capable of division into only two similar parts.
Sympetalous. All the petals coalescent at least basally. See *gamopetalous.*
Synangium. A compound formed by coalescence of *sporangia.*
Syncarp. A structure composed of several fruits, these more or less coalescent, as for example, the "fruit" of a pineapple or a mulberry.
Syncarpous. With coalescent carpels.
Synergids. Two cells lying beside the *egg* in the *megagametophyte* in the *ovule* of a flowering plant.
Syngenesious. With anthers cohering in a circle.
Synonym. A name which is not used because one applied earlier designates the same plant or because the same name was used earlier for a different plant.
Synonymy. See *synonym.* This refers to the series of discarded names applied to a single taxon.
Synonymous. With the same meaning.
Synsepalous. The sepals coalescent.

Tactile. Sensitive to touch.
Tap root. The primary root, when it is larger than the others.
Tap root system. A root system with the primary root markedly larger than the others.
Taxon. A category used in classification, as, for example, a variety, species, genus, family, etc. A living taxon is a reproducing natural population or system of populations of genetically related individuals. Ranks of plant taxa (divisions, classes, orders, families, genera, species, and varieties) depend upon the degree of their differentiation and isolation from each other. Plural, **taxa.** Cf. discussion in Chapter 14.
Taxonomy. The principles of classification.
Tendril. An elongated twining segment of a leaf or a branch, this usually supporting the stem.
Tepal. A term used for sepals and petals which are similar and not easily distinguished from each other; used in the buckwheat family for the sepals of two series, there being no corolla.
Teratological. Abnormal.
Terete. Slender and more or less cylindroidal, approximately circular in any cross section, but of varying diameter.
Terminal. At the end point.
Terminal bud. The bud at the end (apex) of a stem or a branch.
Ternary, ternate. In threes.
Terrestrial. Growing on land.
Tertiary leaflet. A leaflet of the third degree; a leaflet of a secondary leaflet.
Tessellate. Like a cobblestone pavement.
Testa. The seed coat, that is, the hardened mature *integument.*
Tetrad. A group of four.

Tetradynamous. Having four long and two short stamens. A condition found almost throughout the mustard family.
Tetragonal. Four-angled.
Tetragynous. With four pistils.
Tetramerous. With four members; the flower parts in fours.
Tetrandrous. With four stamens.
Tetraploid. With 4n chromosomes per cell.
Thalamus. See *receptacle.*
Thalloid, thallose. Resembling a thallus.
Thallophytes. The algae and fungi.
Thallus. A plant body not differentiated into roots, stems, or leaves. Growth is indeterminate (continuous so long as conditions are favorable), and the thallus does not grow to only a predetermined size.
Thorn. A pointed branch. Sometimes this term is used interchangeably with *spine,* but it is not here.
Throat. The opening of a sympetalous corolla or a synsepalous calyx, that is, the expanding part between the proper tube and the limb (spreading upper portion).
Thyrse. A densely congested panicle, the main axis indeterminate, the lateral typically determinate and therefore cymose. A mixed inflorescence.
Thyrsoid. Of the form of a *thyrse;* thyrselike.
Tomentose. Woolly, that is, densely covered with matted hairs, which usually are not straight.
Tomentulose. With relatively short, fine woolly hairs.
Tomentum. Wool.
Toothed. With minor projections and indentations alternating along the margin.
Torus. See *receptacle.*
Trabeculate. Cross-barred.
Tracheid. An element of the xylem, that is, the cell wall of a single spindle-shaped dead cell.
Trailing. Prostrate but not rooting.
Transverse. Across. See *cross section.*
Trapeziform. In the form of a trapezium, that is, with four sides but unsymmetrical and with no sides parallel.
Tree. A woody plant with a main trunk. Trees in general are larger than shrubs.
Triad. A group of three.
Triandrous. With three stamens.
Trichrome. A plant hair, that is, a slender projection composed of cells usually arranged end-to-end.
Tricolpate. With three furrows (e.g., on a pollen grain).
Trifoliolate. With three leaflets.
Trigonous. Three-angled.
Trimerous. With three members; the flower parts in threes.
Trimorphous. Occurring in three forms.
Tripinnate. Three times pinnate, that is, with primary, secondary, and tertiary leaflets.
Triploid. With 3n chromosomes per cell.
Triquetrous. Having three sharp or projecting angles.
Triradiate. With three structures diverging from a common center.
Truncate. "Chopped off" abruptly; ending abruptly.
Tuber. A thickened short underground branch of the stem serving as a storage organ containing reserve food. An example is a potato.
Tubercle. Diminutive of *tuber,* but not necessarily an underground structure; processes or bumps on a surface.
Tuberculate. With *tubercles,* that is, processes or bumps.
Tuberiferous. With *tubers.*
Tuberoid. Tuberlike; a thick, fleshy root resembling a tuber.
Tuberous. Tuberlike in appearance or character.
Tubular. Forming an elongate hollow cylinder.
Tumid. Swollen.
Tunicate, Tunicated. With a series of membranous coats, as in onions.
Turbinate. Top-shaped, that is, more or less in an inverted cone.
Turgid. Swollen.
Two-lipped. See *bilabiate.*

Ubiquitous. Occurring everywhere.
Umbel. An inflorescence with the pedicels of the flowers arising from approximately the same point. A *compound umbel* is an umbel of umbels.
Umbellate. Like an *umbel,* or in the form of an umbel.
Umbellet. A secondary *umbel* within a compound umbel.
Umbelliform. In the shape of an *umbel.*
Umbellule. A small *umbel* of secondary rank.
Umbilicate. Depressed in the center.
Umbo. With a short projection like the boss of a shield; a projection on the *apophysis* of a gymnosperm cone scale.
Umbonate. With an *umbo.*
Undulate. With the margin irregular and forming a wavy line; i.e. one which winds gently in and out or up and down.
Unguiculate. With the base contracted into a *claw,* that is, a stalk.
Unisexual. Of only one sex. Descriptive of a flower having only stamens or only pistils, not both, or of a gymnosperm producing only pollen or only ovules or of the gametophyte of a pteridophyte bearing the reproductive organs of only one sex.
Urceolate. Urn-shaped.
Utricle. A small 1-seeded more or less indehiscent fruit which appears to be inflated, that is, with a relatively thin pericarp more or less remote from the single seed. At maturity the utricle opens either irregularly or along a horizontal line.
Utricular. Inflated or bladderlike.

Vaginate. Sheathed.
Valvate. Opening by valves, that is, splitting open along regular vertical lines and leaving *valves* in between; applied to sepals when they do not overlap in the bud.
Valve. One of the segments of an open capsular fruit, that is, the area between two lines of dehiscence.
Variety. The smallest taxon usually recognized. See *taxon.* A natural population or population system. The official term is *varietas;* variety is the English version.
Vascular. Containing xylem and phloem (conducting tissues).
Vascular bundle. A threadlike fiber of xylem and phloem in a stem.
Vascular plant. One with the vascular tissues *xylem* and *phloem.*
Vascular trace. A fiber of xylem

and phloem leading to a leaf, branch, or reproductive part.

Vegetable. A vegetative part of a plant; a plant.

Vegetative organ. A root, stem, leaf, or other non-reproductive part of a plant.

Veins. Threads of conducting tissue in a leaf or a flower part.

Velutinous. Velvety, that is, with a pile like that of velvet.

Venation. The type of veining.

Ventral. The side toward the axis as, for example, the upper side of a leaf. See *adaxial* and *dorsal.*

Ventricose. Swelling unequally, that is, more on one side than another.

Vermiform. Worm-shaped.

Vernal. Of the spring.

Vernation. The arrangement of leaves in the bud or the development of a single young leaf.

Vernicose. Shiny; varnished.

Verrucose. With wartlike structures on the surface.

Versatile. Descriptive of an anther attached at the middle.

Verticil. A *whorl* or a *cycle.*

Verticillate. Arranged in a *whorl, cycle,* or *verticil.*

Vesicle. A bladder or air-cavity.

Vesicular, vesiculose. Covered with vesicles or composed of them.

Vespertine. Opening in the evening, as for example, a night-blooming *Cereus.*

Vessel. A tube formed from the walls of a series of dead xylem cells, the end walls dissolved away.

Vestige. A rudiment.

Vestigial. Rudimentary, that is, an undeveloped or poorly developed structure.

Vexillum. See *banner.*

Viatical. Growing by roadsides or along paths.

Villous. With long, soft, more or less interlaced hairs.

Virgate. Wand-shaped, that is, long, slender, straight and erect.

Viscid. Sticky.

Viviparous. The seed remaining attached to the parent plant and germinating there.

Wavy. See *undulate.*

Weed. An introduced plant which grows where it is not wanted.

Whorl. A cycle or a verticil, that is, a group of leaves or other structures at a single node.

Wing. A thin, membranous or leathery expansion on the surface of an organ as, for example, of a fruit; one of the two lateral petals of a *papilionaceous corolla.*

Woolly. Covered with long, matted hairs which are not straight. See *tomentose.*

Xeric. Dry or adapted to dry or desert conditions.

Xerophyte. A plant growing under very dry conditions. See *hydrophyte* and *mesophyte.*

Xylem. The principal cells forming the wood conducting elements, which usually carry water and dissolved salts and sometimes previously stored food upward from the roots to the leaves.

Zygomorphic. *Bilaterally symmetrical.* Often this is described as "irregular," especially with reference to corollas.

Zygote. The fertilized egg, resulting from combining of the male and female gametes (or sex cells).

Index

Page numbers in Roman type (125) indicate that a subject is discussed more than incidentally on the page given. Incidental use of a word or its use as an example ordinarily does not lead to inclusion in the index. Boldface (**125**) indicates the page (or pages) of principal discussion or description. Italics (*125*) indicate a page bearing a figure of the object or an illustration of its parts or qualities or a picture of a person or an illustration of his ideas.

1 2 3 4 5 6 7 8 9 10